THE
OXFORD COMPANION TO
ANIMAL BEHAVIOR

The
Oxford Companion to
Animal Behavior

Edited by
DAVID McFARLAND

Fellow of Balliol College, Reader in Animal Behaviour,
and Head of Animal Behaviour Research Group,
Department of Zoology, Oxford University

Foreword by
NIKO TINBERGEN, F.R.S.

Oxford   New York
OXFORD UNIVERSITY PRESS
1982

© Oxford University Press 1981
Library of Congress Catalog Card Number: 82-80431
ISBN 0-19-866120-7

Printing (last digit): 9 8 7 6 5 4 3 2 1

Printed in the United States of America

FOREWORD

Biologists of my generation still remember the times when the study of animal behaviour was not yet really accepted as a part of modern science, and was practised by no more than a handful of individuals, who indulged in it either as a hobby or as a sideline to their 'real' work. Professional psychologists had begun to study at least a few animals (an influential textbook of those times was dedicated to 'Rattus norvegicus albinus'), but they did so less with the aim of understanding animal behaviour for its own sake than for the purpose of understanding human behaviour, for which they had selected 'subhuman' species to serve as models, or at best as evolutionary 'precursors' of our own species.

Although Darwin was a brilliant student of animal and even of human behaviour, his great contribution to our understanding of evolution channelled the efforts of the next generations of zoologists into the study of animal structure, and Taxonomy and Comparative Anatomy were much more important branches of Zoology than comparative behaviour studies. For a long time, live animals were understood much better by those who hunted or farmed them, or who kept them as pets, than by those who tried to understand them scientifically.

It has been a stroke of great good fortune, and an exhilarating experience for my contemporaries and myself to have grown up just at the time when the scene changed, one could almost say overnight, and the study of animal behaviour became an acceptable and soon even a respectable branch of Zoology (although for many years budgets did not reflect this new status). What happened in the late twenties and the early thirties was that, notably in Central and Western Europe, research on the behaviour of animals in their natural or near-natural environment was started in earnest by a few exceptionally committed individuals—some would call them possessed, others inspired; they were certainly stubbornly concentrating on their subjects rather than on a scientific career—and that the reports of their work, accepted for publication by learned journals, encouraged a growing number of young zoologists to follow their lead (why this revival happened in Europe and why among zoologists rather than among psychologists are still interesting and unsolved problems for the historians of our science).

In the ten years or so before the outbreak of World War II this growing interest expressed itself in the creation of a few minor university posts and in the flocking of enthusiastic and gifted students to the few teachers occupying them. After the war we saw the emergence of what we can now call Comparative Ethology—still in my view the best name for what is now a recognized branch of the Life Sciences; indeed this field became an area of exceptionally vigorous growth.

Two aspects of this truly explosive development have been particularly striking. Inevitably, it has led to growing specialization—on problems; on animal groups or even species; on methods and techniques. But Ethology has avoided the narrowness that characterized for decades the—for a time most prestigious—branch of Zoology called Comparative Anatomy, where in my student years the finding of homologies was still the main aim, and the ways in which the many painstakingly studied structures were used by the live animal were almost totally disregarded ('Functional Anatomy' developed much later). Behaviour students rebelled against following such a narrow course, and also against the prevalent concentration on the pickled and dried carcasses of animals. They aimed almost from the start at understanding the work of, and collaborating with, a variety of colleagues: with the many kinds of psychologists; with physiologists who studied the functions of sense organs, of the nervous system, and of the skeleto-muscular 'effector' system; with ecologists and students of evolution; and later with those working on engineering problems and information science; with human sociologists, and with many other workers in neighbouring fields.

As a consequence, the study of animal (and human) behaviour has now become a vast cooperative effort, in which both specialization and collaboration with related sciences have their parts to play. The resulting progress is fascinating, but it also takes one's breath away. One feels, if I may use a romantic metaphor, as if one is trying to keep up with a team of canoeists, who are steering their way in fast-running, turbulent water, using the current and avoiding wild waves and dangerous rocks, alternately being submerged and coming up to gasp for air, and in great danger of losing touch with the other team members, of being drowned, or of being washed ashore—an 'also paddled'—at the fringe of a quiet eddy.

In this situation it seems to me premature and not conducive to further growth to claim, as some have done, that Comparative Ethology is now a mature science and that we are justified in thinking that we are already attaining a synthesis. In my opinion Ethology is, fortunately, still in very turbulent water, and ethologists have a feeling of being very much in the thick of it. This feeling, I imagine, is what has motivated David McFarland and his large team of able collaborators. Their book seems to express the view that, in the struggle to find their way in this wild water, it helps to pause for a moment in a quiet pool, not only to take stock of how far they have got, but also to try and see more clearly where to go next. This is obviously what they are doing here, and equally obviously they are looking forward to very exciting and challenging tasks ahead. The best I can wish for this splendid *Oxford Companion to Animal Behaviour* is to repeat what I wrote myself, in a rather over-confidently written precursor published some thirty years ago (using a phrase that would, with the addition of only one word, apply here too): '. . . that it should lead to [still] better presentations in the near future'.

NIKO TINBERGEN, F.R.S.

PREFACE

The Oxford Companion to Animal Behaviour has been designed as a non-specialist introduction to the study of animal behaviour. Its scope includes the scientific discipline of ethology and touches upon the related disciplines of ecology, genetics, physiology, and psychology. The *Companion* is a handbook and not a dictionary or encyclopedia. It is intended to provide the reader with a guide to current scientific thought on all aspects of animal behaviour and to aid further study by means of bibliographical references. Numbers occurring at the end of articles refer to the suggested further reading given in the Bibliography.

In order to understand animal behaviour it is necessary to have some background biological knowledge. The *Companion* is designed to meet this requirement by providing the necessary material. This has resulted in articles that are longer than those usually found in Oxford Companions. Each article has been written as an independent entity and readers will obtain the greatest benefit from the *Companion* by reading an article as a whole. To avoid unnecessary duplication, a system of cross-references enables the reader to supplement the information contained in each article. When a subject is mentioned about which there is a separate article in the *Companion*, the relevant words are printed in small capitals. This is normally done only when the subject is mentioned for the first time, but in long articles the cross-references may be repeated to remind the reader of the existence of other articles that are of particular relevance.

The *Companion* has been written as a work of reference for the layman, not for the professional scientist. While every attempt has been made to maintain an up-to-date account of the study of animal behaviour, no attempt has been made to justify assertions by argument or by reference to the scientific literature, as would be the case in a scientific text. Individual scientists are not mentioned except in a historical context.

Animal species

The vast range of species in the animal kingdom precludes articles on particular animals in a book of this size. Where animals are mentioned in the text, their scientific names are provided, and a list of these appears in the Index of Scientific Names of Animals. Readers wishing to obtain information about particular animals can also consult the Index of English Names of Animals, which contains about 1500 entries.

The scientific classification of the animal kingdom is based upon the supposed phylogenetic relationships between animals. Knowledge and opinion about phylogeny change over the years and there may be consequent changes in the classification and naming of animals. These are governed by the International Code of Zoological Nomenclature under the auspices of the International Congress of Zoology. Changes in the scientific names of animals can lead to confusion, especially where the reader is consulting a bibliography that spans a number of years. In order to provide a standard that is accessible to the layman, the scientific names of animals used in this Companion conform, as far as possible, with those used in *Grzimek's Animal Life Encyclopedia*.

Illustrations

Illustrations have been selected to exemplify points in the text and to provide reference material where this can be done only in a pictorial manner. The illustrations are not intended purely as ornamentation and for this reason it has not been thought necessary to use colour. The majority of the illustrations of animals have been prepared by Tim Halliday, from photographs and drawings, and from life. Particular care has been taken to ensure that the illustrations are accurate representations of the animals concerned, because some aspects of animal behaviour can be properly understood only by reference to the detailed morphological features of animals.

Acknowledgements

The author wishes to express his appreciation to all who have given advice during the compilation of this *Companion*. Special thanks go to Dan Davin for his early encouragement, to Bruce Phillips for his invaluable guidance, to Tim Halliday for his splendid illustrations, to Dorothy McCarthy, Jill McFarland, and Joan Pusey for their painstaking work on the copy and proofs, and to Pat Searle for her secretarial assistance.

DAVID MC FARLAND

CONTENTS

LIST OF CONTRIBUTORS

The editor is grateful to the many contributors who have assisted with the compilation of this work. The initials at the end of each entry refer to the main contributor(s) of the article. Where there are no initials the editor is solely responsible for the article. The following list gives the contributors' initials in alphabetical order, followed by their names.

A.C.	Alan Cowey	J.R.K.	John Krebs
A.D.	Anthony Dickinson	J.S.R.	Jay Rosenblatt
A.H.	Andrew Henley	K.L.H.	Karen Hollis
A.R.	Anne Rasa	L.K.H.	Larry Henrickson
B.P.	Brian Partridge	M.D.	Marian Dawkins
B.S.	Bernard Singer	M.E.	Malcolm Edmunds
C.D.	Clifford Davies	M.R.	Mark Ridley
C.E.	Carl Ericson	M.W.	Michael Woolridge
C.G.G.	Carol Gould	N.B.D.	Nicholas Davies
C.J.A.	Charles Amlaner Jr.	N.E.C.	Nicholas Collias
D.F.S.	David Sherry	N.J.M.	Nicholas Mackintosh
D.M.B.	Donald Broom	N.R.L.	Robin Liley
D.M.J.	Duane Jackson	N.S.S.	Stuart Sutherland
D.W.M.	David Macdonald	P.J.B.	Pat Butler
D.W.T.	Donald Thomas	P.P.G.B.	Pat Bateson
E.C.	Eberhard Curio	P.R.	Paul Rozin
E.R.	Elizabeth Rozin	P.W.	Peter Wright
G.B.	Gerard Baerends	R.A.H.	Robert Hinde
G.G.G.	Gordon Gallup Jr.	R.A.M.	Roger Mugford
G.H.	Geoffrey Hall	R.B.	Robert Bolles
H.C.B.C.	Henry Bennet-Clark	R.D.	Richard Dawkins
H.O.B.	Hilary Oldfield-Box	R.J.	Roger Johnson
H.S.	Hermann Schöne	R.J.A.	Richard Andrew
J.B.H.	John Hutchison	R.M.	Ray Meddis
J.D.	Juan Delius	R.M.A.	Neill Alexander
J.E.R.S.	John Staddon	R.M.T.	Roger Tarpy
J.F.E.	John Eisenberg	R.P.	Richard Passingham
J.G.V.	John Vandenbergh	R.W.O.	Ronald Oppenheim
J.H.	Jerry Hirsch	S.D.I.	Sue Iverson
J.H.D.	James Dewson III	S.M.	Susanna Millar
J.L.	John Lazarus	T.H.C.B.	Timothy Clutton-Brock
J.L.G.	James Gould	T.R.H.	Timothy Halliday
J.M.D.	Michael Davis	W.R.A.M.	William Muntz
J.P.H.	Jack Hailman	† W.T.K.	William Keeton
J.R.A.M.H.	Jan van Hooff		

ABBREVIATIONS

cm	centimetre	m	metre
dB	decibel	μg	millionth of a gram
g	gram	MHz	megahertz
h	hour	min	minute
ha	hectare	mm	millimetre
Hz	hertz	ms	millisecond
J	joule	MW	megawatt
kg	kilogram	nm	nanometre
kHz	kilohertz	s	second
km	kilometre	sp.	
kW	kilowatt	(pl. spp.)	species
l	litre	W	watt

A

ABNORMAL BEHAVIOUR may occur as a result of any pathological condition of an animal, but the term is often associated with experimentally induced anxiety, sometimes called experimental neurosis. PAVLOV discovered that, during CONDITIONING experiments, an animal's normal behaviour would suddenly break down, and the animal became very agitated. For example, in one experiment a dog (*Canis l. familiaris*) was required to discriminate between two shapes, an ellipse and a circle. The dog was expected to salivate when the circle was presented, since this was followed by food, but to refrain from salivating when the ellipse was presented, since this was not followed by food. In the early stages of the experiment the DISCRIMINATION was easy, because the ellipse was twice as long as it was wide. As the experiment continued, however, the width of the ellipse was gradually increased until it was almost circular. When the dog could no longer discriminate between the two stimuli, he became excited and showed physiological signs of EMOTION.

Animals subjected to prolonged experimentally induced neurosis may show extreme signs of STRESS, and their behaviour may become pathologically abnormal. They may show great anxiety, loss of appetite, and stereotyped behaviour. Eventually they may develop physical symptoms, such as gastric ulcers. In one experiment, paired monkeys were connected to an electric shock circuit. One member of each pair was required to press a lever to postpone shock. Whenever this monkey failed to make the required response both monkeys received an electric shock. At the end of this experiment, post-mortems revealed that the monkeys that had been required to make decisions had extensive gastric ulcers, but the passive monkeys had no such symptoms of stress, even though they had received the same pattern of electric shock. Abnormal behaviour sometimes occurs in FARM ANIMALS maintained under intensive conditions (see WELFARE).

ACCLIMATIZATION is a form of reversible ADAPTATION, by which an animal is able to alter its TOLERANCE of environmental factors. For example, adult frogs, of the species *Rana temporaria*, maintained in the laboratory at a temperature of either 10 °C, or 30 °C, were tested in an environment of 0 °C. It was found that the group maintained at the higher temperature was inactive at 0 °C, but that the group kept at 10 °C was active at 0 °C. Usually the term *acclimation* is used in laboratory

studies in which adaptation occurs to a single variable, such as temperature. The term *acclimatization* is reserved for the complex of adaptive processes that occurs under natural conditions.

Acclimatization usually involves a number of interacting physiological processes, which occur at different rates. For example, in acclimatizing to high altitude, the first response of human beings is to increase their rate of breathing. After about 40 h, changes have occurred in the oxygen-carrying capacity of the blood, which make it more efficient in extracting oxygen from the rarefied air. As this occurs, the breathing rate returns to normal. True acclimatization is not completed for about 2 weeks, during which there is an increase in the number of red corpuscles circulating in the blood.

Under natural conditions, acclimatization often occurs in response to seasonal changes in CLIMATE. For example, seasonal shifts in the upper lethal temperature of freshwater fish are often directly correlated with monthly changes in their HABITAT temperature. Many fish also show temperature preferences that are related to their state of acclimatization.

Acclimatization can have profound effects upon behaviour, inducing shifts in preferences and in mode of life. For example, the golden hamster (*Mesocricetus auratus*) prepares for HIBERNATION when the environmental temperature drops below about 15 °C. This preparation involves a number of physiological changes amounting to a form of acclimatization to the cold, and makes hibernation physiologically possible when other conditions are propitious. Optimal conditions for hibernation in hamsters include the provision of nesting material, and enough food for the animal to set aside a store. Temperature preference tests, conducted in the laboratory, show that the hamsters develop a marked preference for cold environmental temperatures during the pre-hibernation period of acclimatization. They prefer an 8 °C environment to either a 19 °C or a 24 °C environment. Following AROUSAL from a period of hibernation, the situation is reversed, and the hamsters actively prefer the warmer environments.

ACTIVITY. *General activity* is a crude behavioural category sometimes employed by psychologists engaged in LABORATORY STUDIES of animal behaviour. Different types of activity are sometimes classified in accordance with the methods used to measure the activity. Thus running activity is measured in a circular cage which rotates freely around its axis. This type of apparatus is generally called an activity wheel. The animal, usually a small rodent, runs inside the wheel, causing it to revolve, and the revolutions are counted automatically. Restlessness is often monitored in a cage with a pivoted floor called a stabilimeter. As the animal moves around the cage the tipping movements of the floor are recorded. An alternative method is to pass small beams of light through the cage and to measure activity by interruptions of the light beams caused by the animal's movements. For large animals in a pasture, a pedometer can be used to give an indication of general activity. All these measures are sensitive to particular types of activity, and do not correlate well with each other. Moreover, the notion of general activity encompasses a wide variety of behaviour patterns and is not taken very seriously by students of animal behaviour. Nevertheless, measures of general activity are widely used in psychological experiments, especially in the study of RHYTHMS, SLEEP, and the effects of DRUGS on behaviour. In some cases, the interpretation of the results of such experiments is clear cut, but in many cases the results are very difficult to interpret, due to the arbitrary nature of the measures employed.

ADAPTATION, as used by biologists, has a variety of meanings, all denoting some type of adjustment by the animal with respect to the environment. Firstly, a distinction should be made between *genotypic* adaptation, in which the adjustment is GENETIC and takes place through a process of EVOLUTION, and *phenotypic* adaptation in which adjustment takes place in the individual animal on a non-genetic basis. The *industrial melanism* of the peppered moth (*Biston betularia*) is an example of genotypic adaptation. Originally the majority of these moths were light in colour and well CAMOUFLAGED against their natural background, the lichen-covered, light-coloured trunks of trees. However in regions which became industrialized the tree trunks were darkened by the smoke from factory chimneys, and against this background light-coloured moths would be highly conspicuous. Gradually the peppered moth population in industrial areas became predominantly composed of a dark variety, which was well camouflaged against the dark trees (see NATURAL SELECTION). Thus through a process of evolution, the moths adapted to the change in their normal habitat (see also COADAPTATION).

Phenotypic adaptation may be reversible, or non-reversible, and may involve such processes as maturation, learning, and physiological adjustment. Irreversible adaptation may be observed at different points in the developmental history of an individual (see ONTOGENY). Transient environmental changes may influence the embryo or larva so that a phenotypic characteristic of the animal may become fixed in that individual. For example, in certain bird species the young can become attached to members of another species, notably foster species, at a certain stage in their life history. In later life they often direct their sexual behaviour towards members of the alien species (see IMPRINTING). Adaptation during development enables animals to adjust to the environmental conditions in which they are likely to live their lives. The phenomenon is thought to be widespread, and to include aspects of FOOD SELECTION, HABITAT selection, and MATE SELECTION.

Many animals are able to adapt to environmental change by LEARNING. Not only are animals able to adjust by learning which food is most nutritious, which place gives most shelter, etc., but they are often also able to learn to modify their own behaviour in response to environmental stimuli. For example, the marine worm *Nereis pelagica* normally retreats into its burrow whenever there is a sudden change in illumination, such as a passing shadow. Repetition of the stimulus, however, leads to a gradual diminution of the response, and by this process of HABITUATION the animal is able to adapt to the level of changeability in the environment. Learning is essentially an irreversible process. Although animals may appear to forget, or to extinguish learned behaviour, the internal changes brought about by the learning process are permanent and can be modified only by further learning.

Reversible adaptations generally involve those physiological processes which serve to adjust the animal to local environmental conditions. These range from the slow processes of ACCLIMATIZATION, to the rapid physiological adjustments involved in the maintenance of HOMEOSTASIS. For example, many species are able to adapt to changes in environmental temperature by means of rapid physiological responses, such as sweating in man, and by means of behavioural responses, such as panting, or seeking the shade (see THERMOREGULATION). At the same time, an additional slow process of adaptation generally occurs if the environmental change is long-lasting. For example, a man transported from a cool to a hot environment may decide on a policy of exposing himself to the sun, sweating to keep cool, and gradually acclimatizing to the new environment. On the other hand, he may decide to avoid sunlight, which may restrict his behavioural options and slow down his rate of acclimatization, but will not require such intense short-term physiological adaptation.

In general, the processes involved in physiological adaptation are diverse, have a variety of time-courses, and are complementary with respect to each other. This flexibility of response enables animals to adapt to a wide range of complex environmental situations. Individual animals do not, however, have unlimited capabilities in adapting to changed environmental conditions. The limits of physiological adaptation to a particular aspect of the environment are defined by the degree of TOLERANCE of the animal in that particular respect. Tolerance ranges are generally determined by the genetic make-up of the animal, and are thus characteristic of the species. For example, the stable fly (*Stomoxys calcitrans*) has a thermal tolerance range of 14–32 °C, whereas for the house-fly (*Musca domestica*) the range is 20–40 °C.

The term adaptation is often used for the process of adjustment to environmental change, in a sense different from that outlined above. For example, when we walk into a brightly lit room, the light seems very bright at first and we are dazzled, but after a minute or so the light seems less bright, and we see more clearly. In other words, our eyes have adapted to the change in level of illumination. This type of adaptation is generally called sensory adaptation, and it is widespread in SENSE ORGANS of many types (see SENSES OF ANIMALS). In the case of a bright light, adaptation occurs by constriction of the pupil, so that less light is allowed into the eye, and by photochemical changes within the eye itself (see VISION).

In summary, the term adaptation is used in a variety of ways, and biologists usually distinguish between (i) *evolutionary adaptation*, which concerns the ways in which species adjust genetically to changed environmental conditions in the very long term; (ii) *physiological adaptation*, which has to do with the physiological processes involved in adjustment by the individual to climatic changes and changes in food quality, etc.; (iii) *sensory adaptation*, by which the sense organs adjust to changes in the strength of the particular stimulation which they are designed to detect; and (iv) *adaptation by learning*, which is a process by which animals are able to adjust to a wide variety of different types of environmental change.

ADVERTISEMENT is a form of DISPLAY, usually shown by males holding a TERRITORY. It serves both to attract females and to ward off rival males. To be effective, such displays should be oriented towards likely recipients, and should be conspicuous. The advertising type of bird SONG, for example, should be easily locatable, and should indicate the species, sex, and MOTIVATION of the singer.

The principles of conspicuousness are, in many respects, the opposite of those of CAMOUFLAGE. For example, many animals minimize the conspicuousness of the shadow cast on to the lower part of the body. This may be done by counter-shading, in which the lower parts of the body have paler coloration than the rest of the body. However, reverse counter-shading, in which the ventral parts are darker than the rest of the body, is employed by some species to enhance the conspicuousness of the body shape. Thus the male bobolink (*Dolichonyx oryzivorus*) has counter-shaded coloration in winter, and reverse counter-shading in summer, as illustrated in Fig. A.

Conspicuousness of body outline may also be achieved by uniform coloration, as in the all red male cardinal (*Cardinalis cardinalis*) and the all black common crow (*Corvus brachyrhynchos*), which stand out against a variegated background.

Fig. A. Examples of reverse counter-shading: (**a**) the spectacled eider duck (*Somateria fischeri*), (**b**) the Spanish grunt (*Haemulon macrostomum*), and (**c**) the hooded oriole (*Icterus cucullatus*) are habitually shaded, whereas (**d**) the male bobolink (*Dolichonyx oryzivorus*) is shaded during the summer breeding season, but not in the winter.

Repetitive and geometric patterns are also frequently employed to promote conspicuousness (Fig. B), as are stereotyped movements and the sudden onset of sound or movement.

Advertisement displays that involve conspicuous sounds tend to be easily locatable and to travel considerable distances. The methods of sound location used by animals vary with the type of sound (see HEARING). Low frequency sounds are most easily located by the detection of phase differences, while high frequency sounds are most easily located by the detection of intensity differences between the two ears. The easiest sounds to locate are those that provide a wide range of frequencies and sharp temporal discontinuities. The advertisement sounds of birds generally have these characteristics, and are easy to locate, compared with their ALARM calls.

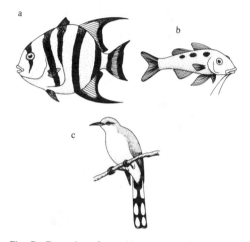

Fig. B. Examples of repetitive patterns that promote conspicuousness. **a.** Atlantic spadefish (*Chaetodipterus faber*). **b.** Spotted goatfish (*Pseudupeneus maculatus*). **c.** Yellow-billed cuckoo (*Coccyzus americanus*).

In some cases an animal may be camouflaged in some respects, and conspicuous in other respects. For example, there are two species of tree frog that appear to be morphologically identical, and have broadly overlapping ranges. Both have drab grey coloration. However, the songs of the two species differ considerably: *Hyla versicolor* sings at a rate of 17 to 35 notes per second, whereas *Hyla chrysoscelis* sings at a rate of 34 to 69 notes per second. That the two types of song really belong to different species, and are not just variants of a single species, can be shown by the fact that the hybrid eggs, resulting from cross-matings of representatives of the two song types, fail to develop, showing that there is genetic incompatibility between the two species.

65

AESTIVATION is a period of dormancy during the summer, in contrast to HIBERNATION, which is a period of winter dormancy. Aestivation enables animals to avoid extremes of CLIMATE, especially excessive heat or drought. It is most common in desert-living species, but is not confined to them. For example, many species of European earthworm, such as *Allolobophora longa* and *Eisenia foetida*, aestivate during the summer months. Each animal hollows out a small chamber deep in the soil, and curls up in a ball. If these worms are maintained in a humid atmosphere aestivation is prevented, suggesting that it normally serves to avoid drought. Similarly, many land snails (Stylommatophora), which inhabit desert and semi-arid places, avoid desiccation by closing the mouth of the shell by a thick diaphragm, which reduces water loss by evaporation. Desert snails have been known to remain in such a state for many years, until dormancy was broken by heavy rain.

Although some species of bird are known to enter periods of torpidity, this is generally thought to be a response to cold. Some tropical birds, such as humming birds (Trochilidae), reduce their body temperature and become torpid at night. The white-throated poorwill (*Phalaenoptilus nuttallii*), which inhabits the California desert, shows torpor at about 18 °C, whereas a typical temperature for mammalian aestivation is 25 °C. It is thought that these birds become torpid at low temperatures as a means of energy conservation, and that aestivation proper does not occur in birds.

Many observers have reported that certain desert rodents, notably ground squirrels (*Citellus* spp.), become inactive during the summer. They enter a torpid state, which is reminiscent of hibernation, in that the body temperature falls and there is a general reduction in the level of most physiological processes. This means that energy expenditure is lowered, and the animal can stay alive for longer periods without feeding. Water losses are also considerably reduced, and this is helped by the habit of retreating into burrows during aestivation. Some rodents store food in their burrows and others maintain a high humidity inside by stopping up the entrance.

AGGREGATIONS. An aggregation is a group of animals which forms as a result of individuals being independently attracted to a particular environmental feature, such as a source of food. Cases in which individuals are attracted to each other, as in FLOCKING in birds and SCHOOLING in fish, are generally regarded as instances of SOCIAL ORGANIZATION.

Aggregations often result from HABITAT preference. For example, woodlice (e.g. *Porcellio scaber*) are usually found in moist places. The animal exhibits a particular form of ORIENTATION, called KINESIS, in which its speed of locomotion is directly related to the humidity. It runs around much more quickly in dry air than in moist, with the consequence that it tends to stay longer in damp places.

Aggregations may occur when animals respond individually to some biological factor such as a

PHEROMONE. For example, butterflies of the genus *Heliconius*, and larvae of the cotton stainer bug (*Dysdercus suturellus*), are distasteful to birds, and have conspicuous warning COLORATION. They aggregate in large numbers in response to chemical attractants. This is a form of DEFENSIVE behaviour, because a predator that samples one quickly learns to avoid others of a similar colour.

AGGRESSION. When one animal attacks another we may say it is being aggressive without really stopping to think what lies behind the behaviour we observe. If we define aggression to include all behaviour patterns which serve to intimidate or damage another organism, then we should not only include physical attack in this list, but also things such as bird song, scent marking, and prey capture. PREDATORY behaviour, however, by general consensus of opinion, is not regarded as aggression *per se*, but rather as a separate MOTIVATION associated with hunger. A cat is not aggressive when catching a mouse any more than a swallow is aggressive when catching a fly.

One question that is raised by this definition is 'Why should an animal want to scare off another member of its own species?' Most animals are specialist feeders, living on a relatively limited food supply. There are exceptions, however, such as the grass-eaters, which usually have food in abundance, at least at certain times of the year. In some species such as the insect-eating birds and mammals, the carnivores, or fruit-eaters, the amount of food available is not unlimited, so it would be of advantage if an individual could secure itself a relatively constant supply of nourishment. One way in which this can be accomplished is to find an area with sufficient food, and prevent other members of the same species from utilizing it. However, there are other things besides food which are just as important to defend, such as special nest sites (holes in trees for example), water holes, a female or a group of females, a place to live (see TERRITORY) or even a position in the rank order of a society (see DOMINANCE).

An organism is designed by NATURAL SELECTION to pass on its GENETIC traits to its offspring and to ensure that as many of them as possible survive. This it can only do if it has an adequate environment: sufficient food, sufficient nest sites, sufficient hiding places, and sufficient potential mates. All of these things are worth defending, and the most efficient way to defend something is to ward off competitors without expending too much energy oneself. So indicating that a territory is occupied, by whichever means the species has at its disposal, be it odour, VOCALIZATIONS, special visual displays or signs left on objects around the perimeter (see SCENT MARKING), or even a combination of these, can only be of advantage to the inhabitant, since such signs have an intimidating effect on intruders, without the owner of the area having to fight to defend it (see SONG). Should an intruder ignore these signals, however, and penetrate the owner's territory, the owner must show aggression against it as a last resort.

This aggression, however, is usually ritualized, the territory owner adopting certain body postures or FACIAL EXPRESSIONS which communicate to the intruder that its presence is unwanted (see RITUALIZATION). Usually these are sufficient to end the interaction with a minimum of expended energy on both sides, but sometimes the situation can escalate until a real fight takes place that can be damaging to both combatants. This whole complex of behaviour is what is commonly termed 'aggression'.

Self-defence and property protection: a basic dichotomy. Let us now return to the concept of defence. We have already seen that there are certain essentials in the environment which an animal should protect against exploitation by rivals. Apart from environmental necessities, there is yet another thing which is essential for an animal to preserve, and that is its own life. Here we can talk of 'self-defence' in contrast to 'property protection'. As we shall see, the behavioural and motivational factors underlying these two basic divisions of aggressive behaviour are fundamentally different, and it is through the lumping of the two under the general heading of aggression that much of the controversy surrounding this concept and its adaptiveness for the species has come about.

Consider the familiar sight of two dogs (*Canis l. familiaris*) meeting on the street outside the garden which is the home of one of them. They approach each other with hackles raised, and move with short, stiff steps. Each sniffs the anal region of the other and they may growl softly. The owner then urinates against a nearby lamp post. The other may sniff the urine mark and urinate over it in turn. Both continue to circle each other, stifflegged and with fur bristling, until the intruder wags its tail slightly and then stalks off purposefully. The other dog then relaxes its stiff stance and returns to its territory, the garden. This is an example of property defence which has progressed in a way adaptive to both combatants. The resident dog has indicated clearly to the other that this area is occupied through his THREAT behaviour, and by marking the site with urine. The intruder has acknowledged the ownership of the resident, and retreated with flags flying, intimidated but not cowed. This pattern of behaviour is familiar to all dog owners and is a daily occurrence. Neither of the combatants gets hurt and yet the question of property ownership is clearly settled.

Now let us consider another case. A dog is chained up in a yard. A rival approaches and both dogs stand stiff-legged with the hackles raised as before. Since the chained dog cannot approach the intruder and meet him on the territory border, the intruder comes to him, thus penetrating deep into the territory and invading the owner's personal space, something which did not happen in the case of the two dogs which met on the street. The dog on the chain cannot retreat or advance and, as a result, resorts to extreme tactics: it tries to drive off the other dog with the most effective weapons at its disposal, its teeth. A typical dog-fight then ensues, both animals biting into the thick neck-ruff of the antagonist and rolling over and over. The chained dog is hampered by being tied up and cannot manœuvre as well as the other dog, which is thus at an advantage. The whole atmosphere of the fight can then change as the chained dog starts attacking desperately, biting its opponent anywhere and hurting it. This is answered by vicious bites on the part of the intruder, and the whole fight escalates until the fur literally flies. Let us consider now the different outcome in the two situations. In both cases there was an aggressive encounter between two dogs, but in one the result was a passive understanding, whereas in the other damage was inflicted. In the second case, owing to external circumstances (one animal being chained up), one of the antagonists was unable to perform the behaviour patterns appropriate to the situation. In addition, since it could not approach the intruder, the situation was made even more critical in that the intruder had to come to him, invading the private sphere or INDIVIDUAL DISTANCE. Usually, this would lead to retreat, but here this was impossible because of the chain. The animal is cornered, the tendency to flee thwarted, and the result is FEAR. As it cannot get away, the dog tries the other extreme, to drive off the stimulus (the intruding dog) which is causing its fear as quickly and effectively as possible, by using the most vicious behaviour patterns at its disposal. This type of behaviour is well known in other animals and

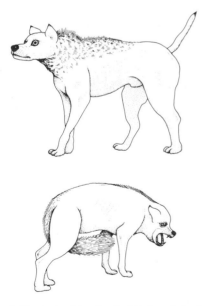

Fig. A. Threatening (above) and self-defence (below) postures of a dog.

reflected in the idiom 'to fight like a cornered rat' (see FIGHTING and DEFENSIVE BEHAVIOUR).

When we look at the dog's external appearance in the two situations we find that, in aggressive threat, the body postures are the antithesis of those shown in self-defence (Fig. A). As previously stated, a threatening dog tries to make itself look as large as possible by erecting its hair and holding its head and tail up, at the same time standing practically on tiptoe. The dog which is fighting out of desperation draws a completely different picture. The body is crouched, the rump drawn in, the tail clamped between the hind legs, and the head lowered. All vulnerable parts are protected. The most vicious weapons, the teeth, are displayed prominently to the antagonist, and the animal usually tries to back up against some object to protect its easily damaged posterior. From this position, darting attacks are made at the other animal, these no longer being restricted to parts of the body, such as the neck-ruff, which are specially protected to receive them, and thus causing damage. Such attacks are usually reciprocated by the rival, with the result that a 'fight to the death' occurs.

Pain and attack. Why should the attacks be reciprocated? To find the answer to this we must look at studies made on PAIN as an instigator of attack. If an animal is severely hurt, but cannot flee, it tries to escape the inflictor of the hurt by driving it away. We have already discussed fear as an instigator of vicious attack, and the effects of pain are the same. Both are stressful to the organism and the organism's response to them is AVOIDANCE. The vicious behaviour shown by the chained dog is understandable in this context but why does such behaviour also occur in the other animal? In all property-protective fights between individuals of the same species a set of rules is laid down, the foremost of these being that attacks are to be directed towards certain body areas which are specially adapted and strengthened to receive them. A certain pain threshold is expected, but when these attacks land on parts of the body which are vulnerable, the pain level exceeds this threshold and the general defensive principle then applies: the thing which causes the pain must be driven off as fast as possible with the most effective means at the animal's disposal.

This situation is especially clearly illustrated in the case of dwarf mongooses *Helogale undulata rufula* raised in isolation. These animals have an inborn tendency to protect their food against rivals, and even tiny babies which still have their eyes closed will growl and try to protect their supposed prey from their litter mates if a little blood is smeared on their tongues. Mongooses raised in isolation from birth onwards, contacting only human caretakers, and with no experience of conspecific rivals, often adopt a part of their body, usually the tail, as a substitute 'rival'. Domestic fowl (*Gallus g. domesticus*) and ducks (Anatinae) do this too. When a mongoose that has been reared in this way is given food, it glances back at its tail and growls, the typical behaviour pattern used to warn off a rival. At the same time it performs a 'hip swing' which, under normal circumstances, blocks the rival's approach. Here it often serves to move the tail out of the line of vision, and the animal then proceeds to feed quietly. If, by accident, the tail becomes trapped, as for instance against the side of the cage, and remains in the animal's field of vision, the growling is intensified and a stabbing bite is delivered to the supposed rival which has not heeded the growling warning. This inflicts pain and the end result is that the animal starts biting the posterior part of the body viciously, and may even mutilate itself by biting pieces off the tail and hind legs. The greater the pain felt, the more vicious the return attack. The vocalizations given here, however, are not the growls directed towards rivals but the piercing screams and spits used to drive off enemies. These usually occur in confrontations with other species and are especially directed towards predators, although they may sometimes be used when a strange mongoose persistently tries to invade a stable group. During the course of events, the tail has changed in stimulus content from 'rival' to 'enemy' through the medium of pain.

Motivational conflict. It is the animal's conceptualization of the antagonist as rival or enemy which directs the behaviour patterns shown. A rival is threatened with ritualized behaviour patterns while an enemy is attacked with no holds barred. We have considered the case where the two dogs fought viciously since one of them could not escape the situation as it was physically cornered. Such fights also occur under conditions where there are no physical restraints. We may call this motivational cornering, since here the animal is unable to escape the situation because it is held there by a motivational CONFLICT between fleeing and staying. A mother hen will not flee if her chicks are in danger. The bond between her and her young is strong enough to hold her near them when, for instance, a fox (*Vulpes*) approaches. The nearer the enemy comes the more critical her response, and she may fly in its face, beat it with her wings, and peck at it. If no chicks were present her response would be to fly up into a tree, or to some place that the predator could not reach. At first, when the predator is at some distance from her young, the bond to the young is sufficiently strong to block flight, although she may try to lead the chicks away from the danger. If the predator approaches too closely, invading the personal space, then the extreme reaction of self-defence is shown, the attachment to the young serving to counterpoise the tendency to flee, the resulting fear setting extreme responses in action.

Similar motivational cornering has been observed in a variety of species ranging from the land iguana (*Conolophus pallidus*) and greylag goose (*Anser anser*) through bank voles (*Clethrionomys glareolus*), bighorn sheep (*Ovis canaden-*

sis), and wild boar (*Sus scrofa*) to dwarf mongooses and rhesus macaques (*Macaca mulatta*). In aggressive interactions in these species it is frequently the subordinate animal which inflicts the first damaging attack, not the dominant one. In all such cases observed, however, the subordinate animal was losing the fight and found itself in a situation from which it could not escape (i) because it was physically trapped, or (ii) because it was held there by a motivational conflict between either a bond to its territory, or a bond to a mate or young which were immobile, or (iii) because the threatening dominant animal was so secure in its dominant role that it practically ignored the attacks of the subordinate and maintained its threatening stance. These observations lend further support to the hypothesis that it is usually fear which motivates vicious attack, not aggression itself.

Special weapons for special tasks. Many animals, such as deer (Cervidae), giraffes (Giraffidae), antelopes, and cattle (Bovidae), have, during EVOLUTION, developed special weapons (horns or antlers) which are only used in property-protective fights. Such weapons are used in a ritualized way, the animals locking antlers or horns and pushing in a trial of strength (Fig. B). It is against the rules to ram a rival in the side, a move which could inflict fatal wounds. In all cases property-protective attacks are made towards protected parts of the rival's body. For example, bighorn sheep charge at each other and collide horn to horn. They have specially thickened skulls which bear the brunt of the shock. Northern elephant seals (*Mirounga angustirostris*) rear up and slash at an opponent's throat and chest with their elongated canine teeth, but the throat and chest area in these animals is protected by a thick layer of blubber and an armour-like skin. There are numerous such examples. (The only exception in the form of property-protective fights seems to be the Rocky Mountain goat (*Oreamnos americanus*), where apparently maladaptive behaviour has been recorded.) When animals with horns or antlers encounter an enemy from which they cannot flee, they normally defend themselves with their sharp hooves, rather than with the weapons used in intra-specific fights. Exceptions are cattle, which charge enemies with their heads lowered, and the oryx (*Oryx gazella*), which uses its spear-like horns against enemies such as the African wild dog (*Lycaon pictus*).

Some animals have gone in the opposite direction, developing special organs of defence against enemies rather than special weapons for use on members of their own species. Some snakes, such as the vipers (Viperidae) and rattlesnakes (*Crotalus*) have a very effective poison, but this is not usually used on rivals. Here a trial of strength is waged, the two animals rearing up and pushing against the opponent's body in an attempt to topple it and pin it to the ground, as if arm-wrestling (Fig. C.). Skunks (Mephitinae) do not use the

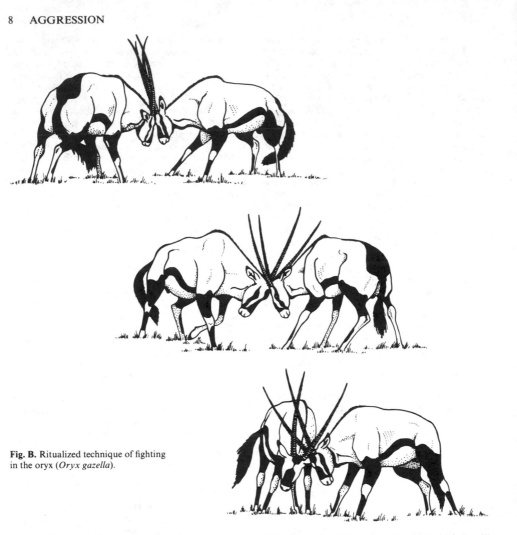

Fig. B. Ritualized technique of fighting in the oryx (*Oryx gazella*).

evil-smelling secretion from their stink glands to blind their rivals, but fight them with their forepaws and teeth. The secretion of their stink glands is only used against enemies. Here again there are numerous examples. In such species, the difference between property-protective and self-defensive behaviour is clearly indicated by the weapons used. In species which do not have such specialized weapons, however, the difference between the two motivations is more difficult to distinguish.

Motivating factors of aggression. When one surveys the literature on aggression there are apparently three main lines of thought. One considers aggression as reactive behaviour, that is, behaviour which only appears when an adequate stimulus for its release is present. Here no internal motivational state leading to APPETITIVE or searching behaviour for an adequate releasing stimulus is postulated. The behaviour appears when the external stimuli elicit it (see RELEASER). Another line of thought considers aggression as something

which the animal is born with, and which it will show even in the absence of an adequate releasing stimulus. The animal may even search for appropriate stimuli to elicit the behaviour when these have been withheld for long periods of time. Yet another point of view is that there is no INNATE disposition to aggress and aggression must be learned during the course of the animal's development. Which one of these is correct? As we shall see, all of them are, depending on the situation in which the animal finds itself.

Some animals, such as certain fish and house mice (*Mus musculus*), if reared or kept for long periods of time in isolation, will show an increase in their aggressiveness and, if no real rival is present for them to aggress against, they will attack objects which, at first glance, do not appear to have adequate stimulus properties. We have described how isolated mongooses and chickens will attack their own tails if no rival is present. A young yellowtail damselfish (*Microspathodon chry-*

surus) (which usually lives on coral reefs where there is an abundance of other fish), when isolated, will carry a piece of food to the surface of its aquarium and display aggressively to it as it sinks slowly downwards, repeating the process time and again. A male white wagtail (*Motacilla alba*) will display to its own distorted reflection in the bumper of a car. A male three-spined stickleback (*Gasterosteus aculeatus*) will attack a piece of red sealing wax suspended in its aquarium. There are several such examples, all occurring under the abnormal conditions of the absence of a conspecific rival. It is important to realize here that the behaviour patterns directed towards these inadequate stimulus objects are of the property-protective kind. Only in cases where the object does not respond by moving away does the attack escalate and self-defensive behaviour is shown, as in the case of the mongoose attacking its own tail.

Experiments to test whether aggression is appetitive, showing an increase when there is no stimulus for its release, have provided conflicting results, however. Some species show an increase in aggression when isolated, others show a decrease (see MOTIVATION). The reason for this discrepancy must be looked for in the mode of life of the animals and the physiological factors underlying aggression in the various cases. An example will illustrate these differences.

Two closely related families of bony fishes, the cichlids (Cichlidae) in fresh water and the damselfish (Pomacentridae) in the sea, have been favourite subjects for the study of aggressive behaviour, since many of their members are territorial and defend their territories vigorously. When cichlids of the species *Pelmatochromis kribensis* are kept in isolation and tested for their aggressive response after long periods of time (up to eight weeks), it is found that they become less and less aggressive with time, and require several days of continual exposure to an antagonist behind a glass partition before they regain their original level of aggressiveness. Yellowtail damselfish, on the other hand, become more and more aggressive the longer they are kept isolated, and will start displaying to very inadequate dummies, such as buttons suspended in the water, or even 'make' their own antagonists, as described above.

To understand why this difference is present we must first look at the functions of aggression in the lives of these animals. As previously mentioned, damselfish live on the reef face where there are a large number of other fish of different species. They take up territories as soon as they metamorphose from the pelagic larval form in the plankton, and become substrate-dwellers. Their territories are feeding-territories and, since they are omnivorous, they utilize the algal growths on the rocks and corals, as well as the tiny animal organisms growing there or brought by the currents. Their territories also include several hiding places to which they can retreat if predators appear. For them, the territory is essential for existence, since adequate living space is hard to come by on the crowded reef face, and loss of the territory is tantamount to a death warrant. Carnivorous fish can easily catch a wandering damselfish if it has no safe 'home hole' to flee into when danger threatens.

Since they feed on all types of food available in the territory, it is important that they not only compete with members of their own species but also with other fish which could plunder their food resources as well. Swimming along a coral reef where such fish live, one can see little 'explosions' of action at these territories, the fish darting about driving off any intruders which pass. This is especially obvious when huge SCHOOLS of young parrotfish (Scaridae) and wrasse (Labridae), which can number thousands of individuals, swim along the reef, grazing off everything in their path. At the damselfish territories the schools scatter, reforming again afterwards, while the territory owners dart here and there, biting and ramming

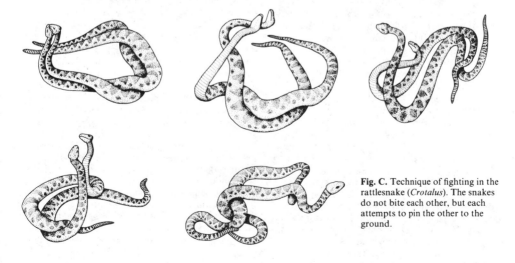

Fig. C. Technique of fighting in the rattlesnake (*Crotalus*). The snakes do not bite each other, but each attempts to pin the other to the ground.

the intruders, so preventing them from grazing the food source clean.

In cichlids, however, the picture is quite different. Young cichlids form schools and only the males take up territories when they are sexually mature. They select a small area on the bottom of a river or lake, chase rival males away, clean a stone or some other hard object as a place to deposit eggs, and then attempt to attract a female to the site to spawn. The territory is a breeding territory, and is often given up again after the breeding season when the young have become independent, the male then returning to the school until the next breeding season. There are usually clusters of these territories on the bottom where conditions are optimal, i.e. the right water depth, temperature, and substrate. The loss of such a territory to a rival would only mean that the fish would have to take up a less optimal breeding area or not breed at all for the season in question. It does not mean almost certain death, as in the case of the damselfish.

There are many species of animals which, like the cichlids, only become aggressive when they are sexually mature; this is especially true amongst male vertebrates. This led scientists to postulate that there must be a connection between male HORMONES, (the androgens, such as testosterone), and aggression. If young male mice are injected with testosterone before their testes are developed and capable of producing the hormone on their own account, the young animals try to mount females and will attack other mice. Domestic chicks and turkey chicks (*Meleagris gallopavo gallopavo*) show the typical threat behaviour of adult males when treated in this way and try to court females. Young cockerels injected with testosterone rise in rank within the flock by becoming more successful fighters. It is more than probable that the territoriality shown by male cichlids is also attributable to the hormone produced by their ripe gonads.

It has been shown in the Barbary dove (*Streptopelia risoria*) that, when no external stimulus is present to which courtship and nest defence behaviour can be directed, the birds will, at first, accept very inadequate substitutes, even a piece of rag tied in their cages, but, with time, the tendency to court and attack decreases. The gonads of these animals were found to have reverted to their non-breeding condition. It is likely that something similar happens in the case of the isolated male cichlid. When no rivals are present against which it can defend its territory, and no females that it can court, it loses breeding condition, the testes revert, and the level of testosterone in the blood returns to its normal level. The animal is no longer territorial and therefore no longer aggressive. A further support for this hypothesis is that it takes several days for an isolated male cichlid to become aggressive again when it can see a rival, something which speaks for a slow build-up of a hormone rather than a purely nervous process.

In the case of the damselfish, however, aggressive behaviour is first displayed when these animals are only about a centimetre in length, still almost transparent larval forms, with the gonads undeveloped. It is maintained for the whole of the animal's life. This suggests that a quite different causal factor than the male hormone is at work here. Apart from this, both males and females show the same level of aggressive behaviour, at least as juveniles, another factor which suggests that testosterone is not involved. Experiments indicate that another hormone, melatonin, may play a key role. When this hormone is added to the water in the aquarium and taken up through the fish's gill filaments, the animals become highly aggressive, suggesting some association between the pineal gland, from which melatonin is secreted, and aggressiveness.

As has been illustrated, even in territorial aggression in closely related fish, which might be thought to have the same causal factors, the physiological and ecological bases can be quite different. It is obvious why scientists have not been able to find a common causal factor for aggression that holds true for all animals. The physiological and ecological factors influencing the defence of a home hole by a mantis shrimp (*Gonodactylus bredini*) may be quite different from those causing a male blackbird (*Turdus merula*) to chase a rival out of its territory, a damselfish to defend its patch of reef, or a male antelope to struggle with a rival over the ownership of a group of females. These were all lumped together under the heading 'aggressive motivation', but more detailed investigation of how animals live and their physiological make-up shows that the set of behaviour patterns we call aggression have probably been developed independently in different animal groups, with different NICHES, and different physiologies, at different times during evolution.

Innate or learned? Is an animal born with a tendency to be aggressive or is this tendency developed during its life history through experience? One indication of whether a behaviour pattern is learned or innate is whether it appears spontaneously at the right developmental stage when suitable stimuli are presented. To exclude all possibility of the experimental animal LEARNING by experience or IMITATION, it is necessary to raise it under the abnormal conditions of being away from its own kind. Then we can discover whether the behaviour patterns are shown either without an adequate stimulus being present, or immediately the appropriate stimulus appears. Siamese fighting fish (*Betta splendens*), raised in isolation from the time they hatch from the egg until adulthood, will display aggressively, with all the behaviour patterns present in a normal fight, to their mirror images, or to another fighting fish, the first time they see them. Here the behaviour is obviously inborn and does not have to be learned. We have already mentioned that baby dwarf mongooses will growl and try to shield their 'prey' the first time a little blood is smeared on their tongues. This also suggests that this type of aggressive behaviour probably has an innate basis and does not depend on experience.

In addition, mongooses raised in isolation from the day of birth onwards show threatening behaviour and growling towards another member of their own species the first time they encounter one. If no rival is present they will show this behaviour towards objects, such as rags, or their own reflections, but this behaviour appears during the course of their development and almost five weeks after that of animals raised in pairs or groups. It seems that the onset of aggressive display behaviour may be retarded in some species when no adequate stimulus is present, but the behaviour appears nevertheless.

These findings suggest that aggressive behaviour probably has an innate basis, but does not exclude the possibility that it may become modified through later experience.

Learning effects. Although aggressive behaviour appears to be pre-programmed in the animal's genetic make-up, its expression in the adult animal varies considerably in both quality and quantity. Some individuals are colloquially termed aggressive, others shy. To determine whether these variations could be attributed to learning effects, scientists conducted experiments in which the type of aggressive experience the animals were exposed to was varied.

An isolated male mouse was allowed to attack an intruder which was dangled by the tail into its cage, and removed as soon as the resident mouse attacked it. Within a few days they found that the resident mouse took a shorter and shorter time to initiate the attack, and concluded from this that aggression was learned. The resident mouse had become a trained fighter, since it was always successful. It was never hurt in the encounter and the effects of its attack were that the intruder immediately disappeared, meaning that the resident had won.

When the behaviour patterns used by mice in a free fight where both animals can behave normally are examined, a balance between threatening behaviour (which is a mixture of attack and fleeing tendencies), aggression, fleeing, and submissive behaviour can be found. The winner tends to show more threat and attack towards the end of the fight, and the loser more fleeing and submission.

If a group of naïve mice is arbitrarily divided into 'winners' and 'losers' and trained as previously described until the winner attacks almost instantaneously when the loser is dangled into its cage, and the animals are then allowed to fight unrestrictedly, a change in the patterning of the events in the fight is found to have taken place. The winners now show no fleeing behaviour and hardly any threat behaviour. They aggress immediately. The losers, on the other hand, flee or submit as soon as they see the winner. The training has resulted in the winner losing all fear of

attacking the antagonist. The actual number of aggressive acts does not increase, but practically all behaviour which has a fear component has been eliminated from the fighting repertoire. The animal has not become more aggressive but has simply lost its fear of the opponent. In the case of the loser mouse, the opposite has occurred. Here, as a result of being attacked as soon as another mouse appears, and being unable to defend itself, it has come to regard fights as a source of pain and stress, and shows no more aggressive behaviour, but submits and flees immediately. When the roles are reversed and the losers are trained to be winners (the process takes much longer than in the case of the naïve mice at the start of the experiment), the previous losers also show no fear of the antagonist and attack immediately, while the previous winners flee or submit.

AGONISTIC behaviour, which is defined as all behaviour shown in the fight situation, both aggressive and fleeing behaviour as well as mixtures of the two, can thus be drastically modified by experience. If an animal only experiences success in its fights, then it will tend to invoke them. If it only experiences being beaten, it will tend to avoid fighting.

Appetite or aversion? With this in mind, we can now answer another question which has caused much controversy: Is aggression an appetite or an aversion? If an animal is able to improve its status in life, that is, win a territory, rise in rank, or capture a group of females from a rival without being badly hurt or stressed in the process, then the tendency to perform such behaviour in the future is increased. If the attempt fails, and the animal is hurt or badly frightened in the process, then the likelihood of it making such an attempt in the near future is reduced. Success makes aggression an appetite, failure an aversion.

Being able to win a fight appears to be rewarding for an animal since it will even perform complicated tasks in order to be able to display aggressively towards an antagonist. Siamese fighting fish will swim mazes or perform complicated swimming patterns through a series of rings suspended in the aquarium in order to gain sight of a model of a male fighting fish to which they can display. Stickleback males quickly learn to bite at the tip of a glass rod which triggers the removal of a partition between the fish's aquarium and a neighbouring one containing a rival male, which they then attack. Damselfish learn on a one-trial basis to swim through a maze into a glass bottle from which they can display to a rival (see MOTIVATION, Fig. J). All this evidence supports the hypothesis that aggressive display in which the animal does not get hurt or stressed is a rewarding experience, and the animals show true appetence for the action. The reverse can can also be true, however. If a damselfish is viciously attacked by a stronger rival when it enters the bottle protruding into the neighbouring half of the aquarium, and so associates entering the bottle with fear, then it will no longer go there. But if the vicious rival is then replaced with one which is not so aggressive, entering the bottle to display aggressively becomes a frequent behavioural event once more.

This association between winning being rewarding and losing a fight being aversive is especially important when the role of young animals in the community is considered. Attempts on the part of a young animal to oust a mature, experienced one usually fail. The animal learns from this experience and will not engage in fights so often, especially with the animal which has vanquished it. Since, on the other hand, it is also likely to encounter individuals as physically weak and inexperienced as itself against which it may win a fight, the tendency to fight at a future date is not completely eliminated. This mechanism ensures that the breeding stock of the future is not destroyed by futile involvement in situations in which they cannot win.

What is important to realize here, though, is that it is not aggression itself which is learned. The behaviour patterns used in aggressive interactions are programmed in the animal's genetic make-up. What can be learned is their effective use in the fight situation, and success or failure during a fight can influence an animal's overt aggressiveness profoundly, enhancing it or decreasing it. Experience may thus play an important role in the level of aggression shown, but not in the form in which it appears.

Social aspects of aggression. Experience in the fight situation is of especial importance for social animals. One of the most essential prerequisites for a social mode of life is the ability to subordinate oneself. Animal societies are built up, in practically all cases, on a system of DOMINANCE which is usually based on the relative strength of their members. Such a system is called a rank order, and the strongest animal in the group, the leader, is called the alpha animal. This individual is then followed by the next strongest, the beta animal, and so on, down to the lowest ranking individual which is called omega. These ranks are usually determined through ritualized property-protective fights. The property protected here is the animal's status in the society. With this status go certain privileges, such as the right to feed first, or to copulate with the females in the group. Lower-ranking individuals attempt to improve their status through challenging animals ranking higher than themselves; if they win they then move up the rank order. The principles behind such rank status are thus not much different from those behind the holding of a territory. The higher the rank, the freer the access to valuable resources, just as the stronger the individual, the better the territory or the quality of the resources available.

Within a society, the role of experience in the aggressive situation is a basic adjunct to stability. Most animal societies are stable with little observable aggression between group members. Usually a low intensity threat posture is all that is neces-

sary to determine ownership of an available resource, the weaker or lower-ranking animal giving up his claim to it without the situation developing into a fight. This balance has come about through learning, weaker individuals having experienced either in their youth through a gradual process of small day-to-day encounters that the higher-ranking animals are stronger than they are, and that attempting to challenge their position would only result in defeat, or through single intense encounters in which the weaker animal has been soundly thrashed and literally put in its place. This experience reduces the number of fights taking place within a society enormously, although occasional dominance strivings occur when a mature young animal attempts to displace an aged one which has become weak.

The system ensures that the most able individuals survive to pass their genetic material on to future offspring, since, as previously mentioned, high rank is usually positively correlated with the right to mate. The system also ensures that younger, weaker animals are protected during the period of their lives when they need this protection most, when they are growing up. Within the group their status is known and, as long as they do not attempt to infringe the right of the higher-ranking individuals, they are not subject to attack. The society is therefore of importance in ensuring the survival of JUVENILES, eliminating extreme competition with conspecifics during a vulnerable stage of development. Animals which have solitary stages in their adult lives, especially adolescent males, very often have to run the gauntlet from resident members of the same species during this stage of development, and many may even die during the process as a result of the stress involved, as has been shown for the tree shrew (*Tupaia belangeri*). In this way, weaker individuals may be weeded out of a population.

Rank within the society is maintained by a balance between the rewarding experience of winning a fight and the punishing one of losing one. Property-protective aggression, in this case, acts as a positive social force which stabilizes the community and ensures maximal benefits for the individuals within it, according to their age and sex. Once initial fights for rank status have been concluded, the society settles down. Experiments with chickens have illustrated this stabilizing function of rank order. Chickens living in a stable flock in which a rank or peck order has been established, show less aggression amongst themselves and a higher egg production than chickens in large groups where no rank order can be established, or in newly formed small groups where the individuals are unknown to one another. Here the presence of a rank order based on aggression and submission between individuals has stabilized the social structure to the point at which the animals indicate the lack of stress by an increase in productivity. Aggression, as a SOCIAL behaviour pattern, may have immediate negative effects from the point of view of an individual within the community but, in the long run, its effects are positive for the community as a whole.

Conclusions. In summary, therefore, when the word aggression is used, the context in which this behaviour pattern occurs must be taken into account if comparisons between different species are to be made. Apart from this, a distinction between property-protective and self-defensive aggression must be made, not only from the point of view of FUNCTION, but also from the causal factors underlying the two behavioural complexes. The former can be rewarding for the individual, but the latter is always a punishment, evoking fear and stress. The property-protective or true aggression appears to be polyphyletic, that is, to have arisen independently in different animal groups for different purposes during the course of evolution. Its function and the physiological bases underlying its expression may be quite different in seemingly similar species under similar environmental conditions. It is thus unrealistic to expect that there is a single entity which encompasses the entire spectrum of property-protective aggression, and that all these different types obey the same basic rules. In contrast to many other aspects of motivation, aggression appears to be extremely sensitive to modification by experience, not in the form in which it is expressed but in the quantity of the behaviour shown. This has far-reaching effects for the social structure of the species concerned, as well as for the individual, optimizing the chance of survival in both cases by regulating relationships between animals. It also serves as a mechanism by which weaker animals are weeded out of the reproductive community, either by being killed or failing to obtain a mate. Aggression is adaptive: through it, the best the environment has to offer is attained by the fittest individuals of the species, thus ensuring that the most successful genetical traits are handed on to the next generation.   A.R.

75, 96

AGONISTIC BEHAVIOUR refers to the complex of AGGRESSION, THREAT, APPEASEMENT, and AVOIDANCE behaviour that often occurs during encounters between members of the same species. Unlike INTERACTIONS among animals of different species, purely aggressive behaviour rarely occurs during SOCIAL INTERACTIONS. Since 'he who fights and runs away, lives to fight another day', recognition of possible defeat is an important aspect of an aggressive encounter. Most animals experience a CONFLICT of aggression and FEAR, when confronting a rival, and this may give rise to a number of forms of agonistic behaviour. In its most simple form the conflict results in an alternation of attack and ESCAPE behaviour, as can often be seen in disputes over a TERRITORY boundary in many birds and fish. One individual attacks an intruder and chases it into the neighbouring territory, whereupon the other individual reciprocates the attack.

At the territory boundary, there is generally an equilibrium between the tendencies to attack and escape, and AMBIVALENT behaviour is often observed. Alternatively, the conflict may be expressed in the form of RITUALIZED threat behaviour, or in DISPLACEMENT ACTIVITY.

Agonistic behaviour may also include activities that are not aroused in direct response to an opponent, but are more a matter of routine. The early morning SONG of many birds, and the SCENT MARKING of territory boundaries in mammals are examples of this. Many forms of ADVERTISEMENT are potentially agonistic: the red breast of the robin (*Erithacus rubecula*), for instance, is conspicuous and also features prominently in the threat display.

ALARM RESPONSES are responses to signs of danger that serve as a warning to other animals. They normally take the form of specific visual, auditory, or olfactory warning signals.

Not all responses to signs of danger should be considered as alarm responses. The house cricket (*Acheta domestica*), for example, simply stops chirping when it senses danger, and there is no evidence that this serves as a warning to others. In those cases in which an alarm response is likely to attract the attention of a predator, the individual that issues the alarm response may be disadvantaged. Such instances of ALTRUISM, in which an individual endangers itself to the benefit of others, are thought to persist in EVOLUTION only when the other animals are likely to be close relatives of the animal giving the alarm response. In social insects alarm responses do benefit related individuals, but among crickets the chief beneficiaries would probably be rival males.

Visual alarm responses are commonly found in species that live in enlarged family groups. Thus rabbits (*Oryctolagus cuniculus*) and the white-tailed deer (*Odocoileus virginianus*) have white tails which are especially conspicuous when the animal is running, and serve as a warning to other members of the group. This type of alarm signal probably does draw the attention of a predator, but this is not always the case. Pigeons (Columbidae), which commonly feed in groups, usually signal by means of flight INTENTION MOVEMENTS when they are about to fly away. A pigeon that signals its intention generally departs without disturbing the others. If a pigeon sees a sign of danger, however, it flies off without giving any intention signals. The other pigeons then immediately take alarm and fly up also. The warning signal, which is the absence of intention movements, obviously does not endanger the initiator of the alarm, since it is the first to take to the air.

Alarm calls occur in many species, and often have auditory characteristics that make it difficult for a predator to locate the calling animal. To be difficult to locate, a sound should begin and end gradually, so that a predator cannot easily compare the times at which the sounds reach his two ears. For the same reason, it should be uniform in pitch. Low-pitched sounds are generally harder to locate than high-pitched sounds, and the alarm calls of many small birds are typically low-pitched continuous calls, which begin and end gradually. In fact, these calls are so similar to each other that they often serve to warn members of more than one species.

Olfactory alarms are given by many animals by releasing special substances that stimulate the CHEMICAL SENSES. If a pike (*Esox lucius*) injures a minnow (*Notropis*), the chemicals released from the broken skin of the minnow include a fright-inducing substance that is avoided by the school of minnows for many hours. This avoidance reaction is not completely specific to members of the same species, since members of other species respond weakly. However, this may be due to similarities in the fright-inducing substances of the different species. Similar alarm substances have been found among aquatic snails (Lymnaeacea), earthworms (*Lumbricus terrestris*), and in a sea urchin *Diadema antillarum*.

Social insects release alarm PHEROMONES that do not depend upon tissue being injured or destroyed. In some ants, such as the cornfield ant (*Lasius alienus*), the alarm substance stimulates flight, but in other species (for example, *Acanthomyops claviger*) it stimulates AGGRESSION. This species lives in large underground colonies, the nests are complex, substantial, and not easily replaced. It is thus adaptive for members of this species to stand and fight in the presence of danger. The alarm substances of the aggressive slave-raider ant *Formica subintegra* not only serve to defend the colony but are also employed in attacks upon colonies of other species. Workers of other species are panicked by the alarm substance, making them more vulnerable to the raiders.

In general, alarm responses are specialized forms of DEFENSIVE behaviour, which have evolved to defend other (usually genetically related) individuals from PREDATION.

ALTRUISM, for our purposes, may be defined as self-destructive behaviour performed for the benefit of others. In contrast to the common usage of the word altruism, the technical definition contains no implication about intentions or motives. We define an altruistic act by reference to its effects, and we do not presume to know the motives or intentions that may have underlain it. We do not deny the possibility that animals may have conscious intentions to be altruistic or selfish, but that has nothing to do with the present definition or the present discussion.

The definition raises the further question of what is meant by benefit. One possibility is to regard the cost to the altruist and the benefit to the recipient as being measured in units of inclusive FITNESS. If we adopted this sensible definition, our discussion of altruism would be mainly a debate over whether it exists at all, other than by

mistake: inclusive fitness is defined in such a way that NATURAL SELECTION would not be expected to favour animals who improved the inclusive fitness of others at the expense of their own. PARENTAL CARE would not qualify as altruism by this definition, for by caring for its young an animal increases its own fitness.

Instead, we shall here define cost and benefit in the simpler currency of individual survival chances. An altruistic act, then, is one which decreases the altruist's chance of surviving, while increasing the survival chances of some other individual, the beneficiary.

By far the commonest and most familiar acts of altruism are directed by parents, especially mothers, towards their children. Parental care definitely increases the survival chance of the offspring: in many cases it would be literally zero without parental care. It is almost as obvious that caring for its young costs a parent some loss of life expectancy. Mothers with young frequently show near-suicidal courage against predators. A mother who consistently denied her offspring food would be better nourished herself and would be likely to live longer as a result.

The Darwinian explanation for parental care has long been clear. Offspring whose lives are saved by parental care tend to inherit the tendency to care for their own young when the time comes. Individuals with a hereditary tendency to starve their offspring would be likely to starve to death before they were old enough to have selfish offspring. Putting the same point in GENETIC terms, genes for parental care tend to be contained in the bodies of the young saved by the parental care. Genes for parental care therefore tend to increase in frequency relative to genes for selfishness towards offspring. Altruism at the individual level is a manifestation of 'selfishness' at the gene level.

Of course this is much too simple an account. It is possible to give too much parental altruism as well as too little. Self-sacrifice to the point of certain suicide is penalized by natural selection just as is selfishness to the point of certain infanticide. There is an optimum balance between individual selfishness and parental altruism. Where the balance is struck depends upon the ecological circumstances of the species. Most mammals, for instance, tend to have rather few offspring and to care for them lavishly. Some fish, on the other hand, have prodigious numbers of offspring, most of whom die because they receive little care from the parents. Neither of these two extremes is superior to the other. Each is suitable to a certain way of life.

Parental care is so obviously favoured by Darwinian selection that the word altruism seems almost superfluous. But a young animal has close relatives other than its parents. Some young birds and mammals receive a substantial proportion of their food from elder brothers or sisters. In Florida scrub jays (*Aphelocoma coerulescens*) about half the nests with young are attended by at least one 'helper' in addition to the mated pair themselves. In most cases the helpers are full siblings of the new brood; in some cases they are half siblings or more distantly related. Of the food that is brought to an average nest, helpers contribute about 30 per cent. The Darwinian advantage of this is in principle no different from that of parental altruism. If an elder sibling has a gene for sibling altrusim, there is a good chance that the same gene has been inherited by the younger sibling who benefits. In fact, for the case of full siblings, the odds are the same as the odds that a gene for parental altruism will be inherited by the offspring. So genes for sibling altruism are really caring for themselves, just as genes for parental altruism are. Far from sibling altruism being hard to explain, we might well wonder why it is not as common as parental altruism. The answer to this is likely to be complex. Perhaps most importantly, siblings often do not have the same opportunities for altruism as parents. Another problem is that in most species there is some probability that they are half siblings rather than full siblings.

Florida scrub jays, and most other vertebrate 'helpers', eventually leave the parental nest and try to have young of their own. Their spell as helpers could be regarded as a brief apprenticeship before they go out to reproduce themselves. It has even been suggested that the Darwinian advantage of helping lies solely in the gaining of useful experience in the difficult art of infant-rearing. But this suggestion misses an important point. However much good experience the helper gains from rearing his younger siblings, and however much this aids his own reproduction, he is also benefiting copies of the helper genes in his relatives' bodies

But there is one type of animal in which such 'helping' has been pushed to its logical conclusion. These are the social insects, the termites (Isoptera) on the one hand, and the wasps, bees, and ants (Hymenoptera) on the other. In the social insects altruism is seen on a grand scale. Most of the individuals in a colony are sterile workers who never reproduce themselves but devote their lives to feeding, caring for, and defending their young siblings. The genes for such altruistic behaviour are passed on in the bodies of that minority of their siblings who are destined to become reproductives (i.e. queens and males).

The altruism shown by social insect workers is legendary, especially their suicidal attacks on nest-robbers. It is epitomized by the barbed hooks on the sting of the honey-bee (*Apis mellifera*). Because of these hooks the bee usually cannot withdraw her sting, and it is torn from her body together with vital internal organs with the result that she dies. It is probable that the sting is thereby made more effective, for it goes on pumping venom into the victim after it has been torn from the bee's body. If this is so, the barb can be regarded as an organ of self-sacrificial altruism towards the rest of the colony.

Most animal altruism works to the benefit of close genetic relatives, and this nepotistic behaviour makes good Darwinian sense, as explained above. But there are other theoretical ways in which natural selection could favour altruism. The best known of these is the theory of reciprocal altruism. This may be summed up as the principle of 'You scratch my back, I'll scratch yours', but once again that would be misleading if taken to imply conscious agreement. No form of COGNITIVE process need be invoked, although something at least crudely equivalent in effect to individual recognition must be. It can be shown mathematically that under some circumstances natural selection could favour a system of grudging reciprocated altruism between non-relatives, even between members of different species. Reciprocal altruism is really just a form of SYMBIOSIS, but with a delay between the first altruistic act and its repayment.

Finally, altruistic behaviour can arise simply by mistake. Among regularly occurring behaviour in nature, the feeding of baby cuckoos (Cuculinae) by their foster parents must qualify as altruism in our definition, and it is obviously not favoured by natural selection acting on the hosts. It is a mistake, a perversion of the host's normal parental behaviour, an aberration engineered by the cuckoo (see PARASITISM).

Although the behavioural phenomena called altruistic are very interesting, and so are the evolutionary theories used to explain them, there is much to be said for dropping the word altruism from the technical vocabulary of animal behaviour. Even if it is not actually misunderstood by biologists, the technical biological usage of the word altruism is not positively very helpful. For instance, when a bee sacrifices herself by using her barbed sting, it is by no means clear to whom she is being altruistic. Is it her fellow workers, the queen, or her younger sisters and brothers who are being reared as future reproductives? The truth is that the question is simply not a useful one; it is the kind of question that is best left unanswered. The important point is that genes tending to make workers develop and use barbed stings will spread through the population if individuals live in close family groups sharing a high proportion of the same genes. Whenever one is tempted to use the word altruism, it is usually possible to convey what is interesting about the phenomenon without using the word at all.   R.D.
34

AMBIVALENT BEHAVIOUR is typical of a CONFLICT situation, in which an animal has simultaneous tendencies to perform two incompatible activities. Sometimes the animal makes INTENTION MOVEMENTS towards both types of behaviour. Thus a half-tame moorhen (*Gallinula chloropus*), when offered food, may make incipient pecks towards the food whilst edging away from the outstretched hand.

Ambivalent behaviour is typical of THREAT situations in which there is generally a conflict between attack and escape. For example, during TERRITORY disputes, herring gulls (*Larus argentatus*) adopt a typical threat posture which has components of both attack and escape, as illustrated in Fig. D, MOTIVATION. The head is held high with the bill pointing down as if to peck an opponent. At the same time the feathers are sleeked, indicating a degree of FEAR, and the bird walks sideways with respect to its opponent, rather than directly towards it as an attacking bird would. Some ambivalent postures have evolved into elaborate DISPLAYS, such as the display of the black-headed gull (*Larus ridibundus*) illustrated in MOTIVATION, Fig. B.

ANTHROPOMORPHISM is the tendency to attribute human characteristics to animals. Human visitors to zoos frequently remark on the humanlike behaviour of the animals they see there, and reports of animal behaviour frequently contain implicit anthropomorphic assumptions. For example, one of the first chimpanzees to endure a space flight returned to earth grinning broadly. This was widely interpreted to mean that he had enjoyed his trip, but in fact this FACIAL EXPRESSION is generally a manifestation of fear in chimpanzees. Anthropomorphism plays a large part in sentimental attitudes towards animals, and such attitudes can easily result in misinterpretation of animal behaviour. When the scientist sees an animal behave in a human-like way, he must guard against immediately attributing human MOTIVATION or EMOTION to the animal. However, in the interests of animal WELFARE, the scientist must also be sensitive to the feelings of animals, in so far as he can judge them. Interpretation of animal behaviour in terms of the emotions experienced by the animal requires a very expert knowledge of the particular species concerned.

Although man is quick to recognize animal behaviour as human-like, he is reluctant to admit that his own behaviour is animal-like. Man's deeprooted tendency to see himself at the centre of the universe makes it difficult for him to realize his insignificant place in nature. The unpalatable scientific revolutions have generally been those which appeared to challenge man's self-esteem. In 1543 Copernicus started a train of scientific thought in which the earth was no longer the centre of the universe, but merely one of many planets moving around the sun. DARWIN's theory of evolution by NATURAL SELECTION, published in 1859, crystallized scientific thought into a pattern in which man is seen as an integral part of the animal kingdom. Both these developments aroused widespread opposition amongst contemporary educated people. Even today, many people are reluctant to recognize their psychological affinity with animals. The scientist studying animal behaviour must always bear in mind the fact that he is an animal

himself, and that his interpretation of the behaviour of other animals may be influenced by this.

Anthropomorphism has its counterpart in the behaviour of other animals. Animals living in close association with man often behave as if men were members of their own species (see INTERACTIONS AMONG ANIMALS). Such cases of mistaken identity among animals generally result from the recognition of particular features of the other animal, which are similar to the sign stimuli by which the animal recognizes members of its own species. For example, Fig. A illustrates a cardinal (*Cardinalis cardinalis*) responding to the stimulus of an open mouth by stuffing some insects into it. This behaviour is characteristic of cardinals, and many other small birds, when confronted with the gaping mouth of a nestling. The gape provides a powerful SIGN STIMULUS which is responded to, even though other aspects of the total stimulus situation are inappropriate. It is probable that the anthropomorphic tendencies of humans result from a similar response to morphological and behavioural patterns which provide a foundation for human SOCIAL behaviour. For example, infant head shape is an important factor in invoking parental responses in human adults.

Fig. A. For several weeks this cardinal (*Cardinalis cardinalis*) regularly fed goldfish (*Carassius auratus*) which had learned to surface to obtain titbits.

Such responses can also be aroused by similar characteristics in young animals, shown in Fig. B, and by cartoons in which these features are exaggerated.

The propensity of animals to respond to but a few of the many possible stimuli in a given situation is sometimes exploited by other species. For example, cuckoos (Cuculinae), which lay their eggs in the nests of other species, often produce eggs which are similar in size and colour to those of the host species. This MIMICRY exploits the egg-recognition mechanism of the host, since it is almost certain that the parent birds are capable of distinguishing their eggs from those of cuckoos. That is, they could be trained to make the DISCRIMINATION, but normally they do not make it. So it is with anthropomorphism. Humans are fully capable of distinguishing themselves from other species, but they do not always do so. They often have to be specially trained to resist the temptation to interpret the behaviour of other species in terms of their normal behaviour-recognition mechanisms.

Fig. B. Common characteristics of these juvenile animals include short faces, high foreheads, and a rounded head shape.

ANTI-PREDATOR behaviour is a form of DEFENSIVE behaviour, which includes any activity that affords protection against PREDATION. Anti-predator behaviour may be passive, as in behaviour associated with CAMOUFLAGE and MIMICRY, or it can be active, as in ALARM RESPONSES and ESCAPE behaviour.

APPEASEMENT behaviour serves to inhibit or reduce AGGRESSION between members of the same species, in situations where escape is impossible or disadvantageous. For example, kittiwakes (*Rissa tridactyla*) nest on narrow cliff ledges, where the avoidance manœuvres characteristic of most other gulls are not possible. The female cannot escape the male's aggressiveness without leaving the TERRITORY. Usually the female turns away her head

and hides her beak in the feathers of her breast, a form of appeasement behaviour.

Appeasement behaviour sometimes takes the form of hiding weapons and other symbols of aggression, as in the above example. In other cases it seems to function by arousing a CONFLICT in the attacker. In monkeys, such as mangabeys (*Cercocebus*), for example, a subordinate individual of either sex may make a sexual invitation to a more dominant animal. The MOTIVATION (sexual) thus induced by the behaviour of the subordinate animal conflicts with the aggressive tendency of the dominant animal. Often the sexual invitation is largely symbolic. The subordinate animal may turn its back and make a *presentation*, and the attacker may briefly mount it, but often accepts the gesture as a token of submission (see DOMINANCE).

In some cases appeasement behaviour takes the form of a stereotyped DISPLAY. The very aggressive cichlid fish (*Tropheus moorii*) appease by presenting a yellow band to the attacker, as illustrated in Fig. A. Appeasement displays often evolve by RITUALIZATION of JUVENILE behaviour. Food begging behaviour in pigeons and doves (Columbidae), for instance, is characterized by a rapid wing vibration combined with a typical begging vocalization. Wing vibrations with a rhythmicity typical of juvenile begging occur during AGONISTIC encounters during COURTSHIP, particularly when the submissive partner is on the nest and is reluctant to leave it. In many birds, such as gulls (Laridae), food begging by the female seems to appease the male's aggressiveness and sometimes results in COURTSHIP FEEDING.

Appeasement gestures often play an important role in greeting ceremonies. Storks (Ciconiidae) greet their mates by placing their heads over their backs and clapping their bills. In this way they turn away their weapon, and behave in exactly the opposite way to their normal THREAT display. Chimpanzees (*Pan troglodytes*) may greet each other by embracing, which is similar to juvenile clasping. Sexual presentation also occurs as a form of greeting. Among humans, greeting behaviour, although very varied, usually contains elements of appeasement. Greeting with the raised open right hand is widespread, even among peoples with no previous

Fig. A. The quiver display of *Tropheus moorii*. The fish on the left is appeasing that on the right.

contact with Europeans. This gesture indicates that no weapons are being carried. When weapons are used in greeting they are held in a non-threatening fashion. Greeting may include elements of submission, as in laying down weapons, baring the head, etc. Human greeting often involves body contact, such as shaking hands, embracing, or kissing. Smiling is basic to a friendly greeting amongst all peoples. These are all aspects of intimate, and sometimes juvenile, behaviour, which are used in a ritualized manner in potentially aggressive situations.

APPETITE is such a common experience in our daily lives that one might expect its relationship to feeding behaviour to be obvious. Yet when examined closely, appetite, like so many familiar concepts, proves to be a complex phenomenon whose origins and role in the control of animal behaviour have long been the subject of serious study and debate.

Attributes of appetite. Today, as in the past, great and general semantic confusion between appetite and HUNGER hinders progress towards understanding how food intake is controlled. The brief characterizations offered here are unlikely to resolve the problem to everyone's satisfaction, but will identify some of the important issues, and serve as a basis for the discussion of appetite which follows.

It is useful to recognize that both concepts have evolved largely on the basis of human experiences, which have, in various ways, also influenced the methods and theories employed by scientists to study and explain the feeding behaviour of other species. With reference to human experience hunger can be described as a generally unpleasant, even painful sensation, which may become so intense that the urgent search for food dominates one's thoughts and actions. Hunger is the normal consequence of prolonged deprivation, and, if food is not found, weakness, emaciation and death inevitably ensue. Thus hunger denotes both a craving for nourishment and the corresponding physiological need.

While appetite is usually the harbinger and congenial companion of hunger, and the boundary between hunger and appetite is blurred, significant distinctions can be made and are important to an understanding of feeding behaviour. Appetite may be described as the comparatively pleasant, though at times compelling, anticipation or relish of certain foods. Appetite is characteristically selective, some foods being eagerly sought while others remain ignored. Selection is guided by the stimulus properties of foods, their flavour, aroma, texture, appearance, qualities which determine their palatability. Specific dietary needs may lead to a greatly enhanced appetite for normally unpalatable foods, and conversely, foods which are normally highly palatable may, under some circumstances, evoke nausea and avoidance.

Appetite is modulated over the course of a day by periodic variations in nutritional need, but may persist and even intensify in the absence of such needs. (All who have savoured a rich dessert at the conclusion of a bountiful meal can attest to this.) Moreover, in some cases the ingestion of food may result in an increased appetite, a phenomenon which led to the custom of appetizers at the beginning of a meal. Just the sight or smell of food may cause one to drool with anticipation. Thus, besides serving as a basis for discrimination among alternative sources of nutrition, the stimulus properties of food give rise to physiological changes and to pleasant experiences which serve as a powerful incentive for eating.

In summary, hunger may be characterized as a set of physiological changes and COGNITIVE events which are the inevitable consequence of prolonged fasting. In the discussion which follows, appetite will refer to those physiological changes and cognitive events that are evoked by the stimulus properties of foods. Implicit is the notion that appetite is labile, the response to a given presentation of a food being determined by a large number of factors, including the present nutritional state and past experience of the individual.

The scientific study of appetite. Although reference to personal experience and verbal reports is helpful in identifying the attributes of appetite, scientists have no access to such information from other species. Nor may they assume that these species are guided in their eating by subjective experiences equivalent to ours (see ANTHROPOMORPHISM). In examining the role of appetite in the lives of animals, scientists must rely entirely upon observations of behaviour and its relation to peripheral stimulation and physiological events. Ideally the animal would be observed selecting and ingesting food in its natural environment, and evidence with respect to appetite would then be distilled from the entire repertoire of natural behaviour and interpreted in that context. Though scientists in the fields of agriculture and ETHOLOGY have accumulated some evidence of this type, FIELD observation of FEEDING behaviour has not been pursued nearly so widely as, for example, the observation of COURTSHIP and AGGRESSION.

The bulk of available information about feeding behaviour in general, and appetite in particular, has been provided by nutritionists, physiologists, and psychologists studying animal behaviour in laboratory experiments. As compared to observations in the field, laboratory experiments enjoy a number of advantages: (i) they are much more convenient, animals in sufficient numbers being readily available at a time and place determined by the scientist; (ii) they permit a wider range of manipulation and more precise control of both the animal and the environment; (iii) many of the physiological techniques and methods of stimulus presentation employed in experimental studies would not be feasible outside the laboratory; and (iv) the carefully controlled and specified circumstances of the experiment facilitate replication and extension of the work by other scientists (see LABORATORY STUDIES).

However, laboratory experiments sometimes suffer from serious shortcomings which must be borne in mind when interpreting their results. Control of the environment is usually achieved through elimination of most of the diverse, ever-changing environmental stimuli which normally guide behaviour. The stimuli which are provided, as well as the tasks presented to the animals, are not always representative of those which are important in the natural habitat. Finally, it should be noted that considerations such as ease of maintenance, economy, and custom have limited the range of animals studied in the laboratory to relatively few species. In fact, most of the laboratory research relevant to appetite, and most of that reported here, has been done on the domesticated rat (*Rattus norvegicus*), a splendid experimental animal but not necessarily a spokesman on all matters for other species, nor even for its wild brethren.

These cautionary notes, applicable to all research in animal behaviour, are offered here as a reminder that, in attempting to understand the role of appetite in animal behaviour, we must evaluate all evidence, that from the laboratory as well as that from the field, with respect to its probable significance to the success of the animal in its natural environment. While a great deal remains to be learned, resourceful observation and ingenious experimentation have provided fascinating clues to the mystery of appetite.

Functions of appetite. Like humans, all animals, except for a few filter feeders, eat their daily ration in a series of discrete meals. At each meal two sets of decisions must be made, one with respect to the selection of food to be eaten (i.e. which food?) and one with respect to the quantity of food ingested (i.e. how much food?). Over the long run the answers to these questions determine the health and ultimately the survival of the animal. Choices must ensure a diet which meets a number of specific nutritional requirements and provides energy appropriate to sustain optimal growth and body composition. While humans, with the aid of intellectual toil and abundant counsel from nutritionists and physicians, achieve only modest success in this endeavour, animals living in their natural habitat appear to regulate dietary selection and energy balance with remarkable skill, guided only by appetite. A large body of scientific evidence confirms this observation and clarifies the status of appetite in animal behaviour.

Appetite as a guide to food selection. Animals typically are very selective in their feeding. Within a given habitat it is relatively easy to observe marked differences between species in the choice of foods. While few animals eat only a single food (are monophagous) most take advantage of relatively few of the sources of energy available in their environment, showing preferences for certain plants or certain of their cohabitants (prey). The necessity for a distribution of sources of nutrition among species sharing a common HABITAT is obvious and has played a fundamental role in the evolution of adaptive specializations, in behaviour as well as morphology, which have been important determinants of the relative success of these species. These ADAPTATIONS impose varying degrees of constraint upon the range of food available to a species, but even animals with a rather restricted set of options are selective in their feeding.

Consider for example the domestic cow (*Bos primigenius taurus*), a herbivore, which normally meets its nutritional needs entirely through grazing. Casual observation might suggest that this grazing is indiscriminate, but upon closer examination this is seen not to be the case. Early in the feeding cycle the cow may ingest clumps of grass, but later will select only the blades, carefully avoiding the stems. The particular patch of grass grazed may vary depending on the season of the year, the weather, and the stimulus properties of the herbage. Chemical analysis of hay and grasses selected by grazing cattle has confirmed that those chosen are the best sources of nutrition among the available alternatives. An unusual manifestation of appetite in the cow is the eating of bones by cattle feeding off phosphorus-poor grazing land, behaviour which reportedly ceased with the addition of phosphorus to their ration. When brought in from the pasture and required to select their diet from a number of alternative foods, such as grains and roughage, cows have chosen diets which provided normal or better than normal nutrition. Considering the low nutritional value and caloric density of the foods usually available to herbivores, the adaptive value of careful dietary selection is apparent.

While the diet and feeding behaviour of predators is very different from that of herbivores, ethological studies show that PREDATION too is selective. Though the relative availability of prey in the habitat is of great importance in determining their contribution to the diet of a predator, it is typical that this contribution is not directly proportional to availability. Studies of selected birds, mammals, and fishes show that as the density of a prey (number of individuals per unit of area) increases, representation of this prey in the diet at first increases rapidly, then rises at a diminished rate, and finally reaches a stable plateau. While the level of this plateau will vary depending upon the predator, the prey, and the alternative foods available, the important point here is that feeding is not limited to a single prey species. Thus, a varied and presumably superior diet is assured (see PREDATORY BEHAVIOUR).

Among all animals the challenge and importance of dietary selection would seem greatest to the omnivores, which must choose a nutritionally adequate diet from among a vast array of animal and vegetable substances which vary widely in composition. In this sense it is appropriate that by far the greatest amount of information concerning FOOD SELECTION has been gathered from the study of one of these animals, the domesticated rat. For the sake of clarity, several aspects of dietary selection by the rat will be considered separately.

Self-selection of a complete diet. In 'cafeteria' studies, rats have been required to select their diet from several containers, each filled with a different purified food. There have been many such experiments and among the pioneering work with this technique were several in which rats chose diets of excellent nutritional quality. In one, for example, rats were offered casein (a source of protein derived from milk), dextrose (a sugar), olive oil, yeast, a mixture of salts, and water. Over the course of forty-five days these animals showed normal growth and good health. They ate fewer calories each day, but gained weight more rapidly than a comparable group of rats which were fed a standard laboratory diet over the same period. Such findings argue for the reliability of appetite as a guide for food selection.

However, in some studies of this type the performance of the rats has been not nearly so impressive. Large individual differences have been noted, and frequently rats have failed to eat sufficient quantities of purified protein, which, though essential, is usually less palatable than the sources of fat and carbohydrate. Those studies where balanced diets have been selected have typically provided two or more sources of protein, a more palatable source of protein, less palatable alternative foods, or some combination of these features.

One conclusion from these studies seems to be that when presented with a cafeteria of purified foods, rats eat what they like (the more palatable foods), and that what they like is not always what they need. In other words, appetite is not an infallible guide to good nutrition. It may be noted, parenthetically, that the devastating effect of refined sugar on the quality of the diet eaten by most children and by many adults leads one to the same conclusion.

On the other hand, in those experiments which provided favourable circumstances, appetite guided rats to excellent diets. This evidence, and the impressive success of the species outside of the laboratory, suggests that in the choice of foods usually available in the natural habitat there must be a general correspondence between the dietary likes and the dietary needs of rats. Cafeteria studies offering a choice of natural foods have demonstrated successful dietary self-selection by several other species, including cows, pigs (*Sus scrofa domestica*), chickens (*Gallus g. domesticus*), and human infants.

Selection of diets which meet nutritional deficiencies. A second experimental approach to the study of food selection by rats has been to create a nutritional deficiency, usually through the feeding of a deficient diet, then to offer the deficient animal a choice between diets which do or do not provide the needed nutrient. If animals show a preference for the enriched diet, or eat enough of it to recover from the deficiency, they are said to have demonstrated an appetite or *specific hunger*.

In a classic experiment of this type, rats deficient in B vitamins failed to discriminate between two diets which differed only in the addition to one of a minute amount of highly concentrated vitamin B. However they quickly demonstrated a preference for the enriched food over several alternatives when the diets offered were distinctively flavoured with basil, lard, cocoa, and 'Bovril', the last-mentioned diet containing the needed vitamin. These results demonstrate both the importance of taste in food selection and the efficacy of appetite as a guide to nutrition under favourable conditions.

But this same series of experiments revealed another striking, and less reassuring phenomenon. Once rats had developed an appetite for the Bovril diet they persisted in their preference for that diet even when the vitamin was switched to the diet flavoured with cocoa. An appetite for the diet now containing the needed vitamin developed only after a few days of 'education', during which rats were offered only this beneficial diet. Subsequent studies have confirmed both the development of appetites (specific hungers) for beneficial foods and the persistence of preferences for those foods after the beneficial ingredient has been removed. Other work has provided evidence of appropriate specific hungers, not only in response to vitamin deficiencies but also to deficiencies in other dietary components, including sodium, calcium, and magnesium, as well as to imbalances in amino-acids. Thus appetite is again seen to be a valuable, though not infallible, guide to nutritionally sound diet selection by animals.

Avoidance of poisons. The wild rat is generally abhorred as vermin because of the threat it poses to the health, pocket, and peace of mind of its human neighbours. Its continuing success is to a significant degree dependent upon its legendary skill in the avoidance of poisons. Laboratory studies suggest the basis of successful poison avoidance by rats to be their usual reluctance to eat novel foods, which following poisoning is greatly intensified and coupled with avoidance of the particular food which was harmful. These tendencies, evident in domestic rats and more pronounced in wild strains, are illustrated in the following experiment. Rats were raised on one diet, poisoned in the presence of a second diet, then offered a choice between these two diets and an entirely new one. Both domestic and half-wild rats demonstrated a strong preference for the safe food they had eaten in the past. Although they ate some of the new safe food as well, they almost completely avoided the diet associated with poisoning. An additional finding, of equal importance to an understanding of dietary selection, was that the pairing of nutritional deficiency, rather than poisoning, with the second diet yielded identical results. Thus, in selecting its foods the rat appears to 'play safe', avoiding harmful foods and turning to untried foods only when no alternative is available which has been eaten safely in the past.

Bases of food selection. The popular belief that animals are capable of choosing foods which are good for them, and avoiding those which are harmful is, in general, confirmed by the results of laboratory research. Conventional wisdom, sustained by concepts such as 'the wisdom of the body' and specific hungers, usually goes on to suggest that this success in food selection is attributable to inherited desires for those foods which satisfy nutritional needs. In some cases, for example the preference for salty foods displayed by animals in need of sodium (sodium appetite), this explanation appears to be valid. But in general, research findings concerning specific hungers and poison avoidance have been difficult to reconcile with the thesis of inherited food preferences. Much of the evidence, including results of the studies cited above, points instead to dietary selection based upon learned aversions to foods which are harmful or fail to meet nutritional needs. This interpretation is not only more consistent with experimental findings, but also more parsimonious than an explanation which envisages animals pre-wired at birth with preferences or aversions for any food they might encounter during their lives. If this view is accepted, the relatively long delay between the ingestion of food and significant post-absorptive effects attributable to it requires that a re-examination of the role of contiguity in the process of learning accompany the reassessment of appetite.

Appetite and food intake. Besides selecting types of food which meet specific nutritional requirements, animals must eat amounts of food commensurate with their energy needs. Too little food will lead to weakness, loss of weight, and stunting of growth, too much will lead to obesity, and either error in food intake will diminish the probability of survival. In general animals meet this challenge with admirable success, and the question is how they manage to do so. In the long history of philosophical speculation and scientific theory regarding the regulation of food intake the roles assigned to appetite have varied considerably.

Ancient speculation placed eating under the control of sensations of hunger which resulted from emptiness of the stomach and gut. Plato mused that man's enjoyment of music and philosophy was made possible by the winding and coiling of his gut, which by slowing the passage of nutrients through the body staved off a constant craving for food that would preclude his intellectual and artistic pursuits. Though not forced into gluttony by the demands of hunger, ancient man was apparently seduced into that fate by the pleasures of the palate. Roman gourmands, having eaten to repletion, used emetics to induce vomiting, which permitted a return of appetite and prolonged the sensual delights of feasting. Thus, early in our history the sensations associated with hunger and appetite were seen to play essential roles in the control of eating.

As the science of physiology emerged in the 18th century, sensory control of feeding remained the prevalent view. In his pioneering textbook of physiology, *Primae lineae physiologie in usum praelectionum academicarum* (1747), Albrecht von Haller noted that 'We are induced to take food, both from the sense of pain which we call hunger, and from that pleasure imparted by the sense of taste.' The sensation of hunger, which Haller attributed to the rubbing together of folds of the empty stomach, became the subject of intense interest and controversy during the following century. A number of competing peripheral theories were proposed, attributing the onset of feeding variously to local sensations of hunger resulting from the activity or lack of activity of the stomach, or to a general hunger based upon cellular needs and sensed throughout the body.

Then in 1912 the matter appeared to be settled. In an experiment to be cited in textbooks for decades to come scientists demonstrated that hunger pangs occur in synchrony with contractions of the empty stomach and that both phenomena are abolished when food is taken into the stomach. In defending this explanation of hunger, these men acknowledged evidence that relatively normal eating persists in animals whose gastrointestinal tract is completely severed from the nervous system, but argued that this may be attributed to appetite rather than hunger: 'Indeed, who eats dessert because he is hungry? Evidently, since hunger is not required for eating, the fact that an animal eats is no testimony whatever that the animal is hungry.' Thus, the presentation of convincing evidence regarding the peripheral origin of hunger was accompanied by reiteration of the importance of appetite in the control of feeding. But already interest in the sensory control of food intake had begun to diminish.

Around the turn of the century the course of scientific thought concerning the control of food intake was altered profoundly by the introduction of Claude Bernard's concept of the constancy of the internal environment (see HOMEOSTASIS). Awareness that the composition and temperature of blood and other fluids bathing body tissue varied only within narrow limits was accompanied by recognition that feeding and DRINKING must be carefully regulated to sustain this stability. The accuracy and significance of this view was forcefully demonstrated by other French scientists, whose studies of animal physiology and behaviour showed that drinking is correlated with changes in the concentration of dissolved substances in the blood, and that eating is precisely adjusted to balance daily caloric intake with energy expenditure and, over the long run, to regulate body weight.

As a consequence the focus of enquiry regarding the control of food intake shifted from peripheral mechanisms which might explain the sensation of hunger to BRAIN mechanisms capable of the precise sensing of cellular needs and nutritional reserves which are reflected in the regulation of energy balance. In succeeding years an enormous amount of research, facilitated by technical advances in neuroanatomy and neurophysiology, has demonstrated the involvement of structures distributed throughout the brain, and the especially important roles played by certain small regions located at its base, in the hypothalamus.

Though the importance of central mechanisms in the regulation of food intake is now well established, any complete explanation of appetite and the control of feeding behaviour must continue to recognize the important contributions of TASTE AND SMELL. Effects of the stimulus properties of foods are not limited to the response of the sensory processes, but contribute significantly to the processes of consumption, digestion, and utilization of foods. Experiments by the great Russian physiologist PAVLOV demonstrated that the sight of food is sufficient to produce salivation and gastric secretion in dogs, and that food passing through the mouth causes much greater gastric secretion and more rapid digestion than food placed directly into the stomach. Pavlov concluded that these effects represent '. . . the digestive value of the passage of food through the mouth, the value of a strong desire for food, the value of appetite.' More recent work indicates that the presence of food in the mouth is also sufficient to release HORMONES involved in the storage and utilization of energy derived through digestion and absorption. The diminution of these and other reflexes to sensory stimulation by food following a meal demonstrates again the interaction of sensory and post-ingestional effects of food and reflects the decline in appetite associated with repletion. Collectively these findings can explain why human beings fed, for medical reasons, through an opening in the abdomen have typically chosen to chew and swallow at least a portion of their food, even though it has to be regurgitated and never reaches the stomach.

Conclusions. Appetite is fundamental to the survival of an animal because of the essential role it plays in the selection of a diet which is nutritionally adequate, and in the normal regulation of energy balance. The bases and functions of appetite are not fully understood, in part because their study falls within the province of no single scientific discipline. In writing of this problem one distinguished scientist observed that psychologists have tended to act as if an animal has no 'body, and physiologists as if it has no head. An ethologist might add that both have been inclined to act as though an animal evolved through successive adaptations geared to survival in their apparatus. A growing recognition of the complexity and importance of appetite together with an increasing awareness of mutual interdependence among the scientific disciplines promises eventual resolution of the many complex and continuing issues suggested by this brief review of the topic.   D.W.T.
106

APPETITIVE BEHAVIOUR is the term that traditionally has been applied to the active, GOAL-seeking, and EXPLORATORY phase of behaviour that precedes the more stereotyped CONSUMMATORY behaviour that an animal exhibits when it reaches its goal. Upon reaching the goal, appetitive behaviour normally ceases. For example, a hungry rat (*Rattus*) shows an increased level of general ACTIVITY, and will search about for possible sources of food in a restless and exploratory manner. When the rat finds food it eats it (the consummatory behaviour), and if there is sufficient food for the rat to achieve satiation there will be no further appetitive behaviour until the rat is hungry again. During the appetitive phase of behaviour LEARNING takes place, and the rat will remember which avenues of exploration led to food and which did not. On future occasions it will repeat those aspects of its previous appetitive behaviour which seemed to be profitable. In LABORATORY STUDIES rats can be induced to learn a particular route through a maze, or to perform a particular response, such as pressing a bar, to obtain food rewards. Such OPERANT behaviour is generally considered to be a form of appetitive behaviour.

The concept of appetitive behaviour is undoubtedly a useful one, but it is not very rigorous. There are philosophical difficulties in separating appetitive goal-seeking behaviour from other forms of behaviour. Consider, for example, the nest-building behaviour of a blackbird (*Turdus merula*). Nest-building begins with the search for large twigs to form a foundation. Smaller twigs are then collected to form the sides. The nest cup is made from mud, and is lined with fine grass and hair. While we may, or may not, wish to regard the complete nest as the blackbird's ultimate goal, there are difficulties in distinguishing between the appetitive and consummatory aspects of the nest-building behaviour. One possibility is to regard the search for each twig as an appetitive episode, and its placement in the nest as the consummatory response. Alternatively, we may wish to regard the search for each twig as appetitive, and finding a suitable twig as consummatory, on the grounds that finding a twig brings searching to a temporary halt. Carrying the twig to the nest would then have to be counted as a separate type of activity.

Appetitive behaviour is not always characterized by active exploratory behaviour. For example, animals that hunt by ambush, such as the praying mantis *Mantis religiosa*, rely on CAMOUFLAGE and stillness to obtain their prey. Only when a potential prey comes within range do they launch an attack, striking with a rapid movement of the fore-legs. A predator that hunts by stealth, such as the chameleon (Chamaeleontidae), can be regarded as an intermediate between an ambush hunter and an active exploratory hunter. HUNTING, therefore, is a category of behaviour that does not comply with the traditional definition of appetitive behaviour. Another type of problem associated with appetitive behaviour concerns those aspects of behaviour to which the concept is of dubious relevance. Clearly, we do not regard appetitive behaviour as relevant to REFLEX activities, such as startle and ALARM RESPONSES. Nor would we normally think of animals as having an appetite for AVOIDANCE of noxious stimuli. Animals show avoidance behaviour when the appropriate situation arises, rather than seeking to confront such situations.

In the study of AGGRESSION it is unclear whether the concept of appetitive behaviour applies or not. Do animals go about actively looking for fights? Scientists are divided on this question. It is true that animals will learn an experimental task where the reward is an opportunity to respond aggressively to a rival. For example, Siamese fighting fish (*Betta splendens*) will learn to swim through a ring to gain an opportunity to attack a rival for a few seconds, or even to display aggressively towards a mirror. Similarly, game cocks (*Gallus g. domesticus*) will learn a response which gives them an opportunity to perform their aggressive display. However, it is known from CONDITIONING studies that inanimate objects can easily become associated with food, sexual partners, or rivals: a pigeon (*Columba livia*) repeatedly shown a red light just before being given access to a mate will start to court the red light, even if the mate is not presented. Similarly, animals can be conditioned to respond to such inanimate objects as if they were food. Thus it is possible that the Siamese fighting fish comes to associate the ring with attack behaviour, rather than simply using the ring as a vehicle of its appetitive MOTIVATION.

100

AROUSAL. It is a matter of common observation that drowsy, tired, ill, or depressed animals are generally unresponsive to stimulation. Such periods of general depression of behaviour are clearly different from effects specific to a particular type of behaviour. For example in a number of male vertebrates copulation remains less likely for some time after a bout of copulatory behaviour, particularly if the same female is used in subsequent tests; other behaviour is depressed less or not at all. A proper understanding of general depressive or facilitatory effects is of considerable importance to the study of the causation of behaviour. By far the most usual approach has been to assume that waking behaviour can be ordered along a continuum of progressively increasing arousal that begins in deep sleep, and presumably culminates in

states of great excitement or even frenzy. The main difficulty in testing, or even in using an approach of this sort, is that there is little agreement on what exactly is meant by arousal.

Five of the most important meanings of arousal will be considered here. These are: (i) the original usage in physiology; (ii) arousal as level of responsiveness or (iii) of activation; (iv) the degree of development of reflex preparation for exertion or immobility, and finally (v) the amount of sensory input which the central nervous system is prepared to accept at a particular time.

The physiological usage was, originally at least, a clear but specialized one. It depends on the fact that it is possible, by recording electrical activity from the cortex of the brain, to distinguish states of SLEEP from states of wakefulness. The transition from sleep, so defined, to waking was described as arousal. Behavioural signs of waking up, such as opening the eyes and bringing the ears to bear on conspicuous stimuli also occur at the transition, and so confirm that it represents true awakening. Problems arise only when physiological arousal ceases to be defined as a transition, and becomes instead a property of the nervous system which increases progressively as the animal becomes more and more awake. Exactly what is meant by becoming more awake, and in particular whether changes of this sort can really be ordered along a single continuum ending in frenzy or fit can be regarded as a behavioural problem, but the question could be framed so as to make physiological investigation proper and profitable by asking what changes follow further activation of the system within the BRAIN which is responsible for arousal (defined as the transition at awakening). However, enquiries of this type are not our concern here.

An important characteristic of behaviour which has sometimes been taken as a measure of arousal is degree of responsiveness. Anyone who has observed domestic dogs (*Canis l. familiaris*), for example, will be familiar with states in which, although awake, they are clearly unwilling to do anything, and conversely states in which they are ready for anything. Strictly speaking an increase in responsiveness can only be deduced to have occurred if an animal can be shown to have become more likely to respond to a range of different stimuli (rather than becoming more likely to respond to only one type of stimulus, such as food). Very few studies have employed this criterion, but there is some evidence that responsiveness may vary systematically during the waking day in the rat (*Rattus norvegicus*). Responsiveness is to some extent correlated in this instance with overt signs of EXPLORATORY behaviour, such as rearing or sniffing, and it may in the future be possible to make useful and precise predictions on this basis.

Another possible way of thinking about arousal is that its level may determine not whether an animal responds, but what type of behaviour it performs. Thus grooming might occur at low levels, feeding at intermediate, and fighting or escape at high levels. This suggestion is effectively a hypothesis about the way in which behaviour is organized in time. Not only does it provide a possible explanation for the way in which different behaviours follow each other during an animal's waking hours, but it also allows specific predictions about what should result if an animal is suddenly startled or aroused in some other way. The term *level of activation* is sometimes reserved for this meaning of arousal.

A great deal of effort has been put into attempts to provide measurements of level of arousal which are reliable and independent of the particular type of behaviour which is being performed. The most popular by far are indices based on various REFLEX adjustments to increased metabolic activity due to exertion, which also occur in anticipation of a vigorous response. Adjustments of heart rate and other physiological variables in response to a stimulus which is likely to call for exertion are familiar from lie-detector tests. Acceleration of heartbeat and increases in the volume of air respired both serve to increase the amount of oxygen reaching the muscles, thus preventing subsequent shortage when effort begins. The beads of sweat on a suspect's forehead can be similarly interpreted as a preparation to dissipate the heat, which will be generated if he begins to fight or flee.

Variables such as heart rate are often held to measure level of arousal. A loose correlation might be expected: startling stimuli are likely both to awake and arouse, and to evoke a *startle response*, which includes preparation for exertion. However, a very large number of studies have shown that there is little justification for the use of this or related measures. The difficulties have been threefold. Firstly, as one would expect, the same changes also appear during exertion, when they serve to correct deficits of oxygen and fuel supply to the muscles caused by the exertion, rather than being commanded in anticipation of it. A proper distinction has rarely been attempted, and has proved difficult in practice. It is generally agreed that it is inappropriate to describe an animal or man, exhausted by sustained effort, unresponsive, and with half-closed eyes, as highly aroused, despite a very rapid heart rate. However, it is impracticable during most behavioural tests to estimate to what extent heart rate is above that justified by current levels of exertion. A second problem is that anticipatory adjustments cannot be sustained if exertion does not develop. In the case of heart rate, it will slow down again. Heart rate may also slow down, at least in dogs and men, during the examination of an interesting novel stimulus. This may represent an anticipation of lower metabolic needs in a period when relative immobility is probable. Thus there are two quite different consequences of the presentation of a novel stimulus, both in the behaviour shown and in the metabolic adjustments: heart rate and similar variables therefore seem to

be of little value as a means of estimating arousal.

Finally, some relationship between arousal and sensory input (see SENSES OF ANIMALS) is widely assumed (and sometimes made the basis of a formal definition); it is, however, exceptionally difficult to set out any such relationship in a generally acceptable form. At its simplest it reduces to the proposition that an aroused animal is more likely to notice stimuli (even if it does not respond to them). This is to some extent evident from overt behaviour: only if the eyes are open, and the animal is prepared to look at objects, or turn its ears to pick up sounds, will it be likely to perceive stimuli. Noticing implies more than perceiving, however; one possible criterion, which obviates the need for immediate response as proof of noticing, is that the animal should subsequently show signs of remembering the presentation of the stimulus (see LEARNING). A number of scientists have postulated that the information relating to a particular stimulus is more likely to penetrate deep into the central nervous system in a highly aroused animal. In functional terms one might imagine such information first reaching mechanisms which could analyse its form, then being recognized on the basis of stored information about previous inputs of similar form, and finally being given significance in terms of past experience and the present state of the animal. Most experiments in this area have been neurophysiological. Some have provided convincing evidence of central competition in waking animals between senses such as HEARING and VISION; only during sleep itself have clear changes been found in the ease with which sensory information proceeds to higher centres.

The concentration of interest in the possibility of central changes as information enters the brain has proved an obstacle to the study of what is probably a more important, certainly a more tractable problem: namely what causes changes in the strategies by means of which animals gather information. The difficulty of ascribing such changes to a variable such as arousal can be illustrated by the following question: is an animal which turns its eyes and head to look at a variety of stimuli in rapid succession more (or less) aroused than an animal which sustains its gaze on a single stimulus, in readiness to respond to it?

The reader may have concluded by this point that studies of general changes in behaviour of the sorts which have just been discussed are best left to the pedantic investigator of esoteric effects. It may help to correct this impression if we consider some particular and striking examples.

A startling or mildly painful stimulus obviously tends to awaken animals and men. Consequently such stimuli are often used as a way to raise arousal in experiments which attempt to study the effects of arousal on some behavioural process. Very striking results have been obtained using a tail pinch or an electric shock in the rat. It seems clear that large and reliable increases in responsiveness occur immediately after the pinch. Thus, if food is present, the rat tends to eat, if water is present it drinks, rat pups induce maternal behaviour, and females evoke sexual behaviour in males. In the absence of the pinch, response to any or all of these stimuli is less likely. The effects are powerful: thus three pinches a day in the presence of food can cause a rat to overeat until it is grossly obese.

Furthermore, a system has been identified in the brain which appears to mediate these changes in responsiveness. Any interference with this system blocks the effect of tail pinches, whilst its total destruction so greatly depresses responsiveness that no behaviour at all can be evoked, even by the most powerful stimuli. The behavioural and the physiological evidence thus agree in identifying level of responsiveness as an important and well-defined variable in the causation of behaviour.

In a number of instances it is possible to make some suggestion as to the function of states of low or high responsiveness. Thus an animal which is ill or injured moves and responds as little as possible, but without necessarily going to sleep or ceasing to notice stimuli. This is adaptive in that it prevents aggravation of the illness or injury by movement and exertion, and conserves energy at a time when it is likely to be difficult to find food. We are so accustomed to this aspect of illness that it may be worth emphasizing that lack of responsiveness at such times does not usually represent a rational decision by the sufferer, but is a state co-ordinated and actively sustained by particular brain mechanisms.

A second example is provided by animals that have been persecuted by a superior. Their *submissive behaviour* often involves immobility coupled with unresponsiveness, even to the approach of the superior (see DOMINANCE). This is adaptive, in that flight is often the surest way to ensure pursuit and attack. Freezing as a response to predators or other sources of danger also involves sustained and impressive unresponsiveness, as anyone who has walked up to a crouching pheasant or hare can testify (see ESCAPE BEHAVIOUR). In a number of species, somewhat similar but more intense states of unresponsiveness can be produced by holding the animal still for a short time. Such animal HYPNOSIS only develops if the animal is at the same time frightened, suggesting some similarity in causation with freezing; it differs from freezing in that the eyes are usually closed and muscular tonus is low.

Another promising approach to some of the problems associated with arousal is to ask how far mechanisms concerned with the initiation and maintenance of sleep also play a part in wakeful behaviour. The occurrence of drowsy periods when the animal, although awake, is relatively unresponsive, is one obvious sign that this is a real possibility.

Even in a constant environment sleep and activity commonly alternate in circadian (i.e. approximately 24 h) and shorter RHYTHMS. These rhythms are known to depend, at least in part, upon physiological changes in the brain; some of the behavioural consequences of these changes can be viewed as changes in one or other aspect of arousal. Firstly, certain HORMONES (glucocorticoids) from the adrenal glands are present only at very low levels during sleep, when, as a result, body fuel reserves tend to be conserved, and metabolic processes come to favour protein synthesis. At waking a great increase in these hormones helps to make stored food reserves available, and so allows the first vigorous exertion of the day. Between this waking peak and the trough of sleep, arousal levels tend to drift downwards through the day. In man, the thresholds at which stimuli can be detected, and at which they can be recognized, change systematically over the same period of time; the changes are almost exactly those which would be expected to be caused by the changes which occur in the amounts of the adrenal hormones in the blood. Extreme variation in these amounts is associated with dramatic disturbances: thus a patient may find it impossible to ignore everyday distractions such as distant sounds or movements, and become, as a result, distressingly agitated.

A second example is provided by changes in a neural system in the brain which is known to play an essential part in sleep. These changes involve the levels of a substance (serotonin) which is necessary for the functioning of the system: when the levels are low, insomnia is likely. Interestingly, these levels depend not only upon time of day, but to some extent upon diet. If the amino-acid from which the crucial substance is made in the body is in short supply, then levels will fall. Both in this case, and in the case of the normal changes through the sleep–wakefulness cycle, a variety of other consequences flow from a drop in level. These include more vigorous startle reactions to painful or startling stimuli, and increased locomotion (perhaps representing increased exploration) when novel stimuli are present. In hamsters (Cricetini) at least, PLAY also is more frequent at such times.

However, perhaps the clearest evidence for the involvement of sleep mechanisms in the organization of wakeful behaviour comes from studies of the periods following copulation and feeding. There is direct evidence in the rat that drowsiness or sleep develops at such times. It is likely from more casual observation that the same is true of many mammals, including man. In the case of copulation in the male rat it has proved possible to delete the drowsy period by specific brain lesions, and to show that the organization and character of copulation are unaffected. The main change is that copulation bouts are now more frequent than in a normal animal. It may be that the drowsy period functions as a means of spacing out copulation and thereby prolonging association between male and female until the brief period of oestrus is more or less over, and the danger that a second male may also copulate and father some of the offspring is avoided. In the case of drowsiness after a satiating meal, the FUNCTION is more obvious, since it is difficult to combine digestion and exertion for a variety of physiological reasons, including the pattern of distribution of blood supply. It is possible that sleep periods function as a means of preventing active behaviour at specific times during hours when the animal would normally be awake and active (see ATTENTION).

States of fatigue are also profitable fields for investigation of effects of this sort. In man, it is well known, and also established by formal experiment, that brief periods of inattention, when even the stimuli on which attention is focused are likely to be missed, occur repeatedly in subjects who are tired or bored. Very little is known of such states in animals, but they are potentially important: if stimuli are not noticed, then even ones whose significance is such that they would ordinarily evoke a response will have no effect.

In conclusion, it must be stressed that the problems which have been lumped together under the heading of arousal are of great importance to studies of behaviour. What most of them have in common is that they involve very general effects which change all behaviour. The first step in any study of such effects is a careful and precise definition of the sort of change which is thought to occur, followed by adequate tests to confirm or disprove this.   R.J.A.

97, 106

ASSOCIATION is a process involved in LEARNING, particularly in the type of situation initially studied by PAVLOV. When an animal responds to a particular stimulus (usually called the *unconditional stimulus*, or UCS), it does so in the presence of other neutral stimuli, such as marks on the ground, background sounds, or any of the normal features of everyday life. If one of these neutral stimuli is consistently paired with the UCS, then it is associated with it, and may eventually come to elicit the response even when the UCS is absent. For example, in a classical Pavlovian experiment, presentations of food (the UCS) were accompanied by the sound of a metronome. The normal, *unconditional response* (UCR) of the dog (*Canis l. familiaris*) was to salivate at the sight of the food. Pavlov measured the degree of salivation. After a number of paired presentations of food and metronome beats, the animal would salivate when presented with the sound alone. The sound had now become a *conditional stimulus* (CS), and salivation had become a *conditional response* (CR). In other words the dog had formed an association between the sight of food and the sound of the metronome, through a process generally termed CONDITIONING.

Classical or Pavlovian conditioning is a very widespread form of learning throughout the animal kingdom, and in the higher mammals it pervades every aspect of life. Thus, associations can be formed with such previously neutral stimuli as the time of day, the physiological state of the body, etc. Conditional responses may include simple REFLEXES, secretion of HORMONES, SEXUAL responses, and FEAR.

O

ATTENTION. We are all familiar with the notion that it is difficult to attend to two things at the same time. Our everyday experience seems to point to this conclusion, and LABORATORY STUDIES on human subjects support it. In a typical experiment a subject is presented with two speakers simultaneously, but asked to listen to only one of them. To ensure that the subject concentrates on only one speaker, he or she may be required to repeat the words after hearing them. When asked to comment on the speaker who was not attended to, the subject can generally recall very little. For example, the subject may be aware that the speaker was a woman, but fail to notice that she switched from the English language to German. This type of experiment demonstrates the operation of a process of selective attention which is commonly used in everyday life.

The phenomenon of selective attention occurs in many animals, but it is more difficult to study than in human beings, because the direct kind of test outlined above cannot be conducted with a non-verbal animal. However, it is possible to devise LEARNING tests which reveal which features of the environment the animal attends to under particular circumstances. As an example, let us consider a laboratory rat (*Rattus norvegicus*) that is required to solve a visual DISCRIMINATION problem.

Suppose that the rat is required to discriminate between a white vertical rectangle and a black horizontal rectangle presented on a grey background, as illustrated in Fig. Aa. In a properly conducted experiment there would be a counterbalanced design involving a number of rats. Some rats would be rewarded for responding to certain combinations of the position, brightness, and orientation of the stimuli, and others to other combinations, so controlling for initial preferences.

To solve this type of problem a rat has to learn which stimulus-object to approach to obtain reward, and which to avoid. However, there are many characteristics of the stimuli with which reward could be associated. Thus the rat might be able to learn to associate reward with the shape, size, position, orientation, colour, or brightness of the stimuli. This particular problem, however, could never be solved on the basis of shape, size, or position, because the rewarded and unrewarded stimuli do not differ in these respects (the positions of the stimuli are randomly changed on each trial). The stimuli do differ in brightness and orientation, but how is the rat to discover which stimulus characteristics are associated with reward?

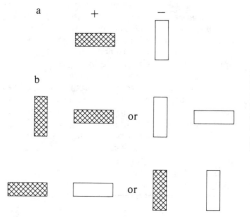

Fig. A. Shapes used in discrimination tests. **a.** The rat is initially required to discriminate between the two rectangles, responses to one (+) being rewarded and to the other (−) not rewarded. **b.** Having learned to discriminate between the two initial rectangles, the rat is required to choose one from each of the four possible pairs of rectangles shown. Each pair differs only in brightness, or only in orientation.

In solving problems of this type, many animals learn two things: (i) which aspects of the stimulus to attend to, and (ii) which of the manifestations of that aspect is rewarded. For example, a rat which is rewarded on a trial in which it attends to the brightness of the stimuli will be more likely to attend to brightness in the future. If the rat happened to choose the black stimulus, it will be more likely to choose black on those future trials during which it attends to brightness. A rat which learned to attend to brightness or orientation, or both, could successfully learn to solve the problem outlined above. On the other hand, a rat which attended to other aspects of the stimuli, such as size or position, would only be able to obtain reward by chance, i.e. on 50 per cent of trials. Therefore the rat that attends to brightness or orientation or both will obtain more rewards.

To determine which aspects of the stimuli our rat has in fact learned to attend to, unrewarded *transfer tests* can be given. During transfer tests the rat is presented with stimuli differing in orientation or brightness, but not both. Thus on half the trials the stimuli are black rectangles or white rectangles differing only in orientation, and on the other half horizontal or vertical rectangles differing only in brightness, as illustrated in Fig. Ab. If the rat had learned to attend to orientation, it would be able to solve the first problem but not the second. If it had learned to attend to brightness, it would be able to solve the second but not the first problem. If the rat had learned to attend to both brightness and orientation, it would be able to solve both types of transfer test.

Experiments of this type, conducted with many types of mammal, bird, and fish, indicate that the more an animal attends to one feature of a situation the less it attends to others. Only when animals are very well trained do they attend to a number of features simultaneously. Attention can be consolidated or disrupted by REINFORCEMENT. Thus an animal that is rewarded while attending to a particular feature is more likely to attend to

that feature in future; an animal that is not rewarded when it expects to be is less likely to attend in the same way in future. The evidence suggests that FRUSTRATION is an important agent in causing switches in attention, and there is also evidence which shows that persistence in attentiveness is influenced by HORMONES.

The phenomenon of selective attention is closely related to the concept of SEARCHING IMAGE. For example, experiments with carrion crows (*Corvus corone*) show that these birds selectively respond to a particular type of prey, even when other equally liked types are available. The birds were trained to search among beach stones for pieces of meat that the experimenter had hidden under mussel shells (Myrtilidae) of various colours. The crows would tend to concentrate on shells of one colour, as if they had an internal 'image' of the object that they were looking for. If the crow turned over a shell of this colour, but found no meat, then it would start searching for shells of another colour. This is clearly reminiscent of the laboratory demonstration (mentioned above) that animals switch attention following non-reward or frustration.

Switch of attention is particularly important in DECISION-MAKING, because it is one of the main ways by which animals cease to pursue unprofitable courses of action and take notice of possible alternatives. An animal that is too persistent in attending to particular environmental features may fail to grasp opportunities (see OPPORTUNISM). 36, 97

AUTONOMIC. The autonomic nervous system is a division of the vertebrate nervous system serving internal organs, such as the heart, blood vessels, lungs, intestines, and also certain glands. It is controlled by the BRAIN and supplies the internal organs with two types of nerve supply which have antagonistic effects. The *sympathetic* nerve pathways have an emergency function, and become active under conditions of exertion or STRESS. They have the effect of accelerating heart rate, dilating air passages to the lungs, increasing the blood supply to the muscles, and reducing intestinal activity. The *parasympathetic* pathways serve a recuperative function, restoring the blood supply to normal and generally counteracting the effects of sympathetic activity.

The autonomic nervous system influences behaviour in a number of important ways: it is responsible for many of the external signs of FEAR, such as feather sleeking in birds, and pallor and sweating in man; it also influences the types of behaviour that are observed in CONFLICT situations. For instance, many species of bunting (Emberizini) show cooling responses during the conflict that arises in COURTSHIP. It seems likely that such THERMOREGULATION occurs as a result of changes in the peripheral blood circulation arising from autonomic activity. In some cases the autonomic activity that is generated during social encounters has undergone RITUALIZATION during evolution, so that it now serves a COMMUNICATION function. For example, many of the feather movements that occur in bird DISPLAY are probably derived from autonomic responses. Blushing in humans has all the characteristics of a ritualized autonomic activity. Normally, in fear situations, the sympathetic pathways act to constrict the peripheral blood vessels, thus diverting the blood to those areas of the body, such as the heart, lungs, brain, and muscles, that are especially active during emergencies. This causes the skin to become pale. However, in some mild fear situations, particularly those in a SOCIAL context, the blood supply to certain parts of the skin is increased. These parts are generally those which are used in communication, such as the face. The change in colour is the opposite of what would normally happen in a fear situation, and this enhances the communicative value of the response. Such reversal of normal function is commonly observed in ritualized activities.

106, 134

AVOIDANCE reactions allow animals to ESCAPE from actual or potential dangers in the environment. They are a form of DEFENCE against circumstances that might be harmful. Physically avoiding areas where predators may lurk, or choosing not to consume a potentially poisonous substance are typical examples. Avoidance behaviour has such obvious SURVIVAL VALUE that it is not surprising that the capacity to perform avoidance responses, based either on INSTINCT or LEARNING, has evolved in virtually all species of animal.

Unlearned avoidance. Much avoidance behaviour is part of the animal's INNATE make-up. For example, some species of young birds (notably field geese, *Anser*) will give distress calls and attempt to hide when the silhouette of a hawk (Accipitridae) is passed over their cage. No experience with predators is required for this to happen;

the birds perform the avoidance behaviour on the first occasion. It is easy to understand why such a response evolved: individuals who could successfully avoid being captured by predators were more likely to survive and thus to endow future offspring with the ability to avoid.

A second example is the finding that some species of animals are thigmotatic, that is, they prefer to stay in close contact with objects rather than venture into open areas. Consider, for example, how adaptive this tendency would be for small rodents such as wood and field mice (*Apodemus*). In avoiding the open areas of their environment, perhaps by remaining close to the base of a tree or bush, they would reduce the risk of being captured by birds of prey.

Learned avoidance. While innate avoidance patterns are widespread amongst animals, the ability to learn new avoidance behaviour is important as well. This is because the environment is never static; new dangers may develop about which an animal must learn if it is to cope with them successfully.

According to one viewpoint, the acquisition of new avoidance patterns is accomplished in two stages. In the first stage an animal learns to FEAR certain stimuli in its environment, because they are consistently associated with an aversive outcome. For example, if the flavour of a new food substance were followed shortly by sickness (see FOOD SELECTION), or if the stimuli associated with a new feeding area indicated the presence of a predator, then an animal would acquire a fear of those stimuli by CONDITIONING.

The second stage of the process, the actual avoidance phase, is where the animal does something about the fear-provoking stimulus; namely it avoids coming into contact with it. Scientists are not entirely certain how this stage operates. For instance, the act of avoiding may or may not cause a reduction in the animal's fear, though it is clear that animals will learn responses that result in the cessation of fear stimuli and the non-occurrence of the noxious event.

There are two general types of learned avoidance reactions. The easiest to train in the LABORATORY, and perhaps the more common in nature, is the passive avoidance response. Here the animal avoids the fear-provoking stimuli by sitting passively some distance away from it. Consider the following example. A laboratory rat (*Rattus norvegicus*) is consistently fed on one side of a two-compartment box. One can easily demonstrate that the animal will always go to that side when hungry and remain there until fed. In the second phase of the experiment, however, the rat is placed near the food source and is given a mild shock. Then, in the final phase, the animal is placed in the non-preferred side with full access to both compartments. Immediately, one observes a dramatic change in behaviour. The rat does not go to the food side as usual, but rather remains motionless on the non-preferred side. Later, as it becomes increasingly hungry, it makes short and tentative forays into the feeding compartment. Traditional theory would claim that the animal first learns to fear the food chamber (where the stimuli are associated with shock) and then later avoids these stimuli by freezing on the non-preferred side. Only when hunger becomes very intense does the animal resume its normal tendency of approaching the food box. This simple form of avoidance learning is influenced by a number of factors, including the intensity of the noxious experience, the discriminability of the fear-eliciting stimuli, and the state of the animal's HORMONE balance.

A second general type of learned avoidance reaction is the active avoidance response. As the name suggests, an animal accomplishes the task of avoiding fear stimuli by becoming active and fleeing. Many studies on fish, birds, and mammals have demonstrated rapid acquisition of active avoidance behaviour. For example, if a laboratory rat is presented with a flashing light followed 10 s later by a mild shock to the paws, it develops a fear of the light. Furthermore, if, during the light, the animal is allowed to avoid receiving the shock by running to the other side of the cage, it will learn to do so quite efficiently. In other words, many animals are well able to learn that actively fleeing from fear-provoking stimuli is also a good defence against the aversive events which inevitably follow those stimuli.

These learned avoidance reactions, both passive and active, are by no means confined to the laboratory or to situations employing lights, tones, and electric shocks. They occur routinely in natural settings as well, and some are commonly observed. Birds, such as tits (Paridae) for instance, soon learn to avoid gardens that have cats (*Felis catus*). The cat does not even have to be visible for the avoidance behaviour to be elicited, once the bird has formed an ASSOCIATION between the garden and the cat. Similarly, a pet dog (*Canis l. familiaris*) will often scurry away with its ears pinned back when its owner speaks sharply or raises a hand. The dog clearly is avoiding a reprimand by responding to a signal—the owner's voice or gesture. Finally, the avoidance of another animal's TERRITORY also occurs as a result of avoidance learning. Previous encounters with the owner make the intruder learn the territory markings, and subsequently keep its distance.

Species-specific defence reactions. The discussion so far has implied that avoidance behaviour is either wholly innate or learned. While this approach serves to highlight these very different origins of avoidance behaviour, most avoidance behaviour is a combination of natural inclinations that are modified by experience of environmental contingencies. Stated somewhat differently, animals have evolved idiosyncratic, or species-specific, behavioural capacities. In some cases these capacities are very inflexible, but in others they are more plastic and are only predispositions to act. Many birds, for instance, are naturally inclined to fly away at the first sign of danger. Certain reptiles, on the other hand, are more inclined to remain motionless (see HYPNOSIS), often to enhance their CAMOUFLAGE, rather than to run when threatened. Where learning comes into play is in the modification of these tendencies. The particular circumstances that signal danger, and so result in the species-specific behaviour, are learned through experience. Similarly, the exact form of the reaction may be a result of learning.

The ideas expressed above are central to an important theory of avoidance learning—the species-specific defence reaction theory. According to this view, an avoidance response may be easily learned to the extent that it is compatible with the animal's natural species-specific defence tendencies. Responses that are not compatible, however, are learned only with great difficulty. So it is that pigeons (Columbidae) learn to flap their wings to avoid shock, but cannot easily learn to peck a plastic disc for the same outcome. Wing flapping is compatible with the species-specific avoidance pattern (flying away) while pecking is not. Similarly, rats rapidly learn to freeze at the onset of a fear signal thereby avoiding shock, but learn to press a lever in the same type of situation only with great difficulty.

Two further examples will help to illustrate how avoidance behaviour is learned. In one study, it was confirmed that mice (*Mus*) are thigmotactic when frightened: they stay in close contact with the wall of a box, rather than venturing into the open areas. It was then shown that the mice are very poor at learning to avoid shock when required to jump on to a platform located in the middle of the box. In contrast, they are very good at learning this simple motor response when the safe platform is located at the periphery. Quite clearly, learning is poor in the former instance because the avoidance response is incompatible with the animal's natural inclination (i.e. to be thigmotactic during times of danger), but good in the latter case because it is compatible.

The second example illustrates more clearly the complexity of the interaction between environmental contingencies and instinctive tendencies. A laboratory rat was confined to a small wire cage placed in the centre of a much larger square box. A cat was then released into the box. The rat, upon seeing the predator, performed its natural defence behaviour, which, in this case, was to remain motionless. Considering the alternatives, freezing was the only form of defence which was effective and, since it was compatible with the rat's natural defence repertoire, it was readily adopted. Later in the experiment the same general procedure was used, except that the door of the small wire cage was left open, and the rat was given access to an escape tunnel. Now, upon seeing the predator, the rat learned to run down the tunnel to safety. That is, an active, as opposed to a passive, avoidance response was adopted, even though the fear or danger signal, the cat, was identical. In other words, the rat was predisposed to adopt either the active or the passive reaction (both acti- vities were consistent with its innate tendencies), but which reaction it, in fact, eventually showed depended upon the particular environmental contingencies which were present at the time.

16, 39

R.M.T.

B

BINOCULAR VISION. If an animal is capable of focusing upon an object with both eyes simultaneously, as we do when we read or write, for example, then we say that the animal possesses a *binocular field* of VISION. If, however, the light from an object reaches only one eye, then the vision of that object is *monocular*. The capacity for binocular vision is restricted to the vertebrates, a group of animals which have evolved a unique eye with special features for the perception of depth and distance. Binocular vision enables an animal to localize objects in space with truly incredible precision. The difference in the ability to judge depth and distance with monocular, as opposed to binocular, vision is easily demonstrated: with your arms partially outstretched and with both eyes open, attempt to bring two pencil-points together. Now close one eye and try again.

When we attempt the pencil trick with both eyes open we are able to depend upon a powerful cue for depth and distance which is not available with monocular vision. The basis of this cue is the extent of *convergence* (or pointing inward) of the two lines of sight. Focusing on closer objects requires more convergence of the eyes than focusing on objects farther away. Quite simply, the direction of the two lines of sight is adjusted, via the muscles of the eyes, until two one-eyed images are unified into a single object. (This process of 'searching' for a single image is analogous to the one in which we adjust the separation of a pair of binoculars until the two separate circular fields of view become a single visual field.) When the object is seen singly, the parallactic angles of convergence of the two eyes are simultaneously registered in the nervous sytem of the animal to yield a perception of distance-judgement.

In man, this PERCEPTION of the singleness of objects in the binocular field is accompanied by, and indeed is inseparable from, a perception of the object's solidity. *Stereopsis*, literally 'seeing solid', is based upon another binocular distance cue termed *retinal disparity*. When two spatially separated eyes look at an object, each retina receives a slightly different view of the object. The disparity of these images varies with the distance between the eye and the object; the closer the object the more disparate are the images. The stereoscope, a device for presenting two flat pictures which, when combined, yield a sensation of depth and stereopsis, makes use of the perceptual outcome of retinal disparity. When separate left-view and right-view photographs of the same scene are presented simultaneously to the left and right eyes, respectively, we see that scene in depth.

But stereopsis is a subjective experience. Do non-human animals see solidly in their binocular fields as well? A recent development has enabled scientists to answer this question in the affirmative. To demonstrate stereopsis, it was necessary to test animals with objects in such a way that binocular disparity was the only cue for depth. Real objects, of course, would provide monocular depth cues like shadow, overlap, and perspective, and these are very effective depth cues, as every artist knows. Hence, real objects would be unsuitable in any test of stereopsis. The advent of the *anaglyph*, or *random dot stereogram*, solved the problem, however. One type of anaglyph appears as a single picture, or field, of randomly-arranged red and green dots on a page. Actually the pattern of red dots is identical to the pattern of green dots; but, when they were printed on the page, one field was slightly displaced, relative to the other. Each of the dot patterns contains a hidden figure, such as the outline of a cube. Viewed normally, each eye sees both the red and green dot patterns. To obtain the stereoscopic effect, however, the anaglyph is viewed with a red filter over one eye and a green filter over the other; thus, each eye sees only one field, and two disparate retinal images are now created. Suddenly, the once-hidden cube pattern jumps from the surface of the dots, appearing to hover in space. The perception is so vivid that we reach to touch it; more importantly we know exactly where its corners are, and we consistently point to precise locations in the empty space above the page. Using the anaglyph and animal LEARNING techniques, scientists have demonstrated stereopsis in the cat (*Felis catus*) and the monkey (Simiae). This is accomplished by first teaching the animal a DISCRIMINATION task; for example, it must depress lever A whenever a square is presented, but must depress lever B whenever a circle is presented. After this task is mastered, the animal is fitted with the red and green filters, and anaglyphs with hidden circles and squares are then presented. If the animal possesses stereopsis it should be able to perform the discrimination task correctly; if not it should perform at chance level. In the LABORATORY monkeys and cats have been tested in this manner, but FIELD observations of the good depth judgement and agility of many mammals, the squirrel (Sciuridae) and the Rocky Mountain goat (*Oreamnos americanus*) for example, strongly imply that stereopsis exists in these species as well.

We have seen that the process underlying stereopsis, retinal disparity, is a concomitant feature of convergence in binocular vision. Converging eyes necessarily create different views of an object. None the less, in many animals, stereopsis may not always accompany binocularity. In mammals, stereopsis depends upon a peculiar anatomical feature of the visual pathway which does not exist in non-mammalian vertebrates. Non-mammalian vertebrates may thus possess binocular vision, that is, singleness of objects in the binocular field, without three-dimensional perception.

In vertebrates, the optic nerves from the two eyes come together at a location beneath the BRAIN, where the nerves cross over, or through, each other. This X-shaped structure is appropriately named the *optic chiasma* (from the Greek, *khiasmos*, a cross). From the chiasma the nerves continue to the brain as the optic tracts. In all non-mammalian vertebrates, all of the fibres from one eye cross to the other side at the chiasma. Thus the left eye is connected to the right hemisphere of the brain, and vice versa. This total crossing over of optic nerve fibres is called *total decussation*. The optic nerve fibres of the mammals, however, only partially decussate; that is, some of the fibres from each retina enter the optic tract on that same side. The neural mechanism of stereopsis is believed to depend upon *partial decussation* of the optic fibres. These crossed and uncrossed retinal fibres eventually converge on cells, located in the *visual cortex* of the brain, which are selectively sensitive to image disparity. It is, in part, the activation of these binocular cells which gives rise to the perception of depth. However, one might argue that, functionally speaking, it should not be important where the decussation takes place. Might it be possible that a decussation of fibres occurs at some subsequent stage in the visual pathway of non-mammalian vertebrates? Indeed, such a supraoptic chiasma has been demonstrated in the owl (Strigiformes) and frog (Anura), and cells which selectively respond to binocular input have been located in a few non-mammalian species, including the pigeon (*Columba livia*). It thus appears probable that some, but perhaps not all, non-mammalian vertebrates see objects as solid in their binocular visual fields.

Binocular vision, with or without stereopsis, would still provide precise object localization and distance judgement, and would seem to afford a clear advantage over monocular vision. However, different vertebrate species do not possess binocularity to the same extent. Some animals, many primates for example, are capable of focusing binocularly on most objects around them. Birds and fish, on the other hand, see much of their visual world with only one eye; that is, their vision is largely monocular. And a few vertebrates possess no binocular field of vision whatsoever. These include lampreys (Petromyzonidae), some large-headed bony fishes (Teleostei), the hellbender (*Cryptobranchus*) which is an amphibian, penguins of the genus *Spheniscus*, and the larger whales (Cetacea). But why is it that not all vertebrates possess large binocular fields? The answer, of course, is that although binocular vision confers certain benefits upon an animal, as we have seen, it also imposes limitations. Further, monocularity is visually advantageous in respects other than depth and distance perception.

The explanation lies partly in the extent of an animal's *panoramic field*. If we imagine that the animal has its head fixed at the centre of a sphere of space, the proportion of that sphere within which the animal can see without moving its head is called the panoramic field (or visual field). The extent of the panoramic field, and the proportion of that field which is binocular, are determined by the shape of the eye, the shape of the head, and the position of the eye in the head.

The visual field of the single eye is roughly constant in vertebrates, and subtends an angle of 170°, although exceptions to this rule do exist. Therefore, the panoramic and binocular fields would vary inversely with each other. Generally speaking, animals with frontal eyes have much larger binocular fields than those with laterally situated eyes. The trend toward *frontality* and the increase in the binocular field at the expense of a large panoramic field are seen in many predatory birds, such as raptors (Falconiformes), with binocular fields of 30–50°, and fishes, such as trout (Salmoninae) and pike (*Esox lucius*), 30–40°, in the ungulates (hoofed mammals) and the larger carnivores, (60–130°), and, maximally, in the higher primates (monkeys, apes, and man), 140°. Animals with laterally placed eyes, which includes most birds (especially song-birds), lizards (Lacertidae) and snakes (Serpentes), the majority of rodents (gnawing animals), and, especially, rabbits and hares (Lagomorpha), have much smaller binocular fields, ranging from 4° in some fish (e.g. the box-fish, *Ostracion tuberculatus*) to rarely more than 60°. The relationship of binocular to panoramic field width is illustrated by a comparison of the strongly laterally-eyed rabbit and the frontally-eyed cat. Because of their frontal position, the eyes of the cat cannot see the half-sphere of space to the rear of its head; that is, its panoramic field is only about 185°. The cat's binocular field, however, is quite large at 99°. The rabbit, on the other hand, possesses a very small binocular field (24°), but can literally see behind its head with a panoramic field of approximately 360°.

The limitation of *laterality* is, of course, that the ability to localize objects in space with the precision of binocular vision is restricted to a very small area. However, the ability to detect movement in a wide field of view (*periscopy*) may sometimes be more important, especially to prey species. The hare, the rabbit, and the rodents can spot a potential enemy coming from any direction and freeze without revealing so much as the movement of an eye. The frontally-eyed pike, known for its lightning-speed predatory attacks, can strike with the accuracy that only binocular vision can afford. Object localization is possible monocularly, of course. If we recall that two views of the same object provide distance cues, it is not surprising that predominantly monocular shore birds (e.g. the sandpiper, *Tringa*) bob their heads while FORAGING, primates with limited binocular cortical areas (marmosets, *Callithrix jacchus*, and squirrel monkeys, *Saimiri sciureus*) cock their heads, and many snakes move their heads from side to side. However, these activities would be inefficient for animals such as the pike. Not only do head movements afford a less precise distance judgement, but they are much more time-consuming than the instantaneous REFLEX of binocular convergence. Indeed, there appears to be a general trend in that predators possess frontal vision while prey carry their eyes on either side of the head. Some exceptions to this rule, for example the PREDATORY crocodile (Crocodylidae) with a binocular field as narrow (25°) as the rabbit, have suggested to some that the predator–prey relationship may not be the only factor determining binocular field width. The greater degree of distance judgement rendered possible by binocular vision may have adaptive significance in many other contexts. In man, for example, the extensive use of the hands, which depends upon precise object localization and eye–hand coordination, is but one example of the diversity of specializations made possible through binocular vision.   K.L.H.

59, 132, 133

BOREDOM. It is not possible for scientists to say whether animals suffer from boredom the way that human beings do. Just as the feelings of other people can be evaluated only by analogy with one's own feelings, the evaluation of the feelings of an animal must rest upon similar analogy, and cannot be a subject of scientific verification. However, on the basis of such analogy, most scientists agree that it is likely that some animals can experience boredom to some degree. The more intelligent animals, which lead more active lives, are most likely to experience boredom akin to human boredom, but one must not assume that the circumstances that bore an animal would bore a human.

Evidence from animals and humans suggests that prolonged exposure to repetitive sights, sounds, etc., leads to impaired performance in tasks which involve ATTENTION to such stimuli. The effects of monotony upon performance suggest that boredom is involved, and it has been demonstrated that animals, when given the opportunity, will work to induce changes in monotonous surroundings. Whether this should be interpreted in terms of a fatigue-like process, or whether change is attractive in itself, is a controversial topic.

It is probable that animals with too little to occupy their time become bored (see LEISURE). Animals are designed by NATURAL SELECTION to make efficient use of their time. In the natural environment certain portions of the day are spent feeding, sleeping, etc., each species filling the day in its characteristic manner, according to the season. An animal will procure food, and other necessary commodities, as quickly and efficiently as it can in the prevailing circumstances. Under domestic conditions such essential tasks may be totally precluded, or performed much more quickly, than in the wild, and this suggests that the unnatural maintenance conditions may lead to boredom. However, it is difficult to be sure that this is the case for any given species. For example, it has been observed that cows (*Bos primigenius taurus*) show much less ABNORMAL behaviour when maintained in stalls for long periods than do pigs (*Sus scrofa domestica*). The fact that cows normally spend 4–9 h per day ruminating may mean that they are not so easily bored by inactivity.

Because of the difficulties of satisfactory investigation, boredom in animals has not been a popular topic for scientific research. The introduction of intensive farming methods, however, has increased concern about animal WELFARE, and problems of boredom are now receiving more attention from scientists. It has been reported that if pigeons (*Columba livia*) or domestic chickens (*Gallus g. domesticus*) are frustrated in getting food, they develop a stereotyped pacing behaviour which is similar to that shown by some animals when confined for long periods. The occurrence of pacing can be reduced if tranquillizing drugs are given early in the animal's development (see FRUSTRATION). Laboratory animals, trained in a simple task to obtain their food, will continue to work even when food is freely available, suggesting that the animals like to have 'something to do'. The behaviour disappears, however, when food is freely available, if the performance of the task no longer results in food rewards. Moreover, there is no evidence that animals do more of the unnecessary work when they are maintained in a barren and unstimulating environment. Thus it seems unlikely that the performance of learned tasks, in the presence of freely available food, is connected with boredom.

To summarize, the question of boredom in animals is complicated, because it cannot be measured directly, and because species differ widely in their reactions to unstimulating situations. In addition, the little research that has been carried out is primarily on domesticated animals, which have been bred specially for FARM or LABORATORY conditions, and are less likely to experience boredom than wild animals kept in CAPTIVITY. Observation of animals in zoos suggests that the more intelligent do experience boredom, which is similar in its behavioural manifestation to that of human beings.

BRAIN. The control of behaviour by the brain remains one of the major unsolved problems in biology. Although much has been learnt about the biochemical basis of life, very little is known of the physical basis of mental activity and of the behaviour which it directs. It is not difficult to see why this should be so when one considers the awesome complexity of the brain, and the crudity of the methods we have for penetrating its mysteries.

The brain is composed of nerve-cells, which are cells specialized for conducting very small electrical discharges termed *nerve impulses* along their length. The structure of these cells can be studied with the light or electron microscope, and their physical and chemical properties have been examined by biophysicists. Our knowledge of what these cells look like and of how they work is fairly advanced. Nerve-cells differ in size and shape, but are alike in possessing a cell body, an *axon*, along which the nerve impulses pass, and *dendrites*, with which the axons of other nerve-cells form connections (Fig. A). Biophysicists have been able to measure the electrochemical changes which occur when a nerve impulse moves down the axon of a cell by making use of certain giant axons found in the squid (*Loligo*). These axons initiate contractions of the mantle, so expelling water and propelling the squid forwards. They are large because nervous transmission is faster the larger the axon, and the squid is thus enabled to make its escape very quickly. Since the diameter of these axons may be up to 0.5 mm, electrodes can be placed inside and outside the cell, and so the electrical potential across the cell membrane can be measured. When a nerve impulse passes down the axon there is a change in this potential, caused by electrochemical changes in the membrane. This shift in potential moves along the axon of the cell, and in its wake processes within the membrane

restore the potential to its normal level. The axon of one cell does not actually touch the dendrite or cell body of another, but comes very close to it at a junction termed a *synapse* (Fig. A). At most of the synapses in the nervous systems of vertebrates one cell affects the activity of another by discharging small quantities of chemicals across the synaptic gap. These change the readiness of that cell to conduct impulses, i.e. to fire. A nerve-cell fires in an all-or-none manner, and there is variation in the rate of firing which depends on the activity of the many other cells which form synaptic junctions on its dendrites or cell body.

Although the mechanisms by which individual cells operate are fairly well understood, in order to understand how the nervous system works it is obviously necessary to know how these cells are connected to each other. Nerve-cells are not connected at random, but during the development of the brain they collect to form various structures. It is the task of the neuroanatomists to describe and name these different parts, and to discover the ways in which cells in the different structures are interconnected. Collections of axons are called *nerves* or *tracts*, and collections of cell bodies are referred to as *ganglia* or *nuclei*. The brains of vertebrates contain many tracts and hundreds of nuclei, but only the connections of the main tracts and the more important nuclei are known, even in those animals which have been most studied such as rats (*Rattus norvegicus*), cats (*Felis catus*), and macaque monkeys (*Macaca*). In invertebrates and many vertebrates much less is known of the connections between the parts of the brain.

Given that there are various structures in the brain it is necessary to ask what functions each performs. These functions can be discovered in various ways. Clues can be derived from the anatomical connections, but these are usually so numerous that little insight is gained in this way.

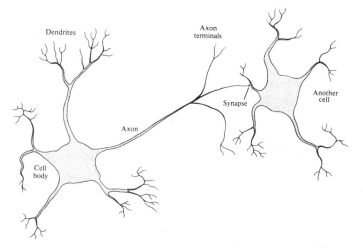

Fig. A. One nerve-cell forming a synaptic connection with another.

Alternatively a structure may be removed, and any interference with normal behaviour noted. This is a crude method because damage to one part of the brain may incidentally disrupt the functions of neighbouring areas. There may also be several alternative explanations of the syndrome produced, and no clear-cut conclusion on the functions of the area in the intact brain.

Neuropsychologists using this technique tend to study the effects of removal of parts of the brain on the behaviour of animals kept in the LABORATORY, because it is only under such conditions that rigorous experiments can be conducted. For convenience they study a limited range of animal species, concentrating their attention in particular on rats, cats, and monkeys, because these are easy to keep and work with in laboratory surroundings. By contrast ethologists investigating the behaviour of animals are often interested in what they do in their normal surroundings (see FIELD STUDIES), and they have studied a wide variety of species to see how they adapt to their chosen environments (see ETHOLOGY and COMPARATIVE STUDIES). Much of the behaviour ethologists have described does not occur or is distorted in CAPTIVITY, because the animals no longer have to work to survive, and are often housed singly, rather than in social groups. As a result laboratory experiments on the functions of particular areas of the brain often fail to suggest what role these areas play in controlling behaviour when the animal is in its natural surroundings.

If the organization of the brain had been well described, and if the functions of at least the main areas had been identified, it would then be feasible to try to give some account of how the firings of cells connected in particular ways carry out the functions of the brain. In fact neurophysiologists have tended to simplify the problem, either by taking for study very simple brains, or by choosing to investigate those areas of more complex brains which have relatively simple known functions. By studying very simple nervous systems it is hoped to find principles which also apply to more complicated brains, the simple brain serving as a basic model. Some threadworms (Nematoda), for example, have less than a hundred nerve-cells, and since their behaviour is very limited it should prove possible to discover how these few cells control what the animal does. However, in the brains of monkeys there are several billion nerve-cells, and in the human brain there are ten billion or more. Matters are made worse for the neurophysiologist by the fact that in a monkey or a person there can be as many as sixty thousand or more cells which form synaptic junctions with a particular cell. The size of the problem for those who try to explain how the brain controls behaviour should now be apparent. Because the techniques which are available allow the recording of the electrical activity of only one or two nerve-cells independently and at the same time, it takes a considerable amount of time to complete recordings on even a hundred or so cells. In spite of this some limited success has been achieved in understanding how the areas of the brain receiving sensory information, and the areas controlling body movements, carry out their functions. Small electrodes are placed near brain cells in animals such as cats or monkeys, which may either be anaesthetized or, if necessary, awake and unrestrained, so that they can perform simple tasks. When patterns are shown to the eye, the changes in the rate of firing of cells can be recorded in the areas of the brain receiving visual information, and an attempt made to specify the functions of the cells in the analysis of patterned light (see VISION). In the same way, if an animal such as a rhesus macaque (*Macaca mulatta*) is trained to move a limb in a particular way, the firing of cells in the areas of the brain controlling movements can be studied for clues as to how they dictate the movements made. But there is little prospect, using such simple methods, of discovering the basis of the more complicated behaviour that ethologists have described in animals living in their natural surroundings. Our ignorance of the way in which the brain controls the behaviour of animals reflects the difficulty of the task.

Invertebrates. A worthwhile approach is to relate the behaviour of different groups of animals to the obvious differences in gross structure between their brains. Since the brain is the organ for behaviour it ought to vary in line with variation in behaviour. The chances of showing such a relationship are best if a very wide range of animals is considered, from the simplest multicellular organisms to the most advanced birds or mammals. Differences in the size of the brain, and thus in the number of cells in the brain, should be reflected in differences in the capacities of the animals, and differences in the specialized parts of the brain should be reflected in the behavioural specializations of the animals. It should be possible to demonstrate relationships of this sort, even though little is known of how any particular area of the brain carries out its functions.

It is most profitable to start with the simplest multicellular animals, because the specializations of the nervous system of more complex animals can be best appreciated in contrast with the unspecialized. To be successful an animal must be able to manipulate the resources available to it without incurring excessive risk to life or progeny. To this end it must have information on the state of the world and some means of controlling the consequences of its own behaviour. Sensory cells are specialized for detecting changes in the environment, whether the change occurs at the body, as when the animal is touched, or when the source of stimulation is at a distance, as when something is smelt or seen (see SENSE ORGANS). In simple animals, such as coelenterates: the sea anemones (Actinaria), hydras (*Hydra*), and jellyfish (*Aurelia*), there are cells which are responsive to touch and others which are sensitive to chemical changes in the water, and in some there are cells which are sensitive to light. In all these animals there are also cells which are partly specialized as muscles, and these allow the animal to move, open its mouth, and so on. But there is also a need for a means by which the activity of the sensory cells can affect the muscles, so that the animal can respond appropriately when there is a change in the environment. Nerve-cells are specialized for just this purpose, to conduct impulses when activated by the sensory cells, and to control the activity of the muscles on which they terminate. In hydras and sea anemones the nervous system forms a diffuse net of nerve-cells round the animal (Fig. B), although in jellyfish the nerves are concentrated into two nerve rings round the bell. Nerve impulses can be conducted either way across the synaptic gaps between the nerve-cells, but it is not known how the activity of one cell influences that of another. The impulses also pass very slowly along the nerve, as compared with nerve impulse transmission in vertebrates. The nerve net coordinates the movements of the animal, as, for example, in *Hydra* when it grasps prey and conveys it to the mouth with its tentacles, but in general, movements are very slow and the repertoire of activities very limited. Behaviour is also stereotyped, and the animals show no evidence of any ability to learn to modify what they do. It is true that if, for example, the mouth of the sea anemone is repeatedly touched it closes reflexly at first, but less readily with time; but if after a rest it is again touched it closes as at first. This happens because the junction between the sensory cells becomes temporarily less efficient with repeated use, and the effect is therefore one which would be attributed to sensory adaptation rather than to LEARNING.

The nervous system of the simplest of coelenterates, such as the freshwater polyp *Hydra*, can be contrasted with that of the simplest of worms. Whereas the coelenterates have radial symmetry, worms and invertebrates, other than starfish (Asteroidea), have bilateral symmetry, possessing a long axis with a head and tail. Particular anatomical terms are used to refer to the front (anterior), back (posterior), side (lateral), upper (dorsal), and lower (ventral) surfaces. In such animals the head meets the world first, and there is therefore a concentration of sensory cells on the head. As a result the nerves are not distributed evenly through the body; instead there is a collection of cell bodies in the head forming the anterior ganglion, as in flatworms (Platyhelminthes) (Fig. C). From the anterior

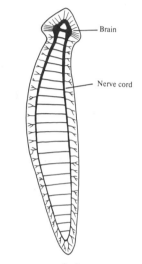

Fig. C. The nervous system of a flatworm.

ganglion two nerve cords run back on either side of the body, and these are linked by nerves in the form of a ladder. From the nerve cords nerve fibres pass to all parts of the body in a net-like pattern, forming the peripheral part of the nervous system. The anterior ganglion may be regarded as a simple brain, and together with the nerve cords it is referred to as the central nervous system. In flatworms such as the planarians (Triclada), the cells of the body are grouped together with cells of similar type to form organs. On the head, for instance, there are two eyes, in which cells sensitive to light are grouped together, and the muscle cells are completely specialized and grouped as muscles. On the head there are sensory cells not only for light, but also for touch, temperature, and the chemical composition of the water. The brain receives fibres from all of these cells, and thus obtains a variety of information about the state of the world. Conduction of nerve impulses along the nerve cords is faster than in nerve nets, so that quicker reactions to events are possible. The behaviour of the planarians is more varied than that of animals with simpler nervous systems. They avoid strong light, rest on the dark undersurfaces of objects in water, glide by movements of hairs (*cilia*) on their undersurface, or crawl by contractions of the body.

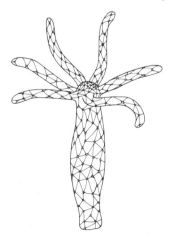

Fig. B. The nerve net of *Hydra*.

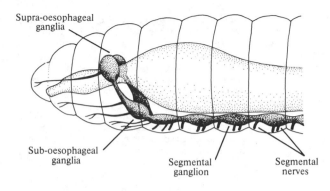

Supra-oesophageal ganglia

Sub-oesophageal ganglia

Segmental ganglion

Segmental nerves

Fig. D. The nervous system of an earthworm.

They are also quick to detect the presence of food in water, and move rapidly towards it. However, flatworms can still move even when the head has been cut off, although they are then poor at recognizing food. Although sensory information from the head is relayed through the brain, and can only influence the rest of the body in this way, the control of movements is less centralized. What flatworms are capable of learning is surrounded by controversy. If a light always comes on 1–2 s before an electric shock is given, planarians appear to be able to learn to anticipate the shock, though it has turned out to be very difficult to prove this. But in other situations the evidence is much more clear. If placed in a simple maze in the form of a T, planarians can definitely learn to turn to one side rather than the other to avoid being tapped with a rod.

Segmented worms (the Annelida) have a more complicated nervous system. It can be illustrated by the earthworm *Lumbricus* in which there is a ganglion above the pharynx (*supra-oesophageal ganglion*) connected to another below it (*sub-oesophageal ganglion*) (Fig. D). The sub-oesophageal ganglion is the first and the largest of many ganglia which are distributed along a central nerve cord, one ganglion for each segment of the worm. From each ganglion nerves pass to the muscles of that segment. Because it burrows in the dark the earthworm has no eyes, unlike other segmented worms, such as the polychaete bristle worms (Errantia), but it has light-sensitive cells on its back, particularly at its front end. If the supra-oesophageal ganglion is removed sensory information from the head can no longer affect the worm's actions, and the worm is poor at feeding and burrowing. If the sub-oesophageal ganglion is removed the worm's muscle tone becomes poor, and the animal is less active. However it is still able to move, since the movements of each segment are determined by the local ganglia, and the coordination of movements depends mainly on the connections between these ganglia. Earthworms are able to learn simple things, and appear to be able to do so even after removal of the anterior ganglia. They can be trained, for instance, to turn down one arm rather than the other of a T-shaped maze to avoid an electric shock. They can learn this task even if the anterior ganglia are taken out, showing that the parts of the nervous system involved in learning can be elsewhere than in the brain.

The brain is very much better developed in insects, partly because they have more specialized sense organs on the head. They have highly efficient compound eyes (eyes divided into units in each of which there are light-sensitive cells and a lens for focusing the light), and antennae which are sensitive to touch and to chemical changes of the body. At the same time insects have jointed legs, and mouths which are often of great complexity, allowing them to manipulate the environment in very varied ways. In the head there is a supra-oesophageal ganglion above the digestive tract, and a sub-oesophageal ganglion below, at the head of a double nerve cord which runs ventrally along the animal. In some insects there are many ganglia on the nerve cord, but in others some of these are fused into larger ganglia, as for example in bees (Apidae) (Fig. E). With this equipment insects dis-

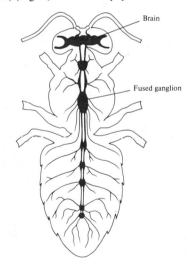

Brain

Fused ganglion

Fig. E. The nervous system of a bee.

play an astonishing range of behaviour, both as individuals and in groups. Consider the female sand wasp (*Ammophila campestris*) which digs shafts in the sand, makes an underground chamber for each shaft, and collects pebbles or earth to close the doors. She then hunts for caterpillars, stings them, places them in the nests, and lays eggs on them. The female is able to bring the right number of caterpillars to each nest even though caring for several nests at a time. Or consider bees constructing a comb, or the waggle dance from which a honey-bee (*Apis mellifera*) can learn the distance and direction of nectar. Insects do not have to learn to do such things, and carry them out in a relatively stereotyped way without necessarily appreciating the reason for doing them (see COGNITION), the nerve cells of their brains being connected in ways which permit such unlearned activities. Of course insects can and do learn. They must learn their way about, just as in the laboratory an insect, such as an ant (Formicidae), can be taught to find its way through a complicated maze. Bees can learn to associate the colours of flowers with the food they contain. If several dishes are put out on different coloured cards bees easily learn which colour is associated with sugar water in the dish. It appears that the very simplest form of learning may not be dependent on the head ganglion; a cockroach (Blattaria) for instance can still learn to lift its leg from water to avoid electric shock even when its head has been cut off. However, in so far as many of the things that insects learn require information from the sense organs of the head, much of the learning almost certainly involves the head ganglion or brain.

The invertebrates with the largest brains are the cephalopods such as the common octopus (*Octopus vulgaris*). The octopus brain is made up of one to two hundred million nerve cells, and can be subdivided into several different lobes (Fig. F). Here the nervous system has become highly centralized, so that instead of having a head ganglion and other ganglia on a nerve cord, the ganglia are collected into a mass of nerve tissue in the head, the sub-oesophageal and supra-oesophageal parts of the brain. The different lobes of the brain are thought to have specialized functions as shown by the different effects of removing them on the animal's behaviour. The sub-oesophageal ganglia coordinate the movements of the arms, and if they are removed the octopus maintains a rigid posture. The optic lobes of the supra-oesophageal brain receive fibres from the eyes, which in the octopus are much like those of vertebrates, and analyse the information received from them. The frontal and vertical lobes appear to play some role in learning. The octopus is easily tested in the laboratory. First it is allowed to collect debris to form a home as it does on the ocean floor, and then it is tempted to emerge from its home to attack a crayfish (Astacidae). If the octopus is then given a small electric shock for an incorrect response, it can be taught to attack a crayfish when shown a black card, but to refrain when shown a white one. By varying the pattern on the cards it is possible to determine which ones the octopus can learn to tell apart. The

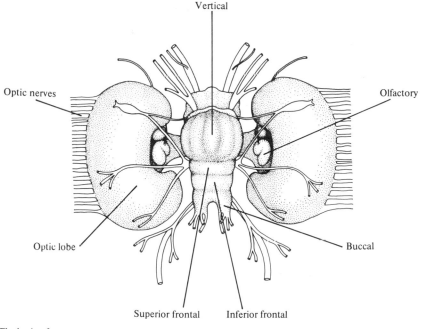

Fig. F. The brain of an octopus.

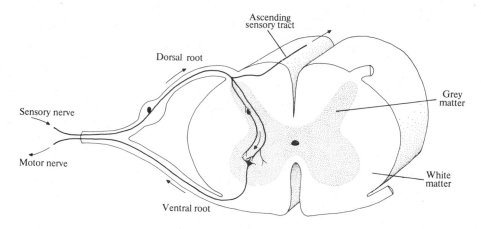

Fig. G. A cross-section through the spinal cord.

same can be done for objects that it can feel but not see. If now the vertical lobes of the brain are removed the octopus is poor at learning these tasks, suggesting that these lobes are involved in learning. It is of particular interest that the octopus's brain also allows it to adapt to repeated changes in the situation, so that if it is taught to attack one object and not another, it can then be trained to attack one which was previously incorrect, and then the first one again, and so on. After many such changes the animal becomes quicker to adapt on each ocasion. Such an animal is well equipped for coping with changes in its environment, in that at least some of the things that it does are not rigid, but can be changed when necessary.

Thus we find that the nervous system varies greatly among the invertebrates from the simplest nerve net to a large brain with specialized areas. In the simplest nervous systems the nerve-cells link the sensory and muscle cells in a peripheral network, but in invertebrates which are elongated rather than radial the head contains a ganglion which collects information from the sense organs on the head. In many invertebrates the head ganglion appears to initiate movements as the result of sensory information, but to play little part in the organization of movements of the parts of the body other than the head. In others the brain takes over control for body movements, and the control of behaviour becomes completely centralized. Complex behaviour is possible if there are a variety of sense organs, a large centralized brain, and complicated limbs. The more information that comes from the senses the better based is any decision that the animal makes. The larger the brain the more numerous the nerve-cells, and in particular the more the influences that can play on any one cell as the result of the activity of those others that form synaptic junctions with it. In general invertebrates are short-lived: for instance, insects have not the time in which to learn how to

feed, build shelters, or play their part in a social group. It is more efficient that those circuits which can organize the appropriate activities should be established on a GENETIC basis. Certainly worms and insects can learn, and may need to learn to find their way about, since it would be difficult to build into the nervous system knowledge of the particular environment in which the animal may find itself. But for the most part the activity of insects is fairly stereotyped, and their success in surviving in such numbers and in so many different environments shows the efficiency with which the appropriate behaviour can be built into the nervous system.

Vertebrates. Unlike the invertebrates, animals with backbones have a single nerve cord along their back, the *spinal cord* Nerves from the sensory organs of the body enter the spinal cord on the dorsal side, that is on the side nearest the back, and those going to the muscle leave the cord on the ventral side, that is on the side facing the stomach (Fig. G). The nerves passing from and to the internal organs, such as the heart or viscera, form part of what is known as the AUTONOMIC part of the nervous system. Outgoing nerves in the *sympathetic* division of this form junctions in a chain of ganglia attached to the spinal cord, whereas the nerves in the *parasympathetic* division enter ganglia near the organs they innervate (Fig. H).

Sensory nerves can form direct connections in the spinal cord with the motor nerves, but more usually form synapses with intervening nerve-cells which directly or indirectly connect with the motor nerves. Relatively simple actions, such as the withdrawal of a limb when contact is made with something causing pain, can be carried out quickly by such circuits in the spinal cord. But, where it is important that more influences should be brought to bear on a decision, information about events must be transmitted to the brain. Sensory nerves make contact with long tracts in the spinal cord which terminate in areas of the brain, and other

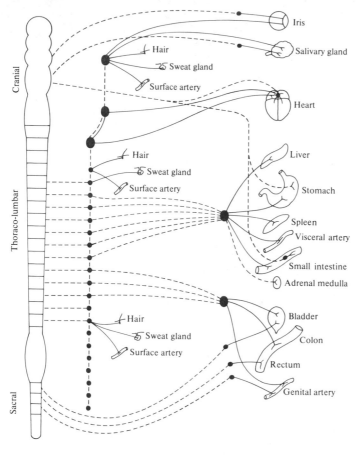

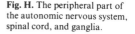

Fig. H. The peripheral part of the autonomic nervous system, spinal cord, and ganglia.

tracts passing from the brain make contact with the motor nerves leaving the spinal cord (Fig. G). At the same time sensory nerves from the nose, face, ear, and eye pass directly to the brain, and together with some other nerves these are referred to as the cranial nerves.

The brain can conveniently be divided into three main parts, a hindbrain, midbrain, and forebrain (Fig. I), and in turn the forebrain is made up of the interbrain and the endbrain. These divisions of the brain are particularly obvious in the developing embryo (Fig. J). At an early stage the brain is like a tube in three segments, not unlike those found in the brain of some worms. Later, however, there is a striking growth of the cerebellum from the roof of the hindbrain, and the *cerebral hemispheres* develop to form the endbrain.

Size of the brain. The brains of some vertebrates are compared in Fig. K, where it will be seen that they differ markedly both in size and shape. Because the brain is easily weighed, the weight of the brain of many vertebrates is known. It might be supposed that the size of the brain was a good guide to the intelligence of the animal, but a glance at Fig. K will show that this is not likely to

be so. Elephants (Elephantidae) and whales (Cetacea) have brains several times larger than our own, but few would want to judge their INTELLIGENCE relative to man's on that basis. In general the organs of the body are bigger the larger the animal; thus the heart must be larger the more work it has to do. Similarly there is a tendency for the size of the brain to increase with the size of the animal, because the larger the body the more sensory fibres that enter the brain, and the more fibres that must leave the brain to control the muscles. It is reasonable to suppose that what determines intelligence is the amount of brain above that required simply for controlling the body. This can be estimated by comparing the brains of animals which are of similar body size.

The brains of fish, amphibians, and reptiles are in general similar to each other but more poorly developed than those of birds and mammals, when account is taken of differences in body size. There is considerable overlap in size between the brains of birds and mammals, although there are some mammals whose brains are developed much further. The rodents such as rats and squirrels (Sciuridae), and the insectivores, such as the shrews

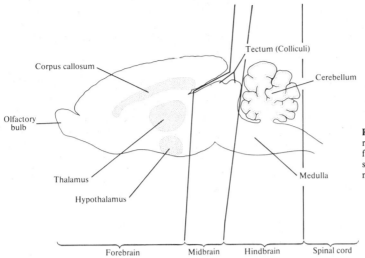

Fig. I. One hemisphere of the rat brain as seen after section from the inside: showing the subdivisions into the forebrain, midbrain, and hindbrain.

(Soricidae) and hedgehogs (Erinaceidae), have relatively small brains, but the brains of ungulates such as horses (Equidae), or carnivores such as dogs (*Canis*) and lions (*Panthera leo*), are very much larger, even when account is taken of differences in body size. The greatest specialization has occurred in the primates. The prosimians (Prosimiae), sometimes referred to as the lower primates, have brains which do not differ much in size from those of some other mammals, but the monkeys and apes (Simiae), sometimes referred to as the higher primates, have relatively larger brains than other land mammals. Some sea mammals, such as the dolphins (Delphinidae), possess brains which are even larger in relation to their body size, but comparisons between animals with bodies adapted for land and those with bodies specialized for sea life may not be valid. Man stands out from all the animals in terms of the size of his brain, which is over three times as large as would be expected for a non-human primate of the same body size. This difference is astounding when it is realized that it is as great as the difference in brain development between a chimpanzee (*Pan*) and a lowly insect-eating mammal such as a hedgehog.

The significance of the size of the brain is that it gives a rough indication of the number of nerve-cells in the brain, as larger brains tend to have more nerve-cells than smaller ones. It is also found that the cells of larger brains are bigger and less densely packed, so that any one cell may have a more complex set of dendrites and be open to influence from a greater number of other cells. The more numerous the influences on each element, the more flexible the system should be. It is widely assumed that the ability of the brain to

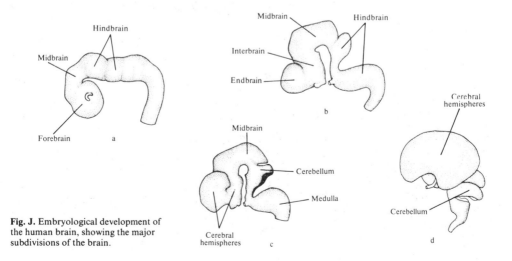

Fig. J. Embryological development of the human brain, showing the major subdivisions of the brain.

solve problems and decide on intelligent action is determined not only by the number of elements or cells that it contains, but also by the number of connections between the cells.

Different groups of nerve-cells in the brain have differing functions, and some account must therefore be taken not only of the number of cells in the brain, but also of the development of specialized regions of the brain, and thus of the number of cells in these regions. Different brains are not just reduced or expanded versions of each other, but may vary greatly in the relative development of particular specialized areas. These areas will now be considered.

Nuclei. There are certain basic functions which must be performed by the brains of all vertebrates,

and, this being so, it is not surprising that the areas which control these functions do not differ greatly across species. The heart and respiratory organs, for instance, must be regulated, and this is done by nuclei in the medulla of the hindbrain (Fig. I). The animal must SLEEP, and nuclei in the central core of the hindbrain and midbrain, called the *reticular formation*, play some part in controlling the level of consciousness. If certain nuclei are destroyed in an experimental animal such as a cat, the animal may fail to wake, whereas if other nuclei of the reticular formation are damaged the animal may remain permanently awake.

All animals must find food and water and must control their intake according to their needs. The nuclei most directly concerned with these func-

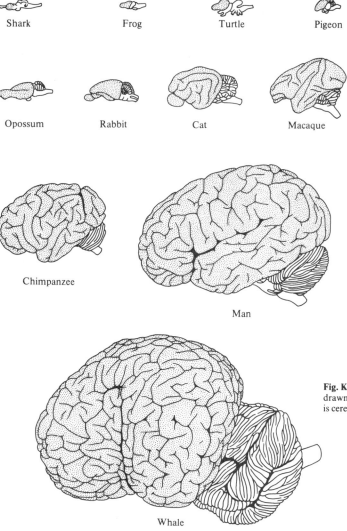

Shark

Frog

Turtle

Pigeon

Opossum

Rabbit

Cat

Macaque

Chimpanzee

Man

Whale

Fig. K. Brain of some vertebrates, drawn to the same scale. Stippled area is cerebral hemisphere.

tions are in the *hypothalamus*, a group of nuclei in the interbrain (Fig. I), although other nuclei, such as those of the *amygdala*, also play some role. If one part of the hypothalamus is destroyed in rats, the animals fail to eat or drink, whereas damage to another area leads to a syndrome in which the animals eat much more than usual, and maintain an abnormally high body weight.

The hypothalamus is also critically concerned in the control of reproduction. It is situated just above the *pituitary gland*, and exerts control over it. The pituitary gland releases HORMONES many of which exert chemical control over other *endocrine glands* causing them to secrete particular hormones. The hypothalamus is itself also sensitive to the levels of some hormones in the blood, and it is therefore in a key position to regulate the endocrine system. Damage to the hypothalamus disrupts the secretion of the sex hormones during the ovarian cycle, and interferes with normal SEXUAL behaviour.

When an animal is in danger it must become alerted and ready for sudden action. In such circumstances the heart beats faster and the rate of breathing increases, the pupil of the eye dilates, sweating occurs as do other changes which prepare the animal so that it can respond quickly and efficiently. When an animal is afraid or is about to attack, a variety of such changes take place, and these may be taken as a crude index of EMOTION. In general it is the sympathetic part of the autonomic nervous system which mobilizes the body for action, and the parasympathetic part which acts to conserve body resources. The effects of sympathetic activation are enhanced by the release by the adrenal glands of the hormone adrenalin into the bloodstream. The release of adrenalin from the adrenal glands is triggered by a hormone produced by the pituitary gland, and the hypothalamus is thus in a central position to initiate and control changes in emotional state. Other areas of the forebrain are also closely involved, such as the nuclei of the amygdala. If the amygdala is removed in monkeys the animals become very tame and placid, and appear to become indifferent to the social advances of their fellows.

It is clear that in the hindbrain, midbrain, and lower forebrain there are nuclei which regulate such basic functions as sleep, eating, drinking, sexual and emotional behaviour. If some of these nuclei are artificially stimulated with electrical pulses an animal, such as a cat, may carry out some specific action, which it will repeat when the stimulation is applied on another occasion. Experiments of this sort have been performed on chickens (*Gallus gallus domesticus*), opossums (Didelphidae), rats, cats, and monkeys, and a great variety of actions have been evoked, especially from stimulation of nuclei in the hypothalamus. Animals can be induced to settle for sleep, to eat food or drink water, even though they have had their fill, to build nests, to retrieve their young, to show fear or to threaten and attack. The

behaviour that an animal shows is usually appropriate for that species, so that, for example, the way a chicken, cat, or monkey displays AGGRESSION is the way that it would normally do so. Even more interesting is the observation that these animals do not carry out the action regardless of the circumstances. An animal will not attempt to eat unless there is food available, to build a nest if there is not suitable material, or to attack in the absence of some appropriate scapegoat. If a rhesus macaque is provoked to attack by electrical stimulation of part of the brain it will only carry out an attack if its partner is one which it is usually able to attack without fear of reprisals. If it is faced with a more aggressive animal it will turn away without a fight.

It seems that the effect of the electrical stimulation is to trigger groups of nerve-cells to discharge in much the way that they typically do, rather than merely to disrupt the normal activity of those cells. It is of especial importance that actions can be evoked in this way, even if the animal has never been exposed to the appropriate circumstances in the environment before, and has never yet shown the behaviour. If a cat is reared in a cage, with no chance to see prey or to practise killing them, it will stalk, paw, and bite a rat in much the same way as an experienced cat when stimulated in the appropriate area of the hypothalamus. The typical way in which a cat hunts and kills prey is built into the animal from birth, although it may be modified by practice later in life. This must be done by the establishment during the embryological development of the brain of circuits of nerve-cells wired up to be ready to control particular actions (see INSTINCT).

Ethologists have shown that much of the behaviour of some vertebrates does not have to be specifically learnt. Many animals know how to build a nest or to court, even if they have never seen it done or had the chance to practise. They do not have to acquire all of the repertoire of behaviour which will serve to adapt them to their environment. It seems that similar areas of the brain control similar functions in different species, as for example the areas in the hypothalamus regulating eating and drinking. But different species carry out such actions in very different ways. Each species has its own way of eating, of copulating, or of defending itself against predators, and if the relevant areas of the brain are electrically stimulated it carries out these actions that way. Although the gross structure of these areas is very much the same in different species, scientists recognize that the detailed circuits of nerve-cells must differ.

Cortex. Although parts of the hindbrain, midbrain, and lower forebrain are quite similar in different vertebrates there are very striking differences in the cerebellum and cerebral hemispheres. These can be seen in Fig. K in which the cerebral hemispheres are indicated by the stippled area, and the cerebellum is shown behind the cerebral hemispheres. Both the cerebellum and the cerebral hemispheres are covered round the outside with cortex (Latin for the bark of a tree), that is, with cell bodies arranged in sheets or layers. The fibres passing to and from the cortex from underneath have a fatty insulating sheath which appears white, whereas the cortical layers have a dull greyish appearance, after death. This has led to the distinction in common speech between grey matter and white matter.

The cerebellum evolved primarily for the regulation of balance and posture. It is more highly developed in birds than in fish and reptiles because of the need for fine coordination of movements to maintain FLIGHT. The mammalian cerebellum differs in having lateral lobes, which are greatly expanded in those mammals with larger brains and which come to form most of the cerebellum. The cerebellum is larger in carnivores and ungulates than in rodents or insectivores, and is best developed in the primates, particularly in man. The lateral lobes appear to be involved in the control of limb movements, and in primates in the regulation of fine movements of the fingers. If the cerebellum is damaged in monkeys, or in people, the smooth performance of skilled tasks is disrupted.

The cortex of the cerebral hemispheres is usually divided into the so-called old cortex, consisting of the *paleocortex* and the *archicortex*, and the new cortex or *neocortex*. The reason for this division is that while fish, reptiles, and birds have paleocortex and archicortex, only mammals have neocortex as well. The distinction between the different types of cortex is made on the basis of the number of layers that can be distinguished using a microscope, there being six layers of cell groups in neocortex and less in other types of cortex. The layers can be roughly distinguished in Fig. L.

The nuclei of the forebrain and the cortex overlying the cerebral hemispheres receive information from each of the sense organs. Fibres from the smell-sensitive tissue of the nose end in the olfactory bulbs of the forebrain, and from the olfactory bulbs other fibres run to certain nearby nuclei and to parts of the paleocortex. The size of the bulbs is a good index of the animal's sensitivity to smell. They are much smaller in birds than in animals such as reptiles, which live on the ground and can make good use of information from smells. The olfactory bulbs are very well developed in most mammals, particularly so in small insectivores and rodents which nose along the ground after food, but they are much smaller in primates which are more dependent on good sight for their life in the trees.

In all vertebrates, fibres from the eye and ear pass to the *tectum*, i.e. the roof of the midbrain, and also to particular nuclei in the collection of nuclei termed the *thalamus* in the interbrain (Fig. I). It has been known for some time that in mammals there were fibres which passed from the thalamus to the neocortex, and it is now known that there are corresponding pathways from the nuclei of the thalamus in reptiles and birds. These fibres terminate in nuclei in what has been termed the *striatum*, and in areas of cortex which are not six-layered. It appears that although these nuclei and cortical areas differ from neocortex in their appearance, they are connected in similar ways to other parts of the brain, and that the cells in these areas in reptiles and birds may have migrated in evolution from the same set of cells in their ancestors as some of the neocortical cells in mammals.

If this were so, it might be expected that the functions carried out by the neocortex in mammals be the responsibility of other areas of the forebrain in fish, reptiles, and birds. The whole of the forebrain of fish such as goldfish (*Carassius auratus*) has been removed, and the fish found to be poor at learning particular tasks, such as learning to anticipate an electric shock and to swim away to avoid it. However, little is known of the particular areas of the forebrain involved with learning. More is known about birds, such as pigeons (Columbidae) and quail (Coturnicini), which can be trained to consistently choose one of two patterns to earn their food, and then to reverse their choice if required. If part of the striatum, the *hyperstriatum*, is damaged (Fig. M) they are then very poor at this task. Though more work is needed to establish the similarities of such areas to the mammalian neocortex it is clear that other areas of the forebrain play some role in learning in vertebrates other than mammals.

In fish, amphibians, and reptiles, e.g. the tortoise (*Testudo*), the old cortex is wrapped round much of the outside of the hemispheres, but in mammals the neocortex pushes these underneath and onto the inner surface (Fig. M). Those mammals with larger brains tend also to have a greater proportion of neocortex relative to the rest of the brain. Because the neocortex is like a sheet, one way of keeping a constant ratio between the area of neocortex and the underlying volume of fibres

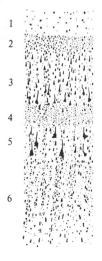

Fig. L. Section through the neocortex as seen through the microscope, showing cell bodies arranged into six layers, and different cell sizes in the different layers.

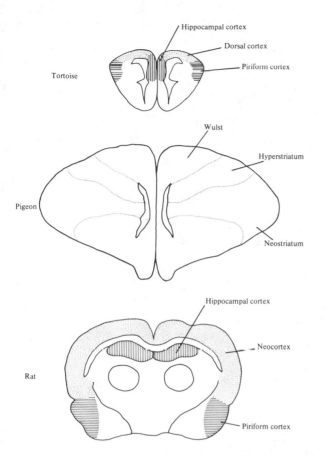

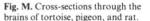

Fig. M. Cross-sections through the brains of tortoise, pigeon, and rat.

is to fold the sheet into creases as the volume gets larger. The folds in the neocortex are called fissures or *sulci*, and there are more of them in larger compared with smaller brains (Fig. K). In small-brained insectivores the neocortex and underlying fibres form less than 30 per cent of the total size of the brain, and their brains are smooth. By contrast up to 70 per cent of the brain in monkeys and around 75 per cent of the brain in the chimpanzee is taken up by the neocortex, and these brains are covered in fissures. Man has a greater proportion of neocortex than any other mammal, around 80 per cent of the entire brain. Even elephants which have much larger brains have proportionately less neocortex in relation to the whole brain.

The neocortex is divided into areas with specialized functions. There are sensory areas, each of which receives information from one of the senses, motor areas which play a part in controlling the movements of the body and limbs, and other areas whose functions are still poorly understood. These last have traditionally been called association areas, because it was supposed by the early neurologists that their role must be to associate together the impressions of the world formed separately in the sensory areas for vision, hearing, and touch.

The areas of the neocortex which receive information from the different senses can be mapped out by recording with small electrodes the electrical activity of large groups of cells in animals that are anaesthetized. In this way it can be shown that the cells are responsive in one area when the animal is touched, in another when sounds are played to the animal, and in yet another if a light or pattern is shown to the animal's eye (Fig. N). The size of the area devoted to a particular sense reflects less the size of the sense organs than the importance of that sense to the animal. Primates such as monkeys are highly dependent on sight for feeding and moving in trees during the day, and they have a large area of neocortex receiving information from the eye, while carnivores such as cats, which are especially active in the evening and at night, depend heavily on sounds to tell them what is happening, and they have an extensive area of neocortex devoted to the analysis of sounds.

In the area for touch, and in the motor area, the body is represented on the surface of the neocortex in a sort of distorted map. Cells in one part of the touch area change their pattern of firing if the paw or hand is stimulated, whereas cells in other parts

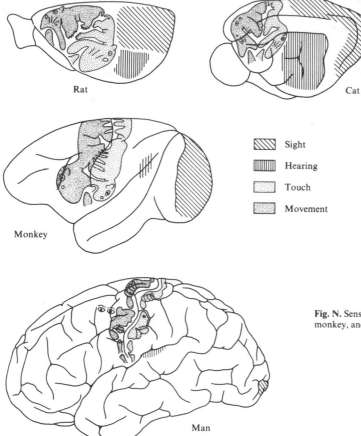

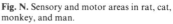

Fig. N. Sensory and motor areas in rat, cat, monkey, and man.

respond if the foot or the face is touched. The map in the motor area (see Fig. N) is charted not by recording from cells, but by electrically stimulating the cells by applying small pulses to the surface of the brain while the animals are anaesthetized. Stimulation of the cells in one area will repeatedly produce a movement in the arm, and in others movements of other parts of the body, showing that different parts of the body are controlled by different parts of the motor area. The main motor area lies in front of the area for touch. The distortions of the maps of the touch and motor areas reflect the importance of particular parts of the body to the animal. Animals such as insectivores and rodents, which use their snouts to snuffle around looking for food, have an extensive representation of the snout and whiskers in the touch and motor areas. Primates such as monkeys, which explore the world mainly with their eyes and hands, have large parts of the touch and motor areas devoted to the hands and feet, and especially to the sensitive and highly mobile fingers and toes. Racoons (*Procyon*), which also have hands, with which they search for and re-

trieve food, have a large representation of the hand in the touch areas. Those New World monkeys, such as the spider monkey (*Ateles*), which use their tail to grasp branches as a sort of fifth limb, have an expanded area devoted to the end of the tail. The map of the body in the touch and motor areas is also very distorted in man. Very large areas are concerned with sensation and movement of the hands and fingers, but the areas for the feet are relatively less extensive than in other primates, presumably because we use our feet for walking upright and have lost the ability that other primates have to manipulate and grasp with the feet and toes.

The proportions of the neocortex taken up by the sensory areas, motor areas, and the association areas are not constant in animals which differ in the size of their brains. In mammals with relatively small brains such as marsupials, e.g. kangaroos (Macropodidae), insectivores, and rodents, the association areas form a small proportion of the total neocortex. In the primates on the other hand, they take up from approximately half the total volume of neocortex in some species, to over 80 per cent in the large-brained chimpanzee. Man's brain seems to be specialized in having a larger motor area and more extensive association areas than would be expected for a primate of his body size. These dramatic differences between different mammals in the development of association cortex can be related to the importance that learning plays in the life of mammals.

Learning. The functions of the association areas of the brain have been examined by observing the effects of removing them on an animal's ability to learn some task. Experiments of this sort have only been performed on a few species of mammal such as rats, cats, and rhesus monkeys. It has been found that removal of any area of association cortex interferes with the animal's ability to learn, but that different areas of association cortex have specialized functions. The main areas are illustrated for the rhesus monkey in Fig. O. It is easy to train the rhesus monkey to choose consistently the one of two objects under which food is always to be found. But after removal of the area marked 'visual learning' the monkey is very bad at learning to do this by sight. Removal of the area marked 'tactile learning' causes the animal to be impaired in learning to choose between objects by the feel of them, and after damage to the area marked 'auditory learning' the monkey has difficulty in learning which of two sounds is the sign for food. These areas seem to play some role in learning to interpret events in the world. The functions of the frontal association cortex marked 'unlearning' are less clear. One way in which monkeys appear to change if this area is damaged is that they become less able to unlearn things that they have been taught to do. For example, they may be poor at switching their choice to an object other than the one they have previously been taught to select. Whatever the correct interpretation of the functions of this area they must be functions which are basic for all mammals, as the area is

found even in marsupials such as opossums (Didelphidae).

The development of neocortex and of association cortex in particular in mammals clearly reflects some specialization for learning compared, for example, with reptiles. Birds show some specializations in the same direction since, although they have no neocortex, they have relatively large brains, and there are areas of their brain which appear to have similar functions to the neocortex of mammals. Basic to the way of life of placental mammals (the higher mammals) is the strategy of having only a few young, which they nurture and protect for a period of development which is a relatively long period in their life. In general the larger an animal the longer it will live, and the longer the period of development before it is mature. During this period there is time to investigate and to learn how to cope with the world, while being protected from the full rigours of the environment by the parents. The longer this period the greater the pressure of NATURAL SELECTION for brains which are efficient at learning. In general, therefore, the mammals have adopted a strategy by which there is a restriction on the amount of the repertoire which is built into the brain at birth, and there is an increase in the brain's capacities for acquiring information and learning new skills. In this way they are better able to adapt to new environments, and to change their behaviour if the world about them changes.

This being so, it should be possible to show that those mammals which have the greater development of neocortex should be the better able to learn and to unlearn. In fact, this proves difficult to test. The test which is needed must compare the performance of different species, and must be fair to all species and not biased in favour of some and against others. In practice, such a test is hard to come by. A test in which the animals are required to learn to choose between objects to obtain food

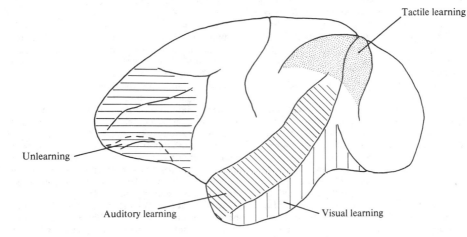

Fig. O. Association areas of the rhesus monkey. Removal of a shaded area impairs the type of learning indicated.

will favour the primates because of their well developed vision, and be biased against such mammals as rodents which are nocturnal, and therefore have poor vision but a good sense of smell. Accepting that there are biases of this sort the results of comparative experiments are still suggestive. In general, those mammals with larger brains and better developed neocortex are more likely to examine and manipulate objects they have not seen before, and they tend to PLAY more as youngsters. On formal learning tasks they are quicker to learn, and also to unlearn something they have been taught and learn something else instead. Some birds, such as pigeons, do as well as many small mammals, as would be expected from the relative development of their brains. Like some mammals, such as rats and monkeys, pigeons are good at exploiting the environment created by man.

The relationship between the development of the brain and intelligence is best appreciated by considering chimpanzees and people. The brains of chimpanzees are very large, and their neocortex is particularly well developed. Like monkeys and some other mammals they can learn TOOL USING, such as using sticks for reaching for food, but they differ in also being able to modify tools as, for example, by fitting two sticks together to give a longer reach. In the wild the chimpanzees of the Gombe Stream Reserve in Tanzania make simple probes with twigs and grasses to be used for getting ants out through the holes in their nests. They learn to do this by watching those who already know how to do it, that is, by the IMITATION of the actions of others. Although many mammals can learn about their environment by watching how others fare in it, the ability to specifically copy the actions of others is well developed only in those mammals with large brains, such as monkeys, apes, elephants, and dolphins. Man stands out from all the other mammals in his capacities both for inventing a technology to ensure survival and for passing on information and skills from one person to another. Man's skills may be attributed to the development of the cerebellum, and the motor area of the neocortex, and his intelligence to the size of his neocortex, and in particular of the association areas.

Cerebral dominance. There is one crucial difference between the brains of man and other animals. While animals have signs and gestures with which their moods and intentions may be communicated, only man has LANGUAGE with which he may talk about the world around him. Dogs and chimpanzees can understand something of what people say but are unable to imitate the sounds of speech, though a Javan hill mynah bird (*Gracula religiosa*) can do this accurately. Not only is an animal's vocal tract limited in the sounds it will produce, but its brain also lacks the mechanism for producing speech. In the human brain there are two speech areas, that is areas with specialized functions for language (Fig. P). If the area towards the front, called Broca's area, is damaged in an accident, or as the result of a stroke, the person has difficulty in talking, whereas if the area further back, called Wernicke's area, is damaged the person may find it hard to understand what is said to him. Removal of the corresponding areas in monkeys does not appear to interfere with their ability to produce calls, so that it looks as if their brains lack specialized mechanisms for speech.

The development of speech areas in man's brain is associated with a further change in the organization of the brain. In all vertebrates the brain is divided into two hemispheres, one on the left and one on the right. Each hemisphere receives information about and directs the movements of the opposite side of the body. The right hemisphere, for example, analyses information about what happens to the left side of the body, and what is seen to the left-hand side of the animal. The result is that if one hemisphere is damaged in an animal or person, it is on the other side of the body that there may be disturbances of feeling or paralysis. That is not to say that each hemisphere functions independently of the other, since there are fibres passing from one hemisphere to another. In placental mammals, but not in other vertebrates, there is a very large bundle of such fibres called the *corpus callosum.*

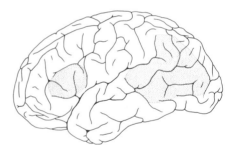

Fig. P. Left hemisphere of the human brain. The shading shows the extent of the two speech areas.

No good evidence has yet been found that in animals the functions carried out by one hemisphere differ from those carried out by the other. Damage to part of one hemisphere seems to produce much the same effects as damage to the corresponding area on the other side. But in man the understanding and production of speech tends to be carried out by one hemisphere alone. In right-handed people the left hemisphere is specialized for language in approximately 96 per cent of cases, and speech is disrupted in these only if damage is incurred to the left hemisphere. In left-handed people the left hemisphere is specialized for speech in only 70 per cent of cases, in 15 per cent the right is specialized, and in the remaining 15 per cent both share the functions equally. At the same time in right handers the right hemisphere tends to be better at carrying out certain perceptual functions than is the left. Remarkable findings have come from studies of patients in which the corpus callosum has been cut in the treatment of severe epilepsy. If an object is placed in the left hand the information about it travels to the right hemisphere. After the operation the speech areas on the left no longer have access to information in the right hemisphere, and the patient is not able to name what he feels, even though he can do so if the object is now placed in the right hand. This is a dramatic demonstration of the specializations in man of the two hemispheres for speech and other functions.

Brain mechanisms. The parts of the brain, and the way the brains of different species vary in size and organization have been described. It will be apparent that we know a little about the functions carried out by some of the main areas of the brain, partly because removing these areas often has dramatic effects on the animal's behaviour. We know rather less about how the many areas work together to produce the normal behaviour of the animal. An attempt has been made here to relate the gross differences between the brains of various animals to their different ways of life, but it is possible only to see crude relationships across animals spread widely over the animal kingdom, such as fishes, birds, and mammals. Examination of the gross structure of the brain provides clues only to such things as the senses on which an animal is most dependent, the agility with which an animal can move, or its abilities at PROBLEM SOLVING.

The differences between the brains of living species have come about as the result of EVOLUTION. For any animal characteristics have been selected which best fit it for survival in the environment it inhabits. To survive in any particular habitat an animal must not only have the appropriate physical structure such as the sense organs and teeth, but must also know the appropriate behaviour for obtaining the necessary resources and avoiding dangers to life. There are therefore selection pressures on the brain so that the animal will be equipped with those capacities which it needs if it is to be adapted for survival in its chosen NICHE.

There are two ways in which mechanisms could be built into the brain to ensure that the animal behaves in a way that makes it likely to survive and to reproduce in the environment in which it lives. Circuits can be built into the brain during its development controlling specific actions which are appropriate for the range of environments in which the animal is likely to live. The animal might, for instance, know how to hunt or how to mate without having to find out by experience in its lifetime. The differences between the way in which two species typically mated would result from differences in their genetic make-up, and therefore in the fine circuitry of their brains. The other way that the brain can promote the adaptation of the animal's behaviour to the environment it chooses is to build in a general capacity to adapt, i.e. by learning the appropriate behaviour during its lifetime. The advantage of doing this

rather than building in a specific repertoire is that the latter may become inappropriate if the environment changes, or if the animal is forced to move into a new habitat; an animal that can learn how to survive in one setting will be better able to change its ways and adapt to new settings. Of course most animals have both certain specific actions that they need not learn, and others that they must incorporate into their repertoire through experience. After all even worms can learn, and however impressive the stereotyped routines of many insects, they too must learn where food is, how to find their way about, and so on. But it is still true that, for example, mammals tend to have to learn more during their lifetime, and that in general their behaviour is less stereotyped and more easy to modify than that of invertebrates or even fishes.

But we know neither how the mechanisms which control specific routines of behaviour, nor how those which underlie the capacity for general adaptability, are built into the brain. Though we can see individual cell bodies and synaptic connections with the electron microscope, the multitude of connections is too great for us to have much hope of describing the intricate circuits which control any particular action. Even if we could identify such circuits we would also need some account of how the building of these circuits is regulated in the developing brain under genetic control. Somehow, in the developing brain, nerve-cells know where to go, and which are the appropriate cells to form synapses on their dendrites. Until we know how this is done we can say little of how the mechanisms controlling the varied patterns of behaviour of animals could be built into the brain. For similar reasons little is understood about the changes which occur in the brain when an animal learns. Most scientists assume that there must be changes in the likelihood with which some cells will fire. But if there are many thousands of nerve-cells which connect with a single cell, and given that there are many millions of nerve-cells in the vertebrate nervous system, it is easy to see why such changes are hard to detect. That is why there is more hope of detecting these events in simple invertebrate nervous systems, such as those of the cockroach and the sea hare (*Aplysia*).

It should not be surprising that, in the years of this century during which attempts have been made to relate the behaviour of animals to the brain, only very basic questions have been answered. We know the general functions of many of the areas of the brain, but much less about how they are carried out. Some exciting questions remain unanswered, how the mechanisms are built into the developing brain, and how they can be altered as a result of the animal's experience. An answer to these would greatly increase our understanding of the brain mechanisms controlling the behaviour of animals.　　　　R.P.

106, 112, 134

BREATHING is, strictly speaking, the movement of air in and out of lungs, and is thus a part of the total function of gas exchange. However, it need not be, and indeed is often not, restricted in its usage to animals with lungs. The cells of most living organisms use oxygen in the final process of converting digested carbohydrates (glucose) into energy. This oxygen consuming process is known as *aerobic metabolism*, and it produces a net total of nineteen times more energy than metabolism of a given amount of glucose in the absence of oxygen by the process known as *anaerobic metabolism*. A waste product of either process is carbon dioxide, so it is clear that, when undergoing aerobic metabolism cells must be provided with a continuous supply of oxygen while carbon dioxide must be continuously removed. In insects and spiders, air is presented directly to the cells via a system of tubes (*tracheae*), but in most other animals a circulatory system carries oxygen in the blood from a part of the body exposed to the environment to the metabolizing cells. Carbon dioxide travels in the opposite direction, from the cells to the external environment.

The part of the body in contact with the environment, across which gas exchange takes place, may be the general body surface, gills, lungs, or tracheae. The means by which oxygen enters and carbon dioxide leaves the body at these gas exchange surfaces is the physical process of diffusion. The gases move from a region of high pressure to a region of low pressure across a thin barrier of tissue. Because the cells are using oxygen and producing carbon dioxide, the pressure of oxygen is low at the cells and the pressure of carbon dioxide high. Thus, when the blood (or gas in insects and spiders) from the cells reaches the gas exchange area, oxygen will diffuse inwards from the environment, and carbon dioxide outwards. This is, of course, only possible if the pressure of oxygen is high on the outside of the animal and the pressure of carbon dioxide low. Breathing, in its widest sense, is the process whereby the external medium is continually renewed at the gas exchange surface, thus maintaining a high external pressure of oxygen and a low external pressure of carbon dioxide. It includes, therefore, the movement of water over the general body surface and across gills, and of air in and out of tracheal tubes, as well as the movement of air in and out of lungs. Breathing is thus a function performed by all animals that utilize oxygen, and not just by those with lungs. Synonyms for breathing are ventilation, which etymologically should be restricted to air breathers, and irrigation which may be applied to water breathers. Respiratory physiologists, in fact, tend to use the word ventilation for all animals.

Air and water are, in terms of their physical properties, vastly different respiratory media. The maximum amount of oxygen that can normally be dissolved in water is approximately twenty times less than is present in an equal volume of air.

Thus, in order to extract the same amount of oxygen as an air breathing animal, a water breather may have to move at least 20 times more of the external medium over its gas exchange organ. At a temperature of 20 °C, water is about 850 times denser and 55 times more viscous than air, so that a water breather has to expend more energy than an air breather per unit volume of the medium that is moved. Also, as water is a far better heat store than air (the heat capacity of 1 litre of water is 3000 times greater than that of 1 litre of air), it is not surprising that no warm blooded animal breathes water.

The pattern of breathing may vary in order to balance the demands of the animal against the oxygen available in the environment, and it can vary in response to a change in either or both of these factors. A change in the external environment may cause a modification of the breathing pattern by stimulating sensory processes (see SENSE ORGANS) which respond to that change (i.e. temperature, movement, or the pressure of oxygen and/or carbon dioxide). On the other hand, the central nervous system (which is where the breathing pattern originates) may initiate changes in breathing activity, either to supplement those produced by the stimulation of sense organs, or perhaps to anticipate an increase in the demand for oxygen. A state of excitement will cause an increase in ventilation, maybe in anticipation of muscular activity. Emotional disturbances in human beings can also elicit changes in the pattern of breathing; the lie detector monitors breathing activity, amongst other variables. As far as animals in general are concerned, most of our knowledge of breathing is related to variations in the external environment and to exercise, and it is apparent that the limitations of the respiratory system may have profound effects upon the behaviour of an animal.

The chemical reactions of the body which involve the use of oxygen, like any chemical reactions, are affected by temperature (a rise in temperature causing an increase in the rate of reaction), and all animals, except birds and mammals, are to a large extent slaves of the environmental temperature (see THERMOREGULATION). Many of these animals breathe intermittently when at rest and when the environmental temperature is low, e.g. crabs, amphibians, reptiles. When the temperature rises and when the demand for oxygen increases, then the duration of the pauses between the periods of breathing may decrease, until ventilation becomes continuous.

Oxygen demand increases for all animals when they raise their level of activity. All animals that have been studied, such as insects, fishes, reptiles, birds, and mammals, increase the rate and often the depth of breathing in response to exercise. In birds and mammals these increases are so abrupt in their onset at the beginning of exercise, that it has been concluded that they must be initiated by the central nervous system. Indeed, there is evidence to suggest that birds increase their breathing frequency slightly in advance of the muscular activity, although for birds and mammals it is also thought that sense organs responding to movement may contribute to the changes in breathing associated with exercise. Often the increased ventilation of the gas exchange organ that is required during increased muscular activity, is, to some extent, produced by the locomotor muscles themselves. In pigeons (Columbidae), for example, there is a one-to-one correspondence between wing-beat frequency and ventilation frequency, and it has been suggested that the movements of the wings cause changes in the volume of the thorax, which assist the activity of the respiratory muscles in filling and emptying the lungs. There is even more convincing evidence that ventilation of the tracheae during flight in insects such as locusts (Acridoidea), dragonflies (Anisoptera) and butterflies (Lepidoptera) is to a large extent dependent upon the activity of the flight muscles. The use of the locomotory muscles to ventilate the gas exchange surface is seen at its ultimate in a beetle, *Petrognatha*, in which air is forced through wide tracheae as the animal flies forward. Such a method is also utilized by very active fish, such as the skipjack tuna (*Katsuwonus pelamus*), which merely keeps its mouth open as it swims so that water is forced over the gills.

The availability of oxygen in the environment can fall under certain circumstances, e.g. in stagnant ponds and at high altitudes. Most animals that have been studied increase their rate of breathing in the face of reduced environmental oxygen. Perhaps the most variable environment is the sea shore, which is periodically covered and uncovered as the tide ebbs and flows. Despite this variability, a number of animals successfully inhabit this region. As the tide retreats some of them may be stranded in rock pools, where they may experience a wide variation in both temperature and oxygen availability. When the water is well aerated, the shore crab (*Carcinus maenas*) stays submerged, pumping water into its gill chamber via openings at the base of its legs. The water leaves the gill chamber via an opening just beneath and between the eyes. If the oxygen pressure in the water falls, it may reach a critical level (which varies with temperature), at which the animal raises itself up so that its anterior region is out of water. Air is then drawn in via the normally exhalant openings and bubbled through the water in the gill chamber. This so-called emersion response allows the crab to obtain sufficient oxygen even though the oxygen pressure in the rock pool may be very low. In fact when keeping *Carcinus*, if the sea water cannot be aerated, it is better for the animals to be in shallow water so that they can emerge, than in deep water where they may suffocate. In tropical and sub-tropical regions there are crabs which are predominantly air breathers, e.g. the coconut crab (*Birgus*). In these animals the gills are reduced and a definite lung is present. If submerged in water for more than 5 h these crabs drown.

The lugworm (*Arenicola marina*) lives in a U-shaped burrow on sandy or muddy shores. When the tide is in the animal irrigates the burrow by undulating its body, thus creating a current of aerated sea water which passes over its gills. Irrigation occurs in bursts, approximately every 40 min. When the tide goes out the worm ceases its irrigation movements, but it does make intermittent testing movements, so that normal activity can be resumed as soon as the tide returns. During low tide, the lugworm, like the shore crab, may use air as a source of oxygen. It has been observed to force a bubble of air down the shaft of its burrow by a modification of the normal irrigation movements. Indeed, the use of air as a source of oxygen is quite common amongst intertidal animals. Barnacles (Balanomorpha), which rhythmically waft their modified limbs through the water at high tide, retract them into their shells when the tide is out, but a small opening is left between the valves of the shell for air to diffuse into the animal. Many intertidal fishes breathe air. Perhaps the best-known example is the eel (*Anguilla*), which at low temperature can survive for many hours in air. Fishes such as the mudsucker (*Callichthys*) gulp air at the surface of the water when the pressure of oxygen in the water is low. Gas exchange takes place across the highly vascularized walls of the mouth.

It was the ability of some fish to breath air and the accompanying development of lungs during the Devonian period, some 300 million years ago, which led to the EVOLUTION of the terrestrial vertebrates, culminating in man. Although the sea shore is a highly variable environment, the cycle of events is fairly regular. Bodies of fresh water, on the other hand, can become deoxygenated if stagnant, or even dry up altogether. Such conditions were thought to be widespread during the Devonian period, which meant that freshwater fishes that could breathe air, and perhaps move from a desiccated pond to a wet one, stood a better chance of survival. The living lungfishes (*Lepidosiren*, *Protopterus*, and *Neoceratodus*) are relatives of those early air breathing fishes. When in water, a lungfish ventilates like any other fish, i.e. by forcing water across its gills as the mouth closes and by sucking water across them as the opercula (gill covers) open. When the lungfish surfaces for a breath of air, the opercula shut, the mouth opens in the air, the glottis opens, and air is forced into the lung as the mouth closes. Thus, the mechanism used in water is modified for use in air. Amphibians, such as frogs and toads (Anura), also fill their lungs by forcing air down as the volume of the mouth is reduced. These animals, as the name amphibian suggests, live in water and on land. When in water they obtain oxygen via the skin, whereas on land they breathe via their lungs. Despite their ability to obtain oxygen from the water, the rate at which this is possible is not always sufficient to satisfy their needs. Thus, at high environmental temperatures, or when the pressure of oxygen in the water is low, the animals have to surface frequently for air. Increased activity of the animal may also necessitate a visit to the surface. The need to breathe sometimes conflicts with another activity. For example, the crested newt (*Triturus cristatus*) breeds under water, and for the male this is a period of high activity. Premature surfacing for a breath of air could lead to an incomplete mating (see COURTSHIP).

Although some reptiles can obtain oxygen from water, the majority, like all birds and mammals, rely solely upon air for their source of oxygen. Despite this fact, a number of reptiles and a few birds and mammals spend a great deal of their lives on or under water. There are changes within the circulatory system which allow these animals to remain under water for extended periods, but in ducks (Anatinae) at least, there is also an increase in lung ventilation which heralds a period of submersion, and is similar to the increased breathing that precedes and accompanies muscular activity in other birds. Of course, as soon as the animal dives, breathing ceases. Compared with mammals, most birds stay submerged for relatively short periods of time (2–3 min), and even the best avian diver, the emperor penguin (*Aptenodytes forsteri*), can only remain under water for 18 min. On the other hand the bottle-nosed whale (*Hyperoodon ampullatus*) has been reported as diving for periods of up to 2 h. However well-adapted a bird or mammal may be for an aquatic existence, it still has to breathe air, and whales (Cetacea) in fact can renew a greater proportion of the air in their lungs at each breath than can human beings.

Insects also have a respiratory system that is primarily designed to function in air. In the larger forms, breathing consists of pumping air through the tracheae. Nevertheless, a number of insects, particularly during their larval stages, live in water. Like air-breathing vertebrates, some insects make periodic visits to the water surface in order to obtain air, e.g. mosquito larvae (Culicidae), but unlike the vertebrates, there are many insects which are totally independent of atmospheric air. The larva of the beetle *Donacia* obtains its oxygen by penetrating the gas spaces in the roots of aquatic plants with a long siphon. The bug *Aphelocheirus aestivalis* has a dense mass of short hairs over its body which hold a thin incompressible layer of air. The tracheae open into this thin layer of air, and as oxygen is removed from there by the animal, more oxygen diffuses into the air bubble from the water. This gas gill is permanent and the animal never has to replace it with atmospheric air. Some insects have developed true gills, i.e. thin outgrowths from the body wall which contain a rich plexus of thin tracheae. Examples are provided by the larval stages of mayflies (Ephemeroptera), damsel flies (Zygoptera), and dragonflies (Anisoptera).

In the larvae of the mayfly *Ephemera*, the gills are held over the top of the abdomen and beat continuously, with the frequency of the rhythm being inversely proportional to the environmental oxygen pressure. These larvae bury themselves into rather coarse sediment, for example the gravel of streams and lakes. Different species of *Ephemera* have a marked preference for different substrate particle size, and oxygen uptake of a particular species is lowest when it is buried in its preferred substrate.

Breathing, then, is a basic activity, and the examples quoted may have given some idea of the variety of breathing mechanisms present in the animal kingdom. The relative importance of breathing in terms of overt behaviour varies from animal to animal, but it is the one overt activity which may remain when other functions have been suppressed. For example, anaesthetics, given in the correct dose, can inactivate the areas of the BRAIN concerned with perception to the point where consciousness is lost, but breathing will continue. Even under more natural conditions breathing can become a dominant part of an animal's behaviour, and there is perhaps no better example to illustrate this than the larva of the midge *Chironomus*, which lives in a U-shaped tube in the soft mud at the bottom of a lake or pond. The behaviour of the larva can be divided into three phases: filter feeding, immobility, and intermittent irrigation. In well aerated water it spends approximately half of its time irrigating the tube, the rest of the time being concerned with feeding or remaining motionless. If the pressure of oxygen in the water falls this general pattern of behaviour does not change until the oxygen pressure falls below 25 per cent air saturation, when there is a progressive decline in the time spent filter feeding and an increase in irrigation. When the oxygen level is below 10 per cent air saturation, feeding ceases altogether and the animal spends all of its time 'breathing'.

P.J.B.

C

CAMOUFLAGE is a form of visual deception, by means of which an animal can elude predators, or a predator may lurk undetected awaiting a suitable prey (see PREDATORY BEHAVIOUR).

The term camouflage or *crypsis* is usually used to denote similarity between the animal and its visual background, as is illustrated by the young ringed plover (*Charadrius hiaticula*) in Fig. A. Where an animal closely resembles an object in the environment, for example, the long-horned grasshopper *Zabilius aridus* looks like a leaf (Fig. A), the term MIMICRY is used.

It is important to remember that in most studies of camouflage judgements as to the effectiveness of concealment are made in terms of human VISION. There have been some attempts to determine a predator's powers of DISCRIMINATION directly by training animals to respond to coloured photographs of camouflaged prey. Blue jays (*Cyanocitta cristata*), for instance, have been trained to respond differentially to the presence or absence of the bark-like underwing moth *Catocala*, presented to the birds as photographic slides. The moths are normally active at night, and spend the day resting on tree trunks. *Catocala cara* and *Catocala retecta* normally rest on oak trees and have brown forewings. *Catocala relicta* has whitish forewings and normally rests on birch trees. The jays had much greater difficulty detecting moths resting on the appropriate background compared with when they were resting on any other background. Detection of moths on their proper substrate was especially difficult when the moths were vertically oriented, but orientation made no difference when the moths were on other backgrounds. Thus the moths achieve maximum camouflage by resting vertically upon a particular species of tree (see DEFENSIVE BEHAVIOUR).

There have also been studies on the effectiveness of camouflage against PREDATION. For example, predation of the green form and the yellow form of the grasshopper *Acrida turnita* by chameleons (Camaeleontidae) is markedly affected by the background. In one study, green grasshoppers had a 30 per cent greater chance of survival on a green background than did yellow grasshoppers, and yellow grasshoppers a 38 per cent greater chance of survival on a yellow background than green grasshoppers. Clearly, the success of camouflage must depend largely on the visual capabilities of both predator and prey.

Techniques of camouflage. An object can easily be detected against a particular background if it has a sharp outline, casts strong shadows, or is a different colour from the background. Most techniques of camouflage are designed to minimize these differences. An animal's body outline can be concealed by disruptive coloration, as in the butterflyfish *Chaetodon* and *Heniochus*, or by morphological projections which break up the body

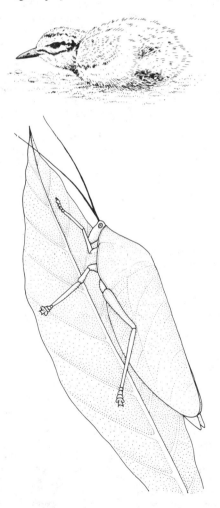

Fig. A. Similarity between the animal and its visual background is shown by the young ringed plover (*Charadrius hiaticula*) and the long-horned grasshopper (*Zabilius aridus*).

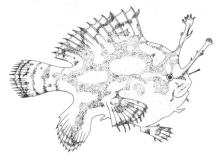

Fig. B. The frogfish (*Histrio histrio*).

outline, as in the frogfish (*Histrio histrio*) which lives among Sargassum seaweed (Fig. B). The coloration of many species of frogs and birds includes an eye-stripe which disrupts the otherwise conspicuous circular eye (Fig. C). Revealing shadows may be suppressed by counter-shading, in which the shadow normally cast on the underside of the body is counteracted by lighter coloration of those parts. Animals which normally rest upside down, such as three-toed sloths (*Bradypus*) and some species of the catfish *Synodontis*, have the lighter coloration on the dorsal side of the body. Counter-shading is most effective when the lighting conditions are strongly directional. Pelagic fish, such as the Atlantic herring (*Clupea harengus*), live at a depth at which there is considerable light scattering so that objects are not strongly illuminated from above. They have a darkly pigmented dorsal

Fig. C. Disruptive coloration in the Pacific tree frog (*Hyla regilla*).

surface which makes them difficult to see from above, and silvery reflecting lateral and ventral surfaces, which reflect light back to a predator viewing from the side, but do not reflect light in an upward direction. This arrangement makes the fish very difficult to see against the background of scattered light, except when the fish is viewed from below. Viewed from above the fish merges into the background, there being no tell-tale flashes of silver. Some fish, such as the glassfish (*Chanda ranga*) have a transparent body, and thus avoid casting strong shadows or presenting a distinctive body outline. Shadows cast onto the ground may give away an animal's presence, and these may be

minimized by adoption of a flattened resting posture. The mantid *Theopompella westwoodi*, whose colour resembles the bark of the tree trunks on which it hunts, flattens itself against the trunk and extends its wings sideways to obliterate any shadow. Shadow minimization can also be achieved by avoidance of directional light, and by venturing only into shaded places and other regions with scattered light conditions. Many animals, for example flatfish, such as flounders (Pleuronectinae) and sole (Soleidae), depend upon the resemblance between their own colour and the background against which they live to provide camouflage. Extremely cryptic patterns can be achieved. Some animals, notably the cuttlefish (*Sepia officinalis*) (DEFENSIVE BEHAVIOUR, Fig. B) are able to alter their coloration to match that of the background, whilst others achieve matching coloration by careful selection of the background against which they rest.

Limitations of camouflage. For camouflage to be effective the animal must remain motionless. Thus there will be times when cryptic behaviour is incompatible with other essential activities, such as courting and foraging. Some species remain motionless during the day, and become active at night, or at dawn and dusk. Others, including many butterflies and moths (Lepidoptera), are active during the day and take up a cryptic position in ALARM. In evolutionary terms, there is often a conflict between the requirements of camouflage and those of COURTSHIP. For example, it is important for many birds to remain inconspicuous during INCUBATION. On the other hand, as a result of SEXUAL SELECTION, conspicuous coloration can make an important contribution to success in courtship. In some species, such as the chaffinch (*Fringilla coelebs*), only the female broods; she is cryptically coloured while the male has conspicuous colours. In the reed warbler (*Acrocephalus scirpaceus*) both sexes are cryptic and both take on incubation duties in an open nest. Among tits (Paridae) both sexes are brightly coloured, but the nest is usually in a hole or crevice, where camouflage is of no advantage.

Many studies have demonstrated the importance of the appropriate background for animals depending on cryptic coloration. The dark coloured moth *Catocala antinympha*, given a choice between a black and a white background, usually rests on the black background, whilst the pale *Campaea perlata* chooses the white background. When these moths are painted to render them conspicuous against their chosen background, their behaviour does not change, suggesting that it is INNATE, and not due to the animal actively matching itself against the background. The yellow moth *Sohinia florida* normally rests on the yellow flowers of the evening primrose (*Oenothera biennis*). It prefers this primrose to other flowers, even when the colour of the flower is hidden by a muslin bag. It comes to rest on the muslin covering the flower, evidently being attracted by the scent of the flower. Many animals only achieve effective camouflage when they align themselves in a particular way in relation to the structure of their chosen background, as described for the moth *Catocala*. The response to the background may not be a purely visual one. The moth *Melanolophia canadaria* normally rests sideways on tree trunks, so that its disruptive markings align with those of the tree bark. Tests have shown that the moths respond in a tactile manner to the grooves in the bark of the tree, and align themselves accordingly.

Changes in the background are responded to by some animals on a short-term basis, as mentioned above. A number of arctic birds and mammals, such as the rock ptarmigan (*Lagopus mutus*) and the snowshoe hare (*Lepus americanus*) show seasonal changes, being white in winter and brown in summer. Some insects change colour when they moult: the pupae of such butterflies as the cabbage white (*Pieris brassicae*), and the African monarch (*Danaus chrysippus*), for example, can be either green or brown, depending on the immediate surroundings of the caterpillar just prior to pupation. Other animals modify their environment in order to improve their camouflage: the black-headed gull (*Larus ridibundus*), for instance, removes conspicuous eggshells from the region of its nest (see FUNCTION). Some web-building spiders rest at the side of their web. Here they are better hidden from

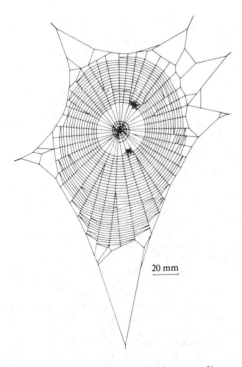

20 mm

Fig. D. The spider *Cyclosa* is resting in the centre of its web, and there are also two dummy spiders made from the remains of prey.

predators, although sometimes they find themselves too far away from prey that become temporarily entangled in the web. Other species rest in the centre of the web, exposed to predation, but nearer the scene of action. Some species of *Cyclosa* and of *Uloborus* add one or two dummy spiders to their webs. These are made of silk and the remains of prey, and are similar in size and shape to the spider itself (Fig. D): a predator will sometimes attack these dummies instead of the spider. Some species of *Theridion* and *Tetragnatha* rest in the middle of the web, but conceal themselves under a leaf or twig that they have taken there. Another ploy is for members of a species to differ in their coloration. There are, for example, five or six colour variants of the banded snail *Cepaea*. The advantage of this polymorphism is that it disrupts the SEARCHING IMAGE strategy of the predator which tends to pay ATTENTION to one colour pattern for a period of time.

39

CANNIBALISM is the eating of members of one's own species, alive or dead. When early Spanish and French explorers in the Antilles discovered that the Carib Indians ate human flesh, that of their slain enemies and captives, they were horrified and adopted their name for the tribe (Canibales) to refer to the practice. Since then, the term has been broadened to include any intraspecific predation or feeding.

Cannibalism can take two forms, active and passive. The first category refers to animals that actively hunt to kill and eat members of their own species. The second category refers to animals that will eat members of their own species that are found dead. Fig. A places cannibalism in the context of FEEDING. Prolicide refers to the killing of offspring. Kronism, after the Greek god Kronos, who ate all but one of his children just after they were born, refers to the killing and eating of offspring. Fratricide is the killing of siblings.

For many people cannibalism is only known from its practice by some of the primitive tribes of the world, where it is often associated with religious or funereal rites, warfare or revenge, but perhaps may occasionally form a nutritionally important part of the diet. However, the practice is by no means restricted to human beings. It occurs in a variety of animal species and may be regarded, not as an aberrant behaviour brought on by stressful or abnormal situations, but rather as a normal response to a variety of environmental conditions. Many of the reported occurrences of cannibalism are anecdotal in nature, or have been casually observed in confined LABORATORY populations. However, sufficient evidence is available from natural populations in the field to indicate that cannibalism is a normal phenomenon in some 138 different animal species, including protozoans, rotifers, copepods, molluscs, arachnids, insects, fish, anurans, birds, and mammals. Most of these 138 species are lower vertebrates or invertebrates,

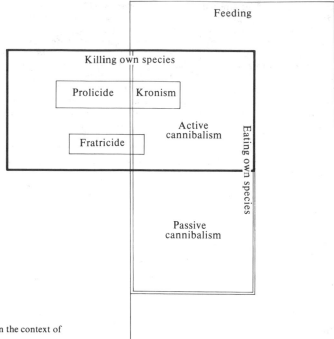

Fig. A. Cannibalism in the context of feeding.

only a few mammals and birds being represented.

It might be expected that the animals best equipped for killing, i.e., the carnivores, would be most likely to be cannibals. On the contrary, in this group of animals, there is usually a strong inhibition against cannibalism. Few of the carnivores will eat the flesh of their own kind or even of other species in the group. In some special cases cannibalism does occur. Male lions (*Panthero leo*), after taking over a new pride, will sometimes kill (and occasionally eat) any cubs in the pride. It has been suggested that this behaviour has the effect of killing off the GENETIC stock of the previous pride owner, and causing the lionesses to come into heat more quickly than they would if they had cubs to look after. The spotted hyena (*Crocuta crocuta*) has been reported to indulge in occasional active and passive cannibalism, but the meat is eaten more slowly than normal prey. Other mammalian examples are found in some of the rodent species. In periods of high population density and numbers, infant mortality becomes very high. This is due to a general lack of parental care, which includes an unquantified amount of cannibalism.

In birds, much of the cannibalism described involves the eating of the young by parents or siblings. This kronism occurs in many of the birds of prey and some storks. The carrion crow (*Corvus corone*) is known to cannibalize eggs and chicks, not for the nutrition they might provide, but in order to prevent a pair from breeding and to obtain a breeding territory for next season. Gulls,

particularly herring gulls (*Larus argentatus*), are also known to eat both eggs and chicks when the opportunity arises. Peculiar to gulls are the specialist cannibals. During the chick season these birds (mostly male herring gulls) exist almost exclusively on the young of their own species. They number less than 1 per cent of the breeding population, but are a major cause of chick mortality, and have been known to take up to 25 per cent of the chicks produced in one season. In contrast to these specialists are the casual cannibals, who will eat their own species when available, every member of the species or population being a potential cannibal. Most cannibalism falls into this category.

There are several factors that are known to influence the practice of cannibalism. In the house mouse (*Mus musculus*), the larvae of the damselfly (*Lestes nympha*), and the fry of pike (*Esox lucius*), cannibalism occurs mainly as a response to the rate of increase in population density. In other species, such as the monarch and queen butterflies (*Danaus plexippus* and *D. gilippus* respectively), the walleye (*Strizostedion vitreum*), and the sea slug (*Aglaja inermis*), cannibalism is regarded as a density-dependent phenomenon, such that the amount of cannibalism is directly proportional to the population density.

Cannibalism may also be related to the availability of alternative food. As sources of alternative food become depleted or unavailable, incidence of cannibalism increases. In the bug, the backswimmer *Notonecta hoffmanni*, the highest incidence of cannibalism occurs after a sharp fall in the abundance of other prey, and declines as abundance of alternative food increases.

In some cases, cannibalism may be induced more by the behaviour of the victim than by the cannibal itself. In viviparous fish of the genus *Poeciliopsis*, large groups of young stimulate females to cannibalize. This behaviour is not affected by the state of hunger of the adults. Small groups of young do not produce the same stimulating effect.

STRESS may also cause the induction of cannibalism. This has been observed mostly in laboratory situations, but there are some examples from the field. The older juveniles and young adults of the Australian grasshopper *Phaulacridium vittatum* undergo mass emigration if food becomes scarce. The migrating animals show signs of physiological stress and cannibalize other grasshoppers. Those remaining at the original site show none of the stress symptoms and do not cannibalize.

In the absence of any of the above inducers, cannibalism may occur purely on the basis of availability of prey: the probability of attack being proportional to the probability of encountering a vulnerable individual.

Much of the information available on the effects of cannibalism comes from studies carried out on the lower vertebrates and invertebrates. A particularly elegant experimental study has been carried out on laboratory populations of flour beetles (*Tribolium*). The results from this study have provided generalities about the effects of cannibalism that are also applicable to field populations. Most of the cannibalism that has been investigated has been found to have some sort of regulatory effect on population abundance. Both population size and age structure are influenced by cannibalism in flour beetles. In some cases, a particular population age structure can result in an unusually large single age class, which then dominates the population structure for a number of generations. This effect has also been found in the European perch (*Perca fluviatilis*), in the midge *Chironomus anthracinus*, and in water-striders (Gerridae).

In the backswimmer, and a species of sheep blowfly larvae *Chrysomyia albiceps*, cannibalism has the effect of increasing population stability and persistence. Under conditions of food scarcity, cannibalism occurs and results in the few surviving members of the population being large enough or sufficiently well fed to produce viable offspring, ensuring the survival of the next generation.

Cannibalism in other species may result in greater reproductive success or individual survival for the cannibal: an increase in growth rate has been found in cannibalistic large-mouth bass (*Aplites salmoides*), and in the plains spadefoot tadpole (*Scaphiopus bombifrons*); increased egg production has been found in the flour beetle *Tribolium castaneum*, and in the aquatic roundworm *Asplanchna sieboldi*.

The occurrence of cannibalism as a normal part of a species's life history, in succeeding generations, suggests the possibility of some sort of genetic influence on the trait. Evidence for such an influence has been found in flour beetles, spadefoot tadpoles and the viviparous fish *Poeciliopsis*. Strains of flour beetles show heritable differences in voracity. In spadefoot tadpoles cannibalism is associated with different morphological forms, and in *Poeciliopsis* hybrids from the crossing of cannibalistic with non-cannibalistic strains show an amount of cannibalism that is proportional to the genetic contribution of cannibalistic traits from the parents. Further evidence for the possible genetic determination of cannibalism comes from observations that the behaviour is often restricted to particular stages in a species's life history, or to only one of the sexes.

Cannibalism is a socially selfish trait, and for NATURAL SELECTION to favour it the advantages to the cannibal in terms of FITNESS must, on balance, outweigh the possible disadvantages. The most obvious advantage of cannibalism is that the cannibal gains a meal. It may also eliminate a potential competitor and sometimes a potential predator as well. Population size will be reduced, leaving a smaller number surviving to utilize the available resources. Selection against cannibalism will occur once individuals become so voracious that they begin eliminating potential mates, or killing off their own genetic stock. The EVOLUTION of a socially selfish trait depends on the degree of genetic relationship between the individuals involved. As the degree of relationship decreases, a smaller fitness advantage is required for the frequency of the trait to increase in the population. If, in the case of cannibalism, the victim is one's sibling or one's own offspring, a large increase in fitness would be required for the trait to be favoured by natural selection. On the other hand, only a small increase in fitness would be required if the victim is a distant cousin (see ALTRUISM). Work on the leaf beetle *Labidomera clivicollis*, where most of the cannibalism is on unhatched eggs by just hatched larvae (i.e. cannibalism of siblings), has indicated that only a small increase in fitness is needed for the cannibalistic trait to be adaptive, when either nutritional benefits are high, or mortality at that particular stage of the life cycle is high anyway. This quite small increase in probability of survival, for an insect that lays large numbers of eggs of which very few will survive to reproduce, serves to point out that selection for cannibalism does not necessarily require large increases in fitness.

To summarize, cannibalism occurs in a variety of animal species, as a normal part of their life. Its occurrence can be related to population density, to the availability of prey or alternative food, to behaviour of victims, or to stress. The behaviour can affect population size and age structure, as well as population stability and persistence. It may also result in increased reproductive success and individual survival. Like other forms of behaviour, cannibalism may be genetically influenced and responds to natural selection.   A.H.

CAPTIVITY. When a wild animal is brought into captivity, it suffers a number of simultaneous changes in its environment, to which it may take some time to adjust. The change in geographical location brings changes in CLIMATE and in apparent TIME of day, which require physiological adjustment and ACCLIMATIZATION. The change in diet and in the distribution and availability of food will induce changes in APPETITE and in FEEDING behaviour. The restricted freedom of movement will induce changes in SOCIAL behaviour, and the proximity to man may induce FEAR and DEFENSIVE behaviour.

Young animals generally adjust to captivity more quickly than older individuals. Animals that are born in captivity do not have to make such adjustments, although they may nevertheless be forced into a life style that is not natural for their species. When we look at the extent to which the conditions of captivity differ from those of the natural environment, we must take account, not only of the magnitude of the differences, but also the degree of ADAPTATION that is characteristic of the species concerned. Some species have rigid requirements whereas others are more flexible. For example, the giant panda (*Ailuropoda melanoleuca*) feeds exclusively upon bamboo, and this food must be provided in captivity. Since bamboo contains little nourishment, tremendous amounts of it must be eaten daily.

The physical environment. Many animals are able to adjust to changes in the physical environment without much difficulty, provided the new conditions are within the animal's TOLERANCE range. Initially, there may be disruption of daily and seasonal RHYTHMS due to desynchronization of the animal's internal CLOCKS, and the local day–night cycle. These are usually transient problems. Similarly many animals, particularly those with efficient THERMOREGULATION, are surprisingly good at adjusting to a new temperature regime. Flamingoes (Phoenicopteridae), for instance, suffer more from the mechanical effects of the ice on their ponds than they do from the cold of a European winter. Reptiles and amphibians, whose metabolic rate is more directly related to the environmental temperature than that of birds and mammals, may be sluggish and fail to grow if maintained at too low a temperature. Giant tortoises (*Testudo elephantopus* and *Testudo gigantea*), for example, are often kept out of doors in European zoos, but they usually fail to grow or to breed under such conditions. Animals in tropical houses are often maintained at too high and too uniform a temperature. Many reptiles regulate their body temperature by shifting their posture or position in relation to the micro-HABITAT. They can do this only if there are sufficient temperature gradients in the environment. Similar considerations apply to the temperature and chemistry of the water in aquaria, and to the humidity and lighting conditions provided for land animals. Some animals thrive on the opportunity to sunbathe, and may require occasional doses of ultraviolet light to aid the synthesis of vitamin D. Other animals are extremely light-shy, and become active only under night-like conditions.

The nature of the ground is often a very important feature of the captive animal's environment. Some arboreal animals, such as the orang-utan (*Pongo pygmaeus*), rarely visit the ground under natural conditions, and consequently never come into contact with their own excreta. Under captive conditions they appear to have no natural avoidance reaction, and may engender disease as a result of handling faeces which they discover on the floor of their cage. A number of primates have fear of the ground, and should be provided with sufficient tree branches to enable them to move about actively at a comfortable height.

Ground-living animals have a variety of requirements concerning the nature of the substratum. Ungulates, such as giraffes (Giraffidae), may develop abnormal hoof growth due to insufficient wear. It is often necessary to provide abrasive material, such as quartz sand, especially when the animals are less active than normal. Soft ground is particularly dangerous for ungulates, because excrement becomes trodden into it, thus encouraging PARASITISM. Burrowing animals must be provided with a suitable substratum, even if this means that they are not often on view. The natural defence reactions of such animals often involve ESCAPE into a burrow, and severe psychological STRESS may occur if this is not permitted.

The biotic environment. The captive animal may make use of vegetation in a variety of ways. Aquatic plants help to oxygenate aquaria, as well as providing food, cover, and nesting material. Many animals require plant material for NEST-BUILDING, or for the provision of a suitable SLEEP site. Perching birds and some mammals prefer to sleep above ground level. Many deer (Cervidae) normally use tree branches for rubbing velvet from their antlers, and if branches are provided in captivity they will immediately be used for this purpose. Similarly, animals such as American bison (*Bison bison*) clean their skin by rubbing against tree trunks. When the first telegraph lines were taken across the North American continent, the wooden posts were often brought down by the rubbing activity of bison.

Indigenous poisonous plants may endanger captive animals that come from a different part of the world. Most herbivores are able to avoid the poisonous plants of their native habitat (see FOOD SELECTION), but may die as a result of eating poisonous plants in a strange environment. When European cattle (*Bos primigenius taurus*) were introduced to the African veld they frequently died from poisoning. There have been cases of death in European zoos, such as the Nubian ibex (*Capra ibex nubiana*) which ate deadly nightshade (*Solanum nigrum*).

Incompatibility may also occur among animals of different species. It is obvious that one should not house predator and prey in the same enclosure, but other relationships may not be so obvious. When a red kangaroo (*Macropus rufus*) and a South American stag (Odocoeleini) were put in the same enclosure, the stag attacked the kangaroo whenever the latter sat up. Stags rear up and hit out with their forefeet when attacking each other, and the typical upright posture of the kangaroo was obviously interpreted by the stag as a sign of AGGRESSION. Hostile encounters between members of different species usually occur when each assumes that the other is a member of its own species, and reacts accordingly (see ANTHROPOMORPHISM). Such misunderstandings may also occur in the relationship between the captive animal and the human keeper. For example, a young king penguin (*Aptenodytes patagonica*) forced itself between a man's feet, just as it would have done with its real parent. During the rutting season roebuck (*Capreolus capreolus*) seem to regard human beings as male rivals, and vigorously attack them. A tame emu (*Dromaius novaehollandiae*) in the Basle Zoo regularly tried to mate with its keeper during the mating season in winter.

Animals in captivity are often visited by members of the indigenous wild population. In European zoos captive animals may be killed by foxes (*Vulpes*), wild cats (*Felis silvestris*), rats (*Rattus*), and raptorial birds of prey (Falconiformes). Their food may be stolen by squirrels (Sciuridae), rats and mice (Murinae), carrion crows (*Corvus corone*), etc. Diseases may be introduced by visiting cats, rats, and birds, and structural damage may be caused by rats gnawing through electric cables etc. In North America similar problems may be caused by racoons (*Procyon*) and in tropical countries by monkeys (Simiae).

Social encounters may be intensified among animals in captivity. Spatial confinement sometimes results in exaggerated DOMINANCE relationships, because the animals cannot avoid each other sufficiently. For example, the inferior of two ibex was unable to obtain food or shelter because these were monopolized by a dominant animal. Overcrowding may lead to heightened aggression due to disputes over TERRITORY, food, or sexual partners. Social harmony will generally be disrupted by scarcity of resources, because of the increased COMPETITION between individuals. Under natural circumstances such conditions can be ameliorated by spacing out, or by the emigration of some individuals. In captivity this type of remedy is in the hands of the human keeper.

The welfare of animals in captivity. Apart from the obvious cruelties, the WELFARE of animals in captivity is difficult to assess. It should not be assumed that the free and natural life is necessarily preferable to that in captivity. Life in the wild can be hazardous and harsh, and there is some evidence that animals prefer the comforts of captivity. It is well known that wild animals retained in captivity for more than a few months are sometimes difficult to re-establish in the wild. They may persistently return to their captors, and even try to re-enter their cages. This behaviour is difficult to interpret, however, because research shows that some animals have a preference for that which is recently familiar. Even domestic fowl (*Gallus g. domesticus*), maintained in battery cages, may initially prefer their cages to the freedom of the farmyard if given a choice. Thus, it is sometimes difficult to disentangle the temporary preferences that are due to familiarity from the more genuine basic preferences.

Animals in zoos often live longer than their wild relatives. For example, the life expectancy of a wild blackbird (*Turdus merula*) is about two years, and the greatest known age is ten years. Aviary blackbirds, on the other hand, have been known to live twenty years. Many mammals live about twice as long as normal when kept in captivity and, surprisingly perhaps, the raptorial birds live more than twice as long, even though they are very inactive in captivity. Many zoologists regard breeding success as an indication that captive animals are being maintained under proper biological conditions. Some species breed readily in captivity, but others require special attention. Research into the breeding requirements of rare animals is of considerable importance, because captive breeding is sometimes the only method of CONSERVATION. In 1979 the large blue butterfly (*Maculinea arion*) was reported to be extinct in the United Kingdom, but it then transpired that some were breeding successfully in captivity. The Javan rhinoceros (*Rhinoceros sondaicus*) is now thought to exist only in the Udjung-Kulon Reserve in Indonesia.

Animals in captivity are much less active than normal, and may suffer from BOREDOM. Activities such as HUNTING and MIGRATION are generally not possible under captive conditions, and many animals spend less time FORAGING and FEEDING than they would in nature. Consequently, captive animals are often not able to fill their time in a natural manner. Some may adjust by increasing the amount of sleep, but others undoubtedly suffer some psychological stress. Many captive animals develop unnaturally stereotyped patterns of movement. They may tread the same route for so long that a path is worn into the ground. SCENT MARKING in the pine marten (*Martes martes*) normally occurs on a regular basis, the marks being renewed after about two weeks when they have been diluted by wind and rain. In captivity the marten may visit his marks every few minutes. In other cases ABNORMAL behaviour may occur as a result of a well-meaning but misguided policy towards the captive animal. For example, it may seem necessary to regularly clean out an animal's cage, but for a slow loris (*Nycticebus coucang*) this can be an unwelcome disturbance to the environment. The loris is comfortable when its cage is thoroughly soaked in urine, and thus provided with adequate scent marks. Every time its cage is cleaned the loris drinks vast quantities of water so that it can systematically re-sprinkle the whole of its cage, so making it habitable once more.

Intimate knowledge of the behaviour and requirements of each species is necessary to ensure that the conditions of captivity are well suited to the animal concerned. Such knowledge must be obtained through detailed FIELD observations and subsequent LABORATORY STUDIES.

71

CHAIN RESPONSES are behaviour sequences in which each movement or activity brings an animal into a situation where the next is evoked. Examples at the REFLEX level are walking and chewing, in which one component of the behaviour, such as stepping out, or opening the mouth, stimulates the next component through the stretching of muscles or pressure on certain parts of the limb.

Chain responses involving the behaviour of the whole animal generally depend upon one activity bringing the animal into a new external situation which leads to the next activity. They may or may not be learned (see OPERANT BEHAVIOUR). For example, the hunting wasp *Philanthus triangulum* flies from flower to flower searching for its prey, members of the bee family (Apidae). Initially, the wasp responds to any moving object of about the correct size, but is indifferent to the scent of potential prey. When a suitable insect is sighted the wasp takes up a hovering position about 10 to 15 cm to leeward. It is now very sensitive to olfactory stimuli, and abandons the insect if it does not detect the bee scent. Upon smelling a bee, the wasp launches a sudden attack and seizes it. Tactile stimuli then trigger the stinging behaviour of the wasp. The HUNTING activity of *Philanthus* is thus made up of a chain of responses, released by visual, olfactory, and tactile stimuli in succession.

Chain responses of a less rigid nature may occur in SOCIAL INTERACTIONS. In the courtship of the three-spined stickleback (*Gasterosteus aculeatus*), for example, stimuli from the female elicit courtship DISPLAY in the male. When the female responds by adopting a head-up posture and swimming towards him, the male turns and leads her to the nest. If the female follows the male, he points his head at the nest entrance, and this causes the female to enter the nest. The male responds by 'trembling' while touching the female, and this induces her to lay her eggs and leave the nest. The male then enters the nest and fertilizes the eggs. At each stage in the chain the response of one partner is necessary for the other to progress to the next stage of the sequence. However, under normal circumstances the sequence rarely progresses quite so smoothly. Thus, if the female delays in responding to the male's courtship, the male may make an inspection visit to the nest. If the nest is disrupted in any way, the male may repair the nest before returning to the female. If the female attempts to follow the male to the nest before the male is ready, he will hold her off by pricking her with his dorsal spines. Thus, although each partner must respond appropriately if the sequence is to be completed, the response chain is not rigid, but acts as a framework for the courtship. In addition to enticing the female to the nest, the male has to guard the nest against raiding male sticklebacks, and he must ensure that the fragile nest is in a fit condition to receive the female and to contain her eggs.

CHEMICAL SENSES. The distinction between a sense of smell (olfaction) and a sense of taste (gustation) cannot easily be made in many animals. *Chemoreception*, or chemical sense, is a useful inclusive term that also indicates the most important property of these two senses: the capability of identifying chemical substances and of detecting their concentration. The term chemoreception has the added advantage that it allows the inclusion of an often forgotten, but important and varied group of receptors within the body: the internal chemoreceptors. These are similar to external chemoreceptors, but are concerned with monitoring the concentration of a number of substances in the body fluids. Strictly speaking, virtually every nerve-cell functions as a chemoreceptor, in that it reacts specifically to substances released by other nerve-cells. While many internal chemoreceptors play a role largely restricted to visceral regulation (for example, carbon dioxide receptors located in the walls of the carotid arteries and aorta are involved in the nervous control mechanisms of blood circulation), others are important in registering levels of HORMONES that reach the BRAIN, and in detecting nutrients in the blood.

Compared with many animals, human beings have a rather reduced sense of TASTE AND SMELL. For this reason we tend to underestimate the role of the chemical senses in determining the behaviour of numerous animals. Chemoreception was certainly the first sense to develop in the course of EVOLUTION. Primitive organisms are critically dependent on the right kind of chemical environment for survival and reproduction. As soon as motility evolved, chemical sensitivity capable of directing such behaviour towards life-supporting surroundings must have conferred powerful SURVIVAL VALUE. The behaviour of present day lower organisms is still often restricted to *chemotaxis*, i.e. movement controlled by chemoreception. The bacterium *Escherichia coli*, for example, is known to be attracted by and to migrate towards sources diffusing oxygen, glucose, galactose, serine or aspartic acid. There are however, GENETIC mutants of the bacterium that are not responsive to one or other of these substances. This indicates that specialized receptors, probably in the form of chemically specific sites on the outer membrane, are responsible for this sensitivity. Similarly the unicellular protozoan *Amoeba proteus*, is selective as to what it engulfs, digests, and incorporates into its cytoplasm. The feeding behaviour is released by certain proteins, basic dyes, and salts.

Chemoreception is similarly widespread among the multicellular invertebrate animals. An example is the much studied sea hare (*Aplysia*, a marine mollusc that feeds on a single species of seaweed). If sea water that has been in contact with this seaweed is injected into a corner of an aquarium containing sea hares, they immediately crawl to that point. Recording the electrical activity of nerve-cells demonstrates that those receiving input from the tentacles are responsive to diluted extract of seaweed: the chemoreceptors are thus located on these appendages. How the chemical recognition of a particular seaweed is achieved is not understood, and several test substances not characteristic of the specific seaweed also lead to excitation of these nerve-cells.

Many species of starfish (Asteroidea) feed on sea urchins (Echinoidea), and many of these move away when a starfish approaches. This response must be chemically elicited. Interestingly, in one such predator–prey pair, the sea urchin *Strongylocentrotus* is not bothered when the predatory starfish *Pycnopoda* is already feeding, and is thus not on the prowl. Whether the starfish does not release the warning substance in this situation, or whether chemicals escaping from the injured prey inhibit the escape response, is not known.

Hydra littoralis, the small freshwater coelenterate, is a classical subject for chemosensory studies. It catches its prey with its tentacles, kills it with its *nematocysts*, veritable miniature poisoned arrows, and then contracts the tentacles towards its open mouth and ingests the prey. This latter part of the feeding behaviour is released quite specifically by a small protein constituent called gluthatione. Other species of coelenterates respond in a similar manner to gluthatione, but some respond specifically to other substances, such as proline and leucine.

Many stationary marine organisms go through a larval phase in which they are mobile, and form part of the plankton. When the time comes to become sessile they choose the permanent site by chemical criteria. The larva of the common barnacle (*Balanus balanoides*) prefers to settle on slate that has been impregnated with an extract of adult sedentary barnacles, as opposed to slate that has not been so treated. The active component of the extract does not lose its potency even when exposed to the extreme heat of 200 °C, or to concentrated sulphuric acid. It is thought that a very stable protein is present in the cuticle of arthropods, and that the receptors suspected of being involved in the recognition of this substance are located on the antennae of the larvae.

In a variety of insects chemoreception has developed to pinnacles of sophistication. Here it is also possible to draw a distinction between olfaction and gustation, in the sense that some of their chemoreceptors are specialized for the detection of very low concentrations of substances, whereas others only respond to high concentrations. The olfactory receptors are generally situated on the antennae, and in some insects these have developed to be veritable scent-molecule sieves. They are studded with thousands of porous sensory hairs, each containing a thin protrusion of a sensory cell which sends its *axon* directly to the insect's brain. A remarkable example is provided by the antennae of the males of many moths, the silk moth (*Bombyx mori*) being one of them. The females have an abdominal gland that secretes a scent. This is carried down-wind, and when detected by the males causes them to move upwind (see TAXES) until they reach the source. By this means the females can attract males over distances of several kilometres. In the silk moth the substance secreted by the female is a polyalcohol called bombykol. Because of the striking effect it has on the behaviour of the male it is called a PHEROMONE, by analogy with hormones circulating in the blood. When it reaches the sensory cells it causes them to become electrically activated. These in turn send nervous messages to the brain, where the detection of the pheromone is registered, and causes the behaviour already described. From experiments it is known that the relevant sensory cells react only to bombykol and not to other related chemicals. In contrast the olfactory sense of other insects is more general. Honey-bees (*Apis mellifera*) for example, can be trained to distinguish a number of different scents. Correspondingly, their antennae carry receptors that are responsive to a variety of substances.

Contact chemoreceptors, analogous to taste receptors, are found on the *tarsi* (feet) and mouthparts of insects. Those located on the tarsi of the blowfly *Phormia regina* are hairs with perforated tips, and contain the processes of several sensory cells. One of these responds to salt solutions, another to sugar solutions, yet another to pure water (a fourth does not respond to chemicals, but is sensitive to the bending of the hair). These receptors control the extension of the fly's *proboscis* (sucking tube).

In vertebrates the olfactory sense is also specialized for the detection of low concentrations of chemical substances, typically ten thousand molecules per millilitre of air or water. It is mediated by sensory cells lining more or less extensive patches of nasal mucous membrane. The area they occupy is a fairly good indicator of the olfactory capabilities of a given species. The sensory cells are ciliated, the *cilia* forming a great tangle embedded in a layer of mucus. A species of average olfactory capabilities possesses some hundred millions of receptor cells, each of which sends off a nerve axon. The axons make up the olfactory nerves, and these terminate in the olfactory bulbs, brain centres in which the neural information about smell undergoes complex processing. The size of the bulbs is a good guide for the assessment of the olfactory capabilities of a species. Secondary nerve fibres carry the higher order olfactory information to other parts of the brain.

Recordings of the electrical activity of olfactory cells in many species of vertebrates have revealed that they are of a generalized type; that is, each cell responds to a wide variety of substances, either by being inhibited or by being activated to various degrees by each of them. Each receptor appears to have its own sensitivity spectrum to scents. In contrast, behavioural studies done in man suggest that there are about seven primary scents: ethereal, camphorous, minty, floral, musky, putrid, and pungent. The results from these two types of research cannot be easily reconciled, and it is fair to say that the mechanism of the sense of smell is not completely understood. This also applies to the processes involved in the chemical recognition of scent molecules as they reach the receptor cells.

As to the multifarious role of olfaction in the behaviour of vertebrates a few examples must suffice. After MIGRATION, Atlantic salmon (*Salmo salar*) return to their native river where they spawn; they are guided by the smell of the local water to which their IMPRINTING occurred as fry. If, at the fork of a stream, they head into water that does not carry the native scent, they let themselves drift downstream until they perceive it again, and try an alternative course. Minnows (*Phoxinus phoxinus*) and many other fish that live in a school scatter hurriedly as soon as they smell an injured member of their own species, perhaps one mauled by a pike (*Esox lucius*). A specific warning substance contained in the skin seeps out through the wounds of the victim and triggers this fleeing response. Some birds, contrary to widespread belief, can smell quite acutely. Vultures (Cathartidae) are thought to locate carrion by smell. Kiwis (Apterygidae), being nocturnal, also seem to locate food by its scent. Voles (*Microtus*) distinguish individuals of their own local variety from those of other varieties by their odour. Male rhesus macaques (*Macaca mulatta*) identify females in heat through a specific sexual pheromone which is secreted by the vagina during the receptive phase.

Some vertebrates, including certain mammals, possess an additional olfactory organ known as *Jakobson's organ*. In lizards (Lacertidae) and snakes (Serpentes) it is located within cavities with two openings to the roof of the mouth. The animal appears to introduce scents into these cavities with the tip of its forked tongue. In other animals, e.g. the golden hamster (*Mesocricetus auratus*), these organs, also known as *vomeronasal organs*, are located within the nasal cavities. If the nerve supplying the organs is cut in male hamsters, their sexual behaviour is markedly impaired. Destruction of the olfactory mucous membrane proper does not have such an effect. This suggests that the male's vomeronasal organ is specially adapted for the reception of sexual smells produced by the female.

The gustatory sense of vertebrates, like that of insects, is sensitive to high concentrations of chemicals (typically billions of molecules per millilitre of water) and is mediated by sensory cells located within the mouth and in some animals also by receptor cells located in the gill cavity, or on the body surface. The receptor cells themselves are clustered within structures known as *taste buds*. Only a fringe of hair-like processes protrudes into the open through the narrow neck of an enveloping structure. The buds, of which a domestic cat (*Felis catus*), for example, has some fifteen hundred, occur in groups within structures known as *papillae* which can be seen on the tongue with the naked eye. Three types of papillae can be distinguished on the human tongue: *folliate*, *fungiform*, and *vallate*. The vallate types are restricted to the base of the tongue, while the two former types predominate respectively on the edges and on the tip of the tongue. Man appears to interpret the taste of substances with reference to four basic taste qualities: salt, sweet, sour, and bitter. The sensitivity to these qualities is not evenly distributed over the surface of the tongue. The tip is more sensitive to sweet, the base to bitter, while areas particularly sensitive to salt and sour lie between. However, the basic tastes are not strictly associated with stimulation of one or the other type of papillae, as experiments in which single papillae are stimulated with small drops of chemical solutions show.

While the taste of some substances relates reasonably well with their chemical nature, for example all sour-tasting substances are acids, it does not do so in other cases. Bitter-tasting substances, for example, are a very heterogeneous group of chemicals. Most sweet-tasting substances are chemically similar, i.e. sugars; but some do not conform to this rule, such as lead and beryllium salts, or the well known artificial sweetener saccharin. Salty taste is generally associated with inorganic salts, but some of them do not fit this rule. Magnesium salts, for example, are perceived as bitter by human beings.

Electrophysiological recordings from single fibres of the lingual nerve of rats give results that do not agree too well with the existence of the four basic tastes deduced from behavioural experiments. A fibre, for example, may respond markedly to one salty substance (sodium chloride), but not to another (potassium chloride). Another one responds to acids as well as to some salty substances, and so forth. How this information is sorted out in the olfactory centres of the brain is not well understood.

Certain human individuals cannot taste very specific substances. A fairly common case is the inability to detect the bitter-tasting compound phenylthiocarbamide and some closely related substances. This inability is inherited, and is interesting because those taste-blind to

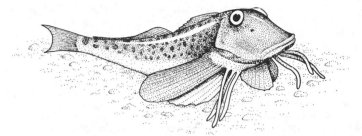

Fig. A. The North American sea-robin (*Prionotus*).

phenylthiocarbamide can still taste other bitter substances. Similarly other animals seem to perceive certain substances differently from us. Saccharin, a substance that tastes sweet to us, seems to taste somewhat bitter to rats (*Rattus norvegicus*). Quinine, a compound that tastes exceedingly bitter to human beings is hardly tasted by pigeons (*Columba livia*), even though they can taste other bitter substances.

The sense of taste is, of course, mainly concerned with FOOD SELECTION and choice of drink. Rats made salt-deficient will preferentially choose food or drink containing sodium chloride. Rats will also learn to avoid food with a specific taste if previously they have been made ill by consuming it. Fish are interesting in that many of them have taste buds on the body surface. The North American sea-robin (*Prionotus*), portrayed in Fig. A, even has specially modified pectoral fin rays, the tips of which are studded with gustatory receptors. The fish samples the substrate with its fin rays, and responds to the taste of food by digging with its mouth. Mouthbreeding cichlid fish (Cichlidae) appear to distinguish their own and foreign fry by taste: the latter may be snapped up by a parent fish, but they are then spat out again.   J.D.

13, 35

CLASSIFICATION OF BEHAVIOUR. A fundamental problem for the ethologist is the description, classification, and measurement of behaviour. These three tasks are so closely interrelated that they are often treated as a single entity. Classification of behaviour depends upon adequate description, and measurement of behaviour depends upon satisfactory classification. The description of behaviour is generally influenced by the types of classification and measurement that are to be employed.

Description. The most fundamental way of describing behaviour patterns is in terms of the exact muscular movements involved. This approach is sometimes used in LABORATORY studies, but is not the one generally employed by students of ETHOLOGY, because it is unnecessarily detailed, complex, and cumbersome for all but the most simple aspects of behaviour.

Behaviour patterns are normally described in terms of some consequences of the muscular movements. Thus, if the main consequence of a particular pattern of muscular contractions is that food is obtained, then the pattern as a whole may be described as FEEDING behaviour. Sometimes the consequences of two behaviour patterns differ only in relation to some aspect of the environment. For example, great tits (*Parus major*) open nuts with a hammering movement that is almost identical to the movements used in attacking rivals. The two behaviour patterns differ only in the speed of movement, and their target. Often, the same consequences result from a variety of patterns of muscular movement, and the ethologist may use a single descriptive term, such as nest-building, or preening. Indeed, the proportion of FIXED ACTION PATTERNS in an animal's behaviour repertoire is usually small, and variability is inevitable in those aspects of behaviour that require COORDINATION with factors in the environment.

The most important rule to abide by when describing behaviour is that the different categories should be mutually exclusive. In other words, the behaviour should be classified in such a way that the members of one class do not also occur in another class. Suppose, for example, that we are observing a pigeon (*Columba livia*) feeding from grain scattered over the ground. The pigeon will walk a few steps, peck at the grain, walk a few steps more, and so on. Suppose we wish to record the pigeon's behaviour once per minute. Our description should represent the behaviour of the bird during the whole of each designated minute. The problem is that we would inevitably find that the pigeon was both pecking and walking during a given minute. If we decide that each minute of observation represents a category of behaviour, then we should not score pecking + walking in one minute and, for instance, walking + scratching in the next minute. We can decide to ignore walking, and score the pecking that occurred in the first minute, and the scratching in the next one. Alternatively we can score pecking-while-walking as a different item of behaviour from pecking-while-stationary, and then simply record the successive periods of time for which each was observed, thus abandoning the one-minute behaviour category.

The way in which behaviour is described is largely a matter of the preferences of the observer, and it will inevitably be influenced by the TECHNIQUES OF STUDY that are available. Whatever form of description is used, it should conform to a *nominal* scale of measurement, in which the different classes should have the same value, and this is only possible if they do not have components in common. Obviously the class pecking + walking cannot be considered as equivalent to the class pecking. Thus if pecking is to be diagnostic of a given class, then all occasions of pecking must be relegated to that class and all occasions must be regarded as equivalent. This is important because the nominal scale is the fundamental basis of all classification and measurement of behaviour, and scientific interpretation of behavioural data is impossible unless a nominal scale has been adhered to in the description of the behaviour.

Classification. The nominal scale is a primitive form of classification that is based upon the criteria by which the observer chooses to label each category of behaviour. These fundamental units of behaviour can then be further classified in accordance with a variety of alternative schemes.

1. *Causal schemes* can be based upon knowledge of the causal factors responsible for the behaviour. For example, many types of behaviour are influenced in strength and intensity by the level of male sex HORMONE. All these activities can be grouped together as male sexual behaviour. This form of classification has certain inherent dangers, which arise from the fact that considerable knowledge of behavioural causation is required. If the classification is based upon partial knowledge, or if opinion about the causation of certain aspects of behaviour changes, then the classificatory scheme is violated.

2. *Functional schemes* of classification can be based upon the adaptive consequences of behaviour patterns (see FUNCTION). Just as in the study of animal morphology parts of the body are classified according to their function (the terms eye and leg are examples), so, in the study of behaviour, the parts may be labelled functionally (hunting and courtship are examples). In some cases there is a correspondence between causal and functional classification of behaviour. This occurs when functionally related activities share causal factors. This is often the case with seasonal activities, such as those involved in MIGRATION and COURTSHIP, but it is by no means a universal rule.

3. *Historical schemes* are sometimes used in the classification of behaviour. These may reflect the evolutionary origin of the behaviour, as with activities that have undergone RITUALIZATION, or they may reflect the ONTOGENY of the behaviour. The latter includes any classificatory scheme that refers to the way in which the behaviour was acquired; the distinction between behaviour that involves LEARNING and behaviour that is INNATE, for example, is essentially an historical one.

Measurement. Scientists recognize different levels of measurement of behaviour, the most fundamental of which is the nominal, or clas-

sificatory, scale already mentioned. This type of scale can be used only for those types of analysis, such as the mode and frequency count, which would be unchanged by a transformation of the symbols used to designate the various classes. To achieve a more useful level of measurement it is necessary to have an *ordinal*, or ranking, scale. This is possible when a comparative relationship exists between the mutually exclusive categories of behaviour. For example, if category A is greater than, or preferred to, category B, then it can be stated that A > B. Often a whole series of categories can be ranked in this way. For example, guppies (*Poecilia reticulata*) undergo progressive changes in COLORATION as they become more sexually motivated. On the basis of an ordinal scale it is possible to perform various limited mathematical operations, such as statistical tests. Many common arithmetical operations are, however, not permissible, because the magnitude of the difference between the categories on an ordinal scale is not specified.

When, in addition to there being a comparative relationship between mutually exclusive classes, the magnitude of the difference between classes is equal, then an *interval* scale is achieved. This is characterized by a constant unit of measurement that is common to all classes. This has the advantage that numbers may be associated with positions on the scale, and the ordinary operations of arithmetic may be meaningfully performed on the differences between numbers. The ratio of any two such differences is independent of the unit of measurement. Thus the unit of measurement on an interval scale is arbitrary, and the scale does not have a true zero. The Celsius and Fahrenheit scales for measuring temperature are examples of interval scales. Interval scaling has the disadvantage that certain arithmetical operations are not permissible. One cannot, for example, say that 20 °C is twice as hot as 10 °C, and the statement is revealed as nonsense when translated into Fahrenheit. These restrictions vanish when it is possible to use a *ratio* scale, which is the most powerful level of measurement. The ratio scale has all the characteristics of an interval scale, and in addition it has a true zero point. Thus the ratio of any two scale points is independent of the unit of measurement. All arithmetical operations can be carried out on ratio scales.

True interval and ratio scaling is hard to establish in animal behaviour studies, and generally has to be justified by elaborate mathematical argument. This difficulty arises because the observer cannot assume that what appears to a human being to be an interval or ratio scale also appears to be so to an animal (see ANTHROPOMORPHISM). For example, male guppies prefer to court larger females. However, although one female may appear to be twice as large as another to the human eye, it does not follow that it is twice as attractive to a male guppy.

CLIMATE refers to the whole range of weather conditions, including temperature, rainfall, wind, etc., that an area receives through all the seasons of the year. Many factors are involved in the determination of climate, particularly latitude, altitude, and position in relation to seas and land masses.

The behaviour of animals is greatly influenced by climatic factors. Many species have special adaptations which enable them to live in areas where the climate is harsh. Others migrate from one region to another, thus avoiding climatic extremes. In such cases the global climatic pattern is relevant to the animal's life, but local climate is generally of prime importance. In studying the behaviour of many species, however, we also need to pay particular attention to micro-climatic factors, which may affect such things as nest temperature and brood survival (see NEST-BUILDING) and distribution of prey (see FORAGING).

The global climatic pattern. Although every point on the earth receives the same amount of daylight each year, the spherical form of the earth results in an uneven distribution of solar energy, as is illustrated in Fig. A. As the angle of incidence

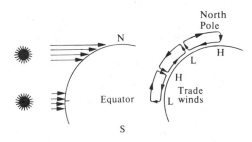

Fig. A. Due to the spherical shape of the earth, polar regions receive less solar energy per unit area than equatorial regions. This gives rise to circulating masses of air, as described in the text.

of the sun's rays approaches 90°, the area of the earth's surface over which the sun's energy is spread is reduced, so there is an increased heating effect. Polar climates are cold because the energy is spread over a wide area at high latitudes. The precise latitude that receives sunlight at 90° at noon varies during the year due to the inclination of the polar axis in relation to the ecliptic (see TIME). It is at the equator during March and September, at the Tropic of Cancer (23·45°N) during June, and at the Tropic of Capricorn (23·45°S) during December.

Another cause of climatic variation with latitude is the pattern of movement of air masses. The heated air near the equator rises, causing a low pressure area. Air from the neighbouring high pressure areas moves towards the equator, causing the trade winds. A similar circulation of air at higher latitudes results from cold currents of air moving from the polar regions. Due to the rotation of the

earth from west to east, a force is induced, called the Coriolis force, which tends to deflect moving objects in the northern hemisphere to the right, and those in the southern hemisphere to the left. For this reason, the trade winds north of the equator blow from the north-east, while those south of the equator blow from the south-east.

The wind pattern greatly influences the world's ocean currents. For example, the trade winds push water into the Caribbean area, causing a pile-up of water, so that the average water level on the Atlantic side of the isthmus of Panama is a few feet higher than that on the Pacific side. This piled-up water escapes northward as the Gulf Stream, which brings warm water to the western shores of Ireland and Scotland. The pattern of wind and current, taken together, largely determine the distribution of rainfall around the world, although the location of oceans and land masses is also of great importance.

Water warms up and cools down more slowly than land masses, and maritime regions therefore have more equable climates, with higher humidity. In summer, land masses heat up quickly, creating a low-pressure area as the warm air rises. In winter, the reverse situation develops, as the land becomes cold more quickly than the ocean, and high-pressure areas develop. In addition to the heating and cooling effects of land masses, climate is also affected by altitude. On average, air temperature falls by 0·6 °C for every 100-m rise in height. There is also an effect of altitude on rainfall. Winds passing over the oceans pick up water, and this water-laden air is forced to rise when it reaches land. The lower temperatures at greater altitudes cause the water to be precipitated as rain, so that the coastal region of a land mass facing prevailing wind generally has high rainfall.

The climate of an area is the result of the many varying factors that affect the locality, although there is a general pattern over the earth as a whole. The climate largely determines the species of plants and animals that can live in an area. Each region has a number of characteristic COMMUNITIES that have evolved and have become adapted to the local climatic conditions. These groups of characteristic communities are called *biomes*, and they are generally classified into six terrestrial types, to which the marine and freshwater biomes should be added. The distribution of terrestrial biomes is illustrated in Fig. B.

Biomes are generally characterized by the most common or dominant types of plant. Thus in Arctic regions the most common plants are dwarf shrubs and other low-growing plants. These have little growth above ground, and carry their buds just below the surface of the soil, where they obtain the maximum protection from cold and wind. Animals also have similar types of adaptation to climate. Thus representatives of a given warm-blooded species living in colder latitudes tend to have greater body bulk than those living in

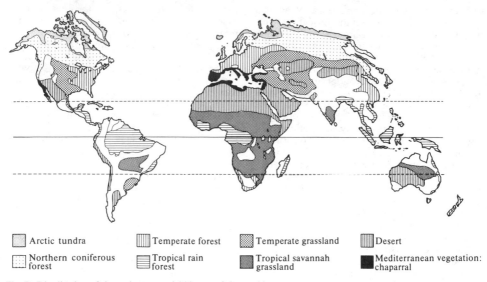

Arctic tundra	Temperate forest	Temperate grassland	Desert
Northern coniferous forest	Tropical rain forest	Tropical savannah grassland	Mediterranean vegetation: chaparral

Fig. B. Distribution of the major terrestrial biomes of the world.

hot regions. This is generally called *Bergmann's Rule*. Similarly, those in cold areas tend to have shorter extremities than those living in warm climates. This is *Allen's Rule*. Both these rules concern the physics of heat loss from a warm body to a usually cooler surrounding atmosphere. A large body has a proportionately smaller heat-dissipating surface than a small one, and the nearer the body comes to being a perfect sphere, the less the heat loss per unit of volume.

In parts of the world where climate is seasonal, animals may have to adjust to prolonged periods of unfavourable climate. Some animals are able to avoid such conditions by MIGRATION. This occurs in many birds, whales, seals, bats, fish, butterflies, and other insects. For birds, the winter environment may provide very different climatic conditions from the breeding area, but this difference is often smaller than that experienced by northern birds which do not migrate. Many birds can survive a wide range of physical conditions, and migration is not so likely to be due to the direct effects of climatic factors, as to the effects of weather on food supply. Other, less mobile, species are able to survive periods of unfavourable climate by entering a prolonged resting stage. Such dormancy is termed AESTIVATION at high temperatures and HIBERNATION at low temperatures. Aestivation is most common in desert-living species, but is not confined to them. For example, many species of European earthworm, such as *Allolobophora longa* and *Eisenia foetida*, aestivate during the summer months. Each animal hollows out a small chamber deep in the soil, and curls up in a ball. The onset of aestivation is associated with low humidity, and it is possible to prevent it by keeping worms in a humid atmosphere. Hibernation occurs in many species in northern latitudes, enabling them to

avoid winter conditions that would otherwise make excessive demands on the energy requirements of the animal. By entering a period of dormancy, hibernating animals are able to lower their energy expenditure to a considerable extent. Where such periods of dormancy occur in response to cold, heat, or drought, as part of the development of the animal, as in many insects, it is generally given the name *diapause*. In many insects diapause is associated with a particular stage of the life cycle. For example, the eggs of the Australian grasshopper *Austroicetes cruciata* are able to withstand extreme heat and desiccation. Eggs in diapause are much more resistant than eggs which have not yet reached that stage of development. In this species, diapause ensures that the animal becomes active only in the season of the year favourable for survival. The hot dry summer is spent as a diapausing egg, which is remarkably resistant to heat and to water loss.

Local climate. The climate of a particular locality is greatly influenced by soil, water, and topography. Different types of soil absorb solar energy at different rates, and the transfer of heat within the soil varies with the conductivity of the soil. This in turn depends upon the density, compaction, wetness, and other physical characteristics of the soil. The structure of soil is also important in determining the distribution of burrowing animals, and in influencing the types of vegetation that can grow in a locality.

Water has a high thermal capacity, so that it warms up and cools down slowly. Thus the proportion of water in a particular environment has a large effect on climatic fluctuations. Deserts cool rapidly at night, and heat up quickly when the sun rises, whereas these processes are relatively slow in wet places. Water has a high heat of vaporization, so a considerable amount of heat can be dissipated by evaporation of water. This is particularly important for animals in hot environments (see THERMOREGULATION). Water in the atmosphere is found in gaseous and in droplet form. The quantity of water found in gaseous form is expressed as vapour pressure. The maximum vapour pressure of the air is called its saturation pressure, and this varies with temperature. The relative humidity is equal to the vapour pressure divided by the saturation pressure, times 100. Relative humidity is thus temperature dependent. Clouds, fog, and mist form by condensation of gaseous water in the atmosphere. Such droplet laden air is the source of precipitation, which may take the form of rain, sleet, snow, dew, and hoarfrost. Although rain and snow provide the greatest quantity of precipitation, other forms, such as fog and dew, can be of particular importance to animals. For example the lizard *Aporosaura anchietae*, which inhabits the coastal desert of Namibia, an area of zero rainfall, is able to obtain water from sea-fog carried inland on the sea breeze.

The topographical features of the environment can have a profound effect upon the weather, and particularly upon local wind patterns. Local winds are found near coastlines and other topographic features under conditions where temperature contrasts result in pressure differences. Consider, for example, the sea-breeze. As we have seen, water warms up more slowly than land, so that a column of air over the land will heat much more rapidly on a sunny morning than a column of air over the water near by. The heated air rises, reducing the local air pressure, so that winds blow from the water to the land, giving the sea-breeze. Similarly, a column of air above mountain slopes that are facing the sun will be heated more rapidly than the air in the valley. This results in air-flow up the slopes from the valley.

The shape of the landscape also influences the flow and accumulation of water, and many other factors which affect the weather indirectly. Indeed, the effects of soil, water, and topography are considerably interrelated, and cannot be considered in isolation from each other. For example, during periods of rainfall some of the water is absorbed into the soil and the rest runs off downhill. The extent of this run-off depends upon the quantity of rain, the slope of the soil surface, the absorption characteristics of the soil, and the resistance to flow on the soil surface. The last factor depends largely upon the nature of the vegetation, which in turn depends partly upon the amount of water absorbed into the soil. So we see that the factors determining the run-off and accumulation of water in a particular locality are inextricably bound together.

Microclimate. Within a particular locality animals are far from evenly distributed. One reason for this is that there may be distinct climatic differences within a very small area. For example,

the surface of a rock in the intertidal region of the sea shore may reach 44 °C when the tide is out. Simultaneously the temperature may be only 25 °C beneath the same rock. Many animals are able to exploit such differences in microclimate.

Microclimate is greatly influenced by vegetation. Tall plants, for instance, provide shade and protection from wind and rain. Areas beneath trees or bushes generally have less variable, warmer, and more humid climates than more exposed places. An important factor in the determination of microclimates is the extent of penetration of sunlight. This can be illustrated by examining the climatic conditions that might be found in a grassland sward including a dense layer of clover, at noon on a sunny day, perhaps after early morning rain. Fig. C shows the conditions of light, temperature, wind speed, and relative humidity that would typically be present at various levels in the vegetation. Bright sunlight does not penetrate far into the vegetation, and half-way down in the grass the intensity of the light is only half that above the vegetation. The amount of light falls off even more sharply once the clover layer is reached. Wind speed is also much reduced by the vegetation, and beneath the clover there is no air movement at all. This factor partly influences the relative humidity of the air. Above the vegetation the water evaporated from the soil and plants is blown away, so the air is only half saturated with water vapour. Where the vegetation is more dense, the water vapour is trapped and humidity is much higher. At soil level the air is fully saturated with water vapour. The distribution of temperature at various levels in the vegetation is the result of interactions between the heating effect of sunlight, cooling by wind, and cooling by the damp surface of the soil. The warmest place is about half-way up the vegetation, where sunlight and wind speed are both about half strength. Above this point conditions are cooler because the wind is stronger, while the temperature falls rapidly below, because of the lack of sunlight.

Variations in microclimate, such as these, have a considerable effect upon the distribution and behaviour of small animals, and even large animals, such as deer (Cervidae), may be influenced by microclimatic factors, as they can gain some protection from cold winds by sitting down (see DECISION-MAKING).

Adaptations of animals to climatic change. There are three main ways by which animals can adapt to climatic changes. These are: (i) rapid physiological adjustment; (ii) changes in behaviour; (iii) ACCLIMATIZATION. Rapid physiological adjustments are generally REFLEX in character. Thus, if a man is suddenly transported to a high altitude, where there is less oxygen in the air, he will immediately start to breathe more heavily. Similarly, if an animal is suddenly exposed to a high air temperature it will show reflex cooling responses, such as panting, sweating, etc. If the climatic change is too extreme the animal will die, but generally the

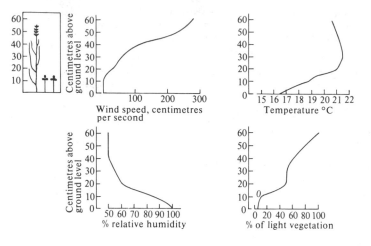

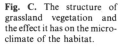

Fig. C. The structure of grassland vegetation and the effect it has on the microclimate of the habitat.

emergency reactions are sufficient to rescue an animal from any sudden change that it is likely to experience in its natural environment. The problem with these reflex responses is that they are costly, and therefore provide only a short-term remedy. For example, heavy breathing utilizes much more energy than normal breathing, and cooling responses generally cause loss of considerable amounts of water from the body. When animals die following a climatic change, it is generally due to secondary factors, such as dehydration and exhaustion.

As an alternative to the physiological response, many animals have the possibility of a suitable behavioural response. For example, a man exposed to the hot sun can seek shade, as an alternative to sweating it out in the sun. Many animals have short-term behavioural responses and postural adjustments which enable them to avoid extremes in the microclimate (see HABITAT). In addition, many animals have behaviour patterns, which enable them to respond to overall changes in the local climate. For example, deer will travel considerable distances to find shelter from cold winds, and sometimes group themselves close together for extra warmth. In the long term, many species avoid climatic extremes by migration, which is usually organized on a seasonal basis. Thus heavy snowstorms cause mule deer (*Odocoileus hemionus*) to migrate to a winter range, whereas the migration back to the summer range is related to plant growth.

In response to climatic change, many animals undergo slow physiological changes, which can be grouped under the heading of acclimatization. In man, for example, the rapid-breathing response to oxygen depletion at high altitude lasts only for a day or so. During this time there are chemical changes in the blood serum which enable it to carry an increased oxygen load. Again, this is only a temporary response, and the body eventually acclimatizes by manufacturing extra red blood corpuscles, which provide the least costly way of increasing the oxygen-carrying capacity of the blood. Many animals have considerable ability to acclimatize, particularly to temperature changes. Extreme examples are the phenomena of hibernation and aestivation, during which profound long-term physiological adjustments occur. Acclimatization enables many species to change their range of TOLERANCE, and thus survive under conditions that would normally be lethal.

In summary, there are a variety of physiological and behavioural mechanisms by which animals can adapt to climatic changes. It is important to realize that these mechanisms are additive in their effects, all serving to maintain viability in the face of a hostile physical environment. This means that the different mechanisms are complementary, and one can be substituted for the other, as in the example of the man sitting in the shade instead of sweating in the hot sunlight. Usually the most rapid behavioural and physiological responses are expensive, and are gradually replaced by slower acting, and less costly, acclimatization mechanisms. The man who sweats in the sun loses a lot of water, the man who sits in the shade curtails his overall freedom of behaviour, but the man who has acclimatized to a high temperature has to make only a small metabolic payment.
27, 107

CLOCKS. Biological clock, physiological clock, or simply clock, are names given to the supposed biochemical mechanism responsible for TIME-keeping in living organisms. Laboratory and field experiments show that many of the annual, lunar, and daily RHYTHMS of behaviour exhibited by animals are maintained by clock mechanisms that are endogenous in the sense that the rhythm persists when the animal is isolated from all possible environmental time cues. However, it is often the case that an external *Zeitgeber*, or time-setter, is responsible for the maintenance of synchrony between the clock rhythm and the rhythm of environmental events. When animals are isolated from the exogenous zeitgeber their clocks drift out of step with the environmental rhythm. For example, the lizard *Lacerta sicula* shows a daily rhythm of activity, even under constant laboratory conditions. This shows that the rhythm is endogenous and not simply a response to 24-h rhythms of environmental temperature or light. However, under constant laboratory conditions the 24-h periodicity drifts slightly from the norm, though very slight fluctuations in ambient temperature are sufficient to prevent this drift. Therefore, it seems that environmental temperature is a zeitgeber in this situation.

The clock mechanism is thought to be a property of individual cells, since rhythmic behaviour is shown by single-celled animals. For example, the protozoan *Euglena* shows a rhythm of swimming activity which is normally synchronized with the motion of the sun. The rhythm persists when the animal is maintained in continuous darkness in the laboratory. Some animals can exhibit more than one rhythm simultaneously, suggesting that an animal may have a number of clock mechanisms.

5, 25, 115

COADAPTATION is one of biology's more versatile words: it has been used in many different senses for ideas that must and can easily be distinguished. The important common idea running through this multiplicity of meanings of the term is that separate structures, or facets of behaviour, are designed (that is, they are ADAPTATIONS) specifically for interaction with each other. Coadaptation is mutual adaptation. The diversity of meanings arises because the term coadaptation is applied indiscriminately, both to different adaptations of a single organism that co-operate for a common FUNCTION, and to different adaptations in separate organisms that interact particularly closely. Furthermore, in the latter case, the term might be used whether the separate organisms belong to the same or different species. The diversity of meanings of coadaptation is not due to confused thinking by biologists, but to the diversity of independent biological disciplines. One of the earlier uses of the term, by DARWIN in *On the Origin of Species* (1859), illustrates several meanings in two sentences: 'How have all these exquisite adaptations of one part of the organisation to another part, and to the conditions of life, and of one distinct organic being to another being, been perfected? We see these beautiful co-adaptations most plainly in the woodpecker and the misseltoe; and only a little less plainly in the humblest parasite which clings to the hairs of a quadruped or feathers of a bird.' The present exposition considers the different meanings of coadaptation in turn, expanding first on coadaptations of traits between different individuals, and then on coadaptations within an individual.

Coadaptations between organisms. Organisms do not live alone in the world; they frequently meet other organisms of their own and of different species. The other organisms might be sources of food, or dangerous predators. Individuals of some species live together with individuals of other species in intimate SYMBIOSIS. Coadaptations are particularly common and noticeable between species that live together exceptionally closely, for example between parasites and hosts, as in the passage from Darwin quoted above. This tendency for mutually specialized species to possess coadaptations is hardly surprising: organisms are adapted to the total NICHE that they live in, so if that niche is mainly, say, the inside of a particular region of another organism's gut, then appropriately specialized adaptations of behaviour and structure are to be expected.

Interspecific coadaptations. The system of plants and their insect pollinators is an example of coadaptation between organisms of different species. In this system, the plant produces nectar and pollen, and the insect imbibes the nectar and has pollen deposited on it while doing so. When the insect visits another flower it will incidentally deposit pollen, so cross-fertilizing the plants (if the two flowers are on different plants). The plant gains cross-fertilization; the pollinating insect gains food. Both partners benefit. In one particularly detailed study of a plant–pollinator system the exclusive pollinators were bumble-bees (*Bombus* spp.) and the plants were of several species of *Delphinium*, *Aconitum*, and *Epilobium* (all of which have their flowers arranged in a spiral, on the type of inflorescence termed a spike). It was observed that the flowers at the bottom of the plant contained more nectar than flowers higher up. For this reason the bumble-bees always land to start FORAGING at the bottom of the plant and work their way upwards. The bumble-bees usually leave the plant before they reach the top, so by starting at the bottom and working upwards the bumble-bee obtains maximum food per unit time. Although the flowers are arranged spirally on the spike the bumble-bee moves vertically. This is because from any one flower the nearest flower is generally that vertically above it, so the bumble-bee can probably obtain food most rapidly by moving vertically. This movement pattern ensures cross-fertilization of the plant by the following mechanism. The flowers low on the spike are receptive to pollen but do not produce it; the flowers higher up produce pollen but are not receptive to it. This virtually guarantees that each plant is fertilized by pollen from another plant, rather than from another flower on its own spike. The bumble-bee's vertical movement results in some flowers being by-passed, which benefits the plant because the arriving bumble-bee does not have sufficient pollen to fertilize all the flowers. The arrangement of the flowers thus ensures that the plant does not waste nectar on a bumble-bee that has no pollen. In sum, in this system of plants and their pollinators the movement pattern of the bumble-bee is coadapted to the arrangement of flowers on the spike, and to the distribution of nectar among them. The reproductive system and the floral arrangement of the plants are coadapted to the movement patterns of the bumble-bee. The plant species and the bumble-bee species form a highly coadapted system. The coadaptation results from a long history of co-evolution of the plants and bumble-bees.

It is a common objection to the theory of EVO-LUTION by NATURAL SELECTION that complex adaptations such as the plant–pollinator system could not evolve in small steps under the influence of natural selection, because, for any change in the plant to be favoured by natural selection, a corre-lated change would have to occur in the pollinator. And it seems unlikely that all the necessary corre-lated variants would appear at the same time. How, it might be asked, could natural selection favour plants that are receptive to pollen in their lower flowers and produce pollen from their higher flowers, if correlated changes did not occur in the bumble-bee? In fact it is usually possible to imagine an evolutionary route along which natural selection could have driven the system. If the bumble-bees already had the vertical movement pattern because the nectar already existed in higher concentration at the bottom, then any small variants of the plants that made them tend to pro-duce pollen only from the higher flowers would be favoured.

It is usual for the various coadapted traits to be disparate: the bumble-bee's movement is co-adapted with the plant's floral structure; and the structures evolved by a parasite differ from their coadapted defensive structures in the host. Thus we can say that coadapted traits are generally asymmetric. An interesting exception to this, where the coadapted structures of different indivi-duals are the same, is MIMICRY. Predators can learn to avoid particular colour patterns if the prey so coloured are customarily unprofitable for some reason (for example, if they are distasteful). Natu-ral selection can then favour other individuals that possess the avoided colour pattern, because these mimics are less preyed upon. Here natural selec-tion favours identity of structure: the coadaptation is symmetrical.

Intraspecific coadaptations. Coadaptation between organisms within the same species is exemplified by nearly all social behaviour. For example, one could often describe signals and response in COMMUNICATION as coadapted. Like-wise the co-ordinated behaviour patterns of ants in the same nest are coadapted, and also beneficial to all practitioners. But as Darwin's example of para-sitism shows, coadaptations between organisms are not always for mutual benefit. Usually, the coadaptations of different organisms have evolved for mutual exploitation, rather than for mutual be-nefit. Indeed, the theory of the evolution of AL-TRUISM tells us that ants are rather exceptional anyway in that genetically the ants of an ant nest are intermediate between normal separate organ-isms and a single organism.

Coadaptations within an organism. The various adaptations of a single organism co-operate to make a coherent whole. Thus, the nervous system and the muscular system, and the muscular system and the skeleton, can be said to be coadapted in the sense that, say, the muscular system of a fish works well with a fish skeleton but it would not work at all with a cat skeleton. The various struc-tures of an organism are said to be coadapted if they interact for a common function; this is, in fact, the original meaning of coadaptation.

Coadaptation, in the sense of the harmonious co-operation of the parts of an organism, was the subject of a considerable debate in the late 19th century among evolutionary biologists; Herbert Spencer and G. J. Romanes thought that natural selection could not explain coadaptation, whilst others, such as August Weismann, A. R. Wallace, and Raphael Meldola, thought that it could. The objection was similar to the one we met above: that because natural selection is constrained to change organisms only by small steps, it could not produce complex adaptations that would have required many correlated changes. Again it may be replied that it is usually possible to imagine a series of small steps, by which any coadaptation could have evolved. For coadaptations within an organism, there is an additional consideration. The GENETIC control of development will often cause the separate traits that are coadapted to vary together. Anatomically separate parts of an organism, such as nerves and muscles, are con-trolled by a single developmental programme (see ONTOGENY) that causes them to coadapt correctly, so any mutation effecting a change in the muscles would tend also to change the nerves, so that the coadaptation would be maintained.

It is not only the structural parts of an organism that are coadapted, but also, behaviour is co-adapted with structure, as is one activity with an-other. Sequences of activities are so arranged that the components are temporally organized and ordered for a common function. More or less any behavioural system would illustrate this; for ex-ample the nesting behaviour of the sand wasp (*Ammophila campestris*) is known to consist of digging a nest, provisionally closing the nest's en-trance, returning with a caterpillar, laying an egg on the caterpillar, closing the entrance, and then reprovisioning the nest at exact intervals until the nest is finally closed. The whole sequence is carried out for the common function of reproduction, and the components are carried out in the correct order and at the correct time intervals. (More re-markably, the wasp can be going through the sequence for several nests at the same time, re-membering both how many times each nest has been provisioned, and the correct time intervals for all nests.) Therefore the activities are described as coadapted for nesting.

Examples also abound for the coadaptation of behaviour and structure or of behaviour and physiology. Again, it suffices to illustrate the principle with a single example. Coadaptation of behaviour and structure is exemplified by CAMOUFLAGE. Many animals are coloured to camouflage themselves, so defending themselves from predation. However, such camouflage would be useless to the animal if its behaviour did not allow the camouflage to be expressed: it must land on an appropriately coloured surface and align itself correctly. Coadaptation of behaviour and physiology is exemplified by THERMOREGULATION; the behaviour of many animals has been found to regulate body temperature at a physiologically optimal level.

One final meaning of the term coadaptation can now be mentioned. If the parts of an organism are coadapted, it can be said that the genetics underlying those structures or activities must also be coadapted. Therefore evolutionary geneticists often refer to 'coadapted gene complexes', by which they mean that the various genes that co-operate in building a body or a part of a body must be coadapted.

Concluding remarks. We have seen two important different meanings of coadaptation: coadaptation of the parts of an organism and coadaptation of traits in different organisms. The process of the natural selection of coadaptation has interesting similarities, but important differences between the two cases. In either case, whether the coadaptation is within or between organisms, it is true that natural selection is favouring those traits that work efficaciously with the other traits that they inevitably interact with. However, despite this similarity, the two cases must be distinguished. When the coadaptation is within an organism this is due to natural selection of a single harmonious developmental programme. When the coadaptation is between different organisms, this is due to the natural selection of one developmental programme that interacts with another independent developmental programme. Because natural selection is fundamentally the selection of different genes within a species, and because genes can only exert their effects in a single developmental programme, our understanding of the very *raisons d'être* of the coadaptation requires that we distinguish between cases where natural selection has been operating in a different way.   M.R.

COGNITION, as used in animal behaviour studies, generally refers to mental processes that are presumed to be occurring within the animal, but which cannot be directly observed. For example, suppose we allow hungry pigeons (*Columba livia*) to see presentations of food that are invariably accompanied by illumination of a small electric light. During the initial observation period the pigeons are not allowed to approach either the food or the light. As a control (see LABORATORY STUDIES) other pigeons observe food and light presentations which are not correlated with each other. At the end of the period of initial observation the pigeons are free to approach the food and light stimuli. All the pigeons tend to peck at the food delivery mechanisms. The difference between the two groups is that the pigeons which observed paired food and light presentations also peck at the light, whereas the control pigeons do not. Clearly, the experimental pigeons have formed some kind of ASSOCIATION between the food and the light. The process of forming an association was not apparent to the observer during the initial phase of the experiment.

Animal psychologists are particularly interested in cognition. The processes involved seem to imply some kind of mental abstraction on the part of the animal. For example, a chimpanzee (*Pan troglodytes*) called Sarah was trained to construct sentences by selecting from an array of coloured plastic shapes (see LANGUAGE). When Sarah was asked to provide a colour token to complete the sentence '. . . is the colour of chocolate', she selected the token for brown. Neither this token, nor the token for chocolate was actually brown in colour, so Sarah must have had some abstract notion of the colour brown. In other words, the remembered colour of chocolate, and the token representing that colour, are separated from each other by a symbolic relationship. The manipulation of such symbols is held by many to be the essence of cognitive processes in animals.

The study of PROBLEM SOLVING, and especially of TOOL USING, in primates has led many scientists to believe that some, such as the chimpanzee, are capable of forming a cognitive map or model of their external environment. FIELD STUDIES show that chimpanzees sometimes take roundabout routes, which suggests that they have a mental picture of the spatial relationships of objects within their environment. For example, a chimpanzee in the forest, coming nearer to a superior (see DOMINANCE), will often approach from overhead, rather than from the same level or beneath, even if this involves taking a longer route. One scientist reported that, in trying to tame a group of infant chimpanzees that had just arrived from the jungle, he went daily into their cage to feed them. Although they would take food from his hand when he was outside the cage, the chimpanzees were too timid to approach him directly within the cage. One infant, however, would make a detour around the scientist so as to take up a position behind a wire partition that partially divided the cage. From this safe position the chimpanzee solicited food from the scientist, who responded by passing titbits through the wire.

Sometimes the cognitive process involves SOCIAL RELATIONSHIPS. For example, if a chimpanzee sees food near a dominant chimpanzee, it usually does not approach the food until the dominant animal has departed. However, if it appears that the dominant animal has not noticed the food, then the other animal may approach stealthily, and surreptitiously obtain it: the non-dominant chimpanzee obviously realizes that the dominant animal constitutes an obstacle to his obtaining the food, only if the dominant animal knows that the other is trying to obtain the food. The many recorded instances of COOPERATIVE behaviour, tool using, and detour negotiation in chimpanzees, leave little doubt that they have some sort of INSIGHT into situations, which is difficult to explain in terms of ordinary LEARNING.

A type of learning experiment that is particularly instructive is called match-to-sample. The animal is shown an object, and then it is shown an array of objects among which the original object is included. The correct solution to the problem is to choose the original object from the array. Children can learn to solve this problem in two or three trials. Pigeons may take hundreds of trials. To determine whether the subject has learned an abstract rule, it is necessary to carry out transfer tests, in which the original objects are replaced by another set of objects. Pigeons virtually have to learn the task all over again, but dolphins (Delphinidae) and chimpanzees perform well on the second task within a few trials. Indeed, there is some evidence that chimpanzees use same–different concepts in their normal lives.

Although pigeons show no evidence of abstraction in match-to-sample tasks, they can readily be taught to DISCRIMINATE between photographs showing, for instance, water versus non-water, person versus non-person, or tree versus non-tree. Pigeons do not respond in any fixed way when they see water, a person, or a tree, so this type of recognition is not like a SIGN STIMULUS, which is characterized by the fact that it elicits a particular type of behaviour, as for example the aggressive behaviour shown by a male stickleback (*Gasterosteus aculeatus*) in response to the red coloration of a rival male. The pigeon can recognize a person, etc., in a variety of contexts and positions. The person may be naked or clothed, alone or in a crowd, etc. To some extent the pigeon must be able to form a concept of person, tree, or water. This type of concept, however, is absolute, whereas the same-as/different-from concept required in match-to-sample tasks is relational. It would seem that the cognitive processes of pigeons are based upon an absolute categorization of the external world, while relational categorization, such as same-as/different-from, or shorter-than/longer-than, plays an important role in the cognitive processes of primates. The whole question of relative INTELLIGENCE of different species is confounded by the fact that each species is designed for a particular ecological NICHE, so that it is difficult to make relative judgements without being unduly ANTHROPOMORPHIC. Nevertheless, it is tempting to think that chimpanzees are cognitively superior to pigeons.

Pigeons are capable of feats of NAVIGATION and of TIME perception that are beyond the ability of human beings. On the other hand, man is far superior at abstract conceptualization and the manipulation of symbols. While we may or may not be justified in concluding that one set of abilities is superior to the other in terms of EVOLUTION, we can recognize that there are differences in cognitive ability, the ability to manipulate abstract concepts in a relational manner.

61, 81, 89, 119

COLORATION in animals is intimately related to their behaviour and way of life. In the course of EVOLUTION animals develop the colour patterns most appropriate to their respective NICHES, and, in particular, to the visual capabilities of other species in the community. Thus, nocturnal animals are not usually brightly coloured because the mechanisms of COLOUR VISION do not work well in conditions of low illumination.

There is often some competition between different pressures of NATURAL SELECTION. Usually the conflict is between the advantages of CAMOUFLAGE and those of conspicuousness. Camouflage is an important aspect of DEFENSIVE behaviour. Many animals have sophisticated colour patterns which help them to hide from predators. Such colour patterns may closely resemble those of the background (see CAMOUFLAGE, Fig. A). Disruptive coloration may so break up the shape of an animal that it is difficult to recognize when motionless (see CAMOUFLAGE, Figs. B and C), and countershading may help to eliminate the tell-tale effects of shadows. Camouflage is also important in PREDATORY behaviour, because many predators need to ensure that they are not detected by their prey, either when stalking or when attempting an ambush.

The advantages of conspicuousness may be connected with warding off predators, or may be the result of sexual competition. Some animals have a distinctive *warning coloration* which is backed up by a noxious taste or poisonous sting. A predator which eats one such animal quickly learns to avoid them in future. Other species adopt warning coloration as a form of MIMICRY. They have no noxious sting or taste, but, by resembling animals which have, they benefit from the avoidance behaviour of predators. For example, the pipe-vine swallowtail butterfly (*Battus philenor*), which is distasteful, is mimicked by the butterfly *Limenitis arthemis* in regions where both species occur. *Limenitis* is differently coloured in regions where *Battus* is absent (see MIMICRY, Fig. A).

There is sometimes a conflict between the advantages of warning coloration and those of concealment, due to the fact that different predators may respond in different ways. Some animals attempt to obtain the best of both worlds by remaining inconspicuous until disturbed, and only then displaying warning coloration. This may be a genuine warning of retaliation, or it may be bluff in animals which have no further defences. An example of such bluff or deimatic display is found in the common cuttlefish (*Sepia officinalis*) which changes the colour pattern of its mantle when disturbed (see DEFENSIVE BEHAVIOUR, Fig. B). It often produces two eye-like spots which help to scare away would-be predators. Eyespots, of varying degrees of realism, are deployed in deimatic display in many insects and amphibia. An example is illustrated in DISPLAYS, Fig. D.

Conspicuous coloration often confers a reproductive advantage. By ADVERTISEMENT animals may attract members of the opposite sex and ward off rivals. Through the process of SEXUAL SELECTION a species may become sexually dimorphic. The males are often highly conspicuous, and may compete in attracting females by means of an elaborate display, as in the peacock (*Pavo cristatus*). Conspicuousness often carries a penalty in that it may attract the attentions of predators. In some species there is an attempt at compromise between the advantages of conspicuousness and those of camouflage. The male bobolink (*Dolichonyx oryzivorus*), for example, is cryptically coloured during the winter, but in the reproductive season the plumage changes to provide a conspicuous reverse counter-shading (see ADVERTISEMENT, Fig. A).

The butterflyfish *Chaetodon lunula* is normally cryptic, but when two individuals are involved in a SOCIAL INTERACTION they change colour and become more conspicuous. At the end of the encounter the colours fade. The males of the three-spined stickleback (*Gasterosteus aculeatus*) usually develop a conspicuous red throat during the reproductive period, and especially during sexual or aggressive encounters. However, at Lake Wapato in north-west America, male sticklebacks are dimorphic in coloration, the throat being either red or black. The red forms are thought to have a reproductive advantage, while the black forms are less prone to PREDATION by trout (Salmoninae).

There are some species which are polymorphic (individuals differ in their coloration); the banded snail *Cepaea*, for example, has several different colours and banding patterns. This can be an advantage for a dispersed, fairly cryptic species that a predator is likely to hunt by SEARCHING IMAGE. The predator tends to pay ATTENTION to one prey type at a time, thus overlooking those with a different colour or pattern. In situations in which the prey is capable of LEARNING the colour characteristics of commonly seen predators, polymorphism may be of advantage to the predator. Open country predators of small mammals and birds, such as buzzards (*Buteo*) and hawks (Accipitridae), tend to be polymorphic in colour, whereas forest buzzards and hawks, whose prey have little chance of seeing the predator before an attack, are usually monomorphic.

An animal's coloration may have consequences for its THERMOREGULATION. Absorption of heat in sunlight is greater for a dark coloured object than for a pale one. A number of lizard species, such as the desert iguana (*Dipsosaurus dorsalis*), are able to change their colour in accordance with their thermal requirements.
39, 65

COLOUR VISION. If we let a beam of sunlight pass through a narrow slit into a dark room, and put a glass prism in its path, the beam will be refracted. If we now put a white screen in the path of the refracted light, a band of colours is formed, known as a spectrum (Fig. Aa). This experiment, which was first done by Sir Isaac Newton in the 17th century, shows that the sunlight consists of a mixture of an indefinite number of differently coloured rays in approximately equal proportions, and that these may be separated from each other by the different extents to which they are refracted. The physical property of light that determines the extent of the refraction is its wavelength (normally measured in nanometres), with short wavelengths being refracted more than long wavelengths. Sunlight contains all wavelengths of light in approximately equal amounts and appears white. If, in an experiment such as that shown in Fig. Ab, we now place a piece of red glass in the beam, it is immediately apparent that the long wavelengths are transmitted more readily than the short wavelengths, and the glass appears coloured by virtue of this fact. The colours of opaque objects similarly depend on the wavelengths of light that they reflect. The fin of the perch (*Perca fluviatilis*) for example, which appears red, preferentially reflects long wavelength light (Fig. Ac). Colour vision may thus be defined as the ability to distinguish and recognize different lights through their wavelength composition.

In Fig. Aa are given the apparent colours of different wavelengths of light when these are presented in isolation, in an otherwise darkened room, to a human observer with normal colour vision. It should, however, be remembered that while the wavelength of a light is a physical property, the colour of a light is a subjective property that depends on the characteristics of the observer's visual and nervous systems. For example, experiments on human subjects who are colourblind in one eye only have shown that a given wavelength can appear quite different in colour to the two eyes. The apparent colour of an object also depends strongly on the wavelength composition of the background against which it is seen, the observer's expectations, and many other factors, which are, however, outside the scope of this article. Although it is an interesting question, it is clearly very difficult, if not impossible, to find out the apparent colour that a given object may have for an animal other than man, and in animal behaviour we are more concerned with the extent to which stimuli can be detected, recognized, and responded to correctly, on the basis of their wavelength composition.

The mechanisms of colour vision. In higher ani-

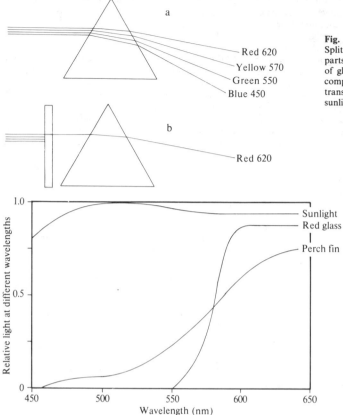

Fig. A. The composition of sunlight. **a.** Splitting a beam into its constituent parts. **b.** The effect of inserting a piece of glass into the beam. **c.** The spectral composition of sunlight, of sunlight transmitted through red glass, and of sunlight reflected off a perch fin.

mals the main visual receptors are the *rods* and *cones* of the eye (see VISION). The former group of receptors are responsible for vision under low levels of illumination (e.g. at night), and the latter for vision under higher levels (e.g. during the day). In general, colour vision is a function of the cones, and is only possible during the day.

In order to achieve colour vision it is clearly necessary that the relevant receptors are differentially sensitive to wavelength. In most cases this is achieved by the receptors having photosensitive pigments that absorb light more readily at some wavelengths than others. The receptors are most sensitive to the wavelengths that are most strongly absorbed by their *photopigments*. Although the technical difficulties involved in measuring the spectral absorption of the photopigments of individual receptors are considerable, reasonably accurate results are available for a number of species. Fig. B shows the approximate characteristics of the receptors of several animals, estimated in this way. In Fig. Ba the receptor contains a photopigment that absorbs most strongly at 500 nm, and it is most sensitive at this wavelength. As we move away from 500 nm, the absorption of the photopigment, and consequently the sensitivity of the receptor, both fall off steadily. However, a

single receptor such as this cannot achieve colour vision, even though it is differentially sensitive to wavelength. Thus, although a stimulus at 500 nm will have a greater effect on the receptor than one at 540 nm, if both stimuli are of equal energy, the latter will be indistinguishable from the former if it is made twice as intense: that is to say, the fact that the receptor is half as sensitive at 540 nm as at 500 nm can be exactly compensated for by providing twice as much energy at the former wavelength. The curve shown in Fig. Ba is correct for human rods, which contain a photopigment known as *rhodopsin*, or visual purple. At night human beings cannot see colours, and all objects appear different shades of grey. Vision of this type is said to be monochromatic.

If however, an animal has two separate types of receptor, with different spectral sensitivities, true colour vision becomes possible. Such an animal is the grey squirrel (*Sciurus carolinensis*), and Fig. Bb shows the spectral sensitivities of its two cone types, maximally sensitive at about 500 nm and 540 nm respectively. In this example, short wavelength stimuli will cause a greater output in the 500 nm receptor, and long wavelength stimuli will have the reverse effect. Each wavelength in the spectrum will in fact result in a specific ratio in the outputs of two receptor types which will be largely unaffected by changes in the intensity of the light. Provided that the BRAIN has some mechanism for recognizing these ratios, colour vision, independent of changes of intensity, is thus achieved.

A simple colour vision system such as that of the squirrel is however, imperfect, in that many stimuli will still be indistinguishable from each other even though their wavelength compositions are very different. The most striking confusion is that a monochromatic light of about 510 nm, known as the neutral point of the spectrum, is indistinguishable from white light. The reason is that this wavelength stimulates both receptors equally. The output of the eye to the brain is therefore identical in the two cases, and the stimuli must inevitably also appear identical. Vision such as that of the squirrel is known as dichromatic vision, and the presence of a neutral point, which has been demonstrated behaviourally in this animal, is diagnostic of such vision.

The number of confusions between spectrally distinct stimuli will be reduced by the presence of a third type of receptor. Three cone types have been detected in a variety of animals, including the goldfish (*Carassius auratus*) and the rhesus macaque (*Macaca mulatta*) (Figs. Bc and Bd), and less accurate data show that human receptors closely resemble those of the monkey. The vision of such animals is known as trichromatic vision. Although trichromatic vision is more effective than dichromatic vision, many physically distinguishable stimuli are still confused. White light, for example, stimulates all three receptors more

or less equally, and it is easy to arrange a mixture of three monochromatic lights at, say, 450 nm, 550 nm, and 620 nm, that will also stimulate the three receptors equally and so appear white. By appropriate manipulations of their intensities, a mixture of three such monochromatic lights can in fact be made to stimulate the three receptor types in almost any ratio we choose. Since the colours of all objects depend on the relative degree to which they stimulate the three receptor types, this means that such mixtures can be made indistinguishable from almost any other stimulus, irrespective of its spectral composition. Colour reproduction with three primaries, such as occurs in colour photography and television, would be impossible if this were not the case.

Clearly, the presence of four or more classes of receptor, each with its own characteristic spectral sensitivity, would further improve colour vision. Such further development has certainly occurred in many birds and some reptiles. The birds and reptiles are of particular interest because of the manner in which the spectral sensitivities of their receptors are determined. As in other animals, the receptors contain different photosensitive pigments, absorbing light preferentially at certain wavelengths. However, many of the receptors also contain brightly coloured oil-droplets. These are usually red, orange, or yellow and they lie between the photosensitive pigment and the incoming light, and thus act as colour filters (Fig. C). The wide variety of receptor types found in these animals is the result of different combinations of oil-droplet and photosensitive pigment. The situation is fur-

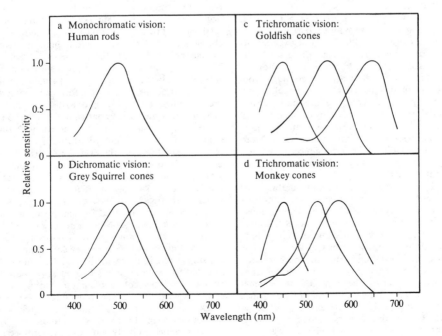

Fig. B. Typical receptor mechanisms for different types of colour vision.

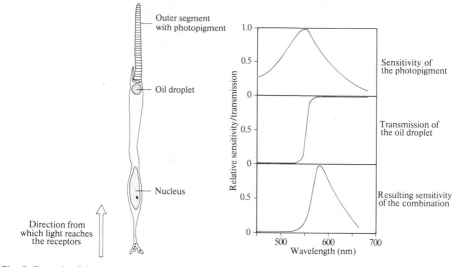

Fig. C. Example of the combination of pigment and oil droplet in the cone of a bird.

ther complicated by the fact that different species have different numbers and types of oil-droplet, and different parts of the retina of a given species may also differ markedly in this respect. For example, in pigeons (*Columba livia*) there is an area in the dorsal retina (the red field) that contains a much higher proportion of red droplets than the rest of the retina, and the eyes of many diving birds, such as gannets (Sulidae), also contain very large numbers of red droplets. Presumably the differences reflect differences in the type of colour vision that occurs.

How can we tell which animals have colour vision? This question can only be answered by behavioural experiments. Physiological experiments can give information, such as that discussed in the previous section, on the types of receptor an animal possesses, and electrical recordings from nerve fibres in various parts of the visual system can show how these nerves are affected by different spatial and temporal patterns of intensity and wavelength. Such data can show that an animal has physiological mechanisms that could in theory give it colour vision. Whether the mechanisms are in fact used for this purpose, however, must be tested behaviourally.

In order to carry out such a behavioural test it is necessary that the animal should respond consistently in some way to visual stimuli. This response can be INNATE or acquired by LEARNING. Innate responses that have been used successfully include the optomotor response, in which the animal is placed inside a rotating vertical cylinder, on the walls of which there are various grey or coloured vertical stripes. The eye, head, or body movements made in response to the movement of these stripes are then measured. The innate tendency for some animals, such as frogs (Anura), to escape towards the light, has also been used with success. It is

generally preferable, however, to use a training technique if possible, since this allows the experimenter considerably more flexibility in his approach (see LABORATORY STUDIES). The animal is usually trained to approach some stimulus for food reward, or to associate a given stimulus with a small electric shock.

A wide variety of behavioural tests, using both innate behaviour patterns and various training procedures, have been employed. One problem common to them all, is to ensure that the animal is really responding to wavelength differences, and not to differences of brightness. The problem may be illustrated by Fig. Ba, which shows the receptor characteristics of a colour-blind animal, namely a human being at night. In spite of being colour-blind, an animal with such receptors might succeed in discriminating a 500 nm stimulus from a 545 nm stimulus, because the former has a much greater effect on the receptors and so appears brighter. In order to demonstrate colour vision it is necessary to show that two spectrally distinct stimuli can be discriminated irrespective of their relative brightness. In the example shown in Fig. Ba we could try to train the animal to discriminate a 500 nm stimulus from a 545 nm stimulus that is precisely twice as bright. This, however, presupposes that we know the spectral characteristics of the receptors involved, which is, of course, generally not the case. More indirect methods of controlling for brightness must therefore be used.

Several methods have been developed. Probably the simplest satisfactory method, and certainly the one that has been most widely used, is to train the animal to select a coloured stimulus from a series of greys, over the whole range from white to black. The white stimulus must look brighter to the animal than the coloured stimulus, and the black stimulus similarly look darker, irrespective of the spectral characteristics of the animal's receptors. If a sufficient range of intermediate greys is used, one of these should appear to the animal to have the same brightness as the coloured stimulus, and we may be sure that a sufficient range of greys has been used if the animal is incapable of discriminating between adjacent greys in the series. If none of the greys is now confused with the coloured stimulus, we may legitimately conclude that the animal has colour vision.

Another problem with behavioural testing is that if an animal fails to perform the discrimination, we cannot necessarily assume that it is colour-blind. It is always possible, for example, that the animal can discriminate the stimuli, but fails to do so because the testing procedure provides insufficient MOTIVATION, or is so constructed that the animal fails to pay attention to the relevant aspects of the stimulus. Thus, although we can show that an animal has colour vision, it is never possible to prove conclusively that it is colour-blind.

In spite of these experimental difficulties, it is clear that colour vision is widespread throughout the animal kingdom. Among the fishes it has been demonstrated in a variety of shallow-water forms, though indirect evidence suggests that it may be lacking in deep-water forms. In some species of fish, colour vision has been shown to be trichromatic. True colour vision has also been shown in a wide variety of amphibians, using the animal's innate tendency to escape towards the light when disturbed. The colour vision of amphibians is apparently not very complex, consisting simply of a tendency to select short wavelength (around 470 nm) lights in preference to longer wavelengths, irrespective of brightness. Among reptiles the evidence indicates the presence of colour vision in some lizards (Lacertidae), crocodiles (Crocodylidae), and tortoises (Testudo). Many birds have been shown to have well-developed colour vision, and even the tawny owl (Strix aluco), a nocturnal species, has been shown to have this ability to some degree. As we have seen above, it is likely that the colour vision of many birds is considerably more developed than it is in man.

There is a prevalent view that, among the mammals, rats, cats, dogs, and cows are colour-blind, and it is in fact often further assumed that the primates are the only mammals to possess colour vision. The available evidence does not, however, support this statement. Experiments have shown convincingly that rats (Rattus) have some degree of colour vision, at least to the extent of being able to distinguish red from other colours. Several experiments have also demonstrated colour vision in both domestic cats (Felis catus) and dogs (Canis lupus familiaris), and in the former case physiological experiments have now revealed the receptors responsible and elucidated some of the neural pathways involved. Colour vision has been demonstrated in the zebu (Bos indicus, the domesticated buffalo of India), the giraffe (Giraffa camelopardalis), the domestic horse (Equus Prze-

walski caballus), the North American opossum (Didelphis marsupialis virginiana), and very many other species. Most of these experiments involved the technique of training the animals to discriminate coloured stimuli from a series of greys, and in some cases the animals learnt the discrimination only slowly, suggesting that their ability to discriminate colours may be only poorly developed. Nevertheless, there is no reason to suppose that colour vision is not widespread, perhaps even universal, among the mammals.

More detailed experiments have been performed on several species of squirrel (Sciuridae). Most squirrels are highly diurnal, with well-developed eyes, and they readily perform colour discriminations. The presence of a neutral point in the spectrum, as well as physiological data on their retinal receptors, shows that they are dichromatic. The primates are known to have well-developed colour vision. Experiments on the rhesus macaque and on the chimpanzee Pan troglodytes have shown that both these species are trichromats, and that their colour vision mechanisms are the same in all measurable respects as those of man. In two New World monkeys (the squirrel monkey, Saimiri sciureus, and the white-throated capuchin monkey, Cebus capucinus) behavioural experiments initially suggested dichromatic vision, for the spectrum appeared to have a neutral point. This neutral point, however, disappeared with further training, and it now appears that these monkeys are trichromatic, but that the red receptors are poorly developed. The vision of the New World monkeys is thus equivalent to a common form of colour defect in human beings, in which the red receptor mechanism is weak: such people are known as protanomalous trichromats.

Finally, colour vision is also known to occur in invertebrates. The best documented data are for the honey-bee (Apis mellifera), which has been shown to have well-developed colour vision that is, however, strikingly different from that of man. In particular, the vision of bees extends into the ultra-violet, where the human eye is totally insensitive. Many flowers have well-developed markings (the honey guides) that are only visible by ultra-violet light, and so are invisible to man, but conspicuous to bees. Among other invertebrates, the common octopus (Octopus vulgaris), which has highly developed eyes and lives in shallow water where colours abound, appears interestingly enough to be colour-blind. Although, as we have seen above, it is never possible to prove conclusively that an animal is colour-blind, careful behavioural experiments on the octopus have failed to reveal any trace of colour vision, and physiological experiments have succeeded in finding only a single receptor type.

Functions of colour vision. The colour of an object is often an important clue to its recognition. The cabbage white butterfly (Pieris brassicae), for example, prefers blues and reds when feeding, but selects green surfaces on which to lay its eggs. The red colour on the belly of the male three-spined stickleback (*Gasterosteus aculeatus*) during breeding is an important stimulus for its behaviour at this time, and the red colour of the breast of the male robin (*Erithacus rubecula*) is similarly a specific SIGN STIMULUS releasing aggressive behaviour in other robins. The wide prevalence of strikingly coloured specific markings in both animals and plants clearly attests to the importance of colour in releasing different aspects of behaviour, and many examples will be found elsewhere in this volume.

It is likely, however, that such specific object recognition through colour is not the only FUNCTION of colour vision, as defined in this article. As we have seen, the ability is widespread throughout the animal kingdom, and it occurs not only in diurnal species, but also, though probably in a less well-developed form, in nocturnal species such as the owl and cat. Nocturnal species are not as a rule brightly coloured, and it seems unlikely that specific colours play an important role in their lives. Nevertheless, colour vision mechanisms may be important for such animals in the more basic task of resolving objects from the background. This task, which involves the detection of any differences there may be on either side of the boundary separating the object from its background, will be performed more successfully by animals having more than one receptor type. For example, if we consider a colour-blind animal with one receptor type alone, an object will be perfectly camouflaged, provided its intensity is such that its effect on the animal's receptors is equal to that of the background, even if its spectral composition is in fact quite different. If two receptor types are present, the object must have an equal effect on both if it is to remain invisible, and this is clearly much less likely to occur. The details of how colour vision evolved will presumably never be known, but it may be that its original function was the simple detection of objects against various backgrounds, with the recognition of objects through their specific colour coming as a later development.   W.R.A.M.

59, 108, 146

COMFORT BEHAVIOUR is the term given to a somewhat heterogeneous group of behaviour patterns, including GROOMING, scratching, shaking, stretching, yawning, etc. These activities all have something to do with body care, whether it be routine maintenance, as in the preening of birds, or removal of parasites as in the REFLEX scratching of mammals. Individuals in a flock of gulls (Laridae), resting on the tideline, often stretch both a wing and a leg on the same side of the body, while balanced precariously on the other leg; blackbirds (*Turdus merula*) 'sun-bathe' in the late summer sun; and, more intriguingly, birds such as the common jay (*Garrulus glandarius*) may engage in the activity of 'anting' (during which they dab ants

on their feathers). These actions, together with water-, sand-, smoke-, and dust-bathing, are all subsumed under the term comfort activity. Activities such as water-bathing or sand-bathing (as frequently shown by desert rodents) are clearly aimed at maintaining the condition of the plumage or pelage, while such actions as yawning and stretching have more obtuse and intangible functions. To recognize the purpose of many grooming actions one simply has to think, by analogy, what steps one would go through in the cleaning of an object. Just as a bird wipes and scratches its bill, and a house mouse (*Mus musculus*) shakes and licks its paws, so we too would shake or bang a brush to remove surface debris, or wash out a comb before use. Only when the cleaning implement is free of dirt can it be used for cleaning. Yawning may achieve a more forceful exchange of gases in the lungs, but is also associated with compression of the salivary glands. Whether this is of significant importance, or a mere corollary of the action, is unknown. Stretching may be concerned with improving muscle tone or muscle function, or increasing circulation to the limbs, etc. Anting behaviour of birds is thought to be a response to ectoparasites (as also may smoke-bathing). The bird may prostrate itself on an ant-hill, fluttering its wings, and occasionally dabbing and rubbing beakfuls of ants over its feathers. It is thought that the formic acid, which the ants produce, may help to cleanse the skin and plumage of ectoparasites.

Other examples of comfort movements abound in the natural behaviour of wild and domesticated animals. Cattle (Bovidae) continually swish their tails to keep attendant flies (Diptera) on the move, and horses (Equidae) may be seen nose-to-tail so that two animals can obtain the reciprocal benefits of the activity of the other. The turning movements of domestic dogs (*Canis lupus familiaris*) prior to lying down is a classic example of a phylogenetically ancient trait (see EVOLUTION), which still appears as no selection has taken place during domestication to eradicate this grass-flattening behaviour pattern. Grooming is used by some animals as a means of regulating their body temperature (see THERMOREGULATION), and actions such as panting are often regarded as comfort behaviour. The vibrating or fluttering of insects on cold days or early in the morning, is the mechanism by which a cold-blooded animal can generate sufficient internal heat to resume normal activity. In addition to the primary FUNCTION of body care, many comfort activities have undergone RITUALIZATION during evolution, and may also have a COMMUNICATION function. Examples are the DISPLACEMENT preening of ducks (Anatinae) (see RITUALIZATION, Fig. A), and the social grooming of monkeys (Simiae) (see SOCIAL RELATIONSHIPS).

COMMUNICATION, in everyday usage, usually means something to do with LANGUAGE. When we communicate with one another we transmit information or convey a meaning. *Language, information*, and *meaning* are words which have been used in specialized senses by students of animal behaviour, as we shall see. But, except in these specialized senses, the words are too closely bound up with subjective feelings to be generally useful in studying animals, and ethologists normally attempt to use the word communication in a strictly objective sense.

In this spirit, one suggested definition is as follows. An animal is said to have communicated with another animal when it can be shown to have influenced its behaviour. Thus a song-bird, seeing a hawk (Accipitridae) overhead, gives a characteristic alarm call, and as an immediate consequence all the other birds in the flock flee for cover. Privately we may speculate that the vigilant bird wished to convey a message with the meaning: 'There is a hawk', and that the other birds understood this information and acted appropriately. But all that we actually know is that one bird gave a certain kind of call which had the effect of changing the behaviour of other birds. We can show that it really is a cause–effect relationship rather than sheer coincidence by experimentally presenting a tape-recording of the call a large number of times, and showing that it is reliably followed by fleeing in birds which hear it. What we cannot discover is the subjective intentions or feelings, if any, of the parties to the communication. We cannot say they do not have subjective feelings; very probably they do, but it is impossible for us to know. Our definition of communication, if it is to be useful, has to recognize this impossibility.

As it stands so far, our crude definition, though undoubtedly objective, allows too much. If we say 'Animal A is said to have communicated with animal B when A's behaviour alters B's behaviour', we are led to the absurd conclusion that A communicates with B when he bites off his foot. B's subsequent behaviour is assuredly changed; for one thing he will now walk with a limp, but this is too far from the everyday meaning of communication to be allowable. Various ways out of this difficulty have been suggested, none of them wholly satisfying. The definition used in this article is as follows. Animal A is said to have communicated with B when A's behaviour manipulates B's sense organs in such a way that B's behaviour is changed. This definition lets in the alarm calls, since they exert their influence via the ears of the responding birds, but excludes the case when A bites off B's foot. B's limp results directly from the absence of his foot (although his SENSE ORGANS presumably register the amputation too). The word 'manipulates' is used rather than merely 'influences', because 'influences' would admit cases where A's effect on B was accidental. For instance it would include as communication the inadvertent movement which betrays a prey animal to its predator. Of course conscious intention to manipulate

another's behaviour is not necessarily implied. An ethologist feels justified in using the word communication, when he thinks that NATURAL SELECTION has acted on the behaviour in question to enhance its power to influence the behaviour of other animals. In practice it is often difficult to be certain of this.

Examples of communication are numerous, and occur in almost every group of animals: SONG in birds, whales, and crickets; flashing in fireflies and some deep-sea fishes; tail-wagging in dogs; hooting in chimpanzees; roaring in lions; hissing in snakes and in cats; rattling of the tail in peacocks and in rattlesnakes; bowing to sexual partners in doves and in monarch butterflies; waving of the arms in jumping spiders and in men. All these behaviour patterns seem to have been adapted to cause some change in the physical environment which is picked up by the senses of another animal, as a consequence of which the behaviour of the receiving animal is changed. The word signal (or, synonymously RELEASER) may be used for any behaviour pattern or structure which communicates.

Notice that there is no reason for restricting the idea of communication to relationships within the same species. A celebrated example of cross-species communication is provided by the so-called cleaner-fish of tropical reefs. These belong to a wide variety of different groups of fish, and even include some shrimps which are not fish at all. All that they have in common is their way of making a living. They pick parasites off the bodies of larger fish. This is good for the large fish too, so a relationship of mutual benefit (see SYMBIOSIS) has arisen. The large fish refrain from eating the cleaners, even though they are of a tempting size. A large fish may let a cleaner go right inside its mouth, pick its teeth, and then swim out unharmed. This relationship of apparent mutual trust is mediated by a system of signals which resembles the courtship gestures normally exchanged between would-be sexual partners. Cleaner-fish have acquired a characteristic stripy 'uniform', and they perform a cleaners' dance in front of a large fish, which induces the large fish to switch off its prey-catching responses, and wait passively to be cleaned. Signals also pass in the reverse direction. When a large fish such as a grouper (*Epinephelus*) wants to move off while it happens to have a cleaner such as the neon cleaner goby (*Gobiosoma oceanops*) in its mouth, it performs a characteristic jerk of the jaw which warns the cleaner that it is time to leave. Impressive as this example is, it remains true that the majority of animal signals which have been studied are exchanged between members of the same species.

The definition does not demand that the response of the recipient of a signal should be immediate. The courtship behaviour of a male Barbary dove (*Streptopelia risoria*) may have no immediate effect on the behaviour of the female. But it does influence her HORMONE production and therefore her behaviour, over a period of days or weeks. In practice it is difficult to spot such long-term effects, and most known examples of animal signals elicit a more immediate response from the recipient.

Who benefits by communication? The central dogma of modern ETHOLOGY is that behaviour, like every other aspect of animal nature, has evolved because it benefits the *genes* of the behaving animal. Communication must be no exception, but we have to be careful here. It takes two to communicate: one to send and one to receive the signal. Are both participants benefited by the transaction, or just the signaller, or just the receiver? The conventional wisdom in ethology has usually been that both parties benefit, and that the whole signalling system evolves because of mutual benefit. Natural selection favours males who signal to influence the behaviour of females; at the same time it favours females who respond to the signals. The joint selection pressure leads to the evolution of more and more effective signals interacting with more and more receptive sensory apparatuses. In some cases this is doubtless a true picture. But the modern theory of EVOLUTION leads us to look at it in a more cynical way.

Except in the rare case of identical twins, no two animals have exactly the same GENETIC interests at heart. Even faithful mates cooperating in the rearing of their joint offspring may have something to gain from deceiving their partners. Even a mother and her child come into conflict over how much milk the child is entitled to suck. Fellow members of the same wolf pack (*Canis lupus*) need each others' help in order to bring down a large moose (*Alces alces*), but once it is killed they are rivals for the tenderest morsels. Perhaps, then, signals should be regarded as the tools by which animals manipulate the behaviour of rivals, in order to exploit them; in order to benefit the genes of the signaller, if necessary at the expense of the receiver.

But a signal can only be effective from the point of view of the signaller, if the receiver responds appropriately. The angler fish (*Lophius piscatorius*) (Fig. A) only succeeds in manipulating the behaviour of his prey as long as the prey respond to his signal (the worm-like bait on the end of the angler's rod-like appendage) by approaching it. The reason a prey fish approaches the lure is that its sense organs relay to its brain a message indicating the presence of a worm; the brain directs the fin- and tail-muscles accordingly, and the little fish chases the lure. On this occasion the consequences of the little fish's action are fatal. Why then does it do it? The answer is that usually small wriggling objects are genuine worms, and natural selection favours fish which chase worms. The angler fish is exploiting an aspect of the sensory-response system of its prey species which usually benefits the prey because it leads them to worms. Of course natural selection will act on the prey fish to sharpen up their discriminatory powers, so that they can successfully distinguish real worms from fake ones projecting from the heads of angler fish. But, no less strongly, natural selection acts on angler fish to improve the quality of their deception. In this case the evolutionary improvement of the signal has proceeded, not because responding to the signal benefits the receiver, but for precisely the opposite reason. The benefit of the communication is all on one side, the signaller's.

In other cases, as we have already seen, it is probably really true that both parties benefit from communication. Even though the benefits may not be evenly shared by the two participants, it may still be the case that the balance of selective forces acting on both sender and receiver is pushing in the same evolutionary direction, i.e. towards the evolution of an efficient mutual signalling system. This is probably true of many courtship signals, signals mediating the parent–child relationship, signals exchanged by fellow members of ant and bee colonies, wolf packs and other forms of SOCIAL ORGANIZATION.

Design features of signals. A signal must travel from sender to receiver via some physical medium. The most important physical media correspond to the four main kinds of sense organ: sight, sound, touch, and chemical (smell and taste). The best way to comprehend the great variety of physical

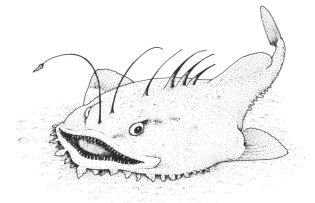

Fig. A. The angler fish (*Lophius piscatorius*).

forms which signals take is in terms of design features which cut across particular examples, and particular physical media. Here we shall consider four such design features. These are the distance of action, the localizability, the time scale, and the specificity of the signals. These design features provide dimensions along which different types of signal can be ranked. For instance auditory signals can be detected at a greater distance than tactile signals. High scoring on one of these dimensions is not necessarily desirable, however: a long range is clearly likely to be a virtue in a mate-attracting call, but this is offset by the dangers of summoning distant predators.

Three of the design features are straightforward, but the fourth, specificity, needs a little elaboration. It overlaps with the idea of information, to be discussed later. Signals may be species-specific. Many courtship signals serve the important function of preventing hybrid matings (see ISOLATING MECHANISMS). For this it is important that they should be as different as possible from signals of other species, and as uniform as possible within the species. But for other purposes it may be important for signals to be individual-specific. Many birds need to recognize the cries of their own mates and their own offspring. This requirement is in partial conflict with the need for species-specificity. The great complexity of bird song allows for individual variety while still retaining species-specificity. No two blackbirds (Turdus merula) sing identical songs, yet all sing recognizably blackbird songs. In the song thrush (Turdus philomelos) the detailed phrases are individual-specific, but certain features of overall patterning, for instance the habit of repeating each phrase twice or thrice, are species-specific. A third kind of specificity may be designated as message-specificity. To the extent that animal signals can be said to have meanings (see below), it is necessary that signals with distinct meanings should not be confused: a sexual call must be distinct from an aggressive call, and so on. An animal must be capable of making a sufficient variety of signals to convey all the different messages which it needs.

Using the four design features as guidelines, we can now look at the different physical media of communication. It is possible that the earliest kind of communication to evolve was chemical. Cells within a body could be said to communicate by chemical means; there is a sense in which developing nerve-cells in an embryo are attracted by chemical signals towards the end organs which they are supposed to innervate. Chemical signals are also used in behavioural interactions between individual animals. The acme of chemical communication is reached in social insects such as ants (Formicidae), which employ a rich variety of ALARM substances, trail-laying substances, and substances for manipulating the sexual development of other individuals.

Chemical substances secreted for signalling purposes are called PHEROMONES. These are either sensed by the receiver making direct contact, as when a worker bee (Apis) licks the body of a queen who is secreting a pheromone; or the pheromone is deposited on a solid object and later contacted by the receiver, as in SCENT MARKING by mammals; or the pheromone is released into the surrounding fluid, the water or air. In the last case the molecules either diffuse outwards until they reach the receiver, or they are carried in a current or wind. Occasionally the signaller manufactures his own current or wind. For instance a male smooth newt (Triturus vulgaris) fans with his tail to waft a current of water over his genital opening towards the female. It has been suggested that when a courting male fruit fly (Drosophila melanogaster) vibrates one wing in the direction of the female, he is blowing a chemical towards her. He is also, by the way, making a sound at the same time.

To the extent that the specificity of a chemical signalling system is limited by the number of distinct substances that can be manufactured, and recognized by sense organs (see CHEMICAL SENSES), it ought to be a very rich medium for communication. Police dogs can be trained to discriminate the odours of any two individual human beings except pairs of identical twins. Of course this is not in itself an example of communication, but it shows the great potential specificity which could be exploited in a communication system.

Pheromones are usually organic molecules with between five and twenty carbon atoms. There seem to be good design reasons for these limits. With fewer than five carbon atoms, the variety of distinct molecules that can be synthesized is too small. Above about twenty, the number of distinct molecules increases astronomically to no good purpose, yet the energy costs of synthesis go up too, and large molecules tend to be less volatile, and so travel less far. A further generalization noticed in insects is that, within this range of molecular sizes, relatively larger molecules tend to be used for sexual attraction, smaller ones as alarm substances. Alarm substances are used, for instance by ants, to alert the rest of the colony to danger. There seems no particular reason why they should have high specificity, so a wide variety of distinct alarm molecules is not necessary. Sex attractant substances on the other hand must be species-specific. The variety of compounds necessary for this is achieved by using larger molecules.

The distance over which a chemical signal can act is partly limited by the sensitivity of the receiver's sense organ. This can be astonishingly acute. The female silk moth (Bombyx mori) attracts the male by secreting a substance called Bombykol, whose exact chemical formula has been analysed. It has also been synthesized and presented to male moths, and also to disembodied moth antennae, whose responses were monitored with electrodes. It turns out that male silk moths and their antennae respond only to Bombykol and, less strongly, to certain very close chemical relatives of Bombykol. This demonstrates very great specificity, but the most remarkable result of this study was that a single molecule of Bombykol was sufficient to trigger a physiological response.

In practice, however, wind is likely to determine the actual range over which a chemical signal is detected. It is also crucial in determining the localizability of a chemical signaller. It is theoretically possible for a receiver to track down a signaller by measuring the concentration of pheromone and moving up the concentration gradient. If there is any wind this will not work, and the male silk moth is probably typical in simply steering upwind. Compared with visual and auditory signals, chemical signals are not very accurately locatable.

The time-scale of action of chemical signals is slow. They cannot be turned rapidly on and off like auditory signals, and they do not fade rapidly. For this reason they are especially suitable for semi-permanently marking out territorial boundaries. A territorial male mammal cannot be everywhere at once around the borders of his territory, but he can do the next best thing which is to leave chemical 'calling cards' for rival neighbours to smell. Some ant alarm substances fade in a matter of 30 s, but even this turnover rate is too slow to allow the kind of intricate temporal patterning which does so much to increase the potential specificity of sound communication systems.

Auditory signals are transmitted at a rate which is, for practical purposes, instantaneous, and they can be turned off instantly too. Sound signals are potentially capable of immense variety. They can vary in pitch (frequency of vibration), in loudness (although this may be confounded with distance), and in temporal pattern. All these degrees of freedom make for great specificity, although they are not always used. For instance male crickets (Gryllidae) of any one species sing all on one note. The pitch of different species varies, but this is very probably largely incidental. What seems to matter is the species-specific Morse-code-like pattern of intervals between pulses of sound.

Vertebrates tend to use both temporal patterning and frequency coding in their sound signalling. Peaks of vocal achievement are reached by birds, whales, and man. The performances of virtuosi such as blackbirds and nightingales (*Luscinia megarhynchos*) are too well known to need further celebration here. Less familiar, but one of the most remarkable sounds on earth, is the song of the humpback whale (*Megaptera novaeangliae*). In pitch it spans the whole range of human hearing and probably beyond, recalling a deep rumbling cathedral organ at one end of the scale, a piercing bat at the other. Human language exhibits the specificity which can be achieved by varying tonal quality. The actual pitch of human speech is fairly uniform, and in any case is sex-specific in an informationally trivial way. What matters is tone quality: different vowel sounds consist of different combinations of pitches, frequencies bearing particular ratios to one another. Consonants employ the same principle in a more complex way. Different temporal sequences of tones convey messages of almost infinite variety, in a large number of different languages.

One obvious limit on the distance of action of sound signals is the loudness which the signaller can achieve. This is directly related to the size of the sound-producing organ, and therefore of the animal. In addition various physical properties of the environment are relevant. Because of a curious interaction between temperature and pressure gradients in the sea, there is an underwater stratum, known as the sound-tunnel, in which sounds, instead of broadcasting uniformly in three dimensions, tend to be reflected in such a way that they propagate approximately in a two-dimensional plane. As a consequence they spread a very great distance horizontally. Sensitive microphones lowered into the sound-tunnel off North America can detect submarines in British waters. The song of the humpback whale is so loud that it could theoretically be heard by another whale right round the world. It is not known whether whales do station themselves in the sound-tunnel, and communication over these prodigious distances has never been demonstrated, but physical calculations allow it as a theoretical possibility. The distance record for a small animal is probably held by the mole cricket (*Gryllotalpa vineae*) (Fig. B). This muscular little creature increases its acoustically effective size by building its own megaphone: it digs a double conical horn-shaped burrow in which it sings. One metre above the megaphone the volume reaches an uncomfortable 92 dB, which, on a still day, can be heard 600 m away.

Bird song too can carry far, the distance depending partly on the terrain. Different kinds of song travel best in different environments, forest, open grassland, etc. We should expect to find that bird species would have songs well tailored to their particular environment. Where this has been looked at, it has been found to be largely true.

How easily sound signals can be localized depends upon the kind of ear possessed by the receiving individual (see HEARING). Small insects tend to have particle displacement ears, whose sensitive part is small enough to be physically buffeted back and forth by the vibrating air molecules. Vertebrates and larger insects tend to go in for pressure receptors, ears which function like high-speed barometers. Like household barometers these are not directionally sensitive. Consequently vertebrates locate sound sources by comparing the input at the two ears. A somewhat complicated physical argument suggests that greatest ease of localizability will be attained with a series of short staccato notes, each including a wide spectrum of frequencies. This is characteristic of the chink-chink mobbing call of the blackbird, and the comparable calls of many other species of song-bird. The mobbing call is used to summon companions to help in harassing a ground-based

Fig. B. The mole cricket (*Gryllotalpa vineae*) in its burrow.

or perched predator, who presents a long-term, but not an immediate, threat. It is of the essence that a mobbing call should be easily located by the companions to whom it is addressed, and the physical properties of mobbing calls seem to meet this requirement. A similar physical calculation shows that if we wished to design a call to be as difficult as possible for a predator to locate, we should make it gradually fade in and fade out; it should be a thin pure tone pitched at an optimum frequency determined by the distance apart of the predator's ears; for a medium sized hawk (Accipitridae) or owl (Strigiformes) it should be around 7 kilocycles per second. This ideal specification turns out to be a good description of the actual alarm call used by a number of small bird species to warn of an approaching aerial predator. In this case the message is 'Danger exists' rather than 'Danger exists over here'. There is nothing to be gained from embodying directional information in the call, and everything to lose, since a predator on the wing is an imminent deadly threat, and he would be likely to use an easily located call as a convenient beacon.

Light has many of the same advantages as sound, as a physical medium for communication. But unlike sound signals, many visual signals can be permanently on display. This is true of static features of external anatomy which serve to label an animal's species, sex, or status: bright feathers and crests, well marked eye rings, garish genitals, specially enlarged limbs. Like sound signals, visual signals can also be rapidly turned on and off, and this potentially increases the variety of messages that can be sent. Male jumping spiders (Salticidae) wave their front limbs in species-characteristic semaphore signals to females. Fiddler crabs (many species of the genus *Uca*) have the claw on one side greatly enlarged, and they jerk this one claw in species-specific rhythms (see DISPLAYS, Fig. A).

Visual signals such as these, which rely on reflected sunlight, cannot be used at night. Some animals have evolved the means to manufacture their own light, using an energy-consuming biochemical reaction. Fireflies (Lampyridae) (really beetles) recognize members of their own species by their Morse-code-like flash patterns. Just as seamen identify lighthouses by their unique flash intervals, so fireflies track down mates of their own species. They are indifferent to colour, and can be readily fooled by an electric torch, but only if the flash interval is correct. The male of the North American firefly (*Photinus pyralis*) for instance, flashes rhythmically with an interval of about 6 s. The female flashes back just 2 s after any one of his flashes. When a male sees a flash 2 s after one of his own he homes in on it (see COURTSHIP, Fig. A).

Potentially, visual signals have a long range, but only over unencumbered terrain. Unlike sound waves, light rays do not travel round corners. The chances are good that a tree or other obstacle will lie between a signaller and a potential receiver of his signal. This, of course, is not always a bad thing: the potential receiver may be a predator, or an interfering rival. It is conceivable that signals may change their effect at different distances. For instance the stripes of a zebra (*Equus hippotigris* spp.), which seem so vivid at close quarters, and which may well serve as social signals, fade into a highly effective camouflage at a distance. If it can be seen at all, a visual signal can be located with extreme accuracy, limited by the distance apart of individual light-sensitive cells in the receiver's eye (see VISION).

Touch-mediated signals can obviously be used only at close quarters. They are important in the final stages of COURTSHIP after physical contact has been established. There is a sense in which a spider's web allows a form of touch-communication at a distance. The web functions primarily as an insect trap, but in species such as the garden spider (*Araneus diadematus*) the male signals to the female by rhythmically twanging one of the outer guy-ropes of her web. This serenade for strings seems to inhibit the female's dangerous tendency to devour anything small that moves. Whether it should be regarded as tactile or low-frequency sound communication is debatable. A similar borderline example is provided by Australian aquatic insects known as water-striders (Gerridae), which send out courtship signals in the form of surface waves in the water. Tactile signals may also be strongly interlinked with surface-acting chemical signals, for instance in the courtship of many insects. The localizability of tactile signals depends on the ability of an animal to know whereabouts in space each of its limbs is. This seems to us a trivial problem, but to the common octopus (*Octopus vulgaris*), for example, it is far from trivial. Tactile signals are not, in general, used for conveying highly specific or complex information, but here a notable exception is the bee dance. This also has auditory and chemical components, but normally does not involve vision as it takes place in the dark interior of the hive. It is such a special case that it will be discussed in detail below.

The evolution of communication. Reconstruction of the remote evolutionary past can only be done by clever detective-work on the basis of indirect evidence. In the case of animal signals, direct fossil evidence is never available and we have to make inferences from surviving animals alone. Ethologists, building on the ideas in DARWIN's *The Expression of the Emotions in Man and Animals* (1872), have shown that signals have evolved from non-signal movements, movements which originally served different, and often more mundane, purposes. For instance the courtship ceremony of the mandarin drake (*Aix galericulata*) includes a delicate pointing with the tip of the bill at a particular bright feather in the wing, called the *speculum*. Comparison of this movement with similar, but less stylized, movements in other, related species has led ethologists to the conclusion that it originated from an ordinary wing-preening movement (see RITUALIZATION, Fig. A). Originally wing-preening served the sole purpose of settling the ruffled wing-feathers, a purpose which is still very important in a species which relies on well-maintained wings for efficient flight. But over evolutionary time natural selection seized on a particular aspect of this useful maintenance activity and modified it as a signal, with the entirely different purpose of manipulating the behaviour of other individuals. The name RITUALIZATION has been given to the evolution of signals from non-signals. The courtship pointing movement of the mandarin drake is referred to as a ritualized preening movement. The word has been used of morphological structures as well as of behaviour patterns. The speculum of the mandarin drake presumably started out as an ordinary feather: its assumption of a bright orange colour and characteristic upward tilt is the result of ritualization.

In many cases the unritualized origin of a signal is clearly betrayed by its form. Signals are sometimes almost indistinguishable in form from their unritualized counterparts still existing in the same animals. Many small birds communicate by means of feather-erection postures, which do not materially differ from the postures used in normal THERMOREGULATION. In other cases the non-signal ancestry of signal movements would be hidden if it were not for the evidence of other living species which exhibit the same kind of signal in a less highly ritualized form. The famous courtship of the Indian peafowl (*Pavo cristatus*) in which the male, the peacock, displays his extravagant fan and pirouettes in front of the female, does not resemble any ordinary non-signal movement. The form of the display on its own gives no clue as to its evolutionary origin. Yet by looking around at other, related birds (various kinds of pheasants and chickens) one can reconstruct a plausible series of intermediates. From the COMPARATIVE evidence we can reasonably guess that the peacock's display has evolved from something like the call *tit-bitting* that the domestic cock (*Gallus g. domesticus*) uses to bring the hen to food. Even this is itself ritualized: in its turn it can be traced back to feeding movements which originally were used only by adults towards chicks. If it were not for the intermediate series of pheasants (Phasianinae), the connection between the peacock's sexual display and the original chick-feeding movement would have been undetectable. Of course nobody is suggesting that peacocks are descended from chickens, or that pheasants are themselves intermediate ancestors. All of them are modern birds, exact contemporaries. What is being suggested is that some species have retained particular aspects of the behaviour of ancestors of varying degrees of antiquity.

If the pheasants had all become extinct, the origin of the peacock's display would have remained shrouded. There must be many cases in which there are no surviving links, and we are left to speculate wildly about the origins of signals before ritualization transformed them and then covered over the traces. An ingenious attempt has been made to follow the evolution of the human smile through a long series of open-mouthed grimaces in monkeys and apes (Simiae), but this will probably never be more than an interesting guess.

What kinds of non-signal movements have been the most eligible candidates for ritualization? One category which ethologists have long recognized is INTENTION MOVEMENTS. These might better be called incomplete movements to avoid the subjective overtones of the word intention. An animal often begins to perform a behaviour pattern from its repertoire, but then stops half-way through. Before a dog (*Canis*) bites a rival it bares its teeth by drawing back the lips. It does not always follow through by actually biting. If an ethologist sees a dog bare its teeth he can say to himself: 'That dog is more likely to bite than he was before'. But if this information is available to a watching ethologist, it is also, potentially, available to another dog. There is no need to ask whether the watching dog goes through the same conscious chain of reasoning as the man. This does not matter for the argument. What matters is that natural selection could, in the past, have favoured dogs who noticed teeth-baring in their companions, and acted accordingly. To act accordingly means to behave as if in anticipation of a bite. Selection would perhaps have favoured dogs who withdrew when a rival bared his teeth.

But now it is time to look at it from the point of view of the signalling individual. If a dog can induce a rival to flee simply by baring his teeth, it could be to his advantage to use his lip muscles as a means of manipulating the behaviour of others. This could have led to the evolution of exaggerated teeth-baring movements, the lips being drawn back further than is strictly necessary to get them out of the way for biting. An ethologist observing the exaggerated grimace so produced would call it a ritualized movement. The evolutionary process of ritualization has taken place.

The exact converse, intention movements with a negative sign, as it were, may also be an evolutionary source of signals, as Charles Darwin long ago recognized in his *Principles of Antithesis*. If baring of the teeth is a fore-runner of outright attack, then an exaggerated covering up of the teeth with the lips becomes a ritual appeasement gesture.

Another category of behaviour which seems to have been a rich source of raw material for ritualization is AUTONOMIC movements. The autonomic nervous system controls a large cluster of important visceral functions, including heart rate, breathing rate, urination and defecation, and temperature control measures such as sweating, hair and feather erection, and surface blood-vessel dilation and contraction. A scientist who wishes to predict a mammal's behaviour can gain useful information if he has instruments to measure such variables as heart rate, breathing rate, and *galvanic skin response* which depends on changes in skin conductivity induced by sweating. The principle of the lie-detector machine is the sensitive measurement of these and other variables. If you can monitor the activity of another individual's autonomic nervous system you are in a good position to predict his future behaviour. If a man's heart suddenly races and his hair stands on end, the chances are that he has become frightened: in behavioural terms he is likely to flee.

Of course animals cannot literally attach instruments to each other to monitor each others' autonomic activity, but they can do the next best thing. They can observe their companions for externally visible (audible, smellable) betrayers of autonomic change. At least some of the tell-tale signs which the lie-detector machine exploits are available to other animals. Both mammals and birds alter the thermal properties of their surface layer by raising and lowering hairs in the case of mammals, and feathers in the case of birds. Since this sometimes

serves as an involuntary betrayer of autonomic activity it is potentially exploitable by other individuals who notice it. It is reasonable to suppose that natural selection would favour the sharpening up of animals' powers to detect subtle changes in the autonomic activity of other individuals.

Once again we now look at it from the point of view of the signaller. If he can influence the behaviour of another individual by controlling his own hair-erecting muscles, natural selection may favour his exploiting this influence. Exaggeratedly ostentatious hair-erection displays may evolve, in association with anatomical changes. Examples include the evolution of special erectile areas such as hackles and crests, porcupine quills and peacock fans. Blushing in human faces and monkey backsides presumably originates from the thermoregulatory use of small muscles controlling the quantity of blood in surface vessels.

Other autonomic manifestations have been ritualized in this way. Urination and defecation are both under autonomic control: spontaneous defecation is a product of extreme fear, as young soldiers discover when they first come under fire. Both urination and defecation have apparently become ritualized in various mammals. Dogs save up their urine and ration it out, one little squirt to each tree or lamp-post. Clearly, urination to a dog is more than just a necessary emptying of its bladder. It has become ritualized, and now has a secondary social or communicative function. It is used for marking out territory. Hippopotamuses (Hippopotamidae) do not defecate merely as a means of voiding digestive waste. They whirl their tails round and round as the faecal material emerges; it hits the fan, as it were, and is sprayed over a wide area as an effective territory marker.

BREATHING movements and their equivalent in fish are under autonomic control. Siamese fighting fish (*Betta splendens*) use an exaggeratedly ritualized erection of their gill covers as part of their fighting display. Emotionally induced changes in breathing rate may have been the starting point for the evolution of the large, inflatable throat sacs which are such a conspicuous part of the social displays of the Ascension Island frigate bird (*Fregata aquila*) and some frogs (Anura) (see CONFLICT, Fig. D). Changes in breathing rate may be heard as well as seen. What was originally a barely detectable sibilant gasp may have been exaggerated over evolutionary time; made louder and more carrying by the development of specialized sound-producing membranes in the respiratory passages culminating in the evolution of mammalian vocal cords, and the equivalent in birds, the *syrinx*.

The other major group of animals to have evolved sound-producing organs, the insects, have followed a different route. Grasshoppers (Acridoidea) sing by rubbing their legs together, crickets by rubbing their wings together, and cicadas (Cicadidae) by noisily buckling a specialized region of the exoskeleton, the *tymbal*, as though it were a tin-lid. In all three cases the unritualized origins are lost in the past and we can only speculate about them. It seems probable that any insect inevitably makes a certain amount of noise as it moves. For instance the legs and wings of the ancestors of grasshoppers and crickets must have made a faint scraping noise as they accidentally rubbed against each other. If ears had already evolved in these insects, perhaps for the detection of predators, it might have been to the advantage of females to use their ears to track down males by listening to these small, accidental scraping movements. It would then have been to the advantage of the males to amplify these tiny noises: hence the evolution of specialized rasping surfaces on wings or legs, and special resonating cavities.

As well as intention movements and autonomic movements, ethologists have emphasized a third general class of behaviour patterns which have commonly served as raw material for ritualization. These go under the general heading of CONFLICT movements.

When an animal is simultaneously motivated to do two or more incompatible things, it is said to be in a conflict. For instance a hungry chaffinch (*Fringilla coelebs*), confronted with a dish of appetizing food on which is perched a stuffed owl, is in a conflict between a tendency to approach and a tendency to flee. How such conflicts are resolved has been a major preoccupation of ethologists. What is interesting for present purposes is that the consequences of motivational conflict seem to have been an especially fertile field for ritualization.

One of the things that an animal in a conflict may do is to alternate rapidly between the two incompatible movements. It may dither, taking a step forwards, then a step back, and so on. It has been suggested that the zig-zag dance of the male three-spined stickleback (*Gasterosteus aculeatus*), performed as part of its courtship ceremony, is a ritualized version of just such a dithering alternation. Sometimes an animal in a conflict may perform mosaic behaviour. It selects from the two alternative behaviour patterns elements which are not wholly incompatible, and combines them simultaneously. Many of the ceremonial postures of gulls have been interpreted as ritualized mosaic or compromise movements (see MOTIVATION).

Sometimes an animal in a conflict between two incompatible behaviour patterns performs neither of the two alternatives but does something quite different: a DISPLACEMENT ACTIVITY. Self-grooming or preening movements are especially common as displacement activities. It may well be that the ritualized preening movements of birds such as the mandarin drake evolved originally from displacement preening.

Intention movements and autonomic movements have been suitable raw material for ritualization because they are in any case tell-tale betrayers of the internal motivational state of the signaller, and hence of his probable future behaviour. Is the same true in general of conflict movements? The answer is probably yes, but the reason is rather less obvious. Motivational conflict will tend to occur at moments of transition from one dominant motivational state to another. For instance when an aggressive fish changes into a fearful fish he will pass through an intermediate period when he is simultaneously torn between aggression and fear. At this moment a conflict movement, perhaps a displacement activity, is likely to occur. A watching fish can use this as a signal that his motivational state is about to change. Such a change is especially significant for a would-be predictor. It is for this reason that motorists signal only when they are about to change their present behaviour. A motorist does not continuously signal: 'I am going to carry on in exactly the same way along this dead straight road'; he only signals when he is about to do something different. An animal in a conflict is an animal who is quite likely to change from one dominant mood to another.

Ritualization sometimes consists of the evolution of one signal into another. Many courting female birds employ a FOOD-BEGGING posture, which often elicits COURTSHIP FEEDING by the male. A glance at the crouched, wing-fluttering stance of a begging female betrays its probable evolutionary origin: it bears a striking resemblance to the begging posture of a juvenile bird. It has probably evolved by the process known as *neotony*: the retention into adult life of juvenile characteristics. It serves to appease the male, to prevent him from attacking the female. Very probably, the reason it was originally effective as an appeasement gesture is that a male would not ordinarily attack a juvenile.

There are cases in which subordinate males appease dominant males by mimicking female behaviour. In many species of monkey, subordinate males present their hindquarters to dominant males, in ritual imitation of the female's sexual presentation gesture. Since in these species a male would normally not attack a female, a male can gain protection by imitating typical female behaviour. In some cases the dominant male seems to acknowledge the submissive gesture by a brief ritual mounting of the subordinate male, sometimes even throwing in a few desultory pelvic thrusts. There is normally no accompanying evidence of genuine sexual excitement, so it is probably incorrect to equate this with human homosexual behaviour.

In general, it seems that a behaviour pattern, in order to qualify as a starting point for ritualization, must pass two tests. Firstly, it must be detectable by the sense organs of other animals right from the start. To be sure, its power to stimulate those sense organs is going to be enhanced and exaggerated by ritualization; but natural selection cannot begin to act in this way unless there is initially some minimal power to stimulate sense organs. The preceding argument leaves us a little surprised that heartbeat, a sensitive autonomic indicator, has apparently not been ritualized. The reason may be that it is initially too quiet for the ears of other individuals to detect it. Therefore natural selection cannot act to amplify it. (Alternatively, of course, it may be that the primary function of the heartbeat, i.e. pumping the blood, is too important for the disruptive effects of a secondary function to have been permitted.)

The second qualification of a suitable starting point for ritualization is that it must, right from the start, be of advantage for a potential receiving animal to detect it and change its behaviour as a consequence. If early mammals had uttered panting gasps at random, in such a way that the hearing of a gasp in no way assisted a listener to predict the subsequent behaviour of the gasper, then it is probable that vocalization would never have evolved.

Meaning is primarily a subjective concept, and one that is difficult to apply objectively to animals who do not use LANGUAGE in our sense. Nevertheless it is often tempting to try to translate an animal signal into a human language, for instance to render the alarm call of a song-bird as: 'There is a hawk'; or 'Look out! Beware!'; or 'I am very frightened and you ought to be too if you know what is good for you'. What objective criteria could we use to justify such translations? Can we really assign meaning to animal signals?

There seem to be two objective ways in which, in principle, we might assign meaning to a signal such as the bird alarm call. Firstly we can do so in terms of the state of the signalling individual. We can observe that this particular call is characteristically given by individuals who have just seen a hawk; or by individuals who show what we judge by other symptoms to be the state known as 'fear'. We can also note the subsequent behaviour of the signaller: birds who have just given an alarm call usually follow it by behaviour appropriate to the evasion of a hawk. This is, indeed, the sole reason why naturalists have called it the hawk alarm call.

The second thing we can do is to observe the response of the receiving individuals. In the case of the hawk alarm call they flee, or show hawk-evading behaviour. In subjective terms this is how we judge what the signal 'means' to the receiver, whereas previously we were concerned with what it meant to the sender.

Sometimes, as in the case of the hawk alarm call, it is possible to regard an animal signal as making reference to something in the outside world. The ability to do this is one of the great powers conferred on man by language, and examples of it in other animals have a special interest for us. The most spectacular example is the famous dance of the honey-bee (*Apis mellifera*).

This is so nearly a human language-like feat of communication that many people find it hard to believe of a 'humble' insect. When a worker bee has found a rich source of food, she 'informs' her colleagues of its whereabouts, using a code. She 'dances' on the vertical surface of a comb inside the hive. Other bees 'read the message' of her

dance and set off towards the food. When the food is very close a simple *round dance* is used, but we are here concerned with the *waggle dance* which is used for more distant sites. It has the form of a figure-of-eight with a straightened central run, during which the bee waggles her abdomen and utters rhythmic piping cries (Fig. C).

Two different aspects of the location of the food are conveyed in a rather precise, quantitative way. These are its distance from the hive, and its direction. The direction is expressed as an angle, using the sun as a reference direction. The dancer has measured the angle subtended at the hive by the food relative to the sun. Inside the hive she tilts her dance on the vertical comb so that the angle of her straight run relative to the vertical is the same as the angle of the food relative to the sun in the horizontal plane. The other bees measure the angle of her dance relative to the vertical. Then when they emerge from the hive, they translate back into the horizontal plane using the sun as the reference direction and fly off towards the food.

The distance of the food from the hive is reflected in various aspects of the rate of dancing, including the rate of turning, the rate of abdomen-waggle, and the rate of piping: the nearer the food, the faster the dance. It is not known which of

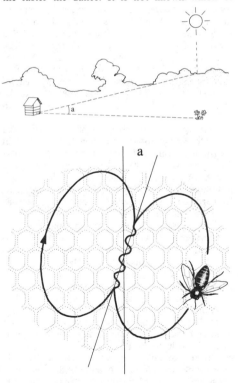

Fig. C. The waggle dance of the honey-bee (*Apis mellifera*). The angle a, between the axis of the dance and the vertical, corresponds to the angle between the sun and the food source (above).

these measures is actually used by the receivers of the information; it could be a mixture of all of them.

In the late 1960s the bee dance became briefly controversial. One group of investigators suggested that it did not in fact serve a communication function. They did not deny the fact that bees dance, nor the fact that it is possible to read from the dance the location of food. They denied that other bees actually read this information. It is true that the evidence, when looked at closely, was not totally convincing. But there is no need to discuss the controversy in detail, because, stimulated by it, experiments have since been done which prove beyond all doubt that other bees do indeed read the information in the dance. Although the sceptics were wrong, they performed a valuable service in goading people into doing the final, definitive experiments.

The bee dance and the hawk alarm call are rather special cases. More usually, the meaning of an animal signal is better thought of as referring to the internal motivational state of the signaller. When a dog raises his hackles he means 'I am angry'. We can twist an alarm call round so that it means 'I am afraid'. Even the bee dance can be regarded as meaning something about the internal state of the dancer. For instance a slow tempo to the dance could be taken as meaning 'I am tired'. A tired bee is likely to be one who has flown a great distance on her most recent foraging trip. Therefore it pays other bees to interpret the dance as if it meant 'The food is a long way off'. This may indeed have been the evolutionary origin of the distance part of the bee code. The origin of the direction part of the code is more difficult to guess.

There seems to be no great harm in using the word 'meaning' for animal signals, nor even in translating them into human language. But there is little positive point in it either, and there is a danger of our being misled by our subjective terminology. It is probably better to stick to straightforward, honest, objective behavioural terms. Then we can reserve the term 'meaning' for discussing our subjective motives and intentions.

Information. If your newspaper headlines consisted of: 'The sun rose this morning'; 'England is in the northern hemisphere'; 'Yesterday lasted 24 hours'; and similar unsurprising facts, you would probably demand your money back. The reason is that you know it all already: it is not news. The facts are all perfectly true but you do not feel informed by them. This idea that a message, in order to be informative, must be at least somewhat surprising to the receiver, has been used by mathematicians to define *information* as a precisely measurable commodity. Although this technical usage of the word information was originally coined for telephonic and other engineers, it has been applied on a number of occasions to animal communication.

The technical definition of information is initially puzzling to laymen because it has nothing to do with the meaning of a message. It is concerned with its surprise value. The statement: 'The sun did not rise this morning' would have very high information content because it would surprise any hearer very much. If he believed it, his prior beliefs would have been shattered. The opposite statement: 'The sun rose this morning' conveys virtually no information: it leaves the hearer in exactly the same state as he was before he received the message.

Mathematically, the information content of a message is measured in terms of the reduction in prior uncertainty caused by the message. Prior uncertainty is measured in terms of probabilities. If the message allows the receiver to decide between two alternatives which had previously been equiprobable, say 'heads' rather than 'tails', or 'boy' rather than 'girl', then one *bit* of information has been conveyed. Just as the calorie is the unit of heat, so the bit is the unit of information. Just as one calorie is arbitrarily defined as the quantity of heat necessary to raise 1 g water through 1 °C, so one bit is defined as the quantity of information needed to halve the receiver's prior *uncertainty*, or, in other words, to enable him to decide between two equiprobable alternatives.

If you know that I have picked a card from an ordinary pack, and I tell you the suit of the card, say 'clubs', the message contains two bits of information. At first sight this is surprising. Since there were four equiprobable alternatives, and my message narrowed down your uncertainty from four to one, why were not four bits of information conveyed? The answer is that it is crucial to the definition of information that it refers to messages which have been recoded in the most economical way possible. Suppose I first tell you my card is black. Since there are two equiprobable alternatives with respect to colour (red or black) I have so far conveyed to you one bit of information. But now consider how much uncertainty remains in your mind. You now know the card is black; therefore it has to be 'spades' or 'clubs'. The one bit of information about the colour has reduced the remaining uncertainty to two equiprobable alternatives; therefore one bit remains. The total uncertainty was two bits not four.

This reasoning is embodied mathematically in the rule that the information content of a message is a function of the logarithm to base two of the number of prior possibilities. In the cases we have so far considered it is in fact equal to this logarithm ($\log_2 2 = 1$; $\log_2 4 = 2$). It is not quite so simple as this when the prior possibilities are not equiprobable, but the principle is the same, and there is no need to go into detail here. If a message has an information content of three bits, this means that it conveys an equivalent amount of information to a message which allows a receiver to choose between eight equiprobable alternatives ($\log_2 8 = 3$), even if this particular message is not concerned with equiprobable alternatives. A message whose information content is two and a half bits has a power to reduce uncertainty which lies somewhere between that of a message allowing a choice between four equiprobable alternatives and one allowing a choice between eight equiprobable alternatives.

In the field of animal communication, *information theory* was first applied to the bee dance. The information content of the dance is best divided into its two components, those of direction and distance. In order to measure the information content of the direction component we need to know the prior uncertainty of the receiving bees. Suppose that the food could be at any point of the compass. We human beings arbitrarily divide the circle into 360 degrees. If bees did the same, the prior uncertainty of a bee as to the direction of the food could be measured as 360 equiprobable alternatives. Then a message which narrowed down the uncertainty from 360 degrees to one would have an information content of $\log_2 360$. But of course the number 360 is entirely arbitrary. To measure the information transmitted from a dancing bee to other bees, we need to measure the accuracy or precision with which a foraging bee is guided to food. It is presumably not a precision of one in 360 degrees, but what then is it? The answer seems to be that bees inform one another about direction with an accuracy slightly greater than the eight points of the compass which we commonly use when we do not have special instruments: north, north-east, east, south-east, and so on. In bits, the information transmitted is slightly more than three. A similar figure has been calculated for the distance information in the dance.

As we have seen, the bee dance is rather a special case, partly because its 'meaning' can be regarded as concerning facts in the outside world rather than, in any simple sense, facts about the internal motivational state of the signaller. Information theory has also been applied to more typical animal signalling systems, for example in hermit crabs (Paguridae).

Here, as in so many cases, the message can be regarded as being 'about' the probable future behaviour of the signaller. Ethologists have developed methods for predicting the behaviour of animals in a statistical sense. They watch an animal for a long period, counting the number of times that it does each behaviour pattern from its repertoire, and also the number of times that each member of that repertoire is followed by each other member. When these sequence counts are collected together in a table and analysed statistically, they enable the ethologist to predict, knowing the present behaviour of an animal, what is the probability that the next behaviour will be each of the possible items from its repertoire. These same probability figures can be used to calculate, in bits of information, the average uncertainty experienced by the ethologist as to the probable future behaviour of the animal. If the average uncertainty of the immediately next behaviour of a single crab

is three bits, what this means is that the ethologist, knowing what the crab is doing at any one moment, can predict what it will do next only rather inaccurately: the ethologist has an uncertainty which is equivalent to a choice between eight equiprobable alternatives. This does not necessarily mean that the crab chooses one from among eight equiprobable alternatives, but the average uncertainty is equivalent to this.

Now if we take into account the behaviour of another crab, we can measure how much our uncertainty is reduced if we know what this other crab is doing. If, through looking at the behaviour of crab A, we are in a better position to predict the behaviour of crab B, then A must be communicating with B, and the information content of the message must be at least as great as the number of bits by which our uncertainty is reduced when we take into account the behaviour of A, as compared with when we ignore the behaviour of A. Using a method basically equivalent to this, although more complicated in detail, the average information transmitted per signal in aggressive communication by hermit crabs has been calculated as between 0·4 bits (*Paguristes grayi*) and 4·4 bits (*Pagurus bonairensis*). The latter is a surprisingly high figure, comparable to a not very concentrated human conversation. It must be emphasized, however, that all estimates of information content, whether in man or in other animals, are necessarily very crude, and involve a host of arbitrary assumptions, so not much weight should be placed on these figures.

Whatever we may think about the usefulness of quantitative measures of information, it is clear that in some cases even the qualitative idea of 'information' is inappropriate. It has been pointed out that a signpost showing the way to Brighton is informative; but a holiday poster saying 'Come to Brighton', with an enticing picture of a sun-baked beach, is not. The advertisement does not contain information; it contains persuasion. Similarly, the display of a peacock may contain a certain amount of information ('I am a sexually aroused male of the species *Pavo cristatus*'); but probably its more important meaning is 'Come and mate with me rather than with my rival'. In our discussions of the information conveyed by signals we must not forget their purely stimulatory or persuasive role.

Towards language. Biologists are interested in communication because it is a part of the way of life of the animals which they study. But no human being can fail to be interested also in uncovering possible clues as to the origin of that distinctively human skill, language. Most of animal communication has rather little in common with true language. If we want to get a subjective idea of what it might be like to communicate in the manner of a non-human animal, we would probably do best to forget about language. A better comparison would be with human non-verbal communication: our involuntary gestures, smiles, frowns, grimaces, gasps, groans of pain, cries of spontaneous delight, screams of terror. Remarkable as it undoubtedly is, even the bee dance is not really language-like. It is an efficient means of conveying precise, quantitative information, but it lacks most of the diagnostic features of language such as flexibility, and hierarchical syntax or infinite extendibility. Despite claims for dolphins (Delphinidae) which are speculative to the point of irresponsibility, there are no naturally occurring cases of animal communication which a linguist would recognize as falling within his domain.

No naturally occurring cases, but some remarkable advances have been made in the teaching of language-like codes to tame chimpanzees (Pan troglodytes). The first serious attempt to teach human language to a chimpanzee failed almost completely. She learned to pronounce some four words of English in barely recognizable grunts. Her pronunciation was greatly inferior to that commonly heard from mynah birds (Gracula) and starlings (Sturnidae). But it seems probable that the chimpanzee simply lacks the vocal apparatus to speak like a person. This is not surprising, and really does not bear on the fundamental question of language itself.

The next attempt, that involving the famous Washoe, was much more successful. She was taught American Sign Language or Ameslan, a standard language involving hand gestures, and used by the deaf in America. She was brought up almost as if she was a human child, as a full member of a family. But her trainers observed a strict rule. They never spoke a word aloud in her presence. They communicated in Ameslan even among themselves. Partly through systematic training, and partly through just 'picking it up', Washoe learned some 160 words of Ameslan. What is more interesting is that she, in some cases spontaneously, put her words together into 'sentences' of two or three words. Her conversations are all of a simple, rather mundane and practical nature: she does not discuss science or philosophy, and is presumably not capable of doing so, although, to be fair, Ameslan is not well-suited to profound abstract discussions.

When Washoe grew up, she was moved from her original home to a primate study centre, and introduced, for the first time, to other chimpanzees. Some of these were also taught Ameslan, and they used it for having simple conversations among themselves. One of Washoe's companions, a young male called Ally, demonstrated good comprehension of spoken English words. Although unable to speak English, he learned to translate a limited vocabulary of English words into Ameslan, and showed by his behaviour that he understood their meaning.

Washoe and her Ameslan-speaking colleagues show evidence of being able to use a real human language, albeit a silent one, in their everyday lives. But some philosophers of language are not impressed. They concede that Washoe has a good vocabulary, and that she uses pairs and triplets of words in sensible combinations. But they suspect that chimpanzees do not use true sentences. 'Tickle Washoe' seems like a sentence, but perhaps she has just learned that this combination of hand movements results in tickling, which she enjoys. Is her achievement anything more than a complex version of OPERANT behaviour?

Further work will be needed before everyone is finally convinced one way or the other, but there are indications that Washoe and her companions are doing much more than a pigeon pecking at a disc for food rewards. They show a facility for understanding new combinations of words which they have not met before. For instance, one of the young chimps, Lucy, was accustomed to being tickled by her human companion, Roger. She frequently requested tickling by signing the Ameslan equivalent of 'Roger tickle Lucy'. On one occasion, Roger tried signalling: 'Lucy tickle Roger'. Lucy showed evidence of confusion at this unusual sentence, but when Roger repeated it she complied, and tickled him.

Washoe was taught the sign for 'dirty' in the context of faeces and soiled clothing. She spontaneously generalized the word so that it became an apparent insult in 'Dirty monkey', applied to a monkey with whom she had had a threatening exchange. She also used it in this insulting way when Roger refused to give her fruit: 'Dirty Roger, dirty Roger, dirty Roger!' Sometimes when Washoe did not know the Ameslan word for an object she would invent her own, putting together two words from her existing vocabulary. She spontaneously invented 'candy drink' for watermelon, and she called a swan a 'water bird', when she saw one for the first time.

It has been suggested that, in the evolution of human language, an Ameslan-like gestural phase preceded the development of vocal speech.

Another famous chimpanzee, Sarah, was taught a very different kind of language, and one which is probably better than Ameslan for getting at the formal logical properties of language. Her 'words' were coloured plastic symbols. She formed them into 'sentences' by placing them on a vertical metal surface to which they would stick because they had small magnets fixed to them. Her language had more in common with reading and writing than with speaking. As in Chinese, but unlike English, her smallest dissociable unit was the 'word': she had no alphabet of letters.

Sarah's vocabulary of about 130 words is slightly smaller than Washoe's: the scientists studying her were more interested in grammatical principle than in vocabulary for its own sake, and they did not teach her more words than they needed for these purposes. The plastic tokens used were arbitrary, and did not physically resemble the object they referred to. For instance, the symbol for apple was a blue triangle, and the symbol for banana was a pink square. In addition to a simple set of everyday nouns, Sarah was cleverly taught a number of verbs: is, take, give, insert, wash. When she had learned to both read and create sentences involving only nouns and verbs, she was taught more abstract logical ideas. For example, she learned to use a form of the conditional 'if–then'. She had the equivalent of a question mark. She had words for 'same' versus 'different'; she had a negative which she would use in conjunction with ordinary sentences to reverse their meaning. She was taught a symbol meaning 'is the name of', and once she had understood this she could be taught new nouns more quickly. After first learning simple adjectives such as red, yellow, small, large, she went on to master the more complex relations 'colour of', 'size of', and 'shape of'. Perhaps most interesting of all, she seemed to understand the rudiments of true syntax. Not only could she understand the fully spelt out sentence: 'Sarah insert apple pail Sarah insert banana dish'. She transferred effortlessly to the abbreviated form: 'Sarah insert apple pail banana dish'. She put the apple in the pail, and the banana in the dish.

Impressive as Sarah's logical achievements are, her language of plastic tokens made great demands on the human trainers' time. An automated training method was used with another young female chimpanzee called Lana. Lana had a large keyboard on the wall of the room where she lived, and she pressed the keys to 'request' a computer to give her the things which she wanted. The keys were illuminated from behind with visual symbols. The computer does not respond to simply one-key 'messages'. It demands sequences of key-presses designated as 'correct' according to the rules of a specially designed artificial language called *Yerkish*. All requests have to begin with 'Please' and end with a full stop. A typical request might be 'Please machine give banana'.

Now, of course, the fact that it is formally possible for a chimpanzee to write out a sequence of five key-presses in a form resembling an English sentence is not, in itself, evidence that the animal is using language. It is easy to teach a pigeon (Columbidae) to peck five different keys in succession, yet this is not taken as evidence that pigeons can use language. Even the fact that Lana showed herself able to recognize sequences of symbols provided by human beings as correct or incorrect is not particularly surprising. But the method has

great potential. Clearly it could be used in very much the same kind of way as Sarah's plastic token language to explore more complex logical ideas which have so far proved beyond the capacities of pigeons. When this is done, the possibilities opened up by the use of an automatic computer are great. The achievements of Washoe and her colleagues may come to have a profound influence on our attitude to animals, and even our attitude to ourselves.   R.D.

75, 76, 100, 122, 135

COMMUNITY, as used by biologists, is an association of animals and plants living in a particular HABITAT. The species making up a community are sometimes classified into *producers, consumers,* and *decomposers.* The producers are green plants that trap solar energy and convert it into chemical energy. The consumers are the animals which eat the plants, or one another, and which thus depend upon plants for energy. The decomposers are generally bacteria and fungi which break down plant and animal material into a form that can be re-used by plants.

Communities can be classified according to the habitats in which they occur. Sometimes fairly sharp dividing lines can be drawn, as between aquatic and terrestrial communities. In other cases, there is a continuous gradation between one habitat and another. This commonly occurs when a climatic factor, such as moisture or temperature, changes progressively from place to place. Such factors have a marked influence on the vegetation, as illustrated in Fig. A, and so determine the nature of the community. A simple example of an ecological community can be found in the desert of Namibia. This desert is so dry that no plants grow there, and the terrain consists largely of sand-dunes. The prevailing easterly winds carry plant detritus into the dunes, where it accumulates at the bottom of the leeward slip-faces. This material consists of fine grass stems and seeds and provides food for a number of species of beetle (especially *Lepidochora argentorisea*) which inhabit the dunes. These animals obtain their water from the occasional sea fogs that invade the dunes. The beetles are preyed upon by the white lady spider (*Carparachne alba*) and by the sand-diving Namib desert lizard (*Aporosaura anchietae*). The lizard also eats

Fig. A. A gradient of increasing aridity from seasonal rain forest to desert.

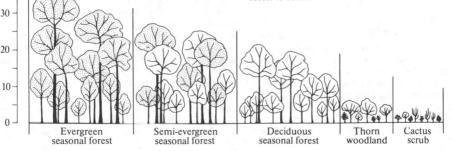

Evergreen seasonal forest	Semi-evergreen seasonal forest	Deciduous seasonal forest	Thorn woodland	Cactus scrub

Fig. B. An example of a food-chain showing the killer whale (*Orcinus orca*), leopard seal (*Hydrurga leptonyx*), king penguins (*Aptenodytes patagonica*), squid (*Histioteuthis*), Antarctic blennies (*Notothenioidei*), krill (*Euphausia*), and plankton.

the spider, various flying insects, and wind-blown grass seeds. In turn, it is preyed upon by the dwarf puff (side-winding) adder (*Bitis peringueyi*), which lies in wait, buried under the sand. Other predators are the pale chanting goshawk (*Melierax musicus*) and the Old World kestrel (*Falco tinnunculus*), which fly in from outside the desert region.

The sand-dune habitat supports a community which can often be described in terms of a food-chain, such as the one illustrated in Fig. B. Each species within the community has its own particular NICHE, or profession. Each is an essential component of the community, necessary for the survival of the other species.
42

COMPARATIVE STUDIES in ethology are those which compare behaviour among two or more species. Often such comparisons allow interpretations that would otherwise be elusive, particularly with regard to evolutionary aspects of behaviour. In fact, comparative study of the EVOLUTION of behaviour was the principal defining characteristic of the rise of modern ETHOLOGY.

Usually comparisons are drawn among closely related species, for example, various species of ducks (Anatinae) or fiddler crabs (*Uca*). Sometimes comparisons are very widely based, as between SOCIAL behaviour of honey-bees (*Apis mellifera*) and man. Occasionally, comparisons are made between populations within a species, particularly within a widely ranging and geographically varying species, or between the wild and domesticated forms of a given species.

Types of behaviour compared. Almost any kind of behaviour may be compared among species or other groups. Closely related species often perform behavioural acts that are fundamentally similar but differ in quantitative aspects. For example, male fiddler crabs extend the enlarged claw in a waving movement that is thought to be a visual signal to other crabs, but the spatial and temporal pattern of waving varies among species (see DISPLAYS, Fig. A). In other cases the behavioural acts compared differ qualitatively, yet appear to be comparable in some specifiable ways. For example, most species of birds scratch their heads in only one of two distinct ways: by directly extending the foot to the lowered head, or by passing the foot between the body and wing such that the leg-joint appears over the top of the wing while the foot touches the tilted-back head (Fig. A).

Comparisons need not be restricted to particular behavioural acts. Sometimes sequences or related groups of behavioural acts are compared, as in the courtship of cichlid fishes (Cichlidae). In other cases entire adaptive complexes of the species' interrelated behaviour and life histories are compared, as in the size and siting of TERRITORY, role of the sexes in PARENTAL CARE, daily movements and *phenology* (periodicity) of the life cycle, and other aspects of behavioural biology among species of ducks.

Finally, comparisons are often made among the results of behavioural acts, whether or not the behaviour itself is known in detail. For example, the nest structure of social wasps (Vespinae), the cases of caddisfly larvae (Trichoptera), the burrow systems of ground squirrels (*Citellus*), the physical characteristics of the alarm cries of birds, and the bowers of bower-birds (Ptilonorhynchinae) have all been subjected to comparative scrutiny.

Homology and analogy. The principal use of comparative studies is to elucidate aspects of the evolution of behaviour. In particular, comparisons often allow one to decide whether behavioural patterns are similar because they serve related functions, or because they are evolved from the same behavioural pattern in a common evolutionary ancestor. Such comparative study of behavioural patterns has its roots in 19th-century comparative anatomy.

1. *Historical roots of comparative study.* When the fact of organic evolution became the cornerstone of biology, two things became clear. First, biology will never secure fossils of all species in the past because fossilization is such a rare process,

Fig. A. Two ways of scratching in birds. The lesser black-backed gull (*Larus fuscus*) brings the leg under the wing. The chaffinch (*Fringilla coelebs*) brings the leg over the wing.

requiring just the right physical and chemical conditions. Therefore, in order to trace probable phylogenetic lineages one must reason from the evidence at hand: the characteristics of contemporary animals themselves which are the end-points of *phylogeny* (evolutionary history). And second, very similar anatomical structures may be found in superficially dissimilar animals. For example, dissection of the forelimb of a bat (Chiroptera) reveals the same set of bones as in the human arm: humerus, radius, ulna, carpals, metacarpals, and phalanges. The bones are shaped somewhat differently in the two species, but they are identifiably similar and have similar spatial relationships with one another in the bat's wing and the man's arm.

Evolutionary relationships between contemporary species and their fossil ancestors were established by structural similarities, and then relationships among contemporary species themselves were also established by structural similarities. Man and bat have similar forelimb structures because they evolved from a common ancestor whose forelimb was built on the same basic plan. The closer the comparative similarities among contemporary species, the more closely related they are in evolutionary history.

The historical development of such comparative reasoning also helped clarify a long-standing problem in pre-evolutionary biology. Anatomists before DARWIN recognized that there were two kinds of similarities among animals: those that were real or fundamental, and those that were only apparent or superficial. Aristotelian thinking dominated pre-evolutionary biology, particularly the notion that all objects observed by man were mere manifestations of ideal types. Being mere copies, images, reflections, or shadows of some theoretical ideal, observed objects were bound to show various differences from the ideal and hence variation among themselves. The anatomist Richard Owen therefore reasoned that real similarities among animals were simply variants of the same ideal or blueprint, whereas apparent similarities were spurious resemblances of things created after different ideals. He called the real similarities *homologues* and the apparent similarities *analogues*.

Evolutionary biology quickly adopted Owen's terms, but with changed causal implications. Homologous structures were those evolved from the same ancestor, so that the ancestor became the underlying blueprint. Analogous structures were more difficult to account for, and only slowly was it recognized that there are in fact two sorts of non-homologous similarities. Sometimes structures are superficially similar because they have evolved under NATURAL SELECTION to serve similar purposes. Thus, the non-homologous wings of dragonflies (Anisoptera) and bats do show certain similarities because they are both structures used in flying, and have been selected to possess certain aerodynamic characteristics. Such structures are now considered analogous. Other structures may be superficially similar by pure accident of construction; thus the horns on the rhinoceros (Rhinocerotidae) and on the rhinoceros beetle (*Diloboderus abderus*) look something alike, even though they are neither evolved from the same ancestral structure nor serve the same function.

One last clarification of comparative anatomy was to emerge. It became clear that one had to specify in what way two structures were considered homologous or analogous. For example, the wings of birds and bats are in one respect homologous, and in another respect analogous, structures. The one-for-one correspondence of bones in the wings can be traced without difficulty, showing that both evolved from some vertebrate ancestor with a similarly patterned forearm structure. As forelimbs, the structures are homologous, but as wings the structures are only analogous, because the common ancestor did not fly. Both bat and bird lineages independently evolved wings from the basic forelimb structure, the differences being reflected in comparative details of the anatomy. The wings of a duck and a house sparrow (*Passer domesticus*), on the other hand, are homologous both as forelimbs and as wings, because both are evolved from a common winged ancestor.

2. *Problems in comparative ethology.* Charles Darwin used comparative reasoning about behaviour, but the development of comparative ethology is usually credited to the German zoo director Oskar Heinroth, who showed that display actions of male dabbling ducks were homologous behaviour patterns. Not only did their component patterns of movement show easily recognizable similarities (comparable to the similarities of bones in the forelimb), but also the sequences of component acts were similar (as the sequence of bones in the forelimb of vertebrates is similar).

The comparative study of behaviour has become one of the cornerstones of evolutionary ethology, but even today there remain a few biologists who resist drawing homologies among behavioural patterns. These scientists see the difficulties of behavioural comparison as outweighing the benefits derived from it, and their viewpoint is useful in maintaining the caution concerning premature interpretations that must be exercised in any science. The chief difficulty is that behaviour has a source of variation not found in anatomy. In both behaviour and structure, individual animals of the same species show some differences. Also, in both behaviour and structure, repeated measurement of the same entity, say length of a bone, or duration of a filmed display, may yield slightly different values because of variation in the measurement process. In behaviour, however, there is a third source of variation: an individual animal may perform differently on different occasions. To take a simple example, consider the head-scratching methods of birds (Fig. A). If a particular individual sometimes scratches under the wing directly, but at other times scratches indirectly over the wing, it shows a kind of variation that is impossible in a bone, which cannot change materially from one instant to the next.

3. *Behavioural homology and analogy.* A common use of the comparative method in ethology is to establish probable homologies and analogies among behavioural patterns. The ideal design of a study involves comparing animal species that show a variety of taxonomic relationships and a variety of ecological situations in which they behave. In the simplest case such a comparative study would involve four groups of species: two taxonomic groups in each of two ecological situations. Suppose the two taxonomic groups are *A* and *B* and the two ecological situations are *1* and *2* so that the four groups of species compared are *A1*, *A2*, *B1*, and *B2*. Depending upon which groups are found to exhibit similarities in behaviour, one may decide from the comparisons whether the behaviour patterns involved are homologous or analogous.

Suppose first that the species in groups *A1* and *A2* show similar behaviour, whereas those of groups *B1* and *B2* differ behaviourally. Then similarities in behaviour correlate with taxonomic relationships, and we judge the behavioural patterns involved to be truly homologous. For example, consider the hundreds of species of birds of the Order Passeriformes (perching birds). These species (groups *A1*, *A2*, etc.) face diverse ecological situations, yet, with only a few exceptions, all scratch consistently over the wing (Fig. A). In contrast, birds of other taxonomic orders (groups *B1*, *B2*, etc., and *C1*, *C2*, etc.) almost all scratch their heads under the wing. Simple scratching behaviour therefore correlates primarily with taxonomic (phylogenetic) relatedness, and we conclude that within the perching birds the similarities are due to having a common ancestor that scratched over the wing. In short, passerine scratching is homologous behaviour among species.

Suppose now that the similarities in behaviour correlate with the ecological situation, rather than with the relatedness of the species involved. When unrelated species show similar behaviour and face similar ecological problems, we conclude that the behaviour is an evolutionary ADAPTATION for solving the problem. For example, most species of gulls (Laridae) build rather shallow nests of vegetation (group *A1*). The kittiwake gull (*Rissa tridactyla*), however, builds a deeply cupped nest using a great deal of mud (group *A2*). Typical gulls nest on relatively flat ground, (ecological situation *1*), but kittiwakes nest on sheer cliffs (situation *2*). Another cliff-nesting gull, the swallow-tailed gull (*Larus furcatus*) of the Galapagos Islands, also nests in cliffs, but builds a nest which is neither like the typical gull's nor the kittiwake's. Instead, its nest is constructed of bits of stone, coral, and sea-urchin spines. In one important respect, however, the nests of the kittiwake and swallow-tailed gull are similar: it is difficult to roll eggs from either kind of nest. The design of the comparative study is completed by the finding that nests of ground-nesting terns (Sternidae) (group *B1*) are much like those of typical gulls, whereas nests of cliff-nesting terns (group *B2*) are much like those of the kittiwake and swallow-tailed gull, in that their construction tends to prevent eggs from rolling out of the nest and being destroyed on the cliffs. Comparative study thus establishes analogous characteristics among nests built by birds, and provides a strong clue as to the functional significance of certain characteristics of nests resulting from nest-building behaviour.

There is a final possibility, too. It is possible that behavioural similarities occur in various species having neither common ancestry reflected by their taxonomic relatedness, nor common ecological conditions. When behaviour cannot be correlated with any other variable, one must, at least provisionally, conclude that behavioural similarities are fortuitous, being neither homologous nor analogous.

No matter how complicated the behaviour, how diverse the taxonomic relationships, or how variable the other factors with which behaviour might be correlated, the comparative method uses similar reasoning in establishing homologous and analogous relationships among behavioural patterns. If comparative study had no other uses, it would retain a prominent place in ethology through its power to deliver evolutionary interpretations so difficult to achieve by any other method

Course of behavioural evolution. Besides the distinction between homologous and analogous behavioural patterns, the comparative method sometimes permits the tracing of possible evolutionary courses (phylogenies) of behaviour. Depending upon the number of extant species and the variety of behaviour they exhibit, inferences about behavioural phylogeny range from strongly to weakly reliable.

1. *Strong phylogenetic inference.* The primary task in reconstructing the probable course of evolution of characters (behavioural or morphological) is to show the gradual changes that might have occurred in past aeons. Strong phylogenetic inference is possible only when a variety of contemporary species show behaviour that forms a spectrum of types along which evolution could have proceeded. Such inference is possible primarily in insects, or other groups of animals in which large numbers of related species still abound.

A typical problem in behavioural phylogeny is represented by the 'symbolic' courtship of the dance fly *Hilara sartor*. The male secretes a ball of silk that is presented to the female, and she manipulates this while copulation takes place. The question is: how could such symbolic gift-giving behaviour arise in evolution? The probable answer is demonstrated by the behaviour of seven related species in the genera *Hilara* and *Empis*, small predacious flies of the family Empididae. As in some other predatory insects, the voracious empidid female is likely to eat anything that comes near her, including the male of the species. *Empis trigramma* males often try to copulate when the female is already occupied with other prey. In some other species (*E. borealis* and *E. scutellata*) the male often catches a prey animal and gives it to the female, who is then occupied, at least for a time, so that he can copulate without so much danger of being eaten. Two other species wrap the prey, thus occupying the female longer: *E. poplita* secretes a large balloon around the prey, whereas *Hilara quadrivittata* wraps the prey with many strands of thread. These species form an easily recognized series that might have been the course of phylogeny in this group, but at this point at least two possible next steps exist. In *H. maura* the male wraps any sort of object, such as a bit of petal or leaf, in the threads, and in *H. thoracica* real prey is so loosely wrapped it often falls out. In either case, manipulating the wrapping occupies the female, even though she receives no prey in the end. It is then a simple step to the 'symbolic' behaviour of *H. sartor*, which presents mere wrapping without any object within.

Comparative study does not tell us why evolution took this particular course in solving the problem of mating in voracious predators. Why it would not be possible for the female simply to become more passive in the presence of the male is unknown. However, the behaviour of contemporary species can be ordered along a roughly linear sequence of types that provides at least one good inference of how evolution may have proceeded. We guess that because *Hilara sartor* has the most unusual and complicated behaviour, it is the derived type, and that evolution has proceeded from the simpler and commoner behaviour, like that of *Empis trigramma*, to the derived type.

2. *Weak phylogenetic inference.* Unfortunately, it is rare that extant species show so many gradations of behaviour that they form a series, possibly resembling the ancestral series leading to a particular end-point. The more usual case is to find a variety of contemporary species, each of which exhibits behaviour that is more or less equally changed from some ancestral type. In this case, only a weaker phylogenetic inference is possible, which may be almost no inference at all. For example, the DISPLAY actions of many of the world's ducks and gulls have been studied, and in neither group can one find a convenient series representing a possible course of evolution, nor even distinctly simple and presumably primitive behaviour in any living species. However, homologous patterns of display can be discerned across species within each group, and the movements, and other aspects that these homologous patterns have in common, is reasoned to be something like the display behaviour of the common ancestor.

3. *Phylogenetic-like series.* Palaeontologists distinguish two kinds of series of fossils: *clades*, which

are true phylogenetic lineages, and *grades*, which are pseudo-lineages. Grades are merely assemblages of unrelated animals of the past that can be ordered in some series to provide clues as to how actual evolutionary trends might have occurred. Just as clades have their counterpart in strong and weak phylogenetic inferences from a comparative study of related contemporary animals, grades have their counterpart in the sequencing of unrelated contemporary animals to form a phylogenetic-like series.

For example, diverse kinds of animals have been compared in order to understand the major factors in the evolution of parental care of the young. From these comparisons one learns that most animal species produce vast numbers of offspring, only a few of which survive, as is typical of many insects, frogs, and fishes. Parental care is rare in such animals, which are said to be under conditions of *r*-selection, where *r* refers to a high reproductive rate characteristic of species living in variable environments where opportunities for survival are unsure. In contrast, the other extreme is represented by the relatively rarer animal species that produce only one to a few offspring at each breeding. These offspring they nurture carefully, and most survive to reproduce themselves. Such animals are said to be under conditions of *K*-selection, where *K* refers to the carrying capacity of a highly competitive and stable environment. Primates, other mammals, and most birds are typically strongly *K*-selected animals, and have elaborate parental care. In between these extremes all sorts of gradations are found: the counterparts of fossil grades of unrelated animals that reveal evolutionary trends through a phylogenetic-like series.

4. *Phylogenetic change*. Sometimes the course of phylogeny cannot be traced with even an educated guess, but comparative study reveals something of the kinds of changes that have occurred during phylogeny. For example, it was once thought that amphibians showed only one of two kinds of responses to coloured light, depending upon the species. One response is to move preferentially towards light only of the part of the visible spectrum we see as the hue blue; the other response is to move preferentially towards lights from the spectral extremes of red or violet. Both frogs and toads (Anura) and newts and salamanders (Urodela) include species that prefer blue and other species that prefer either red or violet.

Comparative study revealed the nature of the difference between species, which proved to be quantitative rather than qualitative. Each species of amphibian tested also moves towards white light, but every species has its own preferred intensity of light, which may differ markedly from species to species. It was then discovered that when the coloured stimuli were dimmer than the species' own preference for white-light intensity, the animals preferred blue; but when the coloured stimuli were brighter than the preferred intensity, the animals preferred the red or violet stimuli. In

other words, all species show both responses to coloured light, depending upon the brightness of the light relative to their own species-specific preference for intensity. In this case comparative study established that an apparent qualitative difference in behaviour, which might have entailed complex evolutionary changes in the photoreceptors and visual pigments, was merely a quantitative change in the preferred intensity of light. In this way comparative study can reveal the kinds of phylogenetic changes that may take place in a lineage of animals.

Behaviour and taxonomy. The comparative study of behaviour can be useful in the classification of animals (taxonomy) in at least two ways. It can help establish phylogenetic relationships among species, and hence become the basis for higher level systematics. And comparative behaviour can help decide which populations of animals belong to the same species.

1. *Systematics* is the science of both naming and classifying animals according to a hierarchical scheme, in which phylogenetically related species are grouped in the same family (such as the cat family, Felidae, and the dog family, Canidae), related families are grouped in the same order (e.g., the dog and cat families are in the order Carnivora), related orders are grouped in the same class (e.g., Mammalia), and related classes are grouped in the same phylum (e.g., Chordata). Because relationships are meant to reflect phylogenetic lineages, they are based on homologous characters: both morphological and behavioural.

The use of behavioural characteristics in taxonomy is not strictly a behavioural discipline, because all taxonomic judgements must be made on as wide a basis as possible. Indeed, morphology still provides the primary basis of taxonomy, and behavioural characteristics are used merely as a part of the additional evidence. There is also a danger of circular reasoning in the use of behavioural characters, because behavioural homologies are established partly on the basis that taxonomically related species show behavioural similarities. Nevertheless, behavioural characteristics often prove a useful adjunct in morphologically based taxonomic decisions.

2. *Species*. Prior to the advent of modern population GENETICS individual animals were judged to be of the same species if they more strongly resembled one type of animal than all others. Thus the species category was a matter of arbitrary judgement, like the categories of genus, family, order, class, and phylum. It is now known that species of sexually reproducing animals are natural taxonomic units, rather than merely arbitrary categories created for convenience.

In the modern view, species are groups of sexually reproducing, and at least potentially interbreeding, animals that do not in nature ordinarily interbreed with any other such groups. It is necessary to say 'ordinarily' because occasional interspecific hybrids do exist in nature, and it is necessary to stipulate 'in nature' because many species that rarely or never interbreed in nature sometimes interbreed in captivity.

Comparative behavioural studies aid in the determination of species boundaries by establishing whether or not two groups of animals actually interbreed under natural conditions, and by describing the behavioural mechanisms that keep species from interbreeding. These mechanisms are called ethological reproductive ISOLATING MECHANISMS, and include species-specific sexual display patterns. For example, it was once believed that most North American fireflies (family Lampyridae) belonged to just one or a few species, as they are very similar morphologically. Comparative behavioural study of these beetles showed, however, that the flashing rhythm of their bioluminescent pulses was a species-specific sexual signal. The studies revealed that fireflies respond to signals of their own species only, and that there are more than twenty different species which had formerly been undistinguished (see COURTSHIP and MATE SELECTION).

Other uses of comparative studies. Although various evolutionary questions are best answered by comparative methods, behavioural study of diverse species has other uses in ethology and psychology. Most of these uses are variants of the traditional research strategy of biologists who search for just the right species in order to study a particular problem. Genetics was greatly facilitated by locating the fruit fly *Drosophila* for inheritance studies, and neurophysiology profited immensely from the discovery of the giant *axon* (long process of a nerve-cell) of the squid (*Loligo*).

The same principle has operated in behavioural studies. For example, early experiments on animal LEARNING used a variety of animals, including racoon (*Procyon*), dog, cat, and so on. By trying various species it became clear that a few types were particularly suitable for maintaining in the LABORATORY and using in studies of learning phenomena, principally the domesticated albino Norwegian rat (*Rattus norvegicus*) and the pigeon (*Columba livia*). For studying hormonal bases of behaviour the canary (*Serinus canaria*), budgerigar (*Melopsittacus undulatus*), and Barbary dove (*Streptopelia risoria*) proved particularly useful. Oskar Heinroth and Konrad LORENZ found ducks to be especially useful animals for studying social behaviour and communicative displays, and, after having worked with many kinds of animals, Niko TINBERGEN devoted special efforts to studying comparative aspects of the behaviour of gulls. For almost any important question that may be asked about animal behaviour, some species will, for one reason or another, be more useful subjects for study than others, and only an initial comparative survey can determine which they are (see FIELD STUDIES). J.P.H.
88, 102

COMPETITION occurs when two or more individuals are using the same resources, and when those resources are in short supply. Food and space are the common essential resources for which animals compete. Competition between members of the same species is termed *intraspecific* competition, and often takes the form of direct interference and AGGRESSION between individuals. *Interspecific* competition is that which occurs between individuals belonging to different species. It most commonly takes the form of *exploitation* of a resource by one species, thus denying the use of the resource to members of other species by reducing its availability.

The introduction of the common starling (*Sturnus vulgaris*) into Central Park, New York, in 1891, provides a good example of interspecific competition. The starling spread widely, and by 1955 it had invaded the whole of the U.S.A. In urban areas it has largely displaced the eastern blue-bird (*Sialia sialis*) and the yellow-shafted flicker (*Colaptes auratus*). These species nest in holes in trees and in buildings, and the starlings can successfully exploit this limited resource, ousting the other species by competitive displacement. In rural areas, flocks of starlings compete for insects and seeds with species of the meadow lark (*Sturnella*), and this is another typical form of competition.

Competition occurs when there is overlap in the NICHES occupied by different individuals. In the case of interspecific competition, it often results in dominance of one species over another, in the sense that the dominant species has priority in the use of resources, such as food, space, and shelter. Subordinate species may be excluded from the use of those resources that the dominant species also uses. Consequently, it is generally true that two species with identical ecologies cannot live together in the same place at the same time. The corollary of this *competitive exclusion principle* is that, if two species coexist, there must be ecological differences between them.

CONDITIONING is a process involved in LEARNING, in which an animal forms an ASSOCIATION between a previously significant stimulus and a previously neutral stimulus or response. Thus in *classical conditioning* (see PAVLOV) an association may be formed between a significant stimulus, such as the sight of food, and a neutral stimulus, such as a flashing light. Initially the animal will respond to the food, for instance by salivating. If the food is presented together with the flashing light on a number of occasions, then the animal will come to associate the light with the food, and it will eventually salivate even if the light is presented alone. The response of the animal is said to have become conditioned to the light.

In OPERANT (instrumental) conditioning, the association is made between a particular response and a particular REINFORCEMENT situation. Thus in a LABORATORY study a rat may associate pressing a bar with delivery of food. In nature, a bird may associate turning over a leaf with the discovery of an insect. In both cases the response, pressing the bar or turning over the leaf, is instrumental in obtaining food, which is said to reinforce the response. Such reinforcement may be negative. Thus a cow that touches an electric fence and receives a shock associates the response with the shock, and is less likely to approach the fence in future. The rat is said to have become conditioned to press the bar, and the cow is said to have become conditioned not to touch the fence. Conditioning is the most prevalent learning process found amongst animals, including man.

86, 98

CONFLICT, as applied in animal behaviour, denotes a state of MOTIVATION in which tendencies to perform more than one activity are simultaneously expressed; it is not normally used to describe aggressive encounters between animals. As a rule, an animal's behaviour is controlled by a single dominant tendency, such as the tendency to feed, or to sleep. At any particular moment, the animal has many incipient tendencies, but by a process of DECISION-MAKING one of these becomes dominant. Generally, only one tendency becomes dominant, but in certain circumstances more than one competes for dominance, and conflict arises.

The dominant tendency represents the activity which the animal has 'decided to do', but the extent to which this tendency manifests itself as observable behaviour depends upon environmental circumstances. For example, an animal may have a dominant tendency to feed, but when it approaches food it finds that its way is blocked by a wire fence. Such situations lead to FRUSTRATION. Conflict, on the other hand, occurs when a dominant tendency is opposed by another tendency, or when two tendencies compete for dominance. Because of INCOMPATIBILITY between most of the activities in an animal's repertoire, conflicting tendencies cannot be simultaneously manifest as overt behaviour, and the behaviour seen during conflict is therefore quite unlike the normal smooth run of activity. For example, if one holds out a piece of bread towards a duck (Anatinae) in the park, the duck will often show a tendency to approach the bread, especially if it is hungry. However, it will also show a tendency to avoid the human being holding the bread. The result is that the duck generally approaches to within a certain distance, and then it may alternate between retreat and approach, or remain stationary, craning its neck towards the bread, while edging away with its feet. This is typical conflict behaviour.

Conflict has traditionally been divided into three main types: (i) approach–approach conflict occurs when the two tendencies in conflict are directed towards different goals. In such a case the animal may reach a point where the two tendencies are in balance, as in the story of the ass

confronted with two bales of straw. However, the tendency to approach a GOAL generally increases with proximity to the goal. This makes approach–approach conflict unstable, because any slight departure from the point of balance towards one goal will result in an increased tendency to approach that goal and a decreased tendency to approach the other, thus resolving the conflict; (ii) avoidance–avoidance conflict occurs when the two tendencies in conflict are directed away from different points. Since the tendency to avoid objects generally increases with proximity to the object (see AVOIDANCE), movement towards either object is likely to result in a return to a point of balance. Such situations are not normally stable, however, because the animal can escape in a direction at right angles to a line between the two objects; (iii) approach–avoidance conflict occurs when one tendency is directed towards a goal, and another away from it. In the case of the duck, mentioned above, we can imagine that the tendency to approach the bread increases as the animal moves nearer to it. Similarly, the tendency to avoid the human being increases as the duck moves nearer to him. Since the human being is holding the bread, the duck can reach the goal only when its approach tendency is greater than its avoidance tendency, at a point from which it can seize the bread. If its avoidance is dominant at this point, then the duck will tend to move away from the goal to a point where its approach and avoidance tendencies are equal. Beyond this equilibrium point, approach is stronger than avoidance, and the duck moves towards the goal. Thus approach–avoidance conflict typically results in a stable situation, in that the animal always tends an equilibrium point, where its approach and avoidance tendencies cross, as illustrated in Fig. A.

Approach–avoidance conflict is by far the most important and the most common form of conflict in animal behaviour. In natural situations its occurrence forms the basis for the evolution of more complex behaviour. For example, in defending its territory against intruders, the male three-spined stickleback (*Gasterosteus aculeatus*) is highly aggressive. If the intruder is another male, the aggression is mixed with fear, particularly near the boundary of the territory. If the intruder is a female, the male will generally show conflicting tendencies of aggression and courtship. So common are such inner conflicts that NATURAL SELECTION has frequently acted upon them, making them more effective as a means of communication. In the case of the stickleback, the position of the territory boundary is established at the point of equilibrium, where the tendencies to attack and flee from an intruder, or neighbour, are equal. In such situations sticklebacks typically adopt a head-down THREAT posture, which is a form of conflict behaviour that has come to serve a COMMUNICATION function through a process of RITUALIZATION.

At points of equilibrium in approach–avoidance

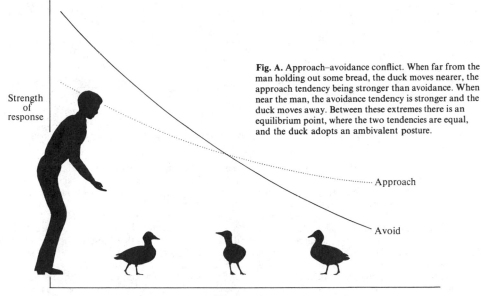

Fig. A. Approach–avoidance conflict. When far from the man holding out some bread, the duck moves nearer, the approach tendency being stronger than avoidance. When near the man, the avoidance tendency is stronger and the duck moves away. Between these extremes there is an equilibrium point, where the two tendencies are equal, and the duck adopts an ambivalent posture.

conflict the animal reaches an impasse, because both approach and avoidance behaviour bring the animal back to the point of equilibrium. Such conflicts can be resolved by a change in the external situation, or by means of some alternative behaviour. Typically, conflict behaviour is characterized by compromise and ambivalence. The duck approaching bread offered by hand may alternate between approach and retreat. This is a form of compromise behaviour, which consists of separate components of the conflicting tendencies. Alternatively, the duck may remain stationary, craning its neck towards the bread while edging away with its feet. This results in a typical AMBIVALENT posture, which compounds elements of the conflicting tendencies.

Conflict behaviour is often replaced by other, seemingly IRRELEVANT, behaviour. For example, sticklebacks in conflict near the boundaries of the territory often indulge in bouts of sand-digging. This behaviour is normally seen as part of nest-site preparation, during which the male makes a depression in the substratum by taking mouthfuls of sand and spitting it out a short distance away. When this behaviour occurs during boundary disputes the nest is often already built, so the behaviour is relevant neither to nest-building, nor to territorial defence. Such apparently irrelevant behaviour is termed DISPLACEMENT ACTIVITY. It has been suggested that the head-down threat DISPLAY seen during territorial disputes has been evolved from displacement sand-digging, through a process of ritualization.

The irrelevant behaviour that occurs during conflict often seems to serve as a means of resolving the conflict. Opinions differ as to the mechanisms responsible for this phenomenon. Some scientists take the view that EMOTION is involved, the emotional tension that builds up during the conflict being released by the irrelevant behaviour. Others believe that the conflicting tendencies cancel each other, allowing a third irrelevant tendency to gain dominance, and to give rise to apparently irrelevant behaviour. A third possibility is that frustration, induced as a result of the conflict, causes a switch in ATTENTION, so that the animal notices and responds to incidental stimuli in the immediate environment. Whatever the mechanisms involved, it seems likely that the irrelevant behaviour typical of conflict situations is a by-product of the resolution of the conflict.

Ritualized conflict behaviour. Animals in conflict situations may, through their ambivalent behaviour, convey information about their motivational state which is significant for another individual. In territorial disputes, for example, the TERRITORY owner often becomes more fearful as he approaches the boundary of the territory. At some point his tendencies to attack and to flee come into conflict. The resulting ambivalent behaviour and displacement activity can be observed by a neighbour or intruder, and taken as an indication of how far the resident is prepared to go in defence of his territory.

If communication of this type is advantageous, then natural selection will tend to make the behaviour more reliable and efficient as a conveyor of information. Through a process of ritualization, conflict behaviour can thus evolve into display behaviour. For example, the *upright threat posture* of the herring gull (*Larus argentatus*) contains elements of both aggression and fear. The

Fig. B. Gradations in the upright position of the herring gull (*Larus argentatus*). With an increasing component of fear, the bill is progressively raised.

raising of the wing carpels (elbows), is characteristic of attack by delivering wing beats with the folded wing; the upright neck with a downward pointing bill is characteristic of pecking at an opponent from above. However, the sleeked feathers, stretched neck, and hesitant walking movement that accompany this display are all characteristic of fearful behaviour. When the birds are slightly more fearful, the head moves back, the bill becomes lifted, the plumage sleeks further, and the bird often turns sideways-on to its opponent, as if preparing to escape. A skilled observer can interpret such gradations in the behaviour of the animal (see Fig. B), and can use these to predict what the bird is likely to do next.

In some species, conflict behaviour has become ritualized into a complex mosaic of postures, such as the threat postures of the greylag goose (*Anser anser*), illustrated in Fig. C. It is sometimes possible to demonstrate experimentally that such postures are indeed derived from conflict situations. In one study of tame Canada geese (*Branta canadensis*) it was found that the geese ignored their keeper if he wore his familiar old clothes, but fled from him if he carried the broom normally used to drive the geese into their house for the night. When the keeper appeared wearing a white coat the geese attacked him uninhibitedly, but if he wore the white coat and also carried the broom, they were thrown into a conflict between attack and escape. On such occasions the typical threat postures of Canada geese were observed, even though the conflict was based entirely on artificial stimuli, far removed from those that would naturally induce threat display.

Other components of displays have been derived from the emotional side-effects of conflict. These include thermoregulatory responses, such as the

Fig. C. Threat postures of the greylag goose (*Anser anser*). From the 'at ease' posture shown in the bottom left-hand picture, components of fear become stronger in the upwards direction, while aggression increases towards the right-hand side.

hair erection seen in the threat postures of cats and dogs. Respiratory side-effects are thought to have led to the raised gill-covers seen in the threat postures of some species of fish, such as the Siamese fighting fish (*Betta splendens*) illustrated in Fig. D. The inflation of air sacs by many amphibians, reptiles, and birds is also thought to have originated in this way (Fig. D).

99

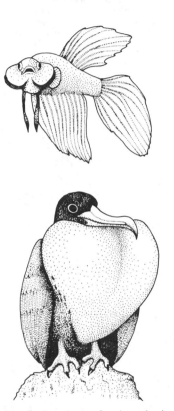

Fig. D. Ritualized respiratory functions: the threat posture of the Siamese fighting fish (*Betta splendens*), with raised gill-covers; inflation of air sacs in the frigate bird (*Fregata magnificens*) during a courtship display.

CONSERVATION is an aspect of WILDLIFE MANAGEMENT which is concerned with the preservation of endangered species, preferably in their natural HABITAT. Many species are threatened with extinction as a result of human activity. Some, such as whales (Cetacea), are excessively hunted by man, others have their habitats destroyed as a result of agricultural or industrial development.

For effective conservation knowledge is required of the species' number, distribution, DISPERSION habits, FEEDING and habitat requirements, and mode of reproduction. This is an area in which the science of ETHOLOGY is of considerable practical importance, both for the research which is carried out in the FIELD and in the LABORATORY.

CONSUMMATORY BEHAVIOUR. Historically, the distinction between a goal-seeking, APPETITIVE phase of behaviour and a goal-finding, consummatory phase was first made by the zoologist Wallace Craig in 1918, although the distinction was anticipated by the physiologist Sherrington in 1906. Craig described how a young male dove (*Streptopelia*) establishes a nest site: 'The first thing the observer sees is that the dove, while standing on his perch, spontaneously assumes the nest-calling attitude, his body tilted forward, head down, as if his head and breast were already touching the hollow of a nest (incipient consummatory action), and in this attitude he sounds the nest-call. But he shows dissatisfaction, as if the bare perch were not a comfortable situation for this nest-dedicating attitude. He shifts about until he finds a corner which more or less fits his body while in the tilted posture; he is seldom satisfied with his first corner, but tries another and another. If now an appropriate nest-box or a ready-made nest is put into his cage, this inexperienced dove does not recognize it as a nest, but sooner or later he tries it, as he has tried all other places, for nest-calling, and in such a trial the nest evidently gives him a strong and satisfying stimulation (the *appeted stimulus*) which no other situation has given him. In the nest his attitude becomes extreme; he abandons himself to an orgy of nest-calling (complete consummatory action), turning now this way and now that in the hollow, palpating the straws with his feet, wings, breast, neck, and beak, and rioting in the wealth of new, luxurious stimuli. He no longer wanders restlessly in search of new nesting situations, but remains satisfied with his present highly stimulating nest' (W. Craig, 'Appetites and aversions as constituents of instincts', *Biological Bulletin*, (1918), **34**, 97–8).

Scientists generally agree that, in some cases, consummatory behaviour brings to an end a period of appetitive SEARCHING behaviour. They often differ, however, in the emphasis given to the role of consummatory behaviour *per se* in inducing such a marked change in the sequence of behaviour. In some cases it seems that simply performing the consummatory behaviour is sufficient. For example, in the SEXUAL behaviour of many animals it appears that ejaculation, or other behaviour associated with fertilization, constitutes the consummatory act of sexual behaviour, because it occurs at the end of the period of sexual activity. In the three-spined stickleback (*Gasterosteus aculeatus*) however, careful study shows that although the behaviour involved in fertilizing the eggs terminates the period of sexual activity, this behaviour is not instrumental in reducing the sexual MOTIVATION of the male. Other aspects of the situation, such as the egg-laying behaviour of the female, seem to be more important. Moreover, in some species, such as rats (*Rattus*), ejaculation occurs a number of times in a bout of sexual behaviour, and it is the cumulative effect of these which is responsible for sexual exhaustion.

Traditionally, FEEDING is regarded as the consummatory behaviour of appetitive food-seeking behaviour, such as HUNTING and FORAGING. Feeding is often followed by a period of quiescence and of postponement of appetitive feeding behaviour. However, feeding has many consequences that influence the future behaviour of the animal. In many animals the sight and smell of food, and the presence of food in the mouth, have the effect of increasing APPETITE, rather than diminishing it. The presence of food in the gut, and its subsequent absorption into the bloodstream, leads to a reduction in HUNGER and an eventual cessation of feeding behaviour (see SATIATION), but food intake may also lead to an increase in THIRST which stimulates DRINKING behaviour. Thus feeding is not simply a consummatory response. It can lead to an initial increase in appetitive behaviour, and it can stimulate other types of behaviour. Rather than thinking of feeding, or any other apparently consummatory behaviour, as terminating the preceding appetitive behaviour, it is perhaps better to think of it as a factor influencing DECISION-MAKING. As a consequence of feeding, the animal may decide to seek more food, to drink, to rest, etc. The factors influencing such decisions are complex, and vary greatly from one species to another.

100

COORDINATION of muscular movements is necessary to accomplish the complex patterns of limb movement used in LOCOMOTION, DISPLAY, and other aspects of behaviour. Coordination is achieved through two main processes. One, called central control, involves a precise series of instructions that are issued by the BRAIN and obeyed by the muscles. The other, called peripheral control, is achieved through SENSE ORGANS in the muscles, which send information to the brain, and thereby influence the instructions issued from the brain to the muscles. In most cases coordination is achieved through a mixture of these two processes.

To determine whether the pattern of instructions from the brain is, or is not, influenced by peripheral factors, it is necessary to sever the nerves leading from the muscles to the brain. Experiments of this type show that simple coordination can result from central control alone. For example, the movements of the sound-producing mechanism of the cicada *Graptopsaltria nigrofuscata* remain unaltered when peripheral factors are de-activated. Similarly, the coordination of REFLEX swallowing in mammals appears to be organized independently of peripheral control. In man, peripheral factors are believed to play only a very minor role in the coordination of many skilled movements, such as playing the piano or swinging a golf club. In novices the peripheral factors are important, but in experts a complex and very precise series of commands is issued by the brain and acted upon without FEEDBACK.

The coordination of walking in insects is accomplished by a combination of central and peripheral

mechanisms. Insects show a variety of gaits, and an individual can modify the pattern of leg movement according to the speed of locomotion, or in response to the loss of one or two legs (see Fig. A). It has been discovered, however, that the coordination of leg movements follows a simple set of rules, which can account for the variety of gaits, even those shown by insects that have lost legs. The insect's brain is responsible for issuing instructions to the legs in accordance with the rules, but information from the legs also exerts an influence. The modification in the gait that occurs when an insect loses one or two legs is due to the

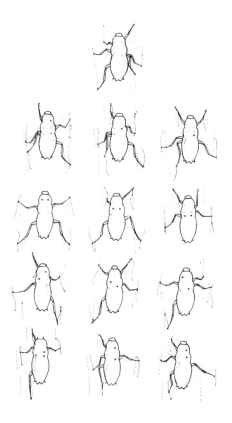

Fig. A. Changes in walking pattern occur when legs are amputated from the oriental cockroach (*Blatta orientalis*). Amputation of a leg is indicated by a cross on the corresponding part of the body. Dashed lines indicate the normal stride of a leg and continuous lines indicate the angular extent of the changed stride pattern.

absence of this information. Central control of muscular coordination is especially important in the generation of FIXED ACTION PATTERNS in insects.

The coordination of swimming movements in fishes is accomplished through an interaction of central and peripheral factors. The brain provides an endogenous RHYTHM that passes down the trunk in waves, coordinating the rhythmic movements of the fins and tail. However, in the dogfish (*Scyliorhinus*) the rhythm disappears if all nerves leading from the muscles to the brain are cut. In dogfish, a member of the Chondrichthyes (cartilaginous fishes) the fins show little independent rhythmic movement, but in the Teleostei (bony fish) the fins have some degree of independence, in that under certain circumstances they beat at different frequencies. The rhythms of different fins influence each other, a feature known as *relative coordination*. Sometimes the rhythm of one fin attracts and dominates that of another, so that they fall into step; in other cases the amplitudes of fin movements summate, so that the movements become smaller when the fins are out of step with each other, and more extensive when they are in step.

Coordination of limb movements in mammals is dependent upon the postural reflexes. These control the length and tension of individual muscles, and coordination between muscles is impossible without them. The coordination of complex and skilled movements involves mechanisms capable of taking account of the animal's initial posture, the physical forces generated by the movement, the effects of gravity, and the relationships of limbs to visually perceived objects. Some of these tasks are believed to be accomplished by REAFFERENCE mechanisms capable of distinguishing between changes in the external environment that are of exogenous origin, and changes which are due to the behaviour of the animal itself. In the control of limb movements, for example, the brain must distinguish between movements of the limb due to outside forces, and movements resulting from the brain's own instructions.

Some animals are able to modify their coordination mechanisms to adjust to changed environmental conditions. Experiments with human beings show that ADAPTATION readily occurs in hand–eye coordination. For example, if a subject wears spectacles fitted with laterally displacing prisms, objects will appear to be displaced either to the left or right of their true position, as shown in Fig. B. Asked to point at a target the subject points incorrectly, especially if he cannot see his hand and so guide it onto the target. If the subject is allowed to see his hand he sees the amount by which his pointing was in error. Under these conditions most subjects rapidly adapt to the situation and correct their error. If, after adaptation, the prisms are removed and the subject immediately points at the target, there is usually an error in pointing, but this time to the side opposite to the original displacement, a phenomenon called the after-effect.

Four stages of adaptation to laterally displacing prisms are illustrated in Fig. C. The whole phenomenon can be demonstrated in less than 5 min, thus showing that there is rapid recalibration somewhere within the visual–motor coordination system. There is evidence that this recalibration takes place in the felt-position of the arm. In other words, the subject who normally feels his arm to be in a certain angular relationship to his body has this felt-relationship changed by repeated observation that there is a discrepancy between the seen and felt-position of the arm. For example, suppose the subject is required to show that he can locate the position of one arm by pointing to it with the other hand, while not able to see the target arm. The subject is then required to use the target arm in pointing at targets while wearing prismatic spectacles, in the manner described above. When the subject has adapted to the displacing prisms and is pointing correctly, his target arm is restrained and he is again required to point to it with his other hand, without being able to see the target arm. The subject now misjudges the position of his target arm by an amount that corresponds to the degree of displacement of the prisms. In other words, his target arm feels to be in a position which is different from its true position, and provided the subject cannot see this difference he cannot correctly locate his own arm in relation to his own body.

The coordination of eye and limb is very important to animals that have to aim at, climb upon, or manipulate objects. Adaptation to prismatic displacement has been demonstrated in the aiming of pecks by chicks (*Gallus g. domesticus*) and in reaching for objects by squirrel monkeys (*Saimiri sciureus*). Indeed it has been shown that

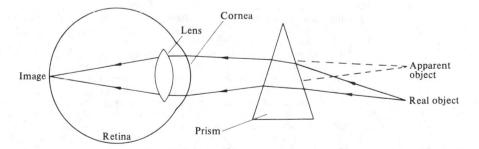

Fig. B. Displacement of the visual field viewed through a prism.

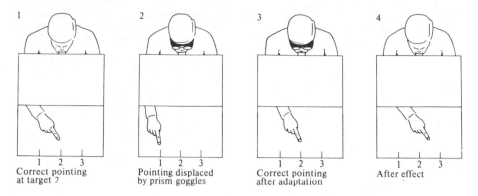

Fig. C. Four stages of adaptation to a laterally displaced visual field.

monkeys can simultaneously adapt to a left visual displacement when in one situation, and a right displacement when in another situation. This illustrates a highly organized ability to recalibrate aspects of the visual–motor coordination mechanisms, which is perhaps not surprising in view of the importance of these mechanisms in the everyday life of tree-living monkeys.
74

COOPERATIVE BEHAVIOUR is behaviour in which members of a species combine in an activity, such as hunting, anti-predator behaviour, or care of the young, and in which there is coordination between individuals to their mutual benefit. Cooperation between members of different species is usually called SYMBIOSIS.

Many examples of cooperative behaviour are known among insects. HABITAT selection by honeybees (*Apis mellifera*) has been well studied and illustrates a number of important features of cooperative behaviour. In the late spring the queen lays an egg in each of several large cells specially prepared by the workers. The larvae are given special nourishment, and develop into fertile queens rather than sterile workers. These virgin queens will fight, and one will become the new queen of the hive. Before this happens the old queen departs from the hive together with about one half of the total population. The swarm of bees forms a compact cluster on a branch of a near-by tree, where it remains for several days. Scout bees fly off and inspect various possible sites for a new hive. They then return and perform a dance on the vertical surface of the swarm. The dance imparts information about the distance, direction, and quality of the potential home. A long and vigorous dance indicates an attractive site. Other workers are stimulated by the dance and fly off to inspect the site. When they return they also dance with a vigour that indicates the quality of the site. This, in turn, stimulates other workers to inspect the site, etc. Usually more than one site is initially reported, and each recruits further scouts. The differences in the quality of the sites become ap-

parent from the number and vigour of the dancing bees. Eventually near-universal agreement is reached as bees from the less good sites cease to advertise their discoveries. The swarm then moves to the chosen site and begins to set up a new hive. Researchers have found that the bees prefer sites that are protected from extremes of CLIMATE. They often choose holes in the ground, or in trees, that are a suitable size in relation to the size of the swarm. The new home should not be too near the old hive, nor too far for the queen to fly to from her temporary resting place.

The remarkable cooperation between the bees is based largely upon their powers of COMMUNICATION. The bee LANGUAGE is particularly sophisticated and capable of imparting precise information in quantitative terms. The collective DECISION-MAKING is made possible because the dance of each worker is influenced both by the quality of the site she visited, and also by the dance of other bees in the swarm. Although each scout is energetic in advertising the merits of a potential home, she does not persist indefinitely, and eventually gives way to the majority opinion. Efficient communication and sensitivity to the actions of other members of the group are important features of cooperative behaviour in general, and are well illustrated by the next example.

Cooperative HUNTING occurs in lions (*Panthera leo*), spotted hyenas (*Crocuta crocuta*), wolves (*Canis lupus*), and wild dogs (*Lycaon pictus*). Wild dogs prey upon animals much larger than themselves, such as the brindled gnu or wildebeeste (*Connochaetus taurinus*) and Burchell's zebra (*Equus burchelli*). They select a single quarry (see PREDATORY BEHAVIOUR) and run it down over a long distance. These communal hunts involve various forms of cooperation. There is a collective choice of the prey. There may be exchange of leaders in a chase, enabling the pack to operate a relay system and so cover great distances. A trailing dog will often cut corners in an attempt to head off the prey. The larger prey are killed by the combined effort of the pack. The food is shared by members of the pack, and the adults often regurgitate food to the young. Sometimes a few adults remain behind with the vulnerable juveniles, and they are also fed when the others return. Cooperative behaviour of this type involves an element of sacrifice, because individuals do not act purely selfishly, but benefit others at some cost to themselves.

Cooperative behaviour often involves some form of ALTRUISM. For example, the Mexican jay (*Aphelocoma ultramarina*) is a communally breeding species that is non-migratory and lives in flocks of five to fifteen individuals. Each flock communally defends a TERRITORY in pine or oak woodland. The nests are built by mated pairs, and the eggs in a particular nest are laid by a single female. There may be between one and four nests per flock. At each nest most of the flock members help to bring food to the nestlings, and about half the food that the nestlings receive is from birds other than their parents. Clearly, these helpers at the nest are behaving in an altruistic manner, and this requires some explanation in terms of evolution. While it is true that the flock as a whole benefits from the activities of the helpers, it is not clear that it is in the interest of the helpers to spend so much time and energy in feeding the offspring of other individuals. In the case of the parents, any incipient evolutionary change, or GENETIC *mutation*, that resulted in failure to feed the young, would not be likely to be passed to the next generation, because the young would probably die. However, it is not clear why a mutation that led to failure to feed another bird's young should not spread through the population and eventually destroy the basis for the cooperative behaviour. In attempting to answer this type of question, scientists have made a number of suggestions. One is that the helpers are sufficiently closely related to the offspring of other members of the flock for the necessary genetic traits to be maintained in the population (see EVOLUTION and ALTRUISM). Another suggestion is that helpers at the nest obtain some benefit from the opportunity to learn the skills required in raising young. In Mexican jays the helpers are often immature birds, or adults who have not yet mated. However, adults who have lost their own eggs or young may contribute more to the young of other nests than do the actual parents. A third possibility is that the helpers contribute to the welfare of others to avoid being evicted from the flock, or that some other form of *reciprocal altruism* is involved. In Mexican jays successful parents are generally about four years old, and they must themselves have been helpers in previous years. Because many birds die before becoming successful breeders, successful parents end up receiving more aid from the flock than they gave to the flock in earlier seasons. Whatever the answer in this particular case, it is important to recognize that cooperative behaviour is potentially unstable in that it is open to cheating by individuals who receive benefits without paying their share of the costs. Scientists are therefore concerned to investigate the mechanisms by which cooperative behaviour remains evolutionarily stable.

28

COPULATION is that part of SEXUAL behaviour which is most closely associated with fertilization of the egg by the sperm. In mammals and birds fertilization takes place inside the body of the female, while in other animals, with a few exceptions, fertilization is external. In frogs and toads (Anura), for example, the male clasps the female from above (see HORMONES, Fig. D) and maintains this position throughout the deposition of the eggs, fertilizing them as they emerge from the female's cloaca. Thus, even this relatively simple form of copulatory behaviour requires a certain amount of COORDINATION between male and female. In the European toad (*Bufo bufo*) the male is fairly indiscriminate in his initial clasping behaviour, and will attempt to clasp any moving object of a suitable size. If the object is another male toad, a few rapid VOCALIZATIONS from him serve as a signal, and the clasping grip is released. When a female is clasped behind her front legs she remains quiet, and eventually arches her back into a typical *lordosis* position. The male responds to this by forming a 'basket' with his hind legs in front of the female's cloacal opening, and in this basket the discharged eggs are collected and fertilized. The copulatory behaviour may be seen as a sequence of CHAIN RESPONSES in which one event leads to another in a stereotyped manner: male clasps object → female forms lordosis position → male forms basket → female discharges eggs → male fertilizes eggs. Any break in the chain causes the male to dismount. Such REFLEX responses to specific stimuli are typical of copulatory behaviour in general.

In most species copulation is preceded by COURTSHIP which is initially fairly flexible, but becomes increasingly stereotyped as progress is made towards coitus. In domestic fowl (*Gallus g. domesticus*), for example, the cock typically takes the initiative, moving among hens as though testing each for sexual receptivity. He may entice a female with a food call, and may allow her to eat food which he has found. He may then perform a waltz-like DISPLAY. If the female responds by crouching, the male will mount her, gripping her comb with his bill, and performing 'treading' movements on her tail. The female responds by moving her tail to one side and everting her cloaca. The male spreads his tail, everts his cloaca, and moves forward so that the vents meet. At this point the male ejaculates, injecting his sperm into the female cloaca. The cock then dismounts and may give a postcopulatory display.

Mammalian copulation is characterized by *intromission*, a term which refers to vaginal penetration by the penis. In some species ejaculation is achieved with a single intromission, while in other species multiple intromissions are necessary. Intromission may or may not be accompanied by *thrusting*. In some species there is a *genital lock* which results in the copulating partners being temporarily unable to separate once intromission is achieved. In dogs (*Canis*), for example, the end of the penis becomes enlarged, and a vaginal sphincter muscle closes around this enlargement. *Ejaculation* refers to the pattern of reflexes which occur at the time of sperm emission. It is normally followed by a *refractory period*, during which the male refrains from copulation. This may be a matter of seconds in some species, or of days in others.

Among the marsupial mammals, e.g. opossums (Didelphidae) and kangaroos (Macropodidae), there is only a single intromission combined with thrusting, and the refractory period is very short, so multiple ejaculations are achieved. In opossums the male clasps the hind legs of the female with his rear feet, and may grip the nape of the female's neck with his jaws. In the larger marsupials, such as the kangaroo, the foot clasp and neck grip are absent. In some other mammals, such as the insectivores (Insectivora), the pattern of copulatory behaviour is similar to that of the marsupials. For example, the tailless tenrec (*Tenrec ecaudatus*) grips the female by the neck while he achieves a single intromission with thrusting and multiple ejaculations. Similar patterns are found among carnivores and ungulates. The genital lock is common among carnivores and the neck grip occurs in some, such as the cats (Felidae).

Rodent copulatory behaviour is characterized by multiple intromissions and ejaculations. Evidence from the laboratory rat (*Rattus norvegicus*) suggests that multiple intromissions facilitate the transport of sperm within the female reproductive organs. A period of quiescence following ejaculation is necessary for sperm transport to be completed, and for successful implantation of the fertilized ova in the uterus. If a second male copulates with the female immediately after the first male has ejaculated, the transport of the first male's sperm may be disrupted, and it may be the second male that succeeds in fathering the offspring.

The copulatory behaviour of primates is varied. Multiple intromission occurs in some monkeys, such as the rhesus macaque (*Macaca mulatta*), but a single intromission is also common, especially among the apes (Hominoidea). Thrusting is universal, multiple ejaculations normal, with the exception of gorillas (*Gorilla gorilla*) and chimpanzees (*Pan*). The genital lock is not found among primates, although the foot clasp is found in some monkeys.

Female mammals are generally quiescent during copulation, and they often adopt the lordosis posture which serves to facilitate access to the vagina. A standing posture is typical, though exceptions have been observed. Bats (Chiroptera) copulate while hanging by their hind feet, and have been reported to copulate in flight.

Ventral–ventral copulation has been observed in gorillas, but the most usual female posture is to rest upon knees and elbows. The success of copulatory behaviour depends largely upon the male in choosing receptive females with which to mate, and in maintaining the pattern of copulation that is typical of the species (see FIXED ACTION PATTERNS). Research shows that deviations from the typical pattern result in a marked drop in breeding success. Many females are dependent upon specific copulatory stimulation to achieve the change in the balance of HORMONES that is necessary for successful pregnancy. It is not surprising, therefore, that copulatory behaviour tends to be one of the most stereotyped patterns of an animal's behavioural repertoire.

82

COURTSHIP is the term used by ethologists to embrace all behaviour that precedes and accompanies the SEXUAL act that leads to the conception of young. Throughout the animal kingdom there is a bewildering variety of courtship behaviour patterns, ranging from those of species that reproduce without any behavioural interaction at all, to species that show very complex courtship sequences lasting for several hours or even days. Courtship behaviour often attracts the attention of the human observer as it frequently involves conspicuous, repetitive, stereotyped movements, the displaying of bright colours and striking patterns, or the production of loud and elaborate sounds. Courtship behaviour has attracted the attention not only of the amateur naturalist but also of the scientist, and many of the early studies in ETHOLOGY were concerned with courtship behaviour. Courtship behaviour raises many interesting questions; for example, why is it that courtship is so conspicuous, so elaborate, and so diverse? Close examination of many courtship behaviour patterns suggests that sexual tendencies are often interacting with other tendencies such as AGGRESSION and FEAR, and this raises questions about the MOTIVATION of behaviour that are relevant not only to courtship, but to behaviour of all kinds. Rather than attempt a survey of all the forms of courtship, we will attempt to establish some general principles which will go some way towards answering these questions.

The EVOLUTION of the wide variety of courtship patterns that can be observed in nature may be understood by consideration of the biological functions that courtship serves. What is meant by the FUNCTION of a behaviour pattern is the contribution that that behaviour makes to the survival of the animal, or of its offspring. In general terms courtship fulfils the very important function of creating the circumstances in which successful mating is possible, but within this general framework are a number of aspects that can be considered separately.

An important function of some courtship behaviour patterns is the initial attraction of a mate, often over a considerable distance. In many species it is the male who attracts the female; for example, the territorial songs of many garden birds serve not only to exclude rival males, but also to attract females (see TERRITORY). In some species the female attracts the male; many moths have evolved wind-borne odours or PHEROMONES which enable males to find the females that produce them. So sensitive are the antennae of the male silk moth (*Bombyx mori*) that he is able to detect the female pheromone several miles from its source. The sensory modality used by a species for mate attraction depends upon the nature of its environment, and on the senses that it has evolved in association with other aspects of its life. Thus birds, which have well developed senses of vision and hearing but a poor sense of smell, generally attract their mates by visual displays and by songs and calls. They do not produce the pheromones which are widely used by insects and mammals, in which the sense of smell is well developed for the detection of food. Birds living in forests where visibility is restricted often use songs and calls or some other sort of sound, such as the drumming of woodpeckers (Picidae). Birds living in more open habitats more often use visual displays. The male lapwing (*Vanellus vanellus*) performs a distinctive soaring and tumbling flight pattern which, together with his black and white plumage, makes him conspicuous from a considerable distance. Animals that are active at night rarely use visual displays, but rely on auditory or olfactory signals to attract mates; the song of the nightingale (*Luscinia megarhynchos*) and the churring of the nightjar (*Caprimulgus europaeus*) both serve to attract females. Nocturnal insects have evolved a variety of methods of mate attraction. As well as moths which produce pheromones, others like cicadas (Cicadidae) and crickets (Gryllidae) make song-like sounds, and male fireflies (Lampyridae) attract females by producing flashes of light as they fly through the air. Some animals modify their environment to make themselves more attractive to potential mates. The male bower-bird (*Amblyornis* spp.) builds a huge nest of twigs which he decorates with flowers and colourful debris. The Egyptian ghost crab (*Ocypode saratan*) digs a burrow in the sand in which he waits for a female, having built the excavated sand into a pyramid nearby which can be seen by females from a distance. To attract a mate from afar an animal may have to make itself conspicuous, not only to potential mates but also to its enemies. The DISPLAYS that we observe represent a compromise between NATURAL SELECTION favouring more effective mate attraction, and selection against behaviour that attracts the attention of predators.

In some species mates do not find one another by long-range attraction, but by aggregating in a particular spot. This is particularly common among amphibians such as frogs and toads (Anura) which migrate to ponds and lakes to mate and lay their eggs. Having come together in large numbers individual toads have still to find a member of the opposite sex. In the European toad (*Bufo bufo*) males will clasp females and other males indiscriminately, but males, when clasped by another male, make a distinctive 'release me' call, in response to which the clasping male relaxes his grip.

To ensure the survival of its offspring it is of paramount importance that an animal mates with a member of its own species since hybrid matings rarely produce any offspring at all, and those that are produced are unlikely to survive or to produce further offspring themselves. The function of many courtship patterns is to minimize the risk of hybrid matings, and to ensure that mating occurs only with an individual of the same species. This phenomenon is called *reproductive isolation*, and any behaviour pattern that achieves it is called an ISOLATING MECHANISM. In many species the isolating mechanism is the behaviour that first brings mating partners together, only members of the same species being attracted by the behaviour. For this reason those behaviour patterns by which animals attract mates tend to be highly distinctive for each species, particularly when one compares closely related species for whom the potential risk of hybrid matings is greatest. For example, two species of grasshopper, *Chorthippus bruneus* and *C. biguttulus*, are almost identical in size and appearance, and, once male and female have come together, they have very similar mating behaviour, but the frequency of the call note they produce by rubbing their hind legs against their abdomen (*stridulation*) is markedly different. Male fence lizards (*Sceloporus* spp.) attract females by bobbing their heads rhythmically, and reproductive isolation is achieved by males of related species bobbing at different and characteristic rates; females will respond only to the bobbing pattern appropriate to their species. Likewise different species of firefly (*Photinus*) have characteristic patterns of light flashes produced by the male, as illustrated in Fig. A. In the Anatinae (ducks and related forms), as in many groups of birds, we find a great diversity in the plumage patterns of the males, whereas the females are rather similar in appearance. This is

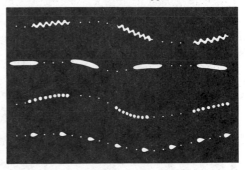

Fig. A. The flash patterns made by male fireflies of four different *Photinus* species as they fly through the air from left to right.

Fig. B. The advertisement display of the male sage grouse (*Centrocercus urophasianus*).

because the male, being the more active partner in courtship, has evolved a number of plumage patterns that enhance his display postures, and to ensure reproductive isolation each species has evolved not only characteristic postures and calls, but also distinctive plumage patterns. Natural selection has favoured those features that make an individual appear different from individuals of closely related species, making it easier for females to recognize males of their own species.

Once a male and female have come together, either by chance encounter or by some form of attraction behaviour, it is often the case that one partner is more ready or willing to mate than the other. In the majority of species the male is initially the more eager partner, and the function of many male courtship displays is to induce a comparable level of sexual arousal in the female. There are two aspects to this process: as well as eliciting a sexual response from the female the male's courtship may also serve to suppress other responses, such as fleeing or aggression, which would otherwise interfere with sexual behaviour and make successful mating less likely. The ways that animals excite their mates are many and varied. Many male birds posture in front of the female in such a way that their plumage patterns are shown off to best advantage; the displays of peacocks (*Pavo cristatus*) and sage grouse (*Centrocercus urophasianus*) are particularly spectacular examples, as illustrated in Fig. B. The male

smooth newt (*Triturus vulgaris*) performs a complex display that includes posturing to show off his brilliant coloration, and a tail-fanning movement that produces a water current directed towards the female which carries his smell to her, and which may also be a vibrational stimulus, as shown in Fig. C. The female newt is thus stimulated through three sensory channels, visual, olfactory, and tactile. This may enhance the effectiveness with which the male induces a sexual response from her, but it also provides her with three sources of information by which to recognize a male as being of her own rather than of a closely-related species.

In some species the male may give the female food as part of his display to her. This COURTSHIP FEEDING may in many species have the important function of providing the female with extra nourishment when she most needs it for the production of eggs or young. It seems also to act as a display, often eliciting a sexual response from the female or decreasing the likelihood that she will attack or flee from the male.

In territorial animals which spend much of their time aggressively excluding other members of their species from their territory, a general antagonism towards potential mates must be overcome if mating is to take place. In the courtship of the black-headed gull (*Larus ridibundus*) there is a display called *facing away* in which both male and female turn their heads away from one another, thereby removing from each other's sight the dark face mask which is normally a stimulus eliciting aggression.

In many spiders (Araneae) the male suitor is faced with a problem of a different kind; the female, who is much larger, is likely to regard him as food. If he is to mate with her he must suppress or circumvent her cannibalistic tendencies, and various species have evolved different techniques for doing this. In some species the male courts the female visually by waving his legs, but does so from a safe distance, approaching her only when she has given a clear sexual response. In some web-making species the male vibrates the web from one corner using a distinctive vibration pattern, quite unlike that which is set up by an insect prey, and he does not venture onto the web until the female has responded. In another species the male ties the female down with silken threads before mating with her; in another he has special appendages with which he can lock her mouth-

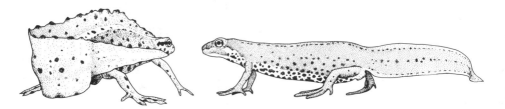

Fig. C. A male smooth newt (*Triturus vulgaris*) (left) fanning to a female.

parts open, rendering them harmless while he mates. In several species the male gives the female a food item, often wrapped in silk, so that he can mate with her while she is busily occupied unwrapping and eating his nuptial gift.

If the actual mating act is to lead to conception it is important that both male and female play their part correctly. In animals that copulate it is usually essential that the female be fully cooperative when the male mounts her. It is thus important that the moment when the male attempts COPULATION coincides with the time at which the female has reached a fully responsive state. A function of some courtship displays is to achieve this synchronization between male and female behaviour. In those species in which fertilization is external (eggs and sperm meet outside the body) this behavioural synchronization is even more crucial, since any delay between the emission of the eggs and of the sperm is likely to cause either or both to become dispersed before they can meet. In the three-spined stickleback (*Gasterosteus aculeatus*) the male courts the female, and, if she responds to him, leads her to his nest where he stimulates her to deposit her eggs; he then swims through the nest and fertilizes the eggs. The stickleback courtship sequence consists of a chain of interactions between male and female, which is such that neither animal will proceed from a given act in its repertoire to the act that normally follows it unless its mate has shown a suitable response to the first act. Thus a male will not usually attempt to lead a female to his nest until she has shown a positive response to his courtship dance. This stimulus–response chain ensures that neither partner can get far ahead of the other, so that the risk that one will produce its *gametes* (reproductive cells) before the other is ready to do so is minimized.

The synchronization aspect of courtship assumes a rather different form in some birds that breed colonially. Courtship displays between the members of a pair stimulate not only the birds in that pair, but also other birds nesting near by, so that in a large dense colony the overall effect of this mutual stimulation is that all the birds become sexually synchronized; this is called the *Fraser Darling effect*. It is known that birds who lay their eggs during the peak of the laying period are less likely to lose eggs or young by PREDATION than birds that lay early or late, so that such synchronization between pairs is clearly highly advantageous.

As we have seen, it is important to an animal that when it mates it does so with a member of its own species, but not all members of its own species will be ideal mates. Those individuals that are stronger and generally healthier than others will tend to have a higher reproductive potential, being able to produce more eggs, to care for more young, or to defend a larger territory, whatever their reproductive role may be. Natural selection will favour an individual who mates with an animal with a high potential, since it itself will thereby leave more offspring. The function of some courtship behaviour seems to be to enable animals to select the best possible mate from a number of potential mates. For example, the male bullhead (*Cottus gobio*) waits in his burrow until a female comes past when he will rush out and seize her by the head in his jaws, as shown in Fig. D. Only females that are fully sexually mature will remain quiescent and allow themselves to be pulled by the head into the male's burrow where spawning will take place. Immature females respond to the male's bite as if it were an attack, and wriggle themselves free. Thus the male tests the sexual state of the female by an apparently aggressive

Fig. D. A male bullhead (*Cottus gobio*) grasping the head of a female.

action, ensuring that he mates only with a fully mature one.

In the stickleback the male tends to alternately court and attack females who enter his territory. This apparently counter-productive behaviour may be advantageous to both partners. His aggression may scare off less sexually motivated females, ensuring that he mates only with a strongly motivated one. In addition his aggression may enable the female to assess his performance as a territorial defender, enabling her to ensure that she lays her eggs in the nest of a male who is likely to look after them well.

The MATE SELECTION aspect of courtship is very much more apparent in those species that mate in groups. Where a number of males and females come together to mate there is great potential for rivalry between members of the same sex to gain the attention of individuals of the opposite sex. To understand how this rivalry affects courtship behaviour we must first consider the evolutionary pressures to which animals are subjected. Natural selection will favour any attribute that increases the number of offspring that an animal produces, but the way that an individual can do this depends on that individual's sex. In many species the female not only produces eggs but also devotes a great deal of time and energy to the protection and rearing of the young. In birds the eggs are large, and must be incubated if they are to hatch. In mammals the young develop within the mother, and after birth are fed on her milk. Thus females generally invest a great deal in terms of time and energy in each of the eggs they produce, and their reproductive potential is accordingly limited. Many female birds, for example, are only able to produce one clutch of eggs each year. In contrast males produce sperms in enormous numbers, and since they are so small very little energy is used in their production. In many species the male plays little part in the care, protection, and rearing of the young, so that his investment of time and energy in his offspring is very small, and his reproductive potential is correspondingly very large. This difference in PARENTAL CARE between the sexes leads to the evolution of rather different reproductive strategies. To ensure the survival of their limited and expensive reproductive potential, females need to ensure that they mate with males of the highest quality. Males, on the other hand, will tend to maximize their reproductive potential by mating with as many females as possible, it being of little account if some of their mates are females of low potential. Thus, in a number of species that mate in groups, the females are slow to respond sexually and the males compete among themselves, often engaging in fierce fights, only the winning males eventually mating with females. Such a social system has a profound effect upon the nature of the males' courtship behaviour, since natural selection will favour displays that do not merely attract the attention of females, but that do so more effectively than the displays of rival males. This SEXUAL SELECTION has led to the evolution in some species of highly conspicuous displays, and very elaborate structures and colour patterns to augment them. The mating display of the peacock, in which the male displays his enormous decorated tail to females, is a classic example of the outcome of sexual selection.

In some birds, such as the ruff (*Philomachus pugnax*) and the black grouse (*Lyrurus tetrix*), mating takes place within a restricted area called a LEK where males compete fiercely with one another for territories, the more central territories being the most strongly contested. Females show a strong preference for certain males which is based on two criteria. They are strongly attracted towards the central territories; these are held by those males which have been most successful in competitive encounters with rivals. Within this group of males they prefer those that show a suitable balance between threatening the female and courting her, thereby inducing in her the ideal compromise between submissive and sexual tendencies that facilitates mating. To be able to achieve this balance a male requires experience of courting females, and it is the older males who attract and mate with the most females. So, by moving to the central territories, females come into contact with the strongest and fittest males in the population, and, by selecting those with a courtship technique which requires experience, they mate with older males who must be fitter than other males to have survived for longer.

In the Pacific tree frog (*Hyla regilla*) males and females congregate in large numbers in ponds where males all call together in a chorus. Some males are able to call for a longer continuous period than others, so that, during a bout of chorusing, males gradually drop out until only one is left calling. The males call in order to attract females, but the latter do not begin to move until the end of the chorus when they start to move towards the solitary caller. Thus females will tend to mate selectively with those males that can call the longest, and who presumably are the fittest.

The idea that courtship behaviour may enable animals to select the best possible mate is a relatively new one, and consequently there are rather few species in which it has been proven. It is possible that, in those displays in which males bring females food items or bits of nest material, they are not only giving the female something of material value, but also demonstrating to her their prowess as a provider, enabling her to select a male who will make a big contribution to the survival of her young. Until more data is available this must remain an intriguing hypothesis.

To summarize this discussion of the biological functions of courtship behaviour, we can say that it ensures that mating occurs under the most ideal conditions possible, in the right place, at the right time, with a member of the opposite sex but of the same species, and that in some species it may enable prospective mates to assess one another's reproductive potential. The conspicuousness of many courtship displays arises from the fact that mates often have to be attracted from a considerable distance; their great diversity is a reflection both of the diversity of ecological conditions under which different species live, and of the fact that many courtship displays have become highly distinctive so as to minimize the risk of hybrid matings. An important feature of many courtship displays is that they are very stereotyped in form; this ensures that there is no ambiguity in the message that they convey. The evolutionary process by which various movements have been changed, so as to become stereotyped, unambiguous displays is called RITUALIZATION.

During a sexual interaction between two animals other behavioural tendencies are frequently apparent. In many fishes, birds, and mammals, tendencies to attack and to flee from the mate are expressed at the same time as the tendency to mate. These activities, which are incompatible with mating, may occur before, during, or after the sexual encounter, or their presence may manifest itself in the form of the courtship displays themselves. When a male chaffinch (*Fringilla coelebs*) meets a rival of either sex in a dispute over food outside the breeding season he will show a head-forward threat posture, as shown in Fig. E. However, when he meets a female in the breeding season and shows sexual interest in her he performs a similar display, but is oriented in such a way that he presents a less threatening lateral view to her (Fig. E); this can be interpreted as a compromise between the tendencies to approach and

Fig. E. A male chaffinch (*Fringilla coelebs*) in the head-forward threat posture (top), the lateral courtship posture (middle), and the upright pre-copulation posture (bottom).

to flee. Gradually he approaches closer and closer to the female to attempt copulation with her, and, as he does so, his display changes from the head-down aggressive posture to a head-up posture characteristic of timid, fearful birds (Fig. E). Immediately after copulation he flies away and gives the same fear motivated call as he gives on seeing a flying predator. Thus, the male's courtship can be interpreted in terms of a shifting balance between the tendencies to attack, to flee from, and to mate with the female. At first he is predominantly aggressive; then, as his sexual tendencies bring him closer to the female, he becomes more fearful of her.

Not all courtship can be interpreted in terms of an underlying motivational conflict involving fleeing and aggression as well as sexual tendencies. During mating a variety of other activities, such as NEST-BUILDING, feeding, or simply looking around,

may be observed. In newts, courtship takes place on the floor of the pond and has to be interrupted from time to time so that either partner may go to the surface for a gulp of fresh air. The phenomenon of simultaneously activated incompatible activities is not limited to courtship situations, but is probably a feature of the greater part of an animal's life (see CONFLICT, DECISION-MAKING).

Many of the general aspects of courtship mentioned above are apparent in the courtship sequence of the smooth newt. Newts are rather unusual animals in that, although fertilization occurs inside the female, the male does not introduce his sperm into her by copulation, but deposits it on the floor of the breeding pond in a capsule called a *spermatophore* which the female then picks up in her genital opening or *cloaca*. Each spring newts migrate to their breeding ponds where they aggregate in large numbers. They show considerable *sexual dimorphism*, the male being more strikingly coloured and patterned than the female, and possessing a dorsal crest which she lacks (Fig C). There is no special behaviour for attracting a mate from a distance; instead courtship encounters are begun by a chance meeting as the newts wander about their pond. When such a meeting occurs the male begins to sniff the female intently, especially around her cloaca, apparently assessing her breeding condition. She usually swims away and the male follows her; thus newts conform to the common pattern in which the male is initially the active partner and the female is unresponsive. During this early phase of courtship the male repeatedly attempts to take up a position in front of the female's head, and across her line of vision. As time goes by she swims away less quickly, allowing the male to display to her from his position in front of her. As already described, the male's display is complex, incorporating different movements that stimulate the female in different ways. The function of this display from the male's point of view is to elicit from the female a positive sexual response. For her it probably provides information by which she can assess his condition, since males carrying a lot of sperm generally display very vigorously, and by which she can check that he belongs to the same species. The smooth newt has a geographical range which overlaps that of a closely-related species, the palmate newt (*Triturus helveticus*). The two species have very similar courtship behaviour sequences, but there are sufficient differences in the details of the males' displays to enable females to differentiate between them. For example, in the fanning display described above, the male smooth newt fans his tail at a very stereotyped frequency of about six beats per second, whereas the male palmate newt fans his tail at about twelve beats per second.

After a period in which the male displays to the stationary female she may begin to show a positive response in the form of a movement towards the male. In response to this the male retreats before her, still displaying to her, and there follows a period in which the female follows the male as he displays and retreats before her. The function of this behaviour is probably that it enables the male to ensure that the female really is positively responsive to him. Provided the female maintains her advance towards him, the male continues this retreating display for about 30 s, after which he turns and creeps slowly away from her. After creeping for 5–10 cm he stops and waves his tail slowly from side to side. The female, who has followed him, touches his tail with her snout, and he responds by raising his tail and extruding a spermatophore onto the pond floor. If she fails to touch his tail he does not deposit a spermatophore, but turns back towards her and reverts to his display. This relationship between the female's tail-touch and spermatophore deposition is a mechanism that ensures that the male does not deposit sperm unless the female is fully responsive to him, an example of the synchronization aspect of courtship.

As soon as he has deposited the spermatophore the male creeps away from it, moving in an arc to left or right until he takes up a position perpendicular to his previous direction of movement, with his tail folded against the side of his body nearest the female. The female continues to follow the male until her nose again touches his tail. She pushes against him, but he resists, and may actually push her backwards by flexing his tail towards her. The male's movement following spermatophore deposition is such that when the female touches his tail for the seond time, she is stopped in a position which brings her cloaca above the point where the spermatophore was deposited. If her cloaca touches the spermatophore the latter becomes attached, and is drawn up into the female's body. This complete sequence is usually repeated two or three times during a courtship encounter, increasing the chance that at least one spermatophore will be successfully transferred in this rather hit-or-miss manner.

The interaction between this courtship behaviour and BREATHING is a complex one. Usually the male is able to complete two or three sequences before he has to go to the surface to breathe, but, if he is forced to breathe earlier, he never does so during that part of the sequence in which the spermatophore is transferred; to do so would mean its almost certain loss. When the male is getting very short of breath he slightly quickens his courtship behaviour, so that he can complete a sequence before breathing.

Newt courtship can thus be seen to have been 'designed' by natural selection to ensure that sperm is transferred to the female as reliably as possible, that the two partners are of the same species, and are in good reproductive condition, and that this important activity is not interrupted at the wrong moment by another essential activity, breathing.   T.R.H.

10, 101, 137

COURTSHIP FEEDING is observed in many species, the male presenting food, or food-like objects, to the female. For example, the initial phase of courtship in the black-headed gull (*Larus ridibundus*) is characterized by a number of mutual displays. The female then starts begging for food, in the manner of a JUVENILE, and the male regurgitates food which the female then eats. Typically, the quantity of food received during courtship feeding is fairly small in relation to the female's normal daily intake. In some cases, however, especially when courtship feeding extends into the incubation period, it may be a substantial contribution to the female's requirements. In the pied flycatcher (*Ficedula hypoleuca*), for instance, it is about half what a nestling of equivalent weight would receive.

Another function of courtship feeding may be the appeasement of the female. The male of the wolf spider (*Pisaura mirabilis*), for example, is smaller than the female and in some danger of being eaten by her. The male captures a suitable prey, such as a fly, wraps it in silk, and presents this parcel to a female during courtship. If the female accepts the food, the male is able to approach her, and copulate unmolested, while she is eating it.

Males of the emphid fly *Hilara sartor* construct empty silk balloons, about the size of their own body. They then join a group of other balloon-carrying males and display collectively to visiting females. The female selects one of the males, and accepts the balloon as a precondition for mating. The behaviour is clearly reminiscent of courtship feeding, even though there is no food involved. Among other species of the emphid family there is a wide range of types of courtship feeding, ranging through the presentation of unwrapped prey, of wrapped prey, of non-nutritious fragments of prey, and of wrappings void of all traces of prey. This COMPARATIVE evidence suggests that the balloon ceremony of the emphid fly evolved from a true courtship-feeding situation, but is now entirely ritualized (see RITUALIZATION).

CULTURAL BEHAVIOUR involves the passing of information from one generation to the next by non-genetic means. Scientists agree that, apart from cultural behaviour, EVOLUTION occurs as a result of NATURAL SELECTION, and that inheritance of acquired characteristics is not possible. In other words, however much an individual animal adapts itself to its environment, whether by physiological ADAPTATION or by LEARNING special skills, etc., these acquired attributes cannot be passed to the offspring by GENETIC means. However, information can be passed from parent to offspring through the processes of IMPRINTING and IMITATION.

Imprinting is a learning process that is characteristic of JUVENILE behaviour. There is often a period during a young animal's life when it is especially sensitive to, and likely to learn about, particular features of the environment. For example, the white-crowned sparrow (*Zonotrichia leucophrys*) will remember its parents' song, provided it hears it when it is between ten and fifty days old. Individuals prevented from hearing the song of their own species during this period never produce a proper white-crowned sparrow song in later life. The juvenile white-crowned sparrow will learn a song that closely resembles the song of its own species, but other birds, such as the bullfinch (*Pyrrhula pyrrhula*), will learn the song of a completely different species, provided it hears it during its sensitive period for song learning. If bullfinches are foster-reared by canaries (*Serinus canaria*) they will adopt canary song, and will even ignore the song of other bullfinches kept in the same room.

In many species, the environment during particular stages of juvenile life may have a profound influence upon later behaviour. Thus, in choosing particular places to breed, and by remaining with their offspring during the early stages of development (see PARENTAL CARE), parents can greatly influence the future behaviour of their offspring. Evidence of this can be seen from the study of DIALECTS in bird song. Although the juvenile white-crowned sparrow will not learn a song very different from its own, it will learn variants of the song of its own species. In the region of San Francisco, populations of white-crowned sparrows, separated by only a few miles, have distinct dialects. Because the sensitive period for song learning occurs before the juveniles are very mobile, they are invariably exposed to the song dialect characteristic of the locality in which they are born. In this way the dialect is passed from one generation to the next by non-genetic means.

Animal dialects represent an elementary form of tradition. Other examples of traditional behaviour include the MIGRATION routes of some birds and mammals. Ducks, geese, and swans (Anseriformes) migrate in flocks of mixed ages, following the same route year after year. The juveniles learn the route that is characteristic of their population, stopping at the same rest places and terminating at the traditional breeding and over-wintering localities. Reindeer (*Rangifer tarandus*) show comparable fidelity to their traditional migration routes and calving grounds. In the case of migrating salmon (Salmoninae), each fish returns to the stream where it was spawned. The juveniles become imprinted on the odour of their native stream, and recognize this when they return to spawn themselves. Imprinting may also be important in HABITAT selection and in MATE SELECTION. Some game trails of deer (Cervidae) and other mammals, and some of the breeding grounds of colonial birds, are known to have been used for centuries. A population of ruffs (*Philomachus pugnax*) in Britain continued to visit their traditional LEK, even after a road had been built through it.

Amongst the primates a number of forms of tra-

ditional behaviour have been observed. Studies of the Japanese macaque (*Macaca fuscata*) show that populations living in different localities have very different FEEDING habits. The monkeys at Minoo Ravine dig out the roots of plants with their hands, while those at Takasakiyama never do this, even though the habitat is similar. The monkeys at Syodosima invade rice paddies to obtain food, but those at Takegoyama do not feed on rice plants, although they pass through the paddies from time to time. Careful studies have shown that many of these differences in feeding behaviour are due to tradition. At Takasakiyama scientists offered caramels to the wild monkeys. Initially the only individuals to accept these were those under three years of age, then mothers began to copy the juveniles, and eventually even the socially distant adult males adopted the habit of eating sweets. On Koshima Island biologists attempted to supplement the diet of the monkeys by scattering sweet potatoes on the beach. The monkeys did not normally frequent the beach, but eventually came out of the forest to feed on the potatoes. A two-year-old monkey called Imo invented the practice of washing the potatoes in water before eating them. Within ten years the habit had been acquired through imitation by the large majority of the population, with the exception of infants less than a year old, and adults more than twelve years old.

Two years later Imo invented another food processing habit. Scientists had been scattering grain on the beach, and the monkeys had been picking the grains up one by one. Imo gathered handfuls of sand and grain and threw them into the sea; the sand sank and the grain could easily be scooped from the surface of the water. The new technique spread through the population in a manner similar to that of potato washing. In each case the new behaviour was first adopted by monkeys of Imo's own age. Mothers learned from the juveniles, and adult males were the last to catch on. One reason for these differences is that juveniles are less conservative than adults and show more EXPLORATORY behaviour; another is that the SOCIAL RELATIONSHIPS of Japanese macaques are such that age peers have greater opportunities to learn from each other. Mothers and juveniles have good opportunities, but social contacts between juveniles and adult males are generally few.

The SURVIVAL VALUE of cultural behaviour is well illustrated by the study of the Japanese macaques. When the sweet potatoes were first provided the monkeys were purely forest dwelling. The food attracted them to a new HABITAT rich with opportunities: the juvenile monkeys learned to enter the water and swim, and a few even learned to dive and bring up seaweed from the bottom; the monkeys not only discovered new sources of food, but also new ways of processing it. These developments are not greatly different in principle from those which can be observed in human beings (see FOOD SELECTION). Many of the seemingly bizarre cultural practices found in primitive human societies are adaptive in the biological sense. For example, most of the traditional methods of cooking maize (*Zea mays*) by the indigenous people of the New World involve some sort of alkali treatment. Usually the maize is boiled for about 40 min in water containing wood ash, lye, or dissolved lime. It is then converted into dough, tortillas, etc., or consumed directly. The alkali treatment is practised for purely traditional reasons, though it is said by some to make the corn more palatable. However, scientific analysis shows that the alkali treatment has important nutritional consequences. Indigenous varieties of maize are characterized by low levels of available lysine, an amino-acid that is nutritionally essential. Most of the lysine occurs as part of an indigestible protein called glutelin. The alkali cooking treatment breaks up the glutelin and thus greatly increases the nutritional value of the maize. Some sort of alkali treatment is found in all indigenous cultures that rely on maize, and it is likely that natural selection, acting through malnutrition, eliminates people who do not follow this tradition.

Cultural behaviour short-circuits biological heredity, in the sense that it can lead to very rapid evolution of behaviour not based upon genetic change. Nevertheless, the products of cultural evolution will still be subject to natural selection, and biologists would expect to find that most long-standing cultural phenomena are also biologically adaptive. However, the cultural aspects of human behaviour with which we are most familiar are much too recent for natural selection to have had much effect upon them.

17, 145

CURIOSITY. Many students of animal behaviour agree that animals often exhibit apparent curiosity, but they find the concept difficult to define. Curiosity implies a VOLUNTARY form of EXPLORATORY behaviour, and the term is not used for REFLEX forms of exploration. For example, the startle response of many vertebrates involves a rapid orientation towards the source of stimulation, which is often followed by subsequent investigation. A startled person first jumps and then quickly looks round. This is a reflex form of exploration which is not normally regarded as curiosity.

Some scientists regard curiosity as the MOTIVATION of all exploratory behaviour, including PLAY. Many mammals investigate novel objects, at first by making numerous cautious approaches and subsequently more confident ones. They may sniff, lick, bite, and manipulate the object for a while, before losing interest. To the casual observer it may look as though the animal is playing with the object, but we have to be very careful in interpreting such behaviour. In the case of novel foods, for example, we know that the initial cautious investigation and sampling is an important part of FOOD SELECTION, and that many animals are thereby able to learn to avoid non-nutritious or poisonous foods. It is questionable, therefore, whether the investigation of novel foods stems from curiosity, or whether it is a more automatic part of normal FEEDING behaviour.

Some animals appear to actively seek out new situations simply for the sake of LEARNING. For instance, rhesus macaques (*Macaca mulatta*) will learn a puzzle game without any reward other than the performance of the task itself. They will also learn to perform a task which enables them to look out of a window in their cage. It is difficult to avoid the conclusion that something akin to human curiosity can be found in animals, but it is equally difficult to understand the role of curiosity in the organization of animal behaviour.

Curiosity and FEAR seem to be related in that animals will sometimes approach objects or other animals of which they are usually frightened. For example, Thomson's gazelle (*Gazella thomsoni*) and wildebeest (*Connochaetes*) often approach and stare at predators such as the spotted hyena (*Crocuta crocuta*), cheetah (*Acinonyx jubatus*), and even lion (*Panthera leo*). Cheetah have been observed to make a sudden dash and to catch and kill an over-curious gazelle, so some danger is involved in this type of behaviour. Maybe there is an advantage in keeping an eye on potential predators, or something to learn from this type of curiosity, which makes the behaviour worthwhile despite the risk.

D

DARWIN, CHARLES ROBERT (1809–82). The founder of the theory of EVOLUTION by NATURAL SELECTION, which received its most famous expression in his book *On the Origin of Species by Natural Selection* (1859). This theory has revolutionized most areas of biological enquiry, with Darwin himself being the main revolutionary. Darwin made important contributions to our understanding of: the physiology and reproduction of plants; the structure and taxonomy of animals (especially in several monographs on barnacles, Balanomorpha); plant–animal relations; animal behaviour; the evolution and psychology of man; the mechanism of heredity. In addition to all this biological work Darwin contributed many important geological ideas, such as his novel theory of the origin of coral reefs. Darwin is historically exceptional not only for the broadness and inventiveness of his work, but also in the extent of his influence on subsequent biology; over a century later his theory of evolution still seems correct, and is the inspiration of much biological research. Outside biology Darwin's ideas have noticeably affected such diverse subjects as philosophy, politics (frequently with disastrous consequences), and even literature.

Brief life. Darwin was born at Shrewsbury, the grandson of Erasmus Darwin (an 18th-century speculative evolutionist) and the son of Robert Darwin, a physician. As a child he was interested in natural history but showed no academic promise. He entered Edinburgh University to read medicine but abandoned the course before graduating. He then went to Cambridge University, officially to study for the ministry. In 1831 he was offered a place on H.M.S. *Beagle* as a naturalist, and he spent five years voyaging around the world, studying its diversity of life and geology. After his return to England Darwin lived for a short time in London but soon moved to Kent where he lived quietly (more or less an invalid), observing and meditating on his flowers and animals, and collecting published information that bore on his ideas.

Evolution by natural selection. Evolution (which until about 1870 Darwin called 'descent with modification') had been discussed by many speculative thinkers before Charles Darwin took up the problem. In the early 19th century the idea of the permanence of species was becoming increasingly difficult to reconcile with discoveries about the fossil record, which apparently revealed extinctions and creations of completely new species. *Catastrophism* (or diluvianism) was the main theory reconciling the permanence of species with the fossil record. Catastrophism postulated a series of divine interventions in the history of life by the agency of floods that caused total extinction of species and were followed by the creation of new species. Secular versions of catastrophism also existed: it was possible that the extinctions and creations were produced by some unknown natural process. When Darwin embarked on the *Beagle* voyage he adhered to some version of catastrophism; but travelling with him on the *Beagle* was Lyell's new book *The Principles of Geology* (3 vols., 1830, 1832, 1833). *The Principles of Geology* was the most important reaction to catastrophism; it urged instead the idea soon called *uniformitarianism*, that the natural forces and laws observable at present have been exclusively responsible for the changes and the periods of stasis that comprise the history of the world. During the voyage Darwin frequently thought on uniformitarianism and catastrophism, with the result that he became converted to Lyell's position. So also Darwin came to doubt the immutability of species (although Lyell himself criticized the idea of evolution), not only because he rejected catastrophism but also because of the facts of biogeography and of the variability of species. Soon after his return Darwin started the notebooks in which he collected the information and ideas about evolution that he was acquiring by systematic reading. This reading led him in 1838 to Malthus's *Essay on Population*, a book that argued that growth in food supply must always lag behind growth in the human population, with starvation and competition for food the inevitable consequence. Darwin was well aware of the profligate fecundity of organisms, and of the consequent 'struggle for existence'. He was also aware of the ubiquity of variation between the individuals of each species, and of the achievements of artificial selection by man in changing domesticated animals and plants. He put all this together to arrive at the idea of natural selection as the cause of evolution; Darwin realized that any variant especially fitted to the environment would leave more offspring than average, and so would come to predominate in the population. If the environment changed, a different variant would be favoured, and thus evolution would occur.

From then on Darwin examined most areas of biology and systematically reorganized the totality of biological knowledge around his new theory. He allowed himself to write preliminary essays on his theory in 1842 and 1844, and was working on a much longer work when, in 1858, he received a letter from A. R. Wallace. Wallace had independently conceived the theory of evolution by natural selection. Darwin, in danger of being scooped, was forced to publish. He contributed to a short joint paper with Wallace, and wrote an abstract of his larger projected work: the abstract was *On the Origin of Species* (1859).

On the Origin of Species falls into two main parts. First, it applies the Malthusian idea to animals and plants in general, and shows how the struggle for existence combined with the known variation of species could cause evolution. Second, it contains a thorough critical review of the evidence for evolution, from biogeography, the fossil record, and taxonomy. Darwin's theory had many consequences for the nature of organic diversity. For example, his theory demanded that organs were changed by modification of previously existing structures, and that each step in the change must be advantageous to the bearer. Thus, for any ADAPTATION, however complex, it must be possible to imagine intermediate stages by which it could have evolved. It was better still for his theory if various existing structures could be arranged in a continuous series showing increasing development of an organ. Darwin had already adopted this approach in his studies of barnacles (published 1851–4), though without drawing the implicit evolutionary moral. In the *Origin* he considered some other evolutionary series, such as the wing of the bat (Chiroptera). Modification in small steps from a common ancestor also explained *homology* (see COMPARATIVE STUDIES), the fact that separate organs such as bat wings, whale flippers, and human hands all have a similar underlying structure. Darwin took this argument furthest in his delightful book *On the Various Contrivances by which British and Foreign Orchids are Fertilised by Insects* (1862), in which he showed that the elaborate reproductive organs of orchids are but modified forms of structures common in other flowers. All of this constituted a total overthrow of the argument from design, which stated that the existence of organic design implied that there must be a 'designer'. Thus did Paley hold up the perfection of the eye as a cure for atheism. Darwin, however, substituted a mundane natural explanation for design, and rightly argued that the 'designer' must have an unbecomingly impoverished imagination if 'he' could only create new organs by refashioning the old.

Ideas about behaviour. Many early theories of evolution postulated behaviour as the driving force causing species to evolve. It is likely that when Darwin started collecting information about behaviour in his notebook on 'Man, Mind and Materialism' he was entertaining the possibility of a similar theory. However, any such theory was dropped after his realization of how natural selection works. Most of the thoughts and information that he had collected reappeared in one of the four main sources for Darwin's ideas about behaviour. These sources are, first, the chapter on INSTINCT in the *Origin*, which was in fact an abstract (emphasizing the parts most relevant to the theory of natural selection) from a longer, posthumously published, essay; second, *The Expression of the Emotions in Man and Animals* (1872); third, some passages in *The Formation of Vegetable Mould Through the Action of Worms* (1881); fourth, *The Power of Movement in Plants* (1880).

Darwin distinguished instinct from intelligent behaviour and, although he explicitly avoided defining instinct, he considered the absence of any knowledge of the end for which an action is performed to be one of its major characteristics. Actions involving INTELLIGENCE could not evolve because each intelligent action is worked out for a particular circumstance; but Darwin did think that an intelligent action or habit, if repeated often enough, could become hereditary and instinctive (by the inheritance of acquired characters, an idea since discredited). After considering this possibility in *The Descent of Man* (1871), Darwin wrote: 'But the greater number of more complex instincts appear to have been gained in a wholly different manner, through the natural selection of variations of simpler instinctive actions. Such variations appear to arise from the same unknown causes acting on the cerebral organisation, which induce slight variations or individual differences in other parts of the body; and these variations, owing to our ignorance, are often said to arise spontaneously. We can, I think, come to no other conclusion with respect to the origin of the more complex instincts, when we reflect on the marvellous instincts of sterile worker-ants and bees, which leave no offspring to inherit the effects of experience and of modified habits.' Instinct, then, evolved just like any organ, by natural selection of undirected variations. Furthermore, Darwin argued that changes in habits were associated with changes in the brain: he was a materialist. Darwin regarded complex behaviour as being the most serious problem for another requirement of his theory, that evolution occurs in small stages, but he did discuss some examples of possible evolutionary routes to complexity.

The Expression of the Emotions has contributed to later ethological theories of COMMUNICATION, though this is not the primary subject of the book. Darwin proposed three principles of the expression of emotion: *serviceable associated habits*, *antithesis*, and the *direct action* of the nervous system. Serviceable associated habits are expressions that are made when an animal is in a particular state of mind, and they are generally adaptive (i.e. serviceable); for example, when a dog (*Canis lupus familiaris*) is in a hostile frame of mind it walks upright, with its back stiff, as is appropriate if it is about to attack. The principle of antithesis is that when an animal is in an opposite frame of mind, it may produce expressions opposite to the serviceable associated habit. Thus a dog greeting its master has its back flexible, and crouches (see AGGRESSION, Fig. A). The antithetic expressions are not themselves adaptive, although they might become so by virtue of the implicit information about the animal's state of mind. Darwin classified a number of inadaptive activities as due to the direct action of the nervous system, rather than to habit or will. He gave trembling as one example; it is an expression of fear but is not serviceable or antithetic. Thus, although Darwin thought some expressions to be adaptive, he also thought that many were the useless by-products of the activities of the nervous or physiological systems.

Darwin patiently studied earthworms (e.g., *Lumbricus*) for about forty years before publishing his book about them. He did a number of experiments on the 'intelligence' of worms, and was led to the conclusion that worms are guided by more than just blind instinct. For example, he observed that worms always drag leaves into their burrows by pulling the leaf by its tip, which is mechanically the most efficient way for a worm to pull a leaf. Darwin then cut out a variety of paper leaves, and offered them to the worms. The results of these experiments showed that the worms were able to judge by which end they could most easily draw the paper leaves into their burrows.

In *The Power of Movement in Plants* Darwin described a number of now famous experiments on how plants respond to stimuli. He knew that a plant shoot will grow towards a source of light. If he covered the tip of a growing shoot, then the shoot did not bend towards the light. Thus the light sensitivity resides in the tip. The bend, however, is caused by differential growth at a region behind the tip, so there must be some communication between the receptor of the stimulus (the tip) and the effector of the response (the region of differential growth). It is apparent from the many analogies between plants and animals in *Movement in Plants* that Darwin was attempting to found a comprehensive theory of the nervous control of movement in relation to stimuli that applied to both plants and animals (Darwin sometimes metaphorically referred to plant nerves). Darwin did not, however, realize this grand theory.

Sexual selection. In many species the sexes look different, with one sex having bizarre anatomical structures more highly developed than the other sex. For instance, the adult male might be more brightly coloured than the female, or the male might possess ornaments or weapons, such as antlers. Darwin put forward his theory of sexual selection to explain these sexually dimorphic structures; this theory was briefly introduced in *On The Origin of Species* and then more thoroughly discussed in *The Descent of Man, and Selection in Relation to Sex*. The structures in question could not be due to simple natural selection to increase the survival of the animal, because in that case it would be expected that the structure would be equally developed in both sexes. Some minor sexual dimorphisms of the reproductive organs can be attributed to natural selection because they increase the efficiency of reproduction; but this cannot be said of organs like antlers, or the tail of the peacock (*Pavo cristatus*). According to Darwin's theory of sexual selection, some structures evolve only because of the advantage they give in obtaining mates. Darwin identified two ways in which this could occur: *male competition* and *female choice*. Structures due to male competition are used by males in fighting or otherwise competing among themselves for females; the males with the most effective weapons would then leave the most offspring, so the weaponry would increase in the population. Structures due to female choice are favoured by causing females to mate preferentially with males bearing the structure. No conscious 'choice' is necessarily implied; all that matters is that females do in fact mate more with some kinds of males.

Darwin's main tests of sexual selection used the COMPARATIVE method. He looked at a wide variety of species for structures that might give a mating advantage, and he showed that such structures tend to occur in animals with more complex nervous systems. This trend is expected because some degree of complexity must be necessary for the females to be able to discriminate among males. Also, sexual selection should work more powerfully in polygamous than in monogamous species, because there would be no advantage to a male that evolved a structure potentially enabling him to mate with many females, if, for some other reason, monogamy were inevitable. Darwin showed that sexual dimorphism is commoner in polygamous than in monogamous species.

Darwin mainly confined his discussion to anatomical differences between the sexes, but there are also sexual differences in behaviour, such as in COURTSHIP and aggression. Later ethologists have explained some sexual differences of behaviour by Darwin's theory. Overall, however, sexual selection has had the most mixed reception of Darwin's ideas. For a long time, the idea of female choice was more or less rejected by evolutionary biologists, mainly because of the lack of evidence for it. Since about 1960 the idea of female choice has become popular again, not so much because of any new evidence, but rather because of a general shift in the attitude of evolutionary biologists.

With the possible exception of Aristotle, Darwin has been the most important biological thinker of all time. He combined a command of a broad array of facts with an ability for profound abstract thought. This combination led him to the discovery of one great system after another, any one of which would guarantee his fame. This is why so many biologists, whatever their speciality, still bask in the light of Darwin's genius.   M.R.

29, 50, 63, 136

DEATH FEIGNING is a form of DEFENSIVE behaviour in which the animal becomes immobile, as if dead. Under natural conditions death feigning usually occurs during the closing stages of a PREDATORY encounter. In response to being grasped and held by the predator the prey becomes immobile, and this behaviour can be simulated in the laboratory by restraining the animal with the hand (see HYPNOSIS). Death feigning has been studied in domestic fowl (*Gallus g. domesticus*), bobwhite quail (*Colinus virginianus*), and ducks (Anatinae), and has been observed in rodents such as rabbits (*Oryctolagus cuniculus*), and in some primates, reptiles, and amphibia. It has been discovered that death feigning in domestic fowl has a marked daily RHYTHM, being least likely to occur around dawn, and most likely to occur at dusk. This rhythmicity is linked to the behaviour of the predators. In a study of the American red fox (*Vulpes v. fulva*), caught in the wild and raised in captivity, it was found that adult foxes ignored ducks which were placed in their pens during the daytime, delaying attack until the onset of their typical nocturnal activities.

The SURVIVAL VALUE of death feigning lies in the fact that predators do not attack dead prey, and do not always eat their prey as soon as it is dead. Foxes may take an opportunity to kill a number of prey, some of which may be subsequently buried. It has been reported that ducks buried in this way have subsequently escaped. Since death feigning is a strategy of last resort, usually occurring after a struggle, the animal has little to lose and has a slight chance of ESCAPE.

DECISION-MAKING. Animals make decisions every minute of their lives. These may be decisions on whether to continue with the current activity, or to change to some other form of behaviour. Alternatively, there may be decisions between alternatives that present themselves anew. In either case, the animal has to assess a variety of factors. Consider an example from the white-tailed deer (*Odocoileus virginianus*). Studies have shown that these animals can reduce heat loss considerably by sitting down. This is largely because winds are less strong near the ground. We can imagine that, in terms of the animal's choice between standing and sitting, there is often considerable advantage in sitting, in that energy is saved, but, on the other hand, the animal's range of vision is much greater in a standing posture, enabling it to detect predators more easily. At any particular point in time, the relative advantages of standing and sitting will depend upon various factors, such as the environmental temperature, the wind speed, the likelihood of predators approaching, and their probable visibility and ease of detection. We can imagine the animal weighing all of these factors in its mind every time it stands up or sits down, but it is unlikely that animals think things out to this extent. Most animals appear to use rules-of-thumb, and to make simple decisions as a matter of routine. Nevertheless, from the scientific point

of view, the principles of decision-making are much the same whether they are inherent in the 'design' of the animal, or whether active thought processes are involved. For example, a child may cross the road using rules that have been instilled into it by its parents and teachers. These rules will have been devised to ensure a proper balance between safety and reasonable progress. One would not expect a child always to wait until there is absolutely no traffic in sight, nor would one encourage it to cross in a hurry regardless of traffic. The proper balance lies between these two extremes. When an adult crosses the road, he or she may ignore the rules and assess each situation on its merits. Such reliance on individual judgement shifts the burden of responsibility onto the individual, who is performing both 'design' and 'execution' aspects of the decision-making process. In the child's case the design is done by the adults who devised the rules, and the execution of the rules is left to the child. The design involves a balance of costs and benefits, and the execution involves evaluation of the situation and performance of procedures designed to fit the situation. This distinction between design and execution is important in the study of decision-making in animals. Generally the design has been carried out by NATURAL SELECTION during the process of evolution, and the execution is the business of the individual animal. In more intelligent species, however, a certain amount of weighing of costs and benefits may be carried out by the individual.

The evaluation of costs and benefits. In considering what an animal ought to do in a given situation, we have to remember that the various possible activities differ in their consequences, and have different costs and benefits attached to them. For instance, many sea-gulls remove the broken egg-shells from the region of the nest soon after the chicks have hatched. Experiments have shown that this behaviour has the advantage of preventing the nest CAMOUFLAGE being spoilt by the conspicuous white insides of the shells. It was found that nests with broken egg-shells nearby were more readily found and destroyed by predatory crows (*Corvus*) and other gulls. Whereas in some species the shells are removed very soon after hatching, black-headed gulls (*Larus ridibundus*) may delay a number of hours before removing the broken egg-shells. This suggests that, despite the obvious benefits attached to early removal of the egg-shells, there must be some counteracting cost attached to their removal. Observation reveals that the wet, newly-hatched chicks are at risk from CANNIBALISM by neighbouring adult gulls to a greater extent than dry chicks. It appears that the cost of leaving a chick alone for the few seconds required for removal of the egg-shell is greater when the chick is wet than when it is dry. The removal of the broken egg-shell has a greater net benefit if it is carried out when the chicks are dry in those species where cannibalism exists, but otherwise it is better carried out promptly.

The decision to remove the egg-shell rests upon a delicate balance of costs and benefits. This need not imply that the individual animal carries out a cost–benefit analysis for itself, although this may occur to some extent in more intelligent animals. The animal is designed by natural selection to behave in such a manner that the greatest net benefit is attained. Removal of the egg-shells after the chicks are dry has greater SURVIVAL VALUE than removal when they are wet. In this case, the evaluation of costs and benefits has been carried out during the process of evolution, and the individual gull has merely to execute the correct behaviour according to some simple INNATE decision-rule.

Almost every activity has both costs and benefits associated with it. Consider, for example, the COURTSHIP behaviour of the smooth newt (*Triturus vulgaris*). This takes place on the substratum of a pond and is occasionally interrupted by ascent to the surface of the water to breathe air. Observations reveal that the male newt breathes less often during courtship than at other times; after he has been to the surface of the pond to breathe he may be unable to find the female again when he returns to the substratum. Clearly, there is a cost associated with breathing during courtship, and it is not surprising that the males make breathing trips less frequently. However, although newts that are prevented from coming to the surface can remain alive by means of the oxygen dissolved in the water, they are forced to become quiescent, and courtship is no longer possible. Thus, there is also a cost associated with postponing breathing. The courtship of the male is designed to induce the female to pick up the *spermatophore*, which he deposits on the substratum. The female can only do this successfully if she follows the male closely. For this reason the courtship should not proceed too quickly, and the most successful courtships tend to be prolonged affairs. So, although the breathing problem can be alleviated by speeding up the courtship, this may make successful fertilization less likely.

The male smooth newt during courtship is faced with a delicate balance of costs and benefits: he is continously faced with the decision to court or to breathe. A well designed decision-strategy is necessary to ensure a satisfactory outcome. We would expect such a strategy to be designed by nature to achieve the most favourable balance of costs and benefits.

Rules and strategies in decision-making. Most animals can 'do only one thing at a time' (see INCOMPATIBILITY). However when we see an animal engaged in one type of behaviour, we cannot assume that its MOTIVATION is towards that type of behaviour alone. In every animal there is an underlying motivational potentiality to indulge in a variety of types of behaviour. In effect, the animal makes a decision about which course of action to pursue, depending upon its internal motivational state, and upon external circumstances.

The most simple type of decision occurs in situations in which an environmental change of overriding importance occurs. In such cases, all current activity ceases as the animal responds to the new situation. This is generally the case with ALARM RESPONSES, but it can also occur whenever a sudden large change in motivation occurs. Decisions that result from such changes in priority of the animal's potential activities are said to be due to motivational competition. The strongest motivational tendency, or urge, takes control of the animal's behaviour, and prevents the other tendencies from gaining behavioural expression. For example, we can imagine that in the courtship of the male newt, discussed above, the courtship tendency is usually sufficiently strong to suppress the tendency to visit the surface of the water to breathe. However, it could happen that the newt was unable to withstand the urge to breathe, and interrupted his courtship to make a breathing trip. In such a case we would say that the tendency to breath became sufficiently strong to oust the courtship tendency by competition.

It may be thought that any activity in which the animal is observed to be engaged is inevitably the victor in a motivational competition. This is by no means so. In the courtship of the smooth newt, for example, the male invariably makes a breathing trip when the courtship sequence is completed, suggesting that the suppressed breathing tendency is released from the suppression by termination of the courtship. Experiments have shown that the male is less likely to breathe as a result of motivational competition during the critical final phase of the courtship than he is at an earlier, and less important, stage. Moreover, the male always breathes if the female is removed, and is more likely to breathe if the courtship is 'going badly'. Thus, although breathing during courtship may occur as a result of competition, its occurrence is more often due to a lull in the courtship routine. Indeed, experiments show that when the tendency to breathe is made artificially stronger than usual (by reducing the oxygen content of the water), the male newt attempts to speed up his courtship, in an attempt to get the sequence finished before he is forced to make a breathing trip. In terms of the costs and benefits involved, it makes good sense for the newt to suppress the tendency to breathe when the likelihood of a successful courtship is strong. It appears that he employs a decision-strategy which maximizes his chances of a successful courtship. Such a strategy is very different from a free-for-all competition between motivational tendencies, and decision-strategies of this nature have proved to be widespread in the animal kingdom.

In some species changes in behaviour occur in a largely pre-programmed manner, so that decision-making becomes a matter of routine. For example, the lugworm (*Arenicola marina*) lives in a U-shaped burrow in muddy sand flats between tide marks. The worm feeds from sand which falls, or is sucked, into the head-end of the burrow. The sand passes through the gut and is deposited on the surface at the tail-end of the burrow at periodic intervals. The feeding behaviour occurs in bursts, which occur at regular 7-min intervals, and are separated by periods of rest. *Arenicola* replenishes its oxygen supply by means of special irrigation behaviour, which replaces the water in the burrow. This behaviour also occurs in a routine cyclic fashion, having a period of about 40 min. When the animal is prevented from obtaining oxygenated water at low tide, the irrigation behaviour is less vigorous, but it still occurs with the normal rhythm. After a prolonged period of oxygen deprivation, the worm may break its rhythm and irrigate for a longer period than usual.

Here we have an example of a series of routine changes in behaviour, as if the animal were driven by clockwork. Many animals exhibit RHYTHMS of this type, particularly those that follow distinctive day–night routines (see TIME). The rhythmic behaviour of *Arenicola* and other marine worms is distinctive in that it seems to be driven directly by a clock-like mechanism, and is modified only in extreme circumstances. This reduces decision-making to a series of very simple rules. However, this clockwork type of decision-making strategy is well suited to animals that live in relative isolation from the outside world.

A preprogrammed decision strategy is suitable for an animal living in a stable and predictable environment, but for an animal living under more changeable conditions it is not sufficiently flexible. Such an animal may have many jobs to do, but opportunities to do them efficiently may fluctuate considerably. Consider, for example, a predatory animal which can exploit a variety of types of prey. The animal may be of the type that employs a sit-and-wait strategy, waiting in one place until a moving prey item comes by, and then ambushing it; or it may have a FORAGING strategy, actively searching out prey. For example, some spiders (Arachnida) spend a considerable amount of time and energy building their webs rather than moving about in search of prey, whereas others do not build webs and forage widely for their food. For the sit-and-wait strategy to be profitable, the prey needs to be sufficiently common, or the prey items sufficiently large, to meet the animal's energy requirements. The advantage of this strategy is that it requires little energy expenditure on the part of the predator, so that its requirements are low compared with those of an actively foraging predator. The main decision that a sit-and-wait strategist has to make is whether to strike when a prey item comes within range. If the prey is too small, the predator may gain little by revealing itself, and may lose the opportunity of obtaining a larger prey. For example, the praying mantis *Mantis religiosa*, a carnivorous insect, lies in wait throughout the day and catches flies (Diptera) with a very rapid movement of its forelegs. It is reluctant, however, to attack prey which is larger or smaller than that which it can handle most efficiently.

Foraging animals have many more complex decisions to make. They have to decide where to search for prey, when to give up in one locality and search in another, which prey to take and which to ignore. Some species have fairly fixed routines; woodpeckers (Picidae), for instance, often exploit crevices in tree trunks in an ascending spiral up to some height, and then fly down to a low point on the trunk of a nearby tree and repeat the process. Other foragers are more adaptable, and are able to adjust their strategy according to the availability of prey, and the energy cost of obtaining it.

In some species the timing of behaviour seems to be an important factor in decision-making. For example, the male three-spined stickleback (*Gasterosteus aculeatus*) builds a flimsy nest of weeds, which contains a tunnel through which the female creeps to deposit her eggs. During courtship the male must move away from the nest to approach the female. The object of the courtship is to induce the female to enter the nest. However, the longer the male spends courting the female, the longer he leaves the nest unguarded. It is essential that the nest be in good condition when he brings the female to it, but there is always a danger that the nest will be disrupted by currents, jealous neighbours, or other animals. The male's solution to this problem appears to be to break off the courtship every so often and pay a quick visit to the nest, as if to check that it is still satisfactory. If the nest has been disturbed, the male will effect some repairs and then dash back to court the female, if necessary leaving the repairs to be continued on his next nest visit. Experimental analysis of this behaviour shows that the courting male allows a certain amount of time for each nest visit. If some of this time is used up because of some external interference, then the nest activities will occur only for the remainder of the time. If the interference lasts longer than the normal duration of a nest visit, then that particular nest visit may be completely missed out, the stickleback returning to court the female. This is somewhat analogous to the situation of a lecturer who typically spends 5 min telling a joke during a lecture. If prevented from telling his joke by some mishap which fills the whole 5 min, he will resume the lecture and postpone or abandon the joke. Thus, at any one time there may be a single major behavioural directive, say nest-building, but time for other activities is shared with the time taken by the major activity, just as a large computer shares out the time between all the various tasks assigned to it. Time-sharing of this type is quite common in animal behaviour, and it enables the animal, like a juggler, to 'keep many balls in the air'.

Choice behaviour. So far we have been discussing decisions between activities directed at different goals. Once an animal has decided upon a course of action, or strategy, it still has a choice of tactics. The aspect of decision-making primarily concerned with choice between various routes to a GOAL is generally called choice behaviour.

An animal's future behaviour depends, to some extent, upon the consequences of its current behaviour. Usually an animal is concerned to attain the most beneficial consequences, in the shortest possible time, without incurring too many risks. Thus a laboratory rat (*Rattus norvegicus*) will generally learn to take the shortest route in a maze, to choose the larger of two food rewards, and to avoid noxious stimuli. To make a choice that is likely to lead to successful LEARNING, the animal must be able to discriminate the situations between which it is choosing. It must both pay ATTENTION to the relevant stimuli, and learn about the consequences of its choice, relating these to the stimuli that are likely to indicate such consequences in the future.

In natural situations it is known that animals may attend selectively to certain food items, apparently ignoring others with which they are known to be familiar (see SEARCHING IMAGE). For example, studies have been made of wood pigeons (*Columba palumbus*) searching for several types of more or less cryptic grain on fields of clover or stubble. The main seeds available were maize (yellow), maple peas (mottled brown), green peas (pale green), and tic beans (dark brown). It appears that most pigeons specialize on one type of food, which is not necessarily the most easily detected type. Moreover, an individual bird might prefer maple peas to green peas on one day, and reverse its preference the next day. Thus, although the animal can discriminate among the various food items, it does not always choose the most conspicuous items, and may even 'overlook' more beneficial items. It seems that such food preferences are determined partly by the animal's internal state, and partly by a process of selective attention, which improves overall searching efficiency, even though it may sometimes result in desirable items being missed.

The choice behaviour of animals is influenced by a number of properties of the external stimuli, such as their conspicuousness, their similarity to other stimuli that the animal has experienced in the past (see GENERALIZATION), and their special features in relation to the animal's innate or learned preferences (see SIGN STIMULUS). Choice behaviour consists of responding to various aspects of the stimulus situation, and of assessing the consequences of such responses. For example, wild Norway rats (*Rattus norvegicus*) normally avoid novel foods, but if they are particularly hungry, or suffering from a dietary deficiency, they will sample novel foods with extreme caution. The rat is particularly careful to eat only a small quantity of a novel food, and then to wait a day or so before touching it again. If, as a consequence of digesting the food, a dietary deficiency is alleviated, then the rat will come to prefer the novel food over its previously normal food. If, on the other hand, the rat becomes sick as a result of eating the novel food, it will avoid it in the future (see FOOD SELECTION).

The consequences of decision-making and choice behaviour can be assessed from two points of view: in terms of the costs and benefits incurred; or in terms of the response of the animal to those consequences. The costs and benefits depend upon the manner in which the consequences are subject to natural selection. Thus, a consequence of feeding in a particular place may be that the animal is not well camouflaged in that place, and is readily detected by a predator. This may be a consequence of feeding that the individual animal is unaware of, and to which it does not respond. To respond to the consequences of its own behaviour, the animal must be able to detect them, through sensory processes. Thus the feeding animal may respond to the taste of the food, to its smell, and to its appearance. It may respond to characteristics of the feeding locality, for example, by seeing, hearing, or smelling signs of danger. All such consequences of decision-making and choice behaviour are likely to influence the future behaviour of the animal.

DEFENSIVE BEHAVIOUR refers to any behaviour which reduces the chances of one animal being harmed by another animal.

We live in an environment that is constantly changing; and to the extent that every living animal is a precision-built piece of machinery which can only function under rather narrow ranges of conditions, environmental changes tend to be hostile to life. Any ADAPTATION which enables an animal to survive these environmental forces can be defined as a protective adaptation: examples include the HIBERNATION of small mammals during winter, and the deep burrowing of earthworms (e.g., *Lumbricus*) to avoid dry soil during drought. Those protective adaptations which protect an animal against other animals can be defined as defensive adaptations.

Defensive adaptations may be static such as the shell of a tortoise (*Testudo*) or the spines of a hedgehog (*Erinaceus*), or active such as running away. But almost all defences have a behavioural component: thus when attacked the tortoise withdraws its legs into its shell, and the hedgehog rolls up into a ball. The simple defensive adaptation of a rabbit (*Oryctolagus cuniculus*) escaping from a fox (*Vulpes*) involves detection of the fox by various sensory processes, translating this information into the response of running fast towards the burrow, and bolting down it to safety. The entire repertoire of defensive responses of the rabbit to the fox and to the other predators it may encounter comprise its defensive system.

It is simplest to think of defensive adaptations in terms of defence against predators, but animals may also be defended against parasites or against other members of their own species. The guenons (*Cercopithecus*), Old World monkeys, groom each other to remove fleas (Siphonaptera) and other skin parasites (grooming also has a SOCIAL function in helping to maintain the bond between members of the same group). Males of many species of birds, mammals, and fish defend a TERRITORY and fight off other males, and have defensive adaptations which protect them from damage during territorial disputes.

Primary and secondary defences. Primary defences operate regardless of whether or not there is a predator in the vicinity, whereas secondary defences operate only after an animal has detected a predator. Primary defences all reduce the probability that a predator will encounter an animal, whilst secondary defences increase the chances that an animal will ESCAPE from such an encounter. There are four types of primary defence: *anachoresis*, *crypsis*, *aposematism*, and *Batesian mimicry*; and six types of secondary defence: withdrawal, flight, *deimatic* behaviour (bluff), *thanatosis* (death feigning), deflection of an attack, and retaliation. These categories are not always clearcut, partly because an animal has to be defended against many different species of predator which hunt in different ways, and so it may be an advantage to be cryptic with respect to one predator, but to rely on flight or retaliation in response to another predator.

The concepts of primary and secondary defence obviously do not apply to defences against members of the same species, but only to defences against predators. But, in general, intraspecific defences are all similar to secondary defences.

Anachoresis is the habit of many animals of living in holes or crevices where they are unlikely to be found by predators. The principal problem encountered by anachoretes is that since they live in holes they may have difficulty in finding food or finding a mate. There are two ways in which animals have overcome these problems. First, animals such as earthworms and moles (*Talpa*) are more or less permanent anachoretes, living their entire lives underground, and as a consequence having very specialized habits and diets. Second, some animals are anachoretic for some of the time, but live at least part of their lives in the open. Thus rabbits feed on the surface of the ground principally at dusk or at night time, but remain hidden during the day when predators that hunt by eyesight could most easily find them. Polychaete fanworms (Sabellidae) live most of their lives in tubes from which they extend tentacles in order to feed, but their larvae are free living. In the soil, leatherjackets (*Tipula*) and other insect larvae are anachoretic as larvae, but the adults fly and reproduce in the typical insect manner.

Crypsis. Cryptic animals are CAMOUFLAGED, so that they harmonize in colour with their background, and are not easily detected by predators: bush crickets (Tettigoninae) are usually green,

planktonic animals in surface waters are transparent, and partridges (*Perdix*) in a ploughed field are brown. Experiments have shown that fish and insects are captured by predatory birds less often when they harmonize in colour with their background than when they are on contrasting backgrounds. But to remain undetected such an animal must remain motionless in the right place. Many moths rest motionless by day on tree trunks where they are cryptic, but fly to feed and reproduce at night-time when visually oriented predators cannot detect them. At dawn they must find a resting place where they will be camouflaged. Experiments have shown that dark brown or blackish moths usually rest on black paper in preference to white, whilst pale coloured moths show a preference for resting on white (Fig. A). Some animals can actually change their colour to make themselves more cryptic. Thus caterpillars of some swallowtails (*Papilio* spp.) and cabbage white butterflies (*Pieris brassicae*) change into green pupae when there are many green leaves present, but into brown pupae when the leaves are dead or absent. The physiological mechanism that brings about this matching response can be very complex, and is not the same in all species.

Mammals and birds of the arctic may change their colour twice a year; thus the rock ptarmigan (*Lagopus mutus*) and the snowshoe hare (*Lepus americanus*) are both brown in summer when they rest and feed on stones and sparse vegetation, but they become white in winter when their background is normally snow, so they are camouflaged at both seasons. Flatfish such as flounders (Pleuronectinae) can change colour to match stony or sandy backgrounds over a period of a few days, whilst the common cuttlefish (*Sepia officinalis*) can do so in a second (Fig. B). Other animals alter the

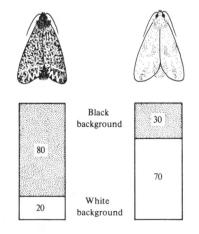

Fig. A. Experiment to demonstrate the percentage of dark and pale moths that choose to rest on black or on white backgrounds.

background in order to conceal themselves; for example, caddis larvae (Trichoptera) live inside cases which they have built from the surrounding debris in streams.

A problem for very abundant cryptic animals is that predators may find one animal, perhaps by chance, and from this learn the characteristic appearance or resting place of the species, so that they can find others more easily. This is called hunting by SEARCHING IMAGE, and there is evidence that birds which specialize in certain types of prey hunt in this way. One defence against this predator strategy is to be polymorphic, that is, for the prey species to have several different colour forms.

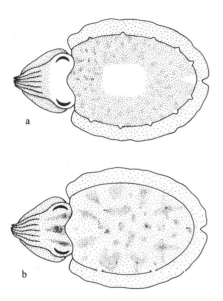

Fig. B. Colour patterns adopted by the cuttlefish *Sepia officinalis* on (**a**) sandy and (**b**) stony substrates.

Then if the predator does learn one colour form and succeeds in finding many of these, the other forms will survive. For example the convolvulus hawkmoth (*Herse convolvuli*) in West Africa has two forms of caterpillar which differ in coloration and behaviour. One form is green and rests on the underside of leaves of the food plant *Ipomaea aquatica*, whilst the other form is striped with yellow and purple, and rests on the leaf stalks and stems which are yellowish green tinged with purple (Fig. C). Both morphs are cryptic, but only because they rest in the appropriate place.

Aposematism. Animals which have dangerous or unpleasant attributes, and which advertise this fact by means of bright (warning) colours, or other signals, are said to be aposematic, and the phenomenon is called aposematism. A typical example is the wasp, genus *Vespula*, which is conspicuously coloured with black and yellow and makes no

attempt to conceal itself from birds or other potential predators. If it is attacked, it is protected by its sting, so warning coloration is the primary defence advertising the fact that the insect is nasty, whilst the sting is a secondary defence which is used if the primary defence fails.

Aposematism can only succeed as a defence if the predators detect and correctly interpret the warning signals. Each individual predator must normally attack and sample at least one prey individual so that it can experience the noxious qualities and then learn from this to avoid similarly coloured prey in future (this is called a conditional avoidance response, see LEARNING). Conditional avoidance responses have been demonstrated in many vertebrates, for example: in birds against black and red ladybirds (*Adalia* and *Coccinella* spp.), orange and black monarch butterflies (*Danaus* spp.), red and black salamanders (*Plethodon jordani*); in toads against bees and bumblebees (*Apis* and *Bombus* spp.); in fish against the poison-fang blenny (*Meiacanthus* spp.). The only invertebrates which appear to be important predators of aposematic prey are cephalopods: the octopus (*Octopus vulgaris*) learns after a few trials not to attack hermit crabs (*Dardanus arrosor*) which have the stinging sea anemone (*Calliactis parasitica*) on their shells, but continues to attack hermit crabs without an anemone. It is possible that other invertebrates, including insects, cannot develop conditional avoidance responses to noxious prey, at least under natural conditions. An animal whose predators are all insects therefore, gains no advantage by being conspicuously coloured, however nasty it may be to eat. This means that aposematic signals can only evolve in animals with vertebrate or cephalopod predators (see SURVIVAL VALUE)

In baboons (*Papio*) one individual can observe

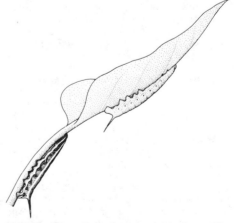

Fig. C. Resting positions of green and striped caterpillars of the convolvulus hawkmoth (*Herse convolvuli*) on *Ipomaea aquatica* in Ghana: green form on lower surface of green leaf; yellow and purple striped form on yellow, green, and purple striped leaf stalk.

another attack and reject a noxious prey, such as a poisonous snake (Serpentes), and then itself learn to avoid similar prey in future. This IMITATION is only likely to occur commonly in social animals.

One problem for an aposematic prey animal is that since the process of sampling by an inexperienced predator can be fatal, the only prey animals that benefit from the warning signals are other individuals in the vicinity. NATURAL SELECTION can only favour the EVOLUTION of warning signals if the prey animals which benefit are genetically related to the individual that was sacrificed (see ALTRUISM). Hence many aposematic animals are gregarious, or at least clumped in distribution, whereas cryptic animals are more evenly, or randomly dispersed. An aposematic animal will also benefit if it has the same warning signal as other aposematic animals in the area, since then predators will have to learn only one pattern instead of two, and the loss of sacrificed individuals whilst the predators learn will be much smaller. Assemblages of distasteful animals sharing the same colour patterns have therefore evolved. This is called Müllerian MIMICRY.

Aposematic animals are often slower moving and easier to capture than palatable but otherwise similar animals. This is probably because it is necessary for predators to capture and sample them, and if an experienced predator frequently sees the same signals it may assist retention of the conditional avoidance response. Furthermore, it is known that for some predators it is the active movement of prey that elicits PREDATORY behaviour, so that a prey that does not move has a good chance of escaping capture. If a predator requires many experiences to develop the conditional avoidance response, or if it requires regular reinforcement to remember it, then the loss of prey individuals may outweigh the advantage of having an aposematic signal. So selection will favour those prey individuals that either are tough enough to survive the experience of sampling and rejection by a predator, or are so nasty that a single experience by a predator is sufficient for it to reject similar prey for a very long time. The toughness of aposematic animals is well known. Butterflies can normally be killed by squeezing the thorax between finger and thumb. However, aposematic monarch butterflies frequently survive this practice, so they may also survive being seized by a bird. It has also been shown that birds and mammals develop a conditional avoidance response more readily to emetic food than to merely distasteful food (see FOOD SELECTION).

But does aposematism really benefit an animal? In Costa Rica the common aposematic butterfly *Heliconius erato* has black forewings on which there is a bright red band. In an experiment wild insects were captured, painted, and then released back into their environment. Some butterflies had the red band painted black, so that the entire wing was black. These insects were then much less conspicuous to a human observer than insects with normal red and black wings. As a control the black part of a number of normal insects' wings was painted black (in case the paint had some effect on their activity, or vulnerability to PREDATION). It was found that the black experimental insects did not survive as long as the more conspicuous controls (Fig. D). Furthermore, examination of surviving insects showed that three times as many experimental as control insects had beak damage to the wings, indicating that they had been seized by a bird and then released. Presumably the wild birds had been conditioned to avoid red and black butterflies, but had not learned to avoid black ones, therefore the blacks suffered more attacks and heavier predation, even though all the insects were equally nasty.

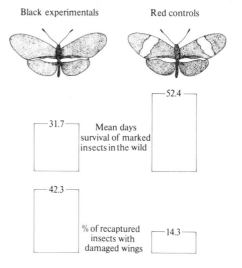

Black experimentals   Red controls

31.7   Mean days survival of marked insects in the wild   52.4

42.3   % of recaptured insects with damaged wings   14.3

Fig. D. Experiment using *Heliconius erato* to demonstrate the selective advantage of warning coloration in the field. Birds have already learned to avoid the normal insects with a red band on the forewing, so these survive well and few have beak-damaged wings. But birds have no previous experience of the more cryptic black insects, so these are attacked more often.

It should benefit an aposematic animal if its predators have an innate AVOIDANCE response to its warning signals and so do not have to sample and perhaps kill it whilst they learn. The yellow-bellied sea snake (*Pelamis platurus*), which occurs on the Pacific but not on the Atlantic side of Central America, sometimes gives a fatal bite when it is attacked by predatory fish. In aquaria, Atlantic fish attack sea snakes readily, but Pacific fish, even those with no prior experience of sea snakes, do not, suggesting that their avoidance response may be INNATE. However, the experience of sampling prey is not usually dangerous, so for most predators there is no advantage to be gained by evolving innate rather than conditional avoidance responses.

Batesian mimicry. Aposematic animals are sometimes mimicked by other non-noxious animals. This type of mimicry is called Batesian. Predators which have learned to avoid the aposematic model are deceived into avoiding the mimic also. This form of defensive behaviour is of advantage to the mimic, but of no advantage to the model. In fact, if there are too many mimics in relation to models present in the population, a predator may eat one and suffer no noxious consequences. It may then continue to eat both mimics and models for longer than it would have done if only noxious models were available. So the presence of mimics in the population may actually be of disadvantage to aposematic animals.

Withdrawal. The most widespread secondary defence of an anachoretic animal is to withdraw into a retreat. Rabbits run into burrows when they detect danger, and tubicolous polychaetes withdraw into their tubes. Animals with a protective shell, such as the tortoise, withdraw into this, whilst animals with spines, such as the hedgehog, roll into a ball or otherwise erect the spines over the softer, more vulnerable parts. A disadvantage of withdrawal is that the animal cannot then feed, and may have no means of knowing when the predator has moved away. So withdrawal in response to any trivial stimulus is likely to be disadvantageous, and this may be one reason why many invertebrates quickly habituate to simple stimuli such as a shadow passing over the tentacles of a polychaete worm (see HABITUATION).

Flight. Most active animals respond to the close presence of a predator by flight; i.e. rapid escape, by running, jumping, swimming, or flying. Sometimes the movement is fast in a straight line away from the predator, but it can also be erratic in direction. Some noctuid and geometrid moths can detect the sonar pulses of a hunting bat (e.g. the little brown bat *Myotis lucifugus*) before the bat can detect them, but they cannot fly as fast as the bat. Consequently if the stimulus from the bat is very strong they have little chance of escaping in direct flight, and so they fly erratically. The bat tries to follow the moth's every turn, but since the direction of movement is unpredictable it may fail to capture the insect. However, if the moth detects the bat from a greater distance it flies directly away from the bat, since then there is a chance of getting out of the bat's range of detection.

Sometimes the prey exposes bright colours as it flees, but when it comes to rest the bright colours are hidden. These *flash colours* probably attract the attention of the pursuing predator, but when the prey stops moving the colours vanish, and the predator may be baffled into assuming that the prey has disappeared. An example is the yellow, blue, or red on the hind wings of insects, such as the red underwing moth (*Catocala nupta*). The forewings of *Catocala* moths are patterned like bark, and the moths rest on tree trunks with the hind wings hidden by the forewings. When disturbed the moths suddenly display their garish hind wings, so startling the attacker and giving the moth time to flee. Conversely, when an underwing moth settles on a tree trunk and hides its hind wings it appears to have suddenly disappeared.

Deimatic behaviour consists of intimidating postures or actions. It occurs in animals which cannot flee very fast, or which have been caught by a pursuing predator. Sometimes the display is a genuine warning that the predator may be harmed if it persists in its attack; for example the arctiid moth *Rhodogastria leucoptera* opens its wings to display yellow or red marks on the abdomen, and it may exude a nauseous yellow fluid from the thorax; skunks (*Spilogale*) rear up on the forelegs when cornered, and may eject a nauseous fluid at the predator. But other deimatic displays are pure bluff. Toads (*Bufo*) inflate their lungs and hence appear to be much bigger than they really are. The praying mantis *Polyspilota aeruginosa* exposes its lateral surface to the predator and raises it wings, thus giving the illusion of increased size. It also turns its head towards the predator and abducts its forelegs, thereby exposing bright colours on the wings, jaws, and forelegs. At the same time it stridulates by rubbing its abdomen between the wings, the noise possibly mimicking the hissing of a snake which would further deter a small bird from attacking. Other deimatic displays involve the exposure of large false eyespots. It has been shown that the sudden exposure of eye-like markings deters yellow buntings (*Emberiza citrinella*) from pecking mealworms, so presumably birds are also deterred from attacking prey that display eyespots in nature. Eyespot displays occur in some moths, for example in the eyed hawkmoth (*Smerinthus ocellatus*), in the target mantis *Pseudocreobotra*, and also in the toad *Physalaemus nattereri*.

Thanatosis is the habit of some animals of DEATH FEIGNING. It is known that some predators only make a killing strike at a moving prey, so that a prey animal that remains motionless may escape. Thanatosis occurs in some beetles, mantids, spiders, and mammals such as opossums (*Didelphidae*). Usually the prey animal remains motionless for only a short time, and then suddenly attempts to escape by flight. It is likely that a predator may relax its attention because of the lack of movement, and so lose its chance of capturing the prey when it eventually darts away (see HYPNOSIS).

Deflection of attack. Here a potential prey may escape from a predator by causing the predator to attack the less vulnerable parts of its body, or as the result of a distraction DISPLAY. Wading birds, such as the ringed plover (*Charadrius hiaticula*), give a distraction display if a predator is approaching their nest or chicks. The parent bird flaps over the ground as if injured, but all the time

% of all pecks which were at painted end

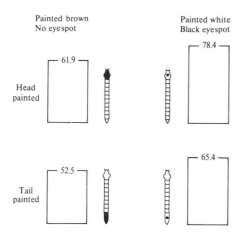

Painted brown
No eyespot

Painted white
Black eyespot

Head
painted

61.9

78.4

Tail
painted

52.5

65.4

Fig. E. Experiment using yellow buntings and mealworm prey to demonstrate the selective advantage of small eyespots. The birds peck more at the head than at the tail, and more at a small eyespot than at a uniformly dark area.

moving away from the nest. The predator is often deceived into pursuing the parent who eventually flies away, whilst the nest or chicks remain unmolested.

Some animals have small eyespots on the body whose function is very different from that of the large deimatic eyespots. For example, some butterflies have one or several eyespots on the wings whose function is probably to divert attacks away

from the head and towards a less vulnerable part of the body. Experiments in which mealworms (beetle larvae) were painted showed that yellow buntings peck more often at the end of the larva with a painted black eyespot than at the unpainted end, and more often at the end with an eyespot than at the end that is entirely brown (Fig. E). But one disadvantage of such an eyespot is that it might attract a predator to a prey that was otherwise well camouflaged. For example, the grayling butterfly (*Hipparchia semele*) has an eyespot close to the tip of each forewing; when it comes to rest it is cryptic apart from the eyespot. However, a few seconds after settling, it lowers the forewing between the cryptically coloured hind wings, so hiding the eyespot. Presumably if a predator had been watching the insect settle it would have attacked at the exposed eyespot, and the butterfly might have been able to escape, albeit with a torn wing. But if, after a few seconds, no attack has occurred, it is reasonable to suppose that there is no predator nearby, and then it is better to be entirely cryptic.

Another type of deflection of attack is to cause the predator to grab hold of some expendable and nasty part of the animal. For example the sea slug *Catriona aurantia* (a mollusc) has bright red papillae. If it is molested these papillae are waved about, and a fish may bite at them (Fig. F). The papillae can be broken off and regenerated, but as they contain stinging cells (*nematocysts*) and glandular secretions the predator may discard them, and refrain from a further attack. *Autotomy* (casting off) of the tail is common in many lizards (Lacertidae), but here the tail is edible, so the predator gets some reward for its attack, even though the prey animal also survives and eventually grows a new tail.

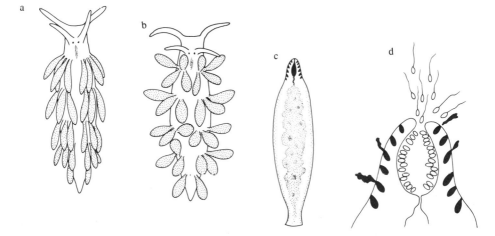

Fig. F. Defensive behaviour of the sea slug *Catriona aurantia*. **a.** Undisturbed animal. **b.** Animal has been attacked and is waving its papillae. **c.** Papilla broken off by a predator. **d.** Tip of papilla showing defensive glandular secretions and sac containing nematocysts, some of which have been ejected and are exploded.

Retaliation. The final defence of an animal when it is attacked by a predator or by a conspecific is to retaliate using whatever weapons it possesses, such as teeth, horns, and claws. Sometimes the attack has very little chance of success, as when a rat (*Rattus norvegicus*) is cornered by a terrier (*Canis l. familiaris*), but sometimes it may succeed in driving off the predator, particularly if the predator is not really hungry. Most animals retaliate if they are seized, and normally retaliation is the final response if the preceding deimatic display fails to deter. One might suppose that deer (Cervidae) use antlers against predators, but in most species the females lack antlers. Yet it seems that this sex should require the better defence because only the female guards the young. Antlers are primarily used for intraspecific encounters relating to status in the social hierarchy, but they are occa-

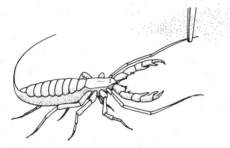

Fig. G. Defensive behaviour of the whip scorpion *Mastigoproctus*. The animal has been seized by the left first leg and responds by rotating the anal gland and ejecting a spray at the aggressor.

sionally used for defence against predators. In some animals however, horns and spines mainly function as a defence against predators. In the three-spined stickleback (*Gasterosteus aculeatus*) the dorsal and lateral spines are normally lowered so that they do not impede swimming, but if the fish is seized by a predator they are erected. The spines then jam in the mouth of the predator which may then reject the stickleback.

Other animals retaliate with chemical defences. The mollusc *Pleurobranchus* responds to being seized by secreting strong sulphuric acid. The grasshopper *Phymateus* secretes a nauseous smelling yellow froth from the thorax. Sometimes the behaviour involved is more complex. When seized with forceps the harvestman *Vonones sayi* responds by regurgitating fluid from the mouth. This fluid accumulates at the edge of the body where a quinonoid secretion is injected into it from glands. The tarsi (part of the leg) are then dipped into the fluid and brushed against the forceps or any other object which is attacking the animal. This particular secretion repels ants (Formicidae), and indeed many chemical defences of arthropods are specifically directed against ants. In the whip scorpion *Mastigoproctus*, the defensive glands open at the tip of the abdomen, which can be rotated so that the secretion can be squirted in the direction from which the attack was launched (Fig. G).

Many caterpillars have irritant hairs which are another defence against ants. In the gold tail (*Euproctis chrysorrhoea*) the caterpillar plucks its irritant hairs at the time of pupation and weaves these into the wall of its cocoon. The female concentrates the hairs at one end of the cocoon, and when the adult emerges she wipes the tip of her abdomen against these larval hairs. Some of them stick and are carried round with her. When she lays eggs she wipes hairs from the tip of her abdomen onto them. Thus every stage of the life history is protected by the larval hairs.

Living in groups and associations. For animals that are cryptic it is of advantage not to be too close to conspecifics, otherwise predators may, by chance, find one individual and then quickly search out and find the rest. This was demonstrated using painted chickens' eggs laid out in 3 × 3 grids with eight of each nine eggs partly hidden under vegetation, and the ninth egg visible to attract carrion crows (*Corvus corone*) to the area: the crows took more eggs when they were close together than when they were far apart (Fig. H). One way in which prey can become commoner without increasing the risk of predation is to be polymorphic for colour: predatory birds tend to concentrate on one or a few types of prey at a time, and to ignore other equally common prey that, by chance, they have not found. By contrast, aposematic animals gain advantage by living in groups, since there is a greater chance that the local predators have sampled and learned to avoid a particular patterned prey if there are many similar prey in the area than if they are scattered widely. Nevertheless there are circumstances where palatable prey also live in groups. One advantage of group living for palatable prey is that it reduces the chances of a predator encountering a prey individual compared with the same number of prey scattered over the same area. Providing that at each encounter with the predator only one or two prey individuals are killed, the group habit may give a greater chance of escaping predation than living in isolation. In a shoal of fish this advantage can be demonstrated: if the shoal is found and one individual killed, the rest will have scattered, re-formed, and swum away before the predator has time to swallow its first capture. In Trinidad, some rivers, such as the Lower Aripo, have predatory fish, whilst others, such as the Paria, do not. When wild guppies (*Poecilia reticulata*) were introduced into a large tank containing a predatory fish, it was found that Paria guppies suffered heavier predation than Lower Aripo guppies. This was because Lower Aripo guppies had experience of predatory fish and formed a shoal, thus increasing their chances of escaping capture, whilst Paria guppies had no experience of predatory fish, and did not show SCHOOLING behaviour. Another possible advantage of group

Eggs close together:

89% eaten by crows

Eggs far apart:

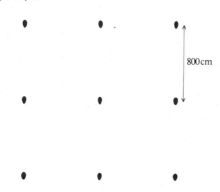

19% eaten by crows

Fig. H. Experiment using painted hen's eggs to demonstrate the survival value of spacing out for palatable prey against predators (carrion crows) that hunt using the searching image method.

living is that a predator may have difficulty in following one particular prey individual in the group, and so fail to make a capture.

Some groups show complex anti-predator behaviour. Thus common starlings (*Sturnus vulgaris*) bunch tightly together when attacked by a peregrine falcon (*Falco peregrinus*) and zigzag erratically, thus making it difficult for the predator to isolate and swoop on one particular bird. A further advantage of group living is that an individual does not need to be constantly alert to danger since others in the group may give the alarm. This is true for schools of fish, herds of gazelle, and flocks of birds. Sometimes groups are composed of several species with SENSE ORGANS of different sensitivities for predator detection. In the East African savannas mixed groups of Burchell's zebra (*Equus burchelli*), white-tailed wildebeest (gnu) (*Connochaetes gnou*), and ostrich (*Struthio camelus*) are common, sometimes with tick birds (*Buphagus erythrorhynchus*) on the backs of the zebra. Any individual can give the alarm, and alert the entire group to danger, so all individuals benefit from the association.

Some animals gain protection from association with other better-protected species, such as those of the Coelenterata (jellyfish, anemones, corals) or Hymenoptera (wasps, bees, ants, etc.). One problem of living close to such formidable animals is that these may attack the harmless *commensal* in mistake for a predator, so it is usual to find complex behavioural mechanisms which enable the two species to live together.

Aphids (Aphidae) and other Homoptera commonly live amongst ants, and the ants ward off predatory insects. The aphids are protected from attack by the ants because when touched by ants they release a sugary fluid that the ants imbibe. Some beetles (Coleoptera) have similar associations with ants, but in others the association is one-sided, with the ant colony suffering predation at the expense of the beetle. Wasps' nests similarly protect various species of birds from attack by nest predators and parasitic flies. For example in Central America nests of icterid finches, the yello-rumped cacique (*Cacicus cela*) and Wagler's oropendola (*Zarhynchus wagleri*), built close to wasp nests, suffer little PARASITISM from botflies (*Philornis*) on nestlings, whereas those built far from wasp nests are heavily parasitized (Fig. I).

In the sea some fish associate with jellyfish or with sea anemones and gain protection from predatory fish by retreating into the tentacles of the

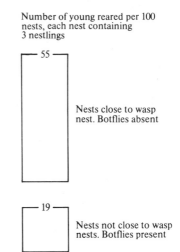

Number of young reared per 100 nests, each nest containing 3 nestlings

55 — Nests close to wasp nest. Botflies absent

19 — Nests not close to wasp nests. Botflies present

Fig. I. Effect of the proximity to a wasp nest on the fledging success of icterid finches in Panama.

coelenterate. When a fish touches a sea anemone this normally causes nematocysts on its tentacles to discharge and injure or kill the fish. But the anemone fish *Amphiprion* is able to swim amongst sea anemone tentacles without causing discharge, and it may thus obtain protection from the anemone. The fish also bring their food into the anemone, which thus benefits from the association (see SYMBIOSIS).

The cost of defensive behaviour. Because selection will favour those prey individuals with the best defences, and those predators best able to detect and capture prey, there will be a constant 'arms race' between predator and prey. Where a prey has just one principal predator it may evolve specific defences directed at that predator, for example the scallop *Pecten* leaps when touched by the predatory starfish *Asterias*, and so gets out of its way quickly, but simply closes its valves and remains motionless when touched by the non-predatory starfish *Porania*.

But evolution of a highly efficient escape behaviour from one predator may make an animal more vulnerable to another, or it may adversely affect some other aspect of its biology. Cryptic animals must remain motionless in order to achieve the full level of protection possible, yet they must also feed, which necessitates some movement. Sticklebacks are best protected from predatory fish if they are grey or black, but the males are most successful in attracting females if they have red bellies. Hence the colour of male sticklebacks in any population will depend on the relative strengths of predation pressure and female preference. In the waterflea *Bosmina longirostris* the defence against predation is by selection of a long-spined form which the predators cannot kill. But the long-spined form has a lower reproductive rate than the short-spined form, so the frequency of long-spined forms in the population depends on the intensity of predation. Thus, the defensive behavioural repertoire of an animal is a compromise between the requirements of defence, and those of other essential systems. And since predators are also evolving counter measures to the prey's defences, the entire defensive system of an animal is in the process of evolutionary change.   M.E.

39

DEVELOPMENT of behaviour is synonymous with ONTOGENY. It is concerned with the history of the individual from conception to death, and with the roles of GENETICS, MATURATION, and LEARNING in shaping the animal's life history.

DIALECT, as defined by ethologists, refers to the VOCALIZATIONS of a population of animals that differ from those of another population of the same species. The occurrence of dialect is widespread in bird SONG, and has also been discovered in the mating calls of the Pacific tree frog (*Hyla regilla*). For example, various populations of white-crowned sparrow (*Zonotrichia leucophrys*), in the region of San Francisco Bay, can be characterized by their song type. When young were taken from nests in three different dialect areas, and raised together, but in isolation from other birds, it was found that they all developed rather similar songs, none of which resembled any of the three home dialects. Results of experiments such as this suggest that the dialect song is acquired by hearing the songs of nearby members of the same species at an early stage of ONTOGENY. Local variations in song would then tend to be magnified, just as in the formation of human speech dialects.

Information can be transmitted from one generation to another by GENETIC means, or by CULTURAL means. In the latter case one individual learns from another. In some birds, such as the European cuckoo (*Cuculus canorus*) and doves (*Streptopelia*), it has been shown that the development of vocalizations is not influenced by early experience. In many other avian species, however, both genetics and learning are of importance. Some birds, including the white-crowned sparrow, can learn modifications of the song of their own species, but they cannot learn parts of the songs of other species. Other birds, such as meadowlarks (*Sturnella*), do not normally imitate songs of other species in nature, but are capable of it under laboratory conditions; while others, including mocking-birds (Mimidae), lyre-birds (Menuridae), some starlings (Sturnidae), and corvids (Corvidae), exhibit true IMITATION of the songs of other species under natural conditions.

The formation of dialects requires some ability to acquire elements of the vocal repertoire through LEARNING. Such vocal learning is most conspicuous in man and birds, although it has been detected in the bottle-nosed dolphin (*Tursiops truncatus*) and in elephant seals (*Mirounga occidentalis*). The formation of dialects is generally seen as a by-product of the process of IMPRINTING, whereby the juvenile animal learns some of the characteristics of its own species. Dialects give a certain distinctiveness to local populations, and these may diverge from each other if interbreeding is impaired by geographical or ecological isolation, perhaps eventually evolving into new species (see ISOLATING MECHANISMS).

73

DISCRIMINATION. When an animal shows differential responsiveness to different stimuli it is said to be discriminating, and since all animals, however rudimentary, show some patterns of behaviour only in certain conditions, discrimination is often said to be a fundamental feature of animal behaviour. Thus, among the coelenterates, the response of the freshwater polyp *Hydra* in stinging and seizing its prey is known to be evoked by the presence of the biochemical glutathione in the surrounding water; *Hydra* may be said to discriminate between the presence and absence of glutathione, which is itself the stimulus that 'controls' the response. This behaviour of *Hydra* gives us a convenient example of *successive discrimination*; successive since the two stimuli being discriminated (the presence and the absence of the chemical) are presented to the animal one after the other. Higher forms of life are capable of quite sophisticated successive discrimination. To take two examples from the behaviour of birds: the herring gull chick (*Larus argentatus*) is fed on food regurgitated by the parent, and begs for this by pecking at the tip of the parent's bill. The discrimination involved in this behaviour is more subtle than a simple sensitivity to the presence or absence of the adult bird, as experiments have shown. The bill of the parent is yellow with a red spot at the end, and the young birds readily peck at a cardboard model of a gull's head which has these features (see FOOD BEGGING, Fig. A). Other models are less effective in eliciting the begging response, and, by using a variety of models, it has been shown that the particular colours used are not especially important; the stimulus crucial in eliciting the response is the presence of a spot which contrasts sharply with the rest of the bill. The chicks can discriminate between a model bill which is grey with a white spot, for example, and a grey bill which lacks the spot, pecking more readily at the former than the latter (see SIGN STIMULUS). A second example of successive discrimination concerns the sensitivity of pigeons (*Columba livia*) to differences in colour. Pigeons can be trained quite easily by OPERANT conditioning techniques to peck steadily (for an occasional food reward) at a response key (a translucent plastic disc) illuminated with light of a particular colour. In one study, pigeons were trained to peck at a key lit with a greenish hue (550 nm). The experimenters then varied the colour of the light, and demonstrated that a change in wavelength of merely 10 nm produced a marked decline in the pigeon's pecking rate (see GENERALIZATION). The birds were clearly capable of discriminating between the successively presented colours; indeed, their ability to discriminate between colours is now known to resemble fairly closely that of the human observer.

The behaviour of any individual animal is constantly changing as it switches from one pattern of responding to another. To some extent these changes in behaviour can be seen as a series of successive discriminations, with the animal discriminating between various states of a changing external environment. But not all changes in behaviour can be interpreted in this way. A laboratory rat (*Rattus norvegicus*) may approach a cup filled with food on one occasion, but fail to do so in a subsequent test. We might say that the rat is performing a successive discrimination, and look for some difference between the stimuli presented in the two trials, but we are unlikely to do this if we know that in the first trial the rat was hungry, while in the second it was satiated; the change in behaviour reflects not a discrimination between different external events, but a change in the MOTIVATION of the animal. In situations which require simultaneous discrimination, however, motivational problems do not arise in the same way. In these, the animal is confronted with two or more objects, and allowed to choose between them. A

consistent choice of one object constitutes discrimination. The choice made may, of course, depend upon the animal's motivational state. For example, a rat may tend to choose food X rather than food Y and reverse its preference when deprived of food Y, but either pattern of behaviour demonstrates its ability to discriminate between the two tastes. Examples of simultaneous discrimination are readily found in the natural behaviour of animals: the duckling (Anatinae) which preferentially follows its mother rather than other adult birds (see IMPRINTING); the gull which returns accurately to its own nest in the middle of a colony; the human being selecting one volume rather than another off a library shelf; all are performing simultaneous discriminations. Less natural discriminations can be demonstrated in LABORATORY STUDIES; for instance, a hungry pigeon can be confronted with two response keys, red and green, and if it is rewarded with food only after pecking at the red key it will come to show a preference for that colour. This is an example of discrimination LEARNING.

Most studies of the processes underlying discrimination have used learning techniques in which external stimuli are differentially associated with reward and non-reward. When the subject performs the response under study only in the presence of the rewarded stimulus it is said to have learned the discrimination. There is a sense in which the animal can be said to discriminate between the two stimuli (e.g. between red and green response keys) before any training is given; presumably the two stimuli produce different states in the animal's nervous system from the outset, and the appearance of differential responding depends upon learning which of these states is associated with reward, and which with non-reward. When the word 'discrimination' is used in this way it refers to unobserved perceptual processes sensitive to differences in the external world, but which might not lead to overt differential responding. Contrast this with the usage favoured so far, in which discrimination has been equated with differences in behaviour controlled by external stimuli. Both usages are current and acceptable, and there is little confusion, since it is always apparent from the context in which sense the word is being used.

Although it is accepted that discrimination learning consists, for the most part, of the formation of associations between discriminated stimuli and certain patterns of behaviour, it still remains possible that, in addition, the animals' perceptual processes are modified as a result of training. In particular, it has been suggested that animals may learn to attend to certain aspects of the stimuli they are required to discriminate between. Most of the evidence in support of this suggestion comes from laboratory studies, but observations in ETHOLOGY underlying the concept of the specific SEARCHING IMAGE point to the same conclusions;

there is evidence that an insect-eating bird may need experience with a certain sort of prey (may need to acquire the appropriate searching image) in order to be able to find it efficiently. The best experimental evidence in favour of a concept of ATTENTION comes from studies of the transfer of learning. Animals (e.g. pigeons) are trained to discriminate, either between different colours, or between different shapes, and are then required to learn a new discrimination in which the stimuli again differ in colour and in shape. All subjects are required to learn to respond to one colour regardless of its shape. It is found that animals which have previously learned a colour discrimination learn the test problem more readily than those pre-trained on the shape discrimination, and it is argued that the former subjects have learned to attend to colour at the expense of shape, while the latter have learned to attend to shape at the expense of colour. Although results of this sort are usually expressed in terms of changes in attentional or perceptual processes it should be stressed that attentional theorists are making no claims about what an animal 'actually sees'. The results show that an animal may learn more or less readily about a particular set of stimuli according to his previous experience with that set. Exactly where in the chain of processes connecting stimulus input to behaviour the change has taken place is not revealed by the experiments.   G.H.

86, 98

DISPERSION. The dispersion pattern of a species is the distribution in space of individuals within their natural HABITAT. It is determined partly by topological factors, and partly by the behaviour of individuals towards each other. Topological factors may include the distribution of nest sites, sleep sites, or food sources, features related to variations in local CLIMATE, and physical barriers to dispersion.

Apart from irregularities induced by topological factors, the patterns of dispersion observed in nature are either random, clumped, or regular. Random dispersion is rare, though it has been observed in various marine molluscs, and in kangaroos (e.g. *Micropus rufus*). Clumped distributions are common, especially in species that tend to form family groups, troops, herds, etc. Regular dispersion often results from the formation of territories, and of schools of fish. Dispersion patterns may change on a seasonal basis, as in many avian species which show FLOCKING behaviour in the winter, and take up territories in the breeding season. There may also be daily changes, as in gelada baboons (*Theropithecus gelada*) and hamadryas baboons (*Papio hamadryas*) which aggregate in large numbers at sleeping sites and disperse to forage in small groups during the day.

Dispersal behaviour usually occurs among non-breeding individuals, either JUVENILES moving from the area of their birth, or adults leaving localities where the population density is high and food is scarce. The Norway lemming (*Lemmus lemmus*) has massive dispersal movements in which thousands of animals may emigrate from areas of high population density, sometimes rushing into the sea, as if in panic. Upon arrival at a new habitat, or breeding area, the members of a species settle into a characteristic dispersion pattern, by means of clumping and spacing behaviour.

Animals that live in groups can be roughly divided into contact species, in which body contact between adults is normal, and distance species, in which individuals normally maintain a small distance between each other, and make contact only in certain situations. Contact species are strongly attracted to each other, and generally sleep or rest in a huddled group. Mutual GROOMING is common in such species. Distance species, although gregarious, maintain an INDIVIDUAL DISTANCE, mainly by AGGRESSION towards individuals that venture within a certain distance. The dispersion pattern that results from spacing behaviour is largely determined by interactions between individuals. Whereas dispersal behaviour is normally non-competitive, the division of a habitat amongst a number of individuals inevitably involves COMPETITION. In many cases the location of a TERRITORY is established on a first-come basis, but the exact boundaries are a matter of dispute. Sometimes, spacing out is achieved by avoidance of proximity with other individuals, but usually it is achieved by aggressive interactions at the boundaries of the territory, HOME RANGE, food supply, or other defended resource.

In general, dispersion patterns are characteristic of the species and maintained through the process of NATURAL SELECTION. Random or scattered dispersal may have evolved as a DEFENSIVE measure against the SEARCHING strategies of predators, and clumped patterns probably evolved as a defence of some resource, such as territory, females, etc. For example, in the rut (mating season), the red deer (*Cervus elaphus*) in Scotland range over large areas of unwooded hills. Each stag attempts to gather together a number of hinds, which he defends against rival stags. This behaviour results in a constantly changing clumped pattern of dispersion. After the rut is over, males and females retire to largely separate home ranges. The females form well-integrated family groups with a hierarchical structure. The mature males band together, but without any stable social structure. Thus the dispersion pattern of the red deer is clumped almost all the year round, even though the groups have a different make-up at different times of year. This dispersion pattern undoubtedly facilitates the detection of predators, and decreases the burden of VIGILANCE on an individual animal. The time taken in watching out for predators can have a considerable effect upon FORAGING efficiency. For example, it has been shown that isolated wood pigeons (*Columba palumbus*) have such low feeding rates due to the time spent looking for predators that they may obtain inadequate amounts of food in situations that could support a flock of wood pigeons. The burden of vigilance for a member of a flock is much reduced, so that more time can be spent foraging.

Clumping may also act as a defence against PREDATION, particularly when the predator is most likely to catch the prey that is nearest. As the predator approaches a group of prey, those on the periphery of the group are most vulnerable, and it is to an individual's advantage to remain close to other members of the group, since this makes it more likely that some other member of the group will be the first to come within range of the predator. In wood pigeons the dominant individuals occupy the safer central portion of the feeding flock. Colonial nesting birds, such as the black-headed gull (*Larus ridibundus*), have been shown to suffer greatest predation around the edges of the colony.

Regular dispersion patterns generally result from situations in which there are both clumping and spacing tendencies, as in SCHOOLING in fish, and some territorial situations. Ideally, a single territory will be circular since this shape has the largest area for the boundary to be defended. When the density of territories is very high, we might expect them to become hexagonal, because hexagons pack together, while retaining a large area for a given boundary length. Hexagonal territories have been observed in the Mozambique cichlid (*Tilapia mossambica*). These fish excavate breeding pits by taking up grains of sand in the mouth and spitting them out at the edge of the territory. In this way sand parapets are built up along territory borders. Territories not in contact with others are usually circular and slightly larger than the densely packed territories. Hexagonal territories have also been observed in colonially nesting royal terns (*Sterna maxima*).

Less regular, though fairly uniform, dispersion patterns occur in many species with exclusive home ranges or territories. In England, for instance, the resident population of red foxes (*Vulpes vulpes*) within a particular locality is made up of family groups, each occupying and defending an area of suitable habitat.

The fox family group generally contains a dog and several vixens, and the territories vary greatly in size, depending upon the type of habitat. There are also a number of non-resident foxes, which disperse from areas where they have been unable to establish themselves in territories. These may travel many miles before successfully integrating themselves in a breeding group. The size of the territories is correlated with food availability and with other features of the habitat. It may range from 30 ha in rich suburban areas to 1300 ha in less productive open country. The density and dispersion pattern of foxes is thus highly correlated with ecological factors, but because of the pool of non-resident foxes it is not greatly affected by predation.

Knowledge of the dispersion patterns and dispersal behaviour of animals is of considerable importance in WILDLIFE MANAGEMENT, both in relation to the spread of diseases, such as rabies, and in relation to CONSERVATION.

DISPLACEMENT ACTIVITY is characterized by its apparent irrelevance to the situation in which it appears. For example, a male three-spined stickleback (*Gasterosteus aculeatus*), in the midst of courting a female, may suddenly swim to his nest and perform the parental fanning activity which normally serves to ventilate the eggs in the nest. However, during the courtship phase of the breeding cycle there are no eggs there, so the fanning is irrelevant with respect to parental care. It is hard to see how the fanning behaviour is relevant to courtship, as it does not appear to influence the behaviour of the female. Such apparently total irrelevance to the current situation is typical of displacement activity, and this type of fanning behaviour is often called displacement fanning.

Displacement activities are common in CONFLICT situations. In the stickleback, for example, it has been found that displacement fanning is most likely to occur when there is an equilibrium between the conflicting tendencies to attack and to court a female which enters the territory. On the one hand, she appears as a potential mate, on the other hand she appears as an intruder into the TERRITORY. Displacement activities often occur in aggressive encounters, when each participant has simultaneous tendencies to attack and avoid the other. Fighting cockerels (*Gallus g. domesticus*), for example, turn aside and peck at the ground as if feeding. They sometimes pick up grain or pebbles, which they generally drop rather than swallow.

Displacement activities often seem to serve as a means of resolving conflict, but opinions differ as to the mechanisms responsible. Some scientists take the view that a tension (see EMOTION) builds up during conflict, and that this is in some way released by the performance of displacement activity. Others believe that the conflicting tendencies cancel each other, allowing a third and irrelevant tendency to gain expression, so giving rise to displacement activity. However, displacement activities are not confined to conflict situations. They can occur whenever an animal is thwarted in attempting to attain the expected consequences of its behaviour. For example, a hungry animal is likely to become frustrated if it is physically prevented from obtaining food which it can see, or if an expected food reward is delayed, or if its food is less palatable than normal. In all these situations displacement activity is likely to occur.

Conflict situations can also be thought of as situations in which the animal is prevented from attaining its goal, through fear or some other conflicting tendency. A possible explanation of displacement activity is that the FRUSTRATION arising in situations of conflict or thwarting causes a switch in ATTENTION, so that the animal notices and responds to incidental stimuli in the immediate environment. These responses appear to be irrelevant to the human observer, and are labelled as displacement activities. Evidence for this view comes from the realization that displacement activities are, to some extent, responses to the external situation. For example, domestic turkey cocks (*Meleagris gallopavo gallopavo*) show bouts of feeding or drinking during fights. If water is available displacement drinking occurs, and if food is available displacement feeding occurs. If neither food nor water is available, displacement feeding may occur, but directed towards inedible food-like particles. Similarly, displacement fanning in sticklebacks can be facilitated by passing carbon dioxode solution through the nest. This has the effect of mimicking the presence of developing eggs. It appears that the nest alone is a sufficient stimulus for displacement fanning, but that the fanning is also influenced by those special factors which normally influence it during the parental phase of the breeding cycle.

Displacement activities often occur in aggressive or sexual encounters between individuals, where conflict is likely to be commonplace. As they generally occur at the equilibrium point in a conflict, they thus provide potential information to the other participant in the encounter. When COMMUNICATION of this type is advantageous, NATURAL SELECTION will tend to make the behaviour more reliable and efficient as a conveyor of information. The evolutionary process by which this comes about is termed RITUALIZATION. By means of ritualization, displacement activities can become incorporated as part of normal aggressive or courtship DISPLAY. For example, in many duck species the male preens at a point in the courtship when he experiences conflicting tendencies to attack and court a female. In the common shelduck (*Tadorna tadorna*) this preening seems to be

a genuine displacement activity, in that it occurs during conflict, but is otherwise similar to normal preening. The northern mallard duck (*Anas platyrhynchos*) preens only a few brightly coloured feathers on the wing. The garganey (*Anas querquedula*) makes incomplete preening movements, directed at a conspicuous part of the wing, while the mandarin (*Aix galericulata*) gives only a 'symbolic' touch to a single conspicuous red feather on the wing. Thus (see RITUALIZATION, Fig. A) there seems to have been a gradual incorporation of preening into the courtship pattern, some species having evolved further in this direction than others. During the process of ritualization, the displacement activity loses much of its original form, so that it may become difficult to tell from what activity it has been derived. The originally irrelevant behaviour has become highly relevant as a means of communication.

Displacement activities are believed by many to occur during social encounters in human beings. Conflict and embarrassment give rise to fidgety movements, such as head scratching, beard stroking, and other forms of grooming behaviour. Sucking on pens, spectacles, etc. is also common, and has been interpreted as displacement feeding behaviour. In the case of man it is not possible to carry out a detailed kind of analysis that separates learned and unlearned aspects of social behaviour, or to conduct satisfactory experiments, so that it is difficult to take such speculations any further.

99, 137

DISPLAYS are stereotyped motor patterns involved in animal COMMUNICATION. Displays are largely genetically determined, and specific to each species. Related species often have similar displays. For example, the courtship displays of fiddler crabs (Fig. A) consist of rhythmic waving of the enlarged claw accompanied by body movements, so that the display looks like a dance. The

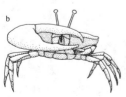
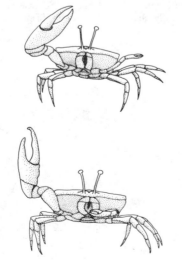

Fig. A. Courtship display of fiddler crabs. **a.** *Uca zamboangana.* **b.** *Uca pugnax rapax.*

Fig. B. Head-up threat display of the great tit (*Parus major*) and head-forward threat display of the blue tit (*Parus caeruleus*).

movement varies in the different species, but it is always stereotyped within each species. In the Philippine species *Uca zamboangana* a series of vertical waves of the claw is accompanied by raising and lowering of the body. In the South American species *Uca pugnax rapax*, the enlarged claw moves outwards and upwards in three jerks and descends smoothly.

Display postures often show off distinctive colour patterns, weapons, or other physical characteristics. For example, the threat display of the great tit (*Parus major*) consists of a distinctive *head-up* posture, which is directed at a rival in disputes over territory (Fig. B). This bird has a broad black stripe on its underside (Fig. C), which is exhibited during the threat posture. The female, which has a thinner stripe, displays less frequently. The coal tit (*Parus ater*), which has a black bib (Fig. C), adopts a head-up posture only momentarily during threat; while the blue tit (*Parus caeruleus*) generally threa-

tens with a *head-forward* posture, and has merely a small black collar (Fig. B). Human beings are no exception in these respects. Comparison of FACIAL EXPRESSIONS in various races of man has revealed that these displays are characteristic of the species as a whole. Smiling, for example, is a universal form of display that is stereotyped, both in its detail and in the situation in which it occurs. Smiling is made more effective as a signal by the distinctive coloration of the lips, which may be regarded as analogous to the coloured markings on the feathers of birds.

The FUNCTION of a display is of considerable interest in understanding animal communication. Experiments can sometimes be devised to demonstrate a particular function of a behaviour pattern (see NATURAL SELECTION), but establishing the function of a display remains largely a matter of guesswork. One of the most obvious functions of courtship display, for example, is to bring the

Fig. C. Black markings on the underside of (from left to right) male and female great tit (*Parus major*), coal tit (*Parus ater*), and blue tit (*Parus caeruleus*).

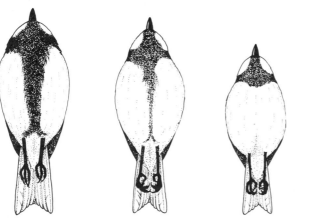

mates together at the right time and place. Many animals breed seasonally, and it is necessary that mating should occur only when both members of a mating pair are physiologically ready. Many conspicuous displays are thought to have the function of attracting a mate, and of bringing potential mates to a state of physiological readiness. These include the SONG of birds, grasshoppers, and frogs, SCENT MARKING in mammals, and other forms of TERRITORY delineation.

The role of displays in synchronizing the behaviour of mating pairs is not only important for successful reproduction, but can also serve as an ANTI-PREDATOR device. For example, mated pairs of colonial birds often display within sight of their neighbours, and the mutually stimulating effect of these displays tends to bring the members of the colony into reproductive condition at the same time. This has the advantage that the eggs and young are produced in large numbers during a short period of time, thus swamping the predators. In the black-headed gull (*Larus ridibundus*), the nests are closely spaced on open sand-dunes, and the majority of the eggs in the colony tend to be laid within a short space of time. The main predators on black-headed gull eggs are crows (*Corvus*), foxes (*Vulpes*), weasels (*Mustela*), and other gulls, and they are able to cope with only a limited amount of food each day. Thus they will take a smaller proportion of the total number of eggs when the period of availability is short than when laying is spread over a long period of time. Studies have shown that the eggs of the black-headed gull have a better chance of survival when they are laid at the peak laying time than if laid early or late in the season.

Another function of displays is to prevent cross-mating between members of different species. Interbreeding is biologically disadvantageous, because the offspring are generally infertile, or less well adapted to the environment than are the parents (see ISOLATING MECHANISMS). Courtship displays often serve as a means of identification, because they tend to be characteristic of each particular species. In many species, discrimination is exercised primarily by the female (see MATE SELECTION), and this helps to prevent cross-mating. Discrimination exercised by males is primarily of advantage in avoiding wasteful courtship of unresponsive females. In situations where there are two species, sufficiently similar to become confused with one another, coyness in females is an advantage (see COURTSHIP). Coyness provokes the male to greater efforts in his display, thus giving the female a better basis for choice between males (see SEXUAL SELECTION).

Displays often function as deterrents. Many species have special THREAT displays distinct from those of courtship. Threat is a form of social interaction, which tends to cause withdrawal without injury on the part of an adversary (see AGGRESSION). It commonly occurs in disputes over territory or quarrels between males over females.

Threat displays may also have a role in courtship, even when females are not directly involved. In deer (Cervidae) and black grouse (*Lyrurus tetrix*) for instance, threats and fights occur exclusively between males, but onlooking females are stimulated by these displays. In other cases, the threat is directed at the female, who withdraws if not receptive. Ripe females are often attracted by threats and attacks. For example, the domestic cocks (*Gallus g. domesticus*) which are most successful in mating are those which chase, peck, waltz, and strut before the female most frequently and vigorously.

In addition to serving as signals in SOCIAL encounters, displays have also evolved as signals between members of different species, particularly between predator and prey. For example, the angler fish (*Lophius piscatorius*) has a worm-like lure with which it attracts prey (see COMMUNICATION, Fig. A). Many insects have eye-like markings which are normally hidden, but which are suddenly revealed when a predator approaches, often frightening it away. There are

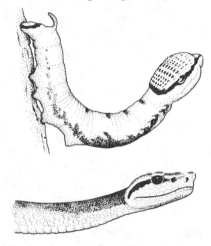

Fig. D. Deimatic display of the hawkmoth caterpillar of *Panacra mydon* compared with the head pattern of the young Wagler's pit viper (*Trimeresurus wagleri*), both from Malaya.

many types of displays which involve MIMICRY of the markings or of the behaviour of other animals. For example, the dorsal side of the head of the caterpillar of the hawkmoth *Panacra mydon* bears false eyes, and resembles that of the young of Wagler's pit viper (*Trimeresurus wagleri*). (See Fig. D.) In the hawkmoth *Leucorampha* the ventral side of the head has two eyespots which are turned towards the predator, while the caterpillar's body curves into a snake-like S shape.

Many animals enhance their CAMOUFLAGE by means of cryptic displays, often mimicking objects regarded as inedible by predators. Twig caterpillars, for example, look like twigs, and attach themselves to plants and remain motionless. If touched, they fall to the ground, like a dead twig. Other animals have markings which imitate the physical structure of leaves, grass, bird-droppings, etc. Many incubating birds run along the ground, feigning injury, when a mammalian predator appears. Such behaviour, which serves to mislead the predator so that it is unlikely to discover the nest, is called a *distraction display*.

The EVOLUTION of displays is of particular interest to biologists, because knowledge of the past history of a behaviour pattern greatly helps in understanding its present-day role. Many displays are complex behaviour patterns, which have evolved from many simple types of behaviour, through the process of RITUALIZATION. Some originated as INTENTION MOVEMENTS or DISPLACEMENT ACTIVITIES, both of which are typical of CONFLICT situations, and form part of the normal pattern of behavioural interaction between individuals of the same species. Other displays appear to have evolved from DEFENSIVE reactions, and from physiological responses associated with mild stress. A typical example is blushing in man. This is an involuntary response which occurs in mildly frightening, or embarrassing, social situations. The normal physiological response to mild FEAR involves a decreased flow of blood in the skin, arising as a result of redirection of the blood supply to areas of the body, such as the muscles, brain, lungs, and heart, where it would be most needed in an emergency. This response appears to have become ritualized, so that a reversal of the process occurs in certain social situations. The reversal is confined to the face and neck, and results in an increased blood flow, which we see as blushing. Such reversal of effect is typical of ritualization, and in this case it appears to have evolved to provide a means of non-verbal communication analogous to the display of animals.

10, 39, 41, 101

DOMESTICATION is the process whereby man has structurally, physiologically, and behaviourally modified certain species of animals by maintaining them in, or near, human habitations, and by breeding from those individuals which seemed best suited for various human objectives. These objectives typically include a variety which relate to economic performance, such as docility, efficient maternal care, high fertility, longevity, maximum production of such products as wool or milk, accelerated growth rates, and efficient food conversion. In other cases, the objectives are those of ornamentation. Well known examples include the development of certain breeds of dogs, birds, and fish, such as canaries (*Serinus canaria*) or Siamese fighting fish (*Betta splendens*). In all cases, however, attempts are also made to make domesticants more suitable .to specific environmental conditions which relate to nutrition, housing, and climate.

A historical note. It is generally assumed that the domestication of animals began at the time of the agrarian revolution, namely that time at which communities of men began to live in settled, as opposed to nomadic hunting and gathering, communities. In these relatively settled communities, species of animals and plants were gradually domesticated. Hence, there was a shift from food collection to food production, with an accompanying increase in population, abundant food, and the gradual development of civilization. The time period of these events is frequently given as between 10000 and 16000 years ago, and with specific reference to a number of geographical locations in the Middle East. The evidence for this latter assumption, however, is not as strong as the common statement would appear to suggest, and it is also clear that the domestication of some animal species preceded the shift to life in settled communities. Further, the evidence for the shift from nomadic to settled communities, which took place within areas of the Middle East, represents one part of the story. Other workers have examined evidence with respect to regions of the New World. There is, for example, evidence for communities of people living in Peru some 20000 years BC.

It also appears that there are marked differences between the Old and New World in the emphasis placed upon animal, as opposed to plant, domestication. Hence, in the Old World people domesticated a wide variety of animal species, and relatively few species of plants, whereas in the New World the situation was the reverse, with hunting remaining the principal way of obtaining meat and fur for a long period. It is interesting, however, that in the New World, as in the Old, the domestication of dogs (*Canis*) apparently pre-dated the domestication of all other animal species.

These few data apart, the reasons why man first domesticated specific animal species remains the subject of speculation. A reasonable suggestion with respect to the evidence is that dogs were used in hunting, perhaps as decoys. There is also evidence that dogs were used as sacrifices in religious ceremonies. Other suggestions include the supposed inclination of people to keep pets. Whatever the initial reasons for animal domestication, however, associations between man and a variety of animal species became more important as man became increasingly organized socially. Some species were domesticated to provide food and other materials, while others were used in work.

Characteristics of potential domesticants. Although a variety of animal species has been domesticated, comparatively few species can be successfully domesticated. Those species which are now domesticated exhibited originally a number of qualities which made them potential domesticants. For example, domesticants should obviously be flexible with regard to the physical and nutritional management of their lives. Promiscuous mating, often within large social groups, such as herds, packs, and flocks is also advantageous. Within such a SOCIAL ORGANIZATION, social stability is established and maintained by differences in rank order or DOMINANCE among members, so that each generally 'knows its place'. With dominance relations established, a group experiences minimum social disruption, even though members live in relatively close proximity. It is also important that strong and permanent bonds may be established early in their lives with members of species other than their own, and not least with man. Dogs provide a good example of this latter point, for they have SENSITIVE PERIODS of socialization during which intraspecific and interspecific bonds are formed and maintained throughout life (see IM-PRINTING). All these behavioural characteristics greatly facilitate the management of particular species.

Physical treatments in management control. Several physical treatments are commonly employed to facilitate the management of domestic animals. For example, some infantile characteristics may be protracted into the life span by administration of HORMONES. Castration, dehorning, and debeaking are all common practices, and have the effect of reducing the full expression of such potentially dangerous patterns of behaviour as FIGHTING. This is particularly important when animals are penned and cannot easily avoid one another; moreover, advantageous dominance relationships may be established without full AG-GRESSION.

As the degrees of management control over animal domesticants become more intensive, there is obviously an increasing emphasis upon the use of artificial environmental factors, such as temperature and lighting regulation, the use of food concentrates, regulation of breeding cycles, and the size and spatial distribution among social groups. The flexibility of domesticants in terms of husbandry regimes is also important. The behavioural effects of various environmental manipulations, however, require detailed evaluation in terms of both the economic performance and the WELFARE of the domesticants.

Some behavioural anomalies of husbandry regimes. The relative behavioural flexibility of domesticants sometimes leads to undesirable behaviour. There are many examples; among the most frequently cited are those observed in large groups of birds, such as domestic fowl (*Gallus g. domesticus*) and domestic turkeys (*Meleagris gallopavo gallopavo*) which may peck extensively at one another, or crowd together in ways which may cause death by suffocation or starvation. A variety of intensive husbandry techniques use small living areas with a high density of animals. Under such conditions one may sometimes observe such anomalies as hyper-sexuality, stereotypy, and damaging aggressive behaviour. On the other hand, keeping animals under various degrees of physical

isolation from one another may also produce a variety of anomalies, not least of which may be the lack of opportunity to learn aspects of their normal reproductive behaviour. They may also become over-reactive to any kind of environmental change, become difficult to handle, and have a low conversion rate of food substances.

There can be no doubt that disruption of normal behaviour, lowering of resistance to disease, and impaired productivity may result from some methods of intensive husbandry. The degree to which they are shown depends upon the species concerned, and sometimes upon different age and sex classes within the species. The amount of space available to captive domesticants may not be, within certain reasonable limits, the crucial factor in the maintenance of normal healthy stock; the quality of the space provided, in terms of relative access to food, warmth, and shelter, is also an important consideration. Aggressive encounters among animals may be reduced, for instance, by the provision of a number of sources of food, rather than a single one. Again, the deleterious effects of some husbandry techniques may be minimized by the breeding of animals that people have referred to as 'utility hybrids'; they are more adaptive because they can be adequately maintained under a number of different methods of husbandry. Conversely, it is the case that over-specialization in breeding may lead to deleterious effects, and especially so in cases where undesirable characteristics are perpetuated on a large scale by the use of artificial insemination techniques.

Selective breeding. Animals within a population are selected for breeding on the basis of characteristics such as their appearance and production records. Hence, it is through selective breeding programmes that the principal effects of the processes of domestication are realized. In some cases, developmental processes are speeded up so that animals mature more quickly; they may then breed earlier. Acceleration of growth may also have effects on adult anatomy, such as a reduction in the proportion of bone to muscle, or in the thickness of the skin. In other cases, the intention of selection may be to slow down the development of the maturing animal, and prolong the infantile and JUVENILE stages of development. This may have the advantages of leaving the animal in a relatively malleable behavioural state, as well as reducing undesirable behaviour such as aggression. These effects may be further accentuated by the physical treatments to domesticants, mentioned above.

Selective breeding may produce dramatic changes in the characteristics of domesticants, but it cannot produce entirely new ones; what is changed is the proportion of relatively desirable traits. Moreover, GENETIC processes are complex, and may result in the long persistence of particular characteristics. Breeding programmes in practice select individuals to mate from groups of animals which show the maximum predominance of a desired characteristic. This is important because the over use of single 'high quality' individuals may give rise to inferior progeny, through the disruption of desirable genetic combinations.

Another aspect of selective breeding programmes is that they sometimes give rise to unintentional and undesirable side effects. For instance, increases in the manifestation of some characteristics may be incompatible with other normal functions, such as resistance to disease. In other cases, apparently undesirable characteristics may be important mediators of advantageous ones. A certain number of undesirable traits will inevitably appear, and their deleterious effects have to be minimized by altering environmental conditions to provide better nutrition, shelter, resistance to diseases, and to avoid stressful SOCIAL RELATIONSHIPS with their own and other species.

Distinguishing indices of the processes of domestication. It is often difficult to decide which characteristics of domesticants are the result of selective breeding, and which are attributable to, and/or accentuated by, particular management regimes. It is sometimes important to have such information. Ways of attempting to examine this issue include comparisons of the behaviour of domesticants with that of contemporary populations of their wild progenitors, and/or by comparing similar domestic stocks in different husbandry conditions. Both areas of enquiry are important in assessing the effects of domestication and in providing useful indices of such, although relatively few studies have been undertaken.

Cases where direct comparisons of wild and domesticated members of the same species can be made, however, indicate at least some of the kinds and extent of change in behaviour that may occur through domestication. Among wild greylag geese (*Anser anser*), for example, monogamous pairs are formed after prolonged courtship involving a variety of patterns of behaviour. Domestic geese, however, may mate relatively quickly, and without remaining together for any length of time. Marked modifications to the courtship behaviour of pigeons (*Columba livia*) have also been observed amongst domesticated varieties. In another example involving care of the young, wild zebra finches (*Taeniopygia guttata*) feed young that show specific markings in their mouths when they exhibit gape responses for food; the parents are highly selective in this respect. Domesticants, however, will respond to gape responses without such specific signals. The domestic turkey is descended from the subspecies *Meleagris gallopavo gallopavo* which lives in the Mexican highlands. There are differences in both physical morphology and behaviour between the wild turkeys and the domesticated ones. Wild turkeys have larger brains and are livelier and more vigilant than the domestic form. In domestic turkeys sexual behaviour occurs at the age of one year, as compared with two years in the wild turkey, an obvious benefit to a turkey

breeder. However, in Missouri, for example, they have their young before the weather is mild enough for the chicks, and so are dependent upon human care in order to raise their offspring. In addition, domestic turkeys do not hide their nests as carefully as their wild relations. Upon hearing the mother's warning-call the chicks of wild turkeys immediately become motionless, whereas chicks of domestic turkeys, or even of crosses between wild and domestic ones, continue moving around, and so may attract a predator's attention. Like the domestic fowl, the domestic turkey has lost the abilities necessary to survive in the wild.

These examples are instructive in indicating the kinds of patterns of behaviour which can be modified through domestication. In some species, all aspects of the biology of the animal may be altered by the processes of domestication, in complex interacting ways. Some investigations have been addressed to these complex interactions; their results are not conclusive in detail, but they do indicate the complexities involved. In this context the Norway rat (*Rattus norvegicus*) has been studied in great detail in the wild and in captivity, and the domesticants are much used as LABORATORY animals in a variety of behavioural and other kinds of biological research. Some specific examples of behavioural comparisons of wild and domestic stocks of this species will now be given in order to demonstrate the complexities involved in making such comparisons.

The Norway rat: a special case for comparison. Some early work on this topic, beginning in the 1940s, compared a number of physiological, anatomical, and behavioural measures between wild subjects and domesticants, and concluded that there was an increase in those activities concerned with the gonadal (sex) hormones, and a decrease in those activities concerned with aggression and FEAR, which depend upon adrenocortical (stress) hormones. As a result of these and other observations, much of the behaviour of domesticants has traditionally been regarded as 'degenerate' when compared with that of their wild counterparts. The situation is far more complex, however, than it might at first sight appear to be. With reference to LEARNING abilities, for instance, a number of potentially confounding variables may hinder the assessment of comparative abilities. For example, wild rats are often more cautious in the artificial conditions of laboratory learning situations than are domesticants, to the extent that they do not learn the task. In such cases, little can be said about their capacity to learn, only that they are too timid to learn. It is also clear that much depends upon the nature of the task when comparing the learning abilities of wild and domestic animals. It is likely to be extremely difficult to show conclusively that there are differences in ability, and little can be concluded from learning-tasks which have been designed for laboratory domesticants.

Another way of approaching this problem is to examine the relative requirements of the animals in their natural environments. Wild rats learn some things very quickly; an excellent example is in learning to avoid areas and food substances associated with poisons (see FOOD SELECTION). Wild rats may sometimes learn this by observing the behaviour of other rats (see IMITATION). Learning is rapid and is clearly advantageous. On the other hand, it is also advantageous that, once learnt, responses of this kind are retained and acted upon in a persistent manner. Stereotyped patterns of daily movements are advantageous in the kind of ecological NICHE in which wild rats live. In contrast, the processes of domestication may well favour relatively flexible behaviour, by initially breeding from animals that are most able to adapt to captive conditions and hence are inherently more flexible. It may also be that freedom from searching for food and avoiding predation, for example, may lead to increased exploration of novel situations. It would certainly seem reasonable to suppose that wild and domestic rats have broadly different constraints upon their learning abilities. Wild rats obviously need to learn, they have also to exhibit behavioural inflexibility; domestic rats may show increased learning abilities through selection for such. In any case, wild rats and domesticants may learn different things and in different ways.

With regard to aspects of their SOCIAL behaviour, comparisons between wild and domestic rats are interesting, if also somewhat difficult to evaluate. Again, and in line with the traditional view of relative degeneracy, much of the social behaviour of domestic rats is thought to be 'watered down'. Investigators have shown that some social behaviour is, in fact, different in wild and domesticated rats. Wild rats, for example, have been found to give more social signals such as threats, to fight more fiercely, and to keep a greater social distance from one another, than domesticants. On the other hand, domesticants show greater general sexual activity than wild rats. Both these findings fit in with the notion of altered hormonal balance in domesticated stocks, as mentioned above. It has been emphasized, however, that there are overt differences in the intensity of the behaviour expressed; hence, they may reflect either genetic changes through the processes of domestication, or that the domesticants respond at higher thresholds. A set of experiments in this context showed that although fighting and death occurred within groups of wild rats, this was not the case among groups of domestic rats. Interestingly, however, when mixed groups of wild and domestic rats were together and competed for water, the aggressive and injurious behaviour of the domesticants increased, albeit to a less marked extent than that of the wild animals; moreover, wild rats were thought to be responsible for the death of some domesticants. Hence, one could argue that social conditions acted to make the domesticants more responsive. On the other hand, the relatively poor repertoire of aggressive and

social behaviour among the domesticants may also argue for a real change in some patterns of behaviour during the processes of domestication. Other experiments have shown that domesticants dominated wild rats which were bred in captivity, in both competitive and non-competitive social conditions. An important difference between the two situations was that in the first case the animals used were taken from the wild, whereas in the second case the subjects were wild animals bred in captivity. Hence, it appeared that the conditions of early experience of wild subjects could make a difference to the quality of their social responsiveness. Moreover, it has been shown in other studies that wild caught male rats are more aggressive than the captive reared male offspring of wild caught animals. Hence, it has been suggested that wild animals bred in captivity are being inadvertently selected for lower levels of aggression. Alternatively, it may well be that the differences between FIELD and laboratory wild rats may be exaggerated by the relatively sudden transition between living in nature, and in captivity.

From even these brief examples, the relative complexity of making comparisons between wild animals and domesticants may be indicated. Many of the contemporary questions about the influences of the processes of domestication obviously involve a variety of interdisciplinary studies. Many pressing problems remain, in terms of both academic interest and economic performance. The history of domestication spans many thousands of years, yet it is within the present century that our ability to modify animal species by selective breeding and other forms of management, has become a highly specialized aspect of scientific endeavour. Moreover, with the increasing human population, and relative lack of food resources and space, the problems of domestication involve us in some of the most economically important questions of our time.   H.O.B
4, 64

DOMINANCE is a feature of SOCIAL ORGANIZATION in which some individuals acquire a high status, usually as a result of AGGRESSION, while other individuals retain a low status. Dominance relationships were first noticed in flocks of domestic fowl (*Gallus gallus domesticus*), in which dominant individuals tend to peck subordinate individuals when they come within range. In a stable flock, the individuals learn to recognize each other, and a *peck-order*, or dominance hierarchy becomes established. Dominant individuals are generally able to use their status to gain priority in access to resources, such as food, roosting sites, etc.

Dominance relationships are widespread in the animal kingdom, and have certain features in common in many species. They are generally established in aggressive encounters. If a particular individual clearly wins a fight, or a ritualized threat contest (see RITUALIZATION), then the encounter may not be repeated on subsequent occasions, but instead the defeated individual may immediately show submission. If the opponents are more evenly matched, then a number of encounters may be necessary to establish a dominance relationship. Once a dominance relationship is established, there may be no further fighting, or overt aggressive DISPLAY. When a dominant animal approaches, a subordinate will often move away, frequently showing typical APPEASEMENT behaviour, but sometimes without any obvious signs of FEAR. Thus, the dominant animal may simply supplant a subordinate at a feeding site, or other desirable position. In FIELD STUDIES the dominance relationships are sometimes so subtle that it is difficult for an observer to establish the relationship between two individuals, unless it is possible to observe an encounter between them. The problem is that dominance relationships are such that encounters are often avoided by the subordinate animal. In LABORATORY STUDIES, staged encounters between pairs of animals are sometimes used to establish dominance relationships. This involves removing the pair from the group and placing them together in a restricted space. In such circumstance it is usually possible to tell which is the dominant animal. However, laboratory studies may not give results that can be reliably translated to the natural situation.

Dominance relationships are generally established by LEARNING, but in some species there may be immediate outward signs of dominance. Dominant individuals may differ in both appearance and behaviour from subordinates. In domestic fowl, for example, dominant males have a larger body size, and a more prominent comb and wattle than subordinates. These features are greatly influenced by male sex HORMONES, and the hormonal state of the individual may, in turn, be affected by its dominance status. In many mammals, body size and adornments, such as horns and manes, are signs of male dominance.

The status of dominance confers many advantages in permitting unchallenged access to limited resources, such as food, space, females, etc. In many social species the dominant males perform most of the matings. For example, in a study of elephant seal (*Mirounga occidentalis*) it was found that the most dominant 6 per cent of the bulls inseminated 88 per cent of the females. The GENETIC contribution of dominant males is so great that it is surprising that subordinate characteristics survive in the population. However, in some species, such as the red deer (*Cervus elaphus*), the dominant males are so busy maintaining their position, that subordinate males are sometimes able to steal an opportunity to mate with one of the females. In other species the mortality rate amongst rival dominant males is fairly high, and subordinate males thus gain a chance to improve their status. For example, among breeding northern fur seals (*Callorhinus ursinus*), the dominant males hold territories that they defend against

rivals day and night for a period of up to two months without food. As a result of fighting, the mortality rate among dominant males is three times that of females. Among red deer, also, the males fight frequently and do not eat during the rut (mating season). They do not often die in combat, but they may be so weakened that they die during the following winter.

Subordinate animals, by employing appeasement and submissive postures, are often able to remain in the vicinity of dominant males, and to profit from opportunities to gain access to food, or to females. They also benefit from membership of the group in terms of defence against predators. In some species, the immature and subordinate males may detach themselves from the group, as in lions (*Panthera leo*), or remain on the periphery of the group, as in baboons (*Papio*).

In many polygynous species, maturity is delayed in the males. As the dominant males age, however, they are challenged by young mature males, and eventually displaced. A successful challenge may result in a long-term dominant status, as in baboons, or it may confer purely seasonal advantages. In red deer, for example, dominance relationships are probably established anew at the beginning of each rut.

In groups of relatively stable composition that travel as a unit, dominance relationships tend to be independent of the location of the group, and to take the form of a stable linear hierarchy. For example, among winter feeding flocks of small birds that range over a fairly extensive area, such as mountain chickadees (*Parus gambeli*), the rank order remains unchanged at different locations within the HOME RANGE of the flock (see FLOCKING). In contrast, some permanent resident species, such as Steller's jay (*Cyanocitta stelleri*), have dominance hierarchies that are highly dependent upon location: the rank of a particular male at a feeding site is dependent upon the distance of the feeding site from the nest of that male. This is true at all seasons, and for all feeding sites within a locality. Thus male A may dominate male B at feeding site I, which is nearer to the nest of male A, but at feeding site II, nearer to male B's nest, male B will dominate male A. Similar site-dependent dominance relationships have been observed in woodchucks (*Marmota monax*), chipmunks (*Tamias striatus*), and octopus (*Octopus cyanea*).

In large groups with changeable composition, as in white-fronted geese (*Anser albifrons*) and Canada geese (*Branta canadensis*), individual recognition is difficult, and stable dominance relationships do not form. Instead, THREAT displays are more widely employed, and fighting may occur. Among these geese there seem to be some generalized rules in that the outcome of encounters usually corresponds to the following order: parents with young > paired adults > yearlings in families > single adults > unattached yearlings. In many species the adult individuals are indistinguishable from each other, and play similar roles. In other species, particularly among social insects and primates, there are marked differences between individuals that are based upon age, rank, and sex. The individuals may play specialized roles, which contribute to a complex social organization.

DREAMING. During human sleep there are phases during which the careful observer may notice rapid movements of the eyes. Physiological measurements indicate that the rates of breathing and heartbeat are elevated during these phases. Moreover the electrical activity of the BRAIN shows a characteristic pattern. Paradoxically, this pattern is similar to that normally associated with wakefulness, even though it occurs during deep SLEEP from which the person is not easily aroused. Scientists distinguish between quiet and active sleep. During quiet sleep, there are no rapid eye movements and the electrical activity of the brain is different from that of the awake person. During active sleep there are rapid eye movements, some physiological AROUSAL, and the activity of the brain resembles that of the awake person. When people are awakened from active sleep they usually report that they were dreaming, whereas they do not usually report dreams when awakened from quiet sleep. It seems, therefore, that the rapid eye movements and other physiological changes are associated with dreams.

Many sleeping animals show rapid eye movements that are accompanied by physiological activity similar to that found in human beings. We cannot wake animals and ask them about their dreams, but scientists can detect many of the phenomena that accompany human dreams. On the basis of the evidence many scientists are willing to agree that animals probably experience dreams that are akin to those of human beings.

In general, young mammals spend a high proportion of their sleep in active sleep, sometimes as much as 100 per cent. In adulthood the percentage of active sleep does not usually exceed 25 per cent. An alternation of quiet sleep and active sleep occurs in many mammals and birds, but not in other vertebrates. Reptiles and fish do not have two distinct types of sleep, but seem to have a single form of sleep that is intermediate between quiet and active sleep. Among the mammals the proportion of active sleep is strongly correlated with metabolic rate, and some scientists believe that the changes in brain activity that occur during sleep are connected with THERMOREGULATION. There is some evidence that mental activity occurs throughout all phases of sleep, but that it is more vivid during active sleep simply because of the greater activity of the brain.

67, 104, 142

DRINKING is an essential activity for most terrestrial animals. The water that is continuously lost through the processes of THERMOREGULATION, excretion etc. must be replaced. Many species are able to conserve water by reducing the rate of loss when they are dehydrated, but this does not make up for water already lost. Some desert-living species, such as the budgerigar (*Melopsittacus undulatus*), kangaroo rat (*Dipodomys*), and gerbil (*Gerbillus*), can live entirely on dry food without drinking water. They eat mainly seeds which contain about 10 per cent free water, in addition to water that is chemically bound and released only in the process of their metabolism. It is because the water conservation mechanisms of these animals are so good that they can survive entirely on the water contained in their food (see THIRST). Some insects, such as the Mediterranean flour moth (*Anagasta kuehniella*) and mealworms (larvae of the darkling beetle, *Tenebrio molitor*), live without any source of water, other than that contained in their food. Indeed, mealworms eat more in dry conditions than in moist conditions. The extra food is taken for its water content, the excess food being passed through the gut undigested.

Some animals are able to absorb water through the skin. Amphibians obtain all their water in this way. The Australian desert lizard, the moloch (*Moloch horridus*), was thought to obtain water in this way. However, it has capillary grooves in the skin which lead to the mouth. When the animal sits in water, the grooves conduct water to the mouth where it is swallowed. If the mouth is closed with adhesive tape no water enters the body. Desert sand-dune beetles, many in the family Tenebrionidae (darkling beetles), and some puff adders (*Bitis*) rely on condensation of fog (see CLIMATE) to form droplets on their body surfaces, which are then taken in by drinking. There are some insects which can even absorb water from moist air. Only larval forms and wingless adults seem to possess this ability. Presumably winged insects can always fly to a source of water. The fire-brat (*Thermobia domestica*) is thought to absorb water through its rectum, and it is probable that other insects employ other parts of the body for water absorption. Sandgrouse (*Pterocles*) live in the desert and may fly 50 km to obtain water. When drinking, they normally stand in shallow water and allow their breast feathers to become saturated with water. A male sandgrouse may then fly to the nest and allow his chicks to suck water from his feathers. Until they can fly, young sandgrouse obtain water in no other way. In the laboratory they will not take drinking water, but will suck water from wet cotton.

Animals which spend long periods without water often have the capacity to drink enormous quantities of water when the opportunity arises. A very thirsty man can drink about a litre of water in a minute, and 2–3 litres in 10 min. A thirsty camel (Camelidae) has a much greater drinking capacity, being able to drink about 30 per cent of its body weight. A male weighing 325 kg was observed to drink 104 litres of water, and a female weighing 201 kg drank 66 litres. Contrary to popular belief, camels do not store water. Members of the pigeon family (Columbidae) are characterized by their method of drinking. They pump water up by suction, whereas most birds take a beakful of water, raise their heads, and rely on gravity to assist in swallowing the water.

Most animals can drink only water with a low salt concentration, but some have special physiological adaptations which enable them to tolerate high concentrations of salt in their drinking water. Some reptiles and birds have special salt glands which secrete salt in a highly concentrated solution. These are found in many desert lizards, and birds such as roadrunners (*Geococcyx*) and sand partridges (*Ammoperdix*). They are also common in marine reptiles, such as turtles (Cheloniidae), sea snakes (Hydrophiidae), and the marine iguana (*Amblyrhynchus cristatus*), as well as in many sea birds. Several freshwater birds, such as ducks (Anatinae) and flamingoes (Phoenicopteridae), also have salt glands. Some animals which do not have salt glands do, nevertheless, have a high salt TOLERANCE. Subspecies of the zebra finch (*Taeniopygia guttata*) which live in salt marshes are able to drink very salt water, and excrete the excess salt in their urine. Many desert rodents have similar abilities. Some animals have special physiological adaptations to particular types of water supply. For example, the pack rat (*Neotoma*) can use cactus as a source of water, but this water is poisonous to other rats. Pack rats are able to metabolize the oxalic acid present in cactus, whereas for most mammals this is toxic.

Although most animals drink in response to various forms of dehydration, this is not the only, or even the most usual, everyday stimulus for drinking. Many animals drink with their meals, FEEDING behaviour being a direct stimulus for drinking. In this way animals drink in anticipation of the possible dehydrating effects of food intake. Thus an animal living on an established eating and drinking daily routine may never become dehydrated. Animals may also drink in direct response to temperature changes. The processes of thermoregulation often involve water loss, and animals can easily become dehydrated in a hot environment. It has been found that both rats (*Rattus*) and pigeons will drink when exposed to a hot environment, in anticipation of dehydration. Although drinking patterns are often in synchrony with those of feeding, the two may become disassociated under certain conditions. Feeding patterns are often driven by the animal's internal CLOCK producing a typical daily pattern of feeding and drinking. If access to food is restricted to certain parts of the day, the daily RHYTHM of drinking may continue. Animals tend to eat and drink in synchrony when both food and water are readily available, but if the cost of changing between feeding and drinking is high, either in terms of distance to be travelled, or because of the presence of predators near the water hole, then the normal pattern may be disrupted. In nature animals cannot afford to drink without taking into account all of its consequences (see DECISION-MAKING).

DRIVE, as a concept, arose out of the idea of INSTINCT as a 'prime mover' responsible for the MOTIVATION of animal behaviour. The influential psychologist William McDougall, in his book *An Introduction to Social Psychology* (1908), held that instincts were irrational and compelling sources of conduct which oriented the organism towards its goals. He postulated a number of instincts, most of which had a corresponding EMOTION. Examples are: flight and the emotion of fear, repulsion and the emotion of disgust, curiosity and the emotion of wonder, pugnacity and the emotion of anger. Robert Woodworth, in his *Dynamic Psychology* (1918), was the first to introduce the term drive. Woodworth distinguished between the energizing and the directing aspects of motivation. He used the term drive to denote the psychological force, which he thought to be necessary to power the behavioural mechanisms into action.

The idea of drive as an energizer has been very influential in psychology, but it has fallen out of scientific use for two main reasons. Firstly, the drive concept involves a misuse of the concept of force. Although the notion of psychological forces can be used legitimately in a descriptive sense, many early psychologists based their views upon a fundamental misconception of the concepts of force, power, and energy, as used in science in general. The problem arises from the fact that, in the physical sciences, ENERGY is not a causal agent, but a descriptive term arising from mathematically formulated laws. The early psychologists were anxious to develop a mechanistic psychology, but their ideas about the energizing aspects of behaviour were based upon conceptions of instinct that were derived from the subjective emotional experiences of man.

The second major problem arising from the concept of drive was empirical. Some early psychologists sought to identify a drive for every aspect of behaviour. Thus they would define a hunger drive responsible for energizing feeding behaviour, a thirst drive, a sex drive, etc. The question of how many drives there should be proved difficult, since there were always some aspects of behaviour that did not seem to fit into the existing CLASSIFICATION of behaviour. Some psychologists advocated a specific drive for every element of behaviour. Thus there might be a tail-biting drive, a thumb-sucking drive, etc. Others claimed that there was a single *general drive* responsible for energizing all behaviour. This theory led to empirical predictions that were easily disproved in LABORATORY STUDIES. Gradually the whole drive concept fell into disrepute, and its place was taken by other approaches to motivation.

16

DRUGS AND BEHAVIOUR. One of the most

important ways in which the study of animal behaviour has proved to be of benefit to man is in the discovery and testing of drugs, which have revolutionized psychiatric medicine in recent decades. In this account of the interactions between drugs, brain chemistry, and behaviour an attempt is made to illustrate the use of behavioural methods for characterizing various groups of psychoactive drugs; to show how brain chemistry is involved in drug action; and to suggest that some ABNORMAL behaviour in man may result from endogenous disorders of the brain.

Chemical coding in the brain. The BRAIN, like all other organs, is made up of cells, the *neurones*, which are organized and connected in various and complex ways in its different functional areas. The brain initiates behaviour, and major advances have been made in understanding how different processes, like SLEEP, PERCEPTION, MEMORY, and movement are controlled.

A drug is a substance administered to change the *interior milieu* (see HOMEOSTASIS). Most drugs distribute throughout the body, including the brain, if the drug molecules are sufficiently small to penetrate the blood–brain barrier. Drugs may alter brain function in many different ways, and result in changed behaviour. For example, by: (i) altering the general metabolism of the neurones; (ii) altering the excitability of the neuronal membrane; (iii) blocking electrical conductance of the neuronal impulse along the neurone *axon* (the process along which a nerve impulse passes) ; (iv) reducing the blood supply to the neurones; (v) altering the distribution of water and ions between the neurones and the surrounding tissue. If the drug influences neuronal function, then changes in behaviour will be observed. If these influences are widespread, the effects on behaviour will tend to be general, rather than specific.

Almost any drug will have effects on one or more of these general aspects of neuronal function, although these effects may be relatively insignificant, and not produce detectable changes in awareness or observed behaviour. These effects may be due to direct influences of the drug on brain tissue, but more often occur indirectly as a secondary response to changes in body functions. Such drug effects are rarely studied, although the situation probably arises more often than in the case of drugs with more specific actions which are taken directly to modify behaviour. However, some drugs, taken for reasons unrelated to the brain function, have a more direct effect on the central nervous system, and alter perception, memory, and even 'mood' in some people. In this article, however, it is the intention to discuss only drugs which act on the brain to influence behaviour directly, and thus are of value in controlling cerebral function in man.

Communication between neurones occurs at *synapses*, gaps in the transmission system across which signals are carried by electrical or chemical messengers. Chemical synaptic transmission in the brain is a relatively recent discovery, although it had been known for many years that it occurred in the nerves serving the body muscles and organs. More important still are discoveries that in the brain several different chemical neurotransmitters exist, and, furthermore, that these chemicals are not randomly distributed throughout the brain. Different brain pathways use different neurochemical messengers, and thus, chemical coding represents an important aspect of brain organization. A variety of methods are used to prove that chemical coding occurs in a given brain pathway. Extraction methods are required to separate the chemical under investigation from related biochemical substrates in the neurone. Biochemical assay methods are used to measure how much of the compound is present in a given brain area, and histological methods are developed to visualize the chemical *in situ*. Using a combination of such techniques, it has been established that the following compounds function as chemical neurotransmitters in the mammalian brain: *acetylcholine* (ACh); *noradrenaline* (NA); *dopamine* (DA); *serotonin* (5HT); *γ amino-butyric acid* (GABA); and also peptide transmitters. The various transmitters are synthesized and stored in well-defined but different anatomical systems, although a given structure may receive input from several different transmitters via different pathways.

Neurotransmitters carry specific commands of an *excitatory* or *inhibitory* nature across the synapse. They may also serve a more general modulatory role in certain areas of the brain, to set levels of activity and responsiveness, rather than to carry specific messages. It is not known for certain what are the roles of a given transmitter in the different brain areas in which it is found. For example, GABA is largely inhibitory where it has been studied; by contrast NA may be excitatory or inhibitory. If a drug alters the function of one or more transmitters it can be appreciated that it may result in widespread and complex changes in the neurochemistry of the brain. It is accordingly difficult to determine the neurochemical basis of a given effect of a drug on behaviour. For example, why do some people feel 'good' after an alcoholic drink, or have frightening visual hallucinations after taking lysergic acid, LSD?

Neurology and psychiatry are the branches of medicine concerned with abnormal brain function and behaviour. Some of the most important discoveries concern the use of drugs to influence nervous activity specifically, and thus promote normal neuronal activity and interaction. Many of these drugs were discovered quite by chance to be of value in treating nervous illness; others were developed by medical chemists who systematically modify a successful molecule in order to find another and more potent compound. In several cases these drugs do not have diffuse influences on neurones throughout the brain, but alter the chemical transmitter substances in specific locations. The discovery that drugs can normalize behaviour by an effect on a chemical system in the brain has resulted in determined efforts to develop highly specific drugs for clinical use, and has reinforced the view that abnormal behaviour in man may have its origins in spontaneous disturbances of the brain's chemistry.

Drugs and animal behaviour. A drug company faced with a new compound of potential use in changing mental function in man uses animal model systems, both pharmacological and behavioural, to characterize the compound. In order to devise controlled testing situations for animals, it is necessary to understand the factors which normally control behaviour. Animal and human behaviour is controlled in a predictable manner by a number of variables in the environment. This is the subject matter of the animal branch of experimental psychology, which attempts to define the variables controlling animal behaviour and their interaction one with another. From these studies there are some general behavioural classifications which have emerged and stood the test of time; one of which is that some behaviour is unlearned, whereas other behaviour has to be learned.

The term INSTINCT is often used in the context of unlearned behaviour. The animal from birth onwards exhibits certain behaviour patterns, peculiar to its species, without the necessity of LEARNING. Most of this behaviour is associated with processes essential to the survival of the individual, and to its reproductive success. For example, at the appropriate time in their development animals breed, the male and the female showing for the first time the relevant behaviour patterns for that purpose.

Learned behaviour on the other hand appears and changes in accordance with the experience of the animal. Two types of learning may be distinguished. One is called classical CONDITIONING, and is typified by PAVLOV's experiments on salivation, in which a dog (*Canis l. familiaris*), having salivated several times to the sight and smell of food presented with a bell, will continue to salivate when the bell is presented alone. The other process, typified by the rat (*Rattus norvegicus*) in a lever pressing situation, is termed OPERANT conditioning. If food is given each time a rat approaches the lever, behaviour is quickly directed to the lever and soon the rat will reliably press the lever to obtain food (the REWARD or REINFORCEMENT). Animals will also learn responses to escape from or avoid unpleasant consequences. Learning is thus dependent on the presentation of a positive or negative reward. SKINNER was the first to realize that the pattern of presentation of the reward determined the pattern of responses made by the animal. REINFORCEMENT SCHEDULES have proved of immense practical value in setting up stable patterns of learned responses in animals and man. If an unpleasant stimulus immediately follows an operant response, the procedure is described as punishment. Mild electric shock is a commonly used noxious stimulus, but the presentation of loud noises and air puffs to the face are also effective in some animals.

The traditional approach in studying drugs and behaviour attempts to classify drugs by their effects on one or other of these categories of response. However, while such an approach has provided an enormous amount of information, it has not in general proved of value in classifying the action of a particular group of drugs. For example, human depression is a complex behavioural response, which, no doubt, cuts broadly across the definitions presented. Drugs which relieve depression are therefore unlikely to act selectively to modify one aspect of behavioural control. The alternative approach is to devise animal models which mimic these complex human disorders, but this is equally difficult, and the animal tests used to characterize a particular group of drugs often bear little resemblance to the behavioural disorders in man which are modified by those drugs.

Responses to drugs in animals and man. With a background of brain chemistry and behaviour it is possible to discuss the major classes of psychoactive drugs used clinically to control brain function in man.

1. *Stimulant drugs.* Amphetamine, the classical stimulant drug, has several effects upon behaviour. In addition to facilitating AROUSAL and behavioural efficiency it induces *anorexia*. Some related compounds are more potent stimulants (e.g. methamphetamine), while others, like fenfluramine, induce anorexia but are weak stimulants, making them valuable for controlling weight in man. Extensive *in vitro* and *in vivo* neuropharmacological studies have been made on these stimulant drugs, and although they have effects on a number of neurotransmitter systems their principal action appears to be on the neurotransmitters NA and DA. These are released by amphetamine-like compounds and their re-uptake in the nerve terminal is inhibited.

Amphetamine stimulates both learned and unlearned behaviour in all species studied, including man. Locomotion is stimulated in rodents and many other species. After higher doses, stereotyped motor behaviour is observed in which unlearned responses peculiar to the species are repeated for long periods of time. Rats, for example, sniff, gnaw, and lick the bars of their cages. In the early stages of intoxication cats (*Felis catus*) show hypermobility, side-to-side-looking movements, and repetitive sniffing. Monkeys (Simiae) frequently show repetitive biting and examination, picking or probing at themselves. Particularly interesting are the descriptions of stereotypy in people taking large doses of stimulant drugs. They analyse details in their environment in a concentrated and repetitive manner. They have a compulsion to take objects apart and put them back together again. Women sort out their handbags again and again, or tidy up their home, whether it needs it or not. Mechanically minded individuals manipulate clocks or car engines.

Amphetamine also has marked effects on learned behaviour. In lever pressing situations, amphetamine facilitates both the acquisition of learned responses, and their performance when animals are already showing stable patterns of responding on particular schedules of *reinforcement*. One such schedule is termed the *fixed-interval schedule*, in which reinforcement is available at regular intervals. This pattern of reinforcement results in scalloped patterns of responding, with the animal pressing the lever infrequently immediately after a reinforcement, but with increasing frequency as the time of the next available reinforcement approaches. Amphetamine produces a marked stimulation of responding in the early parts of the response scallop. The motor stimulation induced by amphetamine is mediated by brain DA pathways. If these neuronal systems are lesioned in the rat, amphetamine is no longer able to stimulate locomotor behaviour or induce stereotyped behaviour.

Amphetamine is used by man to heighten arousal and intellectual function. As in the case of animals, people become more active under the influence of the drug, an effect one supposes is mediated by DA pathways. However, amphetamine also releases NA, and it is possible that the COGNITIVE effects of the drug in man principally involve NA systems. We do not have an equivalent means of assessing cognitive stimulation in animals. Amphetamine is also frequently abused by man, and doses of 400 mg or more may be taken each day. This induces a form of psychosis, which initially may be mis-diagnosed in the clinic as *paranoid schizophrenia*. The drug addicts have olfactory, visual, and auditory hallucinations and pronounced paranoia, although the thought disorder characteristic of schizophrenia is absent. The fact that amphetamine acts on brain DA and NA to induce a condition like paranoid schizophrenia, coupled with the knowledge that NA and DA receptor blocking drugs normalize schizophrenic symptoms, has encouraged the view that overactivity of these brain systems may contribute to the causation of schizophrenia. In support of this hypothesis is the observation that florid psychotic symptoms are intensified when schizophrenic patients receive small doses of amphetamine.

2. *Depressant drugs.* Many different classes of drug depress the central nervous system, disrupt behaviour, and, in sufficient doses, induce sleep or coma. Some drugs like the *neuroleptics* depress brain activity, although this is not the explanation of their effect on psychotic behaviour. Others like alcohol, barbiturates, and large doses of the minor tranquillizers are called *hypnotics*, and have as their principal action a depressant effect on the brain, and accordingly induce, or are used clinically to induce, sleep. All of these drug groups, including the neuroleptics, depress both learned and unlearned behaviour if given in sufficient amounts. The hypnotics however, have one unique behavioural property: notably that at some doses they act to restore behaviour suppressed by the presentation of a noxious stimulus. If, for example, a pigeon (*Columba livia*) has been trained to peck at a key for reward, and is subsequently always given a mild electric shock when it pecks the key for reward, the trained pecking virtually ceases. However, when treated with a barbiturate or a minor tranquillizer, the bird continues to respond for food, despite the presentation of punishing shock.

3. *Antipsychotic drugs.* The development of antipsychotic drugs provides a good example of the scientific approach to drug development. In 1946 reports emerged from a French drug company that certain drugs based on the *phenothiazine* structure showed combined anti-histaminic, and parasympathetic and sympathetic activity on the peripheral autonomic nervous system. One particular derivation, RP 4560, was used initially as an anaesthetic, and it was noted that when given intravenously it produced pronounced sedation without loss of consciousness. It was not until 1961, when the drug was subjected to clinical trial, that it was found to have great value in controlling psychotic symptoms in man, and was named *chlorpromazine*. A variety of phenothiazine derivatives were then synthesized.

The term psychotic is most commonly used in the context of schizophrenic illness, but includes psychopathological states which have in common hypermobility, abnormal initiatives, and increased affective tension. These are the types of severe psychiatric disorders which result in hospitalization. As medical facilities developed after 1945, the number of hospitalized patients in the western world rose dramatically. The marked decline since 1955 is attributed to the growing use of phenothiazine drugs. While it has been argued that these drugs control symptoms rather than cure the schizophrenic illness, the improvement is often so dramatic and long lasting that many patients are able to return to a relatively normal life in society. The dosage depends on the response of the patient, and may be as high as 3600 mg daily. Methods have been introduced by which large doses of antipsychotic drugs are given intramuscularly in an oil suspension, and are released slowly into the body over a two-week period. This method results in sustained rather than fluctuating body levels of the drug, and also alleviates the trauma to the patient of chronic dosage regimes. Unfortunately, the chronic use of phenothiazines results in the development of *tardive dyskinesia*, a motor disorder in which the patient shows uncontrollable movements particularly of the head, mouth, and neck. Efforts are being made to develop complementary therapy to prevent their occurrence.

Studies of the effects of phenothiazines on animals have played an important role in the development and understanding of this drug group, and also provide animal models with which to assess potentially useful new compounds. The phenothiazines and succeeding antipsychotic drugs were classed as neuroleptic drugs, i.e. depressant on the nervous system, and first assumptions were that this effect explained their antipsychotic action. However, this seemed unlikely, as the major class of depressant drugs, the barbiturates, whilst producing sedation were not effective in ameliorating psychotic symptoms. Phenothiazines were originally studied in a shock avoidance task, in which a rat was required to climb a pole when a tone sounded in order to avoid a mild electric shock to the feet. After moderate doses of chlorpromazine, the rat no longer responded to the tone, but did respond to the shock by climbing the pole. This effect on one response, while leaving closely related responses unaffected, demonstrates a selectivity of action of phenothiazines, in contrast with the action of barbiturates, where both avoidance and escape responses are impaired.

4. *Antidepressant drugs.* Depression, together with schizophrenia, represents the major endogenous psychoses. Animal models played a major role in defining the action of antischizophrenic drugs, and continue to be of importance in the search for new drugs, but this is not so for the antidepressant drugs.

Antidepressants were discovered almost by chance. *Iproniazid* was introduced in the early 1950s for the treatment of tuberculosis. It was noted in some patients that a sense of well-being and mood elevation appeared shortly after the start of therapy, and the drug continued to be used for treating depression after it was superseded in the treatment of tuberculosis. Soon afterwards *imipramine*, a drug developed by modifying the structure of chlorpromazine, was reported to have antidepressant activity. Neuropharmacological work on *in vitro* systems revealed that iproniazid inhibited the enzyme monoamine oxidase (MAO) involved in the degradation of neurotransmitters in the synapse. This inhibition results in accumulation of the transmitter substance in the synaptic region, and, it is inferred, increases functional efficiency. Imipramine, one of a class of tricyclic antidepressants, inhibits the re-uptake of released neurotransmitters, and is assumed to have a similar function consequence to MAO inhibitors. These neuropharmacological properties of the antidepressants probably account for the fact that they potentiate the behavioural effects of amphetamine on both learned and unlearned behaviour. As amphetamine acts by releasing neurotransmitters, the antidepressants enhance the effects of the released neurotransmitters by slowing their removal from the synapse. However, no tests are yet documented in which animals have been rendered 'depressed', and the drugs used to reinstate normal behaviour.

Various attempts have been made; for example, rats have been exposed to a schedule of reinforcement where, after a period of low reward, much larger rewards are received and vice versa. The ensuing behaviour of the rat might indicate 'elation' and 'depression', and yet antidepressants do not have an effect on those response swings. Depression in children is very clearly observed after maternal separation or loss. This *anaclitic* depression in infants has been investigated in studies of young monkeys separated from their mothers at birth, but so far it has not been possible to normalize this behaviour with classical antidepressant drugs. An animal model of depression thus presents a real challenge to behavioural pharmacology.

The tricyclic and MAO inhibitors have been of great value in treating *monopolar* depressions. However, they have not been successful in treating *bipolar* depression, a form of the illness in which the patient's mood oscillates between depression and mania. There are no animal models for this condition, and the introduction of treatment with lithium occurred somewhat by chance. It had been noted that lithium had a calming effect on animals, and clinical trials of various classes of psychiatric patients were begun in 1949. However, animals required very high doses to show sedation, and there was a general feeling that any effects in man were due to the toxic effects of the drug. Lithium was not re-evaluated until the 1960s when the antidepressant compounds discussed earlier were introduced and found to be relatively ineffective in treating mania. Further trials in Denmark confirmed that lithium in doses of 250 mg three times daily provided the most effective control of mania yet seen in psychiatry. The neuropharmacological and psychological basis for this action of lithium remains unknown.

5. *Anti-anxiety drugs.* The hypnotic drugs referred to in an earlier section were traditionally used to relieve anxiety and neuroses in man, and to improve the disturbed sleep patterns usually associated with these states. Alcohol is a traditional remedy for anxiety, but excessive alcohol consumption results in hazards to health by interfering with liver function and general metabolism. Barbiturates, which until recently were the only drugs commonly prescribed by doctors for the relief of anxiety and insomnia, carry the associated risk of death by overdose. In the 1950s meprobromate, the first of the drugs which came to be known as minor tranquillizers, was developed. It was noted that the beneficial effect of this drug on anxiety tension was related to its muscle relaxant properties, rather than to the hypnotic sedative effect, as in the case of barbiturates. Some years later another group of drugs with muscle relaxant properties, the benzodiazepines, was developed. Like the hypnotics, these compounds in animals restore behaviour suppressed by punishment. This effect is not attributable to a non-specific depressant or stimulatory action of these drugs. Neuroleptic drugs depress and stimulant drugs enhance brain activity, yet neither of these drug groups re-

stored punished responding. Thus behavioural tests do provide a useful means of identifying and evaluating potentially useful anti-anxiety drugs. Attempts are being made to find other behavioural tests which do not involve the presentation of a noxious stimulus. For example, novelty or a new social interaction are potentially anxiety-inducing to an animal. In both cases uncertainty exists about the nature and the time and place of arrival of stimuli, and tests have been devised in which the drugs are able to relieve the behavioural disruption induced by such manipulations of the environment. If rats are given access to free food in a situation where they are responding for food, the free food is not accepted. Benzodiazepine drugs allow the rat to approach the free food and consume it.

Other workers have explored SOCIAL situations for inducing anxiety. Two male rats placed together in a familiar environment normally show a high level of social interaction. If, however, the test box is unfamiliar or brightly illuminated, social interaction is markedly reduced. Drugs which relieve anxiety, like *chlordiazepoxide*, reverse this dampening of social behaviour. Interestingly, depletion of the brain neurotransmitter *serotonin* blocks the anti-anxiety effect of chlordiazepoxide, suggesting that this brain neurotransmitter system is involved in the processes of anxiety, and in the action of the anti-anxiety drugs.

The effects of drugs on the behaviour of animals are extremely important in medical research. In addition to the direct medical benefits, some knowledge is obtained about the animals' brains and their behaviour. Although this knowledge is not generally the main objective of these LABORATORY STUDIES, it can be of value to students of ETHOLOGY. On the other side of the coin, research in animal behaviour can be of great value to scientists working on drugs, and can sometimes suggest methods of study that are beneficial to the WELFARE of the animals concerned. S.D.I.

134

E

ECHO-LOCATION is a means of ORIENTATION in which animals utter high-pitched pulses of sound, and detect the presence of objects by the echoes produced. It has been studied mostly in bats (Chiroptera), but is also known in other animals that practise orientation in conditions of poor visibility. Echo-location has been demonstrated in dolphins (*Tursiops*) and other marine mammals, in oilbirds (*Steatornis caripensis*), and in the Himalayan cave swiftlet (*Collocalia brevirostris*), which roost and nest in caves that are frequently too dark for vision. These birds can avoid large obstacles while in flight, even in total darkness, but if their ears are plugged they lose this ability. Shrews (*Sorex*, *Blarina*) can locate large objects in darkness by a simple form of echo-location, as can the terrestrial tenrecs (Tenrecidae) of Madagascar.

Echo-location in bats is achieved by means of bursts of ultrasonic sound pulses. These generally have a short duration (5–15 ms) and high frequency (c. 20 000 Hz), and are beyond the hearing sensitivity of man. The brief pulses enable the bats to time the echoes accurately, and so determine the distance of the object producing the echo. The high frequencies can be beamed precisely, allowing resolution of small objects. In natural environments, the sounds produced, for instance, by the wind blowing through the trees, or by other animals, are of relatively low frequency. Thus, by using a high frequency sound for echo-location the bats are unlikely to be subject to interference from other sounds in the environment. LABORATORY experiments have shown that bats subjected to extraneous sound in the 20 000 Hz range begin to become disoriented and collide with objects.

The ultrasonic cries of bats are produced by a highly specialized *larynx*, and are emitted through special horn-shaped nostrils in horseshoe bats (*Rhinolophus*) and leaf-nosed bats (Phyllostomidae), or by the lips, as in naked-backed bats (*Pteronotus*). Since the sound of the loud pulse must be compared with its relatively faint echo, there are many special adaptations which facilitate the detection, timing, and localization of the echoes. Many bats, especially those that feed on insects which they catch on the wing, have relatively large ears, specially shaped to enhance directional sensitivity. The sensitivity of the ear is reduced by special muscles in the inner ear when each loud outgoing pulse is produced. In some bats (the Vespertilionidae) the pulses are so short that there is no overlap between the echo and the end of the outgoing pulse. Since echoes arrive more quickly from nearby objects, the duration of pulses is further shortened as an object is approached. In other bats, overlap between the pulses and the returning echoes does occur, and other means of enhancing echo detection are employed. For example, in the greater horseshoe bat (*Rhinolophus ferrumequinum*) the call is adjusted so that the frequency of the echo falls within the most sensitive part of the HEARING range, while the call itself is emitted at a frequency to which the bat is least sensitive. To achieve this the bat has to alter the frequency of its call in accordance with the distance of detected objects, a technique that utilizes the *Doppler-effect* by which principle the echo-frequency shifts as a consequence of relative motion of the bat and the object.

Many bats feed upon flying insects at night, and are able to capture them on the wing, detecting and tracking them by means of echo-location. Laboratory experiments with the little brown bat (*Myotis lucifugus*) show that it can fly through a fence of vertical wires 1.2 mm in diameter and spaced 24 cm apart. Even in complete darkness the bats could negotiate the fence without touching any of the wires. Bats have also been trained to catch small food particles thrown into the air in pitch darkness, and to discriminate inedible from edible objects on the basis of small differences in shape.

60

ELECTROMAGNETIC SENSES. Many lower organisms are known to orient themselves in artificial electric fields, but the biological significance and the sensory basis of this behaviour is obscure. A number of fish species, however, make use of their electrical sensitivity in the course of their normal life for the purpose of ORIENTATION and COMMUNICATION, and a considerable amount is known about the sensory system that gives them this ability. Two types of electrosensitive fish have to be distinguished: those that detect distortions of the earth's electrical fields, and those that detect distortions of weak electrical fields that are generated by the fish themselves, or by other members of their species by means of special electrical organs. These appear to be modified muscles, and sometimes modified nervous structures. Apart from these so-called weakly electric fish there are also some so-called strongly electric fish, examples being the electric ray (*Torpedo*), and the electric eel (*Electrophorus electricus*). These animals produce shocks capable of stunning prey or even predators. However, they do not seem to possess a specialized electric sense.

An example of electrosensitive fish that are not electric themselves is the dogfish *Scyliorhinus*, which is capable of detecting its prey fish, even when they are buried in the sand, by the local distortion of the geophysical electrical field that they cause. The SENSE ORGANS that are used for this are called the *ampullae Lorenzini*, named after the anatomist who discovered them. Interestingly these organs, which occur widely distributed over the body surface, but especially on the head of sharks (Selachii), were variously suspected of being thermoreceptors, osmoreceptors, or mechanoreceptors until it was established that their function was electroreceptive. They consist of clusters of several sensory cells recessed under the skin and are connected to the outside by a canal filled with modified cells having a low electrical resistance. They are thought to be evolutionary derivatives from lateral line organs (see MECHANICAL SENSES).

The other type of electrosensitive fish, the weakly electric fish, belong to the families Gymnarchidae (with one species, *Gymnarchus niloticus*), Mormyridae, and Gymnotidae (gymnotid eels). They generate their own electric fields, either in the form of continuously and regularly oscillating fields, or of variably spaced electric pulses. There are two types of electrosensitive receptors in these fish: ampulla receptors, rather similar to the ampullae Lorenzini, and tuberous receptors, of a slightly different structure. These receptors differ mainly in that the former respond to static or slowly changing electric fields, while the latter only respond to rapidly changing fields. Some species possess either one or the other type of receptor, others both. Behaviourally the receptors are used to detect, locate, and even discriminate objects in the water on the basis of the distortions that they impose on the electric fields generated by the electric organs. It is, for example, possible to train *Gymnarchus niloticus* to approach an electrically non-conductive stimulus object to obtain a food reward, and to avoid a conductive but otherwise identical stimulus object, the approach to which is punished with a mild electric shock. But these electric fish also respond to electrical discharges of other members of their own species, in the sense that they modify their own discharge rate to be different from that of neighbouring fish. Thus they avoid interference with their electrolocation, and, to some extent, communicate with each other by this means. *Gymnotus*, for example, signals aggressive intentions towards neighbours with bursts of some 250 pulses a second.

Sensitivity to magnetic fields has been demonstrated in a number of animals. Certain bacteria, placed in a drop of water under the microscope, tend to accumulate at its northern edge. If a magnet is brought near to the droplet, they orient according to the magnetic 'north' of its field. Electron-microscopic examination reveals that they contain two chain-like iron-rich structures (similar structures have been found in the abdomen of honey-bees, *Apis mellifera*, and in the retina of pigeons, *Columba livia*) which terminate in two bundles of *flagellae* or hairs, with which they propel themselves. The function of this magnetic orientation behaviour is not understood. It is possible that it helps them to reach the oxygen-poor, deep water in which they thrive: in northern latitudes, where these bacteria are found, the 'north' of the earth's magnetic field not only points northwards, but also downwards. Similar kinds of magnetically oriented behaviour has also been described in lower animals, such as turbellarian flatworms (Turbellaria) and snails (Gastropoda). The sensory organs involved, and the biological function of this behaviour, are unknown.

The earth's magnetic field is also known to cause a slight error in the orientation of the waggle dance of the foraging bee, the dance that it uses to communicate the direction of a food source to hive companions, and which is otherwise determined by the position of the sun and the pull of gravity. It may be that the earth's magnetic field is used in some aspect of NAVIGATION in bees.

The situation is different in birds. HOMING pigeons, with either magnets or electromagnetic coils attached to them, are disoriented when flying back to their home loft when the sky is overcast, so that they are unable to use the sun as an alternative cue for direction. Similarly, experiments with European robins (*Erithacus rubecula*) held in a circular cage revealed that, in the absence of other cues, they correctly orient their escape attempts for MIGRATION, guided only by the earth's magnetic field. If the earth's magnetic field is modified by superimposing an artificial magnetic field with electromagnetic coils, then the animal directs its escape efforts in a manner consonant with the resulting field. Interestingly, robins do not seem to be able to distinguish the 'north' and 'south' of a magnetic field. However they apparently can detect the downward inclination of the earth's magnetic field, which is northward in northern latitudes, and southward in southern latitudes. The sensory structures responsible for this magnetic orientation of birds are not known. J.D.

84

EMOTION is widely accepted as an important part of human experience, and as playing an important part in human behaviour. Nevertheless, there is very little agreement about how emotion works, or even what an emotion is. Part of the confusion arises from the uncertainties inherent in trying to understand the emotions of another person, or of an animal. More confusion derives from the fact that the term emotion means quite different things to different people. It may mean an inner feeling, an aroused state of the body, or the display of certain emotional behaviour. There are difficulties with each of these three approaches.

One difficulty with approaching emotion through subjective feelings is that the rich language of emotion does not correspond very well with the inner world of feelings. Most languages provide a great variety of words that can be applied to almost any situation to describe how we feel, or how we ought to feel. For example, in English we may speak of pity when we are saddened by another's suffering. The word pity refers in part to the particular shade of sadness, but it also refers in part to the relationship we have with the sufferer, such as superiority, and in part to his particular trouble. Our feeling of sadness may be deeply felt and genuine, but the word pity to describe it may be replaced with another word, such as sympathy or compassion or condolence, depending upon what the situation and our relationship with the other person require. The difficulty with the LANGUAGE of emotion, then, is that the words we use to describe our feelings are descriptive of much more than feelings; they refer as well to the SOCIAL situation in which the feelings arise. How then can we know what our feelings, themselves, really are? And how can we know what another person's feelings are? Even if we could surmise that someone feels pity, does that label apply to his feeling or to the social situation? And even if he tells us he feels pity, how can we be sure he is not just applying the emotional label to the situation?

There are difficulties in trying to approach the emotions as though they were objective states of bodily AROUSAL. One might think of each emotion as reflecting a different pattern of arousal, and thereby hope to assess emotion by some objective means. It might be possible to observe an emotional state by measuring respiration, heart rate, etc., as in a lie-detector test. The trouble with this approach is that while such body states as increased heart rate, or changes in HORMONE output, do give evidence of emotional arousal, they do not indicate what emotion the individual is experiencing. This uniformity in response is most apparent in animals. Most animals will react physiologically in essentially the same way whether the arousal is SEXUAL, FEAR provoking, or if there is the anticipation of PLAY or food. The internal body state is much the same in each case, so while the lie detector can reveal that there is emotional arousal, it cannot indicate what the emotion is.

There are also difficulties in approaching the emotions through overt displays of behaviour. The main purpose of using the language of emotion in dealing with behaviour is to provide an explanation of the behaviour. The emotion is assumed to be the underlying cause of the behavioural display. Suppose we see a young boy strike his sister. We may be able to make sense of this act if we assume

that there is in the boy an emotional state of anger. And this assumption may be entirely correct; the boy may have been angry. But without knowing more about the context of the act, without having some further evidence of the anger, the aggressive act by itself provides little justification for any assumption about emotion. It may be that the boy was only settling the score for some earlier act of the sister's. Or he may merely have been showing her who was boss. There are many reasons why one child might strike another besides being in an emotional state. We could only properly conclude that the boy was angry if we knew more about the context, for example if we knew what had provoked the blow, or if we had more behavioural evidence, such as that the boy pursued his sister to hit her again, or swore at her. The problem with inferring emotions from behaviour is therefore one of circularity. It does nothing to promote our understanding of some behaviour to attribute it to an emotion if our only evidence of the emotion is the very behaviour the emotion is supposed to explain.

Apart from these problems of attempting to assess emotion in terms of subjective, physiological, and behavioural manifestations, there is a deeper, more conceptual problem with emotion. While there are certain emotions that are assumed to generate particular behaviour, as anger is assumed to generate AGGRESSION, there is much other behaviour for which there do not seem to be any corresponding emotions. Thus, consider the man at work at his job, or the man busily eating his dinner. We do not have emotional labels to apply to these cases, even when we may be sure that the man dislikes working or enjoys his meal. Indeed, it seems that most of our ordinary activities occur without benefit of any underlying emotion, and that we do not need to refer to any emotion to explain their occurrence. It seems that we refer to emotion, and fall back upon emotional types of explanation, only when some behaviour out of the ordinary occurs. It is mainly the disruption of ordinary behaviour that appears to call for an emotional account. If a bird is busily NEST-BUILDING or FORAGING for food, we are not tempted to speak of emotion. The bird is just going about its business. But if it gives up building to fight with its mate, or if it breaks off foraging to seek cover, then we are inclined to speak of such emotions as aggression in the one case and fear in the other. It is apparent that emotion does not provide a general account of behaviour. Emotions are not the basic source of all behaviour, but more like the occasional disruptors of behaviour. It is therefore not surprising that students of both human behaviour and animal behaviour (ETHO-LOGY) have tended to steer away from emotions as an explanatory concept, and to emphasize instead other kinds of principles of MOTIVATION, such as DRIVE, mood, and TENDENCY.

One of the major differences between human emotion and animal emotion is that while human beings have a great variety of emotions (or at least use many different words to convey how they feel) animals are restricted to just a few basic emotions, as far as we can judge. While we have emotions to fit virtually any kind of social situation, animals probably only have emotions to deal with certain kinds of survival problems, and for which there is some strong adaptive pressure (see SURVIVAL VALUE). For example, we might expect animals to show fear because of the adaptive value of being frightened in a dangerous situation, but there is no reason to expect animals to feel pity because it is not clear that they would have any adaptive advantage if they did. Therefore, we do not have to be concerned with all the subtle shades of difference implied by the language of human emotion. If a domestic dog (*Canis lupus familiaris*) tears up the furniture when it is confined to the house all day, we do not have to suppose that the dog acted destructively because of a spiteful emotion. It is sufficient to suppose that the FRUSTRATION of confinement aroused aggression, and that the aggression was directed at whatever was available.

What is the basic emotional repertoire of animals? Anger probably occurs in a variety of species, if we may take the widespread occurrence of aggression as evidence of anger. Note that the attack of a predator upon its prey does not qualify as aggression, nor does the FIGHTING back of the prey animal at the predator. We can count fighting as aggression only if it occurs among members of the same species. Even then, the occurrence of aggression does not necessarily mean there is any underlying anger emotion. Animals may well contest their status relationships in the absence of any emotion, just as a boy may strike his sister to establish his DOMINANCE in the absence of any anger. In general, it is wiser to study the contests of status among individuals, and to study aggression, than it is to try to get at the underlying emotion.

Fear is unquestionably one of the basic animal emotions. An enormous amount of research has been done on fear. It is the only emotion that has been well studied, it is the only emotion that has been regarded as an emotion, and it is the only one for which the physiological arousal mechanisms are beginning to be worked out.

Joy, or happiness, no doubt occurs widely among the animals. But it is difficult to say whether an animal going about its ordinary business feels any especial emotion. We do not know if the domestic cat (*Felis catus*) enjoys playing with the house mouse (*Mus musculus*); it may be business rather than fun, as far at the cat is concerned. Again, one is well advised to study the behaviour rather than attempting to get at any underlying emotion. Perhaps joy is an emotion that depends upon there being some kind of social context. HOUSEHOLD PETS, the animals we know best, show joy primarily in social situations. At least that is how we interpret the dog's tail-wag and the cat's purr. A kitten purrs to solicit nursing and GROOMING from its mother. A cat purrs because it is infantile; it is soliciting attention from its master. Purring is clearly a social signal that continues to serve the adult cat in much the same way that it serves the kitten. It is a fortunate circumstance that the cat likes to be petted, and that it communicates its pleasure by purring, while at the same time the master likes to hear the purring and communicates his pleasure by petting the cat. This mutual COMMUNICATION of pleasure between animal and human being is partly fortuitous and partly, no doubt, due to selective breeding. The same point can be made regarding the tail-wag of the dog. We recognize this sign of friendliness, and fuss over the dog, while at the same time the dog takes joy in being fussed over and so wags its tail. It is again partly a fortuitous recognition of social signals across species and partly a result of many years of selective breeding (see DOMESTICATION). It should be emphasized that such communication across species is often not effective. If a human smiles at a monkey (Simiae), the monkey may well misinterpret the signal and become aggressive. And when the monkey smiles at the human being, we are most likely to take it as a sign of friendliness rather than for what it is, a THREAT (see FACIAL EXPRESSIONS).

Charles DARWIN, who was one of the first scientists to study animal emotion, always stressed the communicative aspect of emotion. He took the existence of emotion in animals for granted, and devoted most of his research to the question of how the different emotions are expressed. He discussed mainly the three basic emotions that have been noted above, although he noted that more intelligent animals (monkeys) could express a wider range of emotions. Darwin emphasized the communication FUNCTION of the emotions in animals. If an individual animal happened to be in a bad or unsociable mood, it makes sense for this mood to be expressed by some signal that will be recognized by nearby animals, so that they can leave the old grouch alone. Or, if the individual happened to be in a friendly, sociable frame of mind, then it makes sense for that emotional feeling to be expressed and recognized by other members of the group. Thus, we may expect many species of animals to have evolved special social signals to indicate how they would react to a social encounter, just as they have evolved other behavioural signals to indicate danger or to call the young together, and so on.

So Darwin's concept of emotional communication helps to tie up some of the loose ends. It suggests that two of the basic animal emotions, joy and anger, occupy opposite ends of a single dimension of sociability, with expressions of joy over social contact at one end and anger, hostility, and aggression at the other end. The ultimate function of emotional expression might be to promote individual isolation, as may be necessary in defending a TERRITORY or a mate or a FEEDING area, or to promote group and family cohesiveness, as different social circumstances might require. It should be emphasized again that such social signalling should, in general, only be properly recognized by other animals of the same species; only rarely is a human observer able to interpret an animal's social signals correctly. The other basic emotion, fear, would lie on a different dimension from the sociability–unsociability dimension. The concept of emotional communication also helps to explain why the great variety of human emotions depend so much upon the social context: it is because of the richness and subtlety of human social relationships. Animals, such as the primates, that also have complex social systems, might also be expected to show some range of subtly different emotions, but most animals, having simpler social systems, should be expected to reveal only the basic dichotomy of joy and anger.

Charles Darwin's pioneering work on *The Expression of the Emotions in Man and Animals* (1872) is still provocative and instructive.   R.B.
16, 56, 61

ENERGY is the capacity for doing work, and work is a measure of the change of state of a system. The mechanical energy changes that are involved in the LOCOMOTION of animals depend upon muscular work. The capability for this derives ultimately from the food that the animal eats. When food is digested, energy is released as some of the chemical constituents of the food change their state. Some of this energy is lost in the form of heat, and in the excretion of waste products of digestion. Similarly, food chemicals that are absorbed following digestion are further changed in the processes of metabolism. Some energy is again lost in heat and excretion, and some is made available for muscular movement and other bodily processes.

The amount of energy that an animal requires depends upon its way of life, and this in turn is influenced by the energy available in the environment. The food of herbivores has low energy content compared with that of carnivores, with consequent effects upon their life style and behavioural organization. The energy relationships between the animal and its environment play a major role in shaping the animal's NICHE, which together with features of the HABITAT determines the place of the animal in the COMMUNITY.

Animals earn energy by HUNTING, FORAGING, and FEEDING, and spend it upon other aspects of behaviour. Some energy may be saved in the form of fat stores, or by HOARDING food. Scientists can calculate energy budgets similar to the budgets of economists, and can investigate how animals manage their budgets. It has been discovered that the energy economy of animals is often finely attuned to their environment and way of life. Much animal DECISION-MAKING is concerned with the allocation of energy and other resources among the various elements of the animal's behavioural repertoire. For example, pied wagtails (*Motacilla*

alba yarrellii) decide to tolerate or evict intruders into their TERRITORY on the basis of the balance between the energy that they can obtain by foraging on the territory and the energy required to evict the intruders. When food availability is high they are more likely to tolerate intruders than when it is low.

The term energy was sometimes used by early psychologists and students of ETHOLOGY in the context of MOTIVATION. It was thought that motivational energy accumulated during deprivation of food, sexual behaviour, etc., and this energy governed the intensity of the subsequent behaviour. If it was denied its normal outlet it 'sparked over' and energized other activities, or even gave rise to behaviour in the absence of the usual necessary external stimuli (see VACUUM ACTIVITY). It is now realized that since energy is a concept of capacity it cannot itself be a causal agent. Any analogy which implies that energy changes cause things to happen can only be misleading. Behaviour can be caused in such a way that certain energy changes are a consequence, but energy, whether mental or physical, cannot DRIVE behaviour.

ESCAPE is a form of DEFENSIVE behaviour which may occur as soon as a predator is detected, or only after the predator has initiated an attempt to capture the prey. For example, rabbits (*Oryctolagus cuniculus*) run as soon as a predator is detected, whereas the European hare (*Lepus europaeus*) may remain motionless in a CAMOUFLAGED position until detected by a predator, and only then makes its escape.

Many animals have specialized forms of escape behaviour, involving withdrawal to a prepared retreat, such as a burrow, evasive manœuvres, and behaviour designed to confuse the predator. Sometimes precautions are taken against predators, such as the polecat (*Mustela putorius*), which might enter the burrow. Rabbits, for instance, often have a second exit from the burrow, and the African ground squirrel *Xerus erythropus* blocks up the entrance to its burrow as a protection against nocturnal predators. Many species withdraw into a mobile retreat, such as a shell or host animal. For example, hermit crabs (Paguridae) take up empty gastropod (snail) shells into which they withdraw when disturbed. As they grow they discard the shell and select a new one. The anemone fish (*Amphiprion*) retreats into the tentacles of its anemone, from which it gains protection. Many species, including tortoises (Testudinidae), hedgehogs (Erinaceidae), armadillos (Dasypodidae), pangolins (Manidae), and spiny anteaters or echidnas (Tachyglossidae), provide their own protection of spines or armour, into which they can retreat when disturbed. In these cases the escape behaviour consists of adopting a particular body posture.

Evasive manœuvres, sometimes called *protean behaviour* (after Proteus, who escaped from his enemies by assuming various forms), may simply involve unpredictable changes of direction during flight, or may include more specialized means of evasion. For example, hares, snipe (Scolopacidae), and ptarmigan (*Lagopus*) execute sudden changes of direction during flight from predators. Various nocturnal moths (Noctuidae and Geometridae) respond selectively to the ECHO-LOCATION pulses of bats (Chiroptera). A moth can usually detect a bat before it is itself detected; if the bat is far away the moth usually flies as fast as possible in a straight line. However, bats can fly much faster than moths, and if the bat comes close, then the moth executes sharp turns, dives, and other evasive manœuvres.

The more specialized means of escape include the ability to disappear from view as do flying fish (Exocoetidae). These fish are able to take prolonged jumps out of the water, or to skim along the surface of the water propelled by the tail fin. The lizard fishes (Synodontidae) disappear into the sand when danger threatens. They quickly bury themselves by scooping up sand with their pectoral and pelvic fins. *Flash behaviour* is a form of protean behaviour involving apparent disappearance. Various grasshoppers (e.g. *Trilophidia tenuicornis*) and praying mantids (e.g. *Pseudoharpax virescens*) have highly coloured hind wings which are exposed during flight, and concealed when at rest. In evading a predator, the insect jumps or flies a short distance and then suddenly assumes a stationary camouflaged position. Various species of frog (Anura) have bright colours on the inside of their thighs which are conspicuous when they move, but disappear in the rest position. Some grasshoppers (Acridoidea) make a buzzing noise when in flight and become silent upon landing. Although the flash behaviour, whether visual or auditory, may attract the predator during flight, the sudden apparent disappearance of the prey is very confusing. The squid (*Loligo*) has an escape reaction that involves a backward jump, the discharge of a cloud of ink, and a change of colour. Sudden contraction of the muscular mantle ejects water from the mantle cavity through a funnel, causing the squid to become jet propelled. The ink is stored by a gland within the mantle cavity from where it is discharged together with the water jet. The predator is first distracted by the ink cloud, and then confused by the precipitate retreat of the squid, which undergoes a radical colour change, becoming stationary and highly camouflaged.

39

ETHOLOGY is distinguished from other approaches to the study of behaviour in seeking to combine functional and causal types of explanation. Behaviour can be explained in terms of hypotheses which aim to show how NATURAL SELECTION has, in the past, acted as a designing agent in shaping the EVOLUTION of behaviour. Such explanations account for behaviour in terms of its FUNCTION. The alternative form of explana-

tion concerns the way in which proximate causal mechanisms combine to control the behaviour of animals. These mechanistic explanations are usually the province of psychologists who seek to account for behaviour either in physiological terms, or in terms of special behaviouristic concepts, such as those used in accounts of animal LEARNING. Psychologists rarely make use of rigorous argument based upon the theory of natural selection. Functional explanations, on the other hand, are usually offered by evolutionary biologists, and make no reference to proximate causes.

Traditionally, ethologists have sought to combine observations of the form of behaviour, and hypotheses about its causation, with speculation and experiment concerning the function of the behaviour. For example, von FRISCH maintained, in contrast to the prevailing view of the time (1914), that honey-bees (*Apis mellifera*) showed COADAPTATION with respect to the structure of flowers, and were therefore likely to possess COLOUR VISION. It had been established that bees confined in a dark room and presented with two lights of different colour would move towards the brighter of the two lights irrespective of their colour, and it was concluded that bees were therefore colourblind. However, von Frisch showed that when presented with coloured cards in a FEEDING situation, the bees could readily discriminate colours. Differences in the MOTIVATION of the bees in the two types of testing situation were responsible for the different results. A functional view might suggest that colour is important to bees in a feeding situation, but that in attempting to ESCAPE bees should choose the brighter of two light sources.

In the HISTORY of the study of animal behaviour there have been many progenitors of ethology, some of whom were naturalists or ornithologists, others psychologists. Konrad LORENZ and Niko TINBERGEN are generally regarded as the founders of modern ethology. Both zoologists, they sought to divest the study of animal behaviour of its lingering ANTHROPOMORPHISM, and placed great emphasis upon the observation of animals in the natural environment. Together with Karl von Frisch, they were awarded the Nobel Prize for Medicine in 1973, a date when ethology may be said to have come of age.

Modern ethology is loosely organized on a world-wide basis, having an International Committee which oversees the biennial Ethological Conferences. In addition, there are several regional societies, such as the Animal Behavior Society of North America, and the Association for the Study of Animal Behaviour, based in Britain. There are also several learned journals, in which ethologists publish the reports of their FIELD and LABORATORY STUDIES.

41, 72, 136, 137

EVOLUTION. To most of the human race it has always seemed obvious that the immense diversity of life, the uncanny perfection with which living organisms are adapted to survive and propagate, and the bewildering complexity of living structures, can only have come about through divine creation. Yet throughout history the idea has arisen recurrently in the minds of isolated thinkers that there might be an alternative to the creation theory. The notion that species might change into other species was in the air, like so many other good ideas, in ancient Greece. It went into eclipse until the 18th century, when it reappeared in the minds of such advanced thinkers as Pierre de Maupertuis, Erasmus Darwin, and the Chevalier de Lamarck. In the first half of the 19th century it became not uncommon in intellectual circles, especially geological ones, but always in a rather vague form, and without any clear picture of the mechanism by which change might come about. It was Charles DARWIN who, goaded into action by Alfred Russel Wallace's independent discovery of his principle of NATURAL SELECTION, finally established the evolution theory by the publication, in 1859, of his *On the Origin of Species by means of Natural Selection or the Preservation of Favoured Races in the Struggle for Life*. After 1859 it was very difficult for reasonable men to doubt that all living things, man included, had evolved from different forms, and that ultimately all must be traced back to the same simple ancestors.

The facts of the evolution of life on earth, as far as we are able to reconstruct them today from a variety of sources of evidence, are as follows. Nearly 5000 million years ago the earth was formed. By 3000 million years ago life had arisen: we have fossils to prove it, fossils of microscopic bacteria-like creatures. Some time between these two dates, that mysterious event, the origin of life, must have occurred. Perhaps 'event' is the wrong word, because the transition from non-living to living was probably a gradual, imperceptible one, rather than a sudden cataclysmic event. Nobody knows what happened, but most theories agree on certain essentials.

The atmosphere of the early earth probably contained methane, ammonia, carbon dioxide, and other gases still abundant today on other planets in the solar system. Chemists have done experiments reconstructing these primeval conditions in the laboratory. If a plausible mixture of gases is placed in a flask with water, and some energy is released by an electric discharge (simulated primordial lightning), organic substances tend to be spontaneously synthesized. Quite a variety of them is formed, including, most significantly, *amino-acids* (the building blocks of proteins), and purines and pyrimidines (the building blocks of deoxyribonucleic acid, i.e. DNA, the *double helix*, the genetic molecule itself). If, as seems probable, something like this happened on the early earth, the sea would have become a 'soup' of pre-biological organic compounds.

A crucial step must have been the appearance in the primeval soup of molecules capable of assembling copies of themselves. Today the most famous of such replicating molecules is DNA itself, but there could have been others. Once a self-replicating molecule had been formed by chance, something like Darwinian natural selection could have begun: molecules which were especially good at replicating themselves would automatically, by definition, have come to predominate in the primeval soup. Molecules which did not replicate themselves, or which did so inaccurately, would have become relatively less numerous. A kind of molecular natural selection led to ever increasing efficiency among replicating molecules.

As the competition between replicating molecules hotted up, success must have gone to the ones that happened to hit upon special devices for their own self-preservation and their own rapid replication. Such devices probably were made by the manipulation of other molecules, proteins perhaps, in the formation of primitive cell walls to protect the replicator molecules themselves. It may have been at roughly this stage that simple bacteria-like creatures gave rise to the first fossils. The rest of evolution may be regarded as a continuation of the natural selection of replicator molecules, now called *genes*, by virtue of their capacity to build for themselves efficient devices (cells and bodies) for their own preservation and reproduction. Three thousand million years is a long time, and it seems to have been long enough to have produced such astonishingly complex gene-preserving devices as the human body. Genes are often referred to as the means by which bodies reproduce themselves. This is undeniable, but it is a more profound truth that bodies are the means by which genes reproduce themselves.

Fossils were not laid down in abundance until the Cambrian era, under 600 million years ago. By then most of the major animal *phyla* (the large groups into which the animal kingdom is classified), with the notable exception of the vertebrates, had appeared. Behaviour must have arisen long before the Cambrian. Behaviour is movement produced by muscle or its functional equivalent, controlled by a nervous system or its functional equivalent. It is common among single-celled microscopic protozoa today, and it probably arose quite early in evolution. The animals of the Cambrian era had SENSE ORGANS, BRAIN, and muscles. Some of them were specialized as active carnivores, and their prey must have been specialized to escape from them in the same kinds of ways as we see today.

The first vertebrates appear in the fossil record between 300 and 400 million years ago: fish-like creatures, completely covered with heavy armour-plating, perhaps adapted to escape from gigantic crustacean-like predators which infested the seas at that time. The land was first colonized by amphibians about 250 million years ago, and, more effectively, by reptiles slightly later. Birds and mammals arose from two different branches of

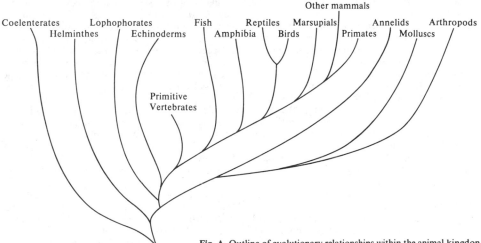

Coelenterates   Lophophorates   Fish   Reptiles Marsupials   Annelids   Arthropods
        Helminthes   Echinoderms   Amphibia   Birds   Primates   Molluscs
                                    Other mammals

Primitive
Vertebrates

Fig. A. Outline of evolutionary relationships within the animal kingdon.

reptiles (see Fig. A). Man evolved from an ape-like primate by an extremely rapid evolutionary spurt during the last few million years.

Logically it is important to distinguish two quite distinct parts of Darwin's contribution. He amassed an overwhelming quantity of evidence for the fact that evolution has occurred, and he provided a workable theory of the reason why it has occurred.

Much of the fossil evidence was known to Darwin, and used by him for demonstrating the fact of evolution, although geologists of his time were not able to place accurate dates on fossils. Indeed, in 1862 the eminent physicist Lord Kelvin greatly worried Darwin by 'proving' that the sun, and therefore the earth, could not possibly be older than 24 million years. Although this estimate was considerably better than the 4004 BC date for the creation then favoured by churchmen, it still did not leave enough time for the evolution which Darwin visualized. Kelvin, a deeply religious man, used his estimate as ammunition against the evolution theory. It was based on the erroneous assumption that the sun burned a fuel equivalent to coal.

Apart from fossil evidence, Darwin used other, more indirect evidence for the fact that evolution had taken place. The rapid alteration of animals and plants under DOMESTICATION was persuasive evidence both for the fact that evolutionary change was possible, and for Darwin's theory of the mechanism of evolution. Darwin himself was particularly persuaded by the evidence from the geographical dispersion of animals. The presence of local island races, for example, is easily explicable by the evolution theory: the creation theory could explain them only by assuming numerous 'foci of creation' dotted around the earth's surface. The hierarchical classification into which animals and plants fall so naturally is strongly suggestive

of a family tree: the creation theory had to make contrived and elaborate assumptions about the creator's mind working out themes and variations. Darwin also used as evidence for his theory the fact that some organs seen in adults and embryos appear to be vestigial: according to the evolution theory such organs as the tiny buried hind-limb bones of whales are remnants of the walking legs of the terrestrial ancestors of whales. The creation theory has trouble explaining them. In general the evidence for the fact that evolution has occurred consists of an enormous number of detailed observations which can all be easily explained by the theory of evolution, but which can be explained by the creation theory only if we assume that the creator deliberately set out to deceive us.

Natural selection, the mechanism of evolution that Darwin and Wallace suggested, had been proposed before. It had had a history almost as long as the theory of evolution itself, although neither Darwin nor Wallace knew this until later. Most previous evolutionists had inclined towards an alternative theory of the mechanism of evolution, now usually associated with Lamarck's name. This was the theory that improvements acquired during an organism's lifetime, such as growth of organs during use and shrinkage of organs during disuse, were inherited. This theory of the inheritance of acquired characteristics has great appeal, but the evidence does not support it. Even if genetic information could somehow travel 'backwards' from body cells into the inherited 'germ-line', it is still almost inconceivable that embryonic development could reverse itself so that bodily improvements acquired during an animal's lifetime encoded themselves in the genes. In Darwin's time the matter was more in doubt, and Darwin himself flirted with Lamarckism when his own theory ran into a difficulty.

The difficulty arose from current views of the nature of heredity. In the 19th century it was almost universally assumed that heredity was a blending process, that is, that offspring are intermediate between their two parents in character and appearance. It can be shown mathematically that if heredity is of this blending type it is almost impossible for Darwinian natural selection to work. This was proved in 1867 and, as we have seen, it worried Darwin enough to drive him in the direction of Lamarckism. It may also have contributed to the odd fact that Darwinism suffered a temporary spell of unfashionableness in the early part of the 20th century. The solution to the problem which so worried Darwin lay in Gregor Mendel's theory of particular inheritance, published in 1865, but unfortunately unread by Darwin, or anyone else until after Darwin's death.

Mendel demonstrated, what Darwin himself had at one time dimly glimpsed, that heredity is particulate, not blending. Offspring are not intermediate between their two parents in all respects. Rather, they inherit discrete hereditary particles— nowadays we call them *genes* (see GENETICS). An individual either definitely inherits a particular gene from a particular parent or he definitely does not. Since the same can be said of his parents, it follows that an individual either inherits a particular gene from a particular grandparent or he does not. This argument can be applied repeatedly for an indefinite number of generations. This leads to the idea of discrete single genes being shuffled through the generations like cards in a pack, rather than being mixed like the ingredients of a pudding.

This makes all the difference to the mathematical plausibility of the theory of natural selection. If heredity is particulate, natural selection really can work. The modern genetic theory of natural selection was worked out in the early 1930s by R. A. Fisher and others. The recent revolution in molecular biology has reinforced and confirmed, rather than changed, the theory which they set out.

The modern genetic theory of natural selection can be summarized as follows. The genes of a population of interbreeding animals or plants constitute a *gene pool*. The genes compete in the gene pool in something like the same way as their ancestors, the early replicating molecules, competed in the primeval soup. In practice genes spend their time either sitting in individual bodies which they helped to build, or travelling from body to body via sperm or egg in the process of sexual reproduction. Sexual reproduction keeps the genes shuffled, and it is in this sense that the long-term habitat of a gene is the gene pool. Any given gene originates in the gene pool as a result of a mutation, a random error in the gene-copying process. Once a new mutation has been formed, it can spread through the gene pool by means of sexual reproduction. This is the origin of genetic variation.

Any given gene in a gene pool is likely to exist in the form of several duplicate copies, either all descended from the same original mutant, or descended from independent parallel mutants. Therefore each gene can be said to have a *frequency* in the gene pool: some genes, such as the albino gene, are rare in the gene pool; others are common. At the genetic level, evolution may be defined as the process by which gene-frequencies change in gene pools.

There are various reasons why gene-frequencies might change: immigration, emigration, random drift, and natural selection. Immigration, emigration, and random drift are not of much interest from the point of view of ADAPTATION, although they may be quite important in practice. It is natural selection which gives rise to progressive evolution, natural selection which accounts for the perfection of adaptation and the complex functional organization of life. Genes in bodies exert an influence on the development of those bodies. Some bodies are better at surviving and reproducing than others. Good bodies, i.e. bodies that are good at surviving and reproducing, will tend to contribute more genes to the gene pools of the future than bodies that are bad at surviving and reproducing: genes that tend to make good bodies will tend to predominate in gene pools. Natural selection is the differential survival and differential reproductive success of bodies: it is important because of its consequences for the differential survival of genes in gene pools.

Not all selective deaths lead to evolutionary change. On the contrary, much natural selection is so-called stabilizing selection, removing genes from the gene pool that tend to cause deviation from an already optimal form. But when environmental conditions change, either through natural catastrophe or through evolutionary improvement of other creatures (predators, prey, parasites, and so on), selection may lead to evolutionary change. The most famous example of this which has actually been observed is the case of the peppered moth (*Biston betularia*) (see NATURAL SELECTION). In a matter of decades a gene for black coloration swept to ascendancy in gene pools of this species in industrial areas. This was because birds could see the lighter moths more easily on industrially blackened trees, and so preferentially ate them. Therefore a larger number of genes for light colour ended up in the bellies of birds than did genes for dark colour. The result was that light forms are now rare in industrially polluted areas. In rural areas, on the other hand, dark moths are more vulnerable to being eaten, and, which is what matters, so are their genes. Dark moths are therefore rare in rural areas. In both rural and industrial gene pools, a minority of the 'wrong' genes are present, probably repeatedly introduced by immigration from neighbouring gene pools.

Genes that make their bodies likely to survive tend to flourish in gene pools. But equally important are genes that tend to make their bodies successful in reproduction. This may include genes for sexual attractiveness. Darwin made a distinction between natural selection and SEXUAL SELECTION. He was impressed with the fact that qualities of sexual attractiveness were often the reverse of qualities leading to individual survival. The gaudy and cumbersome tails of birds of paradise are a notorious example. They must hamper their possessors in flight, and certainly they are conspicuous to predators, but Darwin realized that this could be 'worth it' if the tails also attracted females. A male who manages to persuade a female to mate with him rather than with a rival is likely to contribute his genes to future gene pools. Genes for sexually attractive tails willy-nilly have an advantage to compensate for their admitted disadvantages.

For a time after Darwin's death the theory of sexual selection fell out of fashion. People felt that natural selection would inevitably work on females to change their tastes so that they were no longer attracted by qualities which, when inherited by their sons, could only endanger them. R. A. Fisher resuscitated the theory of sexual selection by means of the following ingenious argument. Given that a majority of females start with the same aesthetic tastes, any female who deviates from the prevailing mode will automatically put her genes at a disadvantage. This is because her sons will tend to inherit from their father qualities which are unattractive to most of the females in the population, and will therefore not provide her with many grandchildren. Fisher showed that even if the prevailing fashion in female taste is for male qualities that are deleterious to male survival, selection can tend to favour sexual attractiveness for its own sake. It is probable that such extravagances as the peacock's tail (*Pavo cristatus*), the bowers of Australian and New Guinea bowerbirds (Ptilonorhynchinae), and the mysterious beauty of bird-song, have evolved by the kind of runaway sexual selection envisaged by Fisher.

Sexual selection of this kind is not, however, the only selective force shaping the COURTSHIP behaviour of animals. Also important is the need to avoid hybridization. Members of widely differing species such as an elephant (Elephantidae) and a hare (Leporidae) would never be tempted to mate with each other but, even if they were, no embryo would be conceived. The same is not true of, say, a horse (*Equus Przewalski caballus*) and a donkey (*Equus Asinus asinus*). The result of their hybridization could be a strong and healthy mule (or hinny): strong and healthy, but, alas for its genes, sterile. As far as a gene is concerned, it might as well be dead as end up in the body of a mule, yet a mule body is just as costly as a fertile one for its parents to make. Genes with a tendency to end up in mule or hinny bodies will not flourish, and this means genes that give donkeys a taste for horses as sexual partners, and genes that gives horses a taste for donkeys. In practice, selection will favour genes that tend to increase fussiness in the choice of conspecific sexual partners, and genes that tend to exaggerate whatever differences there may be in

appearance and courtship behaviour between the two species, which might help them to avoid cross-breeding.

For instance, in the southern United States there are two species of tree-frog, *Microhyla carolinensis* and *Microhyla olivacea*. *Carolinensis* is found in the eastern part of the continent, and *olivacea* in the west, but their ranges overlap in a region of Oklahoma and Texas. In both species the males attract the females by singing to them. The song of *olivacea* is somewhat higher in pitch than that of *carolinensis*. The interesting point is that this difference is exaggerated in the central overlap zone. Presumably selection has penalized genes that make males sing in a way that might be confused by females with the song of the 'wrong' species. The selection against hybridization has occurred in the geographical area where hybridization is a practical possibility.

It seems probable that courtship behaviour may have played a prominent role in the evolution of new species (what Darwin referred to as the 'problem of problems', the origin of species). Evolution obviously does not consist simply of gradual change in a single lineage. The tree of life is a forking, branching tree, giving rise to a vast variety of specialized forms. At every fork, one ancestral species must split into two or more. It is probable that behaviour has played a prominent role in this process. Thus locomotory behaviour and MIGRATION have probably often given rise to geographical isolation between gene pools, which is the usual first step in the evolution of new species (see ISOLATING MECHANISMS). But, more interestingly, as in the case of the tree-frogs, courtship behaviour, evolved under the influence of the selective penalization of hybrids, has reinforced the isolation between races so that they became separate species.

We commonly speak of the evolutionary history of morphological structures such as skulls and limbs. Is it proper to speak of the evolution of behaviour in the same way? At first sight this seems difficult. A particular behaviour pattern seems so abstract: it is not a thing which is there all the time, but is a sequence of muscular contractions that an animal only sometimes does. The difficulty is illusory. The characteristic behaviour of an animal is a product of its nervous system and other organs: it behaves the way it does because of the way in which its nervous system is wired up. Provided at least some of the differences between the behaviour patterns of individuals are inherited differences (see GENETICS), natural selection can work on behaviour patterns as the indirect manifestations of genes, in just the same way as it works on limb-bones as the indirect manifestation of genes. Behaviour does evolve.

However, there are practical difficulties in reconstructing the evolutionary history of behaviour. Skulls and limb-bones fossilize but behaviour does not. To be sure, we can make educated guesses as to how extinct animals such as dinosaurs probably walked and hunted. In some cases we can look at 'frozen behaviour' in the form of footprints. Artefacts such as caddis-larva (Trichoptera) houses are in a very real sense frozen behaviour, and they sometimes fossilize. But in general we can reconstruct the behaviour of ancestors only by indirect inference from COMPARATIVE studies of the behaviour of modern animals.

This leads to some tempting logical pitfalls. Many errors are based on the fallacious assumption that modern animals can be arranged as an ascending series, a *scala naturae* or 'ladder of life'. This notion long predates Darwin; one of its most influential proponents was Aristotle. The theory of evolution is in fact profoundly antithetical to the idea of a *scala naturae*, but this is not always clearly realized, and many people have grafted a pseudo-evolutionary interpretation onto their ladder-of-life ideas, arriving at a so-called 'phylogenetic scale'. Man is at the top of the mythical ladder (of course, that is one of its main appeals). The dubious privilege of occupying the lowest rung is often accorded to the *Amoeba*. All other creatures are neatly perched, one above the other, in between. The hidden assumption behind the phrase 'phylogenetic scale' is that we have only to look around the modern animal kingdom to find all our ancestors, preserved, as though in amber, for our inspection.

If only this were so, the reconstruction of the evolution of behaviour would be ridiculously easy. But it is obviously not so. The 'phylogenetic scale' is one of our sillier preconceptions. If there are grains of truth in it here and there, they need to be advocated with evidence, not assumed as manifest truth. The tree of life is not a vertical ladder but a tree: a branching, spreading, family tree. We, the modern animals, are the tips of the twigs. If some of the twigs are said to be 'higher' than others, we must enquire very critically exactly what 'higher' is supposed to mean. What is a 'higher' animal? Is it one whose bodily structure and behaviour are relatively complex? Is it one that is very efficient at its way of life? (The two often do not go together.) Is a 'lower' animal a simple one, an inefficient one, or one that resembles its remote ancestors? Suppose we could agree upon the latter as a definition of a 'lower' animal, would this necessarily help us to reconstruct the evolution of behaviour? It might, but we must still be careful. We might agree that a spiny anteater (Tachyglossidae) is 'lower' than a rat (*Rattus*), in the sense that it more closely resembles the reptilian common ancestor of both. This is true both of its skeleton (as we can tell from fossils) and of its mode of reproduction (it lays eggs). But does this mean the common ancestor behaved like a spiny anteater? Not at all. The common ancestor probably did not eat ants, and much of the spiny anteater's behaviour is dictated by its dietary habits. In general, it is clearly illogical to infer that because an animal

resembles its ancestors in some respects, therefore it must resemble its ancestors in all respects.

How then can we use comparative evidence to reconstruct the behaviour of long-dead ancestors? We never can with any certainty. But we can make some guesses with reasonable probability (see COMPARATIVE STUDIES). For instance, if we observe that all modern representatives of a particular taxonomic group perform a particular behaviour pattern, it is reasonable to guess that the common ancestor of the group behaved in the same way. All modern pigeons (Columbidae) drink by sucking: it seems probable that the ancestral pigeon did too. This is more parsimonious than the alternative possibility, that the sucking habit evolved independently in more than one pigeon lineage.

It is only slightly more parsimonious, however. Convergent evolution of the same feature several times independently is quite common. ECHOLOCATION has evolved independently in whales (Cetacea), birds (Aves), sealions (Pinnipedia), and more than once in bats (Chiroptera). Social life with sterile female workers has evolved at least eleven times independently in the insect order Hymenoptera. The thylacine (*Thylacinus cynocephalus*), a marsupial carnivore which is probably extinct, but which may possibly still survive in Tasmania, was nearly indistinguishable from a true dog in superficial features: its pouch gives it away as a marsupial, more closely related to a kangaroo than to a dog. It had evolved dog-like features because those features are well adapted to a dog-like hunting way of life. It was the introduction of the dingo (*Canis familiaris dingo*), a true dog, which drove the thylacine into extinction in Australia, by competition. There are also marsupial moles (*Notoryctes*), mice (Phascogalinae), and anteaters (Myrmecobiidae), all of which bear a strong resemblance to their non-marsupial ecological equivalents, both in anatomy and behaviour (see NICHE).

This gives us the clue for another major way in which we can infer the behaviour of extinct ancestors. Clusters of anatomical and behavioural characteristics tend to be correlated with each other in modern animals, on account of their FUNCTION. Carnivores tend to share the same kinds of teeth, guts, sense organs, and behaviour, even if they have evolved them convergently. Herbivores independently share a separate cluster of characteristics. We can use evidence from modern animals to build up a picture of which characteristics tend to cluster together. Then when we come to animals for which our information is incomplete, for example fossils, we can use our picture of correlated clusters of characteristics to fill in the gaps in our knowledge. So far this method has been employed intuitively, but it could be made more systematic and objective.

In some cases it can safely be assumed that a behaviour pattern has evolved from some other behaviour pattern which it resembles. This is particularly so of behaviour involved in COM-MUNICATION. For instance the ritual pointing with the bill at the bright speculum feather of the wing by the mandarin drake (*Aix galericulata*) is sufficiently similar to ordinary wing-preening to betray its own ancestry. It is almost certainly a ritualized preening movement (see RITUALIZATION). In this case one can see the 'ancestral' form of the behaviour preserved in the same individual as the 'evolved' or ritualized form. Sometimes the ancestral form has disappeared from the species itself, but may still be seen in related species. The probable ancestry of the dance of the honey-bee (*Apis mellifera*) can be traced in related bee species. In this case one is not making any assumptions about some modern bee species being 'lower' than others. But there are modern bee species which do things somewhat resembling the honey-bee dance, yet simpler. This provides a plausible model for how the honey-bee dance may have originated, although guesses of this sort may well be wrong.

There are two main reasons why an understanding of evolution is important for the study of animal behaviour. Firstly, and most importantly, animals are created by evolution, their behaviour as much as anything else. Behaviour can only be fully understood in terms of its evolutionary history, and in terms of the role which it plays in the survival and reproduction (more precisely the inclusive FITNESS) of the animal. Secondly, behaviour plays a role in evolution, shaping the course of evolutionary history. It has been rightly said that the whole of biology since 1859 has been a commentary on *The Origin of Species*. The study of animal behaviour is no exception.   R.D.

20, 34, 50, 95

EXPLORATORY BEHAVIOUR is a form of APPETITIVE behaviour which may or may not be aimed at a particular commodity or environmental situation. Exploratory behaviour is shown by animals SEARCHING for food, nest material, etc., and when they search for a nest-site, or some other form of micro-habitat. In all these cases the exploratory behaviour generally ceases when the animal arrives in a particular situation, suggesting that a specific GOAL was involved.

Other forms of exploratory behaviour do not appear to be aimed at a specific goal. This is particularly true of animals' responses to NOVELTY. For example, rats (*Rattus*) routinely explore novel foods, but will usually sample them in only very small quantities. This exploratory behaviour forms part of the normal process of FOOD SELECTION. It is not terminated by the discovery of novel food, or by any other obvious goal.

Many animals seem to explore simply for its own sake. Rats, for instance, tend to enter those parts of a maze that allow the maximum chance of exploration. The opportunity for manual exploration of objects is adequate REWARD for the perfor-

mance of routine tasks by rhesus macaques (*Macaca mulatta*) and chimpanzees (*Pan*). Exploration is closely associated with LEARNING. Rats which have been allowed to explore a maze, or similar novel situation, can be tested to determine what they have learned. If a rat which is not hungry discovers food during the course of such exploration, it may sniff the food, or even appear to ignore it. However, if the rat is subsequently placed in the same situation when it is hungry, it will go straight to the place where food was previously located, showing that it had learned that location during the previous exploration.

Much exploratory behaviour is characterized by a tendency both to approach and to withdraw, motivated by CURIOSITY and FEAR. When we observe an animal exploring a novel object, during PLAY for instance, it is common to see signs of CONFLICT: the strange object may be avoided initially, and subsequently approached only cautiously; there may be a rapid vacillation between approach and retreat, combined with tentative attempts to manipulate, sniff, or bite the object of interest.

EXTINCTION is the process by which learned behaviour patterns cease to be performed when they are no longer appropriate. The evolution of the ability to learn new patterns of behaviour is due to the SURVIVAL VALUE such an ability confers in coping with a changing environment. For many animals important features of their environment, such as sources of food and the nature of predators, may well change in a relatively fortuitous manner from time to time. The ability to learn new behaviour patterns in response to such changes will obviously enhance an animal's chance of reproducing successfully. However, it is also clear that the survival value of learning depends critically upon the ability of animals to give up old patterns of responding when they are no longer appropriate. If an animal is capable of learning the location of new food sources in its environment, the usefulness of such an ability will be severely limited if it cannot also learn to give up visiting this location when the supply is exhausted. Not only would the persistence of the original behaviour consume energy reserves for no return, it would also minimize the likelihood of the animal sampling other areas of its environment, and therefore the possibility of learning about a new food source. Extinction is the condition in which the learned behaviour patterns stop being performed when they are no longer appropriate.

LEARNING occurs when an animal is exposed to a novel relationship between events in its environment. If the relationship between these events is then altered in certain ways which render the response inappropriate, the likelihood of the animal performing that response will decrease. Extinction refers both to the type of change required to produce a response decrement, and to the actual response decrement.

Conditions of extinction. There are two main types of environmental relationship about which animals learn. The first is the relationship between events that occur independently of the animal's own behaviour, and can be studied in the LABORATORY using the *classical conditioning* procedure developed by PAVLOV. In his original experiments Pavlov presented hungry dogs (*Canis lupus familiaris*) with small portions of food and found that if he preceded the presentation of food on each trial by some external stimulus, such as a light, the behaviour of the animal during the light changed systematically in the course of the experiment. Although a number of different responses develop during such conditioning, Pavlov chose to record the salivation elicited by the light, and found that the magnitude of salivation progressively increased with the number of times the light and food were paired. In describing his conditioning procedure, Pavlov referred to the signalling stimulus, the light in this example, as the *conditional stimulus* and the new response, salivation, elicited by the light as a result of conditioning as the *conditional response*. He attributed the development of the salivary conditional response to the strengthening (*reinforcing*) properties of the food, which is itself referred to as a *reinforcer* (see REINFORCEMENT).

In some sense the increase in the salivary response only represents an indicator that the dog has learned that the light signals the occurrence of the food, thereby allowing the animal to take appropriate action before the actual occurrence of this reinforcer. If presentation of the reinforcer after the light is now omitted, the animal should no longer treat the light as a signal for that event. This is in fact what happens. Omitting the food results in a decline in the salivary response elicited by the conditional stimulus (see Fig. A). Withholding the food reinforcer after conditioning represents an example of an extinction condition, and the progressive decline in the magnitude of the salivary response is an example of the extinction of a conditional response. Given that the conditional stimulus is presented alone a sufficient number of times, the conditional response may totally disappear, and the behaviour of the animal in the presence of the light becomes indistinguishable from that produced by the same stimulus prior to conditioning.

In OPERANT or *instrumental* conditioning procedures, the animal learns about a relationship between its own behaviour and some environmental event. For instance, a hungry rat (*Rattus norvegicus*) may be placed in a small chamber with a food hopper and with a lever projecting from one wall. If the experimenter arranges that each press of the lever delivers a food pellet, not surprisingly the animal will rapidly come to press the lever at a fairly high rate. This represents an example of a second type of conditioning in which the food reward acts as a reinforcer for the conditional response of lever pressing. Again, if the condition of extinction is instituted by withholding the food reinforcer, the probability of the conditional response will decrease, and the rat will come to press the lever no more frequently than prior to conditioning. Of course, it is entirely appropriate and not at all surprising that animals should give up learned patterns of behaviour, which are established because they lead to beneficial consequences, when they no longer yield this outcome.

The development of learned responses during both types of conditioning is usually assumed to reflect the fact that the animal has made an ASSOCIATION between two forms of internal representation in the BRAIN: the conditional stimulus and reinforcer in the case of classical conditioning, and the conditional response and the reinforcer in the case of instrumental conditioning. From this point of view, the most obvious explanation of extinction is that withholding the reinforcer in some way decreases the strength of this association or, in other words, the animal unlearns what was learned during conditioning. Let us assume that each reinforcement (a presentation of the rein-

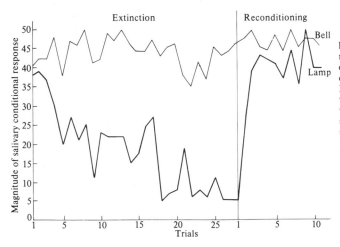

Fig. A. The extinction and reconditioning of a Pavlovian salivary conditional response to a visual conditional stimulus in a dog. Initially both an auditory (bell) and a visual (lamp) conditional stimulus were paired with food to establish the salivary conditional response. During a second stage, illustrated in the figure, the bell continued to be associated with the food while the lamp was presented without food. As a result the conditional response to the lamp progressively decreased or extinguished. In a third stage, when the lamp was again paired with food, the conditional response was rapidly re-established.

forcer) during conditioning increases the strength of this association while each non-reinforcement (an omission of the reinforcer) during extinction decreases its strength. This idea then allows us to understand the well-known fact that the number of responses seen during extinction normally increases with the number of reinforcements administered during conditioning (see Fig. B). However, there are problems with this explanation. For instance, if the conditional stimulus or instrumental response is once again paired with the reinforcer after extinction, the learned response develops much more rapidly than it did during the original conditioning (see Fig. A). It is difficult to see how this could happen if extinction only served to abolish the original learning. The response decrement seen during extinction cannot be solely due to unlearning; other processes must be at work.

Generalization decrement. When an animal learns a new response, this learning usually occurs within a specific environment or context composed of a large number of different stimuli. Although the response will be adaptive within the specific stimulus environment present during learning, it may be totally inappropriate for the animal to perform the same behaviour in another situation. Let us suppose that a fruit-eating animal has learned that the fruit of a particular tree in its domain is over-ripe, and has learned to avoid it. Under these circumstances, it would be clearly inappropriate for the animal to generalize this avoidance behaviour to other fruits, or even to the same fruit in a different situation. The performance of a learned response appears to be restricted to environments which are similar to the situation in which conditioning occurred, and when a stimulus element of this environment is changed or omitted a decrement in the response is observed (see GENERALIZATION and DISCRIMINATION). A response decrement brought about by a change in the stimulus environment is an example of *generalization decrement*.

The relevance of generalization decrement to extinction is that an important stimulus element of the context in which conditioning occurs is the reinforcer itself. During instrumental conditioning a lever press can be reinforced just after the rat has consumed a previously delivered food pellet, so that the presentation of the reinforcer is itself one of the contextual stimuli in which conditioning takes place. Omitting the food during extinction not only removes the reinforcer for the response, but also changes the stimulus environment in which conditioning was established. There is evidence that part of the response reduction seen during extinction is simply due to generalization decrement brought about by omitting the reinforcer. Instead of omitting the food altogether after lever-press conditioning, the experimenter can arrange to deliver food pellets every so often, independently of the animal's pattern of responding. Under these circumstances, one of the conditions of extinction will be in force, for there will no longer be a relationship between the animal's response and the occurrence of the reinforcer. However, food will still be delivered in the experimental situation, so that the stimulus context will be similar to that present during original conditioning. In line with the idea that generalization decrement plays a role in the response reduction during normal extinction, rats do more lever presses under random food delivery than when the reinforcer is completely withheld (see Fig. C).

Although a fairly simple idea, the concept of generalization decrement allows us to understand a striking effect of varying the conditioning procedure upon the persistence of responses during extinction, the so-called *partial-reinforcement* extinction effect. So far we have assumed that the reinforcer is presented on every conditioning trial. However, this is by no means a necessary requirement for stable conditioning; strong Pavlovian and instrumental responses can be established using a partial reinforcement schedule, in which reinforcement is omitted on a certain proportion of trials from the outset of conditioning. We can illustrate the effects of such a schedule by considering an

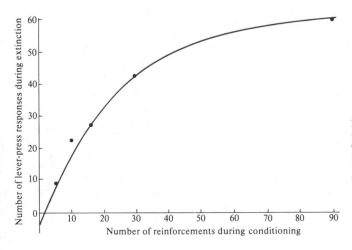

Fig. B. The increase in the number of lever-press responses made in extinction by different groups of rats that received various amounts of food reinforcement during conditioning. As the number of reinforcements received in conditioning increases, so does the number of responses made in extinction.

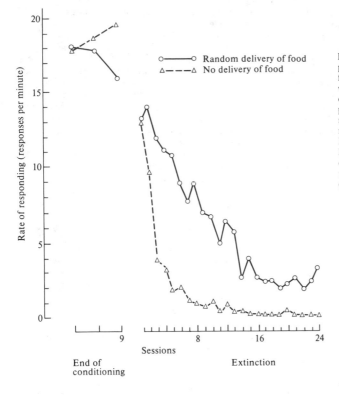

Fig. C. The decline in the rate of lever pressing by two groups of rats when food reinforcement is either totally withdrawn or delivered independently of the animal's response. The left-hand part of the figure shows that the response rates of the two groups did not differ at the end of conditioning. However, when lever presses were no longer reinforced, the response rate declined more rapidly in the group experiencing extinction than in the group exposed to response-independent food.

o———o Random delivery of food
Δ----Δ No delivery of food

experiment in which two groups of hungry rats are trained to run down a straight alley-way in order to receive a food reward at the end. Let us suppose that the rats are given four trials on each of a number of successive days, with a short interval of about 15 s between the successive trials on each day. The difference between the two groups is that one group, the continuously reinforced group, receives food at the end of the alley-way on every trial, while for the other, the partially reinforced group, the food reinforcer is randomly omitted on half the trials. As a result of this conditioning, rats in both groups will progressively decrease the time it takes them to traverse the alley-way over successive trials. If we then extinguish the running response by leaving out the food altogether, we can compare the resistance to extinction generated by the continuous and the partial reinforcement training schedules. A universal finding is that the partially reinforced rats run faster than the continuously reinforced animals throughout extinction.

In order to understand this phenomenon in terms of generalization decrement, we have to recognize that animals can remember events occurring on preceding trials while they are in the apparatus on the next trial (see MEMORY). One of the events they appear to remember is whether or not they received food on the previous trial. As a result, the continuously reinforced rats would have always experienced reinforcement for running in the pre-

sence of the memory that food occurred on the previous trial. Since no food is presented during extinction, the memory of food on the preceding trial will not be aroused after the first extinction trial, and the continuously reinforced animals will experience a change in the stimulus conditions from those in which the running response was acquired. Consequently they show a reduction in responding simply due to a generalization decrement. In contrast, partially reinforced animals will have received reinforcement on some trials during conditioning for running in the presence of the memory of the absence of food (non-reinforcement) on the previous trial. As a result they will not suffer the same generalization decrement when the food reinforcer is totally withdrawn during extinction. This idea argues that the critical feature of partial reinforcement schedules for an enhancement of extinction performance is that non-reinforced trials are followed by reinforced trials during training. It is only on these trials that the rats can be trained to run in the presence of the memory of non-reinforcement. Assuming that the 24-h interval between successive blocks of daily trials is too long for the rat to remember the outcome of the preceding trial, this means that only partial reinforcement schedules which contain trial sequences in which a non-reinforced trial is followed by a reinforced one during the daily block of trials should lead to enhanced extinction. This appears to be so; if the partial reinforcement schedule ensures that all the non-reinforced trials are given after the reinforced trials in each daily block, performance in extinction is similar to that seen after continuous reinforcement.

Inhibition. Even though generalization decrement may control subtle and complex effects in extinction, it cannot constitute a total explanation of the response reduction seen when reinforcement is no longer associated with a Pavlovian conditional stimulus or an instrumental response. The lever-press response of the rat still declined when responses were no longer followed by food, even though similar stimulus conditions were maintained during conditioning and extinction by delivering response-independent food. The obvious explanation of this decrement is that the animal actually learns that the Pavlovian conditional stimulus, or the instrumental response, is now associated with the absence of the reinforcer. This is not the same thing as saying the animal unlearns what was acquired during conditioning; rather it suggests that the animal learns something new during extinction which counteracts the effect of the associations formed during acquisition of the conditional response. Pavlov was the first to put forward this view of extinction when he suggested that pairing a stimulus with non-reinforcement establishes an inhibitory process, which actively opposes the excitatory process that was formed during conditioning. The main implication of this idea is that the underlying tendency to perform the learned response does not disappear during extinction, but is actively inhibited. This means that the conditional response should re-emerge if we could in some way remove the inhibitory process after extinction. The phenomenon of *disinhibition* represents just such an effect. If we present a novel and salient stimulus, such as a loud noise or bright light, after the lever-press response of our rat has been extinguished, the animal will start responding again during the presence of the stimulus (see Fig. D). Even though the reason why such a novel stimulus disrupts the inhibitory process is not fully understood, this type of experiment demonstrates that extinction establishes a process which actively counteracts the tendency to perform the conditional response.

So we now have a view of conditioning and extinction which assumes that, after conditioning, presentation of the conditional stimulus activates excitatory associations, while after extinction it activates both excitatory and inhibitory associations that exactly counterbalance each other. However, it is not clear from the preceding discussion why associating a conditional stimulus or an instrumental response with non-reinforcement should result in the formation of inhibition. Throughout the life of the animal all sorts of stimuli and events occur in the absence of a particular reinforcer and yet do not acquire significance for the animal, let alone inhibitory properties. A crucial requirement for the development of inhibition is that a conditional

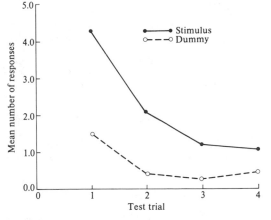

Fig. D. An example of disinhibition. The number of lever-press responses given by a group of rats during a 3-min novel stimulus (stimulus) or an equivalent period without the stimulus (dummy) after extinction. Lever presses were reinforced with food during conditioning. After extinction of the response, presentation of the novel stimulus increases the amount of responding on each of four successive test trials.

stimulus or instrumental response should be paired with non-reinforcement in the presence or context of stimuli that are themselves associated with the reinforcer. In other words, inhibition is only generated by non-reinforcement when it occurs in a context in which the animal might, in some sense, be said to 'expect' the reinforcer.

In the case of simple extinction, the stimulus context causing the animal to expect the reinforcer is provided by the conditional stimulus itself, through its previous association with the reinforcer during conditioning. However, a stimulus can acquire inhibitory properties even though it has never been associated with the reinforcer. This is done by simply pairing it with non-reinforcement in the presence of a second conditional stimulus, which is associated with the reinforcer and maintains the animal's 'expectation' of that event. The development of such a conditional inhibitor can be illustrated by another early experiment from Pavlov's laboratory. Pavlov initially established a rotating object as an excitatory conditional stimulus for salivation in a dog by pairing it with food. He then intermixed such reinforced trials with trials on which the rotating object was presented without the reinforcer. The crucial feature of the experiment was that a second stimulus, tactile stimulation of the skin, was presented at the same time as the rotating object, but only in the non-reinforced trials. Initially the compound stimulus, consisting of the rotating object and tactile stimulation, elicited a salivary response due to the excitatory properties of the rotating object, but with further training the response to this compound stimulus was extinguished. As the rotating object continued to elicit the full salivary response when presented alone, the extinction of the conditional response to the compound stimulus could not have been due to the acquisition of inhibitory properties by the rotating object. The obvious explanation is that the second stimulus, the tactile stimulation, must have acquired the capacity to inhibit the salivary response, even though it had never been paired with the food reinforcer and could not have acquired excitatory properties. It appears that, unlike a simple extinguished conditional stimulus, which has mixed excitatory and inhibitory properties, the tactile stimulus in this experiment had acquired the capacity to activate purely inhibitory associations.

Pavlov demonstrated that the tactile stimulus had, in fact, acquired general inhibitory properties by conducting a final test stage in his experiment. Remembering that the cardinal feature of inhibition is that it opposes excitatory associations that produce the conditional response, presentation of the inhibitor, the tactile stimulus, should be able to inhibit the salivary response elicited by any excitatory stimulus. To test this idea, Pavlov paired a third stimulus, a light flash, with food until a stable salivary response was established, and then for the first time presented the tactile stimulus at the same time as the light flash. The tactile stimulus completely abolished the conditional response, which would have otherwise been elicited by the light flash.

In conclusion it appears that pairing a stimulus with non-reinforcement in a context where an animal expects reinforcement establishes inhibitory associations, just as pairing a stimulus with reinforcement establishes excitatory associations. To the extent that extinction results from the formation of inhibition, the decrease of a conditional response under non-reinforcement depends just as much upon learning, or associative mechanisms, as the initial acquisition of the response during conditioning.

Emotional responses in extinction: frustration and relief. When we wish to establish excitatory associations for a particular response, we associate a Pavlovian conditional stimulus, or an instrumental response, with a motivationally important event, such as presenting food to a hungry dog or rat. In an analogous fashion, the condition for the formation of inhibitory associations, the omission of an expected reinforcer, also appears to be of emotional or motivational significance for the animal. This can be illustrated by considering the effects of omitting a food reinforcer. We can establish an association between apparatus or contextual cues and a food reinforcer, simply by placing a hungry rat in a distinctive chamber, and feeding it on a number of trials. If after a number of such reinforced trials we place the rat in the chamber without any food, the animal will display signs of extreme agitation, suggesting that the omission of

the reinforcer has induced a state of emotional AROUSAL. Such arousal will not be seen in a rat that has never previously received food in the chamber. We can investigate the animal's state further by seeing whether it will learn a new response to escape from the situation that aroused the emotional state. This is usually done by opening a small door leading to a second, distinctively different chamber, on non-reinforced trials. Under these circumstances, rats experiencing non-reinforcement after having previously received the food reinforcer rapidly learn to escape into the second chamber. It is not simply that the animals intrinsically prefer the second chamber, for if we carry out the same experiment with rats that have never experienced food in the first chamber, they do not learn to escape. Clearly non-reinforcement is not a neutral event for an animal. In the case of the omission of a positive reinforcer (reward), such as food, an aversive emotional state is generated, which will motivate the animal to escape from the situation. The emotional state induced by the omission of a positive reinforcer is usually referred to as FRUSTRATION.

So far we have only discussed the extinction of responses to positive reinforcers. However, conditioning can also result from the use of an aversive reinforcer, such as an electric shock (see RE-INFORCEMENT). If we expose a rat to a number of pairings of a light stimulus and a mild electric shock, the light will acquire aversive properties, in that the animal will make a response to escape from it. Presenting the light alone without the aversive reinforcer results in the extinction of these aversive properties. As in the case of an expected positive reinforcer, the omission of an expected aversive reinforcer will generate an emotional state, normally termed relaxation or relief. However, unlike frustration, relief is a pleasant or positively valued state, for animals will actively seek to expose themselves to stimuli paired with omission of an aversive reinforcer when they are in fear-inducing situations.

It is probable that the arousal of these emotional states by non-reinforcement is necessary for the generation of inhibition, and the consequent reduction of the conditional response. If a rat is trained to traverse an alley-way for food, and the response is then extinguished by omitting the food, the persistence of the response during extinction can be increased by administering a minor tranquillizer, such as sodium amytal, to the animal (see DRUGS AND BEHAVIOUR). The drug appears to decrease the capacity of non-reinforcement to arouse the aversive emotional state of frustration, and, as a result, its capacity to generate inhibition of the conditional response.   A.D.

86, 98

F

FACIAL EXPRESSIONS. According to a well-known saying, the face is the mirror of the soul. It is a good illustration of the importance we attribute to the face as a vehicle for the expression of EMOTIONS and attitudes. It is not surprising that facial expression was one of the first modes of non-verbal communication to receive scientific attention. Treatises were written on the subject as long as 200 years ago.

Just over a century ago, in 1872, DARWIN published a major work entitled *The Expression of the Emotions in Man and Animals*, which has become a classic in its field. Darwin was already famous as a result of two previous publications in which he had expressed his belief in the principle of EVOLUTION based on NATURAL SELECTION. He was convinced that the gradual process of evolution affected not only the physical features and morphological structures of organisms, but also their functioning, their behaviour, and the underlying mental processes. According to Darwin, human behaviour and the human mind are also products of evolution. Therefore the 'mirror of the soul' acquired new significance. Facial expressions could be used to demonstrate Darwin's theories. As Darwin pointed out, human beings are not the only species with facial expressions; cats (Felidae) and dogs (Canidae) have them too, as do monkeys and apes (Simiae). In the latter, who are our nearest relatives in the animal kingdom, facial expressions are particularly well developed. Darwin noted that facial displays in different species seemed to be governed by similar principles. He regarded this similarity as strong evidence that the facial expressions of human beings have evolved from those of our non-human primate progenitors. Darwin's contentions were based on limited observational evidence. Much of it was second-hand, and consisted of reports compiled by zoo-keepers and pet-owners. However, in the course of the last few decades a wealth of observational data has been assembled relating to human and non-human primate facial expressions. This new material in fact reinforces and supports Darwin's hypothesis that human facial expressions have evolved from those of man's non-human ancestors.

Investigations have proceeded along different lines and studies have differed in their orientation and emphasis. Psychologists and anthropologists have attempted to understand how man perceives emotions, and to what extent facial expressions provide clues for their recognition. Zoologists have studied facial expressions of various groups of mammals, particularly as an aspect of COMMUNICATION. Let us turn first to the zoological studies.

Facial expressions as communicative displays. Behaviour is not simply a random occurrence of behavioural elements. Elementary patterns are seen to occur in certain sequences or combinations of varying complexity and stereotypy. It follows that any behavioural element is potentially informative, in that it gives an observer, familiar with such behaviour, a clue about what is likely to happen next. There is a message. When a monkey looks at a neighbour and opens its mouth, this may be a sign that he is likely to attack. We know that the receiver has understood the message if he 'takes it into account', i.e., if he adjusts the probabilities of his own behaviour in response to the message, e.g., the neighbour is more likely to leave, or to attack himself than he would have been had he not seen the signal of the other monkey. The behaviour of the sender acts as a signal and the receiver attributes meaning to it.

If it is to the sender's reproductive advantage that other animals take its behaviour into account, natural selection will operate to adapt and modify the informative action, so that it becomes a more effective means of communication. Such adaptive changes may entail the exaggeration and RITUALIZATION of the act, thereby enhancing its conspicuousness and unambiguousness as a signal. Any perceptible movement or posture, even a glandular secretion (e.g., the shedding of tears) can be transformed into a DISPLAY.

The messages of the facial action units. Only in mammals have the structures of the face, such as the lips, the cheeks, a pliable movable facial skin, and the attached musculature differentiated sufficiently to allow the development of a rich repertoire of facial expressions. However, it is only in a few groups of mammals that the face has become an organ of rich expression. The highest differentiation is found in the higher primates: in monkeys, apes, and man. These are the species that have both adopted a more diurnal way of life and evolved comparatively good VISION.

Facial expressions are compound patterns consisting of a number of *action units*, namely the movements and postures of such facial elements as the mouth, the corners of the mouth, the lips, the eyes, the eyelids, the eyebrows, the skin of the forehead, and the ears. A possible way of establishing the signal value of facial expressions is to discover what FUNCTION the action units might have had before they became ritualized, in other words before they were modified to act as displays. Such functions are most probably connected with preparing for action, protecting vital parts of the body, and regulating sensory input. The most important ones are as follows. (i) Openness and receptivity to stimuli, as yet unknown, are increased by opening the eyes, raising the eybrows, and dilating the pupils. These action units are associated with the expectation of unknown or unpredictable changes, on occasions also with FEAR. (ii) When an animal concentrates on a given, known stimulus it steadies its face, eyeballs, and ears, turns them towards the stimulus source, and may frown. (iii) An animal may protect itself against nasty stimuli, or reject them completely, by closing its eyes and flattening its ears. Nasty substances which have got into the mouth can be removed by shaking the head, putting out the tongue, retracting the corners of the mouth and the lips, closing the glottis, or by coughing or sneezing. Similar expressions occur in DEFENSIVE situations (Fig. A). (iv) An animal may make INTENTION MOVEMENTS, preparatory to specific

Fig. A. Expression of slight defence in a tiger (*Panthera tigris*). Note retraction of lips, associated wrinkling of nose, and position of ears.

actions; it may open its mouth to bite (Fig. B), smack its lips in anticipation of grooming, put out its tongue to lick. (v) An animal may make movements that reflect its level of AROUSAL. There are specific responses which are associated with or which anticipate great physical exertion, e.g., the body becomes rigid, the muscles, particularly the throat muscles, are tensed with resultant retraction of the lower lip and the corners of the mouth; the teeth are clenched. (vi) Finally, facial movements may be the result of certain forms of VOCALIZATION. Muscles around the mouth and throat may

contract to support the tissues and vessels of the throat during strong vibration. This too may result in retraction of the lips and the corners of the mouth.

Fig. B. Even in the most primitive mammals one conspicuous facial expression is generally present, the *open-mouth face*. It is a bite threat, though not that of a determined animal. It marks a defensive attitude and is often accompanied by short lunges and by coughing or 'spitting' sounds. (Mongoose, *Herpestes ichneumon*.)

Evidence for ritualization of the facial action units. Undoubtedly these action units do often occur in their original functional contexts. However in many cases, the original functional connection is not evident or it is insufficient to explain the extent of the movements. Human facial–vocal patterns such as laughter, crying, and smiling seem to defy explanation in these terms. This is obviously because these patterns function effectively as SOCIAL signals. They have taken on a new form and operate according to new rules, both of which are in accordance with these new functional demands. Another argument for ritualization is that we find marked interspecific differences in the extent to which the action units are exaggerated. Man raises his eyebrows, macaques (*Macaca*) raise their eyebrows and flatten their ears, but apes hardly do either of these things. Man frowns more deeply than apes or monkeys, but chimpanzees have broader grins. The most bizarre form of lip retraction is seen in the gelada baboon (*Theropithecus gelada*); not only does it raise its upper lip, it turns it back over the nose. Such differences

Fig. C. Exaggerated brightness of the upper eyelids accentuates eyelid and eyebrow movements in the mangabey (*Cercocebus*).

cannot be explained in terms of the original functions of the action units.

An aspect of ritualization is also the development of morphological features that accentuate the display movements. In some species the conspicuousness of eyebrow raising is greatly enhanced, because the skin above the eyes, which is stretched and exposed, is much less pigmented than the surrounding skin (see Fig. C).

The description and analysis of facial displays. The investigator of animal facial expressions must begin by making a detailed description of the patterns of facial movement. In attempting to describe the displays of mammals he meets a problem which is much less acute in the case of fishes, birds, let alone insects. In those groups displays usually are discrete and distinct from each other, because each display consists of a rather fixed combination of movements and postural elements. Mammalian displays, on the other hand, seem to merge or grade into each other almost continuously. Many of the elements constituting the expression even seem to vary independently to a large extent. Because of this an early investigator of the social behaviour of wolves (*Canis lupus*) maintained that it would be impossible to dis-

tinguish the fixed action patterns that we can distinguish in non-mammalian species. LORENZ, however, took a different view. He felt sure that the apparently chaotic variation in patterns could be reduced to a few basic principles. Some facial elements are particularly suitable for the expression of protective and avoidance responses, others are suited to express the inclination to approach and attack. Different combinations of these facial elements yield a great variety of expressions, but in fact all of them can be said to represent degrees of only two basic inclinations, the inclination to attack and the inclination to escape (see MOTIVATION, Figs. E and F).

Categories of facial–vocal displays of primates. There is a great diversity of primates, ranging from primitive prosimians, such as lemurs (Lemuridae), and bushbabies (Galagidae), to monkeys and apes, and finally, to man. Nevertheless, the diversity in facial expressions is comparatively small; the different species appear to use basically the same repertoire. We shall take as our example a macaque-type primate, because it displays characteristics that are common to many primate species. The main facial expressions are depicted in Fig. D.

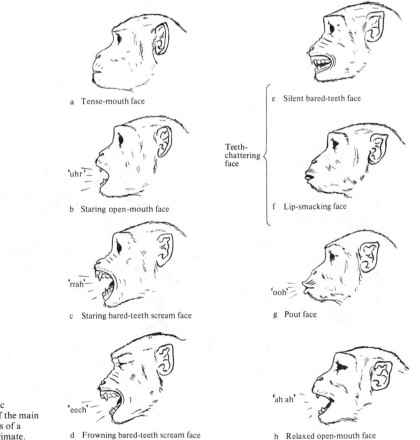

a Tense-mouth face

b Staring open-mouth face
'uhr'

c Staring bared-teeth scream face
'rrah'

d Frowning bared-teeth scream face
'eech'

e Silent bared-teeth face

Teeth-chattering face

f Lip-smacking face

g Pout face
'ooh'

h Relaxed open-mouth face
'ah ah'

Fig. D. Schematic representation of the main facial expressions of a macaque-type primate.

The standard for comparison of expressions is the *neutral* or *relaxed face*, shown when the animal is not engaged in any particular activity. Further relaxation of the facial muscles leads to the expression of drowsiness. The opposite effects are observed in the *alert face*: muscles tense, the movements of the body, head, and eyes become more rapid and abrupt, the eyes open more widely, etc. All these characteristics are associated with general changes in responsiveness, and can be observed either within or outside social contexts. In addition there are ritualized displays that express a more specific attitude.

First let us look at the four agonistic expressions on the left of Fig. D. Roughly speaking the tendency to attack is greatest in the uppermost expression and decreases or is kept in check by an increasing tendency to flee as we go down.

In the *tense-mouth face* (a), the forward tendencies in gaze and body posture are strongest; the jaw muscles are tensed, ready for quick action, the corners of the mouth are pushed forward. The pose signals determined vigour and concentration. If directed towards a fellow animal, this spells AGGRESSION, and clearly functions as a strong THREAT (Fig. E).

The steady stare and the forward tendencies are less pronounced in the *staring open-mouth face* (b).

Sometimes a hesitation to push through is even visible in the body posture (Fig. F). The open mouth can be interpreted as an intention movement to bite. In a confident animal the lips are drawn over the teeth and the corners of the mouth are pulled forward. The more afraid the animal, the more the teeth begin to show. The expression may be emphasized by short harsh grunts, and functions as a warning.

Further retraction of the lips and the corners of the mouth, together with an increased pitch of the vocalizations, leads to the *bared-teeth scream faces* (c and d). These occur in a diversity of situations. Monkeys and apes may show this expression when something they expected to obtain is withheld from them, or when a specific behaviour is thwarted; an important condition is that the animal perceives, rightly or wrongly, that the thwarting is due to the interference of another stronger animal. The expression then operates as a signal of FRUSTRATION and may make the interfering animal give in; it undoubtedly derives from distress responses that infants of many species are known to direct primarily to the mother. Often the expression results from aggression on the part of a fellow animal. The action units that accompany the facial expression reveal whether the animal is inclined to resist and fight (it shows forward ten-

Fig. E. *Tense-mouth face* in non-social context; a chimpanzee concentrates while aiming a ball at a grape.

Fig. F. *Staring open-mouth face* in a crab-eating macaque.

dencies and stares fixedly), or whether it is primarily motivated to withdraw (it flees or crouches and frowns and looks away). The vocalization varies accordingly from short abruptly demarcated shrill barks to longer drawn-out screeches.

The *silent bared-teeth face* (e) might be regarded as a low-intensity phase of the vocalized bared-teeth displays. Indeed it often occurs as a transition phase to the latter. In some of the higher primates, however, this face seems to have acquired an independent specific role in the regulation of social relations. It is often shown in contexts of impending contest; it is then directed to stronger animals and denotes a submissive attitude. The tendency to withdraw may show itself in shrinking movements, averted posture, or evasive glances (Fig. G). There is no obvious tendency to be aggressive. By thus relinquishing possible claims the animal may avoid provoking, or appease, stronger competitors, and may even be allowed to remain in their vicinity, albeit as a subordinate in a compliant role. In a few species the silent bared-teeth face may be shown, on occasion, also by dominant animals to subordinates; the confident animal reassuring a companion about its intentions. There are no furtive aspects to the expression now; the performer moves slowly, so as not to frighten the companion. Its message is one of non-hostility. In some species (e.g. the chimpanzee) it has even become an expression of affection or friendliness, used to reinforce attachment (Fig. H). One can argue that the corresponding expression in man is the *smile*.

To us the *lip-smacking face* (f) that many monkeys use to express friendliness and non-hostility is strange. The central action unit of the pattern is rapid opening and closing of the mouth, often accompanied by rhythmic protrusion of the tongue. Similar movements can be observed during GROOMING when an animal searches

through its own fur or that of a partner and puts the salty particles it finds into its mouth. Social grooming is normally performed in a relaxed atmosphere between animals that have a friendly relationship, such as close relatives or sex-partners. In view of this, it is understandable that lip-smacking, one of the most conspicuous elements, has come to function as a signal which, even when given at a distance, can advertise a positive attitude to another animal, and be an invitation to closer contact.

If the performer is at all uncertain, this is revealed in furtiveness, evasive glances, and the like. With increasing uncertainty, the lips tend to be retracted, leading eventually to the *teeth-chattering face*: it expresses an attitude at once friendly and submissive. In a few other species such as the Barbary macaque (*Macaca sylvana*) or the South-American capuchins (*Cebinae*) the teeth-chattering face has become an intense variant of the affectionate lip-smacking.

The *pout face* (g) is characterized by funnelled lips and plaintive, melodious, drawn-out calls. It is

Fig. G. *Silent bared-teeth face* in a crab-eating macaque.

Fig. H. Reconciliation between two adult male chimpanzees. The animal on the left asks for reconciliation showing a *silent bared-teeth face*. Note also the arm gestures.

commonly shown by young primates when they have become separated from their mother (Fig. I). The pout face seems to signal helplessness, and is very effective in bringing the mother to 'mother' the young one. Upon reunion the juvenile immediately grasps a nipple with its mouth. This suggests that the characteristic mouth posture is a ritualized intention movement of reaching towards something and sucking. The adults of some species, e.g. chimpanzees, pout in this way too, but then it is a general begging display.

The *relaxed open-mouth face* (h) is one of the most fascinating primate displays. Its mouth posture resembles that of the staring open-mouth face, only the corners of the mouth are in the neutral position or are slightly retracted. It also lacks the fixed stare, the rigid face and body posture, and the brusqueness of movement. Everything is relaxed yet lively. The expression occurs during social PLAY or as a prelude to play. Social play is a boisterous affair consisting of chasing, grasping, wrestling, gnawing, etc. (Fig. J). The relaxed open-mouth face can be regarded as a ritualized intention movement of gnawing. It serves to indicate the playful nature of the interaction, emphasizing that it is mock fighting and not the real thing. It may also function as an invitation to play.

These categories are rather crude but include the basic patterns found in the average higher primate. How does man fit into this picture? In order to find out we shall look at facial expressions in our own species.

Human facial expressions and emotions. Whereas in ETHOLOGY the study of animal expressions relies on the analysis of their occurrence in social interactions, the investigation of human facial expressions has always rested heavily on a method which is not applicable to animal behaviour. Man can report on his experiences, on what he perceives to be the motives for his behaviour, and on the meanings he attributes to the behaviour of his fellows.

There is a difference in theoretical emphasis too. Whereas ethologists try to understand facial expressions in terms of their role in communication, psychologists studying human expressions are

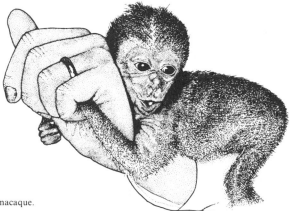

Fig. I. *Pout face* in an infant crab-eating macaque.

interested particularly in the underlying EMOTION. They seek to answer such questions as: Do facial expressions reflect certain emotions? How accurately do facial expressions portray certain emotions? How many categories or dimensions of emotion can be distinguished? Evidence has commonly been based on investigations in which observers are required to judge still photographs or posed behaviour, according to certain preconceived categories or arbitrary scales. The essential difference between the COMPARATIVE ethological and the classical psychological approach can be illustrated, albeit somewhat roughly, by the different interpretations given of smiling. According to the former approach it is an indication of a positive attitude, tending to appease, to reassure, and to increase attachment; in the latter it is regarded as a manifestation of the emotion of happiness. In spite of these differences there seems to be a good case for attempting a comparison of human and non-human primate facial expressions.

Fig. J. *Relaxed open-mouth face* as an intention movement to gnawing during playful gnaw-wrestling in crab-eating macaques.

Although investigators have made many studies of the relation between human facial expressions and the emotions, their conclusions have differed widely. Some sceptics have even questioned whether the face is an efficient vehicle for the expression of the emotions anyway, and have suggested that the emotions said by observers to be associated with certain expressions were inferred from the situational context rather than from the expression itself. Others have argued that the confused picture is due to many methodological differences and shortcomings; that in spite of the numerous contradictions, in all the major studies to date, seven basic categories of emotions consistently emerge; these are unambiguously associated with particular facial expressions. The seven categories are as follows: (i) *Interest:* ATTENTION and receptivity are evident from the directional character of the posture (e.g., craning neck) and gaze. A state of readiness is indicated by slight tension, which is also visible in the face. When a subject is alert and expectantly receptive, he raises his eyebrows slightly; when he is concentrating he frowns. This expression is clearly comparable with the alert face described earlier. (ii) *Surprise:* In its most ritualized, emphatic form the characteristic features are extreme eyebrow raising, eyes open wide and mouth limp and open. This display does not seem to have a ritualized counterpart in non-human primates. (iii) *Anger, rage:* The corresponding facial patterns vary from the tense-mouth face to the patterns described as the staring open-mouth and bared-teeth scream faces (if rage is more intense). The angry sneer with squared lips has not been observed in non-human primates. A morphologically similar expression in the mandrills and drills (*Mandrillus*) has a different origin, and a positive meaning, and should be classified as a variant of the silent bared-teeth face. (iv) *Fear, terror:* The expression of these emotions takes forms that ultimately resemble the tendency to retreat seen in the silent bared-teeth and bared-teeth scream faces. (v) *Disgust, contempt:* These emotions are accompanied by retraction of the mouth-corners, raising of the upper lip combined with wrinkling of the nose, frowning, and screwing up of the eyes, turning away of the face; all these elements were originally aimed at excluding and repelling obnoxious stimuli. Although the same reactions may be shown by non-human primates, for instance if smoke is blown into a chimpanzee's face, it is merely a protective response; the pattern has never been observed in a ritualized communicative form. (vi) *Sadness, anguish:* These seem to have become ritualized. A dejected attitude is accompanied by a 'sad' frown with arched eyebrows, retracted mouth-corners, pulled downward, and outward curling lips. In more intense expressions sobbing, crying and the shedding of tears may be added. The display often induces support and comfort from others. In form and function the expression seems to be a mixture of the infantile begging gesture, the pout face, and some aspects of the bared-teeth face (cf. frustrated screaming). Only in man and in the chimpanzee does this morphological blend occur; in the latter it is known as the *stretched-pout whimper*, which combines the outward curling lips of the pout face with the retracted mouth corners of the bared-teeth yelp face and a plaintive wavering call. (vii) *Happiness, joy:* These emotions are expressed by smiling and by laughter. The whole demeanour may be rather serene and tranquil (during smiling), or more animated if not boisterous (during laughter). The most typical feature is the broadening of the mouth, which involves retracting and turning up the corners of the mouth. This movement may be accompanied by various degrees of teeth-baring and mouth-opening. Many scientists and philosophers have noted the relation between laughter and humour, and have therefore considered laughter to be a uniquely human characteristic. As we shall try to show, there are good reasons for questioning this view.

Are laughter and smiles exclusively human expressions? Any zoo-keeper will readily concur that his chimpanzees laugh. What he refers to is the relaxed open-mouth display which we have come to know as a sign of playfulness. In fact, there is a striking resemblance between this chimpanzee expression and human laughter. However, although the form is similar, it is not the same. The chimpanzee breathes quickly and shallowly, emitting, in quick succession, short staccato grunts that sound like 'ah, ah, ah' (Fig. K); human beings, when they laugh, utter a peal of 'ha's' in one long expiration, then take another breath before the next peal. In the chimpanzee each 'ah' expiration is immediately followed by a short inhalation; the 'ah's' can continue for some time. There is also a difference in the positioning of the lips; in most non-human primates the upper lip remains covering the upper teeth, and the lower lip usually covers the lower teeth entirely or in part. By contrast man's upper teeth and gums are often bared, especially when laughter is ostentatious or particularly intense.

Fig. L. *Relaxed open-mouth face* in a capuchin monkey, while being tickled by a human being.

Fig. K. *Relaxed open-mouth face* ('laughter') in a young chimpanzee being tickled by its mother.

Chimpanzees utter the 'ah-ah' sound especially when there is some sudden change in the play interaction involving a wrestling type of contact or chasing (Fig. L). The essence of play for chimpanzees is that one animal plays tricks on the other, trying to spring a surprise whenever possible. Of course each 'player' expects a tickle, a tackle, or a turn, but the element of surprise is in their precise location and timing. One can test the theory by tickling a tame young chimpanzee and noting which particular move triggers off the 'ah-ah' sounds. Similarly, an adult tickling a child can make the child giggle or laugh (Fig. M). The giggle or laugh is produced when the expected plus unexpected occurs in a playful atmosphere.

In human adulthood laughter is no longer restricted to the context of purely physical mock-fighting. Many distinguished scholars have written essays about the conditions that induce laughter. Not only were they baffled by the form of laughter, which seemed to defy any functional interpretation (why should we shake, uttering peals of laughter, bend double, or slap our thighs when we hear something funny?), they were also intrigued by the essence of 'the comic' which caused the laughter. Clearly laughter was not simply a release of pent-up tension, for that could occur via a sigh of relief. The fact that one laughs less when reading a joke on one's own indicates that laughter calls for a social reference, i.e., someone to laugh with or to. However abstract and intellectual a laughter-provoking event may be, it must have a socially relevant mock-serious element if it is to be really funny. From these considerations we can justifiably conclude that the expression of human laughter is related to the relaxed open-mouth face of the non-human primates.

Can we say the same about smiling? People have always tended to regard smiling as a low intensity, subdued form of laughing. Laughter, it is true, often begins and ends as a smile. However, just as there are diminutive forms of laughter involving slight mouth-corner retraction and weak laughter vocalization, in the form of a few chuckles or snorts, there are also intense, broad smiles, used ostentatiously as a form of greeting, excuse, or reassurance. Obviously it would be impossible to fit all forms of smiling and laughing on one intensity scale: there seem to be at least two dimensions: a broad-smile/greeting dimension, and a laughter/play-joy dimension.

The comparative evidence shows that in most species the silent bared-teeth face is simply a gesture of fear that subordinate animals display to appease stronger animals and to acknowledge their dominance; in some animals however, in the chimpanzee for instance, it serves as a reassuring and attachment-promoting gesture. In the chimpanzee it has even taken over the functional role that in most other primates is performed by the lip-smacking face. It is no longer used as a gesture of reassurance or greeting, and has been retained only in its original context during grooming.

Lip-smacking does not occur in man, unless one regards as a last vestige of this the slight tongue protrusion observed in some contexts, such as courting. We can regard human smiling as a fully emancipated silent bared-teeth display, and laughter as a relaxed open-mouth display. Although laughing and smiling have begun to resemble each other both in appearance (both having baring of teeth) and function (both can be used to 'break the ice'), their typical forms seem to be related to two quite distinct displays in primate phylogeny.

Comparability of human and non-human primate expressions. To summarize, we can say that all non-human primate facial displays, with the notable exception of lip-smacking, are represented in the human repertoire of emotional expression. On the other hand, two human displays (disgust or contempt, and surprise) do not have obvious ritualized counterparts in non-human primates. Both patterns do occur, however, in an unexaggerated form in their original functional context. In contrast, however, expressions such as crying, smiling and laughing, which are often regarded as specifically human, do have ritualized equivalents in non-human primates.

Nature–nurture aspects of facial expression. Elsewhere it has been argued that it is difficult to categorize types of behaviour according to whether they are learned or INNATE (see ONTOGENY). One can, however, try to find out at what stages of individual development particular types of behaviour acquired certain forms, and during what phases and through which influences the development can be modified.

Non-verbal communication depends on (i) encoding, i.e., the production of expressive movement under the appropriate conditions by a sender, and (ii) decoding, i.e., the interpretation of that movement by a receiver. We have fairly convincing evidence that the encoding of facial expressions is largely independent of social LEARNING. Such evidence is provided by studies on non-human primates reared in isolation without social companions. For instance rhesus macaques (*Macaca mulatta*), reared in isolation, came to display the basic forms of expression such as the staring open-mouth face, the various bared-teeth faces, and the lip-smacking face. However, their responses proved to be very crudely programmed; these socially deprived animals did not have that precisely timed responsiveness to social circumstances which enabled them to effectively control and coordinate their social relations via their facial expressions. Evidently it is only as a result of considerable social experience that the application of their expressions becomes sufficiently subtle and adjusted for such purposes (see SOCIAL RELATIONSHIPS).

The decoding of facial expressions is also partly independent of social experience. This has been shown in a LABORATORY experiment in which young rhesus monkeys were housed from birth in isolation cages. Colour slides of various scenes and subjects, including portraits of other monkeys with a variety of expressions, were projected on to one side of the isolation cage. When 2–3 months of age, an isolated monkey was disturbed when shown slides of a staring open-mouth face. The monkey was allowed to shorten the projection time of slides which he disliked; when he saw slides of threatening expressions he terminated these sooner than other slides. At 4 months of age however, he gave up differentiating between the types of slides.

In another classic experiment monkeys were taught to press a lever on hearing a certain sound. If they failed to press the lever they received a mild electric shock. Then a cooperative situation was created. One monkey sat in a compartment where he could hear the sound, but had no lever to press. Another monkey in a different compartment had a lever but could not hear the sound. If the latter failed to press the lever within a certain period after the sound both monkeys received a shock. Of course, the monkey with the lever could not learn to press at the appropriate moment. However, when he was able to observe his partner's face on a television screen he did learn to press correctly, and thus prevented the shocks. This proves that facial expression (in this case an expression of fear) is communicative. When one of the monkeys had been reared in complete isolation the coordination was successful if that monkey was placed in the 'sender' position, but the success rate was con-

Fig. M. Wide-open mouth laughter by a human infant in a tickling game. The wide-open mouth version without retracted lips gets rare in adulthood.

siderably lower than if the sender had been reared normally. So here too we see that social experience has an influence. When the monkey reared in isolation was placed in the 'receiver' position there was no coordination. Clearly he failed to understand his partner's expression. This finding indicates that social experience is of much greater importance for the understanding of expressive behaviour than it is for its performance.

In man too the basic repertoire of emotional expression develops almost normally, even if the social and CULTURAL factors that might have influenced the shaping of such expression are absent. This has been shown in studies of children deaf and blind from birth. Their crying, laughing, and smiling and their expressions of anger, sadness, and fear develop at roughly the same time as in children with normal hearing and sight. Blind children who are able to hear, however, develop more differentiated and complex patterns of expression (coquettish embarrassment has been observed, for instance) than children who have been born deaf and blind. Obviously social experience is again playing a role here. However, it is not the expressive displays themselves which are learned through IMITATION and social experience, but it is the differentiation of attitudes which is promoted; this in turn gives rise to more complex expressions. In the ever dark and silent world of the deaf and the blind such differentiation is not usually instigated.

The species-specific character of expressive displays. According to some anthropologists displays are simply a product of culture, and both their form and meaning are governed by cultural rules. These scholars deny the existence of universal symbols of emotional states, and stress the intercultural differences that exist in communicative behaviour.

It is difficult to reconcile this view with the evidence relating to the deaf and blind children. The conviction that there are no universal expressions has been shaken by comparative studies carried out in a variety of cultures, some very remote. Studies and films of expressions and gestures of people in remote parts of the world show a striking similarity in the form of many facial expressions and gestures. This is in marked contrast to the variability found in other types of expression and in products which are commonly accepted to be culturally based. One of the gestures found to be common to all cultures studied is the *eyebrow flash* which serves as a friendly greeting over a short distance. It is a stereotyped sequence incorporating a smile and the raising of the eyebrows for about one-sixth of a second, followed by a nod of the head. The whole sequence lasts 1–2 s. Most people, it seems, are barely aware that they use this gesture.

In many of the investigations into the recognition of facial expressions, pictures or videotapes of facial expressions are presented to members of different cultures, some literate and civilized, others illiterate and remote. There is a remarkable measure of agreement in the interpretations. This leaves little doubt that both the production and the interpretation of human facial expressions are basically universal. Cultural variation has been found in what are termed the display rules of expression. This is illustrated by an experiment in which Japanese and American students were shown either a stress-inducing and unpleasant film on sinus surgery, or a more neutral film. While the students were watching, and unknown to them, their facial expressions were being filmed. Analysis of their expressions showed that the Japanese and the Americans had similar reactions to the films. However, when later interrogated by compatriots about their experiences the Japanese who had seen the nasty film smiled happily as they commented on it, whereas the Americans clearly showed signs of disgust. This does not mean that the Japanese express disgust by smiling, but what it does mean is that there are conventional rules concerning the appropriateness of expressions of attitudes in certain contexts; even Japanese children have to learn the habit of smiling to mask disgust and contempt.

Can rules for the use of facial expressions be passed on by tradition in animal societies as well? Some evidence suggests that they can. In a study in which three groups of wild Japanese macaques (*Macaca fuscata*) were compared, members of all three groups showed the *lip-quiver* (a very rapid kind of lip-smacking), in which the puckered lips remain together. In two of the groups this expression was frequently used during COURTSHIP, typically by the male towards the female, but it was never observed in this context in the third group.

Every species studied so far has shown universality and uniformity in the basic aspects of facial expressions over its entire geographical range, but not necessarily in the display rules. This even holds for our own species with its great diversity of races and cultures. However, there are considerable differences between species. For example, in crab-eating macaques (*Macaca irus*) teeth-chattering expresses a certain degree of fear, but in Barbary macaques it is a display of intense affection.

It has happened occasionally that an animal has been reared exclusively in the social milieu of another species. Monkeys and apes, for instance, have been raised solely by human beings. In spite of the alien milieu, such animals invariably develop the characteristics of the expressive repertoire of their own species. Such interspecific diversity, therefore, cannot be attributed to accident, tradition, or to environmental factors characteristic for the milieu of the species; the differences must be GENETIC.

The differences in facial expression between species tend to be greater the less the species are judged to be phylogenetically (evolutionarily) related. For example, macaque species all have their own form of lip-smacking, but apes do not engage in lip-smacking at all. This strongly suggests that there is evolutionary continuity, and that the pattern of diversity can be regarded as an instance of evolutionary differentiation. Consequently, the similarities in morphology and in the programming of the expressions can be regarded as *homologous* (see COMPARATIVE STUDIES).

We shall never know for certain what the expressive behaviour of ancestral species was like. Theories about the EVOLUTION of a behaviour pattern can be based solely on the distribution of behavioural variants in extant species. This scheme, of course, should not be taken to mean that man descended from the chimpanzee, and the chimpanzee from the macaques. Instead we must regard these species as representatives of primate types that have existed for different time spans. The essence of the scheme is that, contrary to what we would expect at first sight, smiling and laughing have a different evolutionary background.

Man's use of facial expressions. The sections above have been concerned with non-verbal communication. Verbal communication is a capacity almost unique to man. Although the chimpanzee and the gorilla (*Gorilla gorilla*), man's nearest relatives, seem to have a basic linguistic competence, man's performance by comparison is markedly superior. Human non-verbal communication is closely associated with LANGUAGE, and is greatly influenced by it. Communication in animals is primarily involved with expressing attitudes towards social partners, thereby regulating social relations with these partners, and secondarily with giving information about the non-social environment, such as food sources, predators, nest sites, etc. The reference to these is again for the most part made by expressing a behavioural attitude towards them. Man also uses this type of communication. The information transferred by human language, on the other hand, relates to a much greater extent to features of the environment distant in time and place, and even to abstract concepts. The advent of the verbal mode of expression has greatly influenced the non-verbal mode, be it by posture, gesture, facial displays, or voice.

Man uses emotional expression more extensively than other primates to convey appreciation of the non-social environment. Possibly it is in this context that expressions such as those of surprise and disgust have evolved, expressions which do not have clear, ritualized counterparts in non-human primates. Man also uses the non-verbal mode to support and regulate the process of verbal interaction. Four forms of this can be distinguished. Thus we may accentuate spoken words or phrases, not only by the tempo, intonation, and tonal quality of the vocal utterance, but also by movements of the hands and head, or by movements of the eyebrows and eyelids if the face is involved. Such movements to give emphasis are referred to as *batons*. In conjunction with head movements such as nods, facial expressions can also serve as *regulators*, signalling engagement in conversation, bringing about coordination of verbal interaction, the taking of turns, etc.

Posture, gesture, and facial expression often parallel verbal communication as an additional source of information. First, there are movements or gestures which depict features of the environment, such as spatial relations, forms, and dimensions. Such graphic gestures have come to be known as *illustrators*. (An example might be the gestures of an angler boasting about his catch.) These are performed mainly with the hands, the face having limited capacities in this respect. Facial expressions can, however, be used as illustrators when a subject is reporting on his own or other people's attitudes. Second there are *emblems*. These are closest to verbal communication in that they are actions which can be directly translated by a few words or a short phrase. Winking and sticking out the tongue are instances of emblems in which the face plays a major role. Emblems are used particularly if, in the circumstances, verbal messages are impossible, less efficient, or undesirable. They have a more ostentatious, conscious, and deliberate character than the other three categories of signals. Compared to other categories they lack universality; their meaning is largely a matter of tradition. In every culture or subculture a large variety of emblems are known.

As a rule emblems are compound patterns, involving particularly the arms and the hands, but also the body, the head, and facial elements. Sticking out the tongue is one of the few emblems that are primarily facial. It can have several meanings depending on the context and the manner of execution. In many cultures there is a form with rather brusque protrusion which is intended as an insult or as a provocation: *the rude tongue*. This could be an exaggerated version of the tongue protrusion we have met before as a rejection response. In addition there is the sexual tongue, which may function as an erotic challenge. Its effectiveness as a signal may benefit from the phallic associations it evokes. In courting situations tongue protrusion may also be performed with unconscious spontaneity. Then the action does not really qualify as an emblem.

The smooth, curling movements of the tongue have been interpreted as a vestige of the infantile licking movements that formed part of the mouth-to-mouth feeding process in prehistoric times, when growing youngsters received their first solid food in a pre-chewed state from their mothers. This behaviour, which is still seen in aboriginal cultures, is also thought to be the origin of the human courtship interaction known as the tongue kiss.

This example shows that very diverse meanings can be attributed to an emblematic display in different contexts, and how essential it is to be familiar with a culture or subculture if such a display is to be correctly interpreted.

There is little doubt that the human species has the richest repertoire of facial expressions, with its multitude of blends, even including asymmetrical ones where one half of the face has a slightly different expression from the other. Yet we may be biased in this judgement. By virtue of being human, we have acquired tremendous skill in spotting the slightest changes in human facial expression, whereas we may be very much less skilled in discerning facial changes in non-human primates, or may put an ANTHROPOMORPHIC interpretation upon them.   J.R.A.M.H.
41, 76

FARM ANIMAL BEHAVIOUR. Farm animals, like wild animals, must feed, respond to apparent danger, mate, and maintain their body state. Hence, behavioural components of these essential life processes persist in farm animals and in HOUSEHOLD PETS despite being modified by DOMESTICATION. The frequencies with which many activities are shown in farm conditions may be very different from those shown in the wild, but few activities have completely disappeared from the repertoires of farm animals. Some of the sequences of behaviour serving particular functions are probably little different from those shown by the ancestors. Patterns of movement in grooming, mating, eating, caring for young, or attacking other members of the species have changed little, but the likelihood that a behaviour pattern will be shown in any particular situation and, perhaps also, the frequency of performance of the activity, may well have been changed during domestication. The selective breeding procedures have varied from species to species according to the ways in which man uses the animal. Some behaviour modification has been the result of direct selection for behavioural characteristics, for example docility in the larger mammals, or lack of broodiness in domestic chickens (*Gallus g. domesticus*). Selection for physical characteristics has also had effects on behaviour. For example, larger individuals perform various activities more slowly. The selection of both bulls (*Bos primigenius taurus*) and turkey cocks (*Meleagris g. gallopavo*) for large size and for body forms which give good meat distribution has had adverse effects on mating ability, and on the likelihood that mating behaviour will be shown.

Farm animal breeding and methods of animal husbandry have long been directed towards improving the efficiency with which the food the animal consumes is converted into animal protein. The consequent levels of animal production have been achieved with, until recently, little regard to behavioural factors. A point has now been reached where the rate of improvement in animal production has slowed down, and behavioural factors are assuming a much greater importance. This is especially true because many farm animals are now kept close together in large numbers. Further improvements in production and in the procedures for handling farm animals will depend, in part, on the results of studies of MOTIVATION, FEEDING strategies, PARENTAL CARE, responses to stressful situations, and other behavioural mechanisms.

Feeding behaviour. One major behavioural problem for animals is to find, obtain, and ingest food. Man makes food finding much easier for farm animals than it would be for wild animals (see FORAGING), but FOOD SELECTION is still practised in many farm situations, and avoidance of unpalatable food is always possible. Grazing animals have some choice as to which food they eat, as well as when and how fast they eat. Pasture will always contain several species of plants, and, within a plant species, there will be plants of different ages and with different proportions of stem and leaf. Some plant material will provide a better balanced diet for the animal, and there is some evidence that cattle adjust their feeding so as to compensate for nutritional deficiencies in their diet. If cows are fed on the species and height of grass which they prefer, they increase their body weight faster than they would with less acceptable food. Herbage which is tainted with faeces is also avoided, and this behaviour is important since slurry, the washings from animal houses, is an important grassland fertilizer. The study of grazing preferences and tolerances is therefore important for the efficient production of healthy animals. Feeding preferences are also extensively studied where the animals are fed artificial food. The only choices for the animal, in most circumstances, are whether or not to eat, and how much to eat. Dairy calves are initially fed milk substitute, so that man can drink the cows' milk. In winter many animals are fed concentrated foodstuffs, as well as hay or silage. The composition of the artificial foodstuffs, and the methods used to produce the silage, can be altered so as to change the taste as well as the nutritive value of the food. The farmer must ensure that the animal finds the food palatable enough to eat some of it. Ideally it should eat an amount which results in efficient conversion of food nutrients to animal protein.

Farm animals whose food is presented to them in a trough often have to compete for that food with others housed with them. When the food or the trough space is limited, those individuals which are subordinate in competitive interactions will often get less than their fair share. (See DOMINANCE.) If individuals are unable to consume much food they grow slowly, hence it is useful to study competitive interactions and behaviour when food is presented at different times and in different sorts and numbers of troughs. Animals rapidly learn that fast eating is essential in a competitive situation. They can also learn to open flaps or to press sequences of levers in order to obtain food or some other reward. Pigs (*Sus scrofa domestica*) may grow faster if they are allowed greater control over the timing of food presentation, the nature of the food provided, the temperature, and the lighting in their pens.

Housing conditions and the composition of social groups of farm animals are usually dictated by immediate economic considerations. Observations of farm animals help to detect inefficiencies in husbandry methods, but in trying to create conditions for farm animals which are nearer to the optimum for the species, it is useful to look at closely related wild species, or populations of domestic species which live in a semi-wild (feral) state. Studies of captive jungle fowl (*Gallus*), of feral chickens, and of Soay sheep (*Ovis ammon aries*), have provided valuable information on flock formation, space requirements, and SOCIAL, SEXUAL, and MATERNAL behaviour. Experimental work, in which attempts are made to improve animal production efficiency by varying housing conditions and stocking density (number of animals per unit area), benefits from these insights into the basic requirements of the farm animal species. Such work must also take account of the costs of housing and management. The time during which an animal is engaged in any particular activity is inevitably altered by the restriction imposed by a fence or pen. In many cases differences in behaviour between wild and captive species reflect the degree of ADAPTATION of the animal to changed circumstances, and the results of selection by breeders for animals with certain behavioural characteristics. For example, animals which do not need to spend long periods looking for food, or for a mate, may SLEEP more and grow more. As an example of the influence of behavioural studies on husbandry methods, the use of battery cages for poultry may be considered. Some strains of hens should not be kept in battery cages of a certain design because the hens spend a lot of time pacing up and down and tend to lay when standing, thus cracking some eggs. Preference tests with hens have provided information about the optimum design for nest boxes, and the best floors to use for cages. Other studies have shown that a certain combination of group size, cage size, and cage shape results in the fastest rate of egg production. As economic pressures lead to more intensive farming with greater stocking densities, and larger total numbers of animals in a unit, behavioural factors increase in importance, and more detailed studies become necessary.

Animal husbandry and welfare. The major consideration of farmers in determining which husbandry methods to use has to be economic. Most are also interested in their animals as individuals, so that standards of ethics and aesthetics relating to animal living conditions will affect what they do. If chickens are crowded in cages to such an extent that some of them die or are mutilated, most farmers and most members of the public who consume the poultry products would consider such housing unacceptable on moral grounds. Most people have the same response to the practice of rearing calves for veal in pens which are so small that the animals are unable to walk when taken out of the pens for slaughtering. This is an example of a situation where the amount of meat produced is very small in relation to the feeding costs; the practice is inefficient in production terms, and merely supplies a specialized market. Domestic animals kept in conditions in which there is little variety of sensory input may perform stereotyped movements which are repeated for very long periods. These may have physically deleterious effects, and they certainly result in utilization of energy and consequent reduction in production efficiency. Sometimes the stereotypies can be attributed to specific deficiencies in the surroundings. Calves or lambs removed from their mothers and fed milk from a bucket may suck their pens, and also parts of other individuals, thus causing damage to themselves and to the other animals. This behaviour can be reduced by the provision of artificial teats in the pen, and by early weaning. Isolated calves may eat their own hair, thus forming hair balls in the rumen. This behaviour is drastically reduced if straw is put on the floor of the pens. The presence of straw on the floor also reduces tail-biting in pigs, and both mastitis and trampled teats in cows. Feather-picking is a problem in poultry housed in intensive conditions, but can be reduced by increasing the proportion of fibre in the diet. Many of these ABNORMAL behaviours are not shown if the animals are kept in conditions where there is greater variety in the surroundings, and greater space available for each individual. Since animals which are suffering discomfort, or which are stressed in other ways, often show worse than average efficiency of meat, milk, or egg production, living conditions which are unacceptable on moral grounds are sometimes undesirable on economic grounds. Where the main aim in agriculture is to obtain the maximum efficiency of production from each individual animal, there is little conflict between the economic and moral pressures. Difficulties do arise in relation to problems of capital outlay on accommodation and of labour costs. Cattle kept indoors in large stalls produce more milk and suffer fewer injuries than do those kept in smaller stalls, but the improvement in individual production has to be balanced against the costs of the large buildings. Individual feeding of animals kept in groups might be desirable in order to ensure that all get adequate food, but it would be too time-consuming for farm staff. Research into the behaviour of animals housed in various conditions helps to determine which of the conditions leads to the most efficient production by individual animals, and which result in the least abnormalities of behaviour. Decisions as to which condition to utilize depend upon operating costs, as well as production by individuals. They must be subject to certain minimum requirements giving opportunities for basic activities by the animals, for example those specified by the advisory services of the Ministry of Agriculture, Fisheries and Food in the United Kingdom (see WELFARE).

Social behaviour. The effects of group competition and stocking density on animal production are best understood by studying *stocking behaviour*. Many studies of social behaviour in farm animals have concentrated on competitive interactions, and the rank order which can be compiled as a result of analysing the outcome of a large number of such interactions. Many studies using farm animals have shown that few of these rank orders are linear, and there are examples of situations where A beats B, B beats C, but C beats A; or of more extensive circular relationships. Rank orders can also be compiled on the basis of leadership in movement from place to place, of position in a herd when driven, or of overall levels of activity. Those based on competitive interactions may be different according to the number of individuals in the group, the circumstances of the encounter, and whether attack and victory, or retreat and submission, are the factors used to determine the outcome of an encounter. It is therefore incorrect to assume that there is only one rank order, or to assume that a rank order is the most useful way of describing the social structure of a group of animals (see SOCIAL ORGANIZATION). Nevertheless the subdivision of groups of animals into those which win most of their competitive encounters, those that win some, and those that are subordinate on most occasions has been of some use. Studies of domestic goats (*Capra hircus*) have shown that the most subordinate individuals were last to feed from a limited food supply, and had more parasites than the other animals. As already mentioned, where food is limited in relation to stocking density, the most subordinate cattle also feed least at the favoured feeding times. This may result in poorer growth rates. Attempts to relate a rank order to various measures of size, age, or other characteristics of the animals have given different results according to the composition of the group and the way in which rank is calculated. Certain generalizations can be made where the rank is based upon the number of other individuals which are subordinate in the majority of interactions. Factors which result in higher rank in cows, for example, include greater age, larger size, possession of horns, and group-rearing which results in experience of social interactions from an early age.

When a group of animals is put together for the first time, many competitive interactions occur initially, but their frequency soon declines as the animals learn the characteristics of their neighbours. The rank order thus has the effect of reducing unnecessary FIGHTING. Each individual learns to whom it must give way. The effect of moving animals from herd to herd is an important research topic. Cows or hens put into a new group often

have to take part in a number of fights before reaching a stable position in the group. As a result, frequent movement from group to group may have undesirable effects on production. Growth rates of fowls, pigs, and cattle are worse if frequent changes in social groups are made. Cattle produce less milk, pigs show a slower weight gain, whilst fowls grow more slowly, and show decreased resistance to viral infection. There have been detailed studies of association between pairs, or between small groups of individuals in herds of cattle, and observations on the extent to which sheep position themselves so that they are able to look at other members of their flock. These emphasize the fact that, whilst farm animals show social behaviour towards all members of their herd or flock, their affiliations are principally directed towards certain individuals. If the composition of a group is to be changed it is desirable that individuals should be kept with, or moved with, the other animals with which they have a SOCIAL RELATIONSHIP. If this is not done, the animals may show signs of STRESS: pining, loss of weight, and a decline in general body condition.

Relations with man. Farm animals often also establish social relationships with the people who look after them. It is well known that many horses (*Equus Przewalski caballus*) can be managed much more easily by people with whom they are familiar than by strangers. Studies of herds of milking cows indicate that their familiarity with the stockman, and the way in which he normally behaves towards cows, will alter the ease with which the animal can be driven and yield milk. The benefits conferred by the establishment of good relationships between animals and their stockman are greatest if the herd is not too large. In small herds, more individual treatment of animals is possible. In a milking parlour, milk let-down may be facilitated by a regular routine by a familiar individual. It is likely that some generalized elements of the set of stimuli which would be detected by the cow when its own calf came to suckle must be provided to obtain the best milk yield.

Familiarity with handlers seems important to milking cows, and to other large farm animals, but behaviour which has a calming effect on the animals is valuable for all species. If someone enters a hen-house slamming the door and making much noise, the hens may jump and fly around their cages, and perhaps crack many of the eggs. Quiet, predictable behaviour by farm staff and a similar routine each day result in better production on farms. The responses to unexpected actions by people or other unfamiliar events are part of the mechanism for responding to a possible predator. This mechanism is still present in farm animals, although it has some different characteristics from those of the same or allied species in the wild. The main result of selection by man has been to increase the threshold for the flight response, especially that which is shown when a human being approaches. The selection procedures have not, however, eradicated the HORMONE response shown by mammals and birds in a situation where immediate and energetic flight is required. There are considerable differences between farm animal species in these reactions. Poultry are much more susceptible to disturbance than sheep, and they in turn are more readily disturbed than are cattle. It has been reported that children playing on one side of a hen-house, and running sticks along the galvanized iron wall, inhibited laying on that side of the house, but had much less effect on the hens on the other side. Sheep are easily frightened by dogs, but a larger potential predator would be needed to disturb cattle. It has been shown that intermittent loud noises, like those from a nearby airport, can delay milk let-down in cows, but very occasional loud noises appear to have less effect. Sonic booms cause an initial startle response in various farm animal species, but they adapt readily, and there is no clear evidence of any long-term adverse effect. It is likely that the rate of HABITUATION of the flight response is faster in domesticated than in comparable wild animals. Other types of disturbance, such as movement of animals in vehicles, are traumatic events for farm animals, and may cause temporary declines in production efficiency.

Reproductive behaviour. Animal breeders necessarily take much note of studies on sexual behaviour, for they must be able to ensure that successful mating or successful artificial insemination occurs. Much of the knowledge about the interactions between hormones and behaviour is relevant to farm animals. There are, however, particular problems where the principal production is from female animals, e.g. in milking cows and egg-producing poultry. Small numbers of males are kept separately from the females so that offspring production is confined to a certain time of year. Females kept in single sex groups are less likely to come into *oestrus* and it is important to breeders that oestrus be detected when it occurs, so that a male animal can be introduced or artificial insemination can be carried out. If some cows in a herd fail to come into calf at the optimum time, the delay in calving and lactation may be expensive for the farmer. Behavioural studies have shown that oestrus may be initiated when a 'teaser' male can be seen, heard, and smelled by the females. Sometimes a bullock (castrated male) is used for this purpose. A widely used behavioural method for detecting oestrus in pigs resulted from the observation that the gilt or sow in oestrus would adopt a rigid posture when pressed on the back, especially if the smell or sound of a boar is present. Behavioural methods for synchronizing oestrus in pigs are valuable when pigs can be kept in groups; the detection of oestrus in ewes is difficult unless a ram is present, and the ewe is observed to consort with him. Studies of behaviour have helped to increase the likelihood that most ewes in a flock will be mated without wasting resources on rearing too many rams.

Once fertilization has occurred, changes in behaviour are apparent before egg-laying or parturition. Farmers and veterinarians need to be able to recognize these behavioural symptoms. The interactions between hormones and nesting behaviour in domestic fowl are now well studied and, as a result, conditions which are more favourable to egg-laying can be provided for hens. The selection of hens which do not incubate their eggs has changed behaviour greatly and has been desirable for egg production, but it has resulted in the broody hen becoming a rarity. Selection procedures have had some effect upon egg-laying behaviour in hens, but have not greatly altered the selection of sites for lambing by ewes. Most ewes show a very strong preference for a field site rather than a shed for lambing, even in particularly inclement weather.

Mother mammals must accept their young and allow them to suckle, and the young must be able to find the milk in order to survive. The maternal behaviour of animals is therefore important to farmers. Even in dairy cattle, where the calf is removed in order that milk can be taken from the mother for human consumption, the colostrum (first milk) provides valuable antibodies for the calf, so the behaviour of the mother and the preferences of the young which increase the chances that the colostrum will be ingested are important fields of study. Some young piglets and lambs fail to suckle successfully, and die. This is sometimes due to the litter size which, due to selection by breeders, may be greater than the number of teats on the mother. The behavioural factors which prejudice the chances of successful suckling when there are enough teats are also being studied. Sheep and goats may reject their young if separated from them for more than a few hours immediately after parturition. As a result of this finding, the methods used by ewes for recognizing their lambs have been carefully studied in order to minimize the rejection of lambs by their mothers and to discover the optimum way to persuade a ewe to accept a lamb which is not her own. The rearing conditions of young animals may affect later sexual behaviour. Turkeys and rams reared without contact with potential mates may later address sexual behaviour to inappropriate partners, or may be unable to copulate adequately (see IMPRINTING). Rearing conditions must therefore be altered appropriately.

Abnormal behaviour. Although farmers may not always be aware that they know a great deal about the behaviour of their animals, their knowledge is immediately apparent when they recognize quite small abnormalities of behaviour in sick individuals. The experts on abnormal behaviour are those who see it most frequently, and practising veterinarians rely heavily on behavioural observations when diagnosing diseases. Peculiarities of posture, locomotion, or REFLEXES, and manifestations of localized pain can be used as indicators of various diseases. Much of the knowledge about farm animal behaviour possessed by farmers and veterinarians is not published, but contact between scientists and practitioners can do much to improve the productivity, health, and welfare of animals.   D.M.B.

46, 64, 85

FEAR is a state of MOTIVATION which is aroused by certain specific stimuli and normally gives rise to DEFENSIVE behaviour or ESCAPE. Fear provoking stimuli may be SIGN STIMULI that are responded to without prior experience, or they may be stimuli to which a fear response has become attached through the process of CONDITIONING. A silhouette resembling a hawk (Accipitridae) when passed above ducklings or goslings (Anatidae) induces fear responses when moved in one direction, but not when moved in the opposite direction (Fig. A).

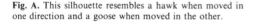

'Goose'   'Hawk'

Fig. A. This silhouette resembles a hawk when moved in one direction and a goose when moved in the other.

The short neck and long tail is characteristic of a hawk, whereas the long neck and short tail resembles a flying goose. The evidence suggests that some birds have an INNATE fear of the hawk-like configuration since they respond appropriately just after hatching.

Animals may learn to fear certain situations as a result of experiencing PAIN or STRESS there, and may subsequently show AVOIDANCE behaviour when they re-encounter the situation. Similarly, situations in which the animal encounters such natural fear-inducing stimuli as the ALARM calls of other animals or the sight of potential predators may induce fear by ASSOCIATION, without the animal ever experiencing pain. Repeated presentation of such stimuli, however, may lead to HABITUATION so that the animal is no longer frightened by them.

Fear is usually accompanied by changes in AUTONOMIC activity, such as increased heartbeat rate and changes in the pattern of blood circulation. Many animals show specific fear DISPLAYS, which involve alarm calls, pilo-erection, or characteristic FACIAL EXPRESSIONS. Some have stereotyped responses involving flight, AGGRESSION, freezing, or even DEATH FEIGNING. The fry of the mouth-breeding cichlid fish, the Mozambique tila-

pia (*Tilapia mossambica*), dash into their mother's mouth when alarmed. Young herring gulls (*Larus argentatus*) run for cover and remain motionless until called by a parent.

The development of fear and associated behaviour in young animals is complicated by a variety of interacting factors (see ONTOGENY). The chicks of jungle fowl (*Gallus*) show escape responses to loud noises and to tactile stimuli immediately after hatching, but visual elicited escape responses develop only later. The escape responses to tactile and visual stimuli show ORIENTATION away from the source of stimulation, whereas the responses to auditory stimuli show no such orientation. Many young animals show distress behaviour if separated from the parents, or if subjected to cold or pain (see MATERNAL BEHAVIOUR). Bodily contact with a parent is often necessary in very young animals, but visual contact may give sufficient reassurance to older offspring. SOCIAL RELATIONSHIPS can also have profound effects upon the development of fear responses. Birds reared in isolation from others are not only more afraid of members of their own species than communally reared individuals, but are also more likely to fear strange objects or situations. Chimpanzees (*Pan*) reared in a moderately restricted environment show enhanced fear of novel objects at about 2 years of age. Rhesus macaque monkeys (*Macaca mulatta*) reared in total isolation for 6 months or a year may never develop normal social behaviour, remaining perpetually fearful of other individuals.

Normally, young animals fear loud or sudden disturbances, but rapidly become habituated to them. Once certain aspects of the world become familiar, then fear of novelty may develop. It has been suggested that there is a SENSITIVE PERIOD during which attachments are made to parents and other common features of the young animal's world. The sensitive period is terminated by the development of fear of novelty. Animals reared in impoverished environments during the sensitive period may never develop normal attachments (see IMPRINTING). Dogs (*Canis lupus familiaris*) reared in isolation from noxious stimuli are slow to learn normal avoidance responses in later life, and it is probable that opportunities for EXPLORATORY behaviour and PLAY are important for the development of normal fear and avoidance behaviour.

In adult animals fear is often mixed up with other aspects of motivation. CONFLICT between fear and approach behaviour may result in DISPLACEMENT ACTIVITIES or in other RITUALIZED behaviour such as MOBBING. Chaffinches (*Fringilla coelebs*) presented with a stuffed tawny owl (*Strix aluco*) approach to a certain distance uttering a characteristic 'chink chink' call. The chaffinches alternately approach and withdraw from the owl, and in the natural setting this mobbing behaviour often succeeds in driving the owl away. It has been shown that the alternating approach and avoidance is the result of a conflict between fear and aggression. This type of conflict forms the basis of THREAT behaviour in many species.

Extreme or prolonged fear can induce STRESS and EMOTION which sometimes leads to ABNORMAL behaviour in the form of anxiety neurosis or depression. Such situations may result from the social DOMINANCE of one animal over another, especially in CAPTIVITY where it is sometimes difficult for a subordinate animal to escape from the attentions of its oppressor. In the natural situation, fear of PREDATION is probably the most common cause of the extreme shyness and VIGILANCE shown by some species. In other species, mild fear may induce CURIOSITY and a tendency to investigate. For example, upon sighting a cheetah (*Acinonyx jubatus*) Thomson's gazelle (*Gazella thomsoni*) may run up from a distance of several hundred metres and approach to within 50 to 80 m. All the gazelle have their ears cocked and stare at the cheetah, occasionally uttering an alarm snort. EXPLORATORY behaviour may also be motivated by fear; animals will explore a novel object or environment, provided their fear is not too great. Black-headed gulls (*Larus ridibundus*) for instance, if near the source of an alarm call, will fly away, but if they are already some distance away they will often approach and investigate the cause of the disturbance.

56

FEEDBACK arises whenever the consequences of behaviour affect future behaviour. For example, when an animal shivers in response to cold, heat is generated by the muscular activity and the animal stops shivering when it no longer feels so cold. As a consequence of shivering there is a feedback effect such that future shivering is affected. When, as in this example, the consequence of an activity tends to diminish future performance of the activity, the feedback is said to be negative. When the outcome tends to enhance the performance of the behaviour, the feedback is positive. The number of animals in a population, for instance, tends to increase as a result of *positive feedback*. If each mated pair produces more than two offspring in a lifetime, then there will be a tendency for the population to increase. When the offspring mature there will be more mated pairs than before, even

more offspring will be produced, and the population will tend to escalate, unless checked by a correspondingly increasing death rate.

Positive feedback processes are comparatively rare in nature, simply because they lead to unstable situations. *Negative feedback* processes, on the other hand, tend to have a stabilizing influence, and to bring a changeable situation into a state of equilibrium. Negative feedback mechanisms are involved in almost every aspect of animal behaviour. They are especially important in HOMEOSTASIS, the regulation of the state of the animal's internal environment. In mammalian THERMOREGULATION, for example, the BRAIN responds to changes in the temperature of the body by activating heating and cooling mechanisms. Thus, if the temperature is too high, REFLEX cooling behaviour, such as sweating and panting, is brought into operation. In human beings and some other mammals an extremely precise regulation of body temperature is achieved by the use of sophisticated negative feedback mechanisms.

A simple negative feedback mechanism is illustrated in Fig. A. A reference value x, such as the desired body temperature, is fed into a *comparator*. This compares the desired value x with a measured value z and produces an *error signal* x−z. The error signal is fed into a *controlling device*, such as the brain together with its heating and cooling mechanisms. The controlling device acts upon the *controlled system*, which in the case of thermoregulation would be the body as a whole. The relevant changes in the controlled system, usually called the *output* y, are monitored by SENSE ORGANS to give the signal z which is fed into the comparator to reduce the error signal. Any deviation in the output will tend to increase the error signal, and this will be automatically corrected by negative feedback from the monitored consequences of the output behaviour.

Negative feedback mechanisms are particularly important in ORIENTATION. Often a visual comparison is made between the position of a target and the position of the animal's body, limb, or other aiming device. In reaching out to grasp an object, for example, the human hand may be visually guided in such a way that the error signal, i.e. the distance between the hand and the object, is continuously diminished. In some cases, however, we can look at an object and then close our eyes

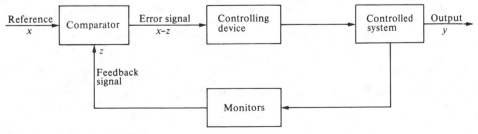

Fig. A. A simple negative feedback system.

and reach for it accurately. Clearly, visual feedback is not being used here. Instead, we compare the felt position of the arm with the remembered position of the object (see COORDINATION). Proprioceptive feedback is then used to guide the movement of the hand.

The feedback mechanisms involved in animal behaviour are often very complex and sophisticated. They enable animals to regulate their internal state, to behave in a controlled manner with respect to the external world, and to distinguish between changes in the environment due to the animal's own behaviour, and changes of external origin (see REAFFERENCE).

103

FEEDING behaviour includes all those activities that are involved in obtaining, handling, and ingesting food. The activities are very varied, ranging from PREDATORY behaviour to grazing. Different species employ different strategies according to their respective ecological NICHES. For example, some predators obtain food by active HUNTING, while others rely on ambush. The prerequisites for a waiting strategy are that the prey should be mobile, and that the predator should be suitably CAMOUFLAGED. Thus the praying mantis *Phyllocrania paradoxa* mimics a brown dead leaf, and it waits in ambush until a suitable prey, such as a fly (Diptera), comes within range. It then strikes at the prey with a very rapid movement of its forelegs. Many ambush predators construct devices that increase the probability of capturing food. The larva of the snipe fly *Vermileo vermileo* constructs a pit in sandy soil, and lies in wait buried at the bottom of the pit. Small insects such as ants (Formicidae) slide into the pit and are seized by the larva. Spiders (Arachnida) are ambush predators *par excellence*. In addition to the familiar web-building spiders, the trap-door spiders (*Tetragnatha*) construct a burrow covered by a trapdoor of silk and soil. The spider waits inside the burrow with the lid closed. When a suitable prey comes near, it springs out of its hiding place and stabs its victim. The spider *Atypus* lurks within a silken tube through which it seizes any fly that alights on the tube. Some ambush predators use lures to attract prey. The tail colour of the green palm viper (*Bothrops bilineata*) is different from that of the rest of its body, and is shaped like a worm or insect larva. The snake lies in ambush and twitches the tip of its tail to attract small lizards and rodents. The alligator snapping turtle (*Macroclemys temminckii*) and the angler fish (*Lophius piscatorius*) both have lures which they display near their wide-open mouths to attract small fish. Many herons (Ardeidae) spread their wings to cast a shadow over water. Small fish are attracted to shaded parts of the water, which usually offer some protection from predators.

Active hunting or FORAGING is necessary when the food is not mobile, or when the predator cannot hide itself from the prey. When the food is patchily distributed, animals have to employ sophisticated SEARCHING and foraging strategies, designed to exploit the food resources with the minimum cost in time and energy. In some species each individual maintains a TERRITORY as a means of protecting its food supply. For example, pied wagtails (*Motacilla alba yarellii*) feed by walking along the river edge and garnering insects which are washed up onto the bank. In winter, when daylight hours are short, they have to spend 90 per cent of their time searching for food. As the bird walks along, it temporarily depletes the food supply, and some time must elapse before a particular stretch of bank can be profitably revisited. Under such circumstances, each wagtail defends a territory along a particular stretch of river, and it exploits the food systematically by walking up one bank to the territory boundary, crossing the river, and then walking back down the other bank to complete the circuit. The length of territory defended is just sufficient to enable the owner to make a complete circuit by the time the food supply has been renewed to a profitable level.

When the renewal rate on the territory is very low, the owner abandons the territory and goes off to feed elsewhere. When the food supply is fairly good, the owner may allow another wagtail, usually a JUVENILE, to share the territory. The lodger helps to defend the territory at a cost of some depletion of the food supply. However, it has been shown that the owner benefits from the association, and when the food supply falls the lodger is evicted. When the food supply is very great the owner gives up all attempts to defend the territory and concentrates on feeding.

COOPERATIVE behaviour is found in species such as wolves (*Canis lupus*) and the African wild dog (*Lycaon pictus*), which hunt animals larger than themselves. An interesting example of cooperative behaviour is found in the brightly coloured painted shrimp (*Hymenocera picta*), which feeds on starfish including the formidable crown-of-thorns (*Acanthaster planci*). The mated male and female cooperate in tickling the tube feet of the starfish so that it loosens its grip on the substratum. They gradually manœuvre the starfish onto its back, and then quickly start eating from the vulnerable underside before the starfish can right itself.

Many animals show a high degree of FOOD SELECTION. When food is plentiful the most profitable food may be selected. Thus deer (Cervidae) may walk through a field of clover to reach another field that has been fertilized, and thus has more luxurious vegetation. Redshank (*Tringa totanus*), foraging for the marine ragworm *Nereis* on the sea shore, ignore the smaller worms and select only the large ones. Grazing and browsing animals generally locate particular food items by smell, and may taste a little before deciding to take larger quantities. Food selection is also partly a result of feeding technique. Features that are important include the shape of the mouth, the thickness and mobility of the lips, the shape of the teeth, and the action of the head in eating. Both grazing and browsing animals lack upper incisor teeth. When browsing the tongue and lip action draw a stem into the mouth, the teeth press the plant against the palate and the leaves are stripped off the stem by a movement of the head. The exact method varies in detail from species to species. In grazers the incisor teeth oppose the palate in a less sharp manner. Grasses are drawn into the mouth by the action of the tongue and lips and then squeezed against the palate. A short pulling swing of the head results in the less strong parts of the plant being separated from the rest. Because of their low fibre content it is generally the leaves that are consumed, the more stemmy material being left behind. African buffalo (*Syncerus caffer*) select their food on the basis of its leaf: stem ratio, preferring species with the most abundant leaf. The speed of grazing depends upon the mechanical properties of the plant species and the width of the mouth, the more selective species having narrower faces and mouths. Herbivores require large amounts of food, and their method of feeding provides them with a much more nutritious diet than would be the case if they were to bite off and eat whole plants.

In general, the feeding mechanisms of each species are designed to enable it to feed in the most profitable manner, without undue waste of energy in handling the food, or in ingesting food that is not worthwhile. At the same time, the feeding animal has to contend with a number of potential hazards and dangers. Poisonous animals and plants sometimes advertise their presence by warning coloration, odour, and taste (see DEFENSIVE BEHAVIOUR). Domestic cattle (*Bos primigenius taurus*) reject food with a bitter taste, and birds quickly learn to avoid conspicuous insects on the basis of their taste (see MIMICRY). Certain colour patterns, such as the eye-like spots on the wings of moths, for example, the eyed hawkmoth (*Smerinthus ocellatus*), will deter even a naïve predator. Many animals exercise caution in sampling novel foods, and may take a little and then wait a few hours before taking more. The sight of other individuals feeding often has a facilitatory effect, and it is probable that some degree of insurance against poisoning is obtained in this way. Feeding animals also must be alert to the possibility of being surprised by predators, and it is notable that many birds and ruminants look up repeatedly while feeding. Research has shown that such interruptions of feeding are more frequent when a predator has been seen in the vicinity.

The MOTIVATION of feeding behaviour is based in part upon the principle of HOMEOSTASIS. As the animal uses up energy and nutrients, various imbalances occur in the animal's internal environment, and these are registered by the BRAIN. Deprivation of food results in a state of HUNGER, which establishes feeding as a high-priority activity. However, APPETITE does not depend on hunger alone, and the circumstances in which an animal chooses to feed depend upon a balance of opposing factors (see DECISION-MAKING).

Feeding behaviour in the natural environment is determined partly by OPPORTUNISM. Animals learn when and where food is likely to be found and distribute their feeding effort accordingly. Appetite may be depressed by high temperature, unpalatable food, or other unfavourable circumstances. Many animals feed on a routine basis, based upon their internal CLOCK, and time of day can be an important stimulus to appetite. In addition, decisions about feeding are influenced by the animal's assessment of the dangers of predation, COMPETITION from other individuals, and the possibility of exploiting a rare opportunity. The latter is considerably influenced by the animal's HOARDING ability; if an animal can store food either in a cache, or in the form of body fat, then it can take advantage of periods of high food abundance. On the other hand, some species of small mammals and birds have a high rate of energy expenditure, and are unable to store food. They are often obliged to forage continuously throughout the daylight hours. Similarly, animals such as herbivores that eat low calorie food have to spend the major part of the day feeding.

FIELD STUDIES. Since DARWIN made his worldwide voyage on HMS *Beagle* from 1831 to 1836 it has been appreciated that field studies are essential for completing our understanding of basic problems in biology. His keen observations of the appearance of animals and plants, and, in particular, his studies of closely related species in different natural environments, made Darwin realize the paramount importance of the phenomenon of ADAPTATION. It became the basis of his theory of EVOLUTION: that each species is adapted to a specific set of factors in its environment which it needs for its maintenance, growth, and reproduction. Jointly these factors are nowadays denoted as the ecological NICHE of the species. Although a HABITAT usually harbours numerous other species, in accordance with what is usually called the *competitive exclusive principle* (see COMPETITION), such species do not compete for the same essential aspects of their ecological niches. In its own niche a species will be the supreme master.

Adaptation encompasses not only the morphological structures and the physiological mechanisms of an animal, but also its behaviour. In fact, only through corresponding behaviour patterns can the anatomy and physiology be adaptive. It is the FUNCTION of behaviour to keep the physiology working, and thus enable the animal to survive and be capable of producing offspring. This implies that behaviour can only be fully understood with respect to the ecological niche of the species. The appropriate knowledge about this relationship has to come from field studies of animal behaviour.

Methods of study. The first phase of a field study should always consist of a thorough description of all aspects of the animal under observation. The activity pattern of the animal should determine the working programme of the observer, who will need all the skill, perseverance, endurance, and sportsmanship of the classical naturalist for maintaining himself in the habitat of his subject while collecting data. Binoculars, of various powers, for work in daylight, as well as infrared devices for watching animals in the dark, will be of assistance, and frequently essential. Although there is much to be said for simply writing notes during observations, the high frequency of events may force the observer to dictate his observations on to tape, and possibly also to use some manual recording device. In particular when more than one individual has to be watched simultaneously, as, for instance, in interactions between animals, it may become desirable to make cinematographic or portable television recordings which can later be carefully analysed in the laboratory. Should the animal vocalize it is usually important to record this too. However, although enormous prospects for TECHNIQUES OF STUDY have been opened up by technological advances, one should not forget that, at least in daylight, the human eye sees more than the camera lens, and straightforward observation is still of fundamental importance and value.

Usually it is important for the observer to ensure that his presence does not influence the behaviour of the animal under observation. The most usual way to achieve this is by the use of hides, and this may sometimes involve considerable ingenuity. For work with larger mammals and birds motor vehicles can often function as suitable mobile hides. In rather exceptional cases (e.g. some of the birds which nest on oceanic islands or in the Antarctic) the presence of an observer seems not to bother the animals. Some observers can carefully and gradually condition their subjects to their visible presence. This is a first step towards taming the animals, which is sometimes a useful technique.

Sooner or later in many field studies the need to distinguish animals individually arises. The best way of doing this is by making use of naturally occurring individual differences, such as the different faces of chimpanzees (Pan), and even of gulls (Laridae), the individual differences in the facial markings of cheetah (Acinonyx jubatus), the irregularities in the horns of antelopes (Bovidae), the patterns of wear of the ears of African elephants (Elephantidae), etc. The second best way is by attaching artificial, individually different markings to the animal. In a few cases this can be done by spraying a dye on the animal from a distance, but usually the animal has to be captured before being marked. Various kinds of nets and traps suitable for different species are in use for this purpose; the larger mammals are tranquillized by means of a dart shot from a gun. Markings vary from coloured and/or numbered tags, rings, or collars, to clippings in toes in small rodents, or in the ears

of hyenas (Hyaena). The presence of toe-clipped rats or mice (Murinae) can be recorded by placing sooted paper on places where they are likely to pass. A very important means of marking animals for tracking over long periods is by attaching radio-transmitters to them. These are available in sizes to suit a range of animals from shrews (Soricidae) to elephants. In addition transmitters can be applied for monitoring physiological phenomena, such as temperature and heart rate. In rather exceptional cases, for animals moving in an easily controllable area over relatively limited distances, radioactive markings detectable with a Geiger-counter have been used, for instance in tracking moles (Talpa).

Markings are not only important for following the behaviour or whereabouts of individual animals. If a known and sufficiently large number of animals in a population has been marked, an assessment of the total population strength can be calculated from the proportion of marked to unmarked individuals encountered during surveys of the area. For counting birds and larger mammals, and for tracking their movements, the use of small planes, usually in combination with aerial photography, can be very successful.

The adaptive organization of behaviour. Persistent observation of individuals of a particular species over a reasonably long period will result in a preliminary picture of its entire life history. For the next stage of the study the observer is likely to concentrate on some particular aspect of the life history. We shall consider here, as examples, three such aspects, differing in the complexity of integration of the behaviour patterns involved.

1. *Protective behaviour.* The first example concerns the relatively simple behaviour patterns involved in DEFENSIVE behaviour against predators. Judged by the human eye a great many species seem to defend themselves through CAMOUFLAGE. A well-known example is the peppered moth (*Biston betularia*). In areas where the bark of trees is covered with lichens the most common form of this moth is light with a mottled appearance. In the industrialized areas, however, where the industrial fall-out has killed the lichens so that trunks and branches are dark, a mutant black form dominates (see NATURAL SELECTION, Fig. A). In a field experiment 473 dark and 496 light marked specimens were released in an unpolluted wood. During the following days 30 (6·3%) dark and 62 (12·5%) light specimens were recaptured. In contrast, of 601 dark and 201 light moths released in a dark, polluted wood, 205 (34%) dark and 34 (17%) light specimens were recaptured. It could be concluded therefore that during the experimental period the mortality of the form matching the background was half that of the contrasting one. Furthermore, direct observations of food-searching birds showed that moths resting on a matching background had less chance of being caught than those on a contrasting one. Consequently the cryptic coloration was indeed found to have a protective effect

against predators hunting visually (see NATURAL SELECTION).

Another strategy for defence is THREAT. This involves the animal making itself, or parts of itself, conspicuous (see ADVERTISEMENT). Whereas the camouflage strategy is only suitable for species which are immobile during the day, the threat strategy is more suited to species moving about in daylight. Often a species is not restricted to either strategy, but can use them in combination. Several species of moth, for instance, possess cryptically coloured forewings, which cover conspicuously coloured hind wings when the moth is resting. When the moth is detected and touched by a predator it exposes its hind wings, which often bear eye-like patterns. This behaviour frightens small birds, delays the resumption of the attack, and thus gives the moth the chance to drop down and reassume its cryptic coloration.

To assess whether, and to what extent, these apparently defensive mechanisms increase the SURVIVAL VALUE of the animal, field studies are needed, preferably combined with experiments. To answer questions about the effectiveness of defensive mechanisms we have to know how often the defence meets with success, and how often not, and how much the effect would be changed by experimentally interfering with the defensive behaviour.

2. *Social systems.* The second example deals with the most complicated behaviour we know: SOCIAL behaviour. Because they occupy the same ecological niche, competition tends to be more severe among members of the same species. Thus it can be important for an animal to keep conspecifics at a distance and this is what AGGRESSION between individuals generally does. On the other hand it is known that predators have difficulty in catching a prey from a dense concentration of animals, such as a herd, a flock, or a school. Furthermore, the larger the number of animals the greater the chance of detecting a predator. Thus, keeping together can be an anti-predator mechanism; it may also facilitate mating, and the discovery of irregularly distributed food sources. Social behaviour makes it possible for the indivduals of a species to compromise between the antagonistic tendencies to cluster and to disperse in a way which best suits their needs. How the social system, or systems, available to a species is adapted to these needs can only be revealed by field studies in the natural habitat. In these studies both the ecology and the behaviour of the species must be investigated. Although we are far from a thorough understanding of the interrelations between the various complex factors involved, the available data permit us to sketch a rough outline (see SOCIAL ORGANIZATION).

If a species' essential resources are evenly distributed over a large area, at least during a sufficiently long period of the year to encompass the reproductive period, we are likely to find that the species has a territorial society. The area is divided into roughly equal territories, each in principle occupied by a family and defended against neighbours and newcomers. Conspecifics who do not manage to establish and maintain a suitable TERRITORY perish or disappear. We find this in several species of song-birds. If, however, no food is available in the territories, as in many sea-birds, the territories tend to be much smaller, and thus the colony denser (which increases the mass protective effect). Such territories may still subserve pair formation, nesting, and possibly the raising of the young, but the food for the chicks has to be brought in from outside the colony. The distribution of the food of such species is often patchy and, to find it, animals benefit from watching the activities of fellow members of the group (see FLOCKING). In animals such as herbivores and fructivores, however, it may be impossible to transport food. In this case the whole family has to move from one patch of food to another. Then both defence against predators and the detection of food will be promoted by the formation of groups of families, though this involves a higher risk of competition within the species. Commonly a hierarchy or rank order exists among the members of a group (see DOMINANCE). Although this usually originates from contests between rivals, it is in general maintained more by threats than by overt fighting. When food gets scarce the lower ranking members are likely to be at a competitive disadvantage.

Thus, even at this level of complexity of behaviour, we find different behavioural strategies being used in an adaptive way to meet the circumstances of an ecological niche. It is especially interesting to study how in taxonomically related species, and thus in different niches, the possible strategies are variously deployed (see COMPARATIVE STUDIES). For instance, primate species living in areas with marginal food resources and few predators tend to have a social system comprised of groups of one male with a number of females on the one hand, and all-male groups on the other. In contrast, in areas with a richer and more constant food supply the species tend to move around in larger groups with numerous males and females. The common black-billed magpie (*Pica pica*) divides its habitat, in which its food is rather evenly distributed, into large family territories. The food of the very closely related yellow-billed magpie (*Pica nuttalli*) of California occurs in irregularly distributed patches, and is often more difficult to find. The individuals of this species nest, like sea-birds, in colonies at short distances from each other, from where they search collectively for food. Sometimes a species, or even an individual, may switch from one social system to another when the ecological circumstances change.

Similarly one may look at the adaptiveness of monogamy, polygamy, promiscuity, and PARENTAL CARE by both female and male, by the female alone, or by the male alone (see MATING SYSTEMS). The last is the most rarely realized system, which seems surprising because one would expect it to give the female the best opportunity to recover from giving birth, and to quickly produce a new batch of young. For the evaluation of the adaptiveness in the natural environment of these social systems of family life, as well as of the social structure of the species as a whole, a careful comparison of the behaviour and of the prevailing ecological factors is the most appropriate approach. Experiments in nature will usually not be feasible, although it may occasionally be possible to use the evidence provided by the occurrence of catastrophes, or by attempts to introduce an endangered species into a new area.

3. *Communication*. Further possibilities for testing experimentally the conclusions derived from comparative observations occur in studies of the function of activities, often referred to as FIXED ACTION PATTERNS, because of their stereotyped form. These are easy to distinguish and are characteristic for the species. Some fixed action patterns subserve COMMUNICATION within a species. From careful observation of the form of these activities, and of their temporal relations with the occurrence of overt attacks or flight, the conclusion has been drawn that a great many of them have originated, in the course of evolution, from the interactions of the tendencies to attack, to flee, and to stay put. It has been proposed that these three tendencies are expressed to different extents and/or in different ratios in different social contexts. To investigate this idea experimentally one should study the consistency of the reactions of an animal, while attempting to manipulate its behaviour by presenting it (successively) with the essential features of different signals. The most useful approach for this research is the use of a model as the donor of the signal.

In the field, much work of this type has been done with gulls; sometimes stuffed birds are presented in different characteristic positions, and sometimes the stuffed birds are 'animated'. One particular component of threat behaviour in gulls, the *upright stretched neck*, is, on the basis of its form, thought to be caused by a tendency to fly up and escape. When this posture is presented (by using a model), as part of a THREAT display, the threat is less intimidating to an intruder than when the model has the neck held horizontally. Moving models have also been used in studies of the courtship of the black grouse (*Lyrurus tetrix*). In this species several males gather on arenas where they have restricted courting territories (see LEK) which they try to defend against each other while also attempting to attract females. Only one or a very few males on an arena manage to copulate with the various females visiting the site. Field observations give the impression that the success of a male very much depends on its tactics. These observations also give some idea what good and bad tactics are in a male black grouse; but only experiments with a model female which is moved about by means of strings can prove, for instance, that to induce approach in a female, the male should

carefully withdraw instead of taking the initiative to approach.

With this type of experiment it is also possible to study the effect of the order in which different signals succeed one another in the interpretation of the signal. Just as the interpretation of words in human communication may vary with the context and syntax, so the interpretation of a signal may be different in the different situations in which it is given. In addition it should be emphasized that the effect of the signal also depends on the identity and MOTIVATION of the receiver. A signal repelling a male or an unripe female may attract a sexually receptive female.

In many animals visual signals are combined with signals in other sensory modalities. Since the development of tape-recorders acoustic signals have become accessible to experimentation. Recorded bird SONGS, for instance, can be played back to individuals in the field, and this can be a very good method for detecting territory owners of the species recorded, as they reply to the playing of the song. With this method it can be shown that, in some cases, a resident bird knows the distinctive features of the songs of its neighbours. When a resident is accustomed to his neighbour's song he will not react as long as the song comes from the usual area, but will become alarmed if the experimenter plays the song to him from a different place. When an acoustic stimulus is combined with a model the effectiveness of the model is increased.

Behaviour not especially evolved to subserve communication may nevertheless have communicative value. For example, plastic models of grazing barnacle geese (*Branta leucopsis*) placed in a group on a meadow can attract wild 'conspecifics'; the attraction is strongest when the models have the posture of a grazing animal. If the models have the alert posture with a stretched neck the real geese stay away.

The examples given above reveal that each species has at its disposal a repertoire of behaviour patterns of different orders of complexity that, when appropriately used, help the animal to adapt itself to its environment. The more related species are, the more similarity can be expected in these patterns. Nevertheless, the behaviour patterns tend to be specifically adapted to the niche of the species concerned. The complex mechanisms enabling animals of a given species to function adaptively may be called the functional organization of the behaviour of the species. These mechanisms include the physiological processes involved in ACCLIMATIZATION and HOMEOSTASIS, as well as the behavioural processes involved in DECISION-MAKING.

The implementation of adaptiveness. The study of the functional organization of behaviour in relation to the ecological niche usually begins with field research. A question that commonly presents itself in this type of research is what mechanisms a species uses to identify in its environment the objects needed for maintenance and reproduction. Firstly, the SENSE ORGANS may be specifically adapted to this end. For example, the hearing capacities of bats (Chiroptera) and dolphins (Delphinidae) are specially designed for orientation and object identification by ultrasound. When it is of positive survival value for the object to be detected by an animal, as in the case of flowers and their pollinators, the adaptation may be mutual (see COADAPTATION). Some flower species have thus exploited the capacity of honey-bees (*Apis mellifera*) to use ultra-violet light, others the sensitivity of birds to the red end of the visible spectrum. Although exact studies of the sensory capacity have to be carried out with equipment only suitable for work in the laboratory, a preliminary phase in the field is useful for narrowing down the statement of the problems. A rough picture of the capacity for colour discrimination of flower-visiting insects, for instance, can be obtained by presenting them with artificial flowers made of coloured paper, among similarly shaped models each of a different shade of grey. As von FRISCH did with bees, it may be necessary to first train the insects to visit such flowers by rewarding a visit with food. Many butterflies, however, tend to visit colour patches spontaneously after having been stimulated by certain odours.

Secondly, mechanisms processing the information received by the sensory organs may be involved in the adaptation. For instance, to analyse which features the herring gull (*Larus argentatus*) uses to identify its eggs, wooden models can be used in field experiments. These can be made in different shapes and sizes, painted in various colours, and with speckled patterns (see SUPERNORMAL STIMULUS, Fig. A). By placing such models on the nest rim in pairs, and watching from a hide which one is first retrieved by the gull, a picture can be obtained of the relative role played by the different features. Shape appears to be little used by the herring gull, but colour and speckling are of considerable importance for identifying the egg as an object for INCUBATION. Although models of all colours are occasionally retrieved, the green and yellow models turn out to evoke the greatest response, and red and blue models the least response. In contrast, when in the same species the response to colour is tested in a feeding situation, the result is the reverse: red scores highest and green lowest. These latter experiments concern food-begging in the herring gull chick, i.e. pecking by the chick directed at the yellow bill of the parent which has a red spot on its end. Cardboard models are used in these tests, and presented to chicks which are 'borrowed' from their real parents for a short while. As it is known from laboratory experiments that the colour sensitivity of the eyes of the chicks is hardly different from that of the adult birds, it can be concluded that a different processing mechanism ('green filter') is used in recognizing an incubation objection from that which is used for feeding ('red filter').

Field experiments with model eggs show that herring gulls attend simultaneously to the different features of the egg, such as its colour, size, and shape. In other cases the animal seems to attend to different features of an object successively. For example, the bee-killing digger wasp (*Philanthus triangulum*) while flying in loops over the heather searching for its prey, the honey-bee, is 'focused' on the movement of small objects. Having spotted one (which may be a small piece of wood tied by string to a stick that the experimenter has planted amongst the heather) it moves downwind of the object, positions itself about 10 cm away from it, and hovers for a while in the air. Only if bee scent is perceived from the direction of the object does the wasp launch a sudden attack. By using two similar models placed carefully in the direction of the wind so that one with bee scent is windward from one without, it can be shown that, following the olfactory control during hovering, the pounce is again visually controlled, and the nearby scentless model is taken. Having contacted the object the wasp further examines it tactually and chemically.

The identification of a particular external situation may elicit and/or orient a behaviour pattern. It may also be essential for continuing a behaviour pattern or even stabilizing a relatively complicated behavioural system. An example can be taken from field studies of the reproductive behaviour of herring gulls. By manipulating the number of eggs in the clutch (the optimal number is three), their size and their shape (by way of wooden models), and their temperature (using copper models with a heating or cooling device), and observing the behaviour of the incubating bird continuously, it can be shown that deviations from normal, perceived while sitting on the eggs, lead to interruptions of the incubation act, and to the performance of other activities in a systematic, predictable, way. Initially the bird is likely to shift its position, and secondly, to rearrange material in and around the nest. Both activities may be functional in restoring the optimal situation. However, if the disturbance continues, activities occur that do not seem to serve a direct function in the prevailing situation. They all belong in the context of body care. It appears that their interruptive appearance is brought about by a conflict between the behavioural mechanism underlying incubation and that underlying escape (see DISPLACEMENT ACTIVITY), the latter being stimulated by the disturbance.

Developmental factors. The functional organization of the behaviour of the adult animal develops during its ONTOGENY. However, one should realize that at any stage during the course of this development the behaviour should always be sufficiently adaptive to enable the young animal to survive, frequently, although not necessarily, without the assistance of adults.

A gradual development of a behavioural mechanism opens up the possibility that the developmental process may depend upon the incorporation of experience. Many experiments under laboratory or semi-natural conditions have shown that the recognition of complex stimulus patterns, essential for normal life (such as characteristics of the parent, of the species, of the adequate partner, or of the home area), may, to a greater or lesser extent, have to be learned. Such LEARNING processes are frequently restricted to particular SENSITIVE PERIODS of development, and some basic characters of the learning situation are often already available before the relevant experience is gained. A striking example is the ability of young salmon (Salmoninae) to learn the characteristic scent of the stream in which they hatch, and to use this scent as a cue for homing after their extensive MIGRATION downstream to the sea. If fry are transferred to differently smelling water before the IMPRINTING is completed, or if the experimenter interferes with sensory perception by cutting the olfactory nerves or blocking the nostrils, the salmon do not return to the stream where they were born. A much studied example of such a programmed and almost irreversible learning process is that by which ducks and geese (Anatidae), and several mammals, obtain the knowledge of the parent and the future sex partner. In some cases the imprinting processes are relatively fixed, so that the developmental processes show little variation in changing circumstances. For instance, in sexually dimorphic ducks, the knowledge that males obtain about the plumage of the females is derived from imprinting on the mother shortly after they hatch. After imprinting on a strange species, subsequent experience with both sexes of their own species, even for several years under almost natural conditions, does not alter the imprinted response. However, the females of most duck species are able to recognize the characteristic conspicuous colour patterns of their male partners without being conditioned to them. This difference in the development programme of the behavioural organization of males and females in these birds seems to be genetically determined.

The above example makes it clear that the possibilities for shaping the behavioural organization through more or less strictly programmed learning processes is determined by the particular characteristics of the species. Thus for gaining insight into how the GENETIC make-up of a species utilizes learning processes and how this may influence the flexibility and adaptiveness of the behavioural organization of a species, field studies are generally necessary.

The type of behaviour generally called PLAY can be expected to be of paramount importance in building up experience. The evidence indicates that the basic coordination of the various species-specific behaviour patterns in the repertoire of animals does not have to be acquired by experience, but that their appropriate deployment does have to be learned. The amount of time spent in play differs greatly between species. As play is likely to be

dangerous because it exposes young animals to predators, the amount of play occurring in a species is likely to be an adapted compromise between the need to use it as a vehicle for learning, and the safety risk and energy expenditure involved. This hypothesis should be checked in the field. Another factor restricting the opportunities for play, and for gaining experience in general, is the time available for development; short-lived insects cannot be expected to have developmental programmes as flexible as those of elephants or man.

The ability to learn from experience may also lead to differences between individuals within a species. These differences can be of great selective value. In territorial males of the great tit (Parus major), for instance, the tendency to behave aggressively is different at different sites in the area, and with respect to different opponents, depending on whether site or opponent is associated with previous losses or victories.

Through the work of experimental psychologists a great deal of information is available on learning processes in the common laboratory animals, in particular on the factors which, in the laboratory, facilitate the formation and EXTINCTION of ASSOCIATIONS. It is interesting to attempt to fit this knowledge into studies of learning as it takes place in the natural life of animals. In this way, for instance, the adaptive value of the different effects of fixed and variable reward rates or intervals can be judged. For example, the development of a SEARCHING IMAGE may enable an animal to find food more quickly than otherwise, because it enables the animal to pay ATTENTION to specific features of the prey. If the prey becomes rare, or the food too monotonous, extinction of the searching image may take place, and it may be succeeded by another one. The degree of consistency of a particular searching image can be studied in songbirds feeding their young. If the birds are nesting in boxes an automatic device can be used to photograph the bird before it enters the box, or the box can be attached to a hide and the birds watched directly.

ORIENTATION by landmarks may also be based on the formation of a cognitive pattern which could be called an image (see COGNITION). Field experiments have revealed that in different species of digger wasp different aspects of landmarks are important for the formation of such images, and relate more to the difference in habits than to differences in vision. The hovering Philanthus wasp, for instance, prefers high, three-dimensional landmarks, whereas the swift flying digger wasp Bembex prefers contrasting flat surfaces.

In these examples we have found a species-specific programming of features that can be learned, and this varies from species to species. The herring gull, for instance, does not learn to identify its eggs. Individuals build elaborate nests at some distance from each other, and the position of the nest is precisely learned by means of landmarks. Additional cues from the eggs are not necessary for distinguishing the owner's nest. Black-headed gulls (*Larus ridibundus*), on the other hand, learn the characteristics of their eggs (or of model eggs) on which they have been sitting for some time; their nests are much closer together. Guillemots (*Uria aalge*) lay only one egg and do not make nests; they recognize their own egg by its colour and speckling pattern, and adjust their image of the egg as it gradually becomes covered with faeces.

The comparative approach. To understand through which mechanisms behavioural organization has evolved, COMPARATIVE STUDIES are needed between a considerable number of closely related species, occupying different but comparable niches. The famous Darwin's finches (the ground finches, *Geospiza*, the vegetarian tree finches, *Platyspiza*, the insectivorous tree finches, *Camarhynchus*, and the warbler finches, *Certhidea*) are such a group, and their structural differences have been extensively compared with their food searching habits. An analogue of the Darwin's finches is found in the cichlid fishes of the great African lakes, for instance those of the genus *Haplochromis* in Lake Victoria. It is estimated that 200 species of *Haplochromis* live in this lake. Their evolution has probably been facilitated by periodic rising and falling of the level of the lake, causing parts to be alternately cut off and re-connected, thus forcing the fish species inhabiting it to adapt to the new conditions, to migrate, or to perish. Adaptation to different kinds of food (plankton, snails, insects, fish, even eggs of other mouth-breeding species) has certainly also played an enormous role in this spectacular speciation (see ISOLATING MECHANISMS)

In several song-birds programmed learning plays a role in the development of the species-specific song. In different geographical areas the members of the population tend to have their own DIALECT of the song of the species, and this variation will be reinforced when young have to acquire this song pattern by listening to their parents. Such processes can only be studied by field work in a variety of localities.

Concluding remarks. A full understanding of the causation, development, function, and evolution of behaviour requires studies of animals living freely in their natural environment, studies of captive animals under semi-natural conditions, and studies of animals and even parts of them in the laboratory. The field studies are essential for understanding what the animal actually does with its behavioural equipment. However, for the sophisticated analysis necessary to understand how nervous and hormonal systems enable the animal to perform its behaviour, the great number of interfering variables usually present in the field is too much of a handicap. This can be reduced by keeping animals in captivity, and, in particular, by using them in LABORATORY STUDIES.   G.B.

107, 138, 139

FIGHTING occurs both in encounters between rivals of the same species and between predator and prey. In AGGRESSION between members of the same species fighting usually shows considerable RITUALIZATION. Thus rattlesnakes (*Crotalus*) and oryx (*Oryx gazella*), both of which have dangerous weapons, do not use them in intraspecific fights (see AGGRESSION, Figs. B and C). Instead they have a ritualized style of fighting in which each tries to push the other backwards, or to the ground. Usually fights between rivals end with the retreat or ESCAPE of one individual, and a single encounter is often sufficient to establish DOMINANCE, or to demarcate the boundary of a TERRITORY. Injury seldom results from such encounters.

Efficient PREDATORY behaviour does not usually involve fighting, because the predator easily overpowers the prey. However, sometimes predator and prey are fairly evenly matched, as is the case in encounters between the ichneumon mongoose (*Herpestes ichneumon*) and the Asian cobra (*Naja naja*) (see HUNTING).

In their DEFENSIVE behaviour against predators animals may make use of whatever weapons are at their disposal, and fighting sometimes results. Although some weapons may have evolved specifically as anti-predator devices, such as the spines of sticklebacks (Gasterosteidae), hedgehogs (Erinaceidae), and sea urchins (Echinoidea), these are rarely used in active fighting. Weapons are often structures which are used primarily for the capture of food. Birds stab with the bill, and mammals bite with their teeth. The scorpion (Scorpiones) uses its sting against a predator, just as it does with its prey. Snakes (Serpentes) use their fangs and venom against both prey and predator.

In some species the weapons have evolved primarily for use against rivals, but may also be used against predators. The antlers of deer (Cervidae) and the horns of antelope (Bovidae) may be deployed when the animal is cornered, or is defending its young against predators. Moose (*Alces alces*) and giraffes (Giraffidae) use their horns for fighting one another, and their hooves for kicking at predators.

FITNESS is a biological and mathematical concept, akin to SURVIVAL VALUE, that indicates the ability of GENETIC material to perpetuate itself in the course of EVOLUTION. The concept of fitness may be applied to single genes, to the genetic make-up (*genotype*) of individual animals, or to animal groups. In animal behaviour, the concept most widely employed is the fitness of a genotype relative to other genotypes in the population. An animal's *individual fitness* is a measure of the relative ability of the animal to leave viable offspring. All factors which affect the animal's fertility and fecundity will affect its individual fitness. These will include the morphological, physiological, and behavioural characteristics of the animal. The process of NATURAL SELECTION determines which characteristics confer greater relative fitness, but the effectiveness of natural selection depends upon the mix of genotypes in the population. Thus the relative fitness of a genotype depends upon the environmental conditions and the other genotypes present in the population.

The *inclusive fitness* of an animal is a measure based upon the number of the animal's genes that are present in subsequent generations, rather than the number of offspring. In assessing the inclusive fitness of an animal it is necessary to take account of the number of its genes that are also present in related individuals. This will depend upon the *coefficient of relationship* between the one individual and another. The coefficient of relationship between parent and child is $\frac{1}{2}$, because a child gains half its genes from each parent. The coefficient of relationship between grandparent and child is $\frac{1}{4}$, between siblings it is $\frac{1}{2}$, since half of the genes from each parent will be the same, on average. Between uncle and nephew it is $\frac{1}{4}$, etc. An animal can increase its inclusive fitness by behaving in a manner that increases the individual fitness of a relative. When such behaviour involves a decrease in the animal's own individual fitness, it is regarded as an instance of ALTRUISM, since the animal is sacrificing its own reproductive benefit for that of other animals. Such behaviour is remarkably common in the animal kingdom.

FIXED ACTION PATTERNS are activities which have a relatively fixed pattern of COORDINATION. Such patterns appear to be stereotyped, although they may show variability in ORIENTATION. A good example is the egg-retrieving behaviour of the greylag goose (*Anser anser*). These birds nest upon the ground, and when an egg has been displaced a short distance from the nest, the incubating bird attempts to roll it back with its bill, as illustrated in Fig. A. If the egg rolls slightly to one side during retrieval, the bill is moved sideways to correct for the displacement. However, if the bill loses contact with the egg, the bird does not immediately seek to re-establish contact, but first completes the retrieval movement.

Fig. A. A greylag goose retrieving an egg into its nest.

The fixed action pattern consists in the bird extending its head to reach the egg initially, and then moving its bill along the ground to a position between its legs, whether or not the bill is still in contact with the egg. The orientation components of this behaviour are flexible in that the bill is initially extended towards the egg, and the exact orientation of the bill with respect to the egg is continuously modified to correct for sideways movements of the egg, but only so long as the bill maintains contact with the egg.

As originally conceived by LORENZ (1932), fixed action patterns were somewhat akin to REFLEXES in that they were characteristic of all the members of the species, or larger group, and thus apparently INNATE. However, Lorenz thought that fixed action patterns differed from reflexes in several important ways: (i) animals show MOTIVATION to perform fixed action patterns, which is not true for reflexes; (ii) fixed action patterns are occasionally performed spontaneously, giving rise to a VACUUM ACTIVITY; (iii) fixed action patterns can be elicited by a variety of environmental stimuli, whereas reflexes only occur in response to specific stimulation. Some of these points would now be disputed; the startle reflex occurs in response to a variety of stimuli, for example, but Lorenz's distinctions were instrumental in breaking the stranglehold of reflexology that prevailed at the time.

It is now recognized that there are many fixed action patterns that are controlled by the BRAIN, without reference to FEEDBACK from the consequences of the behaviour. Examples include the rapid escape responses of squid (*Loligo*) and crayfish (Astacidae), the flight patterns of the locust *Schistocerca gregaria* and the song of the cricket *Gryllus campestris*. These fixed action patterns are almost certainly innate. However, there are other examples of complex behaviour patterns generated by the brain, without reference to feedback, which are certainly not innate. These include many highly skilled movements, acquired by LEARNING such as the movements of playing the piano and of operating a typewriter. Even the skilled swing of the golf club has been shown to continue unaltered if the player is plunged into darkness in midstroke. Because fixed action patterns are often so similar in different individuals of the same species, they appear to be innate. However, such patterns are sometimes influenced by environmental factors during the course of development (see ONTOGENY), and they develop in the same way in different individuals only when those individuals share the same environment. The development of vocalizations in European cuckoos (*Cuculus canorus*) and pigeons (Columbidae) is not at all influenced by early auditory experience, but the calls of many birds are so influenced. For instance, at an early stage of development, the song of the white-crowned sparrow (*Zonotrichia leucophrys*) is greatly influenced by the songs of nearby members of the same species. Normally, this early auditory experience will be provided by the parents.

137

FLIGHT has evolved on only a few occasions. This is because it requires a series of extreme specializations of skeletal design, and of muscular and nervous physiology, which can only be achieved by animals that are already highly evolved for an agile life on land. Because of the specializations that make flight possible, the whole biology of flying animals tends to differ from that of related flightless forms.

Flight is exploited in a wide variety of ways, from its obvious use in ESCAPE, to DISPERSION and MIGRATION. Depending on the way in which flight is exploited, so the mechanism, physiology, and biology differ from group to group, while, for aerodynamic reasons, both the mechanism and the performance of the smaller flying animals differ from that of the larger forms. Indeed, given the properties of skeletal and muscular systems, flight is only possible over a fairly narrow range of body sizes, and, as a result, the ability to fly is lost both in the largest birds and, sporadically, for more complex reasons, in insects.

Aerodynamics and mechanics of flying animals.

1. *Basic principles.* If a plate is moved through the air (Fig. A) the aerodynamic forces which it

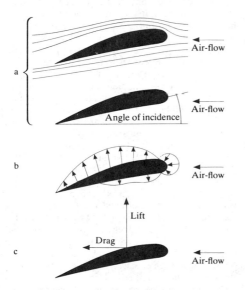

Fig. A. Lift and drag on an aerofoil. **a.** With a small positive angle of incidence, flow lines lift away from the upper surface. **b.** The distribution of pressure around an aerofoil shows that pressure increases at the leading edge and decreases over nearly all the upper surface. These pressures cause the resultant forces shown in Ac. **c.** The resultants of all the pressures can be resolved into an upward lift and a smaller rearward drag.

encounters can be resolved into two components, one acting to slow the plate down, the *drag*, and one which depends on the angle of inclination of the plate to the direction of movement, but which can produce a force which acts upwards, giving *lift*. If the plate is smooth, with a rounded leading edge and a cambered section, the lift forces can be tenfold or more greater than the drag forces, and the plate can become an efficient wing.

The reason why the wing produces lift is that the air-flow over the wing is deflected downwards

both above and below the wing (Fig. A), and so there is a reduction in pressure above, and an increase in pressure below, both of which act together to produce the upward force. The amount of lift that is produced depends in part on the inclination of the wing to the air-flow, i.e., the angle of incidence. As this angle increases, so the amount of lift increases because of the greater downward deflection of the air-flow. At the same time, because the wing presents a greater area to the air-flow, the drag also increases. With large wings at high speeds, the lift does not continue to rise indefinitely, but, at a certain critical angle of incidence, the air-flow separates from the top surface of the wing and becomes turbulent. To set up this turbulent condition requires a lot of energy, and so, when it occurs, the drag forces suddenly rise sharply, and, at the same time, because the air-flow over the wing upper surface has separated, the lift falls drastically (Fig. Ba). This is the stall, so feared by pilots of larger aircraft, as their machine falls out of the sky, and will continue falling unless the wings are brought to angles of incidence at which laminar air-flow can re-start. For large wings, the stall can occur at angles between 5° and 10°, but with smaller wings, and at lower air speeds, turbulence does not set in so easily, and so some insect wings will give useful lift at angles of incidence as high as 45°.

A flying animal must control the lift it produces over a wide range of air speeds. The lift produced by a wing will increase if the angle of incidence to the air stream increases, and vice versa. The lift also increases as the square of the velocity, so at a given angle of incidence a wing will produce four times as much lift when the speed is doubled. At all speeds an animal aims to produce lift equal to the weight of the body; at high speeds this requires a small angle of incidence, but at lower speeds the lift that is required can only be produced by in-

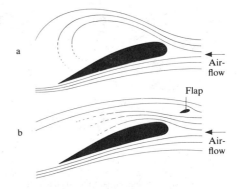

Fig. B. Stalling. **a.** When the angle of incidence of the wing is high, the flow separates from its upper surface and becomes turbulent. The wing does not give lift in this condition. **b.** The stall is delayed by a leading edge flap which delays the breakaway of the air-flow, and so allows the wing to give lift up to higher-than-normal angles of incidence.

creasing the angle of incidence. Ultimately, the lowest flying speed is that at which the wing is not stalled, and is still producing lift equal to the animal's weight. Below this speed, or with higher angles of incidence, a more or less dramatic loss of lift will occur and the animal will fall.

There are various tricks for reducing the stalling speed, such as extending a small leading edge flap (Fig. Bb), the *alula* or bastard wing of birds, lowering the wing's secondary feathers to increase the wing camber, depressing and fanning the tail, or separating the wing primaries so that each forms a tiny wing with a lower stalling speed. The effect of all these tricks is to reduce the speed at which the flow separates from the top of the wing, and to allow higher effective angles of incidence and more lift to be produced.

The stall is used quite deliberately in flight, either to allow very rapid changes of direction or, more commonly, in landing, as when a bird stalls just above its landing spot, folding its wings as it drops on to it.

2. *Gliding*. In considering flight, it is easiest to start with gliding, where the wings are held extended, but the animal loses height continuously, and to proceed from there to flapping flight, where the animal maintains its height as a result of muscular work which overcomes the ENERGY losses of flight.

In gliding, there is a constant lift force from the wings acting upwards, and a drag force acting backwards, while the weight of the animal's body acts downwards. To maintain a steady glide, the resultant of the lift and drag forces must equal the animal's weight (Fig C), and, equally, the glide

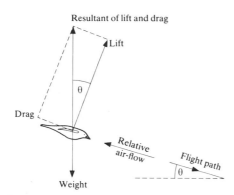

Fig. C. Gliding. An animal in a glide encounters an air-flow along its flight path. The wings produce lift at 90° to the air-flow and drag parallel to it. The resultants of lift and drag act vertically and equal the animal's weight. It loses height continuously.

performance will depend on the aerodynamic properties of the wing and the body; the shallowest glide angle is found when the lift is high and the drag is low. If lift and drag are equal, the glide angle is 45°, and if the drag is very high and the

lift very low as in a stall, the glide is almost vertical. It pays to reduce all forms of drag, which is why the bodies of birds are streamlined, and why they tuck their feet in during flight. Because smaller wings unavoidably produce relatively greater drag than larger wings, smaller birds and insects do not usually glide very well, and tend to concentrate on flapping flight.

The best gliders have long thin wings. In flight a wing has a pressure difference between the upper and lower surfaces which, at the wing tip, produces a circulation around the tip along the direction of flight, giving rise to wing-tip vortices (Fig. D). With longer wings, a smaller proportion of the total lift is lost through wing-tip vortices, which explains the 25 m span of high performance sailplanes, and 3·5 m span of albatrosses (*Diomedea* spp.). In some of the relatively short-winged soaring birds, it is probable that the wing-tip vortices are reduced, and converted into lift and forward *thrust* by the separated primary feathers which are held at an appropriate negative angle of incidence.

3. *Flapping flight*. In flapping flight the animal's problems are different, as it must produce both lift equal to its weight, and thrust equal to its drag. This it achieves by flapping the wing downwards, causing the wing to act as if the animal was in a downward glide, because the resultant wing velocity is the product of the forward movement of the animal and the downward movement of the wing (Fig. Ea). The resultant aerodynamic force on the wing can act in the forward direction, so the downstroke produces thrust. On the upstroke, in a symmetrical system, a reverse thrust is produced (Fig. Eb), but this is minimized in a number of ways; the wing angle of incidence to the air-flow is reduced, so that both lift and drag are reduced, the wing may be partly folded to reduce its area, but the best that many animals can expect is not to lose on the upstroke more than they have gained on the downstroke. Clearly, the up- and downstrokes are not symmetrical and so, for flapping flight, the wing mechanism must include hinges that allow the wing to twist as well as to flap.

4. *Hovering flight*. In hovering flight there is no net forward movement of the wing, and so the wing is usually beaten in a symmetrical horizontal figure-of-eight stroke, with a positive angle of incidence on both the forward and backward strokes (Fig. F). This produces lift on both strokes, but requires considerable twisting of the wing at both wrist and shoulder in birds and, in most insects, a very rapid twisting of the wing at the extremes of the stroke. Because the basic mechanisms tend to be designed to produce an up-and-down stroke, hovering animals of many groups hold the body with the long axis nearly vertical when hovering, and achieve the transition to forward flight by bringing the body round to the horizontal, and changing the wing beat to an asymmetrical up-and-down stroke.

Hovering poses two problems that have led to

Fig. D. Wing-tip vortices. **a.**
Because the pressure above
the wing is reduced, there is
a tendency for there to be a
circulation around the wing
tips. This, with the forward
movement of the animal,
sets up spiral air
movements behind the wing
tips. **b.** By suitably inclining
the feathers at the wing
tips, the circulation can be
exploited and turned into
thrust. This mechanism is
used by the larger soaring
birds.

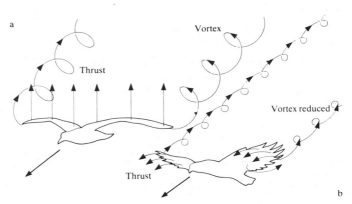

specializations of the flight mechanism. The first is
that because all the lift must be produced by the
flapping strokes, the power required for hovering
tends to be greater than that required for forward
flight, in which the flapping produces thrust which
forces air over the wings at a higher speed and
lower power requirement. The other problem is
that because the wing only executes a beat through
about 120° before reversing, it cannot move far
enough to establish a steady air-flow over its surface and develop as much lift as would be possible
in steady forward flight. In practice, however, the
wings of many hovering animals are clapped together at the top of the stroke, and then rotated
apart before being flapped downwards (Fig. G). It
is thought that this *clap–fling* mechanism estab-

lishes the air-flow around the wing before the
start of the effective stroke, and so allows the wing
to develop relatively large amounts of lift. This
type of wing stroke is seen at take-off in many
birds, and in the hovering of most insects, though
neither dragonflies (Anisoptera) nor hoverflies
(Syrphidae) appear to use exactly this mechanism.

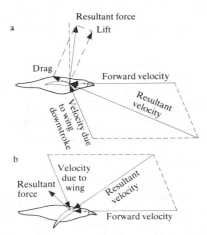

Fig. E. Flapping flight. **a.** The downstroke: the wing beats
down and the animal moves forward, giving a resultant
upward-moving airstream. Due to its high angle of
incidence to the resultant airstream, the wing gives high
lift and some drag, the resultant of which gives an overall
large upward and small forward force. **b.** The upstroke:
the resultant air velocity is now downwards, but the angle
of incidence of the wing is reduced, so the resultant force
is smaller. The wing provides some lift but also drag, not
thrust.

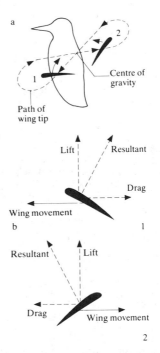

Fig. F. Hovering. **a.** The wings of humming-birds and
many hovering insects beat in a figure-of-eight pattern,
with a sharp wing-twisting at the extremes of the stroke,
so that there is a positive angle of incidence in both halves
of the wing stroke. **b.** At wing position 1, the resultant
force is upwards and towards the centre of gravity (C. of
G.) of animal. The same effect is achieved in the opposite
half of the wing beat (position 2).

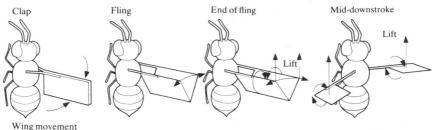

Clap   Fling   End of fling   Mid-downstroke

Lift

Lift

Lift

Wing movement

Fig. G. The clap-fling mechanism. Many hovering animals clap their wings together at the top of the upstroke. As the downstroke starts, the leading edges fling apart, and this sets up the circulation that produces lift at the very top of the wing downstroke. The circulation persists throughout the downstroke, allowing effective use of the whole of the downstroke and giving relatively large amounts of lift.

5. *Animal wings.* For sustained flapping flight various design features are required, and these seem to have evolved in all groups of flying animals. The aerofoil surfaces, the wings, must be large enough to give lift to support the body, and stiff enough to be controllable, both to give the required asymmetry of the up- and downstrokes, and to allow changes of speed, and of angle of the body to the direction of flight. These design requirements have resulted in some sophisticated animal engineering. In many cases there is extensive use of hollow tubular structures and stressed sheets. The wings of a bird have air-filled *humeri* (proximal bones) and even the more distal bones are air-filled in birds of prey. The feathers have a light elastic tubular central beam or *rachis*, from which arise barbs that hook together to form a stiff sheet which may be pre-formed into an appropriate aerofoil (Fig. Ha), while the feathers act together to form a complete streamlined wing. The whole wing folds by bending at the shoulder, elbow, and wrist, and at the same time the feathers rotate to lie back along the bones. When the wing is extended the feathers are automatically extended and linked by a line of tendons that couples the shafts of the primary and secondary feathers; the action can be seen by extending the wing of a bird from which the wing coverts (feathers at the base of the wing) have been plucked. The webbed wing of a bat (Chiroptera) is formed from the hand and the stressed skin which runs between the fingers, and from the fifth finger to the back legs (Fig. Hb). The main stiffness is provided by the arm and finger bones, but the wing membrane is very elastic and is tensioned by extension of the arm and fingers. In the downstroke the bat wing is fully extended, but in the upstroke the wing is both folded and twisted sharply to produce the asymmetric stroke. Bats also fold their wings when at rest, and tend to wrap them around their bodies because their long fingers do not fold. The first two digits, equivalent to our thumb and forefinger, have claws and are used in climbing. Insect wings rely on rather different principles. Arthropods (a large group which includes the insects) have an exoskeleton composed of *chitin* (protein material), not unlike fibreglass in its mechanical properties. This can be deposited in sheets by the insect, but it is not as stiff as bone. Stiffness is provided by forming the wing as a corrugated structure with tubular veins, which tend to resist buckling, at the top and bottom of the corrugations. The principle is the same as with corrugated iron or paper, and, because the wing only operates at low speeds and small sizes, where turbulence is hard to induce, the fact that the wing is not a smooth aerofoil does

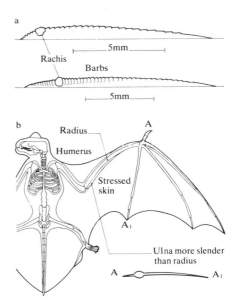

a

Rachis

5mm

Barbs

5mm

b

Radius

Humerus

A

Stressed skin

A₁

Ulna more slender than radius

A ——————— A₁

Fig. H. Wings and wing sections. **a.** The rachis of a feather forms a tubular central beam and the barbs link together to form a continuous aerofoil section. The feathers are: upper, the blackbird (*Turdus merula*); lower, the northern mallard duck (*Anas platyrhynchos*). **b.** The wing of a bat is a stressed skin held between the fingers of the hand and the hind leg. It is curved into an aerofoil section. The finger bones are very long and delicate. The radius and ulna are elongated and the ulna is slender; it fuses with the radius at the wrist.

not affect its aerodynamics. More highly evolved insects fold their wings by rotating them backwards and bending them, so that the leading edge lies facing downwards, pressed against the side of the abdomen, the fore wings covering the hind wings. The fold occurs at a series of small interlocking elements at the wing base, through which, when the wing is extended, the force of the flight muscles is transmitted.

6. *Skeletal structures*. Flapping flight in all cases is achieved by beating the whole wing using muscles in the body. The skeleton of the body must therefore be light but rigid. In birds, considerable economy is achieved by having a single V-shaped girder formed of the *coracoids* (ventral shoulder bones) and *sternum* (breast bone), against which the muscles responsible for both raising and lowering the wings act. The downstroke is produced by the large outer *pectoralis major* muscle, but the wings are raised by the smaller, inner, *pectoralis minor* muscle. The bird sternum is keeled in many species, but the deep keel does not seem to be essential because ducks (Anatinae) and cormorants (Phalacrocoracidae) have shallow sternal keels and fly beautifully. The bat shoulder girdle is typically mammalian, but the clavicle (collar bone) is stout and tends to fuse with the scapula (shoulder-blade) and the sternum (breast bone), which is keeled for the attachment of the large pectoral muscles which power the downstroke. In insects the flight muscles run in two main groups: (i) the dorso-ventrals which bring the top of the thoracic box downwards, and so raise the wings through a lever system at the wing base, and (ii) the dorsal longitudinals which lower the wing by buckling the top of the thorax upwards. The sides of the thorax are ridged to resist buckling, but the top of the thorax may have flexible regions to encourage the essential buckling.

Because the wings both of birds and of bats contain a series of hinged bones, the shape of the up- and downstrokes can be controlled by muscles within the wing. In insects, however, the form of the wing beat is controlled entirely by wing twisting from its basal articulation. In such primitive insects as grasshoppers, crickets, and locusts (Orthoptera) and dragonflies and damselflies (Odonata) this is controlled by direct muscular action, which acts differentially in the up- and downstrokes, but in highly evolved insects, where the wing beat frequency is much higher, the basal articulation of the wing acts differently in the two halves of the wing cycle, and causes an automatic wing twisting at the extremes of the stroke; examples of this are the click mechanism of flies (Diptera) (Fig. I) and the wing coupling and twisting mechanism of wasps, bees, and ants (Hymenoptera). This aspect is explored in the section on the EVOLUTION of flight.

Any structure which is moving has kinetic energy. When a structure flaps up and down, the kinetic energy must be absorbed at the end of one stroke to stop the movement, and energy must

then be provided to produce the movement in the opposite direction. It is partly to minimize this loss of energy that the wings are so light in most flying animals, but in two groups the energy of rotation of the wing is either conserved or used. The feathers of birds are sufficiently elastic to bend under the aerodynamic loads as the wing accelerates at the start of the downstroke, while at the end of the stroke, by straightening, they can decelerate the wing and so make use of the energy of rotation of the wing. Insects have an even more efficient system in which the wing oscillates on a pair of elastic hinges made of a highly resilient rubbery

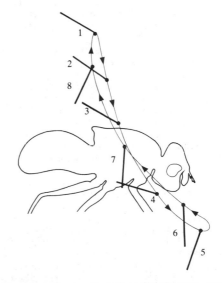

Fig. I. In higher insects, such as flies (Diptera), movements of the thoracic box cause the wing to twist sharply at the top and bottom of the stroke. In the downstroke (1 to 4) the wing travels leading-edge down, in the upstroke (6 to 8) it travels leading-edge upwards, and at the extremes of the stroke (4 to 6 and 8 to 2) it twists rapidly to change its angle of incidence.

protein called *resilin*. By this means, the energy of rotation of the wing is stored at the extreme of the stroke, and then used to re-accelerate the wing in the opposite direction, as if it were bouncing between two springs. Without the resilin hinges, an insect, such as a bee (Apidae) would need to produce about five times as much power as could be expected from even the most powerful muscle.

7. *Muscles, respiration and body temperature*. Driving only one or two pairs of wings, the individual flight muscles tend to be the largest muscles in the animal's body. Their total weight is rarely less than 10% of the body weight, and in hummingbirds (Trochilidae), and some flies and bees, it may approach 30%. The muscles that lower the wings are usually larger than those that raise the wings, but in hovering animals, where both up-

and downstrokes are equally important, the two antagonistic sets of muscles are similar in size.

The typical power output of human muscle is 20 W/kg, but for the flight muscles of humming-birds and bees outputs of over 150 W/kg have been calculated. These remarkable power outputs require specializations: (i) in the muscle to produce both the speed of contraction and the chemical energy required for high power, (ii) of the respiratory system to provide the extremely effective gaseous exchange required, and (iii) of the body surface and circulatory systems to allow high temperature operation and control of loss of the large amounts of waste heat that are produced (see HOMEO-STASIS).

For high power production, flight muscles tend to be specialized, both by having rich internal energy supplies, and by having rapid responses to nervous stimulation. The flight muscles of humming-birds or dragonflies can contract at over 40 cycles per second, which is over five times as fast as the most rapid repetitive movements of man. In some of the most highly evolved insects, the flight muscles do not contract at a rate determined by the nervous stimulation, but at a rate determined by the mechanical properties of the thorax. These, termed *myogenic* muscles, are capable of producing power at otherwise impossible frequencies of between 100 and 1000 cycles per second.

Though the lungs of bats are typically mammalian with a tidal flow of air in and out, the rate of gaseous exchange apppears to be similar to that of birds, where the system is modified to provide a continuous air-flow through the lungs. Flying birds have extensive air sacs both between the mouth and lungs and beyond the lungs (see BREATHING). By a system of tubes which by-pass the lungs, air first moves into the posterior air-sacs. On the next expiratory movement, this air starts to be blown forwards through the lungs, while the anterior air-sacs are emptied to the outside. The next inspiratory movement brings new air into the posterior air-sacs, and sucks air through the lungs into the empty anterior air-sacs (Fig. J). The lungs thus have a virtually continuous forced air circulation which allows a small lung an efficient gas exchange. A rather similar system of air-sacs is found in insects, where air is sucked into thoracic sacs by skeletal movements during flight, and then expelled through the abdomen. The flow is controlled by patterned opening and closing of external holes called the *spiracles*, by which the air-containing system of tubes, the tracheal system, communicates with the outside. The flight muscles are provided with air directly through finely branched tubular tracheoles, which open into the ventilated air-sacs, and along which gaseous diffusion is sufficiently rapid for the respiratory requirements of the muscles.

With the high metabolic rates, flying animals tend to produce sufficient heat to allow body temperatures of 35–45 °C, and this is advantageous as it allows muscles to develop high powers. Most

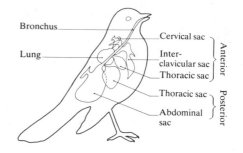

Fig. J. Bird respiration. Birds have extensive air sacs but their lungs are small. The air sacs, bronchus, and lungs form a complex series of passages and chambers. The movement of air is (i) into the posterior air sacs, then (ii) into the lung, then (iii) into the anterior air sacs, and finally (iv) out of the body. This allows a continuous airflow through the lung, and there is no static air as with a tidal lung such as ours.

birds maintain a high body temperature, and can lose excess heat either through the wings in flight, or by panting at rest. Bats, being small and nocturnal mammals, cool down when inactive, and bring themselves up to their flight temperature each evening by shivering. A similar process is used by many insects, notably moths (Lepidoptera) and bumble-bees (Bombini), which maintain temperatures of 30–40 °C in their flight muscles, even when the air temperature is as low as 5 °C. They are well insulated by fine cuticular bristles, just as birds and mammals are insulated by feathers and fur, and are able to control the thoracic temperature by pumping blood from the thorax into the less well insulated abdomen from which heat can be lost (see THERMOREGULATION).

For aerodynamic reasons, not only should the wings act as efficient lift-generating surfaces, but losses due to drag should be minimal. For this reason, the bodies of flying animals are usually fusiform and more or less streamlined. For this purpose feathers are excellent as they mould themselves to the contour of the body, but allow skeletal movement. Many insects are bristly, and this probably serves the same function of smoothing the exterior of the body, and works at that size range because each bristle traps a far larger lump of air around itself.

8. *Flight stability and control.* An additional design problem facing the flying animal is stability. By itself, a body with a pair of wings may not be stable, because it is hard to arrange that the centre of lift of the wings is at the centre of gravity. Instability around the axis of flight, or roll axis, is prevented by placing the wings above the centre of gravity, on the dorsal surface. Stability in both yaw (about the vertical axis) and pitch (about the horizontal transverse axis) are improved by adding a long tail; this was seen in the earliest flying lizards (Pterosauria), early birds (*Archaeopteryx*), and mayflies (Ephemeroptera).

Such a tail, held out straight, can be used in gliding, and can be bent to turn the animal. Exactly this type of control is used in most modern aircraft. As animals became better at both sensing and controlling their flight, so they were able to exploit ever less stable and more manœuvrable designs. The modern sparrow (*Passer domesticus*), and bluebottle (*Calliphora* spp.), fly well without tails and the bluebottle can even land on the ceiling after flying upside down. In both cases the animals rely on elaborate senses of balance and of wind speed.

Birds probably detect changes in the angle and direction of flight both visually and by means of the semicircular canals of the inner ear, while air speed, at least in some birds, is detected through the nostrils. Insects detect air speed either by using special hairs on the head, as in locusts (Acridoidea), or through the deflection of their antennae in the air stream, as in flies. The detection of changes of attitude is less well understood, except in flies; they use only the front wings for flight, the hind wings being modified into club-shaped *halteres* which beat at the same frequency as the wings. If the halteres are removed, the fly can no longer fly straight and level, though stability is restored if a couple of inches of cotton thread are glued to the tip of the abdomen. It appears that the halteres act as a pair of gyroscopes detecting any departures from straight flight in yaw and roll: rotation about either of these axes causes bending of the shaft of the haltere which is detected by strain sensitive receptors at the base of the shaft. Rotation about the pitch axis appears to detected by the antennae.

Flight performance. 1. *Effects of size.* Given a constant proportion of the body devoted to muscle, flying animals will tend to produce constant power per unit mass or specific power. If, as they get larger, the proportions of flying animals remain similar, their body weight will increase as the cube of their length, but their wing area will increase as the square of their length. This means that as animals get larger, their wing loadings get higher and so, to produce the required lift, they must fly faster; it can be shown that the speed should rise as the square root of the length, or that a raven (*Corvus corax*) 60 cm long should fly about 1·4 times as fast as a 30 cm long jackdaw (*Corvus monedula*), and twice as fast as a 15 cm long house sparrow. These rules do not apply precisely, as birds of some groups such as ducks, raptors (Falconiformes), and swifts (Apodiformes) fly particularly fast. Precise measurements of speed are hard to make, but some raptors appear to reach 160 km/h, ducks and pigeons (Columbidae) 95 km/h, but smaller birds, like sparrows (Passerinae), are not capable of more than about 32 km/h. Among the insects, hawkmoths (Sphingidae) and dragonflies can attain between 24 and 40 km/h, bees and horseflies (*Tabanus* sp.) are about half as fast as this, whilst the smallest flies, such as midges (Chironomidae), probably only reach about 1·5–3 km/h.

Similar factors to those that determine the speed of flight determine the wing beat frequency. As animals become larger, so their wings become longer, and to produce a given air velocity they can beat more slowly. Among birds, swans (*Cygnus* spp.) and griffon vultures (*Gyps* spp.) flap at about 2 Hz (beats per second), swifts and sparrows at about 10 Hz, and humming-birds at up to 50 Hz. Among insects, locusts flap at about 20 Hz, bees at over 150 Hz, and mosquitoes (*Aedes* spp.) at 600 Hz. The record seems to be about 1000 Hz for the midge *Forcipomyia*.

The largest birds, the ostriches (Struthionidae) and rheas (Rheidae), are flightless. There are good structural reasons for this. To fly, an animal must produce enough power to overcome the drag of wings and body. With increasing size, and hence flight speed, the power needed for flight increases. This increase is more rapid than the availability of muscle power, which is roughly proportional to weight. The effect of this is twofold: (i) there is a maximum weight at which powered flight is possible, which seems to be about the 10–12 kg of vultures and the giant bustard (*Ardeotis kori*); (ii) because hovering flight requires significantly more power than forward flight, the larger birds can neither take off vertically nor hover. Pigeons, weighing about 0·3 kg, can climb vertically for a brief period, while humming-birds, weighing only 20 g, can hover indefinitely, as can many insects which are even smaller. Thus larger birds must run up to flying speed, or take off into wind, or fly downwards off a tree or cliff to reach a speed at which flight by their muscle power can be sustained. Clearly, for these larger birds, flapping flight is bound to require extreme exertion, and it is for this reason that so many large birds only fly intermittently, or soar using natural air currents.

2. *Exploiting the environment.* Because gliding flight only requires an animal to hold its wings out and bear its weight, it is cheaper, in terms of energy, than flapping the wings, where the muscles have both to produce larger forces and to shorten. For aerodynamic reasons large wings tend to have better lift:drag ratios than small ones, so it pays the largest birds to glide, and make good the consequent loss of height by flying in conditions in which there are natural up-currents.

When a wind strikes a hill, the air is deflected upwards on the windward side (Fig. Ka). On the edges of cliffs the up-current can be a large fraction of the wind speed, and so can be exploited by birds to glide along the cliff. In high winds where the up-current is greater, it may even be possible to fly without flapping close to the bottom of the cliff. One problem is that in really high winds, the bird may be unable to glide fast enough to avoid being swept over the top of the slope; on windy days, gulls (Laridae) can be seen maintaining a constant position above ground in a steeply diving attitude with the wings partly folded. A related type of soaring, seen in seagulls following a ship, makes use of the standing waves that form on the

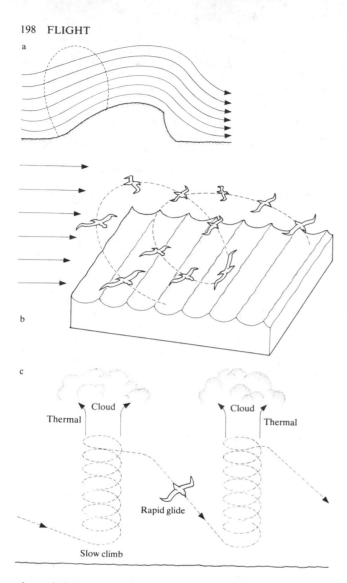

a

b

c

Cloud

Thermal

Cloud

Thermal

Rapid glide

Slow climb

Fig. K. Exploiting natural air currents. **a.** When a wind strikes a hill, it is deflected upwards. Depending on the wind speed, there is a greater or lesser effective lift. A soaring bird can probably fly anywhere within the space shown enclosed by dashes. **b.** A soaring albatross (*Diomedea* spp.) makes use of the wind velocity gradient near the sea. As the animal climbs, it loses ground speed but maintains its air speed because it encounters an ever-increasing wind velocity. At the top of its climb it turns and glides downwind gaining ground speed and kinetic energy; it can only move downwind by this mechanism. **c.** Large soaring birds can progress across country from thermal to thermal, climbing slowly in each thermal till they reach the cloud base and then gliding to the next thermal. Because of the time spent climbing, the average speed over the ground is about half that in level flapping flight, but the energy consumption is many times less. Vultures (*Gyps* spp.) can use this type of soaring to achieve average speeds of about 40 km/h, and make daily round trips of up to 200 km.

downwind side of obstacles; from the gull's point of view, this is a cheap way of crossing large areas of water.

Above the surface of the sea, particularly when there is a strong steady wind and large waves, soaring is possible using one of two tricks. Small to medium sized birds, such as shearwaters (*Puffinus* spp.) and storm petrels (Hydrobatidae), can use the up-currents on the individual waves, and can glide both along the windward side of the wave crest and from the wave crest to the trough, and then take off from the top of the next wave. Albatrosses with wing spans of up to 3·5 m can both exploit this, and make use of the wind gradient that exists above the surface of the sea (Fig. Kb). Near sea level, the wind speed is low, but it rises with increasing height. Albatrosses starting high up glide downwind, losing height, but gaining

speed relative to the sea and, more important, gaining kinetic energy. When they approach the surface, they turn into the weak wind near the sea and start climbing. As they climb, they encounter progressively faster and faster wind and so, because their bodies have considerable inertia, they can climb up the wind gradient until, at a height of 10–30 m, the wind gradient becomes too gradual and the climb flattens out. At this point, the albatross turns downwind again and glides back to the surface. With each vertical cycle, the albatross also moves some distance downwind; these birds tend to be confined to the latitudes between 40° and 60° south, and to circumnavigate the world in a west to east direction.

Over land, differential heating of different regions causes local rising air vortices, or *thermals*, in the centre of which a relatively stable fast up-current may occur. So long as the animal can turn sufficiently rapidly to stay within the thermal, and so long as its rate of sinking in a glide is slower than the speed of the up-current, thermal soaring is attractive. This tends to favour very large birds such as vultures, which can glide with lower sinking speeds than small birds, and, having short wings, can make a tighter turn than is possible for the long thin wings of the albatross for instance, and so can exploit smaller thermals. Vultures appear to be good at detecting thermals at a distance, partly by watching other birds. They make long trips climbing within thermals, and then gliding fast to the next thermal in the line of flight (Fig. Kc). Because they do not need to flap their wings, the energy cost of the journey is between one-tenth and one-fiftieth of that for flapping flight over the same 200 km daily round trip.

Insects sometimes get sucked up into thermals; greenflies (Aphidae) can be seen rising on hot days, and experiments where fine nets have been towed behind aircraft have shown that insects of many different orders can be found at heights up to $\frac{1}{2}$ km. It is known that atmospheric movements are important in insect dispersal; locusts use prevailing winds to invade Africa from breeding sites in the Arabian Peninsula, and then breed and swarm; various butterfly species, such as the painted lady (*Pyrameis cardui*) and the clouded yellows (*Colias* spp.), use southerly winds to invade Britain from the continent of Europe. Very occasionally, the monarch butterly (*Danaus plexippus*), which is a common North American butterfly, manages to cross the Atlantic, but this requires a felicitous combination of wind and weather.

There are, of course, environmental circumstances in which flight is disadvantageous. Flight has been lost both by many groups of island birds and insects, where there is a risk of being blown out to sea, and by animals living in highly localized habitats. In many moth species, such as the mottled umber (*Erannis defoliaria*), only the male has wings. This ensures that the female remains near the food tree, whilst allowing some gene flow as a result of mate-searching by the male. In several parasitic groups, the adult insects are either wingless or use the wings to find the host, and then lose the wings after settling; a similar loss of wings is seen after the nuptial flight and dispersal of both ants (Formicidae) and termites (Isoptera), while the worker castes in both these groups are wingless and forage on the ground from a permanent nest.

3. *Range and fuel for flight.* In many flying animals, the ability to fly long distances permits seasonal migrations. The distances involved may be small, for instance from around the Arctic Circle to southern England or the middle of Europe, or they may be vast, for instance from northern Europe to South Africa. Usually such journeys involve sea passages, which for a small bird can last several days, and therefore require a high level of preparation.

The first requirement is fuel. In normal daily life, the reserves of energy are short term. Birds store a carbohydrate, glycogen, in both the liver and the tissues, and this store is cycled on a day-to-day basis. In the same way, insects use glycogen which is stored in the fat body. For this short-term requirement, glycogen has the attraction that it is speedily mobilized, but suffers from the disadvantage that water binds to it, and so it is relatively heavy. For long-term storage, fat offers a variety of advantages. It can be stored without any weight penalty in bound water, it provides about twice as much energy per unit dry weight as glycogen, and, most important, when metabolized, it produces more than its weight in water as a by-product, whereas glycogen produces only about half its own weight of water. This means that if a flying animal can use fat, it will run less risk of drying out from water evaporation during respiratory exchange, and from water loss during excretion.

Fat is somewhat difficult to use as fuel, and the peak performance may not be as good as when flying with carbohydrate. For the migrating animal this disadvantage is more than offset by the cost-effectiveness of fat, and before long migrations both insects and birds deposit large fat reserves. In some of the smaller passerine birds the weight may double before the start of migration; if later they are caught far from land, they are found to have depleted these reserves. Knowledge of the energy cost of flight for some of these smaller birds makes it possible to calculate the rate of utilization. For some of the passerines with a relatively low energy utilization the endurance is 3–5 days, but it is inconceivable that humming-birds can exceed 30 h. Similar calculations for locusts, which regularly fly in swarms for 5–8 h, suggest that they must be able to re-deposit fat nearly as quickly as they use it in flight. For fat metabolism many flying animals seem to have mechanisms for transporting fatty compounds about the body, and a specially active series of fat metabolizing enzymes within the muscles.

The speed and duration of a migration depends on many factors. Smaller animals naturally fly more slowly and are more affected by winds than larger ones. With higher energy costs they therefore have to stop to feed at shorter intervals. Because the air is more rarefied and offers less drag, it may also pay a bird to climb to heights of some thousands of metres during a long journey, particularly at the start when it has a heavy load of fuel, and so has to fly faster to generate lift. Up at these heights the winds may be more stable.

Flight in the animal's life. 1. *Escape.* Many flying animals operate at close to the limits of muscular and skeletal performance. When flight is used for escape from potential predators it is important to have good acceleration. For this reason, both birds and insects are capable of producing brief bursts of muscle power at moments of crisis, and of exploding into flight. This is clearly demonstrated by two examples: the pheasant (*Phasianus colchicus*) cannot sustain flight because its flight muscles are not capable of rapid respiratory exchange; this results in a quickly built-up oxygen debt and so fatigue; however, the pheasant is capable of very rapid flight for brief periods, and, in fact, uses flight only for escape, otherwise moving by walking. Locusts and grasshoppers also fly when startled, attaining a flying speed of about 3·5 m/s by an escape jump, and using their large back legs to accelerate to this speed.

Escape is often accompanied by visual and acoustic signals which warn other members of a flock and bring about mass flight (see FLOCKING). It is probably more difficult to select one prey individual if many individuals are darting away.

2. *Prey Capture*. The stoop or dive of the peregrine falcon (*Falco peregrinus*) is a dramatic example of the way in which flight can be exploited in prey capture. The peregrine strikes its prey in mid-air at speeds of up to 160 km/h, and kills by stabbing with the claws. Other falconids use similar strikes from the air directed at mammals and birds on the ground. Barn owls (Tytonidae) hunt at night on the wing, and locate their prey by ear; owl flight is particularly silent, and cannot be heard by man beyond a range of 3 m.

Many insectivorous birds, e.g. swallows (Hirundinidae), hunt on the wing (see PREDATORY BEHAVIOUR). In these, the mouth is usually capable of gaping wide, and may be fringed with stiff bristly feathers which increase the size of the trap. Flight of these aerial hunters is usually rapid and they are adapted to be manœuvrable. Insectivorous birds hunt by eye; nightjars (*Caprimulgus europaeus*) for instance, have large eyes for night VISION and are inactive during the day. Bats also hunt insects at night but, according to species and the size of the prey, they may use the tail membrane to form a net in which the prey is caught. The teeth of bats are small and pointed and in this respect analogous to the delicate jaws of insectivorous birds. Prey location in bats is by ECHOLOCATION and, in some, the sound is produced through the nose, leaving the mouth clear for feeding.

Many insects are insectivorous. Some, like robber flies (Asilidae) and wasps (Vespoidea), hunt on the wing, catching their prey with long raptorial legs. Among Odonata, the dragonflies (Anisoptera) hunt on the wing, while the damselflies (Zygoptera) rest on foliage, flying out to attack passing insects. Both groups have long fringed legs, which extend to form a basket-like trap, and sharply pointed mouthparts, the *mandibles*.

The flight season for temperate insects is short, so many insectivores either migrate (swallows, swifts) or hibernate (bats) or phase their life cycle to that of their prey (dragonflies).

3. *Territorial and sexual behaviour*. In the defence of a TERRITORY it is important to be able to patrol the frontier effectively and speedily; for this flight is effective. It may combine with territorial DISPLAYS, such as the singing flight of the skylark (*Alauda arvensis*), and the drumming flight of the common snipe (*Gallinago gallinago*) in which the tail feathers produce a sound during a power dive, or with defence, as when house-flies (*Musca domestica*) chase other flies that intrude into their territories.

There is little division between these types of behaviour and the mating chases of swifts which appear to court on the wing, and then mate in flight. Among insects this type of behaviour is common. Butterflies (Lepidoptera) chase appropriately sized and coloured insects; March flies (Bibionidae) hover above grass waiting for the ascent of females which are then chased; many flies, e.g. midges and the dance fly *Hilara*, form mating swarms in which there is individual display and mating. In honey-bees (*Apis mellifera*) the males chase an ascending queen, and the successful male mates in mid-air, while rather similar nuptial flights occur in both ants and termites.

Evolution of flight. 1. *Insects*. Flight has evolved in various groups of animals, but arthropods were both the first animals to become terrestrial and to fly. The precise course of the evolution of insects is uncertain, because of gaps in the fossil record. By the middle of the Carboniferous era (about 310 million years ago), there were various distinct orders of winged insects, primitive dragonflies, cockroaches (Blattaria), and grasshoppers, and by the end of the epoch, bugs (Hemiptera) had appeared. Beetles (Coleoptera) and flies had emerged by the Triassic era (about 220 million years ago). The explosion of the insects as winged animals was very early, and their later radiation (divergence) has exploited the main adaptive lines that were laid down in the initial radiation (see ADAPTATION).

It is not clear how wings first evolved, but two main theories exist. One, the *gill theory*, suggests that wings evolved from gills which had developed as outgrowths of the thorax and abdomen and which, for ventilation, evolved musculature. The other, the *paranotal theory*, suggests that lobes developed from the side of the thorax and abdomen became gliding planes, and later developed basal hinges and musculature. A series of fossils, the Palaeodictyoptera, have such lobes, but do not antedate other more typical flying insects.

Most of the early flying insects were of medium to large size by present standards. This allowed a wing-beat frequency that could be produced by conventional muscles. Among the Carboniferous insects, some of the dragonflies reached wing spans of about 60 cm. From the fact that neither dragonflies nor mayflies can fold their wings to form a hood over the abdomen, it is thought that one of the early advances was the development of true wing folding, with its attendant advantages of stability when not flying and of wing protection. This grade of evolution, where the wings are powered by direct muscles on the wing bases, and where the muscles are controlled directly by the

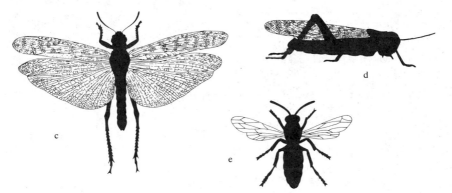

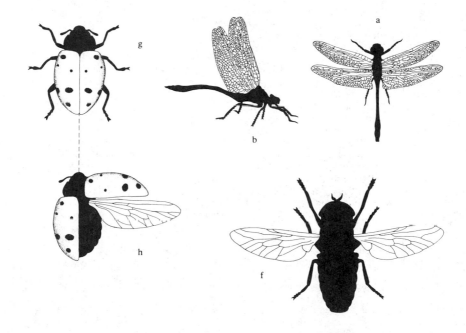

Fig. L. The evolution of insect wings. **a, b.** A dragonfly (Odonata) has four similar wings with a primitive net-like venation. The wings are driven by direct muscles in the thorax, and can only be folded directly above the body. The long body is used as a rudder. **c, d.** A locust (Orthoptera) retains the primitive type of wing, long body, and has partly direct muscles attaching to the wing bases. The wings can be folded neatly along the abdomen with the leading edge downwards. **e.** In a wasp (Hymenoptera) the thorax is fused into a single capsule and flight muscles are all indirect. The wing venation is much reduced and the two wings can be hooked together to act as a single aerofoil. **f.** In flies (Diptera) each metathoracic wing is modified into a stability sensor, the haltere, and the thoracic segments fuse into a single box with large indirect myogenic muscles. Wing-twisting is achieved by the mechanics of the thoracic box. **g, h.** In beetles (Coleoptera) the forewings are modified into elytra which hinge back to cover the abdomen and folded metathoracic wings. To take off, the beetle opens its elytra, unfolds the metathoracic wings, and then starts beating its wings. During flight, the elytra act as passive lifting surfaces.

pattern of nerve impulses, is seen in Orthoptera, Neuroptera (alderflies, snakeflies, lacewings), and other primitive orders. The wings of these primitive insects tend to have complex net-like venation, and both wings beat actively and are similar in size

(Figs. La–d).

Among the evolutionary advances that have occurred in the insects, several have arisen on more than one occasion: among higher insects, it is general to find that the wing venation is reduced and simplified: myogenic (high speed) flight muscles have evolved at least twice in the Hemiptera, Diptera, Coleoptera, and Hymenoptera. This tends to be associated with three advanced features: the thoracic segments tend to fuse into a single stressed-skin box, and the flight muscles run obliquely across this to cause high force but small distance distortions which drive the wings; because of this fusion the wings tend to be coupled together, as in Hymenoptera (Fig. Le), and so to beat in phase, as opposed to the roughly antiphase wing beat that occurs in the two pairs of wings of lower insects; with fusion of the thoracic segments, one pair of wings can be driven and the other pair can assume a different role. The *elytra* (wing covers) of beetles are modified forewings, and do not beat during flight, (Figs. Lg, h). The hind wings of flies beat at the same frequency as the forewings, but are greatly reduced into halteres used to detect course changes in flight (Fig. Lf). A similar modification has occurred to the forewings of the aberrant parasitic order, the Strepsiptera (twisted-winged insects).

Some of the radiation of insect flight must be seen in relation to competition from vertebrates which emerged on to land during the Carboniferous era. By the Jurassic era (170 million years ago), there was an extensive fauna of small reptiles and early mammals, which must have fed on terrestrial arthropods. This will have tended to favour the smaller flying insects. Present-day insects are mostly less than 10 mm long, which is about $\frac{1}{3}$ of the length of their forebears in the Carboniferous era, and insect evolution could be described in relation to advances that allowed decrease in the body size, while retaining the ability to fly.

2. *Pterosaurs.* Soon after their emergence on to the land, the amphibians gave rise to the reptiles which were far better adapted to resist desiccation, and for rapid running. The reptiles radiated extensively, and by the late Permian and early Triassic era (about 240 million years ago) they were the dominant land animals. By the Jurassic era, flying reptiles had appeared and enjoyed a dramatic radiation before disappearing at the end of the Cretaceous (about 70 million years ago). Unlike both birds and bats, pterosaurs flew by a wing membrane supported only at the leading edge by the fourth finger, which was as long as the rest of the arm (Fig. M). There were two main lines, the Rhamphorhynchida which retained the tail, had long delicate teeth, and reached lengths and wing spans of about 1 m, and the Pterodactylida which were tailless and ranged in size from wing spans of about 20 cm up to the toothless but splendid *Pteranodon ingens* of the end of the Cretaceous, which certainly attained a wing span of 8 m and, according to recent fossil evidence, may well have grown even larger.

As with birds and bats, pterosaurs probably flew by raising and lowering the wings, powered for the downstroke by large pectoral muscles (the

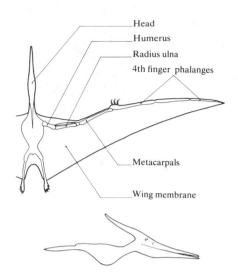

Fig. M. *Pteranodon.* This giant Cretaceous flying reptile lived about 80 million years ago. The wing span exceeded 8 m. The wing was a membrane stretched between the hind legs and the giant hand and fourth digit. The head had a long toothless beak and a posterior keel, which acted as an aerodynamic counterbalance for the forces on the beak as the head was turned.

sternum in some has a keel), but raised by dorsal muscles. The shoulder girdle was fused both to the vertebrae and to the sternum to make a complete ring, and in higher forms the vertebrae were also fused together as in birds. The wing membrane was probably tensioned by a long finger tendon which attached right back to the humerus.

The skeleton of *Pteranodon ingens* has beeen carefully analysed to evaluate its flight performance and its engineering design. The overall weight of an individual with a 7 m span was about 16 kg, somewhat greater than that of a swan of half the span. This low specific weight was achieved by making all wing bones hollow, minimizing the mass of muscles by making the head aerodynamically balanced, and having a lock in the shoulder which allowed the wings to be held extended. Even so, safety factors were probably very low and gliding flight must have been preferred to flapping flight, though the latter was theoretically just possible. *Pteranodon* was essentially a low speed glider, with a far lower wing loading and lower speed and rate of sinking than a vulture; it is hard to envisage it taking off except into a gentle wind, and it would have been very sensitive to turbulent air conditions. It is not known whether any reptiles were warm-blooded, nor why pterosaurs became extinct, but their wing structure must have been more vulnerable and less controllable than that of either birds or bats.

3. *Bats.* Unfortunately, nothing is known about the details of the evolution of flight in bats, which first appear in an essentially present-day form as

long ago as the Eocene era (about 40 million years ago). The skulls of these early bats suggest that they already had the reduced eyes and expanded ear region of modern echo-locating insectivorous bats (Microchiroptera). Unlike birds, the wing bones of bats are not air filled, but the radius and ulna of the arm are fused distally (Fig. H) and the wrist is much reduced. The last four fingers are all hypertrophied, and account for about half the wing span. Bats roost hanging by their hind legs which are very slender, and indeed most bats can crawl only with difficulty, and find it difficult to take off from a horizontal surface.

The other main radiation of bats, the fruit-eating bats of the Old World (Megachiroptera) are larger forms which reach a size of 1·5 m span and 1 kg weight. These are crepuscular forms which rely on large eyes for orientation, though some forms show a primitive low-frequency form of echo-location as well.

4. *Birds.* There is little doubt that the birds evolved from the same reptilian stock as the dinosaurs and crocodiles (Crocodylidae). Features of the organization of skull, limb girdles, and limbs are similar in all these groups.

Although there is argument about whether it was a bird, *Archaeopteryx*, which is represented by only about a dozen fossils from the Jurassic era and was contemporary with the early pterosaurs, had bird-like features, notably feathers, but retained a long tail, had three complete digits in the hand, and only a weakly developed shoulder girdle (Fig. N). The spine was not fused into a girder as in modern birds, and the skull, although lightened,

retained teeth. Calculations suggest that this 60 cm long animal weighed about 500 g, and flew at a speed ranging between 8 and 21 m/s. Because of the weak arm bones, it had a large turning circle and probably had a high landing speed. The three fingers of the hand retained claws and it probably used these to climb trees, from which it took off, as it is unlikely that it could have hovered.

By the Cretaceous, birds showed most modern features, such as reduction of and fusion of the finger bones of the second and third digits, a sternum with a deep keel, the mechanism for wing raising by the pectoralis minor muscle, the long fused bird pelvis, and short tail (Fig. N). They did, however, retain teeth which have been lost in all living birds.

The status of ostriches, kiwis (Apterygidae), and other primitive flightless birds is uncertain. They may have diverged early from the main stock and become large and flightless before the evolution of the typical bird wing and skull, but the fossil record is incomplete. Similar problems exist with the evolution of the typical modern bird orders, fossils of all of which appear patchily from the Eocene onwards, but without obvious forebears. There does not appear to be any obvious trend and both large and small, fast and slow appear more or less simultaneously.

5. *Other flying forms.* Several groups of animals have members that are capable of gliding, but not of sustained flapping flight (Fig. O). Among mammals, there are both flying marsupials, e.g. the gliding possums (*Petaurus*), and placental flying squirrels (*Glaucomys* spp.) which have a skin web

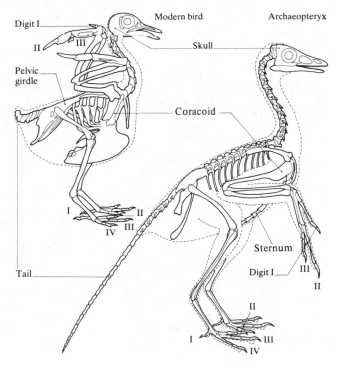

Fig. N. *Archaeopteryx.* By contrast with a modern bird, *Archaeopteryx* retained three free digits in the hand, had only a small sternum and a weak shoulder girdle, and retained a typically reptilian pelvis. The skull still had teeth. It lived in the Jurassic, about 160 million years ago, and is known from about a dozen fossil specimens.

Fig. O. Various flying animals. **a.** Flying squirrels (*Glaucomys* spp.) extend a membrane between the fore and hind limbs and use this as a wing for gliding. **b.** The flying dragon (*Draco*), a lizard, has long ribs which can be unfolded outwards from the sides of the body to support a web of skin which acts as a wing. It makes glides from trees to escape from predators. **c.** Flying fish (Exocoetidae) have greatly enlarged pectoral fins and large pelvic fins which together form wings and elevators. They take off from the surface of the sea, accelerating by beating the enlarged ventral lobe of the tail. **d.** The frog *Rhacophorus* has enormous webbed feet which are used as parachutes during a jump.

connecting fore and hind limbs. Because this is uncontrollable the glide is erratic, and the small wing area and high loading gives a steep glide angle and high landing speed. Mechanisms of this type are interesting because they show how flight could evolve, and it is not surprising that similar mechanisms have evolved in another mammalian order, the Dermoptera (colugos or flying lemurs), in a lizard genus *Draco* (the flying dragons), where the web is supported by elongated ribs and not by the limbs, in a snake genus, *Chrysopelea* (the flying snakes), which flattens its body before dropping and gliding on to its prey, and in the frog genus *Rhacophorus*, for example the Borneo flying frog (*R. pardalis*), which has very long webbed toes which are used as parachutes when the creature jumps from trees. Among fish, two groups have developed long pectoral fins, the flying gurnards (Dactylopteriformes) and the flying fish (Exocoetidae), which are both capable of glides of over 50 m length above the sea. Flying fish have developed a ventral elongation of the tail, and accelerate by a swimming movement with the tip of the tail after the rest of their bodies have left

the water. In these flying fish, there is some wing beating and the flight is controlled, but because of problems with respiration, continuous flight is impossible.

The division between ballistic flight and true gliding is blurred. Animals of many groups jump: fish, frogs, many insects, certain squids (Cephalopoda). True flight must surely be attained only when the animal is capable of prolonging the period it can remain airborne, until ultimately it can remain aloft almost continuously. Such flight has only evolved in insects, bats, and birds.   H.C.B.C.

2, 109, 116

FLOCKING IN BIRDS. (For flocking in other animals, see HERDING.) A flock is a group of birds that remains together as a result of social attraction between individuals. This is in contrast to an AGGREGATION of individuals that arises when each responds independently of the others to some factor in the environment. For example when a blackbird (*Turdus merula*), a robin (*Erithacus rubecula*), and a hedge sparrow (*Prunella modularis*) are seen together at a garden bird table, they are not members of a flock, but of an aggregation in response to the food on the bird table. On the other hand a group of wood pigeons (*Columba palumbus*) feeding on a stubble field in winter constitute a flock. We can recognize that the pigeons are in a flock, and not an aggregation, because when they fly from one field to another they travel in a cohesive group, and do not disperse as soon as they leave a particular feeding place, as do the robin and blackbird when they leave the bird table. In practice, flocks are therefore usually recognized by cohesive movements of the birds.

Flocking is often associated with two other forms of social behaviour involving attraction between individuals: colonial nesting and communal roosting. Approximately half the 8000-odd species of living birds show flocking behaviour at some stage in their life cycle, but nearly all the species that nest in colonies also feed in flocks, so that colonial nesters are more likely to flock than are solitary nesters. Flocking in birds is also associated with diet, so that species that eat seeds, nectar, or fruit are about twice as likely (in terms of per cent of species) to show flocking behaviour as birds that eat mainly animal matter such as insects and fish. This correlation with diet is useful in attempting to understand the advantages of flocking, as is discussed later.

Most flocking birds do not flock throughout their life cycle. The most frequently observed pattern, at least in temperate regions, is solitary nesting (that is, one pair defends a TERRITORY), and flocking outside the breeding season. This seasonal transition is probably under the influence of HORMONES, although little is known about the mechanisms involved.

In addition to seasonal variations, birds often show shorter term fluctuations in flocking behaviour. In the early spring, when great tits (*Parus major*) are just beginning to leave the winter flocks and set up territories, birds may be quite strongly territorial on a warm, sunny day, and revert to flocking if the weather turns cold. In addition to weather effects, food distribution can play a part. Experiments with wintering white wagtails (*Motacilla alba*) have shown that the individuals may be induced to switch from flocking behaviour to territory defence by making the food supply easily defendable and highly localized. This indicates a link between flocking behaviour and the nature of the food supply.

Many non-breeding flocks, in both temperate and tropical regions, contain more than one species. This may also be true in the case of some flocks formed during the breeding season, for example, in some herons (Ardeidae). The different species in a mixed flock usually have similar feeding habits and are often in the same genus. For example, the different species of tits (Paridae), in temperate zones usually form mixed flocks in winter.

In tropical forests, where breeding is less markedly seasonal than in temperate zones, mixed species flocks may contain several breeding pairs that join the flock temporarily as it moves through their breeding territories. In this situation again, there is a rapid switch from flocking to territorial behaviour.

Flocks formed outside the breeding season in temperate zones, and probably also most tropical forest flocks, do not necessarily contain individuals that have hatched from the same clutch (although as mentioned above they may contain pairs). In contrast, there are some birds that live permanently in social groups of perhaps five to fifteen individuals, most or all of whom are close kin, usually parents, children, and grandchildren. These species, which include, among others, babblers (Timaliinae), Florida scrub jays (*Aphelocoma coerulescens*), Australian magpies (*Gymnorhina tibicen*), and long-tailed tits (*Aegithalos caudatus*), defend group territories, nest communally, and always feed in flocks.

Flock cohesion and spatial organization. By definition, the members of a flock show some degree of coordination in their behaviour. The most simple and universal form of group coordination is movement. Although the short hops of individual birds while searching for food in an area are not synchronized, the slightly larger-scale movements, by which, for example, birds move from one group of trees to another, tend to be coordinated so that the whole flock moves at once. These are called integrated flight movements. In practice, one of the clearest ways to recognize the existence of a flock is through these integrated flights. Coordinated movements range from the loose drifting, punctuated by occasional integrated flights, of mixed species of winter flocks of tits, to the remarkably precisely synchronized zig-zag flight manœuvres of sandpipers (*Tringa*).

In flocks of estrildine finches (Estrildidae), for example, in addition to synchrony of movements, other behaviour patterns such as feeding, preening, and sleeping show a strong tendency to be synchronous within a flock. This results from the fact that individuals tend to copy one another (see IMITATION), so that once one or two individuals start preening, for example, the others rapidly follow suit. In the few cases where it has been analysed, it seems that flock synchrony is based either on visual signals or on the contact calls that are continually made by members of many flocks.

In the majority of flocks, individuals do not tolerate close contact. If the birds come within a certain distance, called the INDIVIDUAL DISTANCE, some sort of aggressive interaction takes place and the birds move apart. The minimum distance of tolerance varies from species to species and may be anything from a few centimetres, for example, in chaffinches (Fringillinae), to several metres, for example, in the great blue heron (*Ardea herodias*). Further, the size of the individual distance may vary according to season, or other environmental circumstance. The species that have no individual distance, for example the red avadavat (*Amandava amandava*), are referred to as *contact species*, because the members of a flock often rest huddled together. When huddled together they may preen their neighbours (often a mate), which probably has the effect of reducing AGGRESSION.

Resting in contact with neighbours may well serve as a means of heat conservation. This is especially likely in those species, such as winter wrens (*Troglodytes troglodytes*) and golden-crowned kinglets (*Regulus satrapa*), that form contact groups when roosting in winter. These birds are extremely small, so that heat loss is likely to be a major source of energy expenditure in cold weather.

The maintenance of spacing between individuals in a flock may act to reduce mutual interference in food gathering in some species. Redshank (*Tringa totanus*) hunt visually on mudflats around the British coast for mobile prey, chiefly the crustacean *Corophium*, which tries to escape capture by burrowing into the mud in response to surface vibrations made by the approaching bird's steps. This means that fewer prey are available in an area that a bird has just trodden on. In this situation it pays the redshank in a flock to space out from one another. At night, however, redshank feed in compact flocks, with small inter-individual distances. Instead of hunting visually for prey at the surface of the mud, they now hunt for submerged prey by sweeping their bills back and forth under the mud. Mutual interference is no longer a problem because the prey do not escape by burrowing into the mud; so flock spacing is adjusted according to the degree of interference in feeding. This illustrates the general point that the degree of spacing maintained in a flock is the result of the conflicting selection pressures (see NATURAL SELECTION) favouring clumping and spacing out.

The area over which a flock travels during a day varies considerably from one case to another. At one extreme, the permanent family flocks of babblers defend territories of about 5 ha, while at the other end of the scale, large flocks of hundreds of migrant common starlings (*Sturnus vulgaris*) may travel tens of kilometres in one day. Winter flocks of titmice in temperate zone woodland have a home range of somewhere between 5 and 20 ha.

In mixed species flocks there is usually greater cohesion within species than between species. In temperate zone winter flocks of insect eaters, for example, each species tends to forage in a different NICHE. There are usually clear-cut interspecific DOMINANCE relationships analogous to those in single species flocks; the inter-species dominance relationships are often, but not always, correlated with size (larger species being dominant). Both temperate zone forest flocks and some tropical forest flocks are made up of a few species (perhaps only one) that are represented by several individuals, and several other species that are represented by only one or perhaps a pair of birds. The former are sometimes referred to as *nuclear species* since they constitute the nucleus of the flock. They tend to lead flock movements. The other species are often territorial birds that only join the flock as it moves through their territory; they are *attendant species*. In the mixed flocks of Western Europe, such birds as tits and goldcrests (*Regulus regulus*) are nuclear species, whilst winter wrens and treecreepers (*Certhia familiaris*) are attendant species.

It has been pointed out that in mixed flocks, especially in the tropics, the different species in a group often show some degree of similarity of plumage. It seems likely that this is an adaptation to promote cohesion of mixed flocks. This similarity of plumage means that the different species will react to each other's visual signals associated with flock cohesion, so that any advantage resulting from being in a mixed flock is likely to be enhanced.

The advantages of flocking. Much has been written on the evolutionary forces influencing flocking. When considering the problem it is important to bear in mind that the crucial question is 'How does an individual benefit as a consequence of being in a flock?' There are two ways of collecting evidence for any particular hypothesis. One is COMPARATIVE: to try and correlate the tendency to form flocks in closely related species, with the presence or absence of some environmental factor. Flocks of small birds are rare in Hawaii, where there are relatively few predators, suggesting that PREDATION is an important factor influencing the evolution of flocks. This approach is most powerful if applied to variations within one species. Suppose, for example, that a particular type of bird always formed flocks where hawks (Accipitridae) were present, and never where hawks were absent, this would be good comparative evidence for the

selective influence of predation on flocking. The problem with this approach is that not all flocking species show variation from place to place in the amount of flocking. The second method is to try and measure directly an advantage to the individual that results from being in a flock. In the strict sense, the correct thing to measure is the relative GENETIC contribution to future generations of, for example, individuals in flocks of different sizes, or solitary and flocking individuals. This contribution is called relative FITNESS. In practical terms, it is impossible to measure relative fitness, so one attempts to measure short-term benefit, such as escape from predators, or enhanced feeding success. Even if one demonstrates an advantage of flocking to a species, this does not necessarily tell us what the evolutionary origin of flocking was in that particular species. This evolutionary question is a historical one, and cannot be answered by the study of SURVIVAL VALUE in present day systems.

Flocking and feeding. It has often been suggested that the members of a flock enjoy an increased success in capturing food as a consequence of flocking. The association between flocking and diet, discussed above, supports the notion of a link between flocking and food exploitation. This type of advantage could only apply to those cases where birds actually forage in flocks, and not to roosting or MIGRATION aggregations. How does flocking enhance feeding efficiency? Probably different mechanisms operate in different cases.

Some birds feed on prey that escape from the oncoming predator by rapid movement; for example, many flying insects. During this moment of rapid ESCAPE the prey are relatively conspicuous, so that the approaching bird acts as a 'beater', flushing the prey. If several birds hunt together, each can take advantage of the beating action of the others, so that flock feeding will be an advantage. Some of the large tropical forest flocks composed of several species of insect eaters probably arise as a result of this type of benefit, but the same cannot be true of birds searching for stationary prey, such as seeds or insect pupae. The only direct evidence that flocking birds benefit from flushing of their prey comes from studies of cattle egrets (*Bubulcus ibis*) and anis (*Crotophaga* spp.). These studies show that individuals capture more prey items per minute when associated with cattle, the feet of which flush up the insect prey.

In a few species, integration of flock movements has clearly evolved to take advantage of prey flushing. Double-crested cormorants (*Phalacrocorax auritus*) and pelicans (*Pelecanus* spp.), for example, not only flush their prey but also herd SCHOOLS of fish to prevent them escaping. This involves quite elaborate integrated flock movements.

The correlation between seed eating and flocking suggests a further benefit of flocking. Seeds tend, on the whole, to occur in unpredictable patches of local abundance, and when a bird eats this type of food, the limiting factor for an individual is to find one of these patches, rather than to capture food items once it is in the area of local abundance. The same applies to any bird exploiting a patchily distributed and unpredictable supply of food. In this situation, flocking is of benefit in increasing the likelihood that each individual will find a good feeding area. This is because, with more pairs of eyes looking, the flock is more likely to find the patch, and once one bird in the flock finds a good place, the others immediately join the successful bird. If the patches of food are, on average, large enough for all the flock members to feed, there will be little disadvantage in competing with others within the feeding area, although the size of food patches presumably sets an upper limit to the size of flocks. The tendency to learn where to feed by watching others, sometimes referred to as *local enhancement*, is probably widespread in flocking birds, but direct evidence for the benefit is quite rare. However, results of work on the great blue heron, and on wood pigeons, show that each individual in the flock benefits because the flock is better at finding good patches of food.

Once the flock has found a patch of food, there does not need to be any further beneficial interaction between the birds for flocking to be of advantage. However, in some cases, learning by watching other birds goes on at a much finer level and not simply in the localization of a relatively large patch of food. Studies of captive flocks of tits have shown that when birds are FORAGING close together they continually react to each other's foraging success. If one bird in a flock finds a single piece of food in a particular tree, or in some other typical place, the other birds concentrate their searching effort, for a few seconds, either in the same area as the find, or in a similar place elsewhere. This means that each bird tends to concentrate on searching in places that are most profitable at that particular moment. In flocks of wood pigeons the same effect occurs with respect to choice of food items, so that all the birds in the flock tend to specialize on the most profitable type of food at the same time.

Of course the advantage of learning from others about the location and nature of food will only apply if all the flock members eat the same food, or at least eat foods that occur in very close association. This raises a question about mixed species flocks: can two different species exploiting different food sources acquire information from one another about food places? If the information is of the more general type, such as the whereabouts of rich feeding areas, one could imagine that different species will benefit from following one another. The finer scale learning of the type found in single species tit flocks also occurs in mixed groups of tits. Even though the different species specialize in their own feeding niche, they attend to the foraging success of similar species and acquire information about profitable types of feeding place.

This probably applies only to the species, in mixed groups, that have rather similar feeding niches and not, for example, to such mixed flock members as wrens and woodpeckers (Picidae). In these cases we have to look for an alternative explanation of the advantage of flocks.

Another hypothesis relating to mixed flocks stems from the observation that in some flocks the different species appear to specifically avoid using similar feeding stations. This has led to the suggestion that in a mixed group the various species benefit because they can avoid food COMPETITION with each other, as a direct result of observing each other searching for food. The evidence for this idea is conflicting, some authors reporting that species diverge when in mixed groups, and others that they converge in their feeding habits.

Finally, let us consider the effect of flocking on the SEARCHING path of individuals. Most, if not all, birds do not search for prey at random. Instead, most search with a directional strategy, so that, at any instant, the bird is more likely to continue travelling in roughly the same direction than to turn round and go back in its tracks. An obvious advantage of this is that it reduces the chances of covering the same ground twice in a short period of time. Now some observations suggest that flocks of birds have a greater directional tendency in their search path then do single individuals, so that flocking may act as an adaptation to increase the efficiency of search, for example where food is very thinly scattered. Further, when the flock moves through an area it gleans the food more thoroughly so that when the group does chance to cross its own path, the contrast between the gleaned and untouched areas will be greater, and the birds will be able to rapidly alter their direction of travel.

Anti-predator adaptations. In addition to any advantages that might accrue in terms of feeding, flocking undoubtedly has survival value as an anti-predator device. This could work in any one of several ways.

Just as many pairs of eyes are better than one for food searching, so a flock is likely to benefit in terms of greater awareness of an approaching predator, and be able to take off before the predator comes within striking distance (see VIGILANCE). This assumes that once one bird has detected the predator and reacted in some way (perhaps by giving a special call) the others will recognize the danger virtually at once. Experiments with captive flocks of starlings and tricoloured blackbirds (*Agelaius tricolor*) have shown that a flock reacts more rapidly than a single bird to the approach of a model hawk, and anecdotal evidence indicates that the same effect occurs in natural situations. Not only is a flock more vigilant than a single bird, but the effort that each individual has to put into surveillance, in order to maintain a particular degree of awareness for the flock as a whole, decreases with increasing flock size. Several studies of wild flocks, such as flocks of white-fronted geese (*Anser albifrons*), great blue herons, and wood pigeons, show that, when in a group, each individual spends less time looking around and hence more time searching for food.

If a predator succeeds in getting within attacking range of a flock, there are three ways in which being in a flock may increase the survival chances of an individual. Firstly, the flock may turn round and attack the predator. This is called a MOBBING response, and it generally frightens off the attacker. Black-headed gulls (*Larus ridibundus*) protect their chicks against the carrion crow (*Corvus corone*) by this type of communal mobbing behaviour, although some predators, for example a hawk attacking a flock of tits, attack so rapidly that there would be no time for mobbing. Secondly, the predator may have difficulty in singling out an individual from a rapidly escaping flock, and delay its attack for a vital moment because of a *confusion effect*. There is quantitative evidence for the confusion effect in fish, but in bird flocks little detailed information exists. Thirdly, each individual in the flock might be able to reduce its chances of being selected by the predator if it seeks cover in the middle of the flock, assuming that the attacker will pick off a bird from the edge. The fact that many bird flocks bunch together when a bird of prey approaches may be the result of each individual trying to seek cover in the middle of the group.

Migration and flocking. Many birds migrate in large flocks, even if they are normally solitary. This particular type of flocking may well be concerned with NAVIGATION during migration. If each individual bird is slightly inaccurate in its migratory direction, and if the inaccuracies are distributed about a mean heading which is the correct one, then a cohesive flock, in which outer individuals are attracted to the centre, will have a more accurate heading than any one individual. Homing pigeons have been shown to home more successfully in small flocks than alone, which supports this navigation hypothesis.

Peck-orders and flocking. In most species that have been studied carefully, there is some sort of *dominance hierarchy* or peck-order within the flock. In some birds, such as domestic chickens (*Gallus g. domesticus*), there is a rigid hierarchy in which individuals can be clearly ranked in order of priority in relation to resources such as food, roosting sites, or mates. In other cases there may be one or two dominant individuals in a flock, and many subordinates that cannot easily be distinguished in rank. Some people have argued that the importance of flocking is that it enables the birds to establish stable social organizations and decrease the amount of strife. Evidence shows that strife does decrease once DOMINANCE relationships have been established because each individual has learnt its position, but it is unlikely that peck-order formation is the main function of flocking in birds. It is more likely that formation of peck-

orders is a consequence of being in a flock, rather than the initial factor favouring flocking behaviour. J.R.K.

FOOD BEGGING occurs in animals whose young are neither completely helpless nor completely self-sufficient. Rat pups (*Rattus norvegicus*), for example, are born blind and hairless and are fully dependent upon the MATERNAL behaviour of the mother who ensures that they have no difficulty in obtaining food. The pups suckle at their mother's teats and are initially capable only of simple orientation behaviour towards the nipple. Later in development the pups may demand food by specific VOCALIZATIONS and by food-begging behaviour. The chick of the domestic fowl (*Gallus g. domesticus*), on the other hand, is capable of obtaining its own food very soon after HATCHING. The chicks are dependent upon the mother hen for protection and warmth, but not for food. They start pecking at small food-like objects during their first day, LEARNING what is and what is not edible. The chicks of the herring gull (*Larus argentatus*) are in many respects similar to those of domestic fowl, but they are entirely dependent upon their parents for food. The parents take it in turns to go FORAGING, and upon returning to the nest they regurgitate food for the young. The food-begging behaviour of the chicks consists in pecking at a red spot at the base of the parent's bill, as illustrated in Fig. A. This stimulates the parent to regurgitate

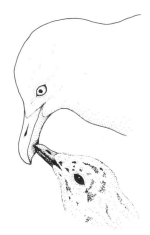

Fig. A. A herring gull chick pecking at the red spot on its parent's bill.

part of its food. When the chicks are no longer hungry they cease pecking at their parent's bill, no further regurgitation occurs, and the parent retains the remaining food.

The food-begging behaviour of juveniles often involves specific SIGN STIMULI to which the parents respond. For example, the nestlings of many passerine birds, such as the blackbird (*Turdus merula*), stretch their necks and gape in response to movement of the nest, and to the sight of rounded objects in the region of the nest. As they become older, this food-begging response becomes more discriminating and oriented towards the parent. The inside of the mouth of the nestling of many passerine species has distinctive markings which serve as sign stimuli for the parent. These markings are usually highly characteristic of the species. However, there are a number of cases of MIMICRY by brood parasites (see PARASITISM). For example among weaver birds the whydahs (Viduinae) are parasitic upon the genera *Estrilda* (waxbills) and *Pytilia* (weaver finches). The nestlings of these birds have distinctive spots and reflecting surfaces on the palate, which are displayed when they beg for food. These are thought to guide the parent in placing food in the gape in the semi-darkness of the closed nest. By mimicking these markings the parasitic nestlings also obtain food from the host parent birds.

Food-begging sometimes occurs amongst adult animals. During COURTSHIP the female may show APPEASEMENT behaviour in response to the male's AGGRESSION. This often takes the form of juvenile food-begging behaviour. In some species, food-begging elicits COURTSHIP FEEDING in which the male offers food to the female.

FOOD SELECTION. Feeding is one of the most essential activities of all animals, so it is no surprise, then, that the lives of most animals are dominated by their never-ending quest for food. Indeed, the structure and the behaviour of most animals are shaped and characterized in very large measure by the nature of the foods consumed, and the ways in which those foods are obtained. Such salient physical features as dentition, shape and size of tongue, and the kind of digestive enzymes produced, are directly related to the problem of feeding; while such behavioural mechanisms as navigational abilities, running speed, stalking activities, SOCIAL cooperation, and FORAGING patterns are also intimately linked with the need to locate and to obtain proper nourishment. The problem of food selection is common to all species, and plays a pivotal role in the evolutionary divergence of many groups of animals.

The world is filled with an immense variety of substances. Some are nutritionally useless, such as stones or sand. Others are poisonous, including a wide array of plants that contain harmful or toxic elements. In addition to these non-nutritive or poisonous substances, the natural world brims with an amazing assortment of potential nutrient sources; there is very little to be found that is edible that is not utilized as food by one species of animal or another.

From this array of potential food resources, both plant and animal, each animal species selects, by means of experience or revolutionary ADAPTATION, foods that allow its kind to mature and to reproduce. Every animal species can be described and characterized in terms of its basic food selection by two related criteria: the types of food selected, and the range of foods consumed. Thus, flesh-eating lions (*Panthera leo*) are distinguished from plant-eating zebras (*Equus burchelli*) primarily in terms of type of food, but the distinction between an animal like the koala bear (*Phascolarctos cinereus*), who subsists on only three types of eucalyptus leaf, and an omnivore like the rat (*Rattus*), who eats almost anything, is most strikingly one of range or variety of foods eaten.

By and large, as we shall see, animals that select a relatively narrow range of foods appear to have, as part of their biological make-up, certain special features or equipment specifically designed to locate or obtain the necessary food item. Their recognition of food is usually INNATE. On the other hand, animals that typically select a wide variety of foods seem to possess fewer of these special evolutionary adaptations, and rely more heavily on experience or LEARNING in the quest for proper nourishment.

Feeding specialists. Animals that consume only one or a very few kinds of food can be described as food specialists: among them are the caterpillar of the monarch butterfly (*Danaus plexippus*), which eats only milkweed, the koala bear, and a number of parasites such as tapeworms (*Taenia* spp.). These specialists survive because they are expert at finding one particular kind of food, and as such they show some of the most impressive food-specific adaptations in the animal kingdom. There are some species of molluscs (Muricidae), for example, that have a special structure called the radula; the radula is a specific adaptation for scraping and boring through the shells of fellow molluscs, on which these creatures feed. Such specialists as anteaters (e g *Myrmecophaga tridactyla*), humming-birds (Trochilidae), tapeworms, and the aforementioned boring molluscs, are uniquely 'designed', with special equipment to obtain the particular kinds of foods on which they thrive.

In addition to equipment for locating or catching its food, the specialist needs a system for recognizing food, and an internal detector system for indicating the need for food. When the internal detector system is activated, the animal searches for things that will trigger its food-recognition system. Since these species have existed successfully for many generations on a very narrow range of similar foods (eucalyptus leaves, ants, flies) the diet must be nutritionally adequate. Thus, there is no real danger of a specific nutritional deficiency: an internal detector that simply arouses the search for food will do the job. This same narrowness in food selection also makes the problem of food recognition relatively simple; it is theoretically not difficult to evolve receptors that will respond uniquely to eucalyptus leaves or to flies.

The same arguments hold for animals that eat a wider variety of similar foods, foods that can be recognized as a category by a specific receptor. Insectivorous animals, like many frogs (Anura), have a visual receptor system that responds to all small moving dark spots, which, in the frog's environment, almost inevitably turn out to be insects. Similarly, carnivorous animals like some snakes (Serpentes) can utilize combinations of visual and chemical information to determine the presence of an appropriate prey, a small moving vertebrate. As with the specialists, in these cases of somewhat broader food selection, food recognition need not depend on experience, but is built into the system, and adult patterns of feeding may commence very early in life.

This picture of the food specialist, with its genetically controlled food recognition systems, and a single HUNGER detector, can be further elaborated. A specialist might consume several different kinds of food and have a special SENSE ORGAN for each one, as well as an internal detector for the specific nutrients contained in each food. The blowfly (*Phormia regina*) appears to be such a specialist. The adult blowfly requires an energy source and water, and the animal is equipped with chemosensory hairs on the legs which contain four to five receptors, one particularly sensitive to sugars and one to water (see CHEMICAL SENSES). When the fly encounters a liquid substance that stimulates the sugar receptor, it extends its *proboscis* (a tube) and sucks up the liquid. If the front part of the gut is filled, an internal detector system signals this to the brain and feeding stops. Sugars provide a completely adequate energy source for the adult flies. The water receptor identifies water, and activates sucking in response to an internal detector that is sensitive to the volume of fluids in the body.

In addition, the adult female blowfly needs a source of protein when her eggs are developing. There is a pattern of response in a few of the receptors that is produced when they are stimulated by proteins, and this seems to activate feeding in gravid females. These females show an increased protein and a decreased carbohydrate selection. In summary, then, the blowfly solves the problem of finding and ingesting the three essential nutrients by possessing a receptor system for each, and an internal detector for each. Each system is inflexible and not susceptible to change by learning.

Another striking example of a food-specific system occurs in a species of flatworm (*Microstomum caudatum*) that consumes the freshwater polyp *Hydra* and a variety of other foods. The *nematocysts* (stinging cells) of the *Hydra* are separated from the rest of the *Hydra*'s tissue in the gut of the flatworm, and the nematocysts are absorbed whole. They are then deposited on the body surface of the worm, where they serve as a kind of pirated defence mechanism. The flatworms consume *Hydra*, which they presumably recognize by a built-in receptor system, only when they have few nematocysts in the body (internal detector), and will continue to consume *Hydra* under these conditions even if they are sated with other energy-rich foods. The assumption is, then, that they possess at least two separate receptor and internal detector systems, one for energy and one for nematocysts.

Feeding generalists. At the opposite extreme from the specialists are the omnivores, or food generalists, among which are rats, some types of bears (Ursidae), and human beings. These animals select their diet from a wide variety of plant and animal foods; indeed, for the generalist, almost anything is regarded as potential food. This ability to eat a great many foods endows the animal with great versatility. Blights, droughts, or seasonal shortages that might devastate the specialist's food supply, have much less effect on the generalists, who can almost always find something else to eat. And although the omnivore is not usually highly specialized to locate or to capture any one kind of food, the wide range of nutrient possibilities open to him generally makes up for this, and in addition allows the omnivore to explore new HABITATS rather easily. In the light of these clear advantages, why aren't all animals omnivores?

First, some environments provide specific foods whose continuing availability is assured, so that animals who are equipped to recognize and obtain these foods can continue to thrive. Second, because omnivores eat such a wide variety of things, there is no simple way to categorize the many foods (leaves, fruits, insects) an omnivore might eat so that a receptor system could clearly identify all such items, but no others, as food. Omnivores cannot have food recognition mechanisms built in genetically; they must learn through experience what is food and what is not. This requires a rather sophisticated system for evaluating potential foods, one that makes considerable demands on INTELLIGENCE. Third, the omnivore gives up the special adaptations (of tongue, claw, digestive system, etc.) that maximize the possibilities of obtaining and utilizing particular foods, and in so doing may find himself at a competitive disadvantage with respect to neighbouring specialists (see COMPETITION).

Fourth, there are dangers to omnivorousness. In exploring the vast potential food resources of its environment, the omnivore is more likely to encounter poisonous substances, or to consume too much of a particular food, so obtaining an imbalance of nutrients. Omnivores can easily become deficient in certain essential nutrients (e.g. vitamins and minerals) because many of the plant foods they eat do not contain all the needed elements. Thus, unlike most specialists, or even carnivores, the omnivore needs much more than one fixed internal detector system; he needs to be instructed not only to eat, but also what to eat.

The conflicting advantages and dangers in omnivorousness can best be described as the omnivore's paradox. New substances must continually be explored and evaluated, but must be treated with suspicion. Old familiar foods are usually accepted and consumed, but there is a countering monotony effect, and a resistance to developing very strong preferences for any particular food. The paradox manifests itself in the omnivore's tendency to explore new potential foods (*neophilia*), while at the same time exhibiting a strong suspicion of them (*neophobia*). The way in which an omnivore works out what is food and what is not can best be demonstrated by the most thoroughly studied omnivore, the common Norway rat (*Rattus norvegicus*) in both its wild and domesticated forms.

That the rat is an enormously successful animal needs no testimony: the species is virtually ubiquitous, and has inspired an entire human industry dedicated to its extermination. The huge and incontrovertible success of the rat is due, in large part, to two basic and essential abilities of the omnivore, the ability to find food almost anywhere, and the ability to detect and avoid dangerous substances. LABORATORY investigations have shown that domesticated rats can choose nutritious combinations of foods when faced with a choice of more than ten individual nutrients (carbohydrate, protein, fat, and solutions of various vitamins and minerals) (see APPETITE). Further, when a deficiency in an essential nutrient is experimentally produced, the rat appropriately increases intake of the needed substance. As the experience of the rat poisoning industry has so vividly shown, rats are extremely skilful at detecting and avoiding man-made poisons, and continue to thwart the efforts of those who would exterminate them by means of poisoning. The question is: how does the rat manage so skilfully to avoid both natural and man-made poisons, and at the same time find enough of the right combination of foods to eat?

There is something of the specialist in this food generalist, in that it has certain built-in receptors for specific substances. As in the blowfly, the rat has oral sugar receptors whose stimulation leads to ingestion, a system that guides the animal toward a variety of nutritive substances. There is also a sodium taste receptor that responds most vigorously to sodium chloride (common table salt). This receptor forms part of a complete 'minispecialist' system, for the rat also has a built-in internal sodium detector. When this detector signals a deficit in body sodium, the positive value of substances stimulating the sodium taste receptor is enhanced, leading to increased sodium intake. This whole specialist system is genetically programmed: rats show an immediate enhanced preference for sodium salts the first time they experience a sodium deficiency. They also possess two oral rejection systems, one for substances that stimulate the bitter taste system, and the other for irritant chemicals that stimulate what is called the *common chemical sense*. Many natural poisons or other harmful substances stimulate one or the other of these receptors, causing the rat to reject or avoid the substance. Rats also possess internal detector systems for water deficit (concentration and volume of body fluids), and for energy balance, though, unlike that for sodium, neither of these systems is tied to a specific receptor. These

taste receptors and internal detector systems are built-in, genetically determined specializations, that enable the rat to detect and obtain certain essential nutrients. But the rat in fact requires more than thirty nutrients (vitamins, minerals, aminoacids, etc.), and it would be cumbersome to have detectors and receptors for each of them. Rather, this food generalist handles the problem of recognizing food and adjusting food intake to correct various internal deficits by a general learning system. This system operates by relating specific flavours to their internal consequences.

There are two difficult problems facing any organism that attempts to evaluate the consequences of ingesting any particular substance. First, the effects of ingestion, whether positive or negative, usually occur many minutes or even hours after eating; while learning typically involves an association of events that occur within seconds of one another. Second, an animal is constantly bombarded with different stimuli (visual, olfactory, gustatory, etc.), and engages in a wide variety of activities such as grooming, exploring, eating, etc. When a new toxic or beneficial symptom appears, how does it associate the symptom with one of the many stimuli or events?

Rats have special learning abilities that deal effectively with these problems, especially with respect to toxic or poisonous substances. As part of their learning system, they take full advantage of a fundamental feature of anatomy, the mouth, through which all food must pass before it gets into the body. However substances are initially perceived, whether, for example, by sight or by smell, they must ultimately be tasted in the mouth; the mouth thus serves as a sort of monitor at the gateway to the body. Since the rat has many taste as well as other receptors in its mouth, it would be of great advantage to have a system that selectively associates tastes with the common consequences of ingested substances, such as nausea. Rats indeed have such a system, one that reflects the real correlation between oral stimulation and visceral after-effects. Rats exposed to lights, sounds, and tastes at the same time will selectively associate the tastes with a subsequent visceral event like nausea, but will associate the lights and sounds with an external event, such as an electric shock to the feet.

Once a food is familiar, it is typically identified before it enters the mouth. Rats rely heavily on odour in this respect, and can learn to associate a particular odour with a particular taste and ultimately, with visceral consequences. Similarly, birds frequently rely on visual stimulation to identify food, and may do so successfully by associating sights with tastes and tastes with subsequent visceral effects.

But how does the rat associate a taste with a visceral event that occurs some half hour or more later, given that most learning ordinarily involves an association of events that occur within seconds of each other? Research on rats has shown that a single experience with a new taste, followed 1 h later by an injection that produces nausea, or some other negative internal effect, will result in a strong avoidance of that taste. The rat apparently has a special ability to learn to associate tastes with visceral consequences that occur some hours later. This ability clearly reflects the natural delay caused by the slow processes of digestion and absorption between eating and the internal metabolic consequences of eating, and contrasts with other learning abilities, which are usually limited to the association of events over much shorter time periods. It would not be very useful, for example, to associate a sudden, sharp pain in the foot with something that happened hours ago.

Within the taste system, rats have another powerful mechanism to aid their selection of nutrient. The rat's foods can be classified as either familiar (previously eaten) or unfamiliar (novel). New substances are potentially dangerous, or potentially valuable sources of nutrients: hence, the omnivore's paradox. Wild rats are extremely suspicious of new foods. They may not sample a new food for some days after its appearance in their environment, and then they do so very tentatively, taking only a small amount and retiring to evaluate the effects.

Minor toxic symptoms from a small sample will result in the rat's rejection of the food, the major problem faced by rat exterminators. (For this reason, exterminators often introduce a new food without poison for a few days; after the rats sample the food and find it acceptable, a, hopefully, tasteless poison is added.) This characteristic sampling behaviour maximizes the rat's ability to evaluate a new food by isolating its taste and visceral effects from those of other foods. If it happens that a rat consumes a familiar (safe) food at about the same time as a new food and then becomes sick, it will subsequently avoid the new food and not the old one. In other words, the rat has learned something; it brings to bear its past experience with food on any new situation.

All these abilities associated with food selection: taste–visceral association, long-delay learning, sampling behaviour, and classifying foods by past experience, enable the rat to recognize and to avoid the many poisonous substances it is likely to encounter. The system for developing preferences for substances that produce positive internal effects seems to operate similarly, but is less powerful. Rats can learn to prefer a food whose ingestion is followed by a reduction in hunger, or by recovery from a specific nutrient deficiency. This learning typically requires a number of experiences, however, and does not result in strong preferences comparable to the strong avoidances of poisons. Still, it appears that the food experience of the rat can be more accurately described by four categories: new (unfamiliar) foods, familiar harmful 'foods', familiar safe foods (no negative consequences), and familiar positive foods.

We can use these categories or principles of food selection to understand one representative problem in *specific hungers*, i.e., the ability of rats deficient in vitamin B$_1$ (thiamine) to select thiamine-rich foods. When the rat is in a situation in which only thiamine-deficient foods are available, it develops an aversion to these foods, because each time the animal eats them it feels worse. In this sense, the deficient foods act like a slow poison. After some exposure to such deficient food, the rat treats it the way it treats a bitter food, spilling or spreading it, and avoiding ingestion. The rat is clearly learning what foods not to eat. In addition, the animal minimizes its food intake, a clearly adaptive behaviour, since the more thiamine-deficient food consumed, the more rapidly the deficiency develops.

At the same time, as deficiency sets in the rat becomes more active, thus enhancing the possibility of encountering a new, potentially thiamine-rich food. In addition, the animal is able to select among the foods in its environment, and to consume those that least exacerbate the deficiency. Since carbohydrates increase the metabolism of thiamine, thiamine-deficient rats show a decided shift in preference to low carbohydrate foods. They exhibit other food-selection shifts as well. Many vitamins are synthesized in the hind gut, but cannot be absorbed into the body from this site. Thiamine-deficient rats show an increased consumption of their own faeces (*coprophagy*), a behaviour that serves to ameliorate the vitamin deficiency.

When the thiamine-deficient rat encounters a new substance, the omnivore's paradox manifests itself, and both neophobic and neophilic tendencies appear in the animal's behaviour. Eventually the new substance is tried and, on the basis of its internal consequences, is rapidly classified as harmful (deficient), safe, or, more slowly, as beneficial. The evaluation is facilitated by the rat's tendency to try only one new food at a time. In this way the rat seems able to cope with a wide variety of potential nutrient deficiencies.

Other strategies of food selection. These behavioural mechanisms for learning about foods should be of great value not only to an omnivore like the rat, but to other animal species whose appropriate foods cannot efficiently be programmed genetically at the level of receptors. Many herbivores, for example, eat a wide variety of plant foods, and face the same basic problems of potential toxicity and nutritional imbalance. Although specific demonstration of poison avoidance has been limited to only a few species of mammals and birds, it should theoretically be widespread in the animal kingdom. Indeed, some forms of MIMICRY, in which a species derives protection from predators by evolving characteristics similar to distasteful or poisonous species, may depend on aversion learning by the predator.

Toxic substances are common in plants and serve presumably as protection for the plant, by deterring animals from eating them. Any general herbivore must learn to deal with this problem in one of two ways: avoidance of the toxic plant, or detoxification of the substance in the gut or liver. Some herbivores use both mechanisms; they avoid the most highly toxic plants, and eat only small amounts of the less toxic plants. By spreading the risk in this way, the animal can maintain a broad spectrum diet, with the advantages of more complete nutrition and minimal dependence on any one food.

The carnivores, as a group, face fewer problems of this sort, since their prey ordinarily contains the full complement of nutrients and a minimum of poisons. In other words, a lion could become vitamin deficient only by consuming vitamin deficient zebras. A preference for sweet foods common in many herbivores and omnivores is absent in at least some carnivores: their natural food is not high in sugars. It is not known whether they possess a bitter avoidance system. Clearly, the big problem for carnivores is finding and catching, rather than selecting, the species of prey.

In the animal kingdom, food selection contrasts strongly with species recognition, in that it is much less frequently determined genetically. While the process of IMPRINTING serves to establish a model for species recognition early in life, such a process rarely occurs in the food system. For omnivores, of course, a special attachment to the food of early life would be disadvantageous, since the clear virtue of omnivorousness is flexibility in choosing among new and changing food sources. And while the appearance of a species remains roughly constant over the lifetime and the geographical range of the animal, food sources can change dramatically over seasons, years, and geographic regions. Mammals provide a striking example of such a change, because the first food of all mammals is milk, a food that, after weaning, will be unavailable in the environment (except to man and his HOUSEHOLD PETS). A strong attachment to this food would serve no useful purpose, and in fact all adult mammals, except for some human beings, are incapable of satisfactorily digesting milk sugar (*lactose*). The production of *lactase*, the enzyme responsible for digesting it, falls to low levels in early life.

An example of flexibility, of adaptability to changes in food selection, is well illustrated by animals who must make certain critical adjustments to a diet consisting primarily of seeds. Many seed-eating birds must learn to extract the seed from its husk. In some cases this involves orienting the seed in the bill so that its line of cleavage is vertical, and parallel to the beak. By this arrangement, the closing of the beak will neatly crack off the husk. It would be most efficient for a bird to select the largest seeds it can handle, because in this way it maximizes nutritional value for work output. Selection of the largest ingestible seeds has been demonstrated for a number of British species, e.g. chaffinch (*Fringilla coelebs*) and greenfinch (*Carduelis chloris*). A moment's thought will indicate why this ability to select is learned and not

genetically determined: these birds increase in size with age, and thus must constantly adjust the optimal size of the seed selected to the size of the growing beak. And this they do.

A final example of the flexibility of certain animals with regard to changing food selection may sound a familiar note to some residents of Great Britain. Great tits (*Parus major*) and blue tits (*Parus caeruleus*) have, within this century, extended their diets to include milk, a new food source that they have discovered with door-step deliveries. Some birds attack this new prey by poking holes in the bottle caps, others by prying them open. Some recognize as food only bottles of a certain shape and size. The habit seems to be spreading, and it is no wonder; all that potential good food left around was too rich a bounty to be ignored (see IMITATION).

Up to this point we have discussed food selection as a uniform characteristic of any given species. But individual specializations based on experience also occur, particularly in many omnivorous species. This ability is especially valuable where there is a large conglomeration of omnivores that might strain local food resources. By adopting new foraging techniques an individual can improve its nutritional yield, with respect both to others of its own species and to other possibly more specialized species. For example, in a colony of herring gulls (*Larus argentatus*) that nest along the British coast, some of the gulls specialize in locating crabs (Brachyura) by finding their breathing holes on the sand flats; others feed primarily on mussels (Myrtilidae) or earthworms (e.g., *Lumbricus* and *Allolobophora*), and still others exploit the human food refuse deposited at the local rubbish dump. Indeed, some of the rubbish specialists come to recognize certain trucks or types of rubbish bags as particularly rich food resources, and know the times when the rubbish is to be dumped.

When these individual foraging specializations are coupled with corresponding food preferences, clear nutritional advantages may result, as can be clearly seen in the British gulls. During the incubation period, one member of the parental pair must always be guarding the eggs. The different foods available to these birds can be obtained only during specific and limited times during the day: crabs and mussels at low tide, rubbish at dumping time, etc. It has been observed that individual pairs of herring gulls seem to have complementary food preferences and foraging specializations, so that feeding time can be more effectively budgeted.

Human food selection. Human beings are animals too, and omnivores at that, though their history as an omnivorous species is complex. Our simian (monkey-like) ancestors in the jungle probably had habits similar to modern chimpanzees (*Pan*), living primarily on fruits, with a variety of other plant foods, insects, and an occasional small vertebrate to round off the diet. These creatures left the jungle habitat in response to dwindling food resources and invaded the open savannah, presumably in search of alternative food supplies, most notably larger game. It was this fundamental change in food selection that is believed to have led to the upright posture, and the remarkable brain development characteristic of man. This primarily carnivorous stage in human food selection was followed, for most humans, by a primarily herbivorous stage, a change initiated by lessening animal food resources and reinforced by the development of agriculture (see DOMESTICATION). Most human beings today eat a diet dominated by bland grain products: rice and wheat account for about 40% of the food eaten in the world.

This pathway from omnivorous but primarily herbivorous habits, through a largely carnivorous stage, and back again to a diet dominated by plant foods, has left its mark on us. The form of our teeth and our intestines reflect both herbivorous and carnivorous feeding. We retain the positive sweet and negative bitter systems of general herbivores and omnivores, but the meat craving of the carnivore as well. In almost all cultures the special or festival foods are meats. The omnivore's paradox is part of our behaviour: we are curious about new foods but also suspicious of them, and that paradox is most clearly expressed when we travel to far and exotic lands.

But what most clearly distinguishes the human from other animals is the social and CULTURAL influences on food selection, the passing of food traditions from one generation to the next. The rat mother incidentally influences her offspring by inadvertently exposing them to the TASTE and odour of the foods she eats. But this is quite different from the deliberate cultural propagation of food traditions that is so characteristic of human behaviour, and to which few genuine parallels are known in the animal kingdom. One well-known example is that of a troop of Japanese macaque monkeys (*Macaca fuscata*) that accepted sweet potatoes as a new food, and developed the habit of washing them in water before eating them. The potatoes were initially dipped in fresh water, but were later dipped in sea water, presumably for the salty taste. This sweet potato dipping and other parallel behaviour diffused through a natural group of monkeys over a period of years, and can now be considered a cultural tradition.

There are in human beings, of course, certain virtually unique food practices. People have a clear tendency to create cuisines: to select certain kinds of foods, to season or flavour them in characteristic ways, and to manipulate or cook them in a specific fashion. Human beings, unlike most other omnivores, also have a decided tendency to develop strong attachments to particular foods. In addition, human beings have an apparently unique ability to develop likings for certain foods that initially trigger their sensory rejection systems, such as the bitter taste or chemical irritation. Widespread preferences for such initially unpalatable substances as alcohol, coffee, and hot chilli peppers are unparalleled in the animal kingdom. Human beings alone seem willing and able to overcome strong, biologically based aversions, and turn them into strong attachments. Here, as in so many other instances, can be seen both our clear ties to our evolutionary forebears in our omnivorous heritage, and the existence of certain uniquely human characteristics.   E.R., P.R.

FORAGING, as used by students of ETHOLOGY, refers to behaviour associated with searching for, subduing, capturing, and consuming food. A full description of the foraging behaviour of a species would include the type of HABITAT used, the substrates from which food is gathered, types of food eaten, and techniques of prey capture. The coal tit (*Parus ater*), for example, searches for food by gleaning insects and seeds from the needles, cones, branches, and trunk of coniferous trees, usually at a height of 2-15 m from the ground. It captures prey by probing into crevices between pine needles or into pine cones, and very occasionally by turning over dead leaves on the ground. It usually hunts by hopping rapidly from branch to branch, or by short flights, and sometimes hovers and catches insects on the wing. This qualitative description of foraging behaviour could be quantified by measuring the relative frequency of use for feeding of different tree types, branch sizes, heights above the ground, searching techniques, and so on.

The term foraging is not restricted to the food gathering behaviour of animals such as the coal tit that actively search for and capture discrete items of food. It can also refer to the FEEDING behaviour of grazing or browsing animals, sit-and-wait predators, such as web-building spiders (Arachnida), and even sessile filter feeders, such as mussels (Mytilidae) and barnacles (Balanomorpha). Obviously the elements of foraging behaviour are very different for a barnacle and a coal tit.

A distinction is sometimes drawn between foraging tactics and foraging strategy. The tactics of a forager are the methods by which it goes about the business of capturing food. For example, hunting groups of spotted hyena (*Crocuta crocuta*) adopt the tactic of running after their prey and overtaking it by superior stamina and by COOPERATIVE hunting manœuvres. The solitary cheetah (*Acinonyx jubatus*), on the other hand, stalks its victim and then overcomes it in a sudden highspeed dash. The term 'strategy' refers to the idea that foraging animals aim to achieve a particular GOAL. This is shorthand for saying that animals which perform best according to some criterion of success are favoured by NATURAL SELECTION. The goal or criterion of success might, for example, be to capture as many items of food as possible each day, or to minimize the time spent hunting each day. It is often fruitful to analyse foraging strategies in terms of DECISION-MAKING. Achieving a goal, such as maximizing capture rate, can be thought of as involving a series of decision rules

(rules about how to search, where to search, which items to eat, and so on) which together constitute the animal's foraging strategy.

Much research on foraging tactics has been aimed at describing how foraging behaviour differs between species living together in the same ecological COMMUNITY. It is an ecological axiom that different species exploit different resources in the environment. Although the term resource refers to any essential requirement of a species, for example shelter, nest sites, water, and shade, the most frequently studied resource is food. A number of studies have shown that species with superficially very similar foraging behaviour, usually referred to as members of the same foraging *guild*, have subtle differences in their patterns of exploitation of food. For example, in a study of several species of similar sized insectivorous warblers living in coniferous forests of Eastern North America, one species, the black-throated green warbler (*Dendroica virens*), preferred to forage in the upper outer branches, while the other, the bay breasted warbler (*D. castanea*), foraged mainly among middle height branches, and a third species, the myrtle warbler (*D. coronata*), used the lower inner branches of the same kind of trees. The significance of these observations is that they suggest that COMPETITION for limited food supplies has led to the EVOLUTION of differences in foraging behaviour. The differences between species are seen as ways of avoiding competition. Competition between species for food is also implicated by other kinds of evidence, much of which is based on the idea of *utilization curves* (Fig. A). A utilization curve is simply a frequency distribution of the use by a species of any resource which can be represented along a linear axis. The frequency of use of different sized seeds by a species of mouse (Murinae) would be an example. Studies of foraging behaviour in relation to interspecific competition have attempted to test whether the utilization curves of a species are limited by the presence of other, competing, species. In one experimental

study of sunfish in North America, it was found that the bluegill sunfish (*Lepomis macrochirus*) forages for planktonic invertebrates at some distance from the shore when it is living in the same ponds as the green sunfish (*L. cyanellus*). However, when the green sunfish, which normally feeds near the shore on the invertebrates living among emergent vegetation, is removed, the bluegill changes its foraging pattern. It moves to the habitat previously occupied by the green sunfish near the edge of the pond, suggesting that the green sunfish normally excludes the bluegill from the edge.

The bluegill sunfish changed its foraging behaviour on an immediate, ecological time scale, but similar effects of competition can also be observed to have occurred over evolutionary time. There are usually morphological differences in the feeding apparatus of closely related species, which reflect differences in their utilization curves. A striking example refers to two species of freshwater snail of the genus *Hydrobia*. Wherever the two species occur together in the same lake they differ in body size, and hence in the size of food items eaten. When either of the two occurs singly, the body size and utilization curve of both are intermediate, so the divergence seems to have been an evolutionary response to competition.

Sometimes it is possible to infer that competition is an important influence on foraging behaviour by showing that the utilization curves of a guild of species in the same community are regularly spaced along a *resource axis*. This is true, for example, of eight species of fruit pigeon belonging to the genera *Ptilinopus* and *Ducula*, which live in the lowland rain forests of New Guinea. Each species weighs approximately 1·5 times the one below it when they are ranked in order of body size, and the size of fruit eaten by a species is closely correlated with body size. On a theoretical level attempts have been made to assess how similar the utilization curves of two species can be, and still permit stable coexistence (assuming that similarity of the curves implies competition). The

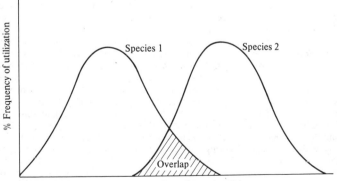

Fig. A. An example of utilization curves. The curves show the frequency of use of different-sized insects by two hypothetical species of birds. The information on which the curve is based could be collected by direct observation, from stomach samples, or from remains in the faeces.

theories have provided useful guidelines, but not much by way of easily testable predictions.

While many studies of vertebrate communities have supported the idea that competition between species places important constraints on the foraging behaviour of a species, the picture is much less clear for invertebrates. For example the limits of the utilization curves of many plant eating insects may be determined by the chemical defences of plants, rather than by competition with other insects.

The ultimate function or SURVIVAL VALUE of foraging behaviour is to acquire food. For many, perhaps most, animals, acquiring food as efficiently as possible is crucial for survival and REPRODUCTION. Small insectivorous birds in Southern England need to find an insect every few seconds throughout the day just to survive during the winter, and even for animals which are obviously pressed for food, efficient foraging may mean more energy stored for reproduction, or more time for mating, ANTI-PREDATOR behaviour, and so on. Perhaps the most staightforward measure of foraging efficiency is the net rate of food intake per unit time. This measure makes biological sense because an individual with a high net rate of food intake builds up reserves rapidly and minimizes its foraging time. It is also justified by the results of a number of experiments which show that the choices made by foraging animals, for example of where to feed and which items to eat, often have the result of maximizing the rate of food intake. The theoretical study of foraging strategies in terms of decisions which maximize the net rate of food intake (or some similar measure of efficiency) is referred to as *optimal foraging theory*. Individual foraging efficiency is only one of many factors influencing survival and reproduction, and the complete study of optimal foraging decisions should take into account such factors as the risk of being attacked by predators while feeding.

A simple example of the optimal foraging approach is the study of preferences for different sized food items. Each prey can be characterized by net food value (E) and a time cost (h), the time taken to subdue and consume one item. Prey with a high value of E/h are more profitable than those ith a low value. If a predator chooses prey so as to maximize its rate of intake, it should prefer more profitable prey, and only pause to consume the less profitable ones when the preferred prey are not readily available. By taking into account the time required to search for the preferred prey, it is possible to predict whether or not the predator should consume an unprofitable item once it has been encountered (Fig. B). This kind of prediction has been tested in a number of studies using birds, arthropods, and amphibians: most of the results agree approximately with the predictions. The choice between qualitatively different kinds of prey is sometimes influenced by requirements of the foraging animal for specific nutrients.

This is especially true of herbivorous vertebrates, whose diet can often be balanced only by taking a variety of different plants. If the various essential nutrients are known, it may be possible to calculate an optimal balance between the different types of food. In one study of a North American herbivorous mammal, the moose (*Alces alces*), it was found that the animals selected a diet in the wild which achieved an optimal balance between energy intake and the requirements for sodium, the major limiting nutrient.

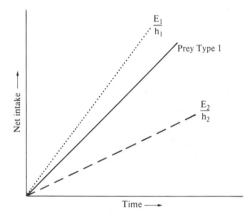

Fig. B. An optimal foraging model of choice of prey. Each type of prey can be characterized by its profitability (net food yield per unit handling time or E/h). The graph shows the profitabilities of two prey as the slopes of the dotted and broken lines. If the predator eats only the more profitable prey type, its intake is given by the solid line. The slope of the solid line is less than the slope of E_1/h_1 because of the time taken to search for prey of type 1. The predator should only eat prey type 2 when the slope of the solid line is equal to or less than that of the broken line. Thus, as prey type 1 becomes rarer, and the solid line rotates to the right, there is a threshold beyond which prey type 2 should be eaten when it is encountered.

Even when one food type is clearly more profitable than others, and when specific nutrients are not a constraint, animals rarely specialize exclusively on one type alone. In a study of bumblebees (Bombini), for example, although some types of flowers yielded more nectar per second of 'handling time' than other species, individual bees occasionally visited the less profitable types. This was seen as an example of sacrificing short-term gain for longer-term acquisition of information about environmental change. As the season progresses, the nectar content of different flower species may change, and only by regular sampling can the bees detect such changes. The same kind of problem must face many foraging animals and sampling is probably important in most foraging decisions.

For many animals food occurs in clumps or patches. Even a seemingly uniform habitat such as a mudflat may be quite heterogeneous. When samples of the mud are examined for invertebrates, some are rich in worms and crustaceans, while other samples contain very few. Shorebirds such as the redshank (*Tringa totanus*), which feed on the invertebrates, tend to aggregate in areas or *patches* of high prey density. The choice of different patches by a foraging animal can be analysed in a similar way to the choice of prey. Each patch can be characterized by its *profitability* (food yield per unit searching time) and an efficient forager should concentrate on the most profitable patches, as do redshank, great tits (*Parus major*), and many other animals. When the first patches visited are depleted by the predator's searching it may pay to switch to a new place. In order to maximize its rate of intake, the forager should switch when its expected gain from moving to the next patch is higher than its gain from the current place. Some experimental studies of birds and insects have supported this prediction.

SEARCHING for food in virtually all animals that have been studied is systematic rather than random. A random search path is inefficient because it often crosses itself, with the result that the same area is searched many times. The search paths of animals usually have a strong forward bias, so that the individual continually explores new ground. The path may become more convoluted immediately after a food item has been found (so called *area restricted search*) as the predator scans the vicinity for other prey. This response is clearly adaptive when prey are clumped, and European blackbirds (*Turdus merula*) show more marked area restricted search when hunting for clumped prey items than when feeding on regularly spaced food. A more complex pattern of systematic search is shown by a nectar feeding bird from Hawaii, the amakihi (*Loxops virens*). After the bird has visited a flower and drunk the nectar it avoids returning to the same spot for some time, allowing a new supply of nectar to accumulate. The amakihi thus moves about its TERRITORY in a systematic way, visiting only flowers which contain an accumulation of nectar.

Efficient foraging skills are often acquired (as a result of experience) during the lifetime of an individual. The bumble-bees referred to earlier learn how to extract the nectar efficiently from each species of flower. Within the same meadow, different individual bees learn the necessary skills for different species of flower, and each concentrates on collecting nectar from its own particular kind of plant. (As mentioned before, each bee also visits other flower species from time to time.) Foraging efficiency in general may improve with age in long-lived animals, particularly those with elaborate hunting techniques. This effect has been clearly shown in birds. Measurements of foraging success of herons (Ardeidae), pelicans (*Pelecanus*), and oystercatchers (*Haematopus ostralegus*), have shown that young birds are appreciably less efficient than

mature adults. Oystercatchers may also acquire foraging preferences for particular kinds of prey as a result of the diet they are fed by their parents during the nestling stage. Differences between species in foraging skills are probably strongly influenced by inheritance (see GENETICS OF BEHAVIOUR). Naïve, hand-reared coal tits and blue tits (*Parus caeruleus*) show similar differences in foraging skills to those observed in wild adults of the two species. Coal tits are more adept at probing into pine needles and narrow crevices for small insects, while blue tits are better at tearing bark and hanging upside down from thin twigs.

In conclusion, a variety of fields of animal behaviour impinge on studies of foraging. LEARNING theory is relevant to the problem of how foragers sample the environment, studies of the regulation of food intake and specific hungers relate to the question of nutrient constraints and choice of diet, and much research on decision-making, *optimality theory*, and cooperative behaviour involves studies of foraging.   J.R.K.

FRISCH, KARL RITTER VON (1886–   ), is one of the most eminent students of animal behaviour. His most famous research is his discovery that honey-bees (*Apis mellifera*) inform each other of the whereabouts of food by a specific dance.

Brief life. Von Frisch was born in Vienna into a family with a considerable academic tradition. He had an enthusiasm for natural history and kept a variety of pets from his childhood, and while still at school published various observations on natural history, such as some experiments on the light sensitivity of sea anemones (Actinaria). He began to study for a medical degree at Vienna University, then changed to studying zoology at Munich University, but finished his degree back in Vienna. In 1910 he returned to the Zoological Institute at Munich as an assistant. During the First World War he worked in a military hospital, and also taught bacteriology to trainee nurses: his lectures were published as *Sechs Vorträge über Bakteriologie für Krankenschwestern* (1918, Six lectures on bacteriology for nurses). After the war he returned to Munich, until he was appointed Professor of Zoology at Rostock University in 1921. He became Professor of Zoology at Breslau in 1923, and then returned, as Professor, to Munich University in 1925. Von Frisch had some trouble from the Nazis (his maternal grandmother was of Jewish extraction), and because of the bombing of Munich he spent much of the Second World War at his Austrian home at Brunnwinkl. From 1946 he was briefly Professor at Graz (Austria); but he soon returned to his chair at Munich, where he remained. In 1973 he shared the Nobel Prize with Konrad LORENZ and Niko TINBERGEN for their contributions to the study of animal behaviour.

The unifying theme runnning through von Frisch's research on animal behaviour, from his childhood studies of the light sensitivity of sea anemones, through his collaboration while a student at Vienna with his uncle Sigmund Exner (himself an eminent sensory physiologist, the discoverer of the 'compound superposition' eye of some arthropods), to his maturer analyses of the sensory physiology of fish and of bee COMMUNI-CATION, is the problem of how animals obtain information about their environment. Although his original research was on a clearly defined group of problems, von Frisch had a broad biological knowledge, and used this in writing several popular books about general biology, notably *Du und das Leben* (1936, translated as *You and Life*, 1940), *Biologie* (1952, translated as *Biology*, 1964) and *Tiere als Baumeister* (1974, translated as *Animal Architecture*, 1974).

Ideas on fish senses. Von Frisch's earliest research at Munich was in reaction to von Hess's claim that fish are colour blind. Von Frisch tested for colour sensitivity by seeing whether he could train fish, by rewarding them with food, to discriminate between colours; and he found that his fish were in fact sensitive to colour. When these results were published von Hess made a somewhat authoritarian attempt to discredit von Frisch's work, and so a feud developed between them which was to embrace von Frisch's subsequent work on colour sensitivity in other animals. In his turn, von Frisch vigorously criticized his senior, so causing more than a few raised eyebrows among the upper scientific echelons; but, as von Frisch subsequently reflected, the controversy probably helped to draw attention to him at the time.

Similarly, in the 1920s there were many biologists who thought that fish were deaf. Von Frisch first tested this on brown bullhead catfish (*Amiurus nebulosus*; generic name now *Ictalurus*) living in opaque tubes, so that when inside its tube the fish could not see an approaching experimenter. He then trained his fish to come out of the tube for food when he whistled. Soon, by conditioning (see LEARNING), the fish came out just for the whistle, proving that they were not deaf. This research was extended to other species of fish, and von Frisch found variability between the different species in their ability to hear. He also worked on the rod-cone theory of COLOUR VISION. In the retina of the eye there are two kinds of light-sensitive cells, called *rods* and *cones*, and it was thought but not proven in 1924 that rods were colour insensitive and concerned with seeing in the dark, while cones were colour sensitive. It was known from microscopical examination that in the dark the rod cells were forward in the retina, whereas in the light the cone cells were forward. Von Frisch trained fish to associate a colour with the receipt of food, and he then reduced the light intensity until the fish started to confuse the colour with grey. He also microscopically examined the retinas of the fish at the different light

intensities, and found that the light intensity at which the fish just confused colour with grey was that light intensity at which the rods and cones were just crossing over in the retina. Thus a combination of behavioural and microscopical work provided support for the rod–cone theory of colour vision. One other significant discovery was of an alarm PHEROMONE in the minnow (*Phoxinus laevis*, now *P. phoxinus*); von Frisch discovered that when the skin of a minnow is damaged a substance is released that causes other nearby minnows to flee the area.

For much of his life, von Frisch worked on fish, usually minnows, during the winter, and on honey-bees at the family retreat in Brunnwinkl during the summer.

Ideas on honey-bees. It is on his research on the sensory capacities and behaviour of honey-bees that von Frisch's scientific reputation securely rests. Although he had been interested in bees for almost all his life, his first important research was begun soon after the First World War. He was studying the ability of bees to detect colour and odour, by methods similar in principle to those used in his work on fish. While conducting this research he noted that initially only occasional 'scouts' appeared at the food dishes, but that once one scout had discovered the food it took only a few minutes for many more bees to arrive at the food source. The bees, von Frisch thought, must have a mechanism of communication through which they learnt that there was a good food source in the vicinity. He noticed that a successful scout performed a recognizable sequence of movements on the vertical honeycombs of the hive; this 'dance', he thought, merely alerted recruits, who then found the food source by scent. The returning scout brings back a sample of food with its distinctive scent, and (von Frisch supposed) the recruits would then fly outwards from the nest in spirals until they could locate the food source. This was the odour theory of honey-bee communication about food sources, formulated in the 1920s. Von Frisch said that he also considered the possibility that the foragers communicated a message about the distance and direction of the food source, but dismissed the hypothesis as too fanciful. He described many of his observations on honey-bee life in *Aus dem Leben der Bienen* (1927, translated as *The Dancing Bees*, 1954), which was written when his friend Richard Goldschmidt invited him to contribute to a series of popular scientific books.

In the 1940s von Frisch did experiments that caused him to doubt the odour theory. He trained a scout to come to a dish containing scented food, and then put another dish containing food with the same scent nearer the hive. The recruited bees should have gone mainly to the nearer dish if they searched by moving outwards from the hive while seeking scent. To von Frisch's surprise the dish nearer the hive was ignored; the recruits went to the dish that the scout bee had originally visited. Similarly, when he put one scented dish to the east of the hive and another the same distance, and with the same scent, to the west, and then trained a scout to visit, say, the east dish, then all the recruits visited the east dish (the odour theory would predict equal numbers of visitors to the east and west dishes). Von Frisch also controlled for the possibility that recruits homed on a scent released by the scout rather than the food by blocking the scout's scent-gland, but this made no difference to the recruits' ability to find the food. The odour theory was manifestly wrong.

Von Frisch then did other experiments in which he observed the dances of the scout bee on her return to the nest. He marked some bees blue and trained them to a food dish near the hive, and marked others red and trained them to a dish far from the hive. When he observed their dances he saw that the blue bees did one kind of dance (a *round dance*) while the red bees did another kind (a *waggle dance*). Here was a mechanism by which the bees signalled the approximate distance to the food source: they do a round dance if it is near, but a waggle dance if it is far. Von Frisch soon noticed more subtle, but more important, variations of the dance. The rate, or speed, of the dance increased if the food was further away. Also, the angle at which the waggle dance was performed with respect to vertical on the honeycomb was equal to the angle of the food source with respect to the sun. Thus, the bee dance signals both the direction and the distance of the food source. A classic series of experiments confirmed that the dance varied in a predictable way as the food source was varied in distance and direction, and that the bees used this information in finding their food. Von Frisch also observed that honey-bee recruits are capable of finding food even when the position of the sun in the sky is obscured by clouds. Yet another sensory capacity of an animal was discovered in investigating this problem: von Frisch showed that when the sun is obscured the honey-bees navigate by the polarization of light that exists in a clouded sky (see NAVIGATION). Von Frisch summarized his experiments and ideas on the bee's dance language in his important book *Tanzsprache und Orientierung der Bienen* (1965, translated as *The Dance Language and Orientation of Bees*, 1967).

The use of symbolic LANGUAGE is often thought to be a trait unique and essential to *Homo sapiens*, and certainly not the kind of ability one would expect in a mere honey-bee. Von Frisch had spent his life discovering new and unexpected SENSES OF ANIMALS; but to many, a symbolic language in a honey-bee was just too extravagant a claim, and so they reacted with bald incredulity. However the theory of a dance language in bees became widely accepted in the 1950s and early 1960s, at least among biologists, only to be seriously challenged in the late 1960s. This time the challenge was sci-entific, not merely prejudiced. It was argued that von Frisch's experiments did not adequately control for the possibility that the bees used odour, rather than the dance, in finding their food. For example, the resinous substance (called shellac) used by von Frisch to block the scent-glands of scout bees was itself attractive to the bees, so even in this experiment the recruited bees could have been following a scent rather than using the information in the dance. And there was also the possibility that the recruits had learned a locality odour (that is, the specific smell of the region near the food source) from the scout, which could in turn be used to find the source of food. Thus an attempt was made to revive von Frisch's older odour theory of honey-bee communication. Von Frisch responded to this latest challenge, but the definitive experiments which controlled against any possible information from odour were done by other scientists. These experiments seem to confirm the dance language theory more strongly than ever.

These celebrated scientific investigations of Karl von Frisch exhibit a fundamental unity. He consistently doubted the commonplace and often anthropocentric bias that animals are severely limited in their sensory abilities. And doubt led to one discovery after another, culminating in the disclosure and translation of the dance language of honey-bees.   M.R.

47, 136

FRUSTRATION is a state of MOTIVATION that arises in situations in which the consequences of behaviour are less than those which the animal has been led to expect on the basis of past experience. For example, a hungry animal is likely to become frustrated if it is physically prevented from obtaining food which it can see, if an expected food reward is delayed, or if its food is less palatable than usual.

It is important to distinguish between frustration and deprivation. Frustration involves an element of EXPECTANCY, whereas deprivation results in a desire or need, but without any associated expectation of attaining the desired state. For example, a domestic dog (*Canis lupus familiaris*) removed from its own home and kept in a strange place without food, would be said to be deprived of food, affection, etc. In a totally strange environment, devoid of any external stimuli associated with food or affection, the dog would have no expectation of attaining these things. We would not say it was frustrated. However, for a dog similarly deprived in its own home, there would be many stimuli associated with meal times. As the dog approached its normal feeding place, or as the normal feeding time drew near, the dog would develop an expectation of food, based upon past experience. If the dog did not obtain the food at the expected time or place, it would become frustrated, in addition to being deprived. The states of frustration and deprivation often merge into each other, especially in situations where associations with external stimuli are weak so that half-expectancies arise, but the distinction is nevertheless important in animal behaviour research.

The first reaction of a frustrated animal is usually to try harder to attain the GOAL. Thus a hungry animal, physically prevented from obtaining food by a wire fence, will try to penetrate the fence with increasingly frantic efforts. Gradually this behaviour is replaced by other, often seemingly irrelevant, behaviour. For example, pigeons (*Columba livia*) often peck at the ground when frustrated in obtaining food. Pecking at small objects on the ground may appear to be part of feeding behaviour, but it is directed at inappropriate and inedible objects. This type of behaviour is termed REDIRECTED behaviour. In social situations, in which one animal is prevented from venting its AGGRESSION on another, it may redirect its aggression to a third 'innocent' bystander.

When frustrated in obtaining food, pigeons often preen, in addition to pecking indiscriminately at the ground. Preening seems totally irrelevant in a situation in which feeding has top priority. It in no way resembles feeding, nor does it help the animal to obtain food. Such totally irrelevant behaviour is common in situations involving frustration or CONFLICT, and is generally termed DISPLACEMENT ACTIVITY. Frustration often leads to aggressive behaviour, particularly in higher animals. In one experiment, for example, chimpanzees (*Pan*) were trained to pull a lever to obtain a small reward. When they had thoroughly learned to expect a reward for each pull on the lever, the experimenter ceased to provide rewards. The first reaction of the chimpanzees was to pull the lever harder and more frequently, but when this did not produce the desired result they then became aggressive towards the apparatus or towards the experimenter.

Frustration in animals has some properties similar to those which human beings normally associate with EMOTION. Physiological changes during frustration are similar to those found in states of FEAR. These, together with some of the behavioural manifestations of frustration, are ameliorated by administration of tranquillizing DRUGS. When animals are repeatedly frustrated in a particular situation, they develop an anticipation of frustration, which they clearly find aversive. When avoidance of anticipated frustration is coupled with a strong tendency to approach to obtain a reward, a conflict may develop. If the conflicting tendencies of approach and avoidance are sufficiently strong, the animal may develop symptoms of acute anxiety, leading to a STRESS syndrome. Alternatively, the animal may learn that the situation no longer embodies rewards, and through a process of EXTINCTION the learned behaviour fades away.

The mechanisms underlying frustration are not well understood. They appear to be similar in a wide variety of animals however, and this suggests that frustration plays an important role in the control of animal behaviour. When an animal has a strong tendency to perform a particular activity, tendencies for other activities are suppressed. When the dominant behaviour is thwarted, there is a danger of the animal reaching an impasse, unable to achieve its goal, but unable to break away from attempting to reach the goal. A classical situation is that of a hen (*Gallus g. domesticus*) persistently trying to penetrate a short wire fence to obtain food on the other side. If the hen could only break away from the most direct route to the goal, it could easily make a short detour around the end of the fence. A dog in a similar situation would do just that. It is possible that frustration functions to disrupt the behaviour that is dominant in a thwarting situation, giving the animal a chance to discover alternative ways of solving the problem.

It has been suggested that frustration acts on mechanisms controlling ATTENTION, causing the frustrated animal to divert its attention to hitherto incidental aspects of the situation. Evidence for this view comes from experiments which show that animals do learn more about incidental stimuli in frustrating situations than in situations where they experience no frustration. This incidental LEARNING is prevented if tranquillizing drugs are administered prior to the experimentally induced frustration, but normal learning remains unaltered. Thus it is possible that frustration acts as a mechanism for diverting attention at times when a behavioural impasse is likely to occur. The switch of attention enables the animal to notice features of its environment to which it can then respond. These responses may help the animal to overcome the frustrating situation (see PROBLEM SOLVING), or they may appear as irrelevant displacement activities.

FUNCTION, as used in everyday language, refers to the job that something is designed to do. We might say, for example, that the function of a radiator is to keep the room warm. The biologist uses function in a more specialized sense, but one which is misleadingly similar. To him, the function of a trait is the increase in reproductive success which it confers on its possessor's GENETIC constitution, or *genotype*. Thus, when the biologist argues that the functions of NEST-BUILDING are to keep the eggs and the incubating adult warm, and to shelter both from predators, he is implying that these consequences of nest-building confer higher reproductive success on the genotype of the individuals which practise it.

When thinking about the functions of traits, it is important to understand the level at which NATURAL SELECTION operates. By definition, natural selection is the differential survival of genotypes, not of individuals or of populations. In most cases, traits that increase the reproductive success of the genotype which produces them also increase the success of their individual carriers, but this is not always true. For example, genes which cause their carriers to assist other individuals sharing the same genotype (i.e. relatives) may be selected for and may spread through the population, even if they lower the reproductive success of their carriers. (This phenomenon is called *kin selection*.) Some biologists have argued that selection may also operate between populations, a process usually referred to as *group selection*. For example, it has been suggested that traits may spread if they increase the reproductive success of the groups in which they occur, even if they also lower that of the individuals carrying them (see ALTRUISM). Behavioural traits cited as products of EVOLUTION at this level include ALARM calls, population regulating mechanisms, territoriality, and inhibition of killing. Though the argument appears plausible at first sight, it faces a fundamental difficulty. If such traits decrease the success of their carriers, they will be selected against by selection operating between members of the same groups, and are thus unlikely to spread. Mathematical formulations of natural selection show that group selection can only operate under very specialized circumstances, and suggest that its importance is slight.

When considering the ways in which traits increase reproductive success, their 'functions' in the biological sense, it is important to remember that these apply to genes rather than to individuals or populations. However, in cases where traits increase the individual success of the animals carrying them (as well as of the latter's genotype) it is usually convenient to ascribe their functions to individuals.

The biological definition of function differs in three other important ways from everyday usage. Firstly, functions are always complex and usually multiple. For instance, INCUBATION might have the following consequences in gulls: (i) the nesting material is compressed and its insulating properties reduced; (ii) the embryos are kept warm; (iii) the egg-shells expand slightly; (iv) the adult is unable to feed and, if it continues incubating indefinitely, will lose weight; (v) the eggs are sheltered from predation; (vi) the adult is more susceptible to attack by predators.

The first and third consequences probably have little effect on survival of either adult or young, and can be regarded as neutral. The second and fifth will increase the adult's reproductive success, and are the functions or benefits of incubating to the adult, while the fourth and sixth will tend to decrease its reproductive success and represent the negative consequences or costs of incubating. The example illustrates two points. If the most important function of incubation is to keep the embryos warm, parents should incubate without interruption from the day the clutch is complete, and the number of changes in the individual incubating (which will affect the temperature of the eggs) should be kept to a minimum. In fact, the en-

ergetic costs of incubating to the adult are sufficiently high to require regular changes, and this may lead to short periods when the nest is uncovered. The situation observed will represent a compromise between behaviour which maximizes the benefits (in this case, keeping the embryos warm and the nest sheltered) and behaviour which minimizes the costs (here, the energetic costs of incubating to the adult). Moreover, the functions of incubating are multiple, though some may be more important than others. The case described here is over-simplified, and the evolutionary processes operating on behaviour are usually far more complex, involving many different costs and benefits. Nevertheless, the principle, that the most advantageous (or optimal) form of behaviour for an individual will be the one which maximizes benefits and minimizes costs, remains the same.

Secondly, biological functions are relative and imply an alternative trait. Thus, if we argue that individuals which build nests enhance their reproductive success, we are implying a comparison with individuals which do not. This point has led some authors to argue that only differences between traits, and not traits themselves, have functions.

Thirdly, where function is used in a biological sense, no *teleology* is implied. To say that the function of a radiator is to keep the room warm suggests both that it is designed for this purpose, and that it may switch itself off as soon as this is achieved. To say that one of the functions of nest-building is to keep eggs and incubating adults warm implies only that the trait has spread because it increases the reproductive success of its carriers; it does not imply at all either that evolution is directed to this end, or that the behaviour of the nest builder itself is GOAL directed.

Although this is the usual way in which function is used in the biological literature, it has often been employed in at least three other senses. Firstly, it has occasionally been used in a mathematical sense to indicate a numerical relationship between two variables, as in the statement that 'behaviour A is a function of behaviour B'. It is usually easy to identify cases where it is used in this sense. Secondly, where it refers to signals given by one animal to another, it is sometimes used synonymously with 'meaning'. Finally, it has been used loosely to refer to any consequences of a trait which are not obviously disadvantageous. In such cases, it is often difficult to be sure what an author means by the term.

Are all traits functional? There has long been dispute as to whether all traits must be functional or whether it is possible to conceive of traits which are neutral or disadvantageous. In general, all consistent differences in behaviour, either between species or between populations, have a functional basis; both because weak but consistent selective advantages are sufficient to cause particular genotypes to spread, and because all traits incur costs which, in the absence of counterbalancing benefits, will lead to selection operating to remove the trait.

However, there are a number of situations in which behaviour which is not selectively advantageous may occur: (i) Extreme environmental conditions may produce non-functional differences between populations or between species. For example, unnaturally high population densities may lead to hyperactivity, and extremely high levels of AGGRESSION, with the result that breeding is interrupted. Such effects are best regarded as pathological products of abnormal situations. (ii) Behavioural ADAPTATION may lag behind ecological changes, with the effect that the behaviour appears in inappropriate contexts: some species of American insectivorous birds apparently still show adaptations to feeding on large migratory insects (e.g. locusts, Acridoidea) although human interference has now removed these traditional food supplies. (iii) Domestic animals, which have been artificially selected, may show traits which would not have been advantageous to their wild ancestors (see DOMESTICATION; FARM ANIMAL BEHAVIOUR). (iv) Animals may show inappropriate behaviour in contexts which share some of the characteristics of situations in which the behaviour would be adaptive. For example, the tendency of blue tits (*Parus caeruleus*) to enter houses and tear paper may be a non-functional expression of a motor pattern adapted to feeding in normal contexts. Similarly, in certain weather conditions when their prey are highly vunerable, carnivores may kill many more than they can possibly eat, expending energy unnecessarily in doing so. Presumably in cases of this kind, the contexts are sufficiently rare for the evolution of suppressing mechanisms to cost more than the rare occurrence of inadaptive behaviour.

How are functions identified? It is seldom possible to demonstrate that a difference in behaviour affects fertility or mortality. In a few studies, mostly involving insects, this has been possible. Females of *Drosophila subobscura* are reluctant to mate with inbred males which cannot perform the courtship dance efficiently. This reluctance is clearly advantageous, for females which mate with these males leave fewer viable offspring than those which mate with normal males. Evidence of this kind is seldom available for wild animals, and removal of populations to the LABORATORY often invalidates experiments. Evidence for function is of two types. It may consist of evidence that a particular trait is associated with a particular ecological situation. For example, among ground-nesting gulls which rely extensively on crypsis to protect their eggs and young from predators, the following traits are common: nests are widely spaced; eggs and young are cryptically coloured; young leave the nest a few days after hatching; young hide and 'freeze' when disturbed; adults learn to recognize their own young a few days after birth; brooding birds carry droppings and egg-shells away from the nest; adults from the same colony cooperatively attack predators.

These traits are not found in cliff-nesting gulls, such as kittiwakes (*Rissa tridactylus*), or Galapagos swallow-tailed gulls (*Creagus furcatus*), which rely on the inaccessibility of their nests to protect their eggs and young from predators. In contrast, these species have developed a number of traits associated with cliff-nesting, including solidly built, mud-based nests with deep cups, peculiar fighting methods, and retention of young in the nest until fledging is almost complete.

This kind of evidence is most reliable where it involves associations running across a number of unrelated species which have evolved the trait independently. Where quantitative evidence for many species is available, complex predictions can be made and tested. For example, many bird species defend extensive territories around the nest site. A variety of functions have been suggested for these territories, but the most likely one is that they provide a reliable food supply for the occupants. In this case, TERRITORY size should be larger in species with bigger body size (and greater food requirements) than in smaller species, and larger in predators than in herbivorous species (which have relatively dense food supplies). Where territory size for different bird species is plotted against body weight both predictions are confirmed: body weight is closely related to territory size, and predators have considerably larger territories than herbivores.

Such correlational evidence has a number of limitations. Firstly, it relies on assumptions about the consequences of particular traits. Secondly, it may give misleading results where similar traits have different functions. For example, in Barbary macaques (*Macaca sylvana*), hamadryas baboons (*Papio hamadryas*), and marmosets (*Callithrix jaccus*) adult males may carry juveniles. The motor patterns involved in this behaviour are virtually identical in all three species, but their functions vary widely. In Barbary macaques, males pick up infants in situations where they may be attacked by dominant individuals; carrying an infant apparently reduces the likelihood that the carrier will be attacked. In contrast, young male hamadryas baboons seek to establish relationships with juvenile females which may subsequently join their harems. Finally, in marmosets, the carrying of infants by the male probably represents an example of paternal care, necessary because the female produces two young at a time and can only carry one.

The second kind of evidence used to support functional arguments consists of observational or experimental evidence that a particular form of behaviour has consequences which are likely to enhance the survival or reproductive success of its carrier. One of the functions of FLOCKING in insectivorous birds may be that it allows individuals to find food by watching other members of the flock. Experimental evidence supports this effect: if great tits (*Parus major*) are released into an aviary where mealworms (larvae of the beetle *Tenebrio molitor*) have been hidden, the rate at which the birds locate the worms is higher if they are released in groups of two or four birds than if they are released singly. After one bird has located a worm in a particular type of hiding place, others tend to search in similar places. Though experiments of this kind provide reliable evidence of the consequences of particular behaviour patterns, they do not show that these are the ones through which selection is acting. For example, even if flocking had evolved for some reason unconnected with food finding, it might still have some effect on feeding rate.

The most satisfactory approach to identifying function is to erect competing hypotheses which can be tested by a mixture of correlational and experimental evidence. For example, black-headed gulls (*Larus ridibundus*) remove old egg-shells from the vicinity of their nests (see SURVIVAL VALUE). Egg-shell removal might have several functions. It might: (i) prevent chicks injuring themselves against the sharp edges of the shell; (ii) remove the possibility that a used shell might slip over an unhatched egg, trapping the chick in a double shell; (iii) reduce the chance that the moist organic material left behind in the shell would provide a breeding ground for bacteria; (iv) reduce the chance that

empty egg-shells lying near the nest would attract predators and thus endanger the brood.

Circumstantial evidence suggests that the first three explanations are incorrect. First, conspicuous objects other than egg-shells are removed if placed close to the nest. Second, the behaviour is performed throughout the incubation period and until around 3 weeks after hatching, and not only when nests contain small chicks. If the function of the trait is to avoid attracting the attention of predators, it should be absent in species nesting in conspicuous places which rely on alternative methods of defence. This is indeed the case, and cliff-nesting species, including kittiwakes, do not remove empty egg-shells from the nest.

To test the last hypothesis, scientists laid out black-headed gulls' eggs with and without egg-shells beside them, and allowed them to be preyed by carrion crows (*Corvus corone*) and herring gulls (*Larus argentatus*). The results showed that more of the eggs laid out with an egg-shell close to them were taken than of those laid out without an egg-shell, providing convincing evidence that the function of egg-shell removal is to prevent nest detection by predators.   T.H.C.B.
88

G

GENERALIZATION. When an animal has been trained to perform a particular pattern of behaviour in response to a particular stimulus, other, previously ineffective stimuli may become capable of evoking the response to some extent. This phenomenon is known as generalization: it is of interest partly because it has been used as an explanation of other features of behaviour, and partly because it can give us some insight into the way in which the animal perceives the world: different stimuli which tend to evoke the same response from an animal are presumably perceived as being in some way similar to each other.

In one of the earliest experiments, a dog (*Canis lupus familiaris*) was classically conditioned to salivate in response to a touch on the left thigh (see LEARNING). Fig. A shows the results of applying this touch stimulus to other parts of the body. It is apparent that stimulus locations other than that used in training are capable of evoking the *conditional response* of salivation: stimulus generalization has occurred. (Some authors prefer to describe this effect as the *irradiation* of a conditional reflex while others use the term *induction*. These words are essentially interchangeable with generalization.)

The effect can also be demonstrated with OPERANT conditioning procedures. In a classic experiment, pigeons (*Columba livia*) were taught to respond for food by pecking at a disc (or response key) illuminated with light of a certain colour. The birds were then presented with new colours of shorter or longer wavelength, and the number of responses given to each stimulus was noted (see COLOUR VISION). The results (Fig. B) show orderly gradients of generalization; the animals clearly discriminate the new colours from the original since they are less ready to respond to them, but they also show generalization in that these new stimuli are capable of evoking at least some responses.

In both the experiments just described, generalization has proved to be a graded effect, with the stimuli that the human observer would judge to be most similar to the original stimulus producing the most responding, while less similar stimuli produce progressively less responding. Generalization effects do not always show this pattern. For example, pigeons trained to respond in the presence of a tone of a given frequency are likely to respond equally readily to new tones differing in frequency, and dogs conditioned to a particular stimulus may, at least early in their training, salivate to sti-

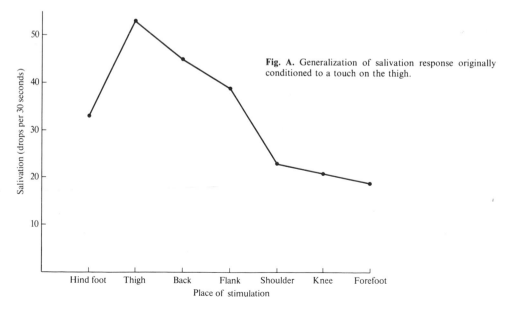

Fig. A. Generalization of salivation response originally conditioned to a touch on the thigh.

muli drawn from quite different sensory modalities. PAVLOV mentions the case of a dog trained to salivate to a given stimulus which also showed the conditional response to the stimulus of Pavlov himself entering the room.

All the examples discussed so far have been concerned with cases of generalization in which the stimuli used tend to excite the response, but it is also possible to demonstrate the generalization of *inhibition*. Thus, pigeons can be trained to respond

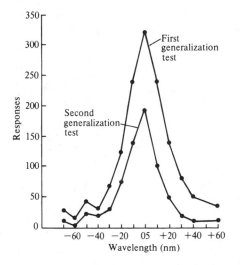

Fig. B. Generalization of response to colour in pigeons.

to a key displaying a colour, but to refrain from responding when the key bears a vertical white line; this is achieved simply by making food available for responses to colour, but withholding it for responses to the line. The generalization test consists of presenting the line stimulus in novel orientations, and usually shows that the birds respond little if at all to the vertical line, but give increasingly more responses as the line diverges from the vertical. It appears that the vertical line has the power to inhibit responding, a power which generalizes to lines of similar orientation.

The concept of generalization has been widely used as an explanation for other phenomena in animal behaviour. Most generally, the point is often made that even in LABORATORY STUDIES it is impossible to present an animal with identical stimuli on different occasions. If a rat (*Rattus*) is trained to run a Y-maze on one day, his perfect performance on the following day might seem surprising if we take into account the fact that the stimulus conditions are likely to have changed: the rat's internal state will probably be different (see MOTIVATION), the lighting may be different, new odours may be present, and so on. But the experimentally demonstrated fact that animals will respond in much the same way to stimuli that are

similar to the training stimulus (i.e. the fact of generalization) will reduce our surprise and explain the *transfer of training* from one day to the next. A second, more specific, example of the use of the notion of generalization in explaining other phenomena is found in its application to *transposition*. A rat, which has been trained to approach a black stimulus object and to avoid a grey one, is said to show transposition when in a subsequent test it chooses the grey stimulus in preference to a white, i.e. when it makes a choice on the relative rather than the absolute values of the stimuli. It is possible to show that if generalization gradients are taken into account, this behaviour can be explained without allowing that rats are able to respond to the relative properties of stimuli. In brief, the generalization of inhibition from the grey stimulus to the white would tend to prevent the animal's making a response to the latter, while the generalization of excitation from the black would be more likely to reach the grey stimulus than the white. Hence a preference for the grey over the white is predicted. This account may be incorrect (there are other aspects of transposition that it cannot easily explain), but it serves to show how effective the concept of generalization can be.

Although generalization has been used as an explanation for other features of behaviour, it nevertheless is itself in need of some explanation. Pavlov's original account which made reference to the spread of excitation and inhibition over the cerebral cortex has been rejected, not least because our ideas about the functioning of the BRAIN have changed since Pavlov's day. Current explanations tend to concentrate not on the underlying brain processes, but on the analysis of the stimuli involved. Their central argument is that the stimulus used in a conditioning experiment can be regarded as consisting of a collection of separate elements; a tone, for example, is an external event, sudden in its onset, of a given frequency, duration, and intensity. All of these elements may become conditioned during training, and a new stimulus that contains some or even just one of them may be capable of evoking the response to some extent. A tone of 1000 Hz may seem to the experimenter to be a very different stimulus from a tone of 300 Hz, and he may think it remarkable that a pigeon trained to respond to the first should generalize to the latter. But although the two tones differ in frequency they hold all other elements in common, and from this point of view it is not surprising that both tend to elicit the conditional response. The same sort of analysis can be applied to the case of the dog which showed generalized salivation to Pavlov's entering the room; here the dog has presumably learned to respond to any sudden change in external stimulation, and the sudden appearance of Pavlov was therefore an effective stimulus.

DISCRIMINATION training is a technique which serves to reduce very widespread generalization. If a pigeon is trained to respond in the presence of the 1000 Hz tone, but is not rewarded for responding in the presence of the 300 Hz tone, then the tendency to generalize to tones of differing frequency is less marked. This sharpening of the generalization gradient can be readily explained in terms of the *common elements theory*, for what discrimination training does is to reduce the number of elements present in the training stimulus that are consistently associated with reward. As the pigeon learns not to respond to the 300 Hz tone he is learning, in effect, that the sudden onset of a tone of given duration and intensity does not ensure that responding will be rewarded; he is being compelled to pay attention to the frequency of the tone. The consequence is that only tones close in frequency to the 1000 Hz tone elicit responding. It should be added that some generalization gradients in some animals are often very sharp without the need for explicit discrimination training (that shown by pigeons to variations in colour is a case in point); these are found when the dimension along which the stimulus is being varied is one to which the animal is particularly sensitive, and to which it tends to attend to without special prompting (see ATTENTION).

There remains the question of why gradients of generalization take the form that they do. It is usual to assume that an animal trained to respond to a yellow light produces some responses to orange, and fewer to red, because it fails to discriminate completely the test stimuli from the training stimulus, and that orange being physically closer to yellow is more likely to be confused than is red. This is undoubtedly largely correct, but it should be added, in conclusion, that there are some examples of generalization which do not follow easily from this analysis. One such is the phenomenon of *octave generalization*; a subject trained to respond to a 1000 Hz tone often shows more generalization to a tone of 2000 Hz (a tone differing from the original by an octave) than to a tone of 1700 Hz, say, despite the fact that the latter is physically closer to the training tone. Why this should happen is not clear, and we must conclude that the extent to which stimuli are perceived as being similar (i.e. the extent of generalization) cannot always be predicted from knowledge of the simple physical properties of the stimuli.   G.H.

86, 98

GENETICS OF BEHAVIOUR. We hear such comments as 'like father, like son', or 'this breed of dog is vicious'. It has been claimed that the genetic make-up of an individual can account for such activities as AGGRESSION, LEARNING, and COURTSHIP. This assumes that individual differences in the manifestation of the activities are caused by genetic differences among individuals. In this article we consider the concepts and methods employed to study such questions.

The genetics of behaviour is normally studied by scientists trained in both genetics and psychology or ETHOLOGY, and the approach used is often termed *behaviour-genetic analysis*. This is simply an approach to the study of organisms and their behaviour which combines the concepts and methods of genetic analysis, based on knowledge or control of ancestry, with those of behavioural analysis from psychology and ethology, based on knowledge or control of experience. The object analysed genetically is a species (or one or more of its component populations). *Population genetics* yields information about the frequencies of genetic factors, and about their distributions in populations throughout a species. With such information one begins to approach the objective of genetic analysis: accumulating knowledge about the genetic endowment of species and of individuals. The ultimate goal would be a complete inventory of the heredity of a species, learning the locus and FUNCTION of every part. Very little of this can be learned from studying an individual and ignoring the family, the population, or the species. Genetic analysis involves the study of individual differences to find genetic factors. Behavioural analysis studies patterns of behaviour and their components, as well as conditions influencing behaviour. The goal of behaviour-genetic analysis is to find relations between genetic factors and behaviour components. A grasp of basic genetic concepts is essential before one can appreciate the hybrid subject called behaviour-genetics.

Basic genetic concepts. The gene can be defined as the basic unit of heredity. The genes are located in long molecules called *deoxyribonucleic acid* (DNA). The DNA protein matrix forms nucleoprotein and becomes organized into structures called *chromosomes*. The chromosomes are found in the nucleus of the cell. Each gene occupies a specific locus on a chromosome. Through *mutation*, a gene may exist in two or more alternative forms called *alleles*.

1. *Phenotype and genotype.* Any organism can be described in terms of a large number of characteristics or *traits*, as many in fact as an observer chooses to analyse. An animal's chemical composition, basal metabolic rate, height and weight, colouring, and sensory acuities are all part of its phenotype: the observable aspects of the individual. Underlying the phenotype is heredity: the developmental potentials transmitted to the individual from its ancestors. An individual's complete genetic endowment is called its genotype. Both phenotype and genotype are of a mosaic nature; the elements of the one bear a definite, but not simple, relationship with the elements of the other. In a complex way, the components of the genotype are precursors of the components of the phenotype.

While the genotype remains constant throughout life, the expression of phenotypic traits may change with the development, MATURATION, and shifting environmental influences (see ONTOGENY). Even when two individuals possess the same hereditary potential for a given character, the character may develop in different ways in these individuals if their environmental backgrounds are sufficiently dissimilar. The phenotype is thus the resultant of the interaction between heredity and environment, and variation in phenotypical characters is a function of both genotypic differences and environmental conditions.

2. *Mendelian genetics*. In 1866, long before the relationship between chromosomes and genes was realized, Mendel formulated the classical concepts of inheritance on the basis of breeding experiments in which the distribution of different phenotypic forms was analysed in the progeny of known matings. Breeding and progeny testing are still the fundamental tools of genetics, but we know now that the inheritance of genes, and ultimately of traits, reflects the pattern of chromosome transmission from one generation to the next. Since the individual receives two sets of chromosomes (one set from each parent) it also receives its genes in duplicate. If both chromosomes in a pair carry the same allele of a gene, the individual is said to be *homozygous* for that gene. It is capable of transmitting only one form (allele) of that particular gene to the next generation. As an illustration, we can consider two alleles, B and b, which in guinea-pigs (*Cavia aperea porcellus*), and many other animals) govern pelage colour. Homozygotes with two alleles, BB, have black coat colour, while those with genotype bb have white coat colour (see Table 1).

Frequently, the *gametes* (sperm and egg) from two parents contain different alleles of a gene. The offspring is then *heterozygous* (Bb). In a mating between BB and bb homozygotes, all of the offspring have the genotype Bb. In this genetic system, B is dominant over b; therefore, the heterozygotes are black and indistinguishable from the BB homozygotes. The activity of the allele B masks

Table 1. The inheritance of a single gene (Black coat–White coat)

Parents (Mating types)	Expected proportion of offspring Genotypes	Phenotypes
BB × BB	All BB	All Black
BB × bb	All Bb	All Black
bb × bb	All bb	All White
BB × Bb	½ BB:	All Black
	½ Bb	
bb × Bb	½ Bb:	½ Black:
	½ bb	½ White
Bb × Bb	¼ BB:	¾ Black:
	½ Bb;	¼ White
	¼ bb	

Table 1 shows the possible types of mating and the progeny ratios for the single-gene coat-colour trait. The relative proportions of genotypes and phenotypes in the offspring are referred to as *segregation ratio*; they represent the probability, not the certainty, that an offspring will fit a given classification. Notice that when one allele is dominant, ¾ of the progeny from the mating of two heterozygous parents are expected to resemble their parents, while the remaining ¼ (the *recessive* homozygotes) are phenotypically unlike either parent.

that of b, preventing the latter from expressing itself in the phenotype. Masking alleles are said to be dominant over the alleles whose effects they cover. The latter are called recessive. With some genes, neither of the two alleles is completely dominant over the other, so that the heterozygote may show a form of the trait more or less intermediate between the homozygous forms. Genotypes, of course, are never seen. They can only be inferred from the phenotypes of individuals and their relatives.

Since homologues segregate to different gametes, a heterozygous individual produces two kinds of gametes, some carrying the maternal chromosome with one allele (e.g., B) and others carrying the paternal homologue with the other allele (b). This is the basis of Mendel's first principle of inheritance: the segregation of alleles.

Table 2. Progeny genotype alternatives in the mating of double heterozygotes for two independent autosomal genes

		BbLl × BbLl		
		BL	Bl	bL	bl
Mating of double	BL	BBLL	BBLl	BbLL	BbLl
heterozygotes	Bl	BBLl	BBll	BbLl	Bbll
Gametes	bL	BbLL	BbLl	bbLL	bbLl
	bl	BbLl	Bbll	bbLl	bbll

Expected distribution of phenotypes if there is dominance at both loci

Ratios of phenotypically homogeneous classes given dominance at both loci

1 BBLL
2 BBLl
2 BbLL :9 : 1 BBll : 1 bbLL : 1 bbll
4 BbLl   2 Bbll   2 bbLl
            :3   :3

9 black short hair: 3 black long hair: 3 white short hair: 1 white long hair

When gametes are formed, all the genes received from one parent are rarely, if ever, distributed together to the same gamete. Whereas the maternal and paternal homologues in every chromosome pair segregate, the segregating pairs *assort* (distribute) their homologues to the gametes independently. Since *genomes* comprising all combinations of maternal and paternal chromosomes are equally likely in the gametes, the individual genes show independent assortment (Mendel's second principle). For example, if a Bb individual is heterozygous for the alleles Ll (allele L is for short hair and l for long hair) of another gene located on a different chromosome, four types of gametes (BL, Bl, bL, bl) can be formed with respect to the two genes. The results of mating between the two double heterozygotes, assuming dominance at both *loci*, are illustrated in Table 2. Most individuals, of course, are homozygous for some genes and heterozygous for others, and are thus capable of producing an enormous array of alternative kinds of gametes.

3. *Heritability and instinct*. The inter-individual or phenotypic variation in behaviour can be parti-

tioned into components assignable to heredity, environment, and their interaction. Heritability (in its broad sense) refers to that part of the total variation which is attributable to genetic differences among organisms. In quantitative genetics, heritability is also defined in a narrower sense.

It used to be common practice to ask what percentage of variation in a behaviour was attributable to heredity, to environment, and to their interaction. Consideration of what is inherited leads us to a critical re-examination of the nature–nurture issue raised by this question. Since the total biological inheritance of the organism is contained in the single-celled *zygote* (fertilized ovum), behaviour, as such, cannot be inherited. Genes are inherited, and these possess certain norms of reaction. For the replicates of a given genotype, a difference in behaviour under two or more sets of environmental conditions is a function of differences between those conditions. For a given set of environmental conditions, a difference in behaviour between two or more genotypes is a function of differences between those genotypes. Heritability therefore cannot be a property of behaviour; it is a property of a population with respect to particular traits measured under certain conditions and not of traits *per se*. The very same behaviour that must have zero heritability in a *clone* (organisms that are genetic duplicates), e.g., a set of identical siblings, might have substantial heritability in a group of heterogenic individuals.

Behavioural traits are properties of organisms that behavioural scientists choose to measure, just as traits, like length and weight, are properties of objects that the physical scientists choose to measure. Heritability is a measure of relationship between a trait and the population in which it is studied, just as weight is a measure of the relationship between the mass of a body and the gravitational field in which it is studied. There will be no confusion about heritability if it is remembered that the methods of genetics only permit the study of differences: Mendel's first concept of segregation defines a gene in terms of an interallelic difference, and Mendel's second concept of independent assortment distinguishes two genes in terms of an interlocular difference. The heritability concept merely summarizes the extent to which phenotypic differences (the trait variance) are related to genotypic differences (the genetic variance) in some population at a given generation under a given set of environmental conditions. Where there are no gene differences between individuals, as in a clone, heritability values must be zero for all their traits. Where there are inter-individual gene differences, as in cross-fertilizing populations, trait heritabilities can only be determined by empirical measurement. For example, aggression, as measured by a given test, may have quite different heritabilities in different populations; it may also have different heritabilities in the same populations under different conditions.

This discussion so far has dealt with measurements at a given time in a given population. Heritability has been represented as the ratio of the genetic variance to the phenotypic variance. In the study of the developing organism over time, it is found that certain behaviour patterns appear with striking regularity: e.g., all or most birds fly at an age that is characteristic of their species and, in the appropriate season, build nests of a kind characteristic of their species. Furthermore, the essential features of these activities do not seem to be acquired as a result of LEARNING by rewarded practice, because they develop even when the immature organism is raised in isolation from conditions in which training or imitation might plausibly be expected to play a role. Such types of behaviour are called instinctive and are frequently said to be the expression of the genetic endowment of a species; although, strictly speaking, no information about their genetic correlates can be known without genetic analysis.

It should be noted that heritability and INSTINCT have different meanings and are studied by different methods. The operations involved in the measurement of the heritability of a behaviour in a population require the measurement of individual differences and of correlations between relatives in the expression of that behaviour. The operations involved in studying the ontogeny of instinctive behaviour usually include the observation of the expression of the behaviour in an organism reared in an isolation experiment.

4. *Strain, species, and populations.* Species can be defined as a group of interbreeding natural populations that are reproductively isolated from other such groups. Defining concepts like *strain* and *population* has been a problem in biology. We shall define a strain (or line) as a group of individuals descended from common ancestors and maintained in reproductive isolation from other such groups. A population can be defined as an interbreeding group of conspecifics occupying the same space at the same time. The distinction between strain and population is not always clear. A small population may closely resemble an unselected strain. The difference between these two concepts at this point becomes a matter of jargon or semantics. Strain (or line) is a term usually employed by the LABORATORY investigator referring to animals that have been reared in an artificial environment, while the term population is usually used by FIELD investigators referring to animals in their natural HABITAT.

5. *Wild type.* The term *wild type* has appeared often in the genetics literature. Unfortunately, behaviour genetics has adopted this term. Wild type is a misnomer at worst, and misleading at best. Wild type implies that some phenotype is representative of a natural population. This is not the case, with the possible exception of a strain being compared with a small natural population. In most instances, the wild type strain is started by

taking a small sample from a large natural population, and bringing it into the laboratory. This can easily lead to *sampling error* in that the sample collected to start the strain is not a true representation of the population in the wild. Also, the selection pressures in the laboratory are more than likely to be very different from those in the wild. An animal with a genotype 'A' may have a selective advantage over an animal with genotype 'a' (of the same species) in the wild. This situation could be reversed in a laboratory environment. It is even possible for a particular genotype to be highly successful in the wild, but to become extinct in the laboratory. Sampling error and different selection pressures in the laboratory can lead domesticated strains to diverge genetically from the natural populations of their origin (see DOMESTICATION). Unselected strain would be a more appropriate term. For, when the behaviour geneticist refers to wild type strains, he is talking about strains in which there is no direct manipulation of the MATING SYSTEM. The strains are merely free-mating.

Genetic methods in behaviour-genetic analysis. There have been two main approaches to behaviour-genetic analysis: the experimental approach and the observational approach. The experimental approach has been primarily restricted to the laboratory and includes selection, strain comparison, segregational analysis, diallel crosses, chromosome and mutant analysis. The observational approach has been concerned with events that occur in nature. This article will be focusing on the experimental approach, though it is appropriate to discuss the advantages and disadvantages of both approaches. The investigator who employs laboratory studies has some degree of control over the subjects being studied. He has a degree of control over both the environment and the heredity of his subjects. This allows that investigator and others to replicate the work. Ironically, it is the laboratory investigator's control that is the basis for criticism of that approach. In nature, there are no perspex mazes or electrical grids. Animals do not live in an electrically controlled day–night cycle at constant temperature and humidity. Although mating combinations are not random in nature, they are nevertheless different from those imposed in the laboratory. Some critics have stated that by creating an artificial environment, over generations, one has created artificial animals. The counter-argument is that there are many variables involved in the study of both behaviour and genetics, and to understand a relationship one must keep all variables constant but one. Through control and replication, one is able to predict. Control, prediction, and replication are the pillars of science. In field studies, the investigator primarily observes animals in their natural environment with little or no manipulation involved. We see that the laboratory investigator sacrifices a portion of reality for control, replica-

tion, and prediction, whereas the field investigator sacrifices these factors for reality. A synthesis of both approaches has become one of the goals of behaviour-genetic analysis.

1. *Breeding studies.* In breeding studies, the investigator controls the system of mating. Selection occurs when animals are mated on the basis of the presence or absence of a particular trait. Inbreeding occurs when closely related individuals are mated.

Artificial selection is a breeding method used by man to change the phenotypic composition of a population. It is useful if there are alleles of additive (non-interacting) gene combinations to be selected. Whether the intention is to arrive at a more productive stock of dairy cattle (*Bos primigenius taurus*) or a fiercer breed of watch-dogs (*Canis lupus familiaris*) or a DDT-susceptible strain of insects, artificial selection involves choosing as parents for each generation only those individuals with the desired expression of the trait or traits (see FARM ANIMAL BEHAVIOUR), and discarding all those that do not exhibit a previously determined level of phenotypic merit. Artificial selection is assortative mating enforced on the basis of phenotypic similarity. Although progress under selection is measured by changes in the phenotypic value of the population, these changes are achieved and maintained through increase in the frequency of appropriate alleles in the gene pool.

When a character is controlled by a single gene with only two alleles, selection for the recessive allele can be completed within one generation, if it is possible to mate only the recessive homozygotes. Selection for the dominant one, on the other hand, proceeds more slowly because inclusion of heterozygotes in the breeding group at each generation retains the recessive allele in the gene pool. The rate of progress depends upon the relative frequencies of the two alleles at the onset of selection: the lower the frequency of the recessive allele, the slower the rate of its elimination.

Many traits of interest in behaviour genetics are undoubtedly controlled by many genes acting in combination, with different combinations of alleles showing different gradations of phenotypic expression. Beginning with a population of diverse behavioural types, artificial selection changes the phenotypic characteristics of the population by redistributing the allelic combination in the gene pools of successive generations. To illustrate, let us consider a population of guinea-pigs which, when scored for speed of running a maze, is found to be phenotypically heterogeneous: some fast, some slow, and some intermediate grades of runners. As is most commonly done in selection experiments, we select for the two extremes of the behavioural distribution. The 'fast' line is initiated by mating the rapid runners, and then, in generation after generation, mating only the fastest guinea-pigs among the progeny; selection of the 'slow' line is carried out in a similar fashion at the opposite end of the scale, and obviously the two lines are never crossed. We might decide also to keep an un-selected, free-mating population as a control that will enable us to estimate the extent of phenotypic changes arising through chance. Now suppose that the running rate of an animal is a heritable trait (otherwise, we could not change the population mean through selection) determined by several genes. We can designate the alleles that contribute to increased speed as 'plus' and those involved in slow running as 'minus'. Under selection, the average scores of the two selected lines should begin to diverge, as more plus alleles accumulate in the fast line, and new combinations of minus alleles build up in the slow line.

Theoretically, selection ceases to have an effect when all the alleles of genes acting additively in one direction have become fixed in a line, i.e. when there is homozygosity at all of the relevant loci, and no further additive genetic variation remains. In practice, selection progress is contingent upon a number of factors: (i) Since permanent phenotypic changes depend upon modifications of the gene pool, we must be able correctly to identify and select for breeding the phenotypes that carry the desired alleles. To the extent that the correspondence between phenotype and genotype is not perfect, and we have seen that one-to-one correspondence is rarely to be expected for behavioural traits, some identifications will be false, and undesirable alleles will be retained in the population. Population changes in the direction of selection will then be delayed. (ii) The intensity of selection pressure that can be applied is important. If, at the start, we have very few extreme scores, we have to include in the breeding group some less extreme phenotypes (presumably with fewer of the appropriate alleles) to get a sufficient number of offspring in the next generation. In doing so, we may be diluting the intensity of selection. (iii) NATURAL SELECTION, too, may operate against extreme types, making them less viable (or less fertile), and thus counteracting the effectiveness of artificial selection. (iv) Many other factors may begin to operate long before genetic variability is dissipated, and all of the alleles have become fixed. For example, phenotypic selection is partially limited by a scoring scale with very few categories, which restricts the possibility of discriminating between phenotypic variants. Physiological and anatomical limitations of the organism may be important considerations; in selecting guinea-pigs for speed of running, how fast could any individual be expected to run?

In behaviour genetics, selection studies are used to estimate trait heritability in a population, and to tailor populations for further, more detailed genetic analysis. The selective breeding method has been used for studying several types of behaviour. Investigators working with populations of rodents have succeeded in developing lines selected for: maze-learning ability and other aspects of rodent INTELLIGENCE; *activity* versus inactivity, scored by the number of revolutions of a running wheel; *emotionality* versus non-emotionality, measured by the amount of urination and defecation within a certain time period (see EMOTION); and susceptibility versus resistance to audiogenic seizures, i.e. convulsive behaviour in response to auditory stimulation. In most studies it has been found that the gains accomplished by selection are lost when selection pressure is relaxed. These findings, together with the rapid response to reverse selection, suggest, though certainly do not prove, that the genetic variance is not simply additive.

Inbreeding is a special form of artificial selection in which closely related individuals are mated. In the laboratory where the mating system can be controlled, inbreeding provides a means of increasing the genetic homogeneity of a population. For cross-fertilizing species, intensively inbred strains offer the best approximation to the 'pure' lines obtainable in species that reproduce by self-fertilization. The intensity of inbreeding, of course, depends upon the genetic relationship between the breeding pairs. Whereas a parent and offspring share almost exactly 50 per cent of their alleles, the degree of genetic relationship between any other two individuals (except identical siblings) expresses the proportion of alleles that are likely on the average to have been inherited from common ancestors. Although the genetic relationship between siblings will vary from zero to unity, it will average 50 per cent.

In rodents intensive inbreeding of brothers and sisters will, theoretically, give homozygosity at most loci within sixty-five generations, i.e. the genetic relationship approaches 100 per cent; with mating of less closely related individuals, the progress is correspondingly slower. In practice, advantages enjoyed by heterozygotes, in terms of superior viability, may sharply limit the rate at which homozygosity increases in a strain. Even if homozygosity is attained, phenotypic variability does not necessarily disappear; in fact, there is evidence suggesting that homozygotes might be more variable than heterozygotes, possibly because the latter are better 'buffered' physiologically against minor fluctuations in their developmental environment. Inbreeding then increases the probability (not the certainty) of genetic similarity at most loci among the individuals of a population.

As experimental tools, inbreeding and artificial selection are similar in several ways. Both are planned assortative mating systems: selection being assortment based on phenotypic merit (together with differential rates of reproduction for the classes of assortment), and inbreeding being assortment based on genetic relationship. After the first few generations, moreover, selection experiments seem to involve some degree of inbreeding within a selected line (especially in very small breeding groups), unless precautions are taken to avoid matings between closely related animals. Without selection, on the other hand, the effect of genetic assortment is unpredictable with respect to fixation of a particular allele at any given locus. Selection and inbreeding are useful techniques primarily because they provide material for further genetic analysis. In artificial selection, matings are based on phenotypic similarity within a line, and further breeding experiments can be performed to study the genetic basis of the phenotypic differences between lines. Different inbred lines known to be genetically unalike can be compared and analysed for phenotypic differences existing between them. Comparisons of inbred strains subjected to a range of environmental treatments yield data on heredity–environmental interactions. Inbred strains thus play an important role in experimental behaviour genetics.

2. *Strain and race comparisons.* Many of the data which have been gathered in behaviour genetics come from comparisons of inbred strains of different *races* occurring in nature. While some studies have been limited to the observation of behavioural differences between strains, other investigations have analysed the genetic correlates of behavioural differences by cross-breeding the strains, and comparing the segregation ratios with those expected under certain genetic hypotheses.

From what we now know about genetics and the properties of populations, we should in fact be very surprised if reproductively isolated populations did not differ in some way on most phenotypic measures (see ISOLATING MECHANISMS). While comparisons have traditionally dealt with average values, other statistics sometimes prove to be more sensitive in the detection of dissimilarities between populations.

It is to be expected, therefore, that behavioural differences will almost always be found when inbred strains are compared. Some of the behavioural variations that have been observed in inbred strains of mice include: fighting tendencies, emotionality or timidity, maze running, and preference for alcohol.

Domestic breeds of dogs, although not as highly inbred as some laboratory strains of rodents, differ in behaviour as well as in the physical characteristics for which they have been selected. In domestic strains of rabbits (*Oryctolagus cuniculus*), differences have been observed in MATERNAL behaviour, i.e. the time of NEST-BUILDING and treatment of young. In nature, reproductively isolated strains or races have developed divergent genetic backgrounds which influence behaviour. For example, the dance by which the light-coloured Italian honey-bee (*Apis mellifera*) communicates the location of a food source to the other bees in the hive has a pattern unlike that used by the dark-coloured Swiss and Austrian strains of the honey-bee.

3. *Segregation analysis.* Changing the distribution of a behavioural phenotype in a population through artificial selection and the measurement of behavioural differences between reproductively isolated groups provide evidence for the existence of genetic differences which may be amenable to further analysis. The nature of the transmission of the genetic correlates of a trait: whether one or many genes are involved, whether some alleles show dominance, etc., can be discovered only through further breeding experiments entailing segregation analysis. The methods and principles of segregation analysis are those of transmission genetics; their application permits the testing of various genetic hypotheses.

Many physical characters and some behavioural traits can be grouped into a few clearly distinguishable phenotypic classes. Very often, the phenotypic differences are found to result from allelic substitutions at a single locus. The single-gene hypothesis is tested by comparing the observed ratios of individuals in the different phenotypic classes with those expected under various possible dominance relations. To illustrate, let us consider two inbred strains of mice, where one consistently displays a behavioural phenotype A, and the other a, a different form of the behaviour. On cross-breeding the two strains, the hybrid (called the F_1) generation behaves like the parental strain A. However, in the F_2 generation, i.e. the progeny obtained by inter-mating the members of the F_1, only three-quarters of the animals have phenotype A, and the remaining quarter resemble strain a. Now we see a familiar pattern emerging: if strain A is homozygous (AA) for a dominant allele, and strain a for the recessive form, the F_1 individuals are all heterozygotes (Aa) and should be like their strain A parents in behaviour. Segregation of alleles in the cross of F_1 heterozygotes (Aa × Aa) would then yield three F_2 genotypes, $\frac{1}{4}$ AA, $\frac{1}{2}$ Aa, and $\frac{1}{4}$ aa, the first two being of the same phenotype. Some of the F_1 mice are mated with strain A and others with strain a. If the hypothesis is correct, the *backcross* to strain A (Aa × AA) should yield only A behaviour, but the opposite backcross to strain a (AA × aa), should give a 50:50 ratio of the two phenotypes. Intermating the a phenotypes of the F_2 of the backcross to strain a should give all a individuals.

The phenotypic segregation ratios for a single gene depend on the dominance relationships among the alleles. When neither allele is dominant, the F_1 individuals are intermediate in behaviour between the two parental strains, as are half of their offspring in the F_2, and half of the members of each kind of backcross. The expectations from the various crosses, under the hypothesis of dominance and under the hypothesis of no dominance, are shown in Table 3.

Some phenotypic characters seem to go together more often than not. The relationship between two characters may result from one of three possible genetic situations. It may be due to the multiple

Table 3. Expected phenotypic and genotypic ratios in experimental crosses under two genetic hypotheses

Hypothesis 1: Allele A dominant over a

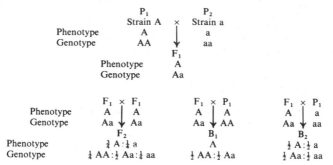

Hypothesis 2: No dominance (I is intermediate between A and a)

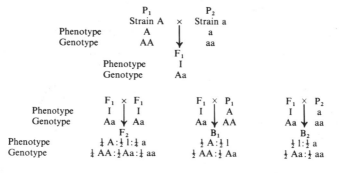

effects (*pleiotropic activity*) of a single gene, in which case segregation analysis will fail to break down the relationship. In one experiment, the floor temperature preference in three mouse strains (*Mus musculus*) showed a correlation between the behaviour and both fur density and the temperature of stomach fur; according to the genetic analysis, these pleiotropic effects appear to be controlled by a single gene with at least three alleles. Another type of association involves traits determined by closely linked genes that occupy nearby loci on the same chromosome. Since linked genes usually remain together in gamete formation, it is often difficult to distinguish between pleiotropy and *linkage*. With sufficiently large numbers of subjects, however, it is possible to detect linkage when crossing-over takes place between homologous chromosomes. For example, if one strain is homozygous for alleles AB on the same chromosome and another for ab, the recombinations Ab and aB are likely to appear in some of the descendents of the crosses between them. The frequency of crossing-over and recombination depends upon the distance between the two genes on the chromosome; the shorter the distance, the tighter the linkage, and the less the likelihood of crossing-over.

In a particular strain, traits determined by genes on different chromosomes may appear to be associated because certain combinations of alleles have

come together in the homozygous state. Crosses between such strains permit segregation and independent assortment of these genes, to be observed in the F_2 and the backcross generations.

Individual differences in the expression of most behaviour patterns do not fit into a few sharply defined classes, but rather show a graded distribution of phenotypes, e.g., maze-learning, emotional reactions, and activity in rats; *geotaxis, phototaxis,* and mating speed in flies (Diptera). These behavioural variations are usually correlated with the accumulated small effects of many genes. *Polygenic* characters may be studied by the same breeding methods used in the analysis of single-gene traits, but the treatment of the data differs. When working with polygenic systems, we summarize data in terms of the *mean* and *variance* of the phenotypic distribution of a continuum of scores, rather than considering the proportions of individuals observed and expected in a few discrete classes.

The simplest illustration of the polygenic model can be provided by considering a trait determined by the combined effects of genes at two unlinked loci. In a cross between two strains, each homozygous for different alleles at the two loci, the F_1 should be uniformly heterozygous, showing no genetic variance, and with a mean score value intermediate between those of the two parents, assuming no dominance or similar conditional genetic effects. Theoretically, segregation in the F_2 produces a range of scores with a mean identical to that of the F_1. The backcrosses of the F_1 to the parental strains are expected to be more variable than the F_1 but less so than the F_2, and to have mean values lying between those of the F_1 and of the parent strain to which the backcross is made. With dominance or similar conditional genetic effects, the expectations regarding the mean value in the various generations are correspondingly altered.

Further analyses of data on crosses between strains that differ in polygenic traits are carried out by means of more sophisticated statistical procedures which entail partitioning of the phenotypic variances into genetic and environmental components. These calculations allow estimations of the number of genes involved.

The majority of behaviour patterns mentioned in the discussions of selection and inbreeding appear to have polygenic correlates. In a few studies, attempts have been made to estimate the number of genes operating in the polygenic system. For example, using two inbred strains of mice, it has been estimated that the exploratory activity, as measured in one type of situation, is controlled by about seven genes. Certain morphological characters are shown to be influenced by many genes. In the fruit fly *Drosophila*, for instance, on one chromosome alone, as many as six genes have been found to influence the number of abdominal bristles formed, and all of the other chromosomes are known to contain genes which are effective in this respect, too. There is good reason to expect that at least part of the variance in almost all behaviour will prove to be related to the polygenic diversity that is characteristic of all bisexual species.

4. *Diallel crosses.* A mating scheme that is of value for the study of quantitative traits provides a means for analysing the existence and degree of additive, dominance and similar conditional effects, maternal effects, and interaction between environmental fluctuations and genotypes. This is the *diallel mating system* which involves all possible mating combinations between several genotypes. By statistical analysis it is possible to isolate non-additive genetic components of variation. For polygenic inheritance, diallel crosses provide a powerful analytic tool yielding data otherwise difficult to obtain. Another advantage of this breeding method is that estimates can be obtained from the F_1 (i.e. first filial) generation without the need for subsequent crosses, although additional confirmation can be provided from F_2 or backcross data.

Mutant genes that are known for their morphological effects may also have pleiotropic effects on behaviour. Breeding techniques that place mutant genes in a uniform genetic background permit the study of their contributions to differences in behaviour. Essentially, the breeding methods are similar to those described in the preceding section, although numerous successive backcrosses are required to ensure an isogenic background on which the effects of allelic substitutions at the locus of interest may be assessed.

Animal behaviour. Next, we shall consider behaviour-genetic analyses of several kinds of animals, involving such behaviour as learning, mating, TAXES, hygiene, and COMMUNICATION.

1. *Learning.* One of the best-known experiments in the behavioural science literature is the Tryon bidirectional selective breeding study of maze-learning in rats, begun in 1925 and continued until 1940. Tryon started with a heterogenous foundation population. After the population had been tested in a multiple T-maze, rats were bred selectively on the basis of their scores (errors and time). This was the beginning of the Tryon 'bright' and 'dull' strains, maintained ever since in reproductive isolation. In succeeding generations, brights were mated with brights, and dulls with dulls. The performance of the two strains gradually diverged and by the eighth generation there was little overlap between them. Many aspects of Tryon's elegant experiment have been replicated for the same and different behaviour patterns in the same and different species.

Later, it was also demonstrated that the brights were not superior to the dulls in all other learning tasks. In fact, sometimes the converse was true. However, brights were superior when the new test resembled the original test. These results suggest two things. First, there is no single 'super-gene' for learning in the rat. Secondly, this INNATE ability of the brights tends to be relatively task specific. Other studies concerned with learning in rodents have involved such measures as: shock avoidance, water escape, preference for or aversion to auditory stimuli, discrimination, and several types of maze-learning.

Much effort has been devoted to demonstrating learning in the fruit fly, *Drosophila melanogaster*, because of its two highly desirable properties: a short generation span and prolific reproductive capacity. Also, its genetics has been studied since 1910. More is known about the genetics of *Drosophila* than about that of any other *metazoan* (many-celled) animal with the possible exception of man.

Although the demonstration of learning in *Drosophila* remains questionable, it is now accepted that there is independently replicated evidence for CONDITIONING in another fly, the blowfly, *Phormia regina*. Blowflies have been classically conditioned by associating their unconditional proboscis-extension response to sucrose stimulation with the presentation of a previously neutral stimulus (water or saline). Also, there are reliable individual differences; some flies do better than others. The analogue of Tryon's rat experiment has been performed with blowflies. After conditioning the free-mating unselected strain, flies scoring six or more correct out of eight possible responses were bred together, and those scoring two or less correct responses were bred together. Repeating this procedure every generation, it was found that, over the generations, the average number of correct responses increased for the brights and decreased for the dulls.

2. *Sexual behaviour.* To understand the relations between heredity and mouse SEXUAL BEHAVIOUR, we need a clear picture of the COPULATION sequence. First, the male mounts the female by grasping her flanks with his forelegs. Then, he begins a series of rapid, short pelvic thrusts. These thrusts lead to intromission, which is the insertion of the penis into the vagina. After a series of deep thrusts, ejaculation occurs. This sequence of behaviour has been analysed into sixteen separate components. By comparing three inbred strains of mice, differences were found in the behavioural components. In such events as time to ejaculation, frequency of intromission, interval between intromissions, time away from the female, etc., the results showed strain differences. Next, the inbred strains were crossed to produce F_1 hybrids. It was possible to assign most of the behavioural components to one of the three modes of inheritance: *dominance*, in which the F_1 hybrid resembled only one parental strain; *intermediacy*, in which the F_1 performance fell midway between both parents on the behavioural scale; *over-dominance*, in which the F_1 out-performed the higher-scoring parent. It was not possible, however, to use knowledge from one experiment to predict the results of another.

The sequence of events in *Drosophila* courtship can be divided into five distinct steps. The first consists of the male's attempts to attract the female's ATTENTION, which he does by ORIENTING around her. Next, he begins tapping the female's abdomen or legs with his legs. In this way he 'tastes' her with the receptors on his legs, and might turn away without courting, if he perceives a 'wrong' taste, e.g. from certain other species. Next, the male vibrates the one wing that is nearest to the female's head. Then, he licks the female's genital region. Finally, the male mounts the female and begins to copulate, if she cooperates by spreading her wings and genitalia.

In another investigation two groups of *D. melanogaster* were used that were genetically alike, except that one carried the yellow gene, an X chromosome recessive. Its phenotypic expression is yellow body colour. Since this mutation is both recessive and *X-linked*, a female will have to be homozygous, i.e. carry the mutation in double dose, to express the yellow body phenotype. But, in the homozygous male, which carries only a single X chromosome, the yellow phenotype is expressed with the mutation in single dose. It was found that yellow were slower than normal males to initiate courtship, and that, once started, they took longer to finish. Comparison of components in the sequence showed that normal males spent more time vibrating and less time orienting than did yellow. The question arose whether females preferred normal males to yellow. If they did, the preference might explain the mating speed dif-ference. By counting the number of the female's rejection movements to both the normals and the yellows, this hypothesis was eliminated. Possible alternative hypotheses are that the yellow mutants have lower sexual MOTIVATION, lower motility, or are less responsive in comparison to the normal males. The relationship between the yellow mutation and courtship speed has not been analysed.

3. *Taxes.* Genetic analyses have been made on the taxes of *Drosophila*. In one investigation starting with unselected (*free-mating*) wild-type fruit flies, the response to gravity was measured by the number of turns toward or away from the pull of gravity in a vertical T-maze. A distribution of scores was obtained, some flies making more positive (or negative) responses than others. From the free-mating foundation population, a positive line was started by mating together flies making the most positive responses (approaching the pull of gravity), and a negative line by mating those making the most negative responses. Henceforth, in each generation, though both lines contained individuals with positive and negative scores, matings were made in the positive line between only the most positively scoring individuals, and in the negative line between only the most negatively scoring individuals. In a few generations the mean scores diverged markedly. But they remain separated only while selection pressure is continued. The changes in the distributions of scores resulting from control of the mating system (the response to selection) indicate the presence of genetic correlates for the trait being studied.

Attempts have been made to do similiar genetic analysis with the *phototaxis* response in *Drosophila*. However, the phototactic response is very sensitive to environmental conditions. Such factors as TIME of day, temperature, sex, wavelength, age, state of dark adaptation, and time since last feeding, tend to have an effect on the phototactic response. Because this response is so sensitive to environmental variables, it is important that investigators interested in relations between genetics and phototaxis in *Drosophila* rigorously define and control environmental variables.

4. *Genetics and nest-cleaning in bees.* One of the problems facing the honey-bee is the American foul brood disease (a bacterial infection) which kills its larvae and pupae. If the dead are not removed, the infection will spread. One investigator worked with two strains of bees. One was resistant to American foul brood disease, and the other was susceptible to it. It was noted that the resistant strain would remove infected larvae while the susceptible strain left most of the infected individuals in the nest. This hygienic behaviour is composed of two steps: uncapping the cell, and removing the infected larva or pupa from the cell. (For simplicity, we shall call the resistant strain hygienic and the susceptible strain non-hygienic.) When the investigator crossed the hygienic strain with the non-hygienic strain, the F_1 from this cross resembled the non-hygienic strain. This indicated

that the gene or genes for hygienic behaviour were recessive. The next question was how many genes were involved in this behaviour. To answer this question, twenty-nine drones from the F_1 generation were backcrossed with queens of the hygienic strain. If one gene were responsible for this behaviour, then among the progeny from this cross one-half would exhibit the hygienic behaviour; if two genes, one-quarter, and, if three genes, one-eighth. The results showed that six colonies were hygienic, nine uncapped the cell but did not remove the dead animals, and fourteen were non-hygienic. From these results (six out of twenty-nine) it was deduced that hygienic behaviour was governed by two genes. Also, it appeared that one gene was responsible for uncapping the cell. Therefore, the other gene might be responsible for removing the dead brood. It was found that, if the experimenter removed the cap, six out of the fourteen non-hygienic colonies removed the dead brood. Here is an example of a behaviour that has been analysed into two components, and for each of which a gene correlate has been found.

5. *Cricket song.* Whereas *Drosophila* males tap females to make sure they are the right species, the cricket female (Gryllidae) listens for the appropriate song to find the right male. The male cricket's song is species specific, and a number of experiments have shown that the female can correctly identify the song of her male conspecific. When two species of crickets were crossed, it was found that the F_1 hybrid sang a song that was intermediate between those of both species. And even more surprisingly, the female hybrid is attracted more to her hybrid brother than to either parental strain.

6. *Relations between evolution and behaviour.* In 1848, most of the moths (Lepidoptera) of the various species found in the Midlands in Britain were light-coloured (see NATURAL SELECTION). The proportion of the once rare dark-coloured forms has increased to the extent that they now constitute 95–98 per cent of the population in some regions. The sixteen unrelated species in which this change has occurred share a common behavioural characteristic and a common peril: they rest on tree trunks and they are the prey of birds. With the spread of industrialism and its sooty by-products, the dark-coloured individuals are afforded protective advantages lacking to the lighter ones against the smoke-blackened background of their resting places. The progress of the *industrial melanics*, as they are often called, represents one of the most striking examples of convergent evolutionary change ever directly observed. We can analyse what has happened. A genetically variable population exists in which most individuals are light-coloured, some carrying genes for melanin. Light coloration provides optimum protection against the light-coloured barks; dark-coloured individuals, when they occur, are more vulnerable to PREDATION, and have a low survival rate, i.e. they are ill-adapted to the environment. Now the

environment changes and the adaptive advantages of the two colours are reversed. The darkening background offers protection to darker organisms. Mutations or gene combinations that produce melanics are favoured; such individuals survive and reproduce. The gene pool is changed (see EVOLUTION). An evolutionary change occurs when the hereditary characteristics of a population have changed over successive generations. In this case, the evolutionary change is adaptive, but it need not always be so.

The ADAPTATION in the moth populations was behavioural as well as morphological. Selection operated on those variants which chose less exposed resting places. There are many illustrations of such behavioural shifts, in the nesting behaviour of various species of birds, for example, that function as adaptations to environmental situations. In fact, the evolutionary changes resulting from adaptive shifts can be initiated by changes in behaviour and followed by changes in structure. Behaviour is a factor in evolution as well as an effect of evolutionary change. The two questions, then, to be asked about the relationship between evolution and behaviour are (i) How have behaviour patterns changed in the course of evolution? and (ii) How does behaviour influence evolutionary processes?

Behavioural changes occurring in the course of evolution are studied by the comparison of related species (see COMPARATIVE STUDIES). An excellent example of behavioural adaptation in *Drosophila* has been analysed. *D. persimilis* occupies the cooler, wetter woodlands, while *D. pseudo-obscura*, a closely related species, is distributed in warmer, drier woods. Experimental evidence reveals that the *D. pseudo-obscura* population has responded to selection stemming from desiccation hazards in a number of complementary ways. Examples are waterproofing of the cuticle, timing of activity, selecting and approaching moisture, and avoiding light (*negative phototaxis*). That the negative phototaxis is part of the adaptive complex in maintaining water balance is shown by the fact that increases in temperature or decreases in moisture intensify the photonegative reaction in *pseudo-obscura*, and produce photonegative tendencies in *persimilis* (which is generally photopositive). It has been suggested that these activities are adaptive consequences of the difference in ecological distribution (see NICHE).

Some of the ways in which behaviour shapes evolutionary changes are through SOCIAL interaction, with its effects on population structure. Between populations the development of differences in habitat selection, and in mating behaviour, is significant for other evolutionary divergences, because of the fostering of reproductive isolation.

Conclusion. Behaviour-genetic analysis is relatively new. The analysis of behavioural components has been limited to a few activities in a small number of species. Knowledge of gene interaction is rudimentary, and very little is known about the biological pathways from genes to behaviour. When we look through the literature, however, we cannot ignore the evidence that individual differences in behaviour are somehow related to genetic differences among individuals.

Behaviour genetics conceives of behaviour as an essentially biological phenomenon, in fact, as a most essential mechanism of species adaptation and survival. Every aspect of behaviour that is studied is intimately involved in the ability of organisms to adapt to their environment. Behaviour of individuals and of social groups is a basic mechanism of population survival.

The study of the relations between heredity and behaviour is of fundamental importance, not only for understanding behaviour itself, but also for a more complete understanding of biology. Behaviour studies occupy a unique position as a bridge between the biological and the social sciences. Through behaviour genetics, animal behaviour is linked with the fundamental biological phenomena of reproduction and evolution. The study of behaviour integrates a spectrum of biological subjects, from the mechanics of nucleotide replication to the dynamics of social systems.   D.M.J., J.H.

40, 77

GOAL-directed behaviour appears to be a universal feature of animal life. This does not mean however, that the principles and mechanisms of goal-seeking are the same for all animals. Scientists recognize a number of different ways in which apparent goal-directedness may be brought about, and a number of different ways in which the term goal can be used in describing behaviour.

Behaviour may appear to be goal-directed simply as a result of its having an obvious FUNCTION. Thus when we see a herring gull (*Larus argentatus*) chick pecking at the red spot on its parent's bill, we may be inclined to say that the chick has the goal of obtaining food. What we really mean is that the function of the behaviour is to induce the parent to regurgitate food. The mechanisms responsible for the behaviour may not have any obvious goal-directed feature. The red spot on the parent's bill is a SIGN STIMULUS to which the chick automatically responds when it is hungry. The response of the parent depends upon its state of MOTIVATION, and upon the efficacy of the FOOD BEGGING of its chick. The process of NATURAL SELECTION ensures that those parents which have the most conspicuous red spots, and respond most appropriately to the begging response of the chicks, are most likely to have chicks that survive to adulthood. Similarly those chicks that respond most vigorously and accurately to the appropriate sign stimuli are most likely to obtain adequate food supplies. A shorthand way of saying all this is that the function of the begging response is to obtain food. Once the function of the behaviour is understood, it may appear that

the behaviour is correspondingly goal-directed, but this is really only an alternative way of labelling the function of the behaviour.

Any behaviour that characteristically results in a particular end-point, or terminal state of affairs, may be said to be goal-directed. Suppose we observed that woodlice (*Porcellio scaber*) run around following a disturbance, and characteristically settle in damp places. It may appear that the animal is actively seeking out a particular HABITAT. However, woodlice move about in an irregular manner, but more rapidly in dry than in moist air, with the result that they spend more time in areas of high humidity. They thus achieve a very simple form of goal-directed behaviour.

Any FEEDBACK situation, in which the consequences of the behaviour influence the subsequent behaviour, is a potential goal-directed system. The consequences of FEEDING behaviour, for example, include food ingestion which reduces the animal's HUNGER and leads to the cessation of feeding. It could be said that the goal of feeding behaviour is satiation, but this is merely another way of characterizing the end-point that typically results from the consequences of behaviour.

Some scientists prefer to reserve the term goal for a state that is represented within the animal's BRAIN, and which corresponds to the state of affairs that the animal seeks to achieve. Such a goal-state may be thought of as a *template*, or internal representation of the desired outcome of the animal's behaviour, which can be compared with the actual outcome. For example, SONG learning in many small birds, such as the white-crowned sparrow (*Zonotrichia leucophrys*), is thought to involve a template of the song that is characteristic of the species. Normally, the naïve nestling hears the song of other individuals of its own species, such as its parents, but in the LABORATORY it may be a song provided by an experimenter to act as a model song. This model may be a modified version of the natural song, or it may be the song of another species.

The development of the template of the naïve bird is influenced by the sounds that it hears during a particular SENSITIVE PERIOD of LEARNING, before the bird itself begins to sing. If a bird is surgically deafened at this stage, it never learns to sing properly. So it seems that it is necessary for the bird to hear itself sing. There is then a progressive improvement, in that the VOCALIZATIONS become more and more similar to the natural song, or to the model song provided by the experimenter. The template, established at an earlier stage of development, is now acting as a goal which the animal strives to attain, by comparing its own vocal behaviour with that represented by the template.

Many scientists think that much of an animal's behaviour is controlled by reference to internal goals similar in principle to the song template. Such goals make it relatively easy for us to envisage (i) how animals distinguish between the consequences of their own behaviour, and events of environmental origin (see REAFFERENCE), (ii) how they regulate the various aspects of their internal environment (see HOMEOSTASIS), and (iii) how they make relative judgements about features in the external world (see COGNITION). Other scientists, however, point to the difficulty in distinguishing between behaviour that is organized on the basis of internal goals or templates, and behaviour that has the appearance of being goal-directed by virtue of the action of feedback mechanisms which do not refer to internal goals. In human behaviour the distinction is often made between intentional, goal-seeking behaviour, and habitual goal-directed behaviour which involves no elements of consciousness or intention. The problem is that it is very difficult to distinguish between these two points of view at the purely behavioural level.

It is often convenient to describe behaviour in goal-directed terms even though the relevant evidence may be lacking. This is acceptable provided that it is understood that the description is by analogy, i.e. as if the behaviour were goal-directed. The danger of this approach is that the behaviour may come to be described in unacceptable teleological terms. For example, it is not good practice to say that an animal adopts a cryptic posture in order to evade predators, since this implies that the evasion of predators is a motivational cause of the behaviour, whereas it is really a function of the behaviour. Similarly, it is not acceptable to suppose that an animal possesses mental capacities in excess of those necessary to account for the observed behaviour. In using goal-directed terminology, therefore, care must be taken to avoid teleological argument.

103

GREGARIOUSNESS is the tendency of animals to form social groups, as in the SCHOOLING of fish, the FLOCKING of birds, and the HERDING of mammals. Such groups should be distinguished from mere AGGREGATIONS which result from the attraction of a particular environmental feature, such as a source of food, rather than from mutual attraction. Gregarious animals have distinct SOCIAL RELATIONSHIPS with one another, and they usually have a type of SOCIAL ORGANIZATION that is attuned to their ecological NICHE.

GROOMING is an activity of primary importance to the survival of an animal, and encompasses all forms of care and attention to the body surface, either by an individual or by conspecifics. For every organism there is an interface (the integument) between itself and the environment, and this must be maintained against attack from environmental factors. Passive protection from physical, climatic factors has evolved in most animals, such as the exoskeleton of insects, the scaly covering of fishes and reptiles, bird plumage, and mammal pelage (fur). These protective integuments must be attended to (e.g., feathers and fur must be re-aligned) in order to maintain their maximal efficiency as protective and insulating barriers. Grooming is also the active means by which the animal defends its integument against attack from biological factors, specifically ectoparasites and other micro-organisms. It is assumed that grooming (mainly as scratching) is the main response to sources of peripheral irritation caused by fleas (Siphonaptera), lice (Mallophaga), and ticks (Ixodidae) (see PARASITISM).

Two other primary functions, related to maintenance of the integument, can be identified for grooming. The removal of surface debris from peripheral SENSE ORGANS (e.g., optical surfaces, auditory membranes, olfactory passages, and vibration sensitive or contact organs) must be carried out so that their functioning is not impaired. Finally, as a further barrier against physical factors (dehydrating effects of air, wetting effects of aqueous media), oils and waxy secretions are often employed to reduce surface evaporation or to maintain the water resistant properties of the integument. These secretions must be actively spread over the body surface.

The extent to which NATURAL SELECTION acts in favour of a hygienic and protective function for grooming can best be illustrated by looking at the sophistication of the system that has evolved in the sea urchins (Echinoidea) to meet these pressures. The spiny, calcareous skeleton of the sea urchin is covered with an outer *epidermis*. The area between the spines would make an ideal NICHE for ectoparasites were it not policed by a unique group of sessile, small stalked spines, called *pedicellariae*. (A terrestrial counterpart of the sea urchin, the hedgehog (*Erinaceus*) is well known for its infestations of fleas.) Each pedicellaria is a tiny, three-jawed grapple, elevated on a stalk and totally flexible at its distal end; and each is triggered by contact with foreign matter. There are at least four classes of pedicellaria, differing in their mode of action: some show maintained closure upon stimulation, while others snap continuously; some cut or crush offending organisms; others inject poison or exude chemical, corrosive agents (*enzymes*). The pedicellariae continually sample the body surface, so that invading micro-organisms face a nightmarish forest of incessantly moving mechanical grabs and chemical clamps, each one a primitive grooming appendage. In this manner they provide a highly effective DEFENSIVE system, protecting the body surface against attack from ectoparasites, thereby fulfilling one of the prime functions of grooming.

The three animal groups in which grooming has been studied to any great extent are the insects, birds, and mammals. Apart from brushing against physical objects, fish show no grooming behaviour; however, they are interesting in that a 'cleaning' service may be provided in the case of some tropical reef fish by SYMBIOSIS with another species. These fish attend specific cleaning stations where they will tolerate the presence of cleaner-fish which remove parasites from the scales, mouth cavity, and gills. The cleaner-fish not only gain a free meal but also some measure of immunity from the larger fish.

The grooming behaviour of insects combines many of the RHYTHMS involved in other insect actions, such as LOCOMOTION. The complex sequencing of grooming activities summoning into use the various body segments and appendages has been investigated from a quantitative viewpoint, and evidence for a complex HIERARCHY of control has been discovered.

Amongst the mammals, most attention has been paid to the grooming of rodents and primates. Here a distinction should be made between grooming of an individual by itself (autogrooming) and that performed by a conspecific (allogrooming); in the latter case, the grooming act is usually reciprocal. Self-grooming has been examined in some detail for rodents, whilst the preponderance of SOCIAL grooming in primate species has made it the subject of extensive study in that group. It is in these two contexts that two secondary functions of grooming have been proposed. Rodent grooming occurs in three typical forms. It may appear as prolonged sequences of care and attention to the pelage; as short bouts of directed scratching and attention to specific regions of the body; or as isolated fragments of grooming at seemingly inappropriate times.

Most types of body grooming take the form of nibbling, biting, or licking of the fur, together with teasing or combing with the forepaws; and are often associated with scratching by the hind legs. Grooming of the head and associated sensory structures (nostrils, vibrissae (whiskers), eyes, and ears) is achieved by synchronous brushing with the forepaws alternating with licking of the forepaws. During bouts of extended grooming the face and head are invariably groomed first, followed by an ordered progression through the other body regions: belly, flank, back, rear legs, genitals, and tail. Face grooming often reappears when the animal changes from one major body region to another, or with repetitions of the whole or parts of the sequence. In consequence, during these more protracted sessions, body grooming, and in particular face washing, appear highly organized and stereotyped.

The different forms of grooming are elicited by varying levels of irritation. If an intense irritant is applied to a certain region of the animal's body it responds directly to that source of irritation with directed body grooming and scratching, after which grooming usually ends. But if a mild irritant is applied, such as water droplets, the animal initiates the complete sequence of grooming starting with the head, passing through other regions, before attending to the irritated area. It would appear, therefore, that the long sequences of grooming function as maintenance routines which are performed over and above immediate needs due to the presence of dirt, dust, or other irritants,

although they can be elicited by the general level of irritation due to such sources.

A fragmentary sequence of grooming is called displacement grooming. DISPLACEMENT ACTIVITIES are a diverse group of behaviour patterns which are generally incomplete or reduced in form. When a dominant behavioural tendency is prevented from being expressed, either by being in equilibrium with an antagonistic and equivalent tendency (CONFLICT between states of MOTIVATION) or through FRUSTRATION due to thwarting, certain behaviour patterns are *disinhibited*. They often appear to be out of context and so are also called IRRELEVANT activities. Grooming (or preening in the case of birds) is one of the most frequently performed displacement activities, and it raises the question: why does grooming appear so consistently?

One theory holds that grooming is a 'low priority activity', occurring only when all other behavioural tendencies have been satisfied, at least momentarily; that is, grooming is normally inhibited by other activities. In the specific event of the behavioural tendencies being blocked, grooming becomes the most probable behaviour to appear. Another theory, not incompatible with the previous one, is that the state of high AROUSAL, normally generated during times of conflict, is associated with peripheral manifestations of AUTONOMIC activity, e.g., sweating, flushing, skin tingling or prickling, or pilo-erection (hair raising or goose pimples), and this produces the necessary relevant stimulation for the elicitation of grooming. Studies of displacement preening in birds have shown that its occurrence is facilitated by the presence of rain, which, presumably, acts as a peripheral irritant.

It has been suggested that a possible FUNCTION of grooming in a high arousal situation is to change the animal's level of arousal. Grooming usually occurs in close temporal association with periods of inactivity, rest, or SLEEP; i.e. at times of low arousal. Its performance at times of high arousal may modify the animal's current state of arousal, and it has been argued that if the slightest shift in state occurs then one or other tendency may again be expressed and the impasse curtailed. The mechanism by which this is brought about may be by a direct effect upon the BRAIN, or it may simply be that the performance of grooming diverts the animal's sensory system briefly away from the arousing stimuli, causing a momentary change in arousal.

Two other specialist functions for grooming have been identified for rodents. At specific environmental temperatures the performance of grooming is an important contributor to THERMO-REGULATION. The spreading of saliva over the relatively hairless parts of the animal, snout, feet, ears, etc., enhances heat loss from the skin through evaporation. Grooming occurs when the ambient temperature exceeds 30 °C and is sufficient to maintain the body temperature at around

38 °C, even with an external temperature of up to 44 °C.

Another finding is that rodents produce a PHE-ROMONE from glands in their nostrils, which, when spread over the fur by grooming, acts as a social arousing substance causing an increase in the attention of conspecifics. Grooming occurs upon awakening from sleep; synchrony in this with other members of the colony minimizes the chances of an animal being taken by a predator by dint of being solitary.

Mutual grooming is one of the commonest forms of behaviour during social encounters among primates, and, with the increased ability to manipulate objects with their hands, these replace the mouth as the primary grooming utensils. From the study of a variety of primate species, there is general acceptance that there are three basic functions for mutual grooming. First and foremost is the biological, hygienic role; ectoparasites are regularly removed from the fur, often with the lips, and eaten. Allogrooming is most commonly directed to regions which the animal being groomed cannot itself reach, or visually inspect. Wounds also receive special attention from grooming animals, and there is evidence from other species that saliva has both disinfectant and healing properties. However, if hygiene were the sole function only the groomed animal would benefit, so grooming would need to be wholly reciprocal for it to persist in the long term. It has been suggested as a possible reward for the groomer, that important quantities of salt and vitamin D, secreted at the skin and spread over the fur, are obtained during mutual grooming. This advantage is offset by the fact that the animal loses water by licking the fur, and fur which is ingested must be eliminated from the body as a fur bolus, an operation which is not without hazard.

Of greater importance is the fact that mutual grooming has acquired two valuable social functions. Firstly, it helps to promote and cement relationships between participating individuals (see SOCIAL RELATIONSHIPS). In the case of a male and a female such a bond may be developed for the purposes of MATING, a *mating consort*, while in the case of two females it may enhance maternal support and cooperation. Secondly, allogrooming has become a conciliatory gesture which reduces tension and AGGRESSION between individuals, and may even forestall or placate potential aggressors. Allied to this, grooming demands imposed by dominant animals are a means of maintaining status within the hierarchy: the linear hierarchy within a group of primates can readily be assessed by observing the direction of grooming interactions (see DOMINANCE). The two social functions of mutual grooming, that of strengthening bonds and reducing tension, are generally fulfilled in different contexts, although both may occur between the same individuals and in close temporal proximity.

Most social grooming in primates is directed towards high ranking animals, and to other 'favoured' individuals; the latter generally being receptive females in the case of grooming by males, and mothers with infants or infants alone in the case of grooming by females. In most primate species the females are essentially less aggressive than the males, and the most frequent and longest episodes of allogrooming occur between high ranking females. The majority of grooming takes place between animals of equal status, or those which are of adjacent rank in the hierarchy, when grooming is invariably initiated by the lower ranked animal. However, it is also the case that many mutual grooming relationships are established in infancy between siblings and other close relatives. These relationships tend to persist so that the individuals often have a similar status.

It is possible that it is the future benefits to be derived from the formation of a coalition between grooming individuals that are important for the groomer. Access to another animal, allowing sufficient proximity for reciprocal grooming, implies that a special relationship exists between the two animals. This is most clearly seen between a male and female which have formed a mating consort. In the case of two females, the grooming may act as an investment for eliciting future support; the benefits of the alliance being realized in such contexts as COOPERATIVE child rearing, defence during agonistic encounters or disputes, and the acquisition of mating partners. Allogrooming would appear to be the currency that is used to maintain the close bond or pact between two partners. If this form of support is, in fact, the real advantage gained from engaging in mutual grooming, then, obviously, it is important to form an alliance with as high ranking an individual as possible. This will create COMPETITION for access to high ranking members, leading to the natural consequence that most coalitions will be observed between animals of equivalent status.

Allogrooming that is seen between males is more often of the type that is considered to be 'tension reducing', and commonly acts as APPEASEMENT following aggressive encounters. Such occurrences of allogrooming tend to be shorter than normal, and are reported to be performed in a rapid and exaggerated manner, consistent with their having undergone RITUALIZATION as an appeasement signal. In some instances the aggressor may return the grooming in a manner which seems to convey the message 'I am no longer angry with you, we can resume amicable relations', indicative of the effectiveness of the tension reducing effect. Studies across a number of primate species show that the incidence of mutual grooming usually correlates with the level of aggression, as would be expected if grooming was frequently employed as an appeasement gesture. The exceptions to this rule, e.g. the chimpanzee (*Pan*), suggest that the hygienic role of allogrooming is still important. This is particularly so, as other features of chimpanzee behaviour (e.g. the construction of a new night sleeping platform each night) suggest that the incidence of ectoparasites may have been of evolutionary importance to this species.

Mutual grooming under peaceful conditions is initiated by the individual who wishes to be groomed approaching another animal and 'presenting' the area to be groomed to that individual. If the invitation is not accepted, the presenter may groom the invitee, and this often has the desired effect of stimulating the approached animal to groom. Being groomed would seem to be a rewarding and pleasurable experience, as the groomed animal often adopts an utterly relaxed posture. In contrast, if the grooming is conciliatory, the groomer remains alert and nervous, and is generally the animal which terminates the act. The special attention given to mothers with infants, and to young animals alone, is difficult to explain without a full knowledge of the degree of relatedness between the individuals concerned. It may be an important aspect of a cooperative infant care system.

It has been argued that the oral forms of grooming are derived from similar patterns found during suckling in the nursing infant. An early precedence is set for grooming between close relatives in that the mother cleans her offspring after birth, and continues to groom them until they are able to do so themselves. Furthermore, mutual grooming by the JUVENILE primate develops fully around the time of weaning. Weaning in many primates is a fairly traumatic process with the mother physically rejecting the attempts of the juvenile to suckle, often because another infant needs to do so. In such circumstances, the young animal continues to seek tactile contact with the mother, and allogrooming is the means by which this is achieved, as it is tolerated by the mother.   M.W.

H

HABITAT is the natural home of an animal or plant: the external environment to which it has become adapted in the course of EVOLUTION. Habitats are usually described in terms of salient physical and chemical features of the environment. Associations of species within a particular environment are generally called COMMUNITIES, and the status of an animal in its community, in terms of relations to food and enemies, is generally called its NICHE.

Habitats are, to a large extent, determined by CLIMATE, which has a profound effect upon vegetation. Thus we can distinguish between forest and desert habitats in terms of climate, or in terms of vegetation. Plant communities depend, first and foremost, upon the physical features of the environment. They provide a variety of possible habitats which animals are able to exploit. However, animals are not always able to fully occupy potential habitats because of COMPETITION from other species. Thus we have to distinguish between the TOLERANCE of animals to environmental factors, and their competitive ability in relation to other species.

Each species has a characteristic ability to tolerate extreme values of environmental factors. For example, for desert lizards (e.g. the Namib desert lizard, *Aposaura anchietae*) the environment in which food can be obtained is usually too cold early in the day, and too hot in the afternoon. Many lizards are able to compensate to some extent by basking in the sun during the morning, and confining their activities to the shade later in the day. At extremes of temperature the lizards usually seek protection beneath stones, or underground. Species differ in their tolerance abilities. Thus nocturnal lizards can generally tolerate lower temperatures than species that are active in the daytime. However, some species have the ability to change their range of tolerance through the processes of ACCLIMATIZATION. For example, the tree lizard (*Urosaurus ornatus*) normally dies at temperatures above 43 °C. By maintaining these animals in the laboratory for a period of 7–9 days at a temperature of 35 °C, compared with a more natural temperature of 22–26 °C, scientists have shown that the lizards could survive at temperatures of up to 44.5 °C.

The tolerance abilities of animals depend largely upon their physiological mechanisms and their ability to adjust to environmental changes by means of appropriate behaviour. Compared with mammals and birds, which can maintain a constant body temperature when the environmental temperature changes, reptiles are much more at the mercy of the environment, because they lack the physiological mechanisms required for THERMOREGULATION. Reptiles often have a well-developed ability to adjust to changes in environmental temperature by means of suitable behaviour, but the necessity for such behaviour leaves them less free to engage in other activities. In general, animals that have evolved the ability to regulate their bodily processes by largely physiological means (see HOMEOSTASIS) have greater freedom of behaviour.

Many animals have some ability to select their habitat, and thus increase their chances of survival. In some species such habitat selection is accomplished by means of simple ORIENTATION mechanisms. For example, when common woodlice (*Porcellio scaber*) are placed in a situation where there is a gradient of humidity, they move about in an irregular manner. These movements are more rapid in dry than in moist air, with the result that the animals spend more time in areas of high humidity, thus achieving a very simple form of habitat selection. This simple form of orientation is also shown by road traffic. It is usual for through traffic to travel more slowly in towns than on main roads in the country. This difference in speed causes cars to aggregate in the towns, even though the drivers may have no intention of lingering there. We might suppose that habitat selection in some animals is similarly fortuitous, in that those animals which happen to disperse to favourable habitats survive in greater numbers than those which find themselves in less suitable habitats. This does happen in plants and some primitive animals, but habitat selection in animals is generally a much more active process (see KINESIS).

In considering more complex forms of habitat selection, it is important to distinguish between evolutionary factors conferring SURVIVAL VALUE on a population, and behavioural factors including the mechanisms by which individuals select their habitats. Observation of the changes in plant and animal life resulting from afforestation of the Breckland heaths in eastern England, showed that the distribution of bird species was largely the result of specific habitat selection which restricted each bird to a habitat which was less in extent than that which it was physically or physiologically capable of occupying. It is generally recognized that two species of animals can coexist in a particular habitat only if they differ in ecology. Such ecological isolation is brought about by competition (see ISOLATING MECHANISMS). For example, among European tits (Paridae) several species often coexist in a habitat. Nevertheless, each species is ecologically isolated from every other; in a few cases by geographical range, and in many cases by habitat, or by differences associated with adaptive differences in overall size and in size of beak. The larger species tend to feed nearer the forest floor, and on larger insects and harder seeds than the smaller species. Likewise, species that live in coniferous forests have longer and narrower beaks than those living in broad-leaved woods.

The behavioural mechanisms by which individuals select their habitats are not well understood. However, the evidence suggests that INNATE factors influence habitat selection. For example, hand-reared ducklings enter the water spontaneously, and seek a suitable habitat. Thus young of the common shoveller (*Anas clypeata*) and of the tufted duck (*Aythya fuligula*) seek reeds, and young eiders (*Somateria mollissima*) seek rocky places, in conformity with the natural habitats of their respective species. The deer mouse (*Peromyscus maniculatus*) can be divided into two ecologically adapted types, each showing a preference for its natural habitat. The short-tailed short-eared prairie form shows little difference in food preferences or temperature requirements, when compared with the long-tailed long-eared woodland form. Nevertheless, the prairie form shows a distinct preference for grassland, when given a choice between woods and grassland. Comparisons of the preferences of adult prairie mice and their offspring reared in grassland or in woodland habitats show that the wild-caught grassland animals have a distinct preference for the grassland habitat. This preference becomes weaker if the mice are bred in the LABORATORY for fifteen or so generations, but early experience of laboratory stock in grassland habitat greatly increases their preference for grassland, although the same exposure in woodland did not cause the laboratory mice to select woodland. The conclusion is that habitat selection in the prairie deer mouse is primarily influenced by heredity.

In another study hand-reared chipping sparrows (*Spizzella passerina*), given a choice between pine or oak branches in an aviary, spent much more time in pine, which is their natural habitat. However, when other hand-reared birds were initially provided with oak branches and foliage in the aviary, and only later were given the choice of pine or oak, the balance was tipped towards a preference for oak. This suggests that a genetically influenced preference for pine was modified, though not eliminated, by the type of vegetation in which the birds lived soon after fledging.

It might be supposed that any innate factor in habitat selection is likely to relate to simple features of the environment, but the evidence suggests that birds respond to a summation of many factors and that habitat selection thus has some variability within a species. The factors which seem to be most important are (i) the characteristics of the terrain, (ii) nesting, singing, feeding, and drinking sites, (iii) food availability, and (iv) other animals. For example, when territory-owning great tits (*Parus major*) were removed from a wood near Oxford, England, their places were taken by individuals owning territories in adjacent hedgerows, where they breed less successfully. Before the experiment the hedgerow birds were evidently excluded from the prime territories by the individuals in occupation, but the rapidity with which the vacated territories were reoccupied (1–2 h) suggests that the hedgerow birds are constantly attuned to the situation. Not only do they appreciate that the woodland territories are better, but they soon discover that the owners have left.

27

HABITUATION is an aspect of LEARNING in which repeated applications of a stimulus result in decreased responsiveness. For example, the ESCAPE response of the guppy (*Poecilia reticulata*) to a shadow passed overhead diminishes progessively if the stimulus is presented every 2 min. Eventually the fish does not respond at all. Different aspects of behaviour may habituate at different rates. For example, the FEAR responses given by Northern mallard ducklings (*Anas platyrhynchos*) to a hawk (Accipitridae) silhouette passed overhead diminish with repeated presentations, but one component of the fear reaction, the ORIENTING RESPONSE, persists long after the other aspects of the response have disappeared.

A common feature of habituation is that the habituated response reappears if the stimulus is withheld for a long period of time. Thus the escape response of the guppy to a moving shadow reappears if no shadows are presented for about a day after the response has habituated. If habituation and subsequent recovery of the response is repeated a number of times, then the habituations tend to become successively more rapid. The European toad (*Bufo bufo*) shows ORIENTATION towards potential prey. This consists of a head-turning response that precedes the extrusion of the tongue, by means of which the prey is captured. The orienting response habituates if nonedible prey-like objects are presented repeatedly, and recovers if the stimuli are withheld. Subsequent habituations to the same stimuli are more rapid. If, however, some characteristics of the stimuli are changed, then the strength of the orienting response may recover and habituation may be less rapid.

In general, a response habituated to one stimulus will show GENERALIZATION to another similar stimulus. That is, the animal will, to some extent, treat the new stimulus as if it had been presented previously. Thus habituation to a new stimulus will be more rapid if it is similar to a stimulus to which the animal has previously been habituated. The recovery of the response upon presentation of a new stimulus is called *dishabituation*. Sometimes dishabituation occurs as a response to a stimulus, such as a novel stimulus, which normally elicits a quite different response. Such effects are usually attributed to an alteration in the animal's level of AROUSAL.

The mechanisms of habituation are usually regarded as akin to those of learning. The waning of a response sometimes occurs as a result of fatigue or of sensory ADAPTATION. It is usually possible to rule out these alternatives experimentally; if after the waning of one response the same muscles can be used in another activity, then the waning cannot have been due to fatigue. Similarly, if after the waning of one response to a stimulus, the same stimulus elicits a different response, then sensory adaptation is unlikely to be involved.

Habituation is a widespread phenomenon in the animal kingdom, and its SURVIVAL VALUE lies in the counterbalancing advantages and disadvantages of responding to stimuli that have uncertain significance. Thus if a toad could always discriminate between edible and non-edible objects, habituation to food-like objects would be unnecessary. If, on the other hand, the toad responded to every presentation of a food-like stimulus, it would waste much time and energy attempting to capture inedible objects. Similarly, an animal that shows AVOIDANCE to every disturbing event will not have much time and energy to devote to other aspects of behaviour. On the other hand no animal can afford to ignore potentially dangerous stimuli. The process of habituation provides a compromise between these conflicting pressures of NATURAL SELECTION.

80, 98

HATCHING is commonly associated with the well-known event of a bird embryo cracking its shell, and the later emergence of the wet and weak chick from the egg. This familiar image overlooks two important facts. First, that the larval or embryonic development of many animals other than birds also culminates in release from an egg or other constraining membranes. Second, that the apparently simple event of an animal escaping from the egg requires a complex series of preparatory mechanisms, often involving an interaction between GENETIC, physiological, and behavioural factors, most of which are hidden from us when we merely observe a chick cracking its way out of the shell. Timing is also crucial, since the hatching event must occur at a rather precise moment in the life history of an animal that has to cope with the environmental differences between the embryonic and the post-hatching existence.

In keeping with the common image most of us have of hatching it will come as no surprise that the first systematic observations and experiments on this phenomenon were done with birds, specifically the domestic chick (*Gallus gallus domesticus*). These were made by the well-known French naturalist and scientist M. de Réaumur in the middle of the eighteenth century. In part because of this historical precedent, and in part because of the inherent technical difficulties in experimenting with many other egg-laying animals, most subsequent work on hatching and its biological mechanisms has tended to concentrate on birds. Nevertheless, there are numerous reports in the literature on hatching in both egg-laying invertebrates and non-avian vertebrates. Consequently, before embarking on a more detailed discussion of avian hatching, it may be informative to summarize briefly what is known about hatching and its mechanisms in non-avian forms, including invertebrates.

Invertebrates. Although there are no entirely satisfactory descriptions of hatching in invertebrate animals there are numerous fragmentary reports, some of the better known and more interesting of which are described below.

In some species of invertebrate the embryos secrete a substance just prior to hatching which serves to digest and weaken the egg *envelopes*, thus allowing the embryo to escape. Muscular activity on the part of the embryo probably also plays a contributory role in some of these cases, for example in the common octopus (*Octopus vulgaris*), although there are reports that eventual escape from the shell may occur even if the embryo is immobilized with an anaesthetic. If true, this might be thought of as a kind of passive hatching. The purple sea urchin (*Strongylocetrotus purpuratus*) and the moth *Antheraea pernyi* also use secreted substances for breaking down the egg envelopes during hatching.

Other types of passive hatching among invertebrates include the imbibing of fluids by the embryo, or the uptake of fluids into the egg from the external environment, both of which serve to increase the intra-egg pressure and thereby lead to a weakening or actual bursting of the egg envelopes, allowing the embryo to escape. This occurs in the American stream planaria (*Dugesia dorotocephala*) for example. In the American lobster (*Homarus americanus*) the female apparently induces or assists hatching by the violent shaking of the eggs through vigorous and rhythmic contractions of her *swimmerets*. Many invertebrate embryos possess *hatching spines*, which are sharp cuticular structures that aid in breaking through the egg membranes, either passively, during the swelling of the egg as described above, or by utilizing embryonic movements which cause the spines to tear open the egg.

The use of specialized active hatching movements, independent of chemical substances, hatching spines, or fluid uptake, is apparently quite rare in invertebrates. On the other hand, there are numerous examples of such specialized behaviour being used in combination with the more passive mechanisms. For example, the yellow-fever mosquito (*Aedes aegypti*) uses a specialized head movement which causes a hatching spine, located on the head, to rupture the egg membrane.

In the common household American cockroach

(*Periplaneta americana*) the embryos develop within a tough, brittle communal egg case which usually contains 12–16 embryos distributed in two opposed rows. As the time of hatching approaches one or two of the embryos become especially active, thereby stimulating the embryos lying next to them, who in turn become active and stimulate the embryos lying next to them, and so on in domino fashion until all the embryos within the egg case are active. Their movements also become synchronized in such a way that by a common effort they are able to pry open the egg case and escape, an event that could not be executed by the movements of only one or a few embryos. Thus, here we have an example of not only a specialized, active hatching mechanism, but a COOPERATIVE one as well.

Among the multitude of mechanisms used by invertebrates for escape from the egg perhaps a favourite example is the somewhat anecdotal report by H. Fabre in the last century that the reduvius bug (Reduviidae), when ready to hatch, fills a membranous bag inside the egg with gas which then explodes, lifting the lid off the egg, so allowing the embryo to escape.

Fish. Hatching in all fish embryos studied to date is aided by the development of specialized glands which secrete a substance which digests the egg membranes. In some cases, the digestion may be complete so that little, if any, muscular activity on the part of the embryo is necessary for its escape (e.g. the rainbow trout, *Salmo gairdneri*), whereas in other species vigorous swimming-like movements serve to rupture the membranes, which have been weakened by prior digestive activity of the hatching substance (e.g. the oyster toadfish, *Opsanus tau*). In both types, however, it is interesting that apparently no specialized movements develop to aid in hatching. Instead, during the hatching process, there are merely swimming-like movements, which are similar to the movements seen at earlier developmental stages in these embryos.

Amphibians. Amphibians have been favourite subjects for embryological investigations since the middle of the last century. Despite this, there have been no detailed studies of hatching in either frogs (Anura) or salamanders (Salamandridae), apart from a few reports of hatching glands, and some incidental observations on embryonic movements occurring around the time of hatching. At least some amphibians (e.g. the African clawed toad, *Xenopus laevis*) are known to possess hatching glands which secrete substances capable of digesting the egg envelope. However, these substances apparently weaken only a restricted region of the membrane; vigorous movements of the embryo then permit exit through this weakened area. Embryos that have been paralysed by an anaesthetic are retarded in hatching, though they may eventually hatch, probably due to the increased mechanical pressure from their continued rapid growth inside the egg. Again, as in fish, the movements used during the hatching process

by amphibians are unspecialized, and are similar to the swimming-like movements seen in earlier stages of embryonic development.

In some terrestrial amphibians (e.g. the robber frog, *Eleutherodactylus matrinicenisis*) the eggs are laid on land, and development is entirely embryonic with no larval or tadpole stage; the young hatch out as fully developed tiny frogs. In the robber frog the embryo has a spine-like projection on the tip of the upper jaw (possibly analogous to the *egg-tooth* of reptiles and birds) which apparently helps to slit open the unusually tough egg capsule.

Reptiles. Turtles (Testudines), lizards (Lacertidae), crocodiles (Crocodylidae), and alligators (Alligatoridae) all lay eggs which have a rather tough, leathery calcareous shell, which differs in some respects from that of lower vertebrates and invertebrates, but is similar to that of birds. And like birds, reptilian embryos also possess an egg-tooth located at the tip of the snout which helps to tear an opening in the shell at the time of hatching. Although there have been no detailed studies of hatching in any reptile the available incidental reports suggest that the initial opening in the shell is achieved by specialized behaviour involving head and snout thrusts, whereas the subsequent escape (hatching) from the egg involves more generalized wriggling or struggling. In the few forms that have been systematically examined it appears that there is an interval between the time the initial opening or tear is made in the shell, and the onset of the final escape. During this interval the embryo is relatively quiescent. In some reptiles this interval is only a few hours in duration (e.g. in the mugger, *Crocodylus palustris*), whereas in others it may last up to several days (e.g. in pythons, Pythoninae).

Monotremes. The monotremes are mammals, and, as is well known, the two existing families, the duck-billed platypuses (Ornithorhynchidae) and the spiny anteaters or echidnas (Tachyglossidae), lay eggs. These are incubated in the maternal abdominal pouch for several days, after which time the young hatch and then attach to a nipple within the pouch, where they continue their development in a manner similar to marsupials. The echidna embryo possesses an egg-tooth which is used to tear an opening in the shell, although the actual process has apparently never been observed. Hatching, after a hole has been torn in the egg, has, however, been observed, and it is reported that to get themselves out of the shell the embryos utilize an ambulatory behaviour pattern involving the head and forelimbs. The ambulatory pattern resembles the locomotor pattern of the new-born opossum (Didelphidae, Marsupialia) as it crawls from the birth canal into the pouch.

Birds. As one closely observes and follows a clutch of eggs being incubated in nature, or in an artificial incubator in the LABORATORY, one gets absolutely no clue as to the extraordinarily interesting series of events involved in hatching, until one day one or more of the eggs in the nest reveals a crack, or perhaps even a small hole at one point on the shell. This is called *pipping*, and is illustrated in Fig. A. In most cases, the beak of the bird is visible through this hole, and if the egg belongs to one of the more *precocial species*, such as chickens, or ducks and geese (Anatidae), VOCALIZATIONS may be heard from within the shell.

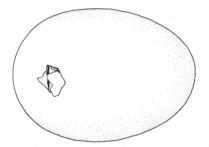

Fig. A. Pipping in a chicken egg. The crack or hole in the shell is at the large end in the region of the air-space.

After an interval varying from several hours to 1–2 days, depending on the species, additional cracks begin to appear in the shell and, starting from the original site of pipping, these always proceed in a counter-clockwise direction around the shell, as illustrated in Fig. B. Within an hour or so the embryo will have broken off the entire shell-cap and escaped from the shell. In some species, such as the domestic chicken, the embryos in a single nest emerge over a period of many hours, whereas in other species like the bobwhite quail (*Colinus*

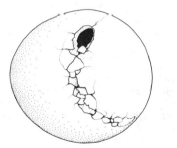

Fig. B. Pipping of the shell during the final hatching process when the shell cap is broken in a counter-clockwise direction around the egg.

virginianus) the embryos are known to synchronize their hatching so that a clutch of ten to fifteen eggs will all hatch within an hour or so of one another. It is this series of events, beginning with pipping and ending with emergence from the shell, which is commonly referred to as hatching in birds.

To students of ETHOLOGY some of the questions that arise concerning this terminal event of hatching are: (i) what are the actual behavioural events involved? (ii) what stimulates hatching? (iii) do all

birds, including the helpless *altricial* song-birds (which need to be cared for for a long period) and the *precocial* ducks (which are ready to leave the nest soon after hatching), hatch in the same way? (iv) do the parents play any role in helping the embryos hatch? (v) are there any special physiological mechanisms involved in hatching? (vi) how is the hatching of birds similar to or different from that of other egg-laying animals? Although at present biologists can provide no final answers to most of these questions, the extent to which the embryo has allowed us to reveal at least some of its secrets can be briefly described.

By making 'windows' in the shell of avian eggs it is possible to observe the embryos without interfering with the normal developmental process. Indeed, by using this technique embryos can be observed and experiments conducted at practically any stage of INCUBATION. In this way it has been shown that during most of the early and middle stages of incubation the embryos are quite active within the egg, and that the movements consist of spontaneous muscular contractions of all parts of the embryo: the wings, legs, trunk, head, etc. are all active. This embryonic behaviour is typically rather lacking in COORDINATION and consists primarily of jerks, twitches and large convulsive-like movements which do not resemble the behaviour patterns of the newly-hatched birds (e.g. pecking, gaping, walking, etc.), and, in fact, if seen in newly-hatched or adult birds would be considered pathological. During the last few days of incubation these uncoordinated movements become more and more infrequent, as indicated in Fig. C, while at the same time a new behaviour pattern begins which is more coordinated and is the first activity specifically related to hatching. The first result of this new pre-hatching behaviour is that the head is raised out of the yolk, and the beak is positioned against the membranes separating the embryo

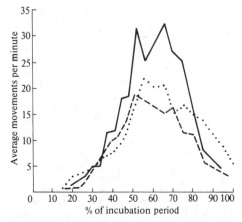

Fig. C. A graph depicting the average embryonic activity levels during incubation for the chick (. . . . .), duck (————), and pigeon (– – – –).

from the air-space, as shown in Figs. D and E. In this position the embryo can pierce through the membranes into the air-space with its beak. There it has access to oxygen and therefore can begin pulmonary respiration, as illustrated in Fig. F. Prior to this time respiration was accomplished by the exchange of gases across the shell and the membranes surrounding the embryo within the egg. A few hours after membrane penetration, the embryo periodically makes vigorous coordinated movements of the entire body, the most prominent component of which is a thrust of the head and beak in a backward direction, causing the beak to strike the shell. It is one of these back-thrust movements which results in pipping (cracking) of the shell (Fig. G); subsequent back-thrusts cause pieces of shell to be broken away, resulting in a pip-hole (Fig. A). In the domestic fowl, where this process has been most carefully studied, the next 15–20 h is a period of relative quiescence; no further cracks appear, and observations of the embryo through windows in the shell reveal that movements are at a low ebb. The same picture emerges from studies of pigeons (Columbidae) and ducks (Anatinae) (see Fig. C). At the end of this quiet interval, however, additional cracks begin spreading counter-clockwise from the site of pipping. Direct observations of the embryo at this time have shown that the embryo has now resumed the same coordinated movements that were seen during pipping; every 10–15 s, on average, one such movement occurs. During each of these movements the entire body of the embryo can be seen to shift slightly in a counter-clockwise direction, resulting in the progressive cracking of the shell around the circumference, as illustrated in Fig. B. After 30–60 min of such stereotyped and periodic behaviour the cap of the shell is sufficiently weakened for the embryo to push it off and wriggle out of the remaining lower half of the shell. With one or two exceptions, all birds which have so far been studied exhibit the same basic behavioural and sequential pattern of hatching. Thus, hatching behaviour is thought to be a rather conservative element in avian EVOLUTION; once an efficient, specialized mechanism was developed it was retained. Furthermore, contrary to common belief, the embryo does not peck its way out of the shell. In fact, movements of the head and beak in just the opposite direction to those of pecking (i.e. back-thrusts) are the critical event in breaking the shell. And finally, there are no known cases in which the parents help the chick by pecking or otherwise breaking the shell open.

Experiments aimed at determining whether avian hatching is triggered or stimulated by HORMONES have so far given negative results. Although hormones, such as *thyroxine*, are known to play an important role in developmental processes in general (see ONTOGENY), in no case has a hormone been shown to selectively or specifically induce or mediate the coordinated behaviour involved in hatching; and, unlike some other egg-

Fig. D. Through an artificial opening ('window') in the shell over the air-space the beak of the embryo (chick) can be seen pushed up under the membranes.

Fig. E. The hatching position of a chick embryo approximately 1–2 days prior to emergence from the shell. m, membrane; rw, right wing; ts, tarsal joint of leg.

laying vertebrates, hatching substances which weaken or completely dissolve the egg-capsule or shell have never been found in birds.

Although nervous mechanisms involved in avian hatching have received little attention from biologists, it is known that certain parts of the BRAIN are more essential than others for normal hatching. For example, whereas embryos lacking both cerebral hemispheres can hatch normally, removal of more primitive areas of the brain retards or completely inhibits hatching in the domestic fowl.

The mechanisms responsible for the rather precise timing of hatching are still largely unknown. If avian eggs are incubated in the laboratory, and thus isolated from many of the environmental events present in nature, they nevertheless hatch at about the same time as they would in the nest, suggesting that there are mechanisms intrinsic to the embryo which are largely responsible for the timing of hatching (see CLOCKS). Unfortunately, there have been very few experiments aimed at this question. It is known, however, that in quail (Coturnicini), which synchronize their hatching time so that all embryos hatch within an hour or two of one another, environmental stimulation between the eggs in the nest plays a crucial role in the synchronizing process. Eggs which are isolated

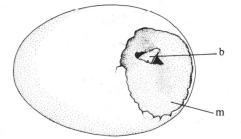

Fig. F. The beak (b) of a chick embryo can be seen to have penetrated through the membranes (m) into the airspace.

Fig. G. Position of a chick embryo at the time when pipping occurs as viewed through a window in the air-space region. The arrow illustrates the direction of head and beak movement (back-thrust) during pipping.

during incubation, or simply prevented from contacting one another mechanically as they would in the nest, hatch over a period of several hours rather than the normal 1–2 h. Sounds ('clicks') produced by the embryos during the last few days of hatching have been shown to be the critical factor here. Even in this example, however, the isolated quail eggs never hatch more than 1 day early or late, suggesting that intrinsic controls are probably never entirely overridden by environmental or extrinsic mechanisms.

In summary, it can be seen that birds have developed a rather complex, specialized behaviour pattern for freeing themselves from the egg-shell, which represents one of the few examples among vertebrates where a behaviour pattern, as opposed to a chemical substance or some other passive mechanical mechanism, has been developed for the apparent sole purpose of hatching. The behaviour is transient, appearing for a day or two, and then disappearing, never to be used again. R.W.O.

113

HEARING, the awareness and appreciation of sound, is used by animals in diverse ways, and is dependent upon both the complexity of their sound-receiving and sound-producing systems,

and upon their specific biological needs. When we examine, successively, moths, crickets, frogs, birds, and mammals, we notice that auditory receiving systems progress from simple two-celled units to structurally elaborate peripheral organs. Moreover, these peripheral mechanisms are complemented by central nervous systems of greatly increasing complexity and capacity. The animals' use of meaningful sounds progresses in like manner from mere PERCEPTION of the environment, through highly complex intraspecific vocal COMMUNICATION to the exotic capability, by some birds and mammals, to hunt and navigate by ECHO-LOCATION.

Sound is defined as rapid, minute changes in pressure within the surrounding medium (e.g. air), which originate from a vibrating source and propagate outwards in waves. Although the simplest sounds are continuous pure notes (such as those produced by tuning forks), natural sounds of biological significance are usually much more complex, and can be best described as either a harmonic series or an aperiodic noise. The strength, or intensity, of a sound is expressed on a logarithmic scale as decibels of sound-pressure level (dB SPL). Decibels thus compare the ratios of acoustic pressures between two sounds (e.g. 6 dB signifies a ratio of 2:1, and 20 dB a ratio of 10:1). Generally, intensity is compared with a standard reference pressure of 0.0002 dyne/cm² which is known as 0 (zero) dB SPL. The use of a logarithmic scale allows one to cover conveniently the great range of pressures involved: the range of human hearing from the threshold of detection to the threshold of pain encompasses a range of 120 dB, which represents a ratio of pressures of 1:1 000 000. The frequency of a sound, e.g. of a pure note, is the number of complete pressure cycles occurring in 1 s. The unit of frequency is the Hertz (Hz) or the kiloHertz (kHz) representing, respectively, one full cycle and one thousand full cycles per second. The lowest note on the piano is 30 Hz; the highest is 4.1 kHz.

Although the auditory systems of animals show considerable variation in their structure, certain universal properties are evident. Among them are a peripheral device for converting sound pressure to vibratory motion, and a specialized SENSE ORGAN for converting this motion into nerve impulses which are coded to represent significant aspects of the original physical signal. This neural code, progressively refined and transmitted to higher levels of the central nervous system, is the true substrate of hearing.

Hearing in invertebrates. Moths (Lepidoptera) have the simplest known peripheral auditory system, yet they are capable of deriving meaning from environmental sounds and responding appropriately. Like all animals discussed in this article, they have two 'ears', a trait which allows for directional hearing. Each ear is composed solely of a *tympanic membrane* and two acoustic receptor cells which are embedded in a strand of tissue attached to it. The nerve fibres of the two receptors, after joining with that of a single non-auditory fibre, form the moth's *auditory nerve*.

Moths are able to discriminate between faint and intense sounds, and, though their peripheral auditory systems are activated by frequencies from below 10 kHz to over 100 kHz, they are apparently 'tone deaf', since they do not exhibit behaviour which would indicate their ability to use this information. Their broadly-tuned systems are, however, capable of responding to rapid temporal changes, such as the very brief pulses of sound in the cry of a bat (Chiroptera). Indeed, the moth uses hearing to detect these cries and is able to determine, in a gross way, both the distance and the direction of the source of the sound. If the bat's cries are weak and distant, the moth will fly directly away from the source. If, however, the cries are close and intense, the moth responds in a somewhat unpredictable manner, either looping or dropping with its wings folded, or making a power-dive towards the ground. The moth's use of hearing for purposes other than escape is not evident.

The peripheral auditory system of crickets (Gryllidae) comprises forty to fifty receptor cells attached to two tympanic membranes on each front leg. The cells are divided into low-frequency receptors which are attached to the larger tympanum, and high-frequency receptors attached to the smaller. The cricket is thus capable not only of intensity analysis, but also of a rudimentary frequency analysis of sounds. Further, and most important, the cricket's auditory *neurones* respond to the temporal properties of cricket SONG, especially to the rate and to the duration of the pulses of sound contained within them. Two groups of secondary neurones are responsible for analysing the temporal characteristics of the acoustic signals. One group responds consistently to the chirps within a given song, while the second group marks each sound pulse within a chirp. Both groups of neurones are apparently connected to the low-frequency receptors, and therefore are able to respond to features within a cricket's calling song and to transmit the information to the central nervous system. Those neurones connected to the high-frequency receptors do not preserve the temporal structure, but do, none the less, respond briefly following the onset of high-frequency sounds.

Crickets produce several types of song, including ones for COURTSHIP, AGGRESSION, and calling (ADVERTISEMENT). The latter is composed of chirps made up of discrete pulses, and also trills produced at a frequency of near 5 kHz. The timing of the pulses (the pulse rate) and the duration of the chirp itself are determined by GENETIC means, while the pulse duration, chirp rate, and sound intensity are a direct function of immediate environmental factors. The female of the species identifies the calling song of a conspecific male and responds by approaching the source of the sound (positive *phonotaxis*). Species recognition is paramount, and depends upon the capability of the crickets' auditory system to process the two invariant properties of song mentioned above. All processing for recognition of the calling song seems to be performed by the low-frequency neurones. The high-frequency neurones are receptive at the frequency of the courtship song (near 15 kHz) which is sung by the male after he has been located by the female, and they are ready to mate. The neurones successfully encode the repetition rate both of chirps and pulses, as well as the duration of each individual pulse. Since the high-frequency receptors are broadly tuned however, they may also signal the existence of bat cries, resulting in negative phonotaxis, or an ESCAPE response by the cricket, somewhat like that of the moth.

Hearing in vertebrates. For convenience of description, the peripheral device of vertebrates is divided into three portions: an outer, a middle, and an inner ear. The outer ear, found especially in mammals, serves to collect and funnel sounds to the middle ear. The middle ear is composed of at least a membrane (the tympanum, or ear-drum) which is set into vibration by sound pressure changes from the environment. In all but the simplest systems, the middle ear includes an additional mechanism consisting of from one to three tiny bones (the *ossicles*) which conduct the vibratory motion to the inner ear. The inner ear is a fluid-filled bony capsule, containing a specialized organ made up of a supporting structure, along which are arrayed numerous receptor cells.

The auditory system of the frog (Anura) illustrates many evolutionary precursors to those of birds and mammals. Frogs have large tympanic membranes which are visible at the surface of each side of the head. Three linked ossicles conduct the vibrations of the tympanum to the inner ear, which consists of two separate auditory organs as well as six *vestibular* (balance) structures. The auditory organs resemble, in a primitive way, those found in mammals, and consist of receptor hair-cells arranged along a supporting structure in such a way that they are in direct contact with a thin membrane lying above them. The entire organ is surrounded by fluid and encased by bone as a closed system. Sound vibrations, transmitted by the ossicles to the fluid of the auditory organs, cause differential movements across the hairs of the receptor cells. These cells are thence in direct contact with the neurones which form the auditory nerve.

The neurones of the frog's auditory nerve have been determined to be of two general types: 'simple' neurones, which respond to frequencies between 1.0 and 1.5 kHz, and 'complex' neurones, which are sensitive to frequencies between 300 Hz and 1.0 kHz. These neurones originate separately from the two auditory organs, and their classification refers principally to the fact that the complex neurones are not only activated by sounds of certain frequencies, but can also be inhibited in their activity by sounds within their particular frequency response range. The two frequency ranges seem well matched to the distributions of acoustic energy in the frogs' mating calls, and the fine tuning of these systems, which are different between different species, facilitates reproductive isolation (see ISOLATING MECHANISMS), since the auditory systems (neurones) of one species do not respond to, nor can they be stimulated by, the sounds in the calls of a different, yet *sympatric* (i.e. having an overlapping HABITAT), species.

Unlike the crickets, the frogs have a rudimentary vocal apparatus with which they produce calls having both harmonic structure and regions of resonance similar to those of human vowel sounds (see VOCALIZATION). These calls include, in addition to one for MATING, calls for both warning and distress, a call for TERRITORY ownership, a release call, and a rain call. A 'model' call, produced artificially in the LABORATORY with acoustic instruments and containing the appropriate low- and high-frequency resonances and a harmonic structure, will consistently evoke answering calls from a listening bullfrog (*Rana catesbeiana*). If just a single one of the frequency components is presented however, no reply can be elicited from the listener.

Birds. While the previously discussed animals possess specialized receptor mechanisms which are limited to discrete frequency bands, the birds exhibit a 'general purpose' peripheral ear. Although its anatomy is not remarkably more complex than that of the frog, the auditory capabilities of birds often approach, and even rival, those of mammals. Birds have a simple external auditory canal covered by feathers. The conductive mechanism resembles that of the frog, consisting of a tympanum connected to the auditory organ by a single ossicle, called a *columella*. Birds also have a single small muscle attached to the columella which, when contracted, decreases sound intensity to the inner ear. The actual organ of hearing, the *auditory papilla*, is like the *basilar papilla* of the frog; it is shorter and wider than the *Organ of Corti* found in the cochlea of mammals, though it does contain a similar number of hair-cells.

Some birds, notably the owls (Strigiformes), use their hearing as an aid in HUNTING. Owls are able to hunt at night not only because of their keen nocturnal VISION, but also because they can accurately determine both the direction and the distance of the source of sounds generated by the prey. This ability stems from the asymmetry in the size and placement of their external ears, and endows the owls with a passive 'sonar' system.

Song-birds have a continuous frequency range of hearing from below 500 Hz up to about 6 kHz, and their sensitivity to sound depends upon its frequency such that their hearing is generally most acute near 3–4 kHz. Their wide frequency band encompasses the range produced in the vocalizations of their own particular species, typically between 2 and 5 kHz in the house sparrow (*Passer domesticus*), for example. Since most song-birds do not themselves vocalize at frequencies below even one kHz, their low-frequency hearing is probably used in the detection of noises, such as those produced by stalking predators. Although birds' auditory sensitivity certainly varies across species, some have hearing which is as sensitive as that of human beings; and while their total range of hearing is not particularly impressive when compared with that of many mammals, the birds' ability to resolve closely-spaced acoustic events is indeed prodigious. This ability is evident whether or not the species in question actively produces rapid sequences of sound in its own vocalizations. Many bird songs must, in fact, be artificially slowed down (thus stretched out in time) in the laboratory before their temporal properties can be fully appreciated by the human listener.

The vocal behaviour of song-birds consists of simple calls and one or two complex songs. The calls are relatively uncomplicated, and have straightforward meanings (an ALARM call simply indicates the presence of a predator), while the songs are structurally elaborate, and are used, especially by the male, in individual and species recognition, in the establishment of territorial boundaries, and in the maintenance of the bond of a mated pair of birds. In many species, the male produces a single complex song which is used for all these purposes, and a truly great variety of songs exists, since they not only serve in species isolation, but also must allow for individual variation within a given species. Beyond this, members of the same species reared in different localities will often produce recognizably different songs, a prime example of vocal DIALECT in animals.

The key feature of bird song is thus the conveyance of information which is relatively rich in meaning, and which, through the sense of hearing, depends upon the birds' capacity for auditory analysis, IMITATION, and LEARNING. Many species show the ability both to learn and to imitate sounds to a degree not found in any animal save man. In some species of tropical birds, a bonded pair will sing, in alternate notes, a duet, which song serves to maintain contact between the singers in areas of limited visibility. When one of the pair initiates that song, the other responds almost instantaneously; each member of the pair is further able to sing the partner's part in the mate's absence, indicating the learning of a complex acoustic pattern seldom, if ever, used before by the singer. Birds such as the grackles (*Quiscalus*) and parrots (Psittacidae) are well known to imitate both human and non-human environmental sounds, an ability which, when coupled with the existence of SENSITIVE PERIODS in the learning of bird song suggests similarities to the development of speech in children.

Mammals. All land mammals share a similar peripheral ear which is far advanced in anatomy and physiology from that of the other animals. It can properly be considered as a wide-band, general purpose receiver. There is a significant outer ear, or *auricle* (also called a *pinna*) which, in many species, is capable of being directed towards a source of sound as an aid in localization. The *middle ear* is more complex than that of the bird or the frog, and contains three small ossicles, attached to the tympanum, arranged and balanced to transmit sound vibrations with high fidelity over a wide frequency range to the inner ear. This transmission is modified by two small muscles which contract reflexly to loud sounds, and which more specifically serve to decrease the intensities of low frequency sounds, such as those generated when the animal is eating. The reflex contraction to very loud sounds is coordinated with muscles of vocalization, so that its net effect is to afford a measure of control of the auditory system in protecting it from the acoustic interference caused by intense, self-generated vocalizations, like those produced by the bat.

The inner ear of mammals has evolved into a three-chambered spiral structure (the *cochlea*) having between two and four-and-a-half turns, depending upon the species of mammal. The middle chamber of the fluid-filled cochlea contains the Organ of Corti which rests upon a basilar membrane and exhibits, along its entire length, two distinct groups of hair-cells. The hairs protruding from these cells, in contact with a different membrane lying immediately above them, are responsible for converting the wave motions transmitted to the hearing organ into the neuroelectric signals which are sent to the BRAIN. These neural events are ultimately interpreted by the brain as auditory sensations.

The mammals' hearing is characterized by high sensitivity and a wide frequency range. For man, the region of highest sensitivity is between 1.0 and 4.0 kHz, and the audible frequency range extends upwards to approximately 20 kHz. The chinchilla (*Chinchilla*) possesses similar sensitivity and range; the range of the guinea-pig (*Cavia aperea porcellus*) extends slightly higher, though its sensitivity is poorer by perhaps 15 dB; cats (Felidae), dogs (Canidae), and monkeys (Simiae) have more sensitive hearing at the higher frequencies than does man, and, additionally, the upper limit of their range of hearing is extended to 30–40 kHz. The bats are exceptional, being noted for both their ultrasonic hearing and their ultrasonic vocalizations. Most bats have a wide range of hearing without any areas of particularly fine frequency sensitivity. The exact range varies; for example, the little brown bat (*Myotis lucifugus*) and big brown bat (*Eptesicus fuscus*) have auditory sensitivity from 12 kHz to 120 kHz, and from 2.5 kHz to 100 kHz, respectively.

Activities which are governed strictly by the sense of hearing are far less often found in mammals than in the other animals mentioned earlier. Phylogenetically (i.e. evolutionarily), behaviour has now come to represent the outcome of increasingly complex syntheses of information derived, over time, from many integrated sensory inputs. Certainly mammals use their hearing to interact closely with their environment especially in prey–predator situations, and for intraspecific COMMUNICATION across distance or when the use of other senses, such as vision, is precluded. However, the mammals' most important use of hearing is within closely-knit SOCIAL frameworks where communication necessarily involves the combinations of several sense modalities, and where acoustic signals in isolation may have no particular meaning to the recipient.

Perhaps the richest examples of intraspecific communication in land mammals are found in the social behaviour of monkeys. A remarkable aspect of their communication is the use of vocalization, and the repertoire of calls of many New World monkeys, e.g. the squirrel monkey (*Saimiri sciureus*), exhibits a wide variety of utterances which are readily distinguishable even to human ears. Variations among these vocalizations are dependent upon social factors, such as the monkeys' status within a group. Vervet monkeys (*Cercopithecus aethiops*) produce as many as thirty-six distinct utterances in numerous and diverse situations. The vocalizations of the gelada baboon (*Theropithecus gelada*) have been studied in a manner similar to that used in the characterization of human speech sounds; these monkeys produce three different voice qualities, and a wide range of both vowel and consonant-like sounds. While direct evidence for the monkeys' use of each of these natural sounds is lacking, we do know that monkeys can be trained to differentiate between many different types of sound including some of those occurring in human speech.

The bats use their highly developed auditory system for ORIENTING themselves in space and for hunting by echo-location, somewhat like the owls. In contrast to the owls however, the bats have an active sonar system with which they themselves produce ultrasonic cries, and subsequently hear the time delays and intensity changes of the reflected sound waves. The bat's echo-location abilities are truly impressive: when required to fly through a laboratory obstacle course composed of fine vertical wires spaced across the middle of a room, the bat can successfully navigate around wires whose diameter is only ⅛ millimetre, being aware of those wires from a distance of at least one metre. They can perform this feat while blindfolded. From observations made in natural surroundings, it is clear that the bats' sonar performance is very resistant to potentially interfering noises as might be expected, for example, in a cave filled with other bats. Some bats are capable, strictly through echo-processing, of drinking while in full-flight, skimming at high speed above the water just near enough to its surface to dip into it with their lower jaw. There are fishing bats which not only fly smoothly and expertly over the water, but which furthermore can detect, and seize with their hind legs, a fish swimming below the surface.

Many bats feed on flying insects, such as the fruit fly (*Drosophila*), mosquitoes (Culicidae), and, as mentioned, moths. Since the moth is keenly aware of the bat through hearing its ultrasonic cries, the bat perforce must contend with the moth's early-warning system in attempting to obtain its food. Such is a perfect example of the use of hearing in two contrasting ways for SUR-VIVAL, and the simple two-celled peripheral ear is a strikingly successful match for the sophisticated sonar system.　　　　　　J.H.D., L.K.H.

21, 66, 100, 122

HERDING in mammals, like SCHOOLING in fish and FLOCKING in birds, is a form of SOCIAL ORGANIZATION which confers advantages, usually in connection with PREDATION. Predators may benefit from HUNTING in groups, particularly when their hunting strategy involves COOPERATION, and their prey may form herds as a means of DEFENSIVE behaviour.

The members of a herd benefit from the VIGIL-ANCE of their companions, in the sense that many pairs of eyes are better than one. This is particularly true when the mode of FEEDING is such that detection of predators is difficult without interrupting feeding. A grazing animal has to look up periodically to scan the environment for possible danger, but if the animal is feeding in a herd it does not have to do this so often, and can thus spend more time feeding.

Members of a herd sometimes cooperate to deter a predator. In response to attacks by wolves (*Canis lupus*), for instance, musk oxen (*Ovibos moschatus*) gather into a defensive formation which presents an array of horned heads to the predator and protects the more vulnerable animals. Similarly, eland (*Taurotragus oryx*) cooperate in protecting their young from the attacks of the spotted hyena (*Crocuta crocuta*).

Lions (*Panthera leo*), hyenas, wolves, and wild dogs (*Lycaon pictus*) usually capture only one member of a herd during a given hunting expedition. The larger the herd the lower the probability that a given individual will be the victim. Moreover, it is generally the very young, old, or infirm animals that are selected. Thus a young healthy animal derives a considerable benefit from being a member of a herd, because the predators' attentions are more likely to be directed against others.

The structure of a herd or group is intimately related to the ecology of the animal. Predators that practise cooperative hunting generally form small groups in which the SOCIAL RELATIONSHIPS are long lasting. The same is true of other species, which may have systems of cooperative defence or FORAGING, as in most primates. A consequence of small stable groups is that there is a tendency for inbreeding, which is genetically deleterious, often resulting in deformed or unviable offspring. In most species, however, incest is relatively uncommon, because young animals tend to leave their natal group and join or found another. In some species, including hyenas, lions, baboons (*Papio*), and most other primates, it is the males

that leave. Among wild dogs and chimpanzees (*Pan*) it is always the females. Young male lions are evicted from the pride by the dominant males, which tend to monopolize mating opportunities. There is no active incest avoidance in lions in CAPTIVITY, and in prides in which young lions have been permitted to remain in the group mating with close relatives does occur. In wolves, however, littermates fail to show sexual interest in each other. In some species which form larger herds there is a tendency for young males or females to leave parents and congregate in subgroups. In wildebeeste (*Connochaetes*) and some deer (Cervidae), for example, juvenile males form separate groups which disperse when mating opportunities arise, but may re-form afterwards.

TERRITORY formation often occurs in large herds of antelope (Bovidae), even though they may be continually on the move. Ugandan kob (*Adenota kob thomasi*), for example, may appear to be scattered over the plain in a disorderly fashion. In fact, there will be several distinct groups as well as a number of solitary animals. The groups may be made up of males of assorted ages, or of females escorted by one or two mature males. The single animals are the territorial males, each holding an area of 500–900 m². The territories are defended against rival males, usually would-be territory holders from the bachelor groups. Female kob are attracted to the territory where the male gives a stiff-legged DISPLAY. After mating takes place the female rejoins her group. Thus the male territories make up an arena or LEK, a form of social organization more commonly found among birds.

Herd organization varies considerably from species to species. Wildebeest herds are constantly on the move, and the bulls stay outside the main herds, defending their immediate surroundings wherever these happen to be. Impala (*Aepyceros melampus*) are not territorial, but the males become strongly attached to a particular HOME RANGE. There are two main kinds of herds, *bachelor herds* made up of males, and *harem herds* made up of females usually under the control of a single male who attempts to prevent females from deserting his harem. Males compete for the possession of females, although the FIGHTING tends to be ritualized.

The new-born antelope is usually quick to reach the stage when it can run with the herd. A young wildebeest may be ready to run alongside its mother within 5 min of its first attempts to stand. In some species, such as topi (*Damaliscus lunatus topi*), females with young calves form separate nursery herds. Female Thomson's gazelle (*Gazella thomsoni*) form small groups with their young and stay in the same locality for about 4 days at a time. They do not rejoin the main herd until about 3 months has passed, and lactation is complete. Bushbuck (*Tragelaphus scriptus*) and duikers (*Cephalophus*) hide their young and leave them in seclusion, and steinbok (*Raphicerus campestris*) keep their fawns safely in underground holes.

Young impala remain in cover for the first few days, with one or two adults standing guard. Even when fully capable of running, the young of many species tend to hide from danger rather than fleeing. Predation on the young can be very severe, and up to three-quarters of all wildebeest calves may be lost to predators, mainly hyena.

When the young are especially vulnerable, one of the best defences against predation is synchronized breeding. This is found, for instance, among wildebeest and Thomson's gazelle. By being born at about the same time of year as others of the species, the young antelope's chances of surviving the critical first few months of life are improved. So many juveniles are produced at once that the predators cannot possibly kill them all. There is thus a strong pressure by NATURAL SELECTION to produce synchrony of mating within the herd, and this in turn affects other aspects of the social organization, such as MIGRATION and choice of HABITAT throughout the year.

HIBERNATION is a form of winter dormancy that is characterized by a slowing of metabolic processes and a marked fall in body temperature. Changes in THERMOREGULATION generally occur on an annual basis (see RHYTHMS) in anticipation of changes in CLIMATE. True hibernation should be distinguished from types of dormancy, such as that shown by the brown bear (*Ursus arctos*) from Europe and the black bear (*Ursus americanus*) from America, in which the body temperature may fall to about 30 °C, and for which body temperatures below 15 °C are lethal. True hibernators are able to allow their body temperature to fall to that of the surrounding air, which may be as low as 2 °C. For this to be possible they have special physiological mechanisms which enable them to survive at body temperatures that would be lethal in other species.

Hibernation, in the sense outlined above, occurs in insects, molluscs, amphibians, and reptiles. However, these animals are cold-blooded and are therefore obliged to enter a state of torpor when the environmental temperature falls. Some scientists question the validity of using the term hibernation in such cases, and prefer to regard them as a form of seasonal ACCLIMATIZATION. Thus terrestrial invertebrate animals of arctic regions often exhibit cold-hardiness as a form of acclimatization to winter conditions. Many insects avoid freezing by *supercooling*, a physical phenomenon by which the body fluids freeze at a temperature well below 0 °C. In some insects the extent of supercooling depends upon the degree of acclimatization, which progressively alters the chemical constitution of the body fluids. Thus the cold-hardiness of the Canadian braconid wasp, *Bracon cephi*, is closely related to the concentration of glycerol in the body fluid, and the freezing point may be lowered to about −46 °C. Some insects, particularly butterfly and moth (Lepidoptera) larvae, are able to withstand being frozen to

the point of brittleness. Such a state is often associated with the *diapause*, a resting stage of the life cycle in which development is arrested, and there is enhanced resistance to heat and cold.

Of the warm-blooded animals true hibernation is primarily characteristic of mammals, though it may be said to occur in some birds. Many birds can enter a state of torpor in which their body temperature falls to low levels. The white-throated poorwill (*Phalaenoptilus nuttallii*) and the nightjar (*Caprimulgus europaeus*) can endure body temperature of about 6 °C for many hours without ill effect. Many species of humming-bird (Trochilidae) show periods of torpidity in which their body temperature falls to that of the environment, although temperatures below 8 °C are generally lethal for humming-birds. In the poorwill the body temperature generally remains slightly above that of the environment, even when the environmental temperature is slightly below 5 °C. However, the lowering of body temperature to dangerous levels does not produce arousal from torpor, as it does in certain hibernating animals. Torpor usually lasts only a few hours in birds, and appears to be confined to the inactive phase of the daily cycle, or to be a response to low food availability. Most birds that can enter torpor, and arouse spontaneously, feed on insects or nectar, both of which may become temporarily unavailable. Energy expenditure is greatly reduced during torpor, and it thus appears to serve as a kind of emergency measure, designed to husband reserves during periods of food shortage. Seasonal dormancy, although well known in mammals, is known in only a single bird family, the goatsuckers (Caprimulgidae). However, our knowledge is very incomplete, due to the few observations that have been made on torpid birds in nature.

Among the mammals, hibernation is found in species belonging to many different groups, including monotremes, marsupials, rodents, insectivores, and bats. True hibernation is found only in the small-sized species although dormancy is found in some large species, such as bears, the North American racoon (*Procyon lotor*), and the European badger (*Meles meles*). Because of their proportionately large surface, small animals cool more readily than large ones; they also warm up more quickly, due to their small thermal capacity.

The monotremes are extremely primitive mammals, which have imperfect thermoregulation. The Australian short-nosed spiny anteater (*Tachyglossus aculeatus*) appears to hibernate for periods of up to three months, with interspersed bouts of activity. The duck-billed platypus (*Ornithorhynchus anatinus*) shows similar periods of dormancy, and both these animals retreat into their burrows during hibernation. Among the marsupials, the North American opposum (*Didelphis marsupialis virginiana*) is thought to be a true hibernator, as is the koala bear (*Phascolarctos cinereus*) and the narrow-footed marsupial mouse (*Sminthopsis crassicaudata*). A large number of hibernators are found among the rodents, and these include the woodchucks (*Marmota monax*), ground squirrels (*Citellus*), dormice (Gliridae), and hamsters (Cricetinae). Two families of insectivores are known to contain hibernators. These are the tenrecs (Tenrecidae) and the hedgehogs (Erinaceidae). The bats (Chiroptera) contain a large number of hibernators, including the serotine bat (*Eptesicus serotinus*), the long-eared bat (*Plecotus auritus*), the noctule bat (*Nyctalus noctula*), and the little brown bat (*Myotis lucifugus*). There are also a number of mammalian species about which there is a difference of opinion among scientists as to the status of their dormancy, or torpor. Many species show some torpor during the night, which is accompanied by a fall in body temperature. Other species show periods of dormancy during which the fall in body temperature is not pronounced, as compared with true hibernators. There may well be some species in which hibernation has yet to be observed.

The characteristics of hibernation. Hibernation is characterized by a SLEEP-like state, in which the rate of heartbeat is lowered, and breathing is slowed. Hibernating animals often adopt a sleeping posture, and choose a typical sleep-site in which to hibernate. During hibernation the body temperature is lowered, and energy expenditure is greatly decreased, often below that of normal sleep. Some hibernators, such as ground squirrels, accumulate body fat prior to hibernation, while others, such as hamsters, do not accumulate fat, but instead store food and build a nest. Many hibernating animals awake periodically, some to eat and drink.

Hibernators differ from non-hibernators in a number of respects. Their thermoregulation is characterized by an ability to set the target body temperature at a low level. Thus hibernators continue to regulate their temperature, but at a point only just above that of the environmental temperature. The body temperature of some hibernators may fall as low as 2 °C, and the body cells of such animals appear to have an enhanced ability to maintain activity at such low temperatures. Hibernators also have deposits of a special brown fat, the primary function of which is the production of heat. This is especially important during the periods of arousal from hibernation.

Some mammals, such as the golden-mantled ground squirrel (*Citellus lateralis*) and the woodchuck, are known to have a marked annual rhythm underlying their seasonal hibernation. Under natural conditions the ground squirrel hibernates for a three- or four-month period, during which its body weight falls considerably. Food consumption increases rapidly after hibernation, reaching a peak in midsummer, and body weight increases up to the onset of hibernation in October. When isolated in the laboratory, this rhythm of hibernation and the associated change in weight may persist for about two years. If serum from a hibernating ground squirrel is injected into a non-hibernating ground squirrel, hibernation is induced. Thus it appears that some basic physiological mechanism is responsible for controlling hibernation, and that this is driven by a biological CLOCK, more or less independently of the prevailing external conditions.

Hibernation in relation to the environment. Hibernation, as the name suggests, normally takes place in climatic conditions that are typical of winter. In the northern hemisphere, this implies reduced hours of sunshine, lower temperatures, and higher rainfall. Although it appears that the hibernation of certain species is based upon an inherent annual rhythm, this does not mean that hibernators are unaffected by climate.

If the golden hamster (*Mesocricetus auratus*) or the fat dormouse (*Glis glis*) is maintained, outside the hibernation period, for a long period at a temperature of about 5–10 °C, which is the temperature normally prevailing during hibernation, acclimatization occurs. Instead of basal metabolism being lowered, as it would be during hibernation, it is elevated in response to the cold, just as it would be in a non-hibernating animal, such as the Norway rat (*Rattus norvegicus*). In other words, the response to cold during the non-hibernating season is the opposite to that which occurs during hibernation.

Ground squirrels will hibernate even when the environmental temperature is as high as 25 °C. Other species, however, will not hibernate at such a high temperature. Thus the maximum for a hamster is about 10 °C, and for a hedgehog (*Erinaceus europaeus*) about 16 °C. It is also known that low temperatures can affect the onset of hibernation: even such a clock-driven species as the ground squirrel will enter hibernation a little earlier and more rapidly if the temperature is 7 °C rather than 20 °C. Thus environmental temperature appears to have some modulating effect upon the basic rhythm of hibernation, although there are considerable differences from one species to another.

Various studies have been made of the effects of light and darkness upon hibernation. From these it appears that hibernation is affected little by level of illumination or by daylength. Thus the Arctic ground squirrel (*Citellus undulatus*) enters hibernation between the fifth and twelfth of October, and arouses between the twentieth and twenty-second of April. Studies of the prevailing ecological and climatic conditions indicate that there is little relationship between the beginning of hibernation and the temperature, the snowfall, or the amount of available food. Although there is some correlation between illumination and the onset of hibernation, attempts to influence hibernation experimentally by manipulation of illumination levels have not been successful.

Animals that hibernate in burrows, such as hamsters, marmots (*Marmota*), and ground squirrels, are well protected against dehydration. For some animals, however, the relative humidity of the air is very important. For example, the northern birch mouse (*Sicista betulina*) may die if forced to hibernate in a dry atmosphere. At the approach of winter, it normally forsakes its dry den for a damp place in which to hibernate. Bats, in particular, are susceptible to dehydration, and often hibernate in damp caves. The serotine bats seek a damp dark place in which to hibernate. They awake every few days and drink, but do not eat. A change in relative humidity from 90% to 80% for 24 h can kill a bat weighing 4 g, whereas a 12-g bat will survive, due to its relatively smaller surface from which evaporation can take place.

The main difference between hibernators and non-hibernators that live in burrows is that the hibernators build separate winter quarters, either by altering their summer residence, or by vacating it in the autumn and acquiring a new one. Most burrowing non-hibernators, such as rabbits (*Oryctolagus cuniculus*) and moles (*Talpa*), retain the same burrow the year round. The common hamster (*Cricetus cricetus*) improves its summer burrow by the addition of storage chambers to hold hoarded food. The woodchuck leaves the alpine meadows in autumn and returns to its previous winter burrow in the woods. The winter burrows of hibernators provide a fairly constant temperature generally above 0 °C, and the closing of the burrow, which many hibernators do, ensures constant conditions and reduces disturbances of the animal's hibernation to a minimum. However, the sealing-up of a burrow means that hibernation takes place in a confined space and this leads to a decrease in available oxygen and a build-up of carbon dioxide. In most animals, including hibernators during the summer time, such conditions have a narcotizing effect, and as oxygen consumption consequently diminishes the animal eventually dies. In the hibernating ground squirrel, on the other hand, the oxygen level at which oxygen consumption starts to decrease is much lower, and the animal is in fact awakened by this decrease, and is able to ventilate its burrow.

Most hibernating bats seek sheltered places in which to hibernate, and this is often associated with an autumn MIGRATION. Hibernating bats are often found in clusters in damp places in caves, in buildings, and in woods. The composition of the clusters varies from one species to another. In the little brown bat mating takes place in the autumn and then the males and females hibernate together. The females are the first to leave the winter shelter, and they gather in nurseries where they give birth and nurse their young. After weaning their young, the females seek out the males in their summer quarter prior to mating again.

Hibernation enables animals to survive during periods in which the climate is a threat, both to the animal itself and to its food supply. To hibernate successfully, the animal must find a safe resting place, and must reduce its energy expenditure so that its food consumption can be reduced to very low levels. In some parts of the world, animals face similar problems in the very hot summer, and some of these respond by a form of dormancy called AESTIVATION. Other animals cope with periods of unfavourable climate by migration, or by acclimatization.

HIERARCHY is a principle of organization that occurs at many levels in the control of behaviour. In a hierarchy the elements are ordered in such a way that higher-ranking elements control lower ones. For example, in the organization of groups of muscles there are groups of muscle fibres which are controlled by a single nerve cell, called a *motor neurone*. All the fibres in such a motor unit contract in synchrony. In a given muscle, there are many such motor units which are controlled by higher-order nerve cells. The motor units in a given muscle may contract simultaneously when a sudden powerful movement is called for, but they may also contract in a more subtly coordinated manner during a graded contraction of the muscle. A yet higher order of control is required for the COORDINATION of the various muscles in a limb. In other words, in the relevant part of the BRAIN instructions may be formulated to move a particular limb in a particular direction. These instructions will be taken up by lower-level motor control centres responsible for the coordination of groups of muscles. More detailed instructions are passed on to individual muscles, to the component motor units, and finally to individual muscle fibres. The organization is hierarchical in the sense that the detailed instructions are not all formulated at the highest level, but at each level commands are issued on the basis of more general instructions from the level above.

Hierarchical principles have often been invoked to account for the organization of large portions of an animal's behavioural repertoire. For example, we may describe the NEST-BUILDING behaviour of the male common cormorant (*Phalacrocorax carbo*) in terms of a hierarchy of GOALS. Thus the overall goal of owning a nest may generate a goal of gathering twigs and another goal of fastening twigs into the nest. The process of fastening may itself be broken down into various subgoals. Students of ETHOLOGY are not agreed as to the usefulness of this type of CLASSIFICATION of behaviour, although it is generally recognized that hierarchy is an important aspect of behaviour organization.

Hierarchies also occur in the SOCIAL ORGANIZATION of animals, though here they are usually linear rather than branching. DOMINANCE hierarchies, are found amongst farmyard chickens (*Gallus gallus domesticus*), for example, and usually involve an α animal which is dominant over all others, a β animal that dominates all but the α animal, etc. Such a linear hierarchy means that there are never members of equal rank as there would be with a branching hierarchy. Human social organizations are often hierarchical, particularly in industrial and military circles. It is generally assumed that a hierarchical structure promotes efficiency, and this argument has also been used with respect to animal behaviour. Thus the hierarchical control of movement may be economical in the material sense, fewer nerve cells being required for a particular job. At the behavioural level a hierarchical organization may make it relatively easy for modifications to be made during ONTOGENY or LEARNING.

33

HISTORY OF THE STUDY OF ANIMAL BEHAVIOUR. Just as the young child first learns about the world before the actual routine of formal education and schooling begins, so early man's first experiences with his surroundings and with animals constituted his first knowledge of species other than himself. It is therefore of great importance to attempt to discover how man first learned about the animal world, and how he reacted to it. The intimacy with and dependence of early man on animals is difficult for us to conceive at the present day. We can obtain some glimpses of it from our observation of those societies or peoples who, for various reasons, depend on certain animals.

The knowledge which the Bedouin has of his camels (Camelidae) and of everything connected with their life cycle would seem extraordinary to the urbanized man or woman. In the same way, the Nuer of the southern Sudan lives a life of almost complete interdependence with his herds of cattle (Bovidae). Evans-Pritchard in *The Nuer* (1940) gives an interesting picture of the nature of this relationship: 'The men wake about dawn at camp in the midst of their cattle and sit contentedly watching them till milking is finished. They then either take them to pasture and spend the day watching them graze, driving them to water, composing songs about them, and bringing them back to camp, or they remain in the kraal to drink their milk, make tethering-cords and ornaments for them, water and in other ways care for their calves, clean their kraal, and dry their dung for fuel. Nuer wash their hands and faces in the urine of the cattle, especially when cows urinate during milking, drink their milk and blood, and sleep on their hides by the side of their smouldering dung. They cover their bodies, dress their hair, and clean their teeth with the ashes of cattle dung, and eat their food with spoons made from their horns. When the cattle return in the evening they tether each beast to its peg with cords made from the skins of their dead companions and sit in the wind-screens to contemplate them and to watch them being milked.'

When men depended upon animals for food, HUNTING, protection, etc., and when their very lives were lost or impaired without their help, one can imagine the nature and degree of knowledge of animals which became essential. One of the few ways in which we can learn about how our ancestors saw animals is to study their surviving paintings on the walls and roofs of caves.

The cave artists. The Upper Palaeolithic period extended from about 34 000 to 10 500 years before the present. At the beginning of this period modern man, known as *Homo sapiens sapiens*, had become dominant, and the cave paintings, known as Palaeolithic art, were his work. Although we know of more than two hundred decorated caves and shelters, the area in which they have been found is rather restricted. Most of them are in the limestone areas of Italy, Spain, and France; particularly in the Pyrenees, Spanish Cantabria, and the Périgord area. In other European areas with limestone caves, no further examples are known, with the single exception of Kapovaia in the Urals. Carved or painted objects, however, as distinct from cave paintings, have been found from southwest Europe to Siberia.

Our present evidence suggests that the Upper Palaeolithic was a period of relative plenty. Game and other food seems to have been abundant, and the image of man involved in a continuing struggle for survival is an inaccurate picture of life during this time. It has been estimated that Europe was inhabited by small nomadic groups of perhaps twenty-five men, women, and children. Since the total population of France did not exceed (say) a few tens of thousands, there was little strain on the natural resources of the environment. Our knowledge of present-day hunters and gatherers, who often live under poorer conditions than their ancestors, indicates that such a life can be far from arduous. Some societies even have two or more days a week in which they are free from essential hunting or gathering, making possible other occupations and interests such as dancing and ritual.

During part of the year, the nomadic bands would perhaps spend some time under the rocky overhangs or in the shallow limestone caves of the valleys. Here they would live, cooking their food and perhaps preparing their tools and decorating some of the shelters with engravings, low-reliefs, and some paintings. It was not in these living-areas, but in nearby deeper caves that the early artists produced their highly painted sanctuaries (Fig. A). These paintings were mostly of animals, a few human forms, and many handprints, with additional signs or symbols known as tectiforms. The most common animal is the horse (Equidae), followed by the bison (*Bison*) and oxen (Bovinae). Of the total number of animals represented, these three groups account for about 60%. The remainder are mostly deer (Cervidae), mammoth (*Mammonteus*), and ibex (*Capra ibex*), with a much smaller proportion of chamois (Rupicaprini), boar (*Sus scrofa*), rhinoceros (Rhinocertidae), and the carnivores: fox (*Vulpes*), wolf (*Canis lupus*), bear (Ursidae), and hyena (*Hyaena*). Birds and fish are much less common, and the

Fig. A. Engraving of cow and bull from the Grotte de la Mairie, Teyjat, Dordogne.

reindeer (*Rangifer tarandus*), considering its importance for meat and bone, is also surprisingly infrequent in cave art.

We know very little indeed about the function or purpose of cave art. The early explanation that it was a form of hunting magic does not tally with our knowledge of the eating habits of the early hunters. Only about 10% of the animals are marked by arrow signs, and the animals most commonly painted were not those most often eaten: the reindeer is an obvious example of this discrepancy. Some have argued that the different parts of the cave-sanctuaries show different animals, with the central positions occupied by horse, bison, aurochs (oxen), the lateral positions showing ibex, hind, and mammoth, and back positions showing the boar, lion (*Panthera leo*), and rhinoceros (Fig. B).

Some theories suggest that the paintings had a religious or ceremonial function and others that the animals had either a symbolic or a totemic meaning. However, our present knowledge is too limited to form a basis for reliable conjecture. Perhaps the application of modern statistical methods would improve the situation and reduce the amount of sheer speculation that is associated with this area of knowledge. There is no doubt on one point, however, which is that the early cave artists were extremely fine and perceptive observers of animal life.

To be able to hunt and kill animals obviously requires a knowledge of the movements and behaviour of the intended prey. Some of the large or ferocious animals must have presented special problems of knowledge and organization, involving techniques acquired over long periods of time. We are still uncertain about the relative amount of animal compared to vegetable food consumed by early modern man. The archaeological evidence would tend to bias our knowledge towards animals, since bones are obviously preserved more often than vegetable products. From fossilized

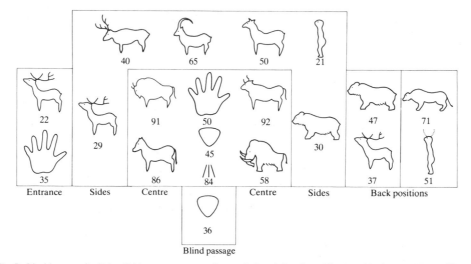

Fig. B. Ideal layout of a Palaeolithic sanctuary according to A. Leroi-Gourhan, *The Art of Prehistoric Man in Western Europe* (1968). The numbers under each drawing are percentages from 865 objects in 62 caves.

kernels and other remains, the evidence is increasing that man did eat large quantities of fruit, nuts, seeds, and cereals. Obviously the type of food consumed would depend upon its availability, and therefore upon climate and geographical area. Modern dietary knowledge suggests that a diet low in animal fats and high in vegetable fibre is beneficial to most individuals. This may give us a clue about the likely diet of early man, but at present we can only speculate.

Beginnings of domestication. The DOMESTICATION of animals and plants perhaps began about 11 500 years ago in the Middle East and in south-western Asia. Dates are frequently revised, but work in the valley of the Euphrates in northern Syria at Abu Hureyra suggests that some form of domestication was taking place at this early period. Here we find a combination of gathering and hunting with simple husbandry. Game, such as rabbits (Lagomorpha) and the onager or wild ass (*Equus hemionus onager*), were eaten, but so were sheep (*Ovis*), goats (*Capra*), and the gazelle (*Gazella*). In addition, the shells of freshwater mussels (Myrtilidae) and the bones of fish have been found. It also appears that cereals and pulses were cultivated.

The distinction between the domestication of animals and the CONSERVATION and herd management of wild animals practised by some peoples in early times is a difficult one. Whatever the complexities of the matter, it is clear that the early farmers needed to acquire much knowledge about the behaviour and the life cycle of animals if they were to be used effectively.

Animals in the early civilizations. By the time of the growth of the great civilizations, much knowledge had been acquired about animal life and behaviour. Deliberate selection and breeding methods were in use, and the ancient Egyptians were already practising the artificial incubation of eggs, a skill not known in Europe until many centuries later. As early as the Fifth Dynasty, over 4000 years ago, encyclopedic lists of animals and plants were being made. Animals were used for hunting and as pets, and already moral tales and fables with animals as the main characters were in circulation. The association between animals and religion in ancient Egypt is well known. A passage from a hymn composed by the pharaoh Akhnaten (reigned *c.*1375–1358 BC), husband of Nefertiti and possibly half-brother of Tutankhamun, is worth quoting:

When the chicklet crieth in the egg-shell
Thou givest him breath therein, to preserve him alive,
When thou hast perfected him
That he may pierce his egg
To chirp with all his might;
He runneth about upon his two feet,
When he hath come forth therefrom.
(From A. Houghton Brodrick, ed., *Animals in Archaeology*, 1972.)

Animals appear frequently in Mesopotamian sculpture, and the menageries of wild animals kept at Babylon are referred to in ancient literature. Perhaps the oldest known book on zoology is the *Har-ra = Hubullu* of Ashur, a Sumero-Akkadian bilingual lexicon dating from the 9th century BC. It contains systematic lists of both domestic and wild animals of the air, water, and land. One section, for example, lists 409 wild land animals. Below we give a short extract (F. S. Bodenheimer, *The History of Biology: an introduction*, 1958).

Probably nowhere do we find a civilization more intimately concerned with animals than that of ancient India. For thousands of years, up to the present day, animals have been portrayed in architecture, literature, and art, and have been involved in almost every aspect of ritual, religion, and domestic life. The belief in reincarnation and the transmigration of souls incorporated the idea of successive rebirths, often in the form of animals. This doctrine became part of Buddhism, and it was believed that even the Buddha had been previously incarnated as different animals, such as the monkey (Simiae), the elephant (Elephantidae), and the horse. The idea of love for all forms of life, no matter how humble, is also part of Buddhism. The Emperor Asoka (*c.*264–227 BC) renounced hunting and flesh-eating after his conversion, and in the 2nd century BC hospices were created for aged and sick animals.

There was a hierarchy of animals, with the more privileged, such as the horse and elephant, near the top. The cow, of course, was ranked highest of

Sumerian	Translation	Akkadian	Translation
lu. lim	deer	lu-lim-mu	deer
si. mul	star horned	a-a-ra	deer
dara	ibex	tu-ra-khu	ibex
dara. bar	foreign ibex	a-a-lu	deer
dara. mash. da	gazelle-like ibex	na-a-lu	? roe-deer
dara. Khal. Khal. la	shy ibex	na-a-lu	? roe-deer
mash	gazelle	ssa-bi-tu	gazelle
mash-da	gazelle	ssa-bi-tu	gazelle
mash-nita	male gazelle	da-ash-shu	male gazelle
amar. mash. da	young gazelle	uz-za-lum	young gazelle
gu. edin. na	young gazelle of the field	an-na-bu	hare

all, and the offence of killing one was as serious as the killing of a high-caste man. It has been suggested that in very early times animals were even regarded as superior to humans, and like deities; certainly the earliest Indian art has many more animal forms than man-like representations.

The god Vishnu, one of three forms of the Hindu Trinity, assumed the shape of different animals, such as a wild boar, or a lion, during his descents to earth. The monkey Hanumâna was part of the cycle of Vishnu, and appears in the famous epic, the *Râmâyana*. Great interest was shown in the behaviour of monkeys, and they are often described in the early folk and religious literature. The horse and the elephant are of particular interest to us in that earlier accounts describe the methods used to train them for their respective occupations during their domestication. Some rulers of ancient India created zoological gardens and reserves where animals could live freely and visitors were allowed to observe them.

Among the Jews of Israel, the Judaic law encouraged the humane treatment of animals. Owners were required to feed their domesticated animals before they themselves sat down to eat. That animals were regarded as sharing with their masters a place in the religious scheme is shown by the obligation to greet animals, as well as humans, at the onset of the sabbath. An understanding of the kinship between animals and men was thought to be a necessary corrective to the arrogance of human kind.

The Bible itself, apart from referring to some 120 mammals, birds, and reptiles, contains many proverbs, riddles, and allegorical passages concerned with animals. Similarly, there was an enormous oral and folk literature of animal stories circulating in early, classical, and medieval times; some of them survive to the present day, and collections for children are often reprinted. The *Panchatantra* is an ancient Indian collection of fables which had enormous influence throughout the world. It shows relationships with the Hebrew stories, and there was also an Arabic version named after the supposed author Bidpai. The *Jatakas* are fables of the Buddha as he passed through various reincarnations in the form of different animals. Even the fables attributed to Aesop of classical Greece show similarities to the other collections just mentioned. Why these tales have had such a wide influence for thousands of years is a matter for speculation; certainly many of them served as sources of moral, religious, and psychological teaching (Fig. C).

The Greek and Roman contribution. The Greeks and Romans had many contacts with animals, often in CAPTIVITY. Pets and animals for show were kept from quite early times. Possibly the most common pets were small dogs (*Canis l. familiaris*), although birds such as starlings (Sturnidae), crows, magpies, and ravens (Corvidac) are often mentioned. Aviaries and vivaries were not uncommon and a mural in the Villa Livia illus-

Fig. C. a. Ganesha (Indian). **b.** Greek Sphinx, 6th century BC.

trates a bird sanctuary. As well as exotic and singing birds, animals such as boar, deer, and antelopes (Bovidae) were kept. Hunting and fishing were considered to be occupations for the nobility, and Roman country gentlemen sometimes kept game reserves. Under the Empire, an enormous number of unusual animals were imported; they were used for the circus or to stock the imperial menageries. Augustus, for example, owned 420 tigers, Nero 400 bears, and Trajan was said to possess 11 000 animals. Travellers to foreign lands would also bring back information: Herodotus (c.425 BC), for example, describes the habits and behaviour of various animals. There is also a rich and interesting literature on country life and agriculture which shows an impressive knowledge of animals and husbandry. The writings of Xenophon (435–354 BC) and Oppian (c.AD 200) on hunting are regarded as classics of their kind, although not all early writers shared the view that it was a noble pursuit.

From very early times interest was shown in the relative differences between man and animals.

Alcmaeon (*c.*520 BC) believed that man had a greater power of comprehension than animals. Anaxagoras (5th century BC) attributed INTELLIGENCE to all animals, but considered man to have the highest degree of sagacity. The Cynics (the name is based on the Greek word for dog) saw animals as superior to man in their simplicity and lack of possessions. Their lack of reason was thus well compensated for by their other virtues. The Stoic logician Chrysippus (*c.*280–207 BC) believed that animals could show reasoning similar to man's. For example, a dog confronted by a forest with three entrances was described as sniffing at the first two openings and then running immediately and without hesitation into the third opening. Not all Stoics felt this way, and Seneca (3 BC–AD 65) objected to the praise of animals and the use of animals as an example for man. Certain writers, known as *physiognomonici*, were interested in the psychology of character, and attempted to understand the characters of human beings by comparing them to the behaviour and features of animals and races. This tendency for ANTHROPOMORPHISM still exists in fable and ordinary life when we compare, say, the rabbit (*Oryctolagus cuniculus*) or the fox with people who share characteristics with those animals.

When we consider animals in terms of CLASSIFICATION into different forms or species, the early writings are somewhat meagre until we come to Aristotle (384–322 BC). Hippocrates (*c.*460 BC) and his associates, however, classified animals according to their diet. Speusippus (*c.*410–339 BC), Plato's nephew, developed some of his uncle's ideas on classification. Surviving extracts name about fifty-five species and genera of animals, birds, fishes, and plants.

With Aristotle begins a new era of observation and description of animal life. The truth is that the direct and simple observation of the behaviour of animals without distortion, and the description of such behaviour without embellishment or bias, is a rare attainment in the entire history of knowledge. Aristotle is a notable exception, and his work abounds with useful observations and insights into animal behaviour. Although, of course, he was sometimes inaccurate, his understanding of the relation between fact and theory remains impressive. For example, he wrote in *De Generatione*, when discussing bees (Apidae): 'Such appears to be the truth about the generation of bees, judging from theory and from what we believed to be the facts about them; the facts, however, have not yet been sufficiently grasped; if ever they are, then credit must be given rather to observation than to theories, and to theories only if what they affirm agrees with the observed facts.'

Aristotle was born in 384 BC at Stagira (Stavró), a small town in Chalcidice in northern Greece. His father, Nicomachus, was a member of the medical guild of the Asclepiadae and was also court physician and friend to Amyntas II, King of Macedon. In 367 BC, when 17 years old, Aristotle entered Plato's Academy in Athens. He remained there for 20 years until Plato's death in 347 BC, when he left Athens for Assos in Mysia, Asia Minor, at the invitation of Hermias, ruler of Atarneus and a former fellow student at the Academy. Aristotle married Pythias, a relative of Hermias, and stayed in Assos until Hermias died in 345 BC. He then moved to the neighbouring island of Lesbos, where he remained for two or three years, chiefly at Mytilene. There is a clear connection between the various places mentioned in his biological works (chiefly the *Historia Animalium*) and the parts of Greece visited during his travels from 347 to 335 BC, such as Troad, Lesbos, and Macedonia. There are detailed descriptions of various marine animals from Lesbos and, in particular, from the lagoon and straits of Pyrrha there.

In 343–342 BC Aristotle was invited to become tutor to the 14-year-old Alexander, son of Philip of Macedon, and in 335–334 BC he returned to Athens after Philip's death. In Athens he rented buildings and began a school outside the city, probably between Mount Lycabettus and the Ilissus to the north-east. Here was a sacred grove dedicated to Apollo Lyceius and the Muses (hence the name Lyceum). At the school, a natural history collection and a library were formed. It is said that Alexander gave Aristotle 800 talents to make the collection and instructed the fishermen, hunters, and fowlers of the Empire to report back to Aristotle anything of biological or scientific interest. In 323 BC Alexander died and a charge of impiety was made against Aristotle. He left the school in the hands of his pupil and friend Theophrastus and moved to Chalcis, his mother's home town, where he died in 322 BC.

In Aristotle's writings, biology and psychology are intimately related; they are not considered to be separate sciences. One has the impression that his COMPARATIVE STUDIES of man in relation to other animals served as a basis for his biological and psychological understanding of man. Here we shall be concerned chiefly with his writings on the behaviour of animals. In all, Aristotle gives an account, in various degrees of detail, of the life and behaviour of some 540 species of animals. The most important source of his ideas is the *Historia Animalium*. Its title in Greek, *Historiai peri ta zoa*, 'Enquiries concerning animals', gives a clearer indication of its purpose. This remarkable work contains numerous observations, either from firsthand knowledge, or from hunters, herdsmen, birdcatchers and the fishermen of the Aegean. Its nine books discuss such topics as the 'ears, nose, and tongue' (Book I); 'of apes and monkeys' (Book II); 'of the pupil of the eye' (Book III); 'of voice and sound' (Book IV); 'of generation' (Book V); 'of the breeding of birds and mammals' (Book VI); 'of pregnancy and birth' (Book VII); 'of the psychology of animals' (Books VIII, IX, and X). The accuracy of the observations can be illustrated by a short extract on bird song: 'Of little birds, some sing a different note from the parent birds, if they have been removed from the nest and have heard other birds singing; and a mother-nightingale has been observed to give lessons in singing to a young bird, from which spectacle we might obviously infer that the song of the bird was not equally congenital with mere voice, but was something capable of modification and of improvement' (Book IV). (See SENSITIVE PERIOD.)

There is an interesting example of an observation on the behaviour of the catfish (glanis or sheat-fish) recorded by Aristotle that was for many years considered to be wrong, but is now known to be a correct account. 'Of river-fish, the male of the sheat-fish is remarkably attentive to the young. The female after parturition goes away; the male stays and keeps on guard where the spawn is most abundant . . . for forty or fifty days. . . . He is so earnest in the performance of his parental duties that the fishermen at times, if the eggs be attached to the roots of water-plants deep in the water, drag them into as shallow a place as possible; the male fish will still keep by the young, and, if it so happen, will be caught by the hook . . .' (Book IX). The common European catfish *Silurus glanis* simply deposits the eggs in a hole and leaves them after fertilization. But the lesser-known *S. aristotelis* behaves in a similar way to the catfish described by Aristotle. This variety lives in southern Greece and Asia Minor.

Aristotle's influence on later thought was enormous: although he was sometimes misinterpreted, for well over a thousand years he was regarded as the primary authority on biological matters. It had clearly not been his intention to serve as a definitive source which could not be questioned, but many treated him in this way, and later exact observation and enquiry were stultified by slavish repetition of his opinions. Two related ideas presented in his writings served as the basis for many later theories of biology and behaviour and even now influence modes of thought. The first is the concept of the 'soul' and the second is the idea of the ladder or scale of nature, 'scala natura', later to be described as the 'Great Chain of Being'.

In discussing the soul, Aristotle used the Greek word *psukhē*. If we translate this by the word 'soul', as is usually done, we shall meet with difficulties and confusion. The word *psukhē* does not mean 'soul' in our modern sense, but has a collection of meanings roughly corresponding to 'life' or 'life principle'. It includes in its wider range all the vital functions such as sensation, reproduction, LOCOMOTION, and reason. Some prefer the term 'faculties' to describe the various vital functions, but others might prefer 'life function' or simply *psukhē* itself. With these qualifications in mind, Aristotle distinguished between plants, animals, and man in their life functions or *psukhē*. Plants had nutritive and reproductive 'functions' which they shared with both animals and man. All men

had sensation and all animals had sensation, or at least touch.

The function of reason or *nous* existed only in man and referred to two different powers: either the power of intuition or the power of understanding (intellect). With the exception of *nous*, the functions cannot exist apart from the body. The *psukhē* is thus the actual functioning or life of the body.

Turning to the ladder or scale of nature, the higher in the scale of perfection, the greater the complexity or structure, the greater the number of 'functions', and the greater the number of aims that can be pursued. Species are fixed, and there is no EVOLUTION in the modern sense. Every organism tends by its behaviour to the form immediately above it in the scale of nature. The scale begins with inanimate matter, passes through the lower and higher plants, then sea creatures such as jellyfish (Coelenterata) and squids (Cephalopoda), and then, after various stages, reaches the mammals and eventually man (see Fig. D). The scale of perfection extends even beyond man. The idea of a scale of nature or being became the fundamental conception of man's place in the universe and remained part of European and even Eastern thought until the 16th century and even later.

In his *De Anima* Aristotle wrote: 'Man, and possibly another order like man or superior to him, [possesses] the power of thinking, i.e. mind' (Book II). It was perhaps such a statement which led later thinkers to believe that he had argued that there was a hard and fast distinction between animals, who lack reason, and man. That he did not hold this rigid view, but believed in a principle of continuity, is shown by the following important passage in the *Historia Animalium*. 'In the great majority of animals there are traces of psychical qualities or attitudes, which qualities are more markedly differentiated in the case of human beings. For just as we pointed out resemblances in the physical organs, so in a number of animals we observe gentleness or fierceness, mildness or cross temper, courage or timidity, fear or confidence, high spirit or low cunning, and, with regard to intelligence, something equivalent to sagacity. Some of these qualities in man, as compared with the corresponding qualities in animals, differ only quantitatively: that is to say, a man has more or less of this quality, and an animal has more or less of some other; other qualities in man are represented by analogous and not identical qualities: for instance, just as in man we find knowledge, wisdom, and sagacity, so in certain animals there exists some other natural potentiality akin to these. The truth of this statement will be the more clearly apprehended if we have regard to the phenomena of childhood: for in children may be observed the traces and seeds of what will one day be settled psychological habits, though psychologically a child hardly differs for the time being from an animal; so that one is quite justified

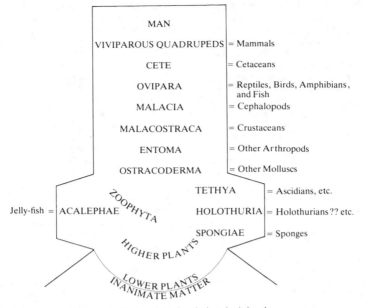

Fig. D. The *Scala Naturae* or 'Ladder of Life' based on descriptions in Aristotle.

in saying that, as regards man and animals, certain psychical qualities are identical with one another, whilst others resemble, and others are analogous to, each other.

'Nature proceeds little by little from things lifeless to animal life in such a way that it is impossible to determine the exact line of demarcation, nor on which side thereof an intermediate form should lie. Thus, next after lifeless things in the upward scale comes the plant, and of plants one will differ from another as to its amount of apparent vitality; and, in a word, the whole genus of plants, whilst it is devoid of life as compared with an animal, is endowed with life as compared with other corporeal entities. Indeed, as we just remarked there is observed in plants a continuous scale of ascent towards the animal. So, in the sea there are certain objects concerning which one would be at a loss to determine whether they be animal or vegetable' (J. A. R. Smith and W. D. Ross, ed., *Aristotle's Historia Animalium*, 1910).

Gaius Plinius Secundus, known as Pliny the Elder (23–79 AD), wrote a *Historia Naturalis* (Natural History) which appeared in AD 77. This was an immense collection of stories, facts, and observations in thirty-seven books based, so Pliny claimed, on 2000 volumes written by a hundred Greek and Roman authors. Books VIII to XI are concerned with animals. Book VIII deals with 'land creatures and their kinds'; Book IX with 'all fishes and creatures of the water'; Book X is 'of flying foules and birds'; and Book XI 'of insects'. Although often criticized, the *Historia Naturalis* is a valuable source of early lore and thought on ani-

mals. For example, the basis of the phrase 'licking into shape' was taken by Pliny from Aristotle, who described the belief that the young bear is born as a shapeless lump which the mother licks into form. The fact that the cub remains blind and hairless for 5 weeks and is constantly licked by the dam probably accounts for the story, which was repeated by many later authors.

The following description of the nesting behaviour of the swallow is more factual: 'Swallows build with clay and strengthen the nest with straw; if ever there is a lack of clay, they wet their wings with a quantity of water and sprinkle it on the dust. The nest itself, however, they carpet with soft feathers and tufts or wool, to warm the eggs and also to prevent it from being hard for the infant chicks. They dole out food in turns among their offspring with extreme fairness. They remove the chicks' droppings with remarkable cleanliness, and teach the older ones to turn round and relieve themselves outside of the nest. There is another kind of swallow that frequents the country and the fields, which seldom nests on houses, and which makes its nest of a different shape though of the same material—entirely turned upward, with orifices projecting to a narrow opening and a capacious interior, and adapted with remarkable skill both to conceal the chicks and to give them a soft bed to lie on' (H. Rackham, ed., *Pliny: Natural History*, vol. III, 1942).

Another great collection of observations, stories, and anecdotes, often with a moralizing purpose, was made by Claudius Aelianus, known as Aelian (*c.* AD 160–235). His *De Natura Ani-*

malium (On the Characteristics of Animals) is a mine of fascinating pieces of curious lore about animals, based mostly on Greek sources, Pliny, and his own limited experience. Famous stories such as the boy on the dolphin and Androcles and the lion are to be found here. An excerpt, possibly showing IMPRINTING, will serve to illustrate Aelian's style and method: 'I am told that a dog fell in love with Glauce the harpist. Some however assert that it was not a dog but a ram, while others say it was a goose. And at Soli in Cilicia a dog loved a boy of the name of Xenophon; at Sparta another boy in the prime of life by reason of his beauty caused a jackdaw to fall sick of love' (A. F. Shoefield, ed., *Aelian: On the Characteristics of Animals*, 1958).

Both Pliny and Aelian shared with the Cynics the view that the greediness, immorality, and unnaturalness of mankind often showed animals in a superior light. This theme was taken up by Plutarch (*c.* AD 50–120), and similar views were put forward by Sextus Empiricus (2nd century AD), and were vigorously supported by Porphyry (*c.* AD 232–84), who used related arguments to justify vegetarianism and to condemn cruelty to animals. However, such opinions were opposed by many Stoics, and apparently even by St. Augustine (AD 354–430). The Stoics in general drew an extreme contrast between animals and men, claiming that animals do not have moral qualities or reason. If animals were treated as rational, life would become impossible for mankind. Since animals had no reason they must also lack rights. Augustine accepted that animals suffer and feel PAIN, but argued that since they did not possess a rational soul they did not share a common nature with mankind. St. Thomas Aquinas (1224–74) also argued that animals suffer, and that man pities such suffering (see WELFARE OF ANIMALS). If he treats animals badly or ignores pain, he is also likely to treat human beings in the same fashion. It is for this reason that animals should be well treated. However, man was still considered to be superior to the brutes since he only possessed a rational soul.

Galen (*c.* AD 129–200) did some experiments demonstrating that certain acts are INNATE. He reared in isolation a young goat that had been taken by Caesarean section. Immediately after the kid was removed from its mother's womb it was put in a room with various kinds of food, such as wine, oil, honey, milk, grains, and fruit. The kid was then observed, and Galen records that 'We observed the kid take its first steps as if it were hearing [*audivisset*] that it had legs; then it shook off the moisture from its mother; the third thing it did was to scratch its side with its foot; next we saw it sniff each of the bowls in the room, and then among all of those, it smelled the milk and lapped it up. And with this everyone gave a yell, seeing realized what Hippocrates had said: "The natures of animals are untutored."' (C. Kühn, ed., *Opera Omnia*, vol. VIII, 1824.)

The Middle Ages and the Renaissance. Isidore of Seville (*c.* AD 560–636) acted as a major link between antiquity and his own time. His *Etymologiae* or *Origines* was an encyclopedia containing much natural history and descriptions of animals taken from many earlier sources.

With the Arab writers and Islam, Aristotle's writings on biology were revived after a period of neglect, although Indo-Persian and pre-Islamic sources were also used. Ishak ibn Husein and his associates translated Aristotle from Syriac into Arabic in the 9th century. The work of Ibn Sina (Avicenna, 980–1037) in his *al-Shifā* and the commentaries of Ibn Rushd (Averroës, 1126–98) on Aristotle were of major significance and became the chief channel for knowledge in the West in the 13th and 14th centuries. Al-Jāhiz of Basra (*c.*767–868) combined a critical appreciation of Aristotle with careful observations obtained from the Bedouin and other sources in his *Kitāb al-Hayawān* (Animal Book). Although it is essentially a religious work, it contains interesting observations, small experiments on subjects ranging from animal COMMUNICATION to the effect of alcohol on animals, further material on animal behaviour, mention of up to 350 animals, and even a doctrine of evolution.

The *Kitāb al-Hayawān* (Animal Book) of Kamal al-Din Al-Damīrī of Cairo (1349–1405) is an alphabetically arranged collection of both traditional lore and actual observations of animal behaviour, with material from Al-Jāhiz and other sources. Many other works on animals were produced in Islam, and in Islamic India the Mogul Emperor Jahāngīr devoted sections of his book *Jahāngīrī-nāmah* (The Book of Jahāngīr) to natural history and animals. The Islamic writers integrated the Aristotelian idea of a scale of being with their own religious conceptions to produce a grand view of the Great Chain of Being involving the 'three kingdoms' (*nawālīd*) of nature: mineral, plant, and animal. Ibn Sina divided each kingdom in relation to its own degree of perfection. Similarly the Ikhwān al-Safā' (Brethren of Purity) produced detailed studies of animals in relation to senses, HABITAT, and their scale of being. There is a large body of Islamic literature concerned with animals in a religious sense, and with their role in spiritual development, for example *The Conference of the Birds* of Attār, the *Mathnawi* of Rumī, the *Garden of Spring* of Jami, and the *Garden of Truth* of Sanā'ī.

Frederick II of Hohenstaufen (1194–1250) was a careful observer and recorder of animal behaviour. His book on falconry and ornithology, *De Arte Venandi cum Avibus* (*c.*1244–8), which he took over thirty years to write, is one of the most remarkable productions of the Middle Ages. Of the surviving six books, the first is a survey of ornithology in general, covering NEST-BUILDING, MATING, FEEDING habits, and MIGRATION. His description and illustration of migration is most impressive. He also viewed with suspicion many of the extravagant accounts of bird behaviour.

Adelard of Bath (12th century) is noteworthy for his extensive travels in the Middle East. He conveyed to the West many ideas from Islam, particularly in his *Quaestiones Naturae*. In a series of discussions with his nephew he examines certain questions, such as whether animals have reason.

Albertus Magnus (*c.*1200–80) played a significant role in the introduction of Aristotle and Greek and Arabic thought to the West. He insisted on the importance of careful observation and basic principles of experimental design, such as repeated results and the methods of agreement and difference. His *De Animalibus* criticized many popular myths and has important sections on REPRODUCTION and embryology. His discussions of insects and their mating behaviour are particularly fine. He was sometimes critical of Aristotle, and modified his taxonomy of the genera of water creatures. At about the same time a number of encyclopedias were produced containing sections on animals. Vincent of Beauvais, a pupil of Albertus, included several books on animals in his *Speculum Maius*, and the *De Proprietatibus Rerum* of Bartholomaeus Anglicus was a source of natural history for Shakespeare and other writers.

Sir Thomas More (1478–1535) actually kept animals, including domestic fowl (*Gallus g. domesticus*), and his *Utopia*, published in 1516–18, contains a clear reference to imprinting: 'They rear a very large number of chicks, by an amazing device. For the hens do not sit on the eggs. Instead they keep a great number of eggs warm with an even heat, and so hatch them. As soon as the chicks come out of the eggs, they follow the men and recognize them as if they were their mothers.' (See E. H. Hess, *Imprinting*, 1973.)

Pierre Belon (1517–64) was one of the naturalist travellers of the 16th century. In 1546–50 he travelled in the Middle East, visiting Arabia, Egypt, Judaea, Greece, and other countries. A careful observer of fish and birds, he is regarded as the founder of comparative anatomy. His *Histoire de la nature des oyseaux* (1555) and his *Histoire naturelle des estranges poissons marins* (1551) are really the first zoological texts based chiefly on observation. Guillaume Rondelet (1507–66) studied the fish and other marine animals of the Mediterranean. Using simple experiments he showed the importance of air for respiration in fish and described over 300 aquatic animals. Konrad Gesner (1516–65), briefly a pupil of Rondelet, was also a keen traveller covering the Adriatic coast and the Alps. His *Historia Animalium* contains 4500 pages, and attempts to present actual observations on animals. The bird which he called *Corvus sylvaticus* has now disappeared from Europe.

Ulisse Aldrovandi (1522–1605), who used to accompany Rondelet to the fish markets to obtain specimens, also collected for his own museum. Although he was sometimes uncritical of earlier writers, his own work on eggs and embryology was most important, and his descriptions of early

Egyptian methods of egg incubation are extremely interesting. Aldrovandi produced an enormous encyclopedia, even longer than Gesner's. Both works were heavily illustrated, and attracted many readers.

The seventeenth and eighteenth centuries. The philosophy of René Descartes (1596–1650) dominated much of 17th-century thought. His ideas of the dualism of body and mind and of the difference between man and animals have influenced our conceptions to the present day. For Descartes, animals did not have abstract reasoning power and self awareness. For these a rational immaterial soul was essential, and animals only possessed a machine-like automaton body. They could carry out the simple mental functions necessary for life, but they could not think and did not have LANGUAGE: '... for the word is the sole sign and the only certain mark of the presence of thought hidden and wrapped up in the body; now all men, the most stupid and the most foolish, those even who are deprived of the organs of speech, make use of signs, whereas the brutes never do anything of the kind; which may be taken for the true distinction between man and brute' (*Letter to Henry More*, 1649). Henry More (1614–87), the Cambridge Platonist and friend of Isaac Newton (1642–1727), was not convinced by such views. He thought that Descartes did not wish animals to have a claim on immortality, since Descartes argued, in keeping with his predecessors Augustine and Aquinas, that immortality necessitated a rational soul. At times, Descartes has been misinterpreted on this point: although he believed that animals were complex machines, he did not say that they were without feeling.

In England a number of thinkers were applying atomistic, Epicurean, and mechanistic ideas to the operations of the mind. Sir Kenelm Digby (1603–65), who introduced Descartes to his contemporaries, proposed in 1644 in his *Treatise on Bodies* an atomistic theory of ASSOCIATION and learning. Thomas Hobbes (1588–1679) similarly put together prevailing ideas on motion and atomism from Pierre Gassendi (1592–1655) and others, to suggest an atomistic theory of the mind. Thomas Willis (1621–75), in his *Souls of Brutes* (1664), presented a detailed associationist account of the thoughts of animals, and his one-time assistant and Secretary of the Royal Society, Robert Hooke (1635–1703), lectured before the Society in 1682 describing a materialistic and physiological theory of the mind and BRAIN.

In Europe, Descartes's admirers continued to extend his views until they came close to explaining the whole of man in terms of mechanisms. Henricus Regius in his *Fundamenta Physices* (1646) emphasized the correlation between psychological processes and their physiological bases. Similarly Rohault in 1671 and Pierre Sylvain Régis in 1690 tried to show a mechanistic basis for even the higher functions in animals. Cureau de la Chambre (1594–1669) paid particular attention to instinctive behaviour, and suggested a physiological basis for INSTINCT and SOCIAL RELATIONSHIPS among animals.

John Locke (1632–1704) not only, perhaps unintentionally, provided a basis for the growth of *associationism* (see VOLUNTARY BEHAVIOUR) but also maintained that there was continuity between man and animals. 'There are some brutes that seem to have as much knowledge and reason as some that are called men...' (J. W. Yolton, ed., *John Locke: An Essay Concerning Human Understanding*, 1965). The associationist ideas of Digby, Hobbes, Willis, and Hooke about the way the mind and body functioned were developed in a remarkable work by David Hartley (1705–57) in *Observations on Man* (1749). Hartley believed that he was using the suggestions of both Locke and Newton. In fact Locke considered that associations interfered with rational thought, but his coinage of the phrase 'the association of ideas' has branded him as one of the founders of the school. Hartley thought that animals, too, worked on the basis of the physiology of association. He found no difficulty in allowing intelligence to animals, and insisted that mankind had a moral obligation not to be unkind to them.

In France, Descartes's ideas on automatism in animals and man came to fruition in J. O. de La Mettrie's *L'Homme machine* (1747). As in England, but with greater confidence and thoroughness, the concept of 'beast-machine' was being replaced by that of 'man-machine'. For La Mettrie, all psychological processes could be based on the mechanisms of the body. The Abbé de Condillac (1715–80), a follower of Locke, extended his ideas on sensations to animals, and elaborated them in his *Traité des animaux* (1754). In France, as in England, the view was becoming accepted that animals and men worked by the same mechanisms.

Among philosophers in France, Britain, and Germany the division between man and animals was rapidly losing hold. Not only Condillac and Leibniz, but also Hume, Voltaire, and Rousseau, declared their views plainly. David Hume (1711–76), in his *Treatise of Human Nature* (1740), wrote: 'Next to the ridicule of denying an evident truth, is that of taking such pains to defend it; and no truth appears to me more evident, than that beasts are endowed with thought and reason as well as men.' Similar ideas are expressed in his *Inquiry* (1748). Voltaire (1694–1778) wrote in his *Philosophical Dictionary*: 'What a pitiful, what a sorry thing to have said that animals are machines bereft of understanding and feeling.' 'It is only in degree that man differs, in this respect [reason], from the brute', Jean-Jacques Rousseau (1712–78) argued in *Origins of Inequality*.

In France there was growing interest in natural history. In 1732 the Abbé Pluche had produced *Le Spectacle de la nature*, a massive nine-volume work in which the wisdom of God in Creation was demonstrated by a series of conversations between a countess, a count, a knight, and a prior. Poorly produced, but with a readable style, it passed through eighteen editions and became a 'best seller'. Pluche's work served as an introduction to the enormous enterprise of Georges Louis Leclerc, Comte de Buffon (1707–88), the *Histoire naturelle, générale et particulière* in thirty-six volumes. The first volume appeared in 1749, and Buffon continued to work on it for eight hours a day for a period of forty years. It was beautifully written, with many engravings, and elegantly printed by the Imprimerie Royale. It described all that was then known about the world of nature. The first printing sold out within weeks, and the entire work was one of the most popular in France for many years.

René Antoine F. de Réaumur (1683–1757) is better known for the thermometer he invented, and some of his work with insects was neglected until republished by Wheeler in 1926. The six folio volumes of his *Mémoires pour servir à l'histoire des insectes* (1737–48) are of major importance for the life history of insects. His range was wide, from ecological observations and the artificial INCUBATION of eggs to studies of the electric ray (*Torpedo*), the digestion of birds, and the behaviour of ants (Formicidae).

Like Réaumur, Charles George Leroy (1723–89) was one of those rare writers on animal behaviour and intelligence who had actual experience of animals in the FIELD. Although an admirer of Condillac and friendly with the intellectuals of the day, Leroy often showed an independence of mind when questions of animal intelligence and behaviour were under discussion. His important work was the *Lettres philosophiques sur l'intelligence et la perfectibilité des animaux* (1764). This series of letters contains many interesting observations on animal life and behaviour, as well as criticisms of earlier philosophical opinions about animals. The views of Descartes and others on automata and instinct are treated with the scorn one expects from a practising naturalist. Leroy himself was a keeper of woods and forests to the court of Louis XV, and from his remarks a keen sportsman. He does not deny that man is different from animals in some ways, but argues that animals act intelligently on the basis of their MEMORY and experience in relation to their needs.

His keen observations on the hunting behaviour of the wolf and the fox, for example, or the fear of the hare (Leporidae) when pursued, are fascinating pieces of writing and illustrate his understanding of the complexities of animal life. At times he anticipates modern ideas in ETHOLOGY. For example, he argues that for every animal a form of *ethogram* should be prepared giving a biography of the species, its individual character, way of life, natural habitat, etc. Observation shows, he writes, that there is much variation in animal behaviour. The seemingly automatic and stereotyped activity of nest-building, for example, varies considerably from one bird to another, particularly among young birds.

In Germany, Johann Pernauer of Rosenau, near Coburg (1660–1731), was a skilled observer of birds, although he is almost unknown today. Pernauer, a member of the nobility, was disgusted by the depravity of his times, and believed that the study of country life and birds would serve as a corrective for children and adults. His book, under the short title of *Agreeable Country Pleasures or Taming and Training Birds*, passed through at least four editions (1702–20). It contains an introduction examining bird behaviour and the main behavioural differences between different species. The main indicators are (i) feeding habits, (ii) ecology, (iii) social behaviour, (iv) nesting, (v) territory, (vi) seasonal colour, (vii) migration, (viii) song, (ix) bathing, (x) feeding the young.

An important technique developed by Pernauer was to train the birds either to remain in the aviary, or to fly out, for the day or for the whole season, into the wild. He carried out many other experiments into the training and general behaviour of birds. The book mentioned contains long chapters on various species of birds, embellished with fascinating observations. For example, 'If one wants to refer several species to one genus, these should at least have a marked similarity of voice and other characteristics.' And, 'The Blackbirds [Amseln, *Turdus merula*] sometimes call each other, but not with the intention of gathering, but rather for inciting each other to fly, or to warn each other, or as a means of threat; for the Blackbirds do chase each other, though not to such a degree as the Nightingales and the Robins.' (E. Stresemann, 'Baron von Pernauer, pioneer student of bird behaviour', *Auk*, 64 (1947).)

In Britain, the ideas of the ancients on the intelligence and superiority of animals were being revived, and Thomasius Jenkin described them in his work *Philosophical Defence of the Souls of Animals* (1713). It will be recalled that Hume and Hartley argued that, for everyday reasoning, there was no difference between man and animals; in addition, it was man's obligation to treat animals well. Some clearly still thought otherwise, and John Granger preaching against cruelty in 1772 aroused insult and hostility. Even the polymath Thomas Young in his *Essay on Humanity to Animals* (1798) felt he would be ridiculed for his book. John Gregory (1724–73), a friend of Hume, Professor of Philosophy at Aberdeen, and later Professor of Medicine at Edinburgh, argued that man has reason but that it is weak and unreliable through his depraved condition. Instead, the instincts which man shares with the animals should be studied.

The Reverend Gilbert White (1720–93), of Selborne in Hampshire, took eighteen years to produce his *Natural History and Antiquities of Selborne* (1789). Despite the fact that it is simply a collection of letters describing his observations of the natural life around his parish, it became one of the most frequently published books in the English language, and has been printed in over two hundred editions and translations. White was not only a classical scholar, well versed in Pliny and Aelian for example, but had read carefully the works of early and contemporary writers on natural history. His most outstanding, and unusual, characteristic was the careful and unprejudiced observation of the bird and animal life of the countryside. The letters are full of interesting remarks and details. For example, like Pernauer, he remarked on the use of birdsong in distinguishing between bird species: 'I make no doubt that there are three species of the willow-wrens: two I know perfectly, but have not been able yet to procure the third. No two birds can differ more in their notes ... one has a joyous, easy, laughing note; the other a harsh loud chirp ...' (*The Natural History of Selborne*, Penguin edn. (1977), p. 46). In one letter he provides a classification of birds based on close observation, according to whether they 'continue in full song till after Midsummer', 'cease to be in full song', or have 'somewhat of a note or song, and yet are hardly to be called singing birds'.

The nineteenth century. *Associationism, behaviour, and transformation of species (1794–1810).* A number of writers, particularly Charles DARWIN's grandfather Erasmus Darwin (1731–1802), Pierre-Jean Cabanis (1757–1809), and Jean-Baptiste Lamarck (1744–1829), attempted to use the new ideas of associationism to explain apparently instinctive behaviour. They regarded the Aristotelian idea of blind, unreasoned instincts as belonging to an outmoded philosophical tradition. In Erasmus Darwin's *Zoonomia* (1794) there is a long chapter on instinct, in which he examines a variety of activities that were customarily described as instincts, and shows how they can be explained as the result of experience and LEARNING. The coordinated pecking of the newly hatched chick, for example, can be attributed to the continual gaping and gulping of amniotic fluid by the foetus. Just as each organism developed by intelligent associative learning, so the same mechanism could explain the transformation of species. For example, attempts at self-preservation lead in different species to the development of wings, or swiftness of foot, or thick protective shells.

Many of Cabanis's views were similar to those of Erasmus Darwin, although he appears not to have known the *Zoonomia*. His important treatment of instinct is in his *Rapports du physique et du moral de l'homme* (1796–1802). He, too, explained activities that appeared shortly after birth as acquired during the time of gestation. However, he stressed that habits were moulded by the anatomical structures of the organism. He divided instincts into two classes, according to whether they were dependent on structures that developed before or after birth. Food-seeking (from swallowing amniotic fluid) and movement (from struggling in the womb), for example, are acquired before birth. Mating instincts, however, depended on structures that develop after birth. Cabanis also wrote that species could be transformed because of new habits acquired in response to felt needs.

Lamarck was greatly interested in Cabanis's book, and there are many similarities (but some differences) in their writings on instinct. Lamarck's most important discussions of instincts are in his *Recherches sur l'organisation des corps vivants* (1802) and his famous *Philosophie zoologique* (1809). He did a great deal of work on invertebrate taxonomy, which is important because he used the nervous system as the primary diagnostic character. This enabled him to arrange animals on a scale according to the development of their nervous system. He also saw, correlated with this, a gradation in the degree of intelligence. The animals lowest in the scale, such as polyps, e.g. the freshwater hydra (*Hydra*), acted only because of the impingements of external fluids. Insects were the first animals (working upwards on Lamarck's scale) to possess a nerve centre, and their behaviour was controlled by internal instincts. A primitive cortex first appeared in fish, and Lamarck supposed that ideas and thought began here. He also built a theory of transformation of species out of his theory of behaviour. At different times he proposed various mechanisms that could cause the transformation of species, the best known of which, although not the most important to Lamarck, was the development of new habits or structures in response to the felt needs of the animal. If the animal's environment changed in some way, the animal would acquire (by a method appropriate to its nervous system) new habits. These habits, by continual repetition, would gradually become inherited, and so the species would be transformed.

Thus, several writers not only applied associationism to explain behaviour, but also developed the theory directly to show how behaviour could act as the cause of the transformation of species.

The rise of experimental methods. Ornithologists such as J. F. M. Dovaston were giving up their guns and using telescopes to study birds; by simple methods, such as luring robins (*Erithacus rubecula*) and then releasing them in order to study their territories, Dovaston obtained valuable information. In America, Thoreau and Emerson were writing about the virtues of the simple life. In Britain, W. H. Hudson fell under Gilbert White's spell, and began to weave his own accounts of nature.

In the 1880s and 1890s a number of books were published on the study of birds, particularly with the use of binoculars, and Edmund Selous (1858–1934) began a series of detailed studies on bird behaviour, sometimes spending weeks in the field. At about the same time some eccentrics began the extraordinary custom of putting out food for the birds. By the end of the century the habit was becoming common. Hudson in Cornwall astonished the locals in 1908 by this curious behaviour, but, according to *Punch*, by 1910 such 'feeding of birds' had become a national pastime. The use of photography and the large-scale appearance of bicycles greatly encouraged trips into the country, and the study of wild life began to grow apace.

In France a division representing two different approaches to the study of biology and nature was coming to a head. On the one hand Georges Cuvier (1769–1832) was arguing for LABORATORY STUDIES, and on the other Étienne Geoffroy Saint-Hilaire (1772–1844) was emphasizing observation of animals under natural conditions. The debate was continued after their deaths by Cuvier's protégé, Pierre Flourens (1794–1867), who employed the term *comparative psychology* in 1864 for the study of animals under laboratory conditions. In 1859 Saint-Hilaire's son, Isidore Geoffroy Saint-Hilaire (1805–61), had introduced the term 'ethology' to describe research in the natural habitat.

In Britain, John Stuart Mill (1806–73) in his *System of Logic* (1843) used the word 'ethology' to mean the science of character and its formation, altering it from its earlier senses of the portrayal of character by mimicry and gestures, or the study of ethics. Mill's usage continued in Britain until very recent times, but in France it was not accepted, and the tradition of Saint-Hilaire was kept alive by Alfred Giard (1846–1908) and J. H. Fabre (1823–1915). In America, between 1902 and 1905, the famous zoologist W. M. Wheeler (1865–1937) was arguing for the use of the term 'ethology' for the study of animals in their natural environment. Perhaps deriving the word from the French writers, Oskar Heinroth (1871–1945) in Germany began using 'ethology' in 1910 and 1911.

The *Origin of Species* (1859) by Charles Darwin (1809–82) had enormous influence on the study of animal behaviour. In his *Descent of Man* (1871) Darwin showed a keen interest in the comparative study of man and animals and in instinctive behaviour: 'We have seen that the senses and intuitions, the various emotions and faculties such as love, memory, attention, curiosity, imitation, reason, etc., of which man boasts, may be found in an incipient or even sometimes in a well developed condition in the lower animals.' Again, his *Expression of the Emotions in Man and Animals* (1872) is a magnificent comparative study of expressive behaviour in animal and man: 'With mankind some expressions, such as the bristling of the hair under the influence of extreme terror, or the uncovering of the teeth under that of furious rage, can hardly be understood, except on the belief that man once existed in a much lower and animal-like condition.'

In America, John James Audubon (1785–1851) was not only making interesting observations on bird life, but was also carrying out simple experiments on bird behaviour, such as the habits of the vulture. In 1843 L. H. Morgan (1818–81) was writing on instinct and learning, and in 1868 he published his major book, *The American Beaver and his Works*, an important study of animal behaviour referred to approvingly by Charles Darwin.

With Douglas Alexander Spalding (*c.*1840–77) begins an era of careful observations and experiments that are often impressive even by today's standards. Although Spalding believed that animals should be observed as far as possible under natural conditions, he carried out a number of interesting experiments using birds, piglets (*Sus scrofa domestica*), and other animals. His work was known to other 19th-century researchers, such as Darwin and William James, but until his 1873 paper in *Macmillan's Magazine* was republished in 1954 he was relatively unknown to modern workers. A friend of J. S. Mill and tutor to the Amberleys, particularly Bertrand Russell's elder brother Frank, he carried out a series of fascinating studies, sometimes alone and sometimes assisted by the Amberley parents and children. His first major talk was delivered before the British Association meeting at Brighton on 19 August 1872 and a summary was reprinted in *Nature* for 10 October under the title 'On Instinct', with an enlarged paper in 1873. Spalding's life was short, and his remaining few papers and reviews were almost entirely published in *Nature* between 1873 and 1875.

Spalding was particularly interested in instinctive behaviour, and in the relative importance of nature and nurture in behaviour. Much of his work was concerned with the problem of how much of our perceptual and intellectual equipment is learned during our lifetime and how much is inherited. He believed in the Lamarckian doctrine of 'Inherited Associations' and that 'Instinct in the present generation of animals is the product of the accumulated experiences of past generations' (*Nature*, 6 (1872)). Using ingenious but simple methods, he made observations on imprinting, ANTI-PREDATOR reactions, and many other topics. For example, from his first brief account: 'Chickens hatched and kept in the dark for a day or two, on being placed in the light nine or ten feet from a box in which a brooding hen was concealed, after standing chirping for a minute or two, uniformly set off straight to the box in answer to the call of the hen which they had never seen and never before heard.' Or again, 'When twelve days old one of my little protégés running about beside me, gave the peculiar chirp whereby they announce the approach of danger, on looking up a sparrow-hawk was seen hovering at a great height overhead' (ibid.).

At about the same period the ever versatile Francis Galton (1822–1911) was writing about DOMESTICATION (1865) and gregariousness (1871) in animals. Galton had travelled a great deal and had observed the gregarious behaviour of the camel and the llama (*Lama*). His understanding of the oxen of the wild parts of western South Africa was based on over a year of travel in their company, spending an entire exploration on the back of one, with others by his side. Galton tries to show that the instincts of gregariousness have developed through NATURAL SELECTION, since separate animals are in great danger from the many beasts of prey around them. If they were too gregarious, they would interfere with each other; less gregarious and they would be too widely scattered to keep a sufficient watch against potential predators.

Sir John Lubbock (1834–1913), first Baron Avebury, was President of the Zoology section of the British Association when Spalding gave his 1872 lecture on instinct. Sir John was already known for his writings on prehistory, and for his interest in insects and plants. In addition, he shared a friendship and correspondence with Francis Galton, who designed some of his apparatus for him. Although Lubbock had relatively little direct influence on British research, his contribution to work in the United States was much greater, and his pioneer laboratory methods of research on insect behaviour were much appreciated by those seeking experimental techniques. His book *Ants, Bees and Wasps* (1882), based on years of research, was keenly studied. Some of his original methods, such as maze learning and PROBLEM SOLVING in the laboratory, were rapidly adopted as standard techniques. Communication between ants and insect COLOUR VISION were among his research interests, and his book on INTELLIGENCE and the SENSES OF ANIMALS (1888) dealt with sensation and instinct both in insects and in the dog. The use of statistical methods, and the care with which he used the reports of others, gave his work an objectivity which was influential on later research.

George J. Romanes (1848–94) was friend and a literary executor to Charles Darwin. Beginning his work relatively late in his short life, he has left a narrow impression of his achievements, since much of what he wrote has not been published. His *Jelly-Fish, Star-Fish and Sea Urchins* (1885) was a major contribution to our understanding of the structure and formation of these animals. In addition, he carried out experiments on the latency and summation of nerve impulses. Romanes' research in animal behaviour was concerned with the behaviour of the white-throated capuchin or cebus monkey (*Cebus capusinus*), the chimpanzee (*Pan*), homing of bees, olfaction in crabs (Brachyura), and direction-finding in cats (Felidae), among other topics. Unfortunately he also wrote some popular material, chiefly his *Animal Intelligence* (1882), in which he was sometimes uncritical and free in his interpretations, and because of this he has never received the credit he deserves for his other researches.

C. Lloyd Morgan (1852–1936), Professor, Principal, and first Vice-Chancellor of the University of Bristol, was a friend and admirer of Romanes. Despite this, he criticized his poor methodology and even that of Spalding. How do we know, he asked, that Spalding's chicks would not move towards any sound? How do we know that his protégés had a specific fear of sparrow-hawks and not a response to unusual ·noises and objects? Although Lloyd Morgan's own research was not large, he carried out valuable experiments on instinctive behaviour with incubated chicks, ducklings (Anatinae), and other birds. In addition, he examined the roles of IMITATION and of learning in animal behaviour. Much of our present-day terminology appears to have originated from his writings, and even 'behaviour' and the extensive use of 'animal behaviour' for this area of research can be ascribed to him. That chicks learned by means of TRIAL AND ERROR, that successful responses were 'reinforced' and unsuccessful ones were 'inhibited' were all terms that he employed.

In his *Introduction to Comparative Psychology* (1894), Lloyd Morgan enunciated his famous canon which has had enormous influence on research in comparative psychology and was the basis for much of his criticism of poor methodology and experimental controls: 'In no case may we interpret an action as the outcome of the exercise of a higher psychical faculty, if it can be interpreted as the outcome of one which stands lower in the psychological scale' (p. 53). Some workers took this principle too seriously and would not allow any interpretation of an advanced process, even if suggested by the evidence, and in 1900 Lloyd Morgan was obliged to add the following rider to his canon: 'To this it may be added—lest the range of the principle be misunderstood—that the canon by no means excludes the interpretation of a particular act as the outcome of the higher mental processes if we already have independent evidence of their occurrence in the agent' (*Animal Behaviour*). Unfortunately, Lloyd Morgan's rider has been almost totally ignored and the errors of the one kind have been replaced by errors of the other. Lloyd Morgan delivered the Lowell lectures in the spring of 1896 at Harvard University and lectured at New York, Chicago, and other universities.

E. L. Thorndike (1874–1949) began his research on animal intelligence at Harvard in the autumn of 1896 after Lloyd Morgan's visit. Thorndike's methods were openly indebted to Lubbock and Lloyd Morgan's. Whereas Lloyd Morgan's dog Tony had learned to use the latch on the garden gate by putting its head through the railings, Thorndike used a small wooden slatted enclosure from which the hungry animal escaped to food by pulling a wire, stepping on a platform, or pressing a lever. The animal, a cat, dog, or monkey, was given repeated trials, and the behaviour changes and decrease in performance time were recorded. From this information a learning curve was plotted to show how the animal learned the problem. Thorndike claimed to demonstrate that the animal would show trial and error until, over a number of trials, the successful behaviour was 'stamped in' and the non-successful 'stamped out'. These experiments had a great influence on re-

search in America and, to a lesser extent, in Europe. In Britain, Thorndike's work was partly repeated, and criticized in detail, by the sociologist L. T. Hobhouse (1864–1929).

Hobhouse, then a writer for the *Manchester Guardian*, was allowed access to many different animals at the Belle Vue Zoo in Manchester. He studied cats, dogs, chimpanzees, monkeys, and even elephants. First he objected to Thorndike's claim that cats learn gradually, since some of Hobhouse's animals learned very quickly, sometimes within minutes. Secondly, Thorndike's cats and dogs had been unable to imitate the successful ESCAPE of other animals, but here again Hobhouse found evidence to the contrary. Further criticisms followed from several other researchers. Lloyd Morgan and Wesley Mills thought the situation in Thorndike's experiments was too unnatural. Small and Kline thought the boxes were too small. Nevertheless, Thorndike's experiments were taken very seriously by later workers, and his research is often considered to mark the beginning of controlled animal experimentation. Small (1900) and Kline (1899) began to use mazes, and were among the first to employ white rats (*Rattus norvegicus*) for their experiments. So, too, was J. B. Watson, who, in 1903, published his study of neurological and behavioural changes in the white rat.

The twentieth century: growth of behaviourism and ethology. Hobhouse (1901) had devised certain TOOL USING tests with monkeys and other animals, in which a stick or box was used to obtain food. Watson (1908) had carried out related experiments with sticks, but had failed to obtain successful results. However, in 1917 Wolfgang Köhler confirmed many of Hobhouse's findings.

Watson had studied under J. Loeb (1859–1924) for a time and was interested in Loeb's extreme mechanistic theory of *tropisms*, in which all behaviour was explained in terms of physicochemical reactions to stimuli. H. S. Jennings (1868–1947) made detailed criticisms of Loeb's ideas, saying that they were artificial and simplistic. Despite this, some students in Germany and America admired this so-called 'objective' approach. At the same time the Russian work on CONDITIONING was becoming known in Europe and America. Wolfson's thesis on conditioned REFLEXES under PAVLOV's supervision was begun in 1899, and Tolochinov first used the term in 1903. Pavlov gave a lecture in Madrid in 1903 and the Huxley lecture in 1906 at Charing Cross Hospital, of which a report was printed in *Science*. A little later, in 1909, R. M. Yerkes and S. Morgulis published a review of Pavlov's work, and this was followed by another review by Watson in 1916. For the previous few years Watson had also been studying Bekhterev's ideas on conditioning.

In the spring of 1907, Watson had spent 3 months on the island of Bird Key in the Dry Tortugas group, where he observed the details of the lives of the noddy terns (*Anous stolidus*) and the sooty terns (*Sterna fuscata*; then known as *S. fuligi-*

nosa) during their nesting season. This resulted in a detailed study of the feeding habits of terns, the mating of the noddies and of the sooties, nesting behaviour, egg laying, mate recognition, recognition of territory by experiment, experiments into distance ORIENTATION, the use of problem boxes with sooties and noddies, development under captivity, and behaviour in mazes. Watson was strongly influenced by many strands of past and present research using experimental and physiological methods, by the introduction of controlled techniques by Lubbock, Lloyd Morgan, and Thorndike, and by the work of Loeb and the Russians. The insistence on phenomenological and introspective studies of animals was clearly distasteful to him, and in 1912 and 1913 he declared his allegiance to the 'new' movement of 'Behaviourism'.

Obviously, the idea of objectivity in studying animals and human beings was not new. Even the definition of psychology as the scientific study of behaviour had originated implicitly with Spalding and Lloyd Morgan, and explicitly with McDougall and Pillsbury years before.

William McDougall (1871–1938) in a number of publications had proposed a theory of instincts of such a rigid and uncritical form that many workers found the topic of instinctive behaviour distasteful. For a variety of reasons, the study of instinct under natural conditions, and the use of anecdotal report, became united in the minds of psychologists as the opposite of objective experimental laboratory work under controlled conditions. In the 1920s and 1930s a number of psychologists began experimental work on learning in animals which was felt to be free from the taint of poor methodology and experimentation; among these were E. R. Guthrie (1886–1959), C. L. Hull (1884–1957), E. C. Tolman (1886–1959), and B. F. SKINNER. The complaints that their work was not truly comparative, that it tended to concentrate on the laboratory rat and the pigeon (*Columba livia*), and, above all, that it was carried out away from the natural habitat and disregarded the animal's life cycle, were either ignored or not heard by many psychologists. Not all psychologists were so inclined. For example, R. M. Yerkes (1876–1956) worked on medusae (free-swimming stage of certain coelenterates), on segmented worms (Annelida), on 'waltzing' mice (Murinae), and extensively on primates. Similarly, F. A. Beach, T. C. Schneirla, D. Lehrman, and K. S. Lashley, for example, did not concentrate on learning, but had wide interests. None the less, the dominant trend in the 1930s, 1940s, and early 1950s was limited in the way just described.

In America also, at the turn of the century, zoologists were interested in animal behaviour and intelligence. Charles Otis Whitman (1842–1910), a distinguished zoologist and founder-director of the Marine Biological Laboratory at Woods Hole, Massachusetts, found that many of the statements of workers in the subject showed an ig-

norance of the actual behaviour of animals. In his very critical *Myths in Animal Psychology* (1899) he wrote: '. . . any attempt to soar to "the nature and development of animal intelligence", except through the aid of long schooling in the study of animal life, is doomed to be an Icarian flight' (*The Monist*, 9). Whitman's most influential work on instinctive behaviour was his study of it in insects and in pigeons, with an analysis of earlier theories. In one of his summaries he wrote: 'Instinct and structure are to be studied from the common standpoint of phyletic descent, and that not the less because we may seldom, if ever, be able to trace the whole development of an instinct.' And, 'Although instincts, like corporeal structures, may be said to have a phylogeny, their manifestation depends upon differentiated organs. We could not, therefore, expect to see phyletic stages repeated in direct ontogenetic development, as are the more fundamental morphological features, according to the biogenetic law. The main reliance in getting at the phyletic history must be comparative study' (*Biological Lectures from the Marine Biological Laboratory of Woods Hole*, 1898).

With such statements we have the beginnings of modern ethology. Whitman's writings were apparently not well known in Europe at the time, but later Konrad LORENZ, among others, was influenced by them. Other observations of Whitman merit attention; for example: 'The clock-like regularity and inflexibility of instinct, like the once popular notion of the 'fixity of species' have been greatly exaggerated' (ibid.). Or again, 'Plasticity of instinct is not intelligence, but it is the open door through which the great educator, experience, comes in and works every wonder of intelligence' (ibid.).

Wallace Craig (1876–1954), a pupil of Whitman, also wrote on instinct and behaviour, and again influenced the thought of Lorenz and other ethologists. Using a large collection of pigeons kept by Whitman, Craig examined the voices of the pigeons as a means of social control, and claimed that it is not possible to explain COOPERATIVE behaviour and SOCIAL behaviour only in terms of a group of social instincts. An individual needs to be guided and regulated by other individuals, and for this the pigeon or dove (*Streptopelia*) uses bowing, strutting, bristling, and many other activities, in addition to voice and SONG. Craig described the effect of song on the behaviour of the dove as a form of control over the mate, in pairing generally, and as a means of proclaiming its species, sex, identity, and territory. In his 1914 paper 'Male Doves Reared in Isolation', he described the behaviour of three out of eight doves reared in isolation after weaning. One dove, Jack, showed all the signs of strong attachment to Craig, and went through the various amorous motions on his hand. The dove Billy also showed a strong attachment to Craig, with 'all the signs of excitement and joy in our presence'. Many of Craig's observations on doves reared in relative isolation led him to observe that

'When a dove performs an instinctive act for the first time, it generally shows some surprise, hesitation, bewilderment, or even fear; and the first performance is in a mechanical, reflex style, whereas the same act after much experience is performed with ease, skill and intelligent adaptation . . .' (*Journal of Animal Behaviour*, 4).

It was Craig's 1918 paper, 'Appetites and Aversions as Constituents of Instincts' (*Biological Bulletin*, 34), which was the most influential of his writings. Here he wrote: 'Instinctive behaviour does not consist of mere chain reflexes . . . this reflex action constitutes only a part of each instinct in which it is present.' At the beginning he makes his famous distinction between APPETITIVE and CONSUMMATORY behaviour: 'An appetite . . . is a state of agitation which continues so long as a certain stimulus, which may be called the appeted stimulus, is absent.' 'The state of agitation . . . is exhibited externally by increased muscular tension; by static and phasic contractions of many skeletal and dermal muscles . . . easily recognized as signs or "expressions" of appetite . . . by restlessness; by activity. . . .' This appetite behaviour 'is accompanied by a certain *readiness to act*. When most fully predetermined, this has the form of a chain reflex. But in the case of most supposedly innate chain reflexes, the reactions of the beginning or middle part of the series are not innate, or not completely innate, but must be learned by trial. The end action of the series, the consummatory action, is always innate.' Independently, Heinroth in 1910 called such actions *arteigene Triebhandlungen* or *species-specific instinctive actions*, and still later they were often referred to as FIXED ACTION PATTERNS. What Craig and Heinroth were saying was that the animal does not simply respond to a stimulus, but searches using appetitive behaviour for a particular stimulus condition which permits the consummatory behaviour to take place; the searching behaviour ends and a state of rest follows.

Interestingly enough, Craig believed that all instinctive activity runs in such cycles, and that, in general, each cycle had four phases. In actual life the cycles and phases 'multiplied and overlapped in very complex ways' (ibid.). The first phase was that of appetite and search; the second, of reception, consummation, and satisfaction; the third phase was surfeit, and aversion to the stimulus, and finally, the fourth phase was freedom from the stimulus and a state of rest. Not only in animals, but also in human behaviour, such cycles are to be found, involving basic needs and even 'higher' activities such as listening to music, levels of AT-TENTION, and feelings of EMOTION. 'The entire behaviour of the human being is, like that of the bird, a vast system of cycles and epicycles, the longest extending through life, the shortest ones being measured in seconds' (ibid.). Craig's 1921 paper 'Why Do Animals Fight?' has relevance to arguments about AGGRESSION in animals and men. From his observations of many species, Craig

argued that if an animal's instincts are not thwarted, or if it is not annoyed, it will not fight. When an animal does fight, it aims only to remove the enemy, not to destroy it; if the enemy submits, the agent ceases FIGHTING. Killing a non-resisting member of one's own species is exceptional. DE-FENSIVE fighting pays but not aggressive fighting, which is wasteful; there is no distinctively 'biological' need for fighting. Finally Craig stated 'The facts of animal behaviour prove that fighting to the death is not necessary for the welfare or for the evolution of the race'.

During the first decades of the 20th century in Britain, the young Julian Huxley was lecturing in zoology at Oxford. In 1912, Huxley published a paper on the COURTSHIP of the redshank (*Tringa totanus*), and this was followed in 1914 by the first of his studies on the courtship DISPLAYS of grebes (Podicepedidae) and related birds. Edmund Selous's dedicated work with birds influenced not only Huxley but also H. Eliot Howard's *Territory in Bird Life* (1929). Howard in turn was in contact both with Huxley and with Lloyd Morgan, while, in the early years of the century, the Irish naturalist C. B. Moffat was already writing on similar topics. In 1936 and 1942 F. H. A. Marshall at Cambridge published studies on the effect of stimulus patterns and displays of the male bird on the SEXUAL and reproductive behaviour of the female.

In Berlin, Oskar Heinroth (1871–1945), an assistant director of the zoological garden, produced his *Ethology of the Anatidae* (1910 and 1911). In this important contribution to ethology, Heinroth made detailed studies of various species of ducks and geese (Anatidae), comparing their movements, anatomy, social behaviour, calls, and reproductive habits. Independently of Whitman, and in more detail, he illustrated how the idea of *homology* (i.e. basic similarity of structure) was as valuable in the study of movement (action) patterns as it was for morphological characters (see COMPARATIVE STUDIES). Many now believe that Heinroth's detailed work on homologies in behaviour sequences marked the actual birth of ethology, and of the comparative study of behaviour. We also owe to Heinroth, who had perhaps borrowed it from Giard, the first substantial modern use of the term 'ethology'. In addition, he had observed imprinting in a greylag gosling (*Anser anser*) after HATCHING in an incubator. Between 1925 and 1933 Heinroth and his wife published their four volumes of *The Birds of Central Europe*, which contained much material on instinctive behaviour and on learning.

The two most significant contributors to the study of ethology in this century, Konrad Lorenz and Niko TINBERGEN, have both expressed their indebtedness to Heinroth's ideas. Lorenz, in fact, was in close contact with Heinroth for at least ten years. We should also mention the important work of Karl von FRISCH, who does not appear to have used the term 'ethology' himself. Von Frisch was born in 1886 in Austria. After graduation he began work on colour changes in fish in relation to the function of the *pineal body* (part of the forebrain). Colour vision in animals became one of his main interests and, despite opposition to the idea, he succeeded in demonstrating colour changes in relation to the background. At his house in Brunnwinkl on Lake Wolfgang, he began to carry out research on colour vision in bees around 1912, and eventually showed that the belief that honey-bees (*Apis mellifera*) were colour-blind was not correct. In 1919, he first noticed the dances of the honeybee when he observed a 'scout' perform a 'round dance' on the honeycomb, exciting the foragers around her to fly to the food source.

Also working early in the century was Jakob von Uexküll (1864–1944), whose ideas on the perception of animals much interested the early ethologists, particularly Lorenz. Von Uexküll proposed the concept of the *Umwelt*, the subjective phenomenal world as the animal itself sees it in contrast with the actual environment. He regarded the first task of research into the *Umwelt* as identification of each animal's perceptual cues among all the stimuli surrounding it. The *Umwelt* contained certain 'key stimuli' which unlocked a releasing mechanism within the animal. A tick (Ixodidae) might wait for years before it perceived the key stimulus of butyric acid on a mammal's body, and then pounce. Lorenz dedicated his famous 'Kumpan' paper to Uexküll on his seventieth birthday.

Konrad Lorenz was born in 1903 in Austria. He first studied medicine in Vienna and later comparative anatomy, philosophy, and psychology. Eventually he became a demonstrator and then a lecturer in comparative anatomy and animal psychology. During this period, until about 1940, he was carrying out naturalistic observations on various HOUSEHOLD PETS and animals at the family home at Altenberg, near the village of Greifenstein. He relates that at the age of five he was given a duckling which grew into 'a fat female Rouen duck ... which lived to a ripe old age.... It is my honest belief that I learned from that bird quite as much as I did from most of my human teachers' (E. H. Hess, *Imprinting*, 1973). In 1940 Lorenz was appointed Professor of Philosophy at Königsberg, but in 1943 he was drafted into the medical service and later into the German army until captured by the Russians. At the end of the war he returned to his home at Altenberg and continued to carry out research under difficult conditions. In 1950 he opened a temporary research station at Buldern, with the assistance of the Max Planck Society, and in 1956 he moved to Seewiesen with the founding of the Max Planck Institute for Behavioural Physiology.

Lorenz's output is enormous and it is only possible to refer to some of his major studies here. From at least as far back as 1927, he was writing about the *Corvidae* (jackdaws, etc.). His famous 'Kumpan' paper of 1935 is usually taken as a landmark in his work. ('Der Kumpan in der Umwelt des Vogels', *Journal für Ornithologie*, 83.

Translated as 'Companionship in Bird Life', in C. H. Schiller, ed., *Instinctive Behaviour*, 1957.) Here he presents most of the ideas that have become associated with 'classical' ethology. Uexküll's ideas on the *Umwelt* are described and the concept of the SIGN STIMULUS (key stimulus) is developed, together with the *innate releasing mechanism* (IRM). The term *releasers* is used for the devices that activate the ready-made IRMs and so set off specific-innate action chains in fellow members of the species. Later Tinbergen was to introduce the term 'social releaser'. The releaser was thought of as a key used to unlock the IRM. 'I have adapted Uexküll's term "releasing mechanism" for the receptor correlate to a releasing stimulus combination, for the readiness to respond specifically to a certain key combination, and thus to activate a certain behaviour pattern' (ibid.). Again, Heinroth's (1910) account of imprinting is described, and Lorenz's own ideas on the topic are developed. Craig's (1918) ideas on appetitive behaviour are briefly mentioned and the concept of VACUUM ACTIVITY is introduced. 'The threshold of releasing stimuli is lowered ... when the reaction is not released over a longer than normal period ... when the threshold value drops to its theoretical minimum, a phenomenon may occur that I have described ... and called "vacuum activity"' (ibid.).

In his 1937 paper on the nature of instinct, Lorenz examines various ideas, including those of Whitman (1899) and Craig (1918). He discusses Craig's distinction between *appetitive* flexible searching behaviour and the innate *consummatory action*. The so-called Craig–Lorenz model of the accumulated *action-specific energy* is presented using the image of a gas constantly pumped into a container until 'under quite specific circumstances, a discharge occurs.... The adequate stimulus, or more precisely, the adequate stimulus combination, corresponds to a single tap, which can reduce the pressure in the container to the level of the pressure outside' (Schiller, ed., *Instinctive Behaviour*). Later, Lorenz used a hydraulic model (1950).

In 1938 Niko Tinbergen came to Altenberg to work with Lorenz. Their work on the egg-rolling movements of the nesting goose was one of their most important studies. Tinbergen was responsible for the design and execution of the experiments, while Lorenz contributed mostly the theoretical arguments in the paper. The year 1949 saw Lorenz's reunion with Tinbergen, who had also suffered in the war, being confined to a type of hostage camp in the Netherlands. Also in 1949 the Society for Experimental Biology and the Association for the Study of Animal Behaviour held a joint conference on 'Physiological Mechanisms in Animal Behaviour' at Cambridge. Lorenz's paper presented, among other things, his general viewpoint on 'action-specific energy', 'innate releasing mechanisms', and fixed-action-patterns (consummatory behaviour), in addition to the hydraulic model of energy discharge. In 1952 Lorenz modified this hydraulic model and suggested other changes.

Lorenz's delightful account of his studies at Altenberg with his animals was published in German in 1949, and in 1952 in English, under the title *King Solomon's Ring*. The whole book is illustrated with his own talented drawings. In 1954 the popular *Man Meets Dog* was published. This was followed by many papers and books, including: *Evolution and Modification of Behaviour* (1965), *On Aggression* (1966), *Studies in Animal and Human Behaviour* (1970, 1971). In 1979 Lorenz published a beautifully illustrated book, *The Year of the Greylag Goose*, containing 150 colour photographs by Sybille and Klaus Kalas.

Niko Tinbergen, born in 1907, was a student of zoology at Leiden in the Netherlands. In 1931 he was appointed assistant in the Department of Zoology at Leiden. His research work, which he began in the early 1930s, was mainly on birds, and his first papers contained studies of the herring gull (*Larus argentatus*) and the common tern (*Sterna hirundo*). The next few years were concerned with a digger wasp *Philanthus triangulum* and its homing behaviour (1932), its hunting behaviour (1935), and its selective learning of landmarks (1938). In 1938 Tinbergen worked with Lorenz on 'the way in which the greylag goose retrieves eggs displaced from the nest' (Schiller, ed., *Instinctive Behaviour*, 1957). In 1939 he published with Kuenen his important study 'Feeding Behaviour in Young Thrushes' (ibid.). Here they carried out major research on sign-stimuli and releasers, using blackbirds (*Turdus merula*) and song thrushes (*T. philomelos*) to examine gaping as a result of jarring the nest, and of the presentation of 'dummy' cardboard heads.

Tinbergen's significant work on DISPLACEMENT ACTIVITIES was published in 1940. He published many other studies at this time, including one on the mating behaviour of the three-spined stickleback. In 1949 he was invited to Oxford University as a Lecturer in Zoology, and in 1950 published an article in the Society for Experimental Biology (S.E.B.) symposium concerned with the hierarchical order of appetitive behaviour and of consummatory acts. The chain of appetitive behaviour found, for example, in the reproductive behaviour of the three-spined stickleback could not be explained simply in terms of the Craig–Lorenz model, and Tinbergen proposed a hierarchical model in which the consummatory act does not end the behaviour, but leads to a further stimulus condition producing additional appetitive behaviour. A chain of behaviour is formed in which the fish goes through a complex sequence of behaviours, beginning with the development of the male stickleback's red belly and the establishment of his territory.

In 1951 Tinbergen's *The Study of Instinct* was published. This was the first major comprehensive text on ethology to be published in English, and its influence was very great indeed. It was followed by *The Herring Gull's World* (1953) and *Social Behaviour of Animals* (1953). Among Tinbergen's later books were: *Curious Naturalists* (1959), *Animal Behaviour* (1965), and, in 1972, a collection of his research studies entitled *The Animal in its World: Explorations of an Ethologist 1932–1972*.

In 1973 Karl von Frisch, Konrad Lorenz, and Niko Tinbergen shared the Nobel Prize for Medicine. In this way the enormous contribution of these three workers to the study of animal behaviour was recognized.

Let us complete this entry with a look at the position at present. What has happened to the difference of attitude between the psychologists concentrating on learning and the ethologists? Clearly there has been a *rapprochement* since the early 1950s. Many ethologists have become aware of the need for greater sophistication in experimental design and statistical analysis, and psychologists have become more aware of the possible artificiality of laboratory conditions compared to field studies. There are some psychologists who generalize far too readily from one species to human behaviour, and on the other hand, some ethologists who generalize from animal behaviour to social questions in human beings with equal facility. Given this awareness of their mutual problems and emphases, much has been gained from the relationship, and perhaps it is a good sign that some researchers appear to be at home in both areas, preferring to consider themselves simply as students of animal behaviour.   B.S.

15, 29, 58, 63, 70, 72, 114, 136, 141

HOARDING is the storage of food, or other things, either in a central cache, or as many concealed items distributed through the HOME RANGE or TERRITORY. Storing food in a single large cache, sometimes called *larder hoarding*, occurs in many small mammals and in some birds. Acorn woodpeckers (*Melanerpes formicivorus*), for example, which live in small groups and are found in Mexico and the south-western United States, prepare trees by drilling hundreds of evenly spaced holes in the trunk and branches, and fill each hole with an acorn. Many social insects collect large hoards of food. Bumble-bees (*Bombus* and other genera) construct separate vessels in their nests for the storage of pollen and nectar. *Scatter hoarding*, the storage of food in many dispersed sites, is common in many birds, e.g. members of the crow family (Corvidae), nuthatch family (Sittidae), and tit family (Paridae), and in mammals such as the red fox (*Vulpes vulpes*), and various squirrels (Sciuridae). The common jay (*Garrulus glandarius*) buries acorns one at a time, and may store several thousand within its home range.

Many types of food can be hoarded. Seeds and nuts are often stored, since they remain edible for long periods, but insects and larger prey are hoarded too. Coal tits (*Parus ater*) prepare insect larvae for storage by removing the head and gut, and hoard aphids (Aphidae) after compacting between twenty and fifty into a small pellet.

There are no simple rules concerning which species hoard food and which do not. Both group-living and solitary animals hoard, though highly mobile species, which rarely return to the same area, do not. Some animals have evolved special structures to transport food for storage, such as the cheek pouches of some rodents, and the sublingual pouches of nutcrackers (*Nucifraga*).

Food is hoarded in a great variety of places. Small rodents often maintain large caches of food in a separate chamber of their burrows. The grey squirrel (*Sciurus carolinensis*) stores food in holes it digs in the ground, which are carefully covered with leaves and grass, and crows, jays, magpies (*Pica pica*), and nutcrackers do the same. Spotted hyenas (*Crocuta crocuta*) cache prey in water up to ½ m deep, and recover it by plunging under the surface. The type of site used for hoarding may be quite important. Clark's nutcracker (*Nucifraga columbiana*) of western North America hoards pine seeds in the ground on south-facing slopes, which accumulate little snow in winter, and thaw earliest in spring. Using a variety of sites, and changing the type of site used, may reduce losses to other animals which exploit stored food.

Food hoarding serves many functions. Periods of low food availability can be survived by eating food stored during periods of abundance. Jays and nutcrackers rear the young with stored food which would otherwise be unavailable when the young are in the nest. Hoarding also has benefits over a shorter period of time. Some animals store food one day and consume it the next if FORAGING conditions are poor. Carnivores such as leopards (*Panthera pardus*), bears (Ursidae), and weasels (*Mustela*), cache prey to prevent its loss to scavengers. When competition for food is intense at a particularly rich source, for example between fox cubs at a carcass, or between tits at a bird table, an animal may be able to garner much more food for its own use by hoarding than by eating immediately and returning later when it is hungry again.

Animals which do not hoard food often exploit food stored by others. Fox squirrels (*Sciurus niger*) and birds attempt to feed at trees used for storage by acorn woodpeckers. Willow tits (*Parus montanus*) will follow food hoarding coal tits and rob their caches. Carrion crows (*Corvus corone*), which sometimes hoard food themselves, will search for food hoarded by jays and magpies. Scatter hoarding and larder hoarding species deal with this PARASITISM of their stored food supply in different ways. Larder hoarders vigorously defend their caches. Acorn woodpeckers attack and drive away any animals, including acorn woodpeckers from other groups, that approach their storage trees. Scatter hoarding animals cannot defend their many storage sites in this way, and instead reduce the likelihood of losing their caches by spacing them out, and using different types of places for storage. If storage sites are sufficiently far apart, any animal finding one by chance will be unlikely to find another by SEARCHING. Fox squirrels and marsh tits (*Parus palustris*) maintain such spacing between storage sites. Varying the type of site used for storage may also reduce the loss to other animals. Tits use a great variety of sites for food storage, including crevices in tree bark, hollow stems, moss, dead leaves, and depressions in the ground. Varying the type of site used may prevent other animals from LEARNING the location of stored food.

While scatter hoarding makes it uneconomical for other animals to exploit the caches, it also presents the hoarding animal with the problem of recovering its own stored food. Although some species may only recover their hoarded food when it is re-encountered during normal foraging, others accurately remember the location of a prodigious number of stored food items. The red fox, the common jay, and the marsh tit all remember the location of hoarded food and readily find it again. Experiments have shown that this is accomplished without relying on the scent of hidden food, but by the MEMORY of its location. Jays are known to store acorns around landmarks, such as trees, which may help them to relocate sites.

Little is known about the immediate causation of food hoarding, or why an animal stores rather than eats a particular item. The amount of food available has some effect, and in animals which hoard seasonally, such as the deer mouse (*Peromyscus maniculatus*) of North America, low temperatures and the short day-length of autumn increase the amount of food hoarded.

Food hoarding may have remarkable effects on the plant producing the food which is hoarded. Trees bearing seed cones or nuts are particularly influenced by food hoarding animals, and CO-ADAPTATION has occurred between them. As not all of the food which is hoarded is later found and eaten, a tree or plant may benefit from having its seeds dispersed and concealed in conditions favourable for germination, safe from rotting, desiccation, or PREDATION by other animals. Food hoarding species may exert NATURAL SELECTION on such traits as the size of seeds or nuts, thickness of the shell, and even the timing and abundance of the seed crop. A food plant may in turn influence the EVOLUTION of the behaviour of food hoarding animals, by varying the properties of its seed or nut crop. Acorns of a certain size and shape are preferred for hoarding by jays. Years in which trees, such as oaks (*Quercus*) and beeches (*Fagus*), produce a particularly heavy crop of mast (acorns and beech nuts) may encourage hoarding.   D.F.S.

18

HOME RANGE is the term given to the area in which an animal normally lives, regardless of whether or not the area is defended as a TERRITORY, and without reference to the home ranges of other animals. Home range does not usually include areas through which MIGRATION or DIS-PERSION occur.

Home range may vary with the age, sex, or breeding condition of the individual. The home range of male tree sparrows (*Spizella arborea*), for instance, varies with the stage of the nesting cycle, being at its smallest just prior to NEST-BUILDING and largest during incubation. Animals do not visit all parts of their home range equally, and among tree sparrows the males have a more extensive range than the females, and both members of a pair frequently visit the core area in the region of the nest. Female tree sparrows do not fight, but the males are territorial, and consequently the home ranges overlap little. Overlapping home ranges are common in species that do not defend their areas, but may also occur in territorial animals if adjacent owners rarely meet. For example, many mammals habitually patrol a system of pathways radiating from a den or core area of the home range. The pathways may penetrate into the next home range, but if they do not cross the neighbours' pathway, then it is likely that the potential rivals will seldom meet. In elephant shrews (Macroscelididae) the territory consists of a burrow, one or more feeding areas, and trails linking them with the burrow. Territories may also interpenetrate in the domestic cat (*Felis catus*). Neighbours may even use the same path, though not at the same time. By keeping to a fairly definite timetable encounters can be avoided and neighbouring cats may not interefere with each other's movements.

In some species, such as the common hamster (*Cricetus cricetus*), all or most of the home range is defended and is therefore the same as the animal's territory. In other species, such as the herring gull (*Larus argentatus*), there is a small defended territory within a vast home range. COURTSHIP, INCUBATION, and PARENTAL CARE occur on the territory and foraging takes place in the rest of the home range. There may be no particular part of the home range that is defended, but DEFENSIVE behaviour may occur in relation to a moving resource. For example, male deer (Cervidae) have a large home range within which they defend a harem (see HERDING). The glaucous gull (*Larus hyperboreus*) may defend a drifting food source within its home range.

Although there is tremendous variation in the size of home ranges among different species, there is a tendency for larger animals to have larger home ranges. As body size increases, daily ENERGY requirements also increase, more food is needed, and a larger area is required for FORAGING. PREDATORY animals seem to have larger home ranges than herbivores of equivalent size, probably because their food is more sparsely distributed.

HOMEOSTASIS refers to the maintenance of an equilibrium between the organism and the environment. The term is usually used to denote the equilibrium of the internal environment of an individual, with respect to body temperature, blood sugar level, oxygen levels, etc. However, it can also be applied to any type of biological regulation, such as that of a population, or COMMUNITY.

It is important for many animals to maintain an internal equilibrium despite environmental change. This idea was first propounded by Claude Bernard in 1859. Bernard had observed a constant level of sugar in the blood of animals which had just consumed meat, or food containing sugar, or which had been deprived of food. He recognized that this constancy of blood sugar implied that processes of regulation and control must have been designed to maintain the constancy of the internal environment. Bernard's famous dictum, 'La fixité du milieu intérieur est la condition de la vie libre' (*Leçons sur les phénomènes de la vie communs aux animaux et aux végétaux*, 1878), refers to the fact that an animal which is able to regulate its internal environment, despite fluctuations in the external environment, has greater freedom to exploit a variety of potential HABITATS. For example, mammals have many mechanisms of THERMOREGULATION, and are able to maintain their body temperature within 5 °C of the normal value of about 37 °C, despite a range of environmental temperature from about −20 °C to 40 °C. Mammals can thus remain active over a wide range of environmental temperatures, whereas lizards (Lacertidae) are generally able to remain active only in the 25–35 °C range. Thermoregulation in lizards is confined to a few primitive physiological responses plus behavioural measures, such as sun-basking, by which they are able to alter their body temperature indirectly. Compared with the warm-blooded mammals and birds the geographical distribution of lizards is very much restricted by their TOLERANCE of environmental temperature. Likewise, their behaviour is curtailed by the fact that they can become active only at certain times of day, when the temperature is neither too high nor too low.

The term homeostasis was first used by the American physiologist Walter Cannon, who wrote as follows: 'The coordinated physiological processes which maintain most of the steady states in the organism are so complex and so peculiar to living beings—involving, as they may, the brain and nerves, the heart, lungs, kidneys and spleen, all working cooperatively—that I have suggested a special designation for these states, *homeostasis*' (*The Wisdom of the Body*, 1939). Cannon envisaged a situation in which sensory processes, monitoring the internal state of the body, initiated appropriate action whenever the internal state deviated from a pre-set, or optimal, state (see SENSES OF ANIMALS). For example, when human body temperature rises above 37 °C cooling mechanisms, such as flushing and sweating, are brought into action. When the temperature falls below the optimal level, warming mechanisms such as shivering come into play. By employing a number of such finely tuned mechanisms, humans are able to achieve a precise thermoregulation and thermal homeostasis.

Although the regulatory mechanisms of the type envisaged by Cannon are now known to be widespread in the animal kingdom, and to involve a great variety of physiological processes, they are not the only type of process involved in the control of the internal environment. Until recently, it was thought that the increased DRINKING by animals in response to high environmental temperatures was a result of dehydration arising from the increased water losses involved in cooling responses, such as sweating and panting. Such a view is entirely consistent with a theory of the maintenance of homeostasis by regulation, as outlined above. Evaporation of water is necessary to maintain thermal homeostasis in a hot environment, and this upsets the fluid balance of the body, the restoration of which requires increased drinking. However, it is now known that in species such as the rat (*Rattus*) and pigeon (Columbidae), the drinking occurs as a direct response to the temperature change, in anticipation of any change in fluid balance arising from thermoregulation, and not in response to it as the regulatory theory would require. In other words, the animals drink in order to have water available for thermoregulation.

The role played by behaviour in the maintenance of homeostasis varies considerably with species and with circumstances. Drinking behaviour, for instance, is essential for the maintenance of homeostasis in many species, the physiological mechanisms of water conservation being unable to prevent lethal dehydration after prolonged water deprivation. However, some species, such as the mongolian gerbil (*Meriones unguiculatus*) and the budgerigar (*Melopsittacus undulatus*), are able to survive indefinitely without water, because of the great efficiency of their water conservation mechanisms. Others, such as aquatic species, need no special behaviour to obtain the water necessary for the maintenance of homeostasis.

In many aspects of homeostasis the role of behaviour is normally negligible. However, experiments involving surgical interference with the physiological mechanisms responsible for homeostasis show that animals often have the capacity for appropriate behaviour, even though it is not manifested in their normal lives. For example, when the thermal homeostasis of rats is interfered with by removal of the thyroid gland, they respond, if given the opportunity by the experimenter, by building warmer nests and by indulging in other forms of behavioural thermoregulation. Similarly, removal of the adrenal gland, which is involved in salt balance, causes rats to shift their preferences towards more salty food and water. In other cases there may be no special organs involved. Thus rats fed a vitamin deficient diet are able to select food containing the required vitamins, even though they cannot taste the presence of the vitamins in the food. It seems that they are able to learn which food made them feel better (see FOOD SELECTION).

The concept of homeostasis has had great influence upon the development of thought in animal behaviour research. This is partly because an animal's physiological requirements are bound to have a great effect upon its MOTIVATION, and partly because homeostasis is a good example of the FEEDBACK principle, by which the behaviour of a system is partly controlled by its own consequences.

103, 106

HOMING occurs in a wide variety of species and ranges from the relatively simple task of returning home after FORAGING to the complexities of a MIGRATION from one home to another. Adelie penguins (*Pygoscelis adeliae*) may leave their young chicks for up to 2 weeks while they forage far out at sea. They are able to return to their nest even if displaced by storms, or by a student of ETHOLOGY. Animals that travel in the air or water always risk being carried off course by wind or current (see ORIENTATION) and many possess remarkable homing abilities. A shearwater (*Puffinus*), which was removed from its burrow in Britain and transported to the USA, returned home within 13 days of being released, a journey of over 4500 km. The green turtle (*Chelonia mydas*), which breeds on Ascension Island, feeds off the coast of Brazil about 1500 km away. The adult female turtles visit the island, which is only 7.5 km wide and is in the middle of the Atlantic ocean, and lay their eggs in the sand of the beaches. They then return to their feeding grounds where they remain for several years, before returning to the same beach to lay another clutch of eggs. Even more remarkable is the homing behaviour of the Californian newt (*Taricha torosa*). These newts spend the summer months underground, forage widely in the forests in the autumn and winter, and migrate to their particular breeding grounds in the spring. They breed in pools in mountain streams, and return to the same pool each year. Of a group of newts that were taken 4.5 km to a deep canyon with a 300-m high ridge between them and their home pool, a number were found back home the following year. Such studies illustrate not only remarkable homing ability, but also a very persistent MOTIVATION to return home despite the obstacles and difficulties.

To return home successfully across an ocean, or a mountain barrier, requires a considerable feat of NAVIGATION. The homing animal has not only to set off in the correct direction, but has also to maintain a correct course despite changing weather conditions. A number of navigational cues are known to be used by animals, although the way in which they are actually deployed is not fully understood. An important factor is TIME. Apparent time changes with one's position on the earth, and an indication of position can sometimes be gained by comparing the apparent local time with one's personal time. Many animals are able to use their internal CLOCK to keep track of their personal time. The internal clock will be synchronized to the apparent time at the place where the animal last spent a long period of time. Thus the internal clock of a pigeon (*Columba livia*) living in London will be set at London time. If the pigeon were to be transported to New York then the apparent time in New York would be about 5 h earlier than that given by the pigeon's internal clock. By comparing this personal time with the local apparent time, the pigeon would have the information required to indicate that it was about 4500 km west of London. In fact, pigeons and many other birds are known to use their internal clocks in this way. If the internal clock is experimentally reset, then the animal makes characteristic navigational errors.

Apparent time, and some positional clues, can be read from the sun and stars. Many animals are known to use this type of information, but some can also navigate when the sky is overcast. Adelie penguins removed 1800 km from their rookery set off walking in a NNE direction, which is the direction that they would normally take to the sea from their rookery. They were able to maintain this course, provided that the sun was visible. When the sun was obscured the birds' movements were random. However, pigeons and many other birds are able to navigate when the sun is overcast. Many SENSE ORGANS are known to be implicated in navigation and homing, including ELECTRO-MAGNETIC SENSES, CHEMICAL SENSES, and various aspects of VISION. Honey-bees (*Apis mellifera*) use the position of the sun as a navigational aid. When the sky is overcast they obtain equivalent information from the pattern of polarization of the light in the sky. Homing salmon (Salmoninae) are known to be able to discriminate some chemical characteristics of the river in which they were born, even when they are several kilometres out to sea. Salmon migrate great distances and return to the same river for each spawning. Visual landmarks are used by a number of animals in locating the exact position of their home when they are already near. If the burrow of the bee-killing digger wasp (*Philanthus triangulum*) is ringed by pine cones, the wasp will use these as a landmark. If the ring is removed a metre or so away from the burrow, then the wasp returning from a foraging trip will home in on the centre of the ring.

The complexity and sophistication of the sensory mechanisms used in homing suggest that there has been considerable pressure of NATURAL SELECTION, and that homing is very important in the life of many animals. Most animals show specific ADAPTATION to a particular HABITAT and if they are displaced from their normal HOME RANGE it will be important for them to be able to return. A well-developed homing ability is particularly valuable in saving time and energy when large distances have to be covered, and when the starting point is unfamiliar. Many HUNTING animals, such as the African wild dog (*Lycaon pictus*), the spotted hyena (*Crocuta crocuta*), leave their young in the den and travel up to 30 km in a night. They bring food home to the young from what is sometimes unfamiliar territory. Similarly, foraging birds may become lost in a storm and have to find their way back to the nest. When a particular habitat becomes unfavourable, due to seasonal changes in CLIMATE, some animals migrate to an alternative residence, sometimes thousands of kilometres away. It is amongst these migratory animals that the most impressive homing abilities are found.

24, 37, 69, 123

HORMONES are well known to play a vital part in the control of our physiology. In medicine, the use of synthetic hormone preparations has revolutionized treatment of diseases which were thought to be incurable. As a consequence, the structure of our society, in terms of population growth, has been influenced by these blood-borne chemical messengers. While the physiological effects of hormones prescribed for therapeutic purposes are usually well understood, it is no overstatement to say that our knowledge of the effects of hormones on human behaviour has shown few advances since the discovery of these substances at the beginning of the century. We have to look to research on animals for any comprehensive view of the relationships between hormones and behaviour.

To attempt to describe these relationships it is necessary to consider how hormones exert their effects on behaviour. One of the few generalizations that can be made with any confidence is that hormones by their own action do not normally initiate the performance of behaviour. Hormones ensure the correct physiological background which will permit the occurrence of behaviour, given certain conditions of external stimulation, and depending upon the prior behavioural experience of the individual. Hormones act, therefore, in association with other factors which are equally important in producing the behaviour. For example, the sexually inexperienced male cat (*Felis catus*) is unlikely to display normal SEXUAL behaviour despite high blood levels of *androgen* and adequate stimulation by the female; sexual experience during critical phases of development is as vital to sexual behaviour as hormone level. The fact that hormones have effects on behaviour that are difficult to isolate from other contributory factors obviously complicates the search for answers to questions about the mode of action of hormones on behaviour. However, due to improvements in methods of describing behaviour as endpoints of hormonal action, and also to a better understanding of the physiological actions of the hormones themselves, we know which hormones are likely to influence behaviour, and whether fluctuations in the level of hormones in the blood can be correlated with particular properties of behaviour specifically associated with hormones. Given these facts, we can begin to look into the actual hormone-sensitive mechanisms in the BRAIN which are the physiological intermediaries between hormone action and behavioural expression.

An insight into these problems requires some understanding of what the word 'hormone' means. This is probably best gained by viewing, in a historical perspective, the discovery of hormones, and the attempts to elucidate their structure and functional relationships.

Historical background. The origins of the scientific study of behavioural endocrinology can be traced back with some certainty to classical Greece where, some 300 years before Christ, Aristotle recorded (Book 9, volume 4 of the *Historia Animalium*) that when the immature cockerel (*Gallus g. domesticus*) was castrated, the sexual characteristics such as colouring of the comb, mating attempts, and the characteristic call of the male never developed in adulthood. He went so far as to compare the changes in the cockerel to those occurring after castration in man. Early experiments of this nature demonstrated that the testes were related to the reproductive powers of the individual, and built upon knowledge, probably pre-dating recorded history, of the use of castration to produce docile slaves and animals such as oxen (*Bos primigenius taurus*), and also to improve the quality of meat. In the Middle Ages attention was focused not only on castration as a means of modifying behaviour, but also on the use of hormonal extracts to restore sexual potency. For example, the medieval Chinese associated the testes with beard growth and male virility, and capitalized upon this knowledge by using testicular preparations, taken either desiccated or raw, for male sexual debility, hypogonadism, and impotence, for which testicular hormones (androgens) would be prescribed today. At the end of the fifteenth century, females who were suffering from miscarriages, or who were barren, were given 'Ta-Tsao Wan' pills, consisting of extracts of placental tissue, now known to be rich in female hormones (*oestrogens*), to improve the 'Yin' force in the body, including sexual function. It seems unlikely that the early Chinese pharmacologists were able to purify either androgens or oestrogens in effective concentrations, although it has been claimed that they were capable of preparing crude extracts of steroid hormones. However, despite a lack of any real theoretical knowledge of hormones or their action, an association was made between endocrine tissue and reproductive processes, including behaviour. It was not until the last decade of the nineteenth century that Berthold of the University of Göttingen, in a series of classic experiments, satisfied scientific requirements for the existence of a hormone. Berthold showed very clearly not only that castration eliminated normal masculine behaviour and comb development in young cockerels, but also that removal of one testis and transplantation of the other into a new site, a different part of the abdominal cavity, or transplantation of a testis from a normal sexually active bird to the abdominal cavity of a recently castrated bird, maintained male behaviour and comb development. Berthold's experiments represented the first clear-cut demonstration that there must be blood-borne substances that not only control reproductive processes, but are essential for the occurrence of sexual behaviour. Berthold concluded that the testicular substances secreted into the blood must influence nervous mechanisms underlying behaviour; this was remarkably predictive, because the action of hormones on the brain was not demonstrated convincingly until well over half a century later.

In 1905, Starling proposed the word 'hormone', derived from the Greek ορμαν (arouse to activity), for the internal secretions which occur in small quantities in the blood, as chemical messengers, and exert their effects at a distance from their source. During the early part of the twentieth century, knowledge of the types of hormones and their secretory glands increased enormously. The generally recognized endocrine glands, the pituitary, pineal, thyroid, parathyroids, thymus, pancreas, intestinal mucosa, adrenals, gonads, and placenta, are well known and widely studied. The student of ETHOLOGY is particularly interested in specific hormonal influences on the performance of particular types of behaviour. Most studies have concentrated on the gonadal or sex hormones from the testis and ovary which are better understood than the other hormonal systems. Hormones from other endocrine glands do affect behaviour, but often in a more general way; by influencing metabolic rate, for example.

Hormone physiology. Before embarking on a detailed survey of the relationships between gonadal hormones and behaviour, it is necessary to understand something of the way in which these hormones and their endocrine glands work physiologically. An important principle in the regulation of hormonal secretion is that endocrine glands must be able to turn their secretions on and off at appropriate times. Control of hormonal secretion is achieved by the self-regulating circular series of events which can be described as negative FEEDBACK, i.e. a hormone produces a biological effect that, on attaining sufficient magnitude, inhibits further secretion. Such a system contains: a detector of hormonal imbalance, secretory cells, an organ capable of an appropriate response to the hormone, and a means of promptly shutting off secretion when the response is adequate to restore HOMEOSTASIS. Hormones are available in very small amounts in the blood (in some cases less than a 50–100 millionth of a gramme), generally measured by chemical *assays*. However, the requirements of receptor tissues are so small that the hormones have to be inactivated continuously and excreted if their presence is to act as an effective signal. Most hormone molecules have a life of less than 1 h in the blood.

From modern knowledge of the endocrine system, we know that the hormone involved in Berthold's experiments on the cockerel was *testosterone*, a member of an important class of hormones, the *steroid* sex hormones, which are all derived via complex metabolic pathways from a parent substance called cholesterol. The reproductive endocrine glands form targets for a stimulatory chain of hormonal actions involving two quite different classes of hormone, the protein *trophic* hormones of the pituitary, and the structurally simple releasing hormones which are secreted by the brain and effectively link it with the endocrine glands. It is of historical interest that the duration of flow in this latter link was predicted by Vesalius of the School of Anatomy in Padua (1547) who proposed that 'pituita' or 'phlegm' drained from the brain to the pituitary. In the case of testosterone, the stimulatory route involves a hormone called *gonadotrophin releasing hormone* secreted by cells in the ventral hypothalamic areas of the brain which acts, via a special blood system in the brain (the hypophyseal portal system), on the adenohypophysis of the pituitary, to stimulate the secretion into the general vascular system of the gonadotrophic hormone called the interstitial cells stimulating hormone. This hormone in turn induces the secretion and release of the male hormone, testosterone, from the interstitial cells of the testis. The negative feedback system controls levels of testosterone in the blood; a rise in blood level of testosterone is monitored by special cells in the hypothalamus, and quickly results in a depression of secretion of first the gonadotrophin releasing hormone, second the interstitial cell stimulating hormone, and finally a diminished secretion of testosterone.

A second important principle in endocrinology is the specificity of hormone action. Since all hormones travel in the blood from their glands of origin to their 'target' tissues, all cells must be exposed to all hormones. Yet under normal circumstances, tissues respond only to their appropriate hormone. The specificity of hormonal action appears to depend on the capacity of receptors in the target tissue to recognize only their own hormone. The receptors are proteins in the cell cytoplasm, which react with the hormone in a specific manner, transfer the hormone to the cell nucleus, and ultimately initiate a response in the cell which leads to a further chain of biochemical reactions characterizing the response of the target to the hormone.

Hormones act primarily as signals to initiate or terminate specific responses in target tissues. Hormones may also affect the sensitivity of target tissue to other physiological signals, such as other hormones, hormonal products, or nervous impulses. In addition, the receptor tissue may vary in sensitivity to its own specific hormone.

The association between hormones and behaviour.

A great deal has been learnt about the relationships between hormones and behaviour by studying the way in which normal variations in concentration of a hormone in the blood-stream are correlated with behaviour. This is best seen in relation to seasonal cycles, where the female, particularly in lower mammals such as rodents, will only accept the male during a limited period of her reproductive cycle known as *oestrus*, derived from the Latin meaning a gadfly. The term is thought to come from the observation that periods of 'frenzy' occurred in cattle attacked by the gadfly (*Oestrus*), so the term oestrum came to mean any condition of periodic frenzy. Oestrus was later used to describe recurrent periods of sexual heat behaviour seen in many female mammals, which ensure that COPULATION occurs at the time of ovulation to maximize chances of fertilization. This particular type of behaviour can be correlated with a peak in blood concentrations of the ovarian hormones oestrogen and progesterone. However, the correlation may be far more complex than it appears at first sight, because the female is not only willing to cooperate in sexual activity, she may also behave in a way that makes her sexually attractive to the male. She may, for instance, actively seek out males for herself. Bitches, free to visit dogs (*Canis lupus familiaris*) of either sex, show a strong preference for males during oestrus, but show no preference at other times. Aspects of oestrous behaviour may require different effects of oestrogen, and may involve the action of other hormones. For example, oestrus in some female primates is correlated not with oestrogen and progesterone, but with oestrogen and androgen. The latter hormone is usually regarded as testicular in origin, but in female primates it is derived from the adrenal glands. Approximately half-way through the MENSTRUAL CYCLE, the attractiveness of the female rhesus macaque (*Macaca mulatta*), in terms of her sexual signal value to the male, and receptivity in terms of willingness to engage in copulatory activity, are both high, and depend upon the concurrent actions of oestrogen and androgen acting upon different target organs. This is implicated by the rather paradoxical finding that appropriate hormonal manipulation, consisting of removal of the ovaries to eliminate ovarian oestrogen, and androgen therapy, can produce an unattractive but receptive female. Conversely, an unreceptive but attractive female is produced by removal of the adrenals to eliminate adrenal androgen, and oestrogen therapy. Oestrogen appears to act directly on the female genitalia to facilitate the production of olfactory signals or PHEROMONES, which act as sexual stimuli to the male. In the latter case, the behavioural action of the female hormone is on the male, because it is he who is made more responsive to the female. The behavioural action of androgen on the other hand, is directly on the female, where it increases receptive behaviour.

In other species, mating behaviour and the

endocrine systems of the male and female are coordinated to ensure fertilization. An example is provided by induced ovulators, such as the rabbit (*Oryctolagus cuniculus*), or ferret (*Mustela putorius furo*), where the female, influenced by the hormones oestrogen and progesterone, is receptive to the male. The tactile stimulation from coitus initiates a series of neural and hormonal events which permit ovulation to occur during the period when the spermatozoa are likely to be in the genital tract of the female.

The relationship between hormones and behaviour can seldom be seen as a simple positive correlation between the availability of a hormone and the occurrence of a particular type of behaviour. In the reproductive cycle of the Barbary dove (*Streptopelia risoria*), for example, not only is there a relationship between the endocrine system and the environment, but also, within an individual, the changing relationships between hormones and behaviour must be synchronized with both the behaviour and endocrine system of the partner. Thus participation in COURTSHIP induces the secretion of hormones which facilitate collaboration between male and female in

the building of a nest; participation in NEST-BUILDING under these conditions stimulates the secretion of the hormones which play a part in inducing the birds to sit on eggs. Stimulation from the act of sitting on the eggs induces the further secretion of hormones which help to maintain INCUBATION, and also brings the birds into a condition where they will feed the young when they hatch. Synchrony in reproductive development depends upon a mutual interaction of stimulation which influences the endocrine systems of each partner, and which in turn contributes to sequential changes in their behaviour.

Long-term hormonal changes and behaviour. In many species of mammals and birds living in temperate areas of the world, the patterning of the production of reproductive hormones is regulated by seasonal changes in day-length. For example, peaks in testicular activity and hormone production in the common starling (*Sturnus vulgaris*) are synchronized very closely with the spring-time DISPLAYS on the TERRITORY and sexual interactions. The disappearance of reproductive behaviour coincides with testicular atrophy and a lowering of androgen levels in the autumn (Fig. A). Animals

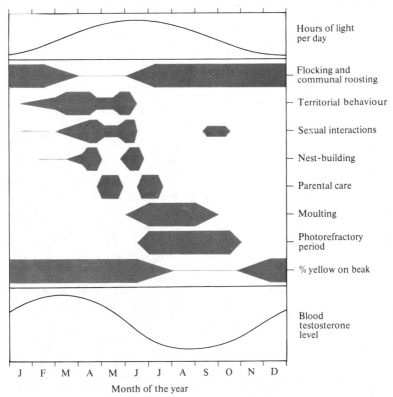

Fig. A. Upper figure: Events in the annual reproductive cycle of starlings. The width of each horizontal bar is roughly proportional to the frequency and intensity of each event. Lower figure: Annual cycle of the steroid hormone, testosterone, in the blood of male starlings. Note the close synchrony between elevated testosterone levels and the breeding season.

whose reproductive behaviour depends on the regulation of the endocrine system by day-length are adapted to make best use of the available stimuli. Male Japanese quail (*Coturnix coturnix japonica*) for example, whose testes are undeveloped because they have been kept on a short day-length, may show a thousandfold increase in testicular weight within a few days of being exposed to a longer day-length. Thus, the endocrine response to such environmental stimuli, which initiates reproductive activity, can be shown within a very short period. A different species of quail, the Californian quail (*Lophortyx californica*), which breeds irregularly in arid areas, does not reproduce during periods of continuous drought, at which times it feeds on the leaves of stunted desert forbs which produce plant oestrogens (phytoestrogens, similar to the female oestrogenic sex hormone in structure). The natural plant oestrogens appear to have the effect of suppressing the endocrine system of the female, so preventing reproduction in conditions too hostile to support the rearing of young. In a wet year, the forbs grow prolifically, contain little phytoestrogen, and produce abundant seed crops for the quail (see COADAPTATION).

SOCIAL stimuli also influence the endocrine system to induce long-term changes in the patterning of hormone secretion which either suppress or activate behaviour. For example, the tendency in the male Barbary dove to show high levels of aggressive behaviour (Fig. B) is correlated with

Fig. B. Aggressiveness during courtship in the male Barbary dove. The male launches himself at the female who is turning away from him.

low blood testosterone levels and small testes, both of which indicate an early stage of reproductive development. This aggressiveness has a marked effect on the female, suppressing her reproductive cycle until the male's condition has advanced to a stage where the synchrony in male and female cycles, so essential for reproductive success, can be maintained.

The importance of the social environment in determining increases in blood level of androgen has been demonstrated in studies of the talepoin monkey, the dwarf guenon (*Cercopithecus talapoin*). In this species, males which are dominant in the male social HIERARCHY show a marked elevation in blood testosterone level after mating interactions, whereas subordinate males show no such increase. This raises the interesting possibility that the social status of the DOMINANCE hierarchy of the group influences behaviour by an effect on the hormonal condition of the individual. The sexual performance of the dominant male, and his attractiveness to females, may be enhanced by elevated testosterone levels. A more general conclusion from this work is that the organization of a social group, and an individual's role in that group, modify the effects of hormones on behaviour. Social stimuli are no less important in their regulatory action on endocrine systems associated with behaviour than other environmental stimuli, such as day-length.

Short-term hormonal changes and behaviour. Biochemical assay methods allow the identification of hormones in the blood. This means that, in the LABORATORY, accurate information is easily obtained on the rapid fluctuations in hormone level which occur within minutes of the perception of external stimuli that are important in evoking sexual behaviour. Whether such stimulus-induced fluctuations in hormone level actually cause changes in behaviour is open to question. Certainly some of the evidence suggests a close relationship between sexual stimuli and sex hormone level. For example, significant increases in blood levels of the sex hormones, interstitial cell stimulating hormone, and testosterone are apparently associated with the sight and smell or both of a 'teaser' cow being led past the bull on her way to his covering yard. Similarly the sight or smell of a female rabbit elevates blood androgen levels in the male within minutes. The same sort of phenomenon has been observed in male hamsters (Cricetini) where exposure to vaginal odour alone increases testosterone level to the same degree as that occurring after pairing with a female. Male house mice (*Mus musculus*) which have been paired with a female for 1 week show an elevation in blood testosterone within 30 min of the resident female being replaced by another female. This elevation appears to be a specific response to a strange female, as it does not occur if the resident female is replaced by a male. Even anticipation of a sexual encounter increases hormone levels. Blood androgen rises rapidly in male rats (*Rattus norvegicus*) alone in a cage in which they are used to being tested for sexual behaviour. A similar, though necessarily more anecdotal, account of an anticipatory effect has been reported in a human male who lived alone on an island, but sought female company at regular intervals. Increased beard growth, which is mediated by testosterone, occurred in the days immediately preceding his visits to the mainland, suggesting that preoccupation with the sexual activity lying ahead raised blood testosterone level, which in turn accelerated his beard growth. More recently data have become available from volunteers who have sampled their own blood during sexual AROUSAL and following intercourse. Blood levels of testosterone are predictably higher during these periods than during other sample periods in the day. These results, suggesting that the actual performance of sexual behaviour causes a rapid elevation in testosterone level are confirmed by studies in sub-primate mammals.

It would appear that changes in blood hormone level not only precede sexual behaviour, but also follow immediately upon such activity. Although it is tempting to speculate that hormonal elevation directly influences the tendency to display sexual behaviour, there is insufficient evidence to make any positive statement.

The action of hormones on behavioural mechanisms. Theoretically, there are at least three ways in which a hormone can influence the mechanisms underlying behaviour. The hormone can influence (i) *effectors*, such as special structures involved in the performance of behaviour; (ii) the peripheral sensory systems to modify brain functioning indirectly, by altering sensory input to the brain; and (iii) behavioural mechanisms directly in the brain itself. All of these routes are involved in hormonal action.

Actions not involving the nervous system. There are many examples of the effector action of hormones. For example, on the forelimbs of the aquatic African clawed toad (*Xenopus laevis*), the male develops specialized nuptial pads consisting of a thick mass of small spines set into the skin. These increase friction and adhesion of the male's forelimbs as he clasps the smooth, mucus-laden skin of the female's pelvic area during mating, when the pair swim about together (Fig. C). The appearance of the pads and their progressive development up the forelimb depend on the testicular hormone, testosterone, as does the development of the muscles of the forelimbs, which is also essential for maintaining the male's grip on the female.

Hormones may also influence the development of specialized structures that have value in sociosexual encounters. For red deer stags (*Cervus elaphus*) access to hinds depends on the use of antlers

in aggressive encounters during the mating season. Castrated males, who lose their antlers, show normal AGGRESSION when treated with testosterone, but in the absence of antlers they do not attain the social position which would normally enable them to attract hinds.

Actions involving the nervous system. Hormonal action which influences behaviour by modifying the sensitivity of sensory receptors involved in processing stimuli relevant to a particular type of behaviour is not easy to demonstrate, but there are a few examples. Thus the skin covering the glans penis of the male rat becomes thinner under the influence of testosterone, with the result that the tactile receptors are more exposed to sensory stimulation. The influence of the pituitary hormone *prolactin* on the FEEDING behaviour of the Barbary dove is also partly due to a peripheral effect. Prolactin induces engorgement of the crop of both the male and female with a 'milk' formed by sloughed epithelium cells. The sensory stimulation provided by the distended crop has been shown to be essential for PARENTAL feeding of the young; desensitization, using a local anaesthetic, reduces this feeding. Another example of the peripheral effects of hormones on behaviour comes from detailed studies of the nesting behaviour of the canary (*Serinus canaria*). In this species, the nest, which the female has built, provides an important source of stimulation for subsequent events in the reproductive cycle (e.g. egg-laying). As the female builds her nest she loses feathers from her breast region, and develops a patch of bare, highly vascular skin called the *brood patch*. The loss of feathers brings the nest into closer contact with the skin, and thereby increases the effectiveness of stimuli emanating from the nest. The latter process is aided by the skin's increasing sensitivity to tactile stimulation. This stimulation, derived from the nest cup, is important in determining the female's preference for feathers rather than grass as a nesting material just before the eggs are laid, and in bringing nest-building to an end. All of the concurrent changes in the brood patch skin are under the control of the hormones, oestrogen, progesterone, and prolactin. Whether the change in tactile sensitivity is due to a direct effect of these hormones on skin mechanoreceptors is still a matter for speculation. It can be argued that the hormones influence the PERCEPTION of sensory input

Fig. C. A mating male African clawed toad lying on top of the female and clasping her. During mating the pair swim about actively in water, diving deeply and swimming in perfect synchrony. After fertilization of the female, the pair separate.

by the brain, rather than necessarily influencing the tactile receptors themselves. The recording of neuronal impulses from the nerves supplying the skin suggests that the tactile receptors in the skin of the brood patch are extremely sensitive, irrespective of hormonal condition during the breeding cycle. An effect of oestrogen on cutaneous sensitivity of the perineal skin area (near the urinogenital openings) of the female rat has been established with some certainty, however. Oestrogen acts in this case both to sensitize the skin and to increase the sensory area of the pudendal nerve, which innervates the areas adjacent to the vaginal opening. This area is extensively stimulated by the penis of the male prior to and during intromission; stimulation which is known to enhance the female's sexual response to the male.

While the peripheral actions of hormones must not be ignored, there is little doubt that the most important target organ is the brain. Birds are particularly useful for the study of hormone-sensitive brain mechanisms, because of their elaborate courtship displays which are highly dependent upon hormones. In the case of the male, castration results in the rapid decline and disappearance of all courtship displays. Intramuscular therapy with androgen rapidly restores them. To determine which areas of the brain are androgen-sensitive, minute amounts of hormone have been implanted in different areas with an accuracy of fractions of a millimetre. Normally the tip of a very fine needle is coated with crystalline hormone and, using a special instrument, it is introduced into the brain area to be studied. The hormone then diffuses into the relevant area of the brain without interfering with the animal's normal behaviour. The surface area of the hormonal implant, and consequently the rate at which diffusion of hormone from the implant occurs, can be controlled, and is limited to the area of tissue surrounding the implant. These experiments are carried out in castrated animals which have no sex hormones, and are therefore in a sexually quiescent condition. In doves, qualitatively normal courtship behaviour can be restored by testosterone implants in only a limited area of the brain (the *anterior hypothalamic nuclei*). Implants of testosterone in other areas are ineffective. These experiments cannot in themselves be considered as evidence that testosterone normally reaches the relevant sensitive areas of the brain involved in the effects of hormones on sexual behaviour. For it could be argued that introduction of hormones into the brain breaks down natural barriers that normally prevent blood-borne steroid hormones from influencing the brain. However, tracer studies employing testosterone in which radioactivity is incorporated into the testosterone molecule, have shown that radioactivity associated with the hormone is taken up by nuclei of cells in the hypothalamus within 1 hour of the labelled hormone being introduced into the blood. Clearly such hormones do find their way very rapidly to target cells in the brain, and in general the brain

area from which the most complete behavioural response can be obtained corresponds well with the area that contains receptors for testosterone. Other areas of the brain are also sensitive to hormones. In the canary, tracer studies with radioactive testosterone have shown that various midbrain areas take up the hormone. The limits of these areas compare very well with the extent of experimental lesions, which disrupt androgendependent VOCALIZATIONS involved in courtship song. Evidently components of the vocal control system are androgen-sensitive.

Some apparent anomalies have come to light from studies of the action of hormones upon the brain. For example, in the male dove, the action of testosterone implanted into the hypothalamus in restoring male courtship behaviour can be duplicated to some extent by oestrogen, the female hormone. Specificity of action, one of the fundamental principles of hormonal physiology would appear to be invalidated. However, biochemical work shows that testosterone is converted to oestrogens by enzymes in the brain target cells. The male hormone, testosterone, can be regarded, therefore, as a pre-hormone whose effects on male behaviour are, rather paradoxically, mediated by a metabolic product identical to the female hormone oestrogen. Several instances of this phenomenon have been recorded in mammals as well as in birds. For example, red deer stags show rutting behaviour, consisting of roaring, *flehmen* (i.e. lip curl), wallowing in peat bogs, thrashing about in the herbage with the antlers, repeated flicking of the penile sheath, and protrusion of the penis with emission of spurts of urine which impart a characteristic rutting odour. Castration eliminates every component of the stag's sexual behaviour. However, virtually complete restoration of rutting can be achieved with subcutaneous implants of oestrogen; just as it can with testosterone.

While we tend to think of hormonal action on the brain in terms of single systems, i.e. one hormone specifically influencing mechanisms underlying one type of behaviour, there is no doubt that hormones may affect more than one system, or may act on more general brain mechanisms. One such influence is that affecting the animal's ATTENTION to external stimuli. Androgen, for example, appears to increase the persistence with which a behavioural response is carried out, possibly by maintaining attention to the specific stimulus which evokes the behaviour pattern.

The developmental action of hormones. In the foregoing discussion, emphasis has been placed on the role of hormones in increasing or decreasing the probability that behaviour patterns in adult animals will occur in conditions of appropriate external stimulation. However, hormones are also important in determining the type of adult behaviour by acting during the period when the animal is developing, either just before or after birth, and also later when behaviour is being modified by experience. The best example of hormonal actions of this type concerns the sexual differentiation of behaviour. Much research has been carried out on rodents, particularly the rat, where mechanisms underlying masculine and feminine sexual behaviour remain undifferentiated during embryonic life. Each GENETIC sex appears to have the potential to develop in either a masculine or feminine direction. However, during a brief period either before birth in species with a long gestation period, or immediately following birth in species with a short gestation period (the first five postnatal days), the potential for male sexual behaviour becomes irreversibly established by the action of the male hormone testosterone. So the behaviour of adult females (genetically female) can be modified to resemble the male type by treating them with testosterone during very early development. This means that in adulthood, despite adequate hormone treatment, the tendency of such females to show feminine behaviour is diminished, whereas their tendency to show male behaviour is increased. Conversely, the elimination of testosterone in males, either by using chemicals which antagonize the action of the male hormone, or by castration in early development, results, at adulthood, in males which have little potential for male behaviour, and an increased potential for female behaviour. The differentiating hormone is not necessarily an androgen. New-born female rats can be masculinized as effectively with oestrogen as with androgen, suggesting either a lack of hormonal specificity in the mechanisms underlying behaviour, or that oestrogen may be involved in the process of differentiation.

One of the most challenging problems in this field is the degree to which the brain is involved in the early actions of sex hormones. It has been suggested, for example, that the differentiating action of testosterone in rodents may not involve the brain at all, because castration early in development deprives the male genitalia of the testosterone required for normal growth of the penis. Deficits in male sexual behaviour may be due to an inadequate genital response in adulthood. However, very small amounts of testosterone applied locally within the brain of female rats increases the potential for male sexual behaviour, strongly implicating the brain in the developmental action of testosterone.

The differentiating actions of sex hormones are best viewed as being exerted on a range of target organs of which the brain ranks as the most important. Exactly how these hormones exert their developmental effects is not well understood. But we can say with some degree of certainty that the plasticity of the system involved in the differentiation of sexual behaviour, at least in mammals, is limited to a short SENSITIVE PERIOD in early ONTOGENY. Furthermore, differentiation early in development to the behavioural type characteristic of the male is due to the influence of the hormone itself, and probably does not depend to any extent on environmental stimuli. During later stages of development, social stimuli and LEARNING processes assume great importance in moulding adult behaviour.　　　　　　　　　　　　　　J.B.H.

11, 82, 93

HOUSEHOLD PETS. The term pet animal indicates that a special relationship exists between an animal and its human owner, a relationship where the quality of social interactions between pet and owner are more important than the animal's economic or utility value. In particular, the pet is a source of companionship and pleasure to its owner, and as in any SOCIAL RELATIONSHIP the interactions tend to be reciprocal. Pet animals are entirely valid as subjects for those with an interest in animal behaviour; indeed they have a number of advantages which are not to be found in either the LABORATORY or FIELD situation. For instance, artificial selection over some 14 000 years has produced several hundred distinct breeds of dogs which differ both anatomically and behaviourally from one another (see DOMESTICATION). They provide ample confirmation of a GENETIC basis for behavioural variability, but the moulding influence of the environment upon behavioural development can also be clearly seen in pet animals.

The objective of this article is to present a general description of the behaviour of the most popular and accessible species of pet animals, the dog (Canis lupus familiaris) and the cat (Felis catus), followed by a section on social interactions between pets and people. Information about the behaviour of dogs and cats comes from four different sources. They have been studied in everyday home settings as pet animals, in the laboratory, in the feral state, or as pre-domesticated species in the wild. The dog and cat are, therefore, of special scientific interest because their biology and behaviour can be compared across these very different ecological conditions.

The dog. It is now generally accepted that all known breeds of dogs were derived from the wolf (Canis lupus). However, domestication of the wolf was probably not a single historical event but rather was repeated on several occasions in different parts of the world, and from different geographical races of the wolf. The SOCIAL ORGANIZATION, feeding habits, and general biology of the wolf have been systematically studied in recent decades and it is clear that it provides the key to understanding dog behaviour. In particular, wolves are a highly sociable species, possessing a complex communicatory repertoire of auditory, visual, and chemical signals. Their social organization ranges from large, cooperative, hierarchical packs to a relatively solitary existence. The terrain, and the type and abundance of prey are the critical factors determining group size. Wolves are generally carnivorous, but they also eat significant quantities of plant materials when mammalian and invertebrate prey are in short supply.

Social behaviour. The development of social responses in the dog has been intensively investigated, following the common observation that early social experience of the puppy could determine whether it grew into a fearful and withdrawn, or an outgoing and sociable adult. Scientists have demonstrated that JUVENILE social experiences, such as exposure to other dogs, or to human beings, are essential prerequisites for the development of subsequent social responsiveness. This does not mean that social contact after the age of 13–15 weeks is irrelevant, or that the effects of mild social deprivation within this SENSITIVE PERIOD of socialization cannot be reversed by experiences later in life. Nevertheless, the concept of a sensitive period of social development is of great practical importance to those engaged in breeding and rearing well-adjusted dogs.

Between 3 and 5 weeks of age the young puppy's social relationships are mostly centred upon the bitch and its littermates, but thereafter other objects and animals assume significance. The development of social order or precedence can be observed in very young litters of puppies, as evidenced by teat choice, differential growth rates, and sleeping positions relative to the bitch. PLAY is the primary vehicle for development of social skills, and in particular of the *dominance–subordinance* relationships that characterize all aspects of social behaviour in the adult dog (see DOMINANCE).

As with most of the wild canids, the primary unit of social organization in the dog is the pack. Stray, feral, and wild dogs readily form loose associations similar to those observed in wolves, and packs of up to twenty individuals have been recorded. However, the bitch and her juvenile offspring, or pairs and trios of unrelated dogs, are more permanently associated with one another than with such large packs. Canine society is closely regulated by dominance–subordinance interactions between its members. The 'top dog' has precedence to food, is the leader on expeditions, and is generally responsible for initiating activity within the pack. The most dominant dog within a mixed-sex group is usually a male, but as in wolf packs leadership is sometimes taken by a bitch.

Dominance–subordinance is expressed by a complex series of communicatory gestures. In particular, the tail, hair on the ridge of the back, facial expression, eye contact, and general body stance indicate the dominance relationship between two familiar individuals. FIGHTING is a mechanism for establishing dominance between strange dogs, and if it is allowed to continue until a clear winner has emerged, is unlikely to recur between those individuals. Dogs utilize a number of very effective affiliative gestures which serve to reduce the likelihood of fighting. The most extreme such gesture is the 'roll-over' or *inguinal display*, but a low wagging tail and sideways approach is also used by a subordinate dog in relation to a dominant. The outcome of the AGGRESSION between two strange dogs is greatly influenced by the proximity of their respective home territories, and by

whether or not one is accompanied by its pack leader. Owners of pet dogs can readily observe how social dominance is relative to territory and proximity of the pack by noting that a dog usually wins a fight on its own territory and in its owner's presence.

Dogs are a prime example of a territorial species, although usually only a small home area is actively defended against intrusions. Under wild or feral conditions the HOME RANGE is centred upon a den or sleeping and rest area, whereas pet dogs usually identify their home along the boundaries of the owners' property. The borders of the TERRITORY of male dogs are usually marked regularly by urine. Territorial marking by urination in bitches and defecation by both sexes also occurs.

Sexual maturation of the dog has been accelerated during the domestication of the wolf, and male puppies often show sexual responses as early as 5 weeks of age. Sexual responses are obviously an important feature of social behaviour, and their normal development in the dog is greatly influenced by early social experience of the puppy. The bitch is quite influential in the selection of her mate (see MATE SELECTION). One indicator of social dominance in the male is that he usually gains preferential access to the females in heat, maybe because he exerts dominance over the bitch during coitus.

Communication. Dogs are such ubiquitous animals that most people have a fairly good intuitive knowledge of their communicatory repertoire. Because man is a highly vocal species himself, there is a tendency to over-emphasize the behavioural significance of the dog's bark, howl, or whine at the expense of other aspects of COMMUNICATION. The tendency to bark has been increased in most breeds of dog by artificial selection; and this VOCALIZATION is used only infrequently by the wolf. Certain breeds, for instance the Basenji, do not bark, and there are other breeds which do not whine or howl. Such variations indicate marked genetic control over canine vocalization, and it is difficult to determine precisely the communicatory significance of sounds such as barking. Puppies of all breeds have vocal repertoires which are different from those of adults. The most characteristic is a piercing yelp made at times of STRESS or social isolation. Dogs also use a rich variety of body postures and chemical signals to communicate with one another. The visual quality of postural DISPLAYS in dogs has been markedly affected by selective breeding; the face of a chow or the tail of a Bedlington terrier must surely carry a different signal than the same anatomical regions of a German shepherd dog.

Chemical cues are used by dogs in a variety of behavioural situations; urine marking by male dogs is probably the best known. The urine of bitches on heat is the main source of an attractant for male dogs, and such bitches increase the frequency of their markings as they come into heat. Another source of odours is the anal glands, whose viscous and acidic contents can be released during stress and at other times to coat the faeces. Various sebaceous glands upon the skin, secretions from the external ear, and sweat glands on the foot pads are all presumed to have some communicatory significance, though little experimental data is available to establish precisely what this might be. All of the chemical signals mentioned consist of complex mixtures whose composition varies between individuals by sex, diet, age, and other characteristics. With their remarkable olfactory powers, dogs are probably able to gain fairly accurate knowledge of an individual from the chemicals it deposited upon the environment on some earlier occasion (see CHEMICAL SENSES). As in other species this well-developed chemical communicatory repertoire allows the social life of dogs to span both space and time.

Feeding behaviour. Under natural circumstances the FEEDING behaviour of a carnivore such as the dog is closely associated with HUNTING and prey killing. The dogs of TV commercials for pet food obviously have very different ecological constraints upon the composition, quantity, and timing of their meals. The ability of pet dogs to manipulate the purchasing habits of their owners is an interesting topic in its own right, but we will be concerned here with more normal aspects of feeding behaviour.

The transition from mother's milk to solid food normally occurs when puppies are between 4 and 6 weeks of age. In the wild the mother or other pack member usually carries food in the stomach to the puppies from the site of a kill. The partially-digested contents are then regurgitated in response to begging and pestering by the puppies. This behavioural sequence is called *et-epimeletic* vomiting, and can be observed in domestic breeding bitches if the puppies are not provided with solid food by the owner, and so have not been weaned at the conventional 6 weeks of age. Feeding by puppies and indeed by some adult dogs is augmented when they are within a group. Social facilitation of feeding is, presumably, an adaptation to pack living where individuals must gorge the kill as quickly as possible before it is taken by other individuals. There are large individual and breed differences in the speed and quantity of food consumption by dogs. Individual Labradors and beagles can ingest 10% of their own bodyweight within a few minutes, whereas small dogs like miniature poodles generally eat more slowly.

When given a choice between two or more nutritionally-balanced diets, pet dogs usually take a novel or unfamiliar diet, and are also very sensitive to the moisture content of foods (wetter foods being preferred). It has been shown that dogs, like rats (*Rattus*), quickly learn to avoid diets which have provoked illness or nutritional distress. This ability to form *conditioned aversions* to harmful diet items is obviously a useful protective device against natural toxins (see FOOD SELECTION).

The spacing or timing of meals in the domestic dog may vary immensely. For instance, under experimental conditions dogs can maintain satisfactory body weights when fed *ad libitum*, once a day or even once every 2 days. However, when they are permitted to regulate voluntarily, dogs generally adopt a nibbling feeding strategy of numerous small meals equally spaced throughout the daylight hours. Under feral conditions where dogs are keen to avoid human contact, nocturnal hunting and feeding is more common, and the animals SLEEP during the day. Feeding in dogs is also disrupted by hot weather, with meals being taken at the coolest time of day.

Learning ability. Dogs have been trained to perform a multitude of different tasks throughout human history, and dogs would probably not be kept as pets today if they were not amenable to some degree of training. Formal experiments by PAVLOV and other workers have also amply confirmed the remarkable LEARNING ability of dogs.

Dogs are probably unique amongst animals in their responsiveness to social REWARDS, such as petting or praise from a human being. Conventional rewards, such as food, light, warmth, and water, can also be used to modify behaviour, but social contact linked to verbal commands or gestures is particularly effective in dogs. The mere presence of a human being is very encouraging for a dog, particularly if the person has clearly established dominance over it. Variations between breeds in learning ability have often been suggested, but the evidence points to these being based upon temperamental factors rather than differences in INTELLIGENCE. In particular, the degree of social attachment to people varies considerably between breeds, and as a result some dogs will work harder 'to please' than others.

The artificial selection of several hundreds of dog breeds from the wolf must be the longest-running experiment ever performed in the genetics of behaviour, and the remarkable diversity of temperament and social behaviour that has resulted from this historical process provides a telling insight into how the individual animal is a product of both heredity and experience.

The cat. *Felis catus* is a much more homogeneous species than the dog, and its historical and contemporary relationship with man is also quite different. The historical record of the domestication of the cat is by no means complete, but its association with man is thought to have begun considerably before the Egyptian dynasties of 4000–5000 years ago. The likeliest ancestral species is the Kaffir cat (*Felis libyca*) of North Africa, though other small felids in other parts of the Old World might also have contributed to the genetic make-up of the present-day domestic cat. Like dogs, cats now enjoy a world-wide distribution, and exploit HABITATS ranging from sub-Antarctic islands to hot deserts, from fireside chairs to a virtually wild existence well away from human settlement.

Social behaviour. Cat owners will know very well that their pet is capable of very affectionate displays on some occasions, but on others seems intent on avoiding human or feline company. Cats are usually less sociable than dogs, but it is misleading to describe them as an anti-social species for that reason alone. The precise form of social organization adopted by cats depends upon local circumstances. If food is scarce and the habitat favourable, there may be little social contact between cats in adjacent territories, but in more favourable urban areas and around farms quite large integrated communities of cats can be observed.

Early social experiences of the kitten within the first 3–4 months of life have been shown experimentally to be as important for normal development of social behaviour in the cat as in the dog. Contact and handling by human beings during this juvenile period is an essential prerequisite of later tolerance and social relationships with people; without such socialization kittens quickly become wild and unmanageable. A measure of the plasticity of feline social responsiveness can be gauged from the ease with which kittens of small wild felids can be tamed to almost domestic manners by early handling. Domestication of the cat no doubt was much less demanding than that of the wolf, and probably succeeded with a single generation of hand-reared kittens taken from the wild.

Cats are territorial animals and under free-living conditions utilize three different types of space. The first is a nest or den area which might be anything from a hole in a tree, a deserted rabbit burrow, a false roof, to a favourite chair. Around the nest is a home range where most of the cat's time is spent in rest, sunbathing, play, etc. In female cats the home range tends to be fairly compact and vigorously defended against intrusions by strange cats of either sex. However, males have less distinct home ranges, and are less territorial than females. Castrated males, however, tend towards the female territorial pattern. Beyond the home range and linked to it by a series of paths and runs is the hunting range. The hunting range is not patrolled or defended as an exclusive area, and indeed is likely to be shared with several other cats. It might be a hedgerow, cornfield, or series of backyards and rubbish bins.

Household or pet cats are obviously very much affected by the timetable and architectural features provided by their owners. Female and castrated cats tend to adopt a home range along their owners' property limits, whereas tom-cats traverse the whole neighbourhood. The home range territory may be shared by several co-existing cats, particularly if they are related or have grown up together. Multiple cat-owning households and feral cats in urban areas generally form themselves into such cooperative clans. Regular nocturnal gatherings of cats from a city neighbourhood can also be observed. The participants in these gatherings would normally be intolerant of one another on their own territories, but away from their territories they associate as though members of a feline tribe.

Cats develop a clear social organization that is based upon dominance and established by the outcome of fights. Actual fighting within a settled group of cats is rare; a single confrontation between a pair usually establishes their relationship for ensuing weeks or months. Dominance-subordinance is mostly relative to geographical location, but it seems that dominance relationships within the clan produce a linear rank order. The normal territorial configuration of cat society does not usually develop in densely housed populations, such as breeding catteries and laboratories. Under the latter conditions, a single despot usually emerges which dominates the rest, and all of the colony may persecute one or two bottom-ranking pariah cats.

Cats are prolific breeders, and their numbers would quickly outstrip available food resources unless checked by disease and predators. Females are generally fairly promiscuous and allow a succession of males to mount them, though more selective sexual partnerships have also been recorded. COURTSHIP and mating represent a substantial change in the female's behaviour, the females only allowing close approaches by strange males when in heat. The ambivalent nature of the female cat's social attitudes are very well illustrated by her reactions to the male before and after copulation has taken place. Sexual invitation by the oestrous female is highly explicit, consisting of loud vocalizations, treading, and ready adoption of the *lordosis* posture following a touch on the back. However, at the moment of ejaculation the female erupts into a dramatic rage and attempts to attack the male, who has to leap off her in order to escape injury. Ovulation is induced by mating, and the HORMONES involved in sexual receptivity have been intensively studied in the cat.

Communication. Vocal communication is important in certain social contexts, namely during CONFLICT, courtship, and MATERNAL care. Some sixteen characteristic sounds have been identified in the cat, but mostly it is not as vocal as the dog (the Siamese cat being an exception). Chemical and visual or postural signals are of great importance in social contacts between cats, and between the pet cat and its owner.

The most obvious and certainly least forgettable form of SCENT MARKING is the spraying of urine. Spraying is most commonly performed by intact male cats, though females and neuters also spray at the margin of their home range and along pathways. The posture during spraying is quite characteristic in all cats; the tail is held high and as it quivers a fine mist of urine is ejected against bushes or other vertical objects. The social context of spraying and the reactions of conspecifics to these marks suggest that they provide a mechanism for communicating the individual identity, sexual condition, and time of passage of a cat across its environment.

Scratching with the claws is also performed in a social context and leaves both a visual and a chemical signal upon prominent objects such as tree stumps or posts. Other distinctive marking displays in the cat occur during rubbing against objects or conspecifics. In particular, the skin along the forehead and around the lips is richly supplied with sebaceous glands, and an oily secretion is deposited when the fur on these regions is touched. The normal greeting sequence of two friendly cats usually includes rubbing of the head, flank, and tail one against the other as though they were exchanging body odours, and of course pet cats like to rub and to be stroked at these points by people.

Visual displays by the cat involve whole-body postures that are used in finely graded sequences to indicate mood and intention. Many of the displays incorporate components characteristic of both FEAR and aggression, and it is possible to link the two in models which account for most social situations (see MOTIVATION). These sequences are repetitively exercised during PLAY and for the adult they comprise a 'language' that generally obviates the need for physical fighting. It is questionable whether cats have a posture indicating active submission to a potential aggressor, such as is expressed in dogs by the inguinal display. However, the greeting response of cats in which they approach with tail held high probably has a submissive component, but it is not used at close quarters when a fight is in the offing.

The lordosis and treading behaviour of female cats during sexual courtship was previously mentioned. Another component of sexual behaviour in the cat is the so-called *flehmen* (i.e. lip curl) reaction of the male when he investigates the female's urine and vaginal secretions. A quite characteristic facial expression follows licking and investigation of sexual secretions, and flehmen is probably a visual indication of tongue movements as chemical stimuli are transferred to the secondary olfactory apparatus located on the roof of the mouth. Dogs also display the flehmen response, which is often accompanied by tooth-chattering movements of the jaws.

Ingestive behaviour. The cat is exclusively carnivorous, and a substantial portion of its waking time has to be expended in hunting under most wild conditions, but of course the pet cat has an alternative food supply in its owner's larder. PREDATION by cats upon rodents and small birds has been shown to have a significant effect upon populations of the latter in meadows, barns, and hedgerows. Hunting by cats is almost always a solitary activity except where the mother is accompanied by her young during their training. Considerable skill is employed to stalk and catch prey, and one of the functions proposed for play in felids is that it exercises skills used for hunting. Whether or not that is so, a considerable amount of play and baiting is usually directed towards the prey. The habit of many pet cats to return with live prey to their owner's house or to leave the corpse upon the door-step is an illustration of the inherent sociability of cats. Feral and farm cats frequently pool their kills with young or friendly members of the same social groups, so sharing prey with people is a natural extension of the same cooperative tendency.

When given free access to food, cats are very efficient at regulating intake to match their caloric requirements (see HUNGER). Numerous small meals are taken throughout the 24 h, typically about twelve meals per day. This nibbling strategy has an interesting and possibly coincidental implication for meal size, in that each meal provides about the same energy as would a house mouse (*Mus musculus*).

Like dogs, most cats maintained upon commercially prepared diets prefer novel or regularly changed varieties. However, some individuals develop an exclusive APPETITE for a particular diet, or a single ingredient such as liver. Food additions of this sort sometimes run counter to nutritional wisdom, and serious metabolic imbalances can result.

Learning ability. Cats are not renowned for their obedience in response to commands from their owners, and so it is popularly supposed that cats are either stupid or exceptionally artful. In fact, cats are very fast learners under circumstances where their natural response tendencies are exploited, and in particular where food, warmth, or social contact are used as rewards. The training of cats cannot proceed in the same way as with dogs, because their relationship with people is not centred upon dominance–subordinance interactions. Rather than compel compliance through using physical or psychological punishments, cat owners should design reward-oriented learning situations which strengthen their social relationship with the pet. Cats have been used in laboratory experiments for many years, where they perform extremely well on most REINFORCEMENT schedules. They are able to learn behavioural sequences or CHAIN RESPONSES in order to escape from a confined situation, and transfer the same general learning strategy to related situations. Another feature of the behaviour of cats in learning situations is that they tend to be rather indolent compared to industrious species such as the rat or pigeon (*Columba livia*). For instance, whereas the rat and pigeon actually prefer food obtained by OPERANT behaviour and continue to work in the presence of food, cats generally only work when hungry. Negative reinforcers which deliver pain, such as electric shock or air puffs, often elicit bizarre behavioural reactions in cats, and they tend not to be good subjects in Pavlovian CONDITIONING experiments.

Pets and people. The main reference point for any discussion of the behaviour of pet animals must be human behaviour, including our social history and the psychological needs which are satisfied by interactions with animals (see ANTHROPOMORPHISM). Until recently, there has been little formal interest in the underlying motives for pet ownership and the effects of pets upon people, but the scale or popularity of pet-keeping amongst all sections of society is a constant reminder to social scientists and to the medical and veterinary professions of the importance of this social phenomenon. The final section of this article will consider why people keep pets, and the effects of pet animals upon human behaviour.

The majority of people in any western country cite companionship as their predominant motive for having a pet. For instance, recent surveys conducted in Great Britain, Australia, and Sweden each found that over 80% of dog owners mentioned companionship as the predominant motive and reward for owning a dog. Detailed analysis of the word companionship reveals that it is being used as a term to describe a variety of social needs. The most important needs satisfied by pets are those of emotional and physical security, followed by needs for social affiliation or belongingness and self-esteem. Each of these dimensions has been experimentally investigated in the context of pet ownership, and the general conclusion is that relationships which people form with pet animals are largely an expression of normal social behaviour.

People certainly behave differently in the presence of a furry, non-threatening animal. It generally becomes a focus for conversation between people, reduces anxiety, and thereby strengthens their capacity to interact with one another. There are several practical implications of the effects of pets upon human behaviour, ranging from a specialized therapeutic role in penal and psychiatric institutions to the everyday and informal setting of family life. We have seen that the behaviour of pet animals can be studied from many vantage points; for instance the historical, evolutionary, ecological, and social perspectives are all valid approaches to the subject. Only the dog and cat have been considered here, but the domestic horse (*Equus Przewalski caballus*), budgerigar (*Melopsittacus undulatus*), and numerous species of rodent are also part of the same human tradition. The word 'pet' is an unfortunate misnomer that has discouraged academic interest in the behaviour of these animals, and the term 'companion animal' is probably more appropriate. There is no need for an elaborate laboratory or uncomfortable and expensive travel to study the behaviour of companion animals; they are ubiquitous and easily accessible and yet can be used to explore fundamental scientific issues by anyone who is curious about the natural world.   R.A.M.

4, 12, 44

HUNGER. In order to survive all animals must fulfil certain bodily needs, for food and water, for oxygen, and for regulating body temperature. Hunger, the bodily sensation which appears under conditions of food deprivation, and which results in the desire and craving for food, has been extensively studied by physiological psychologists as the prototypical example of a primary state of MOTIVATION. Unlike the need for oxygen, which is highly specific, our needs for food seem to be unspecific. At least for omnivorous animals such as man, foodstuffs are interchangeable, but there are always certain specific needs for salt and essential vitamins, which, if deficient, produce a selective search for these nutrients resulting in *specific hungers* (see FOOD SELECTION). Man and other animals normally adapt their food intake to changing needs depending on the amount of work done, the CLIMATE, and the nutritional value of the food. This short-term regulation of food intake is superimposed on a long-term regulation, which counteracts temporary inadequacies in the diet, and ensures a return to normal bodyweight; for example, if animals are force-fed to make them fat, and then returned to a normal diet, they will eat less than before until they return to their original bodyweight. When adult animals are deprived of food and become hungry, once food is restored they consume not only appropriate substances, but also just enough food to regulate their bodyweight within fairly close limits.

Much of our understanding of the underlying mechanisms of hunger comes from studies which examine factors influencing the control of food intake. How an organism regulates its requirements for food involves three separate questions: What initiates eating? What determines how much is eaten? What stops eating? The answers to such questions come from examination of the way in which animals eat, and, in particular, by examining the pattern of eating.

Historically, the discipline of physiology has influenced the study of food intake and produced an enormous literature on the relation between BRAIN mechanisms and FEEDING. The sensation of hunger was thought to arise from detection of some underlying physiological need, and it was Carl Richter in the 1920s who first applied the concept of HOMEOSTASIS to behaviour. When Norway rats (*Rattus norvegicus*) are allowed to select their own diet from a variety of foodstuffs in cafeteria-style experiments, they develop motivated behaviour appropriate to the correction of their underlying physiological need and to the maintenance of the internal environment of the body. The popular theoretical view was that food acts to reduce an underlying state of hunger DRIVE, which builds up as a result of food deprivation. However, much feeding behaviour appears to be anticipatory; animals are not responding to current needs but to future need. A simple LABORATORY demonstration illustrating this provides a rat with access to food for three distinct 1-h meal periods a day, separated by 7-h intervals. If the middle meal is withdrawn, the rat will initially increase its food intake in the meal which follows this 15-h gap, but within a few days food intake increases for the meal which precedes the long interval, i.e. the rat has learned to increase its meal size in anticipation of future needs. Study of the factors involved in determining the size of meals has shown that both short- and long-term factors are important.

The blowfly (*Phormia regina*) engages in apparently random flight patterns until it encounters a suitable smell, a TAXIS based on the CHEMICAL SENSES then introduces directionality into its flight pattern which begins to look GOAL oriented (see ORIENTATION). The blowfly eventually lands near a source of food, such as drops of honeydew secreted by aphids (Aphidae) on the surface of a plant. As it walks it may step into the sugary solution, and the simple TASTE receptors within the hairs covering its legs signal detection of the sugar solution to the central nervous system, and cause a REFLEX uncoiling of its *proboscis* (sucking tube). Taste receptors on the tip of the proboscis, when stimulated by the sugar, reflexly open the channel into its gut, and the blowfly commences to pump up the sugar solution. The vigour with which the fly pumps depends on the food's sweetness, which, of course, depends on the concentration of the sugar. Like all sensory receptors, the blowfly's taste receptors adapt quite quickly, and as they do so the sweetness of the solution in effect decreases, and the rate of pumping slows down. Eventually they adapt so much that the sweetness falls below some critical value and the pumping stops, and the proboscis is coiled back up. The fly may then wander away from the sugar, and as it does so the receptors begin to recover. When the recovered sensitivity raises the sweetness above a critical value, then the siphoning resumes and the sensory ADAPTATION starts all over again. If that were all to its eating, the fly would cycle rapidly between feeding and stopping according to the moment by moment state of its sugar receptors. In fact, these successive episodes of eating become briefer until the proboscis does not become extruded and eating ceases. In common-sense terms the fly has eaten enough to satisfy its hunger. If the sugar solution is very sweet, the blowfly will consume more solution in one bout of sucking than if it is less sweet: the higher the critical concentration the longer in time it takes to fall down to this level, and so the fly eats more of a rich nutritious solution than of a poor one. Its meal size has been determined by a peripheral factor.

In addition to the usual digestive tract, the blowfly has a blind alley called the crop which stores food while eating is in progress. After each meal a train of reflexes shunts food from the crop into the main gut. The rate at which this occurs is a function of the distention of the crop and the level of blood nutrients. The sweetness criterion, which determines whether the fly eats or not, itself depends on the activity of the crop: the more rapidly food is being transferred to the gut, the sweeter a substance has to be to elicit eating. After the fly has eaten a large amount, the rate of transfer is maximal and the fly will only respond to a sweetness so intense that it is bound to find nothing, and therefore the meal ends.

As the activity of the crop subsides, so the sweetness criterion relaxes. So the common-sense statement that the fly eats because it needs to becomes: the blowfly eats because its crop is not passing material to its gut, and there is something adequately sweet in the vicinity. The cessation of feeding is not caused by there being sufficient glucose in the gut or in the bloodstream; the blowfly begins its search for food when its blood sugar is high and its crop empty. Food takes time to pass from the crop to the gut and the blowfly cannot afford to delay its search for food until its blood glucose is low. As it begins the search when blood glucose is high and the crop empty, it has both available energy for foraging and available space for storing the food. If the blowfly is very active and uses a lot of energy, then food is used up more quickly and the interval between its meals becomes shorter. The frequency of its eating, i.e. the number of meals eaten, depends on the fly's energy expenditure, but the size of its meals depends on the nature of the food encountered. Food intake depends on the two variables of meal size and inter-meal interval which are themselves controlled by different factors.

Foods vary enormously in the amount of nutrient they can supply for a given bulk. Grass requires a great deal of ingestive digestion before it supplies the nutrient requiremennts of the herbivore, whereas relatively small quantities of meat will provide the same amount of nourishment. The type of food to which the animal is adapted will largely determine its typical pattern of consumption. Horses (Equidae) on poor pasture may spend 22 h of the day eating, and the need for food largely governs behaviour, whereas a lion (*Panthera leo*) may kill and eat only once every 3–4 days, and so its behaviour will be regulated by many other factors.

In the case of human beings, although we eat every day, the time we eat and the type of food are influenced largely by social and CULTURAL factors. As infants, however, this is clearly not the case, and food requirements will in part be a function of the nature of the milk produced. Analysis of the milk composition in many animals shows that the composition is not similar by virtue of species relatedness, for example the brown bear (*Ursus arctos*) and the kangaroo (*Macropus rufus*) have virtually the same milk composition, as do the reindeer (*Rangifer tarandus*) and the lion (*Panthera leo*). The correlation appears to be between ecology and nursing behaviour rather than relatedness. Marsupials, such as the kangaroo, and animals which rear their young in hibernation have virtually identical milk composition. In these animals, the mother is available at all times and her milk is relatively dilute. Animals which are born in a relatively mature state, and which follow or are carried by the mother at all times, also have rather dilute milk. They show the nursing pattern of a continuous feeder; examples are the chimpanzee (*Pan troglodytes*), man (*Homo sapiens*), and the domestic pig (*Sus scrofa domestica*). Animals such as the lion, the rabbit (*Oryctolagus cuniculus*), and the red deer (*Cervus elaphus*), which leave their young in secluded places and return to nurse at widely spaced intervals, have a very concentrated milk with a high fat content; the young are intermittent feeders. Human breast milk on a volume per volume basis contains less protein and calories than commercial formulae, and, assuming that an infant's intake is limited by volume, a bottle-fed baby will get more food per feed and is likely to SLEEP longer. The breast-fed infant lies closer to the continuous feeder end of this continuum, with the bottle-fed infant closer to the intermittent feeder.

Returning to an examination of food intake in terms of its constituent variables of meal size and inter-meal interval, we can show, with a series of simple demonstrations in the omnivorous rat, how they are influenced by different parameters. Altering the palatability of food influences meal size over short time intervals, and both size and frequency of meals over a longer time period. The rat presented with a highly palatable food for only 1 h each day will eat and gain weight, but with *ad libitum* access to such food it will compensate for the increased meal size by decreasing its meal frequency. With foods of low palatability, or which are unfamiliar, the rat takes smaller meals but eats more often. Altering the nutritional value of the food, by dilution with a non-nutritive filler such as cellulose, will initially lead to little alteration in meal size but increase the meal frequency to maintain the same calorie intake. After 3 or 4 days the rat will increase its meal size and decrease the frequency, and the initial compensation is altered and the balance restored. Therefore, mechanisms which affect alterations in meal size would seem to require longer to come into operation than those affecting meal frequency. Reduction of nutritional need by infusion of glucose intravenously after a meal delays the start of the next meal, i.e. meal frequency decreases. Increasing energy requirements by lowering the ambient temperature increases the meal frequency, but does not affect the size of meals. A strange environment reduces meal size, but this reduction is compensated for by increases in meal frequency. A sexually receptive female house mouse (*Mus musculus*) reduces its meal size, and this reduction is not compensated for by increases in meal frequency. Examination of the relationship between meal size and meal intervals in the rat indicates that the amount eaten at a particular meal determines the length of time the rat waits before initiating a new meal, and not, as common sense at first would suggest, that more is eaten the longer the time since the previous meal. Factors such as the taste of food, or the amount of food already in the stomach, determine how much is eaten. But if a large meal is eaten, then the rat waits a long time before initiating a new meal: the onset of a meal is determined by a physiological hunger state. Such findings are important as they suggest we must look for different physiological mechanisms underlying our initial questions: What initiates and what stops eating?

Local theories. Subjectively hunger seems to be a general sensation localized in the stomach region, appearing when the stomach is empty and disappearing or changing to feelings of satiety when the stomach fills with food. In 1882, a hunter on the American frontier was injured by a gunshot wound leaving an opening in his side (a *fistula*) through to the stomach. His physician made the important discovery that food placed directly into the stomach relieved the sensations of hunger. From experiments in which subjects swallowed a balloon, the physiologist Walter Cannon found subjective reports of hunger were correlated with contractions of the balloon, and this led to *local theories* of hunger, i.e. the sensation of hunger is due to the contraction of muscles in the stomach wall. However, such contractions are in large part an artefact of placing a balloon in the stomach, and when the stomachs of animals are experimentally denervated or removed altogether (a not uncommon operation in man) eating behaviour is remarkably unaffected. So although such contractions may be one factor leading to the sensation of hunger, they are not essential.

Again on common-sense grounds, we might expect that the act of eating, the chewing and swallowing movements, will contribute to the control of food intake. The combination of TASTE AND SMELL, proprioceptive, and MECHANICAL stimuli which arise from the execution of these movements are collectively known as *oropharyngeal* (mouth and throat) stimuli. One of the first experiments separating the influence of oropharyngeal and gastrointestinal factors on feeding was performed over a hundred years ago by Claude Bernard. He found that if water is prevented from entering the stomach of a horse while it is DRINKING, the horse will drink for much longer than usual, and then resume drinking after a short interval. That the horse drank more than normal suggests a role for the stomach in suppressing drinking, and that it did stop drinking, indicates that oropharyngeal factors do have at least a temporary satiating effect. Similarly, a dog (*Canis lupus familiaris*) with an oesophageal fistula which prevents food from entering the stomach will eat for much longer than an intact animal, but eventually appears to be satisfied and stops eating.

If we examine the fine structure of eating within a meal, it is found that, up to a point, the first few bouts of feeding become successively longer, i.e. in the early stages of a meal some form of positive

FEEDBACK is in operation which facilitates the act of eating. Increasing the palatability of a foodstuff maximizes this effect, and is familiar to most of us as the 'salted peanut phenomenon', i.e. we feel hungriest as we begin to eat; so, the hors-d'œuvre served just before the meal increases our appetite and interest in the main course. If we compared the OPERANT response of a rat working for food which was either delivered directly to the stomach, or into the mouth, the rat would work harder and press for a longer period of time for the food in the mouth. This is probably because the positive feedback effects just described are operating for the oral delivery, whereas only negative feedback occurs with intragastric feeding. But it is also the case that intragastric feeding is less satisfactory than feeding by mouth; human subjects report strong cravings for chewing, tasting, and swallowing after meals delivered directly into the stomach. Purely local gastric and oral factors cannot adequately account for feelings of hunger and satiety, and it seems that such peripheral stimuli are best viewed as things which accompany rather than lead to the initiation of feelings of hunger (see APPETITE).

Brain mechanisms. Several theories have been proposed which postulate the existence of some receptor within the brain which is sensitive to peripherally determined physiological conditions. One such popular candidate is the detection of blood glucose, which is a prime source of energy and is especially vital as a fuel for brain cells. This *glucostatic* hypothesis states that when the blood glucose is low, hunger is experienced, and when high, hunger and feeding are inhibited. However, it is well known that people suffering from diabetes and who have abnormally high blood sugar levels conversely often feel very hungry. This led some scientists to modify the glucostatic hypothesis, and to suggest that absolute levels of blood glucose are not important, but rather the availability of that glucose for use by cells. It has been shown experimentally that a decreasing availability of glucose is very well correlated with feelings of hunger, and there is evidence for the existence of both central glucoreceptors in the brain, and peripheral glucoreceptors in the liver.

At one time the role of the brain in controlling food intake seemed very clear. Whereas peripheral, oropharyngeal, and gastric factors had their role in the initiation and cessation of eating, this was thought to be primarily to monitor intake; the more fundamental regulation was thought to depend on structures located within the hypothalamus, a small region found at the base of the brain and to which is attached the pituitary gland. Experiments showed that destruction of an area in the middle and basal region of the hypothalamus resulted in massive overeating and consequent obesity in rats, cats (Felidae), and monkeys (Simiae). This was conceptualized as a *satiety centre*, in that its destruction seemed to lead to an inability to detect when to stop eating. Conversely,

damage to more lateral areas of the hypothalamus resulted in the opposite syndrome, i.e. cessation of both eating and drinking. Electrical stimulation of this same area resulted in satiated animals resuming the act of eating. Furthermore this electrically elicited eating seemed to have all the characteristics of normal feeding behaviour in the deprived animal, and in particular animals would work to obtain food REWARD and learn new tasks for food reward under its influence. Accordingly, the lateral hypothalamus was designated a feeding centre, and the two regions were thought to jointly control food intake according to the balance of activity in the inhibitory 'satiety' centre and the excitatory 'feeding' centre. More recently it has become obvious that this extreme localizationist position is almost certainly incorrect, and that such experiments which attempt to find the underlying mechanisms operating under extreme conditions of food deprivation may not apply under the more normal conditions of continuous access to food. In particular, the anticipatory nature of feeding and the importance of learned influences requires the participation of higher levels of the brain, such as the *limbic system* and the *association cortex*.

Hunger is therefore a complex sensation, and subject to a wide variety of both internal and external influences. The special conditions which result in an animal starting to eat do not have to be eliminated in order for that eating to cease, and it would seem that what starts and what stops eating are indeed separate questions. How much food is eaten at a particular meal is determined by factors such as the taste of food, and the animal's previous history. If that food is highly palatable, then in man and other omnivores this may lead to excessive intake and consequent obesity. The spontaneous feeding behaviour of animals is not likely to be understood simply in terms of eating in response to deficit signals of hunger, as much of their eating is of a non-regulatory and often anticipatory nature. In circumstances where hunger does result from periods of food deprivation it is unlikely that there is any one signal which is of prime importance and acts as a hunger stimulus, but rather a number of influences, both peripheral and central, which are acting together.   P.W.

106

HUNTING is an aspect of PREDATORY behaviour and involves an animal of one species, the predator, capturing a member of another species, the prey. In this way hunting may be distinguished from the more general SEARCHING and EXPLORATORY behaviour.

Hunting may or may not involve an active exploratory phase. A predator may choose a suitable ambush site and lie in wait, or it may move about searching for prey, or it may attempt to flush prey from its place of concealment. For example, the barn owl (*Tyto alba*) sometimes searches for prey visually, while at other times it sits motionless in the dark, listening for a wood mouse (*Apodemus*) or other small animal to reveal itself. Barn owls can catch mice in complete darkness, guided by accurate auditory location. They also beat upon branches of trees to flush small birds from their roosts. The barn owl therefore, is an example of a predator that employs a variety of methods of hunting. Other predators, however, specialize in a particular technique designed to catch a particular type of prey.

Scavenging. Many carnivores, including lions (*Panthera leo*), spotted hyenas (*Crocuta crocuta*), and jackals (*Canis mesomelas*), scavenge on carrion and steal directly from other predators, in addition to hunting and killing on their own account. Animals that steal prey from others are sometimes called *kleptoparasites*. Some, such as Arctic skuas (*Stercorarius parasiticus*), habitually attack birds, such as puffins (*Fratercula arctica*), and steal their food (see PARASITISM).

Some marine fishes and birds take prey flushed by members of other species. For example, cattle egrets (*Bubulcus ibis*) follow herbivorous mammals, or farm machines, and catch insects as they are flushed from the ground. Many small fish are attracted to the cloud of sand raised by feeding goatfish (*Mulloidichthys dentatus*), which dig in the substratum with their barbels, disturbing small organisms in the process.

Stalking. Predators such as lion and cheetah (*Acinonyx jubatus*) are able to outrun their prey over short distances, and have a good chance of capturing Thomson's gazelle (*Gazella thomsoni*) or Burchell's zebra (*Equus burchelli*), provided the chase is initiated within a certain striking distance. The chase is often preceded by a stealthy approach.

Successful stalking requires inconspicuous manœuvring. Some predators rely on CAMOUFLAGE. Thus, the common cuttlefish (*Sepia officinalis*) adopts a cryptic COLORATION and approaches its intended victim by inconspicuous undulations of its transparent swimming membranes. Other predators, such as pike (*Esox lucius*), lion, and cheetah, make skilful use of cover, moving from one clump of vegetation to the next when the victim is looking away or is feeding with head down.

Stalking animals often avoid detection by adopting an inconspicuous mode of locomotion. Some owls have special feather structures which allow them to fly silently. The carnivorous mammals remain quiet during a stalk, and reduce their apparent body size by crouching and approaching the prey head-on. They also avoid sudden movements which might attract attention. Many predators are able to make detours which enable them to take advantage of wind direction or cover and so approach more closely to the prey before having to reveal themselves. Chameleons (Chamaeleontidae), which stalk insects among the branches and twigs of trees, are able to make extensive detours when they see a possible victim that is out of reach. The animal manœuvres itself

into a position from which it can strike at the insect with its long extensible tongue.

Loss of visual contact with the prey often occurs during the initial phase of a detour, yet many predators seem able to judge the correct subsequent approach path. This suggests that they possess a particular kind of PROBLEM SOLVING ability.

Pursuit. Two modes of pursuit can be distinguished. One may be called *guided pursuit*, and the other *ballistic attack*. During guided pursuit the predator continually modifies the direction of its approach by FEEDBACK from monitoring the position of the victim. For example, cheetah attempting to follow the zigzag evasions of a fleeing gazelle do not always run at their maximum speed. They may sprint during an early phase of the pursuit to bring themselves within striking range, and then closely follow the zigzags of the fleeing animal. They attack with a final sprint when the victim is running directly away from them. During a ballistic attack the predator launches itself at the victim, having first estimated the future path or position of the prey. No attempt is made to modify the direction of the attack once it has been launched, and indeed this is often not possible because of the speed of the attack. The success of ballistic attack therefore depends entirely upon the initial aiming. For example, both large-mouth bass (*Micropterus salmoides*) and the common octopus (*Octopus vulgaris*) approach their prey under visual guidance, but the final phase of the attack, which is characterized by rapid acceleration, is not affected by visual feedback. The predator takes aim and then launches itself at the prey. If the prey moves in an unexpected manner, once the final strike is under way, then the target will be missed.

Predators often attempt to take short cuts during a pursuit, or to intercept the flight path of the prey. Whenever the line of flight is not directly away from the predator, then the possibility of a short cut exists. Thus a prey that adopts a zigzag dodging manœuvre when the predator is still some distance away gives the predator an opportunity to draw closer. Dodging is a worthwhile strategy only when the predator is very close. Predators may attempt to induce dodging by means of mock attacks. A golden eagle (*Aquila chrysaetos*), for example, may swoop on a hare (*Lepus europaeus*) in a series of mock attacks, and then launch a real attack at a time when it is difficult for the hare to dodge, such as when it is about to enter a patch of thick vegetation.

Mock attacks may also be used to exhaust a potentially dangerous prey. For example, the ichneumon mongoose (*Herpestes ichneumon*) may launch a series of mock attacks in tackling the Asian cobra (*Naja naja*) or the mountain viper (*Vipera xanthina*). Each feint induces the snake to strike, and when the snake becomes tired and inaccurate in its counter-attacks, the mongoose delivers a deadly bite at the back of the snake's neck.

Capture. The exact mode of prey capture is

often tailored to the species of prey. For example, the pygmy owl (*Glaucidium passerinum*) kills small birds with the powerful grip of its talons, but kills mice by repeated pecks into the head. The orb-web spider *Argiope argentata* discriminates between moths and butterflies (Lepidoptera), and other insects which become caught in its web. Most insects are wrapped in silk and then bitten. Moths, however, are always bitten before being wrapped in silk. This difference is probably connected with the fact that the loose wing scales often enable a moth to escape from a spider's web.

Prey capture is often preceded by an accelerated approach to the prey, and in addition to seizing the prey, the predator must take care not to overshoot the mark, or to loose the prey as a result of impact with inanimate objects. Open water predators are not impeded in this way, and are often carried by their momentum beyond the point of seizure. The same is true for the mid-air attacks of birds of prey. When the prey is near the substrate it is necessary for the predator to decelerate to protect itself from injury. Both octopus and large-mouth bass apply rapid braking movements just before prey capture. Falcons (e.g. the prairie falcon *Falco mexicanus* and the peregrine falcon *Falco peregrinus*) maintain their grip when they attack a pigeon (Columbidae) during level flight, but when they stoop from a point high above the prey, they knock it to the ground with a glancing blow from the talons.

The final phase of an attack requires considerable COORDINATION and accurate estimation of the relative velocities of predator and prey. Remembering that the final attack phase is often preprogrammed and essentially ballistic, it is interesting to note that research studies of cuttlefish, bass, osprey (*Pandion haliaëtus*), and puma (*Puma concolor*) attributed only about a 10% failure rate to miscalculations on the part of the predator. When account is taken of other causes of failure, such as evasive action by the prey, interference from other predators, etc., then the failure rate may be considerably higher. The spotted hyena are successful on about 32% of occasions that they attack wildebeest calves (*Connochaetes*), and Forster's tern (*Sterna forsteri*) was observed to attain a success rate of only 24% in diving for fish. Success rate is undoubtedly influenced by whether or not the predator assesses the chances of success before embarking on a hunt. The evidence suggests that some predators do this. Thus spotted hyena adopt different hunting techniques for different prey, and from their mode of setting out for a hunt it is often clear which prey they are after. For example, they hunt zebra in packs of about twenty, but gazelle singly. Sometimes they communally run down the prey, sometimes they stalk, and sometimes they steal from other predators. Despite the diversity of methods the success rate is surprisingly similar for different types of prey.

Communal hunting. Communal hunting requires a degree of SOCIAL ORGANIZATION, and the fortuitous arrival of more than one predator at a potential prey should not be counted as communal hunting. This means that FLOCKING behaviour, when combined with feeding, should not be considered as communal hunting. In such situations each predator is usually capable of dispatching a prey animal unaided, whereas in true communal hunting the concerted effort of at least two individuals is required to hunt down the prey.

A number of predatory species, including killer whales (*Orcinus orca*), porpoises (*Phocaena vomerina*), and wolves (*Canis lupus*), herd their prey by approaching with the pack fanned out, and eventually encircling the prey or driving it into a place from which there is no escape. Wolves and lions are also known to drive prey: a few individuals manœuvre themselves around the other side of the prey and drive them towards their colleagues waiting in ambush. In some species, different roles are adopted by the members of the hunting group. In a mated pair of lanner falcons (*Falco biarmicus*), for instance, the heavier female may chase jackdaws (*Corvus monedula*) and pigeons (*Columba livia*) away from a sea cliff, by flying in and out of gullies and caves. Meanwhile the male waits on the wing, some distance from the cliff, and attacks the birds flushed by his mate.

Communal hunting greatly increases the chances of success. Thus lions hunting in a group kill about twice as many prey per hunt as do single lions. Laughing gulls (*Larus atricilla*), either singly, or in groups of up to eight, chase the common tern (*Sterna hirundo*) and the Arctic tern (*Sterna paradisaea*) that are carrying fish to feed to their brood. The gulls try to steal the fish from the tern by attacking it in mid-air. These attacks are much more successful when there are a number of gulls involved. However, only one gull gets the fish, and for any individual gull the chances of getting a fish are lower if it joins an attacking group than if it attempts to attack the tern alone. When an individual gull initiates an attack he has approximately a 15% chance of success, but if he is joined by other gulls his chance of success falls. Moreover, gulls that join the attacking group later are more likely to get the fish, because they are less exhausted than the gulls that initiated the chase. Thus it appears that it is to a gull's advantage to initiate an attack, provided he is not joined by too many others. On the other hand, it is even more advantageous to join a group of attacking gulls at a late stage.

When spotted hyenas are hunting wildebeest calves they have the greatest individual advantage from hunting in pairs; a single hyena has no chance of killing the calf if its mother attacks. However, one of a pair of hyenas can keep the wildebeest cow preoccupied, while the other deals with the calf. The chances of success are only slightly increased if other hyenas join the hunting pack. Ostensibly, two hyenas will do better to share the dead calf between them than between members of a larger hunting pack. However, this advantage is often offset by the arrival of other hyenas after the kill.

Another advantage of communal hunting is that it enables predators to obtain larger prey than they could otherwise do. The communal spiders *Agelena consociata* live in large webs, up to 1 m in diameter, along borders of the Gabon forest. Small prey are captured and eaten by single individuals, but larger prey are killed by groups of spiders. Killer whales move in groups, but individually attack small prey. When they encounter a large baleen whale (Mystacoceti) they attack in a cooperative group. A similar response is seen when wolves attack moose (*Alces alces*), when cheetah attack zebra, and when Eskimos hunt bearded seal (*Erignathus barbatus*) as opposed to the smaller ringed seal (*Pusa hispida*).

The obvious advantages of communal hunting are partially offset by the fact that the carcass is generally shared between the members of the group, so that the individual may obtain less than if he had hunted alone. However, it is likely that the probability of hunting success is related to the availability of the various prey. Thus, when small game are plentiful it may be better to hunt alone, but when there are relatively more large prey it may be of advantage to the individual to join a hunting pack.

28

HYPNOSIS. An awareness of animal hypnosis dates back at least to the Old Testament, but the phenomenon was popularized by a report 325 years ago by an Austrian monk who demonstrated the reaction in domestic chickens (*Gallus gallus domesticus*). His procedure was quite simple. He took a stick and drew a straight line on the ground. Next, he held a chicken so that its head was on the ground in order to get the animal to 'fixate' on the line, much as a hypnotist often requires his subjects to fixate on an object dangling in front of their eyes. Much to his surprise the bird, after a few frantic moments, stopped struggling and went into a frozen posture, a state of paralysis which persisted long after he removed his hands.

Fig. A depicts a chicken in this so-called hypnotic state. Notice, however, the absence of a line. Fixation is irrelevant to the onset of the hypnotic condition, as evidenced by the fact that blind animals still show the reaction. Many other superstitious techniques for inducing animal hypnosis have been devised over the years, such as the alligator wrestler who rubs his 'opponent's' stomach or the magician stroking a rabbit into acquiescence. The common denominator of most procedures designed to produce hypnosis in animals involves some form of physical restraint. In practice, all one has to do is hold an animal down in any stable position on a flat surface for a few seconds. Virtually any animal will do, as long as it has not previously been exposed to much human contact. Some HOUSEHOLD PETS are probably too tame to show the reaction; however, a caged canary (*Serinus canaria*) or lizard (Lacertidae) will demonstrate the condition well. Hypnotic reactions have been observed in such diverse animal groups as insects, crustaceans, fish, amphibia, reptiles, birds, lower mammals, and primates. In fact, one of the most impressive things about the phenomenon is its generality throughout the animal kingdom.

In addition to the profound state of unresponsiveness which accompanies animal hypnosis, other characteristics include muscular rigidity, Parkinsonian-like tremors, changes in heart and respiration rate, altered BRAIN electrical patterns, and diminished responsiveness to external stimulation. Occasionally animals may also close their eyes during the hypnotic episode, giving the impression of SLEEP or even death. Termination of hypnosis is usually quite abrupt; the animal stands and then often attempts to escape. The duration of the hypnotic condition varies both between and within species. In chickens, which have been by far the most popular LABORATORY subjects, the reaction typically lasts for about 10 min, but may continue for as long as 3 h or more.

Theories of animal hypnosis. Historically, interest in animal hypnosis was prompted by the notion that it might represent the prototype or precursor of hypnotic phenomena in man. Even today there are advocates of this view. However, there have been a number of other theories, including (i) PAVLOV's notion of *cortical* inhibition, (ii) sleep, (iii) spatial disorientation, (iv) fear, and (v) DARWIN's concept of DEATH FEIGNING.

Of the various interpretations only a few have been conducive to much systematic research. The hypnotic approach, while intuitively appealing,

Fig. A. A chicken in the state of hypnosis.

has never succeeded in generating much scientific data. The problem with hypnosis is that, although it inspires the imagination, it lacks objectivity. Even hypnosis in man remains elusive to scientific investigation. One could argue that the biggest impediment to research on animal hypnosis has been the word 'hypnosis'. Not only has it tended to attract an unobjective mystical type of investigator, but it has probably deterred many reputable scientists. Thus, most contemporary investigators prefer *tonic immobility* as a more neutral and descriptive designation.

Both sleep and cortical inhibition, while also attractive, fail to provide adequate explanations. It has been established, for example, that in spite of the animal's apparent lack of responsiveness during tonic immobility, it is awake and alert, and remains capable of monitoring the environment and processing information. Rabbits can undergo CONDITIONING while in the hypnotic state, and show evidence of long-term retention of what was learned.

Pavlov thought that tonic immobility was due to a generalized inhibition of higher (cortical) brain centres. It appears, however, that rather than causing tonic immobility the cortex antagonizes the reaction. Rats (*Rattus norvegicus*), which have either been surgically deprived of their cortex, or had their cortex deactivated with chemicals, are actually more susceptible to the hypnosis reaction and show much longer durations of immobility. Moreover, many animals lacking any naturally occurring cortex (e.g. reptiles and insects) still show robust immobility reactions.

The role of fear. As early as 1878 it was proposed that animal hypnosis might represent a FEAR reaction. After all there are many anecdotal accounts of people who have been 'scared stiff', which perhaps is comparable to tonic immobility. Unlike many of the other interpretations, the fear hypothesis lends itself more readily to experimentation.

What does one do to an animal to produce fear? The abrupt onset of a novel, pain-producing stimuli, such as mild electric shock, often elicits responses with overtones of EMOTION and fear. What would happen then, if we gave a chicken a mild shock prior to inducing tonic immobility? If it is a fear reaction, one would expect shock to exaggerate the immobility episode. This is precisely what happens. Not only do animals given a brief shock show longer durations of tonic immobility, but the stronger the shock the more prolonged is the subsequent duration of immobility. On the other hand, perhaps there is something about electric shock which might affect tonic immobility irrespective of fear, such as the tetanizing effects of shock. After all, shock has a considerable physiological effect.

So, how does one scare an animal but at the same time preclude the confounding muscular effects of a painful aversive stimulus? Abrupt exposure to loud noise would seem to be relatively free of any tetanizing effects. In support of the fear hypothesis, chickens confronted with a loud sound prior to being manually restrained show greatly accentuated durations of tonic immobility. Unfortunately, however, some animals show a tendency towards seizures when exposed to loud noise, so this method also has undesirable side-effects.

Perhaps the best experimental means of inducing fear is to employ the *aversive conditioning* technique, in which a neutral stimulus, such as a light or tone, is used as a signal for something unpleasant, such as a mild electric shock. As a result of repeated paired presentations of the neutral cue and shock, the cue eventually becomes a predictor of shock, and begins to elicit a conditioned or anticipatory kind of fear in its own right. When chickens are presented with such a cue prior to the induction of immobility, they remain immobile about six times longer than appropriate control birds which had received the same number of unpredictable shocks during training. Moreover, the stronger the shock paired with such a cue, the more effective is the cue in prolonging the duration of tonic immobility.

Immobility in such diverse animals as the chicken, the Carolina lizard (*Anolis carolinensis*), and the leopard frog (*Rana pipiens*) is accentuated by the injection of adrenalin, which increases fear responses in a variety of situations. Moreover, the induction of tonic immobility can be used as a punishment. If chickens have been taught to make some simple response for a reward, the induction of immobility contingent upon making this learned response greatly weakens such behaviour in the future, much as other aversive stimuli would do. Thus, many conditions, all designed to increase fear, share in common the ability to intensify the immobility reaction.

On the other hand, if animal hypnosis is really a fear reaction then procedures designed to reduce fear should antagonize or abbreviate the response. In support of this expectation, handling, taming, and repeated testing reduce the immobility response, which is the reason it rarely occurs among household pets. The ingestion of tranquillizers in chickens has the effect of reducing both the susceptibility to immobilization, and the duration of the reaction. Certain anti-depressant DRUGS also make tonic immobility much more difficult to obtain.

One could still argue, however, that the evidence which might seem to support the fear hypothesis remains uninterpretable. All of the manipulations in question could conceivably be affecting processes other than just fear. Electric shock, loud noise, and tranquillizers might, for example, be confounded with changes in a generalized state of AROUSAL, which in turn could be what underlies the hypnotic state. In other words, it may be that tonic immobility reflects changes in an organism's overall level of arousal, and not just fear *per se*.

However, contrary to this arousal hypothesis,

attempts to manipulate arousal in a manner unrelated to fear have had no effect on tonic immobility. Animals deprived of food for a sufficient period of time to produce hunger and corresponding signs of arousal show immobility reactions which are identical to those found in non-deprived, satiated animals. Similarly, FRUSTRATION, produced by omitting an expected food reward, leaves immobility unaffected. Amphetamine drugs, which are normally thought to profoundly alter arousal states, actually diminish immobility reactions rather than enhance them. Thus, non-specific changes in arousal fail to yield predictable effects on tonic immobility, and we appear to be dealing with something more than mere arousal.

The role of predation. Knowing that fear enhances the immobility response, we must ask why animals exhibit it. So far we have only been dealing with contrived laboratory manipulations; now we must ask whether so-called animal hypnosis has any evolutionary or ecological significance?

Darwin thought that the reaction represents a form of death feigning, and that it may have SURVIVAL VALUE. His idea was based on the observation that at least some predators may refrain from eating dead prey, so that by simulating death the animal might be able to evade PREDATION. This is certainly not to imply that the potential victim knows, anticipates, or appreciates the ultimate consequences of his behaviour; on the contrary, hypnotic reactions are not learned.

A more modern view of the possible role of predation in tonic immobility is that many prey can be seen to progress through four fairly distinct phases during a PREDATORY encounter. First, at appreciable distances, when detection first occurs, the most common reaction, especially for CAMOUFLAGED animals, is to *freeze* (sit tight), and this serves to minimize detectability. As the distance decreases, however, ESCAPE becomes the next most likely response, followed by FIGHTING at close quarters. Finally, if there is more than momentary contact with the predator, tonic immobility ensues. Thus, hypnotic reactions in animals may, in fact, represent a type of DEFENSIVE reaction against predators. A major implication of such a view would be that tonic immobility might have survival value, and could have evolved as a result of selective pressure due to the predatory action of other animals (see NATURAL SELECTION).

To substantiate such an evolutionary interpretation, however, requires evidence on a number of points. First, it would be helpful to demonstrate some GENETIC involvement, because EVOLUTION is based upon inheritable traits. Second, if the predation hypothesis is correct, it should be possible to show that the threat of predation affects tonic immobility. And finally, it would have to be shown that under natural conditions the reaction provides for a reproductive advantage, by presumably decreasing the likelihood of succumbing to predation.

The existence of a genetic influence on tonic immobility is now well documented. Not only are there substantial differences between species, but within species there are strain differences as well. There are at least a hundred strains of the domestic fowl, and each is distinguished from the next as a result of selective breeding for a variety of characteristics. Such breed differences among animals raised under comparable conditions can be attributed to genetic factors (see DOMESTICATION). Tonic immobility varies widely among the different strains of domestic fowl, White Leghorns being one of the most susceptible. It also varies between different strains of rats. Perhaps more compelling in a genetic sense is the fact that both chickens and rats have been selectively bred for no reason other than to show changes in tonic immobility. In just a few generations substantial hypnotic differences can be obtained between animals bred to show long reactions, and those bred to show short reactions. So tonic immobility does seem to be subject to a genetic influence, and therefore could have evolved in response to natural selection. But this still leaves the nature of such selection unspecified.

If tonic immobility is based on a fear of predation, then it should be possible to increase the duration of immobility by merely simulating a predatory episode. In support of this expectation, laboratory raised chickens, confronted with a stuffed Cooper's hawk (*Accipiter cooperi*) as shown in Fig. B, show greatly accentuated immobility reactions which last for inordinately long periods of time, even though they have never seen a hawk before. But what is it about a stuffed hawk that causes chickens to react in this way? It seems to have something to do with the hawk's facial features, since obscuring the hawk's head with a hood greatly diminishes the effect. In fact, the enhancement seems to be almost entirely dependent on the hawk's eyes. If his glass eyes are covered with small pieces of black tape, chickens react as if the hawk were no longer there. Indeed, it has been shown that artificial eyes suspended on wooden dowels are just as effective as an intact hawk for prolonging the immobility response.

Could it be that when a person grabs an animal and holds it down on the ground, he is in effect simulating a predatory episode? Animals tested behind opaque partitions concealing the experimenter's presence show much briefer reactions. Staring at an immobilized animal prolongs the response in much the same way as eye contact with a stuffed hawk. Even immobilizing a chicken in front of a mirror is sufficient to greatly lengthen the reaction. Why? Because each time the bird looks at his reflection he makes eye contact with the image. Dead birds in the testing situation with their eyelids sewed open have the same effect as a mirror, but the presence of a dead bird with his eyes closed leaves the reaction unaffected.

Now for the last and most important question. To view tonic immobility as an evolved predator defence presumes that the response has survival

Fig. B. A chicken showing the hypnotic reaction in the presence of a stuffed hawk.

value. In fact, many predators are very sensitive to prey movement, and without such stimulation they may lose interest or become distracted. Take the familiar case of a cat (*Felis catus*) 'playing' with a house mouse (*Mus musculus*). Following the initial capture, if the mouse remains motionless the probability of further attack is often diminished. Recent field as well as laboratory investigations have shown unequivocally that tonic immobility in chickens, lizards, ducks (*Anatinae*), and mealworms (*Tenebrio molitor* larvae) reduces the chance of being killed and eaten by a number of predators.

Human implications. Although it should be clear by now that the analogy to human hypnosis is grossly misleading, tonic immobility does show some striking similarities to certain psychotic states in man. Specifically, catatonic schizophrenia and tonic immobility share a number of symptoms in common. As a component of schizophrenia, catatonia is characterized by a profound state of physical immobility, often referred to as a catatonic trance. Comparable to tonically immobile animals, catatonics often show muscular rigidity, a hypnotic or stuporous gaze, waxy flexibility (*catalepsy*), and Parkinsonian tremors.

Catatonia, like tonic immobility, is often precipitated by or accompanied by an acute emotional crisis. While in the catatonic state patients give the impression of being overwhelmed with fear, and as a consequence it has been characterized as a panic reaction. Indeed, upon recovery, many patients report that they had in fact been afraid to move, or were even paralysed with fear. In terms elements, animals typically show intense escape reactions following termination of tonic immobility, and under conditions of simulated predation often attempt to attack the experimenter as soon as immobility is over. Similarly, catatonic people frequently go into manic, hyper-aggressive states following a catatonic episode. The abrupt onset of such behaviour, and the frenzy of the attacks can make such patients particularly dangerous.

Finally, while the catatonic can give the impression of being detached and totally unaware of what is going on around him, subsequently he frequently describes what was happening in surprisingly vivid detail. As we have already seen, tonic immobility is comparably deceptive. In spite of their physical immobility and lack of responsiveness, animals remain capable of monitoring the environment during the immobile state.

Although speculative at this point, maybe there are elements contained in catatonia which represent fragments of primitive predator defences that now appear inappropriately under conditions of exaggerated STRESS. After all, during his evolution man was probably not immune to the effects of predation. When viewed in this light some so-called ABNORMAL behaviour may not be pathological as such, but rather may have a basis in normality and evolution. G.G.G.

I

IMITATION. When Charles DARWIN helped to launch the study of animal behaviour as an empirical science in the latter half of the 1800s, he also indirectly helped focus attention on the topic of imitation. One of the implications of Darwin's theory of evolution by NATURAL SELECTION was that the property of human beings we refer to as 'mind' could be found in varying degrees in animals. Man was not unique in his ability to think intelligently, to reason, or to infer, for animals also manifested such capacities, albeit to a lesser degree than man. In particular, the ability of animals to imitate was often cited as evidence to support this notion that mind, like other characteristics, had evolved.

Whether one accepts the above line of reasoning or not, it must be conceded that many people find imitative behaviour interesting simply because it does seem to suggest that animals may be, to some extent, intelligent, rational creatures. Perhaps, too, we think we see something of ourselves, or possibly parallels to human culture, in the imitative behaviour of animals. Consider, for example, the washing of sweet potatoes in water to remove sand and grit which has been observed in a troop of Japanese monkeys (*Macaca fuscata*) from its first accidental discovery by a juvenile female, through its adoption by the youngster's playmates and her mother, and its eventual propagation through the troop. These intensively studied primates of southern Japan have provided many other examples of the origin and propagation of novel behaviour patterns, ranging from eating candy and wheat to learning how to separate wheat grains from sand by tossing the mixture into the water and scooping up the grains of wheat as they float on the surface. It is interesting to note that SOCIAL RELATIONSHIPS have been found to be important in the transfer of these habits. Mother–infant and dominant–subordinate relationships are particularly potent in this respect. Moreover, youngsters generally are more open to adopting new modes of behaviour than their elders, although this is not invariably the case. These are obviously important findings for anyone interested in the origins of human culture and society.

One need not limit considerations to primates, however, to find imitative behaviour and suggestions of CULTURE in animals. DIALECTS have been found in certain bird SONGS, much as in human LANGUAGE, that indicate that variations in species-typical patterns of VOCALIZATION can be learned socially and passed on from generation to genera-tion to within discrete social groupings or geographical areas. Similarly, young rats (*Rattus*) are known to learn from their parents which substances in their environment are safe to eat and which are not.

On the other hand, not every case of imitative behaviour necessarily reflects higher INTELLIGENCE or rational inference. It would be difficult, for example, to interpret as a cogitative act the imitative pecking response of a small chick to a tapping pencil. Nor is there any reason to suppose that sheep (*Ovis*) are behaving rationally when they blindly follow one another in single file. Similarly, whatever it is that causes many animals to eat more when fed together than when fed apart, it would hardly appear to require any more intellectual functioning than a group of people in a small room exhibit when they start yawning contagiously.

Since some instances of imitative behaviour seem intelligent and some not, early students of the problem came to distinguish two forms of imitation, one 'intelligent' and the other 'instinctive'. Instinctive imitation was thought to reflect merely a primitive copying tendency for simple actions, whereas intelligent imitation seemed to require conscious intent to profit from the example of another individual, especially as evidenced in novel or atypical patterns of behaviour. This basic distinction has persisted ever since, despite changes in the identifying labels, although scientific views on behaviour have evolved. Today the term *social facilitation* has replaced instinctive imitation, because the word INSTINCT created more problems than it solved. Yet both terms have denoted basically the same examples of imitative behaviour. For a while, there was an effort to reserve the word imitation itself, or 'true imitation', for use in place of intelligent imitation, the implication being that real imitation was much more than merely copying another individual. This seems rather pretentious in view of the fact that as will become evident, we are blatantly ignorant of just what is responsible for the vast majority of imitative behaviour, however one wishes to classify it. Certainly, without some means of reading animals' minds, if indeed they even have minds, it is not very helpful to have to decide whether or not one 'consciously intends to profit from another's experience'. So of late, reflecting the dominant interest in LEARNING processes within the field of animal psychology, the term *observational learning* has largely supplanted the older concept of intelligent imitation, while still suggesting that the behaviour in question involves circumstances for which the species' evolutionary history has not specifically 'programmed' it. Rather than contribute further to the futile debates over what is really meant by imitation, let us simply examine some of the variety of ways in which animals are thought to imitate one another. It is hoped that the range and diversity of phenomena illustrated will make it self-evident that there is no single, unitary process or phenomenon that can be called imitation, but rather that the word is only a crude abstraction, a label for our ignorance. By placing these examples in a roughly historical format, it may also be possible to appreciate some of the reasons that so little is known about what, how, and why animals imitate.

Anecdotal reports. As noted above, Charles Darwin provides an appropriate starting point for a history of imitation. For example, in one case taken from his notes, the bumble-bee (*Bombus bombax*) was observed to cut holes through the underside of the calyx of kidney bean plants, and thereby obtain nectar more easily than by entering the mouth of the flower. Honey-bees (*Apis mellifera*), which had until then only been seen to feed from the flowers' mouth, thereafter were observed to use the holes by the bumble-bees. Discounting the possibility of their being guided by the scent of the nectar escaping through the holes, Darwin concluded that the hive bees probably saw the bumble-bees at work, understood what they were doing, and imitated them in taking advantage of the shorter path to the nectar.

Another example of such anecdotal reporting is provided by reports of pet dogs (*Canis l. familiaris*) learning certain habits by imitation during their foster rearing with domestic cats (*Felis catus*). Among the feline habits claimed to have been acquired in this way were washing of the paws and face, bounding after a ball and rolling it over with the fore-paws, watching a mouse-hole for mice (*Mus musculus*), chasing them, and having caught the prey, playing with them by allowing them to run a short distance before again pouncing on them.

Behavioural scientists quickly became sceptical about the value of these anecdotes. The basic problem was that any phenomenon that is to be examined scientifically should be reproducible and verifiable. Many naturalists observed what were admittedly improbable, even unique, events, which were difficult to substantiate. Another problem with the anecdotal method was that it was generally unknown whether the same behaviour might not have occurred merely by chance in the absence of any model. How many dogs, for example, show feline traits without ever having been in contact with a cat? An answer to this question would be a prerequisite for evaluating the claims that dogs imitate cats.

Naturalistic observations. Observing behaviour under natural conditions nevertheless has its value as a legitimate scientific tool and, particularly in

the hands of dedicated naturalists and students of ETHOLOGY, has produced much better documented reports of imitative phenomena. One example, well known to British bird-watchers, concerns the habit of opening milk bottle caps by species of tits (Paridae). With the introduction in Great Britain of cardboard and metal-foil capped milk bottles placed on doorsteps, a new food source became available for exploitation in the rich cream floating on top of the milk. In themselves, the techniques employed to open the caps were not unusual, resembling the birds' normal approaches to pecking open nuts and tearing at loose bark. But responses to questionnaires in ornithological periodicals indicated that the habit of stealing cream was largely confined to blue tits (*Parus caeruleus*) and great tits (*Parus major*), and that it spread progressively from various centres of origin, which suggests some degree of social influence in the acquisition of the adaptive behaviour.

Evidence of the gradual diffusion of new feeding habits has also been reported for other species and food sources, including: great tits eating peanuts in shells suspended on a string; a group of wild northern mallard ducks (*Anas platyrhynchos*) learning to eat dry corn meal by scooping up the meal in their beaks and then walking over to a pond for water to wash it down; the recently acquired and economically distressing feeding habit of the redpoll finch (*Carduelis flammea*) in New Zealand that has resulted in more and more damage to fruit blossoms; and the adoption of birdseed as a food source by neighbouring flocks of common linnets (*Acanthis cannabina*). In each of these cases the behaviour pattern was essentially unknown prior to its first recorded occurrence, after which it was found to take place with increasing frequency, and usually with expanding geographical distribution.

Not only may the selection of a food item be socially influenced, but the actual techniques used by a number of PREDATORY species to obtain food apparently develop through social interaction with more experienced hunters, normally the parents. There are many reports that young birds of prey show little indication of knowing instinctively how to kill prey, and that they are actually quite inept at learning how to hunt without assistance, some would say guidance, from their parents, or trainer, in the case of falconry. During the comparatively long period of association with the parents, the young have ample time to develop the required skills by observing their parents. But it could be a *shaping* process that accounts for this apparent learning by imitation, for the sequence of development of feeding habits in many predators that eat mobile prey generally starts with one or both parents bringing already dead prey to the young to eat. Then, as the young mature, injured prey may be brought for them to PLAY with, attack, and kill, as their abilities allow. As their locomotor and other skills develop, the young can go out to hunt in the company of the parents, and, eventually, separate from the parents to fend for themselves. Thus, the parents may not teach so much by example as by providing appropriate opportunities in which developing skills can be practised and refined.

Other cases of social influence on feeding habits seem less amenable to this sort of hypothesis. Naturalistic observations of various species have suggested that certain noxious food items may be avoided as a result of another individual's unpleasant experience with them. Calves (*Bos primigenius taurus*) and young lambs (*Ovis ammon aries*) have been said to eat poisonous herbs unless in the company of more experienced individuals. Attempts to kill large flocks of the common crow (*Corvus brachyrhynchos*) in the United States of America by providing poisoned bait were unsuccessful because, after a few individuals had been poisoned, the others avoided the bait. Similarly, the reaction of just one rat to a novel food has been said to determine the reactions of other members of the group to the same substance. If the food is acceptable, other rats will join in eating it. But if the first individual sniffs the food or bait and rejects it, the rat will sometimes defecate or urinate on the bait, thus warning others to reject it also. Even without explicit marking, however, young rat pups are known to avoid poisoned bait as a result of their parents' and other adults' past experience with it. A rather similar phenomenon has been suggested by some avian species' tendencies to take or avoid conspicuously coloured or otherwise distinctive prey. One report said that a missel-thrush (*Turdus viscivorus*) was observed on two occasions by a starling (*Sturnus vulgaris*) in an adjacent cage to take an unfamiliar and unpalatable species of grasshopper (Acridoidea), drop it, and repeatedly wipe its beak on the perch, a response often associated with unpalatable food. When the same grasshopper was offered to the starling, it too repeatedly beak-wiped, moved away from the prey, and made no attempt to attack it.

In addition to feeding behaviour, observations sometimes indicate that behaviour patterns show imitative influences. Various aspects of reproductive behaviour, for example, can apparently be facilitated through social interactions with individuals other than one's own mate. Originally it was F. Fraser Darling who noticed that gregariously nesting herring gulls (*Larus argentatus*) closely synchronized their breeding cycles. Darling inferred that a social influence was operating because larger flocks synchronized their breeding more closely than smaller ones, and because lesser black-backed gulls (*Larus fuscus*) that happened to nest in herring gull colonies accelerated their reproductive cycles to match the herring gulls. The *Darling effect* has since been confirmed in a variety of species. The facilitating effect of the presence of other individuals has also been noted for copulation in ring-billed gulls (*Larus delawarensis*), egg-laying in pigeons (*Columba*

livia), calving in the brindled gnu (wildebeest) (*Connochaetes taurinus*), NEST-BUILDING in black-headed village weaverbirds (*Textor cucullatus*), pair formation in Californian quail (*Lophortyx californica*), and copulation in Norway rats (*Rattus norvegicus*). Conversely, it has been speculated that the failure of some species, such as the razor-billed auk (*Alca torda*) and the Northern gannet (*Morus bassanus*), to reproduce successfully in small groups could be due to the lack of sufficient social stimulation. It is thought that such a deficiency could also have been a factor in the precipitous decline and extinction of the passenger pigeon (*Ectopistes migratorius*).

There are a number of miscellaneous categories of behaviour that could be mentioned here. The consistency from generation to generation of MIGRATION routes used by such birds as the Canada goose (*Branta canadensis*) has prompted speculation on whether such behaviour amounts to a tradition learned by each new generation of geese by travelling in association with experienced adults, notwithstanding the fact that other factors such as prevailing winds might also contribute to the observed uniformity of behaviour. The 'transfer of mood', for want of a more precise phrase, is another example of socially induced behaviour, described by various ethologists as a contagious tendency within certain FLOCKING birds and for that matter many ungulate herds, to be easily panicked into taking flight or stampeding, due to the influence of one or more frightened individuals. Conversely, a calming influence on excitable members of a group has also been noted at times, as in the case of 'leader' steers used by cattle drivers.

The rather frustrating characteristic common to many of the above reports is their failure to go beyond the level of merely describing the behaviour involved. An understanding of imitative phenomena requires knowledge of the necessary and sufficient conditions under which the behaviour occurs. Because of the inherent limitations of naturalistic observation, experimental analysis is necessary to answer questions about the causes and functions of behaviour. Thus, we turn now to consider some of the experimental studies that have been conducted on imitative phenomena and their theoretical interpretations.

Experimental studies. There have been numerous experimental studies of various forms of imitative behaviour, a few of which will be described in order to illustrate the type of research that has been done. One of the better-known phenomena under the broad heading of imitation is the tendency of many species to eat more when fed in groups than when fed alone, and this has been demonstrated experimentally in more than one way. For instance, if a chicken (*Gallus g. domesticus*) is allowed to eat until completely satiated and is then introduced to another chicken that is still hungry and feeding, the satiated bird will resume eating. Alternatively, if, for example, puppies are fed either alone or together on alternating days, they show a pattern of eating more on the days of social feeding. Similar work has been done with rats, fish, opossums (Didelphidae), and other species. In some cases, however, a species may show socially facilitated eating by one procedure but not the other. The reasons for these differences are not clear. 'Competitiveness' has occasionally been mentioned as a basis for socially facilitated eating, the idea being that animals tend to eat more when in the presence of others because of a generalized tendency to compete for what is often a limited resource. But this is really only begging the question, since it provides no explanation of what causes the animals to be competitive in some cases but not others. Besides, some studies have made an effort to eliminate any form of direct or active competition, but this has not resulted in any reduction in the facilitation of eating.

Related perhaps to this work is the phenomenon of socially facilitated pecking in chicks. Using live chicks and hens, as well as mechanical models, a number of studies have tried to discover what factors influence a chick's tendency to peck when another animal pecks. An interesting instance involves a study using a simulated hen which showed that not only did the chicks peck more frequently in the presence of the mechanical model, but they even chose to peck at the same coloured grains as the model. Results of a similar type were obtained using chaffinches (*Fringilla coelebs*) and house sparrows (*Passer domesticus*): these birds were more inclined to consume an unfamiliar food (e.g. green pastry dough) if another member of their species first did so.

Although experimental demonstrations of such phenomena are valuable, there still remains the task of finding out what underlies the imitative act. A good illustration of this point is provided by a series of experiments investigating what has been termed the *social transmission* of learned taste aversions (see CULTURAL BEHAVIOUR). Rats and a number of other species, including man, are known to associate ill effects (e.g. nausea) with a particular taste and thus learn to avoid ingesting substances having that taste. This FOOD SELECTION ability lies behind the problem of bait-shyness that often plagues efforts to eradicate vermin with poisoned baits. What makes this phenomenon relevant to the topic of imitation is that young rat pups have been found to acquire these aversions from their parents and other adults, even though the bait in question was entirely safe and palatable during the pups' exposure to it. Moreover, they continue to maintain their socially acquired avoidance tendencies even after being separated from the older rats. Thus, it is clear that the youngsters learn something from their parents about which foods to eat, but the questions then arise as to exactly what they learn, and how they learn it.

Many experiments on observational learning or socially facilitated learning fail to separate adequately the effects of observation from the effects of normal individual learning. This is because the observation periods and the experimenter's test periods are usually interspersed, and may even be simultaneous. Of course, control groups that learn the task without exposure to a model's performance are always used for comparison with the experimental subjects. But this approach still does not allow one to conclude that purely observational learning occurred. In order to justify such a conclusion, it would be necessary that all observation time preceded the test period. Relatively few studies using non-primate species have used such a procedure successfully. Among the small number of examples of this sort is a report that rats can learn a discriminative avoidance task by observing pre-trained models. The experimental procedure in this case required the rats to leave an enclosed area by means of a trapdoor, depending upon which visual pattern was present. For example, if vertical black and white stripes were displayed on the wall the rat should run from that area within 5 s of the trapdoor opening, otherwise it would receive a shock to the feet. Conversely, if the stripes were horizontal it should remain there when the door was opened to avoid being shocked. Rats quickly learn to perform the responses appropriate to the cues provided, and so generally avoid receiving any shock. This particular study used two rats simultaneously as models in the task, because they seemed to facilitate one another to perform better and more consistently. Observer subjects were either placed directly in the test box with the models, or were kept in an observation box from which they could observe the model's responses without actually being able to take part themselves. It is important to note, however, that the experiment provided that even the observer subjects that participated in the models' trials did not receive any shock. Thus, none of the observers ever experienced any direct aversive stimulation at any time. Yet, when placed in the test box alone, the observer subjects that had had either six or twelve observation trials (their control groups would have had no trials or only two trials, respectively) performed significantly better than would be expected by chance. In other words, these animals, including even those that had only observed without participating, had evidently learned whether or not to flee from each of the stimuli, based solely on the response of other individuals to those patterns.

Bird vocalization. It is fitting to conclude this article with a discussion of avian vocal imitation, for, apart from its own intrinsic interest, this topic provides a convenient recapitulation of many of the themes already presented. First, there has been the usual concern about terminology and distinctions. *Vocal imitation* generally has been the broad heading for the copying of one's own species, as may occur when young birds learn their song from others around them. *Vocal mimicry*, on the other hand, has usually been reserved for the sort of copying that includes apparently non-functional sounds, for example, the so-called talking birds, such as the Javan hill mynah (*Gracula religiosa*). Although convenient for descriptive purposes, the distinction is actually quite arbitrary, and should not be taken too seriously.

There has been a long history of natural observation, some of it quite dilettante, but a great deal also of a very careful and precise nature, especially since the advent of high-fidelity recording instruments, and the sound spectrograph (see TECHNIQUES OF STUDY). The latter device provides a graphic description of three physical dimensions of sounds: their temporal patterning and duration (even as brief as a few thousandths of a second); their frequency (pitch); and their intensity (volume). Descriptions of avian vocalizations that were formerly limited to onomatopoeic renditions such as 'tseet', 'ah-ooo-ah', and 'whip-poor-will', have now been supplanted by internationally understandable graphic representations that have many advantages over musical notation. Hand in hand with these technical advances in description have come experimental studies on SONG learning that have provided fascinating insights into what birds learn and how they learn.

Scientists have discovered not only that young chaffinches need to hear the song of other chaffinches in order to develop a normally structured song, but that the first year of life is critical for such learning (see ONTOGENY). During their first 90 days after hatching the young birds assimilate the basic structure of their song, although at the time they themselves of course do not sing. The following spring they appear to develop the finer details of the song through exposure to and competition with their territorial neighbours. Thus there are actually two separate periods during the first year which are critical for the young chaffinch's song development. Although details vary considerably for different species, the importance of early auditory experience has been experimentally established for a number of other birds. One long-suspected result of the fact that certain birds learn their songs socially is that DIALECTS may be established which characteristically differentiate individuals of the same species inhabiting different geographical areas. White-crowned sparrows (*Zonotrichia leucophrys*) of the San Francisco Bay region of California, for example, are known to show such local geographical differences in vocalization patterns. By hand-rearing young sparrows in sound-proof chambers, and then exposing different individuals at various ages to recorded sounds of their own and other dialects, as well as other species' song, it was found that nestlings taken, for instance, from the Sunset Beach area could eventually adopt the characteristic variations they heard in the song of adult males from the Marin Beach area near San Francisco. It is also important to note that the young birds were not completely subject to modification by their auditory experiences. First of all, they had a limited period of sensitivity to the songs of their species, whatever the dialect. Thus, their usual period of learning lasted only from about day 10 to day 50, after which they showed considerably less or no ability to acquire a normal song. Secondly, their plasticity during this SENSITIVE PERIOD did not extend to the songs of other species such as the song sparrow (*Melospiza melodia*). So even though certain properties of vocalizations may be learned socially, constraints can nevertheless exist on the degree of variance that can be assimilated in the songs of some species.

Not all cases of vocal imitation develop at an early age. One striking example is provided in the learning of duets by certain tropical birds. FIELD and LABORATORY STUDIES of the bou-bou shrike (*Laniarius aethiopicus*), for instance, have revealed how mated pairs have the ability to sing in rapid alternation and remarkable coordination the notes of an antiphonal song. Moreover, the birds are quite flexible in their vocal performances, for either bird can sing the whole song alone if the other is absent, and both birds can simultaneously and perfectly duplicate each other, as well as sing in alternation. It is also known from careful observations of aviary-housed bou-bou shrikes that when a new pair is formed, they develop their own particular patterns of singing over a period of time, so the songs are clearly learned socially.

Another example of imitative vocalization is the sort of matching of song that sometimes occurs among TERRITORY neighbours of certain species, such as the great tit and the blackbird (*Turdus merula*). Thus, while some birds learn to imitate the vocalizations of their mates, others may imitate their neighbours.

Given that birds variously imitate the songs of their parents, their mates, or their neighbours, it may not seem too surprising that some learn the vocalizations of other species. In fact, there are quite a few birds known for this sort of copying, some of the best-known examples of which are the common starling, the North American mockingbird (*Mimus polyglottos*), the marsh warbler (*Acrocephalus palustris*), various members of the crow family (Corvidae), the Australian lyre-birds (Menuridae) and bower-birds (Ptilonorhynchinae), and various parrots (Psittacidae). Reports indicate that as much as 80 per cent of the vocalization of the superb lyre-bird (*Menura novaehollandiae*) consists of (i) calls borrowed from other species (including at times the sounds of barking dogs), (ii) the sounds of musical instruments, and (iii) other noises, such as the blows of an axe.

It has been hypothesized that such behaviour is learned because it has been associated, even if only indirectly and secondarily, with rewards. Thus, birds that are physically capable of producing the same sounds as their trainers come to do so with increasing frequency because the sounds themselves have taken on rewarding properties through association with the trainers, who in turn have provided the birds with food and other basic requirements. A number of problems have arisen within the field of LEARNING theory that bring into question the plausibility and validity of this hypothesis on vocal mimicry. For example, how does one account for the first occurrence of the 'learned' response? The theoretician's answer has been that the bird just happens upon the sounds during random vocalizing. But surely our credulity is stretched to breaking point if we must believe that a bird just happens to whistle the particular combination of notes that comprises 'Yankee Doodle Dandy' or the motif from Beethoven's choral symphony. Obviously, there is some serious question-begging involved here. Even more damaging though, is the fact that an experiment that was conducted to support the theory produced results instead that argued against the necessity of reward for the occurrence of vocal mimicry. In essence, the study paired one tone with the presentation of food and another tone without, and found that 8-month-old mynahs exposed to these two sounds reproduced them both about equally often. Clearly, there must be more to vocal imitation than learning by reward.

It has been suggested that imitation abilities have evolved to promote individual recognition. What then about the birds that copy other species? First of all, there is no evidence that either parrots or mynahs regularly imitate any other species in the wild. In fact, the adult Javan hill mynah's imitativeness in its natural environment is reported to be limited to matching certain vocalizations of neighbours. Thus, if we were not already aware of the mynah's mimicking abilities in captivity, we would probably class it with the great tit or the blackbird as vocal imitators go. Obviously, there is a significant difference when it comes to the mynah, but the basis for this difference remains unknown. J.M.D.

IMMOBILITY generally occurs as part of DEFENSIVE behaviour although it can sometimes be induced by a form of HYPNOSIS. Many animals become immobile as a result of FEAR, and this is an important aspect of CAMOUFLAGE in some species. Immobility can also result from DEATH FEIGNING, a strategy used by some animals to foil predators.

IMPRINTING is an aspect of LEARNING which takes place during the early stages of an animal's life. It is well known that a lamb (*Ovis ammon aries*) will follow the person who has reared it on a bottle. Even after the lamb has been weaned and joined the flock it will approach its former keeper and try to stay near by. Thus imprinting has both short-term and long-term aspects: as a JUVENILE the lamb follows, and as an adult it shows some attachment to, its keeper.

The following response. The young of many *precocial* species, which can run around soon after birth, show a fairly indiscriminate attachment to moving objects in their immediate environment. For example, recently hatched northern mallard ducklings (*Anas platyrhynchos*), separated from

their mother, will follow a crude model duck, a slowly walking person, or even a cardboard box that is moved slowly away from them. Some stimuli are more effective than others in eliciting the following response. Mallard ducklings prefer yellow-green objects, while domestic chicks (*Gallus gallus domesticus*) more readily follow blue or orange objects, and will even approach a flashing light. The following response of some species, such as mallard ducklings, is enhanced by appropriate auditory stimulation, and young wood ducks (*Aix sponsa*) will move towards the source of an intermittent sound in the absence of any visual stimuli. At their first exposure they respond to a wide range of sounds, but subsequently discriminate against those which are unfamiliar. Wood ducks nest in holes in trees, and the young are normally called out of the nest by the mother from some distance away.

In general, the more an animal develops an attachment to one object the less interested it is in others. The attachment can be enhanced in chicks by rewarding their approach responses with food. In the natural environment the most effective stimuli will normally be provided by the mother, and approaches to the mother will often be rewarded by body contact and warmth, or by food that the mother scratches up. Imprinting thus involves a process of learning through which attachment to the mother develops (see ONTOGENY).

The sensitive period. Mallard ducklings will form an attachment to a moving object most readily between 10 and 15 days of hatching. They will follow a moving object upon first exposure, for the first 2 months of life, but the tendency to do this becomes progressively weaker. There seems to be a SENSITIVE PERIOD during which the young animal is most susceptible to the imprinting process. The period varies with both species and circumstances. Domestic chicks kept in groups cease to follow moving objects 3 days after hatching, but chicks reared in isolation remain responsive for much longer. In natural conditions the young birds would become imprinted upon one another; chicks and ducklings tend to stay close together, even in the absence of a parent.

The decline of sensitivity with age may be a result of the development of EXPLORATION and FEAR. Newly hatched birds of many species, including domestic fowl, turkeys (*Meleagris gallopavo gallopavo*), ducks and geese (Anatidae), and pheasants (Phasianinae) do not show AVOIDANCE of novel objects, but instead tend to approach and explore them. After a few days, however, the young birds become more timid, and show signs of fear when exposed to anything unfamiliar.

For the newly hatched bird, nothing is familiar and nothing strange. Only as the animal becomes familiar with some stimuli will other novel stimuli be differentiated. It is not surprising that a developing fear of strange objects should interfere with the imprinting process, and it has been suggested that the increasing tendency to avoid such stimuli is responsible for the termination of the sensitive period. Animals reared in isolation develop avoidance responses more slowly than socially reared animals, which suggests that the development of fear depends upon experience rather than MATURATION. They also imprint at an age when socially reared animals will no longer do so, supporting the view that the process of imprinting is terminated by the development of fear.

Long-term effects of imprinting. In addition to its effects on parent–offspring relationships, imprinting can have marked effects on the SOCIAL RELATIONSHIPS of adults, also on other aspects of behaviour, such as FOOD SELECTION and HABITAT selection.

In many mammals early experience affects subsequent social adjustment. For example, among dogs (*Canis lupus familiaris*) there is a sensitive period from about 3 to 10 weeks of age, during which normal social contacts develop. If a puppy is reared in isolation beyond 14 weeks of age, its subsequent social behaviour will be ABNORMAL. Dogs, like some bird species, readily accept human beings as social partners (see HOUSEHOLD PETS). A short acquaintance at the height of the sensitive period is sufficient for a puppy to form a lasting social relationship with its owner.

The sexual preferences of many birds are influenced by early experience, a phenomenon termed *sexual imprinting*. Among domestic species of fowl, ducks (Anatinae), pigeons and doves (Columbidae), and finches (Fringillidae), individuals of one breed with a distinctive coloration can be reared by parents of a different breed and colour. It is generally found that, when mature, such individuals prefer to mate with birds of their foster parents' colour, rather than of their own colour.

Cross-fostering experiments can also be carried out with birds of different species, provided that they are fairly closely related. Experiments involving pigeons and doves, ducks and geese, jungle fowl (*Gallus*) and domestic fowl, house and tree sparrows (*Passer domesticus* and *Passer montanus* respectively), herring and lesser black-backed gulls (*Larus argentatus* and *Larus fuscus* respectively), and various species of finches usually result in a strong sexual attachment to the species of the foster parent. For example, if male zebra finches (*Taeniopygia guttata*) are raised by Bengalese finches (*Lonchura striata*), they court Bengalese finch females when adult. Even when provided with the two types of female simultaneously, and when the zebra finch females show obvious readiness for courtship, whereas the Bengalese females are unenthusiastic, the zebra finch males attempt to court the Bengalese finch females.

Birds which are hand-reared often become sexually imprinted upon people (see LORENZ). This phenomenon has been reliably reported for more than twenty-five different species of birds. Mammals which are hand-reared and maintained in CAPTIVITY can also develop long-lasting social relationships with people, but these are rarely sexual. Cross-feeding among mammals rarely results in altered sexual preferences.

Sexual preferences for the foster parent species are not restricted to particular individuals, but show GENERALIZATION to all members of that species. This is not due to an inability to discriminate between individuals. Among zebra finches, for example, preferences for individuals may develop after pair formation, and these are obviously not due to the initial imprinting experience. Thus sexual imprinting seems to involve a preference for the type of female that played the PARENTAL role. This is in contrast to *filial imprinting* which is much more likely to result in attachment to particular characteristics of the parent or parent substitute. Interestingly, among several species of sexually dimorphic ducks, only males show sexual imprinting as a result of cross-fostering. They prefer partners similar to the female that reared them, whereas females prefer to mate with males of their own species. Of course, the females of a sexually dimorphic species would not normally be greatly exposed to the male colour pattern during early life, because INCUBATION and subsequent parental care is carried out exclusively by females. In the monomorphic Chilean teal (*Anas flavirostris*), on the other hand, both males and females show sexual imprinting. In this case the female ducklings are normally exposed to the colour pattern characteristics of both parents.

Sexual imprinting often persists for a number of years. Mallard ducks foster-reared by other species of ducks and geese persist in their attempts to court their foster species, even though they do not get much cooperation. In one experiment, zebra and Bengalese finches who had been cross-fostered were then maintained with members of their own species, and carefully isolated from their foster species for a number of years. Most of these finches bred successfully with members of their own species, but when eventually they were given a choice between their own and their foster species, they strongly preferred to court the latter.

Sexual imprinting occurs most readily to members of the animal's own species, less readily to closely related species, and least readily to inappropriate species such as human beings. The rigidity and persistence of sexual imprinting also conform to these general rules. When finches are raised by a mixed pair of zebra and Bengalese finches, they always imprint upon their own species. Similar results have been obtained with ducks. Cross-fostered finches will sometimes court members of their own species in later life, but finches raised by their real parents never court other species. Hand-reared finches and other animals do show sexual behaviour towards human beings, but they usually also court members of their own species. Thus sexual imprinting is not absolute, but seems to be innately biased towards members of an animal's own species.

The biological significance of imprinting. Imprinting provides an alternative to an INNATE recognition of members of one's own species. Both filial and sexual imprinting normally involve learning the characteristics of one's parents. This may be important in species in which the COLORATION and other characteristics of the parents are inconspicuous, and designed to facilitate CAMOUFLAGE. For innate recognition to be reliable, conspicuous SIGN STIMULI are probably necessary. Imprinting enables the young animal to establish precise and detailed recognition of its parents and of other members of its species. In sexually dimorphic species the male is usually more conspicuous, having a distinctive colour pattern or SONG. Among such species the females are generally less susceptible to imprinting, relying more on innate recognition of members of the opposite sex.

From the point of view of EVOLUTION, it is usually important that parents should care only for their own offspring, and that mating should occur only between members of the same species (see ISOLATING MECHANISMS). Among ducks and geese the sensitive periods for sexual imprinting correspond closely to the duration of parental care, being more prolonged in geese than in ducks. Thus the young bird is susceptible to this type of learning only while it is a member of the family group. In species that breed in close association with other species, as in the mixed breeding colonies of gulls, the sensitive period is generally short. In the ring-billed gull (*Larus delawarensis*), for example, the sensitive period is over before the chicks leave the nest and mix with members of other species. Many animals have specific mechanisms for ensuring that strange juveniles are not accidentally adopted. Thus many gull chicks develop a specific recognition of their parents' call and *vice versa*. These *contact calls* occur after periods of separation, and ensure that the integrity of the family unit is maintained.

Shortly after she gives birth a mother goat (*Capra*) is sensitive to the smell of her kid for about 1 h. During this period a 5-min contact with any kid is sufficient for it to be accepted as her own. If there is no such contact the kid will be rejected and will not be allowed to suckle.

Imprinting tends to occur in those species in which attachment to parents, to the family group, and to members of the opposite sex are an important aspect of their SOCIAL ORGANIZATION. Imprinting seems to have evolved in those species in which there is a danger of such attachments being misplaced.

69, 99, 130

INCENTIVE is a type of MOTIVATION that derives from expectation of reward or punishment. When an animal has learned a particular task to obtain food REWARD, it will perform more strongly when its expectation of reward is high. Suppose an animal is given large rewards in situation A and small rewards in situation B. A DISCRIMINATION between the two situations will generally develop,

and the animal will learn to expect a larger reward in situation A. On the basis of this expectation it will respond more strongly in situation A than in situation B. Situation A is said to provide more incentive than situation B.

Incentive is thus based upon LEARNING. It is a property of external stimuli which have a motivational effect, because the animal has learned that certain REINFORCEMENT properties are associated with the stimuli. The reinforcement may be positive or negative. Thus expectation of food reinforcement gives rise to a positive incentive, whereas expectation of PAIN gives rise to a negative incentive. The incentive motivation influences the animal's behaviour independently of other aspects of motivation. Thus an animal may be very hungry but have little incentive to obtain food, if external cues indicate that the likelihood of obtaining food is low. The behaviour of the animal will then depend upon the balance of hunger and incentive. Similarly, an animal may be hungry, but its assessment of the situation is such that it expects there to be some pain involved in obtaining food. The combination of high HUNGER and high negative incentive may result in CONFLICT behaviour.

16

INCOMPATIBILITY. Behaviour patterns are said to be incompatible when they cannot occur simultaneously. Some activities are logically incompatible: no animal can simultaneously stand up and sit down, or move forwards and backwards. However, there are other types of behaviour which can be seen to be incompatible only through knowledge of the muscles and REFLEXES involved. Pairs of reflexes that have some muscular movements in common are neurologically incompatible in that stimulation of one reflex inhibits performance of the other. Such *inhibition* is often reciprocal, so that one activity completely suppresses the other or is completely suppressed by the other. *Reciprocal inhibition* of this type is typical of walking and other aspects of LOCOMOTION.

Psychological incompatibility can arise in situations in which the animal's ATTENTION is divided between different features of the environment, or where different stimuli give rise to incompatible states of MOTIVATION. For example, a duck (Anatinae) offered bread may be in CONFLICT between incompatible TENDENCIES to take the food and to avoid the person holding it.

Incompatibility by definition can arise in cases in which mutually exclusive classes of activities are important in the CLASSIFICATION of behaviour. Scientists often define activities in such a way that they fall into mutually exclusive classes, because this aids subsequent description and analysis of the behaviour.

INCUBATION. The eggs of all birds require particular conditions of heat, humidity, oxygen, and carbon dioxide pressure to develop successfully. Most birds satisfy these requirements by sitting over their eggs, and maintaining them in a fairly stable thermal and gaseous environment. Some bumble-bees (Bombini) incubate their eggs in a similar way, as do some pythons (Pythoninae), which shiver to generate the necessary heat. Not all birds incubate their eggs, however. Some birds of the mound-builder family (Megapodiidae) build a mound of decaying vegetation in which they lay their eggs. The decomposition of the mound provides heat for development of the egg, and the adult tends the mound until the eggs hatch. Other megapodes lay their eggs where they can be heated by the sun, or by underground volcanic activity.

Incubation has long held a special fascination for students of animal behaviour (ETHOLOGY). The behaviour used by the incubating greylag goose (*Anser anser*) to retrieve eggs that have rolled out of the nest was the subject of one of the first ethological experiments by LORENZ and TINBERGEN in 1938, and has remained the classic example of a FIXED ACTION PATTERN (see also HISTORY). Research on the colonially nesting herring gull (*Larus argentatus*) and the black-headed gull (*Larus ridibundus*) has examined the FUNCTION of incubation and the general question of the organization of behaviour. The attention paid to incubation by ethologists has been due to its relative accessibility, the birds returning to the same place day after day. In addition, there are few other activities for which the consequences, in terms of reproductive success and FITNESS, can be seen so directly.

The optimum temperature for development of the embryo inside the egg varies from species to species in birds, but is usually in the range of 34–39 °C. Incubation below this temperature may lead to developmental abnormalities. However, eggs can remain for several days without ill effect below the temperature at which no development occurs, which is around 25 °C. The duration of incubation also varies from species to species, from a short incubation period of 12–15 days in many small *passerine* (perching) birds (Passeriformes), to incubation periods of 50 days or more in some sea-birds, including penguins (Spheniscidae). The length of the incubation period is, in general, longer for species with larger eggs.

The eggs of many species require regular turning to prevent premature adhesion of the embryonic membranes to the shell, and to allow the developing embryo to remain uppermost within the egg. Incubating birds adjust the position of the eggs, especially when beginning a bout of incubation, by moving the bill among them. The incubating parent also protects the eggs from PREDATION, and some remain on the nest much more than would seem to be required to regulate the temperature of the eggs. This *attentiveness*, as spending time on the eggs is called, allows the parent to repel or distract predators, may help to conceal the nest, and prevents CANNIBALISM by neighbours in colonially nesting species. Nest attentiveness may also be important in species which are invaded by brood parasites, such as the European cuckoo (*Cuculus canorus*) and the brown-headed cowbird (*Molothrus ater*), which do not build a nest but instead lay their eggs in nests of other birds, to be incubated by the host (see PARASITISM).

How does the incubating parent maintain the environment essential for development of the embryo? Adult birds maintain their body temperature within a narrow range, despite wide fluctuations in environmental temperature, by producing enough heat to equal their heat losses (see THERMOREGULATION). At high temperatures they are able to increase their loss of heat to prevent the body temperature from rising. For the temperature of the egg to remain stable its gain and loss of heat must also be equal, but it is the parent and not the embryo that must control this. When the environmental temperature is low and the egg is losing heat, the incubating parent provides the heat necessary to balance the heat loss. At moderate environmental temperatures, when the heat loss of the egg is low, the amount of heat the parent must provide will be low too. It has been shown in a number of small passerine birds that the amount of time the parent spends incubating decreases as air temperature increases.

Though the developing embryo cannot regulate its own temperature, it does produce some heat metabolically. The amount of heat produced by the egg increases during development, and in large eggs this may be sufficient to reduce the amount of heat the parent must provide, especially in the late stages of incubation. This heat production by the embryo cannot, however, be varied to compensate for changes in heat loss. Instead, it increases and decreases as the environmental temperature rises and falls, in the same way that metabolism varies with temperature in fish, reptiles, and other cold-blooded animals (see HOMEOSTASIS).

Parent birds can adjust the amount of heat they transfer to their eggs in other ways besides getting on and off the eggs. Experiments in which the nest or eggs were artificially cooled have shown that a number of species adjust the amount of contact between the eggs and the body, and thus vary the amount of heat transferred. When the temperature of the experimental eggs was very low, the birds incubated very closely, shivering to keep both themselves and the eggs warm. The incubating bird can also prevent the egg from overheating by shading it from the direct rays of the sun. In climates where the air temperature rises above the optimum for incubation, the birds incubate closely to prevent the egg temperature rising above the body temperature of the bird.

The nest can have a major effect on gain and loss of heat by the egg. A well-insulated nest, such as that built by the hole-nesting great tit (*Parus major*), can reduce heat loss by the eggs, and reduce the amount of heat the parent must supply. The location of the nest can also be of thermal significance (see NEST-BUILDING). The rufous humming-bird (*Selasphorus rufus*) builds its nest at greater heights, and in different types of trees, as the summer progresses, to take advantage of favourable microclimates (see CLIMATE).

Although most eggs require particular conditions of humidity, oxygen, and carbon dioxide pressure to develop successfully, less is known about the regulation of these factors during incubation than about temperature regulation. It is not clear whether they too are regulated by the incubating parent, or whether they are kept within a tolerable range as a side effect of the maintenance of a stable temperature.

What causes a bird that has just laid a clutch of eggs, or whose partner has just done so, to carefully settle down on them and begin to incubate? Experiments with the Barbary dove (*Streptopelia risoria*) have shown the importance of HORMONES in this process. One of the effects of COURTSHIP in female doves is to raise the level of hormones circulating in the blood-stream. The levels of two hormones produced by the ovary rise following courtship, first *oestrogen*, then *progesterone*. Experimental doves in which the ovaries had been removed, and which therefore were producing no ovarian hormones themselves, would begin to build a nest after treatment with oestrogen and progesterone in sequence. The appearance of the nest and the eggs then produces further hormonal changes which both stimulate and maintain incubation.

The hormonal system has important effects not only on the behaviour of incubating birds, but on their morphology as well. In many species a patch of de-feathered skin appears on the breast, often with an enriched blood supply. In gulls (Laridae) three such brood patches appear, one for each egg. This structure facilitates the exchange of heat between the adult and the egg, and may, to some extent, control incubation behaviour. The tactile sensitivity of the area increases during incubation, and it has been shown that local anaesthesia of the incubation patch causes an increase in attentiveness, and thus an abnormally high egg temperature.

The timing of the onset of incubation is important for a number of reasons. It is crucial in birds breeding at temperate latitudes that HATCHING of the young occurs when suitable food is abundant. This means that the beginning of incubation must be timed appropriately. Environmental factors, such as temperature and day-length, which can be predictive of food availability several weeks later, are thought to control the initiation of breeding and incubation. It is also important in *nidifugous*, i.e. *precocial* birds (which hatch well developed and leave the nest immediately) that hatching of all young be synchronized. Hatched young can wait only a short time for their siblings to emerge before they must be led away by their parent. In many species the parent synchronizes the hatching of the eggs by delaying the onset of incubation until the full clutch has been laid. The young of some species, notably bobwhite quail (*Colinus virginianus*), have the ability to synchronize hatching themselves, by communicating while still in the shell (see VOCALIZATION). Eggs will hatch up to 24 h earlier than usual if they are in contact with an egg which is in the process of hatching. Sounds made by the hatching bird are responsible for this synchrony.

Incubation of the eggs can be costly, in both TIME and ENERGY. It has been suggested that incubation would require very little additional energy expenditure if the incubating bird were able to use heat that would normally be lost as part of normal thermoregulation. If this heat could be directed through the brood patch to the eggs, and the loss of heat by other routes reduced by fluffing the feathers and adopting a compact posture, the eggs could be incubated with little additional expenditure of energy. Although this may be possible in some species, it appears to be so only within a favourable range of environmental temperatures.

In species of birds in which both parents incubate there is sufficient time for both individuals to feed frequently and still attend the nest. In many species in which only one parent incubates, as occurs in most gallinaceous birds (Galliformes): ducks, geese (Anatidae), pheasants (Phasianinae), penguins, and others, incubation and frequent feeding are incompatible. Incubation may occur at a time when little food is available, for example in geese which breed in the Arctic in early spring, or at great distance from food, for example in penguins that nest far from the sea. Adequate protection of the nest, especially in ground-nesting ducks and pheasants, may leave little time for FEEDING. In many of these birds energy is stored in the form of fat before breeding commences, and is used up as incubation proceeds. In some species the amount of energy in reserve is carefully regulated to prevent exhausting the supply before incubation is completed. Another way to ensure an adequate supply of nutrients during incubation is for one partner to incubate while the other gathers food, and then feeds its mate on the nest. The hornbills (Bucerotidae) carry this to an extreme. The male uses mud to wall the female into a nesting cavity in a tree, leaving only a small aperture, through which he feeds her for the duration of incubation.

The division of incubation duties between the sexes shows a great deal of variation among birds. It has been estimated that in 50 per cent of bird families both parents incubate, either relieving each other several times a day, or at longer intervals, sometimes after many days in penguins and other sea-birds. In other species, such as the cockatiel (*Nymphicus hollandicus*) of Australia, the male incubates during the day, and the female overnight. Because incubation by both sexes is widespread and occurs in families thought to be evolutionarily older, it is regarded as the original form of incubation in birds. In a further 25 per cent of bird families, the female incubates alone, and in about 6 per cent the male alone. In the remaining families different species have different ways of incubating their eggs.

The SOCIAL ORGANIZATION of a species may be strongly influenced by whether one parent can successfully incubate and rear the young alone. When one sex does all the incubating the partner freed from incubation is free to mate again, if it is not obliged to feed its mate or help raise the young (see PARENTAL CARE). The harem organization of some pheasants is possible because a male can have, at one time, several females incubating clutches of eggs, unassisted. In birds such as the American jacana (*Jacana spinosa*) and the spotted sandpiper (*Arctitis macularia*) the usual roles of the sexes are reversed. Males incubate a clutch of eggs laid by their mate, which then moves on to produce additional clutches which are incubated by other males. D.F.S.

38, 48

INDIVIDUAL DISTANCE is the distance from an individual at which another of the same species provokes AGGRESSION or AVOIDANCE behaviour. Some species show virtually zero individual distance. For example, striped mullet (*Mugil cephalus*) swim together with their bodies repeatedly touching. Many other species have a strong tendency for GREGARIOUSNESS, yet maintain characteristic individual distances. Thus the black-headed gulls (*Larus ridibundus*) maintain a distance of about 30 cm between individuals. Amongst flocks of the greater flamingo (*Phoenicopterus ruber*) individual distances of about 60 cm are maintained. There may be differences between the sexes in their individual distances. Male chaffinches (*Fringilla coelebs*), for example, maintain much greater individual distances than do females. Experiments in which females are dyed to resemble males show that plumage colour is a major factor, such females being treated as males by both males and females. Interestingly, in a group of eight females, all painted to look like males, the individual distances were typical of males. It appears that if one's neighbour looks like a male one does not approach too closely.

INNATE BEHAVIOUR is a concept that has been the subject of much controversy. One extreme view is that GENETIC influences upon behaviour are minimal, and that most behaviour develops as a result of LEARNING and IMITATION. At the other extreme there are those who believe that many behaviour patterns are innate, in the sense that they develop without example or practice. However, most scientists recognize that all behaviour is influenced to some extent by the animal's genetic make-up, and, at the same time, by the environmental conditions which exist during development. The extent to which the two influences, nature and nurture, determine the outcome varies greatly from species to species and from activity to activity within a species. From the

study of bird SONG, for example, it is apparent that many species-characteristic calls, such as those of pigeons and doves (Columbidae), develop independently of any experience of the calls of other individuals, whilst other species fail to sing normally unless the song is heard at a certain stage of development. Between the two extremes of perfect imitation and complete absence of imitation, are many species which are able to learn only a restricted range of sound patterns. Generalizations about the development of behaviour patterns can be based on a comparison of species which differ markedly in their performance. These generalizations may be tentatively summarized as follows: the naïve bird has a *template* of the species-characteristic song, which may, or may not, be modified by experience of the song produced by other individuals. Such modification may be restricted to a SENSITIVE PERIOD, which generally occurs before the bird itself begins to sing. This initial learning cannot, therefore, be accounted for by classical learning theory. The extent to which the template can be modified may be limited. In the chaffinch (*Fringilla coelebs*), for example, the limitation seems to be set by the resemblance to normal song. In the second stage of song learning there appears to be a comparison between the template and the sound produced by the bird. Singing the 'correct' song appears to be self-rewarding, in that it reinforces the establishment of the song pattern in the animal's behavioural repertoire. The progressive improvement in VOCALIZATION, which marks this stage of song learning, does not occur in birds deprived of auditory FEEDBACK by deafening. However, the song pattern is little affected by deafening after the vocalization has reached its final form.

From the study of bird song it can be seen that activation of particular genes may be a necessary condition for the development of a behaviour pattern, but is never a sufficient condition. Even the stereotyped postural REFLEXES of animals, which, like the vocalizations of pigeons, appear independently of experience, require a suitable embryonic environment in which to develop. Normally, the conditions inside the egg or the mother are those that provide the correct medium for the growth of appropriate nerve connections, etc. This provision of an environment suitable for the development of the embryo is little different in principle from the provision of an environment suitable for the development of a JUVENILE. Thus a chaffinch normally develops the song characteristic of its own species, because it is reared in an environment in which it will normally hear the song of other chaffinches.

The term innate is sometimes used for behaviour which was adapted for its present FUNCTION by NATURAL SELECTION during the course of EVOLUTION. An activity can be innate, in this usage, even though it is learned. For example, the juveniles of many species learn the characteristics of their own species members, and of their HAB-

ITAT, at a particular stage of development. For this IMPRINTING to occur in the proper manner, the PARENTAL CARE must follow its normal pattern, so that the juveniles are exposed to those environmental stimuli which will lead them to learn the behaviour that is characteristic of the species. Thus the action of natural selection upon the behaviour of the parents may be just as important as that upon the juveniles.

Much animal behaviour is innate, in the sense that it inevitably appears as part of an animal's repertoire under normal and natural circumstances. Genetic factors are often necessary for such behaviour, but the species-characteristic processes of MATURATION and learning may be just as important. The naïve dichotomy between nature and nurture, which has characterized so much of the debate about innate behaviour, bears no relation to the complex and subtle interactions between various processes that are responsible for the development of behaviour patterns (see ONTOGENY).

1, 95

INSIGHT is a term used to denote the apprehension of relations between stimuli or events. In some animals, such as chimpanzees (*Pan*), relational LEARNING comes readily to hand, whereas in others, such as pigeons (Columbidae), it is difficult to establish. The correct solution to a task involving relationships, to choose the larger of two objects, for example, requires a degree of abstraction on the part of the animal (see COGNITION). A pigeon can easily learn to choose the larger of two particular objects, but has great difficulty learning to select the larger of any two objects presented by the experimenter. Chimpanzees are good at PROBLEM SOLVING, especially when the task requires the appreciation of relationships. For example, they readily learn to use a stick to rake in a banana that is out of reach behind bars (see INTELLIGENCE). To do this the animal has to see the relationship between the space to be bridged and the size of the stick.

Some scientists regard insight learning as different from ordinary forms of learning, because it involves the sudden production of a new response which is not arrived at through a process of trial and error. One of the difficulties with this view is that it is very hard to know whether the response is genuinely new. The chimpanzee may have developed a facility to use sticks through PLAY, may have observed others using sticks (see IMITATION), or may have an INNATE tendency to manipulate objects (see TOOL USING). In one study, six chimpanzees were given tests involving reaching for food with sticks. The animals had been born in captivity and raised under controlled nursery conditions, and records of their early development and experience were available. After the initial tests, some of the chimpanzees were given sticks to play with for 3 days, after which all of the animals were tested again. The results showed that the chimpanzees which were allowed to play with sticks improved greatly in the reaching tasks.

It appears that much apparent insightful behaviour is the result of collating past experiences and deploying appropriate responses in a new situation. Most scientists agree that this does not require any special theory of learning. Insight learning differs from conventional learning primarily in that the animal can improve its performance on the basis of experience gained outside a test situation.

89, 119

INSTINCT was regarded by early writers as the natural origin of biologically important motives. Thus Thomas Aquinas wrote that animal judgement is not free but implanted by nature. Descartes regarded instinct as the source of the forces that govern behaviour, being designed by God in such a way as to make the behaviour adaptable. The associationists appeared to reject all notions of instinct, although Locke did write of '... an uneasiness of the mind for want of some absent good. ... God has put into man the uneasiness of hunger and thirst, and other natural desires ... to move and determine their wills for the preservation of themselves and the continuation of the species.' (*An essay concerning human understanding*, 1690.)

While the associationists insisted that human behaviour is maintained by the knowledge of and desire for the consequences of behaviour, others such as Hutcheson (1728) argued that instinct produces action prior to any thought of the consequences. Whereas instinct had previously been regarded as the source of motivational forces, Hutcheson made instinct the force itself. This concept of instinct was seized upon by the new rationalists, such as Reid (1785), Hamilton (1858), and James (1890), as a convenient vehicle for the non-rational elements of behaviour. Thus man's nature was seen as a combination of blind instinct and rational thought (see VOLUNTARY BEHAVIOUR).

The idea of instinct as a 'prime mover' was taken up by psychologists such as Freud (1915) and McDougall (1908). Freud developed a motivational theory of neuroses and psychoses, emphasizing the irrational forces in man's nature. He saw behaviour as the outcome of two basic energies: a life force underlying man's life-maintaining and life-continuing activities, and a death force underlying his aggressive and destructive activities. Freud thought of the life and death forces as instincts, the energy of which was seen as requiring expression or discharge. McDougall thought of the instincts as irrational and compelling sources of conduct, which oriented the organism towards its GOALS. He postulated a number of instincts, most of which had a corresponding EMOTION. Examples are: flight and the emotion of fear; repulsion and the emotion of disgust; curiosity and the emotion of wonder; pugnacity and the emotion of anger.

These various conceptions of instinct were derived from the subjective emotional experiences of man. This essentially unscientific practice involves difficulties of interpretation, of agreement between different psychologists, and of determining the number of instincts that should be allowed or recognized. DARWIN (1859) was the first to propose an objective definition of instinct in terms of animal behaviour. He treated instincts as complex REFLEXES, which were made up of units that were compatible with the mechanisms of inheritance. Instincts were thus the product of NATURAL SELECTION, and evolved together with other aspects of the life of the animal. This is very much the view propounded by the early students of ETHOLOGY. LORENZ (1937) claimed that there were many FIXED ACTION PATTERNS that were determined by an animal's GENETIC make-up and were characteristic of species. Lorenz identified instincts with CONSUMMATORY behaviour, recognizing that APPETITIVE behaviour was modifiable by LEARNING. Initially, the ethological view was largely structural, little attention being given to MOTIVATION in relation to instinct. Subsequently, it was postulated that each instinct or fixed action pattern was motivated by an *action-specific energy*. Lorenz (1950) likened the source of energy for each instinct to liquid in a reservoir. The energy appropriate to each instinct accumulated in the reservoir and was discharged when the appropriate RELEASER stimulus was presented. TINBERGEN (1951) proposed that the different sources of action-specific energy were arranged in a HIERARCHY so that the energy pertinent to one class of activity, such as reproduction, would motivate a number of subordinate activities, such as NEST-BUILDING, COURTSHIP, PARENTAL CARE, etc.

The classical ethological view of instinct does not find favour amongst many behavioural scientists for two main reasons. The first concerns the idea that there are instinctive forces that propel an animal into action. As in the case of the concept of DRIVE, behavioural scientists no longer believe that it is necessary or helpful to account separately for the energizing aspects of behaviour. The concept of ENERGY was misused by early psychologists and ethologists, and has been replaced by a different view of motivation.

The second objection to the classical instinct concept concerns the idea that certain aspects of behaviour are INNATE, in the sense that they develop independently of environmental influences. Most scientists recognize that behavioural development involves a complex interaction of genetic predispositions and experience (see ONTOGENY). It may well be misleading to say that an animal behaves instinctively, because this implies that the behaviour has not been influenced by experience. Even when the behaviour of new-born animals is highly characteristic of the species, careful analysis is needed before it can be concluded that the form of the behaviour is predominantly genetically determined, or is not modifiable by experience. A particular example may serve to illustrate this point.

Newly hatched chicks of herring gulls (*Larus argentatus*) and laughing gulls (*Larus atricilla*) peck at the tip of the parent's bill, and this stimulus induces the adult to regurgitate food for the chick. This looks like a classic case of instinctive behaviour, because it is characteristic of all newly hatched chicks and is performed in an apparently stereotyped manner. However, research shows that there is considerable variation in the form of the first pecks of different individuals. Initially the angle of pecking, the rapidity of pecking, etc. differ considerably from individual to individual, but as the chicks gain experience their pecking accuracy improves, they become more selective in what they peck at, and their pecking behaviour becomes less variable in force and reach (distance between the chick and the target). Some of these changes are probably due to MATURATION: as the muscles develop the chicks become more stable on their feet and are able to peck in a more co-ordinated manner. Some of the changes are due to learning: initially the chicks will peck at any elongated object of a suitable size, but when they do not receive food they learn to exercise greater DISCRIMINATION. In nature, all the chicks are confronted with a similar situation, and it is not surprising that their behaviour develops along similar lines. The pecking behaviour of the older chick is more stereotyped and more like that of its peers than is that of younger chicks. Practice and experience in similar situations lead to similar results, and it is misleading to regard such behaviour as purely innate.

The concept of instinct has undergone many changes in its HISTORY. In its early usage, instinctive behaviour was seen to be inborn, reflex-like, and driven from within. Modern research has separated out the innate, the reflex, and the motivational aspects of apparently instinctive behaviour, and regards these as separate issues. This is probably the main reason for the diminution in the importance of the concept of instinct.

1, 16, 137

INTELLIGENCE. Before the time of DARWIN it was usually assumed that the behaviour of animals was under the control of blind INSTINCT. This was flattering to man, who prided himself on his possession of reason. The sharp divide between man and the brutes was challenged by Darwin's theory of EVOLUTION by NATURAL SELECTION. In his book *The Descent of Man* he argued that 'animals possess some power of reasoning'. He was convinced that 'the difference in mind between man and the higher animals, great as it is, certainly is one of degree and not kind'.

Once it is admitted that animals may be guided by more than instinct there is a danger that their intelligence be exaggerated. By the end of the nineteenth century the literature was full of anecdotes purporting to show the great powers of reasoning of HOUSEHOLD PETS and other animals. A pet animal had only to open the latch of a door and it was assumed that it must be making an intelligent attempt to imitate the people seen using the door. Nobody disputes that some animals do marvellous things, but experiments rather than anecdotes are needed to establish whether these feats are more dependent on LEARNING or on GENETIC mechanisms for their control. The nineteenth-century entomologist, Henri Fabre, was able to show that many of the remarkable achievements of insects could be explained without attributing great intelligence to them, and that insects could often be tricked into stupidly repeating an activity even though Fabre had now rendered the action pointless.

What is required is some test or set of tests which might be given to animals so as to judge their relative intelligence. It was not until 1911 that Thorndike, an American psychologist, published a book called *Animal Intelligence*, in which he proposed a series of tests and described how animals of different groups from fishes to monkeys fared on them. He set the animals tasks in which the animal must learn how to obtain food or to escape from a confined space. Domestic cats (*Felis catus*), for example, were placed in cages or puzzle boxes, from which they could only ESCAPE by learning to press a lever, or to pull at a piece of string. Using such tests, not only can the animal's performance be compared with that of animals of other species, but the mode of solution can also provide useful clues as to the animal's level of intelligence. Thorndike concluded from his use of puzzle boxes that animals learnt by a process of TRIAL AND ERROR rather than by the development of any INSIGHT into the problem. He supposed that their random attempts to escape occasionally met with success, and that the successful actions were more likely to be tried in the future.

It might be thought that little progress would be made in devising valid tests of animal intelligence until some clear idea was formed of what intelligence might be. Yet intelligence tests for people, first devised by Binet in 1905, have proved highly successful, although there is still controversy over the nature of intelligence. They can be said to be successful because they are reasonably accurate in predicting how well a person will do in his education or career, and this success may be used to support their validity as tests of intelligence. A modern intelligence test usually consists of several sub-tests, each of which tests some aspect of intelligence. Rote memory, knowledge, arithmetic, and spatial abilities may all be tested, but the items which have proved particularly valuable as measures of general intelligence are those which test the ability to form concepts, and those which require the person to reason. Items of the first sort may require the person to say in what way two things are alike, or to pick the odd man out from five words. Tests of reasoning may present a series of letters or designs with instructions that the person is to generate the next item in the series. To solve such problems it is necessary to discover the rule obeyed in the series. In general, intelligence tests for people may be said to test the ability to appreciate relations, and to make valid inferences.

Tests of animal intelligence. The simplest tests of an animal's ability to appreciate relations are those in which it is required to learn that one event follows another in TIME, or that an action is followed by a particular consequence. In the procedure referred to as classical CONDITIONING an animal such as a dog (*Canis l. familiaris*) might be fed only after a tone has sounded to see how quickly it learns the temporal relation between the sound of the tone and the presentation of food. Similarly in *instrumental learning*, a Norway rat (*Rattus norvegicus*), or some other animal, might be set the task of learning that food can be obtained only by pressing a lever, so testing the animal's ability to appreciate the relationship between its actions and their consequences. The animal's task may be made more difficult by restricting the conditions under which food is given, for example by specifying that it will follow one tone but not another, so that the animal must infer the precise conditions under which it can predict food. Simple tests of the sort described here have not proved useful in revealing differences in intelligence between different species. There are no consistent differences on such tasks between the performances of goldfish (*Carassius auratus*), pigeons (*Columba livia*), and chimpanzees (*Pan troglodytes*).

It has proved more profitable to compare animals on tasks which require them to learn some more general rule. An animal, for example, can be given a series of similar PROBLEMS such that it can benefit if it sees the similarity. If it improves over a set of such problems it is said to show a *learning set*. For example the animal can be given a series of DISCRIMINATION problems (Fig. Aa) on each of which it must learn under which of two different objects food is hidden. Though the objects change from problem to problem, the problems are similar in that food is placed under only one of the objects, irrespective of whether that object is placed on the right or the left, and it is to be found there on each of the trials for which those objects are presented. The speed with which an animal acquires a learning set for such discrimination problems is an index of the animal's ability to learn the rule that applies across the different problems. Children have been given sets of discrimination problems in much the same way, and their performance on these problems turns out to be related to their level of intelligence measured by conventional intelligence tests.

Another test that has proved to be of value is the *reversal task* (Fig. Ab). An animal is first taught to find food under one of two objects, and then the food is placed under the other object until the animal learns to reverse its original choice. If the position of food is then repeatedly switched between the two objects the animal has the opportunity to improve its performance over a series of reversals by learning to switch more quickly each time it finds that the food is no longer under the object which had previously covered it. Alternatively, the animal may be given a series of problems, each with a different pair of objects. On each problem the animal is first taught to choose one of the objects for food and then required to learn to choose the other. If the animal shows an improvement over a series of such problems it must be that it has benefited from the similarity between the reversal problems, and learnt the rule that is common to all.

In the reversal task the animal is given no warning that the position of the food is to be changed. It learns of this only when it continues to choose an object and finds that the food is no longer there. A task which is closely related to the reversal problem is the *conditional task* (Fig. Ac). On this the animal must also learn to choose sometimes one object and sometimes the other, but which is correct is indicated by some external signal. For example, the animal may be required to choose one object if the two objects are red and the other if they are blue. The rule that determines this choice is an arbitrary one.

The conditional task is in turn closely related to

the problem referred to as *matching* (Fig. Ad). Here, too, the animal is taught to respond to one object on some occasions, and to the other on other trials, but the rule that determines which must be chosen is not an arbitrary one. Instead, which object is correct is determined by the presence of a third object referred to as the *sample*. The sample is identical to one of the two objects between which the animal must choose. On some trials the sample is the same as one of these, and on others it is the same as the other. If the animal can see the similarity between the sample and one of the objects it can learn to match by choosing that object, since food is placed under whichever of the two objects is identical to the sample. The problem can be given the other way round, so that the animal must choose the object which is not the same as the sample, and it is then referred to as the *oddity task* (Fig. Ae), since the animal must choose the odd man out. If a series of matching or oddity problems are given, with new pairs of objects for each problem, it can be seen whether the animal shows improvement over the course of the problems. If it does, it suggests that it can appreciate the rule common to all problems, that the correct choice is governed by the principle of iden-

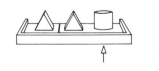

a

Fig. A. Series of discrimination problems used to test learning sets.

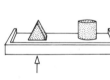 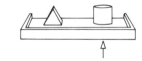

b

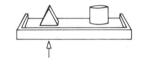 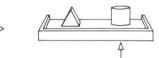

c

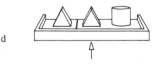 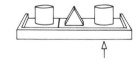

d

  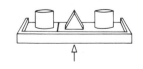

e

Fig. B. Chimpanzee using a stick to rake in food from outside its cage.

tity in matching, and non-identity in oddity, problems (see COGNITION).

These tasks, discrimination learning sets, reversals, conditional problems, and matching, all set the animal the problem of discovering some general rule. In all cases that rule determines under which of two objects food is to be found, i.e. which of two signs is associated with food. There is another set of problems in which the animal must find out not where food is, but what act it must carry out to obtain it. The best tried of such tests are those which present the animal with food which can be reached only by the use of some implement. It was tasks of this sort that were used by Wolfgang Köhler in a series of classical experiments carried out between 1913 and 1920 to test the abilities of a group of chimpanzees. In the rake problem the animal is shown food outside its cage which it can reach only by raking it in with a stick (Fig. B). In a more difficult version the animal is given two sticks, such that the food can only be reached if the end of one stick is inserted into the end of the other to make a longer implement. The box-stacking task (Fig. C) is formally similar, food being hung from the ceiling which can be reached only by standing on a box or even by piling the boxes up. Köhler chose these problems because they differed from those previously devised by Thorndike in one very important respect. He noted that the animals in Thorndike's puzzle boxes had no way of understanding the mechanism that opened the door because it was hidden, and that they therefore had little alternative but to work by trial and error. In the rake and box-stacking problems, by contrast, all the elements necessary for the solution are visible. The animal has a chance to develop insight into how the food could be obtained, and can start with hypotheses concerning the solution which are not randomly generated. Köhler concluded that some of the solutions

to these problems that were produced by his chimpanzees were indeed evidence of insight. The difference between the views of Thorndike and Köhler was noted with amusement by Bertrand Russell. 'All the animals that have been carefully observed . . . have all displayed the national characteristics of the observer. Animals studied by Americans rush about frantically, with an incredible display of hustle and pep, and at last achieve the desired result by chance. Animals observed by Germans sit still and think, and at last evolve the situation out of their inner consciousness.'

There has been much interest in how Köhler's chimpanzees did solve these tasks. It has become apparent that the previous experience of the animals is crucial. If chimpanzees are set the rake problem without having had prior access to sticks they fail to use them as rakes, whereas they succeed if they have been allowed to play with sticks earlier. It is necessary that the animal should become familiar with the properties of sticks if it is later to appreciate how they might be used to rake in food. The solution of problems of this sort requires that the animal see the relevance of its past experience, and use knowledge previously acquired when faced with a new problem. The answer comes not out of the blue, but rather from putting together elements already in the animal's repertoire.

The validity of tests of animal intelligence. The validity of intelligence tests for people is sometimes questioned. It is argued that the tests are biased in favour of one group or another. The items may be thought to presuppose knowledge which is available to members of some cultures but not others. To answer this criticism attempts are made to produce tests which are 'fair' from a CULTURAL point of view. For example, test material with abstract designs may be used as these may

Fig. C. The chimpanzee has stacked up boxes to reach the bananas.

be no more familiar to Americans or Europeans than they are to people from other countries.

The validity of tests of animal intelligence has been questioned in the same way. It is pointed out that animals of different species differ widely in the complexity of their sensory equipment, and in the movements they can make. There is therefore the danger that a test might favour one species more than another, not because of differences in intelligence but because of physical differences between them.

It might have been hoped that tasks such as discrimination learning sets might be less affected by such factors, because they test not how well an animal learns any one task, but rather how much improvement is shown across a series of tasks. But in fact how well an animal does on such a test depends on the SENSE ORGANS that the animal must use in making the discrimination. Monkeys (Simiae) improve greatly over a series of visual discrimination problems, whereas rats (*Rattus*) show little if any improvement, but monkeys possess

acute eyesight and rats have poor VISION for detail. If, on the other hand, rats are given a series of problems in which they must discriminate between different smells, they turn out to perform remarkably well, and to learn later problems much more quickly than earlier ones in the series; their sense of smell is very well developed, and with it they can show a much greater improvement over a set of problems than when the problems are presented visually.

It is even more difficult to devise fair tests of the ability of animals to learn what action they must carry out to obtain food. Manipulating levers or other gadgets is much easier for some species than others, and the use of sticks and other implements is sometimes possible only for those with hands. For this reason studies are often confined to a comparison of primates.

There is a further difficulty which is less obvious but quite as serious. Attempts to compare different animals in intelligence often assume that they differ only in general intelligence. But some animals clearly have special abilities directed towards particular problems which those animals face in their natural HABITAT. Chaffinches (*Fringilla coelebs*), for example, like some other song-birds, have the ability to learn to imitate the songs of other chaffinches, and the young learn their full song in this way (see IMITATION). Yet there is no reason to suspect that they are also able to copy actions they see. There is a danger that where such special abilities exist the animal will be taken to be more generally intelligent than is warranted.

When all these considerations have been taken into account the validity of tests of animal intelligence can be evaluated only by seeing to what extent the performance of different species agrees with some other criterion. With intelligence tests for people the criterion may be success at school or work, but there is no good independent test for animals. Recourse has sometimes been taken to some phylogenetic scale in the hope that those species higher on this scale (i.e. more highly evolved) might do better on the tests than those lower on the scale. But such scales are themselves very dubious. It is usual on such a scale to rank mammals higher than birds, but there is no justification for this since mammals did not evolve from birds, but rather both mammals and birds evolved from ancestral reptiles. Again it is sometimes suggested that there is a phylogenetic scale such that monkeys rank higher than cats which rank higher than rats, but in fact primates, carnivores, and rodents all diverged at the same time from the ancestral stock of placental mammals.

Rather than trying to relate the results of comparisons of tests of intelligence to the ranking of the animals on such an arbitrary scale it would be better to see how the results relate to a ranking of the species on a scale of BRAIN development. On such a scale the monkey is indeed higher than the cat, and the cat higher than the rat. On the other hand the brains of many birds are as well developed as those of many mammals, so that there is no reason to rate many mammals above birds. If the performance of an animal in a LABORATORY task can be predicted with reasonable accuracy, given the knowledge of the brain development of that species, then the validity of the tests will be supported.

Another obvious criterion would be the animal's performance in the wild. It would be expected that an animal that did very well on laboratory tasks should show evidence of its abilities in its home environment (NICHE). Macaque monkeys (*Macaca*), for example, which tend to perform as well on such tasks as any animals other than chimpanzees, do indeed show an impressive resourcefulness in exploiting their native habitats. They adapt quickly to different foods, learn to raid farms and settlements, and pass traditions of behaviour from one generation to another (see CULTURAL BEHAVIOUR). However, for many animals we still have too little information on the problems the animals face in their natural environment, and the way they solve them, and so there is too little evidence of this sort to use when considering the validity of laboratory tests.

One intriguing problem raised by the comparison of the performance of animals in the laboratory and in the wild is that it is not always obvious how animals might use their ability to solve some laboratory task in their native environment. In the laboratory chimpanzees easily solve oddity tasks or conditional problems, although it seems unlikely that they would meet such problems in the wild. They have also been shown to be able to recognize themselves in mirrors, and to be able to learn to use the sign language developed for the deaf and dumb (see COMMUNICATION). It is by no means clear why chimpanzees should have such abilities if they are not needed in their ordinary life. Of course, the chimpanzee may succeed on these and other tasks because they can be solved given a high level of general intelligence, but it is still puzzling why the chimpanzee should be so intelligent, since it does not appear to need such a high level of intelligence in the wild. Perhaps it needs to draw on its true abilities only in rare emergencies, or perhaps it does indeed meet with problems which truly test it, whether in devising new ways of obtaining food, or in its dealings with its fellows. It would clearly be worth studying the chimpanzee in the wild with the aim of determining exactly how it is tested by its natural environment.

Comparison of different species. In spite of the difficulties referred to above it has proved possible to make rough estimates of the relative intelligence of different animals. Comparisons will be safest where the animals are closely related and therefore share similar sensory and motor equipment.

On discrimination learning sets there is overlap between the performance of birds and mammals, the pigeon for example succeeding as well as the ferret (*Mustela putorius furo*), and this is not sur-

prising given the overlap in brain development for birds and mammals. The rat and grey squirrel (*Sciurus carolinensis*) hardly improve over a series of visual discrimination problems, but, for the rat at least, the reason may lie in poor vision. A carnivore, such as the cat, may perform as well as a primate such as the squirrel monkey (*Saimiri sciureus*), no doubt a reflection of the well-developed brain of many carnivores. However, no non-primate mammal has approached the level of the rhesus macaque (*Macaca mulatta*) or the chimpanzee, which quickly learn to solve new problems in only one trial. Indeed chimpanzees improve at the same rate as do young children.

There is also evidence that rhesus macaques and chimpanzees learn more general strategies in solving such problems than do cats. If these primates are tested first on a series of reversal problems their performance on discrimination learning sets is enhanced. This must be because of features common to the two sorts of problem, and especially the fact that on both the animal must learn to repeat a successful choice and change to the other object if unsuccessful. By contrast, cats are not helped by experience on reversals, seeming not to see the relevance of one type of problem to the other, nor to appreciate the similarity in the rules that apply to both series of problems.

A wide variety of animals have been tested on reversal problems. Fish learn to improve across a series of reversals only with difficulty, and more slowly than reptiles, birds, or mammals, but it has proved difficult to make the conditions equally favourable to fish as to the other groups. Again some birds, such as the pigeon, but not others, such as the chicken (*Gallus g. domesticus*), show an improvement comparable with that of mammals, and even of the squirrel monkey, a primate. Within the mammals, rats and squirrels improve more slowly than do cats or rhesus monkeys. In general the great apes (Pongidae) perform better than monkeys.

Conditional tasks can be learnt by both birds and mammals. However, there appears to be a difference in the way that these problems are solved by different species. Whereas neither rats nor cats show any improvement in their performance over a series of such problems, rhesus monkeys clearly perceive the similarity between the different problems and do much better on later problems than they do on the first problems they meet. The same difference is found between primates and other animals on matching or oddity tasks. Pigeons, rats, cats, monkeys, and chimpanzees can all learn single problems of this sort, but only the primates tested are consistent in improving their performance when given a set of these problems. It appears that only they appreciate the general rule that applies, and can use it when faced with new items to compare. Rhesus monkeys are also capable of more sophisticated judgements, and can be trained to pick from a group of different objects all those which are the same colour, or even to choose on the basis of similarity in shape or colour depending on some other cue which specifies which characteristic is relevant in this trial. The most impressive feat of all was that of a chimpanzee called Viki, who was brought up by two psychologists in their own home. She was given objects to sort into dishes, and objects were deliberately chosen which could be sorted on the basis of form, colour, or size. She was able to sort them by any one of these features, and, surprisingly, she spontaneously changed the basis of her sorting from one run through to the next, sorting now into piles by colour and on the next occasion by shape or size. This flexible handling of abstract concepts attests to the high level of intelligence reached by the chimpanzee.

One further concept that we have reason to be interested in is that of number, since we greatly value our arithmetic abilities. It is important to distinguish the ability to perceive the similarity between sets of things which are alike in number, and the ability to count out how many things there are in a set. These abilities have been found in some birds and primates. A jackdaw (*Corvus monedula*) has been taught to choose the lid which has the same number of spots (up to five) as a card, even when the spots differed in size, shape, or distribution. A grey parrot (*Psittacus erithacus*) has been trained to take the number of pieces of food that corresponds to a number of light flashes, or to the number of notes played on a flute. Similarly a rhesus monkey can learn to choose the card bearing a particular number of symbols, even if the shape, size, and distribution of these is altered from trial to trial, and monkeys can press a lever the number of times that corresponds to the number of signs shown. Chimpanzees have even been taught to use the binary notation for numbers, by pressing up to three switches to turn on lights forming the same number in binary as the number of objects seen. While it is intriguing that there are some birds and some primates which are able to appreciate number, it remains to be shown whether they use this ability in the wild, and whether they are alone in possessing it.

A further similarity between birds and mammals is that TOOL USING occurs in a few species from each group. The woodpecker finch (*Cactospiza pallida*) picks cactus spines in its beak and uses them to winkle out insects from crevices in bark (Fig. D), and the Egyptian vulture (*Neophron percnopterus*) drops stones onto eggs to break them open. In the laboratory, blue jays (*Cyanocitta cristata*) have been found to tear strips of newspaper for use as probes to draw in food from outside the cage. Among mammals, the sea otter (*Enhydra lutris*) retrieves stones from the sea-bed, and uses them to open abalone shells (Haliotidae). There are more reports of tool use in the wild for primates, particularly chimpanzees, than for other mammals. Monkeys and apes (Simiae) sometimes throw objects, such as stones, to intimidate predators, and chimpanzees have even been reported

Fig. D. The woodpecker finch uses cactus spines to probe for insects in the crevices of bark.

to use sticks as clubs with which to beat a stuffed leopard (*Panthera pardus*) which was presented to them in an experiment. Chackma baboons (*Papio ursinus*) have been observed using stones to crack open tough fruit, and chimpanzees in the Gombe Stream Reserve in Tanzania poke grasses or twigs down termite (Isoptera) holes and then withdraw the implement so that they can eat the insects clinging to it. In the laboratory brown capuchin monkeys (*Cebus apella*) and chimpanzees can learn to use sticks to rake in food, although rhesus monkeys appear not to be able to learn this.

It is noticeable that, although there are isolated instances of tool use in animals, it is rare for any one species to use more than one sort of tool, or to use a tool for more than one purpose. More information is needed about the general adaptations of tool-using animals before concluding that tool use in the species mentioned is indicative of high intelligence. We need to know in particular whether these animals could learn new techniques or make new inventions when faced with novel problems in the laboratory. It is clear that certain primates, such as the capuchin monkey and the chimpanzee, can indeed solve such problems, and appear to have a general ability to appreciate the various possible uses of objects for achieving their ends. We also know that chimpanzees not only use twigs for fishing out termites, but can even modify the twigs by paring them down, and thus fashion a tool which suits their purpose better. The ability to make a tool is one which had previously been supposed to be unique to mankind. The number of different uses to which chimpanzees can put tools, and the fact that they can modify them, suggests an intellectual pre-eminence compared with other mammals or birds.

Intelligent animals. Since it has proved possible to find some way of ranking animals in abilities, two questions of interest arise. First, why are some animals more intelligent than others; that is, what are the pressures of NATURAL SELECTION that have led to the development of high intelligence in some animal groups and not others? Second, why is man in particular so intellectually superior to other animals; that is, what were the conditions in our history that were responsible for producing such a remarkable animal?

In answering the first question, it helps to consider why any animal should have the capacity to learn. During their lives animals face many problems, but the solution to many of these can be built into the animal by genetic mechanisms. A cat does not have to learn to kill a mouse, nor do all birds need to learn their route on their first MIGRATION. There are, however, problems of a type for which it would be very difficult to equip an animal with ready solutions, simply because the environment is not always predictable, so there is no one solution which will always apply. For example, the genes can know nothing of the look of one's parents, and a *precocial* bird (i.e. one which leaves the nest soon after hatching) must learn which adult to follow. The types of food available and the local predators have to be learnt where there is great variation within the area occupied by a species, because under such circumstances details of what to approach and what to avoid cannot be written into the genetic instructions of the species.

The shorter the life of an animal, the faster the species can adapt genetically to environmental change. For such animals it may pay to rely mainly on solutions to problems that are handed down genetically.

On the other hand, for animals which are long-lived it is more important that the animal adapt to environmental changes which occur during its own lifetime. Larger animals tend to live longer than do small ones, and thus to have a longer period when the young are immature. During this period, if the young are adequately fed and protected, they have time in which to learn about the world into which they have been born. In mammals, in particular, the mother tends to have few young, especially in larger species, and these are well cared for during the period in which they are not able to fend for themselves (see PARENTAL CARE). In big animals such as lions (*Panthera leo*), chimpanzees, elephants (Elephantidae), or dolphins (Delphinidae) this period lasts many years. The strategy in such animals is to produce only a few young and ensure their survival to adulthood by elaborate MATERNAL care. By the time they are adult they will have acquired considerable knowledge about their environment, and developed skills for dealing with it. In such animals the brain is also very well developed, endowing them with considerable capacities for learning.

It is of interest that PLAY is seen only in those animals to which there is reason to attribute high intelligence. It has not been observed in fish or reptiles. There are some birds, such as the jackdaw, which have been reported to play, but in general it is amongst mammals that play is most commonly seen. On tests during which they are presented with various objects, rodents show much less interest than do carnivores or primates, and young chimpanzees excel over all other animals. Compared with monkeys, the great apes are much more creative in the ways in which they manipulate objects, doing so in very varied ways. By thoroughly testing the properties of things in this way the animal is better prepared for coping with its environment in the future.

Our closest relatives are the great apes, and especially the two African apes, the gorilla (*Gorilla gorilla*) and the chimpanzee. Observations of the abilities of these animals give some indication of the level of intelligence possessed by the ancestors we share in common with them. By bringing up a chimpanzee, for example, in a human home, it is possible to directly compare the chimpanzee with a child, and thus to identify the characteristics that are peculiar to man. In such surroundings chimpanzees show a remarkable ability to learn to use household equipment such as kettles, doors, cutlery, and so on. Though they do not naturally acquire human speech they can be taught to use a symbolic means of communication such as American sign language or 'Ameslan'. In this system manual gestures correspond to words. Chimpanzees can learn a large vocabulary of signs, and even put them together into short sentences three signs long. In their natural surroundings, however, chimpanzees have not invented a complex technology or new means of communication, and they still lead what is really a very simple life, spending much of the day just picking fruit, and moving on to new fruiting trees.

What then were the pressures of natural selection that led to the widespread production of tools by our ancestors, and the invention of LANGUAGE? We know that a few million years ago our ancestors in Africa moved from woodlands into open country, and that they changed from a vegetarian diet to a mixed diet including meat. It seems plausible that it was this change of habitat that led to the rapid advance in technology, because tools were now invaluable both as weapons and for skinning carcasses. One may suppose that the pressures for improved methods of communication were then increased, since success in HUNTING depended on the close COOPERATIVE behaviour of the hunters and the formulation of complex strategies. Furthermore, language provided an efficient means for transmitting cultural traditions in technology from one generation to the next. We differ from animals not because we alone possess reason, but because historical circumstances led our ancestors to rely increasingly on their intelligence for survival. We became specialized as brainy animals.   R.P.

81, 89, 119

INTENTION MOVEMENTS are incomplete initial phases of behaviour patterns. For example, when a bird on the ground is about to take off in FLIGHT, it first crouches, raises its tail, and withdraws its head. It then extends its head and neck and springs off. The crouching part of flight preparation may occur a number of times before the bird actually takes off, or it may not be followed by flight at all. Such movements provide potential information to other animals, and it is not surprising that they play an important role in COMMUNICATION. For example, pigeons (Columbidae) in a flock are alert to each other's flight intention movements. A pigeon leaving the flock does not disturb the others, provided the normal flight-intention movements are given. If a member of the flock flies away suddenly, without showing the normal intention movements, then all the birds take flight. Thus the absence of the flight-intention movement is a type of ALARM signal. Intention movements have sometimes undergone RITUALIZATION and evolved into stereotyped DISPLAYS. The displays of the common golden-eye duck (*Bucephala clangula*) and the American green heron (*Butorides virescens*) are derived from flight-intention movements (see RITUALIZATION, Figs. B and C). Many THREAT displays are derived from intention attack behaviour, and some mammalian FACIAL EXPRESSIONS are thought to be derived from intention biting, or from DEFENSIVE behaviour.

INTERACTIONS AMONG ANIMALS. Every animal interacts with other animals during the course of its lifetime. The interactions between members of different animal species can be classified according to the cost or benefit that each receives as a consequence of the interaction. This CLASSIFICATION is summarized in Table 1, and it provides a useful framework within which to discuss relationships between animals. Interactions between members of the same species are normally considered as SOCIAL INTERACTIONS, and will not be considered here.

Table 1. Interactions between animals

Type of interaction	Participants A	B	Nature of the interaction
Competition	–	–	Both competitors are affected adversely
Predation	+	–	The predator benefits at the expense of the prey
Parasitism	+	–	The parasite benefits at the expense of the host
Amensalism	0	–	One participant is affected adversely, while the other remains unaffected
Neutralism	0	0	Neither participant is affected
Commensalism	0	+	The commensal benefits without affecting the host
Mutualism	+	+	The relationship benefits both participants

+ indicates a beneficial effect
0 indicates a neutral effect
– indicates an adverse effect

Whenever two or more individuals are using the same resources, such as food or space, and when the resources are in short supply, then COMPETITION occurs. Competition between species is characterized by the fact that members of both species are affected adversely. For example, the red-cheeked salamander (*Plethodon jordani*) and the slimy salamander (*Plethodon glutinosus*) occur on mountains in the eastern USA. When the two species occur on the same mountain, *jordani* occurs at a higher altitude than *glutinosus*, and there is very little overlap in their ranges. In the absence of *jordani*, however, *glutinosus* occurs at the higher elevations. Hence *jordani* excludes *glutinosus* from HABITATS that it would otherwise use. Similarly, the presence of *glutinosus* restricts the lower range of *jordani*. Thus competition between the two species affects both adversely.

PREDATION is a relationship between animals, in which the predator benefits while the prey is affected adversely. Predation implies that the prey is killed, and usually eaten, by the predator. It thus differs from PARASITISM in which the parasite exploits the host, but generally without killing it. Many animals have evolved elaborate forms of DEFENSIVE BEHAVIOUR against both predators and parasites. Parasites that infest the internal organs may be countered by chemical means, by the production of antibodies, for example. External parasites may be removed by special behaviour patterns, such as the mutual GROOMING of monkeys (Simiae). Sometimes, the same behaviour is used as a defence against both predator and parasite. For example, the bark mantis (*Tarachodes atzelli*) guards her young from predatory ants in exactly the same way that she guards her eggs from attacks by parasitic wasps (Hymenoptera) and flies (Diptera).

Relationships in which one animal is affected adversely, while the other is unaffected, are given the name *amensalism*. As it is virtually impossible for an animal to remain unaffected in its relations with another, pure amensalism probably does not occur in nature. For the same reason, true *neutralism*, in which neither animal is affected, is generally considered to be very rare or non-existent. Amensalism may be approximated to in cases of competition, where one species is dominant over another, in the sense that the dominant species has priority in the use of resources. Absolute priority would imply that the dominant animal was unaffected by the competition.

SYMBIOSIS is the living together of animals of different species, to their mutual benefit. It can occur as a relationship between individuals, between individuals and societies, and even between entire societies. There are some, probably rare, instances of *commensalism* in which one species benefits from a symbiotic relationship, while the other remains unaffected. A possible case is that of the cattle egret (*Bubulcus ibis*), a bird which often lives in close association with cattle or buffalo (Bovidae) and feeds on insects disturbed by them. It is known that cattle egrets associating with cattle obtain more food than do those feeding apart from cattle. The birds do not remove parasites from the cattle themselves, as do tick birds (*Buphagus erythrorhynchus*), so it would seem that the cattle do not benefit from the relationship with cattle egrets. It is possible, however, that the egrets give warning of approaching danger. An example of social commensalism is that of the trumpet fish or pipefish (*Aulostomus*), which sometimes joins schools of yellow sturgeon fish (*Zebrasoma flavescens*). It takes advantage of this CAMOUFLAGE to approach smaller fish, which it darts at and seizes as prey. The sturgeon fish seem to remain unaffected by the association.

Mutualism, sometimes called true symbiosis, is characterized by relationships in which both participants benefit. It is widespread in the animal kingdom. In many cases simple signals develop between the participants, and these improve the COORDINATION of behaviour and COOPERATIVE BEHAVIOUR. For example, the cleaner wrasse (*Labroides dimidiatus*) lives off parasites that infest the bodies of larger fish species. It entices a host to permit itself to be cleaned by means of a special form of swimming, the *cleaner dance*. It butts its snout against the fins and gill covers to signal to the host to spread them so that they can be cleaned. Similarly, it induces the host to open its mouth, so that it can enter and take parasites from the mouth cavity. While the cleaner fish is going about its work it continually vibrates its ventral fins, so that they tap against the host's body. Thus the host knows where it is being cleaned, and reacts by holding that part immobile. Host fish generally signal to the cleaner fish when they are about to move. They invite the cleaners to enter their mouth by opening it wide, and signal them to leave by jerking the mouth half shut and then opening it again. The cleaner leaves the mouth following this signal. Many different species of fish allow themselves to be cleaned in this way, and they all behave in the same manner when being attended to. Some cleaner fish take up a station at a particular place, and their host fish congregate and wait to be serviced. There are many species of cleaner fish, and they generally have similar distinct stripy markings, which act as SIGN STIMULI and facilitate recognition by large host fish, which might otherwise eat them. The sabre-toothed blenny (*Aspidontus taeniatus*) imitates cleaner fish, and approaches a host fish with a cleaner dance. If the host is deceived, the blenny approaches and bites chunks from the fins and gills (see MIMICRY).

The honey-badger (*Mellivora capensis*) lives in symbiosis with a small bird called the black-throated honey guide (*Indicator indicator*). When the bird discovers a hive of wild bees, it searches for a badger and guides it to the hive by means of a special DISPLAY. The badger opens the hive with its large CLAWS, being protected from the bees by its thick skin. It then feeds upon the honeycombs, while the bird gains access to the bee larvae and wax, which it could not have done unaided. If the honey guide cannot find a badger, it transfers its ATTENTION to the next best alternative, which often happens to be man. In accordance with old tradition, the natives understand the bird's behaviour, and are able to follow it to the hive. It is an unwritten law that the bird is allowed to take the bee larvae. Thus the symbiotic relationship is transferred from badger to man.

This example raises the question of relationships between man and other animals. Hunters, farmers, and fishermen benefit greatly from understanding the habits of the animals from which they gain their livelihood. Urban man, on the other hand, has no necessity for such contact, and yet continues to express interest and pleasure in the company of animals of many kinds. The fact that so many people keep HOUSEHOLD PETS, visit zoos, and indulge in other purely recreational activities in association with animals, suggests that interest in animals is deeply embedded in man's nature. Animals living in close association with man often behave as if men were members of their own species. In the zoo, a male kangaroo (Macropodidae) will often behave as if the upright posture of the keeper were a challenge to fight. The kangaroo's AGGRESSION can be appeased if the keeper adopts the bowed posture characteristic of peaceful kangaroos. Similarly, many people treat animals as if they were human beings. The poodle with its human name, nail varnish, and hair style is an obvious example. Human visitors to zoos constantly remark on the human-like behaviour of the animals they see there. This tendency to attribute human characteristics to animals is called ANTHROPOMORPHISM. It has its counterpart in many species, which tend to treat other species as if they were members of their own (see IMPRINTING).

IRRELEVANT BEHAVIOUR. It is a fairly common observation that an animal engaged in one activity may suddenly show some apparently irrelevant activity before resuming its former behaviour. For example, a courting pigeon (Columbidae) may suddenly start to preen, or a ring-necked pheasant (*Phasianus colchicus*) may show a little feeding behaviour in the middle of a COURTSHIP bout. Such behaviour appears to be out of context, and is often referred to as DISPLACEMENT ACTIVITY. However, considerable care must be taken in the interpretation of such observations, and in labelling behaviour as irrelevant.

It makes little scientific sense to use terms like 'incongruous', 'out of context', or 'irrelevant', without reference to some theory or generally agreed standard of relevance. One such theory is that animal behaviour normally has a particular GOAL or obvious biological FUNCTION. For example, FEEDING behaviour normally leads to food intake, but out-of-context feeding behaviour is often incomplete in this respect. For instance, pigeons in a CONFLICT situation often peck at particles of food, pick them up in the bill, but do not swallow them. This type of feeding activity is functionally irrelevant, in that it does not fulfil the normal goal of food intake.

An activity may also be said to be functionally irrelevant when it has no apparent SURVIVAL VALUE. Male three-spined sticklebacks (*Gasterosteus aculeatus*) often visit the nest in the midst of courting a female, and this appears to be functionally irrelevant, because there is no apparent survival value to be gained from visiting the nest before the female has laid any eggs there. However, there is some evidence that it is necessary for the male to check that the nest is in good condition before inviting the female to enter it and deposit her eggs. Here we can see that to an observer, who assumes that there is no point in visiting the nest during courtship, the behaviour seems irrelevant, but to another observer it may seem not only relevant but adaptive (see ADAPTATION). In general, it is impossible to show that an activity has no survival value, and mistakes in interpretation are easily made; thus many apparent instances of displacement activity have subsequently been shown to be important features in animal COMMUNICATION.

An activity may be considered irrelevant with respect to the stimuli that normally cause it, when it occurs in the absence of such stimuli. Such behaviour has been described as VACUUM ACTIVITY and as REDIRECTED behaviour. The most common situation is one in which the animal is observed to direct its behaviour to an irrelevant object in the environment. For example, fighting herring gulls (*Larus argentatus*) may direct attacks at clumps of grass and pull at them with the bill in a manner which is typical of their fighting technique. Such redirected AGGRESSION is very common in animals, but may appear quite incongruous to a human onlooker. A difficulty with the interpretation of this type of situation, however, is that of stimulus GENERALIZATION. It is difficult to show that the object to which the apparently irrelevant behaviour is directed does not, in fact, in some way remind the animal of the object to which it would normally direct such behaviour. Studies of redirected behaviour have shown that it is greatly influenced by the presence of stimuli which are similar in some way to the stimuli that normally elicit the behaviour.

In summary, it should be emphasized that the words irrelevant, incongruous, etc. should only be used with reference to some formulation with respect to which the behaviour can be considered relevant. This formulation may concern either the functional or the causal aspects of behaviour. When an activity is observed in circumstances which are not in agreement with the accepting formulation, it may be called irrelevant. However, a detailed study of such a situation generally comes to one of the following conclusions: (i) That the activity in question was not, in fact, irrelevant, but appeared to be so because the observer was mistaken in drawing up his criteria for relevance. Thus the observer might be mistaken about the functional context of the behaviour, or might have overlooked some features of the stimulus situation. (ii) That the activity should properly be considered merely as unusual, but apparently irrelevant with respect to what is usual. (ii) That the activity appeared to be irrelevant, because the observer was led to make a mistake in CLASSIFICATION, by the similarity of the activity to some other type of behaviour. Such misidentifications of the nature, function, and causes of behaviour are easily made, especially by the untrained observer, and they often lead to apparent paradoxes which are only resolved by subsequent research.

ISOLATING MECHANISMS. Behavioural (or sexual) isolation is one of several isolating mechanisms which prevent interbreeding and gene exchange between animal species (see Table 1). In maintaining reproductive isolation, isolating mechanisms play an essential role in protecting the genetic integrity of a species' adaptations to its environment.

Table 1. Isolating mechanisms

Biological properties of individuals that prevent the interbreeding of populations that are actually or potentially sympatric.

1. Pre-mating mechanisms: mechanisms which prevent interbreeding.
   (a) Ecological (habitat) isolation: closely related species live in the same general area but breed in different habitats.
   (b) Temporal isolation: related species breed at different times of day or in different seasons.
   (c) Behavioural (sexual) isolation: differences in courtship behaviour and sexual preferences prevent interspecific mating.
   (d) Mechanical isolation: morphological differences, as in the structure of genitals, prevent completion of interspecific mating.

2. Post-mating mechanisms: mechanisms that prevent full success of interspecific crosses.
   (a) Gametic mortality: sperm transfer takes place but egg is not fertilized.
   (b) Hybrid inviability: the F_1 (first filial generation) and subsequent generations have a reduced viability.
   (c) Hybrid sterility: the F_1 hybrids and/or subsequent generations are viable but partially or completely sterile.

Geographic or spatial barriers prevent any possibility of interbreeding between related species, and it is generally recognized that such obstacles are instrumental in the formation of new species. But where such extrinsic barriers no longer exist isolation must be maintained by intrinsic mechanisms which depend upon the biological properties of the species concerned. Differences in ecology, HABITAT preference, and daily or annual RHYTHMS of breeding may be sufficient to ensure that individuals of closely related species do not interact at the time of breeding, even though they occupy the same general area. In the absence of such obstacles, morphological incompatibility, or differences in COURTSHIP behaviour, may result in a failure to complete pairing or mating procedures. Because these mechanisms operate before gametes (reproductive cells) are shed or transferred they are referred to as pre-mating mechanisms.

If, in spite of pre-mating mechanisms, interspecific mating does occur, reproductive isolation may be maintained by post-mating mechanisms. Gametes may fail to survive in the reproductive tracts of another species. If fertilization does occur the GENETIC incompatibility of the two species may result in hybrids which are inviable or sterile.

One isolating mechanism or several in concert may act to maintain reproductive isolation. Even if post-mating mechanisms are highly effective and no viable or fertile hybrids are produced we should still expect to find that NATURAL SELECTION has favoured the EVOLUTION of effective premating mechanisms. This is because pre-mating barriers provide a more efficient first line of defence against dilution of the *gene pool*, in that they prevent the wastage of gametes and other costs incurred by mismating attempts.

Although habitat preference and temporal responses to the environment are in part mediated by behaviour, the term behavioural isolation (or sexual isolation) is generally reserved for those mechanisms based upon differences in courtship and mating behaviour. The remainder of this article will be concerned mainly with behavioural isolating mechanisms and their significance in the evolution of behaviour.

Courtship has been defined as the heterosexual reproductive COMMUNICATION system leading up to the CONSUMMATORY sexual act. Courtship is frequently initiated by an active searching or advertising by one sex, usually the male. A potential partner may be attracted to the advertising individual and an exchange of signals follows. Only rarely do animals follow a rigid sequence of signal exchange (response or reaction chain) and then usually only in those steps immediately preceding mating or spawning. Nevertheless, as many courtship signals are specific to the species, if individuals of two species meet, the failure of one or both individuals to provide the other species' signals and responses is likely to result in a disruption of the courtship exchange, and thus terminate pairing before mating can occur.

Studies of the SOCIAL signals of animals have revealed a spectacular diversity of special movements, postures, calls, and chemicals which play a role in courtship as well as other forms of social interaction. Movements and postures are often associated with distinctive features of morphology and colour pattern to form particularly striking DISPLAYS. Many of these signals are highly distinctive and specific to a species or group of species, so much so that they may be valuable as taxonomic characters.

However, although it is tempting to identify differences in the signals of closely related species, and to infer that these differences function as barriers to interbreeding, conclusions arrived at in this way may be misleading unless there has been some attempt to demonstrate that these features are actually utilized in the isolation of species. The difficulty arises from the fact that not only does courtship advertise the identity and location of an individual, but it must also serve a number of other functions. Courtship displays may also serve as THREATS towards competitors for territory or mates. Other signals may be required to suppress non-sexual aggressive or PREDATORY responses in a potential mate. Some or all of these signals may be important in synchronizing the reproductive condition of the partners, in maintaining a pair bond, and in facilitating the COORDINATION of movements required by the act of COPULATION or spawning. In addition the signalling system must be adapted to the particular social and physical environment occupied by an individual of a species. In fulfilling these numerous FUNCTIONS it is likely, especially in species with extensive signal repertoires, that displays have evolved under a variety of pressures of natural selection, which are, to some extent, independent of each other. Thus it is perhaps not surprising that it is often difficult to identify with certainty those features of courtship which play a particular role in maintaining behavioural isolation.

Survey of behavioural isolation. Numerous careful FIELD and LABORATORY STUDIES provide examples of the operation of behavioural isolating mechanisms. The type of signal an animal uses in its courtship behaviour depends in large measure on the environmental context in which an animal breeds: conspicuous visual and vocal displays may be highly effective in gulls (Laridae) nesting on open dunes or cliffs, whereas rhythmic electrical discharges may serve a comparable role in electric fish occupying dark or turbid waters (see ELECTROMAGNETIC SENSES).

Just as we are able to distinguish very many animal species on the basis of morphological features and pattern of COLORATION alone, so it seems likely that at least among vertebrates active during the day, individuals recognize and respond preferentially to potential partners of their own species on the basis of these same features. Quite fine details of pattern may provide the basis for species recognition. One investigator examined the role of behavioural isolating mechanisms in four species of gull which breed in the Canadian Arctic. Although the species differ somewhat in habitat preference, there may be as many as three and in some cases all four species nesting in the same colonies without evidence of interbreeding. In all four species the body is white and the back and wings grey. They differ in the pattern of wing tip marking, and in the degree of contrast between the white head and the iris and fleshy ring which surrounds the iris. For example, Thayer's gull (*Larus thayeri*) has a brown iris and reddish purple eye ring which provides a more marked contrast with the white head than does the yellow iris and brighter yellow eye ring of the glaucous gull (*L. hyperboreus*).

In a series of field experiments in which the eye ring of captured birds was painted over, it was established that eye–head contrast is an important factor in species recognition for both male and female. Females rarely chose males of their own species given the 'wrong' pattern of eye–head contrast. However, if a pair was allowed to form before the male's eye ring was altered the birds remained paired and bred successfully. In contrast males would initially accept and pair with females with the 'wrong' eye colour, but in spite of vigorous soliciting by the females, males refused to mount and the pairs broke up after 1–2 weeks. Evidently the female must present the correct eye-colour stimulus for copulation to occur. Thus the same relatively subtle visual feature plays an important role in maintaining behavioural isolation, but works in different ways and at different stages in the courtship of males and females. By giving female glaucous gulls a dark eye ring after they had already paired with male Thayer's gulls previously given a light eye ring the investigator was able to induce fifty-five pairs to persist at least until eggs were laid.

The four species also differ in the patterning of the back, the wings, and the wing tips. Experimental manipulation of wing tip pattern suggests that this feature does function in species recognition, but is perhaps of secondary importance to the part played by the eye ring colour in maintaining reproductive isolation.

Almost invariably, distinctive colour patterns are associated with special movements or postures which render the display more conspicuous and distinctive. The display movements of live-bearing fish provide an effective combination of colour pattern and movements which have been shown to play a role in maintaining reproductive isolation. A number of species of live-bearing fish occur in different combinations in north-eastern South America and Trinidad. Two of these, the guppy (*Poecilia reticulata*) and *P. picta* are similar in size and habits and frequently occur together without interbreeding. Males of the two species differ markedly in appearance and each performs an elaborate courtship dance (Fig. A). The guppy orientates in front of the female, twists his body into sigmoid shape, and quivers, while holding this position for several seconds before attempting to mate. The males of *P. picta* literally dance in a circle around the females before attempting to mate. While circling a female's head the male spreads his fins and displays distinctive markings on the

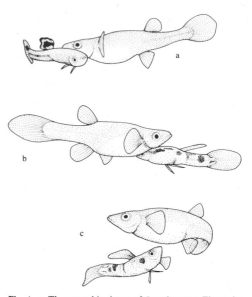

Fig. A. a. The courtship dance of *Poecilia picta*. The male circles repeatedly around the head of the female displaying the distinctive markings on his caudal and dorsal fins as he does so.

b. The sigmoid display of *Poecilia reticulata*. The male holds a position in front of the female and quivers for several seconds. Males of this species have conspicuous but variable markings, mainly on the body.

c. A mating attempt by a *P. reticulata* male. After displaying the male swings around and attempts to insert his gonopodium (a modified anal fin) into the genital pore of the female. If the male is successful packets of sperm are transferred to the female. Fertilization is internal.

dorsal and caudal fins. In laboratory aquaria males of both species frequently court and attempt mating with females of either species. In contrast receptive females respond by cooperating with the mating attempts of their own males, but ignore or evade the attentions of males of the other species. It appears that coloration and pattern of movement combine to provide the cue used by the female to select her mate.

In this example female choice provides the principal obstacle to interbreeding. That this selectivity is present in a female without any previous exposure to male courtship indicates that her selectivity is inherited. Males, on the other hand, initially fail to discriminate between females of different species. But after experiences of females of their own and other species the number of 'mistakes' decreases. Evidently the males learn, as a result of occasional acceptance by their own females and rejection by females of other species, to distinguish between females, perhaps on the basis of the relatively minor differences in colour pattern. Thus, although behavioural isolation depends ultimately upon inherited female selectivity, acquired selectivity in males must, in the context of the highly promiscuous MATING SYSTEM of these fish, make the search for a conspecific mate more efficient and simultaneously strengthen the isolating mechanism considerably.

The distinctive SONGS of birds, calls of frogs, and the stridulations of many insects are familiar to us all and can readily be used to identify the calling species. In some cases VOCALIZATIONS may allow us to distinguish species which are difficult to distinguish morphologically. Experimental studies making use of playback of recordings have confirmed that for many species these calls may be used in MATE SELECTION.

Although there is considerable diversity in chemical signals (PHEROMONES) there have been relatively few demonstrations that such signals serve as isolating mechanisms. In one investigation males of two morphologically similar species of leaf-rolling moth, *Archips mortuanus* and *A. argyrospilus*, were found to be active at the same time and in the same orchards. Males were attracted almost exclusively to traps containing the extract of the scent-producing gland of females of their own species. A similar specificity was found in males of another pair of morphologically similar moths. In this case males were attracted to synthetic chemical attractions believed to be identical to the natural sex pheromones of females of the two species. The investigators concluded that sex pheromone specificity is the only obvious isolating mechanism in the two pairs of species.

Other studies reveal less specificity in pheromone mechanisms. Electrically recorded responses in the antennae of giant silkworm moths (Saturniidae) are not specific to airborne sex pheromones released by conspecific females: males respond to pheromones from related species of the same genus, or in some cases other genera within the same subfamily. As hybridization does not occur it is probable that other isolating mechanisms are effective, for example many of these species are active at different times of day. Perhaps in these and other species of invertebrates and vertebrates pheromonal mechanisms supplement other isolating mechanisms rather than provide the principal barrier.

In this brief survey of behavioural isolating mechanisms it is readily apparent that behavioural isolation is unlikely to depend upon a single mechanism acting through a single sensory system. Isolation may depend on several behavioural components acting simultaneously or in sequence. Visual displays of birds are usually accompanied by song or distinct calls. In species of the sunfish genus *Lepomis* visual cues are evidently involved in species recognition, but the courtship is accompanied by auditory signals which also differ from species to species. Male fruit flies of the genus *Drosophila* produce distinctive songs by vibrating the wings during courtship. Contact before or after wing vibration transfers chemical stimuli which are known to serve in species and sex recognition.

The development of sexual preference. It is generally assumed that species-specific signals and the responses to them are INNATE. Although this is probably true of most invertebrates it is unlikely to apply to all vertebrates. The study of live-bearing fish showed clearly that although species recognition in the female is inherited, in that the female does not have to have had previous experience of male courtship for her to show the 'correct' response, males learn to distinguish their own females from those of other species. In effect females teach the males how to recognize their own species.

There are numerous anecdotal accounts of misdirected sexual behaviour in animals fostered by another species or by a human keeper. Most of these accounts refer to birds, but there are a few reports of acquired SEXUAL preferences in mammals. This early LEARNING which results in the acquisition of sexual preferences is referred to as sexual IMPRINTING. Of course under normal circumstances sexual imprinting, where it occurs, will result in an individual directing its sexual attentions towards individuals bearing the same characteristics as the natural parent.

In one experimental study ducklings (subfamily, Anatinae) of one species were raised with a foster mother and ducklings of a different species. As adults, males of several sexually dimorphic species would court and attempt to pair with females of the species with which the male ducklings had been raised. On the other hand, females were not affected by *cross fostering*, and responded as adults to the distinctive display and plumage of conspecific males. An interesting exception is provided by the Chilean teal (*Anas flavirostris*), a monomorphic species of duck in which both male and female have the same relatively inconspicuous 'female' plumage. In this case it proved possible to imprint females on foster parents. Evidently in highly dimorphic species sexual imprinting results in an acquired sexual preference in which males distinguish the relatively fine details of plumage which identify females of a closely related species of duck. On the other hand, because of the absence of a male model at the nest, and because of the distinctive male plumage and display, female young do not normally have to learn the male characteristics, except in a species in which the male lacks the highly distinctive plumage characteristic of dimorphic species.

Experimental studies with a variety of other species including doves (*Streptopelia*), gulls, and estrildine finches (Estrildidae) demonstrate that in the early learning of species characteristics of morphology, visual and auditory signals may play a role in species recognition. However, even in species which undergo sexual imprinting a capacity to acquire a sexual preference for another species is not unlimited. Just as in the case of the acquisition of song during ONTOGENY, there appear to be inherited predispositions in the ability to imprint upon another species. Zebra finches (*Taenopygia guttata*) imprint most easily on their own species, and sexual preferences acquired in this way are more rigid than those of a male finch raised by another species. The males of the northern mallard duck (*Anas platyrhynchos*) are predisposed to imprint upon their own or closely related species, and are less readily sexually imprinted on more distantly related species, such as coots (*Fulica*) and chickens (*Gallus g. domesticus*). Indeed it seems unlikely that behavioural isolation could ever depend entirely upon an acquired sexual preference. Learning provides a 'fine adjustment' and perhaps allows the use of more subtle cues as to species identity.

Evolutionary considerations. It is generally agreed that the initiating factor in *speciation* (the formation of new species) is geographic separation. An ancestral species may become fragmented as a result of a physical change in the environment or an unusual MIGRATION. Because geographically separated descendant populations are genetically isolated they are able to evolve independently of each other. Environmental conditions in the areas occupied by descendant populations are almost certain to differ, and thus, as each population becomes progressively better adapted to its environment, there will be an evolutionary divergence in some or all morphological, behavioural, ecological, and physiological traits. But even if populations remain similar in external features they are likely to become increasingly genetically incompatible as a result of the accumulation, through natural selection, of different mutations, gene combinations, and perhaps changes in gene function. If the geographic barrier should be overcome or disappear the outcome of the meeting of the descendant populations will depend largely upon the degree of divergence which took place while they were separated. If only minor differences emerged, the descendant populations are likely to interbreed freely and merge. Alternatively, the populations may show such profound differences that no interbreeding is attempted, or if mating does occur any hybrids produced are inviable or sterile. In this case the populations are reproductively isolated and would be considered separate species. In practice biologists find every intermediate between these two extremes.

In this picture of geographic speciation it is apparent that isolating mechanisms must arise initially as by-products of general evolutionary divergence. Differences in ADAPTATIONS to the biological and physical environment may appear while the populations remain geographically isolated. One or several of these differences may be sufficient to ensure that mating is not attempted, or, if it occurs, that hybrids produced are poorly adapted to the local conditions.

However, it can be argued that although isolating mechanisms must arise initially in geographically isolated populations, intrinsic mechanisms do not develop fully until the populations become *sympatric* (i.e. live in the same place). Once the ranges of previously separate populations overlap, natural selection will strengthen the isolating mech-anisms by eliminating those genotypes which form hybrids which are less fit than either of the parental genotypes. In other words, selection will favour those individuals which discriminate in favour of, or are selected by, members of their own population. Individuals which mate with members of the other population may give rise to hybrids, but if hybrids are less viable or perhaps sterile then those individuals will make a smaller genetic contribution to subsequent generations than those individuals which mate with their own kind. The process of developing and strengthening isolating mechanisms under conditions of sympatry can only occur if hybrids are at some disadvantage. Furthermore, it must apply mainly to pre-mating mechanisms as it is these that prevent the wastage of gametes.

Indirect support for this view may be derived from the observation that many examples of natural hybridization occur in situations where there are reasons to suspect that previously separated populations of two distinct, but evidently related, species have been brought into contact by recent natural or man-made changes in the environment. Hybridization occurs because differences acquired while the species were geographically separated are not fully effective as isolating mechanisms.

Examination of the isolating mechanisms themselves should reveal evidence of the part played by natural selection for reproductive isolation. However, because courtship serves numerous functions, the form of a behavioural isolating mechanism must inevitably reflect a wide variety of factors in addition to those favouring reproductive isolation. The mating system and the accompanying intensity of SEXUAL SELECTION is likely to exert a particularly profound influence.

Females invariably produce relatively fewer but energetically more expensive gametes than males, and thus we should expect females to be more selective in choosing mates; errors in mate selection are more serious and could result in the loss of a year's or even a lifetime's gametes. On the other hand, for males with their plentiful supply of gametes errors are less costly, and selection may be expected to favour behaviour which results in an increase in the number of offspring by mating with more than one female. Thus, as a general rule females will be expected to exercise a strong preference, whereas males may be expected to be less selective. The difference in degree of selectivity exercised by the two sexes will also be influenced by the type of mating system. In a species without PARENTAL CARE or in which only the female cares for young a *polygynous* or *promiscuous* mating pattern is likely to prevail. Under these circumstances sexual dimorphism may be extreme, and males compete vigorously for the possession or attention of females. In such species, selection is likely to reside almost entirely with the females, and males may be quite unselective. In species of grouse (Tetraoninae) that form LEKS the males will readily attempt to mount dummy females, or other inappropriate objects. In monogamous species with persistent pair bonds the sexes are frequently monomorphic, and both sexes are involved in the selection process. For example, female gulls evidently make the initial choice, but at a later stage selection by the male becomes a factor in the maintenance of the pair bond.

If natural selection for reproductive isolation has played a role in shaping courtship behaviour, we should expect to find that mate selection is most marked at the start of courtship, thus ensuring that wasteful inter-specific encounters are terminated at an early stage. This appears to be the case. The initial advertising phase is usually the most conspicuous species-specific phase of courtship. Indeed, when these initial barriers are overcome artificially, as in the close confinement of CAPTIVITY, it is frequently found that differences in the subsequent stages of courtship are insufficient to prevent mating. Thus, in spite of obvious differences in song, courtship, and plumage, interspecific pairings of caged European goldfinch (*Carduelis carduelis*), greenfinch (*C. chloris*), and domestic canaries (*Serinus canaria*), result in a roughly similar output of eggs per pair and probably as many young as conspecific pairs.

Moreover, there are studies which suggest that related species differ more in ecological and behavioural characteristics in regions where their ranges overlap than outside the region of overlap. The enhancement of differences which serve (i) to minimize ecological competition, and (ii) to maintain reproductive isolation is referred to as *character displacement*.

One of the best-documented studies of character displacement affecting a behavioural isolating mechanism is a study of two species of tree frogs whose ranges overlap in south-eastern Australia. Only *Hyla ewingi* is found to the west of a region of overlap; *H. verreauxi* occurs to the east and north of the overlap. Males of both species have calls which consist of a series of notes broken into pulses. There are differences between the species in both number of pulses per note, and in the pulse repetition frequency. However, the calls of males of the two species from outside the region of overlap differ far less than do the calls of males taken in the overlap zone. In experiments in which recordings of calls were played back to gravid females, females taken from the overlap zone approached the loudspeaker transmitting the call of a conspecific male. However, females of *H. ewingi* failed to discriminate between the calls of males of the two species taken from outside the overlap. In spite of the 'call barrier' 5% of pairs observed in the field in the overlap zone proved to be mixed species pairs.

The arctic gulls referred to earlier were examined for evidence of an enhancement of isolating mechanisms which could be related to the degree of range overlap of the four species. Both Kumlien's gull (*L. glaucoides*) and Thayer's gull have a dark eye ring, but the amount of pigmentation in the iris varies, particularly in Kumlien's gull. In areas which are also occupied by herring gulls (*L. argentatus*) and glaucous gulls with their relatively low eye–head contrast, Kumlien's gulls have a dark iris. But where Kumlien's gull occurs alongside Thayer's and glaucous gulls, the iris is mainly light. This indicates that selection has favoured dark-eyed individuals of Kumlien's gulls where they nest with light-eyed herring gulls, whereas in the presence of dark-eyed Thayer's gulls light-eyed Kumlien's have been selected for.

A number of laboratory studies with the fruit fly *Drosophila* have provided a clear demonstration that it is possible to select for reproductive isolation between two closely related species, or even between two genetically identifiable strains of the same species. One investigator selected for reproductive isolation between ebony and vestigial winged strains of *Drosophila* by removing all hybrids. After more than forty generations of selection the proportion of ebony flies attempting to mate with vestigial winged flies (and vice versa) had fallen significantly. Observations revealed that the changes in behaviour were as predicted: females had become more selective and repelled males of the other strain.

While studies of this nature cannot prove that selection for reproduction does occur in nature, they do provide clear evidence that it is possible to obtain an increase in behavioural isolation as a result of selection. Furthermore, the changes which occur are of the type which would be predicted on the basis of a consideration of the mating system of the species concerned.

Conclusion. It is generally recognized that courtship is but one component in the coadapted complex of responses to an individual's physical and social environment. Furthermore, the maintenance of reproductive isolation is only one of many functions of courtship. Thus it has proved difficult, particularly in COMPARATIVE STUDIES, to investigate behavioural isolating mechanisms without reference to other adaptive features of behaviour. In spite of these difficulties and the relative paucity of direct evidence, most biologists accept the idea that selection for reproductive isolation has been a potent force in directing and shaping the evolution of courtship behaviour, and that the demands of behavioural isolation have been in large measure responsible for the spectacular diversity and species-specific characteristics of courtship in so many animal groups. N.R.L.

20, 131

J

JUVENILE BEHAVIOUR, like adult behaviour, is subject to NATURAL SELECTION. Although some aspects of juvenile behaviour are precursors of adult behaviour, other aspects are specifically adapted to the survival of the young animal. For instance, the ALARM responses of herring gull chicks (*Larus argentatus*) are quite different from those of the parents. When alarmed, the chicks move a short distance from the nest and crouch silent and motionless amongst the vegetation, whereas the adults fly away from the nest uttering alarm calls. Chicks which ran around during an alarm would be highly vulnerable to PREDATION from other gulls; by remaining motionless they obtain some protection from their cryptic coloration (see CAMOUFLAGE).

Typical juvenile behaviour includes FOOD BEGGING, PLAY, and specialized behaviour such as HATCHING. There are also various processes, such as IMPRINTING, which are confined to the early stages of life. PARENTAL behaviour often involves some sort of protection for the young, but the juvenile animal has, nevertheless, to fulfil certain immediate requirements, and to prepare for life as an adult.

The development of juvenile behaviour is the result of an interplay of GENETIC predispositions, MATURATION, and LEARNING (see ONTOGENY). The emphasis given to these different processes depends upon the HABITAT and life style typical of the species. Juvenile animals occupy a NICHE that is both typical of the species, and typical of the particular stage of the life cycle. Sometimes, as in the larval forms of frogs (Anura) and butterflies (Lepidoptera), the life of the juvenile is quite distinct from that of the adult. Their tadpoles and caterpillars have to fend for themselves and occupy a niche which is quite different from that of the adult form. In other species, such as the wildebeest (or gnu) (*Connochaetes*), where the young animal runs with the herd within a few minutes of being born, the juvenile life style is virtually identical to that of its parents.

K

KINESIS is a form of ORIENTATION in which the animal's response is proportional to the intensity of stimulation, and is independent of the spatial properties of the stimulus. For example, common woodlice (*Porcellio scaber*) are very active at low levels of humidity and less active at high humidities. Consequently, they spend more time in dry parts of the environment, and their rapid locomotion in moist places increases the probability that they will discover drier conditions. Woodlice tend to form AGGREGATIONS in damp places beneath rocks and fallen logs, and their selection of this HABITAT is based entirely upon kinesis.

The velocity of locomotion of the flatworm *Planaria* is proportional to the intensity of light, and this kinesis has the consequence that the flatworm tends to spend its time in dark places. The darkness acts as a trap, slowing or stopping moving individuals. The type of kinesis in which there is a relationship between intensity of stimulus and speed of locomotion is called *ortho-kinesis*. Another flatworm, *Dendrocoelum lacteum*, also aggregates in dark places, but its rate of movement is not influenced by light. However, the rate of change of direction is related to the light intensity, and this is called *klino-kinesis*. Upon entering an area with a high level of illumination the rate of turning of *Dendrocoelum* increases considerably, but if the animal remains in light conditions there is some ADAPTATION, and the turning rate gradually falls. This adaptation has been shown to be necessary to enable the animal to seek darkness in conditions where there is a gradient from high to low levels of illumination.

L

LABORATORY STUDIES. Studies of animal behaviour are of two main types. There are those in which the animal is left much to its own devices, interference with its environment is minimized, and the investigator confines his activities to simple observation of behaviour, or to observation following manipulation of a few carefully selected aspects of the environment. This type of work is generally carried out in the field (see FIELD STUDIES), or with animals kept in semi-natural conditions in the laboratory. In contrast, there is the type of study in which an attempt is made to control the animal's environment as rigorously as possible, and to record selected aspects of behaviour which change in response to manipulation of the environment. This type of work is invariably carried out in the laboratory.

Each approach has its advantages and disadvantages. The rationale of the former approach may be summarized as follows: Much animal behaviour is specific to the species, and highly integrated with the animal's normal environment. It is therefore appropriate to study animals in their natural environment as we cannot expect to observe an animal's full range of behaviour in the laboratory. Although it is possible to employ a rigorous approach in the experimental study of behaviour in the field, this should be preceded by a thorough descriptive analysis of the species concerned. And finally, investigation of the way in which an animal is adapted to its environment is important for a proper understanding of the behaviour observed (see ADAPTATION).

Straightforward description of behaviour observed in the field provides information about the stimuli that an animal responds to, and the full range of its normal behavioural repertoire. It is a valuable preliminary to experimental work in the field, and to more detailed studies in the laboratory. Some behaviour patterns in the laboratory would be difficult to understand if they had not also been observed in their natural context. For example, ducks (Anatinae) and gulls (Laridae) sometimes show foot trampling on a damp cage floor. In the wild, this is part of their normal FEEDING behaviour, perhaps serving to stir up food in shallow water. Some behaviour patterns which normally occur in nature are completely absent in the laboratory. For instance, many pigeons and doves (Columbidae) breed perfectly well in the laboratory without displaying the complicated aerial flights which are part of their breeding behaviour in the natural environment. Laboratory conditions may produce ABNORMAL behaviour, that is, behaviour rarely or never observed in the field. For instance, animals kept in too small a cage are often much more aggressive than under natural conditions. Animals bred in the laboratory for many generations may develop a behaviour repertoire which is quite different from that of their wild relatives (see DOMESTICATION). For example, the exploratory behaviour of laboratory rats (*Rattus*) differs from that of their wild relations. While the former show little hesitation in investigating unfamiliar objects, the latter are extremely cautious and generally avoid regions of their TERRITORY which are suddenly changed.

The first, observational stage of a field study often leads to the formation of a preliminary hypothesis, which then has to be tested. There are three main ways in which this can be done in the field. These are: (i) by statistical analysis of the observations, (ii) by interference with the animal's environment, and (iii) by the use of models which more or less resemble natural stimuli. Statistical analysis can be used to determine whether different activities tend to occur together, or periodically, and what reliability may be attached to such conclusions. Selective interference with the natural environment, and the use of models, are often employed when laboratory studies are not possible or appropriate.

Although the study of animals in their natural environment provides important information about the range and variety of animal behaviour, laboratory studies are of prime importance in investigating the mechanisms of behaviour. The variable conditions in the field make properly controlled investigations extremely difficult, while in the laboratory much of this variability can be excluded. In bringing animals into the laboratory it is necessary to ensure that they are not disturbed by the alien environment. Some animals, such as fish, can be studied in semi-natural conditions in the laboratory. This is largely a matter of convenience, as it makes selective interference with the environment much easier. However, to conduct a fully controlled experiment it is usually necessary to study the animal in completely artificial conditions. Wild animals can be brought into the laboratory and gradually accustomed to the new conditions, but a number of advantages are gained by breeding experimental animals in CAPTIVITY: the animals grow up under similar conditions to those in which they will be studied; the history of each individual is known from birth; an important source of variation can be eliminated by ensuring that all animals to be used in a particular study are of the same age and have been reared under the same conditions. In general, the investigation of animal behaviour in the laboratory is necessary for a science in which hypotheses must be tested experimentally. Although a certain amount of progress can be made in the field, the more sophisticated techniques and the fully controlled conditions which are available in the laboratory are essential for this purpose.

Direct observation and measurement of behaviour. The development of adequate technique in the recording of animal behaviour should always involve a period of observation and familiarization on the part of the observer. Too often, studies using automatic recording methods are undertaken without a proper appreciation of the behaviour of the species concerned. Direct observation of behaviour involves problems of interpretation and CLASSIFICATION, which need to be well thought out before routine recording methods are embarked upon. For example, the use of both check lists and keyboard operated event-recorders presupposes the formulation of well-developed behaviour categories, and an adequate system of time-scaling.

Recording of preliminary observations may take the form of simple pencil-and-paper notes, a method long used by field workers. Recently there has been great development of the use of tape-recorders. The recording of spoken commentary onto magnetic tape has the advantage that the observer need not take his eyes off the animal to record information, and can thus record in greater detail, speech being generally faster than writing. The method has the disadvantage that long hours are required for transcribing the record, and methods that provide a record directly onto paper are more suitable for routine work (see TECHNIQUES OF STUDY).

Photographic recording on film or video-tape has certain advantages over the spoken word. It provides a permanent and detailed record that can be analysed at leisure, and the recording process is less affected by problems of observer fatigue. However, film or video-tape recording only postpones the tasks of categorizing and transcribing the data. Although the facility to slow down and to repeat the record is extremely useful for detailed work, the disadvantages for routine work remain.

The analysis of behaviour requires, not only that the behaviour patterns be adequately categorized, but also that a record be obtained in a form that is suitable for subsequent mathematical treatment. For this purpose, a record is generally made on paper, although the development of digital computer techniques sometimes obviates this requirement. A suitable paper record may be obtained as a result of transcribing notes, tape recordings, film or video-tape recordings, but it can also be made directly as part of the initial recording procedure. To some extent, the check list

fulfils this function, and in simple studies this method can be satisfactory. In recording complex behaviour patterns, however, there is a limit to the detail that the observer can record efficiently. In such situations it has become common practice to use an *event-recorder*.

The purpose of an event-recorder is to register sequences of events continuously, for a period of time. Most modern event-recorders register such events by means of pens operating on separate channels, and record time by moving the recording paper at a controlled rate. For recording direct observation, event-recorders are generally used in conjunction with a keyboard. Different behaviour categories are each assigned a key, and the operator depresses appropriate keys, while observing the behaviour.

The main disadvantage of moving-paper event-recorders is that the records are extremely tedious to transcribe, and many hours must be spent in preparing the data for statistical analysis. In this respect, there has been a significant improvement with the advent of small laboratory digital computers, and a number of laboratories have developed systems for recording events in a form suitable for direct computer analysis. Although it is possible to attach a keyboard directly to a small laboratory computer and record 'on-line', it is much more economical on computer time to make the recording 'off-line'. For off-line recording it is necessary to retain the data in a form that is readily transportable and available for computer analysis. Frequently the investigator uses a portable keyboard tape-recorder system at the site chosen for the behaviour study, and the data is stored, in coded form, on magnetic tape. These tapes can be fed into a specially designed decoding system, which is interfaced with the laboratory computer. Such systems have the advantage that they are portable, and can be used for recording observations in the field.

Automatic methods of measuring behaviour. Direct observation, as a stage in the measurement of behaviour, has two distinct drawbacks. Firstly, it inevitably involves an element of subjectivity on the part of the observer, who is required to make numerous judgements between behavioural categories. Secondly, it is costly in terms of man-hours. The former difficulty can, to some extent, be mitigated by cross-checking between different observers, but this procedure aggravates the second problem. A solution adopted in many laboratories is to employ automatic methods of measuring selected aspects of behaviour. This solution can suffer from a certain lack of generality, and from misinterpretation of the information obtained. However, the use of direct observation as a checking procedure does much to overcome these drawbacks.

The most common method of measuring behaviour automatically is the operant method (see OPERANT BEHAVIOUR). Operant conditioning consists essentially of training an animal to perform a task to obtain a REWARD, or to avoid a punishment. A rat, for example, may be trained to press a bar, or a pigeon to peck a key, by means of a method called *shaping*. Consider the training of a pigeon to peck an illuminated disc (the key) to obtain a food reward. After 1 or 2 days of food deprivation in its home cage, the pigeon is placed in a small cage equipped with a mechanism for delivering grain, and a key at about head height (Fig. A). This apparatus is called a Skinner box,

Fig. A. A pigeon in a 'Skinner box' pecks one of the square panels, upon which distinctive patterns are often presented by the experimenter. Rewards are delivered into the square hopper below.

after **B. F. SKINNER**, who developed the method. Delivery of food is normally signalled by a small light which illuminates the grain. Pigeons soon learn to associate the switching on of the light with the delivery of food, and approach the food mechanism whenever the light comes on. The next stage of shaping is to make food delivery contingent upon some aspect of the animal's behaviour. A pecking response is frequently used, but pigeons can also be trained to preen, or to turn in small circles, for example. Pecking is shaped by limiting rewards to movements which become progressively more similar to a peck at the key. So when it has learnt to approach the key, the pigeon is then rewarded only if it stands upright with its head near the key, then when it makes pecking movements towards the key, and finally only when it pecks the key. Up to this point the reward delivery is under manual control, but once the pigeon has started to peck the key the rewards can be delivered automatically, because when the pigeon pecks the key it closes a small switch in an electric circuit. The animal is now ready for use in an experiment.

Many types of experiment utilize this operant conditioning procedure. For example, discrimination LEARNING can be studied by rewarding the animal for responding only when a certain colour or pattern is presented, or by allowing the animal to choose between two keys which are visually differentiated. The technique has proved particularly useful for studying the effect of different types or patterns of reward (see REINFORCEMENT). Thus, instead of rewarding a pigeon for every peck, it can be rewarded for every nth peck, so that there is a fixed ratio between the number of pecks and the number of rewards. This procedure is called a *fixed-ratio schedule*. Other common schedules include *variable ratio*, *fixed interval*, and *variable interval*. On an interval schedule, reward is given at intervals of time specified by the experimenter. The animal is rewarded for the first response made during a given interval. Different schedules of reward have been found to have different effects upon an animal's performance. For example, a variable interval schedule produces a very uniform rate of responding, and for this reason it is often used to test the effects of DRUGS, reward substances, etc.

The operant method has been adapted for use with a wide variety of species, in situations ranging from the analysis of meal patterns, to the monitoring of performance during space flight. As a method of measuring a variety of aspects of behaviour, such as degree of MOTIVATION, perceptual ability (see PERCEPTION), or the effects of drugs on behaviour, it has considerable advantages. In particular, the standardized nature of the apparatus facilitates control of environmental variables, and permits comparison between species. In addition, the key, or other mechanism that the animal is required to operate, is a convenient transducer between the behaviour and the monitoring apparatus.

The problems connected with non-operant methods are primarily those of finding suitable transducers, by means of which various aspects of behaviour may be monitored. The methods used tend to follow advances in technology, and touch-sensitive devices, pressure transducers, strain gauges, temperature and light sensitive devices are all used in converting measurements of behaviour into an electrical form suitable for automatic recording. In addition, a wide variety of techniques are employed in measuring physiological changes in freely moving animals. The main considerations in the development of measuring devices are much the same as those in other sciences. Reliability, freedom from noise and other undesirable artefacts, and the ability to calibrate the monitoring apparatus in relation to the variable being measured, are all important. Considerations particularly relevant to behavioural studies are that the apparatus should be tailored to the experimental situation, and should be checked against other methods of measurement.

The methods used can vary considerably with the type of behaviour being investigated. For example, in recording the position of an animal within a cage, maze, or other apparatus, microswitches, or strain gauges, may be placed under panels set into the floor, or the animal may break a light beam when passing certain points in the apparatus. In all such cases, information as to which detector was operated at a particular time is important to the investigator, and must be recorded as a distinct event. In recording general ACTIVITY, on the other hand, similar methods may be used, but the number of detectors activated within a certain time-period is the variable of prime importance. In this case, the information from the different detectors can be pooled.

In recording behaviour such as meal patterns, general activity, etc., it is important to check one method against another, at least during the pilot stage of a study. In one study, for example, an 'eatometer' was devised in which food pellets were automatically delivered into a small trough. The presence of a pellet in the trough interrupted a light beam, generated by a small bulb and detected by a photoresistor. Removal of the pellet activated the photoresistor, and caused another pellet to be automatically delivered. This apparatus served the double function of making food freely available, and of monitoring the removal of pellets by the animal. The apparatus was used in an experiment in which the feeding pattern of rats with food freely available was compared with that of rats required to press a bar to obtain a pellet of food. The results showed differences in the pattern of feeding as measured by the two methods. When the rats were required to press a bar for food, they increased the number of short pauses in feeding.

Having devised suitable methods of measuring the behaviour to be studied, the next stage is to develop methods of recording the data automatically. In principle, this problem is much the same as that for recording direct observations, discussed above. One extra requirement, in the case of automatic recording, is that in addition to obtaining a permanent record in the form of a pen recording, or on computer tape, it is generally useful to be able to spot-check on the current progress of the experiment. In the case of direct observation this is not necessary, but with automatic recording methods much time and effort can be saved if the experimenter is occasionally able to check that the experiment is in working order, and that the animal is in satisfactory condition.

Control of the experimental situation. A controlled experiment is based on the rationale that all variables should be kept constant, or their effects known, except those that the experimenter is deliberately manipulating and recording. The necessity for such controls in behavioural studies is well illustrated by the example of 'Clever Hans', the horse that amazed the German public at the beginning of this century by his ability to answer questions on mathematics, spelling, and other aspects of the school curriculum. Hans answered questions by tapping his right forefoot to indicate numbered letters or words on a blackboard, or as a direct answer to mathematical questions. Although some people suspected trickery, Hans's owner was quite sincere, and Hans had a high success rate when questioned in the owner's absence. In scientific circles Hans's achievements caused quite a stir, and psychologists from the University of Berlin undertook to investigate the matter. They found that Hans could answer questions put to him in French, but had difficulty in answering questions to which the answer was one, and always failed when the questioner did not know the answer himself. Thus, if one whispered a number, and another questioner, not knowing the first number, whispered another number to be added to the first, then Hans could not give the correct answer. As a result of the investigations it was concluded that Hans watched his questioner very carefully, and stopped tapping when the person gave some involuntary signal, such as a small muscle twitch or change of breathing. Even questioners who knew which cues the horse used often had difficulty in suppressing them.

The controls used in behavioural studies fall into two general categories: control by experimental design, and control of the environment. The most straightforward experimental design is that in which an experimental group and a control group of animals are given treatments that are identical, with the exception of the 'key' treatment. This type of design may suffer from confounding variables within the key treatment. For example, studies in which animals are visually deprived in early life, and then compared with undeprived controls, in an experiment testing some aspect of VISION, have been criticized on this account. Thus, the visually deprived animals may perform less well than the controls for a number of reasons: (i) The animals may have been deprived of the opportunity to learn to organize their perceptual abilities, a possibility that this type of study is often designed to test. (ii) The animals may perform less well than the controls as a result of degeneration in the visual system, through lack of use. Such degeneration has been demonstrated in certain cases. (iii) There may be a decrement in performance due to the emotional shock of receiving visual stimulation for the first time, or (iv) due to lack of opportunity to practise the appropriate responses. (v) The visually deprived animal may come to rely on other sense modalities, as a blind man comes to rely on touch and hearing. In view of the numerous possibilities, the finding that visually deprived animals perform less well than controls in visual tasks is very difficult to interpret.

Problems inherent in the type of experimental design outlined above can sometimes be overcome by use of 'yoked' controls. The experimental animal and control animal are tested simultaneously, and when the experimental animal responds the same consequence is delivered to both animals. For example, a monkey (Simiae) which received electric shocks for making incorrect responses in a simple task developed gastric ulcers, whilst its partner, which received the same shocks but did not have to make decisions, remained healthy. Confounding experimental treatments can also be controlled by the use of a counterbalanced experimental design, in which groups of animals are given a number of treatments in an order which makes it possible to test each treatment combination. The effects of the treatments and their interactions can be subjected to statistical tests aimed at detecting sources of variability in the data.

Much of the variability in experimental data can be reduced by careful control of environmental factors. This is true, not only for testing conditions, but also for housing animals between tests. Many experimental designs require that animals be tested for a specified period each day. Between tests they should be maintained under conditions designed to ensure that the animal is in an appropriate physiological state at the start of each experimental session. For example, experiments involving food or water reward require that the animal be appropriately motivated during the testing period. This requirement is usually met by depriving the animal of food or water beforehand. As degree of motivation affects performance, the motivational state of experimental subjects must be controlled as far as possible. This can be done by subjecting the animals to a routine deprivation schedule, such as 22 h of food deprivation prior to each daily testing session. The deprivation schedule must be tailored to the species concerned, and a check kept on the condition of each animal. This is generally done by measuring bodyweights prior to each testing session. In addition to deprivation of the reward commodity, a large number of other factors affect an animal's motivational state. Thus, energy balance, water balance, and body temperature interact at the physiological level, and one cannot be controlled without control of the others. It is therefore important that food and water availability, and ambient temperature should all be controlled in the home environment. Similarly, motivational state is much affected by daily routines (see RHYTHMS), and care should be taken to maintain constant lighting conditions, and to test animals at the same time each day. It is often convenient to maintain nocturnally active animals on a reversed day–night cycle, so that they are most active at testing time. Animals subjected to frightening, sexual, or other disturbing stimuli in their home cages are unlikely to perform well in the experimental situation, and it is often wise to keep animals visually isolated from each other in the home conditions.

The above considerations also apply when transporting animals from their home cages to the testing laboratory. Sudden changes in temperature, illumination, or background noise must be avoided, and care must be taken in the manner in which the animals are handled. In experiments with monkeys, for example, it is common practice to transport each animal in its home cage, and even to allow it to remain in this cage while it performs the experimental task.

Control of the experimental environment itself requires considerable forethought. However much care is taken, the experimenter can never be absolutely certain that the animal is not being affected by outside factors. Thus the smell of a previous occupant of the apparatus, the barometric pressure, and the existence of slight draughts, have all been shown to influence the performance of animals in a prejudicial manner. In addition to controlling environmental variables, such as temperature, level of illumination, background noise, etc., the presentation of stimuli and of rewards must also be carefully controlled. Rewards and stimuli are often made contingent upon some aspect of the animal's behaviour, and much effort has been put into the automation of such procedures. The presentation of a stimulus or reward may be made directly consequent upon some automatically measured behavioural change. In an operant situation, for example, it is common practice to provide visual or auditory signals to indicate to the animal that it has made an effective response. Thus, when a rat presses a bar there may be a distinct click, or brief illumination of a lamp, that would not occur if it merely touched the bar. Frequently, the presentation of a stimulus or reward is made jointly contingent upon a response and some pre-programmed factor. For example, rewards may be consequent upon both a correct response, and a pre-programmed irregular sequence of reward and non-reward. Similarly, the pairs of stimuli used in choice tests may change their relative positions in accordance with some pre-arranged schedule.

To illustrate some of the control procedures outlined above, it may be helpful to consider a particular example. Suppose we wish to determine whether rats can distinguish between red and green hues. A suitable procedure would be to arrange for the rat to choose between two coloured lights. The apparatus could consist of a Y-shaped box with the entrance at the base of the Y, and a coloured light at the end of each arm of the fork. But before lights are introduced into the box it is necessary to allow the rats to become accustomed to the apparatus and experimental procedure. The usual training is to deprive the rats of food or water in their home cages and to reward them appropriately for each response in the experimental box. The coloured lights are then introduced and the choice of only one of the colours will be followed by a reward.

Before starting the experiment we must ensure that any other aspects of the rat's environment, which might be used as a cue to reward, are controlled. The experimenter must make sure that he cannot be seen by the animal, or heard placing the reward on the correct side. The reward itself must not be visible or provide olfactory cues. If the choice is to be between two hues, care must be taken to ensure that the animal is not able to respond on the basis of brightness or saturation differences between the lights used. This cannot be done by making the brightness and saturation physically the same for the two stimuli, because a colour-blind animal's brightness sensitivity is not the same over the whole spectral range (see COLOUR VISION). For this reason the brightness and saturation of the stimuli have to be varied so that no one property except hue can be associated with reward. The two stimuli must not always be at the end of the same arms, otherwise the rat could simply learn to go to the right or to the left every time. Changes in position, brightness, etc. should be made in a random order to prevent the animals responding to a pattern of changes. Rats can learn, for example, to run to the right on alternate trials.

Having eliminated all other cues, we are now in a position to test whether rats can distinguish between red and green hues. If the choice of red is followed by reward and, after a number of trials, the rats consistently run towards the red light, we can conclude that they are capable of the DISCRIMINATION. Should the rat not learn, however, the experimenter would not know whether the rat could not discriminate between the stimuli, or simply had not sufficient INTELLIGENCE to learn this type of task. In practice this difficulty is often avoided by choosing a task appropriate to the animal's known learning capabilities. For example, it is almost impossible to train frogs (Anura) to choose between coloured lights to obtain a food reward. But these animals will demonstrate certain colour preferences in an escape situation, and this method has been successfully used in the study of frog colour vision.

125

LANGUAGE is considered by many to be a purely human attribute. The possibility that other animals may have language has driven some linguists, philosophers, and other students of language to take refuge in definition, the last sanctuary of a threatened idea. Lists of the features required of language are drawn up and constantly amended, specifically to exclude interloper species. However, each year scientists are uncovering new and ever more astonishing abilities of animals, abilities which we certainly never suspected, and many of which we ourselves lack. Similarly, one by one the special features dear to the hearts of linguists are being invalidated.

Some of the linguistic criteria offered to date seem to reflect a real attempt to come to grips with language. To qualify as language, according to some, a COMMUNICATION system must be symbolic; to others, it must be open, i.e. capable of accommodating infinite creative constructions. Still others require that the information it encodes must be broken down into discrete *bits*, and must be able to denote abstract ideas, things, and events distant in time and place. These criteria seem reasonable, on the whole. Other specifications, however, seem to be required merely for purposes of exclusion. An exhaustive definition of language in a standard linguistic text, for instance, considers language to be an exclusively human attribute.

One supposed feature of language seems more resistant than others to disproof: the requirement that language be learned and transmitted by CULTURAL means, rather than being INNATE. This feature makes language available for use in new structures, to be applied in intellectually creative ways to novel situations. But we must determine whether these distinctions are real, and basic to our intuitive sense of language, or merely convenient and clever strategies for excluding animal communication. Should we accept this one *ad hoc* criterion on faith, we must apply it with our minds open, even to possibilities that students of language have been traditionally unwilling to accept. For example, there is evidence of cultural transmission of bird SONG, and recent discoveries have revealed innate tendencies in human infants towards consonant recognition, and other linguistic abilities. The time has come to explore the ramifications of recent research in animal behaviour, and to attempt a more sensible definition of language; one which would satisfy our intuitive feeling that there is something distinctive about human language, but one which would include the possibility that this very distinction is the result of a difference in degree in some sort of communication continuum throughout the animal world, rather than one necessarily of kind; one which would not high-handedly discard good strong distinctions between language and non-language, but would make use of those which ring true, whether or not they might break down the exclusive species barriers we have clung to so desperately.

The most reasonable course to follow in constructing a definition, then, begins with asking how language might be more than, rather than different from, simple communication. In looking over the endless examples of animal communication in the natural world, we see it repeatedly serving two general FUNCTIONS. By far the most common is the transmission of relevant SEXUAL information which allows the species to propagate, to persist from generation to generation. This sort of communication occurs in virtually every species, SOCIAL or solitary, from the use of common chemicals in single-celled organisms to specific PHEROMONES in moths (Lepidoptera); from the simple flashing code of fireflies (Lampyridae) to the complex visual signals of courting ducks (Anatinae); from the patterned rasping of crickets (Gryllidae) to the singing of birds. Both introspection and cross-cultural studies reveal that our species too has its own collection of such signals. In large part, these examples of communication can be thought of as signals of sexual status or physiological or emotional state, which, despite their often arbitrary and abstract character, hardly satisfy our intuitive sense of language. In the more social animals, a new range of communication opens up, and the lines separating language from mere communication begin to become less clear. Though the signals these creatures use may be chemical, as in ants (Formicidae), or physical, or acoustic, the information they convey may range from concepts as elementary as age or social status, to realities such as the presence of danger or of food. They may serve the purposes of facilitating social intercourse, or they may be used to manipulate the behaviour of others in the group, rousing them to attack, or recruiting them to a new source of food. Again, though the form of the signals may often seem arbitrary and unrelated in any way to the behavioural situation itself, or even abstract (the by-word of the linguistic philosopher), they are still basically statements of present physiological condition, or present mood.

The logical distinction that we are looking for, then, between communication and language, could lie in a creature's ability to communicate subtly different aspects of MOTIVATION. For example, when it behoves us to do so we can convey to others of our species whether we are hungry for bread or for meat, whether we are afraid of them, or of a thunder-storm. A second possible distinction for language could be its ability to allow an animal to escape the personal and immediate present, and to communicate information about events distant in both space and time. But we must be prepared to face the probability that both of these intuitive distinctions may fail to exclude quite all of the rest of the animal world. Many creatures, for example, have more than one way of communicating FEAR, and use ALARM calls to distinguish between classes of predators. Birds typically have both a whistle which is difficult to localize and which serves to spread the warning about airborne predators, and an easily located set of chirps to rally forces for MOBBING in other situations. Ground squirrels (*Citellus*) distinguish two classes of predator, aerial and terrestrial, and flee from both, but the behaviour of other individuals on HEARING the calls of a fleeing squirrel is thought to reflect the difference in the danger that each group poses. While any burrow will do to ESCAPE from a hawk (Accipitridae), only one with a second entrance will yield safety from badgers (Melinae), or other predators capable of digging their prey out. Hence, squirrels alerted by the terrestrial predator alarm scurry past burrows which would shelter them from birds of prey, and make for the nearest one with a back door.

Vervet monkeys (*Cercopithecus aethiops*) employ a more elaborate system. Their alarm vocabulary is now known to consist of at least three and probably four discrete signals, each attached to a particular class of predator. In one troop being studied a 'bark' from males or a 'chirp' from females or JUVENILES signals the presence of a leopard (*Panthera pardus*) or other carnivore, the 'rraup' call indicates a martial eagle (*Polemaetus bellicosus*) or other hunting bird overhead, while the 'chutter' means a python (*Python*) or other large snake is near by. There is a fourth call that seems to be elicited by the presence of baboons (*Papio*), or human beings. The reactions of other, non-calling members of the troop is usually immediate, and appropriate to the pre-

dator. Eagle calls cause monkeys in the tops of trees to drop like stones into the dense and inaccessible centre layers of branches, while those on the ground look up or take cover. Leopard warnings send monkeys on the ground up into the trees, while those already there go still higher. Snake calls elicit standing and visual search of the ground for this class of slow-moving predator. Juveniles seem to give alarm calls less accurately than adults, and respond to the warnings of others less reliably. These observations are consistent with either a cultural interpretation based strictly on LEARNING, or an instinctive one, in which calls and accompanying RELEASERS are innate and a selective learning process sharpens behavioural DISCRIMINATION. A cursory comparison with a second vervet troop from a different area has revealed differences in the calls, an observation consistent with the cultural hypothesis. On the other hand, regional variations in communication signals based strictly on GENETICS exist in frogs (Anura), and learned DIALECTS or adumbrations of innate, species-specific bird songs are well known. In either case, this alarm system of vervet monkeys seems to defy facile CLASSIFICATION either as language or simple communication.

Perhaps the most famous example of animal communication that has consistently challenged linguists with its perverse similarity to 'real' language is the dance language of honey-bees (*Apis mellifera*). A forager bee which has found a good source of food will return to the hive and begin a dance which excites the interest of the other bees. These bees will then fly out and search for the food in a non-random fashion. Von FRISCH showed that the odour of the food is carried back to the hive on the waxy hairs of the forager's body, and that it supplies recruited bees with information which they will then use in their search. He also showed, however, that the direction and duration of part of the dance specify the food's location; both abstract and arbitrary conventions which are shared within the group clearly communicate concrete information. Dances are normally performed on vertical sheets of comb in the darkness of the hive. The bees consider 'up' to be the direction of the sun outside the hive, and the angle of the waggle phase of the dance with respect to vertical corresponds to the angle between the sun's azimuth and the food. Hence, a dance with a waggle run directed 90° to the left of vertical corresponds to a food source 90° to the left of the sun (see COMMUNICATION, Fig. C). As the sun moves west, the dances rotate slowly counter-clockwise.

The distance to the food is coded in the duration of the waggle run, and there is considerable 'cultural' variation in this distance component. Each of the two dozen or so geographic races seems to have its own dialect, so that one waggle is about 5 m to an Egyptian honey-bee, 25 m to an Italian bee, and 75 m to German ones. Of course, this variation is perfectly acceptable so long as the convention for encoding and decoding distance in dances is shared by all the members of the hive (or linguistic community). Then, too, defining 'up' as the direction of the sun is another arbitrary convention. The bees could have chosen down, 13° to the right, or defined up as, say, geographic north. The element in this communication system that gives it its linguistic-like quality is that all the bees in a colony must agree on the convention. Indeed, the tropical honey-bees from whom our bee evolved seem to lack the up-is-the-sun convention, and rely instead on dancing in the open, with at least a partial view of the sky through the trees for a reference. Under these circumstances the bees generally point their dances directly at the food, as will our temperate-zone honey-bees if forced to dance on a horizontal surface. In ORIENTING their dances the bees rely entirely on the direction of the sun if it is visible through the leaves and clouds. Otherwise, they seem to infer its position from the polarized light in the sky. The latter ability is difficult to understand, since most polarization patterns exist at two places in the sky at once. Bees must distinguish between the two, which they do with arbitrary, linguistic-like rules. If the two possible locations for the pattern are at different distances from the sun, bees take what they see to be the alternative further from the sun. If the two are at equal distances, the bees take the pattern they can see to be the one to the right of the sun. Although these rules result in the bees' misidentifying the pattern half of the time, the universal acceptance of the conventions ensures that the message is transmitted none the less. Although the dancer is signalling the direction of the food relative to an arbitrary interpretation of the patch of sky it sees, the recruits are interpreting the dance in accordance with the same rules. Since all the bees involved agree to be consistent in such mistakes their errors are cancelled, and the recruits find the food.

That honey-bees use the information coded in the dance has been challenged. Bees, like other social insects, have such a wealth of pheromones and such a good sense of smell that essentially all of von Frisch's results could be explained on the basis of the primitive communication system to be derived from an 'olfactory map'. Indeed, many experiments have shown that bees are capable of well-directed recruitment in the absence of well-oriented dances, and can suffer from poorly directed recruitment when apparently normal dances are available. The question of whether honey-bees can use the dance information was resolved by tricking foragers and recruits into using two different direction conventions. Even bees dancing on a vertical surface will abandon gravity and orient their dances directly to the sun if they can see it. If the simple eyes of a bee (the *ocelli*) are painted over, however, the bee becomes about eight times less sensitive to light. Such bees ignore the sun when dancing on vertical comb, and orient their dances to gravity, as if the sun were not visible. By painting over the ocelli of foragers, then,

and controlling the angle of an artificial sun to which the unpainted recruits are orienting their interpretations of the foragers' dances, researchers are able to 'aim' the recruits in any arbitrary direction. This shows that the bees can and do use the information encoded in their dances.

But is the dance language, strictly speaking, a language? In addition to its transmission of abstract information by means of demonstrably arbitrary conventions and the existence of numerous dialects, it is also used under 'novel' circumstances: (i) to direct recruits to water when the hive is over-heating (bees use an evaporative cooling system); (ii) to direct them to propolis, tree sap which is used to caulk the hive, smooth its interior, and seal off dead animals too large or too unwieldy to remove; or (iii) to direct recruits to new nest possibilities at swarming time. A colony may go a year or more (about six forager generations) between occasions for communicating about any of these three goals, and there is no reason to suppose that bees would be unable to adapt the dance system to other goals should the need arise, or should some artificial reward be offered for doing so.

The most language-like characteristic of the honey-bee dance, though, is its undisputed ability to communicate information about referents which are distant in both space and time. The food or nest cavity being advertised may be 10 km or more away, and have been visited some minutes or even hours previously (during which time the forager has kept 'mental' track of the sun's movement to the west). Then too, other bees seem to evaluate the information in seemingly logical fashion, weighing the distance to be travelled, in the case of food, against the current needs of the colony.

The argument against the dance system's claim to be a language, beyond mere matters of degree, boils down to its necessarily innate basis: it cannot compare with the human cultural achievement called language because it is inborn, rather than contrived through sheer intellectual force. The difficulty we see with this admittedly aesthetically appealing standard is that it assumes both that human language is entirely cultural, and that animal communication systems are entirely innate, and our reasons to believe in these previously compelling criteria are rapidly being undercut, reducing scientific fact to articles of faith.

As an outstanding example of communication which is assuredly not innate, for instance, there are the systems being taught in various LABORATORIES to primates, particularly to chimpanzees (*Pan*). These animals, though they communicate very effectively among themselves in the wild, have no abstract language as we know it. However, when given the appropriate linguistic tools, which consist in their case of manual referents (the American sign language Ameslan, plastic symbols, or keyboards) rather than verbal ones, thus taking advantage of their digital expertise and by-passing their inadequate vocal apparatus, they are able to master impressive vocabularies. They demonstrate a clear ability to categorize and to generalize; they are able to master at least English grammatical sentence structure, as well as the classes of noun, verb, adjective, adverb, even linguistic social graces like 'please' and 'thank you'. And it has been shown that they can use their acquired linguistic abilities to communicate with each other both spontaneously and in controlled situations.

Must we then consider chimpanzees to have language, merely on the grounds that they are, like us, capable of being taught it? There are limits to their capacity, though whether these limits are imposed by GENETIC means, or by our incompetence as teachers, remains in doubt. On the other side of the coin, must human beings be denied language if our ability is found to have a strong genetic basis?

Across the world, linguists have traditionally argued, language systems differ radically, thus illustrating their arbitrary, and therefore non-genetic, origin. English and Chinese, for instance, are completely different languages. Or are they? Perhaps, as preliminary evidence is beginning to suggest, the myriad human tongues are only magnificently ingenious elaborations on an innate basis which we, as human beings, are driven instinctively to complete. The eminent linguist Noam Chomsky has discovered a clear, basic, abstract, cross-cultural core to language which he calls its *deep structure*. He has formulated a theory by which human beings come equipped with a template for this structure, a sort of innate language acquisition program.

Solid evidence for this sort of BRAIN structure is difficult of access, but increasingly significant data indicate that human infants come programmed to recognize and to respond to a variety of stimuli. By far the most interesting of these genetically predetermined systems is the one associated with language acquisition. Consider an adult's difficulty in learning a language, even badly, compared with the ease with which even flighty, distracted, and uncooperative children manage it, regardless of intellect. It is difficult to see how this phenomenon could be accomplished without the aid and guidance of some internal monitor. Considering vocabulary size as an index, the task appears still more Herculean. Beginning at about the age of one, children acquire passive vocabulary, that is, words which they recognize without being able to articulate, at a rate of about seven words per day. This increases to more than twenty a day by age eight. A well-educated adult could have a passive vocabulary in excess of 40 000 words, a quarter of the *Shorter Oxford English Dictionary*. We are able to acquire and manipulate this vast internal lexicon because we learn rules by which a far smaller number of root words may be made to generate other words in various ways, transforming themselves into verbs, nouns, adjectives, adverbs.

Children, though, seem to be programmed to look for these rules, and to generalize them. It is this phenomenon which causes them, at the height of their language acquisition, to begin making studied errors in irregular forms which they had previously used correctly, such as changing 'I brought' in mid-sentence to 'I bringed'.

Our world is further organized and simplified by our strong drive to break things down into general concepts and classifications, and language reflects these basic abilities. Linguists see three basic patterns of CLASSIFICATION in language: *superordinate*, *basic*, and *subordinate*. The categories correspond to, say, the classes of vehicles, automobiles, and Volkswagens. Once a higher-level mental category is created, a person's ability to acquire particular names, and to use them creatively to generate new names and meanings is dramatically enhanced. This universal linguistic strategy may be merely a reflection of our species' supreme cleverness, but it seems that animals no more intellectually impressive than pigeons (Columbidae) do the same thing. When shown a series of slides and rewarded after seeing a few examples of, say, trees, incubator-hatched and laboratory-reared birds that have never seen anything more tree-like than a perch inside a cage are able to generalize to trees in general, including trees seen from odd angles or under marginal viewing conditions, unusual species of tree in leaf and out, shrubs, twigs, and branches. At the same time, they categorically reject television antennae and telephone poles.

All three thousand human languages are generated out of forty phonemes (units of significant sound), spawned in turn from perhaps a dozen vocal 'gestures', though many other combinations are physically possible. Syllables are composed of vowels and consonants, which convey their identities by means of the intensity, relative frequency, and modulation of two frequencies of sound emitted during speech. Vowels are the consequence of the size and thus the resonance of the vocal apparatus, while consonants result from specific movements of the lips and tongue. Linguists arrange consonants into a matrix of voice-onset times (VOT), and site of origin in the mouth. The VOT is the time between the beginning of the strictly mechanical part of a syllable and the 'voiced' part, when the vocal cords come into play. In 'ba', for example, the so-called 'plosive' release of air which has been held in by the lips is followed almost immediately by the 'ah', or voiced part of the sound, whereas in 'pa' there is a perceptible delay between the two. Thus the major difference between the sounds, which are both made by the lips, is the VOT.

A continuum of sounds with VOTs running from zero to as much as a second can be synthesized, and experiments run to determine what VOTs mean to human beings. On the ba–pa continuum, for instance, humans the world over, it turns out, hear either 'ba' or 'pa', depending on whether the VOT is longer or shorter than about

40 ms. This astoundingly discrete perceptual boundary exists for other consonants as well, and even 6-week-old infants recognize it. Is this a case, perhaps, of extraordinarily rapid conditioning? Apparently not, as we can see from languages in which some consonants are missing. For example, English lacks a consonant on the ba–pa continuum which has a negative VOT: the 'ah' precedes the plosive. In fact we as English speakers cannot distinguish these from 'ba's', but our newborn children can. Several such examples exist, and serve to show that we are to some extent preprogrammed to recognize and pick out salient elements of human speech from all the confusing noise that we hear. Since we lose some of this ability if we do not hear the appropriate sounds, there is probably a SENSITIVE PERIOD during which our perceptual abilities are determined. Native Oriental speakers may be baffled by the liquid English 'r', while English-speaking people may find the subtleties of the soft non-plosive French 'p' difficult to master.

The emerging picture of human language is one of continual internal guidance and help: help in recognizing and classifying sounds, organizing concepts, formulating rules for generating words, and arranging those words meaningfully into spoken or written phrases to communicate our thoughts. This is a disturbing turn of events. According to strict traditional definitions, the innate component of human language disqualifies it as 'real' language on the same grounds as it disregards the honey-bee dance language. Perhaps only vervet monkeys and trained chimpanzees now survive this well-worn linguistic touchstone. In fact, we are forced to conclude that language, like behaviour in general, is part of an evolutionary continuum. Human language is not discernibly difficult in kind, though it is very different in degree. It is clearly far more open to creative construction, less instinctively determined, and capable of transmitting far more information than the systems evolved by other species. And yet, our language is, in all likelihood, only an elegant elaboration of a set of genetically established and maintained opportunities for which our species was unique only in its degree of need. The FITNESS to be derived from exploiting its linguistic opportunities occurs at the undoubted sacrifice of the efficiency and clarity that are the obvious correlates of more primitive communication systems. The traditional linguistic ploy of 'defining animals out' is simply a covert, highly intellectualized tactic to deny evolution where our species is concerned.   J.L.G., C.G.G.

53, 76, 81

LEARNING. The behaviour of many animals is often rigid and unvarying. The same stimulus elicits much the same response, whether the response in question is as simple as flexing the leg in response to a pinprick to the foot, or as complex as the sequence of coordinated movements whereby a greylag goose (*Anser anser*) retrieves an egg that

has rolled out of her nest. This is not particularly surprising: an animal living in an unvarying environment can rely on equally invariant patterns of behaviour to ensure survival, and individual deviations will tend to be eliminated by NATURAL SELECTION. Not all aspects of all environments, however, are perfectly stable. If the environment changes in some crucial respect, persistence in a fixed pattern of behaviour may be maladaptive. In many cases, there is little or no regularity in the relevant aspects of the environment at any time: an omnivorous animal, for example, must find its food where it can, rather than rely on a set of fixed responses. When we observe that an animal's behaviour changes in response to some change in its environment, we may attribute this change to learning. Just as the INNATE behaviour patterns of a species may change over many generations in response to slow environmental changes, so individually learned patterns of behaviour may change in response to abrupt or transient environmental changes. Although this analogy between the natural selection of advantages patterns of behaviour and the modification of learned behaviour should not be pressed too far, it is helpful if it reminds us that the ability to learn would not have evolved had it not been of some adaptive significance, and that particular instances of learning can therefore normally be expected to have advantageous consequences for the learner.

Learning is a familiar enough phenomenon, but as is often the way, one not easily captured by the scientist's definition. Experimental psychologists have sometimes identified learning with observable changes in behaviour, but this clearly does not accord with any usual meaning of the term. Rather we infer (or may infer) that an animal has learned something when we observe a particular change in its behaviour. We must then distinguish between those changes which we wish to attribute to learning and those which we would attribute to other causes. A thirsty animal will drink, although 6 h ago it refused water. Its behaviour has changed, but we do not necessarily suppose it has learned anything in the interval: its state of MOTIVATION may have changed. A male puppy urinates in the same way as a female (by squatting), but an adult dog (*Canis lupus familiaris*) cocks his hind leg. The change is not a consequence of learning, but of increasing sexual maturity (if the puppy is injected with male HORMONES it will promptly behave like the adult), and we should probably regard it as an instance of MATURATION.

Although there is often no difficulty in deciding whether or not a particular change in behaviour is due to learning, at other times careful experimental analysis may be necessary. Young birds, for example, cannot fly. As they grow older, they can be seen practising, and after a few weeks they can fly quite proficiently. It seems reasonable to suppose that they must learn to fly, and that the period of practice is necessary for the development of this skill. Both pigeons (Columbidae) and swallows (Hirundinidae), however, have been brought up under restricted conditions which completely prevented their practising flight movements. When they were released at the age at which normally reared birds were first flying successfully, there was essentially no difference in the proficiency of those that had practised and those that had not.

Since the ability to fly develops without the opportunity for practice, we are inclined to say that it does not depend on learning. This suggests that we may define learning by reference to the particular set of circumstances that are responsible for the observed changes in behaviour. If motivational changes are those produced by one set of conditions (e.g., by depriving an animal of food or water), then changes attributable to learning are those produced by circumstances which provide certain other experiences. The definition will be of limited value, however, unless those circumstances can be specified. It is not very much use to say that learning is produced by practice. Now in particular cases, sufficiently precise definitions can be provided: a useful definition of classical CONDITIONING, for example, would refer to changes in an animal's behaviour to a particular stimulus brought about by exposing the animal to a correlation between that stimulus and another. However, animals can, it seems, learn a large number of different things, and psychologists and ethologists have devised a large number of different procedures for studying learning. The variety is such that it may be difficult or impossible to formulate a definition of the conditions producing learning that is not either vacuous or so restrictive that it excludes certain cases which we should certainly want to regard as instances of learning.

It may be foolish, therefore, to waste too much time attempting to provide a precise, all-embracing definition of learning at this point. For the time being, it will be more useful to provide particular descriptions of some of the particular situations in which learning has been studied. The definitions so provided will in effect specify the sorts of changes in behaviour that an experimenter may observe, and the conditions or experimental operations responsible for those changes. This procedure has the further advantage that it does not, as would a single overall definition, prejudge the issue whether there is one variety or many different varieties of learning. It is certainly not obvious that the same processes are involved in learning to read music, play tennis, and solve differential equations. We shall be in a better position to judge how much is common to the various situations in which animals learn after we have seen what those situations are.

Habituation. If the snail *Helix albolabris* is moving along a wooden platform, it will immediately withdraw into its shell when one taps on the wooden surface. After a pause, it will emerge and continue on its way, but will again withdraw if one taps again. This time, however, it will re-emerge more rapidly, and a third and fourth tap

may elicit only a brief and perfunctory withdrawal response. After half a dozen trials, the stimulus which initially elicited an immediate response may have no detectable effect on the animal.

What is the cause of this change in behaviour? Perhaps sensory ADAPTATION has reduced the animal's sensitivity to the stimulus, in much the same way as we rapidly adapt to bright light after emerging from the dark. Or perhaps muscular fatigue reduces the probability of the response occurring. Both these interpretations, however, can often be ruled out, as experiments with the marine worm *Nereis pelagica* have shown. This worm lives in a burrow or tube from which it emerges to eat. A variety of sudden stimuli, including either an increase or a decrease in illumination, a passing shadow, or a touch from a small rod, will cause immediate withdrawal into the tube, but with repeated presentation any one stimulus is less and less likely to elicit this response. At this point, however, a different stimulus (even a weaker one) is still able to elicit complete withdrawal. The decline in responding, therefore, can hardly be due to muscular fatigue. Moreover, although the worm will no longer withdraw upon presentation of a particular stimulus (such as being touched by a rod), it is clearly still able to detect the stimulus, since it may now respond by turning towards it. Thus the change in behaviour can hardly be due to sensory adaptation.

When adaptation and fatigue can be ruled out in this way, a decline in responsiveness to a repeated stimulus is referred to as HABITUATION. According to the generally accepted definition, habituation is the learning process whereby the experimental operation of repeatedly presenting the same stimulus to a subject results in a decline in the probability of the response or responses initially elicited by that stimulus. Physiological analyses have been able to locate some of the changes at the cellular level which underlie this learning process. Studies of the habituation of the gill-withdrawal reflex in the sea hare (*Aplysia*), for example, have implicated certain specific electrical changes in the nerve cells controlling the gills.

Evidence of habituation has been obtained throughout most of the animal kingdom, from the simplest animals to human infants. It is particularly noteworthy that habituation has been reliably documented in animals, such as the freshwater polyp *Hydra*, which have shown no good evidence of being capable of associative learning (see ASSOCIATION). This suggests that the processes of habituation cannot be the same as those required for successful conditioning. There is, of course, no guarantee that the processes involved in habituation are the same in all animals. Indeed, since even protozoa (single-celled animals) may show a decline in responsiveness to a repeated stimulus, and since in this case there can be no question of distinguishing between sensory adaptation, muscular fatigue, or central habituation, there is good reason to suppose that the processes responsible for similar behaviour at different levels cannot always be the same.

The generality of habituation implies that it is of considerable biological significance. This is not hard to see. A snail (Gastropoda) that failed to respond to sudden, novel stimuli by withdrawing into its shell would forfeit much of the advantage of having a shell. But a snail that continued to withdraw into its shell in response to every change in stimulation would rarely be out for more than a second at a time. Novel stimuli may signify danger, and an animal must react to them by appropriate DEFENSIVE behaviour; but a stimulus which has no further consequences on its first few occurrences will probably continue to be safe, and regularly repeating stimuli are as likely as not to form part of the general background against which more significant events occur. To the marine worm *Nereis*, for example, an overhead moving shadow may signify the approach of a predator, but if it occurs sufficiently often, it is more likely to be caused by some waving seaweed. Thus the function of habituation is to discriminate between novel and familiar events, and to ensure that the animal's behaviour is more or less appropriate to each.

Sensitization. Habituation may be regarded as a mechanism for eliminating unnecessary responses. Learning may also, however, result in the appearance of new responses rather than the disappearance of old ones. Indeed, the belief that learning always produces an increase in the probability of a particular response, or set of responses, has been one reason for the prevalence of *stimulus–response* (S–R) *theories*, which attempt to reduce learning to the establishment of connections between stimuli and responses. The analysis is not easily applied to habituation, nor even to what appears to be one rather simple form of learning which does actually result in an increase in the probability of responding.

The common octopus (*Octopus vulgaris*) usually lives in a small cave or opening between rocks, sitting at the entrance to this home, emerging to attack its prey, and retreating inside when danger threatens. Octopuses may be kept in the LABORATORY in tanks of sea water equipped with a 'home' made of a few bricks at one end, and can then be trained to swim down the tank to attack a food source at the far end, or to retreat if such an attack is met with a mild electric shock. The probability of emerging to attack a neutral stimulus, such as a white plastic disc suspended on a rod, is increased if the octopus has recently been given food, and decreased if it has recently been shocked. These effects do not depend upon rewarding or punishing the octopus for attacking, for they can be obtained by feeding or shocking it in its home. It seems rather that feeding or shocking the animal increases the probability of the responses (of approach or withdrawal) elicited by food or shock, so that they may temporarily be elicited by a neutral stimulus.

Such an increase in the probability of a response elicited by a biologically significant stimulus consequent upon repeated presentation of that stimulus is termed *sensitization*. The astute reader may perceive a certain conflict between the principles of habituation and sensitization. We have just been considering cases where the repeated presentation of a stimulus results in a decline in the probability of the response elicited by that stimulus. Are we really to suppose that the same set of operations also increases the probability of responding? Part of the solution to this apparent paradox is to note the difference between those stimuli whose presentation leads to habituation, and those whose presentation results in sensitization. Habituation occurs to relatively insignificant stimuli such as changes in illumination or loud noises, or perhaps models of possible predators. Repeated presentation of an electric shock, however, may result in sensitization, and certainly will not produce habituation. Similarly, the only reason why responsiveness to food declines with repeated feeding is because the animal becomes satiated (a change in motivation rather than in learning). It must also be recognized, however, that this distinction between stimuli that habituate and those that produce sensitization is unlikely to be rigid. There is, in fact, good reason to believe that in many cases both processes actually occur simultaneously, with the balance between the two determined by the strength or significance of the stimulus. Thus the overall decline in responsiveness observed in an experimental study of habituation will often be the result of two opposed processes, one decreasing, and the other increasing, the probability of the response.

Recognition of the importance of sensitization, therefore, is necessary for a proper understanding of habituation. It is also relevant to the analysis of associative conditioning, for as the experiment with octopus showed, repeated presentation of a significant stimulus may increase the animal's responsiveness to the point where a normally inadequate stimulus temporarily elicits the response in question. Such a change in responsiveness may easily be mistaken for conditioning. Another experiment with marine worms *Nereis diversicolor* illustrates the point. The worms were placed in long horizontal tubes and either fed or shocked at intervals. Interspersed between these presentations of food or shock was a series of test trials on which the level of illumination was changed. Over the course of the experiment, the shocked worms became progressively more likely to withdraw in response to this change in illumination, while the fed worms showed a progressive tendency to advance along the tube in response to the light. Both groups, in other words, showed sensitization. The experiment also contained groups of worms for whom the change in illumination was actually paired with food or shock, and these groups also showed the same significant changes in behaviour. In the absence of the first pair of groups one might have attributed these latter changes in behaviour to the pairing of light with shock or food, i.e. to a process of conditioning. Since, however, there was no difference between those animals for whom the light was paired with food or shock, and those for whom it was not paired, it seems more reasonable to attribute all the behavioural changes observed to a process of sensitization.

Sensitization can thus produce changes in behaviour that, in the absence of careful experimental analysis, may be mistaken for instances of associative learning or conditioning. This has proved somewhat unfortunate, for it has meant that sensitization is usually treated simply as an inconvenient nuisance which has to be controlled for in more interesting studies of conditioning. There has, therefore, been very little systematic investigation of the range of conditions under which sensitization occurs, the sorts of behavioural changes it may produce, and the underlying processes responsible for those changes. It seems likely, however, that it represents a fairly widespread form of learning, at least among invertebrates, and may profitably be thought of as a necessary precursor of associative learning or conditioning. It is not difficult to see how such a form of learning could be of adaptive significance. Provided there is some regularity in the environment, an animal which shows an increased propensity to leave its home and explore when food has recently been obtained will have a better chance of obtaining more of the same food; conversely, one that shows an increasing tendency to withdraw to its home following recent exposure to severe danger stands a better chance of escaping the lurking predator. The process of sensitization thus enables an animal to take advantage of any statistical regularity in its environment, without requiring it to learn what are the specific events correlated with food or danger. But sensitization can be regarded as only a precursor of associative learning, for it seems obvious that associative learning must be the prime method by which an animal discovers what is correlated with events of such significance as food or danger.

Associative learning. Associationism has a long, largely philosophical tradition behind it. British associationist philosophers from Hartley to Mill assumed that the complex ideas which form our picture of the world were built up by associating simple ideas. An understanding of the laws of association was, therefore, central to an understanding of the human mind. Associationism has always provoked opposition, both in philosophy and in psychology. Against the empiricists' view that the mind is a *tabula rasa* on which experience writes its impressions, the European rationalist philosophers insisted that the mind imposes its own categories on experience, and that without the structure that we bring to our experiences, those experiences would be meaningless. This philosophical tradition has also had its psychological heirs. *Gestalt* psychologists, such as Wolfgang

Köhler, argued vigorously that associationist theory is totally inadequate as an explanation of both human and animal behaviour. Modern students of PERCEPTION and COGNITION in human beings might well accept this argument. Many of the more interesting problems in experimental psychology are probably not greatly illuminated by simple associative theories. Even when psychologists attempt to force their human experimental subjects to learn simple associations by rote, as when they require them to memorize lists of 'nonsense syllables', most people probably solve the task by imposing some meaning or structure onto the list. The fact that a human adult's knowledge of the world is not based entirely on simple associative processes, however, does not imply that associative learning never occurs in humans, let alone in animals. Indeed, if the operation of such a process is often obscured in humans by more complex processes, this is all the more reason for using animal subjects to study it.

The study of learning in animals has, indeed, for a long time been dominated by the study of associative learning or conditioning. In a typical conditioning experiment, the experimenter arranges correlations or contingencies between certain events, and observes whether the subject's behaviour changes as a result of exposure to this contingency. In an experiment on classical conditioning, of the type pioneered by PAVLOV, the contingency is between a neutral stimulus, such as a light or buzzer, and a motivationally significant event, such as food or electric shock: the illumination of the light signals the delivery of food, and the experimenter records changes in the subject's behaviour to the light. In Pavlov's terminology, the light is the *conditional stimulus* (CS), and the food the *unconditional stimulus* (UCS). The food unconditionally elicits a set of CONSUMMATORY responses, one of which is recorded by the experimenter and designated the *unconditional response* (UCR). In Pavlov's case, of course, the UCR recorded was the response of salivating. After a number of pairings of CS and UCS, the CS also starts eliciting salivation, and this is then termed the *conditional response* or CR. The presentation of the UCS following the CS is said to reinforce this conditional reflex of salivating to the CS. By extension, therefore, the UCS is often referred to as a reinforcing event or *reinforcer*.

Classical conditioning therefore occurs when a CS starts to elicit a CR and this change in behaviour can be attributed to the experimental operation of arranging a contingency between CS and UCS. This latter requirement, of course, implies the need for various control procedures, to ensure, for example, that the change in behaviour is not a consequence of sensitization.

In an experiment on instrumental or OPERANT conditioning of the type pioneered by Thorndike and extensively studied by SKINNER, the occurrence of a motivationally significant event (again termed a reinforcer) is contingent on the subject's performing some designated response. In Thorndike's puzzle box, the animal obtained food only if it depressed a particular catch or panel. In its modern equivalent, the Skinner box, the rat (*Rattus norvegicus*) obtains food only if it presses a lever that protrudes from one wall of the box: the depression of the lever automatically results in the delivery of a small pellet of food into a magazine. In another version of the Skinner box, designed for birds, a domestic pigeon (*Columba livia*) is required to peck an illuminated disc mounted on one wall, and successful pecks result in the automatic delivery of grain to a magazine situated below the disc. In all these cases the experimenter records any change in the probability of the designated instrumental response, and *instrumental conditioning* is said to have occurred if such a change can be attributed to the contingency between that response and the reinforcer.

In both varieties of conditioning experiment, therefore, the experimenter arranges a particular set of contingencies, in one case between a CS and reinforcer, in the other case between a response and reinforcer. Perhaps the simplest interpretation of conditioning, therefore, is that the subject associates these pairs of events, and the changes in behaviour are a consequence of an association between CS and reinforcer in one case, and response and reinforcer in the other. In order to span the gap between association and overt change in behaviour, theorists have proposed certain principles of REINFORCEMENT. Pavlov proposed that REFLEX responses elicited by one stimulus would come to be elicited by other stimuli associated with it. Thus food reflexly elicits salivation in a hungry dog, and if the dog, as a result of the contingency between light and food, associates these two events, the light will also come to elicit salivation. The principle is intended to apply only to those responses reflexly elicited by a stimulus, and thus may not apply to what we should ordinarily regard as voluntary responses. To explain changes in responses such as lever pressing, Thorndike proposed his *law of effect*, which stated that a response followed by a rewarding or satisfying state of affairs would increase in probability, while one followed by an annoying or aversive consequence would decrease in probability.

The adaptive significance of instrumental conditioning is so obvious as to need little comment. If the only way for a hungry rat to obtain food is to perform a particular response such as pressing a lever, then an increase in the probability of lever pressing becomes a prerequisite for survival. The law of effect, indeed, has often been viewed as an analogue of NATURAL SELECTION as a means of shaping the individual's behaviour to the requirements of the environment. The adaptiveness of classical conditioning, however, has sometimes been queried. Since the dog will get food on each trial regardless of its behaviour, what is the point of salivating just before the delivery of food? One suggestion has been that salivating at just this

moment is in fact rewarding: dry food is relatively unpalatable, and it is made more palatable for the dog if moistened by anticipatory salivation. The argument amounts to applying the law of effect to classical conditioning, for what is being said is that the animal performs a particular response because that response has rewarding consequences. Although this may seem a reasonable account, it fails to recognize the importance of the distinction between VOLUNTARY and involuntary responses. Voluntary responses may be readily modified by their consequences in the manner implied by the law of effect, but involuntary responses are precisely those which are not readily modified in this manner. In fact, if we require a dog to control its rate of salivation in order to obtain food it will perform rather inefficiently. The experimenter can arrange that a light signals the delivery of food, but only provided that the dog does not salivate when the light is turned on. On early trials, before conditioning has progressed very far, the light is turned on, the as yet unconditioned dog does not salivate, and receives food. In due course, however, the dog starts salivating on some trials, and of course gets no food on those trials. The efficient solution (predicted by the law of effect) is for the dog now to refrain from salivating until the food is actually delivered, but this the dog fails to do. Instead the dog continues to salivate on more than half the trials, thus forfeiting more than half the food it might have obtained. The result is very much what Pavlov's theory of reinforcement would lead one to expect: salivation, as a response reflexly elicited by food, will be elicited by a stimulus correlated with the delivery of food, regardless of the consequences of salivating, and despite the fact that in this particular case salivating leads to a net loss of food.

Classically conditioned responses, therefore, appear to differ from instrumentally conditioned responses precisely in not readily modified by their consequences. But this does not mean that they are normally without beneficial consequence. The mistake is to suppose that the particular, discrete responses typically recorded by experimenters, such as salivation in response to food, flexion of the leg in response to shock to the paw, eyeblink in response to a puff of air, exhaust the range of classically conditioned behaviour. As Pavlov fully realized, a dog will do very much more to a CS signalling the delivery of food than just salivate: for example it will show excitement as soon as the CS is turned on, and approach the place where either the CS is presented or the food delivered. For purposes of scientific analysis, Pavlov chose to ignore these other patterns of behaviour, and rely exclusively on conditioned salivation (among its virtues, it was easily measured and quantified, and its very arbitrariness discouraged the investigator from making ANTHROPOMORPHIC assumptions about the dog's thoughts and expectations). But to appreciate the significance of classical conditioning, one must forgo this concentration on one arbitrary aspect of conditioned behaviour, and take a wider view of the range of responses that are actually conditioned.

Reinforcing stimuli, such as food, water, or sexual partners, elicit a wide variety of APPETITIVE and consummatory responses in hungry, thirsty, or sexually motivated animals, ranging from general approach responses to specific patterns of COURTSHIP behaviour, and characteristic consummatory behaviour. Similarly, aversive reinforcers such as electric shocks, very loud noises, attacks by predators, elicit a characteristic array of DEFENSIVE reactions, which in different species may range from escaping, fleeing, freezing, or 'playing possum', to mobbing and attacking. It does not require great imagination to see that if it is advantageous for a particular animal to curl up into a ball, or to flee when a predator attacks, it will be even more advantageous to do so in response to a stimulus that has regularly preceded that attack. Similarly, just as animals must approach food in order to survive, so it must be advantageous to approach places which have been associated with finding food.

Classical conditioning, therefore, by ensuring that animals respond to stimuli correlated with the occurrence of reinforcing events in the same way that they respond to those events themselves, may enable animals to behave appropriately in anticipation of a reinforcer, and thus increase the probability of obtaining appetitive reinforcers and evading aversive reinforcers. In this sense, the process of classical conditioning, requiring only the capacity to detect correlations between external events, may be an effective means of producing adaptive modifications of behaviour. But it does not guarantee success. A general tendency to approach the location of food may not be sufficient actually to obtain food. The predator that merely approaches its prey will find the prey disappearing. Even when obtained, the prey may be inedible without fairly intricate manipulation. Having caught a mussel (Myrtilidae), for example, the Californian sea otter (*Enhydra lutris*) brings it to the surface, together with a stone. Floating on its back, it places the stone on its chest as an anvil, and cracks the mussel open by repeatedly banging it against this anvil. In many cases, of course, the requisite set of consummatory responses forms part of the species' natural repertoire of behaviour, which develops without the need for learning. When the blowfly *Phormia regina*, for example, lands on a solution of sugar and water, chemical stimulation of receptors on its legs elicits extension of the proboscis; stimulation of the tip of the proboscis now elicits a pumping action which leads to ingestion of the solution. In many invertebrates, at least, intricate consummatory responses are released by the appropriate stimulus characteristics of food, mates, etc. Little or no learning is thus required. But many vertebrates and perhaps most mammals have a less rigidly specified set of consummatory responses, and may

have to learn how actually to obtain their food. A common observation is that a rough approximation to the required behaviour appears in the absence of any opportunity for learning, but that reinforced practice may be necessary to perfect the necessary skill. Thus several species of passerine birds, such as great tits (*Parus major*), hold down their food with one or both feet while tearing bits off with their beak. A young tit will hold down a meal-worm (larva of the darkling beetle, *Tenebrio molitor*) as soon as it is given one for the first time, but the behaviour is clumsy and inefficient, and is only perfected by specific practice. Young red squirrels (*Sciurus vulgaris*) are able to open hazel nuts at the first opportunity, but they gnaw inefficiently and at random until the nut breaks open by chance, and it is only with practice that they learn to gnaw a long furrow down one side, wedge their teeth into this crack, and break the nut in half.

In cases such as these, what we are seeing is a process of instrumental conditioning serving to modify the animal's natural appetitive and consummatory behaviour so as to maximize efficiency in obtaining the appropriate reinforcer. The sight of a stimulus associated with food elicits, by a process of classical conditioning, an appropriate set of consummatory responses, but aspects of these responses may then be modified as the animal learns the correlation between variations in its behaviour and variations in its success. In the laboratory, of course, instrumental conditioning may be able to produce behaviour that bears little or no relation to the natural behaviour that is classically conditioned to stimuli associated with reinforcers. Mammals, at least, can sometimes be taught relatively arbitrary responses to obtain food. Domestic dogs can be trained to lift one paw (as well as to sit up and beg), rats to press a lever, guinea-pigs (*Cavia apera porcellus*) to turn their heads 1 cm to the left or the right, and monkeys (Simiae) to pull levers, undo catches, and turn handles. But it must be emphasized that under natural conditions the more normal process will be the instrumental modification of classically conditioned behaviour. Moreover, the attempt to produce instrumental conditioning of totally arbitrary responses in the laboratory may founder on the incompatibility between these arbitrary responses and the consummatory responses related to the animal's motivational state and reinforcer. These consummatory responses may become classically conditioned to stimuli associated with the reinforcer and then interfere with the performance of the arbitrary instrumental response. Thus racoons (*Procyon*), trained to pick up a metal coin and deposit it in the slot of a vending machine in order to obtain food, have no difficulty in learning to pick up the coin, but instead of then depositing it they may spend several minutes vigorously rubbing it back and forth in their paws. This behaviour is in fact one of the racoon's normal responses to prey such as crayfish (Astacidae). Its intrusion in the animal's sequence of instrumental behaviour might not seem particularly surprising, except that the racoon will sometimes persist in rubbing the token, thus postponing the delivery of food, for minutes on end. The required instrumental behaviour of releasing the token is incompatible with the classically conditioned consummatory response of rubbing it back and forth, and performance is accordingly inefficient.

As defined here, therefore, instrumental conditioning involves the modification of an animal's responses in accordance with the law of effect, as a consequence of exposure to a specific set of contingencies between a response and its consequence. Just as an experimenter may not conclude that he is studying an instance of classical conditioning unless he has controlled for sensitization (i.e. unless he has shown that the change in the subject's behaviour is produced by the contingency between stimulus and reinforcer), so he may not conclude that instrumental conditioning has occurred unless he has shown that the contingency between response and reinforcer is responsible for the change in behaviour. This is not always as simple as it may seem. Over the years, for example, thousands of pigeons have been trained in hundreds of experiments to peck a small illuminated disc on the wall of a Skinner box in order to obtain food. These experiments were regarded as almost prototypical studies of instrumental conditioning: the delivery of food was contingent on the response of pecking the illuminated disc, and the probability of pecking changed appropriately. It turns out, however, that pigeons will peck in Skinner boxes without any such instrumental contingency. If the experimenter simply illuminates the disc for a few seconds just before each delivery of food, but delivers food automatically regardless of the pigeon's behaviour, pigeons will end up pecking just as rapidly as they would if there were an instrumental contingency. The only contingency now is one between a stimulus (the illumination of the disc) and a reinforcer, but this classical conditioning contingency is sufficient to generate a high rate of pecking. When one then sees that pecking is the most obvious component of the pigeon's consummatory response to food, it becomes apparent that a process of classical conditioning is probably responsible for at least much of the pecking that experimenters have been recording and categorizing as instrumental for the past 20–30 years.

Perceptual learning. The distinction between classical conditioning and sensitization is that a conditioned animal will, for example, approach only those stimuli specifically associated with food, while a sensitized animal may approach any stimulus shortly after being fed. Unlike sensitization, classical conditioning requires of the subject a sufficiently well-organized sensory apparatus to enable it to discriminate between those events that have been correlated with a reinforcer and those that have not. A dog conditioned to salivate to a tone will not salivate when a light is turned on; a pigeon conditioned to peck a disc illuminated with red light will not peck when a white triangle is displayed on the disc. Animals do not, of course, always discriminate perfectly: the dog that did not salivate to a light will probably show some tendency to salivate when a tone differing only in frequency from the original CS is turned on; the pigeon may not peck at a white triangle, but will almost certainly peck, albeit at a lower rate, when the disc is illuminated with orange or yellow light. Conditioned responses generalize to stimuli similar to the original CS (see GENERALIZATION).

How do animals discriminate between different stimuli, and why should responses conditioned to one stimulus generalize to similar stimuli? The general assumption is that the nervous system of most animals is automatically organized so as to detect differences between tones differing in amplitude or frequency, and lights differing in intensity or wavelength, and is capable of ordering stimuli along these simple physical continua. This is plausible enough, but it is hardly possible to suppose that, at least in mammals, there is no development of the ability to perform perceptual discriminations. Before the age of 4 months, the human infant does not discriminate between its mother and other adult faces; the medical student must be taught to discriminate what he sees through a microscope; and the educated palate of the professional wine taster enables him to make discriminations which the rest of us can only pay for. We can learn, in other words, to perceive the difference between stimuli which we were initially unable to discriminate, and such learning, we can hardly doubt, depends upon the opportunity to inspect those or related stimuli.

PERCEPTUAL LEARNING has been studied in laboratory experiments with animals, either by providing specific experience with a particular set of stimuli before testing the animal's ability to discriminate them, or by depriving animals of, for example, all visual experience before testing their ability to learn a particular visual discrimination. The testing situation might be the same in both types of experiment: a hungry subject is presented with two stimuli on each trial; a response to one stimulus produces food, while a response to the other goes unrewarded. DISCRIMINATION is measured by the rate at which the animal learns to respond only to the rewarded stimulus. If rats are brought up with circles and triangles painted on the walls of their cages, they may later learn to discriminate between circles and triangles faster than a normally reared control group. Conversely, rats brought up in total darkness for the first 2–3 months of their life may learn such a discrimination more slowly than a normally reared control group. Such visual deprivation, however, has no effect on the learning of a simple brightness discrimination, or on the ability to discriminate between lines shown in different orientations (vertical, horizontal, or oblique). This is important, for it shows that the disruption of the more complex visual discrimination is not a sign of some general deficit resulting from being brought up in the dark, and it also confirms that it is only the capacity to perform relatively complex discriminations that depends on prior experience, while simpler discriminative abilities require no such experience.

Just as we must sometimes learn to discriminate between two or more similar stimuli, so we must sometimes learn to recognize the same stimulus on its reappearance. The two processes are obviously related. To discriminate one stimulus from another we must, in some sense, have a description of it that differs from our description of the other, and some such stored description is necessary if we are to recognize that stimulus again. Perceptual learning, therefore, may be thought of as partly a matter of building up a set of descriptions of the relevant, differentiating features of particular stimuli and situations.

One of the most interesting cases where such a process occurs is that of SONG learning in birds. Although certain simple call notes develop normally in most birds in the absence of specific experience, the complete development of the full adult song nearly always depends on some learning. Chaffinches (*Fringilla coelebs*), for example, reared in isolation will sing only a crude and simplified version of the normal chaffinch song; the development of the full song requires both exposure to the song, and the opportunity to practise singing it. There is thus both a phase of perceptual learning, normally occurring in the first few months of life, when the young bird stores a description of the complete song, and a later phase when the year-old bird learns to reproduce a song matching this stored description.

This long gap between the occasion when the young bird normally learns the characteristics of the species' song, and the occasion when it shows what it has learned by starting to produce the song itself, is found not only in the case of song learning. In many species, young animals learn the perceptual characteristics of other members of the species, to whom they later direct various aspects of their social behaviour (see IMPRINTING). The most familiar example of imprinting is the following response shown by many young birds to the first moving object that they see. This particular case has proved relatively easy to study, and provides nice evidence of the extraordinarily wide range of objects (which includes rubber balls, balloons, painted cubes, and flashing lights) which young ducklings and goslings (Anatidae) will accept as substitute mothers to be followed in times of danger. But the process of imprinting affects other aspects of SOCIAL behaviour, and is not confined to birds. There is good evidence of imprinting in a number of mammals; and an adult bird's choice of sexual partner may also be determined by the stimuli to which it was exposed during a relatively brief period of infancy. It is obvious, moreover, that under natural conditions

imprinting may involve a much greater degree of perceptual learning (in the sense of ensuring finer perceptual differentiations) than is displayed when a young duckling is imprinted on a green cube and tested for its discrimination between the green cube and an orange sphere. In animals that live in large groups or herds, one of the functions of imprinting, as far as it affects the following response, is to ensure that mother and young can discriminate one another from a large number of similar alternatives. Imprinting, therefore, is a case, like song learning, where the young animal may learn to formulate a relatively precise description of a relatively complex stimulus situation.

Insight and intelligence. Classical and instrumental conditioning are usually said to be instances of simple associative learning. Popularly, indeed, they may be held in even lower esteem: Pavlov's dogs helplessly salivating at the sound of the bell, and white rats finding their way through mazes by blind trial and error are often regarded as epitomes of a passive, mechanical process of stamping in associative connections, the complete antithesis of intelligent learning. This is to misunderstand the nature both of conditioning and of scientific method. The typical conditioning experiment may seem to place subjects in an impoverished environment and to record only trivial aspects of their behaviour. This represents, however, a deliberate policy of abstracting certain features from what is recognized to be a more complex reality, justified by the common scientific criterion of needing to exercise precise control over an experimental situation in order to determine the relevant causal factors and their mode of action. It is moreover quite misleading to describe Pavlov's dogs as helplessly salivating at the sound of the bell. They are more accurately described as detecting a particular set of contingencies in their environment, and reacting accordingly. The processes whereby they do so are more subtle and complex than has so far been suggested, too complex indeed for most existing theories of conditioning. By a process which is far from understood, for example, conditioning occurs selectively to stimuli better correlated with the delivery of a reinforcer at the expense of stimuli worse correlated, and to stimuli occurring closer in time to the reinforcer at the expense of stimuli further removed from the reinforcer. This is not because a poorly correlated stimulus does not get conditioned under any circumstances: for example, in the absence of any better predictor of food, a bell that signals food on only 50% of trials will be conditioned very rapidly. The presence of another stimulus perfectly correlated with the delivery of food, however, may completely prevent the bell being associated with food. Animals may thus be said to be selectively ignoring less valid signals of reinforcers in favour of more reliable signals. Quite how this happens is the subject of much doubt and some dispute, but it is surely a reasonably intelligent way to behave.

Nevertheless, the prevalence of the term 'simple

conditioning' attests to the prevalence of the belief that there are other, more complex forms of learning. As we have seen, one of the central theses of *Gestalt* psychologists, such as Köhler, was that simple associative theories were inadequate to deal with the complexities of behaviour. Although the argument was most often applied to perception, Köhler extended it to the case of learning and problem solving by insisting that learning was largely a matter of perceptual reorganization: a subject learning to solve a particular problem was said to be really learning to see the situation in a new light. These arguments were supported by Köhler's observations on a captive group of the chimpanzee *Pan troglodytes* kept on an anthropological research station on the island of Tenerife during the First World War. The chimpanzees learned to use a bamboo pole to rake in a banana lying beyond their reach outside their cage, and when one pole was too short they might fit two together to form a single long pole; they climbed on boxes to reach bananas hanging from the ceiling, and would stack one box on another to gain a higher platform. According to Köhler, they were perceiving objects in their environment in a new way. In a flash of INSIGHT the sticks were no longer seen as sticks, they became extensions of their arms; the boxes were no longer just boxes, they were seen as a crude ladder.

Although Köhler's chimpanzees were obviously solving problems, the evidence of insight and INTELLIGENCE which he found so impressive, and which he communicated so persuasively in his classic book, *The Mentality of Apes*, becomes rather less impressive in the light of further information about the natural behaviour of chimpanzees. Given the opportunity, laboratory-reared chimpanzees will play with sticks and boxes; they will bite the end off one bamboo pole and fit another into the hollow tube so formed; they will pile up boxes and climb up the rickety structure so built. All these activities have been observed in the course of PLAY, in chimpanzees who were neither solving problems nor attempting to obtain food, and who had never been trained to obtain food by these methods. Thus, although Köhler's chimpanzees were learning what responses were instrumental in obtaining bananas, the responses in question formed part of their natural repertoire. There appears to be no need to appeal to a process more complex than that of instrumental conditioning to account for their behaviour, and little to be gained by talking of their intelligent manipulation of arbitrary objects, or their insight in perceiving that those objects could be transformed into useful tools.

We are entitled to question Köhler's conclusions because he had not observed the natural behaviour of his subjects closely enough, a criticism often levelled by students of ETHOLOGY against psychologists. Observations of chimpanzees in the wild have provided other examples of their use of tools, and their ability to adapt raw materials into useful implements. Adult chimpanzees use a twig as a probe to stick into termite (Isoptera) nests; the twig is withdrawn, and any termites clinging to it are eaten. A good probe, of course, is one without numerous side branches or leaves, and chimpanzees can be seen selecting appropriate twigs and stripping them of such excess leaves. They have also been observed using larger branches both to beat on the ground and wave about as part of a display, and as a weapon with which to attack a stuffed leopard (*Panthera pardus*) left in their path by a curious ethologist. One interesting observation to come out of this experiment was that not all chimps chose particularly appropriate weapons or used them with any particular skill. Chimpanzees living in forest areas were especially unskilful, in some cases, for example, merely bending over a growing sapling in the vicinity of the leopard.

Relatively little is yet known about the limits of the chimpanzee's capacity to use and manufacture tools, and even less about the stages of learning necessary for the perfection of such performances (see TOOL USING). Such information as we have, however, and casual observation of a young child's first clumsy and maladroit attempts to master such skills as hitting a ball with a bat, suggest that Köhler's expectations for his chimpanzees were extraordinarily optimistic. There is little reason to suppose that any animal is capable of instantaneously discovering a new solution to a problem, when that solution requires the complicated manipulation of strange objects to form novel tools. All that we know about the development of the young, both animal and human, suggests that extensive practice (albeit in a different context such as play) is usually necessary for the appearance of skilled behaviour. And all that we know about the life of young mammals reveals numerous opportunities to learn, both by play and by observation and IMITATION of others, those necessary skills.

One might argue, of course, that to appeal to the possibility of learning by observation or imitation in order to explain skilled behaviour is to concede a role to processes notably more complex than those of associative conditioning. But the argument does not seem wholly convincing. An animal learning something by observation of the behaviour of another may be simply associating together certain classes of events, i.e. stimuli provided by the behaviour of the other animal with the consequences of that behaviour. This may not be apparent without detailed experimental analysis. For example, long-term FIELD STUDIES of the behaviour of groups of Japanese macaques (*Macaca fuscata*) under semi-natural conditions in Japan has revealed the spread of certain feeding habits in particular groups. These include the practice of accepting paper-wrapped caramels from visitors, and washing the sand off sweet potatoes and wheat in the edge of the sea, or in streams running down to the sea. In each case, the behaviour originated with one or two individuals (the two latter cases with the same juvenile female monkey), and spread gradually through the group, beginning with the close companions of the inventor, and only slowly reaching such conservative members of the group as the adult males. There can be little doubt that the monkeys were imitating the actions of individuals with whom they came in contact, but the field study does not tell us what processes are involved in imitative learning. Laboratory experiments have shown that, in at least some cases, an observer monkey is better described as learning what are the consequences of the agent monkey's actions than as imitating the agent. An observer monkey can learn the solution to a discrimination problem in which one object is correlated with food and another is not, by observing for a single trial the performance of another monkey. The observer, however, is not in fact imitating the action of the agent, since if the latter makes a mistake, choosing the unrewarded object, the observer is actually slightly more likely to show that it knows the solution than if the agent had chosen the rewarded object. By seeing the consequences of the agent's choice, therefore, the observer learns which of the two objects is correlated with reinforcement, and when given the opportunity displays this knowledge by choosing appropriately.

Like studies of song learning or imprinting, therefore, experiments on observational learning keep distinct the occasion on which learning occurs from the occasion on which that learning produces a change in behaviour. In this respect they are also similar to typical experiments on *latent learning*, in which a well-fed rat is permitted to explore a complex maze at the end of which food is available. Since the rat is not hungry, it does not eat the food, and there may be little change in its behaviour at the time; but when later made hungry, it shows what it has learned during this period of exploration by choosing the correct path through the maze as soon as it is placed in the starting compartment. The phenomenon of latent learning posed problems for a restricted version of S–R learning theory, which assumed that learning may be reduced to the stamping-in of stimulus–response connections, but this is a view which would find few modern adherents. If we take the view that conditioning is a matter of associating events that are correlated in time, no new problems are posed by the phenomena of observational and latent learning.

Although imitation has sometimes been regarded by animal psychologists as an extraordinarily complex and mysterious process, evidence of great intelligence and insight, it is frequently dismissed by others as a mark of feeble intellect, unfavourably compared with creativity and originality. When John Stuart Mill wrote of the 'ape-like faculty of imitation', he did not intend this as a compliment: the implied contrast was not with yet simpler modes of behaviour, but with innovation and initiative. The contrast is surely valid: the inventor shows the way, the slavish herd of imitators follows. Animals too can behave in an original manner: consider the young female monkey who invented potato and wheat washing. We may understand little of the processes involved in creativity, but it does not seem likely that associative conditioning will be found to exhaust the possibilities for intelligent behaviour.

One way of approaching this problem it by use of COMPARATIVE STUDIES. Suppose that it was claimed that a particular form of apparently intelligent behaviour depended only on a process of simple associative conditioning. We might question the truth of this claim if we found that although chimpanzees, dogs, rats, and pigeons all showed rather similar behaviour in experiments on simple conditioning, only the chimpanzee displayed this particular form of intelligent behaviour. Implicit use was made of this type of argument earlier: the fact that habituation occurs virtually throughout the animal kingdom, whereas there is little satisfactory evidence of conditioning in many simple animals, suggests that habituation and conditioning constitute rather different forms of learning. A similar conclusion is suggested by the observation that chimpanzees have been successfully trained to manipulate symbols in a primitive form of sign LANGUAGE, while attempts to teach pigeons and dogs the sign language have foundered on the animals' failure to master the earliest stages of the training regime. The implication is that at least some of the processes involved in this simplified form of language are more complex than those involved in experiments on conditioning, on which chimpanzee and pigeon perform in a similar manner. This tortuously expressed and conservative conclusion will surely come as no surprise. Nobody, one is inclined to say, has ever supposed that the ability to use a language could be reduced to a process of conditioning. The answer, of course, is that several eminent psychologists have supposed exactly this. They may be misguided, but it is the task of science to find simple order in what appears to be great complexity, and, if nothing else, the attempt may help to clarify the nature of language and other forms of intelligent behaviour.

Conclusion. If nothing else, the reader should now have some appreciation of the variety of situations in which scientists have studied learning in animals. This variety is sufficient to make one pause before attempting to propose a single, monolithic principle to encompass all cases of learning. It seems improbable, for example, that there is very much overlap between the processes of sensitization and those of perceptual learning, or between habituation and language learning. In other cases, however, there may be more similarity between the different categories of learning than their separate treatment here might suggest: if habituation is a matter of recognizing familiar stimuli and discriminating them from novel stimuli, then there must be considerable overlap between the processes of habituation and those involved in more complex cases of perceptual learning.

In general, however, it is easier to be impressed by the difference between the various things animals can learn than by their similarities. All the examples considered here have conformed to the very general definition of learning noted at the outset: we infer that learning has occurred when we observe a change in an animal's behaviour and can attribute that change to a particular set of circumstances. But neither the changes in behaviour nor the circumstances are readily reduced to any simple, neat formula. The circumstances responsible for perceptual learning, for example, are quite different from those that produce associative conditioning. Although the set of operations that produces habituation is the same as that which produces sensitization, the behavioural changes are quite different. The term learning, we may therefore conclude, refers to a variety of processes whereby normally adaptive changes in individual behaviour may occur in response to a variety of environmental changes.   N.J.M.

80, 86, 95, 98

LEISURE. Animals can often be seen resting, or apparently doing nothing. Herring gulls (*Larus argentatus*), for instance, generally preen after feeding, and then stand around in groups, apparently resting, for 1 or 2 h. Such observations seem to challenge the general belief of biologists, that animals are designed by NATURAL SELECTION to make efficient use of their TIME.

An animal normally spends a certain amount of time each day FEEDING, sleeping, defending its TERRITORY, etc. The amount of time it needs to complete each task depends upon the prevailing environmental circumstances, and upon the animal's state of MOTIVATION. For example, when food is scarce, an animal will have to spend more time FORAGING than when food is plentiful. When an animal is reproductively motivated it may have to spend some time gathering nest material, and when it has young offspring it may have to spend some time in PARENTAL care. When the demands upon an animal's time are severe, certain aspects of its behaviour will be curtailed. These will generally be the less essential activities, such as SLEEP, GROOMING, and PLAY. When time is very short, some of these less important activities may be abandoned altogether. We can conveniently define a leisure activity as an activity which disappears from an animal's repertoire when the demands on the animal's time are very severe. These activities will vary from species to species, but they will generally include those aspects of the behaviour which have relatively low SURVIVAL VALUE. Thus we would not expect feeding to be a leisure activity, but we might expect play to drop out of the animal's repertoire in times of hardship. Animals normally procure food, and other necessities of life, as quickly and efficiently as possible in the prevail-

ing circumstances. In nature, COMPETITION tends to push animals towards the limits of their efficiency. However, in artificial environments, such as zoos and other forms of CAPTIVITY, the animal may have too little to occupy its time. The animal will generally have its food provided, and some normal activities, such as territorial defence, may be totally precluded. When an animal's daily activities take much less time than is natural it will tend to sleep more, and may develop symptoms of BOREDOM.

LEK is a type of TERRITORY held by males of certain species, and used solely as a communal mating ground. For example, the sage grouse (*Centrocercus urophasianus*) inhabits the high sage brush plains of western North America. The female has a brownish cryptic coloration while the male, when in breeding condition, is conspicuous, being larger than the female and having yellow air sacs, a white breast, and spiky tail feathers. Large numbers of males aggregate in the breeding season and take up breeding territories or leks, in which there is usually a dominant male and several subordinate males (see DOMINANCE). It appears that the females are choosing a suitable mate when they move around and between leks, prior to copulation. The majority of the copulations are performed by the dominant males. The males and females live apart most of the year and only the females care for the young. It seems likely, therefore, that the females are choosing mates largely on the basis of their apparent ability to hold leks (see SEXUAL SELECTION).

Another typical lek species is the ruff (*Philomachus pugnax*), in which there are two types of male bird. *Independent* males are aggressive, defend a small mating territory, and are generally dark in colour. *Satellite* males do not defend a mating territory, are not aggressive, and often have a conspicuous white ruff around the neck. It has been suggested that the plumage of satellite males attracts females to the lek, and so it is advantageous for independent males to tolerate their presence. The independent males compete for mating opportunities. The satellite males attach themselves to particular independent males, and steal copulations while the independent male is busy defending his territory. Thus both types of male perpetuate themselves genetically.

Leks are found in many species of grouse (Tetraoninae); also among pheasants (Phasianinae), hummingbirds (Trochilidae), some bowerbirds (Ptilonorhynchinae), some manakins (Pipridae), and in some weavers, such as the village weaver bird (*Textor cucullatus*). Many mammals have communal mating grounds, but the lek type, with an element of female choice, is rare, although it has been observed in the Ugandan kob (*Adenota kob thomasi*). In polygamous species leks serve the important function of ADVERTISEMENT. They are often situated at prominent sites, and contain large numbers of conspicuous displaying and vocalizing males. It has been observed that there are more females per male in large colonies of the village weaver than in small colonies, and it is likely that females are generally attracted to larger colonies, in which they have a greater choice of mates.

LOCOMOTION. The study of locomotion is generally concerned with VOLUNTARY movements which displace the whole body. Some animals are restricted to one particular type of locomotion, while others have alternative possibilities, such as walking and FLIGHT. Students of behaviour are usually concerned with such questions as why particular patterns of movement are used, whereas the details of the COORDINATION of muscles are generally the province of physiologists.

Movement in water. 1. *Passive movement.* Many small organisms which float free in water, rather than resting on the bottom, have little or no control of their movements, but are carried hither and thither as the water moves. A great many of the small animals of the plankton, even those which have means of locomotion, are moved more by water movements than by their own efforts.

2. *Undulating swimming.* A great many aquatic animals propel themselves through water by eel-like movements. The body, or part of it, is thrown into waves which then move backwards along it, propelling the body forwards. The diagram (Fig. Aa) shows a protozoan less than 0.1 mm long being propelled by an undulating *flagellum*, but a sequence of pictures from a film of a tadpole swimming would look very similar. Undulating movements propel animals, because a bar moving through water experiences less resistance if it is moved lengthwise than if it is moved broadside on. When the pictures from Fig. Aa are superimposed (Fig. Ab) it is apparent that any section of the flagellum (such as the section shown thicker

than the rest) is moved transversely while inclined at an angle. Since it can move more easily lengthwise than broadside on, it moves forward as shown by the continuous arrow in Fig. Ac, rather than directly transversely as shown by the broken arrow. This is the principle which propels animals with smooth undulating flagella or tails, from small Protozoa (unicellular organisms) to large eels (Anguilliformes) and water snakes (*Natrix*). However, the forces which the water exerts on the body are mainly due to its viscosity in the case of the Protozoa, and mainly due to its inertia in the case of the larger animals. The difference is due to the differences of size and speed.

Some Protozoa have 'hairy' flagella instead of smooth ones (Fig. Ae). When these are undulated the animal is propelled in the same direction as the waves: smooth flagella propel the animal in the opposite direction. The difference arises because the hairs are long and numerous enough to have more influence on the movement of the flagellum than its main strand: when the flagellum is moved transversely inclined at an angle, it moves as indicated by the continuous arrow in Fig. Ad. Many polychaete worms, such as the ragworm *Nereis*, swim by undulation, and have structures called *parapodia* projecting from the sides of the body which make them travel in the same direction as the waves.

Eels and many other fish depend on undulating movements of the whole body for fast swimming, but in many fish the main propulsive effect is due to the hydrofoil action of the tail, which is explained in a later section. Undulation of individual fins is also important, particularly for slower movements. Many bony fishes (Teleostei) have about the same density as the water they live in, and can manœuvre in remarkably varied and precise ways by undulating their fins. Some of the

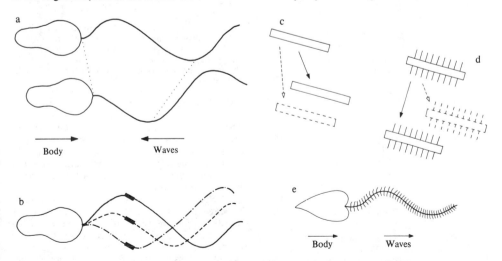

Fig. A. Diagrams illustrating swimming by flagellate protozoa. **a**, **b**, **c**. A flagellate with a smooth flagellum, such as *Strigomonas*. **d**, **e**. *Ochromonas*.

possibilities are illustrated in Fig. B. Seahorses (*Hippocampus*) and a few other fish do not use body undulations but depend entirely on fin undulation, even for the fastest swimming they can manage. Cuttlefish (*Sepia*, which are not true fish but molluscs related to the octopus) also swim by undulating fins.

3. *Rowing*. In rowing, an oar blade is moved backwards through the water, driving the boat forwards, and is then lifted clear of the water for the return stroke. The same effect could be achieved (less efficiently) without removing the oars from the water, if the blades were held broadside on to their movement during the backward stroke, and edgeways on (feathered) during the return stroke. Animals which row vary in size from ciliated protozoans to swans (*Cygnus*). One of the ciliated protozoans is shown in Fig. Ca. It is covered by *cilia*, tiny hair-like structures which have the same internal structure as flagella. They beat in the manner shown in Fig. Cb. In the backward stroke (1–3) the cilium is more or less straight, moving broadside on through the water and exerting relatively large forces on it. In the forward stroke (4–7) it bends so as to move mainly lengthwise through the water and exert as little force on it as possible. This action tends to drive the animal towards the left of the diagram.

Fig. Cc and d shows the rowing action of a water

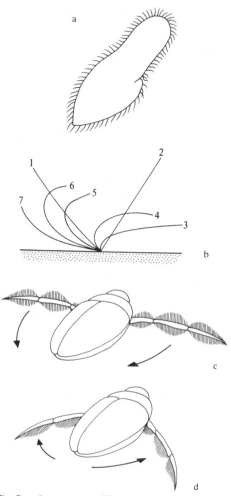

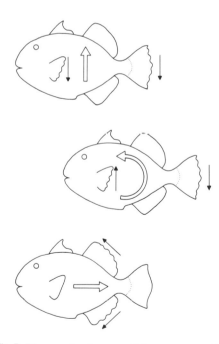

Fig. B. Diagrams showing some of the movements which can be made by teleost fish (especially perch-like fish, Perciformes) by undulating their fins. Thin arrows show movement of the waves along the fins, and thick ones show the resulting movement of the fish.

Fig. C. a. *Paramecium*, a ciliate protozoan about 0.2 mm long. **b.** A diagram showing successive positions of a beating cilium. **c, d.** Diagrams showing a water beetle swimming. The first two pairs of legs have been omitted. Cc shows the third pair of legs making their backward stroke and Cd shows the forward stroke.

beetle (Dytiscidae). The third pair of legs (the only legs shown in the diagram) are the principal oars. They are fringed by long hairs which are hinged to them in such a way as to spread out in the back stroke (c) and trail behind in the return stroke (d). They are broadside on in the backward stroke, but they move lengthwise in the forward stroke, so the beetle is propelled forwards.

4. *Hydrofoil swimming*. Imagine a flat plate moving through water, tilted at an angle to its direction of motion (Fig. Da. The angle α is called the *angle of attack*). The force which the water exerts on it has a component called *drag*, which acts backwards along the direction of motion, and a component called *lift*, which acts at right angles to the direction of motion. In the diagram the

plate moving towards the right drives water downwards so the lift on it acts upwards. The lift is zero when the angle of attack is zero, and has its maximum value, for a given speed of movement, when the angle of attack is about 20°. At higher angles of attack 'the phenomenon called stalling occurs, and the lift is less than it would otherwise be. The lift is zero again when the angle of attack is 90°.

Lift on the wings of an aeroplane keeps the aeroplane airborne, and lift on the blades of a propeller can be used to provide the thrust which drives it forward through the air. Similarly, lift on propeller blades can be used to drive ships through water, and lift on fins or flukes can be used to drive animals through water. Structures designed to produce as much lift as possible for as little drag as possible are called aerofoils if they are intended for use in air, and hydrofoils if they are intended for use in water. They are generally streamlined and asymmetrical in cross-section, as shown in Fig. Db.

Tunnies (*Thunnus*) and many other fast-swimming fish are propelled by hydrofoil tails (Fig. Dc and d). The tail beats from side to side as the fish moves forward, so it takes a sinuous path through the water (Fig. Dd). Its angle of attack is adjusted so that in every stroke the resultant of lift and drag acts somewhat forward. As the tail moves to the right, the resultant acts forward and to the left: as it moves to the left, the resultant acts forward and to the right. The components to left and right cancel each other out, and the net effect is a forward force, driving the fish through the water.

Whales (Cetacea) swim in the same way, but their tail flukes are horizontal and are moved up and down, not from side to side. The flippers of marine turtles (Chelloniidae) also serve as hydrofoils; they beat up and down with appropriate adjustments of angle of attack, so as to propel the turtle forwards. Penguins (Spheniscidae) and auks (Alcidae) swim with their wings in the same way. No really small animals depend on hydrofoils for swimming.

5. *Jet propulsion.* Squids and octopods (Cephalopoda) swim by jet propulsion. They draw water through a wide opening into a cavity which has muscular walls, and then squirt it out again through a much smaller opening, the funnel. They can point the funnel forwards or backwards: when water is squirted forwards from the funnel the animal is propelled backwards, and when it is squirted backwards the animal is propelled forwards. Dragonfly larvae (Anisoptera) also use jet propulsion in their escape movements: they squirt water from the anus to propel themselves forward.

6. *Buoyancy.* Many of the materials of which animals are made are denser than either fresh water or sea water. Many aquatic animals, for instance most sharks (Selachii), are denser than water, and can only prevent themselves from sinking by swimming upwards, or by means of hydrofoils, such as the pectoral fins of sharks, which extend

laterally like aeroplane wings and serve a similar function. Many other aquatic animals have buoyancy organs which adjust their density to that of the water. Most teleost fish have gas-filled bags (*swim-bladders*) in the body cavity. The cuttlebone of cuttlefish is another buoyancy organ, filled largely with gas. Some other aquatic organisms gain buoyancy from low-density organic compounds, such as the hydrocarbon squalene in the livers of some deep-sea sharks. Others have light ions replacing some of the heavier ions in their body fluids, or in the jelly of jellyfish (*Aurelia*).

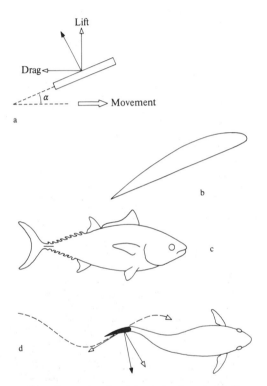

Fig. D. a. Forces on a flat plate moving through a fluid. The plate is seen in section. **b.** A section through a typical hydrofoil. **c.** A tunny. **d.** A diagram of a tunny swimming, seen from above.

Movement at the surface of water. There are a number of animals, including ducks (Anatinae) and men, which swim at the surface of water, with only part of the body submerged. Ducks row themselves with their feet. The 'dog paddle' used by terrestrial quadrupeds and the breast stroke of man are rowing techniques in which all four limbs are used as oars. Ducks float very high in the water because of the air trapped between skin and feathers, and in the large air-sacs connected to their lungs.

Some insects walk on the surface of water, supported by surface tension, but only very small animals can do this.

Movement on solid surfaces, without legs. 1. *Amoeboid movement.* Fig. E shows how *Amoeba* crawls. It is a single cell, so there are no partitions across the body. The outer parts of the cell contents are jelly-like, but the core is fluid. The fluid

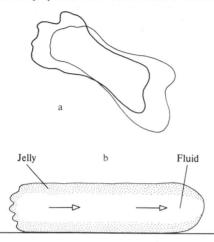

Jelly   b   Fluid

Fig. E. a. Two successive positions of a crawling *Amoeba*. **b.** A diagrammatic section through a crawling *Amoeba*.

flows forward while the jelly remains stationary. (The nature of the forces which propel the fluid is not known with certainty.) Fluid is converted to jelly at the leading end of the body, and jelly is converted to fluid at the other end, and so the whole animal moves forward.

2. *Ciliary crawling.* Small flatworms (Platyhelminthes) glide over submerged surfaces by the action of cilia on the underside of the body. The cilia beat in the manner illustrated in Fig. Cb, and propel the worm.

3. *Concertina crawling.* This section is about various techniques of crawling used by segmented worms (Annelida) and gastropod molluscs (Gastropoda), and various other animals. In all of them the body (or at least the part of it which rests on the ground), alternately lengthens and shortens. In the simplest technique, the whole body lengthens and shortens. This is how leeches (Hirudinea) crawl (Fig. Fa). There is a sucker at each end of the body. The suckers attach by turns as the body lengthens and shortens. The body lengthens while the posterior sucker is attached and shortens while the anterior one is attached, and so the animal progresses.

The same principle, of lengthening in front of an anchored region and shortening behind one, is used by many worms. Fig. Fb shows the technique used by earthworms (e.g., *Lumbricus*). The body is divided into a large number of segments, one behind the other. The body cavities of successive segments are separated by partitions. Each segment can be made long and thin, or short and fat, but its volume remains constant. At any instant

during crawling, alternate groups of segments are contracted and elongated. At the front of each contracted group the segments are elongating, and at the back they are contracting, so waves of muscle action travel backwards along the body. The contracted regions are effectively anchored, partly because being fat they support most of the weight, and partly because a few bristles (chaetae) project obliquely from each segment, so as to allow forward sliding but prevent backward sliding.

Fig. Fc shows a different technique used by a few marine worms such as *Polyphysia*. These worms have no partitions across the body cavity, so fluid can be displaced from one region to another. Regions of the body are alternately short and thin, or long and fat. The fat regions are effec-

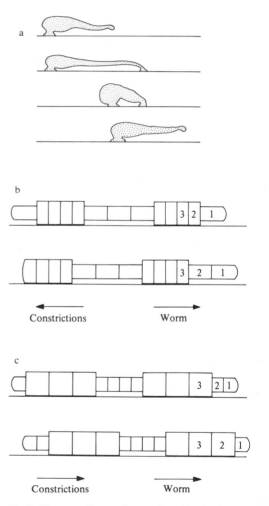

Fig. F. Diagrams of concertina crawling: (**a**) a leech, (**b**) an earthworm, and (**c**) a worm such as *Polyphysia*. In each case, two or more successive positions are shown.

tively anchored. Lengthening occurs immediately in front of them, and shortening immediately behind, so the animal moves forward, but the waves of muscle action also travel forward.

Snails and other gastropod molluscs have a large flat 'foot' which rests on the ground. Waves of muscle contraction travel along it, throwing the sole into ripples. Forward movement is caused in some species by ripples which travel backwards and in others by ripples which travel forwards. In the first case the longitudinal muscles are shortest at the crests of the ripples (which are stationary in contact with the ground). In the second they are shortest in the troughs (which are raised off the ground). Though the waves of muscle action are restricted to the foot the principles are precisely the same as in the two types of crawling used by worms, shown in Fig. Fb and c. The larger flatworms also crawl on moving ripples, and so (very slowly) do sea anemones (Actinaria).

4. *Serpentine crawling.* Snakes (Serpentes) use various techniques of crawling of which serpentine crawling, shown in Fig. G, is the most usual. The body is thrown into waves which travel backwards along the body. Stones, tussocks of grass, and other irregularities in the ground prevent sections of the body from sliding broadside on across the ground, but allow them to slide lengthwise, and so the body moves forward. The principle is closely related to the principle of undulating swimming as used by eels, which involves very similar movements. *Sidewinding* is a variant of serpentine crawling used by rattlesnakes (*Crotalus*) and some other snakes, especially on loose sand.

Movement on legs. Most terrestrial vertebrates run on legs, and so do most arthropods. The number of legs used for running varies from two (in birds, man, kangaroos, and a few other mammals) to more than a hundred (in some millepedes). It will be convenient to consider first two-legged animals, and then animals with larger numbers of legs.

1. *Two legs.* A three-legged stool is stable but a two-legged stool is not. Three is the smallest number of small feet which will give stability. Resting kangaroos (Macropodidae) set the tail on the ground to serve as a third support. Men and birds can stand on two feet, or even on one, without performing any remarkable feat of balancing, because their feet are large.

A vehicle on wheels travelling at constant velocity over level ground has constant potential energy and constant kinetic energy. The engine need only deliver enough power to overcome the drag which the air exerts on the moving vehicle, the rolling friction (due to distortion of the tyres), and friction in the transmission system. An animal moving on legs, however, inevitably rises and falls (gaining and losing potential energy) or accelerates and decelerates (gaining and losing kinetic energy) or both. These fluctuations occur in every stride. If the total energy of the animal falls at one stage in

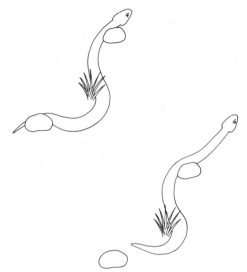

Fig. G. Two successive positions of a crawling snake.

the stride it must be replenished at another, by muscular work. The power required for this purpose is substantial, which is why running uses more power than cycling. A running man uses oxygen (and energy) twice as fast as a man cycling at the same speed.

There are two basic types of gait illustrated by the human walk and run. Fig. Ha is a simplified and idealized representation of walking. The right foot is set on the ground when the hip is at A. The leg is kept straight, so the hip moves in a circular arc through B to C. When it reaches C the right foot is lifted and the left foot is set down. The centre of mass of the body rises as the hip moves from A to B, and falls as it moves on to C. Hence the potential energy is greatest when the hip is at B, immediately over the foot. However, the force on the foot keeps in line with the leg: it slows the body down while the hip is behind the foot between A and B, and speeds it up while the hip is in front of the foot, between B and C. Hence the body is moving most slowly and has minimum kinetic energy when the hip is at B. The kinetic energy is least while the potential energy is greatest, and vice versa. Energy is converted from the kinetic to the potential form and back again in the course of the step. There is little change in total energy, so the muscles do not have to do much work. Energy is similarly converted back and forth between kinetic and potential forms as a pendulum swings: the pendulum bob is highest and has most potential energy at the extremities of its swing, but travels fastest and has most kinetic energy in the middle of the swing.

The style of walking illustrated in Fig. Ha is very economical of energy at low speeds, but it is impossible at speeds greater than $\sqrt{gh}$, where g is the acceleration of free fall, and h is the length of the leg. Walking at higher speeds would involve the impossible feat of falling with acceleration greater than g. Adult men have legs about 93 cm long, so for them $\sqrt{gh}$ is about 3 m/s. Athletes achieve speeds up to 3·7 m/s in walking races by using a technique significantly different from the one which has been described. They move their hips in a way which reduces the vertical movements of the centre of mass.

Even ordinary walking is a little different from the idealized gait represented in Fig. Ha. The leg is not kept absolutely straight while the foot is on the ground. One foot is set down a little before the other is lifted.

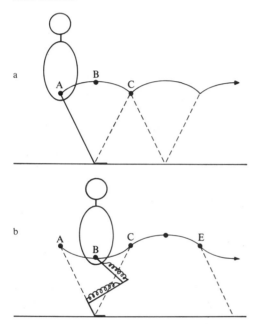

Fig. H. Diagrams of (**a**) walking and (**b**) running.

Running is illustrated in Fig. Hb. In walking there is always at least one foot on the ground, but in running there are times when neither foot is on the ground. In the diagram the right foot is set down on the ground when the hip is at A, and remains on the ground till the hip reaches C. At that stage it leaves the ground, and there is no foot on the ground until the left foot is set down when the hip reaches E. The centre of mass is lowest when the hip is at B, and highest when the hip is at D. The velocity is lowest when the hip is over the foot at B, and highest when there is no foot on the ground. Thus potential and kinetic energy both have their minimum values when the hip is at B, and there is no possibility of saving energy by converting it back and forth between the potential and kinetic forms. Instead, potential and kinetic energy are converted to elastic strain energy, which is later recovered in an elastic recoil. The diagram shows

springs in the legs which are stretched as the leg bends. They are most stretched, and have most energy stored in them, when the hip is at B. A man running is like a child bouncing along on a pogo stick. The diagram shows springs, but the elastic structures which are stretched in a running man are the tendons of the leg muscles.

Running is possible at any speed, but it is only at speeds above about 0·8 $\sqrt{gh}$ (2·5 m/s for adult men) that it is more economical of energy than walking. Men usually change from walking to running at about 2·5 m/s, though it is possible to walk a little faster than this.

Ducks walk in essentially the same way as men, but they rock from side to side (waddle) because their hips are wide apart. The slow run of quail (*Coturnix*) is different: there are stages when both feet are on the ground, but the centre of mass is lowest as it passes over the foot. Some lizards (*Lacerta*) when running fast are on their hind feet alone.

Bipedal hopping is just like running except that the feet are set on the ground simultaneously. It is used by many birds and also by kangaroos and a few other mammals.

2. *Four legs: mammals.* If an animal with four small feet is to be stable at all stages in its stride, it must always have at least three feet on the ground. Moreover, it must move its feet in appropriate sequence, so that a vertical line through its centre of mass is always within the triangle formed by the supporting feet: otherwise it will overbalance. The only sequence which allows stability when each foot is off the ground for a quarter of the time is the one shown for walking in Fig. I. Walking is the slowest quadrupedal gait, and all faster gaits involve unstable phases when there are two, one, or no feet on the ground.

All quadrupeds which have been observed move their legs, in walking, in the sequence shown in Fig. I. In at least the larger quadrupedal mammals, such as dogs (*Canis*) and horses (Equidae), walking seems to be mechanically equivalent to the human walk. Energy is converted in the same way between the kinetic and potential forms. A horse walking is like two men walking one behind the other, a quarter of a cycle out of step.

The trot is like two men running one behind the other, half a cycle out of step. The *rack* is like two men running in step. In each case, as with running, energy is saved by elastic storage in tendons. Most mammals walk at speeds up to about 0·8 $\sqrt{gh}$, trot at speeds from about 0·8 $\sqrt{gh}$ to about 1·5 $\sqrt{gh}$, canter at slightly higher speeds, and gallop at still higher speeds. Giraffes (Giraffidae) and camels (Camelidae) rack instead of trotting, and horses can be trained to rack.

In galloping, the two forefeet are set down together (or in rapid sequence), and similarly the two hind feet. The back bends at one stage of the stride and straightens at another (Fig. J), lengthening the stride. Thus the muscles and tendons of the back have an important role in galloping, as well

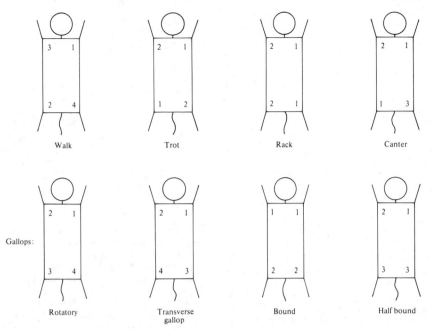

Fig. I. Diagrams of quadrupedal gaits. The numbers show the order in which the feet are moved. The gaits in the lower line are all versions of the gallop.

as the muscles and tendons of the legs. The four variants of the gallop shown in the lower line of Fig. I are characteristic of different groups of mammals. Antelopes (Bovidae) tend to use the rotatory gallop, horses use the transverse gallop, rabbits (Leporidae) use the half bound, and squirrels (Sciuridae) use the bound.

Dogs, horses, and other medium to large quadrupedal mammals stand with their legs rather straight, but rodents and other small mammals stand and move with their legs much more bent (Fig. Ka and b).

3. *Four legs: amphibians and reptiles.* However much they bend their legs, mammals stand with their knees and elbows close to the body, and their left and right feet close together. Newts (*Triturus*), lizards, and tortoises (*Testudo*) stand very dif-

ferently, with their feet much further apart (Fig. Kc).

Because their feet are so far to the side, newts and lizards are able to extend their stride considerably, by bending the body from side to side (Fig. La). The legs must move in diagonally opposite pairs, as in trotting. Note that the bending has the form of a standing wave (Fig. Lc), not a travelling wave such as is used by swimming eels and crawling snakes (Fig. Lb).

4. *More than four legs.* Insects stand with their feet far to either side of the body (Fig. Kd). They have to do this, because they are small: a breeze which causes no inconvenience to a dog would bowl an insect over, if it stood with its feet close together like a dog. Lobsters (*Homarus*) are relatively large but stand in similar fashion: if they

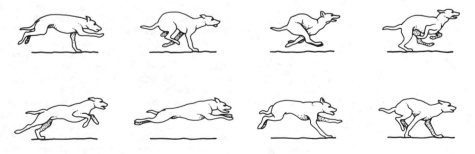

Fig. J. Outlines traced from cine film of a dog galloping.

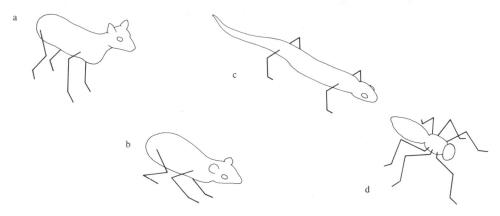

Fig. K. Diagrams showing the stance of (**a**) a large mammal, (**b**) a small mammal, (**c**) a newt or lizard, and (**d**) an insect.

did not, they would be apt to be bowled over by water currents.

Insects have six legs which are generally moved in sets of three, as shown in Fig. Ma. There are always three feet on the ground and the insect is always stable. Lobsters have eight legs (in addition to the pincers) and move them in rather irregular sequence. The left and right legs of each pair move in turn, but the order in which successive pairs move varies. Centipedes (Chilopoda) and millepedes (Myriapoda), with far more legs, move them with beautiful regularity. Waves of leg movement seem to travel along the animal, either backwards or forwards. The appearance is due to the legs of a segment moving immediately after (Fig. Mb and c) or immediately before (Fig. Md) those of the preceding segment. The difference between Fig. Mb and Fig. Mc is that in the former the legs of a pair move simultaneously, but in the latter

alternately. Some polychaete worms crawl with their bodies more or less straight, moving their parapodia as shown in Fig. Mc.

5. *Jumping.* Most jumping animals are either vertebrates or insects. Most of them jump, from a standing start, in the same way as men do: they bend their legs and then extend them rapidly. However, click beetles (Elateridae) jump by a jack-knife bend of the body.

Vertebrate leg muscles can do work in a single contraction up to about 150 J/kg of muscle. Locust (Acridoidea) leg muscle can probably do rather more. Imagine an animal with jumping muscles amounting to 10% of its body mass and capable of doing 150 J/kg. These muscles could in principle accelerate it from rest to a take-off velocity of 5·5 m/s, and make the centre of mass rise (in a vertical jump) 1·5 m. Bush babies (*Galago*) have very large leg muscles and can jump even higher than this, up to 2·26 m. Men cannot raise their centres of mass as much as 1·5 m, even in a running jump (an athlete need only raise his centre of mass about 1 m to get over a bar 2 m from the ground).

An animal which jumps by extending its legs must accelerate from rest to its take-off velocity over a distance rather less than the length of the extended legs. To reach the same take-off velocity a small animal must accelerate over a shorter distance, and so in a shorter time. Even to jump to its modest maximum height of about 4 cm, a rabbit flea (*Spilopsyllus*) must extend its tiny legs in only 0·001 s. This is made possible by a catapult mechanism. The leg muscles contract quite slowly, distorting blocks of elastic protein. Then a trigger mechanism is released and the protein makes a rapid elastic recoil which extends the legs. Locusts have a different catapult mechanism and can jump to a maximum height of about 30 cm.

Movement in trees. Squirrels and most monkeys (Simiae) run along the tops of branches in essentially the same way as they run on the ground. Perching birds hop along branches in the same way as on the ground. Apes (Hominoidea), especially gibbons (Hylobatidae), and spider mon-

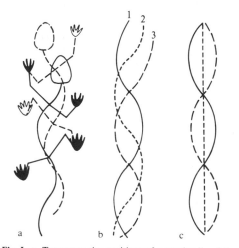

Fig. L. a. Two successive positions of a running lizard. **b.** A travelling wave. **c.** A standing wave.

keys (Atelinae) hang by their arms from branches, travelling through trees by swinging alternately from their left and right hands. This is called *brachiation*. Spider monkeys use their tails as well as their arms.

Vertical tree trunks present special problems. Fig. Na shows a woodpecker (Picidae) on a vertical trunk. Since the weight W acts vertically downwards the claws must be used to obtain an upward force W from the tree. The two forces W are not in line, but tend to rotate the bird clockwise. To counteract this the horizontal forces N are needed; the claws must exert a pull towards the trunk, and the tail must push on the trunk. The bird cannot climb the trunk unless the bark is rough enough to enable the claws to obtain the necessary pull. Fig. Nb shows the climbing technique of some other birds, such as nuthatches (Sittidae). The force on the upper foot has a component tending to pull the body towards the trunk, and the force on the lower one has a component tending to push it from the trunk. The climbing ability of squirrels depends on the same principle. To descend a trunk head first a squirrel must rotate its hind feet so that their toes point posteriorly, and their claws can exert the necessary pull. Cats (Felidae) cannot rotate their feet like this, and are therefore incapable of controlled head-first descent of vertical surfaces.

Movement underground. Earthworms burrow by the same technique as they crawl by (Fig. Fb).

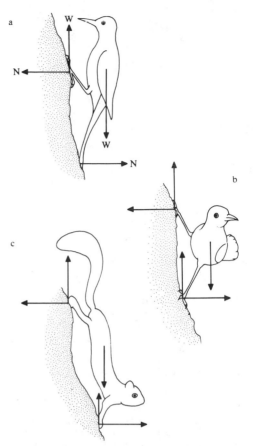

Fig. N. Diagrams showing the forces which act on animals resting on vertical tree-trunks: (**a**) a woodpecker, (**b**) another bird such as a nuthatch, and (**c**) a squirrel.

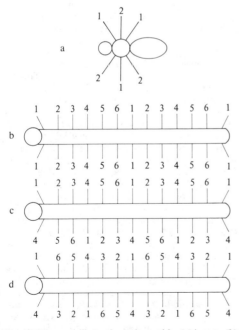

Fig. M. Diagrams illustrating gaits used by (**a**) insects, (**b**) millepedes, (**c**) a long-legged centipede, and (**d**) most centipedes. The numbers show the order in which the feet are moved.

Each segment advances while it is thin, so the anterior end can be insinuated into narrow crevices which are enlarged when the segments thicken. The thick segments are firmly anchored, by being jammed in the burrow. Bivalve molluscs (Bivalvia) burrow by means of a muscular foot which can be inflated with blood. The shell is allowed to open fairly wide so that it is jammed in the sand, and the foot is protruded (Fig. Oa). Then the foot is inflated so that it is anchored, the shell closes a little so that it is released, and the shell is pulled towards the foot (Fig. Ob). Thus the animal burrows, advancing the foot while the shell is anchored, and advancing the shell while the foot is anchored.

Comparisons. Flight is the subject of a separate article in this volume, but it seems appropriate to make a comparison here between flight and running. Many measurements have been made of the oxygen consumption of birds in flight, and of mammals running. It has been found for instance that budgerigars (*Melopsittacus undulatus*) in flight

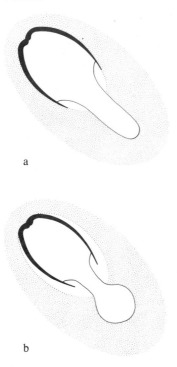

a

b

Fig. O. Diagrams of a bivalve mollusc burrowing.

use oxygen (and so energy) three times as fast as kangaroo rats (Dipodomys) of the same weight running at the highest speed at which they would run steadily. The budgerigars flew at about 10 m/s, but the kangaroo rats ran at no more than 0·5 m/s, so the energy used by a kangaroo rat running 1 km is almost enough for a budgerigar to fly 7 km. It seems to be a general rule that flying animals use much less energy per unit distance than running animals of similar size. This is why long MIGRATIONS are feasible for birds but not for mammals. Large birds such as storks (Ciconiidae) migrate by soaring, which is particularly economical, instead of flapping flight. Swimming fish travel even further for given energy cost than flying birds, but they travel as slowly as running mammals. Trout (Salmoninae) of the same weight as budgerigars cannot maintain speeds above about 0·5 m/s.   R.M.A.
2, 23, 49, 55, 83, 140

LORENZ, KONRAD ZACHARIAS (1903–   ), was one of the founders of ETHOLOGY. His most important practical research has been on the behaviour of various species of birds, and through this work he formulated many of the problems that have come to dominate ethology. Lorenz also developed a grand theory to explain behaviour, his theory of INSTINCT, which has been the focus of the most important controversy in the history of ethology. He is known to a wider audience

through his popular books, some of which contain controversial applications of Lorenzian ethology to human behaviour.

Brief life. Konrad Lorenz was born in Altenberg in Austria, the son of a famous surgeon, Adolf Lorenz, who had invented a method of curing a congenital hip disorder and had accumulated a small fortune by conducting the operation. For this discovery, Adolf Lorenz was proposed for, but did not receive, the Nobel Prize for medicine. From his childhood, Konrad Lorenz kept and reared many animals and was curious about their behaviour. After leaving school, he studied briefly at Columbia University (USA) in 1922, but soon returned to Austria. He then joined the anatomy department of Vienna University and obtained a medical degree, after which he remained attached to the department and also took time to observe the behaviour of jackdaws (*Corvus monedula*). Lorenz did his most important ethological work during the 1930s at the family home in Altenberg, while unattached to any institute. In 1940 he was appointed Professor of Philosophy at Königsberg, but soon afterwards he was conscripted as an army doctor. He was taken prisoner of war by the Russians in 1944, and released in 1948. He then worked in a variety of places, but mainly at Buldern (attached to the University of Münster), until establishing himself at Seewiesen in 1956. He remained at Seewiesen until he retired to Altenberg in 1973. In the same year he shared the Nobel Prize with Niko TINBERGEN and Karl von FRISCH.

Lorenz's theory of instinct. Ever since his earliest papers in the 1930s Lorenz has maintained that there is a fundamental distinction between the INNATE and the learned; however, his conception of the distinction has undergone several developments.

At university, Lorenz studied post-Darwinian comparative anatomy: his doctorate, for example, was a COMPARATIVE STUDY of the mechanisms of FLIGHT, and the ADAPTATIONS of wing form in a variety of bird species. (This was later published as *Der Vogelflug*, 1965.) Comparative anatomy aims at description of how structures develop, how they work, and how they evolve; its method is the comparison of structures in different species. Lorenz realized that activities could be treated like any anatomical organ, and he was soon investigating behaviour by the methods of comparative anatomy. This insight was a crucial step in bringing the study of behaviour into the Darwinian synthesis that was being forged in the 1930s. The influence of comparative anatomy is particularly overt in his early definition of instinct, taken from H. E. Ziegler: 'I have defined the difference between instinctive and intelligent actions in the following way: the former are based on *inherited pathways*, the latter on *individually acquired pathways*. Thus a *histological definition* replaces a psychological definition' (*Journal für Ornithologie*, 80, 1932). At this stage, Lorenz regarded instincts as REFLEXES. Instincts were to be thought of as inherited neuronal structures which were unmodified by the environment during development. In about 1933 he was encouraged by his teacher, Karl Bühler, to study previous authors on the subject of instinct. The resulting analysis, a paper of 1937, contained a minor addition to his theory of instinct: the idea of APPETITIVE behaviour. He now thought of instincts as a particular kind of reflex: those preceded by a period during which the animal actively seeks out the conditions for the instinct to be evoked. An animal does not just show feeding movements in the presence of food, but also seeks places where it can feed. This SEARCHING behaviour would be called appetitive. Lorenz used his theory of instinct in his masterpiece on the SOCIAL behaviour of birds, 'Der Kumpan in der Umwelt des Vogels'. (*Journal für Ornithologie*, 83, 1935. Translated as 'Companionship in Bird Life', in C. H. Schiller, ed., *Instinctive Behaviour*, 1957.) This paper was based on his personal observations of thirty species of birds, and it analysed the functions of, and conditions evoking, behaviour shown by parents, offspring, SEXUAL companions, and other social companions. Lorenz also tried to explain the differences in behaviour of the various species, making this paper a fine example of the comparative method. Elsewhere, he gave examples of behavioural characters that could be used in taxonomy, thus completing his analogy between comparative anatomy and the comparative study of behaviour. His best example of behavioural CLASSIFICATION comes from pigeons (family Columbidae): there is no anatomical character that defines this family, but they all make sucking movements while drinking.

An important component of Lorenz's method of studying avian behaviour was the IMPRINTING of the birds on to himself. In many bird species, the newly hatched young directs its filial responses to the first moving object that it sees after HATCHING. Usually this object would be the bird's mother; but if the egg is artificially incubated, the young bird can be imprinted on to a person. Lorenz imprinted birds on to himself so that he could study their behaviour more closely. It also gave him an opportunity to observe differences between species in the process and consequences of imprinting. For example, he found that goslings of the greylag goose (*Anser anser*) become imprinted on the first moving object that they see, whereas newly hatched mallard ducklings (*Anas platyrhynchos*) will only imprint on an object that makes the appropriate duck-like noises. Imprinting of some species, such as jackdaws, on human beings disturbs the bird's subsequent sexual behaviour (it becomes directed at humans, not jackdaws). In greylag geese, however, imprinting on humans does not alter their sexual behaviour from that normally practised by the species. One of the main reasons why greylag geese have featured in so many of Lorenz's studies is that they can be easily imprinted on human beings without pathological side effects.

The next development in Lorenz's theory of instinct came when he was inspired by some of von Holst's work. Von Holst had demonstrated the existence of spontaneously occurring RHYTHMS of activity in the nervous system of animals. In a classic paper (1938) written in collaboration with Tinbergen, on the egg retrieval response of the greylag goose, Lorenz incorporated von Holst's discovery into the theory of instinct. If the animal continually produces internal nervous activity, this could explain two observations about behaviour: first, if a stimulus is withheld, then the level of that stimulus necessary to evoke a response gradually falls, and second, a VACUUM ACTIVITY occurs (i.e. an activity produced in the absence of its normal goal, such as pecking in the absence of food). Both these observations could be explained by postulating a continual build-up of ENERGY or DRIVE for that activity; as the energy built up, it became progressively easier for it to be released. Initially Lorenz supposed that this energy was *action-specific*; but his observation of DISPLACEMENT ACTIVITIES led him to reject this view. A displacement activity is one that appears to be functionally inappropriate in its context, and which appears when the animal has conflicting behavioural tendencies. For example, in an aggressive encounter different factors may be stimulating an animal both to approach and to retreat from its opponent; in this case a bird might wipe its beak, or preen. If the internally produced nervous 'energy' was action-specific, it would not be expected to cause other actions, such as displacement activities. (Lorenz tinkered with his model to account for displacement activities, but subsequently abandoned, or, at least, became less enthusiastic about energy models of drive.) The other addition to the theory during the late 1930s was to point out that instinctive behaviour, once stimulated, is completed without any further FEEDBACK from the environment.

So, from initially conceiving of instincts as simple reflexes, Lorenz built up a theory of instinct that increasingly emphasized internal causes of behaviour. This contrasted with the then common view that behaviour should be understood as sequences of stimuli and responses, with environmental feedback between each activity. However, Lorenz's theory of instinct was not only intended to explain the EVOLUTION and causation of behaviour. It could also help to explain DEVELOPMENT (see also ONTOGENY). Instincts were said to be inherited, and to develop independently of the environment. Lorenz regarded isolation experiments (in which an animal is reared in isolation from some environmental factor) as the best source of evidence for whether a particular activity was innate or learned: if the activity developed even when the animal was reared in isolation, then that activity was probably innate. However, Lorenz himself hardly ever did carefully controlled isolation experiments, but relied instead on less direct evidence. For instance, he argued that instinctive behaviour tends to be stereotyped and similar in all members of a species, and that instincts tend to be evoked by relatively simple stimuli.

Lorenz has been strongly criticized for the developmental connotations of his theory of instinct, especially his statements that behaviour itself is, in some cases, inherited and unmodified by the environment. Several animal psychologists (most notably Daniel Lehrman in 1953) pointed out that Lorenz's idea of instinctiveness can be very obstructive. The inheritance of an organism only expresses itself as behaviour, or anything else, by its interaction with the environment. It can be misleading, therefore, to describe an activity as unmodified by the environment, because environmental effects are the focus of interest in some studies of development. Lorenz's response to these criticisms can be detected in changes of emphasis; but he has never admitted that his original theory of instinct contains any major errors. In his small but important book *Evolution and Modification of Behaviour* (1965), he no longer describes behaviour as innate or learned. What the animal inherits is not an activity, but a 'limited range of possible forms in which an identical genetic blueprint can find its expression in phenogeny' (ibid.). Furthermore, 'No biologist in his right senses will forget that the blueprint contained in the genome requires innumerable environmental factors in order to be realized in the phenogeny of structures and functions' (ibid.). Thus Lorenz denies that describing something as inherited has any implications for how the environment mediates during its development. He is not even interested in analysing the various environmental influences on development, but what he does want to understand is the fact of adaptedness. Adaptedness implies that the animal has some kind of information about the environment. (In this language one would say that the fact that a bird pecks at food is due to its having information about what food looks like.) Now, Lorenz argues that there is a dichotomy between the sources from which an animal acquires information about its environment. On the one hand, there is phylogenetically (evolutionarily) acquired information, which is due to NATURAL SELECTION operating over generations. On the other, there is individually acquired information, which could be obtained by LEARNING, for example. Learning itself, however, is not haphazard or arbitrary, but adaptive, a fact that can only be explained by innate teaching mechanisms that are 'in existence before all learning in order to make learning possible' (ibid.). So demonstrating an environmental influence, such as learning, on development does not mean that the trait is unadaptive, or that it is independent of heredity (see GENETICS).

If the word 'innate' is taken only to refer to phylogenetically acquired information, then it is a valid consequence of our explanation of adaptation by natural selection. The dichotomy of 'sources of information' is, however, importantly different from the innate–learned dichotomy as it was understood by Lorenz's earlier critics.

Popular and human ethology. Lorenz's first, and most enduringly popular, book was *King Solomon's Ring* (1952, a translation of *Er redete mit dem Vieh, den Vögeln und den Fischen*, 1949). It describes in simple language many of his earlier observations on birds and other animals.

From about 1960 Lorenz has become increasingly interested in human ethology. He has pursued this theme in several popular books, and his ideas have reached a wide audience, usually provoking impassioned controversy. He believes that writers on human behaviour have hitherto overlooked the importance of instincts. *On Aggression* (1966, a translation of *Die Sogenannte Böse*, 1963) notoriously argues that there is an aggressive instinct in human beings as well as in other animals. AGGRESSION will therefore occur spontaneously unless the instinct can be redirected into a less bloodthirsty substitute. These speculations have been criticized, not least because of their ideological content; but the debate has tended to be unscientific and misdirected. In *Civilized Man's Eight Deadly Sins* (1974) Lorenz repeated among other things his fear that civilization is dysgenic. (He had written on this in a paper of 1940 that contained subsequently retracted Nazi jargon; a mistranslation from this paper led to a false accusation of racism.) There are, however, no scientific arguments to justify this fear of genetic deterioration. In addition, Lorenz's claims that instinctive behaviour will inevitably and spontaneously appear are in marked contrast with the analysis of innateness presented in *Evolution and Modification of Behavior* (1965).

Lorenz wrote his first paper on epistemology (theory of knowledge) in 1941, and has always been interested in what Darwinism implies for the theory of human knowledge. For example, he identifies his idea of innate with Kant's notion of *a priori*. *Behind the Mirror* (1977, a translation of *Die Rückseite des Spiegels*, 1973) is a review of all the mechanisms by which animals acquire information about their environments, and also discusses how people are able to be knowledgeable.

110, 136   M.R.

M

MATERNAL BEHAVIOUR. It is characteristic of mammals that the females suckle their young from specially developed mammary glands, which produce a milk sufficiently nutritious to sustain the young during the early stage of life. This gives the female a special role in PARENTAL CARE that is not found in other groups of animals. It is, therefore, appropriate to restrict the term 'maternal behaviour' to mammals, and to use the general heading 'parental behaviour' when considering other animals.

Maternal behaviour is that behaviour, exhibited by mothers towards their young, which is presumed to aid the young in their survival, growth, and development, both physically and behaviourally. This is a functional definition of maternal behaviour, and one which is general enough to cover all mammals. The specific maternal behaviour exhibited by different species is quite varied. It depends upon a long history of EVOLUTION in different kinds of HABITATS, to which each species has become adapted in all of its behaviour, including its maternal behaviour. Seasonal differences in temperature, food supply, presence of predators, and availability of nesting places determine many features of maternal behaviour in each species, and give rise to common breeding seasons among species living within the same ecological zone.

The variety of patterns of maternal behaviour can be grouped into three types with a number of species falling into each type, and some species being difficult to classify. These types are determined mainly by the nature of the young at birth, their rate of development, and by the principal mode of food-gathering of the species.

One large group of mammals give birth to *altricial* young. These are helpless at birth, their eyes and ears are sealed, they cannot walk or even maintain their body temperature or eliminate (excrete) by themselves. They are entirely dependent upon the mother for food, THERMOREGULATION, elimination, and protection from predators. The mother takes care of these needs by providing milk and nursing the young, building a nest in which they can huddle together for warmth, retrieving them to the nest if they accidentally get out of it, licking them to stimulate elimination, and attacking any predator, or even any other member of the same species, that might harm the young. Rats and mice (Murinae), the golden hamster (*Mesocricetus auratus*), cat (*Felis catus*), dog (*Canis lupus familiaris*), and rabbit (*Oryctolagus cuniculus*) all exhibit this type of maternal behaviour.

Another large group of mammals, generally herd animals such as sheep (*Ovis*), goats (*Capra*), cows (Bovidae), deer (Cervidae), and horses (Equidae), give birth to *precocial* young. These can stand and walk at birth, can regulate their body temperature and eliminate by themselves, and their eyes and ears are open. These species are grazing animals, and so must constantly move from place to place to obtain food. Most of them live in large SOCIAL groups that move together from one feeding area to another (see HERDING). After birth it is dangerous for the mother and her young to remain apart from the herd, so there is only a short period of maternal care at a nest site. The mother and young then rejoin the group, and the young must follow the mother as she moves with the herd and grazes. In these species the mother learns to recognize her young very shortly after birth, and the young learn to recognize their own mothers. So, when they rejoin the herd they are not separated from one another, as each mother will care for only her own offspring. Maternal behaviour in these species initially consists of clearing a nest site, hiding the young, suckling, and defending the young against predators. Later it includes leading them and watching that they remain close to the mother when the herd is moving. The young begin to graze for themselves at quite an early age, but they continue to supplement this with suckling for some months after birth.

Among the primates, who exhibit the third type of maternal behaviour, the young remain clinging to the mothers's body from birth until several months of age, depending upon the species. The young are generally quite helpless at birth, but they can grasp the mother's fur and cling to her body, their eyes and ears are open, and they regulate their body temperature to a large extent. However, they are entirely dependent upon the mother for food. Most of these species live in social groups that FORAGE for food, and therefore must move from place to place, generally on a daily basis. Some species of primates live as individual family groups which stake out HOME RANGES, where they take up permanent residence, and within which they are able to collect sufficient food to raise their young.

In contrast to the clinging of most primate young, the human infant needs to be cradled during feeding, and requires other maternal care. Maternal behaviour among human beings alternates between cradling the infant and depositing it in a nest (the crib or some other type of 'nest'), although some cultures have adopted a means for the mother to carry her infant in a sling when she is not cradling it. In primates maternal behaviour consists largely of suckling the infant, cleaning it, transporting it from place to place, protecting it from predators, and sometimes from fellow members of the species, and keeping an eye on it so that it does not engage in dangerous activities.

The course of maternal behaviour in relation to the development of the young. Maternal behaviour changes as the young develop, declining as the young become capable of taking care of their own needs. In species that live in large social groups the young males generally join a JUVENILE group of the same sex, while the young females remain in association with the mother. In species that live semi-socially, or in small family groups, the young usually disperse and establish their own living areas. The males usually leave the mother earlier than the females and go further away. In some species, such as wolves (*Canis lupus*), the young remain part of the social group, which is essentially an extended family group, and join in HUNTING, but as the group gets too large to be supported by the local food supply it splits up into two smaller groups that separate and hunt in different areas.

Changes in maternal behaviour between birth and weaning of the young are synchronized with the behavioural and physical development of the young, and with their growing ability to function independently. In many species (e.g. rats and mice) this process may take place in a few weeks, but in others it may last several years, as in several species of primates. In the domestic cat the course of maternal behaviour may be divided into three main periods leading to weaning. During the first 3 weeks the mother takes the initiative in approaching the kittens to feed them, but during the second 3–4 weeks, the young take the initiative and approach the mother to suckle, and she assists them. As weaning begins in the sixth or seventh week the mother begins to avoid the kittens when they approach her to suckle and when they follow her around trying to suckle. She often leaves them and they then begin to find other sources of food, and thus they are gradually weaned. In the rat the same three periods are seen, but after the twenty-first day, in addition to avoiding the pups, the mother begins to actively reject them by pushing them away, and by picking them up and tossing them to one side. This grows more frequent after the fourth week, and the young begin to leave the mother shortly afterwards.

Among monkeys (Simiae), chimpanzees (*Pan*), and other primates, similar changes take place in the mother's behaviour towards her young. However, after the mother begins to reject the suckling approaches of the young she still allows them to remain near her, within arm's reach, and this type of association continues for a long time, until the young begin to leave the mother for long periods to explore places at a distance from her, and to join other young of the same age.

Maternal responses and the stimuli that elicit them. Although mothers are prepared to respond maternally to their young at parturition, it is the stimuli which the young provide that actually elicit specific maternal responses. These stimuli differ in different species, and the responses to them by mothers also vary in different species. Here we can describe only a few of the better-known stimuli and the maternal responses which they elicit (see RELEASER).

At birth the young are covered with birth fluids which elicit licking and cleaning by the mother. The umbilical cord is also licked and, in most mammals, when the placenta is delivered the mother begins to eat it, in the course of which she severs the umbilical cord and frees the new-born. These actions by the mother are believed to initiate the breathing of the new-born. In addition, several investigators have suggested that the mother begins to form her attachment to the young during parturition as a result of licking them. In the goat it has been shown that the licking of the kid during the first 5 min after birth, when the young are covered with birth fluids and need to be freed of birth membranes, stimulates the mother to become attached to her new-born, and, in addition, helps her to learn to identify it, and to distinguish it from other kids. If mothers are prevented from licking their new-born during these first 5 min, they will then reject their own kids, and any others that are offered to them, even if tested only 2 to 3 h after the birth. Stimuli other than the odours and taste of the birth fluids must also stimulate the goat mother, since mothers whose ability to smell is blocked and who are also prevented from seeing their new-born during parturition nevertheless become attached to them. Also, mothers who are prevented from smelling their kids when they are returned to them 3 h after birth will accept them and mother them if they have had 5 min contact with them at birth. Since there is a good deal of calling between mother and new-born at birth it is possible that the mother is stimulated by the new-born's calls, and distinguishes them from the calls of other kids.

Parturition in the cat has been studied in great detail, and these studies have revealed how the mother gradually shifts her ATTENTION from the many kinds of sensory stimuli she received from her own body during the birth (i.e. birth fluids, muscle contractions, emerging foetus) to the new-born as it emerges from the birth canal and is finally delivered. These studies suggest that initially the mother treats the emerging foetus as an extension of her own body, as the hamster mother also appears to do. The foetus possesses many stimuli in common with the mother's own body: the same birth fluids, wet fur, movement, etc., and during parturition the mother probably does not distinguish it from her own body. After delivery, however, it becomes more distinct, and the mother alternates between licking her own body and licking the new-born. Gradually her attention focuses on the new-born, and she begins to lick it and manipulate it with her forelimbs. With parturition over, her own body ceases to provide the unusual stimuli associated with parturition, but the new-born kitten continues to be a source of unusual stimulation: it is active and it retains many of the stimuli that were previously so attractive to the mother. In rats it has been shown that if foetuses with placentas attached are presented to mothers they are more likely to be adopted and reared than foetuses without placentas that have been washed and dried before they are presented to the mothers.

Among the new-born's activities which are most stimulating to mothers are those which are aimed at suckling. All mammalian new-born have nipple exploratory movements (i.e. nuzzling patterns) which they exhibit immediately after birth, and sometimes even during delivery. In altricial young this involves forward crawling and a side-to-side head movement which brings the nose and mouth into contact with the mother's nipples. In precocial young the nuzzling movement is usually an up-and-down head butting against the udder, or some other part of the mother's body. Of course these activities can only be performed by the new-born if the mother remains in contact with them after birth, which for most mammals is the case. The cat mother is exhausted after parturition, and lies down to rest with her body and forelegs encircling the litter. This gives the new-born the opportunity to search out her nipples, attach to them, and suckle for the first time, a process which is usually accomplished within an hour of parturition. The goat mother usually stands with her new-born, licking it, and moving her body into position for suckling, while the new-born begins nuzzling against her hind leg, then her belly, and finally reaches her udder, where it locates a nipple and soon attaches to it and suckles. This all occurs within an hour of the birth.

The mother receives many kinds of stimuli during the suckling of the new-born: there is tactual stimulation of the belly and of the nipples, thermal stimulation from contact with the new-born, and olfactory and taste stimuli when she sniffs and licks them. Moreover, the mother usually licks the anogenital region of the new-born during suckling, and consumes the urine and faeces, substances which appear to be attractive to her. The mother's attachment to the new-born, initiated during parturition, is certainly strengthened in the early hours after birth. It has been shown in the rat, in fact, that these stimuli are sufficient to establish attachment to the young. Females which were prevented from having contact with their young during parturition (by delivering the young surgically, under anaesthesia, by Caesarean section), upon recovery from the anaesthetic responded maternally towards the young, and established an attachment to them. Rats, and perhaps other mammals with altricial young, retain the ability to respond positively to new-born for a longer period after parturition than, for instance, goats.

Suckling is not the only stimulation provided by the young. In rats, mice, hamsters, and other species with altricial young, the new-born emit both audible VOCALIZATIONS and what are, to us, inaudible vocalizations (i.e. ultrasonic) when they are distressed either by being cooled or by painful or uncomfortable tactual stimulation. Depending upon the circumstances the mother may either pick up the pups in her mouth and carry them to the nest, or she may add to the existing nest; if she is out of the nest she may return to it, and suckling may follow. Goat young when they are distressed also emit audible sounds which the mother responds to either by calling in return, thereby signalling her position if the young is lost, or by approaching the kid and comforting it. Monkey mothers come to the aid of their young when they cry, as do human mothers, and they offer their bodies for comfort, or cradle the infant in their arms.

Many of the stimuli from the young which the mother responds to are not easily classified according to the sensory system, since it is the meaning of the stimulating situation, rather than the sensory system through which this meaning is conveyed, that determines the mother's response (see COMMUNICATION). The stimuli range from very subtle movements made by the young, to the mother's PERCEPTION of impending situations of danger to the young, and include interactions through gestures, and through FACIAL EXPRESSIONS.

There is a second type of maternal behaviour which consists of protective responses that are elicited by animals other than the young, but which are nevertheless maternal. Predators elicit aggressive responses from mothers, and these are basically maternal protective responses. The mouse mother will attack males that encroach upon the nest and litter; when she is not attending a litter she will not attack males. Maternal AGGRESSION is greatest at parturition, and declines afterwards in the mouse and in a number of other mammalian species. In several species of monkey the mother is very protective of her new-born, not permitting any other animal to handle it except a privileged 'aunt'. The aggressiveness of the newly maternal cat and dog is well known, even when the mothers were formerly HOUSEHOLD PETS with gentle natures.

It is not yet clear what changes occur in the young that elicit rejection responses from mothers of many species in the period preceding weaning. Not all species exhibit this change; among chimpanzees the transition to weaning is gradual and non-aggressive, but in most species, at a certain age, young elicit mild or even strong negative responses from their mothers. Among kittens, play activities are associated with the onset of maternal rejection, especially when the play is directed at the mother's face. Among primates it may be that the increased size, greater weight and strength of the infant, and its greater initiative in relation to the mother, as well as other changes in the physical appearance of the young, make some of the earlier activities more difficult or burdensome to perform (e.g. carrying the infant, allowing it to climb on her, suckling, etc.).

What makes the mother attractive to the young? To understand maternal behaviour we must know not only what features of the young attract the mother, but also what features of the mother attract the young. Among species with altricial young, the young are without VISION and HEARING at birth, but they are sensitive to tactual, thermal, olfactory, and taste stimuli. Even though precocial young have vision and hearing they are no less sensitive to these other kinds of stimuli. Artificial mothers have often been used to study the characteristics of the mother to which the young respond, since the experimenter can arrange for the artificial mother to include specific characteristics of the real mother and to exclude others. The newborn rat responds to a warm, moist, pulsating wide plastic tube much as it does to its own mother, except for suckling. The kitten requires a warm, soft surface with a belly-like shape to exhibit normal suckling responses, and if the artificial mother is equipped with a nipple it will often suckle from it. Puppies and rabbits require a similarly constructed artificial mother; warmth, or a soft insulating surface, is especially important. Monkeys cling to an artificial mother consisting of a cylinder mounted with a prominent 'head' with facial features, provided the cylinder is covered with a soft material. They prefer this to a wire-surfaced artificial mother. Warmth and a slight rocking movement increase the attractiveness of the artificial mother, and, in one species, the model must have the proper odour for the infant to be entirely comfortable with it. A protruding nipple supplying milk again adds to the attractiveness of the 'mother'. The requirements for kids, lambs, and calves are quite simple: a bag suspended at a certain height and filled with warm milk with a protruding nipple serves quite well to elicit suckling.

Initially, therefore, it is mainly the physical characteristics of the mother to which the young respond: the shape of her body and her nipples, her body warmth and soft furry surface, the slight movements she makes, and her odour are all important. Of course these are made available to the young through the mother's behaviour; in a sense she presents these features of her body to the young.

Artificial mothers begin to fail in comparison with the real mother as the importance of the mother's own behavioural responses to the young become more important than her physical characteristics. An artificial mother cannot retrieve a rat pup back to the nest, thereby relieving its distress; the pup must find its way back to the artificial mother. It cannot adjust itself to the young, nor can it respond to subtle communications that the natural mother normally responds to. Perhaps most important, it does not behave in a manner which encourages the developing young to become more independent. Therefore the young remain at a relatively simple level of social responsiveness compared to their littermates who are reared with the real mother. In the same vein, the artificial mother never rejects the young, therefore the young remain attached to it and do not wean themselves. This often has harmful consequences for the later development of the infant (see SOCIAL RELATIONSHIPS).

There is another kind of stimulus in rat mothers that may account in part for the frequent approaches to the mother seen in older pups around 16 to 27 days of age. At a time when the mother begins to decline in her approaches to the young, the young exhibit an increase in their approaches to her. Testing pups in an air chamber that carried odours to the pups from their own mother and from a strange, non-lactating female, it was found that each pup could locate its own mother through her odour. But pups could not distinguish their own mother from other mothers who had pups of the same age. Further study showed that the mother's odour emanated from her faeces, but only from one volatile portion of it. This portion is produced in a small region of the digestive tract (the caecum) in which digestive bacteria reside, and is in fact the product of the digestive processes of these bacteria. Between 16 and 27 days after parturition the mother begins to produce an excess of this odorous material (i.e. PHEROMONE). The young ingest faeces of the mother, thereby establishing their own digestive bacteria. They come to associate the odour with the mother and can sense it at a distance. Most interesting was the discovery that if two groups of mothers are fed different diets, the pheromones produced by each group have slightly different odours. Tested in the air chamber with the odours alone, the young chose the odour associated with their own mothers, and ignored the odour of the other group of mothers. The association of the pheromone with the mother is therefore learned by the pup during the second or third week of life.

Lactation and maternal behaviour. Lactation was formerly thought to be the basis for maternal behaviour: it was believed that internal pressure created by accumulated milk motivated the mother to suckle and to exhibit all the other maternal activities. It is now known that at least in rats and mice this is not the case: rats whose mammary glands are removed even before they become pregnant nevertheless show perfectly normal maternal behaviour when they give birth, although of course they do not secrete milk. They adopt the suckling position over the young, and engage in all the normal maternal activities. Moreover, since virgin males and females among mice and rats, and even among primates, can exhibit maternal behaviour when their mammary glands remain undeveloped, it is unlikely that the mammary glands play any direct role in causing maternal behaviour.

In the rabbit, in which a short period of suckling occurs only once a day, and at the same time each day, it has been shown that it is not the empty mammary gland that causes the female to wait 24 h before returning to suckle her young again. If the mother is anaesthetized, and the young are allowed to suckle her until her mammary glands have been emptied of milk, the mother, after recovering from the anaesthetic, returns to the young to suckle them with empty mammary glands. It is not therefore the emptying of the glands, but rather the experience of suckling which the mother counts as suckling, and while under the anaesthetic she does not receive this experience.

Suckling by the young in all species is an important stimulus for lactation. Stimulation of the nipple and its surrounding alveolar region causes the release of HORMONES from the pituitary gland, which act both on the mammary glands to cause milk production and on the smooth muscles within the glands to cause them to contract and squeeze the milk into ducts from which they can be withdrawn by the suckling of the young. If the mother is not suckled her mammary glands begin to lose the capacity to produce milk, because no hormone is released, and gradually they deteriorate. The modern view, therefore, is that lactation is dependent upon maternal behaviour, rather than the reverse which earlier scientists believed.

The rat has been studied very intensively with respect to the relationship between maternal behaviour and lactation. It has been found that, although suckling is a necessary stimulus for lactation during the first week and a half after parturition, after that time a mother needs only to smell her pups for the secretion of a sufficient amount of hormone to maintain lactation, at least for a short time. This is referred to as *exteroceptive* stimulation of the secretion of hormones which cause lactation, and it is undoubtedly learned in some manner. This phenomenon would mean, however, that lactation would not cease even if the young stopped suckling, provided they remained near the mother and she could smell them. This is not the case. Pups of around 21 days of age, although they continue to stimulate hormone secretion exteroceptively, also have another effect on the mother. The maternal hormone which is secreted in response to their stimulation no longer acts on the mammary gland to cause milk production. Some other hormone, perhaps adrenalin, blocks the action of the hormone on the mammary gland. It is probable that the sight of the now larger juvenile causes this hormonal change in the mother.

Hormonal basis of maternal behaviour. It has long been held that maternal behaviour is based upon the hormonal secretions at the end of pregnancy, even though there was no direct evidence for this, nor were the hormones identified. Studies in the rat have now established that this is true: blood was drawn from rat mothers shortly after they gave birth and began to show maternal behaviour, and the plasma was injected into non-pregnant females. Within 2 days the non-pregnant females began to show maternal behaviour towards foster pups that had been placed in their cages. Blood transfused from new mothers to non-pregnant virgins is even more effective, and stimulates maternal behaviour in about 14 h. Later studies showed that three hormones might be involved: oestrogen and progesterone from the ovaries, and prolactin from the pituitary gland. In species other than the rat, only NEST-BUILDING has been stimulated by hormone injections: in the rabbit prolactin seems to be effective, and in the hamster and the mouse oestrogen and progesterone are effective.

Maternal behaviour without hormones. Non-pregnant adult rats and mice exhibit maternal behaviour when they are exposed to pups: in mice maternal behaviour is exhibited within 5 min, but in rats it takes anywhere from 2 days to 5–7 days. Females and males are equally responsive to pups, and removal of the gonads does not affect the response; nor, in females, does removal of the pituitary gland. The response is therefore not under hormonal control. Of course, since lactation is dependent upon hormones these females and males do not lactate, but they do show nursing behaviour.

In several species, the young can stimulate maternal behaviour in females (and sometimes males) that have never been pregnant and given birth. Hamster females are maternal to pups of 6 days of age and older. Female cats and dogs exhibit maternal-like behaviour in an incomplete form to the young of their species, and monkey females often act as 'aunts' alongside lactating mothers who are rearing infants. Among monkeys, older brothers and sisters can be recruited as 'mothers' by their younger siblings when the real mother has, for one reason or another, abandoned the infant.

As indicated above, in rats it requires 5–7 days of exposure to foster pups before non-pregnant adult females begin to exhibit maternal behaviour. There is evidence that the female is at first repelled by the odours of the pups, and she avoids them. Gradually, however, she overcomes this aversion and begins to respond to the other stimuli from the pups. If she is prevented from smelling the pups by the desensitization of her olfactory system, or the cutting of her olfactory nerves, the female becomes maternal to the pups within 24 h. She will remain maternal for as long as she is given young pups to respond to, but of course she does not lactate. In contrast, if non-pregnant female mice are prevented from smelling the pups, they kill and eat them. Thus, there is no general rule about the role of olfactory stimuli in the maternal behaviour of non-pregnant animals.

Development of maternal behaviour. It is generally assumed that maternal behaviour is adult behaviour, and this is true under natural conditions. However, young rats, mice, and hamsters have been observed to exhibit parts of the pattern of maternal behaviour even before they have themselves been weaned. This is not ordinarily observed, because younger pups are not available to them. If they are given younger pups, however, or if the mother gives birth to a second litter before she has weaned the first litter, which is often the situation among rats and mice, then maternal behaviour may be shown by these young animals.

Young female and male rats of 24 days of age begin to show maternal behaviour after less than 3 days of exposure to 3- to 5-day-old pups, which is more rapid than in adults. At this age they have not yet reached puberty, and it is doubtful that significant amounts of gonadal hormones are being secreted. Around puberty (36–42 days) or even earlier, they become less responsive to pups, and require about the same duration of exposure as adult animals. It is not know what causes this change, but it may be the result of the hormonal changes normally associated with the onset of puberty in both males and females.

In some strains of rats there are differences in the behaviour exhibited by adult males and non-pregnant females when exposed to pups. In one strain the males do not retrieve pups, although they show nursing, licking, and nest-building. If males of this strain are castrated as infants (before their fifth day), when they become adults they exhibit retrieving, as females do, when they are exposed to young pups. Male hormone (testosterone) secreted early in life therefore produces a difference in the behaviour of male and female rats of this strain, and this difference disappears if the hormone is removed early in life. Among rabbits, males are almost totally unresponsive to young, even when they are injected with prolactin, which makes females exhibit maternal behaviour. Females born of mothers that have been injected with testosterone during pregnancy, are unresponsive to the hormones that would normally make them maternal when they become adults. The male hormone therefore completely eliminates the maternal behaviour of female rabbits.

The maternal behaviour of mammals is an example of parental care in which the greater investment is made by the female. The mammalian form of reproduction, involving live birth and suckling of the young, makes this inevitable. It is not surprising, therefore, that sexual dimorphism and SEXUAL SELECTION are common features of mammalian behaviour.   J.S.R.

MATE SELECTION. In those animal species that reproduce sexually, the quality of the mate is a critical determinant of reproductive success and the GENETIC constitution of the offspring (see NATURAL SELECTION). Hence it is not surprising to find that animals seldom mate indiscriminately. Various mechanisms ensure some selectivity in the SEXUAL process, and in many species this selectivity depends not merely upon the mechanical compatibility of the sexual organs but upon the perceptual discriminations and selective sexual arousal of the animals involved.

A survey of animal mating patterns reveals a remarkable variety in the frequency and form of mate selection. While some animals take but a single mate in their lifetime, others may acquire many, either successively or simultaneously. Polygyny, polyandry, monogamy, and promiscuity are only general labels for the many different MATING SYSTEMS to be found in the animal kingdom (see SOCIAL ORGANIZATION).

When does mate selection occur? Commonly, mate selection takes place when animals are ready to engage in reproductive activity. Thus, it is most frequently observed when animals are sexually mature and approaching the breeding season. Mate selection is less likely to occur when females are pregnant or caring for young offspring. In short, the selection of a mate usually constitutes the initial step in the reproductive cycle, and its appearance generally reflects the physiological readiness for reproductive activity.

There are some notable exceptions to this generalization, however. In some bird species heterosexual relationships may form long before reproductive activity commences. For example, among some of the lovebirds (*Agapornis* spp.) heterosexual pairs form when the birds are about 2 months old and still have their JUVENILE plumage. In the African violet-eared waxbill (*Uraeginthus granatinus*) monogamous pairs are established before the partners are 35 days old. At this time they are still being fed by their parents and are far from reproductive age.

In some primate species sexual relationships emerge from juvenile associations that are not themselves sexual in nature. For example, the male hamadryas baboon (*Papio hamadryas*) acquires a harem of several females long before attaining adulthood. Young males kidnap infant baboons from their mothers, and spend much of their time grooming, holding, and playing with them. As these males become older, they more frequently select female infants which then become members of the harem. Often these associations develop 2–3 years before the females achieve sexual maturity. Therefore, there is again no contiguity between mate selection and sexual interaction.

In those species in which mate selection is a direct prelude to sexual activity, it is not necessarily the case that every breeding cycle is preceded by the selection of a new mate. Many animals, especially birds, retain the same mate from one breeding cycle to another, and in some cases they remain together for many years. For example, kittiwakes (*Rissa tridactyla*) in the north of England have been observed breeding together each year for 16 years, while a pair of Buller's albatrosses, 'mollymawks' (*Diomedea bulleri*), that were found breeding together in 1948 were still paired and actively producing young in 1971. Among the thousands of Bewick's swans (*Cygnus bewickii*) wintering at The Slimbridge Wildfowl Trust in Gloucestershire, England each year, there has been no recorded case of 'divorce'. Although individuals will select new mates if one dies or is lost, there has been no occasion during a 10-year period of intensive observation when both partners have returned with new mates. Thus, in many of these species the selection of new mates occurs infrequently during their lifetime.

Selecting the right species and sex. When parents are of different species, the offspring are generally infertile and incapable of adapting to the ecological NICHE of either parent. Hence, hybridization in nature is invariably disadvantageous, and the pressures of natural selection generate mechanisms that ensure the selection of mates from among members of the same species. The barriers to hybridization, whether geographic, climatic, mechanical, or behavioural, are known as reproductive ISOLATING MECHANISMS. An interesting example of the effectiveness of these mechanisms is provided by the crickets of the eastern United States. Male crickets attract females through songs produced when they rub their specialized forewings together. In some species the song consists of a series of rapid pulses which continue without interruption for minutes at a time. Two species of the same genus of tree cricket, *Oecanthus quadripunctatus* and *O. nigricornis*, are often found together in weedy fields, and both reproduce at approximately the same time each summer. If females are placed in a chamber where tape-recorded songs are played from a loudspeaker, a female *O. nigricornis* will approach the sound source when songs of her own species are played, but not when those of other species are presented. This is also true of female *O. quadripunctatus*. If artificial songs of the proper pulse rate are played to the females, they will also be attracted to the sound. The pulse rate of male *O. nigricornis* is much more rapid than that of male *O. quadripunctatus*, and it is this difference that allows the females to identify the males of their own species.

It is interesting to note that the pulse rate of both species increases with a rise in environmental temperature. Females make the appropriate choice of mates at all temperatures however, for a change in temperature produces a corresponding shift in the female's preferred pulse rate, and this shift precisely parallels the change in pulse rate sung by males of her species.

Intermittent signals are also used by fireflies (Lampyridae) for identifying and attracting mates. Males fly about in their HABITAT emitting a flashing pattern which is specific to the species, while females remain perched and flash answers to the males. When the flash of the female is given after a fixed and appropriate delay, and when this flash is of a specified duration, the male approaches. A male and female engage in several flash exchanges until the mate reaches the female for mating. For the females of some species the flash pattern serves a second function. In *Photuris versicolor*, for example, the female is a predator upon the males of at least four different firefly species. These *femmes fatales* mimic the female flash patterns of these other species and lure the males to their deaths. Clearly, hybridization is not the only unfortunate consequence of improper mate identification.

Mates must be selected according to sex as well as species. While in many species males and females differ in a number of structural and behavioural dimensions, sexual identification often depends upon the presence or absence of relatively few features. For example, the two sexes of the flicker (*Colaptes auratus luteus*) are similar, except that the male of this bird species bears a black stripe resembling a moustache at the corner of his beak. If the female of a pair is captured and painted with a similar moustache, she will be attacked by her own mate as if she were a male.

In those species in which males and females are similar in appearance, sexual identification usually depends upon differences in odour, VOCALIZATION, or behavioural DISPLAYS. For example, in many species of pigeons and doves (Columbidae), the male performs a bowing display when he encounters a member of the same species. It should be noted that animals may discriminate these same features in identifying both the species and the sex of a potential mate.

Selection of mates according to their physiological state. Animals commonly select mates only when they are physiologically prepared to do so. Similarly, the attractiveness of animals as mates often depends upon their physiological state as manifested in a variety of morphological and behavioural features. The seasonal variations in nuptial plumage and vocalization that are so widespread among male birds are directly linked to changes in their internal state, and COURTSHIP displays are especially dependent upon HORMONES secreted from the sex glands. As the breeding season approaches, these glands secrete increasing amounts of steroid hormones and induce the display which attracts mates.

In mammalian species, visual and auditory stimulation is generally less effective than odour in arousing sexual attraction. The odours used in COMMUNICATION are called PHEROMONES, and they depend upon the hormonal state of the animals emitting them. For example, female rats and mice (Murinae) ovulate every few days, and during the hours prior to ovulation the females become receptive to the sexual advances of males. Moreover, at this time the odour of the female changes and becomes more attractive to the males. Both the odour and the behavioural receptivity are dependent upon hormones from the female's ovaries.

The attractiveness of female rhesus macaques (*Macaca mulatta*) depends upon similar factors. During their MENSTRUAL CYCLE female monkeys are more attractive to males in the phase prior to ovulation than during the interval immediately afterwards. High levels of the hormone, oestrogen, secreted by the ovary as ovulation approaches, induce changes in the lining of the vagina, and these, in turn, create an odour which attracts males. If oestrogen is applied locally to the vaginal lining, a male will show increased interest in a female even if she has had her ovaries removed, and is not motivated to accept the advances of the male.

The context of mate selection. Although several individuals may be of the same species, sex, and physiological readiness for breeding, their attractiveness as mates often depends upon some associated feature and not upon any behavioural or morphological characteristic intrinsic to the individual animal itself. For example, the quality of a TERRITORY held by a male bird may be an important determinant of his attractiveness as a mate. Clearly, it is advantageous for a female to select a male whose territory is richly supplied with food and adequate nest sites. Indeed, this can be so important that it may account for the EVOLUTION of some polygynous mating systems. A polygynous mating system is one in which a single male maintains a breeding relationship with several females. Most avian species are monogamous and in these instances the male commonly aids the female in caring for the eggs and young. If the quality of territories varies greatly, however, it could be of greater advantage for a female to mate with a polygynous male holding a superior territory than to pair monogamously with one possessing an area of inferior quality. Although she may have to sacrifice the aid of her male mate in parental care, a female in a polygynous system can achieve a net gain if the male possesses an area containing good nesting and feeding sites. This explanation has been most useful in understanding the evolution of polygyny among those American birds that breed in marshlands. Marshlands provide a marked variety of territory quality, and the areas held by neighbouring males often yield striking differences in reproductive success. This may account, then, for the high incidence of polygyny among bird species that breed in these areas.

The bower-birds of Australia and New Guinea provide yet another example of mate selection through associated features. The spotted bower-bird (*Chlamydera maculata*) uses colourful objects to attract females. The male constructs an elaborate edifice of thin twigs and grass stems to be used, not as a nest, but as a bower for courtship of females. Two parallel walls, 12–22 cm thick, are constructed with an avenue between them. These walls are 15–22 cm apart, up to 0·5 m high, and often as much as 0·75 m long. When the walls are complete, the male collects conspicuous objects and strews them along the avenue and in the open

area at each end of the structure. It is particularly common for the spotted bower-bird to collect the bones of small mammals, a behaviour pattern which accounts for the bird's common name, 'sepulchre bird'. More than 1300 bones have been found at a single bower. Near urban areas spotted bower-birds will collect a variety of glittering or colourful objects, and they have been known to take the keys from automobiles. In one instance a spotted bower-bird took the glass eye from a bushman who kept it in a cup while he slept. As in some other bower-bird species, the male spotted bower-bird paints the inner walls of his bower by mixing charcoal or vegetation with saliva in his beak, and wiping the paste on the dry twigs and straw. This gives the interior of the bower a dark-stained appearance. Females appear to be strongly attracted by the structure and the display of the male; they approach and allow the male to mount and copulate, sometimes within the bower itself.

Mate selection may be influenced by the constitution of the mating population. In the fruit fly (*Drosophila pseudo-obscura*) unusual individuals are more successful in obtaining mates than are those that occur in greater abundance. For example, pairs of fruit flies were collected from two geographical areas of the United States and placed together in an arena for mating. Those from Texas were very frequently selected as mates when their numbers were low relative to those from California, but this advantage disappeared when their numbers were approximately equal to those of the Californian strain. When the Californian strain was relatively infrequent, its mating success increased as well. Thus, both strains exhibited a mating advantage when they were relatively rare in the population. Observations of their behaviour indicated that male fruit flies are quite indiscriminate in their mating choices, but females are highly discriminating and seem to be capable of identifying the relative frequency of various males prior to mating. In a LABORATORY experiment fruit flies of two kinds were placed in an arena containing a cheesecloth floor. Below the floor were placed many males of the type that were rare in the main arena. The presence of the males in the lower chamber eliminated the mating advantage of rare males in the upper area. Fatty extracts from males in the lower chamber were also effective in reducing this *rare male effect*, suggesting that the preference of females for rare males depends upon the prevalence of odours of the various types.

Selection of familiar mates. We have already seen that many animals return to their former mates in each breeding season, and in many cases the same partner is retained for life. This suggests that the partners recognize one another as individuals and exhibit a preference for one another as mates. Such conclusions must be drawn with caution, however. For instance, it is very common for birds to return to the same nest site in successive breeding periods, and it is possible that the continuing relationship of a pair is due as much to their individual attachments to the nest site as to an attachment of one animal to the other. If attachment is to the nest site, there may be, in fact, no particular preference for the mate or even any recognition of the animal as an individual.

None the less, many pairs do develop complex modes of interaction which suggest a basic attachment of one to the other. For example, a male may become aggressive towards all females other than his mate. Conversely, his mate alone may be the object of his courtship displays, his COPULATION, and his protection. Her disappearance or death may produce symptoms of STRESS.

One can discern the contribution of individual recognition and preference through experimental techniques. An experiment on the Barbary dove (*Streptopelia risoria*) is illustrative. A large number of female doves were assigned as mates to an equal number of males. Each pair was placed in a separate laboratory cage and allowed to mate and raise young twice in succession. At the completion of the second breeding cycle all birds were individually isolated for up to 8 months, and at the end of this period they were tested for mate recognition and preference. This was accomplished by observing the animals in small groups. For example, three males and three females were taken from isolation and transported to large outdoor cages in a nearby forest. All birds in each group had previously mated with one of the animals of opposite sex in the group. The birds were then observed, to determine which of the previously mated doves located themselves in close proximity to one another, and which of them copulated, built nests together, and incubated the same eggs. This procedure continued until all of the original pairs had been examined. The results left no doubt that Barbary doves could identify their former mates and would do so at an unfamiliar nest site. Without exception all birds paired with the same individuals with whom they had mated in the laboratory. It should be noted that mates were not selected originally by the birds themselves but were assigned by the investigators. Nevertheless, it was these assigned mates that were chosen again as partners. This indicates that, as a consequence of mating, the doves had acquired a preference for these particular mates. Moreover, the possible influence of a common attachment to a nest site was eliminated by testing the animals in an unfamiliar environment. This study demonstrates that Barbary doves learn to recognize their own mates as individuals and that they acquire a preference for them as partners in subsequent breeding periods.

The retention of a partner from one breeding season to another may provide several advantages. For instance, if animals recognize one another as individuals, the time and effort spent in identifying a partner's species and sex can be reduced or eliminated altogether. This may provide a considerable advantage in terms of breeding efficiency, especially in extreme latitudes where the breeding season is very brief and the young must be produced quickly in order to avoid exposure to inclement conditions. Mate retention may provide an additional advantage if reproduction requires an extended and complex series of interactions between the partners. If the two individuals improve their reproductive success by learning the idiosyncrasies of the other, loss of the mate can constitute a serious setback with regard to their production of offspring.

On the other hand, it would appear to be biologically adaptive for animals to part and then select new mates when young are not produced successfully. A study of kittiwakes indicated that although the majority of individuals returned to the mate of the previous breeding season, about one-fourth of the animals selected new mates even though their former mates (presumably familiar) were present in the breeding colony. An examination of their breeding histories revealed that a high proportion of those animals choosing new mates had reproduced unsuccessfully in the previous season. In most cases the birds' eggs had failed to hatch. Although it is possible that both the failure to remate and the failure to breed successfully were due to a deficiency in the initial relationship of the partners, it seems likely that retention of a mate depends somewhat on the outcome of their breeding effort.

Which sex selects? DARWIN suggested that it is more often the female than the male who selects from among potential mates (see SEXUAL SELECTION). Although this generalization appears to be broadly correct, it remained unclear for nearly a century why this should be the case. Biologists now seem to have reached some agreement in their explanation, however. In most species the costs to the female of producing offspring are relatively great. Typically, she must provide the embryo with yolk as a nutrient reserve, or she must nourish a foetus throughout an extended period of gestation. In many animals the provision of post-natal care falls to the female alone. One expects, then, that females will commit themselves to such an investment only when conditions are optimal for reproductive success. Since the quality of the male mate may represent a very important condition in this regard, one expects females to be highly discriminating with respect to potential mates.

By comparison, the costs of reproduction to the male are usually low. The males of most species can produce millions of sperm cells and can copulate with the females at relatively little expense in terms of time and energy. Males may attempt to increase their production of offspring, if only marginally, by mating with inferior females in poor environmental conditions. Even such very slight chances of success are likely to warrant their reproductive efforts when the costs of these efforts are low. Such a view is consistent with the general observation that males will court and copulate with a wide variety of females including, if given the opportunity, those of other species. On the other hand, females are generally 'coy' and slower to be aroused sexually, and they will permit copulation only after a more extended exposure to males.

Not in all species, however, is there such a marked disparity in the contributions of male and female to the reproductive effort. For instance, in a great many birds the male and female share in constructing the nest, in INCUBATION, and in brooding and feeding the young. When the reproductive investments of both sexes are great, one finds that both are discriminating in their mate selection (see PARENTAL CARE).

Although the features of mate selection vary widely among animal species, they invariably reflect the best reproductive strategy for those exhibiting them. In many, if not most, species the overall strategy is only partially understood, but clearly the quality of a mate is a critical determinant of reproductive success.   C.E.

20, 91, 101

MATING SYSTEMS. The mating system that is typical of a species provides a fundamental aspect of its SOCIAL ORGANIZATION. The two basic types of mating system are monogamy and polygamy. In a monogamous system each breeding adult mates with only one member of the opposite sex. Over 90% of the birds of the world are monogamous, but monogamy is rare in other animals, including mammals. In perennial monogamy, an individual mates with the same partner for life, or until a 'divorce' takes place as a result of the failure of one partner to fulfil its duties. The partners often maintain some association even outside the breeding season. Such relationships are found amongst swans (*Cygnus*), field geese (*Anser*), cranes (*Grus*), and gibbons (Hylobatidae). Seasonal monogamy is characteristic of species that are monogamous during the breeding season, but lead separate lives for the rest of the year. Many migrant birds come into this category, and they usually have a strong tendency to return to their exact nesting locality, where they join up with their partner each season.

In polygamous species an individual generally has two or more mates, either successively or simultaneously. The most common form is polygyny, in which one male mates with a number of females. In polyandry one female mates with two or more males. An example of polyandry is provided by the American jacana (*Jacana spinosa*), in which the female is more conspicuous, dominant, and territorial than the male. Having successfully courted, the female lays the eggs, which are incubated solely by the male, while the female attempts to gain more males to incubate successive clutches. In promiscuous species there are no pair bonds, and both males and females mate with more than one member of the opposite sex.

The formation of *pair bonds*, and the type of mating system that characterizes a species, has developed by EVOLUTION in relation to the animal's ecological NICHE, and in response to the genetic interests of the sexes. Within a species, the sexes often differ considerably in their PARENTAL CARE. Usually, the female invests much more than the male in terms of energy expenditure, risk of PREDATION, etc. As a result of this imbalance, the consequences of making a mistake in MATE SELECTION are much more severe for females. For example, if the male is a carrier of a GENETIC abnormality that results in high mortality amongst his offspring, he wastes a certain amount of energy and time at each COURTSHIP and mating. The female that mates with him, however, loses much more, in that she has the costs of bearing the offspring and may miss a whole season's breeding opportunity.

We should expect the pressure of NATURAL SELECTION for females to exercise care in choosing a mate to be much greater than it is for males. Indeed, it is a general rule that, where the female invests more in the offspring than the male, she also exercises greater choice during courtship. Normally, the male actively woos the female, and is aggressive towards rival males. In those species, such as the three-spined stickleback (*Gasterosteus aculeatus*), in which the male carries a greater burden of parental investment, the females play a less coy role in the courtship.

The physical characteristics of a potential mate may or may not be a good indication of its parental contribution (see SEXUAL SELECTION), but an animal's reproductive success will be greatly influenced by the genetic make-up of the mate, in so far as it is manifest in the offspring; and by the behaviour of the mate in so far as it affects the survival of the offspring. The evolutionary implications of this situation provide a good basis for the understanding of mating systems. A male may maximize his FITNESS, in terms of his genetic contribution to future generations, by mating with many females, and thus fathering a large number of offspring. However, this polygyny is a good strategy only if the offspring themselves survive to reproductive age. The survival of the offspring depends upon the degree of parental investment of the female, and of the male, and upon the ecological circumstances. A polygynous male can have little time for caring for his offspring. Therefore, polygyny is likely to be successful only in those species in which the female can support her young unaided. This may be possible in species which have few young, and in which the female is specialized for parenthood.

If we compare mammals and birds, for example, we see that polygyny is much more common among mammals. Female mammals are specially equipped to feed the new-born, providing milk and specialized MATERNAL care. Male mammals inevitably have less of a role to play in caring for the young, although they can often help by providing food etc. Male birds are just as well equipped to look after offspring as are female birds, and their contribution is often necessary. Polygyny is not common in birds, but it occurs more frequently in birds with *precocial* young, which are ready to leave the nest soon after hatching, than among those with *altricial* offspring, which are helpless for a relatively long period. Thus the occurrence of polygyny is ultimately related to the necessity for parental care.

When a female mates with more than one male, her parental behaviour is devoted to her own offspring, but any parental contribution by a male might benefit other males. For this reason males should refuse to mate with females that have recently mated with other males. Therefore it is not surprising that polyandry is rare in the animal kingdom. Among the few species of polyandrous birds, the female often leaves a different clutch with each male and the male usually incubates only those eggs that he has fathered.

To some extent, a female suffers a disadvantage in mating with a male that has already mated, since he is unlikely to care especially for her young. However, this factor will be more important among birds than among mammals, because male mammals are not specially adapted for parental care. A female bird will try to choose a mate who is likely to help rear the offspring, especially if this enables her to support a larger clutch of eggs. A male who can demonstrate his ability to support a family, perhaps by securing a TERRITORY before mating, is likely to improve his mating success. However, a male that devotes all his time to securing a territory, courting a female, and helping to incubate and care for the young, has little time for other females, and is likely to be monogamous.

Monogamy is much more common among birds than among mammals, and this is partly due to the fact that female birds are not especially equipped to care for the young, in the way that female mammals are. Females of all species exert a choice of mate that is likely to maximize their reproductive output. In general, birds are more likely to choose on the basis of the male's degree of parental investment, and to avoid choosing a male that has already mated. For female mammals, the parental abilities of the male are not so important, and there may be some advantage in mating with a male that has already mated, especially if he is a dominant male that has demonstrated his ability to compete with other males.

20

MATURATION of behaviour is an irreversible process of change that occurs as part of the total ONTOGENY of the animal. Maturation does not depend upon experience, and is thus distinct from LEARNING and practice. Early experimenters kept JUVENILE pigeons (*Columba livia*) and swallows (Hirundinidae) under confined conditions so that they could not move their wings. When released they could fly as well as other birds of the same age. Although young birds can be seen flapping their wings whilst remaining on the ground, and human parents support their children during their first tentative steps, this apparent practice has no effect upon the development of the behaviour. Walking and flying depend entirely on the growth of the nervous system and its coordination with the muscles. The development is one of maturation. Growing frog tadpoles (Anura) flex their tails in an increasingly coordinated manner, and this occurs whilst the animal is still in the jelly-like spawn. These incipient swimming movements can be inhibited by application of a light anaesthetic which permits the tadpoles to continue growing. If, when they have reached the size of normal swimming tadpoles, the anaesthetic is removed, they swim normally. Thus swimming does not depend on practice, but upon growth and maturation.

In some cases, however, practice is important. Cuttlefish (*Sepia*) attack the small shrimp *Mysis*. First, they fixate the shrimp with one eye, then turn to face it with both eyes, and finally seize it with their long tentacles. Newly hatched cuttlefish take as long as 2 min to execute this manœuvre, but the time required rapidly diminishes with practice, and reaches 5 s after about ten trials. The improvement is the same whether or not the cuttlefish is hungry, and irrespective of its success in obtaining shrimps. However, the improvement depends upon the experience of attempting to catch shrimps, and does not simply mature with time. Since the improvement occurs whether or not the cuttlefish is successful in catching shrimps, it cannot be due to learning, but simply to practice. Practice depends upon experience, but not upon specific consequences of the behaviour, in contrast to learning.

The pecking behaviour of domestic chicks (*Gallus g. domesticus*) improves as a result of a combination of maturation, practice, and learning. The accuracy of pecking at food particles improves with age, even if the chicks are kept in the dark and fed on powdered food so that they have no pecking experience; it improves with practice, even if the grains are stuck to the floor, so that the chick cannot eat them, and it also improves as a result of learning, particularly the DISCRIMINATION between edible and inedible objects.

99

MECHANICAL SENSES. Mechanoreceptors are receptors that are sensitive to mechanical deformation of one kind or another. They provide the basic element of a variety of senses, depending upon (i) where in the body they are situated, and (ii) which accessory structures are associated with them. In vertebrates, for example, two different senses, that of HEARING and that of balance, have their receptive organs in the inner ear. Essential to both of these are almost identical sensory cells that respond to the bending of hairs, called *cilia*, that protrude from them. However, in each case the cells are associated with very different accessory structures, the cochlear organ and the vestibular organs respectively. These structures are sensitive to sound-waves, in the case of hearing, to position with respect to gravity, and to rotational acceleration, in the case of the sense of balance.

Touch and pressure. Let us first consider a sense in which the sensitivity of mechanorecentors is put to a more direct use: the sense of touch. The sense of balance is dealt with later, and that of hearing is treated separately. A variety of touch receptors have been identified in the skin of vertebrates. These consist of nerve fibre terminals simply embedded in the skin, or enveloped by more or less complex non-neural structures. In some cases the cells making up these accessory structures aid the translation of the mechanical stimuli into neural messages. Often these receptors carry the name of their discoverer, e.g., Ruffini, Merkel, and Grandry corpuscles. Some of these only occur in a particular vertebrate species, often in a special region of the body. For instance, Grandry corpuscles only occur in the beaks of birds that probe with the bill in the mud for food (ducks, Anatinae and plovers, Charadriidae). Sometimes clusters of different corpuscles occur in association, such as at the base of the whiskers of many mammals. Here the hair acts as a lever mediating mechanical disturbances that are clearly of importance for animals that habitually operate in narrow constrained spaces, such as burrows, crevices, and thick undergrowth.

To discover what sort of mechanical stimulus each of the receptors described by anatomists responds to, one must record from the nerve fibres coming from them. In this way it has been discovered that Merkel corpuscles react mainly to rapidly changing deformations of the skin, while Meissner corpuscles respond to a steady pressure on the skin. Of course the detailed sensitivity characteristics of a corpuscle also depend on the mechanical properties of the tissue in which it is embedded, which varies from the stiff skin of a foot pad to the soft skin of a lip. In other cases identification is less certain. It is assumed that free nerve endings mediate the sensation of PAIN, and respond to strong mechanical, extreme thermal, and irritative chemical stimulation, but positive evidence for this assumption is lacking.

In mammals information coming from the touch receptors is relayed by a series of neuronal pathways to higher centres of the BRAIN, notably the cortex of the cerebellum and of the forebrain. It is interesting that the two cortical forebrain areas receiving such information are laid out so that there is a map-like representation of the body surface. Evidence from physiological studies of the behaviour of nerve cells in these areas indicates that the information coming from the individual receptors is processed here in such a way that complex stimulus configurations, such as stroking or tickling, can be recognized.

Most lower animals also possess a sense of touch. Many for example show withdrawal behaviour upon being touched: segmented worms (Annelida), sea-anemones (Antinaria), and wood-lice (Isopoda), to name but a few. Touch is also important for mating behaviour in virtually all bisexual species, and for obvious reasons. Even brief observations of honey-bees (*Apis mellifera*) in their hive, or ants (Formicidae) at their nest, reveal the important communicatory function that touch stimuli must have for them. In many instances we do not know which sensory structures mediate these activities. In insects virtually the whole body surface is covered with sensory hairs that function as touch mechanoreceptors, or, at least, are suspected to do so. Typical mechanoreceptive hairs are the *sensillae trichoidea*, which are stiff hairs set in a membranous socket, and innervated by one or more nerve cell protrusions. As the hair is bent the nerve cell fires repeatedly, and thus codes the information about the mechanical disturbance which is transmitted to the central nervous system. Similar arrangements are found in the joints of invertebrates. Sometimes they serve special purposes in conjunction with rather specialized appendages. For instance, a joint in the antenna of the dung beetle *Geotrupes* is equipped with mechanoreceptors known as *campaniform sensillae*. They sense the deflection of the antenna caused by air currents. This enables the beetle to move upwind when stimulated by the smell of dung, and thus to find the cowpats on which it lives. Water striders (Gerridae) sense water surface waves with leg-joint mechanoreceptors, and are thus guided to find food and avoid predators.

Fish and some amphibians, for example the clawed toad *Xenopus*, possess a system of channels under the skin and along the side of the body and head that are studded with groups of hair cells, the cilia of which are embedded in a gelatinous mass. They form the lateral line sensory system. While these receptors certainly are highly sensitive to movements of the surrounding water, the overall function of the lateral line system is not completely clear; it may serve to mediate ORIENTATION, guided by water pressure waves, during such activities as searching for food or SCHOOLING.

Pigeons (*Columba livia*) are highly sensitive to changes of barometric pressure; they can easily detect a pressure change equivalent to a 30-m altitude difference. The sensory organ responsible seems to be a small closed vesicle in the middle ear which contains a number of sensory cells. It occurs only in birds and some sharks (Salachii), and is well developed in flying birds, but only poorly so in ground-dwelling species. Interestingly it is also large in some diving birds, for example penguins (Spheniscidae), and might be used by these to assess depth on the basis of hydrostatic pressure alterations. Sharks might use it in a similar way. In bony fishes (Teleostei) stretch receptors associated with the swim bladder serve the same function.

Muscle contractions are monitored by stretch receptors embedded in the muscles and tendons. Those present in muscles are associated with accessory structures constituting what is known as a muscle spindle. They embody a principle called *efferent control*, which is also found elsewhere among sensory organs. Each spindle is embedded within the main muscle, and consists of a small muscle with stretch receptors. This small muscle is controlled independently of the main muscle by a special set of nerve cells and fibres. Thus the sensitivity of the spindle stretch receptors to the tension of the main muscle can be altered by neural commands emanating from the central nervous system. Stretch receptors, arranged in a somewhat different manner, are found associated with the intersegmental muscles of many crustaceans, such as the lobster, *Homarus*. Their precise role is not fully understood, but it seems likely that they are important in the COORDINATION of swimming movements.

Mechanoreceptors also occur amongst the viscera of vertebrates. A well studied example is the Pacinian corpuscles. These consist of nerve endings enveloped in an ovoid, layered capsule, and are found embedded in the mesenteries, i.e. the membranes that support the guts of vertebrates. The capsules are deformed by pressure. A capsule deformation of less than a thousandth of a millimetre is sufficient to stimulate these nerve endings. Pacinian corpuscles serve to monitor displacements and pressures resulting from gut movements associated with digestion.

Balance. Mobile animals can, and usually do, maintain a stable position with respect to gravity (the pull of the earth's mass). This requires a sense of equilibrium. Mechanoreceptors, associated with complex structures provide the relevant information in most animals. In vertebrates one finds them in a cavity of the inner ear, close to, but still separate from, the hearing organ proper, the cochlea. They cluster in two or three small patches known as *maculae*. The tips of the hairs of the sensory cells are embedded in a gelatinous concretion of calcareous particles (statoliths). When the head is tilted the weight of these statoliths shears the hairs sideways, and this causes the receptor cells, depending on the direction of the shear, either to increase or decrease their electrical potential with respect to the surrounding tissue. The nerve fibres that make contact with the hair-cells translate these potentials into nervous messages which are relayed to the cerebellum, and other parts of the BRAIN, where they contribute to the coordination of motor responses that maintain the spatial orientation of the animal.

Many lower animals have similar arrangements. The common octopus (*Octopus vulgaris*), for example, has a pair of statocysts that are analogous to the maculae of vertebrates. Information coming from these controls, among other things, the position of the eyes. These have slit pupils which always remain horizontal irrespective of the tilt of the animal. If the statocysts are disturbed experimentally, then this response disappears. In crustaceans such as lobsters and crayfish (Astacidae), the statoliths (sand particles contained within the statocysts) are lost with each moult, and the animals have to use their claws to put grains of sand into the statocysts at the base of the antennula. If forced to use iron filings instead of sand they behave towards a strong magnet as they normally do towards gravity. In honey-bees orientation with respect to gravity is very accurate when workers communicate the direction of foraging flowers to their hive mates. A bee returning to the hive performs a dance in which the angle of the food source direction with respect to the sun is translated into a dance path with the same angle but with respect to anti-gravity (see COMMUNICATION). In the bee the whole head functions as a statolith: mechanoreceptive hair-plates at the head–thorax joint monitor the direction of the pull of its weight.

Mechanoreceptors also monitor the angular accelerations of the head in vertebrates. Each member of the paired sensory organs consists of a horizontal, a transverse, and a sagittal *semicircular canal* connected with the inner ear cavity. The canals are filled with a liquid. At one end of each canal there is a cluster of hair cells, the hairs of which are crowned with a gelatinous mass that almost blocks the cross section of the canals. When the head is rotated the inertia of the liquid deflects the cupula, and the hairs of the sensory cells are bent. As in the case of the macula this is translated into nervous messages that are transmitted to the brain-stem via the vestibular nerve. From there they are relayed to a variety of brain areas, and contribute to the coordination of balance reflexes. However, before the information of the vestibular organs can be utilized for balancing the body it must be corrected for any bending of the neck. The information for this is provided by mechanoreceptors located between the neck vertebrae. For birds this introduces an unacceptable delay since they are critically dependent on very accurate balancing of the body when flying and perching. Accordingly, they have evolved an additional sense of balance that monitors the position of the body directly. It consists of stretch receptors situated among the membranes that support the viscera, and which are stimulated when the body of the animal is rotated and the viscera lag behind. This information reaches the spinal cord via visceral nerves, and there it controls the balancing movements of wings, legs, and tail.   J.D.

MEMORY, as used in everyday conversation, is usually thought of as a person's ability to express or perform some previously learned bit of information or habit; very rarely is there any misunderstanding about its meaning. A precise definition of memory, however, is more difficult to formulate, particularly for the scientist who studies memory processes in infra-human animals. One reason is that memory processes (i.e. those mechanisms that cause the information or habit to persist over time) are so intertwined with LEARNING processes (the mechanisms that cause the information or habit to come into existence in the first place) that the two are easily confused. Consider, for example, a very simple case of animal learning/memory. A pigeon (*Columba livia*) is placed at the centre of a large cage, on one side of which are a feeding trough and a light. The pigeon's task is to learn that food is available in the trough whenever the light is illuminated. On the first few occasions, the bird takes some time to approach the feeder when the light is presented, but after repeated exposures to this situation, it approaches immediately. Are we observing memory or merely learning? The answer, of course, is both: learning and memory processes are contributing to the animal's behaviour, such that on each trial the animal is remembering something from its immediate past experience (or else the bird would never show any improvement with training), and yet is having the strength or 'status' of the habit constantly altered by learning. In short, we can say that virtually all animal studies involve the expression of memory, but how the memory processes are to be distinguished from the learning processes (or even, say, from MATURATION, fatigue, HABITUATION, and the like) is not always obvious. Fortunately, problems of the type just described are often resolved by careful research design.

For most scientists, and for our purposes here, memory refers to the persistence of a learned response over time. This means that studies on memory involve the deliberate use of an interval of time between the learning, and the retention or test phases. Given the above definition of a memory experiment it is useful to remember that studies on memory contain the same three phases that characterize the course of memory itself: (i) habit formation or learning, (ii) storage for some specified period of time called the *retention interval*, and (iii) the later expression of the habit on a retention test. In other words, a habit is learned during the first stage of an experiment. During the second stage it persists or is 'stored' for some period of time; here the investigator often introduces certain variables in an effort to manipulate and understand the physiological and behavioural dynamics of the storage phase. Finally, the third phase of a memory experiment involves the utilization of the stored habit, the actual expression of the remembered response on a *recall*, *recognition*, or *relearning* test. (Any of these three traditional measures of retention could be used, although each probably reflects distinct facets of the memory process.)

In summary, the basic stages of memory, and consequently the basic components in all experiments on memory, are learning, storage, and retention. Careful consideration must be given to the experimental design, so that the behavioural outcome observed by the investigator can be interpreted in the light of the memory processes, and not the learning processes. More specifically, variables or conditions that are thought to affect memory *per se* must be shown to operate after the learning phase has been completed, not during the learning phase when the habit is still being formed.

Basic theories of memory. Given the extensive research and the complexity of the issues, it is not surprising that a very large number of theories of memory have been proposed. Most of these theories, however, fall into one of two broad categories, either the *decay* or the *interference* theory. Decay theorists postulate that memories gradually dissipate or decay with time. Learning creates a basic impression, but it fades like a footprint in sand if not maintained by periodic rehearsal.

In contrast to this view, interference theories focus on the competition between memories. Any given habit is not in itself weakened by disuse, but the storage of the habit as a discrete memory trace, or its expression on a retention test, is subject to interference by other memory traces. The formation of a recent trace, for example, may inhibit or interfere with a previously formed memory by displacing it, or by creating a form of confusion during recall. Note that the forgetting of habits due to either decay or interference would be an adaptive characteristic for an animal to possess, since forgetting would protect the organism from retaining useless or outdated information.

A brief example may help to illustrate these different approaches to memory. Consider your retention of the contents of the first paragraph of this article. The decay theorists would claim that the strength of your memory has declined simply because you have not been rehearsing the material; the unaltered memory trace has dissipated over time. Interference theorists, on the other hand, would say that either the material in subsequent paragraphs has displaced or interfered with your memory for the first paragraph, or perhaps some other material read previously has caused confusion and interference. Either way, forgetting would be a by-product of learning something else, an indirect consequence of the formation of other memories. Naturally much of the research on memory, even on the infra-human level, has been directed at resolving these two major types of theory.

Types of memory. In human memory research it has been useful to distinguish between three basic types of memory. *Immediate memories* are those bits of information that last only a few milliseconds. In a real sense, immediate memory refers to the input or encoding stage of the memory process, and thus is occasionally referred to as the *short-term sensory store*. Because it is so difficult to study this type of memory in non-verbal animals, little research has been completed.

The second type of memory, which is discussed below in some detail, is *short-term memory*. Here, information is retained over a period of seconds or even minutes before it is forgotten. The exact length over which a short-term memory may persist depends, of course, on many factors, but in every case, the time scale is in seconds rather than hours or days. A good example of short-term memory occurs when you look up a new telephone number in the directory. You manage to remember it long enough to dial the number, but usually forget it almost immediately afterwards. It appears that short-term memory obeys a special set of rules, which are distinct from those of both immediate and *long-term memory*. For example, the absolute amount of information capable of being held in short-term memory at any one time is quite limited; this is not true of long-term memory.

The third basic type of memory is long-term memory. When we mention memory in our everyday conversation, it is usually long-term memory to which we refer. Examples, such as remembering how to ride a bicycle or the significance of a historical date, are good illustrations of this. As the name suggests, long-term memories last considerably longer than short-term memories. (For our purposes here, retention for longer than 24 h will be used as an operational definition of long-term memory.) But long-term memory differs in other ways too. For example, the number of long-term memories that human beings are capable of possessing is thought to be infinite. Although such a notion would be impossible to verify in the extreme, it is clear that long-term memory is virtually unlimited for all practical purposes. Finally, long-term memory is not always as accurate as short-term memory; people appear to change or forget many of the specific details, while retaining the essential meaning of the information.

Short-term memory. The problems of interpreting studies on animal memory are considerable, primarily because there may be important differences between what an animal is capable of remembering, and what sort of memory it ends up demonstrating. For example, an animal may indeed remember where food is located in a cage, but, for one reason or another, not choose to demonstrate that fact. Unlike research with people, where it is possible to clarify the instructions through direct COMMUNICATION, animal investigators may often be deceived about the 'real' nature of the animal's behaviour. Nevertheless, a stylized methodology has evolved which appears to be generally adequate for investigating many memory phenomena in animals.

One such technique involves standard classical CONDITIONING procedures (see LEARNING). Here, the experimenter presents a relatively neutral *conditional stimulus* followed shortly by a more biologically powerful *unconditional stimulus* which reliably elicits a strong reflexive-like response. After repeated presentations of these stimuli, the conditional stimulus acquires the power to elicit a conditioned form of the REFLEX. The simplest example of this type of learning is PAVLOV's work with hungry dogs. In his experiments, the sound of a bell (the conditional stimulus) was paired with food presentations (the unconditional stimulus) which in turn caused the animals to salivate (the unconditional response). After sufficient training,

Pavlov found that the bell alone was capable of eliciting the salivation reaction (the conditional response).

This method provides us with a simple, albeit crude, measure of short-term memory. If the subject remembers the conditional stimulus over a short period of time, then its memory trace will be contiguous with the unconditional stimulus presentation, and a learned association (the conditional response) will be formed. If, on the other hand, the animal forgets the conditional stimulus soon after it has been presented (whatever the cause of the forgetting) then clearly the stimulus cannot become associated with the unconditional stimulus, meaning that the conditional reaction will be weak or non-existent. The main question, therefore, concerns the extent to which the unconditional stimulus presentation may be delayed without precluding the formation of a conditional response.

Many experiments focusing on this question have confirmed that the maximum permissible delay depends upon a variety of factors, including the type of stimuli used, and the species of animal being studied. In laboratory rats (*Rattus norvegicus*), for example, external conditional stimuli, such as lights and tones, are remembered anywhere from a few seconds to half a minute or so, when they are followed by typical Pavlovian unconditional stimuli, such as a mild electric shock or food. In contrast, flavour or odour stimuli are remembered over a very much longer period of time if they are followed by an unconditional stimulus that induces a mild gastrointestinal illness; in fact, the unconditional stimulus presentation can be delayed up to 12 h or more, and learning will still take place (see FOOD SELECTION). This extraordinary ability to associate flavours and illness, even when separated by many hours, appears to be an evolutionary feeding specialization in many animals, such as rodents, that have a highly developed CHEMICAL SENSE. The capacity to remember the taste of a novel food hours later when experiencing illness helps the animal to identify and, therefore, avoid consuming poisons a second time. Accordingly, this special memory capacity has been selected during the animal's evolutionary history.

Unlike rodents, birds have a highly developed visual sense which they use for identifying food. Not surprisingly then, birds have been shown to have an excellent short-term memory for visual as opposed to taste cues. The colour of a fluid, but not its taste, will become strongly conditioned to a poison even if the presentation of the poison is delayed several hours. The specialized ability of birds to associate the visual features of food and later illness has been subject to NATURAL SELECTION during EVOLUTION because, in their particular NICHE, birds depend strongly upon their VISION for procuring food.

Short-term memory has also been tested using *instrumental conditioning* techniques. Instrumental (or OPERANT) conditioning involves the presentation of a potent unconditional stimulus (or in some cases its withdrawal) contingent upon some designated response. For example, in a typical reward procedure, food is presented to an animal once the animal has pressed a lever or run through a maze. In an escape/avoidance situation, a mildly painful shock is terminated (ESCAPE) or not presented at all (AVOIDANCE) if the animal performs the requisite behaviour. In each case the animal is an active participant in the experiment, in that the presentation of REINFORCEMENT is entirely dependent on its behaviour.

Operant conditioning techniques have been used extensively to investigate animal memory. In the *delayed response* test, for example, rats are exposed to three transluscent doors. The 'correct' door on any given trial, i.e. the door behind which the animal can obtain food, is signalled by a light located behind the door. However, there is a delay between the light signal and the time when the animal can actually open the door and get food. This means that the rat has to remember which door is correct on that trial (which light had been turned on) for at least the length of the delay period. Rats are surprisingly poor at this task unless they are allowed to physically orient their bodies towards the appropriate door. Additional research shows that dogs (*Canis lupus familiaris*) also depend to a large degree on body orientation, but racoons (*Procyon*) and children do not.

Evidence from LABORATORY STUDIES does not always agree with results obtained in the FIELD, and because of this fact, important misunderstandings can arise. In terms of delayed responding, for example, it has been shown that chimpanzees (*Pan*) are very adept at locating hidden pieces of banana, and dogs at finding bones, in a field situation, but when given the same test in a laboratory environment they performed quite poorly. In each case there was a delay between observing where the rewards were placed and being given the opportunity to find them. Why such a discrepancy should exist is not clear, but it illustrates that an animal's memory capacity cannot be divorced from the context in which it is assessed (see INTELLIGENCE).

The delayed response procedure can be modified in order to improve its precision as a test of memory. The modern version is termed *delayed matching to sample*. Here a sample stimulus is presented followed by a delay (or retention) interval. Then, two comparison stimuli are given, one of which matches the original sample while the other does not. The animal's task is to respond to whichever comparison matches the sample. If the match is correct, then a reward is given. If, however, the animal chooses the comparison that differs from the original stimulus, then reward is withheld. Clearly, this situation is a good test of short-term memory, because the animal must remember the exact features of the sample over the length of the delay interval if it is to match it successfully with one of the comparison stimuli. If the animal does not recall the sample when asked to match, its success will be no greater than chance.

A great deal of information on short-term memory, particularly in pigeons and monkeys (Simiae), has been derived from delayed-matching-to-sample studies. For example, memory is better if the animal experiences darkness during the retention interval. Although this finding suggests that the improvement results from a decrease in competition from distracting cues, such a hypothesis is far from proved. Physically restraining a subject, which also ought to decrease distractions, or deliberately presenting noise distractions during the retention interval, has little effect on performance.

There are important differences between species in delayed-matching-to-sample studies. For example, the maximum delay that pigeons can tolerate is about 10–15s, whereas dolphins (Delphinidae) can match with intervals of 2 min or longer. Primates do even better. One rhesus macaque monkey (*Macaca mulatta*) improved his performance over many sessions of training until he could successfully match with a delay of 9 min. Even on a conditional matching task, where the stimuli are highly arbitrary, and bear no physical relationship to each other, capuchin monkeys (Cebinae) can successfully match with delays of over 100 s.

Another species difference is that monkeys require only a brief presentation of the sample stimulus for effective matching, whereas pigeons, and paradoxically human beings, find that matching becomes easier as the duration of the sample presentation increases. Moreover, if the sample is shown twice, performance in human beings and monkeys is better when the two presentations are separated by a fairly long time interval. In contrast, pigeons cannot tolerate such a spacing procedure; improved matching occurs when the two presentations occur in rapid succession.

Medium-term memory. Scientists who study memory processes in human beings do not normally use the category 'medium-term memory', but it is convenient to do so when discussing the animal literature. The main reason is that several important research programmes have focused on retention phenomena that appear to be neither unequivocally short- nor long-term in nature. For our purposes, 'medium' means from several hours to about 24 h, i.e. longer than the short intervals used in delayed-matching-to-sample tasks, but shorter than the long retention intervals that characterize the long-term memory research.

One main area of study in this category is the work on the retention of AVOIDANCE responses in rats. In the paradigm study the animals are trained to a moderate degree to avoid a shock, and then retested at various time intervals. It is found that performance of the avoidance response is a U-shaped function of time since learning: no decrement is observed immediately after training or 24 h later, but at intermediate times avoidance performance declines, suggesting that the animals have forgotten what they have been taught only a few hours earlier.

It is possible to interpret these effects in terms of memory and claim that animals have trouble retrieving the memory at intermediate intervals, and therefore perform poorly on the retention task. Why should a rat be unable to retrieve a habit 3 h after learning it, but easily retrieve that same habit 24 h later? The reason is that the appropriate external and internal stimuli which were needed to elicit the response are not present 3 h after training, but they reappear after 24 h. For memory to be expressed, the salient features of the original complex of conditional stimuli must be present. This is because memory is no more than the elicitation of a previously learned behaviour by the appropriate set of conditions (or stimuli). If the important features of such a complex of stimuli are absent, then the memory is not retrieved, i.e. the behaviour is not elicited. Shock distress initiates a massive time-dependent hormonal change in the animal, such that, after about 3 h, the salient features of the animal's combined internal and external stimulus environment have changed dramatically. The consequence is that the subject cannot retrieve the memory because the stimulus environment that was present during original learning (the complex of cues that had become 'bonded' to the habit at that time) was no longer present.

The evidence for this memory theory is impressive. If, for example, rats are taught to avoid shock by running vigorously from one side of a box to the other, they readily learn the opposite behaviour (i.e. to avoid shock by sitting passively) if the training is given 3 h later. No confusion arises because they have forgotten all about their active avoidance learning. However, animals cannot easily learn the new passive response if training takes place immediately or 24 h later. Here, the remembered tendency to perform the active response interferes with their learning to do the opposite behaviour, namely to be passive. The reverse experiment has also been done. Passive avoidance learning interferes with learning an active response when training for the latter is given 0 or 24 h later, but not when given 3 or 4 h later.

There is a great deal of additional evidence supporting the general proposition that memories of aversively motivated behaviour are poorly retrieved after an intermediate retention interval due to a marked change in the internal environment of the animal. The generic term for this phenomenon, which occurs in many other species and testing situations, is *state-dependent memory*. Proper retrieval of habits or information depends to a large degree upon whether the animal, even a human being, is in the same 'state' (meaning its combined external and internal environment) as when it ori-

ginally learned the habit or information. If the states do correspond, whatever they happen to be, then good memory is found; if they do not, forgetting is observed.

This retrieval theory has been strongly supported by research on an altogether different phenomenon, that of retrograde amnesia. If an animal experiences electric shock treatment of the sort given to people for the treatment of depression, the events (responses or stimuli) that happened just seconds prior to the electric shock are forgotten; the animal is quite unable to demonstrate any memory for them when tested at a later time. This retention loss was traditionally believed to support a decay-type theory of memory. According to the argument, if memories are to become long-term they must consolidate for some period after they are formed. If the memory is not allowed enough time to complete this consolidation process, then it will fail to become fixed, and will be forever unavailable to the subject. Electric shock therapy, therefore, is said to interfere with consolidation, to disrupt the creation of a long-term memory. Recent evidence, however, favours a retrieval (or interference) type theory, by showing that a supposedly destroyed (i.e. unconsolidated) memory can be performed on the retention test provided the animal is given a reminder trial just beforehand. Such a reminder serves to jog the memory by restoring the internal and external stimuli to their original condition (i.e. the condition that existed before the electric shock treatment). In general, then, much research has demonstrated that an experience which changes the physical or psychological state of the organism to a sufficient degree may later cause interference with the retrieval of memories that had been learned in a different state.

An animal's medium-term memory may also reflect important biological constraints. A good illustration of this is the rat's learned anticipation of a diurnal meal. Several different investigators have shown that rats will learn to anticipate their daily meal by becoming active just prior to feeding, provided that food is presented at the same TIME every day. This phenomenon is not at all uncommon among zoo animals. The way they do this is by monitoring their internal cue state at the time of feeding; when the appropriate cues are present, perhaps a certain level of blood sugar, then the anticipatory behaviour occurs. This process, of course, requires a highly accurate memory for the internal cues over a 24-h period.

This memory phenomenon assumes even greater significance when one notes that rats are quite poor at anticipating other diurnal stimuli that are not biologically meaningful. For example, if an animal is given a brief electric shock at the same time each day, no anticipation, in the form of incipient avoidance behaviour, is ever acquired: the rat apparently cannot remember the internal stimuli that precede shock from one day to the next.

Long-term memory. It should be clear from our previous discussion that the duration or permanence of any given animal memory over the short and medium term depends on a great number of factors. While our knowledge of these incredibly complex processes is very incomplete, we can specify some of the general principles of animal memory. This is the case for long-term memory as well.

Some research, for example, has investigated long-term memory of several different types of behaviour in adult rats. In one study, subjects were trained to press a lever to obtain food according to a fixed interval schedule, i.e. a lever press would be rewarded if a constant and fixed amount of time had elapsed since the last food pellet (see REINFORCEMENT SCHEDULES). Under these conditions, rats, and most other species, normally pause for a few seconds immediately after receiving food; then later they begin pressing again, but the rate of responding gradually increases until it reaches a maximum just prior to the next scheduled reward. The investigators argued that this post-reinforcement pause is an indication that the rats remember that once they receive food they will not receive any more for the next few seconds. In testing the endurance of this memory-for-no-reward, the experimenters found that after 24 days the animals had forgotten the behaviour; when put back into the apparatus and allowed to obtain food, the rats failed to show the normal post-reinforcement pause, suggesting that they had forgotten that food was never delivered at that time.

Other studies of memory for rewarding behaviour produce essentially the same results. For example, 66 days after being trained in a maze, rats appear to have forgotten the visual characteristics of the maze (changing the colour did not affect behaviour the way it did when carried out with a 1-day retention interval). Similarly, rats forget how large a reward was during original learning. When, as in one experiment, they were accustomed to receiving twenty pellets for a response, switching them after 1 day to two pellets produced a very marked decrease in responding, but switching them after a 68-day interval did not produce such a reaction.

The fact that these well-learned habits were subject to forgetting after a few months does not suggest that all memories follow the same time-course. In particular, two sorts of habits appear to be highly resistant to forgetting. First, a rat's long-term memory for very simple spatial ORIENTATION behaviour (e.g. choosing the right-hand lever) is undiminished after $1\frac{1}{2}$ months. Although we must be very cautious in our interpretation of this finding (it is almost certain that such behaviour is conditioned more strongly to begin with) it makes sense from the point of view of evolution for easily learned behaviour to be easily remembered as well. Spatial orientation, smell, taste, etc. are important sensory dimensions for rats; and hence tasks which involve those dimensions seem to be both learned and remembered better than certain other tasks, such as those based on vision. Conversely, birds, for whom visual tasks are 'easy', but spatial or olfactory ones are not, are better at remembering visually oriented tasks. In short, it appears that in this context, where 'simple' is used to mean 'biologically meaningful', simple tasks are retained better in long-term memory than more difficult, or less biologically relevant habits.

The second class of activity which is resistant to long-term forgetting in adults is fear-based behaviour. More specifically, when a conditional stimulus is followed by an aversive cue such as a shock, the animal learns to fear the new stimulus. Because this fear reaction is maintained at full strength for many months following original fear conditions, many investigators believe that long-term memory for aversive stimuli (those signalling potential or actual harm) is especially good.

The unique facility for remembering aversive events is, perhaps paradoxically, not found in young animals. Indeed, the research on this so-called *age-retention* (or infantile amnesia) phenomenon is clear and consistent. Memory is exceedingly poor during the very early stages of development, but improves gradually, reaching a maximum in early adulthood. If the animals are very young, forgetting occurs within days; at somewhat older ages, retention may be observed for a longer time. In all cases, however, infantile memories are more transient than those found in adult animals.

The age-retention phenomenon has been thoroughly investigated in a number of different contexts, and while it is true that memory improves with age, the rate of improvement depends upon the nature of the response. In one study, for example, juvenile rats trained at 15, 17, or 20 days of age showed greater forgetting of an escape response after 1 week than subjects who had been trained at 25 or 35 days of age (rats are weaned at about 21 days and reach sexual maturity around 35 days). Other studies have shown that rats who were given tone/shock pairings at 17–19 days retained their fear reaction for 42 days, whereas animals trained at 15–17 days of age completely forgot the response within 4 weeks. Finally, studies of memory of aversive flavours show that appreciable forgetting occurs only if the subject is trained at an even earlier age (i.e. 10–20 days); animals who had been trained at 18 days of age demonstrated good memory for the aversive flavour 56 days later. The main difference in these results, therefore, is the youngest age at which nearly complete forgetting occurs. Although young animals tend to forget all aversive responses and stimuli, the rate of forgetting for simpler and/or more biologically relevant behaviour does not seem to be quite as rapid as that for other responses.

It should be emphasized that considerable attention has been given to the possibility that young animals simply learn less well than their mature counterparts; if such were the case, it would, of course, be impossible to draw any conclusions about the strength of memory as a function of age. Fortunately, scientists do not believe that this is a problem. On the one hand, most of the behaviour, although admittedly not all, is acquired at the same rate or to the same degree by all age groups. Therefore, the performance levels do not suggest any deficit in learning for the juvenile subjects. Second, perceptual and muscular limitations, which change as a function of age, do not account for the forgetting of particular stimuli. For example, young rabbits (*Oryctolagus cuniculus*) which had apparently 'forgotten' a conditioned head-lift response to an odour, in fact showed retention for the conditional stimulus by performing a gross locomotor response instead; their perceptual and movement COORDINATION had changed during the first few days of life, but the strength of the memory for the odour was unaltered. This finding does not bear directly on long-term memories, but it strongly suggests that forgetting by juvenile animals may occur quite independently of other maturational changes.

Another important conclusion about the age-retention effect is that it is adaptive; forgetting, especially in young animals, is a useful trait, one that has evolved because it endowed the individual with selective advantages. If young animals were unable to forget past traumas, they would be forever haunted by outmoded threats. As they mature and improve in their defences and motor abilities, the need to remain fearful of previously conditioned noxious stimuli is diminished. To continue fearing such stimuli would not only be unnecessary, but indeed might preclude the learning of new and more relevant information concerning those or similar stimuli. Furthermore, in most of the mammalian species studied, a good memory is not really very important early in life because sufficient MATERNAL care is afforded to the young animals at a time when they are most vulnerable. A good memory for trauma only becomes imperative once the animal has matured and has left the mother's care.

The age-retention phenomenon poses a serious question: how can we explain the fact that certain animals do seem to remember important traumas even though they were learned early in life? The answer is *reinstatement*. This term refers to a procedure whereby animals are given 'reminders' during the retention interval which result in the preservation of the infantile memory. In other words, when the original conditions (those under which learning took place) are periodically restored, the animals show perfect retention; otherwise, they forget.

In the typical reinstatement procedure, only a portion of the original training experience is given. For example, consider a learned fear reaction: if conditioned at an early age, it is forgotten within a week or so. However, if a rat is given a single electric shock once a week, even without the other stimuli involved in original learning, such as the

same apparatus, the memory of the aversive experience remains for an indefinite period of time.

It should be noted that the reinstatement procedure does not produce additional learning. Rather, it is merely a reminder of past learning.

The reinstatement phenomenon helps us to understand the functional value of the age-retention effect because it provides a mechanism for explaining why certain infantile memories may persist while other, perhaps less 'relevant', memories are forgotten. In human terms, early traumas may indeed come to dominate adult behaviour, but only when the aversive stimuli were consistently reinstated during childhood.

The study of reinstatement provides strong support for the retrieval theory of forgetting because reinstatement, by its very nature, involves the restoration of original training cues, hence the maintenance of an effective memory retrieval system. As discussed earlier, effective memory retrieval occurs when those stimuli that were 'bonded' to the behaviour during training are also present during testing. If they are not present, the behaviour is not elicited. From the experimenter's point of view, the habit is forgotten.   R.M.T.

78, 119

MENSTRUAL CYCLE refers to the repeated pattern of periodic bleeding from the uterus and vagina which occurs in sexually mature women, in all species of apes, and in some kinds of monkeys (all classified as Simiae). The bodily changes are in response to fluctuations in the level of circulating HORMONES, and the latter may also give rise to changes in behaviour, especially SEXUAL behaviour in the case of sub-human primates. In human beings the changing hormone levels influence more global aspects of behaviour, and in particular produce variations in mood and temperament which in Western culture is recognized as a distinctive syndrome. The menstrual cycle in non-human primates has a considerable similarity to the *oestrous* cycle in other female mammals. The oestrous cycle is so named because of the presence of a period of oestrus or heat, meaning a sharply delimited time when the female shows intense APPETITIVE behaviour for COPULATION. Oestrous cycles are not accompanied by any external blood loss.

In order to follow the changes in behaviour over the course of the menstrual cycle it is necessary to describe the fluctuations in the circulating levels of hormones, which are a result of complex interactions between the BRAIN and the ovaries. The cycle is under the control of the pituitary gland situated at the base of the brain. The anterior part of the pituitary gland secretes a number of hormones which have an effect only on their target organ, the ovary. In fact, the activity of the anterior pituitary is directly controlled by the part of the brain called the *hypothalamus*, and when the blood supply connecting the anterior pituitary with the hypothalamus is severed, the anterior pituitary becomes almost functionless, as the hypothalamus itself effectively acts as an endocrine gland producing substances which trigger the release of hormones from the anterior pituitary. The sequence of events in the human menstrual cycle is shown in Fig. A.

The ovary contains many thousands of egg cells, but the release of eggs for potential fertilization must await the age of sexual maturity or menarche, i.e. when the first menstruation occurs. The exact conditions which result in menarche are unknown, but undoubtedly nutritional influences play an important role in human beings. The age of menarche in women is almost 5 years earlier today than it was a hundred years ago, and may range from an average of 12·3 years in Cuba to 18·8 years for the Bundi tribe of New Guinea. There is good evidence that the critical factor for menarche is reaching a particular bodyweight in relation to height. In polyoestrous mammals such as the laboratory rat (*Rattus norvegicus*), the exposure of immature animals to extra illumination leads to an earlier onset of sexual maturation as indexed by earlier spontaneous vaginal opening, and if kept in constant darkness, sexual maturity is delayed. In view of this observation, it is surprising that blindness in human beings seems to accelerate sexual maturity, and the more total is the visual loss, the earlier the age of menarche. The important environmental signal in the normal control of the oestrous cycle in the rat is the diurnal light–dark cycle, and if sexually mature rats are kept in either continuous darkness or continuous light, then they will stop cycling altogether, and cease to ovulate (see RHYTHMS). It is unlikely that the control of the human menstrual cycle is dependent on light, although some women report that their cycles may be a few days longer in the winter months, when the daylight hours are shorter, than during the summer months. As 'menses'

means a lunar month, it is worth noting that one hypothesis for the timing of menstrual cycles is that they are dependent on monthly variations in moonlight (see TIME). In one recent study of women with variable onset of menstrual periods, it was found that artificial illumination of the bedroom from the 14th to the 17th nights following the onset of menstruation resulted in the cycle length becoming a regular 29·5 days (the natural synodic month). That this is a biologically significant period for human beings is suggested by the fact that the average duration of pregnancy is a precise 9 × 29·53-day synodic months.

Although it is conventional to consider the first day of bleeding to be day 1 of the menstrual cycle, it is more useful to describe the hormonal control of the cycle by commencing with the development of *ovarian follicles*. These follicles contain the egg attached to the lining wall, and they develop under the influence of follicle-stimulating hormone (FSH) released into the blood-stream by the anterior pituitary. For unknown reasons, it is most unusual for more than one follicle to ripen at a time, and only one of the ovaries is active in each menstrual cycle. The egg is surrounded by a group of smaller granulose cells and as these multiply they secrete an oestrogenic hormone called oestradiol. The latter has a positive FEEDBACK influence on the hypothalamus and the anterior pituitary, causing the release of more FSH. In this early *follicular* phase of the cycle, the lining walls of the vagina and uterus begin to thicken, and the uterus in particular develops a rich blood supply. In women, after approximately 2 weeks the follicle is ripe and the egg becomes detached from the wall of the follicle. The ripe follicle is now about the size of a large pea, and has a very thin outer wall; it bulges out of the surface of the ovary. The level of circulating oestradiol has now reached a peak and

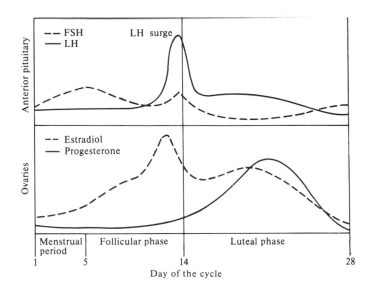

Fig. A. Changes in hormone levels during the menstrual cycle.

begun to fall, and shortly after its decline the pituitary releases a massive amount of a second hormone, luteinizing hormone (LH). This surge of luteinizing hormone probably causes the follicle to burst, and ovulation occurs with the release of the egg into the Fallopian tube. When the follicular wall ruptures there may be slight bleeding and some blood loss into the abdominal cavity. This irritates the lining of the abdomen and causes abdominal pain such that some women know when they ovulate by the occurrence of this midcycle pain (known as *Mittelschmerz*, from the German). There may also be some very slight blood loss externally at ovulation. In some mammals, e.g. the rabbit (*Oryctolagus cuniculus*), ovulation is triggered by the act of copulation, and this mechanism has been used to explain cases in which human pregnancy apparently follows copulation early or late in the cycle, but there is no definite evidence for such a stimulus in man. The only reliable way of detecting the time of ovulation is to record body temperature daily; in a normal cycle there is a sudden rise in temperature by about 0·5 °C on the day of ovulation, which is maintained for the rest of the cycle. If the cycle is anovulatory, no temperature change is detected.

The cells of the ruptured follicle now multiply and develop into the *corpus luteum* which has a characteristic yellow appearance. This corpus luteum secretes a second ovarian hormone called progesterone which has a negative FEEDBACK effect on the hypothalamus and pituitary, resulting in a reduction in the secretion of FSH. Progesterone has additional effects on the uterine wall, preparing it for the implantation of a fertilized egg; in particular it makes the uterus more secretory. The corpus luteum continues to secrete oestradiol, so whereas the early follicular phase of the cycle is dominated by oestradiol, the later *luteal* phase is characterized by high levels of both oestradiol and progesterone. If fertilization of the egg does not occur, then the lining of the uterus which has been built up by oestradiol and further stimulated by progesterone disintegrates and is shed into the uterine cavity. The blood capillaries in the uterus also burst and this blood together with the sloughed-off uterine wall constitutes the menstrual flow, the external and distinguishing sign of menstruation. Rhythmic contractions of the uterine wall gradually expel the discharge and may be sufficiently strong to be experienced as pain. Variability in the length of the cycle is due to the follicular rather than the luteal phase, and in woman the likely time of ovulation is most accurately placed with respect to the first day of menstrual bleeding, and usually occurs 14 days prior to this event. The period of bleeding is highly variable and ranges from 2 to 7 days. Anovulatory cycles are not uncommon, particularly for the first few cycles after the menarche and after about 30 years in woman, when the ovaries cease to function and menstruation ceases at the menopause.

Apes and Old World monkeys have cycles characterized by the same physiological events as women, whereas New World monkeys do not bleed externally, but have instead a regular rhythm of slight internal blood flow; in the very strict sense of the term therefore, they do not exhibit menstrual cycles. However, even in some Old World monkeys, e.g. the blue monkey (*Cercopithecus mitis*), the bleeding is very slight and not visible externally. In baboons (*Papio*), chimpanzees (*Pan troglodytes*), and rhesus macaque monkeys (*Macaca mulatta*) the time from onset of menstruation to ovulation is variable, mostly in the direction of lengthening rather than shortening the cycle. STRESS and SOCIAL factors are almost certainly responsible for prolonging cycle lengths in these species, as group-living animals, either free or in CAPTIVITY, have longer mean cycle lengths than isolated females. Otherwise the cycle length, like that of human beings, ranges from 20 to 35 days, whereas in a New World species, such as the squirrel monkey (*Saimiri sciureus*), it is only 1 week in duration.

Whereas in human beings menstruation is the norm, in other primates regular menstrual cycles rarely occur, usually being replaced by prolonged periods of pregnancy or lactational amenorrhoea. The distinctive external changes of the non-human primate menstrual cycle are not so much the occurrence of blood loss, but external skin changes around the genital area at mid-cycle. The difficulty of observing menstruation in the wild often makes it impossible to be certain of the exact phase of the menstrual cycle, and complicates any attempt to match behaviour with the underlying physiology. The presence of bright red hindquarters in baboons and other Old World monkeys usually signifies that the female is receptive, and she may in addition develop a sexual swelling sometimes reaching enormous proportions. This sexual skin is a specialized area adjacent to the external genitalia, flesh-coloured to bright red, and characterized by a thick dermis, a marked oedema (swelling) of the subcutaneous connective tissue, and by having a very rich blood supply. Swelling and colour changes roughly coincide with the follicular phase of the cycle. The prominence of these changes depends on the species and on the individual, and may also vary because of social factors. For example, when a female talapoin monkey, the dwarf guenon (*Cercopithecus talapoin*), is caged with a male, her swelling becomes redder; isolated female baboons have much larger and paler swellings than those living in groups. Generally, in baboon species, the colour and shape of the sexual skin is highly individual, and it is claimed that a baboon is easier to recognize from her rear than from her front view. The chackma baboon (*Papio ursinus*) has an enormous swelling, much bigger than the other baboon species, the yellow baboon (*Papio cynocephalus*) has a heart-shaped swelling, and the olive baboon (*Papio anubis*) a less clearly defined lumpy swelling. Macaque monkeys (*Macaca*) have wide varieties of sexual skin, and in the rhesus macaque and the Japanese macaque (*M. fuscata*) there is reddening but no swelling of the rump; in response to oestrogen, reddening appears also on the face, nipples, and belly. In the bonnet macaque (*M. radiata*) there is no sexual swelling and no sexual skin change. The colour changes of pregnancy are in fact more dramatic: all the above areas are brilliant red by the end of gestation, but immediately after parturition they pale again.

The swelling and distinctive coloration of the sexual skin magnify the effect of females presenting their hind-quarters to the male as an invitation to copulation. But presenting is not only shown by females on heat, but by females of all ages and at all stages of the cycle, and also by males and very young animals. It is, in fact, a submissive gesture, broadly distributed among Old World monkeys, and functioning as an APPEASEMENT or 'greeting' gesture (see DOMINANCE). Females on heat present most intensely, and presumably these conspicuous œstrous symptoms have a high stimulus value to the male. In view of the changes in EMOTION reported by women: feelings of elation and well-being mid-cycle, and irritability, fatigue, and tension premenstrually, it is interesting that female chimpanzees present without any sign of fear in mid-cycle, and rhesus monkeys are particularly aggressive.

It is very difficult to decide from FIELD STUDIES whether at ovulation and at other stages in the menstrual cycle, females become more attractive to the male, or are more receptive and solicitous of male advances. Actual measures of copulatory activity in LABORATORY situations in rhesus monkeys indicate that there is a significantly higher frequency of successful copulation in the follicular compared to the luteal phase. Whereas it is characteristic of sub-primate species to mate only during the restricted period of oestrus in the female, in female apes and Old World monkeys at least, as in human beings, copulation occurs throughout the menstrual cycle, but in some species is far more frequent in the follicular and mid-cycle phase. Whereas the laboratory studies indicate a clear period of oestrous behaviour in rhesus monkeys, in field and group studies whether a female mates with a particular male is determined not only by her reproductive condition, but also by the dominance relationships among the males themselves. In the laboratory, there is evidence that the increase in attractiveness of females at mid-cycle is due to some breakdown product of oestrogen present in the vaginal odour, which functions as a PHEROMONE. However, if males are offered the choice of two female rhesus monkeys in an experimental situation, it is not possible to predict the pattern of interaction simply from a knowledge of the hormonal state of the females.

In human beings, there are obvious difficulties in obtaining accurate accounts of reproductive activity in relation to phase of the menstrual cycle. Some of the older questionnaire studies revealed peaks in women's desire for copulation before and immediately after the menses, but not mid-cycle. More recent studies in which women kept daily diaries of their sexual activity revealed peaks in the frequency of copulation at about the time of ovulation, and a secondary peak just before menstruation. How far this reflects hormonal influence rather than CULTURAL influences is impossible to assess. Certainly the time of the menses is associated with a variety of taboos and regulations in many different cultures. Semitic races in the Middle East consider the woman to be impure and unclean at this time, and require her to leave home and live in a menstruation hut or tent. Sexual intercourse is proscribed during the menses, and the woman may not prepare food, or enter the home of a sick person or of a woman in labour. As indicated earlier, in Western culture there is recognition of a premenstrual syndrome characterized by irritability, headaches, abdominal pain, insomnia, and tearfulness. Recognition that behaviour is determined by endocrine status receives formal recognition in France, where by law a woman who commits a crime during her premenstrual period may use this fact in her defence, and claim temporary impairment of sanity. Careful measurements in laboratory situations indicate that there are slight changes in the sensory thresholds for olfaction, VISION, and tolerance of pain in the course of the menstrual cycle. Most of these changes are not, however, apparent to most women, and have no significant influence on behaviour. As with the primate field studies, the exact phase of the cycle is often not known with certainty, and, to be truly reliable, human studies need either to follow the same subject over several consecutive cycles, or, alternatively, to show a correlation between behaviour and the level of circulating hormones.   P.W.

MIGRATION. In the minds of many people the word migration is closely associated with the seasonal movement of certain species of birds. Since biblical times it has been observed that 'the stork in the heaven knoweth her appointed times; and the turtle and the crane and the swallow observe the time of their coming' (Jeremiah 8:7). Such migration involves regular, cyclic journeys between summer breeding ranges and winter resting areas. There is also another type of migratory phenomenon with which the biblical authors were familiar: the plague of locusts (Acridoidea) in Egypt that were brought by 'an east wind upon the land all that day, and all that night' and which 'covered the face of the whole earth' (Exodus 10:13, 15). Many insects make long journeys at particular stages in their life cycle. This is often a means of DISPERSION. Similarly, there are human population movements which serve a dispersal function. The exodus of the Israelites from Egypt and their 40-year journey to the land of Canaan provides a biblical example of such a migration.

The problem is to frame a definition of migration which is wide enough to include the various patterns of animal movement that are generally considered as migratory, while excluding trivial movements from the category. In order to understand this problem it may help if we review some of the FUNCTIONAL reasons which drive animals to move about within their environment.

Functional aspects of migration. In order to stay alive and grow an animal needs to feed at more or less regular intervals. FORAGING may involve active SEARCHING and HUNTING in a suitable HABITAT. SEXUAL activities also often require active search, and many animals travel long distances during the breeding season.

Most authorities would not include movements made in search of food or sexual partners as examples of migration. However, the spatial distribution of these resources in the environment may be very wide indeed, involving the animal in substantial journeys.

The position of the earth in the solar system imposes a number of cyclic rhythms on all life on the planet. The rotation of the earth on its axis produces a 24-h periodicity of light and darkness which places temporal restrictions on the TIME at which an animal can exploit the resources in its environment. In adapting to this external periodicity many, if not all, animals exhibit endogenously timed daily cycles of activity. The rotation of the moon around the earth produces a 28-day periodicity of variation in tidal height (and also in brightness of moonlight). This rhythm restricts the times at which the resources of certain regions of the inter-tidal area of the sea-shore are available to many marine and maritime animals, and is reflected by the endogenous lunar- and semi-lunar rhythms of LOCOMOTION and sexual activity shown by these species. Superimposed upon the daily and monthly rhythms of life on earth is the annual periodicity caused by the rotation of the earth around the sun. The seasonal changes caused by this annual cycle are most pronounced in their environmental effects at the poles, have intermediate effects in the temperate latitudes, and minimal effects on the equator. Many animals have

adapted to these seasonal changes in CLIMATE either by changing their behaviour as the seasons change, or by moving regularly between different environments, vacating a particular area when the season becomes inhospitable.

The regular annual and lunar movements of animals, either as individuals, or as populations, are normally accepted as instances of migration. Regular movements on a daily cycle, however, are not often considered as migrations, except when exhibited by animals with a relatively short life cycle.

An animal's life cycle passes through a series of stages, each of which may require the exploitation of different environmental resources. Food requirements may be quite different for immature and adult individuals of the same species; the JUVENILES may be more prone to PREDATION, and hence require an environment that affords them better opportunities for evading detection; while adults will be concerned with finding conspecifics of the opposite sex for reproduction. These changes throughout ONTOGENY may well cause an animal to make one or several journeys during its lifetime.

A possible way of classifying migrations is illustrated in Fig. A. The first distinction is between species migration, where an entire species moves its geographical range, and individual migration, where only the movements of a single individual during its lifetime are considered. If migration is assumed to be of SURVIVAL VALUE to the individual, its initiation must normally be non-accidental; purely accidental migrations do, however, sometimes occur and may be of great significance in the colonization of new HABITATS. Of animals that are intentionally on the move, some may know where they are going (*calculated migrants*) while others will be venturing into uncharted land (*non-calculated migrants*). Non-calculated migrations may be EXPLORATORY, serving to extend the familiar area within which the animal is living, or may involve the permanent removal of the animal to a new and unfamiliar spatial unit. Calculated migrations may also involve one or a series of

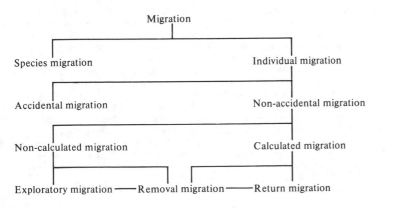

Fig. A. Classification of migrations.

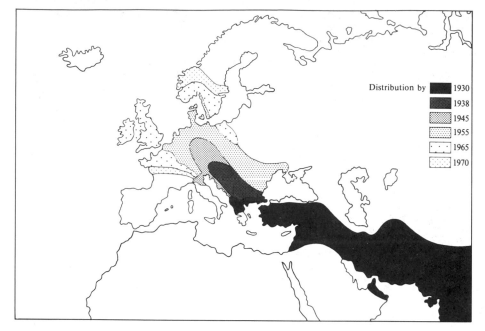

Fig. B. Map showing the expansion in range of the collared dove (*Streptopelia decaocto*) this century.

removals between spatial units; furthermore, these removals may eventually bring the animal back to the place from whence it came, thereby constituting a return migration. Thus, both exploratory and return migrations can be considered as series of removal migrations. This CLASSIFICATION provides a useful framework within which to consider specific examples of animal migration.

Species migration. Animal migration at the individual level describes the movements of a single animal as it traces out its lifetime track, and the summation of these tracks comprises the HOME RANGE of that species. Extensions of this range can only be made by individuals wandering further afield than any other past or present members of the species.

A good example of such range expansion is provided by the recent spread of the collared dove (*Streptopelia decaocto*). Prior to 1930 the species was primarily an Asiatic bird, not extending further into Europe than the Balkans, where it was largely resident. It then began a dramatic north-westerly extension of its range, colonizing the whole of central Europe and southern Scandinavia (see Fig. B) and reaching Britain as a breeding bird in 1955, since when it has spread to become common throughout the country. Its success can probably be attributed to the existence of an under-populated ecological NICHE that it was able to exploit as it spread. However, its expansion has been remarkably uni-directional, and it seems likely that the spread was triggered by a GENETIC mutation (sudden change in a gene) in certain individuals of the original western peripheral population, which caused the majority of all juvenile dispersions to be north-westerly. The spread of the collared dove has now been temporarily arrested at the North Atlantic seaboard, but birds have been observed flying out over the ocean in a westerly direction, so the colonization of North America remains a possibility.

A similar, though less dramatic, extension of range has also been seen in the brown-headed cowbird (*Molothrus ater*). This is a North American song-bird, which lays its eggs parasitically in the nests of other species (see PARASITISM). It typically feeds by picking up insects disturbed by herds of browsing cattle (Bovidae) and, before the European colonization of America, was restricted to the great plains where it associated with the American bison (*Bison bison*). The introduction of domestic cattle (*Bos primigenius taurus*), and the clearance of the eastern forests, has allowed the cowbird to extend its range across to the eastern seaboard, and in so doing it has benefited additionally by being able to parasitize new host species not adapted for dealing with nest-parasites. Thus, as the bison declined in numbers, the cowbird flourished (see Fig. C). In this example there is no evidence that the range extension was genetically mediated; major habitat changes will have simply opened up new FEEDING and nesting opportunities for the species.

Accidental migration. For a migration to be classified as accidental its initiation and progress must be due to the failure of that animal's normal station-keeping mechanisms. The migration need not, however, be disadvantagous to the animal concerned. For instance, a pregnant female lizard (Lacertidae) could be washed out to sea on a piece of driftwood, land on an oceanic island, and there lay its eggs and found a new colony. It is argued that much island colonization must have occurred in just this way. Again, a few rats could board a cargo boat and be transported to another continent; both the black rat (*Rattus rattus*) and the Norway or brown rat (*Rattus norvegicus*) first entered Britain in this manner. On the other hand, neither a ballooning (or gossamer) spider (Linyphiidae) that spins its parachute, takes flight, and is then carried by the wind well beyond the limits of its normal range, nor a migrating bird that gets caught up in adverse weather conditions and is blown off-course, is an accidental migrant, because in both cases the initiation of the movement was deliberate. The occasional records of vagrant American song-birds that arrive most autumns somewhere along the European Atlantic seaboard (the Isles of Scilly being a particularly popular landfall) constitute an intermediate case. It is clear that at least some of these birds are assisted in their transatlantic journey by resting on ships, where they may obtain food, shelter, and passive transport. These birds certainly 'meant' to begin their migratory journeys, and their FLIGHT paths may have been based on expectations of particular weather conditions over the ocean, but it seems unlikely that any species has yet evolved the strategy of deliberately exploiting ships as a means of transport.

Exploratory migration. An exploratory migration is one which takes an animal outside the limits of its familiar area without precluding the possibility of subsequent return. It is the means whereby an animal can assess the resource properties of unfamiliar, possibly even distant, habitats before deciding whether to move permanently to a new location.

To keep track of its position during these explorations, an animal must possess a well-developed spatial MEMORY. For this reason, examples of exploratory migration in invertebrates are rare. Honey-bee (*Apis mellifera*) workers may forage in areas around the hive that are unfamiliar to them and to other workers from the hive. When a good food source is located, the foragers will return to the hive and communicate the position of the food by means of their 'waggle dance'. When searching for possible locations at which to establish a new hive, foraging workers use the same COMMUNICATION system to spread information about the location and characteristics of possible sites that

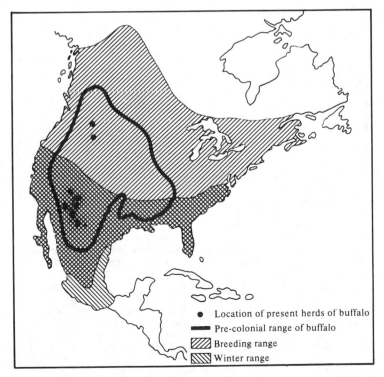

● Location of present herds of buffalo
▬ Pre-colonial range of buffalo
▨ Breeding range
▧ Winter range

Fig. C. Current world distribution of the brown-headed cowbird (*Molothrus ater*) compared with the original and current range of the American buffalo (*Bison bison*).

they have visited. In this way the exploratory migrations of individual workers can, through SOCIAL communication, extend the effective size of the familiar area of all the closely related members of the hive.

Social communication of the location of newly discovered food sources is accomplished by ant (Formicidae) and termite (Isoptera) foragers by laying a SCENT trail during their return to the colony. Other colony members can then follow this trail to the food. There is, however, no evidence that the foraging ants or termites lay scent trails on their outward journeys to help them find their own way back, although slugs (Arionidae) certainly use this mechanism to retrace their steps.

Amongst vertebrates, however, it seems that exploratory migration is a widespread, if not universal, phenomenon, being a particular characteristic of the movement patterns of immature animals. Immaturity is the period of an animal's life cycle in which it comes to command sufficient resources to be a successful reproducing individual. These resources will have been available to the animal's parents in its natal area, but to remain close to the PARENTAL home would be to compete for resources with one's own family, and possibly also to run the risk of inbreeding. For these reasons, many vertebrates, once they are independent of their parents, enter a period of exploratory migration, often called post-juvenile dispersion.

The barn swallow (*Hirundo rustica*) provides us with a good example of such dispersion. Ringing recoveries have revealed that young swallows may, after fledging, move as far as 200 km in any direction from the natal site, although most movements are less than 50 km in distance. Later in the autumn the ringing recoveries indicate that the young birds are making an oriented emigration to the south-east, along with adults, who do not show such dispersive movements. It is not known exactly what the juvenile swallows do on their dispersive journeys, but there is observational and circumstantial evidence that they are exploring the areas around their natal sites, and so building up detailed knowledge of the resource characteristics of their expanding familiar area. In particular it seems that young male swallows may be searching for suitable NEST-BUILDING sites in which to breed the following season. Young birds have been observed entering barns where swallows were still rearing second clutches, and ringing evidence shows that birds who were hatched early in the season, and who have therefore had plenty of time to explore their environment before emigrating south, are much more likely to establish themselves as breeding adults the following year than are their later-hatched second-clutch siblings. Similar dispersion patterns are characteristic of many bird species in which the males establish the breeding TERRITORY, to which they attract a female, and where their food resources show a consistently patterned distribution in space.

Where food resources are less reliable, animals

may have to make exploratory migrations to establish the location of new, adequate supplies. The red crossbill (*Loxia curvirostra*) feeds largely on the seeds of spruce, which are set in the June of one year, and remain available to the birds until the following June, when they fall. Spruce seed crops in any one area are, however, highly variable, so that any particular area is unlikely to sustain a crossbill breeding population for two successive years. When the new crop of spruce seeds is forming in early June the crossbills are most likely to initiate exploratory migrations, which may take them anything from 100 to 4000 km away. The birds, having found a suitable new feeding area, will settle to breed early the following year. However, ringing recoveries indicate that they retain the ability to return to their natal area in subsequent years, which distinguishes this exploratory migration from the non-calculated removal migrations (also known as *irruptions*) which will be described in the next section.

Exploratory migration also seems to be a characteristic feature of the behaviour of many Pinnipedia (sea-lions, seals, etc.). In adult life these animals show impressive colony fidelity, and there is evidence from at least some species that juveniles may return to their natal colonies to breed as adults. However, marking studies have also indicated that juveniles may visit and sometimes settle to breed in colonies far distant from that in which they were born, and it seems likely that these eventual removal migrations are being made within a large familiar area that was established by their juvenile wanderings.

In the three examples above, the exploratory migrations described are *facultative* rather than *obligatory*; the animals are free to choose when to leave, although their DECISION may be based on their assessment of a dwindling environmental resource. Obligatory exploratory migrations seem to be rare, although post-fledging or post-weaning conflict between parent and offspring might determine the time at which the juveniles of some species begin their explorations. As will be seen in the next section, obligatory migration is more likely to lead to the removal of the animal to a new habitat altogether.

Removal migration. Within an established familiar area, animals may make removal migrations to areas of higher resource availability by using information detected by their SENSES, by relying on a memory of the distribution of resources within that area, or by LEARNING from conspecifics where the currently most profitable part of the area is.

Migrating brindled gnu or wildebeest (*Connochaetes taurinus*) on the plains of East Africa use their eyes and ears (but not, apparently, their noses) to sense nearby areas where rain has produced, or will soon produce, a fresh growth of grass. The herds move steadily towards the sight and sound of storms, but only so long as the rain is falling within their familiar area.

Hibernating bats (Chiroptera) in temperate regions have very precise requirements for their THERMOREGULATION if they are to survive the winter on their accumulated fat reserves, and must select their HIBERNATION places with care. Ambient temperatures near or below freezing will be lethal unless the bats increase their metabolic rate to generate body heat, and, in the process, use up energy reserves; alternatively, high ambient temperatures will again increase the metabolic rate of the animal, also leading to energy loss. To minimize energy utilization during the winter, the bats may need to migrate between caves, and there is evidence that at least one species, the greater horseshoe bat (*Rhinolophus ferrumequinum*), does just that. The distance between caves may be as much as 30 km, and it is believed that the bats use a memory of the seasonal characteristics of a range of potential hibernating places in order to choose their optimal cave for each period of the winter.

These examples provide instances of calculated removal migration. However, most insect migrations are non-calculated (i.e. the animal has no direct knowledge of its destination at the time that the migration is initiated), although some are none the less spectacular for that.

The monarch butterfly (*Danaus plexippus*) is a large, boldly patterned butterfly found in Canada, the U.S.A., and Mexico. In autumn most of the population makes a southward migration to the southern states and Mexico, where it may overwinter as a free-flying individual or, when the weather turns cold, aggregate in dense roosts. The autumn migration is fairly rapid, the butterflies averaging 120 km a day, but in the spring their return progress is much more leisurely. This is in part due to the fact that mating and egg-laying both occur *en route*, and the offspring so produced may themselves also mate and lay eggs before they reach the northern limits of their range in mid-July. Thus, the butterflies that return may be the first-, or even the second-generation offspring of those that set out the previous autumn, and it is uncertain whether any of the original migrants survive the entire round trip. The picture is further complicated by the fact that some (about 30 per

cent) of the northern butterflies do not migrate in the autumn, but hibernate, and it is the offspring of these individuals that are first seen in the north the following May.

Thus, what looks at first glance like a simple case of periodic return migration in a butterfly turns out to be a series of non-calculated removal migrations, involving two or even three generations of the species. Similar patterns of migration are found in Europe in the red admiral (*Vanessa atalanta*) and in the painted lady (*V. cardui*) butterflies. Both, like the monarch, are strong fliers, and are capable of sustaining rectilinear (straight-line) flight in a preferred direction, using the sun as a cue to ORIENTATION.

Apart from butterflies, the other group of insects that are widely recognized as migratory are the locusts. Because of the economic damage caused by locust swarms, the locust *Schistocerca gregaria* may lay claim to being the most-studied migratory species in the world. It is an inhabitant of seasonally arid areas, and the emerging hopper instars (stages in larval development) may expect to have to make an early removal migration in a relatively predictable direction if they are to breed successfully. The conditions of crowding under which the locusts have developed determine whether they develop into individuals of the *gregaria*, *solitaria*, or *transiens* phase. When they have become adult and the cuticle (the tough outer covering) has hardened, all three phases show a low migration threshold, but it is only the *gregaria* individuals that then form up into dense, day-flying swarms. The *solitaria* individuals migrate alone, at night. Although locusts can orient themselves with respect to the sun, the swarms tend to migrate downwind into areas of low pressure, where they are most likely to encounter rain. Wet conditions stimulate the locusts to stop migrating, and COPULATION and deposition of eggs occurs as soon as sexual MATURATION is complete. Locust swarms seem to describe seasonal circuits, which are either circular, or to and fro across the same territory (see Fig. D). However, the generation time of the *S. gregaria* is too short for individual

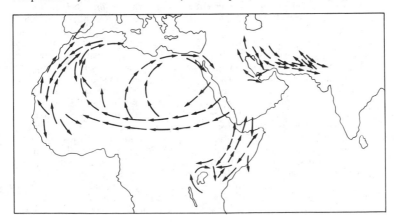

Fig. D. The migratory circuits of the locust *Schistocerca gregaria*.

animals to migrate round the entire circuit, although some seasonal changes of preferred direction are not associated with a generation change. Thus, locust movements must be considered as removal, and not true return, migration.

A final example of insect migration will illustrate the important distinction between facultative and obligatory initiation of migration in a species whose means of movement is largely passive. The bean aphid (*Aphis fabae*) spends the winter as a fertilized egg on the spindle tree, its primary host. In the spring wingless females that are both *parthenogenetic* (eggs develop without fertilization) and *viviparous* (eggs develop internally) hatch, and after a couple of generations these produce winged females that are obligatory migrants, flying out into the airstream in the hope of finding a secondary host plant, such as the broad bean. During the summer a number of generations of females are produced, which may be winged or wingless and are facultative migrants, only moving under conditions of low food availability and high population density. In the autumn winged males are produced that are obligatory migrants, and they fly off in search of a spindle tree; from the same generation winged females that are obligatory migrants are also produced, and these also fly off to the spindle trees, where they produce a generation of wingless *oviparous* (egg-laying) females who copulate with the males and produce eggs that overwinter. Within this cycle, no individual performs a return migration, and the extent to which the flights, obligatory or facultative, are directional is severely limited. None the less, there can be no doubt that the bean aphid is a migrant species.

One vertebrate that seems sometimes to exhibit non-calculated removal migration similar in nature to that of the migrant insects is the Norway lemming (*Lemmus lemmus*). These animals inhabit the alpine zone of Scandinavian forests, above the tree-line, and at times of food abundance their numbers can increase dramatically, partly as a result of young animals reaching reproductive maturity and breeding much earlier than is normal. These population explosions are then followed by a removal migration of a large section of the population, with immature males predominating. The streams of lemmings usually head down the mountain sides and into the valleys. Individual animals move in a straight line, giving no suggestion that they are either constructing or retracing a familiar path. They will not normally enter water unless they can see across it, although lemmings are quite strong swimmers. However, the build-up of lemming numbers along the edge of large expanses of open water may cause some individuals to set off into the open sea, thereby creating the myth that the species is essentially suicidal. It is assumed that the function of these emigrations is to colonize new habitats that are either vacant, or under-populated. The likelihood of achieving this aim will decline as the length of the migration in-

creases, but, if the natal area was full to capacity, the migrant lemmings have no alternative but to continue their search.

It can therefore be seen that lemming emigration differs fundamentally from crossbill exploratory irruptions. Insufficient data are available to decide whether the migrations of other irruptive species (e.g. the nutcracker *Nucifraga caryocatactes*) fit either of these two patterns.

Return migration. Return migration, in which an animal migrates back to a habitat that it has visited previously, is the most widely recognized form of migration. The most commonly identified return migrations are those with a periodicity of 1 year, like the return of the European cuckoo (*Cuculus canorus*) to the United Kingdom every spring. However, many return migrations have much shorter periodicities, while others may only occur at particular stages in the ontogenetic DEVELOPMENT of the individual.

Most animals of the oceanic plankton show a diurnal pattern of vertical movements. It is probable that these migrations are initiated in response to changes in light intensity and water temperature. In addition to the diurnal vertical movements, most species of zooplankton also show seasonal migrations within the surface waters. Vertical return migrations of daily, lunar, and seasonal periodicity can also be found in the movements of many inhabitants of sand, soil, and leaf-litter.

Return migrations with a lunar or semi-lunar periodicity are found in many animals that inhabit the inter-tidal region of the sea-shore, or the waters immediately off shore. The rock crab (*Cancer pagurus*) feeds in the inter-tidal zone, spending the low-tide period well buried in the sand. At spring tides sub-adult crabs move high up the beach, only to be trapped there by the retreating water, and certain herring gulls (*Larus argentatus*) at Walney Island in Cumbria, England, have learned to hunt for the buried crabs by SEARCHING for signs of the buried crab shells.

Return migrations of longer periodicities are, however, particularly characteristic of vertebrate movements. Fowler's toad (*Bufo fowleri*), for example, performs a regular DRINKING migration down to the shores of Lake Michigan, with a periodicity of 5 days. These migrations may involve a journey of up to 1·5 km each way and take place at night. They are presumably triggered by THIRST.

The exploratory migrations of swallows have already been described, but their annual return migrations are far more famous. Many farmers will insist that it is the same pair of birds that arrives every spring to nest high under the roof of his barn or cowshed, and while an average adult mortality rate of 67% argues against this regularly being the case, it is none the less true that an adult male swallow will continue to return faithfully to his nest-site as long as he survives, while his mate will, if she survives, return faithfully to him. The pattern of returns for first-year birds is less accurate, young males returning, on average, to within 5 km of their natal site, while females may settle 50 km or more away. This pattern of dispersion might well be taken to argue against the hypothesis that young swallows do indeed perform an accurate return migration in their second summer, but it is possible, indeed likely, that their future nest-site was selected during their post-fledging dispersion the previous year, so that all swallows young and old may be considered capable of returning to their familiar area of the previous summer.

British swallows migrate to Cape Province in South Africa for the northern hemisphere winter, and there is some evidence that they and other migratory birds return regularly to the same wintering area year after year. It is generally accepted that, to complete such a journey, a migratory bird must be capable of complex feats of NAVIGATION. The first journey can be considered as a special kind of exploratory migration, with the bird remembering the landmarks and other salient cues as it proceeds, perhaps willing and able to retrace its steps if it gets lost.

Long-distance migration also occurs in other vertebrate groups. Atlantic salmon (*Salmo salar*) and Pacific salmon (*Oncorhynchus* spp.), which spend most of their lives at sea, relocate their natal rivers when they return to spawn by discriminating the specific olfactory characteristics of that river. They initially become acquainted with this smell during their time spent as fry, prior to their migration down the river to the sea. As their olfactory sense is good enough to identify the water from a particular river, it probably also permits the animal to learn, as it moves out into the ocean, the olfactory characteristics of the currents in which it is swimming. It has been shown that plaice (*Pleuronectes platessa*) in the southern North Sea move round their migratory circuit by swimming up into mid-water, and then maintaining themselves in a favourable tidal stream until the next slack water, when they drop to the bottom again. It has been suggested that the progress gained in this way may be monitored with reference to the TASTE AND SMELL of water from submarine freshwater springs seeping out on the sea-bed. It is difficult for us to imagine what life within such an olfactory world would be like, but we have no reasons for believing such a navigational system to be impossible.

A great many vertebrate return migrations have been studied, involving animals moving by air, land, and sea. Whales (Cetacea) swim thousands of kilometres to return to the same traditional mating and calving areas; birds, such as the wandering albatross (*Diomedea exulans*), circle the southern oceans in a perpetual easterly direction, returning to their nesting island every 2 years; eels (*Anguilla*) hatch out in the Sargasso Sea on the western side of the North Atlantic, yet migrate as elvers across to the river systems of Europe, where they grow, before returning to the Sargasso to spawn and die; and the moose (*Alces alces*), though often considered to be a purely nomadic species, can be shown to make regular altitudinal return migrations within its northern forest habitat.

For most temperate species it has long been known that changes in the length of daylight as the seasons change may trigger many functions of an animal's annual cycle, including migration. Some animals, especially those migratory birds that spend some part of the year on or near the equator, where seasonal day-length changes are very small, may have their own internal *circannual* CLOCK which tells them when to moult, when to begin and stop migrating, when to breed, etc. It has been established that a number of European warblers, including the willow warbler (*Phylloscopus trochilus*), and the garden warbler (*Sylvia borin*), possess such clocks, and it is likely that they also occur in many more long-distance migrants, especially those that migrate to or across the equator.   C.D.

7, 24, 37, 123, 144

MIMICRY is the resemblance of one animal (the mimic) to another animal (the model) such that the two animals are confused. For a more precise definition it is necessary to specify a particular type of mimicry, the most important being Batesian, Müllerian, aggressive, and intraspecific mimicry.

Batesian mimicry. In this type of mimicry a predator, which avoids a noxious animal producing a particular signal (the model), is deceived into avoiding an edible mimic which produces a similar signal. It is of advantage to the mimic (see DEFENSIVE BEHAVIOUR), but is of no advantage to the model. The model is normally an animal with a nasty taste, a sting, or spines, but plants and inanimate objects can also be mimicked. For example stick insects (Phasmida) are elongated, and closely resemble twigs which are inedible to the predators that are likely to prey on the mimics.

In the early part of this century there was considerable controversy as to whether or not Batesian mimicry is really of SURVIVAL VALUE to an animal. In the LABORATORY it has been shown that birds that have been conditioned to avoid wasps of the genus *Vespula*, or the salamander *Notophthalmus viridescens*, both of which have warning coloration, will subsequently also avoid at least some mimics, for example hoverflies of the genus *Syrphus* and the red salamander *Pseudotriton ruber* respectively. Control birds that were offered only the mimics ate them, so it was only the prior experience of the noxious model that caused them to reject the mimics. In the same way it has been shown that Batesian mimicry can be of survival value against predatory fish and amphibia.

To demonstrate the effectiveness of Batesian mimicry in a natural situation one needs to demonstrate that a population of mimics suffers greater PREDATION in the absence of its model than with the model present, but this is not easy to do.

However, it is possible to predict that if a species is deriving protection from mimicry of an *aposematic* animal (which is an animal with warning coloration), then in the absence of the model it should exhibit different signals, otherwise these would attract inexperienced predators to the edible mimic instead of to the noxious model.

This prediction can be tested using the aposematic pipe-vine swallowtail butterfly (*Battus philenor*) and its mimic *Limenitis arthemis*. *Battus* occurs in North America from the Gulf of Mexico northwards to the Great Lakes. Throughout this region the population of *Limenitis* is black, like *Battus*, but in Canada where *Battus* is absent *Limenitis* is cryptically coloured black and white. In the narrow region where *Battus* is scarce both forms of *Limenitis* occur together with intermediates. Clearly where the model is present it is of advantage to *Limenitis* to be black, but where it is absent it is of greater advantage for it to be cryptically coloured (Fig. A).

Typically Batesian mimicry involves imitation of one species, the model, by another species, the mimic, and although the relevant predators may often be deceived into confusing the two, it is normally easy for a biologist to distinguish them. However, it is possible to have visually indistinguishable distasteful, aposematic individuals and edible, mimetic individuals in the same population. This is called *automimicry*. For example, caterpillars of the American monarch butterfly (*Danaus plexippus*) feed on milkweeds (Asclepiadaceae). Some species of milkweed are highly toxic whilst others are palatable. If a monarch butterfly was reared on a toxic milkweed (such as *Asclepias curassavica*) then it will be very poisonous and cause birds that

eat it to vomit, whereas a butterfly that was reared on a non-toxic milkweed (such as *Asclepias tuberosa*) will not make a bird ill. Birds that vomit after eating a toxic butterfly tend to refuse to eat similar coloured butterflies in future, so the insects reared on non-toxic plants are perfect mimics, both visually and in their behaviour, of those fed on toxic plants.

One limitation of Batesian mimicry is that the mimic species cannot become too common relative to the model, otherwise the predators will sample the mimics first and then develop a SEARCHING IMAGE for the mimics instead of a conditioned AVOIDANCE response to the models. Experiments have shown that the critical frequency of models and mimics that still gives some protection to mimics depends on the nastiness of the models and the behavioural characteristics of the predator, for example on how long it can retain the conditioned avoidance response and how hungry it is. It is well established that predators that normally refuse a particular prey may eat it if they are sufficiently hungry.

One way in which the population of a mimetic species can be increased above this critical mimic:model ratio is by limiting the mimetic form to one sex, as occurs in the tiger swallowtail butterfly (*Papilio glaucus*). Since the female can live long and lay many clutches of eggs from a single mating, whilst the male has only to fertilize the female, it is normally the female that is mimetic and the male that is non-mimetic.

Finally, some predators learn to discriminate mimics from models, and clearly the selective advantage of such DISCRIMINATION will be greater the higher the frequency of mimics in the population.

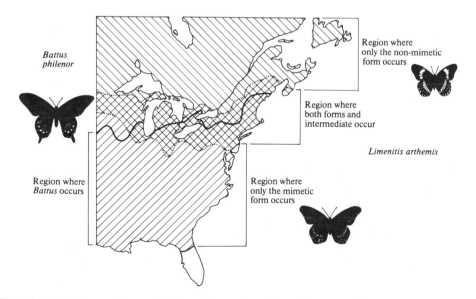

Fig. A. Geographical range of aposematic *Battus philenor* and two forms of *Limenitis arthemis*.

Hence there will be a corresponding pressure of NATURAL SELECTION on the mimics favouring those individuals that bear the closest resemblance to the model. In fact it has been shown, using artificial prey made of pastry with birds as predators, that even a very slight colour resemblance gives some protection to a mimic, but the greater the resemblance to the model the greater the level of protection achieved. Hence one can see how Batesian mimicry has evolved.

Müllerian mimicry. In this type of mimicry a group of noxious species share the same warning signals. European examples are the black and yellow social wasps of the genus *Vespula*, solitary wasps of the genera *Bembex* (digger wasp), *Eumenes* (potter wasp), and *Crabro*, and the cinnabar moth caterpillar (*Callimorpha jacobaea*). Birds that have learned to avoid cinnabar caterpillars will not attack wasps, but birds with no experience of either wasps or cinnabar caterpillars will attack wasps. Since predators need only learn one pattern in order to avoid all species in the assemblage, Müllerian mimicry is of benefit to all individuals and no deception is involved. Because all species in the assemblage are nasty, and are evolving similar warning signals, it is not always possible to say which species is the model and which are the mimics. In theory this is quite different from Batesian mimicry where models and mimics are easily distinguished, but in practice it is not always easy to decide if a species is a Batesian or a Müllerian mimic. This may be because a species varies in its palatability, or because it may be noxious to one predator but palatable to another, or because it may be avoided by predators when they are well fed, but eaten when they are starved.

Aggressive mimicry. In this type of mimicry one animal (the mimic) preys on or otherwise exploits a second animal because it mimics a signal that is normally attractive to (or at least is not avoided by) the second animal. For example the Malayan praying mantis *Hymenopus bicornis* is pink and rests on the flowers of *Melastoma polyanthum*, and closely resembles them in colour and shape. Insects attracted to the flower are caught by the mantid.

In Pacific coral reefs there is a small black and white striped fish, the cleaner wrasse (*Labroides dimidiatus*), which removes parasites from the bodies of larger 'customer' (or 'host') fish. Customers recognize the pattern of the cleaner and do not attack it, but adopt a characteristic posture which enables it to search their bodies and remove parasites, (see INTERACTIONS AMONG ANIMALS). However, the sabre-toothed blenny (*Aspidontus taeniatus*) is another black and white striped fish that occurs in the same area. Customer fish may approach *Aspidontus* in mistake for *Labroides*, and wait to be cleaned, but *Aspidontus* attacks and bites chunks out of their fins (Fig. B). *Aspidontus* is therefore an aggressive mimic of *Labroides*, and natural selection will favour those *Aspidontus*

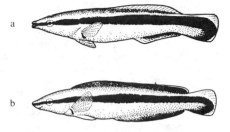

Fig. B. (**a**) The cleaner wrasse (*Labroides dimidiatus*) and (**b**) its mimic the sabre-toothed blenny (*Aspidontus taeniatus*).

which mimic the colour and behaviour of *Labroides* the closest, but also those customer fish which can discriminate *Labroides* from *Aspidontus*, and so avoid being attacked.

Intraspecific mimicry. In this type of mimicry the model, the mimic, and the animal perceiving the mimicked signal all belong to the same species. In the African cichlid fish *Haplochromis burtoni*, the female normally broods the eggs in her mouth until they hatch, and thus protects them from being eaten by predators. When ready to copulate the male digs a small scrape in the sand and attracts a female to the place. She spawns here and immediately takes all the eggs into her mouth. The male then ejects semen in the scrape, at the same time expanding his anal fin which has orange circles on it of the same size and colour as the eggs. The female attempts to suck these egg dummies into her mouth, and in so doing takes in semen which fertilizes her eggs. Obviously this mimicry is of benefit to both male and female, since it increases the chances of successful fertilization occurring (Fig. C).   M.E.

39

Fig. C. Male *Haplochromis burtoni* (right) ejecting semen whilst female (left) attempts to pick up dummy eggs on his anal fin, and in so doing takes up the semen.

MOBBING is a form of harassment directed at predators by potential prey. For example, owls (Strigiformes) that appear in the daytime are often mobbed by small birds. The birds make mock attacks upon the owl, alternately approaching and retreating, and uttering a characteristic VOCALI-

ZATION. For example, Brewer's blackbirds (*Euphagus cyanocephalus*) have been observed circling around a tree containing an owl, repeating a harsh call. Red-winged blackbirds (*Agelaius phoeniceus*) make hit-and-run attacks on owls, while wren tits (*Chamaea fasciata*) remain in dense shrubbery uttering ratchet-like calls which they may keep up for 2–3 h. The mobbing calls of birds are characteristically easy to locate, and serve to recruit would-be attackers from far afield. Mobbing is also found amongst mammals, particularly the primates. Baboons (*Papio*) and chimpanzees (*Pan*) launch mock attacks on leopards (*Panthera pardus*), screaming, charging, and retreating. Chimpanzees may even throw branches and sticks at the leopard (see TOOL USING). California ground squirrels (*Citellus beecheyi*) mob snakes (Serpentes), as do agoutis (*Dasyprocta*). Axis deer (*Axis axis*) may follow tigers (*Panthera tigris*) and leopards, barking at them from a safe distance.

Mobbing is characterized by CONFLICT between AGGRESSION and FEAR. In addition to the typical alternation of approach and retreat, DISPLACEMENT ACTIVITIES and other signs of EMOTION are common. Chimpanzees, for instance, frequently scratch themselves, hug and reassure each other, and show signs of diarrhoea during their encounters with leopards. Mobbing is also related to CURIOSITY and EXPLORATORY behaviour. An owl-like object will be mobbed by chaffinches (Fringillinae), but if some of the owl features are removed, the object will be explored. Chimpanzees presented with a stuffed leopard initially mob it, but later investigate it.

The SURVIVAL VALUE of mobbing lies partly in the warning given about the presence of a predator, for which the typical loud vocalizations are particularly effective. Mobbing often succeeds in driving a predator away, and thus increases the probability that it will eventually hunt elsewhere. A predator that is mobbed has lost all chance of a surprise attack. Breeding pairs of pied flycatchers (*Ficedula hypoleuca*) increase their frequency and intensity of mobbing during the course of their nesting period. It reaches a peak just before fledging, and may serve to distract predators from the nest. Mobbing is less prevalent in unmated birds, and in birds that have lost their brood. Flycatchers in Germany which live within the range of the red-backed shrike (*Lanius collurio*) will mob shrike dummies, but the Spanish subspecies living beyond the shrike's geographical range will not mob dummy shrike. Thus, in flycatchers at least, mobbing is related to the vulnerability of offspring, and is a form of DEFENSIVE behaviour against the predators typical of the vicinity.

90

MOTIVATION. In everyday language, the term motivation is used to describe the experience of desiring to act in particular ways in order to achieve certain ends. Thus if a man wants SEXUAL satisfaction, we say he is sexually motivated.

Desires may be classified along a continuum, ranging between impulsive irrational urges and calculated rational desires. During impulsive and irrationally motivated behaviour the individual feels himself to be driven by an internal urge, such as HUNGER, SEX, or FEAR. During rationally motivated behaviour the individual is usually conscious of a definite GOAL towards the attainment of which his actions are directed. This goal-directed behaviour may be deliberate and VOLUNTARY, or may, to a certain extent, be habitual. People motivated in this way can usually give coherent reasons for their actions. Other types of motivated behaviour appear to lie between these two extremes. For example, the motivation of SOCIAL behaviour is often a mixture of deep-seated urges and deliberate intentions, woven with learned conventions and habits.

The task facing the scientific student of motivation is a formidable one. He has to account for observed changes in behaviour in an objective manner, without recourse to his subjective feelings and experience, and without invoking any axioms, assumptions, or doctrine, other than a belief in the laws of causality. To the untrained observer, the changes in behaviour appear to be much the same, whether they are caused by 'blind instinct' or by 'rational thought'. For example, when an incubating herring gull (*Larus argentatus*) retrieves an egg that has rolled out of the nest, is this merely an automatic response to a particular stimulus, or does the gull deliberately retrieve the egg in order that it should not become cold, knowing that it will not hatch if it cools too much? William of Occam, the 13th-century philosopher, in pointing out the merits of simplicity stated that a plurality must not be asserted without necessity. So far as behaviour is concerned, this principle is embodied in Lloyd Morgan's canon, which states that one should not attribute to an animal a mental faculty higher than the simplest possible to explain its behaviour, unless one has independent evidence for the occurrence of the higher faculties (see HISTORY).

Most behavioural scientists hope that it will eventually be possible to account for behaviour in material terms; that is, in terms of the physiology of the animal and of events in the BRAIN. They recognize, however, that the day when a material account of behaviour will be possible is far away, and some prefer to define laws of behaviour that are independent of the material and mechanisms that govern the behaviour. This empirical determinism developed as a protest against the assumption that causal laws must always be couched in material terms. Like the meteorologist, the behavioural scientist has to account for the observed phenomena, without reducing his explanation to the level of chemistry and physics. Empirical determinists ask, not whether the terms and laws of psychology can be reduced to those of physiology, but whether they could be deduced from them.

Rather than starting from preconceived ideas about the nature of motivational forces, the scientist prefers to proceed cautiously, and begins by attempting to create some order from his observations. It is a common observation that animals change their behaviour in response to changes in their external environment. For example, a sleeping dog (*Canis l. familiaris*) may awake at the sound of its master approaching; a feeding pigeon (Columbidae) may fly off when it sees a hawk (Accipitridae). Thus external stimuli can be classified as one type of causal factor influencing behaviour. However, it is also commonly observed that animals change their behaviour when there is no change in the external situation. Therefore the cause of the change must be sought within the animal. It is now generally recognized that changes in behaviour may be due to any one of, or any combination of, the following five factors: (i) external stimuli, (ii) maturation, (iii) injury, (iv) learning, and (v) motivation.

Changes due to MATURATION occur during the development of the individual, and are not due to experience or to injury (see ONTOGENY). Thus a bird flies, and a child walks, at a certain stage of its development, without the necessity for LEARNING or practice. Changes due to injury may result from disease, or from physical damage. Thus a dog may limp if its foot becomes infected, or has been caught in a trap. Many insects adopt a new pattern of walking after one or two limbs have been amputated. Changes in behaviour due to learning must result from external events experienced by the individual concerned. These external events often have immediate consequences, though this is not essential for learning. For example, a dog may learn to avoid fire as a result of being burned, but it may also learn to avoid a particular food, the consumption of which had made the animal sick a number of hours later. Some songbirds learn the song of their parents when they are JUVENILE, even though they do not themselves sing or show any response to song until they become mature weeks later.

Changes in behaviour in an unchanging environment are very often due to changes in the internal motivational state of the animal. Thus a domestic hen (*Gallus g. domesticus*), presented with an egg, may eat it on one occasion and sit on it on another occasion, even though the external conditions are unchanged. We account for this difference by saying that the hen was hungry on the former occasion and broody on the latter. This implies a change in motivation, for the terms hungry and broody are convenient labels for the respective motivational states, the nature of which is a matter for experimental investigation. The characteristic of changes in behaviour that are due to motivation is that the underlying change in internal state is reversible. For example, an animal kept without food will become increasingly hungry. If it is then allowed to eat, it will revert to its original non-hungry state. Provided that there

has been no maturation, learning, or injury in the meantime, the animal will revert to exactly the same condition.

The property of reversibility does not apply to the processes of maturation, injury, or learning. Maturation is part of the process of growth and development, which is irreversible. Permanent injury leads to irreversible changes in behaviour. Learning is also an irreversible process. Although animals may appear to forget, or to extinguish learned behaviour, they are no longer the same as animals that have never learned the behaviour (see EXTINCTION). The internal changes brought about by the learning process are permanent, and can be cancelled only by further learning.

Motivation can thus be defined as that class of reversible internal processes responsible for changes in behaviour. This definition distinguishes motivation from the other four types of process responsible for behavioural change. It is purely classificatory in nature, invoking no material or mechanistic explanation of the processes involved. These are empirical matters, the subject of experimental investigation by physiologists and psychologists. In defining motivation in this manner, the scientist is not committing himself to any doctrine, or set of preconceived notions about the nature of motivational processes. Nothing need be said about animals being driven (see DRIVE), organized, energized, or directed by their motivations, which is not the result of observation and experiment. For the scientist, the term motivation stands for a class of phenomena, to be investigated and pronounced upon with caution.

The scientific perspective. The scientific study of motivation is diverse in its origins and methods of approach. The study of human motivation has developed more or less independently from that of animal motivation. This is partly because students of human behaviour do not have at their disposal the variety of tools and TECHNIQUES OF STUDY available to students of ETHOLOGY. Although LABORATORY STUDIES are sometimes possible, research into human motivation relies heavily on the use of questionnaires and social surveys. Lessons learned from animal research have, however, been applied to human beings with some success; nevertheless, the study of human motivation remains a separate discipline, and one which is outside the scope of this article.

Observation of behaviour. The study of animal motivation must begin with direct observation of the behaviour of animals, preferably in their natural environment (see FIELD STUDIES). Only by studying an animal in its natural environment can one observe the full range of its behaviour, and gain an appreciation of the organization and role of the behaviour in adapting the animal to its environment. An animal's way of life has a marked influence on its behavioural organization.

In addition to what can be said about the essential details of the behaviour of particular species, there are some important generalizations which come from direct observation of normal and natural behaviour. The first of these is that animals 'can do only one thing at a time'. More correctly, certain activities within the repertoire of any given species will be incompatible with other activities (see INCOMPATIBILITY). For example, BREATHING is perfectly compatible with COURTSHIP in most terrestrial animals, but in newts which conduct their courtship under water, this is not so. In the smooth newt (*Triturus vulgaris*), courtship takes place on the substratum of the pond, which forms the natural HABITAT of these animals. The male has to remain on the substratum to court the female, but he has to swim to the surface of the water to breathe. Breathing and courtship are therefore incompatible in this species.

The fact that animals can do only one thing at a time means that, when an animal is engaged in one type of behaviour, such as FEEDING, it is not performing other types for which it has the motivational potential. For example, a hungry dog is aroused from SLEEP by the arrival of food. If the food had not arrived until later, the dog might have gone on sleeping longer (see AROUSAL). The dog, therefore, is still motivated to sleep when awakened by the arrival of food. Presumably, the dog's hunger took priority when the food arrived. In addition to being hungry, the dog might have been motivated to go for a walk, but this had lower priority than either sleep or hunger. If such were the case, then the dog would be likely to go for a walk when it had satisfied both its hunger and its sleepiness.

Suppose that, when the dog was asleep, it had a motivational potential for eating and for going for a walk. The former would not be expressed in behaviour, because the appropriate external stimuli had not yet arrived. The latter would not be expressed in behaviour, because it had a lower priority than sleeping. Clearly, when we see an animal engaged in one type of activity, we cannot assume that it is motivated towards that type of behaviour alone. We must recognize that in every animal there is an underlying motivational potentiality for a variety of types of behaviour. In effect, the animal makes a DECISION about which course of action to pursue, depending upon its internal motivational state, and upon the external circumstances.

The motivational potential to perform various types of activity is sometimes called the *primary* aspect of motivation, in contradistinction to the *secondary* aspect, which is characteristic of the behaviour that the animal is involved in at any particular time. For example, primary THIRST is brought about by dehydration of the body tissues, and is represented in the brain as a motivational potential for DRINKING. But in addition to being affected by primary thirst, which reflects the basic bodily needs, drinking is affected by secondary thirst, which is contributed to by such factors as the palatability of the liquid, the recency of feeding, and the environmental temperature. Many animals drink in association with meals, even though their primary thirst is negligible. Such drinking is anticipatory of the fact that the food eaten will induce a primary thirst, once the food has been digested. Although the necessity for a distinction between primary and secondary aspects of motivation can be recognized by direct observation of behaviour, the details of the mechanisms involved, and the manner in which the various aspects of motivation interact, can be unravelled only by experiment.

A second generalization is that behaviour can be divided into APPETITIVE and CONSUMMATORY aspects. Observation of behaviour often indicates that certain activities, for example running, nosing around, sniffing, scratching at the ground, tend to occur in temporal association. In the case of running, sniffing, etc., we may like to call this SEARCHING behaviour. Observation may also reveal that the searching behaviour tends to occur in bouts which are terminated by eating. It seems reasonable to conclude that the animal was searching for food, so that the observed behavioural bouts might be called 'feeding behaviour'. Such behavioural bouts can therefore be divided into an appetitive phase, characterized by searching, and a consummatory phase, which terminates each bout of feeding behaviour. In general, appetitive behaviour takes many forms, but it is generally flexible, repetitive, and interspersed with other activities. Consummatory behaviour, on the other hand, tends to be stereotyped and less readily interrupted.

The recognition that bouts of behaviour may be terminated by a consummatory activity suggests that the behaviour patterns making up the bouts share common causal factors and serve a common FUNCTION. In the case of feeding behaviour, the common causal factor would be hunger which activates the various behaviour patterns involved in searching for food, as well as in eating. The common function, served by all aspects of feeding behaviour, is that of obtaining food. Thus, observation of behaviour can suggest various ways in which behaviour can be classified, an important step in a scientific analysis (see CLASSIFICATION OF BEHAVIOUR).

Fig. A. Oblique display of the black-headed gull (*Larus ridibundus*).

Although the generalization that animals can do only one thing at a time has important theoretical implications, in practice there is often a CONFLICT between tendencies to perform two different types of behaviour, so that the animal resorts to some sort of compromise behaviour. For example, if one holds out a piece of bread towards a duck (Anatinae) in the park, the duck will often show a tendency to approach the bread, especially if it is hungry (see CONFLICT, Fig. A). However, it will also show a tendency to avoid the human holding the bread. The result is that the duck generally approaches to within a certain distance, and then it may alternate between retreat and approach, or it may remain stationary, craning its neck towards the bread, while edging away with its feet. This will result in a typical compromise posture, combining elements of both approach and avoidance. Such postures are commonly observed in animal behaviour because internal conflicts of this type frequently occur. For example, in defending its TERRITORY against intruders, the male three-spined stickleback (*Gasterosteus aculeatus*) is highly aggressive. If the intruder is another male, the AGGRESSION is mixed with FEAR, particularly near the boundary of the territory. If the intruder is a female, the male will generally show conflicting tendencies towards aggression and courtship. So common are such inner conflicts that NATURAL SELECTION has frequently acted upon them, making

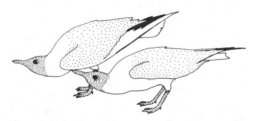

Fig. B. Forward display by a pair of black-headed gulls.

them more effective as a means of COMMUNICATION. In such cases the behaviour evolves into a stereotyped form, and becomes incorporated into the animal's repertoire as if it were a ritual (see RITUALIZATION). For example, the black-headed gull (*Larus ridibundus*) has a ritualized meeting ceremony which develops during pair formation. When the male bird on the ground sees the female flying towards him, he adopts an *oblique posture* and utters a *long call* (Fig. A). The female then alights next to him, and both birds adopt a *forward posture*, usually placing themselves parallel to each other (Fig. B). Then they change suddenly to an *upright posture*, and *face away* from each other (Fig. C).

In analysing the motivation behind this type of DISPLAY, the scientist in the field has five main methods at his disposal: (i) He can study the situation in which the behaviour occurs. In territorial fighting, for example, ritualized threat postures

Fig. C. Facing away during the greeting ceremony of the black-headed gull.

occur primarily near the boundaries, where there is good reason to suppose that there will be a balance between the tendencies to attack and flee from a rival. (ii) The behaviour that accompanies the display can be studied. Displays which are accompanied by incipient attack and FLIGHT movements, betraying the presence of a conflict between aggression and fear, are likely to be THREAT displays. (iii) The behaviour occurring before or after the display may give clues to the motivation of the display. The greeting ceremony of the black-headed gull is followed by amicable behaviour, whereas elements of aggression are usually observed when two males meet. If a display is preceded or followed by overt aggression, it is likely to be a threat display. Such displays often result in retreat of one of the participants, implying victory for the other. (iv) The nature of the display itself may give important clues to its motivation. Some postures are composed of elements of recognizable behaviour patterns. The upright posture of the herring gull (*Larus argentatus*) is a good example (Fig. D). The raising of the wing carpels, or 'elbow', is characteristic of attack by delivering wing beats with the folded wing; the upright neck with a downward pointing bill is characteristic of pecking at an opponent from above. However, the sleeked feathers, stretched neck, and hesitant walking movements that accompany this display are all characteristic of fearful behaviour. Thus the form of the display suggests that it has elements of both aggression and fear. The upright posture is generally regarded as a threat display.

Fig. D. The upright posture of the herring gull (*Larus argentatus*).

(v) COMPARATIVE STUDIES of different species can also give clues about the motivation of behaviour patterns. For example, most gulls, when simultaneously aggressive and frightened, adopt a threat posture similar to that of the herring gull (Fig. D), having an upright carriage with the neck held vertical, and the bill pointing downwards. The wing carpels are held away from the body and the feathers are sleeked as is usual in frightened birds. The exposed carpels could indicate readiness to fly away (fear) or to beat an opponent with the wing (aggression). Carpel exposure is common, however, in the aggressive postures of many species which use wing-beating during fighting, whereas in a species which does not use wing-beating, the great skua (*Stercorarius skua*), for example, the carpels are not raised during threat. The evidence obtained by comparison of species, therefore, suggests that raising the carpels during threat has a primarily aggressive motivation.

Displays have evolved mainly as a means of communication. As Charles DARWIN recognized, however, the display is an expression of EMOTION, and can thus provide a useful indicator to the human observer interested in motivation. The study of displays, FACIAL EXPRESSIONS, VOCALIZATIONS, and COLORATION of animals has contributed considerably to our understanding of animal motivation. For example, the balance between fear and aggression in cats (Felidae) can be gauged both from body posture (Fig. E) and from facial expression (Fig. F). In the cichlid fish, *Hemichromis fasciatus*, as in many other animals, changes in motivational state are mirrored by changes in external coloration. Colour patterns characteristic of aggressive, sexual, PARENTAL, and fearful states can be distinguished, although these normally grade into one another.

The study of animals in their natural environment is important for understanding behaviour as part of the total biology of the animal. The range and variety of behaviour that animals show in their natural environment is often restricted when they are kept in CAPTIVITY. However, observational studies can only give a preliminary picture, and laboratory studies are of great importance for investigating the mechanisms of motivation in detail. Carefully controlled experiments are necessary to unravel the complex motivational interactions that govern behaviour. The variable conditions in the field make this type of experiment extremely difficult, but in the laboratory much of this variability can be excluded.

Experiments on motivation. Hypotheses tentatively formulated on the basis of observation of behaviour can often be substantiated by experiment. As a first step, selective interference with the animal's environment can provide valuable information prior to a fully controlled laboratory experiment. The aim of this selective interference is to change aspects of the environment in such a way that any change in the observed behaviour can be attributed to the environmental change.

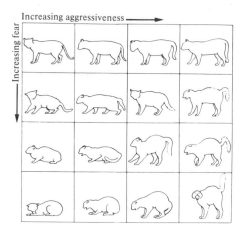

Fig. E. Simultaneous expression of aggression and fear in the body posture of the domestic cat (*Felis catus*).

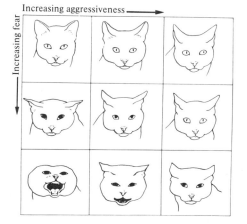

Fig. F. Simultaneous expression of aggression and fear in the facial expression of the domestic cat.

For example, experiments with the great water beetle (*Dytiscus marginalis*), which is carnivorous, and has well-developed eyes and good visual capabilities, have shown that VISION is not used for catching prey. Tadpoles placed in a test-tube were ignored by the beetle, whereas those in a muslin bag were immediately attacked. It is known that *Dytiscus* has a well-developed sense of smell, and the experimental results suggest that the beetle uses CHEMICAL stimuli to detect prey under water, although it is known to respond to visual stimuli in other situations. Preliminary investigations of this type are often possible in the field, but it is usually necessary to bring the animal into the laboratory to study its behaviour in detail, so that adequate controls can be introduced into the experiment.

A controlled experiment is based upon the rationale that all variables should be kept constant, or their effects known, except those that the experimenter is deliberately manipulating and/or recording. Experiments on motivation generally fall into two categories: (i) those in which determination of the primary motivational state of the animal is the main aim; (ii) those designed to study the motivational changes that take place in conjunction with the observed behaviour (secondary motivational changes). In investigating primary motivational states, the experimenter maintains his animals on a strict regime, with maintenance conditions such as day-length, temperature, etc., kept constant. He then introduces a change into a single known aspect of the maintenance conditions. For example, in investigating the effects of ambient temperature upon drinking in pigeons, the experimenter might maintain the pigeons for 2 days without water at a particular temperature, and then administer a standard test to see how much they drink. He would then repeat the experiment with a different maintenance temperature. Thus, in studying the effects of various maintenance temperatures on drinking, the experimenter is comparing differences in treatment that occur prior to the observation of the behaviour. The standard drinking tests are always conducted in the same way, the birds being offered water from a standard drinking bowl, in a standard cage, at a standard temperature, etc. Any differences in the amount drunk during such tests must be due to the differences in maintenance temperature, since this is the only aspect of the repeated experiments that has been allowed to vary. The variation in maintenance temperature can only induce changes in primary motivation (the animal's potential for drinking), because the conditions under which drinking behaviour takes place are always the same.

In studying secondary aspects of thirst, the maintenance conditions are held constant, and a selected testing condition is varied. For example, the birds may be deprived of water for 2 days, always at the same temperature. The drinking tests are carried out in a standard cage, with a standard drinking bowl, etc., but with the ambient temperature varied from one occasion to the next. Any differences in the amount drunk during such tests must be due to differences in temperature at the time of drinking, since this is the only aspect of the repeated experiments that has been allowed to vary. There is no difference in primary motivation from one experiment to the next, because the conditions under which the animals are deprived of water are always the same. Therefore, any differences in the amount drunk must be due to changes in secondary motivation.

Experiments of this type have been carried out on the Barbary dove (*Streptopelia risoria*). The results of varying the ambient temperature during water deprivation are summarized in Fig. Ga. From this figure it can be seen that maintenance temperature has no detectable effect on the amount drunk. This result is somewhat surprising, because doves, like many animals, lose water in

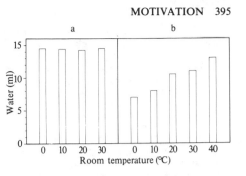

Fig. G. Effects of room temperature upon drinking in doves. **a.** The effect of temperature during water deprivation upon subsequent drinking. **b.** The effects of temperature during drinking.

keeping themselves cool, and might be expected to be more thirsty when deprived of water at higher temperatures. The result of varying ambient temperature during drinking tests is summarized in Fig. Gb. From this figure it can be seen that the temperature has a marked effect on the amount drunk, more water being drunk at higher temperatures. It appears from these results that the birds drink in response to a given ambient temperature, in anticipation of the water that they are likely to need for cooling themselves, rather than drinking to make up the water that they have already lost in cooling. Similar mechanisms are known to operate in other species, and it is recognized that providing for one's needs prior to expenditure is a more sophisticated form of behaviour than making up for losses after they have occurred (see THERMOREGULATION).

In general, studies of primary motivation are concerned with the relation between the animal's physiological state and its behaviour. Questions relating to the effects of starvation on hunger, the effects of water deprivation on thirst, the effects of HORMONES on sexual behaviour, have received a great deal of attention from scientists. Also important are the interactions between different types of motivation. For example, hunger reduces sexual motivation in some species, and thirst has an inhibitory effect on eating.

The maintenance of a stable internal physiological condition is an important aspect of motivation (see HOMEOSTASIS). Thus hunger and eating are instrumental in maintaining a satisfactory ENERGY balance, and the water balance of the body is maintained partly through the mechanisms of thirst and drinking. Many animals have the ability to compensate for physiological malfunction by means of appropriate behaviour. For example, rats (*Rattus*) can learn to compensate for digestive deficiencies by altering their diet. The study of the mechanisms by which physiological functions control behaviour, and vice versa, is generally undertaken by scientists with a physiological training, whose prime interest is in the physiological basis of behaviour rather than in the behaviour itself.

Studies of secondary motivation are generally concerned with the more psychological aspects of motivation. The consequences that result from an animal's own behaviour have marked effects on its motivation. For example, highly palatable food increases the feeding incentive of hungry animals. On the other hand, if the consequences are less than the animal has been led to expect, the motivational state of the animal will be considerably altered as a result of FRUSTRATION. For example, a hungry animal is likely to become frustrated if it is physically prevented from obtaining food which it can see, if an expected food reward is delayed, or if its food is less palatable than usual. Other aspects of motivation that are particularly affected by the prevailing circumstances are fear, CURIOSITY, aggression, and EXPLORATORY behaviour.

An important question in motivational studies is whether or not depriving animals of the opportunity to perform a particular activity causes a build-up of motivation and results in an increase in the relevant behaviour when the deprivation is over. In some cases the answer is obvious. Deprivation of food results in increased feeding behaviour when food is made available again, and this is clearly due to a build-up of hunger. Sleep deprivation results in increased sleep, and this is generally attributed to accumulated tiredness. In other cases it is equally obvious that deprivation does not involve a motivational build-up. For example, avoidance of particular stimuli, which is motivated by fear, is not enhanced by deprivation of those stimuli, nor does the level of fear build up inside the animal, in the way that hunger does. Rather, it is exposure to the frightening stimuli that causes a build-up of fear.

In some cases it is not clear whether deprivation has an enhancing effect or not. Aggression is a case in point. Some scientists believe that there is a build-up of aggressive motivation when opportunities for aggressive behaviour are denied. For example, male yellow damselfish (*Microspathodon chrysurus*) behave aggressively towards males of their own species presented to them behind a glass barrier. In one study, each male was allowed one such opportunity per day, during which its aggressive behaviour was observed and recorded. After being deprived of the opportunity for a period of 7 days, a fivefold increase in aggressive behaviour was observed. This result suggests that there had been a build-up of aggressive motivation during the deprivation period. However, other scientists disagree with this interpretation, arguing that the level of aggression was not directly measured during the deprivation period. For example, in another study, using the cichlid fish *Haplochromis burtoni*, single males kept in a tank with numerous juveniles were exposed to another male of the same species, presented behind glass for periods lasting 30 s. Aggression was measured after each exposure, by counting the number of times the male attacked the juvenile

fish. During deprivation of the opportunity to attack another male, (i.e. there was no other male behind the glass) it was found that the level of attacks on the juvenile fish actually declined, but there was an increase in aggression when the deprivation period was terminated. The most plausible explanation of this result is that a degree of familiarity with the rival male was established before the deprivation period. During the deprivation period aggressive motivation declined and familiarity with the other fish also declined. The rise in aggressive behaviour observed after the deprivation is attributed to this loss of familiarity. In other words, although the level of aggressive motivation was lower, the familiarity of the other male had worn off, making it a more powerful stimulus. Familiarity effects of this type are known to occur in a variety of situations involving fear or aggression (see HABITUATION).

The question of motivational build-up is in many ways related to the question of whether animals actively seek out situations appropriate to their motivational state. Animals deprived of food not only experience an increasing hunger, but actively seek out food and will work to obtain it. This appetitive aspect of behaviour is not found in association with REFLEX activities. For example, many animals have reflex and stereotyped AVOIDANCE and ESCAPE responses, but they do not accumulate a desire to perform such responses, nor do they actively seek out situations which provoke such behaviour. The demonstration that animals will actively seek particular situations is often taken to show that the animal's response to the situation is not purely reflex. For example, the male damselfish will show aggressive behaviour towards another male. This behaviour may seem to be merely a reflex response to the stimuli provided by the other male. However, it has been shown that damselfish can learn to enter a 'bottle' to gain the opportunity to see and behave aggressively towards another male (Fig. H). When the fish has learned

Fig. H. A male yellowtail damselfish (*Microspathodon chrysurus*) entering a bottle in order to display aggressively towards another male.

this behaviour, it will periodically enter the bottle, even though the other fish is not present on every occasion. This suggests that the fish is actively seeking an aggressive encounter with another individual.

The ability of animals to learn new forms of appetitive behaviour is frequently exploited by scientists in devising measures of motivation, and also in investigating other aspects of behaviour, such as learning, PERCEPTION, and PROBLEM SOLVING. The most common method is the OPERANT procedure, in which the animal is allowed to indulge in some learned behaviour at its own pace. Rats and pigeons have been chosen most frequently for this type of experiment, though many other animals, including human beings, have been used. Operant CONDITIONING consists essentially of training an animal to perform a task to obtain a reward. A rat, for example, may be required to press a bar, or a pigeon to peck an illuminated disc. One method of training is called *shaping*.

Let us consider the training of a pigeon, which has to peck an illuminated disc, called a 'key', to obtain a food reward. After 1–2 days of food deprivation in its home cage the pigeon is placed in a small cage equipped with a mechanism for delivering grain, and a key at about head height (see LABORATORY STUDIES, Fig A). Delivery of food is normally signalled by a small light which illuminates the grain. Pigeons soon make an ASSOCIATION between the switching on of the light and the delivery of food, and approach the food mechanism whenever the light comes on. The next stage of shaping is to make food delivery contingent upon some aspect of the animal's behaviour. A pecking response is frequently used, but pigeons can also be taught, for example, to preen or turn in small circles to obtain a reward. Pecking is shaped by limiting the rewards to movements which become progressively more similar to a peck at the key. So when the pigeon has learned to approach the key for reward, it is then rewarded if it stands upright with its head near the key. At this stage the pigeon usually pecks at the key spontaneously; slow learners can be encouraged to peck if a grain of wheat is glued temporarily to the key. When the pigeon pecks the key it closes a sensitive switch in an electric circuit, which causes the food to be delivered automatically. From this point the pigeon is rewarded only when it pecks the key, and manual control of reward is no longer required. The animal is now ready for use in an experiment.

This training technique depends upon the fact that the animal's appetitive behaviour is flexible and easily modified by learning. For the pigeon, pecking the key becomes a new means of seeking food. For the experimenter, it is a convenient way of measuring the pigeon's motivation to seek food, because each peck at the key operates an electric circuit, and can be recorded automatically. There are other ways of training animals to modify their appetitive behaviour, and situations can even be devised in which the animal trains itself. The essential features are that the animal should be motivated to perform and modify the appetitive behaviour, and that the behaviour should be capable of being modified. Consummatory behaviour, and some forms of appetitive behaviour, are not readily modified by learning, or can only be modified by certain types of REWARD.

Many types of experiment utilize the operant procedure. For example, DISCRIMINATION learning can be studied by rewarding the animals for responding only when a certain colour or pattern is presented, or by allowing the animal to choose between two keys that are visually differentiated. The technique has proved particularly useful for studying the effect of different types or patterns of reward (see REINFORCEMENT SCHEDULES). Thus instead of rewarding the pigeon for every peck, it can be rewarded for every nth peck, so that there is a fixed ratio between number of pecks and number of rewards. Thus animals can be made to work harder to obtain the rewards. This procedure is called a *fixed ratio reward schedule*. Other common schedules include *variable ratio*, *fixed interval*, and *variable interval*. On an interval schedule, reward is given at intervals of time specified by the experimenter. The animal is rewarded for the first response made during a given interval. Different schedules of reward have been found to have different effects on the animal's performance. For example, a variable interval schedule produces a very uniform rate of responding, and for this reason it is often used in studies of motivation. Variations in hunger, or some other aspect of motivation, show up clearly against a background of uniform response rate.

Animals have been trained by operant methods to obtain many and various types of reward. In addition to the obvious rewards, such as food, water, warmth, etc., animals have been trained to work for the opportunity to attack a rival, and for the opportunity to court a member of the opposite sex. Some song-birds have been trained to work to hear snatches of the SONG of their own species. A particularly interesting example, which illustrates many of the motivational principles discussed above, concerns the pair-bond in zebra finches (*Taeniopygia guttata*).

The zebra finch inhabits the semi-arid plains of Australia, and has a nomadic and highly SOCIAL way of life. These finches are ideal laboratory animals, being easy to maintain and breed. The male and female pair for life, and a number of experiments have been conducted on this aspect of their behaviour. Experiments involving separation of the male and female have shown that the male is strongly motivated towards reunion with the female, and will actively search for her, even if he is in the presence of other females. Upon reunion, there is an upsurge of sexual activity between the pair, and this is more prolonged the longer the period of separation, suggesting that separation from the mate causes a build-up of sexual motivation. The sexual behaviour is directed towards the

Fig. I. A male zebra finch (*Taeniopygia guttata*) which has been trained to hop from perch to perch to be allowed to see its mate.

mate, and very seldom towards any other individual substituted by the experimenter. Similarly, mated females separated from their own males avoid or act aggressively towards substituted males.

A male zebra finch, which has been separated from its mate, can be trained to jump on and off a special perch, to operate a mechanism which allows it to see its mate for a 10-s period (see Fig. I). Isolated males will also work in this way to see other zebra finches, but will work much harder to see their mates. They will work even harder if the mate is visually presented together with another male. These experiments suggest that the *pairbond*, which develops between male and female after an initial period of courtship and MATE SELECTION, has strong motivational components, as well as components of learning and mate recognition. There are three main characteristics of the male behaviour that are generally taken to be typical of motivation. Firstly, the males appear to show a build-up of sexual desire during periods of separation from their mates. However, further work would be required to establish this with certainty, because the upsurge of sexual behaviour following

separation may be due to loss of familiarity. This is unlikely in this case because the periods of separation never exceeded 6 h. Secondly, males separated from their mates exhibit typical appetitive behaviour. This involves calling and searching, which tend to increase the likelihood of reunion. The females also call, and it is probable that the male can recognize the call of his own mate, and home-in on it. Thirdly, isolated males will learn an operant response, which is rewarded by the sight of another finch, especially its own mate.

The degree to which animals will work for certain rewards is often taken as a measure of motivation, though this approach is not without its difficulties. Thus the extent to which an animal will work may be counteracted by fatigue, competing motivations, or poor ability to learn the correct, or most efficient, responses. Motivational processes are complex, and interwoven with the processes of learning and maturation. Only by shedding his ANTHROPOMORPHISM, and by careful observation and experiment, can the scientist hope to achieve a good understanding of motivation.

16, 74, 103

N

NATURAL SELECTION. Modern evolutionary theory can be said to begin with the publication, in 1859, of Charles DARWIN's book entitled *On the Origin of Species by Means of Natural Selection, or the Preservation of Favoured Races in the Struggle for Survival.* The elements of Darwin's theory can be stated as follows. Within any population of animals of the same species, there is considerable variation among individuals. Much of this variation is due to heredity. Many more individuals are born in each generation than survive to maturity. Therefore the likelihood of an individual surviving to maturity will be affected by its particular traits, especially those that it has inherited from its parents. If the individual survives to maturity and reproduces successfully, its offspring will tend to perpetuate the inherited traits within the population. In Darwin's own words, 'Now can it be doubted, from the struggle each animal has to obtain subsistence, that any minute variation in structure, habits or instincts, adapting that individual better to the new conditions, would tell upon its vigour and health? In the struggle it would have a better chance of surviving; and those of its offspring which inherited the variation, be it ever so slight, would have a better chance. Yearly more are bred than can survive; the smallest gain in the balance, in the long run, must tell on which death must fall, and which shall survive. Let this work of selection on the one hand, and death on the other, go on for a thousand generations. Who will pretend to affirm that it would produce no effect, when we remember what, in a few years, Bakewell effected in cattle, and Western in sheep, by this identical principle of selection?'

Thus the SURVIVAL VALUE of a trait is determined by natural selection. That is, the extent to which a trait is passed from one generation to the next, in a wild population, is determined by the breeding success of the parent generation, and the value of the trait in enabling the animals to survive natural hazards, such as food shortage, predators, and sexual rivals. Such environmental pressures can be looked upon as selecting those inheritable variations which best fit or adapt the animal to its environment. This is what Darwin meant by 'survival of the fittest'.

The expressions 'struggle for survival' and 'survival of the fittest' are somewhat unfortunate, because they tend to focus attention on avoidance of death and perpetuation of the individual. In fact, natural selection operates only by differential reproductive success, and differential mortality can be selective only to the extent that it differentiates between individuals in respect of the number of progeny they produce. The FITNESS of an individual is often measured in terms of the individual's contribution to the genetic make-up of the population. In this context, the expression 'survival of the fittest' refers to the perpetuation of the genetic characteristics of the individual (see GENETICS). This use of the term 'fitness' differs slightly from that implied by Darwin, and the term 'Darwinian fitness' is nowadays used to indicate the contribution of an individual to the population as a whole, in terms of number of progeny, rather than in terms of a genetic contribution.

Although Darwin spent 26 years collecting evidence for his theory of EVOLUTION by natural selection, there were two fundamental gaps in his chain of evidence. Firstly, Darwin had no knowledge of the mechanisms of heredity. Secondly, he had no example of evolution at work in nature. Since Darwin's time the industrial revolution has provided a natural 'experiment', which has been studied intensively by scientists. The peppered moth (*Biston betularia*) is widely distributed throughout Britain. Originally the majority of these moths were light in colour with small dark markings, as illustrated in Fig. A. However, in

Fig. A. The peppered moth (*Biston betularia*): normal form above, and melanic form below.

industrial parts of Britain the majority of individuals of this species became almost completely black, a phenomenon called *industrial melanism*. In unpolluted parts of the country, the moths commonly alight on lichen-covered, light-coloured tree trunks, where the pale form is well camouflaged. In industrial regions smoke particles accumulate on the leaves of trees, and are washed by rain into the bark on the boughs and trunk. This kills the lichens, and makes the tree trunks bare and black. Against this background the light-coloured moths are highly conspicuous, whereas the dark form is almost invisible. As a result of experiments designed to test the rate of survival of the two forms in these contrasting types of woodland, it was found that PREDATION by birds removes a high proportion of the conspicuous moths. That is, the pale moth in industrial areas, and the dark moth in rural areas, are more frequently taken by birds.

Here we have natural selection in action. The dark form was first reported near Manchester in 1848, and probably arose as a result of a genetic mutation. In the decades that followed the dark form was subject to predation to a lesser extent than the light form, and so it was at a selective advantage. Consequently more dark forms survived to reproduce, and their offspring were likely to be dark, because the colour character is inherited. Gradually the peppered moth population in industrial areas became predominantly composed of the dark variety. In rural parts of Britain the dark variety is quickly picked off by birds, and has no chance to establish itself. A large number of other species of moth, in industrial areas of Britain, Europe, Canada, and the U.S.A., are known to have become darker in colour during the past century. In more recent years, as pollution control has begun to become effective, the environmental conditions that have until recently placed light-coloured moths at a disadvantage are returning to the natural state. There is evidence that industrial melanism in moths is beginning to reverse itself.

Natural selection operates on the external characteristics, or *phenotype* of an individual, so that an animal's immediate fitness is determined by its total phenotype. However, the effectiveness of natural selection in changing the composition of a population depends upon the degree to which the phenotypic characteristics are inheritable. Thus the effectiveness of natural selection depends upon the genetic influence that an individual can exert upon the population as a whole. This fact raises important problems concerning the phenotypic unit upon which natural selection acts. For example, we might recognize that the tendency for a rabbit (*Oryctolagus cuniculus*) to run when it sees a fox (*Vulpes*) is a genetically influenced trait, which benefits the individual. The white tail of the rabbit, which becomes conspicuous when it runs, does not help the rabbit to escape from the fox, and may even aid the fox in keeping track of the rabbit. However, the white tail could serve as a warning to other rabbits, and thus be of benefit to the group. Similarly, the tendency of rabbits to thump upon the ground with a hind leg when frightened probably draws the attention of the predator, but also serves to warn other rabbits.

There is a difficulty in accounting for the evolution of such apparently altruistic behaviour. It would appear that a rabbit which did not thump upon seeing a fox, and had no white tail, but merely ran for cover, would be at an advantage over ordinary rabbits. More offspring would be left by this rabbit, and the population would gradually become composed of a non-warning variety of rabbit. A possible way out of this paradox is to suppose that an individual could better maintain its genetic characteristics in the population by sacrificing its own life, if by so doing it could save the lives of other individuals with a genetic make-up very similar to its own. If such were the case, then a process known as *kin selection* could occur. Generally, those individuals most likely to have a genetic constitution very similar to that of a given individual will be its brothers, sisters, and children. This kin selection hypothesis accounts for the fact that PARENTAL care often involves elements of ALTRUISM, and for the warning given by members of a close-knit community, of which a large proportion are likely to be interrelated.

The question of whether natural selection can operate to the benefit of a group is a controversial one. Most scientists agree that certain cases can be satisfactorily explained on the basis of kin selection, but deny that explanation in terms of a more general *group selection* is possible, except under very special circumstances.

29, 34

NAVIGATION, as a term used by biologists, denotes the most advanced and complex form of long-distance ORIENTATION by animals. Thus the most widely used CLASSIFICATION of long-distance orientation behaviour recognizes three different categories. *Pilotage* (or type I orientation) is steering a course using familiar landmarks. *Compass orientation* (type II orientation) is the ability to head in a given compass direction without reference to landmarks. And *true navigation* (type III orientation) is the capacity to orient toward a GOAL (e.g. home, or a breeding or overwintering area), regardless of its direction, by means other than recognition of landmarks. For example, HOMING pigeons (Columbidae) that promptly take up a homeward course when released hundreds of miles from home in unfamiliar territory to the east, and then also take up a homeward course a few days later when released hundreds of miles away in unfamiliar territory in some other direction, are doubtless performing true navigation; they are clearly not relying on familiar landmarks, nor are they merely taking up a constant compass bearing.

The reader will recognize that it is not always easy in practice to distinguish between navigation and the other two types of orientation. Consider, for example, long-distance MIGRATION by birds. Numerous studies have yielded data consistent with the notion that for much of their migratory journey many birds merely fly in a species-characteristic compass direction night after night, rather than carrying out goal orientation. Does this mean that true navigation is not an aspect of avian migration behaviour? To investigate this question, scientists captured young starlings (*Sturnus vulgaris*) as they passed through Holland on their first autumnal migration from their natal area around the Baltic Sea to their wintering grounds in southern England, Belgium, and northern France. After ringing (banding) the starlings, they took them to Switzerland, so displacing them approximately 750 km south-south-east, and released them. Almost all the reported recaptures of these birds were from the west-south-west, indicating that the young starlings had not corrected for the displacement but had, instead, continued to orient in the direction that would have been appropriate in Holland, even though this direction now led many of them into Spain, an area where they would not normally have gone (Fig. A). But the recapture results for adult starlings displaced at the same time were very different. Most of these birds moved north-westerly, the proper direction to their normal wintering grounds. Thus it appeared from these experiments that the young starlings relied on simple compass orientation but

that the adults used true navigation to correct for the displacement, and orient towards the appropriate migratory goal.

In bird species where no such correction for displacement *en route* has been found, even in adults, there is nevertheless often rich evidence from ringing studies indicating that individual birds regularly return to the same breeding spot in spring, and to the same overwintering spot in autumn. Such accuracy of return, often across thousands of miles, strongly suggests that at least towards the end of their migratory flights, if not before, the birds are navigating towards a goal. In short, depending on the age or experience of the birds, or on the stage of the migratory journey (or perhaps even on weather conditions or other local factors), birds may sometimes perform true navigation, and at other times use simpler types of orientation, such as compass orientation or pilotage. The same sort of behavioural flexibility probably holds for other animals as well.

Bicoordinate navigation. One of the most obvious ways to carry out long-distance navigation is to use a grid formed by two coordinates. Thus, if we know the latitude and longitude of our present position and also the latitude and longitude of the goal, we can plot a course from the one to the other. To determine their latitude and longitude and to plot their course, present-day human navigators utilize a variety of aids, including com-

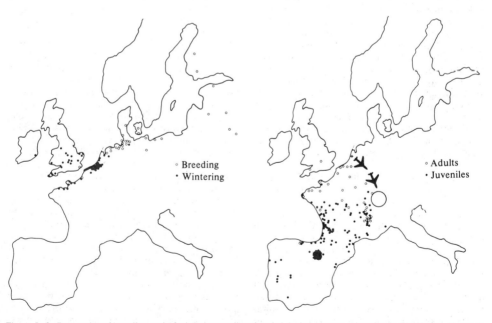

Fig. A. Left: Recoveries of breeding and of wintering starlings banded during autumn migration in The Hague, Holland. The birds migrate along a south-westerly course. Right: Recoveries of adult and juvenile starlings displaced to and released from three airports in Switzerland during autumn migration. The juveniles flew in the usual south-westerly direction, which took them into areas of southern France and Spain which they would not normally visit. Most adults chose directions that would lead them to their normal wintering areas.

passes, sextants, maps, and elaborate tables that show the positions of the sun, the moon, the stars, and the planets for every minute of every day of the year. But what environmental variables might enable other animals to perform bicoordinate navigation?

One possibility is that an animal, such as a homing pigeon (*Columba livia*), released at an unfamiliar site far from home might determine its position relative to home by observing the sun. According to the *sun-arc hypothesis*, the pigeon could determine its latitudinal displacement by observing the sun's movement in arc, extrapolating that arc to its noon position, and then comparing the altitude of the sun's noon position at the release site with the sun's noon altitude at home, as remembered. A noon sun lower than that at home would indicate that the release site is north of home, and a noon sun higher than that at home would indicate a position south of home. The magnitude of the difference between the noon altitudes of the sun at the release site and at home would indicate the distance of displacement along the north–south axis. Longitudinal displacement could be determined by, in effect, comparing local sun time at the release site (as indicated by the sun's position on its arc) with home time as indicated by the bird's internal sense of time (internal CLOCK). A local time ahead of that at home would mean displacement toward the east, and a local time behind that at home would mean displacement toward the west. In summary, the pigeon would determine its latitudinal displacement by a comparison of sun altitudes, and its longitudinal displacement by a comparison of times. By vectorially combining the latitudinal and longitudinal information thus obtained, the bird could calculate the direction it must fly to get home.

Unfortunately, the results of many kinds of experiments conducted during the last two decades seem to indicate that birds do not navigate according to the sun-arc method. They and many other kinds of animals do, however, regularly use the sun as a compass, as can be demonstrated by artificially shifting their internal clocks 6 h out of phase with true sun time by holding them for a few days in a closed room where the lights are turned on and off 6 h earlier than sunrise and sunset. Animals with such a shifted time sense usually misread the sun compass, and orient approximately 90° to the left of control animals whose internal clocks are on normal time (Fig. B). But the animals do not appear to use the sun as a source of latitudinal or longitudinal information.

Like the sun, the stars could potentially provide sufficient information for bicoordinate navigation, but there is no convincing evidence that animals use them in this way either. Many species of birds that migrate at night do, however, derive compass information from star patterns. It seems, then, that animals often possess two celestial compasses, a sun compass and a star compass, but that they do not carry out celestial navigation (see TIME).

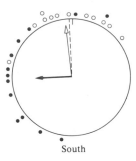

South

Fig. B. Effects of a 6-h-fast clock shift on the initial bearings of pigeons released south of home. The bearings chosen by the experimental birds are shown as solid symbols, those of control birds (on normal time) as open symbols. The mean bearing of the experimental birds (arrow with solid head) points approximately 90° to the left of the mean bearing of the controls (arrow with open head). The home direction is indicated by the dashed line. These results are consistent with the predictions of the map-and-compass model. According to this model, the birds would consult their navigational map and determine home to be northward. They would then use the sun compass to locate north. The test release was performed in the early morning, under an easterly sun, so north was roughly 90° to the left of the sun. The control birds, with a correct sense of time, read this relationship properly and oriented homeward. By contrast, the clock-shifted birds read the time as midday, when the sun would be in the south. They therefore determined north to be roughly opposite the sun's direction, and they flew in that direction. But since their internal clocks were wrong, and the sun was really in the east, flying opposite the sun's direction took them in a westerly direction.

Now, a compass alone, no matter what kind, is not sufficient for navigation to a goal. A compass can tell an animal where north, south, east, and west are, but it cannot say which is the direction to the goal. An animal first needs an analogue of a map from which to determine its position, and to calculate the direction to the goal; it could then use a compass to locate the calculated direction in order to begin its journey. Though the map-and-compass idea has proved useful in describing much of what is known about animal navigation, and in predicting the results of experiments (e.g. ones in which an animal's internal clock is phase-shifted so that it will misread the sun compass; see Fig. B), the nature of the 'map' component remains unknown. The task of elucidating what positional cues animals use in concert with their sun or star compass remains a challenge.

There are, of course, other environmental variables besides celestial ones that could potentially provide coordinates for a navigational grid. Thus, it was suggested that homing pigeons might use as one coordinate the intensity of the vertical component of the earth's magnetic field, and as the other component the Coriolis force resulting from the earth's rotation. This idea was quickly rejected because the experimental evidence did not confirm the predictions it makes, but later investigators did find evidence that homing pigeons, many migratory birds, and a variety of other animals (including some insects, elasmobranch fishes, and even certain mud-dwelling bacteria) can detect the earth's magnetic field (probably not just the vertical component) and use it in their orientation (see ELECTROMAGNETIC SENSES). Though the bulk of current evidence can be interpreted as indicating that this magnetic sensitivity merely provides the animals with another compass (a magnetic compass) it is too soon to rule out completely the possibility that some parameter of the magnetic field may provide information for one coordinate in a bicoordinate navigation system.

Another environmental variable that could potentially provide one coordinate for grid navigation is the north–south gradient in the strength of gravity. This gradient is centred on the pole of the earth's rotational axis, and hence forms a pattern that differs from that of such magnetic parameters as magnetic inclination and magnetic vertical intensity, which are centred on the magnetic pole. The possibility that animals might use gravitational cues in navigating has only very recently come under investigation, however, and there is not yet any convincing evidence that they can detect such tiny differences in gravity as would be required to read the gradient, let alone use them as part of a bicoordinate system.

Navigation by non-bicoordinate methods. There is a variety of procedures animals can follow to achieve true navigation without the use of a bicoordinate grid. Some of these can function only in return navigation to a location previously experienced, because they depend on information gathered during the outward journey. Others, which are independent of outward-journey information, permit not only return navigation but also the possibility of navigation to locations not previously visited.

1. *Methods that require information from the outward journey.* Perhaps the most obvious of the navigational procedures that depend on outward-journey information is the use of an inertial guidance system. If an animal could detect all the accelerations (both linear and angular) of the outward journey and double-integrate them to calculate the bearing to the starting point, then such an animal could return to that starting point (e.g. home) even if it had been displaced under conditions that precluded any view of landmarks or or celestial cues. Appealing as such a system may be to the human mind, however, no convincing evidence of a true inertial navigation system for long-distance movements by animals has ever been found.

A system superficially resembling inertial guidance but not dependent on detection and integration of accelerations is seen in honey-bees (*Apis mellifera*). When a bee flying out from the hive in search of a new food source follows a very circuit-

ous course, perhaps making detours round obstacles or visiting a series of different possible FEEDING localities, it is able to integrate the many legs of its outward journey and fly home by a direct route; it need not retrace its outward path. It seems that the bee keeps track of the directions of the legs of its outward flight by measuring the angle of each leg relative to the sun (or sometimes relative to landmarks), and that it keeps track of distances by measuring the energy expended in flight; it is these quantities rather than inertial data that the bee integrates to determine the homeward course. Since neither angles to the sun nor energy expended in flight can be measured if the bee is displaced in a closed container by an experimenter, this system, unlike an inertial guidance system, does not enable a bee to determine a direct homeward course after experimental displacement.

There is a variety of other cues, in addition to inertial ones, that conceivably could be detected by an animal during artificial displacement, and that could provide information for return navigation. An animal might, for example, detect changes in the magnetic field during the outward journey. It might then use this information in either of two ways. (i) It might previously have learned that the detected pattern of change in the magnetic field occurs, let us say, to the north. Upon release it would therefore locate south by one of its compasses, and head off in that direction. (ii) Alternatively, the animal might upon release choose the course that would most rapidly reverse the magnetic changes detected during the outward journey and thus bring the magnetic field values back to those characteristic of the goal locality; such a procedure would require no prior knowledge of the geographic pattern of magnetic change. At the present state of our knowledge it is impossible to say whether any animals can actually use either of these methods in return navigation. There is some preliminary evidence that when pigeons are exposed to an artificially altered magnetic field (e.g. one in which magnetic north points toward geographic south) during the outward journey to a release site, their orientation is often disturbed. But even if this preliminary evidence is authenticated, it could be that exposure to the abnormal field has merely altered the birds' later sensitivity to normal magnetic cues, rather than giving them false information about the outward journey. Only further research can clarify this point.

According to one hypothesis of pigeon homing, the outward journey is important in providing the birds with olfactory information about the direction of displacement. Thus, if the pigeons have previously learned that odour *A* characterizes the region to the north of home and odour *B* the region to the east, then during an outward journey in one of those directions the birds would detect the odour associated with that direction, and even if the release site itself were beyond the region of their familiarity with odours the birds would none the less be able to act on the information about the direction of the first part of the journey. For example, if during the first part of the outward journey the birds detected a strong odour of *B*, meaning displacement to the east, then even if no familiar odour were evident at the release site a reasonable strategy for homing would be to fly west. Experimental evidence relevant to this hypothesis is mixed, with some strongly positive and some equally strongly negative; we therefore do not have a definitive judgement.

2. *Methods that do not require information from the outward journey*. One of the most extensively investigated feats of animal navigation is the return of salmon to their natal stream to spawn. Young Pacific salmon (*Oncorhynchus*), for example, hatch in streams of the western United States and western Canada and, after passing through an important developmental transformation called *smolting*, they swim downstream to the Pacific Ocean. They spend 2–3 years at sea, often ranging northward into the Gulf of Alaska. Then when they are sexually mature the salmon migrate back to the shore, probably using, among other orientational cues, a sun compass similar to that already discussed for birds. They locate the mouth of the river of their origin and then swim upstream, past fork after fork, often travelling many hundreds of miles until they reach the same tributary stream where they began life. There they spawn, and shortly thereafter die. How do these salmon so accurately choose the correct course at each branching point of the stream? How do they identify the particular locality of their birth?

After many years of work it has been convincingly demonstrated that odours provide the principal cue used by salmon during the upstream phase of their migration. During the brief period (as short as 2 days) of their smolting, IMPRINTING of the young salmon on the odours of their native stream occurs. Their upstream journey as adults 2 or 3 years later is guided by their ability to discriminate olfactorily between water from the native stream and water from other tributaries of the same river system. Thus the choice of direction at each fork depends on detecting which fork contains water from that native stream. Though it is possible that some orientationally useful information is obtained by young salmon during their seaward migration, it is clear that this outward-journey information is not essential for the return migration, because if salmon are held in their native stream through the crucial smolting stage and then transferred to a different stream, they return at spawning time to their native stream, not to the stream from which they descended to the sea.

Olfaction also clearly plays a central role in the orientation of many other kinds of animals, though often only within a region with which the animals are familiar, not from the great distances that characterize salmon migrations. It has been suggested, however, that olfaction is the basis for the remarkable long-distance homing feats reported for the red-bellied newt (*Taricha rivularis*) in California. Scientists displaced large numbers of these small slow-moving amphibians distances (up to 8 km) that for them were very great indeed. Moreover, some of the displacements were across steep mountain ridges. A high percentage of the newts (which were marked) were later recaptured in their native stream. The researchers went on to show that even blindfolded newts could successfully return, but that newts with severed olfactory nerves could not. From these results they concluded that olfactory cues provide the principal information used by the newts in homing. However, a cautionary note is in order here: as so much has been learned since these experiments were done about the variety of environmental cues animals can use in orienting, the experiments need to be repeated with proper controls, for such factors as magnetic cues (which some research suggests amphibians can detect) and cues from polarized skylight (which amphibians are now known to detect). It seems possible that one or more of these other cues may be used by the newts to orient homeward while they are at such a distance that olfactory cues from the home stream would be minimal, and that it is only after they have reached the immediate vicinity of their goal that olfactory cues become essential.

A recent hypothesis of olfactory navigation by birds, especially homing pigeons, has gained wide attention and deserves mention here. The proposal (in its original form) is that a young pigeon growing up at its home loft would learn to associate various odours with wind directions. For example, odour *A* might be especially strong when the wind is blowing from the north, odour *B* when the wind is from the east, and odour *C* when the wind is from the south. (The wind direction would be determined by the bird from its sun or magnetic compass.) When displaced to a distant release site, the bird would detect the odour of the site (let us say *A*) and establish the home direction as opposite to the one from which the odour had usually come to the loft (since *A* came from the north, the home direction in this example would be determined as south). The bird would then locate the deduced direction by means of one of its compass systems and begin its homeward flight. In this system, then, olfactory cues would provide the 'map' component envisaged in the map-and-compass hypothesis of navigation. (Note that in its original formulation the olfactory navigation hypothesis of bird navigation assigned no important role to information gained during the outward journey; later, however, the hypothesis was expanded to include the gathering of olfactory information during displacement, as discussed above.) The proponents of this hypothesis have performed many ingenious experiments, mostly in Italy, that have nearly always yielded supporting results, but unfortunately attempted replications elsewhere have often yielded either negative or ambiguous results; we are therefore unable to pass judgement on the hypothesis.

In our discussion of navigation thus far, our only mention of human beings has been to refer to their use of sextants and compasses and complicated charts. But many primitive sea-going peoples long ago developed a system of navigation that relies heavily on the analysis of wave patterns. A skilled navigator from Puluwat in the Caroline Islands of the south-west Pacific, for example, sits on a special seat in the canoe he is guiding in a voyage to some tiny distant island, and feels 'through the seat of his pants' the pattern of waves in the sea beneath him. From this pattern, which is the resultant of a host of interacting waves of varying direction, frequency, and magnitude, he can resolve certain important constituent waves whose characteristics he has learned during years of intensive training. These waves tell him compass directions and also, at least sometimes, the directions toward major islands or reefs, or other objects that alter the pattern of the water's movement. Navigation by waves, especially when combined with the use of stars, as it often is, permits remarkably accurate voyages over vast expanses of open ocean. It is important to point out, however, that this system requires prior knowledge, indeed extensive prior LEARNING, of the locations of islands and reefs and the appropriate courses between them. In a sense, then, the system only permits navigation in a 'familiar' region, even if the familiarity is based not on having been there before but on having been told about it.　w.t.k.

51, 84, 94, 123, 124

NEST-BUILDING. A nest is a place where eggs are laid and hatched. The term is also used for places where young animals are born and undergo at least part of their development, as in the brood nests of rats and mice (Murinae). Sometimes it is used to refer to a place of rest, retreat, or lodging, such as the special sleeping nests which some birds build outside the breeding season. The emphasis of this article is on the more widely used meaning, which refers to breeding nests.

Nests are of widespread occurrence in the animal kingdom and worthy of study as important natural phenomena. In addition they deserve special attention in an ecological sense because nests help to bring to a focus many of the more significant HABITAT requirements of the species concerned. Furthermore, nests are of particular interest to the student of ETHOLOGY because they are more or less permanent and diagrammatic records of behaviour made by the animal itself. The nests of birds and of the highly social insects have attracted most attention. This article will deal with birds' nests.

Why birds build different kinds of nests. In the nest-building of birds we have a classic example of species-specific behaviour. There are some species of birds that are virtually indistinguishable in appearance, but can easily be told apart by marked differences in their nests and songs. This is true, for example, of certain small flycatchers of the genus *Empidonax* in North America, especially *E. trailli* and *E. brewsteri* which for many years were considered to belong to the same species.

1. *Evolution of nest diversity.* The building of a nest often requires considerable energy. Many species of birds must make a thousand or more trips in gathering the materials needed to build the nest. NATURAL SELECTION will therefore often favour any behaviour that tends to economize on effort, so leading to increased efficiency in building, as well as to diversification of nests between species.

Nests are closely related to differences in HABITAT and behaviour. There has been, therefore, a tremendous amount of ADAPTATION as well as convergent and parallel EVOLUTION, of different nest types in birds. It is of interest to examine the ecological conditions under which different kinds of nests and building behaviour have evolved. Reproduction and survival depend on the total biology of the species, and to understand fully the forces of evolution of the nest of any species one must often be familiar with all aspects of its life history.

COMPETITION between different species of birds has led to specialization and to differences in the habitat and nest sites occupied by related species. In turn, differences in nest site have imposed special requirements with regard to nest placement, materials, structure, and form of the nest, as well as leading to profound differences in building behaviour. A familiar example is the nests of various swallows. The bank swallow or sand martin (*Riparia riparia*) lays its eggs on a litter of straw or grass and feathers at the end of a long tunnel which the bird digs into a sand, clay, or gravel bank. The barn swallow (*Hirundo rustica*) builds its nest on a beam, joist, or ledge in a barn, shed, or stable, constructing a saucer of mud which it lines with straw or grass and feathers. Occasionally it places its nest against a wall or beam, and then the nest is a half saucer which often falls. The house martin (*Delichon urbica*) builds an enclosed mud nest, with entrance hole at the top, under eaves of houses and barns, strengthened by attachment above. It sometimes nests on cliff faces, undoubtedly its original habitat, usually building below some overhanging rock. The mud is added in successive layers and is collected from puddles, ponds, and streams. The American cliff swallow (*Petrochelidon pyrrhonota*) has very similar nest sites, nests, and building behaviour, but the mud nest is retort-shaped, with a short entrance tunnel opening below the roof. The nest of the tree swallow (*Iridoprocne bicolor*) of North America is a feather-lined cup in a hole in a tree or nest box. The various species of swifts (Apodiformes), which despite their somewhat similar life and appearance are not at all related to the swallows, show convergent evolution of nest site with virtually all the types of swallow nests mentioned above. In fact, some species of swifts may take over and breed in nests built by swallows.

2. *Mound-builders and the origin of nest-building.* The origin of the nest of birds can be traced back to the nests of their reptile ancestors which were the evolutionary inventors of the land egg. The eggs of reptiles are often buried in the ground, and generally take much longer to hatch than do those of most birds. Some modern birds like the megapodes or mound-builders (Megapodiidae) of Australia and adjacent islands also bury their eggs in the ground, and depend on the heat from the sun or from decaying vegetation to provide warmth for the incubation of the eggs. It has been shown that one of these megapodes, the mallee fowl (*Leipoa ocellata*), is far more efficient than any known reptile in regulating the temperature around the eggs in its mound nest. This it accomplishes largely by scratching sand on or off its mound according to the temperature conditions of the environment.

Development of the eggs in birds was greatly accelerated when the parent began to sit on the eggs. Such direct INCUBATION no doubt arose in close association with the evolution of warm-bloodedness and the ability to fly. As birds developed a high and relatively constant body temperature, direct incubation of the eggs conferred a tremendous advantage. It became possible to keep eggs warm during the cool nights or days. It also made it possible for birds to invade cool climates where they would have few reptilian competitors.

Why a few modern birds, like the megapodes, do not sit on their eggs is something of a mystery. Perhaps, because of the great efficiency these birds have developed in regulating the temperature of their mounds, direct incubation of the eggs would not necessarily confer any superior advantage to them in the environmental conditions under which most megapodes exist.

It has been suggested that the habit of incubation by sitting on the eggs probably evolved from a tendency of the parent bird to conceal the eggs with its body from predators. In the early history of birds the danger of PREDATION, especially from contemporary mammals which were small and probably often nocturnal, would give value to staying with the eggs and defending them if necessary.

3. *Cavity-nesting birds.* Many birds nest in cavities, including whole orders of birds such as parrots (Psittaciformes), trogons (Trogoniiformes), kingfishers (Coraciiformes), and woodpeckers (Piciformes). Birds nesting in cavities or enclosed nests commonly have a higher fledging success than do birds that have open nests. Studies of fledging success in birds have shown that only about half of some 22 000 eggs of various species with open nests fledged young, whereas fully two-thirds of 94 000 eggs of birds with enclosed or cavity nests were successful.

The primary functions of a nest are to help the parents furnish warmth and protection to the developing eggs and young. By substituting for these functions, nesting in cavities tends to block further evolution of nests built by addition of materials except as mere filling for the cavity.

Regressive evolution of nests may follow adoption of nesting in tree holes. All degrees of simplification to reduction of the nest to a mere pad are to be seen in the nests of various species of sparrows (Passerinae) of the Old World which nest inside holes or cavities in trees.

The climax in the evolution of excavated nests by birds is the construction of nesting cavities inside the nests of social insects. The distribution of the orange-fronted parakeet (*Aratinga canicularis*) in Mexico and Central America closely approximates that of a colonial termite (*Nasutitermes nigriceps*) in the nests of which the parakeet breeds. Certain species of kingfishers, parrots, trogons, puffbirds (Bucconidae), jacamars (Galbulidae), and a cotinga (Cotingidae) are known to breed in termite nests. As excavation by the birds progresses, the termites seal off the exposed portions of their nests.

4. *Evolution of open nests.* After direct parental incubation evolved it was no longer necessary to build a pit for the eggs. However, most modern ground-nesting birds begin their nest by making a circular scrape with the feet, while crouching low and rotating the body in a horizontal plane. This hollow may then be lined with various materials that help protect the eggs from the ground. The same sort of movements used in making the initial scrape help make a rim of insulating materials around the body of the incubating bird. In addition, many ground-nesting birds, such as the Canada goose (*Branta canadensis*), build up the surrounding rim of nest materials and keep it from becoming flattened down by a characteristic act of building in which the bird, while sitting on the nest, reaches out with the bill and draws nest materials to its breast, or passes them back along one side of its body before dropping the material on the rim. The next step in evolution of building would be to walk or fly to the nest while carrying materials in the bill, as cormorants (Phalacrocoracidae) and gulls (Laridae) do.

Some birds such as the common stone curlew (*Burhinus oedicnemus*) build no nest, except for a scratched hollow on stony ground. The colour patterns of the eggs, young, and parents closely match the surroundings. Presumably natural selection has resulted in the virtual disappearance of the nest because it might be conspicuous to predators in the open habitat. The nest is also virtually absent in some cases where eggs or nestlings are exposed to strong tropical or subtropical sun in open situations, and often need to be shaded by the body and wings of the parents. The nest of the sooty tern (*Sterna fuscata*) of Midway Island is merely a scrape in the coral sand.

In the perpetual search for safety from terrestrial predators some birds have become adapted by evolution to nest over water. It was found during one study in Manitoba that losses to predators were less for nests of diving ducks (Anatinae) which are built on platforms over water in a marsh than for dabbling ducks which nest in the uplands.

The dangers of ground nesting and the intense competition for tree holes have provided a strong natural selection leading to the evolution of nests placed amongst branches. In the prairie country of Oklahoma one study found that nests of mourning doves (*Zenaidura macroura*) built in trees were almost twice as successful as were nests of the same species when built on the ground.

Species of birds with *precocial* young (which run about and feed themselves on the day of HATCHING) are, as a rule, ground nesters, whereas species with *altricial* young (which are blind and helpless at hatching, and dependent on the parents for food) often nest in bushes and trees. The latter include the passerine birds (Passeriformes), which comprise about half the world's species of birds. With their perching foot structure, and relatively helpless nestlings, it seems probable that they evolved first in relation to arboreal life, and then some species secondarily re-invaded ground habitats where they generally continue to make well-constructed nests.

The nature of the materials used in arboreal nests varies with the body size of the bird and its lifting power. Large birds, such as herons (Ardeidae) and eagles (Accipitridae), use twigs and branches which are not readily blown out of the tree by wind. Medium-sized birds use twigs or grasses, or both, sometimes adding mud to help attach and bind the nest materials, as does the European blackbird (*Turdus merula*). A great many small birds use spider silk or insect silk as a binding material to attach the nest to the substrate, and for fastening together various plant materials in the nest.

The building movements of birds that make open nests in trees are basically similar to those of ground nesters, particularly in shaping the nest concavity by means of scraping movements of the feet, combined with rotation of the body and pushing movements of the breast. Additional characteristic movements are the thrusting of twigs or grasses into the nest mass with trembling movements of the bill. Very small birds, such as the icterine warbler (*Hippolaius icterina*), will gather cobwebs which they wipe from the bill onto the growing rim of the nest, thus helping to fasten the plant materials of the nest together.

Nests attached to vertical faces of cliffs offer protection from non-avian predators, but pose special problems for nest attachment. Swifts have generally developed an adhesive saliva, while many swallows have evolved toward the use of mud, probably with some admixture of saliva.

Different species of cave swiftlets (*Collocalia*) can be arranged in a graded series from those like *C. francica* which makes nests of pure saliva (source of the ideal bird's nest soup of the Chinese), through various admixtures with plant materials, to more conventional types of bird nests. The nests of *C. francica* can be glued to vertical faces in a cave, but the nest cement of *C. fuciphaga* (= *C. salangana*) is sparse and soft, and its nest of moss and other plant materials is placed on some irregularity in the cave wall that will take all or a good part of the weight of the nest.

5. *Evolution of roofed nests.* Small birds in particular require protection from cold, rain, and predators, such as is furnished by enclosed nests. Building of a roofed nest by the successive addition of materials is rare among non-passerine birds, whereas about half of some eighty-two families and distinctive subfamilies of passerine birds build roofed nests, or contain representatives that do so. At the same time such roofed nests are unusual among passerine birds of the north temperate zone, the long-tailed tit (*Aegithalos caudatus*) being a notable exception. Roofed nests are very common among small tropical birds, being typical of many families and genera. The Galapagos or Darwin's finches (Geospizini) have an equatorial habitat, and unlike most finches of the family Fringillidae, which build an open nest, they build roofed nests.

Convergent evolution, resulting from natural selection in similar environments, has led to evolution of roofed nests that are often composed of very different materials in different families or genera of tropical birds. Roofed nests may be woven or thatched grasses in true weaverbirds (Ploceinae), made of a mass of short heterogeneous plant materials bound together by spider or insect silk in sunbirds (Nectariniidae), or built of mud in some of the swallows (Hirundinidae). In the rufous ovenbird (*Furnarius rufus*) of South America the solid, two-chambered enclosed nest is made of a mortar of sand and cow-dung. The Cape penduline tit (*Anthoscopus minutus*) of South Africa combines cobweb and felting to bind together a nest of cottony fibres of the kapok tree, and of wool or hairs. The African broadbill (*Smithornis*), like a number of other birds of tropical forests, uses black fungus fibres (*Marasmius*) for a binding material. The white-headed buffalo weaver (*Dinemellia dinemelli*) of East Africa places a roof of thorny twigs over its grassy nest, and even places thorny twigs along the boughs leading to its nest, presumably to discourage mammalian predators.

The roofed nest among birds reaches its evolutionary climax in the pendulous hanging nest, and in the compound nest. The coherence and firm binding of materials in roofed nests has predisposed them to evolve a pendulous attachment. In turn, the enhanced safety of a hanging position of the nest at or near the tips of branches may

have been an important component of natural
selection leading, for example, to the evolution of
weaving by the true weaverbirds. These birds
generally tear long flexible strips from fresh green
leaves of grasses or palms for use in weaving. The
nests of different species can be arranged in a
series leading from loose, crude, irregular weaving
to the close, neat, regular pattern especially found
in those species that build pendulous nests with
long entrance tubes. For example, Cassin's
weaver (*Malimbus cassini*) of the tropical rain
forests of central Africa builds what may be the
most skilfully constructed nest of any bird. From
the brood chamber there hangs down an entrance
tube with a finely woven warp-and-woof pattern.
This tube may be over 60 cm long with the opening at the bottom, and is assumed to enhance
protection from such enemies as tree snakes
(*Boiga*).

In compound nests birds of the same species
occupy separate compartments in the same nest
mass, which also has some communal feature,
such as a common roof shared by all. Instances of
species where different pairs of birds build their
nests in physical contact are not uncommon, and
may illustrate initial stages in the evolution of the
compound nest.

Compound nests are made by relatively few
birds, the best-known example being the sociable
weaver (*Philetairus socius*) from the deserts of
south-western Africa. This species builds what is
probably the most spectacular nest of any bird.
The small, sparrow-like birds construct an immense nest of straws, grass tops, and fine twigs,
usually in camel-thorn acacia trees. Each nest
mass is often a metre thick, of irregular extent and
may reach over 7 m in its greatest dimension. The
common roof, on which all of the birds may work,
is dome-shaped, while the underside of the nest is
riddled with up to sixty or more separate nest-
chambers. The birds live in their nest all the year
round.

During the non-breeding season sociable weaverbirds sleep several to one chamber, thus helping
each other to keep warm during the cold winter;
some of the chambers are therefore not occupied.
During the breeding season, which seems to be
governed by the erratic rains, the birds tend to
occupy separate chambers as pairs or families.
Careful measurements have indicated that the
greater the number of birds present and the larger
the nest mass the more stable the temperature is
inside the occupied chambers during the winter. It
therefore appears that natural selection has
favoured increase in size of nest mass, in response
to the climatic conditions in which these birds
live.

Brood parasites such as the European cuckoo
(*Cuculus canorus*) and most species of American
cowbirds (*Molothrus*) do not build a nest, but instead lay their eggs in other birds' nests, leaving
their young to be raised by the foster parents. Possibly this trait arose from the habit of sometimes

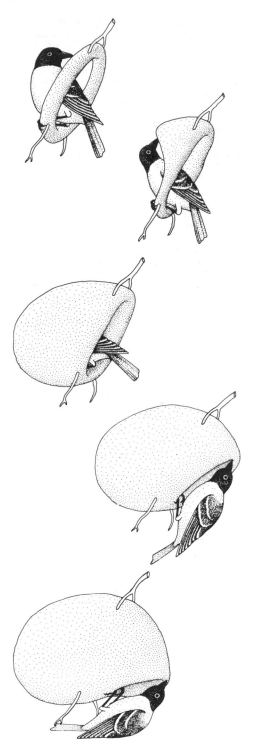

Fig. A. Stages of nest-building in the common village
weaverbird (*Ploceus cucullatus*).

laying eggs in the nests of other birds. If the nest were fresh and the host happened to be of a suitable species, the foster young might be reared successfully. The bay-winged cowbird (*Molothrus badius*) of South America may build its own nest, incubate its eggs, and raise its young, but often it also lays eggs in the nests of other species. A related species (*M. rufoaxillaris*) parasitizes solely its relative, *M. badius*, while the brown-headed cowbird of North America (*M. ater*) parasitizes a great variety of passerine species.

How a weaverbird builds its nest. As an example of just how a bird builds its nest a weaverbird will be selected. The common village weaverbird (*Ploceus cucullatus*) breeds over much of Africa south of the Sahara. Usually its nesting colonies are found in trees near human habitations, where it is easily observed. The species can be successfully maintained in captivity, and when breeding in aviaries its building behaviour is essentially the same as in the wild. The nest is ovoid with its long axis horizontal and with the entrance opening downwards at one end. The outer shell is woven by the male using long strips torn from fresh green leaves of tall grasses or from palm fronds. The different stages in weaving a nest are shown in Fig. A. Each stage of nest-building automatically provides the external stimuli for its own termination and for the starting of the next stage. Any portion of a fresh nest can be removed and will be replaced by the male, so long as the bird is left with the part where it normally perches, i.e. the bottom half of the initial ring.

After the male has woven his nest he DISPLAYS it to visiting, unmated females. In this display he hangs upside down at the entrance and vigorously flaps his wings, at the same time uttering special call notes. This display attracts the female, who may enter and inspect the interior of the nest. If she accepts it, she lines the nest with soft grass-heads and feathers, mates with the male, lays and incubates eggs in the nest, and does all or most of the work of feeding the nestlings. The male of this polygynous species then builds a new nest and attempts to attract additional mates. If all females reject a male's nest, the nest becomes even less attractive as it ages and fades from a bright green to a dull brown. Generally, before the rejected nest is more than a week old the male tears it down and builds a fresh nest in its place.

Weaving consists of interlocking loops and requires flexible materials. Experiments demonstrate that the male prefers green nest materials to any other colour and to different shades of grey. This preference for the green colour of fresh grass or palm leaves helps ensure the necessary flexibility of the materials used in building.

Fig. B (lower left figure) shows the very first strip for the start of a nest. This strip is first partly coiled about a twig, and is kept in place by then winding and threading the strip back and forth between the twig and the doubled-back opposite half of the strip which is held in place by the feet

as the male weaves with his bill. The initial attachment of strips for a nest results from the tendency of the male to hold the strip at one end in his beak, while he then pokes and vibrates this end into or alongside twigs and winds it about twigs. The male also tends to reverse the direction of winding a strip between adjacent twigs or strips, and he also tends to push and pull ends of strips through holes made by the act of weaving itself. The tendency to alternate the direction of winding is particularly clear in the weaving of the brood chamber, and in the case of weaving on an artificial frame (Fig. B, right-hand figures).

Construction of the ring stage results from the tendency of the male to weave around himself and along twigs, while generally perching in the fork of a twig keeping each foot more or less in the same place. Usually the bottom half of the ring is the last part to be closed, and if the male is given only strips shorter than the length of his body he may be unable to close the ring beneath himself. Fig. C illustrates how a male village weaverbird actually weaves a strip into his ring.

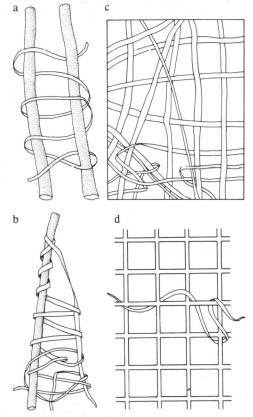

a c

b d

Fig. B. Details of weaving technique in the common village weaverbird. **a.** Alternately reversed winding between twigs. **b.** Coiling about a single twig, with alternately reversed winding and split strip. **c.** Details of weaving from egg chamber. **d.** Strip woven into wire frame.

The normal polarity of the nest, with the brood chamber on one side of the ring and the antechamber and entrance on the other, is a consequence of the plane of tilt of the basic ring frame. The male sits in the ring and faces so that the ring tilts towards him. He then builds the brood chamber out in front of himself. If the ring is tilted in the opposite direction by the experimenter and fastened in this new position, the male promptly reverses the direction in which he faces, and therefore reverses the polarity of the nest he builds.

The globular form and normal size of the brood chamber result from the fact that the male faces one way and always stands in much the same place as he works. He perches on the bottom half of the ring while he weaves and repeatedly pushes out the developing brood chamber with his bill as far as he can reach in all directions from his fixed location.

The male shows a gradient of weaving tendencies, and he works especially over his head. When the meshwork of the roof becomes too fine to permit easy weaving, he puts in a ceiling just under the roof. This ceiling is not woven but is a thatch of overlapping, short, broad pieces of grass leaf or of dicotyledonous leaves. Its function

is presumably to help shed rain, since like weaverbirds generally, this species breeds during the rainy season. It appears that the ceiling is terminated when it becomes opaque and blocks off the entrance of light through the roof. If a piece of green cloth is sewn over the brood chamber and the ceiling strips beneath it are removed, the male will not replace the ceiling beneath the cloth, although he will put in an abundance of ceiling strips just beneath the roof of the antechamber where penetrating light is not blocked by the cloth cover.

The antechamber is woven out from the back of the ring (Fig. A). The male, keeping his feet on the bottom part of his ring, weaves over his head, gradually leaning over backwards farther and farther, until he has built the entrance down to a horizontal level. This level is the cue to stop. If the nest is then tilted back 90° until its long axis is vertical and is fastened in this position, the male continues to weave backwards over his head at the entrance, until he has built the entrance down to the new horizontal level at right angles to its former and normal position.

The male adds a short entrance tube after a female has accepted the nest. The female prevents the male from entering, and he continues building on the nest by adding to the rim of the entrance,

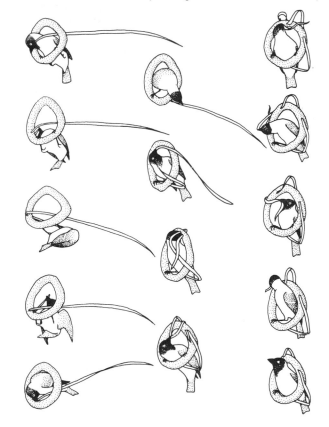

Fig. C. How a male village weaverbird weaves a strip of grass into his ring.

and by reinforcing the outer attachments of the nest. The smoothing of the rim as the male works in loose ends tends to terminate growth of the entrance tube. If the experimenter adds a strip to the rim, loosely threading in one end, and leaving the other end dangling, the male instead of removing this strip weaves it in all around the rim at the entrance. By continuing to add such strips one can cause the male to extend the entrance tube 30 cm or more beyond the normal length.

The lining put in the bottom of the egg and brood chamber by the female village weaverbird is thatched and not woven. First, she puts in a few strips torn from grass leaves, then a thick layer of soft grass tops and finally, if available, a layer of feathers. As in the canary (*Serinus canaria*) and many other birds, the peak of nest-building and of copulation comes 1–2 days before the first egg is laid. In the case of the open nest of the canary, the shift from putting in grasses to feathers results in part from the reduced inside diameter of the nest cup. If this diameter is artificially reduced by adding an inner cup, the female canary begins carrying a higher proportion of feathers into the nest.

Development of the ability to build. Nest-building in birds has often been cited as an excellent example of an INSTINCT, but few detailed and systematic studies have been made of how ability to build develops in young birds (see ONTOGENY).

Building does not require a teacher. Domestic canaries and village weaverbirds reared in isolation are able to construct the nest typical of their species when furnished with suitable building materials. Such experiments, although they rule out need for social example or teaching, do not really rule out need for practice, since even during construction of its first nest it is possible that a young bird may be LEARNING some of the techniques involved. This is more likely to be true of birds that make relatively complex nests, such as the village weaverbird, than of birds, like the canary, that make a simple open nest in a cup. The nest of the canary is built by the female, and females reared without access to normal nest materials will, when they reach breeding age and are given suitable materials, build nests as neat in appearance as those of normally experienced birds.

Male village weaverbirds do not mature until their second year, but soon after fledging they start to actively manipulate all sorts of materials. They are able to build crude nests long before attaining sexual maturity. Nests built by yearling males are more loosely constructed, and are woven of shorter and more varied materials than are the nests of experienced adult males.

The normally crude appearance of nests of yearling males is caused by insufficient practice in selecting and preparing nest materials. Young males first exposed to normal nest materials when almost a year old are scarcely able to tear off

strips the first day or so. Even after they learn to tear strips correctly from leaves of tall grasses, these strips are abnormally short and not so suitable for weaving as are later strips. Performance improves with practice and there is a marked difference in average length of strips in nests built by the same individual male in his first and second year.

When young male village weavers are hand-reared in the absence of nest materials they often manipulate or try to 'weave' their own feathers or those of their cage mates. When given a choice of different coloured nest materials, these deprived young birds prefer green to yellow, blue, red, black, or white; their preference is the same as that of 'control' birds reared with access to normal nest materials. In an experiment in which the birds were exposed to variously coloured artificial nest materials for one hour each day, the hand-reared birds interestingly showed only a slight preference for green materials during the first 2 days' exposure, but this preference doubled after 4 days. When tested with vinyl plastic strips these hand-reared young weavers also had the normal preference for flexible as against stiff nest materials.

When they were about 1 year old, the three survivors of this experiment were given their first normal nest materials, and although they often picked up and carried these materials about, they wove a significantly smaller percentage of such strips than did three control males of about the same age. But this difference greatly diminished with 3 months of practice in handling strips, and two of these hand-reared males managed to weave two nests. During the same period the three control males wove eleven nests. The most subordinate of the three deprived males never learned to build a nest, although he developed normal adult breeding coloration, held a territory, and lived for 9 years.

Even partial deprivation of nest materials, starting a month after the young have left the nest, may have a significantly retarding effect on the ability to weave by yearling male village weaverbirds. It would appear that opportunities for practice are important even during the first month out of the nest. However, if male village weavers have built nests during their first 2 years of life, this ability is not lost, even if later on they are deprived of nest materials, and not allowed to build nests for prolonged periods (over 2 years).   N.E.C.

NICHE. The status of an animal in its community, in terms of its relations to food and enemies, is generally called its niche. Animals are commonly referred to in terms of their feeding habits, terms such as carnivore, herbivore, and insectivore being widely used. The concept of the niche is simply an extension of this idea. For instance, there is the niche which is filled by birds of prey which eat small mammals, such as shrews (Soricidae) and wood and field mice (*Apodemus*). In an oak wood this niche is filled by tawny owls (*Strix aluco*), while in the open grassland it is occupied by the Old-World kestrel (*Falco tinnunculus*). We can think of the niche as the way in which an animal earns its living, i.e. its *profession*.

Just as we find the same professions represented in different human communities, so we find a close parallelism between niches in widely separated animal communities. There is a mouse niche, filled by various species in different parts of the world; and a rabbit niche, filled with herbivores, such as hares and rabbits (Leporidae) in the northern temperate zone, the agouti (*Dasyprocta*) and viscacha (*Lagostomus maximus*) in South America, wallabies (*Wallabia*) in Australia, and the hyrax (Hyracoidea) and mouse deer (Tragulidae) in Africa. These examples result from evolutionary *convergence* (see EVOLUTION), and they are generally known as ecological equivalents. Further examples are illustrated in Fig. A. One of these, the African yellow-throated longclaw (*Macronyx croceus*), inhabits some African prairies and grasslands, and looks and acts so much like the Eastern meadowlark (*Sturnella magna*) of America that a competent birdwatcher might easily confuse the two species, although they belong to different families. Such similarities are due to the action of NATURAL SELECTION operating in similar environ-ments upon species which have a very similar way of life.

42

NOVELTY. An animal's response to novelty depends partly upon the nature of the novel stimulus or situation, and partly upon the animal's internal state. However, certain generalizations can be made, particularly with respect to the reactions of birds and mammals towards novel objects. When presented with a novel object the animal typically shows an ORIENTING RESPONSE which consists in a sudden turning of the head or body so that the eyes and ears are focused on the object. The animal pays ATTENTION to the object and this is usually accompanied by a certain amount of AUTONOMIC activity and of AROUSAL. If the novel stimulus is repeatedly presented, HABITUATION occurs and the generalized orienting response gives way to a localized orienting response, which either develops into a specific adaptive response, or wanes to the point that the stimulus no longer evokes a response. For example, a domestic cat (*Felis catus*), presented with a clockwork mouse for the first time, will initially show a generalized orienting response, turning to face the mouse and stare at it. There will be an increased heartbeat rate and other signs of autonomic arousal. Upon subsequent presentations the cat will initially glance at

Fig. A. Pairs of independently evolved, but ecologically similar, species which occupy similar niches are known as ecological equivalents. The examples shown here are (**a**) the eastern meadowlark (*Sturnella magna*) from America, and (**b**) the yellow-throated longclaw (*Macronyx croceus*) from Africa; (**c**) an Australian wombat (*Phascolonus ursinus*) with its skull, and (**d**) an American woodchuck (*Marmota monax*) with its skull.

the mouse, but show no heightened arousal. Eventually it will either ignore the mouse, or it will start to show EXPLORATORY behaviour and PLAY, the specific adaptive responses.

When released into a novel situation, such as a new cage, animals initially show signs of FEAR, and birds especially may remain motionless for a long period of time. The initial fear response is often followed by exploratory behaviour and those particular activities, such as SCENT MARKING, by which animals in CAPTIVITY make themselves at home. The probability that a given situation will elicit exploration rather than fear depends partly upon the animal's internal state. Animals that have recently had frightening experiences, or have been reared in isolation, are more likely to be wary of novel situations. New-born animals will investigate novel situations without apparent signs of fear, but this tendency decreases with experience, and as some situations become familiar the response to novelty becomes increasingly fearful (see IMPRINTING).

Certain types of novel stimuli are treated in a specific manner. For example, when a new member is introduced to a group of domestic fowl (*Gallus g. domesticus*) AGGRESSION is usually the first reaction of the resident birds. If a novel food is made available to a rat (*Rattus*), it will tend to ignore it if its diet is satisfactory. If it is suffering from a dietary deficiency, the rat will cautiously sample the novel food, and will generally wait a day before sampling it again. This aspect of FOOD SELECTION acts as an insurance against poisoning, and is illustrative of the CONFLICT between fear and exploration that is characteristic of responses to novelty.

O

ONTOGENY. In many respects the behaviour of the young animal is less complete, less complicated, and less competent than that of an adult. The ontogeny of behaviour is commonly seen as the elaboration and perfection of behaviour as an individual grows up. The ways in which these things happen and the conditions which have to be fulfilled for them to occur are central problems in the study of behavioural DEVELOPMENT. Nevertheless, it is worth noting at the outset that much of a developing animal's behaviour is required for the specific problems that beset it while it is young. Its ecology can change dramatically during the course of its life, and such a change can be associated with wholesale reorganization of body and behaviour, as in the metamorphosis of a caterpillar into a butterfly (Lepidoptera). It would be absurd to regard the caterpillar as an inadequate butterfly, or the suckling of the young mammal as an incompetent version of adult eating. Descriptions of what happens during ontogeny can, therefore, be a misleading guide to the way in which adult behaviour is assembled.

Another general point is that in many cases behavioural development may not cease until death. This is particularly obvious in animals whose behaviour is enriched and differentiated by LEARNING throughout life. So, the developmental processes responsible for many behaviour patterns may not have a clearly defined end-point which is typical of adults of that species. In seeking ways to understand how a given pattern of behaviour has developed, two things can be done. First, we can look for the starting ingredients without which development of that behaviour would be impossible. Second, we can examine the 'cooking' processes which transform the ingredients and give rise to the end product. These approaches will be examined in turn.

Origins of behaviour. It has proved convenient to distinguish between the factors that control behaviour from moment to moment, and those that are responsible for its development. The distinction may not always be easily drawn in practice, since a factor responsible for the development of a behaviour pattern may lie close in time to the occurrence of that behaviour. In general, though, sources of behavioural distinctiveness usually lie considerably further back in time from the behaviour they affect than do the controlling conditions. Developmental determinants are agents that lastingly give the behaviour pattern its peculiar characteristics, differentiating it from other types of behaviour; they represent necessary, though not sufficient, conditions for development. Once embarked on tracing back through the web of historical events that preceded the emergence of an adult behaviour pattern, there might seem no obvious stopping point. However, what is usually meant by a developmental determinant of an individual's behaviour is a factor that was responsible for the distinctiveness of the individual's behaviour, and which operated at some point in the life of that individual.

The cut-off point in the historical analysis is obviously arbitrary, but the distinction between ontogeny and *phylogeny* is reflected in the ways in which ideas can be tested. It is no easier to test the predictions of a hypothesis about the evolution of an animal species than it is to replay the social history of humans with some supposedly crucial factor changed. However, it is possible to test hypotheses about ontogeny, because the starting point of a new individual's development can be accessible in practice as well as in principle.

Few people would disagree that some of the initial determinants of behaviour are already present in latent form within the fertilized eggs; some determinants are perhaps present as cytoplasmic factors, but most are represented in the nucleus of the *zygote*, presumably in genetically coded form. Furthermore, few people would doubt that environmental factors are crucial in determining the way internal factors express themselves, and that very often they determine the detailed patterning of behaviour. How is an external or an internal factor shown to contribute to the distinctiveness of an individual's behaviour? What is needed is an analysis of the sources of difference between individuals when their behaviour is measured in a particular way.

Sources of differences between individuals. The most direct way to demonstrate that something is responsible for one individual behaving in a different way from another is to vary that factor. Meanwhile, other things are kept constant or randomized so that they cannot contribute systematically to differences between individuals. The experimenter's art lies in being sensitive to closely intertwined sources of variation among individuals, and in disentangling them (see LABORATORY STUDIES). In general, this procedure does not pose any major problems for studies of external factors. Indeed, many (but not all) studies of learning rely on this approach. For instance, when day-old domestic chicks (*Gallus g. domesticus*) hatch out they will peck at a variety of small objects. If a particular object is coated with a bitter-tasting substance like methylanthranilate, a chick that pecks at it is likely to do so only once. On subsequent occasions it withholds its pecks from that particular object.

It must be emphasized that the effects of a particular treatment need not necessarily be specific as they are in this instance. The point is that animals treated in one way behave differently from those treated in another way.

Usually the internal factors involved in determining the characteristics of behaviour are genes in the cell nuclei of the nervous system. Even with simple systems it is not easy to specify the precise moment in development when a gene is activated. And it is even more difficult to activate genes experimentally. Consequently, the analysis of GENETIC sources of variation is nearly always less direct than that of external sources of variation.

The first step is to establish that a distinctive feature of behaviour is inherited, by studying related and unrelated individuals brought up under identical conditions. The second step is to perform breeding and cross-fostering experiments, in order to show that the inheritance is not SOCIAL or non-genetic. This is necessary, because behaviour can be transmitted from one generation to the next in a variety of ways. Especially convincing evidence for a genetic influence on behaviour will be when a behavioural character is shared by grandparents and grandchildren, but is not expressed in the intervening generation; the presumption would be that the behaviour is influenced by a recessive gene.

An interesting example of genetic analysis comes from work on hygienic behaviour in the honey-bee (*Apis mellifera*). Honey-bees frequently re-use the wax cells in their hives for rearing the young. If the brood is infected, the wax cells gradually fill up with dead larvae and pupae. In most strains of honey-bee this is just one of the facts of life. However, in some strains the workers uncap cells with infected larvae in them and remove the larvae. These bees are called 'hygienic'. The behaviour was studied first of all in true-breeding lines of hygienic and non-hygienic bees. Half the cells were inoculated with American foul brood, a bacillus which attacks the larvae. As a control the other half were injected with water. The non-hygienic strain removed a small proportion of the infected larvae, and by the end of the experiment 20% of the cells contained larvae killed by American foul brood. By contrast, the hygienic strain of bees removed all the larvae killed by American foul brood, doing so to the greatest extent on the 9th day of larval life.

In order to elucidate the genetics of the honey-bee's hygienic behaviour, a virgin queen of one strain was artificially inseminated with the semen of the other strain. When the hygienic strain was crossed with the non-hygienic strain all the off-

spring were non-hygienic. When these non-hygienic hybrids were crossed with the original parental strain of hygienic bees the offspring consisted of three times as many non-hygienic bees as hygienic. This suggested that two different genes were involved. Examination of the behaviour of the non-hygienic offspring of the back-cross revealed a remarkable and quite unexpected thing. One third were like the original parental non-hygienic strain, one third would uncap the cells of the infected larvae but would not remove them, and one third would remove the dead larvae of the cells which had been uncapped. This suggests that one gene controls uncapping and one controls removal, and both are necessary for the hygienic behaviour to be shown. The behaviour of the non-hygienic strain suggests how the genes are acting. Despite their name, the bees of this strain do remove the major proportion of the infected larvae. So it looks as though the genes in question do not directly control the motor patterns of uncapping and removing the dead larvae, but are affecting the readiness with which these patterns are elicited. In other words, in the hygienic strain the *thresholds* for elicitation of the behaviour patterns are lowered.

Sometimes clear evidence of genetic transmission from one generation to the next can be obtained by hybridizing different species. For instance, different species of dove (*Streptopelia*) have similar bow-cooing DISPLAYS, but the amplitude of the bow is characteristic of the species. Hybrids between two species have bows that are different from, and sometimes mid-way in amplitude between, the two parental species. Unfortunately, further genetic analysis is impeded because the hybrids are sterile, or have markedly reduced fertility.

When related individuals behave more like each other than unrelated individuals, it is easy to conclude that the differences are transmitted genetically from one generation to the next. One way of checking to see whether or not this is the case is to take the young of one line and foster them on the parents of the other line. For instance, half the pups of one strain of house mouse (*Mus domesticus*) are fostered on the parents of the other strain. When the young grow up they are first given a test in which the frequency with which they press a hinged panel is measured. Secondly, measurements are made of the frequency with which the mice press the same panel when doing so causes a light to be turned on for 1 s. While the first measure of panel-pressing is unaffected by the foster parents, and is influenced by the genetic strain of the pups, the frequencies of turning on the light are strongly influenced by the foster parents. Clearly some non-genetic factor is involved.

While it is commonly accepted that environmental and genetic sources of variation can both play a part, it is tempting to ask how much of the variation can be attributed to internal factors. An estimate of 'heritability' is an attempt to provide an answer. The technique is based on a set of assumptions about the ways in which the *genotype* and the environment produce variation in the *phenotype*. Heritability is defined as the estimated variance in a phenotypic character due to genetic factors, expressed as a percentage of the total observed variance in that character. Such an estimate has a clear value in artificial selection experiments carried out in laboratory conditions. However, heritability estimates are also subject to great abuse. If, as usually happens, only a limited range of environments have been used, then the heritability appears larger than would have been the case if individuals were obtained from a wider range of environments. Conversely, if only a limited range of genotypes have been sampled, then the heritability would seem smaller than it would have been if a larger range of genotypes had been used. A more serious point is that an absolute value for heritability assumes, quite unreasonably, that no interaction takes place between environmental conditions and gene expression. So an absolute value for heritability can lend a spurious air of precision to the question of how much the development of behaviour is dependent on inherited factors.

Analysing sources of variation is an important part of investigating the origins of behaviour. The conclusions are that, if animals known to differ genetically are reared in identical environments, any differences in their behaviour must ultimately have genetic origins. Similarly, differences in genetically identical animals reared in different environments must be attributed ultimately to the environmental conditions. However, identification of a particular source of a difference does not preclude the possibility of others. Moreover, it is only the first step in unravelling the dynamic processes of development. Genetically based differences may be mediated through the external environment, and environmentally based differences may rely on the induction of particular genes. For instance, a mutant form of the fruit fly (*Drosophila melanogaster*) has been found which differs from other flies in one respect. It is unable to learn an ASSOCIATION between an odour and the delivery of an electric shock, and is appropriately called *dunce*. Undoubtedly the mutation may influence the learning capacity of the flies in a number of different ways. Nevertheless, the mutation of a single gene does influence the way in which a fly picks up information from its environment. On the other hand, an environmental factor, such as overcrowding, can cause the offspring of the non-migratory form of *Locusta migratoria* to grow up differently from their parents. By degrees the population becomes migratory. The solitary and migratory forms of the locusts are so different that they were once classified as different species. It would seem that the induction of a large number of the locusts' genes depends on external conditions. These two examples serve to remind us that in understanding the process of behavioural de-velopment, we have to look at the interplay between many internal and external factors.

Evolution and development. The problem of how an individual animal develops is distinct from the problem of how its ancestors changed during the course of EVOLUTION. So it is necessary to tread carefully when evolutionary and functional arguments are brought to bear on developmental issues. One way of thinking about evolution is not in terms of the need of species, groups, or individuals, but in terms of the requirements of genes, defined as the agents which are necessary for the expression of a character, and which are transmitted to the next generation. Different genes combine in each generation to form a temporary federation. The alliance is an individual organism. By reproducing, individuals serve to perpetuate the genes which in the next generation recombine in some other kind of alliance. Therefore, genes are sometimes thought of as being selfishly bent on replicating themselves by the best possible means. While this style of thought is plainly teleological, it is easier for many people to think of a complex system in terms of the best ways for it to reach a specific end-state. It certainly seems to be a helpful way of thinking about evolution. Confusion arises when this language is treated as being virtually equivalent to the language of what genes do. What is implied is some simple correspondence between teleological and causal explanation. This leads inexorably to the belief that if it is valuable to view genes as selfish, it is also valuable to suppose that they uniquely bring into being the phenotypic character of the whole animal. It cannot be stated too strongly that the activation of a particular gene may be a necessary condition for the development of a given pattern of behaviour without it being a sufficient condition. Furthermore, the gene may exclusively 'program' a particular bit of behaviour without being the only agent to do so. So the switch from a useful way of thinking about evolution to a misleading way of thinking about development merely encourages a crude form of preformationism—the doctrine that a recently fertilized egg is a miniature adult (see also INSTINCT).

A much more subtle and pervasive confusion of evolutionary and developmental arguments arises in the use of the term INNATE. It was originally used for behaviour which developed without example or practice, and that usage is strictly developmental. More recently 'innate' has been used for behaviour which was adapted for its present FUNCTION by NATURAL SELECTION during the course of evolution. Because of the two usages, it is very easy to switch from one to the other without being aware that a change in definition has occurred. The problem is set out in Fig. A. It is easy to see that behaviour can be adapted to a particular function in three separate ways: (i) by natural selection operating during evolution; (ii) by trial-and-error learning during ontogeny; or

Source of adaptedness

	Evolutionary history	Individual history	Cultural history
No learning involved in development	Direction of migration in garden warblers	No examples	No examples
Learning involved in development	Kin recognition in ducks	Avoidance of bitter-tasting food in domestic chicks	Opening of milk bottles in blue tits

Fig. A. Behaviour adapted during evolutionary history by natural selection may involve learning.

(iii) by selection of appropriate habits during CUL-TURAL history.

The last of these possibilities is by no means confined to human beings. Increasingly examples are being found in mammals and birds of highly adaptive habits that are transmitted from one generation to the next by social means. One of the first to be discovered was the opening of milk bottles by blue tits (*Parus caeruleus*). Other famous examples include the skilful methods used by oystercatchers (*Haemotopus ostralegus*) for opening mussels (Myrtilidae) and the potato washing by Japanese macaques (*Macaca fuscata*). When the individual learns the habit by copying other members of its own species, it does not need to make any errors, in other words the process of selecting an appropriate skill went on in a previous generation. In both this form of ADAPTATION and in the second, the rules for learning have almost certainly been subject to natural selection during the course of evolution, but the specific form of behaviour emerging from the learning processes has not. In the second form of adaptation the individual itself has to tune its own behaviour to the environment, as in the case of chicks suppressing their pecks at bitter-tasting objects.

Clearly, learning need not be involved in the first case. Hand-reared European garden warblers (*Sylvia borin*) kept in cages all their lives will start to show migratory restlessness at what would normally be the appropriate time in the autumn (see MIGRATION). They attempt to fly in a south-westerly direction, and after about a month make a course correction which under natural conditions would prevent them from flying out into the Atlantic. It is difficult to believe that this behaviour is learned, but dramatic examples like this do not mean that all cases of behaviour which have been subject to natural selection during evolution develop without specific instruction from the environment. Learning may be involved in case (i), as it must be in (ii) and (iii). The recognition of kin, for instance, is presumably a form of be-

haviour which has been subject to natural selection during the course of evolution. However, by the process of IMPRINTING, kin recognition in many birds and mammals involves a learning process in the course of development. So if natural selection is the source of adaptedness it hardly follows that no learning is involved in development. On the other hand, if no learning is involved in the course of development the strong implication is that the adaptedness of the behaviour was achieved solely by natural selection. In other words, developmental evidence can be used to draw conclusions about evolution. In principle there is nothing wrong with doing this, providing we are clear what we are up to. In practice, there are a number of difficulties about excluding learning as a possibility. This is a controversial area, and it is worthwhile, therefore, to consider carefully both the character of the arguments and the character of the evidence.

The isolation experiment. Konrad LORENZ strongly argued for an experimental approach, in which it would be possible to identify internal mechanisms responsible for the adaptedness of behaviour, by systematically excluding likely sources of environmental 'information'. The isolation experiment, as it is called, clearly has been of service in eliminating possible explanations for the determination of some behaviour patterns. For instance, the young of many species of bird that are feathered and active at hatching can be taken out of the incubator in which they hatched and placed on a wooden platform running across a clear piece of glass (see Fig. B). Many young animals will avoid walking over what appears to be a cliff. Young birds such as chicks of the ring-necked pheasant (*Phasianus colchicus*), which under natural conditions nests on the ground, nearly always walk off the wooden platform on the side that looks shallow. By contrast, the young of species such as the mandarin duck (*Aix galericulata*) behave differently on the two sides. When they leave the platform on the shallow side they simply

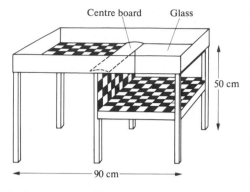

Centre board   Glass

50 cm

90 cm

Fig. B. Apparatus used in testing a recently hatched bird's response to a visual cliff.

walk off. When they leave on the deep-looking side they launch themselves with a little jump. Such a jump would be what a recently hatched mandarin duckling would do as it came down from the hole high up in a tree where its mother had made her nest. It would be difficult to argue that the adapted behaviour was learned in any obvious way. Of course, neither this experiment nor any other kind of isolation experiment can show precisely how the behaviour developed. Even so, as a strategy, Lorenz's approach has the great merit of being positive and directed. Rather than bother about possible unknown sources of variation, the advice to the experimenter is straightforward: if you consider something as the source of variation, then remove it. Difficulties in interpretation arise for a number of quite separate reasons.

(1) *Specific–general continuum.* Konrad Lorenz originally proposed the metaphor of a blueprint for the origin of behaviour patterns which he thought were coded in the gene. The metaphor was helpful in the sense that nobody would suppose that blueprints were sufficient for a building. Clearly to raise a building a work force is required along with bricks, mortar, and so forth. However, a sharp distinction had been made by Lorenz between the information on which the detailed characteristics of the finished building depend, and the conditions necessary for translating that blueprint into a building. The distinction is between a determinant with a specific and qualitatively distinct effect on behaviour, and a determinant which has a general effect on all behaviour. In practice this distinction does create problems. It is extremely difficult to know what to do when considering a spectrum of environmental conditions ranging from those that exert a highly specific effect on behaviour, such as those required for learning, through to those that produce general effects on behaviour and, indeed, on anatomy, such as a low protein diet. Where do we draw the line and say from here on the experiences are no longer providing relevant information? For instance, simple

exposure to patterned light can speed up the process by which chicks peck accurately at seed, approach potential foster-parents, and the rate at which they learn about visual targets. In Lorenz's sense, are these environmental factors supplementary to the blueprint, or are they part of the work force? It really does not matter, and if we insist on an answer we have been trapped by the metaphor. Nature is not going to package herself conveniently to match our distinctions. Undoubted aids to thought at one stage of analysis can easily become shackles at the next.

(2) *Equivalence.* Even when considering experiences that have a specific effect on behaviour, it may be very difficult to know in advance when an animal is likely to generalize the effects from one kind of training to a novel situation. Can we really be so certain that we know what are equivalent types of experience for an animal? We might, for example, be inclined to treat tactile input as being so different from visual information that experience of an object in one modality would not help recognition of that object when using the other. In the rhesus macaque (*Macaca mulatta*) opportunities to discriminate between potential pieces of food in the dark, using tactile cues, make it easier for them to discriminate between the same objects in the light when they have to choose on the basis of visual cues. It is very difficult to have useful intuitions about these kinds of equivalences in animals living in very different perceptual worlds from ourselves.

(3) *Equifinality.* It is possible for a given pattern of behaviour to develop by several different routes. The term *equifinality* is used for cases in which a given, end-point can be reached in more than one way. An isolation experiment that deprives an animal of a particular kind of experience may force it to develop the pattern of behaviour, which normally depends on that experience, in another way. While this result would be very interesting, it would not show that the excluded environmental factor had no influence on development when it was normally available. To argue like that would be like arguing that travellers who are forced to use bicycles because of a fuel shortage do not need petrol to run their cars.

(4) *Self-stimulation.* Even though an animal is isolated from relevant experience in its environment, it may do things to itself which enable it to perform an adaptive response later on. Normally treated northern mallard ducklings (*Anas platyrhynchos*) are able to respond preferentially to the maternal call of their own species. However, if they are devocalized in the egg so that they do not make sounds and thereby stimulate themselves, they do not show the same ability to recognize the calls of their own species. In other words, FEEDBACK from their own activity is an integral part of normal development. In many cases it may be difficult to cut such feedback.

It can be seen, then, that the evidence obtained by an isolation experiment can usually be interpreted in a variety of ways. It does not follow, though, that because it is possible to think of an objection, the objection is necessarily valid. There is no problem in principle in conceiving of highly complex unlearned behaviour that owes its adaptedness to processes of natural selection occurring in the remote past. We have no difficulty in accepting that an extraordinarily complex adaptive organ like the kidney can arise in this way; the special devices used, say, by desert-living animals for separating and re-absorbing virtually all the water from the unwanted excretory products develop whether or not these animals grow up in the desert. We should be equally willing to accept the principle that behaviour adapted for a particular environment by natural selection may develop, even though the animal has not grown up in that environment (as a result of an accident or deliberate experimental intervention). That having been said, we also have to be aware that behaviour which does not require learning processes for its development may still be greatly dependent on environmental conditions which are normally constant. If these conditions are changed then behaviour may be dramatically altered.

Multiple determination of behaviour. When thinking about the origins of behaviour it is easy to focus on the factors that have qualitatively distinct effects on behaviour. However, many things have quantitative effects. For instance, the accuracy with which domestic chicks peck at seeds is improved by their being exposed to light before they see the seeds. Even in the dark-reared animals, the accuracy improves with age (see Fig. C). So what light does is facilitate certain aspects of behavioural development. Eventually, however,

rearing in the dark leads to a regression in the ability of the animals to see. So as the chicks get older light seems to serve a maintenance function. A very similar effect is found in the development of depth perception in hooded rats (*Rattus domesticus*). The performance of dark-reared rats on the visual cliff initially improves with age, although the rate of improvement is not as rapid as in the light-reared rats. Up to around 60 days of age light seems to have a facilitatory effect on development. However, after 80 days of age, the performance of the dark-reared rats sharply deteriorates, whereas the performance of the light-reared rats remains stable; in the older animals light appears to serve a maintenance function.

The conditions necessary for the development of a behaviour pattern can be seen, then, to act in a number of quite different ways. They may initiate developmental processes, they may facilitate processes that are already in operation, or they may maintain the end products of the behavioural repertoire. The distinctions are somewhat arbitrary since it is difficult, for instance, to draw a sharp line between something which has a qualitative effect and something which has a quantitative effect. Nevertheless, in thinking about a problem as complex as this, it is convenient to make some arbitrary distinctions.

It should be clear that the distinctions apply not only to the external factors influencing behavioural development (experience) but also to the internal ones. A further point is that both the external and internal developmental determinants of behaviour range from those that have specific effects to those that have general effects. Here again a distinction between specific and general cuts across a continuum. What we end up with, then, are several different kinds of determinant. It seems reasonably likely that many patterns of behaviour of complex long-living animals, such as human beings, will be determined by a combination of conditions from all categories. This is a daunting prospect for anybody who wants to understand development. It is not surprising, therefore, that many people would prefer to focus exclusively on determinants that have specific and qualitatively distinct effects. Sometimes attempts are made to classify behaviour in terms of this category of determinant. The classification is usually in terms of learned and unlearned components of behaviour. Just as some people continue serenely to use this classification, others vehemently deny its usefulness. As we have already seen, some of the difficulties arise because of the problems in practice of interpreting the isolation experiment. Also, the arbitrariness in defining a specific effect or one that is qualitatively distinct from another excites heated debate.

Despite the controversy, let us follow through the logic of classification based solely on the agents that give rise to specific and qualitatively distinct patterns of behaviour. We deduce that some patterns will be influenced by external fac-

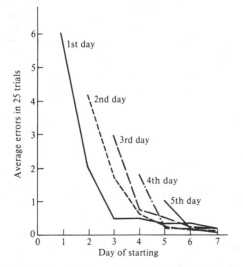

Fig. C. Changes in the accuracy of pecking by domestic chicks. Groups were given the first experience of pecking on different days after hatching.

tors alone, and some will be influenced only by internal agents. But we also expect some to be affected by both internal and external factors. It has sometimes been argued that such cases can invariably be decomposed into components that reflect respectively the internal and external determinants. Sometimes such analysis may be possible, but it does not seem very plausible that the song of a chaffinch (*Fringilla coelebs*) can be subdivided in this way. A more interesting point is that some forms of distinct behaviour may arise from the interaction of agents none of which have specific effects. For instance, the social status of its parents and its genetic sex may determine the specific ways in which a rhesus monkey responds to its peers.

If we were omniscient and were able to quantify all the determinants exclusively affecting any given behaviour pattern occurring at a particular stage of development, it would be possible to build up a scatter diagram, as shown in Fig. D. (It is of course impossible to justify the relative positions of the four dots placed on the scatter diagram.) The diagram makes the point that there is no particular reason why the dots should be scattered round the edge, or in any one part of the space in the middle. A different point is that such a diagram would represent a snapshot of ontogeny. The positions of some behaviour patterns would doubtless move more during development than others. Many would move to the right on the diagram, as the behaviour patterns became increasingly enriched and differentiated by experience. Some might move upwards or diagonally, as fresh genes affecting the details of already established behaviour patterns became activated during development.

Processes of development. The major thrust of the research on the ontogeny of behaviour has been to show the large variety of ways in which a particular pattern of behaviour can be determined. What this amounts to in terms of the analogy with

cooking is an analysis of the major ingredients, and also some of the conditions under which the ingredients are thrown together. If behavioural development is to be properly understood, it is necessary to analyse the ways in which the various conditions that are necessary for development of behaviour act together to produce their effects. The sharp polarization of opinion already encountered when considering the origins of behaviour is also found in the theories about the processes of behavioural development. On the one hand there are those who think that straightforward correspondence can be found between the starting-points and the end-points of development. On this view components of behaviour influenced by external conditions are intercalated among those components influenced by internal conditions. An advocate of this idea has been Lorenz. His major opponents have argued that the interaction between internal and external factors is complex and continuous. It is absurd, they argued, to parcel adult behaviour into genetic and environmental components.

Lorenz's views have the virtue of simplicity. Those of his opponents, though probably correct, have often seemed to lack lucidity. Of course, it is not enough simply to refer vaguely to interactions. It is essential to consider actual ways in which internal and external factors work together. Before doing so, however, it is worth considering some of the legacies of setting internal and external factors in opposition to each other.

Internal–external alternatives. 1. *Maturation and experience*. It is plainly the case that many forms of behaviour do not occur until a particular stage in development. Fully developed sexual behaviour is characteristic of adults, for instance. Many such examples seem to develop without any obvious practice. For instance, pigeons (*Columba livia*) start to flap their wings and fly uncertainly about at a certain age, and the naïve inference is that they are learning to fly. However, in a classic ex-

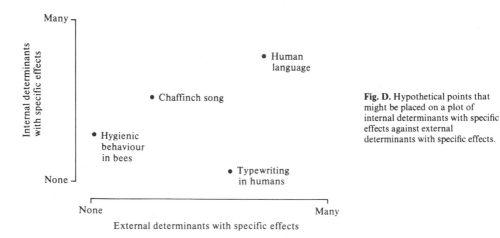

Fig. D. Hypothetical points that might be placed on a plot of internal determinants with specific effects against external determinants with specific effects.

periment one group of pigeons were placed in tubes so that they could not move their wings, while another group were left without restraint. When the unrestrained birds were able to fly adequately, those that had been kept in tubes were released and were also found to be able to fly. On the face of it this result looks like a convincing demonstration of internal changes, but there are some difficulties of interpretation. How does one know what is relevant experience? Furthermore, when it is possible to exclude with some plausibility external factors having specific and qualitatively distinct effects, other external factors may also have an important influence on development. For instance, in a cold-blooded animal external temperature is likely to have a profound influence on the general rate of development; and in human beings the onset of puberty is greatly influenced by the nutritional state of the child.

A change in behaviour with age does not necessarily imply that the change was dependent on internal factors alone. For instance, as a result of an internal change in responsiveness, the developing animal starts to learn about features of its environment; as a consequence of learning, the animal's behaviour changes. Similarly, the experiences associated with birth, and the changes that follow afterwards, may be responsible for the appearance of behaviour patterns after birth. One way of examining this possibility is to use different reference points for calculating age, such as the time of conception, and the time of birth. This was first done in the duckling, in which there is considerable variation in the embryonic age at which HATCHING occurs. So, if some event occurring after hatching such as the time of filial imprinting is examined, it is possible to separate the effects of hatching, and the change in environment which occurs as a result of hatching, from those of general development. It turns out that both the variables associated with hatching and those associated with general growth and development play a part in determining when imprinting can occur most readily. The result serves to demonstrate how misleading are the either/or formulations of research problems springing from the opposition of internal to external control.

2. *Stability and lability.* When the left and right forelimb rudiments of salamander (Salamandridae) embryos are interchanged, normal limbs develop, but they face backwards instead of forwards. When the nerve connections are established, the grafted limbs move just as they would have done if they had been left in the original position, working to move the animal backwards, whilst the movements of the rest of the body produce forward movement. A year's experience does not change the inappropriate movement of the grafted legs. What this shows is that certain aspects of the underlying machinery once developed are extremely stable and cannot easily be changed by experience. However, a striking finding such as this one does not justify the conclusion

that the development of stable behaviour is uninfluenced by external factors, nor that labile behaviour is uninfluenced by internal factors.

Learned behaviour can be extremely stable, and behaviour which is not learned can be modified by experience. For instance, if nestling zebra finches (*Taeniopygia guttata*) are reared by striated finch (*Lonchura striata*) foster-parents, and then put with their own species, they will breed with their own species quite successfully. Even so, if, much later in life, males are given a choice between a female of their own species and a female striated finch, they prefer to mate with the striated finch. Preferences for the striated finch acquired early in life have shown astonishing stability in the face of subsequent experience that might have been expected to modify those preferences. On the other hand, with no experience of objects to peck at, young laughing gulls (*Larus atricilla*) will peck appropriately at an object similar to the beak of the natural parents. However, the accuracy of the orientation of the response towards the bill, or the bill-like object, improves as a result of practice. This kind of interaction, between a predisposition to behave in a particular kind of way and learning, is very common.

The idea that the internal and external origins of behaviour could be neatly attributed to processes that were well buffered from the environment, and those that were open to environmental instruction, is obviously false. It is necessary to move towards more elaborate ideas of developmental process.

3. *Closed and open programs.* An appropriate metaphor for much of the development of behaviour is that it represents the expression of a program. However, the metaphor can easily lead the unwary into supposing that all the instructions for development reside in the same place, and are coded in the same way. This deduction is particularly common when the emerging behaviour is concerned with long-term propagation of the animal's genes, and seems not to involve any kind of learning. In such circumstances, many people will immediately state that behaviour is 'genetically programmed'. An explicit distinction is made between these closed genetic programs and open programs that are subject to environmental change. Two kinds of confusion flourish on this conceptual diet. First, it is easy to muddle the neural machinery necessary for the expression of behaviour with the genetic and environmental conditions necessary for the formation of that neural machinery. In terms of programming metaphors the program for behaviour control (see MOTIVATION) is not the same as the program for writing the motivational routines in the first place.

Secondly, the distinction between open and closed programs encourages misunderstanding of exactly how phylogenetic theory relates to ideas about ontogeny. It is argued that natural selection leads to evolutionary changes in the phenotypic characters of organisms. If these changes occur in stable environmental conditions, then they must result from changes in gene frequency. Therefore (and this is the part of the argument that does not follow from what went before), the expression of phenotypic character is exclusively under genetic control. If we are to be strict about this, all the conditions that are necessary for the expression of the phenotype are involved in controlling the development of that phenotype.

However stable they may be, from one generation to the next, environmental conditions that are required for normal development exert some degree of control over the final outcome, and, therefore, should be likened to a part of the program in the computing metaphor. Of course, a change in the part of the program, whether it be environmental or genetic, can have subtle and specific effects or gross and general ones. As we have already seen when considering the specificity of a developmental determinant, interest tends to be focused on the fine control, and commonly a sharp and arbitrary line is drawn across the continuum from fine to gross control.

The dogma that all information flows out from the genes has had a powerful influence on thinking about the development of behaviour, and has probably done more damage than anything else to an understanding of the integrated way in which genes and external conditions work together. Many genes are part of control systems and can be likened to the switch thrown by a thermostat. When the value of external conditions falls below a certain point, the discrepancy is detected and a signal is fed to a particular gene in a particular set of cells. The gene is activated, and its products set in motion a string of chemical and neutral processes which generate the behavioural analogue of heat. In such examples of negative feedback, and many others in which gene action is part of a system, instruction of a kind clearly flows into as well as out from the gene.

Stages in development. When a house is built, the roof is not put on until the walls have been raised. Similarly, it is logical to suppose that one kind of behaviour must develop before another if the second kind of behaviour is to function properly. For example, exploration of the environment may be disastrous if it occurs before the young animal has established some standards of what things are familiar, and where it has a secure base. To take another example, the song of the male white-crowned sparrow (*Zonotrichia leucophrys*) is greatly influenced by the bird's early experience with songs of its own species. As the bird develops its own song it matches up the sounds it makes with those that it heard when it was very young. When the match is perfect it need no longer make the comparison each time it sings, and therefore can speed up its singing rate. This is a bit like a musician learning a new part. While he has to monitor the individual sounds he is making to ensure their accuracy, he must allow at least a tenth of a second between each note. However, in the final performance, when such control is no longer needed, the gaps between notes can be greatly reduced. Examples like this suggest that behaviour may develop in stages.

Descriptive stages in development have often been identified. However, they do not necessarily carry any implications about developmental process. The young animal may have a very different ecology from the adult. It may have special problems to overcome at birth or hatching, as in the case of the European cuckoo (*Cuculus canorus*) that removes its hosts' young from the nest. Such juvenile specializations may account for sharp discontinuities in development. A quite different point is that a distinction between one stage and another may merely represent a convenient classificatory device in which the classifier is drawing a sharp line across the continuum. Also, an apparent discontinuity in development may arise for rather trivial reasons connected with the CLASSIFICATION and measurement of behaviour. Nevertheless, when a stage in development has been recognized, it may represent reorganization of behaviour arising from the operation of a fresh internal or external source of variation.

Regulation of development. If a developing animal or, indeed, a young child is deprived of food or falls ill, it stops growing for a while. If the period of starvation or illness persists the young animal may weigh considerably less than others that have had no such misfortune. However, as can be seen in Fig. E, the weight of the animal can quickly recover to what it should be if it is put back on a normal diet, or if it recovers from illness. In other words, the animal seems to 'know' how heavy it should be at a given stage in development. Some kind of internal self-regulatory process is clearly implied. It is not difficult to suggest internal control mechanisms which would ensure that the animal would regulate its feeding to match the requirements that are normal for a given stage of development. As yet, very little is known about the actual nature of the regulatory mechanisms. In some, the machinery must reside within the animal itself. In other cases, it may reside within the parent, so if it detects that the behaviour of its offspring is different from what it should be, the parent behaves in such a way as to make good the discrepancy. For instance, a domestic goat (*Capra hircus*) with two offspring will, unless one of them is very weak, prevent the more vigorous one from suckling first. Presumably if she did not do this the more vigorous of the kids would get most milk and grow at the expense of the other, and the net benefit for the mother in terms of producing grandchildren would be reduced. Fig. F shows the manner in which the kids could be controlled. The extent to which the mother compensates for slight deficiencies in the young is particularly obvious if the stronger of the two kids is stupefied with an anaesthetizing drug. Under natural conditions her behaviour would act to keep the offspring on a normal development pathway.

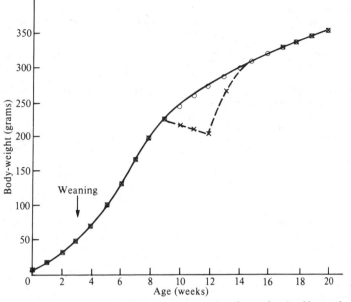

Fig. E. Effect of malnutrition on rats between the ages of nine and twelve weeks. At this age the rats rapidly recover in weight when returned to the normal diet.

Compensating for shortcomings may have diminishing returns. In the case of the goat, the weaker kid will be neglected by the mother if it has been stupefied, suggesting that there may come a point when she settles for one strong offspring. Similarly, if animals are deprived of food at stages when they are growing very rapidly, they may be permanently stunted. One functional explanation for stunting is that the animal does not endlessly attempt to reach a state which may never be achievable in the particular conditions in which it is developing. Deprivation of optimal conditions for one system does not necessarily imply that conditions are bad all round. Normal development of the animal's other systems may still be possible. Although the animal may be handicapped, its chances of surviving and leaving offspring may not be reduced to zero. A behavioural example of

settling for less than the best is provided by the nest-site selection of great tits (*Parus major*). In the early spring tits will visit a large variety of holes and crannies, many of which are obviously unsuitable. If a site does not match up to the characteristics of an optimal nest site they keep on SEARCHING, at least to begin with. However, if optimal sites are unavailable or already occupied, the birds ultimately nest in places which they would have previously rejected. It would make good sense if they were equipped with a rule for gradually relaxing the conditions under which searching for a nest site was brought to an end, and NEST-BUILDING began. It might seem like a glimpse of the obvious that, metaphorically speaking, a starving man is not fussy about what he eats. Nevertheless, the relevance of the great tit example to a discussion of development is that once the bird has

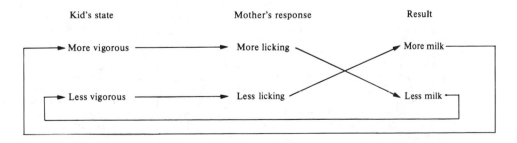

Fig. F. The way in which the mother goat ensures that both her kids grow at the same rate. Licking by the mother makes it more difficult for the licked kid to get milk.

selected a sub-optimal site it may, for that breeding season at least, prefer it, even if an optimal site should subsequently become available.

Sensitive periods. It has frequently been found that an individual's characteristics can be more strongly influenced by given events at one stage of development than at other stages. These phenomena are referred to variously as sensitive phases, critical moments, optimal periods, vulnerable points, crucial stages, susceptible periods, and so forth. In behavioural studies they were first recognized in the case of imprinting in birds and socialization in the dog. The ideas were borrowed from the embryologists. It is doubtless from embryology that the SENSITIVE PERIOD notion has spread into many other disciplines as well. For instance, the masculinizing influence of androgens in rodents and monkeys is only found at certain stages of development; starvation only has a stunting effect on the size and other features of adult rats when it has occurred early in life; handling in infant rats only influences their behaviour as adults when they are still being reared by the mother. The use of the sensitive period concept in the imprinting literature influenced the thinking on development of bird song. It may also be the source of its current usage in discussions about the influence of experience on the mammalian visual system.

In some cases it may be quite misleading to suggest that some descriptively defined sensitive periods have any function whatsoever. They may, for instance, represent a time of rapid reorganization when the developing animal is more easily destabilized by deprivation or environmental insult. In other cases, though, it does seem likely that the sensitive period represents a time when specific information from the environment instructs the developing animal. Even in these cases it seems inconceivable that all the phenomena that attract the name 'sensitive period', or one of its synonyms, arise in the same way. Unfortunately, much of the discussion has rested on the implicit assumption that one explanation will do for the lot. A rather extreme view can be represented by an analogy. Think of the developing animal as a train travelling one way from a place called 'Conception' to some indeterminate place where it disappears off the tracks. The theory has it that all the windows are closed for the first part of the journey. Then at a particular moment all the windows are thrown open and passengers are exposed to the outside world. A little later all the windows are shut again. A more refined thought is that the train is divided up into compartments and different windows open and shut at different stages during the journey. Each compartment in the train, with its occupants, represents a particular behavioural system, sensitive to the environment at a certain stage in development. Once embarked on an analogy of this kind, a large number of possibilities become apparent. For instance, one could argue that at a particular stage of development the window to a compartment opens and never shuts

till the train stops running. On this view the apparent end to the sensitive period results from changes to the occupants and not from the shutting of the window. This is probably the case with filial imprinting in birds.

IMPRINTING is a pre-emptive self-terminating process in the sense that it narrows social preferences to that which is familiar, and therefore tends to prevent fresh experience from further modifying those social preferences. Moreover, while some stimuli are much more effective in restricting filial preferences than others, birds such as domestic chicks and mallard ducklings can eventually form social preferences for sub-optimal stimuli, such as the static cages in which they have been isolated. It follows that if some birds are reared with optimal stimuli such as their siblings, and some are reared in isolation, it should be possible to imprint the isolated birds with a novel object at an age when the socially reared birds escape from, or are unresponsive to, new things. This is the case with both domestic chicks and mallard ducklings.

What emerges is that the bird will start to learn when it is ready to learn, and when it is ready to learn is determined, in part, by internal changes. In terms of the developmental train analogy, the window of the filial compartment opens at a particular stage in the journey and then stays open. As a result of what they learn about the outside world, the occupants subsequently avert their gaze from anything strange. Since they can learn nothing until the window does open, the timing of the ending of the sensitive period is also dependent on the internal processes responsible for opening the window in the first place. If this analogy can be pursued a little further, it looks as though the occupants can, under certain circumstances, be persuaded to study strange things outside the train later in the journey, and when they do so they are influenced by what they see.

An important lesson which arises from the work on imprinting is that an end of a sensitive period can arise not so much through an incapacity to learn as through an unwillingness to do so. If the various systems that make an adult animal unready to respond to novelty are silenced, by HABITUATION for example, the animal may then learn about new things once again. It should be obvious that the more powerful are the protective mechanisms, the more stable will be the consequences of the initial experience and the more robust will be the sensitive period phenomenon. It must be emphasized that this need not invariably be the case. It is perfectly possible that sometimes the state of motivation necessary to pick up information from the environment only persists for a limited period of the animal's life, and there is no way in which the behavioural system involved can be subsequently modified by experience.

The well-analysed examples of sensitive periods show as clearly as anything else the kinds of interactions that can take place between internal processes and external influences during the course of development. It seems likely that this kind of thing is the rule rather than the exception in development.

Learning. The processes of learning have been frequently alluded to in this article and it should be obvious that they can play a very important role in the course of development. By learning, an animal can perceive causal links between different events in its environment and so make predictions. Similarly, it can perceive a causal link between something it has done and a subsequent change in the environment. Learning is required when becoming familiar with companions and particular features of the environment; it is involved in dropping from the behavioural repertoire acts that no longer serve a useful function; and it plays a major role in the transmission of information from one generation to the next. It should be clear that the list of uses to which learning is put is a long one.

Learning processes may range from very simple types of habituation through to the most elaborate kinds of IMITATION and reasoning. The process may involve association of independent events, or, at other times, a MEMORY for a particular event may be formed merely as the result of exposure to that event, as though a photograph were being taken of the outside world.

As in the study of any other kind of biological process, it is helpful to consider the context in which learning takes place, since animals live in very different HABITATS and in one animal's life the rules for detecting the causal structure of the environment may vary greatly from one context to another. For instance, in the rat (*Rattus norvegicus*) the rules for detecting a causal link between movement of a leg and a change in the environment (the tolerable delay in reinforcement) appear to be very different from the rules for detecting a causal link between eating poisoned food and its adverse consequences (see FOOD SELECTION). The importance of adding the developmental perspective is that the ecology of the young animal may be very different from that of the adult, and, of course, the learning processes of the adult must themselves develop. Furthermore, the young animal may have behaviour patterns, such as PLAY, that are needed to gather information from the environment, or to acquire particular kinds of skills. Consequently, the rules for gathering information may change as the animal gets older.

Learning processes are, of course, inferred. They are not directly observed and relatively little is known about exactly how they work. It is usually supposed that they involve the implantation of fresh information in the nervous system. However, it need not always be quite like that. We have already seen the case of the locust which becomes solitary or migratory depending on whether or not it was crowded when it was young. In this case it seems unlikely that the migratory behaviour is in any sense learned in response to the crowded conditions. It is more as though the animal is like a juke-box in which a particular record can be played as a result of a specific button being pressed from outside. The finger that presses the button does not carry the information which is on the record. In this case the environment selects an appropriate form of behaviour. Similarly, some of the examples of learning may represent the selection of a particular pattern of behaviour which is already, in some sense, latent within the animal. The conventional view would be that the learning animal is like a tape-recorder; if a song is performed, then the animal was instructed by the environment. Since the underlying processes are inferred, it is extremely difficult in practice to know whether or not selection or instruction has taken place. However, a reasonable starting point is to see whether the new form of behaviour elicited by a particular environmental event bears some correspondence to the conditions that gave rise to it; in other words, was the analogue of the performed song available to the animal as a song earlier in its life?

Conclusion. The study of ontogeny has in the past attracted some of the most bitter and protracted controversies in the whole field of animal behaviour, largely because the arguments have reflected more general ideological battles about nature and nurture. Consequently, much research work was concerned merely to establish that a particular kind of internal or external factor could be important, or that evidence can sometimes be obtained for a certain logical possibility. Such activity abated with the growing acceptance that both internal and external factors play important roles in the development of any one pattern of behaviour. Also, the air was cleared by the realization that an interest in how behaviour has been adapted to its present uses is not the same as an interest in what makes one individual animal different from another one.

As actual processes were studied, it became increasingly obvious that what was needed was an approach that would cope with (i) the multiple and varied nature of developmental determination and (ii) the interactions that take place between the determinants. As a consequence the focus of the search has shifted towards coherent explanations of what actually happens during development. Major sources of variation operating in an animal's lifetime can be systematically located, the ways in which they interact can be modelled, and the various models can be set in opposition to each other and tested. The changes in research style did not, of course, mark the end of all arguments. Nevertheless, the scientific controversies about ontogeny lost some of their perennial and befuddled character. The subject had settled into a progressive phase with widely accepted rules for settling disputes. Where will it go? It would be pleasant (but doubtless over-optimistic) to suppose that the complexities of behavioural development will miraculously disappear with the emergence of

a set of general explanatory principles.   P.P.G.B.
8, 9, 54, 74, 75

OPERANT BEHAVIOUR. Many of the things that animals do can be considered as operant behaviour: a wild rat (*Rattus*) searching for food; antelope (Bovidae) fleeing from a lion (*Panthera leo*); a cat (*Felis catus*) exploring a novel environment. The distinguishing feature of operant behaviour is its modifiability: the rat returns to the place where it has found food and abandons an empty site; antelope learn not to run from a caged lion, or from one at a safe distance; the cat ceases to explore once the environment has become familiar. Studies of LEARNING in animals are largely concerned with operant behaviour, its dependence on REWARD and punishment, on the history of the individual animal, and on the animal's *phylogenetic* (evolutionary) inheritance.

Historical background. The term 'operant' was coined in 1937 by the American behaviourist B. F. SKINNER. He applied it to a kind of behaviour which he defined by two properties. (a) It is 'emitted' by the animal, in the sense that no obvious external stimulus can be found for it; (b) it has effects (operates) on the environment, and is, in turn, affected by these environmental effects: operant behaviour is modified by its consequences. Skinner worked within a very circumscribed experimental arrangement (the *Skinner box*, described below) but applied the term operant widely to what had earlier been termed VOLUNTARY or instrumental behaviour.

The operant concept was in part a reaction against the simple stimulus–response (reflex) view of human and animal behaviour then in vogue. In its simplest terms, a REFLEX is a cause–effect relation between a physical stimulus, such as a light, a blow, or a sound, and a stereotyped muscular or glandular response that invariably follows it, such as pupillary contraction, salivation, the knee jerk, or startle. The reflex idea has venerable origins, beginning perhaps with the mechanistic philosophy of Descartes, and later elaborated by Whytt, Prochaska, and others. More recently, the concept of the reflex was brought to its most refined form by the English physiologist C. S. Sherrington in his book *The Integrative Action of the Nervous System*, published in 1906. Sherrington worked with 'spinal' animals, i.e. animals whose higher nervous centres, such as the cortex and midbrain, are surgically isolated from the spinal cord. These preparations allowed him to study the stimulus–response properties of the spinal cord, unaffected by the more complex behaviour of the higher centres.

Spinal reflexes are not modifiable by learning, but the concept of the reflex was applied to learned behaviour by the Russian physiologist I. P. PAVLOV. Pavlov's researches achieved wide recognition in the West following the translation of his book *Conditioned Reflexes* in 1927. This work provided a target for humanists such as George Bernard Shaw (in *The Adventures of the Black Girl in Her Search for God*), who despised the mechanistic view of man, and a weapon for mechanists who sought to promote that view. Among the mechanists was the influential American psychologist John B. Watson, the originator of the term *behaviourism*. Combining Pavlov's conditioned reflex with the concept of tropism (forced movement) developed by the Chicago biologist J. Loeb, Watson promoted a view of psychology that emphasized determinism, and the total dependence of behaviour on the stimulating environment.

Skinner's concept of operant behaviour is partly within and partly outside this mechanistic tradition. Skinner shares with Watson a firm belief in determinism and the power of the environment to mould behaviour. He differs from Watson in his view of reflex action and its relevance to voluntary behaviour. Watson's behaviourism was of the stimulus–response variety. For every response there is a stimulus, which is its total explanation. Skinner's concept of 'emitted' operant behaviour is a departure from this view, and a return to earlier notions of essentially spontaneous behaviour that is nevertheless attuned to its environment. Skinner's determinism meant that the spontaneity of emitted behaviour was accepted as only apparent: it is not that no stimulus exists for such behaviour, but that '. . . no correlated stimulus can be detected upon occasions when it is observed to occur' (*The Behaviour of Organisms*, 1938). In short, the spontaneity of operant behaviour reflects the limited knowledge of the observer, not a real spontaneity on the part of the animal.

Skinner retained a place for reflexive behaviour in his sytem, denoting it by the contrasting term *respondent* behaviour. Respondents are 'elicited' in a reflex fashion, as salivation by food, or eye-blink by a puff of air.

Operant behaviour may be brought under the control of stimuli by the operation of REINFORCEMENT. However, no matter how automatic such relations may become, Skinner always describes the 'discriminative stimuli' that produce an operant response as 'setting the occasion for' the response. In this way, the appearance of spontaneity is preserved. This distinction between reflex or respondent behaviour and operant behaviour is, of course, blurred in common speech: when a child runs in front of a car, most people would describe the driver's response of swerving or hitting the brake as reflexive, in the sense that it is automatic, not conscious until after the fact, and very rapid. Yet, since this behaviour is learned, Skinner would have to refer to the child as merely setting the occasion for the response, as if the driver in leisurely fashion mulls over the possibilities, and then spontaneously emits the appropriate response.

Despite such flaws, the concept of emitted behaviour was a very fruitful one, because it allowed Skinner and his students to study recurrent activities, such as pecking or lever pressing, without worrying about identifying a separate stimulus for each response instance. Situations of this kind, in which the response can occur freely at any time, have come to be known as *free-operant* situations, in contrast to the *discrete-trial* situations favoured by stimulus–response theorists, where the animal is only periodically presented with the opportunity to respond to a stimulus. Thus, in a typical free-operant situation an animal, such as a pigeon (*Columba livia*), is trained to peck a lighted disc which is continuously available, and occasionally receives food reward (reinforcement) for doing so. The process of DISCRIMINATION can easily be studied in such a situation by changing the colour of the light illuminating the disc every minute or so, and arranging for pecks to produce food only when a particular coloured light is on. If the rewarded light is discriminably different from the others, the pigeon will soon learn to peck only when that light is on. However, if the rewarded light is not easily discriminated from the others, the pigeon may merely peck rather more (i.e. at a higher rate, measured as pecks per minute) on the rewarded stimulus than on the others. In a comparable discrete-trial experiment, the animal might be presented with one or the other stimulus on each trial, and be required either to respond or not respond, depending on the stimulus (a go/no-go discrimination). The required response might be pecking a key, for a pigeon, or lever pressing or running down an alley, for a rat.

In experimental science, the greatest advances in understanding often follow on advances in technique, such as inventions like the microscope or the oscilloscope. Few would dispute Skinner's technical contributions to the study of learning and MOTIVATION in animals. In his well-known essay, 'A Case-History in Scientific Method', he entertainingly describes the sequence of steps, a mixture of expediency and accident, that led him to develop the Skinner box, starting with a conventional runway apparatus. At the same time he devised the technique of intermittent reinforcement, when he discovered that his rats would continue to press a lever for food reward, even if food was not forthcoming for every press. The lever-press apparatus, which allows for the automatic delivery of food to a rat who can respond at any time, is associated with Skinner's name, although a similar apparatus was devised at about the same time by Grindley in England. However, an invention of Skinner's that has had almost as much influence is the cumulative recorder. In a cumulative record, time is on the horizontal axis, and the vertical axis represents cumulated responses, so that the slope of the record corresponds to rate (responses per unit time) of responding.

The Skinner box allows the instrumental behaviour of an animal to be recorded entirely automatically (see TECHNIQUES OF STUDY), while the cumulative recorder provides a moment-by-moment record of its behaviour in an easily readable form. These factors pushed operant conditioning in the direction of the exhaustive study of the behaviour of individual animals, rather than groups of animals, the method more typical of users of discrete-trial procedures. Skinner and his students soon found, moreover, that maintained behaviour, i.e. the pattern of lever presses or key pecks in time, under various schedules of reinforcement, is much more reliable than an animal's pattern of behaviour when he is first exposed to a situation. 'Learning' is, of course, our name for the transition between the animal's first fumbling efforts to solve a simple problem, and its final smooth performance. Thus, by ignoring transitions and concentrating on steady-state behaviour, operant conditioners ceased to deal directly with learning. Moreover, learning is usually thought of as an essentially irreversible process. An animal that has learned to make some response for food reward, and then learns not to when the reward is no longer forthcoming, is not the same animal he was at the beginning. For example, he will relearn the task much more quickly a second time. Consequently, the study of learning, conventionally defined, requires the use of 'control' groups (see LABORATORY STUDIES). To compare the effects of some experimental treatment on the speed with which an animal can acquire some skill, it is necessary to compare treated and untreated groups of animals. It is not possible to train individual animals to learn the task, extinguish their performance by withholding reward, apply the treatment, then have them relearn. Thus, Skinner's emphasis on the study of individual animals, while it produced a great gain in precision and experimental control, inevitably led to a change in the focus of interest. Operant conditioners of the Skinnerian school traditionally stood somewhat apart from other students of conditioning phenomena, partly because of their determined avoidance of the topic of learning.

Skinner's idiosyncratic philosophical views on the nature of scientific method and the place of theory in science have also helped to isolate his disciples. Perhaps discouraged by his own early efforts at developing quantitative measures of conditioning, Skinner persuasively argued that much more experimental analysis was necessary before any theorizing about the mechanisms of learned behaviour could be fruitful. At certain times his followers interpreted his positivistic strictures against certain kinds of theories as implying that theory may be acceptable, but only at some date in the far future.

If not theory, then surely induction, in the Baconian sense, is an acceptable strategy. First we need to gather facts and classify them (see CLASSIFICATION), and then it may be permissible to theorize. But no; in an oft-cited passage Skinner warns that '... the number of possible [conditioned] reflexes is for all practical purposes infinite, and that what one might call the botanizing of reflexes will be a thankless task' (*The Behaviour of Organisms*). This injunction against collecting sets of stimulus–response relations has been widely interpreted as a prohibition of inductive research generally, even though, as will become clear below, the major contribution of studies of operant behaviour is a sort of 'natural history' of animal–environment interactions.

Experimental procedures. Circus trainers have known for generations that to teach an animal a complicated trick it is necessary to proceed in stages, building up each element separately, and by degrees, before putting them all together. Skinner was the first scientific student of animal behaviour explicitly to use this technique and to analyse the reasons for its effectiveness. He called it 'shaping by successive approximations'. For example, if the trainer wishes to teach a pair of pigeons to play ping-pong, he begins with a hungry pigeon (deprived of food for a day or two, or, in later work, maintained at 80% of its weight when given food *ad libitum*). In this condition, the animal will be responsive to food as a reward (technically, a reinforcer). Training then proceeds in three steps. First, the animal is given intermittent, brief access to food (delivered by an automatic tray mechanism) until he learns to approach the feeder as soon as it operates. This training allows the sound of the feeder (which the animal obviously hears before he can actually eat the food) to act as a reward in itself (a *conditional reinforcer*). Next, some response that the animal makes without explicit training is selected as a likely precursor of the behaviour of interest. Head movements, ground pecking, or the sideways sweeping movements that pigeons make to push aside fallen leaves to look for food, are all possibilities. The trainer then operates the feeder for a second or two immediately after each instance of the selected response. Very soon the response begins to increase in frequency and, more important, new responses, more closely related to the desired final behaviour, also begin to occur.

This initiates the third stage. At this point the contingencies of reinforcement, i.e. the rule that the trainer is using to dispense food, can be shifted so that food now only follows a response closer to the desired goal than before. As this second response begins to increase in frequency, still other responses begin to occur, some of which will be even better approximations to the response desired. By this process of selection, almost any pattern that is within the limits of the animal's muscular apparatus can be increased in frequency. This same sequence of progressive approximations to a desired response is then repeated for other elements of the task; a ball may be introduced and only head movements directed at it rewarded. Next, reward may be made contingent on the ball's being driven for some distance. Finally, the ball may be thrown toward the animal by the experimenter. Once two pigeons have been trained in this way, with each having learned to hit a moving ball to the opposite side of an enclosure, the system becomes more or less self-sustaining. When the two animals and the ball are placed together they appear to be engaging in a passable game of ping-pong.

Even more complicated sequences can be built up by *shaping* each element separately, and then linking them up one at a time to form a chain. This is done by making use of discriminative stimuli, i.e. stimuli whose presentation is necessary for the conditional response to occur. For example, suppose a teacher wishes to demonstrate the power of conditioning principles by training a rat successively to climb some stairs, run in a wheel, and finally pull on a rope which raises a flag. He first establishes the last link in this behavioural chain by shaping the hungry animal to pull on the rope for food reward, in the presence of a light. If pulling on the rope with the light off is ineffective, the light soon comes to 'control' the pulling response, in the sense that the animal will not pull on the rope if the light is off, but will do so at once when it comes on. Next, wheel running is shaped by arranging for a few turns of the wheel to switch on the light. The rat will soon learn to run until the light comes on and then go to the rope and pull on it until food appears. The final link in the chain is easily established by allowing the rat to run up some stairs to get to the running wheel. Here the contingency (dependence) between running and availability of the wheel is due simply to their spatial separation, and the sight of the wheel itself is the discriminative stimulus controlling approach.

From these simple beginnings Skinner developed a whole philosophy of human instruction. The most tangible result has been the programmed instruction movement, which has resulted in numerous programmed textbooks and teaching machine programs. The two basic notions behind all these educational efforts are (a) immediate reinforcement for responses, and (b) breaking down the task into simple steps. A programmed text typically consists of a sequence of simple fill-in-the-blanks questions which guide the reader through the subject matter in very small steps. Reinforcement for correct responses (answers) is supposed to be provided by prompt FEEDBACK that the answer is correct. However, if the questions are indeed simple, then the feedback is often redundant (the student knows he is right), so that the concept of reinforcement does not seem to apply in any obvious way. The programmed instruction idea has had the beneficial effect of forcing textbook writers to think carefully about the organization of their material, and it has aided the training of personnel in large organizations, such as the armed forces, where the tasks are well defined and teachers are in limited supply. The method has worked well in training people on operating procedures and routines of all sorts and, to some degree, in teaching skills like languages and mathematics. It is less useful where the subject matter is not intrinsically structured, and where it requires the exercise of judgement or the active participation of the student. Moreover, intelligent students find most teaching machine programs just plain boring. (This is less true of computer-controlled instruction, but this movement does not derive directly from Skinnerian ideas.)

From a theoretical point of view, the most interesting issues raised by the shaping procedure, and programmed instruction, concern the origins of novel behaviour, and the relation between a desired target behaviour and its precursors. For example, if we wish to teach a child differential calculus, what prior skills should be taught, and in what order? Is the best sequence the same for all individuals? The difference between more and less creative people seems to lie in the range of behavioural variation that they show: the creative person is a 'man of many parts'; he knows, and can do (hence can learn), more things than the less creative person. What affects the range of behavioural variation? In shaping, for example, occasional free food deliveries sometimes help to break up an undesired rigid pattern, allowing new behaviour, closer to the desired response, to emerge. How does this process work, and what is the analogous process in human learning? Unfortunately, these questions have received relatively little direct study.   J.E.R.S.
79, 126

OPPORTUNISM. An opportunity is a favourable set of circumstances that is open to exploitation. Some species are particularly adapted to exploit opportunities in colonizing new HABITATS. To be able to make the most of such an opportunity, a species must be able to produce many individuals with good dispersal abilities. A good example is the colonization of Australia by rabbits (*Oryctolagus cuniculus*). There are also many examples among PARASITIC animals. Such species are capable of a rapid rate of reproduction when circumstances are favourable.

In individual animals, opportunism often takes the form of rapid exploitation of new sources of food. This is particularly common amongst omnivores (see FOOD SELECTION). For example, in many areas of urban development herring gulls (*Larus argentatus*) have been quick to exploit the food that is available at refuse dumps, and gull populations have grown considerably in these areas.

Many omnivorous animals are adept at exploiting short-lived daily, tidal, or seasonal feeding opportunities. Thus herring gulls on the coast of Cumbria, England, rarely miss the opportunity presented at the spring tides, when the water is sufficiently low for starfish (*Asterias*) to be obtained. Primitive man is often an opportunistic feeder, taking advantage of the seasonal or occasional abundance of fruit and game.

ORIENTATION is an essential element underlying all behaviour: a hunted mouse (*Mus musculus*), for example, darts to its hole; the giraffe (*Giraffa camelopardalis*) stretches up for high foliage; the

migrating garden warbler (*Sylvia borin*) heads for the south, the pigeon (*Columba livia*) for its loft; the male hamadryas baboon (*Papio hamadryas*) turns to confront an approaching rival, or offers his back to a female for delousing. All these behavioural events contain orientation components. The movements are related to specific spatial requisites. A behaviour pattern consists of a sequence of muscle actions, ordered in a specific way in time and space. In so far as the spatial order is under the animal's control, we call it orientation. What is oriented? On the face of it, the question seems simple enough. The migrating starling (*Sturnus vulgaris*) orients its whole body with respect to the sun. But the praying mantis, *Mantis religiosa*, moves only its predatory legs towards the fly. Thus an animal can orient its body as a whole, or merely part of its body (Fig. A). Even inanimate objects distinct from the body may be oriented. The bird builds its nest with the opening upwards, that is, with respect to gravity; and, on his walls, man adjusts the pictures to the vertical, coinciding with the direction of gravity.

Any orientational event can be examined from several aspects, including (1) inspecting the geometry of the spatial structure, (2) investigating the achievements of the orientation systems, (3) analysing the mechanisms involved.

1. The geometrical arrangement of orientation. Points, straight lines, angles, and their spatial relations are the geometrical basis serving to explain some of the terms often used to describe orientational events. Many butterflies (Lepidoptera) flutter from one flower to another. After landing, the butterfly turns on the spot, exposing its extended wings to the sun. This simple behaviour demonstrates the two basic forms of spatial alteration: translation, i.e. moving from one locus in space to another, and rotation, i.e. turning. The spatial control of a translatory direction is called *course orientation*. The sitting and turning butterfly changes the angle of its body axis with respect to the sun's direction. This is called *angular orientation*. A course direction with respect to the sun is called a *sun-compass* course. The term *compass orientation* was invented for the orientation first observed in ants (e.g. the garden ant, *Lasius niger*), which was compared with our use of a magnetic compass for direction finding. Like the magnetic compass the sun offers the reference for maintaining a distinct heading direction. Although we may think it simple for an ant or a honey-bee (*Apis mellifera*) to maintain a sun-related course, we have to remember that there are two complicating factors. One is the movement of the sun, and the other, for a flying animal, is the wind. Animals have developed compensatory mechanisms to overcome these difficulties. The sun's daily movement is from east to west across the heavens, and the angle between north and the sun's geographical direction is called the *azimuth angle*. Bees, ants, and all animals using a sun compass have a 'compass-shifting' device which alters

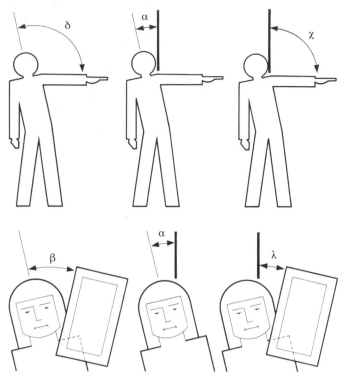

Fig. A. Frames of reference in the orientation of body parts and of 'non-body' objects. The upper row shows the orientation of body parts to an external frame of reference (right); body orientation with respect to the external frame (middle); and body orientation with respect to an internal frame (left). The lower row shows the same for the orientation of a 'non-body' object. (The vertical bar indicates the direction of gravity.)

their sun-compass angle continuously in the opposite direction by an amount corresponding to the change in the sun's azimuth angle and according to the same TIME schedule. By this continuous compensatory adjustment of compass direction, a constant geographical direction in the animal's course is maintained. The shift system depends upon the animal's internal CLOCK, and there have been many studies showing that animals use clock-compensated orientation during NAVIGATION.

An animal capable of FLIGHT is in quite a different situation from one moving on the ground. The locomotory substrate of the pedestrian is the ground, so his direction of movement refers directly to the ground. But the airborne animal's locomotory efforts are related to the surrounding air; the substrate here is the air. The course of a flying animal is therefore its locomotory direction with reference to the air. The animal's translation with respect to the ground may differ from this, as the air may move with respect to the ground. Thus, in the air we have two motions to consider: one due to the animal's own locomotory actions, the other depending on the air's shift with respect to the ground. Together they result in the animal's path with respect to ground, which is called its *track* (R, in Fig. B). It is evident from the figure that the animal's heading direction through the air (F) does not coincide with its shifting direction over ground (R). As seen from the ground, the air

movement is called wind, but experienced from the air it has quite a different implication and effect. The airborne creature senses no wind, but sees only a ground pattern flowing in the opposite direction to its track. The visually perceived pattern movement is called *opto-kinetic* movement of the ground's pattern flow below the animal. This visual impact is an important cue for orientation in the air. It indicates the track, and that is the crucial orientation direction. The angle between the track and the animal's heading direction in the air is called *drift angle* (β_0) as it depends upon the animal's drift due to the wind. The angle between the track and the wind direction is called the *track angle* (β_{ow}). It has been shown that animals try to maintain a constant track angle. When wind speed is experimentally changed, the moth *Plodia interpunctella* cruising upwind is found to keep its track angle (β_{ow}) and its speed along the track (R) constant by adjusting its flight velocity (F) and drift angle (β_0). This type of orientation, therefore, serves to stabilize the animal's track with respect to the wind. But the animal has no information about the wind's direction. If the wind changes, the animal will be taken with and turned with the air mass without sensing it. The animal sees the opto-kinetic movement flowing in the same direction with respect to its body as before. In other words, it senses a track, but it does not 'know' which way the track is heading. The opto-kinetic movement alone is not sufficient, because the animal records only the flow direction of an anonymous pattern. In order to orient this pattern flow (i.e. the track), the animal needs geographical references, that is, particulars of the pattern (landmarks). Another possible reference is the sun. Thus the animal can maintain, for instance, a compass angle between its track (= pattern flow) and the sun's direction.

2. Achievement and performance of orientation. There are two main classes of orientation achievement, which may be designated as *positional orientation* and *goal orientation*, serving to maintain a position in space, or to reach a particular GOAL, respectively. A third class is characterized as *stabilization*. A person heading for a goal maintains his upright position in space. Thus positional orientation is a sub-system of goal orientation. Because a particular, normal position is a prime requisite of most behaviour patterns, positional orientation has also been called primary orientation, whereas goal orientation, operating only on specific demand, has been termed secondary orientation. Stabilization may serve these two as a sub-system, improving or completing them by preventing deviations or oscillations.

Positional orientation. Perhaps the best-known positional orientation is the maintenance of the body in a state of physical equilibrium. It is of particular importance on land in long-legged animals forced to balance the body mass high above the substrate. Equilibrium orientation is often based on the function of the gravity SENSE ORGANS (*statocysts*), but the eyes may also be involved, as well as MECHANICAL receptors which signal the body position relative to the substrate.

The maintenance of equilibrium is a type of orientation of preference position. Preference position may refer only to the substrate, irrespective of gravity: thus a climbing animal prefers always to

have his ventral side close to the substrate, for obvious reasons. The same holds true for substrate-bound animals in the water, crabs (Brachyura), for instance, which may cling to vertical walls or even hang upside-down. A butterfly in the sun's rays may adopt a preference position called a sunning posture. Another preferred position is the resting position adopted for rest or SLEEP. Many animals demonstrate preference positions with respect to SOCIAL partners. The individual members of SCHOOLING fishes or FLOCKING birds keep to specific positions and at specific mutual distances.

Stabilization. The difference between stabilization and positional orientation is that stabilization cannot guide the animal into a particular posture, whereas positional orientation does. Stabilization serves to maintain a particular positional orientation. Opto-kinetic orientation may be mentioned here. A particular kind is found in man. If a patterned cylinder rotates around a person sitting motionless, he soon has the strange experience that he himself is turning, and not the cylinder. He feels himself moving in the opposite direction to the cylinder's movement. This illusory perception has been named *circularvection*. To overcome this unwanted 'motion', the person has to turn himself with the drum. This is exactly how an animal behaves when performing the well-known opto-motor reaction. It tends to turn with a rotating cylinder. We can assume that the animal, too, perceives itself as being moved, and reacts in response to this unintentional 'movement' in the same way as a person does. This response to PERCEPTION of changes in the pattern surround serves to stabilize the animal's position with respect to the surroundings.

Positional changes also require stabilization. Systems serving to control the transition from one position to another supervise the expected 'shift' of the visual pattern of the surroundings. This con-

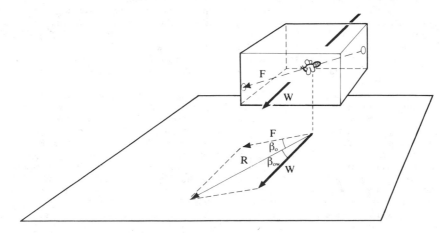

Fig. B. The orientation of an animal in space. The flight path F of the animal is affected by the direction of the wind W, giving a resultant track R. The angles β_o and β_{ow} are the drift angle and the track angle, respectively.

trol can be demonstrated in a simple experiment on the drone fly (*Eristalis tenax*). The head of a fly is carefully rotated by 180°, a reversible operation which does not damage the animal, as the articular membrane between head and trunk is elastic. The fly's left eye now looks to the right and the right eye to the left of its body. If such a fly walks straight ahead, the visual pattern flows from front to back of each eye as in the normal head position, and nothing happens. But if it begins a turn it will perform circling manœuvres of increasingly smaller diameter until it whirls round on the spot. By the turning, the eyes are stimulated in the opposite way to that in the normal animal. They signal, for instance, a left turn when the body turns right. This elicits increasing efforts to turn right. In walking straight, however, the opto-kinetic shift is normal on both eyes from front to back. Thus the opto-kinetic signals serve to supervise the turning manœuvres (see REAFFERENCE).

The semicircular canals of the labyrinth of the mammalian inner ear have a similar stabilizing function. Ring-shaped tubes are filled with a fluid and equipped with sensory hairs. When the animal turns the fluid lags behind because of its inertia, and the hairs are stimulated by this. The signals stemming from this stimulation serve to control both changes of position and maintenance of position. When we move, the spatial arrangement between the body and its surroundings changes. The positional and stabilizing systems record these alterations; the respective sensory devices are stimulated. But we perceive the surroundings as immobile and ourselves as moving, and not vice versa. The space around is perceived as being stable. This is called space constancy. Without the mechanisms of space constancy, the world around us would appear to move and tumble about at every spatial manœuvre we undertook, making an orderly performance of behaviour almost impossible. Space constancy systems are found in animals and man. The circular self-movement (circularvection) described above can be interpreted as a consequence of a space constancy system, i.e. the perceptual mechanism 'evaluates' environmental cues which fill large parts of the sensory field as being stable in space. Therefore, the visually perceived shift of the environmental pattern is interpreted as self motion. Another mechanism of space constancy is that sensory changes are prevented by compensatory movements of the sensor. The eyes, for instance, counteract a body tilt, so that the position of the eyes with respect to the surround remains constant. A third mechanism depends upon nervous compensation, relying upon the input of a second sensor. A body tilt, for instance, is recorded by the statocysts and by the eyes. The latter sense an angular shift of the surroundings. The statocyst signal about the body tilt compensates for the signal of the visual shift, resulting in space constancy of perception of the surroundings.

Goal orientation leads the animal towards a spa-tially distinct end. It is usual to distinguish between proximate and distant orientations. Proximate orientation implies direct sensory access to the cues marking the target, whereas distant orientation lacks direct sensory contact with the goal. Mechanisms of proximate orientation are involved, for instance, if a leopard (*Panthera pardus*) leaps at an oryx (*Oryx gazella*), or a barn owl (*Tyto alba*) pounces on a rustling wood mouse (*Apodemus*) in complete darkness. These animals throw their whole body onto the target, guided by visual or acoustic cues respectively. But the much less dramatic actions of animals steering for a nest entrance or springing from tree to tree have also to be mentioned here. In other cases only parts or appendages of the body are aimed at the target; chameleons (Chamaeleontidae), frogs (*Rana temporaria*), and toads (*Bufo bufo*) use their tongues, birds their beaks, mantids (Mantida) and mantis shrimps (*Squilla mantis*) hit their prey with their specialized forelegs, and many mammals also use their forelegs to reach for food and other objects.

Naturalists and students of ETHOLOGY will agree that animal MIGRATION is one of the most fascinating occurrences in the animal kingdom. Familiar examples include migrating birds which fly thousands of kilometres to reach their destination; migrating locusts, *Locusta migratoria*, which move in clouds that darken the sky as they follow suitable air currents across North Africa and the Middle East regions; also HOMING pigeons and wandering barren ground caribou (*Rangifer tarandus arcticus*) in Canada. But butterfles also migrate, as do bats (Chiroptera), whales (Cetacea), walruses (Odobenidae), and spiny lobsters (*Panulirus argus*), which walk one behind the other in long files. All of these operations are long-range migrations controlled by systems of distant orientation. But the term 'distant orientation' is equally applicable to the behaviour of an ant, which marches only a few millimetres following a sun-compass course until it finally reaches the nest. The honey-bee also performs distant orientation, using sun compass and landmarks, and arriving at the food source at a distance of a few hundred metres from the hive (see COMMUNICATION, Fig C).

Sun-compass and landmark orientation are intervening systems, as are the star and moon compasses. Migrating birds may orient by the stars at night, as shown by experiments under the artificial sky of a planetarium. During the period of migration, birds show continuous locomotory activity: migratory restlessnness. They hop and flutter, predominantly in a specific geographical direction, the migratory direction. If the planetarium star pattern is the same as the natural one, the birds show preference for the normal migratory direction, demonstrating their capacity to orient by the stars.

Animals can also orient with respect to the earth's magnetic field by using a magnetic compass. This has been shown by experiments on the migratory restlessness of robins (*Erithacus rube-*

cula) and warblers in rooms shielded from the earth's magnetic field, and equipped with a device producing artificial magnetic fields (Helmholtz coils). In the absence of any field, the birds showed randomly distributed migratory direction. But when exposed to an artificial magnetic field copying that of the earth, the birds were oriented in the same way as in control experiments under natural magnetic conditions, preferring the migratory direction (see ELECTROMAGNETIC SENSES). The compass mechanism may be only one of several systems for distant orientation. The migrating animal uses other additional navigational aids. A compass by itself is of limited usefulness, because it does not tell about the geographical site. In addition to a compass, a human hiker, for instance, needs a map. And indeed, the paths of migrating birds often follow coastlines, river systems, mountain chains, or valleys. Local landmarks may also play a role if the bird, at the end of its journey, recognizes its goal visually. But there are other ways in which migration may be successfully terminated. In warblers an internally controlled timing mechanism has been found which delimits the migratory restlessness. Migratory speed and the duration of migratory restlessness are adjusted to each other, resulting, under normal conditions, in the conclusion of the travel at the destination area of the species. Desert woodlice (*Hemilepistus reaumuri*) use a sun compass to find the way home after a foraging excursion; yet they do not rush straight to the nest, but stop a few millimetres short of the entrance and start a zigzag searching approach. This is the normal strategy, which is apparently a safer system than reliance upon a compass mechanism alone, which might not be sufficiently precise to hit upon the nest spot from distances of many metres.

3. Mechanisms of orientation. Environmental stimuli are important in orienting behaviour as well as in eliciting changes from one activity to another. For example, a flock of ducks or geese (Anatidae) rushes away at the appearance of a hawk (Accipitridae). Their flight behaviour is oriented to the nearest cover. Evidently the hawk presents stimuli eliciting the ESCAPE behaviour, while the cover provides the stimuli directing it. In this case, releasing and directing signals come from different sources. In other cases they may originate in one object. For instance, the pounce of a predator is both released and directed by cues of the prey object.

Changes in behaviour may be based on internal processes, without any external stimulus being necessary. The turtle, the pond slider (*Chrysemys scripta*), in an aquarium, for instance, swims or rests under water, but after a time it orients upwards, rising to the surface when BREATHING is necessary. The upward orientation is obviously released by physiological processes related to oxygen consumption. The same observation can be made with the air-breathing larvae of some water beetles, for instance *Acilius sulcatus* and *Dytiscus marginalis*. A deficit of oxygen induces a water beetle larva to take a course angle of about 15° to the direction of incident light. Thus internal processes bring about a specific directional value, the reference value of the orientation.

Any spatial cue is designated by two essential properties, which we can illustrate by asking 'What is it?' and 'Where is it?' The first question refers to the quality, serving to identify the cue. For instance, if bees are given a choice of differently coloured hives, they aim for the blue one as blue marks the 'home' hive in this case: it is the 'What'. 'Where' deals with the spatial coordinates, e.g. of the hive in relation to the bee. Thus two physiological mechanisms are needed to record a spatial cue, one for identification, the other for localization. The cues used in localization may be directional, such as those provided by light, gravity, etc., or they may take the form of a field of graded stimuli, of temperature or chemical stimuli, for example.

Orientation without external cues. How can we find a direction in space without an external reference to it? In familiar surroundings, we can move about in the dark even with our ears covered. We may know exactly how to get to the door, or how to find the refrigerator in the kitchen. This capability is often called *kinaesthetic orientation*. The movement pattern consists of a sequence of single actions, and the whole set is unrolled according to a particular learned program of motions. Every spatial manœuvre is based on the spatial situation left by the previous manœuvre. Another example of such a system is the correcting behaviour shown by a millepede (Myriapoda) when an obstacle obstructs its course. After a detour the millepede resumes its previous course, obviously using information registered when deviating from the original course to negotiate the obstacle. This endogenous spatial cue, referring to the previous orientational state of the body, is called *idiothetic* information, while information gained from external spatial cues is *allothetic*.

Orientation based on external cues. In systems depending upon external cues, we may differentiate between systems composed of many sensors, called *rasters*, systems based on two sensors (bi-sensor systems), and systems working with only one sensor (uni-sensor systems) (Fig. C). Examples of rasters are the eyes of human beings and other vertebrates, and also those of arthropods. The sensors may be sensitive to stimulus strength only, or they may also be sensitive to stimulus direction.

Orientation based on the direction of stimuli. In most animals the direction of light is recorded by rasters, for instance, by the eyes of vertebrates or by the complex eyes of insects. Those sensors of the eye which are impinged upon by the stimulus give information about its direction (see VISION). The directions of sound and gravity are usually recorded by bi-sensor systems, i.e. by a pair of symmetrically arranged ears or statocysts re-

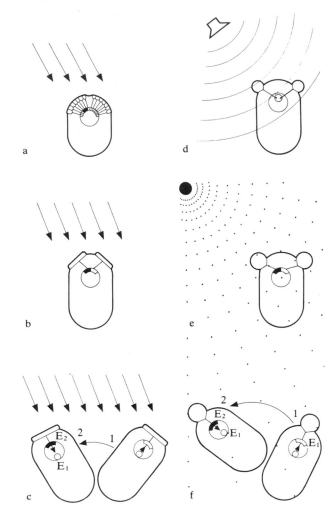

Fig. C. Schematic representation of some of the basic principles of sensory orientation. **a.** The direction of stimulation (e.g. light) is registered by a raster of sensory receptors. **b.** Direction registered by simultaneous comparison between two receptors. **c.** Only one receptor is available and the animal makes successive comparisons by moving its body. **d.** Time of arrival of stimulation (e.g. sound waves) is compared by two receptors. **e.** A gradient of stimulation (e.g. chemical) is registered by comparison between two receptors. **f.** A gradient is registered by a single receptor as the animal moves to sample different localities.

spectively. The two organs are differently stimulated, depending on the direction of the stimulus in each of them. For instance, the ear directed towards the sound source receives a higher intensity than the opposite one. The difference is transformed into information about the stimulus direction. Sound direction can also be recorded with only one sensor. Many animals have rather mobile pinnae (outer ears). The direction of highest sound intensity, which indicates the direction of the sound source, is found by searching, the pinna being turned from side to side. A similar system of successive measuring with one sensor is found in fly larvae (e.g. *Calliphora* spp.). These maggots crawl away from light by moving the head part of the body alternately from side to side. By this movement the sensors in the head will be successively stimulated by the light. In HEARING, the mechanism is based upon time difference evalua-

tion (Fig. Cd). Sound waves reach the nearer eardrum first, the further one a fraction of a second later. Some mammals, including man, are capable of utilizing this time lag in ear stimulation to obtain directional information.

Searching the gradient. Odour particles are molecules of a volatile substance, and their density is highest at the source of evaporation. The field of decreasing strength around the source is called a *gradient*. To find the source, the gradient must be searched. One way is to sample the strength of the scent at different positions on the gradient. A bee does this, for example, with its two antennae. It then turns to the side of the more stimulated antenna. This is gradient searching with a bi-sensor system. If, however, the gradient is too slight for the antennae to record differences in concentration, the bee will search it by moving its body from side to side, sampling the scent conccntration at gradient points further apart than was possible with both antennae simultaneously. Under these conditions the two antennae act as if they were a single sensor. The bi-sensor and uni-sensor procedures are quite different, however, one recording the direction of the source by means of direction-sensitive sensors, the other recording different stimuli at different sites along a gradient with sensors sensitive to stimulus strength only, not to direction. A third method of finding the source of a gradient is by indirect orientation. The stimuli affect the orientation process indirectly. A yellow-fever mosquito (*Aedes aegypti*), for instance, may happen to move from an area having attractive humidity, temperature, and CO_2 content into a less suitable, unattractive area. This change of stimuli causes TRIAL AND ERROR turning behaviour, which usually leads the animal back to the preferred environment. The direction of the turns depends upon idiothetic information, and is not directly influenced by environmental stimuli.

Orientation with respect to landmarks. Landmarks can play a role in both proximate and distant orientation. In the first case, the animal uses the landmark as a target. In distant orientation, landmarks serve as NAVIGATION aids and are very often used in combination with a compass. For localizing visual landmarks, raster sensors are used, so that an orientation movement brings the image into a position where it can be fixated. Insects such as digger wasps (e.g. *Bembex rostrata*) rely predominantly on landmarks. The female excavates a nesting burrow and supplies her progeny with living prey which she has paralysed by stinging the central nervous system. The burrow entrance is often carefully closed and camouflaged before the wasp leaves to go FORAGING. Nevertheless, she finds the opening unerringly. She has learned its spatial coordinates with respect to the surrounding pattern of landmarks. If these are shifted experimentally, her approach flight shifts accordingly. Such experiments show that the three-dimensional shape of the landmark is important.

Trail following. A domestic dog (*Canis lupus familiaris*) follows its master's trail by sniffing, nose to ground, and continually taking scent samples. He tracks the line of the strongest concentration of his master's scent. This is the principal technique employed by all trail followers. The best-known trail specialists are the ants (Formicidae). Their paths are marked by a scent which is species-specific: it contains either hind gut content, or the secretion of the poison gland or other glands, or a mixture of these. The trail follower touches the ground with both its antennae, taking samples alternately, with the left antenna from the left of the trail, and with the right antenna from the right of the trail (bi-sensor procedure). If it has only one antenna at its disposal it takes the samples with the one antenna, alternating left and right along the trail (uni-sensor procedure). Unisensorial and bi-sensorial probing may work together in the ant's trail-following mechanism.

Airborne scent trackers zigzag against the scent-carrying wind. This is how male moths (e.g. *Plodia interpunctella*) find their way to a female, which releases a sex-attractive scent (see PHEROMONES). This method of approach has also been observed in carrion beetles (*Nicrophorus* spp.), led to a carcass by smell. The mechanism of this aerial pursuit is quite different from that of a ground trail, however. The scent is windborne, and fliers must orient with respect to the wind, the scent serving merely as a RELEASER for this orientation. The animal cruises upwind. Experiments with moths have shown that the track angle (Fig. B) changes as a function of the scent. Without the sex-attractant scent the track angle is almost 90°, i.e. the male moth zigzags without progressing. If scent is added to the wind, the angle increases and the moth pursues its zigzag track upwind. Other experiments have shown that the onset of the zigs and zags is related to the borders of the scent trail. When the scent concentration drops below a certain threshold, the flier turns back. It seems that the direction of this turn is not related to the stimulus, but to the previous direction and/or the previous turns; for example, a left turn is always followed by a right turn. In other words, the turning direction depends upon idiothetic information. Thus the system resembles indirect orientation: chemical stimulation releases idiothetically directed reactions.

Taxes and kineses. The foregoing CLASSIFICATION of orientations by incident stimuli and gradient stimuli was based mainly upon the manner in which the animal obtains the information about the course or position to be established. TAXES and KINESES deal with the orientation process as a whole. That is, they describe the spatial manœuvre which is released by the stimulus. Some terms merely denote the spatial arrangement with respect to the stimulus, as for instance positive, negative, and transverse reactions. Similar denotations are light compass reaction (which has been discussed already), dorsal-light reaction, and ventral-gravity reaction. Other terms indicate the stimulus modality: photo-, geo-, chemo-, phono-, and anemotaxis, referring to light, gravity, chemical, acoustic, and wind stimuli respectively. Only a few terms also characterize the essentials of the mechanism of the manœuvre. These are *telotaxis*, *tropotaxis*, and *klinotaxis*. In telotaxis the stimulus strikes a spot in a composite organ, eliciting a turning manœuvre transferring the stimulus to the fixation area, which is usually frontal. Evidently this procedure requires a raster system. In tropotaxis two organs are differently stimulated, the difference causing a reaction bringing the animal in line with the stimulus, where the organs are equally stimulated. The underlying hypothesis assumes a tendency of the system to attain a central balance of excitations. Obviously tropotaxis requires a bi-sensor system. There is no differentiation between reactions to incident stimuli and to gradient stimuli, as in the above classification. The term klinotaxis refers to a successive sampling and comparing of the stimuli, resulting in orientation towards, or away from, the stimulus direction. For this a uni-sensor organ suffices. Kineses designate manœuvres which show no direct spatial relation to the releasing stimulus. The stimuli affect the locomotor activity of the animal. When the velocity of the movement is altered, this is called *orthokinesis*. If the rate or size of the turns is changed, it is *klinokinesis*. Although not directly related to the stimuli, these reactions result in a translocation of the animal along a gradient.

Quantitative relations between stimulus and reaction. The experimental investigation of orientation reactions presupposes a constant relation between stimulus and response, in the sense that the same stimulation always produces the same reaction. In other words, the animal aims always at the same orientational value: for instance, it turns towards a light source until the light stimulus is received from the front. If the stimulus is deviating from this direction, a reaction occurs which is a function of the deviation. It is often measured in terms of the turning tendency, which is defined, for instance, as the measurement of the physical force expended in the turning of an animal towards the stimulus (the torque). In one experiment, a shrimp *Palaemonetes varians* in a harness is connected to a tiny watch spring. It is free to turn by swimming movements about an axis parallel to the long axis of its body. The torque exerted by the shrimp in turning can then be read in numbers of spring extension. For instance, the shrimp will try to turn its back to a light source, as this is its normal position. The mathematical relationship between the turning torque and the light direction can thus be systematically measured.

In nature an animal is often exposed to more than one stimulus. Let us first examine the simplest situation in which two stimuli of the same modality are present. We may distinguish between reactions of positional orientation, as in the shrimp's back-to-light behaviour, and reactions of target orientation, in which the animal aims at the stimulus. In positional orientation the animal usually adopts a compromise position when exposed to two stimuli. This also happens in compass course orientation. But in target orientations the animal often decides to aim at one or other of the two stimuli. An insect larva, for instance, crawling in the field of two light beams crossed at right angles, heads away from the lights (negative phototaxis). In doing so, it takes both lights into account and steers a course exactly midway between the two, provided they are equal in intensity. If not, the larva's course lies nearer the direction of the weaker light. This indicates that the light intensities are weighted, and a course is steered between the two directions according to the weighted ratio of the two stimuli. From a mathematical point of view, the weighted average

corresponds in this case, as in many others, to a vector addition. That means that the stimulus–response situation can be presented as a triangle of forces. The two lights resemble two forces, producing two turning tendencies, which counteract each other until they are in counterbalance. It has been suggested that this is a facsimile of the central balance of excitation of the nervous system as postulated by the tropotaxis hypothesis. But the description in terms of forces is a metaphorical one, serving to explain a process whereby two input data are averaged, their weights being assessed in a specific manner. The situation with respect to gravity is quite different from the one with two lights. Two gravity forces, one produced for instance by centrifugation, the other by referring to earth's gravity, interact physically to produce a resultant force acting as a single stimulus. So for the animal's gravity SENSE ORGANS, only one stimulus exists.

Multimodal orientation. The normal position of many fishes is related to both light and gravity. If these reference cues diverge, for instance, if the light impinges upon a guppy (*Poecilia reticulata*) from one side, the fish adopts an intermediate position. Experiments revealed that the effects of directions and intensities of these two reference stimuli upon the positional orientation of the fish can also be described in terms of a vector addition.

Similarly, many insects walking on a tilted plane with illumination from one side will adopt a course between light direction and gravity. Object orientation in relation to spatial cues may also be placed in this category of multimodal orientations. When we evaluate the orientation of a picture hanging on the wall with respect to the vertical (Fig. A), sensory processes of two modalities are involved: vision and gravity sensing. The gravity sensors indicate the position of the body with respect to gravity (angle α), while the eyes signal the position of the picture with respect to the body (angle β). The integration of these two delivers the information about the position of the picture referring to gravity (angle δ).

The visual sense may be involved in another way in perception of the vertical orientation of the figure. It may mediate the reference for the vertical separately from the one given by gravity. It has been found that we perceive shapes and contours which are predominant in our environment as vertical. In a normal environment vertical (or horizontal) shapes and contours dominate the scene: trees, the edges of buildings, windows, doors, etc. This 'optical vertical' may interfere with the gravity vertical, if their directions diverge (Fig. D). In an experiment, a person looks into a fully equipped dummy room which is tilted to one side. If asked to adjust a movable rod inside the room to the vertical, he will give it an intermediate position between optical and gravity vertical, but rather favouring the optical vertical, i.e. closer to the room's contours. This response changes if the visual surround is reduced to a minimum, for instance to a rectangular frame. The test person will adjust the rod more closely to the gravity vertical than to the frame direction. Obviously the weighting ratio of gravity and optical reference has changed as a function of the size and structuring of the visual surroundings. This experiment is called the rod and frame test.

Weighting the stimuli. The relationship between stimulus and response may not be a fixed one. The animal's state of MOTIVATION may affect the orientation mechanism and alter this relationship. This alteration may take the form of changes in the weighting given to different stimuli. For instance it has been found that in man the weighting ratio of the rod and frame test depends upon personal characteristics of the subject. In women, for instance, the weighting ratio is shifted in favour of the frame reference. That is, in women the rod's position deviates more from the gravity vertical than in men. These findings can be interpreted in terms of the subject's susceptibility to environmental influences. A fish's position between gravity and side-light, for instance, is affected by the hydrostatic pressure. Pressure increase which corresponds to a descent into greater depths induces the fish to close up to the gravity direction, that is, to change the weighting ratio of gravity to light in favour of gravity.

Active change of orientation. The stimulus–response relationship can also change as a result of changes in the goal of the orientation. An animal may respond in a REFLEX manner to an environmental stimulus, such as a movement in the visual field. If, however, that same stimulation is a consequence of the animal's own behaviour, no response may occur. In other words, if the animal's goal or intention is to turn in a particular direction, and this behaviour results in apparent movement of the visual field, there will be no reflex reaction to this apparent movement. To illustrate

this reafference principle, let us consider a particular example. An ant moves along a course at an angle of 40° to the left of the sun; that is, its eyes record the light direction at 40° from the right side. The *afferent* (incoming) signal from the eyes matches the central reference value (output copy) corresponding to 40°. If, due to some disturbance, the ant is forced to deviate from this direction by about 5° to the right, then the eyes would record a compass angle of 35°. There would therefore be a discrepancy between the reference value and the incoming signal. This discrepancy acts through the orientation control mechanism to turn the animal until the signal from the eyes again matches the reference value. Here we have an example of FEEDBACK correction in response to disturbance. If the ant were to initiate a new course, the reference value (output copy) would be altered, and a discrepancy would again arise. In this case, however, the feedback would serve to steer the animal in the new direction. Suppose the ant intended to steer a compass angle of 50° to the left of the sun. The visual signal would still indicate 40°, a difference of 10°. The ant would then turn so as to reduce this 10° to zero, and it would then be heading in the new direction.

The tendency of an animal to turn (and therefore the mechanism of altering the reference value) can be investigated experimentally. For example, the larvae of the water beetle *Acilius sulcatus* use a light compass to orient their swimming behaviour, i.e. to steer up to the surface for breathing, and for swimming downwards. These two compass courses correspond to two reference values of light incidence. They can be manipulated experimentally. Preventing a larva from reaching the surface for several minutes produces a reference value for swimming upwards. A downward course is observed if a larva starts swimming away from the

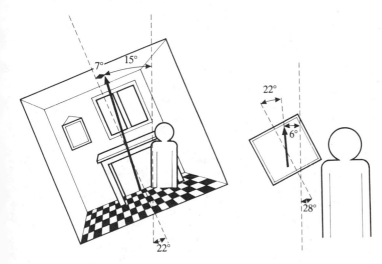

Fig. D. Left: A subject viewing a tilted room is asked to adjust a rod to the subjective vertical. When the room is tilted by 22°, the subject usually positions the rod at 15° to the true vertical. Right: When the subject is viewing a tilted frame, the deviation of the rod from the vertical is much less.

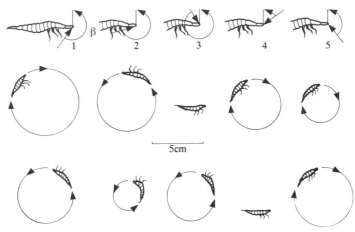

Fig. E. Turning manœuvres of water beetle larvae. The top line shows the angle β of vision permitted after blacking out some of the animal's visual field. The middle line shows the magnitude of the swimming circles obtained when the animal is diving, and the bottom line shows the magnitudes obtained when the animal is surfacing.

surface. Each of these reference directions is based on a set of turning tendencies, responsible for regulating the direction of swimming. Turning tendencies are measured by fixing the impinging light in relation to the animal, and thus causing continuous turning. This is done by covering the animal's eyes with a light-proof lacquer, except those open to a particular direction (the larvae have six eyes on each side). Swimming in a vessel illuminated from all sides, the animals receive light continuously from the direction of the uncovered eyes. If this direction deviates from the animal's desired direction (i.e. the *reference value*), the larva performs turning manœuvres. Because the feedback loop is interrupted, the movement of the animal can no longer affect the sensory input. Therefore the turning manœuvres result in circles. The diameter of the circle can be taken as a measure of the turning tendency (large diameters refer to weak turning tendencies, small diameters to strong ones), as illustrated in Fig. E.

This type of experiment, which can be made highly quantitative and precise, provides the basis for rigorous theories of orientation. The general consensus from these theories is that orientation involves both action and reaction. Changes in orientation behaviour may be reactions to disturbances of environmental origin; they may be responses to major environmental changes, such as the movement of the sun; they may be actions initiated from within the animal itself, and they will often involve a mixture of such action and reaction.   H.S.

45, 84, 124

ORIENTING RESPONSE. When an animal is suddenly presented with a stimulus, particularly a novel stimulus, it quickly turns its body or head so that its eyes and ears can be brought to bear on the source of stimulation. This orienting response was first described by PAVLOV and is REFLEX in character. The orienting response which occurs when an animal or a person is startled is involuntary. It is often accompanied by increased heartbeat rate and other signs of AUTONOMIC arousal.

The orienting response has been called a 'what is it?' reaction, which serves to focus the animal's ATTENTION on the stimulus. The reflex, autonomic, and psychological AROUSAL prepares the animal for an emergency response. The behaviour that follows the orienting response depends upon the nature of the stimulus. A potentially dangerous stimulus may elicit ESCAPE, DEFENSIVE behaviour, or AGGRESSION. A novel stimulus may elicit initial caution followed by EXPLORATORY behaviour or other signs of CURIOSITY (see NOVELTY).

Repeated presentation of a stimulus leads to HABITUATION of the orienting response. The autonomic aspects of the response usually wane more quickly than the behavioural aspects. Habituation of the orienting response shows a high degree of *stimulus specificity*. That is, if the stimulus is changed in intensity or in quality then the habituated orienting response reappears. After the response has disappeared it can often be elicited by presenting the stimulus when the subject is drowsy. This suggests that the waning is due to an active inhibitory process which does not function at low levels of arousal. Suppression of the orienting response thus appears to be a type of economy measure introduced when the response is evidently unnecessary.

74

P

PAIN cannot be defined in an objective scientific manner. Suppose we wish to know whether animals experience pain. We can present a stimulus which would be painful to a human being and observe the animal's response. We might find that the animal withdraws from the stimulus and learns to avoid such situations in future. We can see that the stimulus is instrumental in inducing FEAR responses and other outward signs of EMOTION. However, as we cannot ask the animal about its feelings we have no means of knowing whether it consciously experiences pain. The withdrawal and AVOIDANCE behaviour could be shown by a man-made robot. Can we be sure that animals are not like such machines?

This question is philosophically controversial. Because of the anatomical and physiological similarities between man and a number of animals, some people argue that we have no right to assume that animals do not feel pain, particularly when we are responsible for their WELFARE. On the other hand, we have no empirical means of verifying the theory that animals do have conscious experience similar to ours. Indeed, some philosophers argue that there is no logically valid way of determining whether people have similar private experiences to each other. If a person says 'I know I am in pain', we cannot argue that he may be mistaken in the way that we could argue if he had said 'I know that it is raining'. A person's knowledge and feelings about pain are entirely private and inaccessible to outside observers. Human beings can express the fact that they are in pain by means of LANGUAGE and other signs of distress. The latter also occur amongst animals. However, distress calls in animals have clearly evolved by virtue of their effect upon other animals. A chimpanzee (*Pan*) with a thorn in its foot will scream and seek aid from its fellows. The screams serve to attract the ATTENTION of other members of the group. A wildebeest (*Connochaetes*), however, which is being torn to pieces by African wild dogs (*Lycaon pictus*), or other predators, will suffer silently. Distress calls in this situation would serve only to endanger other members of the group that were attracted by the calls. Even in our own homes we learn to discriminate between genuine distress and the sham version practised by both children and HOUSEHOLD PETS. We learn that VOCALIZATIONS alone are not a reliable guide to pain, and we usually seek further evidence.

Within our own bodies, however, pain clearly serves to draw attention to sites of injury, disease, or other malfunction. A painful stimulus to the exterior of the body elicits a REFLEX withdrawal response, usually followed by an ORIENTING RESPONSE, by which the person or animal turns to examine the source of stimulation. The subsequent behaviour depends upon the nature of the stimulus. If another animal is responsible, the reaction may be one of AGGRESSION or of DEFENSIVE behaviour. If there are clear signs of danger, the animal will usually attempt to ESCAPE.

PARASITISM is an INTERACTION among members of different species, in which the parasite exploits the host, but generally does not kill it. Many parasites are fully dependent upon the host at some stage of their life cycle, and NATURAL SELECTION therefore exerts a strong pressure for them to become highly adapted to their NICHE. At the same time many host species have evolved DEFENSIVE mechanisms, so that there may be an evolutionary 'arms race' between parasite and host.

The life cycles of parasites are often complex, involving a number of ecologically distinct stages. For example, the parasitic flatworm *Cryptocotyle lingua* has three such stages. The adult lives in the intestine of a suitable vertebrate, such as the herring gull (*Larus argentatus*). The eggs of the parasite are released into the intestine and pass out with the faeces, thus passing from a stable, warm environment to a cold and variable one, such as the waters of an estuary. Each egg hatches into a free-swimming *miracidium* (Fig. A), which generally infects a mollusc, such as the periwinkle *Littorina littorea*. In the mollusc the miracidium develops into another free-swimming stage, the *cercaria*. When this animal comes into contact with a suitable species of inshore fish it penetrates the tissues and encysts. The cyst is resistant to the digestive processes of the vertebrate host, so that if the fish is eaten by a herring gull the cyst survives, and is able to grow into an adult worm within the herring gull's intestine.

A parasite which is dependent upon its host throughout its life is called an *obligate* parasite. For these parasites the host provides the complete living environment. Obligate parasites may be *endoparasites*, living inside their hosts, as do tapeworms (Cestoda); they may be *ectoparasites*, such as the feather lice (Mallophaga), which live on the bodies of birds and feed upon blood and feathers; or they may be social parasites, such as the beetle *Atemeles pubicollis*, which is discussed below. Other parasites are dependent upon the host

Fig. A. Life cycle of the herring gull fluke *Cryptocotyle lingua*.

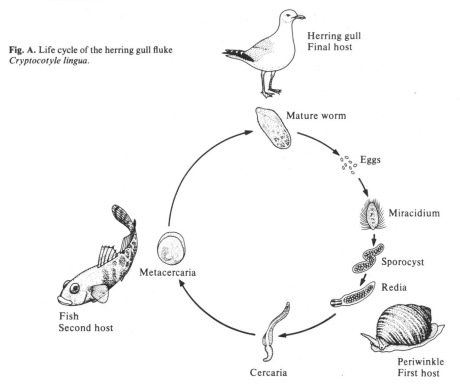

Herring gull
Final host

Mature worm

Eggs

Miracidium

Sporocyst

Redia

Metacercaria

Fish
Second host

Cercaria

Periwinkle
First host

during only part of the life cycle. For instance, the adult of the fly *Protocalliphora azurea* lives off the nectar and dew of flowers, but its larva lives by sucking the blood of nestling birds. There are also many *periodical* parasites, such as ticks (Ixodidae), which gorge themselves with blood from a bird or mammal and then drop onto the ground to digest their meal. Similarly, leeches (Hirudinea) drop back into the water after engorging themselves from the mouths of horses (Equidae) and cattle (Bovidae) drinking in ponds and streams. *Facultative* parasites can live, and complete their life cycles, as free animals, but resort to parasitism when circumstances are favourable. Among these we should include many feral animals, such as members of the cat family (Felidae) and the dog family (Canidae), which roam freely in many countries, intermittently parasitizing themselves on human beings (see SYMBIOSIS).

Birds and mammals are often infested by many different types of parasite (see Fig. B), against which they have little or no defence. Preening and scratching may serve to remove ectoparasites to some extent, but some internal parasites can only be avoided by gross changes in behaviour, such as a change in diet or nest site, which may help to prevent infection. For example, digger wasps (Sphecidae) lay their eggs at the end of long burrows dug into sand. Their larvae are often parasitized by larvae of bee flies (Bombyliidae), blow-

flies (Calliphoridae), cuckoo wasps (Chrysididae), and velvet ants (Mutillidae). Some digger wasps temporarily close their burrows after each visit, and this probably deters parasites to some extent. Other species dig a false burrow near the entrance of the true nesting burrow. This reduces the incidence of parasitization, as the adult parasite often enters and lays its eggs in the false burrow, instead of the true burrow. Digger wasps such as *Sphex argentatus* and the hunting digger wasp (*Philanthus coronatus*) not only dig false burrows but also rebuild them if they are destroyed. In some cases the parasite can influence the host to its own advantage. For example, the amphipod 'shrimp' *Gammarus lacustris* is often parasitized by the roundworm *Polymorphus*. This parasitization induces *Gammarus* to take on a blue coloration which makes it more conspicuous to predators. The parasite benefits from this, because *Gammarus* is commonly preyed upon by ducks (Anatinae), which are also the final host in the *Polymorphus* life cycle.

The defensive behaviour of host against parasite inevitably has an influence upon the numbers of host and parasite within a given population. The giant cowbird (*Scaphidura oryzivora*) of Panama is a *brood parasite*, i.e. an animal which puts its eggs into the nests of other species and allows its young to be raised by the host species. It parasitizes species of oropendola (*Zarhynchus*) and cacique

(*Cacicus*). These birds build dangling oriole-like nests in mixed-species colonies, often with a hundred pairs of birds nesting together. In some colonies the cowbirds lay mimetic eggs that closely match the visual characteristics of the host eggs (see MIMICRY), whereas in other colonies the cowbirds lay non-mimetic eggs. Mimetic females always lay a single egg in a host's nest in which the host has already deposited eggs. Non-mimetic females deposit their eggs in empty nests as well as in full ones, and often lay two or three eggs in a nest. Mimetic cowbirds are very cryptic and try to avoid being seen by the host birds, whereas non-mimetic cowbirds often lay their eggs conspicuously, and in full view of the host birds. The hosts in colonies with mimetic cowbirds discriminate between eggs and throw mis-matched ones out of their nests. However the hosts in colonies with non-mimetic cowbirds do not discriminate in this way, and tolerate even mis-matched cowbird eggs. Many more cowbird eggs are hatched in the nests of host birds that do not discriminate, and it would appear that the strategy of dumping non-mimetic eggs is more successful than that of cryptic laying of mimetic eggs.

However, the host species are also parasitized by botflies (*Philornis* spp.), which lay their eggs on the chicks. Upon hatching, the botfly larvae burrow into the chick's body and feed upon its tissues. Chicks with more than about seven maggots normally die. Some host colonies are protected from botfly by being situated near nests of wasps and bees (Hymenoptera); such colonies are invariably made up of mimetic cowbirds and discriminating hosts. However, even away from the protection of bees and wasps, nests containing cowbird chicks are rarely infested with botfly, even though nests in the same colony, but containing no cowbird chicks, are heavily infested. It appears that the cowbird chicks protect the host from botfly infestation. So in one case the relationship between host and cowbird is parasitic, as the host is protected from botfly by the proximity of bees' and wasps' nests. The mimetic strategy of the cowbird is typical of brood parasites. In the other case, however, the relationship is symbiotic, in that the cowbirds gain by the nest care of the host, and the host species gain by being protected from botfly by cowbird chicks.

A less complicated type of brood parasitism occurs in the European cuckoo (*Cuculus canorus*). When the female returns from wintering in Africa she sets up a TERRITORY, preferably in fairly open country. She defends the territory against all other

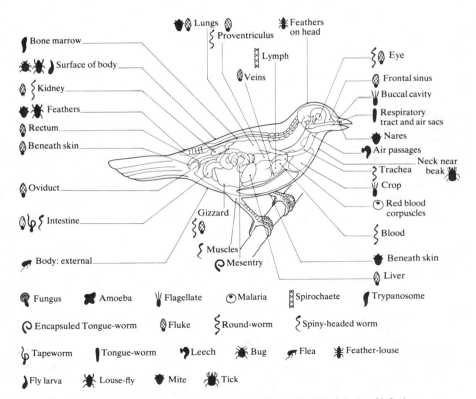

Fig. B. Diagram of a passerine bird illustrating the main groups of parasites with their site of infection.

females that are rivals for the same host species. The male cuckoo also establishes a territory, but this rarely coincides with that of any particular female. The males mate with several females, and females may also mate with more than one male.

The female cuckoo searches for the nests of a particular host species, generally a small passerine bird. Individual females specialize on one particular host species, and when they find a suitable nest, they surreptitiously observe the behaviour of the owners from a point of vantage. The female cuckoo lays her egg in the nest of the host, shortly after the foster mother has begun to lay. She lays her egg quickly and stealthily, generally in the afternoon when the nest owners are likely to be absent. She may destroy some of the host eggs, either by throwing them out of the nest, or by eating them. The eggs of the female cuckoo are distributed singly around a number of nests of the favoured host species.

Most female cuckoos parasitize only one host species throughout their lives, and this preference is largely genetically determined. The eggs of a particular female usually resemble those of the favoured host species. However, this mimicry is by no means perfect, and the nest owners may remove the cuckoo's egg, or may desert the nest. They will also attack the adult cuckoo, should they chance to find it near the nest.

The cuckoo embryo develops more quickly than the host embryos and, provided brooding had not started before the cuckoo's egg was laid in the nest, it hatches a few days before the host's eggs. When about 10 h old the blind nestling cuckoo attempts to remove any remaining eggs from the nest. It tries to roll the egg onto its own back, where it has a specially developed sensitive hollow. Having manœuvred the egg into this hollow, the nestling cuckoo hoists it backwards over the rim of the nest. This is repeated until the young cuckoo is the sole occupant of the nest. The foster parents do not attempt to retrieve their eggs or nestlings from outside the nest, and appear to accept the young cuckoo as their own. They feed it, even though it develops very differently from, and usually becomes much larger than, a member of their own species. It is difficult to believe that the foster parents are not capable of the DISCRIMINATION required to distinguish between a nestling cuckoo and a nestling of their own species. It is also difficult to understand why, during the course of EVOLUTION, the host species have not developed more effective counter measures against such brood parasitism. Some scientists believe that there may be some, hitherto undiscovered, compensating advantage, similar to that seen in cowbirds.

Brood parasitism is a special case of temporary social parasitism. Another example is *trophic parasitism*, in which one species obtains food from another. The spotted hyena (*Crocuta crocuta*) parasitizes the African wild dog (*Lycaon pictus*) by following the dog packs during hunting, and stealing the prey from them. The arctic skua (*Stercorarius parasiticus*) habitually attacks less powerful gulls and other sea birds, and steals their food. Similar piracy is practised by ants of the genus *Crematogaster*; they ambush worker ants of the genus *Monomorium* as they return home along the foraging trails, and steal the seeds that the foragers have collected. Some small ant species of the genus *Solenopsis* live in the walls of large nests built by other ants, and stealthily enter the chambers to prey upon the inhabitants.

Temporary social parasitism is very common in insects, and takes a wide variety of forms. For example, the mound-building ant *Formica excectoides* is a facultative temporary parasite. While the majority of new colonies are founded by the adoption of queens by workers of their own species, after they have been inseminated during the nuptial flights, some are founded by parasitizing other species. The queens of *F. excectoides* may approach a nest of *F. fusca*, and attempt to enter the nest by stealth, or by inducing the host workers to carry them into the nest. This they do by DEATH FEIGNING, in which they pull their appendages into the body in a typical pupal posture. In this position they are picked up by the host workers and carried down into the nest, where they somehow manage to displace the host queen and take over her reproductive role.

The most highly evolved form of social parasitism in insects is *iniquilism*, in which members of one species spend their entire lives as parasites within the society of another species (see SOCIAL ORGANIZATION). To varying degrees the individuals of the parasitic species attach themselves to host colonies, gain acceptance as members, and induce host workers to feed them. For example, the beetle *Atemeles pubicollis* parasitizes ant colonies in Europe. In summer it lives with *Formica polyctena* and in winter with a species of *Myrmica*. In each case it spends its time where the food is most abundant, *Myrmica* being active throughout the winter, whereas *Formica* is quiescent. Migrating beetles, moving from one host colony to another, are guided by the odour of the ants. After reaching a nest entrance, a beetle wanders around until it encounters a worker. It then presents the worker with secretions of its special appeasement gland, and the worker feeds upon this fluid. In effect, the beetle dopes the worker ant, and then induces it to sample secretions of a second gland, the adoption gland. After tasting this substance, the worker usually transports the beetle into the nest. However, should the ruse fail, and the beetle is attacked, it is able to use secretions from a special defensive gland which repel the ants. Once inside the nest, the beetle induces each recently fed ant to regurgitate food from its elastic crop by copying the normal begging behaviour of ants. This consists of lightly and repeatedly tapping part of the mouth of the donor ant. Thus by copying the chemical and tactile codes of the ants, the beetles are able to exploit them. The beetles lay their eggs in

the ants' nest, and the beetle larvae which hatch out induce the worker ants to regurgigate food in the same way as the ant larvae. The beetle larvae are actually more successful than ant larvae at eliciting regurgitation of food, so presumably their mimetic behaviour is a SUPERNORMAL STIMULUS for the ants: beetle larvae accumulate more than their share of food compared with ant larvae when workers are fed with radioactive labelled food. *Atemeles* larvae are straightforward parasites on the ants and are also CANNIBALISTIC, eating both fellow beetle larvae and ant larvae.
120, 145

PARENTAL CARE includes all the activities of animals which are directed towards the maintenance and support of their own offspring or offspring of near relatives. The popular conception of parental care is generally restricted to activities by parents toward young, but in a strict definition parental care includes all energy expenditure commencing at the initiation of the developmental phase of any embryo. In sexually reproducing forms, parental care usually commences at the time of fertilization, and includes such disparate forms of energy expenditure as the building of a nest, transfer of nutrients via the maternal circulatory system, and the provisioning of a natal chamber.

Evolution of sex and asymmetry of parental investment. Regardless of how individualistic an animal may be in its food-seeking behaviour, if the species in question reproduces sexually, then some sort of SOCIAL grouping for exchanging gametes must arise during the process of reproduction. NATURAL SELECTION has shaped the COMMUNICATION system of species in various ways for the synchronization of reproductive behaviour. Generally certain environmental cues serve to prepare the sexes for reproduction. Courtship DISPLAYS serve not only to attract mates, but also to ensure species recognition during mating. The synchronization of partners' activities is an important outcome of so-called COURTSHIP interaction. A survey of the animal kingdom reveals that courtship and mating can take place in a bewildering variety of social groupings.

Parental care is one aspect of the investment process an organism contributes to its progeny. In sexually reproducing organisms GENETIC material is halved to produce sex cells, which then unite through the process of fertilization to form a zygote or egg, which will develop into a new individual. The process of producing sex cells or gametes is referred to as *gametogenesis*. In general, sexually reproducing organisms have two types of gametes which define, on the one hand, the male sex and, on the other, the female sex. Female gametes tend to be larger and to have stores of nutritive material for sustaining the early development of the young. Why, in the EVOLUTION of sex, gametes tended to evolve in this asymmetrical fashion is a mystery. The fact remains that for

most sexually reproducing organisms the gametes are of unequal size, with the male contributing less energy to the production of each individual gamete than the female. Even before fertilization, therefore, the investment by a female in her expected progeny tends to be higher than that of the male. The male's effort in reproduction may equal or exceed the female's, since he has the option of producing many more gametes than does the female, and thereby has the possibility of fertilizing gametes from more than one female. This asymmetry in contribution between the two sexes has often led to the main burden of parental care falling on the female, in cases where parental care extends beyond the deposition of the fertilized egg.

The timing of investment in eggs or young. When the entire animal kingdom is surveyed, the vast majority of invertebrates and vertebrates reproduce by depositing eggs. The internal development of young (viviparity) occurs in all placental mammals, and sporadically in other groups.

Parental behaviour may be shown prior to the fertilization of eggs, or prior to the deposition of eggs (fertilization and deposition in the case of external fertilization may be coincident), or only after the deposition of eggs or young. When parental care occurs prior to egg deposition, it may fall upon the female, the male, or both. If it falls upon both, the relative energy expenditure may be asymmetrical. In general, if there is an asymmetry, the bulk of the parental care falls to the females, with the exception of a few invertebrates, fishes, amphibians, and birds. Cases where the male participates in parental care exclusively, while the female goes on to lay additional clutches, are rare in the animal kingdom.

The vast majority of invertebrates and vertebrates which breed in an aqueous medium fertilize the eggs externally, and, after fertilization, the eggs are given very little attention. Unattended eggs usually float to the surface and the larvae, upon hatching, spend part of their existence as members of the plankton. This permits a wide DISPERSION of the progeny. Of course, eggs may be deposited in sheltered places or in elaborately constructed nests, both by aquatic and terrestrial animals. Eggs may be incubated to the point of HATCHING in the oviduct of the female, a condition known as ovoviviparity. In some cases a special pouch or compartment on the body of the male or the female serves as a repository where development to the point of hatching can proceed.

Eggs may undergo development in the oviduct with a true exchange of materials between the female circulatory system and the larval system giving rise to a true gestation and viviparity (birth of live young). In such cases an even higher investment is contributed by the female to the development of the young. Species exhibiting this form of early parental care only exceptionally show participation on the part of the male in the care of the young after birth.

From this introduction it is clear that parental

care may involve the participation of both sexes, often with the main burden falling on the female; it may be the prerogative of the female or, in the rarest cases, it may be performed exclusively by the male.

Some patterns of parental care. 1. *Invertebrates.* Systems of parental care have evolved independently within all the major groups of the animal kingdom which reproduce sexually. In invertebrates rudimentary forms of parental care can involve simply placing the eggs in a sheltered or secure spot. A more advanced form would involve not only placing the eggs in a sheltered spot, but also placing the necessary foodstuffs for the newly emergent young, so that feeding can commence upon hatching. This is a stage of parental care often employed by certain species of insects, such as the solitary hunting wasps of the genus *Bembex*. The female digger wasp *Bembex rostrata* lays an egg in a burrow which she constructed earlier, and then provisions the egg with the carcass of a paralysed insect. Thus, not only is shelter provided, but also a food supply for the emergent larvae. Associated activites include guarding the locus of the egg, or egg and food supply. A further refinement is for the young to be fed for a time after hatching.

It has been proposed that the higher social insects such as ants, wasps, and bees (all members of the Hymenoptera) passed through an evolutionary stage of parental care such as that exemplified by *Bembex*. To pass beyond this stage the female must feed the larvae directly, and be able to retard the SEXUAL development of her first brood, so that they can be recruited to assist her in rearing a second brood. This sets the stage for the development of permanent non-productive castes, and the further evolution of complex insect societies.

2. *Fishes.* Among the fishes one can see an entire spectrum of parental care, from none, to great and complex forms. Let us consider first of all the modes of reproduction. The common cod (*Gadus*) breeds in a school with males and females ejecting eggs and sperm into the open water. The egg receives no parental care whatsoever. A more complex process is found among trout and salmon (Salmoninae) where the female and male pair off prior to spawning and fertilization, and dig a gravel pit on the floor of the stream for receiving the eggs. Most of the digging is done by the female.

This simple ejection of unfertilized eggs with external fertilization is less advanced than internal fertilization and the ejection of fertilized eggs. The latter evolved in many members of the Cypriniformes (carp) and usually involves a corresponding evolution of a lower egg number per female. Viviparity has occurred in many lines of bony fishes (Teleostei), as well as in sharks (Selachii). Embryonic nutrition has been achieved by a variety of mechanisms in sharks, attesting to the parallel evolution of the systems. Nutrition may be provided within the follicle of the egg itself as well as within the uterus of the female, as occurs in the

mud shark, *Squalus*. Live bearing is always accompanied by a reduction in the total number of young produced during any spawning season.

Care of the young after hatching or birth occurs in a variety of ways. It may involve simply guarding the eggs, as shown by the blenny, *Pholis*, or the construction of a nest cavity as in the Salmoninae, or the construction of complicated nests out of bits of plant material, as in the three-spined stickleback (*Gasterosteus aculeatus*), or the construction of a floating bubble nest as in the Siamese fighting fish (*Betta splendens*).

Complex methods for transporting eggs during development have evolved independently in several families of fish. Certain species of the cichlid *Tilapia* transport the eggs in the mouth, and are properly referred to as mouth brooders. The bottom-dwelling fish, *Aspredo*, rolls in its egg mass, and the eggs attach to the skin where they are overgrown by a bit of epidermis; this is referred to as cutaneous incubation. The male of the sea horse (*Hippocampus*) develops a temporary pouch into which the eggs are deposited by the female, and there they are carried until they hatch.

Trends in the evolution of parental care shown by fishes can be traced within family groups. For example, in a study of parental care among fourteen species of darters (*Etheostoma*), it has been demonstrated that the number of eggs laid is inversely correlated with the degree of parental care shown. Parental care involves tending the eggs and aerating them, and is always restricted to the male.

3. *Amphibians* have also developed varied and complex modes of parental care. The history of amphibian evolution involves coping with an egg that is vulnerable to desiccation. Their eggs must be kept moist, because the amphibians did not evolve a hard protective shell for their eggs when they evolved mechanisms for BREATHING air and living on the land. Thus, for most amphibians a source of permanent water is necessary for successful reproduction. On the other hand, in all of the major families of living amphibians, mechanisms have evolved for making use of temporary water, or for keeping the eggs moist in other ways. Some completely aquatic species, such as the salamander genera *Hynobius* and *Cryptobranchus*, still employ external fertilization in water, but other salamanders and some frogs (Anura) have developed mechanisms for successfully producing young on land.

The care of the eggs by amphibians is quite varied. Guarding the eggs during their early stage of development occurs in some species of salamanders and in the limbless caecilians (Gymnophiona). In *Hynobius* and *Cryptobranchus* it is the male that typically guards the eggs, whereas in the arboreal salamander (*Aneides lugubrus*) the female coils around the newly laid clutch, so reducing the chances of the eggs being attacked by fungi. The carrying of eggs has evolved many times within terrestrial breeding frogs. The midwife toad (*Alytes*) lays its eggs on land. The eggs are extruded by the female into a triangular-shaped space formed by her hind legs. The male fertilizes externally, and then inserts his hind legs into the egg mass and transports them around his waist throughout the developmental period. During this time, the male very carefully selects optimum humidity and temperature conditions to ensure that the egg mass does not dry out. Just before hatching, the male returns to a pond and places his hind legs into the water, whereupon the tadpoles hatch. The toad *Pipa* has a curious method of egg transport, in which the eggs are pushed into the tissue of the female's back by the male, and there they develop in skin pockets. In the South American frog (*Rhinoderma darwini*) the eggs are also laid on land and the male guards the eggs for varying periods of time; but, before hatching, the male will take each egg into his mouth and by appropriate movements force the egg into his vocal sac. The eggs are carried in the vocal sac until the young hatch, whereupon they emerge fully developed from the male's mouth. In one amphibian genus, the toad *Nectophrynoides*, true viviparity has evolved, the young developing completely within the reproductive tract of the female. At the tadpole stage the blood vessels of the tail permit nutrient transfer between the circulation of the oviduct and that of the tadpole. This arrangement is similar in function to the mammalian placenta, and is the most extreme ADAPTATION for live-bearing amongst amphibians. Some salamanders of the genus *Salamandra* are also live-bearing, but do not show such an elaborate adaptation for gaseous exchange between maternal and foetal circulations.

4. *Reptiles*. In so far as we understand parental care in reptiles, several generalizations may be made. There is little or no *pair bonding*, and the holding of a TERRITORY appears to be weakly developed, with the exception of the lizards (Lacertidae) and crocodilians (crocodiles, Crocodylidae and alligators, Alligatoridae). Internal fertilization is nearly universal, and all degrees of the retention of the eggs within the female tract may be demonstrable within the group. Although most reptiles are oviparous (i.e. they lay eggs), ovoviviparity (i.e. the embryo develops in the female, from whom it derives nutrients, though it is separated from the female by egg membranes) and viviparity have evolved. The most singular advance in the reproduction of the reptiles was the evolution of the *cleidoic egg*, i.e. an egg which is protected by a shell and a series of membranes, so that it can be laid on land and will not dry out as readily as the membranous eggs of fishes and amphibians.

Parental care is highly developed in crocodilians. Elaborate nests may be built, and the type of construction varies with the HABITAT. In the American alligator (*Alligator mississippiensis*), a nest of rotting vegetation is constructed by the female, and this lifts the clutch of eggs above the water line and provides it with a warm and humid environment. The female guards the nest and uncovers the eggs when the young utter peeping calls just prior to emergence. Even more elaborate parental care has been described for the Nile crocodile (*Crocodilus niloticus*). Here the eggs are deposited in a hole dug by the female on a sandbank. The female remains in the vicinity of the nest until the eggs are pipped (i.e. have a crack or a hole). The calls of the young again attract the female and she opens the nest. Furthermore, the female may transport the young in her mouth to the water's edge, and she stays in the vicinity of the young for many weeks during their early development.

In Testudines (tortoises, turtles, and relations) there is essentially no parental care other than the selection of a nest site, the digging of a nest cavity, and the covering of the eggs. A similar situation pertains for the primitive form, the tuatara of New Zealand (*Sphenodon punctatus*). In the Squamata (snakes and lizards) different systems have been developed by the lizards and the snakes. Among snakes two major solutions have evolved: (i) many young are brought forth alive after an internal incubation period, and there is negligible posthatching parental care; (ii) many eggs are laid, and, apart from the selection of a nest site and the covering of the eggs, little parental care is shown. There are exceptions to this general account, however. For example, females of the genus *Python* coil around the eggs and actually assist in incubating them by raising the temperature of the clutch. This is accomplished through rhythmic muscular contractions on the part of the female. Guarding the eggs is a well-known phenomenon in some species of cobra (*Naja*), the female coiling above the clutch which she has placed in loose leaves.

Among the lizards, guarding the eggs has been described for the American glass snake (*Ophisaurus*) and the skink (*Eumeces*). In the latter case, the female will even collect scattered eggs, and redeposit them into a consolidated clutch. In the desert night lizard (*Xantusia vigilus*) the egg develops within the female's body, and the association between the egg membrane and the wall of the oviduct suggests a primitive form of placentation. At the time of birth the female seizes and breaks the egg membranes, and pulls at the emerging young. This is one of the most extreme cases of viviparity to be found within the reptiles.

5. *Birds*, like reptiles, have a cleidoic egg, but no bird has evolved viviparity. Parental care by one or, most often, both parents is almost the universal rule. Care of the young is often complex and lasts until they reach maturity. Parental care involves the construction of a nest, an INCUBATION phase where the warmth of the parent's body is employed to maintain a constant temperature during embryonic development, and finally a posthatching phase, involving feeding and/or guarding the young. During incubation, male birds often feed females if the female does most of the incubation, and a concomitant strong pair bond develops.

A survey of the trends of reproduction in birds

in relation to their taxonomic position is instructive; in such a survey one can almost read the steps in the evolution of parental care (see COMPARATIVE STUDIES). In the closely related families, the ostriches (Struthionidae), rheas (Rheidae), cassowaries (Casuariidae), and tinamous (Tinamidae), the male assumes most of the parental care after the female lays the eggs. However, some variation on this theme occurs. For example, in the tinamous the male incubates and guides the precocial young after they hatch. In this group the females are generally brightly coloured, hold territories, and actively court males. A given female may court more than one male. Here we see almost a reversal of the more usual sex roles. In the ostriches the male incubates during the night, and the female incubates during the day. Although more than one female can deposit eggs in a nest, generally a dominant female shares the incubating with the father.

An intense involvement of the male in parental care has evolved independently in other families of birds. In the Turnicidae (button quail), brooding and rearing of the young is done by the males only. The jacanas (Jacanidae) and the painted snipes (Rostratulidae) almost universally have male parental care, and in the family Phalaropodidae (phalaropes) the females do the courting, and the males incubate the eggs and care for the young. There is some tendency for exclusive male parental care in members of the woodpecker family (Picidae), but in the Passeriformes (passerine, or perching, birds) only the family Artamidae (wood swallows) of Australia shows complete male parental care. (All passerines are altricial, i.e. have helpless young.)

On the other hand, the female assumes the major role in incubation and care of the young in many other families of birds. Examining different orders reveals a trend for increasing involvement of the female in incubating and feeding the young. This is demonstrated in the Anseriformes (screamers, ducks and geese), in the Galliformes (gallinaceous birds, poultry, pheasants, etc.), and in the Trochiliformes (humming-birds, with one family, the Trochilidae). Most parrot-like birds (Psittaciformes), owls (Strigiformes) and passerines have a parental care system in which the female does most if not all of the incubation, but the male helps tend and feed the young. In some passerines, as an adjunct to this process, the first brood of the year remains around the nest and assists the parents in rearing a second brood.

One can attempt a phylogenetic reconstruction of the evolution of parental care in birds. The ancient trend was probably for the production of large eggs and large young, which were precocial and could begin to feed themselves fairly soon after hatching. Thus there was little or no necessity for parental feeding. On the other hand, such a process required long incubation periods, and long periods of attentiveness to the nest. It is difficult to tell what the primitive incubation process may have been like. Perhaps it did not involve direct transfer of bodily heat from the parent to the egg, but instead involved placing the eggs in a mound of rotting vegetation, and controlling the heat by removing vegetation or replacing it, in much the same way as do the mound-builders (Megapodidae) of the present day. Interestingly enough, in the megapods the male not only constructs the mound, but attends it throughout the incubation period.

Departures from the ancient process are apparently specializations for a smaller size on the part of the adult bird. This is especially true of the evolution of the Passeriformes. Evolution of small size and greater mobility was probably in response to the occupancy of numerous arboreal feeding NICHES, but smaller size has a price. It means that one must lay smaller eggs relative to the size of the female. If the female does not have the energy resources to lay eggs that contain a large amount of yolk, or if natural selection favours a large clutch size, then the net result will be more altricial young, which are dependent on external feeding. Therefore, feeding becomes mandatory, although the overall rearing time may be less than is the case with the incubation of large eggs producing precocial young. Once the energetic drain became extreme on the female, in terms of incubation time and feeding the young, natural selection could then favour participation by the male in either sharing incubation and/or sharing the feeding of the newly hatched young. Selection favouring the retention of helpers at the nest could be seen as a further refinement of this process.

6. *Mammals.* In mammals the sex involved most often in parental care is the female, by virtue of the fact that one of the defining characteristics of a mammal is the possession of functional mammary glands in the female sex. These glands of course evolved as a method for providing neonates with nutrition. The integration of a male mammal into the family unit is rare, but does occur with monogamous pair bonding in approximately 4% of all mammalian species.

Among the living mammals, we find three different reproductive strategies. The monotremes, such as the duck-billed platypuses (Ornithorhynchidae) and the spiny anteaters or echidnas (Tachyglossidae), lay eggs. The echidna's egg is laid in a pouch, whereas the platypus places her two eggs in a nest. Monotreme females feed their young on milk after they hatch. The young echidna remains in the pouch until it becomes so large that it has to stay in a nest, to which the mother returns at intervals to suckle. All the parental care is done by the female.

The marsupials, for example kangaroos (Macropodidae), give birth to young in a extremely altricial state. The young crawl from the opening of the urogenital sinus to a teat area (often enclosed in a pouch) where they attach to a nipple and undergo a prolonged teat-attachment phase. This may be followed by a nest phase prior to weaning. In all marsupials studied, parental care is restricted to the female.

Eutherians (higher mammals) give birth to young which are much more advanced in their development than marsupial young, although there is a great range in the degree of dependence of the young. There is no prolonged teat-attachment phase, but rather successive bouts of suckling until weaning. The early nutrition of the young is provided solely by the female. In many mammals the mother is the sole support system for the young, but varying degrees of male and family participation have evolved in different groups.

There is considerable variation in the form of parental care shown by different mammalian species. When we compare the various evolutionary lines of mammals, we note the formation of more complicated SOCIAL RELATIONSHIPS occurring independently many times. Mechanisms have evolved for communication that permit the control of inter-individual behaviour within a social context, and without pathological results. The reproductive unit in mammals will become more complex, in terms of the numbers, sexes, and age classes involved, if any or all of the following occur: (i) the male is involved in parental care; (ii) some or all of previous offspring are retained in the group during subsequent rearing phases; (iii) extended family groups derive from (i) and (ii) involving either males and/or females. Why these more complex social patterns for rearing evolved can only be discovered by analysing the overall adaptation of the species for the exploitation of a particular habitat at a particular point in time. Complex social patterns generally correlate with long life and slow maturation rates (see MATERNAL BEHAVIOUR).

Parental care and the number of young produced. Parental care is well developed in some species of insects, crustaceans, and vertebrates. Indeed, the parent or parents, and the young or egg clutch, can form an interacting system involving a mutual exchange of stimuli which coordinate their activities. It would appear that sexually reproducing animal species have two broad avenues open to them in the process of reproduction. Firstly, the species may be selected to produce a great many young or eggs, which are placed in an advantageous spot and then left to develop without further contribution from the parents. Natural selection has refined such forms so that a minimum of LEARNING by IMITATION is required. FEEDING behaviour is pre-programmed, shelter-seeking and mate identification likewise seem to require little prior experience.

On the other hand, only a small number of eggs or young may be produced by a female. If, however, the male takes over sole parental duties after laying or birth, the female may lay subsequent clutches with other males, and so still produce a respectable total number of young for a given breeding season. If, however, the female alone tends the eggs or young, or if both parents are occupied with rearing a brood, then the number of young produced per breeding season is usually low. The young of species with prolonged parental care usually require intense early experiences with the parents in order to learn to fend for themselves. Increased potential longevity and small litter sizes tend to go together in the mammals. Small litter sizes and relatively prolonged parental investment also go together, with relatively large BRAIN size and presumably an increased learning capacity. This association of characters appears to define an adaptive set which also characterizes the origins of human evolution.   J.F.E.

19, 54, 87, 111, 118

PAVLOV, IVAN PETROVICH (1849–1936), Russian physiologist. His importance for the study of animal behaviour is mainly due to his work on conditional REFLEXES; this work provided the basis for much of the subsequent research on LEARNING. Pavlov also conducted significant research on the physiology of digestion and on neurosis.

Brief life. Pavlov was born in Ryazan, Russia, and attended the religious school and seminary there, where he studied natural science. He did not complete his studies, but entered St. Petersburg University in 1870, where he continued to study natural science, and decided to make his career as a physiologist. After graduation in 1875, he went to the Military Medical Academy to pursue his research. He completed his doctorate there in 1883, and then went to Germany (1884–6), where he studied in Leipzig with Carl Ludwig, and in Breslau. In 1890 he was appointed Professor in the Department of Pharmacology in the Military Medical Academy, and in 1895 he moved to the Department of Physiology. In 1904 he received the Nobel Prize for his work on the physiology of digestion. He remained Professor of Physiology until 1925, when he resigned in protest against the expulsion of sons of priests from the Academy (he himself was the son of a priest, but would not have been expelled). Initially Pavlov was outspokenly opposed to the Bolsheviks, even though they supported his research. In 1922 he asked Lenin's permission to transfer his research abroad, but was refused: Lenin wanted prestigious scientists. However, during the last few years before his death (in Leningrad) Pavlov increasingly accepted and approved of the Bolsheviks. From 1925 to 1936 he worked mainly in three laboratories: the Institute of Physiology of the Soviet Academy of Sciences (which is now named after him), the Institute of Experimental Medicine, and the biological laboratory at Koltushy (now Pavlovo), near Leningrad.

Physiology of circulation and digestion. Pavlov held that physiologists should study 'the actual course of particular physiological processes in a whole and normal organism'. He also held that the main problems for experimental research were the mutual interactions of organs within the body, and the relation of the organism to its environment. The method of working on the whole, healthy body of an animal contrasted with the mainstream of physiology in the latter half of the 19th century; most investigations then were on isolated organs and prepared specimens.

Pavlov's work on the physiology of circulation (c. 1874–88) was mainly concerned with the mechanisms that regulate blood pressure. His experimental animal for this, and for most subsequent research, was the dog (Canis l. familiaris). He was mainly interested in nervous mechanisms. He discovered, for instance, that the vagus nerve controls blood pressure, and that there are four nerves controlling the heartbeat, which can vary the heartbeat's rhythm and intensity (work on the nervous control of the heart had formed his doctorate).

Pavlov's work on the physiology of digestion began in about 1879, and culminated in his book The Work of the Digestive Glands (1902, a translation of Lektsii o rabotie glavnych pishchevaritel'nykh zhelez, 1897). He investigated the nervous mechanisms controlling the secretions of the various digestive glands, and how these nervous mechanisms were stimulated by food. He had to expose the structures of interest surgically, and work on them in a healthy animal, so it was crucial to his success that he was also a brilliant surgeon. (Similar experiments had been attempted in the laboratory in Breslau that he visited in the mid 1880s, but had failed because the experimenters lacked Pavlov's surgical skill.) Once he had exposed part of the gut, Pavlov could directly insert food or chemicals, and observe the effects on the activity of the digestive glands. The method of sham FEEDING was a related development. A slit is made in the animal's throat so that food entering through the mouth falls out through the neck before reaching the stomach. The animal can be fed through a second opening made into the stomach. By sham feeding, Pavlov could observe the effect of food in the mouth on the secretion of digestive juices elsewhere in the gut; he found that the TASTE of food in the mouth causes the release of gastric juices in the stomach. A smaller quantity of juice is released if food is put directly into the stomach. Sham feeding has been used and developed extensively by later workers.

Pavlov's own theory for the control of digestive secretions postulated control exclusively by nervous mechanisms. Subsequent research has shown this theory to be incomplete: control by HORMONES also occurs. He made many other important discoveries while working on digestion. Two of the most important were the enzyme enterokinase which controls the activity of another digestive enzyme, and the connection between the properties of the saliva and the type of food being eaten (the Pavlovian curves of salivary secretion). N. P. Shepovalnikov and Pavlov were co-discoverers of enterokinase.

Pavlov's work on the physiology of digestion is important for understanding his work on animal behaviour, for his explanations of animal behaviour are similar to those of the control of digestion. In addition, his methods were similar in all his studies.

Conditional reflexes. While at the seminary in Ryazan Pavlov had read I. M. Sechenov's Refleksy golovnago mozga (Reflexes of the Brain), which argued that mental events were reflexes. Then, while working on the physiology of digestion, he had noticed 'psychic' salivation: when the dog was confronted by a stimulus that customarily preceded feeding, it salivated even before being fed. This psychic salivation could be induced, for example by the animal's food container, or by the presence of the attendant who normally fed the animal, or even by the sound of the attendant's approach. Armed with these incidental observations and with the reflexology of Sechenov, and stimulated by DARWIN's evolutionary arguments about animal behaviour (which encouraged materialistic analyses of mental events), Pavlov set out to investigate the psychic salivation of dogs. This was a simple extension of his earlier studies on the control of digestive secretions, and once again it was the nervous reflex that he looked to for his explanation. He worked on conditional reflexes from about 1902 until his death.

A typical Pavlovian experiment on psychic salivation would be as follows. On several occasions a bell is rung just before the dog is fed, and the dog salivates on receiving its food. Then the bell is rung without presentation of food. It is observed that the dog salivates in response to the bell's ringing. Pavlov termed the food the unconditional stimulus, the sound of the bell the conditional stimulus, the salivation to the food the unconditional reflex, and the salivation to the bell alone the conditional reflex. ('Conditional' is what Pavlov actually wrote, but the early mistranslation of 'conditioned' is now widespread in the psychological literature.)

Many of the fine details of the conditional reflex were studied by Pavlov and his collaborators. First, there is the temporal sequence of stimuli. Pavlov found that it is much easier to form a conditional reflex if the unconditional stimulus (food) follows the conditional one (bell) than if they are simultaneous, or if the conditional stimulus follows the unconditional. Second, there is the time delay between stimuli. Here he found that a discrete conditional stimulus is more effective in forming a conditional reflex if it occurs near in time to the food than if it occurs a long time before. However, if the conditional stimulus starts a long time before, but continues right up to when the unconditional stimulus is presented, then it is as effective as a conditional stimulus which starts just before the food is given. Third, there is the intensity of the stimuli. A dog salivates more if it is trained on bigger pieces of food; and similarly, it salivates more to a louder bell. Fourth, Pavlov studied GENERALIZATION of the conditional stimulus. If the animal has been trained on a stimulus of one pitch, it can then be tested for a response to a stimulus of another pitch. This leads to a method of investigating the animal's powers of sensory DISCRIMINATION, which again was originated and developed by Pavlov.

Pavlov was not only interested in how conditional reflexes were gained; he also studied how they were lost. He classified factors causing loss of a conditional reflex into cases of either *external inhibition* or *internal inhibition*. If an animal, conditioned in one way, is moved into a new environment, or is exposed to new stimuli before being fed, it loses its original conditional reflex; this is called external inhibition. There are several forms of internal inhibition. The most straightforward is the gradual loss of the conditional reflex if the food is withheld after the conditional stimulus; the conditional reflex requires regular REINFORCEMENT (to use Pavlov's term) by the unconditional stimulus.

Pavlov thought of the conditional reflex as similar to any other kind of reflex. The flow of digestive juices is stimulated by the MECHANICAL and CHEMICAL properties of food, through the mediation of a nervous (unconditional) reflex. Similarly, salivation could be induced by some environmental indicator of food, again by a nervous (conditional, in this case) reflex. The conditional reflex, however, is easily modifiable by the environment, according to whatever the local indicators of food happen to be. So, Pavlov regarded the formation of conditional reflexes as an ADAPTATION whereby the animal could survive better in a changing environment.

Pavlov also speculated on the fine details of the formation of a conditional reflex. He suggested that the cells of the central nervous system must change structurally and chemically when a conditional reflex is formed: 'the locking in, the formation of new connections, we attribute to the functioning of the separating membrane, should it exist, or simply to the branching between neurones'. This idea has subsequently been confirmed.

Although nearly all his research was on dogs, Pavlov also showed that conditional reflexes can be formed in mice (Murinae) and monkeys (Simiae), and he was in no doubt that they occur in man, and in all other animals. He wrote: 'A temporary nervous connection is a universal physiological phenomenon in the animal world and exists in us ourselves.' He also showed that more or less any environmental factor can act as a conditional stimulus (though this conclusion has been slightly modified by later research). Pavlov was aiming at truly universal laws of learning, and that is why his discoveries are so fundamental to modern theories of associative learning. One branch of psychology, termed *behaviourism* (see OPERANT BEHAVIOUR; SKINNER), went so far as to attribute all human behaviour to conditioning and reinforcement. Pavlov, however, made no such extravagant claims for the conditional reflex, and

ridiculed the claims of behaviourism, to scientific status. The most famous expression in English of Pavlov's work on conditioning is his book *Conditioned Reflexes* (1927). In psychiatry, conditioning is used in *behaviour therapy*.

Experimental neurosis and personality. In a famous experiment (1921) by Shenger-Krestovnika a circle was used as a conditional stimulus before feeding, and the dog was also trained to associate an ellipse with not being fed. By small steps the ellipse was then made more and more like a circle. When the ellipse was almost round, initially the dog could usually distinguish it from a circle. But after a few weeks the dog became neurotic: it ceased to be able to recognize obvious ellipses and circles, became very excited, and was no longer calm during experiments. Pavlov termed the animal's ABNORMAL condition *experimental neurosis*, and he attributed it to a disturbance of the balance between excitatory and inhibitory processes in the nervous system. This explanation of experimental neurosis is grounded in Pavlov's theory of personality, He explained personality by variation in the excitation of the nervous system. He did not, however, attribute neurosis solely to external factors, such as contradictory stimuli. His experiments on experimental neuroses showed that dogs with different 'personalities' were differentially susceptible to the treatment: the same treatment on different dogs could produce quite different neuroses. In the 1930s Pavlov decided to work on the GENETICS of behaviour, and his government built him the biological station at Koltushy for this research.

There is an underlying unity to all Pavlov's work. In his earliest work on blood circulation he established his method (intervention in unanaesthetized, whole dogs) and his paradigm explanation: nervous control. The same procedure was used in his work on the control of digestive secretion, and he started research on conditional reflexes as just another kind of nervous control of digestion. Late in his life, he saw how conditioning could be used to analyse personality and neurosis, and once again he was able to carry over his theoretical framework into a new field.   M.R.

57, 62

PERCEPTION, the appreciation of the world through the senses, depends upon the SENSE ORGANS possessed by the animal, and the interpretation that is placed upon the incoming sensations by the BRAIN. It is difficult for us to realize the extent to which the perceptual world of some animals differs from our own.

In describing the perceptual world of the tick (Ixodidae), Jakob von Uexküll wrote in his book *Umwelt und Innenwelt der Tiere* (1909): 'After mating, the female climbs to the tip of a twig on some bush. There she clings at such a height that she can drop upon small mammals that may run under her, or be brushed off by larger animals. The eyeless tick is directed to this watchtower by a general photosensitivity of her skin. The approaching prey is revealed to the blind and deaf highway woman by her sense of smell. The odor of butyric acid, that emanates from the skin glands of all mammals, acts on the tick as a signal to leave her watchtower and hurl herself downwards. If, in doing so, she lands on something warm—a fine sense of temperature betrays this to her—she has reached her prey, the warm-blooded creature. It only remains for her to find a hairless spot. There she burrows deep into the skin of her prey and slowly pumps herself full of warm blood. Experiments with artificial membranes and fluids other than blood have proved that the tick lacks all sense of taste. Once the membrane is perforated, she will drink any fluid of the right temperature.'

The tick's behaviour is complex and shows a high degree of ADAPTATION, but it is guided by a primitive sensory system responding to just three stimuli: diffuse light, the smell of butyric acid, and heat. It is a nice illustration of what can be done with a simple perceptual apparatus, and of how animals' sensory systems are adapted to their needs. The perceptual world of the tick may be contrasted with that of higher animals, such as birds and mammals, who use perception to form an accurate representation of the world, so that they can find their way around and satisfy their elaborate needs within it.

In order to survive, all animals must be able to detect and locate food, shelter, and mates, and they must recognize and avoid predators. Perception is the means by which this wealth of information is acquired. All perception is initiated by the stimulation of specialized nerve cells known as *receptors*, which are sensitive to specific physical events. These receptors, together with ancillary structures, make up the SENSE ORGANS which provide the basis for HEARING, VISION, TASTE AND SMELL. In addition there are the CHEMICAL, ELECTROMAGNETIC, and MECHANICAL SENSES which depend upon less elaborate sense organs. Perception of PAIN, temperature (see THERMOREGULATION), and internal states associated with HUNGER and THIRST also depends upon specialized receptors, often located within the brain itself.

Stimulation of receptors will only be of use to animals in so far as they can interpret it in terms of the objects in the world which it signifies. Brightly illuminated coal sends much more light to the eye than does a sheet of white paper in dim light: nevertheless, many species see the coal as black and the paper as white. The interpretation of incoming signals depends on the context in which they occur, and the animal reconstructs in its brain a representation of its world. Human beings do not see a flat image varying in brightness and colour; they see three-dimensional surfaces, houses, trees, and tables, and they see them at their correct distances, and in their correct relationships to one another.

The mechanisms that allow us to reconstruct our environment from the fragmentary information arriving at the SENSES are subtle and complex: perception in higher animals may well involve considerable INTELLIGENCE. Perceptual processing has evolved over billions of years, whereas the manipulation of mathematical symbols, for instance, is a comparatively recent human activity that is carried out slowly and painfully: contrast the speed at which the meaning of a scene can be registered at a glance.

The animal's perceptual world is bounded by physical factors. For example, most higher animals rely heavily on sensitivity to light to provide information about the external world. In physical terms, visible light is electromagnetic energy, and lies on a continuum with wireless waves, X-rays, and cosmic rays. Different kinds of electromagnetic energy differ in wavelength, and all members of the animal kingdom that can see are sensitive to approximately the same narrow band of electromagnetic wavelengths ranging from about 350 nm (violet) to 750 nm (red). One may wonder why there should not be some species with eyes sensitive to wireless waves, and others with eyes sensitive to X-rays. Part of the answer is that the energy radiated by the sun is maximal in the visible spectrum, but there are other limiting factors (see VISION).

Electromagnetic energy of a slightly longer wavelength than red light is known as infra-red and is felt as heat: the bodies of all warm-blooded animals radiate energy of this wavelength and this constant background radiation would make it impossible for a warm-blooded animal to detect variations in infra-red energy actually caused by its reflection from surfaces. In fact, some snakes (Serpentes), which are of course cold-blooded, do have directionally sensitive infra-red receptors buried in pits on their heads, and they use these receptors to detect the presence of a warm-blooded animal within striking distance of their fangs (SENSES OF ANIMALS, Fig. A).

If we move in the other direction from the visible spectrum, to light of shorter wavelength than violet, we come to ultra-violet. Most naturally occurring surfaces reflect little ultra-violet, and it is perhaps for this reason that vertebrates have not developed any sensitivity to it. Many insects, however, are sensitive to ultra-violet: some flowers reflect such light strongly, and have developed patterns of ultra-violet that serve to guide bees (Apidae) and other insects to their nectar and pollen. *Pollen guides* picked out in ultra-violet can be readily detected by bees, but are, of course, invisible to man. Beyond the ultra-violet part of the spectrum are X-rays: they are much more penetrating than light rays and tend to damage the organic molecules of which all life is composed. Fortunately little energy of this wavelength reaches the earth, or life as we know it could not have evolved. Because of their penetrating power, X-rays are not readily reflected from surfaces, and would not therefore in any case be a suitable medium for vision. Just as there are organisms, like bees, that are sensitive to a slightly greater range of wavelengths than ourselves, so there are some animals that are sensitive to a reduced range. Red light does not penetrate far below the surface of water, and the eyes of many deep-sea fishes are sensitive only to the ghostly blue-green light that pervades the ocean.

The perceptual world is also influenced by the sensory medium that is utilized by the animal. For example, if organisms are to use light to provide detailed information about the world, it is necessary to form an image. An animal lacking this ability would be sensitive only to fluctuations in the average amount of light present in its environment, and would not be able to detect differences in the brightness of light reaching it from different directions. In fact, some simple animals, like the earthworm *Lumbricus*, have light detectors scattered over the whole body and do not form images: they make withdrawal movements when a part of their body is stimulated by light, and hence avoid bright sunlight. Three different ways of forming images have been evolved. Some simple organisms, for example *Nautilus* (a primitive member of the Cephalopoda, squids and octopuses), use the principle of the pin-hole camera and admit light to the eye through a tiny hole (see VISION, Fig. E). This ensures that an image will be formed on the receptive surface, but the device is inefficient since it admits very little light. A second method of image formation is used by many invertebrates, including insects. Their eyes consist of a series of tubes (*ommatidia*) with opaque sides (VISION, Fig. E). Each ommatidium points in a different direction and they are arranged radially with the outward ends on the surface of a sphere and the inner ends pointing to the sphere's centre. Each will therefore capture light from a single direction in space, since light not aligned directly with the axis of the tube is trapped by its opaque sides and fails to reach the light-sensitive elements at its base. Eyes based on this principle are known as compound eyes. In practice, each ommatidium has its own individual lens in front of it to help in the collection of light. There is a popular fallacy that flies (Diptera) having such eyes with multiple facets see hundreds of different pictures of the world: in fact their receptors are wired to the brain in a way not unlike those of vertebrates, to yield a single coherent image.

Vertebrates (and some invertebrates) have evolved an even more sophisticated method of image formation based on the principle of a single lens. When light passes from one transparent medium to another, its rays are bent, and by interposing a lens of the right convexity between the receptive surface and the external world it is possible to ensure that all rays emanating from a given point in space are bent in such a way as to fall on the same point on the receptive surface, and hence form an image. As VISION, Fig. G, illustrates, a lens of a given convexity will only form a sharp image for objects at a given distance away: to form

sharp images of objects at different distances, it is necessary to alter the lens in some way. Animals have invented two solutions to this problem: mammals change the convexity of the lens in accordance with the distance of the object they are looking at, making it more convex for near objects and less convex for far objects. Many fish have a lens of fixed convexity, but achieve a sharp image for objects at different distances by varying the distance of the lens from the retina.

Most animals having a single lens in front of the eye have a receptive surface (retina) behind the lens in the shape of a hemisphere, ensuring that for a given lens curvature all points in space at the same distance from the animal will be in focus on the retina. Some animals, however, have evolved retinas of a more complicated shape: for example, the upper half of the retina of horses (Equidae) is further from the lens than is the lower half. The effect of this arrangement is that nearby objects lying below the level of the eye will be in focus, while a sharp image of higher objects at a greater distance will simultaneously be formed. The horse can therefore get a clear picture of the grass it is cropping whilst simultaneously keeping a look-out for predators appearing on the horizon.

Light is not always the best medium through which information about objects in the outside world can be gained. Animals which are active at night, or in murky water, have solved the problem in other ways. Many fish have lateral line organs that are specialized to detect small changes in the flow of water round them, and certain fishes that live in the turgid waters of the Amazon and other South American rivers have organs in the skin that are sensitive to changes in the electric field around them. These fish actually produce an electric field around themselves, and it has been shown that they can detect changes in this field caused by the presence of nearby objects: such fishes have been trained to select a cylinder full of water in preference to one full of a non-conducting material (paraffin wax).

Another approach to the problem is ECHO-LOCATION, the best-known examples of which are found among the nocturnal bats (Chiroptera). A bat emits cries of a very high frequency (above 20 000 Hz), and for the most part beyond the range of human hearing. They may use these high frequencies because such sounds are rarely produced in nature, and therefore the reception of echoes from their own cries will not suffer interference from other noises. The wind rustling the trees for example, produces sound of considerably lower frequency.

Bats can not only locate their prey on the wing by listening to the echoes received back in the silent intervals between cries, but can also use echo-location to identify prey and to avoid obstacles. Bats can identify a moth (Lepidoptera) from about 2·5 m away, and the fruit fly (*Drosophila*) from 60 cm. Some fish-eating bats, such as the Mexican bulldog bat (*Noctilio leporinus*), use sonar to detect the presence of fish by echoes from their fins sticking just out of the water. Porpoises (Phocaenidae) also use echo-location and emit short bursts of high-frequency sound. They have, of course, good vision, but echo-location could be useful to them in turbid water.

Perceptual organization. Many features of the external world can be recognized by animals only through complex organization and interpretation of the messages entering the brain from the various sense organs. For instance, any organism that uses vision to guide its way around the world must extract information about the distance of objects and surfaces, in order to construct a three-dimensional representation of the arrangement of objects in its environment. The two-dimensional images falling on each eye contain a number of cues known to be used by certain animals as follows.

(1) *Motion parallax.* If the animal moves its head from side to side or up and down, nearby objects are swept across the retina faster than distant ones: the further away the object, the more slowly it moves across the retina. Most animals are thought to use this cue. For example, when locusts (Orthoptera) prepare to jump across a gap, they characteristically move their heads from side to side a few times before leaping. In an ingenious experiment it was arranged that as they moved their heads the image was moved either too quickly or too slowly across the retina: if the motion was too fast, they did not jump far enough, whereas if it was too slow they overshot the point at which they were aiming.

2. *Stereoscopic vision.* Since each eye has a slightly different view of the world, the visual images of objects at different distances do not coincide exactly on the retina. If one forefinger is held close to the eyes and the other at arm's length and the eyes are opened in succession one at a time, it will be seen that the far finger jumps to the right when seen with the right eye and to the left when the left eye is opened. This disparity in the two images can be used to gauge the distance away of different objects in the field of view. It is known that the praying mantis *Mantis religiosa*, and some birds and primates, use stereoscopic vision as a cue to depth, and it is likely that many other species have this capacity.

3. *Looming.* When an object approaches, its retinal image increases in size. If a disc of light is cast on a screen and made to expand suddenly in front of a monkey (Simiae), it flinches, as do human infants. The monkey interprets the expansion of the disc as meaning that the disc is moving towards it, and accordingly ducks out of the way. Several other animals, including crabs (Brachyura), chicks (*Gallus g. domesticus*), frogs (Anura), and kittens (*Felis catus*), have been found to respond to the cue of looming by running away.

4. *Shadows.* Man can use the direction in which a shadow falls to provide information about the shape of an object: concave objects cast a shadow in the region nearest the light source, whilst convex objects cast a shadow in the region furthest from the light source. Chickens also use information from shadows as a cue to depth. A picture was taken of a surface covered with grains of corn; chickens were shown two copies of the picture, one the correct way up, the other upside-down. They pecked only at the corn in the picture shown the right way up, presumably because in the upside-down picture the shadows signal concavities rather than convex objects.

Man is known to use many other cues to depth. For example, the extent to which the eyes are converged can yield information about the distance away of the object being fixated, there are many cues associated with perspective effects, and where the image of one object is interrupted by the image of another the first object is seen as lying further away than the second. It is likely that many species of vertebrates make use of such additional depth cues, but this has not been proved.

What is certain is that many animals have very accurate depth vision. Consider the common frog (*Rana temporaria*) shooting out its tongue to the correct distance to catch an insect on the wing, or a squirrel (Sciuridae) leaping unerringly from bough to bough, or running along a fence made of stakes. Moreover, it is known that in many species at least some of the mechanisms for detecting depth are INNATE, and do not depend on LEARNING. In a typical experiment, a shallow platform is laid on top of a sheet of glass: on one side of the platform the glass is supported by a surface immediately beneath it, but on the other side there is a sheer drop below the glass to the floor some distance away. The animal is placed on the platform and eventually steps off onto the glass: the experimenter notes whether it comes down on the side of the shallow surface or on the side where there is a visual drop. Animals using depth vision would be expected to come down consistently on the shallow side, and this is precisely what has been found for a great range of animals, including crabs, rats (*Rattus*), lambs (*Ovis*), and human infants. To discover whether animals can use depth cues without previous learning, individuals are kept in the dark until old enough to move around, and then placed in the *visual cliff* apparatus. Such animals will have had no previous visual experience, so if they come down consistently on the shallow side they must be able to detect depth innately. Using this procedure many animals, including rats and lambs, have been shown to have an innate appreciation of visual depth. Others, such as kittens, perform well on the visual cliff after only a few hours of exposure to light.

A further ability possessed by all animals that have been tested is the correct judgement of the size of objects despite variations in their distance. When a tall man is seen at 1·5 m, the retinal image produced is four times as large as that of the same man at 6 m, and yet we see the man as the same size in both cases. This phenomenon is known as *size constancy*. To discover whether animals have size constancy they can be trained to select the larger of two cubes placed equidistant from them. They are then tested with the large cube placed further away and the small cube nearer. Even when the large cube has a smaller retinal image than the physically smaller cube, all animals tested continue to select the cube that is physically bigger. In experiments of this sort, it has been found that monkeys, chickens, cats, fishes, and the common octopus (*Octopus vulgaris*) all select the object that is the same real size as the one they were originally rewarded for taking: they do not respond to the changes in retinal size.

In addition to the perceptual organization of the external world, animals have to cope with the information from the sense organs that detect changes in their own body. Muscles and tendons have receptors that signal changes in contraction or tension, and blood vessels have receptors signalling pressure and distension. In addition, most species from jellyfish (*Aurelia*) to man have organs of balance that enable them to maintain an upright ORIENTATION. Mammals detect changes in head position through the semicircular canals set in three different planes approximately at right angles to one another: any acceleration in the movement of the head will set in motion the fluid in these canals, and this motion deflects hair cells which in turn send signals to the brain. Also, many species have sacs containing small crystals, which, being heavier than the fluid in which they are immersed, will exert pressure on different hair cells depending on how the head is oriented with respect to gravity. Most species manufacture the crystals themselves out of calcium, but the lobster (*Homarus*) actually uses grains of sand. Some insects have balance receptors located at the base of the wings that work on a rather different principle: these receptors are differentially stretched when the insect makes yawing or pitching movements, and hence help to keep it on a level FLIGHT path. Flies have developed a secondary pair of wings known as *halteres*: these oscillate very rapidly and actually function like gyroscopes. They tend to remain in the same position relative to external space when the fly changes direction, and hence exert mechanical forces on the fly that are detected and used to maintain a straight and level flight path.

The organ of balance in the octopus is known as a *statocyst*: as well as enabling the octopus to remain roughly upright, it controls the positions of the eyes which rotate in the head, so that no matter what its head position the eyes are always correctly oriented. If the statocysts are removed, the octopus walks unsteadily like a drunk person, and it can no longer maintain a straight course when swimming, but tends to spiral around helplessly. If an octopus is trained to seize a vertical rectangle and to avoid a horizontal rectangle, and then the statocysts are removed, further responses to the rectangles are determined solely by their position on the eyes, not by their position with re-

spect to gravity. Its eyes no longer rotate when its head is tilted, and if it is sitting with its eyes at 45°, it will seize a rectangle at 45° to gravity but vertical on the eyes, and avoid one that is horizontal on the eyes. The eyes of most mammals only make small rotatory movements, and when the head is tilted the visual system itself corrects for the unusual angle at which objects in the real world fall on the eye. This is yet another example of how different species can attain the same end by different means.

In general, changes in the pattern of stimulation are of more significance for survival than continuous stimulation, and all sensory systems are designed to respond mainly to change (see ADAPTATION). Continuous sounds or smells cease to be noticed, and the horse, no doubt, is as little aware of its saddle and bridle as we are of our own clothes. The visual system responds mainly to changes in brightness and colour across the retina. The absolute strength of stimulation is less important than the ratios between different amounts of stimulation present at the same time: coal continues to look black and paper white despite changes in incident light, and animals can recognize sounds by relative changes in intensity and frequency rather than by the absolute value of any one component. In vision and hearing it is necessary to respond to a considerable range of intensities, and there are special devices to make this possible. Because of the importance of ratios, a sound or light that is double the physical energy of another is not perceived as twice as intense. The loudness of a sound increases in proportion to the logarithm of its intensity, so that for one sound to appear twice as loud as another, the energy must be squared.

The senses of different species are finely adapted to help their survival in the ecological NICHE in which they find themselves. Quite simple organisms such as ticks or moths may achieve highly adaptive responses with sensory systems that are primitive in comparison to those of most mammals. In the course of EVOLUTION many different ways of achieving the same end have been invented. We have seen how three different ways of forming an image have been evolved, and different animals obtain information about their own position relative to gravity either by using suspended weights, or by the principle of the gyroscope. Higher animals use their senses to construct in their brains a representation of the world around them. We have seen how it is possible to extract from the stimulation received cues to the distance of objects, and how objects can be represented in the brain in their true sizes, shapes, and colours despite variations in the retinal input. Nevertheless, the details of how such representations are constructed, and of how they are manipulated in order for the animal to find its way around the world, are largely unknown, and this problem remains one of the most interesting challenges facing the experimental psychologists, or the student of ETHOLOGY working in the LABORATORY.   N.S.S. 59, 100

PERCEPTUAL LEARNING. Do animals have to learn to organize their perceptual world, or is their recognition of external stimuli INNATE? This question is a controversial one, with regard both to the way in which investigations are carried out, and to the interpretation of the results of such investigations.

The easiest way to investigate the ONTOGENY of perceptual processes is to rear young animals in the partial or total absence of sensory input. Pigeons (*Columba livia*) raised in the dark, or with translucent goggles which admit only diffuse light, subsequently take longer in LEARNING to distinguish between differently shaped objects than normally reared pigeons. Kittens (*Felis catus*) reared in darkness or diffuse light between the ages of 18 and 28 days show deficiencies in the *placing response*. Normally, when a kitten is held in the hands and lowered towards a surface, the forepaws are extended before contact with the surface occurs. Visually deprived kittens do not show this response at 28 days of age, the time of its appearance in normal kittens. However, only 5 h experience of patterned light is necessary for the kittens to show the correct response.

The interpretation of sensory deprivation experiments is open to criticism on a number of grounds. Firstly, animals allowed to see normally for the first time in their lives may experience some EMOTION that impairs their performance in experiments. When rats (*Rattus norvegicus*) which have learned to run a maze in the dark are tested in the light, their performance deteriorates. Similarly, rats trained in the light perform less well in the dark. Secondly, an animal deprived of visual experience may come to rely more on other senses, such as HEARING and touch. When allowed visual experience it may be difficult for the animal to change its previous habits. Blind people become very skilled in the use of non-visual senses, and may retain these abilities if they subsequently become sighted. Thirdly, an animal deprived of normal visual input may suffer deterioration of those parts of the BRAIN responsible for the interpretation of visual information. When eventually tested in a visual task, its poor performance may be due to nervous degeneration rather than lack of opportunity to learn. In one experiment, a number of kittens were reared in darkness from birth to 5 months of age. From 2 weeks, when their eyes opened, they spent 5 h each day in an environment so arranged that they could see only horizontal or vertical stripes. At 5 months the kittens were put into normal surroundings, and, after an initial period of disorientation, they were able to cope with the change. However, their VISION remained abnormal. When a kitten raised in a vertically striped environment was put together with another raised in a horizontal environment, marked differences could be seen. If a rod was

held vertically, and moved about, the vertically raised kitten would play with it, but the 'horizontal' kitten would ignore it. When the rod was held horizontally the vertical kitten behaved as though the rod had suddenly disappeared, while the horizontal kitten now played with the rod. Subsequent physiological experiments showed that there were differences in the brains of kittens raised in the two environments, which could be correlated with the type of visual environment that the kittens experienced during their early life.

The deficiencies of sensory deprivation experiments can be partially overcome by providing animals with abnormal sensory experience, or by simply observing how much ATTENTION young animals give to various features of their environment. Rats reared with three-dimensional objects which they could explore both visually and tactually performed better in two-dimensional visual DISCRIMINATION tests than rats reared with two-dimensional shapes that they could explore only visually.

The evidence suggests that EXPLORATION and PLAY are important in enabling animals to learn to refine their perception and appreciation of the visual world, and, in particular, to correlate different sensory modalities. For example, kittens that are not allowed to move in relation to their visual world, and so have only passive visual experience, do not develop the normal paw-placing response. On the other hand, studies of SIGN STIMULI and of the ATTENTION that young animals pay to features of their environment show that the ability to discriminate between, and respond to, particular features of the external environment is often present from birth. When visual preference is assessed by measuring the relative duration of visual fixation on various visual targets, it is found that young animals often have marked preferences which imply that they can discriminate between the objects. Infant rhesus macaques (*Macaca mulatta*), chimpanzees (*Pan*), and human beings can discriminate certain patterns from birth, without the opportunity for perceptual learning. They show preferences which change with age, and which sometimes correspond to natural stimuli. For example, at about 4 months of age human babies prefer to look at a simple picture of a human face, compared with the same visual features rearranged into a haphazard pattern (Fig. A).

99

Fig. A. Stimuli used for testing face-recognition in human infants.

PHEROMONES are CHEMICAL messengers which provide what is perhaps the most primitive form of COMMUNICATION. Bacteria and unicellular organisms utilize chemicals to detect sources of food in the way that their ancestors presumably did when the first motile cells appeared on earth. As complexity increased and SEXUAL reproduction emerged, chemicals were utilized to enhance the chances of contact between the sexes. From such primitive systems in free-living single cells, aggregations developed, and the original metazoans made their appearance with the beginnings of cellular specialization. The existing slime moulds are an excellent example. The slime mould *Dictyostelium discoideum* exists as free-living unicellular amoebae in moist soil. Under certain environmental conditions, such as a decline in the bacteria upon which they feed, the individual amoeba secretes a specific chemical, which attracts neighbouring amoebae. The amoebae congregate to form a multicellular slime mould which shows rudimentary cellular differentiation, with a basal portion and a stalk containing a fruiting body. As EVOLUTION proceeded, chemicals took on important roles as messengers acting to attract, repel, and to identify the sex and status of individuals, and also to control another individual's physiological processes.

The term 'pheromone' was first applied to such chemical messengers in 1959. Its Greek roots mean 'to transfer excitement'. Considerable controversy has surrounded the definition of 'pheromone' and debate has occurred over the types of chemical signals to be included under this term. The definition currently accepted in the scientific community is a chemical or mixture of chemicals that is released into the environment by an organism, and that causes a specific behavioural or physiological reaction in a receiving organism of the same species. Pheromones have been subdivided into two types: *signalling pheromones* refer to substances that induce a behavioural response, usually of an immediate nature, and *priming pheromones* refer to substances that induce a physiological change, usually of long duration.

Signalling pheromones are widespread in the animal kingdom and serve as (i) sex attractants, (ii) coordinators of copulatory behaviour, (iii) aggregation promoters, (iv) trail substances, (v) ALARM substances, and (vi) advertisements of individual identity, SOCIAL status, and territorial boundaries. Much of our knowledge of signalling pheromones comes from detailed studies of insects but investigations have revealed that they play an important role in other animals as well.

Sex pheromones are chemicals that are specific to the species, disperse readily, yet have a long fade-out time. *Bombykol* from silkworm moths (*Bombyx mori*) was the first chemically defined sex pheromone, and others that followed were hydrocarbons with similar characteristics. The elegant complexity of how moths are drawn together to copulate is exemplified by the red-banded leaf-rolling moth, *Argyrotaenia velutinana*. The sex pheromone from the female effluvium in this moth elicits long-distance ORIENTATION to the female, enhancing the frequency of landing, wing fanning, and walking to the chemical source by the male.

The quantity of a pheromone present is important in inducing a response. In attracting the cabbage looper (caterpillar), *Trichoplusia ni*, to traps, the response is maximal at a release rate of 60 μg/h. Responses decrease at rates of 180 μg/h and 2 μg/h. Some species respond best to only very low release rates. For example, the red bollworm (moth), *Diparopsis castanea*, responds optimally at only 0·7 μg/h. Responses of insect sensory receptors are remarkably sensitive, especially to sex pheromones. Dilution in the air can be so great that only a few molecules of a pheromone reach a receptor, yet a perceptible effect can be detected.

Sex pheromones are not limited to insects. Observations of mammals in the wild often suggest that pheromones are involved in attracting males to receptive females, but only a few cases have been verified experimentally. Rats and mice (Murinae), and gerbils (Gerbillinae), have the ability to discriminate between males and females, and the male rat can determine whether the female is receptive. Olfactory stimuli also play a role in species identification in several rodents, and may be important as ISOLATING MECHANISMS. An animal's preference for its own odour appears to be the result of olfactory experiences encountered in early DEVELOPMENT, rather than the result of INNATE preferences.

Reactions to pheromones are probably most complex among primates, and LEARNING plays an important role. The female rhesus macaque (*Macaca mulatta*) emits a chemical signal at the time of ovulation that attracts the male's interest, and leads to mounting behaviour. Suggestions have also been made that chemical communication plays an important role in human sexual attraction and COPULATION. Perfumes have been used throughout history, suggesting that there are rewards for their use. As yet, however, no substance meeting the definition of a pheromone has been identified in human beings.

REPRODUCTION is only one outcome of pheromonal mechanisms bringing animals together. AGGREGATION (assemblage) of many social insects is achieved by chemical communication. The termite *Kalotermes flavicollis* congregates as a result of a pheromone, which is produced by the action of an intestinal protozoan (unicellular animal) living in the termite's hind gut. In the honey-bee (*Apis mellifera*) assemblage occurs as a result of a secretion from the *Nasanov gland*. This pheromone is emitted by a bee that has found a new food source, or if it has lost contact with its companions for a time.

Disturbed honey-bees emit an ALARM pheromone, which excites and attracts other bees. If the disturbance becomes so great that the bee stings its enemy, the sting is torn from the bee's body exposing the poison gland containing more pheromone. This sudden increase in pheromone excites additional bees to enter the fray, resulting in a swarm of bees around the enemy. Somewhat more complex alarm systems exist in many species of ants. For example, when the fire ant *Solenopsis saevissima* is injured or restrained, it simultaneously secretes an alarm substance from the head and trail substances from *Dufour's gland* in the abdomen. Other ants in the area are excited by the alarm substance and attracted to the location of the threat by the trail substances.

In mammals there are many examples of the use of pheromones in recognizing individuals, and their age or social status. Secretions from the tarsal gland of the black-tailed deer (*Odocoileus hemionus columbianus*) are used in this species for individual recognition, and in the pronghorn (*Antilocapra americana*) a pheromone is used to advertise territorial boundaries (see SCENT MARKING). Although we remain ignorant of the chemicals involved, it is apparent from behavioural testing that many species of rodents use pheromones to detect the social status of potential opponents. Urine from adult, gonadally intact male mice elicits attack, whereas that of castrated males or females is either inactive, or may actually be able to inhibit AGGRESSION. It is clear from these results that the emission of pheromones controlling aggression is controlled by the levels of the HORMONE testosterone secreted by the testes. There is an intimate relationship between hormones and pheromones, with each of these classes of chemical messengers able to affect the other.

Priming pheromones differ from signalling pheromones in that they act upon an organism to modify development or adult physiology. The response upon presentation of the pheromone extends over a relatively long period of time, and any behavioural consequences that result occur secondarily in response to hormonal or other changes.

Priming pheromones play an important role in the control of reproductive development in a variety of species. Among social or aggregating insects reproductive synchrony is mediated by chemical signals. An unknown substance produced by mature males of the locust *Schistocerca gregaria* hastens the MATURATION of both sexes, thus maximizing the number of reproductively mature individuals present. Such reproductive synchrony may be partly responsible for the plague-like outbreaks seen in this species. Similar priming pheromone effects have been noted in the darkling beetle (*Tenebrio molitor*), and it is likely that priming pheromones play an important role in synchronizing reproduction in many species that aggregate in dense populations. In the truly social insects such as bees, ants, and termites, the queen produces a substance, appropriately named *queen substance*, which inhibits ovarian development among members of the group. Queen substance results in the attenuation of reproductive development of the workers, and forms the basis of the caste system.

In addition it has a variety of behavioural effects. The effect of queen substance appears to be mediated by olfactory responses to the airborne molecules, as well as by mechanical transmission and absorption directly into the bodies of the workers. This brings up an important point about priming pheromones; they may act directly on the physiology of the recipient after absorption or ingestion in addition to acting through the olfactory system (see TASTE AND SMELL). In other words, pheromones need not be odours but can also be chemicals that have communicatory functions through being absorbed or being ingested.

Priming pheromones also play an important role in the lives of many mammals. Much of our understanding of the importance of priming pheromones in mammals comes from studies on the reproduction of the LABORATORY mouse (strains of *Mus musculus*). Three priming effects have been discovered: control of puberty, stimulation of oestrus, and blockage of pregnancy.

SEXUAL maturation in the mouse is an inexorable process given normal nutrition and the absence of disease. However, the chronological age at which a mouse attains puberty is dependent, within limits, upon social stimuli. The presence of adult male mice or their urine accelerates puberty, and the presence of females or their urine delays puberty. Acceleration of female puberty by the male has been extensively examined, and the results of this work provide an insight into the complex interplay between hormones, pheromones, and behaviour in regulating reproductive functions. The urine of adult males hastens the onset of puberty in JUVENILE females, but if the males are castrated their urine rapidly becomes ineffective. Injection of the normal amount of testosterone into a castrated male restores his ability to hasten puberty in juvenile females, indicating that the male's ability to control the female's sexual maturation is dependent upon his testosterone level. Testosterone level in the male can be increased by social DOMINANCE, and it has been found that males losing fights with dominant males also lost their ability to hasten puberty in females. Thus, SOCIAL INTERACTIONS between males can have an effect on the rate at which females in a population mature. The male pheromone acts upon the female to induce an elevation of luteinizing hormone from the pituitary gland. This hormone, in turn, stimulates ovarian activity, which brings about the changes associated with puberty.

Working to balance these acceleratory factors is a urinary cue from females in dense populations that can inhibit the onset of puberty in females. The urine of a female living alone in a cage, or living in a sparse natural population, is without effect on the sexual maturation of other females. When, however, females are crowded, either in a laboratory cage or in a dense natural population, their urine changes and now contains a pheromone that delays puberty in other females (i.e. the first time they come into oestrus or 'heat'). It is apparent from these findings that the precise timing of the onset of puberty in mice can be controlled by accelerator factors from the male, and inhibitory factors from the female. A balance between these factors, both influenced by social conditions, can set the average age of puberty in a mouse population, and thereby have a strong effect on population.

Priming pheromones can also influence the oestrous cycles of adult female mice. Crowding adult females induces a period of anoestrus in which the females' ovaries may remain quiescent for 40 or more days rather than showing the typical 4- or 5-day cycle. The suppression of oestrous cycles in crowded females can be quickly overcome by isolation from the group, or by exposing the females to an adult male or male urine. Synchronization of oestrus by male stimuli is not limited to the mouse. Similar effects have been noted in sheep (*Ovis*), goats (*Capra*), and cows (*Bos primigenius taurus*). There is also evidence that the human female's MENSTRUAL CYCLE can become synchronized among women living in close proximity, as in college dormitories. The cues responsible for this synchrony are not known, but, from the work done on animals, it seems quite likely that a pheromone is involved.

One of the most interesting of the priming effects on mouse physiology is that of pregnancy blockage. Female mice are typically very fertile. When mated, 90–100% of females become pregnant in laboratory strains. However, when a strange male, i.e. a male other than the stud, is placed with the female within 4 days of insemination, as many as 40–80% of the females will fail to continue their pregnancies. This blockage of pregnancy occurs because the fertilized eggs fail to implant in the uterine wall. Exposing newly inseminated females to the urine of a strange male induces the same effect, indicating that a chemical cue mediates this phenomenon.

It is clear that pheromones are potent conveyors of messages and are used extensively in the animal world to modify behaviour as well as physiology. However, upon analysis, it is clear that chemical cues have some major disadvantages as messengers. They are relatively slow and dependent upon currents of air or water or, in some cases, upon mechanical transfer of the chemical. Another disadvantage is that they cannot be transmitted as abrupt on or off signals. If a message cannot be turned off, it may be left on too long, resulting in confusion. For certain forms of communication, however, pheromones have clear advantages over other sensory cues. They are active in very small amounts, and are energetically cheap to manufacture; in fact they are very often the 'debris' of metabolic processes which otherwise might have gone to waste. Pheromones can convey highly specific messages, which in many cases trigger behaviour specific to the species, or physiological changes essential to the survival of the species. Lastly, pheromones can convey a lasting message,

such as a territorial boundary, or the presence of a reproductively active female in the area. Given these advantages, it is little surprise that pheromones play such an important role in the lives of animals.   J.G.V.
13

PLAY. Leaping, jumping, bucking, or running when there is no obstacle to overcome, no enemy to flee, or object to attain; sniffing, licking, pawing, and manipulating familiar rather than novel objects; sex without coition; fighting in friendly rather than aggressive encounters which avoids injuring or routing the partner: these are instances of activities typically called 'play'. They occur in most mammalian species, mainly, although not exclusively, in JUVENILES, and are more frequent and more varied in the more highly evolved species.

Problems of definition and theories of play. Play activities have only recently begun to receive systematic study. The long neglect stems partly from the difficulty in defining them. In ordinary LANGUAGE the term refers to all activities that apparently have no use or FUNCTION, and appear to be undertaken for pleasure. But these are not workable criteria. Almost any behaviour of which an animal is capable may appear in play; no common activity is unique to it. Quite apart from the practical difficulty in accurately identifying an animal's mood, it is still very much open to question whether a specific emotion, 'pleasure or joy', is unique to and accompanies even all human play. However, the main problem has been that play activities fit badly into the system of CLASSIFICATION that has been useful for the more obvious behaviour patterns of adult animals. Traditionally behaviour sequences are defined in terms of specific GOALS and states of MOTIVATION which are assumed to propel or 'drive' the sequences to their end-states. Advances in studying the energy exchanges between the organism and its environment, and the complex external and internal factors that govern behaviour, have led to the demise of the concepts of DRIVE and INSTINCT. Nevertheless, the traditional method of classification has been both pervasive and persistent, because it seemed to answer the important question: why do organisms behave as they do?

Because of the difficulty in defining play in terms of obvious goals, there have been a great many theories, each attempting a global explanation but in reality deriving plausibility only from specific aspects or categories of play. Several variants of Herbert Spencer's theory of 'Surplus energy' suggest that play results from overflow of too much ENERGY or leisure. A more modern version has linked play to unpleasantly low levels of AROUSAL. It implies that animals actively seek stimulation and information in conditions of boredom. Another notion favoured by some LEARNING theories regards play as a repetition of activities that have been found to be rewarding in other situations. Another solution has been to assume special instincts or drives for EXPLORATORY behaviour, manipulation, and play, and to argue that these are, in all respects, similar to fighting, feeding, and sex. Perhaps most popular have been various restatements of the practice theory originally advanced by Karl Groos in the late 19th century. This held that play is necessary for adaptable animals with few completely inborn activities, so that they learn to adjust to other animals and to assimilate information and perfect skills for later use. This learning occurs during a period of protected infancy, which shields them from harmful consequences of their actions. The frequency of apparently IRRELEVANT activities in widely divergent species is an argument for their having some function.

In behavioural sciences as in other sciences, the causal question 'why' can only be answered by a series of detailed studies of 'how' given phenomena operate, and what conditions are necessary and sufficient for their occurrence or change. *A priori* definition and global theories are largely irrelevant. Specific, testable hypotheses are required for this approach. A major impetus to study activities that are incomplete or appear out of their normal sequence has come from ETHOLOGY. This favours detailed observation of animals in their natural environment, as well as in more restricted conditions (see FIELD STUDIES; LABORATORY STUDIES).

Characteristics and categories of play activities. It is not easy to recognize play in animals. Incomplete or out-of-sequence behaviour has often been called play, but may more usefully be studied separately. DISPLACEMENT ACTIVITIES, such as the feeding that domestic cockerels (*Gallus g. domesticus*) do during a fighting bout, may intervene when the ongoing activity has received some check, but generally require the presence of appropriate stimuli. Parts of a recognizable sequence may occur when some, but not all, of the external or internal conditions for it are present. When sufficiently satiated, a domestic cat (*Felis catus*) will 'toy' with a house mouse (*Mus musculus*) without eating or even killing it. NEST-BUILDING movements may occur when an animal is in the appropriate condition, but has no access to materials. An activity may be misinterpreted due to the ignorance of the observer. For example, at one time fish seen leaping over floating objects were considered to be having 'fun'; later, however, it was found that the activity actually scrapes off parasites, and is therefore more accurately classified as GROOMING behaviour.

Several characteristics have been noted as symptomatic of play activity. These do not constitute a definition, but may serve to distinguish play observationally. They may be summarized briefly as follows: (i) Play is VOLUNTARY behaviour. A simple REFLEX cannot be called play, and the term cannot apply to activities under external constraint. For instance, a lion (*Panthera leo*) made

to jump through a hoop cannot be said to be playing. The notions of pleasure and fun implied by the term relate to its meaning as activity which is both voluntary and free from constraints. Not all voluntary activity is play. The following criteria delimit it further. (ii) Play is apparently paradoxical behaviour. The activity is not only dissociated from the behaviour sequence and goal that normally define it, but occurs in a context that is inappropriate, incongruous, or contradictory in relation to that goal. For example, aggressive actions like chasing, shaking, hitting, and biting are used in FIGHTING to hurt, incapacitate, and rout an enemy or competitor. They are labelled play only when they occur in the paradoxical context of friendly SOCIAL encounters and keep the animals together. Sex is considered as play only if mating is impossible or excluded. To say that an object is played with typically implies that the object is used inappropriately, and is not put to its normal use. (iii) Play bouts are often apparently random sequences in which a number of activities from functionally different adult behavour systems follow each other rapidly. (iv) Certain activities may be executed repeatedly; the behaviour may have undergone RITUALIZATION, or it may have rules that are apparently specific to a given bout. (v) Exaggerated forms of normal movements are frequent. (vi) Play bouts in which two or more animals interact are typically preceded or accompanied by special signals.

Not all of these characteristics of play necessarily occur together, and the question whether they cover a unitary type of behaviour is not important. The fact that there is one word for it in ordinary language is not sufficient to decide whether all play is governed by the same mechanisms and conditions. Classification is a matter of convenience depending on facts uncovered by study. At least six main subdivisions may be considered.

Categories of play activities. 1. *Superfluous activity.* Prancing, frisking, leaping, gambolling, and other exaggerated movements have been observed in cattle (Bovidae), horses (Equidae), sheep (*Ovis*), goats (*Capra*), some wild animals, monkeys and apes (Simiae), and human children. They are associated with moderate environmental changes, such as release from short periods of confinement, someone running alongside, and other factors favouring heightened arousal. They are more common in the young, and vary with SEXUAL state, health, and the weather. Lack of both stimulation and exercise are known to have adverse effects on health and development. It is not yet clear how much play relates to physiological systems concerned in the maintenance of muscle tone and motor development. FLIGHT frolics of flocks of birds also depend on atmospheric changes to which birds are sensitive, but this is apparently not related to maturity.

2. *Aimless exploration, manipulation, and object play (diversive exploration).* Novel stimuli and objects typically elicit approach, touching, snif-

fing, mouthing, and other manipulation, provided the animal is not frightened. The fact that this behaviour declines with familiarity over short periods (see HABITUATION) has served to distinguish specific 'what is it?' exploration experimentally from 'diversive' exploration, in which the animal actively seeks stimulation or information by engaging in a variety of activities with familiar objects, sometimes incorporating them into repeated games. This is most varied and persistent in young primates, and is associated with the complexity of the SENSE ORGANS and of the BRAIN mechanisms involved in detecting, selecting, and storing information about the environment. Baboons (*Papio*) may carry bags and baskets, chimpanzees (*Pan*) may punch holes in leaves and look through them, or bang hollow tins and make a noise. The extent to which new uses of objects are discovered probably depends on chance and INTELLIGENCE. The now famous chimpanzee Sultan fitted small sticks together to make a TOOL that could rake in a banana (see PROBLEM SOLVING). A few such examples have been observed in the wild, but object play is more frequent in CAPTIVITY. Some objects seem to derive their popularity from being familiar: one European badger (*Meles meles*) played endlessly with a brush and slippers he had SCENT marked; a juvenile panda (Ailuridae) took a toy hoop to bed with him.

3. *Practice play.* The popularity of this notion probably stems from observations that some animals whose skill in movement is not perfect at birth seem to repeat and elaborate newly acquired and chance actions. This may relate to the establishment of the control and FEEDBACK systems governing such skills, rather than to systems involved in providing stimulation or information. It is not clear how far play is a necessary rather than a facilitating condition for the smooth execution of complex learned movement patterns. LEARNING is, of course, involved in the more complicated games of the higher primates.

4. *Responses to the wrong object.* INNATE stereotyped movements to inappropriate objects are often found in young animals. Such objects usually have general characteristics, such as movement, size, or brightness, to which the animal is innately sensitive, and which have to be narrowed down to appropriate objects as a result of experience. For example, chasing and pouncing upon a rolling ball of knitting wool is typical play in kittens.

5. *Social play.* Play between young and parents is less common than play amongst coevals, and its frequency varies between species and with living conditions. In CAPTIVITY, or when isolated from other animals, chimpanzee mothers may initiate play with an infant by hugging, grasping, or tickling it. Play with adults initiated by the young has been observed in howler monkeys (*Alouatta*), langurs (*Presbytis*), and baboons. Adult males tolerate or respond to such approaches from juveniles. However, play between young animals is by

far the most frequent category of social play, and is most often reported as play-fighting. Its form and amount varies in different species. Pursuit and fighting movements occur mainly in PREDATORY animals. Movements from different adult sequences follow each other rapidly. Bull-calves will suddenly frisk, shake their heads, butt, and attempt to mount each other. DEFENSIVE actions, DISPLAYS, and sexual fighting are shown in rapid succession by young mongoose (Herpestinae). The frequency of different types of movement varies in male and female Steller sea lions (*Eumetopias jubata*). Parts of mating patterns occur in the young of most species, but less frequently than wrestling, hitting, biting, or other fighting movements. Young monkeys spend most of their time playing with each other, the games varying from simple wrestling and chasing by young howler monkeys, to the far more elaborate games of chimpanzees. Indeed, some COOPERATIVE games with simple rules like 'taking turns' have been observed in gorillas (*Gorilla gorilla*).

One exciting discovery is that play-fighting is preceded by special signals, not used in aggressive encounters. Infants of the howler monkey and the gibbon (Hylobatidae) chirp at each other, rhesus macaques (*Macaca mulatta*) look at their partners upside-down through their legs. Exaggerated or clumsy movements are common. Most widespread is the play-face, a relaxed, open-mouthed expression (see FACIAL EXPRESSIONS) used by chimpanzees and human children. But the black bear (*Ursus americanus*) and other mammals also have special postures for soliciting play. Signals such as these rarely occur in solitary play. Some may be specific to play-fighting, indicating that what follows will be a friendly bout (see COMMUNICATION).

It has been suggested that social play may serve to establish DOMINANCE relationships, or control AGGRESSION between group members, or develop group cohesion. The evidence so far suggests that this differs between species. Dominance hierarchies develop differently even among different breeds of domestic dog (*Canis l. familiaris*). Predatory, carnivorous animals who hunt in packs usually have recognizable dominance patterns. These may be established in infancy in play-fighting while canine teeth are still absent, and then persist throughout adulthood. However, in free-ranging baboons the mother's rank is also important in determining an infant's position in the HIERARCHY. In gorillas, dominance rank during play can be different from that for food. It seems possible that play-fighting which is delimited by signals that show the context to be friendly has different functions from play-fighting which falls short of actual harm only because the protagonists do not as yet have an adult set of teeth. Rough-and-tumble play, and games that ensure physical contact between animals, possibly constitute yet another subdivision, and cooperative games with rules, as in gorillas, may need to be distinguished further. Social play is probably important in the development of SOCIAL RELATIONSHIPS. However, the precise relation of various types of social play to MATURATION levels, environmental conditions, and adult activity for different species can only be determined empirically.

6. *Pretend play*. It has been suggested that play is behaviour in the simulative mode, and that it exhibits the non-literal use of resources. This is over-inclusive: frisking, gambolling, and contact play are hardly pretence. But some animal play does look like pretence: for instance, attacks that are signalled as friendly, or the inappropriate use of objects, such as the kitten that pounces on a ball of wool instead of on a mouse. For animals other than human beings this does not seem to require a separate category. The category is needed, however, for the play of human young, and seems to relate to the development and use of symbols. It has been shown that in children bouts of social play with imagined events are delimited from periods in which information is exchanged by tone of voice, gestures, and amount of repetition. Play with symbols, such as substituting a dish-cloth for a baby, or re-enacting changed versions of past events in either solitary or cooperative games, is typical of children. Whether similar pretend games occur in other species is difficult to tell.

The study of play. Methods of investigation necessarily depend to a large extent on the questions asked. The study of play has been somewhat hampered by the apparent need to demonstrate the function or usefulness of play. The usual method of experimentally depriving the animal and assessing effects is hardly possible. Young animals cannot be deprived of play without depriving them also of other activity, social contact, and stimulation. However, the young animal's environment can be modified. The great majority of investigations of play have been observational studies, for instance providing an enriched environment and comparing the play activity with that observed in an impoverished environment. Most of these studies have concentrated on social play, and some of the more sophisticated use check-lists of specific movements, and observe the kind of controls that are used in ethological studies.

A great deal of information is being accumulated about play in different species, although observational studies have also suffered from uncertainties about the criteria that distinguish social play from other forms of social interactions between young animals. This has led to difficulties in the interpretation of results. Observations geared to relatively restricted questions have yielded some unexpected results. For instance, social cohesion was observed within a troop of squirrel monkeys (*Saimiri sciureus*) during a period of food scarcity, despite the fact that the young did not engage in social play under these conditions. This apparently contradicts an often-supposed correlation between social play and social cohesion.

In the light of our current knowledge, the most reasonable assumption about play activities is that they are related in some manner to the development and acquisition of the control systems that regulate different types of behaviour. This view would account for the difficulty in describing play activities in terms of a unique behaviour, or a common goal. It enables us to explain the fact that play is most obvious in species whose behavioural repertoire is not fully established at birth, and who have the most complex nervous system, sensory apparatus, and behavioural mechanisms, and that although play occurs most frequently during infancy, it is not confined to young animals. Even mature animals may have to adapt to new situations or learn complex new skills. Precisely how play activities relate to the various control systems is not known, but this hypothesis allows us to link, for instance, superfluous activity with ONTOGENY, practice play with feedback in sequencing complex skills, pretence with COGNITION, diversive exploration with the acquisition of information, and social play with the structure of social relationships in a given species. It may be that play subserves many functions that are important in the development of behaviour in species that have a flexible behaviour repertoire at birth.   S.M.

105

PREDATION is a relationship between animals, in which one, the predator, benefits, while the other, the prey, is affected adversely. This situation is in contrast to other types of species INTER-ACTION in which the relationship is symmetrical, for example in COMPETITION, where both participants are affected adversely. Predation implies that the prey is killed, and usually eaten, by the predator, and it thus differs from PARASITISM, in which one animal exploits another but generally does not kill it.

Although predation usually occurs between members of different species, it may take the form of CANNIBALISM in which an animal kills and eats a member of its own species. For example, herring gulls (*Larus argentatus*) prey upon the eggs and chicks of their neighbours during the breeding season. Some individual gulls specialize in this method of obtaining food. However, the benefit gained from predation upon members of one's own species often takes the form of increased reproductive success, in contradistinction to preying upon other species, which is normally undertaken purely as a means of obtaining food. For example, when a male lion (*Panthera leo*) takes over a pride, he generally kills all the cubs. He then mates with the females, having created a situation in which his own offspring will have an enhanced chance of survival. Infanticide of this type also occurs in the entellus langur (*Presbytis entellus*). Marauding bands of nomadic males raid a troop, drive off the resident males, kill all the juveniles, and quickly mate with the females.

For many species, preying upon other species is the only means of obtaining food. Thus NATURAL SELECTION will tend to increase the predator's efficiency at finding, capturing, and eating its prey, especially when the abundance of suitable prey is low in relation to the size of the predator population. However, the greater the level of predation, the greater the pressure of natural selection upon the prey, which respond by evolving more effective DEFENSIVE behaviour. Thus there can develop a kind of escalating evolutionary 'arms race', which can lead to very refined ADAPTATION on the part of both predator and prey. For example, the little brown bat (*Myotis lucifugus*) feeds by catching insects while flying in the air. The bat detects a flying insect by ECHO-LOCATION (see also ORIENTATION) from a distance of about 80 cm, and attempts to intercept the flight path. Some insects, such as the moth *Prodenia eridania*, have specially sensitive organs of HEARING, that are able to detect the ultrasonic cries used for echo-location by bats, and when they hear a bat coming close to them they are able to take evasive action. However, the bats have developed a counter-tactic. Instead of merely attempting to catch the insect in the mouth, they make use of their FLIGHT membranes to effectively enlarge the area of the open mouth. The insect is scooped up by the membranes of the tail or wing, and transferred to the mouth by means of an intricate mid-air manœuvre.

The PREDATORY BEHAVIOUR of animals is often finely attuned to the characteristics of a particular prey. For example, the ichneumon mongoose (*Herpestes ichneumon*) has special tactics for killing poisonous snakes, such as the cobra (*Naja*). It lunges at the prey time and time again, and elicits retaliatory strikes which tend to exhaust the prey. Many of the lunges are mock-attacks carried out with no true biting intent. However, when the snake begins to tire, the mongoose dodges about repeatedly to confuse the aim of the snake, while it attempts to manœuvre into a position from which it can deliver its deadly bite at the base of the snake's head.

The defensive behaviour of the prey can, equally, be tailored to the habits of particular predators. For example, Thomson's gazelle (*Gazella thomsoni*) has different *flight distances* for different predators, which seem to correspond to their respective pursuit abilities. The gazelle flee from African wild dogs (*Lycaon pictus*) when they are about 500–1000 m away, but they will not flee from cheetah (*Acinonyx jubatus*) or lion until they are at a distance of about 100–300 m. The flight distance is 50–100 m in response to the spotted hyena (*Crocuta crocuta*) and only 5–50 m in response to jackals (*Canis mesomelas*). Although there is some variation in flight distance, according to the age, sex, and SOCIAL status of the gazelle, there does seem to be a good correlation between the hunting capabilities of the predators and the response of the prey. Thus, while the cheetah is the fastest of all land animals over distances up to about 500 m, wild dogs are noted for their tremendous stamina, and ability to maintain a chase for many kilometres. In the Serengeti National Park, East Africa, Thomson's gazelle is a common prey of cheetah, but the victims are usually juveniles. Jackals probably cannot run down an adult gazelle, and rarely prey upon them. Hyenas take a fair number, often by surprising animals at rest.

Predation can have profound effects upon population size. For example, the population of the Californian vole (*Microtus californicus*) is kept in check by feral house cats (*Felis catus*) in the hills behind Berkeley and Oakland, California. The voles can breed very fast, and their population rises even though they are preyed upon by the cats. However, they are forced to stop breeding for about 6 months of every year due to regular seasonal fluctuations in CLIMATE. At this time the predation by cats reduces the vole population considerably. The voles are then harder to find, and the predators tend to switch to other prey, such as wood rats (*Neotoma*) and gophers (*Thomomys*). When the voles start to breed and increase their numbers again, the cats turn back to them.

Clearly some kind of balance must be achieved in the complex interactions between predator and prey. In the more simple situations, such as in the Arctic, a predator may depend almost exclusively upon one species of prey. Thus, in Alaska and Northern Canada, wolves (*Canis lupus*) prey primarily upon the herds of the barren ground caribou (*Rangifer tarandus arcticus*). The wolf population is probably controlled by the fact that, although they successfully prey upon the juvenile, sick, and old caribou, most of the healthy adults escape. The situation is much more complex in the tropics, where there is usually more than one species of predator, and numerous alternative prey. In the Serengeti there are six or more important predatory mammals, which are in competition with each other for a similar number of common prey species. Such situations generally lead to SPECIALIZATION, in which each predator becomes adapted to a particular method of HUNTING. Thus the lion and hyena are specialized for taking large prey, such as white-tailed gnu or wildebeest (*Connochaetes gnou*) and Burchell's zebra (*Equus burchelli*), the lion by stalking and the hyena by coursing. Cheetah are the coursing specialists *par excellence*, but they are unable to kill larger animals, such as wildebeest or the African buffalo (*Syncerus caffer*). The somewhat similar leopard (*Panthera pardus*) is a stalker which also specializes in smaller prey. Wild dogs are highly specialized as coursing predators, and are sometimes able to tackle large prey by virtue of their socially well-coordinated predatory behaviour. The jackals are primarily scavengers, and generalized predators on smaller game.

The effect of predators upon prey populations is not entirely deleterious. For example, the puma (*Puma concolor*) of the Americas helps to regulate the numbers of deer (Cervidae). In the Kaibab Forest of Arizona, where pumas were eradicated by man, the deer population grew so large that all food supplies were used up, and thousands of deer starved to death. Predators usually take a higher proportion of the lame and the sick than of healthy animals. Thus, by removing diseased individuals, they help to reduce infection by parasites. By removing the congenitally deformed, they may prevent these individuals from breeding, and passing their undesirable traits into the population. It is sometimes suggested that the predator should crop its prey selectively, so as to maximize the reproductive rate of the prey, and increase the corresponding yield to the predator. Although man has the capability for such 'managed' predation, it is not generally thought to occur in animals. This is because the individual predators are always in competition with each other, so that any individual that evolves a trait for ALTRUISM, that leads him to pass over certain prey opportunities for the future benefit of his fellows, would be less successful than other individuals. Thus any animal who 'cheated' by grasping all opportunities would subsequently produce more offspring than its altruistic rivals.

28

PREDATORY BEHAVIOUR is the behaviour by means of which an animal of one species, the predator, kills and eats a member of another species, the prey. It can be distinguished from CANNIBALISM in which an animal kills and eats a member of its own species, and from PARASITISM in which one animal exploits another, but generally without killing it.

The motivation of predatory behaviour. Generally, HUNGER provides the MOTIVATION for predatory behaviour, and the speed and efficiency of prey capture increase with hunger. However, this is not always so. In some species, such as the praying mantis *Hierodula crassa* and the jumping spider *Epiblemum scenicum*, the stereotyped movements of prey capture remain relatively unaffected by hunger. Other aspects of predatory behaviour, such as VIGILANCE, are affected by hunger in these species.

Many species may exhibit an APPETITE for a particular prey species, which has no very obvious relationship with normal hunger. There are a number of different aspects of predatory behaviour which need to be considered with this in mind.

1. *Choice of prey.* Choice of prey is usually dictated by the availability of the prey, but many species will feed for a period upon a particular prey, and then suddenly switch to another type of prey, even though both prey types are available. For example, the redshank (*Tringa totanus*) is a wading bird whose FORAGING upon the shore is somewhat selective. When feeding upon the marine polychaete worm *Nereis* it tends to maximize its rate of food intake by taking large worms whenever possible, and passing over small worms. When feeding upon the crustacean *Corophium*, a type of shrimp, it moves up and down the beach selecting areas in which the prey are abundant. When both worms and shrimps are available, it appears that redshank prefer shrimps, in that they are increasingly unlikely to take those worms which they encounter as the availability of shrimps increases. This is surprising because it has been shown that redshank gain energy more quickly by feeding upon worms. Individual redshank may switch from one prey to the other, and, within a particular area, some redshank may be seen feeding upon worms while the majority are feeding upon shrimps. In general, however, shrimps are preferred, even though they are less profitable in terms of energy. It seems likely that the shrimp contain particular nutrients that the redshank require, and that their predatory behaviour is adjusted to this. Studies of FOOD SELECTION show that many species are able to learn to recognize and select those foods which suit their physiological requirements. Having fed upon one prey for a period of time, many predators then switch to another type of prey. For example, a clan of spotted hyenas (*Crocuta crocuta*) may hunt the brindled gnu or wildebeest (*Connochaetes taurinus*) for many days, and then suddenly switch to Burchell's zebra (*Equus burchelli*). This kind of sudden change is not simply due to changes in the chance encounter of prey, but seems to be due to a change in preference. Hyenas intent on hunting Thomson's gazelle (*Gazella thomsoni*), or wildebeest, tend to form a pack of two or three, whereas they form packs of up to twenty-five individuals when they are hunting zebras. Moreover, prior to hunting zebra, the hyenas go through elaborate greeting ceremonies as the pack assembles, and they may pass by numerous wildebeest and other potential prey without paying any attention to them. When a wildebeest is hunted down, the pack size rapidly increases, as other members of the clan join in to share the booty. This does not happen when a zebra is killed, the pack having already formed. Thus the number of hyenas that share the carcass is about the same for wildebeest and zebra. However, as a zebra weighs much more than a wildebeest, each hyena obtains more meat: the return for the greater social cooperation required to hunt zebra.

2. *Storage of prey.* Many birds and mammals kill more than they eat and store the surplus. Carnivores, such as the leopard (*Panthera pardus*), the puma or mountain lion (*Profelis concolor*), and the tiger (*Panthera tigris*), tend to hide the carcass of the prey animal only if they are disturbed, and therefore unable to consume it all immediately after the kill. They do not store several prey at a

time. The red fox (*Vulpes vulpes*), on the other hand, often buries a number of prey and is able to remember the location of such caches. It will move a cache to a new location if its hiding place is discovered by another animal. Caches may be made while the fox is still hungry, as a way of protecting food from competitors, or as a way of increasing the time available for the capture of prey.

Ravens (*Corvus corax*) are more likely to store food when they are hungry, whereas shrikes (*Lanius*) store more when they are satiated. This difference may be connected with the fact that ravens store large quantities of prey at the time they are feeding their young. In other words, the ravens store in response to demand, whereas shrikes store in relation to the abundance of available prey. Such different strategies show that the relationship between hunger and prey storage is complex, differs from species to species, and probably serves different functions in different species, according to their ecological NICHE.

3. *Feeding the young.* Many predatory species obtain food which they do not eat themselves, but give to their young. The motivation of such behaviour may appear to be similar to that of ordinary feeding, except that it has a different endpoint. However, in many species there are distinct differences in the pattern of predatory behaviour in the two cases.

It is usually the case that the process of obtaining food for offspring is under the control of stimuli from the young, rather than under the control of the hunger of the parents. Birds of many species adjust their parental foraging to the size of their brood. The begging behaviour of the young is usually the prime stimulus by which parents adjust to their food requirements. The foraging or HUNTING animal often kills its prey soon after capture, but this aspect of predatory behaviour may be absent when the predatory behaviour is for the benefit of the young. Domestic cats (*Felis catus*) and cheetah (*Acinonyx jubatus*), for example, carry live prey to the litter, and release it in the presence of the kittens. If they do not succeed in chasing and killing the prey, the mother recaptures it and presents it again to the young. When live food is delivered to a brood, it is often disabled, or presented in a special manner. For instance, the Eurasian kingfisher (*Alcedo atthis*) normally holds fish with the tail protruding from the bill, prior to swallowing it head first. When the fish is intended for the young, it is held the other way round, and offered head first to a nestling. In many birds of prey the captured animal is torn to pieces of a size suitable for the nestlings to swallow. Other birds eat the food themselves, and then regurgitate it to the young. Some birds, when foraging for the young, select prey items different in size from those they would eat themselves. Thus great tits (*Parus major*) provide older broods with larger caterpillars than the ones they consume themselves. This is probably because a foraging trip

made for the benefit of the young is more efficient if a large prey item is selected, whereas a small item can be more quickly eaten at the point of capture by a bird foraging for itself.

4. *Rhythms of predation.* Many predators show rhythms of predatory behaviour. In some cases these are clearly synchronized with the availability of prey, but in other cases the relationship is not so clear cut.

Seasonal RHYTHMS have been observed in predators, but little is known about how these correlate with prey movements. Circumstantial evidence in birds indicates that hunting intensity and prey selection vary on an annual basis, even under LABORATORY conditions. Great bustards (*Otis tarda*) have been reported to accept field mice (*Apodemus*) in the summer, but not in the winter. The northern shrike (*Lanius excubitor*) usually hunts both insects and lizards, but in the autumn the percentage of lizards in the diet drops to zero, even though the availability of lizards remains the same all the year round.

Many predators have a marked *circadian rhythm.* Thus, in some nocturnal carnivores, such as the poto or kinkajou (*Potos flavus*) and common genet (*Genetta genetta*), the onset of daily activity is at dusk, although it terminates somewhat variably before dawn. The same pattern has been observed in nocturnal birds. Laboratory studies of the circadian activity of domestic cats, which were given food at irregular times throughout the day, showed that the cats had a pattern of nocturnal hunting activity that exactly matched the pattern of activity of their characteristic prey, the long-tailed field mouse (*Apodemus sylvaticus*).

Similarly, the predatory pattern of the pygmy owl (*Glaucidium passerinum*) in Finland closely matches that of its prey, the bank vole (*Clethrionomys glareolus*). In general, the evidence suggests that the hunting motivation of many predators is influenced by the activity of endogenous CLOCKS.

Searching and hunting strategies. The DEFENSIVE behaviour of many animals involves CAMOUFLAGE, hiding, and particular patterns of spatial or temporal distribution, all of which tend to hinder detection by predators. For example, many marine SCHOOLING fish have a pattern of daily activity that seems designed to inconvenience predators. Diurnally active species seek shelter for the night just before sunset, and nocturnal fish depart from their inshore nesting places and head for offshore feeding grounds at the end of twilight. Similarly, many fish exclude the morning twilight period from their long-range movements. Twilight is the best time for a predator fish to approach to within striking distance of prey without being detected. So by avoiding activity during twilight, the prey fish make it more difficult for predators to catch them unawares.

In response to the defensive strategies of their prey many predators have evolved efficient patterns of SEARCHING behaviour, often designed for the capture of particular prey. For example, honey-bees (*Apis mellifera*) can learn that particular species of food plant produce nectar only during a certain period of the day, and they visit such flowers only at the appropriate time of day. As we have seen, hyenas may form different types of hunting packs according to the type of prey being hunted. Many predators behave as if they have a particular SEARCHING IMAGE, which corresponds to the prey they are hunting for. For example, carrion crows (*Corvus corone*) can be trained to search for bait under variously coloured mussel shells (Myrtilidae), spaced out upon the ground by the experimenter. When the bait was placed under some shells and not under others, the crows tended to concentrate upon shells of the same colour as the one under which they first found bait. In other words, if the crow found bait under a black shell, upon a particular day, it would turn over all black shells, ignoring shells of other colours. The following day, the crow might concentrate its search upon another type of shell, and ignore black shells, even though they might contain food. It may be that, by concentrating its ATTENTION upon a particular type of prey, or on a particular place or time of day, a predator is able to improve its searching efficiency.

When a searching predator captures a small prey, it tends to search for others in the vicinity of the capture. This strategy is likely to increase the predator's chances of finding another prey, especially if the prey is of the type which tends to have a patchy distribution, due to factors, such as microclimate, which may influence its preferred HABITAT. For camouflaged prey, however, a good counter-strategy is to avoid each other, so that their distribution is scattered rather than patchy. Provided that they cannot easily be detected at a distance, the survival chances of prey are greater when they are spaced out.

The means by which predators recognize their prey vary greatly from one species to another, but it is nevertheless possible to make some generalizations. Most predators feed upon a variety of prey. There are some exceptions, such as the snail kite (*Rostrhamus sociabilis*), which preys exclusively upon apple snails of the genus *Pomacea*, but there can be few predators that can depend solely upon specific SIGN STIMULI in recognizing their prey. Most predators rely upon more general features, such as size, colour, movement, and configurational cues, such as bilateral symmetry and the presence or absence of legs. For example, experiments with a highly insectivorous primate, the rufous-naped tamarin (*Saguinus geoffroyi*), showed that they recognize their prey, stick-insects (Phasmida) and mantids (Mantida), by the presence of head and legs. Insects with head and legs removed were not generally detected. Stick-insects, such as *Metriotes diocles*, normally adopt a cryptic position in which the legs are pressed close to the body. As soon as only one pair of legs is made to

stand out from the body, the prey is easily detected.

Many predators are attracted to particular patterns of movement. For example, the LOCOMOTION of many small crustaceans and of mosquito larvae (Culicidae) consists of alternate bouts of upward wriggling and sinking. So powerful is this movement stimulus to predators that it is copied by the free-swimming stage of many fish parasites (see PARASITISM and MIMICRY).

The piranha fish (*Serrasalmus nattereri*) recognizes its prey primarily by VISION, although the threshold of attack is lowered by MECHANICAL and CHEMICAL stimuli. Piranhas tend to attack all fish shapes that are more than four times as long as they are wide. Any fish that is less than three times as long as it is wide is exempt from attack, and this shape category includes the piranhas themselves.

Another example of a capture-inhibiting stimulus is provided by the special coloration and movement pattern of cleaner fish, such as the cleaner wrasse (*Labroides dimidiatus*), and some cleaner shrimps (see SYMBIOSIS). These animals are usually boldly striped, and approach their hosts by means of a characteristic 'dancing' motion. Although the host fish are often large carnivores, such as groupers (*Epinephelus*), they allow the cleaner fish to approach and enter their mouths unharmed. The cleaner fish remove ectoparasites and food particles from around the teeth, but they are never themselves eaten, even though they are the same size as the host's normal prey.

In addition to recognition of prey, the circumstances of a hunt may affect its success, and may give valuable information to the predator. For example, the condition of the water surface is of considerable importance to birds that capture fish by plunge-diving. In calm water fish can see quite well what is happening above the surface and can take evasive action if they see a predatory bird. Observations of the brown pelican (*Pelecanus occidentalis*), the common tern (*Sterna hirundo*), and the sandwich tern (*Sterna sandvicensis*) have revealed that when the surface of the water is rough, not only do these predators have greater hunting success but they also hunt more frequently. This shows that the birds take the weather into account in deciding when to hunt.

In addition to being attracted towards prey by such semi-direct cues as odour trails and giveaway sounds, such as distress or contact calls, some predators rely on information from other members of their own species. For example, by FLOCKING around a food source and uttering food calls, feeding gulls (Laridae) attract others. Similarly, when spotted hyenas make a kill, their VOCALIZATIONS often attract lions (*Panthera leo*) and hooded vultures (*Necrosyrtes monachus*).

A number of predators hunt by speculation. Thus a lion that is familiar with the locality may run up a slope and look down the other side of the ridge, in the hope of surprising a suitable prey animal. *Octopus cyanea* frequently pounces on coral rocks, or clumps of seaweed, which may happen to conceal crabs. Herons (Ardeidae) use a foot to stir up debris in shallow water, and then, using the bill, they spear any fish flushed from its hiding place. Some marine fishes and birds take prey flushed by other animals. Hornbills (Bucerotidae) in Africa, drongos (Dicruridae) in Asia, trogons (Trogonidae) in Panama, and a number of other tropical birds, regularly feed upon insects disturbed by monkeys (Simiae). Cattle egrets (*Bubulcus ibis*) habitually follow cattle and other large animals, and feed upon insects flushed from the ground.

Ambushing is another form of hunting by speculation that is practised by many predators. The predator usually lurks in a camouflaged fashion in a place that potential prey are likely to frequent. Only when a prey comes within range is an attack launched. For example, the mantid *Parastagmatoptera unipunctata* is a carnivorous insect that lies in wait throughout the day and catches flies which come within reach. It sits motionless among vegetation which resembles its own form and coloration. The prey is detected by the well-developed compound eyes, and is faced by an aiming movement of the head. The mantid then strikes the prey with a very rapid movement of its forelegs.

Some ambush predators position themselves so as to obtain a good view of potential prey. Thus grouper fish (*Mycteroperca* and *Epinephelus* spp.) lurk low down in the water, and attack prey fish that swim above them, silhouetted against the light sky.

Modes of attack. The three main modes of hunting used by predators are lying in wait, stalking, and searching for relatively immobile prey, as described above. Once the predator comes within striking range of the prey it launches an attack, which generally consists in aiming and then striking at the prey. The strike is often launched without any further guidance from sensory processes. For example, kingfishers close their eyes as they hit the water surface in their dive after a fish, and owls (Strigiformes) do the same prior to impact with the prey animal. In the laboratory it can be shown that the attack of the common octopus (*Octopus vulgaris*), and of the common cuttlefish (*Sepia officinalis*) is not affected, after a certain point, by switching off the lights during an attack. A similar result can be obtained with the stroke of a competent golf player. If the lights are extinguished during the back-swing, the rest of the stroke, including hitting the ball, remains unaltered. This type of experimental result shows that the attack movements are pre-programmed during aiming, and are then executed regardless of further changes in the situation. Usually, however, the strike is so fast that there is little time for the prey to move and spoil the aim, once the attack is launched. Sometimes, an attack involves pursuit, during which the predator must launch the strike.

Thus a cheetah, in pursuit of prey, often makes a final sprint which is terminated by a leap at the prey. Birds of prey, such as owls and the northern goshawk (*Accipiter gentilis*), thrust their feet forward and strike in mid-air.

A pursuing predator may attempt to intercept the prey on the basis of a prediction about the future flight path of the prey. The dwarf chameleon (*Microsaurus pumilis*) shoots out its prehensile tongue just ahead of a moving prey, and cuttlefish often strike, with their specialized prehensile tentacles, just behind or to the side of a crab, anticipating its most likely ESCAPE direction. By dodging, a prey can make it more difficult for a pursuing predator to aim its attack. Some predators, such as the ichneumon mongoose (*Herpestes ichneumon*) and the golden eagle (*Aquila chrysaëtos*), make mock attacks which confuse and tire the prey. African wild dogs (*Lycaon pictus*) spread out during a chase, so that a dodging quarry can often be intercepted by animals to one side of the main line of flight. Wild dogs also make short cuts in situations where the prey heads off in a new direction. The dogs immediately behind the quarry continue the chase, while others cut across to intercept it.

A mode of attack used by some predators is disguise, sometimes called aggressive mimicry. The predator relies on his own appearance and behaviour to approach a prey without being noticed, or to induce the prey to mistake the identity of the predator. Both piranha fish and Norway rats (*Rattus norvegicus*) have been observed close to their prey, 'pretending' not to be interested, and not showing any sign of stealth. The predator then launches a sudden attack upon the unsuspecting prey. Similarly, trumpetfish or pipefish (*Aulostomus*) ride upon the backs of parrotfish (Scaridae), as illustrated in Fig. A. Small fish are normally in-

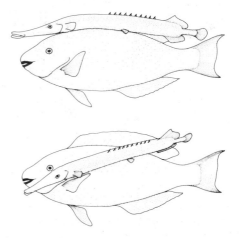

Fig. A. The trumpetfish rides on the back of a harmless parrotfish, and slides off to attack its prey.

different to the harmless parrotfish, and they seem not to notice the pipefish, which slides sideways off its mount to attack small fish that come within range. A true aggressive mimic is the cleaner mimic, the sabre-toothed blenny (*Aspidontus taeniatus*), which imitates the cleaner wrasse, both in its coloration and in its behaviour. As described above, the cleaner fish approaches large fish by means of a special 'dance', and is usually permitted to enter the mouth and remove food particles and parasites. The cleaner mimic has the special markings typical of cleaner fish, as illustrated in Fig. B, and it also copies the approach dance.

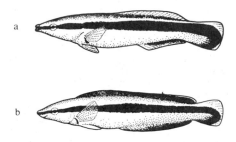

Fig. B. The cleaner fish (**a**) and its mimic (**b**).

When the mimic gets close to the large fish that is inviting cleaning, it suddenly bites a piece from the body of its prey and quickly escapes. Since cleaner fish are normally not attacked by the large host fish, the mimic enjoys the double advantage of its own predatory behaviour and relative immunity from predation from larger fish.

Man appears to be the only aggressive mimic amongst mammals. North-American Indians used to use wolf-hides when approaching bison (*Bison bison*), and Australian aborigines use kangaroo skins when hunting kangaroos (Macropodidae). The Hausa bowmen of Northern Nigeria hunt prey under a head-dress made from the head of the Abyssinian ground hornbill (*Bucorvus abyssinicus*), as illustrated in Fig. C. These are

Fig. C. Hausa bowman disguised as a hornbill.

common birds, which would normally be ignored by game animals. Illustrations of this technique appear in cave paintings, and are thought to be at least 10 000 years old.

28

PREENING is a form of GROOMING behaviour performed by birds as part of feather maintenance. It is often seen in conjunction with other COMFORT behaviour such as bathing, dusting, and sunning. Preening consists in the cleansing and arrangement of feathers by the bill. This is done both in response to dirt and wetness, and as part of LEISURE activities. Typically, attention is given to individual feathers as well as to the arrangement of the feathers as a whole. Both mandibles (the two parts of the bill) are employed during nibbling movements directed at particular feathers. The feather may also be drawn through the bill in one movement. This is especially common for the primary feathers.

In arranging the feathers the bird may employ a stroking action with a closed bill, which is moved rapidly down the feather near to the shaft. There may also be a quivering or trembling action of the bill as it moves down the feather. During preening the feathers may be raised or ruffled to permit better access. The head is usually groomed by scratching with the foot, or by rubbing it against other parts of the body. Mutual preening, especially of the head, occurs in herons (Ardeidae), pigeons (Columbidae), parrots (Psittacidae), and some passerines (perching birds, Passeriformes). In addition to its cleansing function, mutual preening is thought to be important in the maintenance of SOCIAL RELATIONSHIPS.

Preening is an essential activity for maintaining the feathers in the proper condition for FLIGHT and thermal insulation (see THERMOREGULATION). Most birds have a preen gland near the tail that secretes an oily substance. The bird stimulates the flow of oil with its bill, and then applies it to the feathers with stroking and quivering movements. This is frequently done if the feathers are wet. However, dry waders (Charadriiformes) and grebes (Podicipedidae) have been observed to wet themselves prior to oiling, so it may be that water on the feathers helps to spread the oil. Exposure to sunlight increases the vitamin D content of preen oil, which is sometimes ingested. The preen oil helps to keep the feathers supple and water resistant. Some birds anoint their feathers with other substances. Parrots have been observed to use eucalyptus oil in this way.

PROBLEM SOLVING ability is often regarded as an indication of INTELLIGENCE in animals, including human beings. Animals differ widely in their abilities to solve problems, but this does not necessarily mean that they can be ranked along a single scale of intelligence. An animal may be especially good at certain kinds of problem, but poor at others. For example, compared with primates and some other mammals, birds tend to be poor at detour problems. They tend to persist in taking the apparent direct route to a GOAL, when an indirect route is required. If a pigeon (*Columba livia*) or chicken (*Gallus g. domesticus*) is presented with food behind a wire fence it will persist in attempting to reach the food through the fence. A domestic cat (*Felis catus*), dog (*Canis l. familiaris*), or chimpanzee (*Pan*) may make an initial attempt to reach the food directly, but will quickly see that a detour around the end of the fence is required. When an animal fails to reach the goal by a direct route it experiences FRUSTRATION. This causes it to switch its ATTENTION to other aspects of the situation, and creates an opportunity for the animal to break away from the unprofitable course of action. This is an essential step in solving a detour problem. Although it can be demonstrated that pigeons and chickens are capable of LEARNING about aspects of the situation other than the direct route to the goal, they do not seem to have the ability to recognize the significance of alternative routes. A bird may eventually solve the problem by TRIAL AND ERROR, and will thereafter learn the correct route. Some mammals, however, seem to have INSIGHT into the situation, in that they quickly realize that the end of the fence provides a solution to the problem.

Most detour problems require the animal at some stage to move away from the goal. This is what many animals find difficult, even when the problem is fairly abstract. For example, a chimpanzee which had learnt to fit two sticks together to make a pole long enough to reach food outside its cage failed to do so when particularly appetizing food was presented. Instead the chimpanzee became so excited that it persisted in attempting to reach the food with a short stick. To take time to fit two sticks together involves a mental detour that is counteracted by the animal's strong MOTIVATION.

The fact that pigeons are poor at detour problems does not necessarily mean that they lack intelligence. Their feats of NAVIGATION would tax the mental ability of any primate. Reptiles, which show little sign of intelligence in many types of test, are, nevertheless, good at some detour problems. Chameleons (Chamaeleontidae), for example, are frequently required to make detours among the branches of trees to come within striking distance of their prey. In LABORATORY tests the chameleon is particularly good at spatial detour problems and at negotiating complex mazes. The common wood turtle (*Clemmys insculpta*) can learn to negotiate a complex maze in just a few trials. It progresses in a systematic manner, visually scrutinizing each choice point. By contrast, a rat (*Rattus norvegicus*) will subject the maze to a thorough EXPLORATION, but may not learn any faster.

Animals vary considerably in their mental abilities. Pigeons seem to be good at forming concepts, but relatively poor at tasks involving matching to

sample, or recognition of oddity (see COGNITION). The complexity of a problem can be measured in various ways.

Abstraction is one form of complexity, and the number of features that require attention is another. A chimpanzee can be trained to choose the larger of two white squares, and then learn that if the two squares are black the smaller is the correct choice. He then has to learn that if the shapes are triangles instead of squares the previous relationships are reversed, so that the smaller of the white triangles and the larger of the black triangles is the correct choice. The problem can be further complicated, until the chimpanzee has to take account of five different factors or cues in order to solve the problem. Experiments show that rhesus macaques (*Macaca mulatta*) can handle four cues in this type of problem, but rodents and carnivores only two.

Conditioning versus insight. Many psychologists subscribe to Lloyd Morgan's canon: Never attribute animal behaviour to a high mental function if the behaviour can be explained in terms of a more simple process (see HISTORY). From the beginning of the 20th century psychologists have sought to account for the problem-solving abilities of animals in terms of conventional CONDITIONING. An alternative view, known as the *Gestalt* school, was that animals gained insight into problems through an INNATE tendency to perceive the situation as a whole. A protagonist of this view, Wolfgang Köhler, carried out a series of experiments on the island of Tenerife, where he was isolated throughout the First World War.

In his book *The Mentality of Apes* Köhler claimed that the TOOL USING abilities of chimpanzees when solving the type of problem in which food is in an inaccessible place, such as outside the bars of the cage, or suspended from a high beam, demonstrated that they had insight into the problem. The chimpanzees learned to use a stick to rake in food (see INTELLIGENCE, Fig. B), or to pile up boxes to form an improvised ladder to reach food from a high place (see INTELLIGENCE, Fig. C.). One of Köhler's chimpanzees was given two bamboo poles that would fit together to make a longer pole. Fruit was presented outside the cage and out of reach of either of the short bamboo poles. After trying many times to reach the food with a single pole, without success, the chimpanzee accidentally joined the two poles together by pushing the narrower one inside the hollow end of the other. This was apparently done during idle PLAY. At once the chimpanzee jumped up, rushed to the bars of the cage, and used the now long pole to retrieve the fruit. In the *Gestalt* view this is an instance of insightful behaviour.

The *Gestalt* interpretation has been criticized by other psychologists on two main grounds. Firstly, it is argued that the experiments are set up to discover whether the animals behave with insight under conditions which are supposed to require such behaviour. There is no independent evidence that a given task does require insight, but if the animal succeeds in the task it is said to have demonstrated insight. Many scientists do not find this to be an acceptable line of reasoning. The second objection concerns the interpretation of the way in which the chimpanzees behave in arriving at a solution to the problem. All observers agree that the apes engage in much IRRELEVANT behaviour, play, and abortive attempts to obtain the food or other REWARD. Some take the view that the chimpanzees arrive at a solution to the problem through a cumulative process of trial and error, arriving first at a partial solution to the problem and later at a complete solution. Thus, a chimpanzee may learn, by accidentally standing on a box, that this behaviour brings him nearer to fruit suspended from the roof of his cage. He may then learn that moving the box under the fruit brings him even nearer. Later he may learn to place one box upon another during play, the action being unrelated to a particular problem, etc. A strength of this argument is that it can be tested experimentally. It has now been demonstrated repeatedly that experience with the string, sticks, and other objects used in the tests is a necessary prerequisite for successful problem solving. Moreover, many of the activities, which are regarded as insightful in the context of a particular problem, occur during play when there is no problem to be solved. For example, in one study forty-eight chimpanzees were given sticks that could be fitted together. There was no problem to solve. Of these, thirty-two fitted the sticks together within 1 h. The older chimpanzees generally achieved this more quickly than the younger ones. Careful studies have shown that the ability to manipulate objects in ways that are relevant to problem solving is largely a matter of MATURATION. The younger animals simply do not have the manipulative skills that would enable them to perform some of the required tasks. Once an individual discovers the ability to carry out a par-

ticular manipulation, it tends to be repeated over and over again. When given boxes to play with for the first time, 6-year-old chimpanzees dragged them across the floor, sat on them, stood on them, stacked them on top of each other, and climbed up them as if to reach something. Once manipulation of objects has become firmly established as part of an animal's repertoire, it is not surprising that they are used in problem solving.

Although most psychologists accept that prior experience and innate tendencies play a large part in problem solving, insight has not been ruled out as a possible component of problem-solving behaviour.

What does emerge from this type of study, however, is that the ways in which chimpanzees solve problems are not very different from those of human beings. For example, in one experiment people were given the problem of hanging a loop on a hook in the wall, without approaching closer than 2 m. When provided with two sticks and a piece of string hanging on a nail, most tied the sticks with the string, so that they could then use this enlarged stick as a tool to hang the loop on the hook. However, when the string was supporting a picture on the same nail, few people thought to use it to tie the two sticks together. This experiment illustrates that much depends upon the exact context in which the task is set. One difficulty in understanding the problem-solving abilities of animals is that the human investigator may not have a clear idea of the relevance of objects as perceived by the animal (see ANTHROPOMORPHISM). Clearly, it is unfair to set a problem that involves behaviour beyond the animal's manipulative ability, or behaviour that is obviously against the animal's natural inclination. Do we regard intelligence as ability to go against the natural inclination, or do we regard intelligence as a measure of an animal's ability to solve those problems to which it is suited by nature?

89, 92, 119

R

REAFFERENCE theory offers an explanation of certain phenomena that are widespread in animal behaviour. Imagine that you are holding on to the branch of a tree with one hand. On a calm day you may shake the branch, and the pattern of movement of the branch will be determined solely by the pattern of movement of your arm, which in turn will be determined by instructions issued by your BRAIN. Moreover, you will feel that you are moving the branch in a VOLUNTARY manner. Now imagine that you are holding the branch in the same way, but that it is a windy day. Your arm is completely relaxed and the pattern of movement of the branch is caused by the wind. The movement of the branch causes your arm to move passively, and this movement is detected by your brain. You will now feel that your arm is being moved by the branch and not that you are voluntarily moving the branch. The pattern of movement of the branch could, by chance, be the same in the two cases, yet you would easily be able to tell the difference between the two situations. How does the brain distinguish between the movements of external objects that are a result of its own instructions, and the movements that are the result of external forces?

Reafference theory seeks to answer this question. According to the theory, any animal that is capable of bodily ORIENTATION must be capable of distinguishing between *exafferent* stimulation, which results solely from factors outside the animal, and *reafferent* stimulation, which occurs as a result of the animal's own bodily movements. The idea is that the motor commands, issued by the brain, not only cause patterns of muscular movement, but also set up an *output copy*, which corresponds to the expected input from the sensory processes that monitor the limb movements of the animal. The brain then makes a comparison between the output copy, i.e. the pattern of signals sent out by the brain, and the incoming or *afferent* information, i.e. the pattern of signals transmitted to the brain by these sensory processes. A schematic representation of this situation is shown in Fig. A. If, as a result of comparing the output copy and the incoming information, the brain finds no differences, then it concludes that all the incoming information was reafferent. This means that all the instructions given by the brain were matched from the monitoring of the animal's voluntary movement, indicating that the instructions were carried out. If, however, the comparison shows a discrepancy between the output copy and the incoming information, then, either the instructions were not carried out due to some malfunction of the muscles, or exafferent information was also received by the sensory processes involved.

A good example of this phenomenon is the control of eye movements in human beings. Normally the movement of objects in the outside world results in a corresponding movement of the image of the object on the retina of the eye (see VISION), and this is perceived by the brain as an actual movement of the object. However, when the eyes are moved voluntarily, there is also a movement of the image on the retina, but this is not perceived as movement of objects in the outside world. According to reafference theory, the brain makes an output copy of its instructions to the eye muscles, and the output copy is translated into the expected retinal movement. If the real retinal image movement is the same as the expected, then no movement is perceived. However, any movement in the outside world will result in movement of the retinal image which has no counterpart in the output copy, and this is perceived as movement. The theory can be tested by interfering with the instructions from the brain to the eye, or by moving

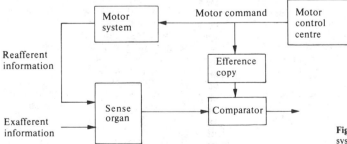

Fig. A. Outline of a basic reafference system.

the eyeball in a way that is not expected by the brain. If you apply gentle pressure from your finger to your eyelid, you can make your eyeball move in its socket, whilst keeping the eye open. Such movement is perceived as a movement of the outside world. Good evidence for reafference theory comes from experiments in which a person's eye muscles are temporarily paralysed by means of an injection, and the person is then requested to attempt to move his eye to the right. Most people report that the objects in the outside world appear to move to the right whenever they attempt to move their eyes to the right. In reality the brain commands the eye to move and issues an output copy, but the eye does not move because the muscles are paralysed. There is no movement of the image on the retina, but movement is perceived because the output copy is not cancelled by the incoming information.

In animal behaviour, reafference theory has proved to be particularly important in accounting for features of the orientation of animals. For example, a dronefly (Eristalis tenax) placed inside a cylinder painted with vertical stripes shows a typical optomotor REFLEX, turning in the direction of the stripes when the cylinder is rotated. Such reflexes do not occur when the fly moves of its own accord, although the visual stimulation is similar. When the head of the fly is twisted through 180° by an experimenter, the optomotor reflex is reversed as expected, but when the fly attempts to move on its own accord its movements appear to be self-exciting, and it tends to go into a spin. The results of this experiment show that the fly distinguishes between exafferent and reafferent movement stimuli, and does not merely block the optomotor reflex during self-initiated movement. According to reafference theory, the outcome of the comparison between output copy and incoming information induces movement to the right or the left according to its sign. When the outcome is zero, movement ceases and the target condition is achieved. When the head of the fly is reversed, the sign of the reafference signal is reversed, resulting in a positive FEEDBACK situation. The unstable spinning behaviour is thus a consequence of the fact that bodily movements are determined directly as a consequence of the comparison between output copy and incoming information.

Reafference theory has also been invoked to account for swimming reflexes in fish, limb movements of land vertebrates, and postural reflexes of birds. Studies of INCUBATION in birds' show that the tendency of a bird to incubate in an undisturbed manner is partly determined by the satisfactoriness of the nest situation. If the nest is deficient in size or shape, or in containing more or fewer eggs than normal, then the incubating bird is often restless and may interrupt incubation to attend to the nest. In accordance with reafference theory, it has been postulated that the bird has an expectancy, in the form of a template or output copy, of the feel of the nest and its contents. The

less the tactile information conforms to the expected, the less the bird's tendency to continue incubating. Experiments involving manipulation of egg number, size, shape, and temperature have tended to support this theory.

When applied to the behaviour of the whole animal, reafference theory becomes similar to other expectancy theories of behaviour, such as those which maintain that an animal experiences FRUSTRATION when its expectancies are not confirmed.

103

REDIRECTED BEHAVIOUR is the term applied to activities that are directed at external stimuli which are IRRELEVANT to the current situation. For example, a thirsty pigeon (Columba livia) that is denied access to water at an accustomed place or time may show drinking movements directed at smooth surfaces or shiny pebbles. Redirected activities occur in situations of FRUSTRATION or CONFLICT. For example, a herring gull (Larus argentatus) foiled in a dispute for a TERRITORY will often redirect its AGGRESSION towards an innocent bystander, or even towards an inanimate object. In winter flocks of small birds, when one is supplanted at a food source by a superior, it may attack an inferior bird instead of retaliating. Redirected behaviour may undergo RITUALIZATION during EVOLUTION, as seen in the FIGHTING of herring gulls: they frequently seize and pull at a clump of grass, as if it were the opponent's wing. This grass-pulling behaviour tends to be stereotyped in form, as if the animal were performing a ritual.

99

REFLEX behaviour is the most simple form of reaction to external stimulation. Stimuli such as a sudden change of tension in a muscle, a sudden change in the level of illumination, or a touch on some part of the body, induce an automatic, involuntary, and stereotyped response. The SENSE ORGANS involved, such as muscle spindles or light detectors, are usually directly linked to specific reflex mechanisms. They send messages to the BRAIN, or to more peripheral parts of the central nervous system, such as the spinal cord of vertebrates. Instructions are then sent directly to the effector organs, usually muscles or glands. The reflex action is not entirely automatic, because there is generally a possibility of interference from within the central nervous system. There may be inhibition from other INCOMPATIBLE reflex mechanisms, or modifying influences from the brain itself. For example, the pupillary light reflex, a feature of the vertebrate eye, acts to control the level of illumination falling on the retina: the diameter of the pupil is adjusted in response to changes in the amount of light on the retina. Thus, if we are dazzled upon entering a brightly lit room, our pupils contract and cut down the amount of light entering the eyes. After a few seconds we are no longer dazzled and can see more clearly. Although

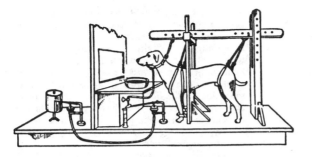

Fig. A. A diagrammatic illustration of a Pavlovian salivary conditioning apparatus showing the dog restrained in a harness, the fistula system for recording the amount of salivation produced by presenting a conditional stimulus, and the bowl used for presenting the food.

adjustment of the pupil is reliable and automatic, the magnitude of the response may be altered by the brain. Many mammals, including human beings, show marked changes in pupil size in situations involving EMOTION. The pupil expands in positive or pleasant circumstances and contracts in response to negative emotion.

Reflex behaviour often involves parts of the body, such as limbs or other specific groups of muscles. Reflexes are involved in the COORDINATION of limb movements and in all aspects of LOCOMOTION. In some cases, especially in startle responses, the whole animal is involved in the reflex response. This occurs in the ESCAPE behaviour of the squid (*Loligo*) and in the withdrawal reflexes of many invertebrates, such as snails (Gastropoda) and polychaete 'bristle' worms (Errantia). The escape reflex is triggered by giant nerve fibres which carry very fast messages to all the muscles involved, so that they contract suddenly and simultaneously.

REINFORCEMENT refers to the process by which certain responses become strengthened as an animal learns. Associative LEARNING occurs when an animal is exposed to a consistent relationship between different events, either in its natural environment or in the LABORATORY. Such learning is usually revealed by the fact that the behaviour of the animal changes in a way that reflects this relationship. However, in the absence of such a change, we cannot necessarily conclude that the animal failed to learn about the relationship between the events, for learning will only be manifest in overt behaviour if at least one of the events has MOTIVATIONAL significance. The process of reinforcement reflects the way in which associative and motivational mechanisms interact to change behaviour during the course of learning.

There are at least two classes of environmental relationships or contingencies about which some animals can learn, and each may be studied in the laboratory using different CONDITIONING procedures. Classical conditioning experiments, first performed by the Russian physiologist PAVLOV, investigate learning when a relationship is arranged between two environmental events or stimuli that occur independently of the animal's behaviour. In contrast, an instrumental or OPERANT conditioning procedure can be used to study learning arising

from a contingency between an animal's behaviour and some environmental event. As the learned ASSOCIATIONS formed in these two procedures produce different behavioural changes, the process of reinforcement has to be considered separately for each type of conditioning.

Classical or Pavlovian conditioning. In the original classical conditioning experiments Pavlov presented small portions of food at irregular intervals to a hungry dog (*Canis l. familiaris*) restrained in a harness (see Fig. A). When he repeatedly signalled the delivery of food by turning on an external stimulus, such as a bell or light, some seconds before the food was given, the behaviour of the dog during the stimulus changed progressively in the course of the experiment. Most noticeably, the animal began ORIENTING towards the stimulus, licking its lips, and salivating. Pavlov chose to record salivation systematically by placing a small collecting tube in the salivary gland. As a result, he found that the magnitude of the salivary response elicited by the external stimulus increased as the animal experienced more pairings of the stimulus and food (see Fig. B). Pavlov referred to the external, signalling stimulus as the *conditional stimulus*, and the food or signalled stimulus as the *unconditional stimulus*. Unconditional stimuli, such as food, consistently elicit a set of CONSUMMATORY responses, which are referred to as the *unconditional responses*, while the conditioned stimulus gradually comes to elicit a new response, the *conditional response*, as a result of the conditioning procedure. It is the development of this conditional response (salivation) which is taken as evidence that the animal has learned something about the relationship between the conditional stimulus (bell or light) and the unconditional stimulus (food).

Conditioning experiments such as Pavlov's which employ a pleasant or positively valued event as the unconditional stimulus are examples of positive conditioning. However, a noxious or unpleasant stimulus can be successfully used to study defensive or aversive conditioning. For instance, if the onset of a tone precedes the delivery of a puff of air to the eye of a rabbit (*Oryctolagus cuniculus*) by a second or so, this tone will come to elicit a blink of the eyelid as a conditional response.

Since the rapidity with which a conditional re-

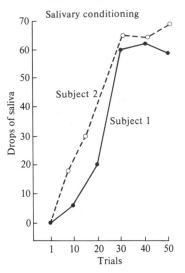

Salivary conditioning

Fig. B. The development of an appetitive salivary conditional response in two dogs. The number of drops of saliva elicited by the presentation of the conditional stimulus progressively increases with the number of trials in which the conditional stimulus and the food are paired.

sponse develops and its size at the end of conditioning seem to depend largely on the potency, intensity, or magnitude of the unconditional stimulus, this event is usually regarded as the agent responsible for strengthening or reinforcing the conditional response. So in Pavlov's salivary conditioning procedure, the food serves as a positive (sometimes called APPETITIVE) reinforcer, while in rabbit eyelid conditioning the air puff or shock acts in an analogous fashion as an aversive reinforcer. The defining characteristic of a reinforcer, namely that it will support the development of a conditional response during learning, is often not an intrinsic property of the unconditional stimulus, but arises by virtue of the relationship of that stimulus to the current motivational state of the animal. Food will only reinforce salivation if the dog is hungry at the time of conditioning. In a similar fashion, unconditional stimuli will only support defensive conditioning if they have motivational significance by being noxious or unpleasant for the animal. Even though a weak puff of air produces a reliable eyeblink, such a stimulus will not necessarily lead to the formation of a conditional response unless it is actually aversive for the animal. Generally, a stimulus will only have reinforcing effects if it has motivational or EMOTIONAL significance for the animal.

Although most effective reinforcers, such as nutritive substances and stimuli which potentially lead to tissue damage, appear to exert their strengthening property through INNATE mechanisms, this is by no means a necessity. Pavlov clearly demonstrated that a conditional stimulus itself could act as an effective reinforcer. Suppose that a bell has already been established as a positive conditional stimulus for salivation in a dog by pairing it with food. If a second conditional stimulus, a light, is then paired with the bell in the absence of food, we might well observe the development of salivation to the light during the initial pairings, even though the light itself has never been directly associated with food. This phenomenon, which Pavlov called *higher-order conditioning*, demonstrates that reinforcing properties can actually be acquired through conditioning.

Instrumental or operant conditioning. In Pavlovian or classical conditioning, either the experimenter or the natural environment arranges a relationship between two events which occur independently of the animal's behaviour. In contrast, the cardinal feature of instrumental or operant conditioning is that a relationship holds between an aspect of the animal's own behaviour and the occurrence of an external event. So, following a procedure developed by the American psychologist SKINNER, a hungry rat (*Rattus norvegicus*) may be placed in a small chamber containing a lever projecting from one wall, and a small food-well into which pellets can be delivered. If a contingency is set up such that depression of the lever causes delivery of a food pellet, we shall not be surprised to find that the animal starts to press the lever, and that the frequency of lever pressing rapidly increases with the number of previous responses that have been rewarded with a pellet. Such an experiment can be regarded as a demonstration of reinforcement in instrumental conditioning, in which the presentation of the food as a positive reinforcer strengthens the conditional response of lever pressing. There is nothing special about the lever-press response, and the experimenter could make the positive reinforcer contingent upon a number of different aspects of the animal's behavioural repertoire. Perhaps the most bizarre demonstration of the power of positive reinforcement comes from a legendary laboratory rat who was prepared to climb to the top of a spiral staircase, 'bow' to the audience, push down and cross a drawbridge, climb a ladder, use a chain to pull in a model railway car, pedal the car through a tunnel, climb a flight of stairs, run through a tube, and descend in a lift, all to receive a single food pellet.

Like positive Pavlovian reinforcers, effective positive instrumental reinforcers have to be pleasant and positively valued events, such as food, and often the term REWARD is used interchangeably with positive reinforcement. The parallel between positive Pavlovian reinforcers and instrumental positive reinforcers extends to the fact that a positive Pavlovian conditional stimulus, such as a light paired with food, may not only act as a reinforcer in higher-order classical conditioning, but will also reinforce instrumental responses. A hungry rat in a conditioning chamber will come to press the lever if, instead of following a response by food, we present a light which has previously been paired with food. Under these circumstances the light is referred to as a *secondary* or *conditional reinforcer*.

On the basis of this parallel one might expect that aversive Pavlovian reinforcers would also exert reinforcing properties in instrumental conditioning, but this is not so. In fact, on a simple level of analysis, they appear to have exactly the opposite effect to that of an instrumental reinforcer. Remembering that an aversive Pavlovian reinforcer is an unpleasant or noxious event, it is not surprising that presenting such an event after the animal has performed a response tends to decrease the likelihood that the response will be given again. Such suppression, of course, reflects the effect of punishment.

However, it is possible to arrange a relationship between aversive events and the behaviour of animals in such a way that a particular response becomes strengthened. Suppose that instead of reinforcing the rat's lever presses with food pellets, we occasionally present an aversive stimulus such as a mild electric shock. If, in addition, it is arranged that a lever press stops the shock while it is on, we shall again not be surprised to find that the animal learns to ESCAPE from the shock by pressing the lever. Another refinement to this procedure can be added by allowing responses in between shocks to postpone or omit the next shock which would have occurred. Under these circumstances rats can succeed in avoiding the large majority of aversive stimuli which would have occurred if they did not press the lever. From the point of view of EVOLUTION, the development of the ability to acquire a new response in order to escape or avoid noxious or potentially harmful events in the environment is of just as much adaptive importance (see ADAPTATION) as the ability to learn new ways of getting rewards. However, if our interest lies in isolating the reinforcing event, then AVOIDANCE learning presents problems; for an animal that is successfully avoiding all the aversive stimuli continues to respond in the absence of any external stimulus change that we might identify as the reinforcing agent. By its very nature a simple avoidance situation arranges that nothing happens after the response. The event that appears to be of importance in strengthening responding is either the termination (escape) or omission (avoidance) of a stimulus that would otherwise have occurred, and the concept of *negative reinforcement* is taken as referring to such an event.

So the parallel between Pavlovian and instrumental reinforcers is not complete. Although it is true that most positive Pavlovian reinforcers can act as positive reinforcers for instrumental responding, negative reinforcement arises from the omission or termination of a stimulus which is likely to be an aversive Pavlovian reinforcer. In instrumental conditioning the concepts of positive and negative reinforcement do not refer to the emotive value of the reinforcers, but to whether the reinforcement process is activated by the presentation or termination (or omission) of a stimulus.

Principles of reinforcement. So far we have just identified the events that are of importance in strengthening responding during conditioning, but said nothing about the way in which these reinforcers bring about the specific behavioural changes observed. Often the most notable feature of classically conditioned responses is their similarity to the unconditional responses elicited by the Pavlovian reinforcer. Pavlov observed that dogs began to salivate to a conditional stimulus paired with food, a REFLEX response that is of course elicited by the food itself. Similarly a tone paired with a puff of air to the eye of a rabbit elicits an eyeblink that is similar to the defensive reflex elicited by the air puff itself. It is as though pairing a conditional stimulus with a Pavlovian reinforcer endows it with the same behavioural properties as the reinforcer itself possesses. This simple principle of reinforcement is usually referred to as *stimulus substitution*, because in some sense the conditional stimulus comes to substitute or stand for the unconditional stimulus. As a result, classical conditional responses are often assumed to be involuntary responses to the conditional stimulus, much in the same way as unconditional responses are reflexively elicited by the unconditional stimulus or reinforcer.

At first sight, the principle of stimulus substitution seems to be totally unable to cope with the behavioural changes seen during instrumental learning. Since the relationship arranged by the experimenter is between the animal's own behaviour and a reinforcer, it is not obvious that any particular stimulus will be paired with that reinforcer, and so come to substitute for it. A second point is that, at least on a simple level of analysis, there is often no obvious similarity between the response strengthened by the instrumental reinforcer, such as lever pressing by a rat, and the response elicited by a positive reinforcer, such as consumption of a food pellet and its attendant visceral changes.

However, the principle of stimulus substitution cannot be dismissed so easily as an account of instrumental conditioning. The fact that an experimenter or the environment arranges a certain contingency between the animal's behaviour and a reinforcer does not ensure that this is the relationship which controls the animal's behaviour. When a lever press is reinforced, the sight of the lever will be paired with food. So it may well be that the lever itself becomes a Pavlovian conditional stimulus paired with a positive reinforcer, food. If this is so, according to the principle of stimulus substitution, the lever should take on some of the properties of food, and the fact that the animal presses the lever may be a fortuitous consequence of the fact that hungry animals tend to approach and manipulate Pavlovian conditional stimuli for food in the same way as they approach food itself.

A crucial feature of instrumental conditioning, which makes a complete account in terms of stimulus substitution difficult, is that highly specific patterns of behaviour, rather than general approach and withdrawal tendencies, can be reinforced. For instance, rats can be successfully trained to press levers with a specific force or duration if only responses with these characteristics are reinforced. It is difficult to see how these specific response characteristics could arise from the fact that particular stimuli in the environment had become Pavlovian conditional stimuli. So a second principle of reinforcement, the *law of effect*, is required. The law of effect was originally formulated by the American psychologist Thorndike, and in its most general form maintains that following a particular response by an instrumental reinforcer directly strengthens this response, and increases the probability that it will subsequently be given by the animal. Under the law of effect, the success of instrumental conditioning is specifically attributed to the fact that behaviour can be directly modified by its consequences for the animal.

Having accepted the need for an instrumental principle of reinforcement, we might well reverse our question and ask whether the principle of stimulus substitution is required to explain classical conditioning. Perhaps the growth of classically conditioned responses can be explained in terms of the law of effect. The main strength of the principle of stimulus substitution is its ability to explain the similarity of the conditional response to the unconditional response. However, if the responses elicited by a particular unconditional stimulus are usually determined by innate neural mechanisms, one may ask why evolution has brought about the development of these specific responses. A usual reply is that these responses enhance the ability of the animal to cope with that particular stimulus. Salivation in the dog aids mastication and swallowing of food, while eyelid closure in the rabbit potentially defends the eye against damage by noxious stimuli presented in the orbital region. Given this perspective, perhaps the reason why the dog salivates to a positive Pavlovian conditional stimulus is because the salivation is in reality an instrumental response which is positively reinforced by the fact that it enhances the value of the food. Similarly, an eyeblink to an aversive Pavlovian conditional stimulus might be negatively reinforced by the fact that it reduces the aversive consequences of the shock. This is tantamount to saying that the law of effect will often select conditional responses in a classical conditioning situation which are similar to the unconditional responses because this principle is in some way analogous to the laws governing the selection of the unconditional responses by evolutionary mechanisms; namely that these responses enhance the ability of the animal to cope with that type of stimulus in nature.

As in the case of instrumental conditioning, it is impossible to come to a general conclusion about whether classical conditional responses are strengthened according to the principle of stimulus substitution, or to the law of effect. The best we can hope to do is to show that stimulus substitution is required to explain the reinforcement of a particular response. The way we do this is by investigating whether the instrumental contingency we think might be operating will in fact strengthen the response to the same extent as a simple classical conditioning procedure. Consider the conditioning of the rabbit's eyeblink. An explanation in terms of the law of effect maintains that the conditional eyeblink is reinforced because it reduces the noxious effect of the air puff, while stimulus substitution maintains that the eyeblink occurs because the conditional stimulus acquires similar properties to the air puff. Suppose that we arrange that every time the animal blinks to the conditional stimulus, the air puff is omitted; or, in other words, establish an avoidance contingency with the eyeblink as the avoidance response. If the eyeblink can be instrumentally reinforced this should, presumably, increase the magnitude of instrumental reinforcement available. We can then compare the strength of conditioning in animals receiving this avoidance procedure with that in animals simply given pairings of the conditional stimulus and air puff in a Pavlovian manner with no avoidance contingency. Experiments of this type show that in some cases stronger conditioning occurs under the simple classical conditioning procedure than in the avoidance situation, in spite of the potential source of instrumental reinforcement available with this latter procedure. Clearly, the fact that the eyeblink is stronger in the classical conditioning situation in spite of the absence of a potential source of instrumental reinforcement shows that this particular conditional response is not reinforced according to the law of effect.

In conclusion we need both the principle of stimulus substitution and the law of effect to provide a general account of reinforcement. However, in the absence of experimental analysis, we often cannot determine which principle is operative simply on the basis of the type of procedure (classical versus instrumental) used to establish conditioning. Of course, the same caution is required in determining the principles of reinforcement at work in the natural environment of the animal.

Selectivity of reinforcement. So far we have assumed that any stimulus to which the animal is sensitive can come to substitute for a Pavlovian reinforcer, and that any response in the animal's behavioural repertoire can be strengthened by an instrumental reinforcer. This does not appear to be so. The effectiveness of a given reinforcer may depend upon the particular conditional stimulus or response employed.

Rats will show conditioned aversion to a flavoured solution if this solution is ingested before gastric upset is produced by exposure to X-rays or by the injection of a poisonous substance. This in itself is not surprising. More notable is the fact that a marked aversion can be seen after a single pairing of the flavour and the poison, even if the interval between the two events is an hour or so. Nothing like the same strength of conditioning will be seen if either a shock is used as the aversive reinforcer, or an external stimulus, such as a light or tone, as the conditional stimulus. Whether the difference is due to the fact that rats, as omnivorous animals, are 'biologically prepared' to develop aversions to tastes associated with illness, or to the past experience of the animal with these two classes of events is not known. (see FOOD SELECTION). However, it is clear that for a given species Pavlovian conditioning develops more readily with the conjunction of certain classes of reinforcers and conditional stimuli.

A comparable situation also holds for instrumental reinforcement. Sometimes it is claimed that only so-called VOLUNTARY responses can be instrumentally reinforced, although the problem of the instrumental reinforcement of responses mediated by the AUTONOMIC nervous system is a hotly disputed topic. Even so, the responses within an animal's behavioural repertoire which are mediated by the striate musculature (which moves the skeletal parts) do not appear to be equally modifiable by instrumental reinforcement. For example, if different response patterns of the golden hamster (*Mesocricetus auratus*) are instrumentally reinforced by food, certain responses, such as scrabbling, digging, and rearing, show much larger increases in the time the animal spends performing them than other responses, such as face washing, scratching, and SCENT MARKING. What is particularly significant is that those responses which are susceptible to instrumental reinforcement by food are just those behaviour patterns which show a spontaneous increase when the animal is deprived of food in the absence of any reinforcement. It appears that a response can be more easily reinforced by a given reinforcer if the response and reinforcer are related to the same motivational system.

However, it is not always obvious whether the difficulty in establishing a certain type of instrumental conditioning arises from the inherent ineffectiveness of a particular reinforcer for a particular response. Often instrumental conditioning can be masked by competition from other factors. This can be illustrated by an experiment on the male three-spined stickleback (*Gasterosteus aculeatus*). Although the response of swimming through a ring can be easily reinforced by the opportunity to court a female, it is remarkably difficult to instrumentally condition the fish to bite a rod for the same reward. Observation reveals that the fish tends to direct its COURTSHIP to the rod. When viewed in terms of stimulus substitution, this finding makes sense: the instrumental conditioning procedure ensures that the sight of the rod is followed by the presentation of the female, thereby embedding a classical conditioning relationship within the instrumental contingency arranged by the experimenter. By the principle of stimulus substitution, the rod should gain some of the properties of the female. As stimuli arising from the female probably exert an inhibitory effect on the biting response, the rod as a Pavlovian conditional stimulus should have a similar effect. So it can be seen that the difficulty in instrumentally reinforcing the biting response does not arise from the ineffectiveness of the female as an instrumental reinforcer for that response but from the presence of a competing inhibitory tendency established by classical conditioning.

Relationship between reinforcement and learning. Obviously the processes of reinforcement and learning are closely interrelated, for reinforcement directly refers to the strengthening of learned responses. However, the nature of this relationship is rarely made explicit.

Although psychologists tend to use the concepts of reinforcement and learning as though they are synonymous, there is no doubt that associative learning can occur in the absence of reinforcement. In a sensory pre-conditioning experiment, two neutral stimuli, such as a tone and light, are paired in a classical conditioning procedure with, for example, the tone as the conditional stimulus and light as the unconditional stimulus. The fact that no obvious conditional response develops to the tone shows that a reinforcement process is not engaged by this procedure. The animal will, however, often learn about the association between the tone and light. This can be demonstrated by then, in a second stage, making the light a conditional stimulus for an effective Pavlovian reinforcer, such as food. If we now present the tone to the animal it will elicit a conditional response even though it has never been directly paired with the reinforcer. The only way this could happen is if the animal learnt about the association between the tone and light in the first stage when there was no effective reinforcer present. Given that associative learning can take place in the absence of a reinforcer, the mechanisms underlying learning cannot be reduced to those of reinforcement.

A second point is that the principles of reinforcement, stimulus substitution and the law of effect, can be stated in a general way which does not specify exactly what is learned during conditioning. Obviously the principle of stimulus substitution is most easily reconciled with a theory of learning which assumes that the animal forms associations between internal representations of the conditional and unconditional stimulus. However, it is also compatible with the idea that the animal forms a direct association between an internal representation of the conditional stimulus and the mechanisms responsible for the unconditional responses, and also probably with a number of other accounts of learning. Similarly, the general law of effect can be linked with at least two

different accounts of the associations formed during instrumental conditioning. Stimulus–response theorists have argued that instrumental behaviour is mediated by an association between the response-producing mechanisms and the stimulus environment in which the response is reinforced. From this viewpoint, our rat forms an association between the stimuli of the experimental chamber and the mechanisms generating the lever-press response. However, some scientists have maintained a position more in line with common sense and argued that instrumental learning results from the formation of an association between the response and its consequences, namely the instrumental reinforcer. Whatever the merits of these different theories of learning, both are compatible with the law of effect as a principle of reinforcement. From this perspective the process of reinforcement is best regarded as the mechanism by which learning and motivation interact to control learned behaviour.   A.D.
98

REINFORCEMENT SCHEDULES. Everyone knows that animals, like people, tend to learn things that lead to reward and to give up punished behaviour; hence the principle of REINFORCEMENT, the strengthening or weakening of behaviour depending on its consequences. Historically, the term arose from the so-called *law of effect*, which summarized the results of the American psychologist E. L. Thorndike's experiments on cats (*Felis catus*) escaping from puzzle boxes. Although he was later to revise his position substantially, Thorndike's first statement of the law—that animals tend to repeat actions that have satisfying consequences and desist from actions that have unsatisfying ones—was the most influential.

The Harvard behaviourist B. F. SKINNER extended Thorndike's law of effect. He simply made it a definition of reinforcement: a reinforcer is anything which, if made contingent upon some aspect of behaviour, causes that behaviour to increase in frequency. Of course, this definition implies that all, or at least most, reinforcers affect behaviour in approximately similar ways and through similar mechanisms. Skinner also assumed that the response, the reinforcer, and the stimulus in the presence of which the reinforcement occurs, are all more or less arbitrary. Any reinforcer could strengthen any response in the presence of any stimulus, provided the response was within the animal's motor capacity and the stimulus within his sensory range. Subsequent research shows that this kind of generality cannot by any means be assumed. For example, the Norway rat (*Rattus norvegicus*) will much more readily learn to avoid tastes associated with sickness than sights or sounds similarly associated. Pigeons (*Columba livia*) learn about visual stimuli in food situations much more readily than about sounds, and so on.

Let us consider a pigeon which has been trained to peck at an illuminated disc, called a key, to obtain a food reward. When the pigeon pecks the key it closes a sensitive switch in an electronic circuit, which causes food to be delivered automatically. This experimental technique is widely used in the study of OPERANT behaviour, and it has proved particularly useful for studying the effect of various types or patterns of reward. Thus instead of rewarding a pigeon for every peck, it can be rewarded for every *n*th peck, so that there is a specified relationship between the number of pecks delivered by the pigeon and the number of rewards obtained. Such a relationship is called a *reinforcement schedule*.

In nature, every response made by an animal has consequences which are detected by the animal and are capable of modifying its behaviour. A reinforcement schedule merely makes responses and their consequences, and the relationship between the two, explicit in a LABORATORY setting. The experimenter decides which aspect of the animal's behaviour, such as the key-pecking response, is to be followed by consequences of importance to the animal. The experimenter decides upon the relationship between the response and the reinforcement in that he designs the reinforcement schedule. Of course, the exact relationship is determined by the animal, since its response necessarily determines the exact timing of the reinforcement.

Much of the interest of reinforcement schedules derives from the fact that many species of animals show similar, predictable adjustments to them. Four simple schedules have been widely studied. These are *fixed-* and *variable-ratio schedules*, and *fixed-* and *variable-interval schedules*. On a fixed-ratio 20 schedule, food delivery follows immediately on every 20th response, independently of its time of occurrence. On a fixed-interval T seconds schedule, food follows immediately on the first response T seconds or more after food delivery. On fixed-ratio schedules, animals typically pause briefly, and then respond at essentially their maximum rate until food is delivered; whereas on fixed-interval schedules, they pause for about two-thirds of the interval, and then (even though only a single response is required for each food delivery) respond at an accelerating rate until food is delivered.

The number of responses for each food delivery on ratio schedules, or the time that must elapse before a response is effective on interval schedules, need not be fixed, but can vary from one food delivery to the next in some fashion. Such schedules are termed variable-ratio or variable-interval. If the exact moment of food delivery is completely unpredictable, animals tend to respond at a more or less steady rate, with no pauses, corresponding to the fact that an opportunity for food can occur at any time. Experiments have shown that animals' rate of responding is closely related to the distribution of opportunities for food: responding ceases only when there is no opportunity for food. For example, on both fixed-interval and fixed-ratio schedules, animals do not make the instrumental response in the period just after each food delivery, and this corresponds to the fact that, in their experience with the schedule, food is never delivered during that time.

Numerous more complex reinforcement schedules have been studied: for example, schedules that require the animal to space his responses in time; schedules whose intervals or ratios adjust, depending on the animal's past performance; schedules in which brief neutral stimuli are sometimes presented in place of food; chained schedules, in which responding sometimes produces food and sometimes produces a change of stimulus; multiple schedules, in which different simple schedules are associated with different, successively presented stimuli, so that an animal may show quite different patterns of responding, depending on the prevailing stimulus; and many others.

The extreme reliability of schedule performances, and the automatic nature of the procedures, has made operant conditioning a useful tool for exploring the effects of DRUGS. Many pharmaceutical companies now maintain large operant conditioning laboratories studying the effects of drugs on the behaviour of animals. Interesting quantitative relations have emerged from this work, most notably the rate-dependence of drugs such as amphetamine. This drug, normally thought of as a stimulant, in fact has different effects depending on the rate of the baseline behaviour: if the animal is on a schedule (such as fixed-ratio) which produces a high rate of responding, amphetamine injections will generally reduce the rate; however, if the schedule is one that produces a low rate, amphetamine will generally increase it. This property of amphetamine may be related to its effectiveness in moderating the behaviour of so-called hyperkinetic (overactive) children.

A reinforcement schedule usually arranges for a three-term relation between an instrumental response, a stimulus (or time), and reinforcement. This three-term relation can be broken down into three aspects. Consider, for example, a fixed-interval schedule. Normally, the experimental subject— a pigeon, say—receives most of its food in the experimental apparatus. Hence the stimuli associated with the apparatus predict food better (are associated with a higher rate of reinforcement) than the other stimuli he sees through a 24-h day. This kind of predictive relation between a stimulus and the occurrence of a reinforcer is called a stimulus contingency. Once the animal is in the Skinner box, however (see OPERANT BEHAVIOUR), reinforcement is more probable at some times than at others: at post-food times greater than the interval value, the probability of food is one, but at other times it is zero. This is a temporal contingency, a predictive relation between time and the delivery of food. Finally, if the animal does not make the

instrumental response, the probability of reinforcement is zero, but if he does respond the probability is finite (albeit small on most intermittent schedules). This is a response contingency.

Much recent work on operant conditioning is concerned with how these three kinds of predictive relation jointly affect the animal's behaviour. Several conclusions have emerged. (i) Stimulus and temporal contingencies alone are sufficient to produce behaviour, even behaviour conventionally regarded as operant, such as key pecking by pigeons. (ii) Habitual behaviour is often maintained almost entirely by stimulus and temporal contingencies, so that abolition of the response contingency alone (e.g. presenting food periodically, independently of behaviour, to an animal already trained on a fixed-interval schedule) often has little effect: animals (and people) continue to respond even though responding is now unnecessary. (iii) Stimuli that strongly predict reinforcers induce characteristic behaviour, the terminal response, that is usually related to the reinforcer, e.g. salivation or pecking for a food reinforcer, courtship behaviour for a sexual reinforcer, and so on (see LEARNING). Stimuli that strongly predict the absence of a reinforcer usually induce IRRELEVANT activities. The most striking of these is the excessive drinking (polydipsia) shown by rats and several other species on food schedules, at times when food delivery is improbable. (iv) The effects of stimulus and temporal contingencies, which depend on the species and the nature of the reinforcer, interact with the effects of the response contingency in ways yet to be explored fully.   J.E.R.S.

126

RELEASER is an alternative term for SIGN STIMULUS. The term developed in conjunction with the idea that motivational ENERGY was released by the action of a specific external stimulus upon an INNATE releasing mechanism. The appropriate behaviour was then discharged. Thus the red belly of the male three-spined stickleback (Gasterosteus aculeatus) may act as a releaser for AGGRESSION in a rival male. LORENZ pointed out that sign stimuli and the specific response that they release become mutually adapted to each other during the course of EVOLUTION. This idea is generally accepted, but the mechanisms of PERCEPTION and their effect upon MOTIVATION are thought to be more complex than those postulated by early students of ETHOLOGY.

137

REWARDS are obtained by animals during the process of LEARNING. Although in everyday language rewards are obtained for duties performed or general good conduct, in animal psychology the term is restricted to the learning context.

Rewards strengthen learning if they are positive and weaken it if they are negative. Thus a pigeon

(*Columba livia*) that is rewarded with food for pecking at a disc is likely to repeat the response, whereas a pigeon that is rewarded by a mild electric shock is less likely to peck the disc in future. Some psychologists employ the term REINFORCEMENT in place of reward. Others, however, reserve 'reinforcement' for the process of strengthening and weakening that occurs as a result of rewards being received. Thus the pigeon that obtained food would experience positive reinforcement, while the pigeon that received a shock would experience negative reinforcement. The situation is complicated by the fact that the question of whether reinforcement is necessary for learning is a controversial one. It is evident that learning can occur in the absence of overt reward, as in the cases of IMPRINTING and the SONG learning of birds. Some psychologists argue, however, that all learning requires reinforcement, since this is the process of strengthening and weakening stimulus–response connections.

Many students of ETHOLOGY and others not specialized in the field of animal learning use the term 'reward' in a general sense that is not far removed from its everyday usage.

RHYTHMS occur commonly in animal behaviour. Some rhythms, such as those involved in animal LOCOMOTION, are fairly readily observed, whilst others, such as daily or annual rhythms, require routine observation before they become apparent. In general, there are three main ways in which rhythms can arise: (i) they may be driven by a biochemical CLOCK, which is *endogenous* in the sense that the rhythmic mechanism is independent of external events; (ii) they may be *exogenous*, entrained to rhythmic external stimuli, such as daily fluctuations in temperature or light intensity; or (iii) they may arise as a result of the physiological or mechanical organization of the animal, in the way that a steam engine exhibits rhythmic movement. This third type of rhythm is called a *relaxation oscillation*, because it involves an alternating build-up and release of energy, as with the steam in a piston cylinder.

Rhythms involve wave-like variations in behaviour, which can be characterized by their period, frequency, and amplitude, as illustrated in Fig. A. The *period* of a rhythm is the interval of time between successive peaks in behaviour. For example, tidal rhythms have a period of about $12\frac{1}{2}$ h. The *frequency* of a rhythm is expressed as the number of peaks in a given period of time. Thus we may speak of a heartbeat frequency of sixty beats per minute. The *amplitude* of a rhythm is a measure of the magnitude of the change in behaviour that occurs between a peak and a trough. For example, in normal human walking the head moves rhythmically up and down with an amplitude of about 2 cm. In discussing the rhythms typical of animal behaviour, it is convenient to start with those with the longest period, and progress towards those of short periodicity.

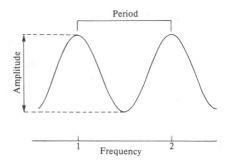

Fig. A. Standard terminology for rhythmic behaviour.

Circannual rhythms. In many parts of the world the seasonal CLIMATE means that animals have to adjust by organizing much of their behaviour on an annual basis: seasonal periods of unfavourable climate can be avoided by annual MIGRATION or by HIBERNATION; animals may take advantage of regular favourable seasons by organizing their reproductive behaviour on an annual basis. In some species it has been shown that such annual changes in behaviour are based upon an endogenous biological clock. For example, the marine polyp *Campanularia flexuosa* has an annual cycle of growth, development, and longevity, as illustrated in Fig. B. Under constant laboratory conditions the rhythm persists, although the period is slightly longer than 1 year. The term *circannual* is used to refer to a rhythm which has a period of about 1 year under constant laboratory conditions, where all exogenous influences are excluded. The persistence of rhythms under such conditions is evidence that the clock is endogenous.

In birds, endogenous circannual rhythms have been established in European warblers. Some of the warblers, such as the garden warbler (*Sylvia borin*), the sub-alpine warbler (*Sylvia cantillans*), and the willow warbler (*Phylloscopus trochilus*), are long-distance migrants which show marked seasonal changes in body weight, moult, testis size, nocturnal restlessness, and food preferences. The European populations of these species winter in Africa and migrate across the Sahara desert. When they are maintained under constant laboratory conditions in Europe, having been hand-reared from a few days after hatching, the normal seasonal changes associated with migration still occur. It appears that there is an INNATE endogenous circannual clock, which is responsible for the timing of the physiological and behavioural changes associated with migration. Warbler species that migrate only short distances, such as the blackcap (*Sylvia atricapilla*) and the chiff-chaff (*Phylloscopus collybita*), or partial migrants, such as the Sardinian warbler (*Sylvia melanocephaia*), also show circannual changes when reared in the laboratory, but these are not as marked as those of the long-distance migrants.

There have been many studies of the seasonal rhythms of reproductive activity in birds, and some of these suggest that an endogenous process is involved. Most of these studies concern species that breed exclusively in the temperate zone, and at these latitudes an annual breeding cycle is likely to have high SURVIVAL VALUE because of the alternation between seasons favourable and unfavourable for the survival of the offspring. At equatorial latitudes, however, where there are seasonally stable food supplies, and less marked seasonal changes in climate, it might be of advantage to maintain reproductive condition continuously. Nevertheless, some equatorial species do show a marked breeding rhythm that appears to be unconnected with seasonal changes. Several sea-birds, including the sooty tern (*Sterna fuscata*), the brown booby (*Sula leucogaster*), and the lesser noddy tern (*Anous tenuirostris*), breed at intervals of 8–10 months under equatorial conditions. The progressive shift between such breeding cycles and the seasonal cycle suggests that no one season is preferable, but that there is some benefit inherent in the breeding rhythm itself. Perhaps the energy required for continuous reproductive activity is prohibitive, and a periodic rest during which moult can occur is beneficial.

In mammals, there is evidence for endogenous circannual rhythms in a few species. In the Western European hedgehog (*Erinaceus europaeus*), and in man too, there are marked physiological changes, which are correlated with the seasons, and may well be endogenous. In the golden-mantled ground squirrel (*Citellus lateralis*) there is a marked circannual rhythm, which is evident even when the animals are maintained under constant laboratory conditions. This small rodent is found in western North America at altitudes of 1500–3600 m, ranging from northern British Columbia to southern California. Under natural conditions they hibernate for a 3- or 4-month period, during which their body weight falls considerably. Food consumption increases rapidly after hibernation, reaching a peak in midsummer, and body weight increases up to the onset of hibernation in Octo-

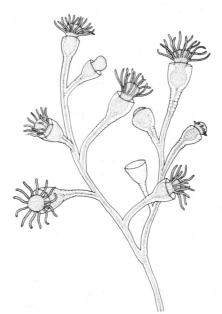

Fig. B. *Campanularia flexuosa* (above) consists of a number of bell-shaped polyps attached to a branching stalk. In the graph (below) the average life span of a polyp cultured in the laboratory at 10 °C is plotted against the water temperature of the natural environment. A clear circannual rhythm is evident.

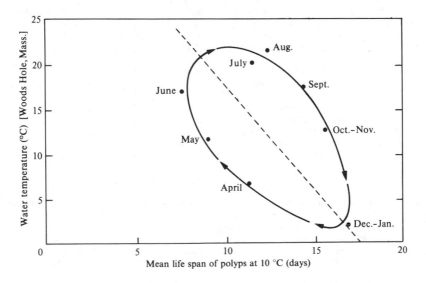

ber. In laboratory experiments on these animals all external factors are maintained as constant as possible, and all disturbances kept to a minimum. The rhythm of hibernation has been observed to continue for over 2 years under these conditions, clearly showing that a circannual clock is involved. The period of the rhythm under laboratory conditions is actually a little shorter than the natural period. Normally the clock is reset to the rhythm of external events by some particular *Zeitgeber*, or time-setter, such as the duration of the daylight period. This has been clearly demonstrated in the woodchuck (*Marmota monax*), which is also known to exhibit a circannual rhythm of hibernation. Woodchucks from the eastern USA have an annual rhythm of deposition of fat which is maintained under laboratory conditions. When transported to Australia and exposed to the natural light and temperature conditions there, but with food and water freely available, woodchucks were found to reverse their normal rhythm within 2 years, bringing them into line with local conditions.

Circannual rhythms are evidently widespread in the animal kingdom, and it is generally assumed that they enable animals to anticipate seasonal changes in environmental conditions. Such a mechanism clearly facilitates the accurate timing of events, especially under conditions where climatic factors may vary from year to year. Such precision is particularly important in the timing of migration, hibernation, and various aspects of reproductive activity.

Lunar and tidal rhythms. A number of animals are known to exhibit rhythms of behaviour which correspond to the lunar cycle of 29.5 days (see TIME). The most famous is the palolo worm (*Eunice*), in which reproductive activity reportedly occurs only during the neap tides of the last quarter moon in October and November. Laboratory studies indicate that rhythms related to the lunar cycle are, in some cases, based upon an endogenous clock. In the sea-hare *Aplysia*, a marine mollusc, nerve cells have been found which have a rhythm of activity with a period exactly half that of the lunar cycle. Although most laboratory studies have been carried out with marine animals, other cases are known. The freshwater guppy (*Poecilia reticulata*) has a rhythm of change of spectral-sensitivity (see COLOUR VISION) which corresponds to the lunar cycle, and the terrestrial beetle *Calandra granaria* has cycles of phototaxic responsiveness (see TAXES) which also correspond to the lunar cycle. Most laboratory studies, however, have been concerned with tidal rhythms in animals.

The tides result from changes in the combined gravitational pull of the sun and moon. The tidal cycle is repeated twice in the lunar cycle. Thus for about 7 days after the new or full moon, the sun and moon move progressively out of line, and the tides progressively diminish until the sun and moon are at right angles relative to the earth.

After this neap tide, the tide range increases for 7 days to the spring tide. Because the earth moves progressively around the sun, an annual rhythm is superimposed on the lunar cycle. As the equinoxes (about 21 March and 21 September) are approached, the spring tides become progressively larger, because it is at these times that the earth, sun, and moon come most nearly into line. Conversely, the spring tides are smallest at the time of the summer and winter solstice (about 21 June and 21 December), when the alignment is least. Although the general pattern of tidal rhythms is the same around the world, it can be modified considerably by local conditions.

Numerous marine animals show rhythms of behaviour that coincide with the tidal cycle, and continue under constant laboratory conditions: the shanny (*Blennius pholis*), a littoral fish, shows a rhythm of swimming activity with an approximately 12-h period; fiddler crabs (*Uca*) emerge from their burrows at low tide and become very active, foraging, courting, etc., creeping back to their burrows with each flood tide. The crab's rhythm of activity can persist for up to 5 weeks under constant laboratory conditions. There is a considerable amount of evidence that the endogenous clock controlling the tidal rhythms of many marine animals is of a biochemical nature. In some cases it is clear that the behaviour of the animal is controlled by both a tidal clock and a 24-h clock. The shore crab (*Carcinus maenas*), for example, often shows a daily rhythm of activity which is superimposed upon a tidal rhythm. It appears that the superimposition of two rhythms can allow the animal to adapt to the irregular changes of high and low tide that occur in some parts of the world. On the Californian coast, for example, a time interval of 13.80 h between high tides is followed by one of 10.43 h, and these two periods alternate with each other, as illustrated in Fig. C. The intertidal crustacean *Synchelidium* has a pattern of swimming activity which closely follows this tidal pattern (Fig. C), and this swimming rhythm continues for several days in the laboratory.

Circadian rhythms. Changes in environmental conditions between night and day are a feature of the world of most animals. Changes in climatic factors, such as temperature and light intensity, affect animals directly, but there are also indirect effects of climate, such as fluctuations in food availability and in numbers of predators, which make the difference between night and day an important aspect of animal life.

In adjusting to the differences between night and day, an animal organizes its behaviour according to a daily routine or rhythm. The daily rhythms of animals have been the subject of considerable study, and there is now little doubt that they are generally controlled by an endogenous clock which is capable of being entrained to exogenous factors; the daily rhythms are thus synchronized with local conditions. For example, the lizard *Lacerta sicula*, when hatched in an in-

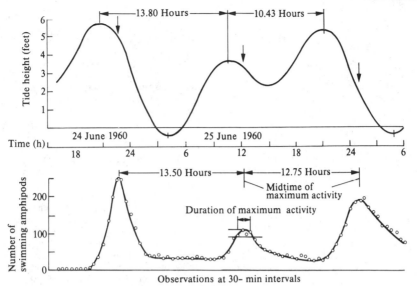

Fig. C. Changes in tide height at La Jolla, California (above), compared with the swimming activity of *Synchelidium*, under laboratory conditions.

cubator under a temperature and light regime designed to correspond to a day length of 16 and 36 h respectively, exhibited the normal 24-h rhythms of activity when tested under constant conditions. This shows that the 24-h rhythm is endogenous and does not depend upon individual experience of the normal day–night cycle. However, under constant laboratory conditions the 24-h periodicity drifts slightly from the norm. Such patterns, with a period of about one day, are called *circadian* rhythms. Under natural conditions a 24-h rhythmicity is maintained by some exogenous *Zeitgeber*, which serves to continually reset the endogenous clock and prevent it from drifting out of phase with the cycle of environmental change. In the case of the lizards mentioned above, it has been found that synchronization can be maintained by quite small daily fluctuations in environmental temperature. In laboratory experiments, a 24-h temperature cycle with an amplitude of 1.6 °C was sufficient to maintain 75% of the animals in synchrony, while at an amplitude of 0.9 °C only 25% were synchronized.

The most important daily changes in the external environment are those in temperature and light intensity. In a hot climate it is advantageous for a small animal to be nocturnal, and so to avoid activity in the heat of the day. In cold climates also, it can be of advantage for small mammals to be active at night when temperatures are low. This is because their period of greatest heat production coincides with the colder part of the day, and the period of activity is thus harnessed as a means of THERMOREGULATION. In other species it is advantageous to be active in the daytime, especially if they are able to sleep in a warm nest at night.

Rhythms of rest and activity are widespread in the animal kingdom, and offer a number of benefits to animals which possess them. Thus animals specialized for daytime VISION may be disadvantaged at night, because they cannot forage efficiently and may be in danger from predators. For such species there are advantages in sleeping at night: SLEEP provides an opportunity for saving energy and for avoiding predators. Although it is often difficult, and sometimes impossible, to determine whether the rest period of a particular species constitutes true sleep, it is generally true that rest and sleep are the activities with the lowest energy expenditure. When it is positively disadvantageous to be active at night, the best policy is often to sit tight and save as much energy as possible. Moreover, in many species the rest period of the day is also the most secure: by choosing particular places to sleep, animals can avoid predators and inclement aspects of the weather.

So we see that daily rhythms in the physical environment make some activities advantageous at one time and disadvantageous at another. An animal which is adapted to its environment will inevitably settle into a daily routine, designed to maximize the survival value of its various activities. It appears that, in many species, the daily routine is controlled to some extent by an endogenous circadian clock, and in the laboratory animals sometimes retain the daily routine that they exhibit in the natural environment. Daily activity patterns may be altered by LEARNING, and may change in characteristic ways at different seasons of the year. For example, in nocturnal animals, such as bats (Chiroptera) and deer mice (*Peromyscus maniculatus*), which become active at sunset, total activity time tends to be longer in winter than in summer. In diurnal animals, on the other hand, it tends to be shorter in winter. However, jackdaws (*Corvus monedula*) depart from their roosts later and return earlier, in relation to twilight, in summer than they do during winter. Thus they counteract the seasonal effect of daylight on activity time. The great tit (*Parus major*) also tends to compensate for seasonal and geographic differences in daylight. The rest period during the summer at latitude 68° N begins at the same time as that at 52° N, despite the differences in daylength. The northern birds therefore have a shorter rest period and are more active at lower light levels during winter than during summer. This adjustment is necessary to enable the animals to obtain sufficient food in winter.

Short-term rhythms. Rhythms of activity of relatively short duration are known in a number of species. Rhythms of feeding, drinking, preening, etc. have been studied in the Barbary dove (*Streptopelia risoria*), zebra finch (*Taeniopygia guttata*), and skylark (*Alauda arvensis*). The period of these rhythms ranges from a few minutes to about 40 min. In some cases it has been possible to establish that the dominant rhythm is that of feeding, and other activities, such as grooming, or drinking, tend to fit into the gaps between meals. It might thus appear that the feeding rhythm is established by an alternate waxing and waning of HUNGER, such that a build-up of hunger induces feeding, which then temporarily reduces hunger. Such a situation would constitute a relaxation oscillation totally dependent upon the nutritional state of the animal. However, the mathematical analysis of meal patterns in animals tends to suggest that some independent timing factor is also involved, so that the animal normally takes up an endogen-

ous FEEDING routine which is merely modified by its degree of hunger. A similar type of organization seems to exist in the case of locomotory rhythms.

The rhythmical characteristics of locomotion are usually generated endogenously by the central nervous system (see COORDINATION). In the intact animal, the pattern is modulated by input from SENSE ORGANS, particularly those involved in sensing the movement of limbs. In many species of fish, the fins move rhythmically with a frequency that is primarily determined endogenously. However, each fin can beat with a rhythm of its own, and the rhythms of different fins can influence each other. Two main principles are involved in this *relative coordination*. First, the frequency of one rhythm may attract that of another, resulting in the so-called *magnet effect*. Thus, where two fins initially are beating with independent rhythms, one rhythm may dominate the other, which gradually falls into step. Second, the amplitudes of different rhythms may interact, giving a *superposition* effect in which, for example, the amplitudes of different fins may add or subtract from each other at different parts of the cycle. These phenomena have also been observed in the locomotion of domestic chicks (*Gallus g. dómesticus*), as illustrated in Fig. D. As the chick walks the head nods so that the eye remains stationary for a moment, while the rhythms of head and leg movement may fall into complete coordination in which the pattern of leg movement is usually dominant. Sometimes relative coordination occurs, so that the rhythmic relationship between head and leg resembles that of two different fins on a fish.

The rhythmic aspects of animal behaviour seem to conform to a common pattern with an endogenous clock or pacemaker, modulated by an ex-

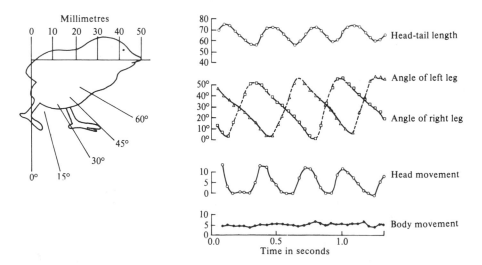

Fig. D. Locomotion in a 3-h-old domestic chick, showing rhythmic coordination of head and leg movement.

ternal *Zeitgeber* or by the consequences of the animal's own behaviour. This pattern is very widespread in the animal kingdom: it governs the swimming movements of crayfish *Procambarus clarkii*; it provides the basic principle in the control of insect FLIGHT; it appears in the locomotion of vertebrates, in the circadian rhythms of many animals, and even in the timing of seasonal events. 25, 100, 115

RITUALIZATION is an evolutionary process by which behaviour patterns become modified to serve a COMMUNICATION function. Evidence that such a process of evolution has occurred comes primarily from COMPARATIVE STUDIES of animal behaviour. For example, male zebra finches (*Taeniopygia guttata*) often perform beak-wiping during COURTSHIP, and this is thought to be a DISPLACEMENT ACTIVITY. In the related striated finch (*Lonchura striata*) and spice finch (*Lonchura punctulata*) the male performs a bow, and remains with head lowered for a few seconds. No beak-wiping occurs, but the similarity between the bow and beak-wiping in the zebra finch, and the context in which the bow occurs, strongly suggest that the bow is a ritualized form of beak-wiping. Many other instances of ritualization have been carefully studied by students of animal behaviour. In general, whenever it is of advantage for an animal that some incidental aspect of its behaviour be understood by another animal, NATURAL SELECTION operates to transform the behaviour pattern into a more reliable and conspicuous form of communication.

There are many ways in which a behaviour pattern can be changed during the process of ritualization. Frequently the behaviour becomes stereotyped in form, and incomplete in its execution. For example, ritualized grooming is often restricted to certain parts of the body, and the movements themselves become curtailed, and are sometimes merely token movements. In other cases movements 'freeze' into postures; many threat postures are derived in this way. A further characteristic of ritualization is that an activity which previously varied in its intensity comes to have a stereotyped and typical intensity, which makes the behaviour less ambiguous. For example, the black woodpecker (*Dryocopus martius*) drums against dry branches to indicate to other males that the TERRITORY is occupied, and to attract females. This drumming has a characteristic rhythm, which is stereotyped compared with the sound made by chipping out a nest cavity, from which it is clearly derived. During the excavation of the nest cavity, the partner doing the work will fly to the entrance of the cavity and drum with a very slow rhythm at the edge of the entrance hole. This is a signal for the other partner to relieve the bird and continue the work. Here we have two examples of the use of a typical intensity of behaviour to make a signal unambiguously different from the normal behaviour.

Ritualization often involves a change in MOTI-

VATION. A well-known example is the courtship feeding of birds. During the late stages of courtship the females of some species beg from their mates with a posture that is otherwise found only in juveniles begging for food from their parents. The females do this only during a restricted period of the breeding cycle, and they are not especially hungry at the time. Clearly, the motivation for food-begging during courtship must be very different from that of the begging JUVENILE.

The ritualization of behaviour patterns is often accompanied by the EVOLUTION of special markings or structures, serving to enhance the conspicuousness of the display. For example, in many duck species the male preens during courtship, as illustrated in Fig. A. In the common shelduck (*Tadorna tadorna*) this seems to be a genuine displacement activity, in that it occurs during con-

a

b

c

d

Fig. A. Displacement preening in various duck species: (**a**) the shelduck *Tadorna tadorna*; (**b**) the mallard *Anas platyrhynchos*; (**c**) the garganey *Anas querquedula*; (**d**) the mandarin *Aix galericulata*.

flict, but is otherwise similar to normal PREENING. The northern mallard (*Anas platyrhynchos*) lifts one wing and reveals brightly coloured feathers to which its preening is confined. Here we have the beginnings of ritualization: the preening movements are restricted in range and are directed at the conspicuous feathers. In the garganey (*Anas querquedula*) incomplete preening movements are directed at the outside of the wing, thus conspicuously displaying the light blue wing coverts. The courtship preening of the garganey drake is really sham preening, in that it is ritualized to the extent that it no longer serves any GROOMING function. The courtship preening of the mandarin drake (*Aix galericulata*) is more extensively ritualized than in any other duck species. The sham preening is merely a touch directed at an enlarged conspicuous rust-red feather, which becomes erected like a sail during courtship. The preening is further enhanced by the crest on the back of the head, which is raised so that the preening movement is emphasized.

Ritualization often involves a change of function. For example, in the black-headed gull (*Larus ridibundus*) copulation is preceded by a head-tossing DISPLAY, in which both birds walk together jerking their heads up repeatedly while uttering a soft call. This behaviour pattern is very similar to the FOOD BEGGING behaviour of juveniles. As such it would trigger PARENTAL responses and acceptance of the begging bird by the parent. In the process of ritualization the food begging has become stereotyped and exaggerated, and has come to indicate a readiness to copulate on the part of both birds. Ritualized forms of juvenile behaviour are quite common as a type of APPEASEMENT during courtship, when the two animals are initially wary of each other.

The origin of ritualized behaviour. The evolutionary origin of ritualized behaviour can be traced to a variety of types of everyday activity. However, these usually have the common feature that they are already potential sources of information to other animals. Such activities include CONFLICT behaviour, INTENTION MOVEMENTS, and displacement activities. They are especially suitable as a means of COMMUNICATION, either because they characterize a particular state of motivation, or because the form of the behaviour is itself an indication of the mood of the animal. For example, the take-off leap of a bird before flying consists of a crouching phase, in which the bird bends its legs, withdraws its head, and raises its tail, followed by a leap phase, in which it extends its head and legs and depresses its tail as it springs off. The first phase of the take-off leap may appear alone, as an intention movement, and may be repeated a number of times if the bird is 'undecided' whether to fly. Ritualized intention take-off is fairly common in birds, examples being the displays of the common golden-eye duck (*Bucephala clangula*), illustrated in Fig. B, and the American green heron (*Butorides virescens*), illustrated in Fig. C.

A common starting point for ritualization is conflict behaviour. An animal in a conflict situation has the tendency to perform more than one activity simultaneously, and may, through its AMBIVALENT behaviour, convey information about its motivational state which is significant for another individual. Thus many THREAT postures are derived from ambivalence between attack and escape, and many other displays have evolved from by-products of conflict behaviour, such as displacement activity and 'emotional' responses.

As an example, let us consider the courtship behaviour of the green heron. These birds have been studied in a salt-marsh HABITAT, where they nest in the tops of dead trees. The males arrive before the females in spring and defend a nest tree by means of a distinctive 'skow' call, which serves to ward off other males. Intruders are challenged by a *forward display*, in which the male adopts a horizontal posture with the bill pointing towards the opponent. The feathers are fully erected and the tail vibrates up and down. The heron may lunge at its opponent, waving its wings and uttering a harsh call. This display is a threat display,

Fig. B. Courtship in the golden-eye duck *Bucephala clangula*. The display of the male (left) is oriented towards the female. This display is derived from intention flight behaviour.

Fig. C. Displays of the green heron *Butorides virescens*. **a.** Forward threat display. **b.** Snap display. **c.** Stretch display.

which is thought to have been derived from intention attack behaviour.

The females are attracted by the male 'skow' call, but their attempts to approach nest sites are initially rebuffed by the forward display of the male. Despite this the females persist, and the male behaviour gradually changes. The readiness of the male to accept a female is signalled by the *snap display*. This is similar to the forward display, but the beak is pointed diagonally downwards and the mandibles snap together. This display is thought to be a ritualized form of displacement nest-material gathering. It is similar to the behaviour by which the males break twigs from trees to build the nest. Originally the gathering behaviour would have appeared as a displacement activity, resulting from the male's conflict between aggressive and sexual behaviour towards the female. It may have indicated that the male has a site suitable for a nest, but in its present highly ritualized form it signifies readiness to accept the female.

The *stretch display* of the male green heron is clearly a ritualized form of intention flight take-off, similar to that found in many other birds. The bill is turned upward, the feathers depressed, and a soft 'aaroo' call is given. In these respects the stretch display is the antithesis of the forward display, and indicates the opposite of aggressive tendency on the part of the male. The snap and stretch displays gradually give way to mutual displays in which the male and female fly around the territory together. This is followed by contact displays, such as mutual billing and preening. Copulation follows shortly after the contact displays.

The displays of the male green heron, shown in Fig. C, exhibit many typical features of ritualized behaviour. They are derived from activities which in themselves signify the 'mood' of the male: intention attack and intention take-off signify opposite moods, while displacement nest-material gathering indicates that the male is in conflict with respect to his behaviour towards the female. In the process of ritualization these activities have become modified to make them more effective as signals. In particular, the snap display and the stretch display are 'frozen' versions of twig-gathering and flight take-off respectively. In addition, their signal value is enhanced by the development of special VOCALIZATIONS and by markings associated with the postures. For example, the bright red lining of the mouth is exhibited during the forward display as the beak is directed toward the opponent and a harsh call is given. The antithesis between the forward display and the stretch display is typical of ritualized behaviour. In the forward threat display the beak, which is the male's main weapon, is directed towards the opponent, whereas it is directed away from the female in the stretch display. The forward display is accompanied by a harsh call, whereas the stretch display is accompanied by a soft call. In the forward display the feathers are ruffled, increasing the male's apparent size, whereas the feathers are sleeked in the stretch display. The ruffling of the feathers can be regarded as a ritualized form of the 'emotional' side-effects of conflict, somewhat similar to blushing in human beings (see AUTONOMIC).

S

SATIATION is a state of MOTIVATION, usually, but not always, referred to in the context of FEEDING. When feeding undisturbed from a source of freely available food, an animal will eventually stop; it is said to have reached its satiation point, or to be in a state of satiation. This is not necessarily the point at which it is incapable of eating any more food. Nor is it necessarily the state at which zero HUNGER is achieved (see HOMEOSTASIS). The factors which determine the state of satiation are complex, and may include the animal's state of ACCLIMATIZATION, its degree of THIRST, the extent of COMPETITION with other animals, and the animal's wariness concerning the proximity of predators.

During the process of satiation, the rate of feeding gradually lessens until it reaches zero. Sometimes animals stop feeding abruptly, but a gradual deceleration is more common. This phenomenon is thought to be due to a shift in the balance of the costs and benefits of feeding. When the animal is very hungry it is important to reduce the food deficit quickly before the opportunity is lost. In this sense a high degree of hunger carries a high cost, or risk of death through starvation. When the animal is less hungry the relative costs associated with the feeding behaviour itself become more pronounced. When an animal is feeding at a high rate it pays little ATTENTION to other aspects of the environment, and runs some risk of being surprised by a predator. While this risk may be worth taking when hunger is high, it becomes relatively less worth taking as the satiation point is approached. Many animals spend a progressively greater amount of time looking up from feeding and scanning the environment as the state of satiation approaches. Even animals with few predators, such as lions (*Panthera leo*) and elephants (Elephantidae), behave in this way. This may be due to the complex SOCIAL RELATIONSHIPS that occur among such species, rather than to FEAR of PREDATION. It is commonly observed that social DOMINANCE is associated with a relaxed style of eating, compared with the more furtive feeding of subordinate individuals.

Although the typical decelerating nature of the satiation process can be accounted for by the need to exercise VIGILANCE during feeding, scientists do not assume that animals weigh up the relative merits of feeding and pausing in a COGNITIVE manner. They assume that satiation mechanisms have been so designed by NATURAL SELECTION that an optimal balance of costs and benefits is achieved. The mechanisms by which the state of satiation is achieved require a separate type of investigation.

Many mechanisms are known to be involved in the process of satiation. In the blowfly (*Phormia regina*) feeding stops as a result of sensory ADAPTATION of TASTE receptors (see HUNGER). In vertebrates, SENSE ORGANS in the mouth, gut, and circulatory system provide FEEDBACK to the BRAIN about the nature of the food that has been ingested. Cognitive factors are also known to be involved in satiation. For example, if pigeons (Columbidae) which usually eat dry grain are fed wet grain, they show the normal pattern of feeding and satiation, even though wet grain produces a much greater distension of the crop. Experiments show that the pigeons behave as if they are counting the number of grains eaten. Human beings are capable of accurate FOOD SELECTION (e.g. in a cafeteria), taking the amount required for satiation before they even start eating. Considering the many different types of food involved in this selection, this is an impressive feat of predictive assessment of food requirements. Studies of FORAGING show that many animals are capable of similar judgements.

SCENT MARKING. Our own sense of smell is sufficient for us to know that animals, such as the boar (*Sus scrofa domestica*), polecat (*Putorius*), or house mouse (*Mus musculus*), have distinctive odours, but our perception of odour must be crude compared with that of the many animals which clearly have an acute sense of smell with fine discriminatory powers (see CHEMICAL SENSES). One has only to consider the ability of a domestic dog (*Canis l. familiaris*) to identify and follow the trail left by one of two human twins, to verify this.

Many animals, including ourselves, have specialized glands on the body surface which produce scented secretions (see PHEROMONES). Scent may be released in a volatile form into the air, or the secretion of the gland may be actively deposited by an animal onto the ground or some other surface. The latter type of behaviour is known as scent marking, and it appears to be a method of COMMUNICATION between individuals. The advantage of depositing scent, rather than releasing it into the air, is that the scent will dissipate at a slow rate to form a relatively permanent signal, which will persist for many hours in the absence of the animal that produced it. This makes scent marks particularly effective in certain situations, such as marking trails or pathways, or maintaining communication between individuals in a HABITAT where encounters are infrequent, due, perhaps, to the size of the HOME RANGE, or the density of the vegetation.

It is probable that all mammals possess scent glands, and they are found also in fish, reptiles, amphibians, and in insects and other invertebrates. In birds the sense of smell may be fairly well developed (particularly in water birds), but it seems that birds have no scent glands, and there is no evidence that olfaction is important in communication. Deposition of the products of the scent glands by marking behaviour has developed particularly in the mammals, and, to a lesser extent, in the insects.

Examples of scent marking in insects are found in the honey-bee (*Apis mellifera*), and the wasp, *Vespula vulgaris*, both of which deposit scent at the entrance of the nest. This behaviour appears to be important in the recognition of the entrance. The scent may be produced from glands on the feet or elsewhere on the body. Bees scent mark in a similar way at food sources, leaving a deposit which is attractive to other bees and probably helps them to locate the food. Several species of insect lay scent trails from glands on the thorax or in the gut. For example, the fire ant *Solenopsis saevissima* lays a trail back to the nest after finding food, and this induces other ants to leave the nest and helps them to locate the food. The solitary bumble-bee *Bombus* has been observed to leave spots of scent on vegetation, with the result that a network of scent marks is formed, but here the function of the behaviour is uncertain. They may serve as ORIENTATION marks for the bee, or perhaps they play some role in the attraction of mates.

The development of scent marking behaviour is greatest in the mammals, and it is in this group that has been most extensively studied. By examining what is known for a wide variety of species some generalizations about marking can be made, although it is its diversity that is striking. Over forty different locations for scent glands have been described in various species. For example, in rodents, glands may be found on the ventral, lateral, or dorsal surface of the abdomen, on the feet, at the base of the tail, or on the face. In deer (Cervidae) and other ruminants, glands may occur between the toes, on the first or second joint of the leg, on the forehead, or immediately in front of the eye. Those of bats (Chiroptera) may be on the throat, lips, and wings. Many species have scent glands in the genital region, and some use urine and faeces as scent carriers, in which case the scent comes from associated glands: for example, scent may be added to urine from glands located in the penis, and to the faeces from anal glands. The use of faeces as a scent mark is common in the carnivores, but also occurs in various other species, such as the rabbit (*Oryctolagus cuniculus*), and the black rhinoceros (*Diceros bicornis*). Many species use urine in scent marking—in particular the carnivores, and many of the rodents and marsupials.

The use of faeces or urine in scent marking can usually be distinguished from a normal act of elimination by the frequency with which it occurs, and often by a distinctive behaviour pattern associated with deposition. This is illustrated by the frequent urinations of the dog which are characteristically carried out with one leg cocked. If scent marks are to be set frequently by such methods, then it is to be expected that adaptations in excretory function will be found. Thus, the rabbit has been reported to produce as many as eight hundred faecal pellets in 24 h, and is probably aided in this rate of production by its habit of eating its own faeces. In the bank vole (*Clethrionomys glareolus*) urination may be frequent, but the size of the one or two drops released each time is so small that it is unlikely that the bladder ever becomes empty.

The behaviour pattern used in marking is inevitably related to the position of the scent gland on the body. The golden hamster (*Mesocricetus auratus*), which has glands on its flank, marks by rubbing its side on tufts of grass or similar objects, whereas the Mongolian gerbil (*Meriones unguiculatus*) marks with a gland on the ventral surface by depressing its abdomen as it passes over small objects. Some species go to more trouble to make their scent marks: lorises (Lorisidae) and some New World monkeys (Platyrrhina) urinate on their feet, and thus spread the urine onto the branches they travel along. The dwarf mongoose (*Helogale undulata rufula*) and several closely related species adopt a handstand position to mark with their anal glands, so that the mark is placed at the maximum possible height. Some animals have been observed to prepare a site before marking. The oribi (*Ourebia ourebia*) bites off the tops of the tall grass stems of its habitat to a suitable height for marking with a gland situated in front of the eye, as illustrated in Fig. A. The grey squirrel (*Sciurus carolinensis*) will strip the bark off a branch at its marking points, possibly with the effect of adding a visual signal indicating the location of the scent mark. Bears (Ursidae) and cats (Felidae) also scratch bark from trees, and deer will strip the bark from bushes and saplings by thrashing them with their antlers. These acts may also be part of a marking behaviour. In some cases scent marks are placed without a specific act of deposition. Rodents with glands on the flank probably leave scent on the walls of burrows and runways as they pass along them, and deer and other species with scent glands on their feet may leave a trail of scent as they move about.

Why do animals scent mark? It is assumed that it is a form of communication because of the interest and behaviour shown by an animal encountering a scent mark. But animals often make no obvious response to scent marks other than to show apparent interest. When a more specific response can be seen it remains difficult to deduce what information has been transmitted. Many of the scientists who first described the scent glands of

Fig. A. Adult male oribi, (**a**) biting intact grass flowering stem; (**b**) marking bitten-off stem with its antorbital gland; (**c**) biting the tip off a previously marked stem prior to re-marking; (**d**) normal grazing.

various species avoided the question of FUNCTION. Secretions of the glands were thought to maintain the condition of the skin. This is no doubt true to some extent, but does not account for the addition of scent to the secretion. Historically, SEXUAL attraction has been a frequently suggested function of scent glands; however, sometimes more curious ideas were proposed. The European badger (*Meles meles*) has been said to lick and apparently ingest the products of its scent glands as a tonic to maintain its health. The scent of peccaries (*Tayassu*) has been thought to act as a mosquito repellent, and that of the short-tailed shrew (*Blarina brevicauda*) to make the animal unpalatable to predators. In fact, the function of scent marking remains uncertain. A common suggestion is that scent marks are territorial, warning other animals to 'keep out' of occupied TERRITORY. There is not much evidence to support this, however. If scent marks delineated boundaries they might be expected to occur particularly at the edges of the territory, but observations on the rabbit, the beaver (*Castor fiber*), the common hamster (*Cricetus cricetus*), the European water

vole (*Arvicola terrestris*), and other animals have shown that they occur throughout the home range. Moreover, experimental studies and field observations on many species do not provide any evidence to suggest that an alien scent mark is avoided. Generally such marks are approached and investigated. It is clear that in many cases marking is not related to territorial possession, in that it occurs in individuals that do not hold territories, and in species believed not to be territorial at all, such as the rhinoceros. Furthermore, if an animal is removed from its territory to a strange place, marking may continue.

Marking has been observed to occur when animals are sexually or aggressively motivated. These motivations are related at a physiological level by the fact that both marking and AGGRESSION are dependent on the sex HORMONES. Generally, the males of a species mark more than the females. If a male is castrated, the size and secretory activity of the scent gland, and the frequency of marking, are reduced. If the castrated animal is then injected with male hormones, marking activity is restored.

Marking behaviour has been observed to occur in association with aggression in many species, and it seems that it can be part of a threat DISPLAY. Animals of high DOMINANCE status are often found to mark more frequently. This is true, for example, in the rabbit, the gerbil, and the common marmoset (*Callithrix jacchus*). In the hamster, the guinea-pig (*Cavia aperea porcellus*), and other species, the victor in a fight marks after the encounter, but the loser does not. In various species, such as the guinea-pig, Maxwell's duiker (*Cephalophus maxwelli*), and the white-nosed coati (*Nasua narica*), marking may be carried out by two rivals as a preliminary to FIGHTING, apparently as a form of THREAT. A threatening element in scent marking is indicated also by the observation that threat display may be observed in response to a scent mark. Thus, the hamster may sometimes grind its teeth in a characteristic expression of threat when an alien mark is found. Similar behaviour occurs in the beaver, the mongoose, and other animals.

Although marking is clearly associated with aggression and threat in many species, its function in this context is not clear. Scent is important in aggression, in that the scent characteristic of a female, a juvenile, or a fellow colony member will inhibit aggression, whereas that of a strange male may elicit it. Thus, marking may be a means of displaying identity. Alternatively, marking may occur in aggressive encounters, with the effect of self reassurance, and may increase the animal's aggressiveness. This may be similar to the marking which occurs in response to an encounter with a predator, or other signs of danger. A captive ring-tailed lemur (*Lemur catta*) may mark when somebody approaches its cage, and in animals such as skunks (Mephitinae) scent release has become a powerful means of defence. In other species such as mice, dogs, and the black-tailed deer (*Odocoileus hemionus columbianus*) odour may be deposited when the animal is alarmed, and it seems that other animals will avoid these scents. For these situations, it can be said that marking occurs as a result of the animal itself being threatened.

Marking appears to be related to several factors which suggest that it might have some sexual function. The development of the scent glands and the frequency of marking behaviour are generally found to be greater in the male than the female. In some species (the hamster, the rabbit, the water vole) the frequency of marking by males follows an annual cycle in which the peak of activity coincides with the mating season, but in others, such as the Eurasian river otter (*Lutra lutra*), the squirrel, and the gerbil, this relationship is not found. The marking behaviour of females is believed to show variation related to their sexual cycle, such that when the animal is receptive a high frequency of marking occurs. It has been suggested that in, for example, dogs, squirrels, and the slow loris (*Nycticebus coucang*), females leave scent trails when sexually receptive and males locate them by following these.

It has been demonstrated that in the hamster and gerbil the scent glands are not essential to mating. The surgical removal of the glands appears to have no influence on breeding success. However, in some species, such as the rabbit, the orange-rumped agouti (*Dasyprocta aguti*), and several of the lemurs, scent marking has been seen during sexual excitement. In a number of species, particularly rodents, animals have been observed to urinate on each other. This often occurs during COURTSHIP and it is suggested that making the potential mate smell of 'self' has the effect of resolving the approach–avoidance CONFLICT that may be involved in such an encounter. The relationship between marking and sexual behaviour is not clear, but it appears that marking is related to sexual status, though not necessarily directly associated with sexual MOTIVATION.

So far, the examples of marking that have been given relate to SOCIAL behaviour, but this is not true of all marking. If, for example, a tree shrew (*Tupaia belangeri*) is placed by itself in a clean cage, a considerable amount of marking is observed, but as the animal becomes used to the cage, the frequency of marking declines. If an object such as a new piece of branch is then introduced, marking will be particularly directed towards this. It seems that the animal seeks to maintain its own smell throughout its surroundings. Experimental studies on various species have shown this phenomenon to be widespread. Scent may make an area familiar to an animal, in the same way that our own sense of familiarity is largely based on sight.

The importance of scent marks in communication must not be underestimated, however. In a study of the otter it was observed that animals made journeys to certain specific sites which were jointly used for scent marking by individuals from several neighbouring territories. A visiting animal would investigate the site and then add its own mark. Communal marking sites occur in many species. Access may be restricted to colony members in colonial species, such as the rabbit or the southern mountain cavy (*Microcavia australis*), or be open to individuals from neighbouring territories, as in the otter, in the cheetah (*Acinonyx jubatus*), Verreaux's lemur (*Propithecus verreauxi*), and others. The attractive properties of an alien scent mark were common knowledge to the Canadian fur trappers of the last century, who used the secretions of the beaver's scent gland as bait to trap other beavers.

The scent marks of other animals appear to be a strong stimulus in eliciting marking. For instance, the honey glider (*Petaurus breviceps*), which is a gliding possum, the hamster, and the mongoose, are all found to exhibit a higher frequency of marking when presented with an object which has been scent marked by another animal of the same species, compared with the same object presented unmarked. Similar behaviour has been observed in many other species.

Thus, marking sites may be established to which all individuals in a neighbourhood contribute, apparently as points for a general exchange of information. Experiments have shown that rats (*Rattus*) and mice and some other species can recognize the sex, breeding condition, or individual identity of other animals from their scent alone. These odour cues could be contained in a scent mark. When two animals meet it is of interest to note that as they sniff each other they pay particular attention to the region of the scent gland. Presumably, whatever information is gained about identity in this way is also present in a scent mark. In a study on the rabbit it was discovered that human observers were able to discriminate between anal gland secretions on the basis of the intensity of the odour with respect to age, sex, breeding condition, and dominance. Thus, marking sites probably contain information about the individuals who are present in the area.

There is no doubt that marking has become functionally specialized in various ways in different species. More than one scent gland is commonly found in a species, and the secretion of each may have a separate function. In the mongoose, for example, the scent from the anal gland identifies the individual, while the scent from the glands on the cheek does not have this property, and is used primarily in aggressive contexts. The effect of a scent mark may also depend on who encounters it. A mark might signal 'home' to the animal that deposited it, act as an identity check for another member of the colony, but signal threat to a foreign male, or the presence of a potential mate to the female. Marking thus appears to be functionally diverse.

Marking may be associated with responses made to frightening stimuli. This is suggested by the fact that it commonly occurs when an animal is exposed to novelty, and, sometimes, as a reaction to a predator or other danger. From this level of response, marking could have developed into an alarm signal, or into a method of defence, as in the skunk. Marking when another animal is encountered may also derive from an element of stress or fear associated with the meeting. In these situations the scent produced might act to increase the confidence of the animal producing it. In encounters, marking may have become a form of threat, or enhance sexual attraction, or it may be a means whereby animals display their identity. In all these situations scent would be equally effective whether it was deposited as a scent mark or released in volatile form. The adaptive advantages of specialized behaviour patterns for the deposition of scent are more obvious where scent marks serve to maintain communication amongst a community in lieu of encounters. It is evident that in the insects, as well as in the mammals, marking has developed specifically in those situations, such as the identification of food sources, where the release of volatile odour would be ineffective.

It may be argued that a primary function of scent marking is the advertisement of an animal's identity, but its importance in SOCIAL ORGANIZATION remains unclear. It is difficult to determine the signals carried in a scent, and the nature of their effect, when we are almost totally insensitive to these scents ourselves, and have no means of recording or measuring odour.   R.J.

43

SCHOOLING. The formation of schools, or shoals, by fish provides one of the most familiar examples of animal SOCIAL behaviour. Despite this, remarkably little is known about them. What constitutes a school? Why do fish school? How do fish school? Only in recent years have attempts been made to answer these questions.

In addition to the difficulty of observing species which live in the seas, there is the complicating fact that different species appear to school to a greater or lesser degree. At one extreme, some species form highly polarized assemblages. Individuals maintain a constant distance from and orientation to their neighbours, and all swim at a constant pace. Other species only occasionally form groups, and even then the individuals, although attracted to one another, do not show the characteristic polarization.

The extreme examples of this are fish AGGREGATIONS, where the group is formed not as a result of social attraction but merely in response to some environmental stimulus, such as a localized food source. Aggregations are not really schools at all. None the less, even discounting those species which form groups only in response to external stimuli, one can still count over 4000 different species of fish which school.

Anti-predator aspects of schooling. The fact that so many species of fish have developed the schooling habit leads one to surmise that it must have overriding SURVIVAL VALUE. Although it is possible that schooling in different species might serve different functions, most scientists agree that the primary FUNCTION of schooling is as an anti-predator device. As evidence for this, most schooling species are relatively small, and few PREDATORY fish school (with the notable exceptions of the Pacific barracuda, *Sphyraena argentea*, and tuna, *Euthynnus pelamis*). Tuna school when small, but as they grow the schooling breaks down, and large animals are solitary.

Schooling protects fish from PREDATION in a number of ways. First, a predator is less likely to encounter fish of a prey species if they stay clumped than if they disperse randomly. The mathematical proof of this is not complex, and we can understand it as follows. The maximum distance over which an object can be seen depends on the size of the object, hence the angle it subtends on the viewer's retina, and on the contrast of the object with its background (see VISION). Warships, for example, are painted grey to reduce contrast. In water, even of exceptional clarity, backscattering and absorption greatly reduce contrast, with the result that the maximum distance that an object of any size can be seen is about 200 m, and in most bodies of water a great deal less. We can imagine around each fish a sphere, the radius of which is the maximum distance at which it can be seen by a predator (the maximum detection distance). If two fish swim close together, the spheres surrounding them overlap to a large degree, so the chance of a predator finding the two of them is only slightly greater than his chance of finding either of them, had they been swimming alone, and the chance of his finding a single fish is roughly half what it would have been. If, upon discovering the hypothetical two-fish school, a predator were to eat both members, the advantage would be lost. But if, on the other hand, the school had more members than the predator could possibly eat at once, the advantage would remain. For this to provide an advantage to an individual fish requires either that members of a school be close relatives (see EVOLUTION), or that reducing the chance that a predator find any food increases an individual's survival prospects, by, for example decreasing the total number of predators.

Because of the limits to vision in aquatic environments the advantage to individuals increases as the size of the school increases, so that with 1000 fish the chance of a predator finding a member of the school is only about 1/1000 as great as if they had been randomly dispersed, and similarly the risk of an individual being eaten, should the school be discovered, is only about 1/1000 as great as it would have been had he been alone.

It is easy to see that schooling is disadvantageous for fish when dealing with human predators fishing with trawls or seines which capture all or a large part of every school they locate, and which use sophisticated electronic aids, as well as aircraft, to find the schools.

The same considerations regarding maximum detection distance and clumping, although operating in only two dimensions, led to the formation of convoys of supply ships during wartime. The chance of a ship being detected by a predator such as a submarine is reduced, and, if the convoy were detected, relatively few ships would be lost to a single submarine. Of course, instead of relying on immense numbers of ships to keep losses low, the convoys could actively defend themselves. They could also take evasive action when detected, in the form of sharp turns and, in some cases, the actual disbanding of the convoy. Fish schools, as we shall see later, take similar action when under attack by predators. In both cases, the aim is to make it more difficult for the predator to predict where an individual will be in the future.

The 'evolutionary' response of submarines to the 'schooling' of supply ships was the same as that made by tuna and barracuda: they hunted in loose schools of their own. Had they kept very close together then the total area they searched would have been very much smaller than that of the same number of single submarines. But, by staying just inside radio range of each other, they kept the total area they searched constant, while allowing each individual to benefit from the detection of a school or convoy of prey. Schooling fish do not communicate by radio, of course, but a similar result is obtained, since each predator will notice if his neighbour finds food.

Although schooling predatory fish might hunt cooperatively, like wolves (Canis lupus) or lions (Panthera leo), there is no evidence that they do so. The advantage that they gain over solitary predation is, therefore, due entirely to the optical properties of water. Schooling increases the effective SEARCHING area of each member of the school, since each can take advantage of prey located by any of the others. In air, where vision is possible over kilometres rather than metres, one would not expect flying predators such as hawks (Accipitridae) to school (or flock). This is because for the predators to increase their search area they would have to stay so far apart that it is unlikely that one could capitalize upon the success of the others. Vultures (Aegypiinae and Cathartidae) provide an interesting exception to this. Their well-known habit of HOMING in on each other's finds, however, depends upon the fact that they prey upon dead animals, which are most unlikely to move. Because of the optical properties of air, groups of birds are more conspicuous than individuals. This difference in visual range between air and water may be responsible for the fact that while 4000 species of fish school, a much smaller number of birds flock.

In order to demonstrate the advantage of schooling, it has been convenient to think in terms of a sphere of maximum detection distance. Unfortunately, in real life the situation is slightly more complicated. The reason for this is that, in water, objects are more visible from above and below than from the sides, due to the increased background contrast provided by the depths in one case and the water surface in the other. This should result in the imaginary sphere of maximum detection distance becoming elongated at the top and bottom until it resembled a Rugby football standing on end. Most open water species of fish, however, have developed special coloration called counter-shading which counteracts this (see CAMOUFLAGE). By evolving lighter bellies and darker backs they counter this increased contrast, since from above they look dark, like the depths below them, and from below they appear white like the water's surface.

Coloration of this sort is nearly universal among open water, or pelagic, fishes. Perhaps the only exception to the light belly/dark back rule is the upside-down catfish (Synodontis batensoda) which has its colouring reversed, its ventral surface being darker than its dorsal one. The apparent contradiction is solved by watching the animal feeding on surface plankton in the wild. Because, like other catfish, its mouth is located on its ventral side, it must swim upside-down to feed at the water's surface. The result is that, like other fish, it appears dark above and light below.

Fish in schools may protect themselves from predators in more ways than simply reducing the chance of detection. One of the first scientists to study schooling quantitatively suggested that in open water fish might respond to the lack of cover by 'hiding behind' each other, and thus forming schools. European minnows (*Phoxinus phoxinus*), for example, form schools which are more highly polarized and much denser than normal when under attack by a predator. This hypothesis was later termed the *selfish herd* effect, since a mathematical model based upon it dealt with land animals rather than fish.

Another suggestion is that a predator might perceive a dense group of small prey as a single large (and one assumes) frightening object. The most convincing example of this is seen in the *ball schools* which form under docks and have sometimes been observed in open water. A factor of great importance is that a predator confronted with a large number of prey simultaneously may have difficulty choosing a single individual to attack. This phenomenon is called the *confusion effect*, but actually the term may refer to two quite separate processes, one operating in the BRAIN, and the other at the periphery. In the first, the predator simply cannot make up his mind between the various prey members. This may be likened to the dilemma faced by a child in a sweet-shop: everything looks so good, he is unable to make his choice. The second case, that of peripheral or sensory confusion, can be better compared with the difficulty a man has hitting one of two tennis balls thrown at him simultaneously. Movement by one is likely to distract him from hitting the other.

There is good experimental evidence that schools do have a confusing effect on predators. The success of the common cuttlefish (*Sepia officinalis*) and the common squid (*Loligo vulgaris*) hunting schools of small silversides (*Atherina* spp.), and that of pike (*Esox lucius*) and perch (*Perca fluviatilis*) attacking schools of young cyprinids, bleak (*Alburnus alburnus*), and dace (*Leuciscus leuciscus*), has been investigated experimentally. With each of the four predatory species, increasing the number of fish in a school from one to six to twenty decreased the predators' success rate, with a concomitant benefit to the individuals in the school.

Much of the confusion effect seems to stem from the identical appearance of the different prey individuals, since it seems that predators prefer to take fish which look different from the rest, or are behaving unusually. Even small differences are sufficient to overcome the predator's inability to single out individual prey. Minnows marked with Indian ink are taken by pike attacking schools.

Similarly, anchovies (*Cetengraulis mysticetus*), whose gill covers flash when they are feeding, are taken from mixed schools in the wild. The effect is also seen when hawks prey upon pigeon flocks (Columbidae). A single black pigeon in a white flock is likely to be attacked, as is a single white pigeon in a black flock.

If minnows are placed one at a time into a large tank containing a small pike, it will devour six or seven before SATIATION, capturing each in a single lunge. If, on the other hand, a school of minnows is placed with the pike, it may take 2–3 h and eight or nine lunges to capture a single prey. Between strikes, the pike can be seen ORIENTING itself towards individual after individual, apparently unable to decide between them. However, the confusion effect may not be all that is operating here, since, once a school notices a predator, it begins DEFENSIVE manœuvres, much like the ship convoys discussed earlier, which may further reduce the predator's chance of success.

A number of these manœuvres are now well documented. When under attack, a school may respond by leaving a gaping hole, or vacuole, around the predator (as illustrated in Fig. A). More often, the school splits in half, and the two halves turn outwards, eventually swimming back around the predator and rejoining. The result of this tactic, which is known as the *fountain effect*, is that the predator is left with the school behind it. Each time the predator turns, the school splits and rejoins behind him. The advantage of this is that while the prey are generally more agile than their predators, they cannot swim as fast. By a succession of quick turns, they out-manœuvre a faster-swimming predator.

The most spectacular of the schools' defences is the *flash expansion*, so called because on film it looks much like a bomb bursting as each fish simultaneously darts away from the centre of the school as it is attacked. The reason for its great interest to students of schooling behaviour is that, although the entire expansion may occur in 1/50 s, and they may accelerate to ten to twenty body-lengths per second within that time, the fish do not collide. Not only does each fish know in advance where it will swim if attacked, but it must also know where each of its neighbours will swim.

In addition to anti-predator functions, shoals may have a number of social functions. They may ensure breeding by bringing members of both sexes together; for example, in the straits of Dover, during the herring breeding season schools have been observed which stretch for 25 km. (Such a school must contain millions of individuals.) Schooling allows individuals to copy one another. If, for instance, one tuna or barracuda finds food, the others notice and begin FEEDING as well. In the same vein, many pairs of eyes are better than one pair in spotting predators. For many years it has been known that fish in groups learn more quickly than when alone, but whether the effect is due to increased ATTENTION or to IMITATION is unclear.

It has even been suggested that during MIGRATION the pooled information of many individuals may result in a more accurate determination of ORIENTATION. If the distribution of individual estimates of the correct direction in which to travel was *normal* about the true direction, then the 'consensus' of the group would indeed point the way.

Fig. A. Formation of an empty space around an immobile predatory fish.

Hydrodynamic aspects of schooling. Scientists studying fish propulsion, in an effort to build faster submarines, were surprised to discover that fish could travel faster in schools than alone, and that individual fish use up more energy in swimming than do fish in schools. One suggestion is that the difference is merely due to solitary fish being 'upset' at being alone, or expending energy searching for a school to join. Alternatively, several scientists have proposed attractive hydrodynamic models, whereby fish in schools could gain an advantage by making use of the vortices, or eddies, and areas of turbulence left by fish swimming ahead and behind them. If fish maintained the hypothetical spacing, they would benefit from induced water velocity relative to the ground, since the water in the vortices is moving in the same direction as individual fish in succeeding rows. The resultant 'tailwind' might give all the fish, except for the very front one, an advantage over fish swimming alone. Although fish do tend to avoid areas of high turbulence, such as that directly behind the tail of the fish in front, there is, as yet, little evidence that they actually take up the positions required to obtain the postulated hydrodynamic advantage. Studies on the Atlantic spadefish (*Chaetodipterus faber*), however, lead to the related suggestion that schooling may allow fish to take advantage of minute quantities of slime given off by fish in front of them. The slime has the remarkable property of reducing the surface tension of the water, thereby reducing drag. In the case of the species tested the savings may be as much as 60%.

Sensory basis of schooling. In order for a fish to school, it must first locate and join a moving body of fish, and second, adjust its position and velocity in relation to them. The way in which a fish does this was thought for a long time to be exclusively by vision, since fish usually stop schooling at low light intensities. For instance, in one study, ten species all stopped schooling at a level of illumination close to the limits of human perception.

One question which has long puzzled scientists is how fish, scattered in the darkness, regroup when it becomes light. Tuna seem to have solved the problem in a rather striking way. In the early morning, they often exhibit a marked LEARNING behaviour, and it has been suggested that this allows them to use reflections of the sun off their silvery sides to aggregate from extreme visual ranges.

That fish use vision to school is clear. A number of species, including European roach (*Rutilus rutilus*), the thick-lipped grey mullet (*Mugil chelo*), and tuna, will school with a species companion behind a glass partition, or, less willingly, with their mirror image.

While vision plays a major part, it is now known that schooling, like homing by birds, involves several different sensory processes. Two carangids (*Caranx hippos*) separated by a clear partition swim closer together than when no partition is present. This suggests that pressure waves set up by the swimming fish are normally important in inter-fish spacing.

Blinded Atlantic silverside fish, such as *Menidia menidia*, will maintain position with and swim parallel to an appropriate school for several seconds at a time. Similarly, blind mackerel (*Pneumatophorus grex*) will turn towards a school of conspecifics when it passes. Proof that fish use other SENSES as well as vision to school comes from experiments on saithe (*Pollachius virens*). The fish were fitted with opaque contact lenses, which effectively eliminated vision. In the large circular tank where they were tested, the temporarily blinded saithe were still capable of joining and maintaining position indefinitely within a school of normal fish. The experimenters guessed that the blind fish were judging the positions of their neighbours by special pressure-sensitive organs, known as the *lateral lines*, which run along the length of the animals. However, fish with their lateral lines severed at the gill-covers schooled normally. Only if both vision and lateral lines were eliminated was schooling prevented.

It is not surprising that the lateral line system should play a part in inter-fish spacing, especially if fish are trying to monitor the pressure waves set up by their neighbours in order to avoid areas of turbulence, or to obtain the hydrodynamic advantage predicted by the mathematical models mentioned above.

Another reason why one might expect lateral lines to be involved in inter-fish spacing is that most fish have BINOCULAR VISION over only a small part of their visual range, and depth PER-CEPTION, necessary for maintaining inter-fish spacing, is difficult with monocular vision in an aquatic environment.

A question posed by the demonstration that, at least in saithe, vision is not required for individuals to school, is why schools should normally break up in darkness. Actually, not all species stop schooling at night. For those species which do cease schooling in the dark, the reason may be one of MOTIVATION rather than physiology; that is, although they do not school, they may be perfectly capable of doing so. In the case of the experiments on saithe, even though blindfolded so that they could not see their neighbours, the fish could still perceive light/dark via the *pineal organ* (sometimes known as the third eye) on top of the brain. If the primary function of schooling is to reduce predation, then it may be less important to school tightly in the dark.

In addition to vision and the lateral line, several other senses have been implicated in schooling, or at least in assisting the aggregation of the fish in the first place. CHEMICAL SENSES, for instance, have been shown to be important in many aspects of fish behaviour: the remarkable homing by salmon (*Salmo salar* and *Oncorhynchus* spp.), individual and species recognition, location of prey, and the transmission of the ALARM RESPONSE in fish such as the minnow. Chemical senses may play an important role in maintaining school cohesiveness, especially at night. Species which do not disperse at night, such as the roach, none the less school more loosely. This has been taken to suggest that olfaction may be keeping them together in the absence of visual cues.

Many fish have been shown to recognize species-specific odours, but their power of discrimination may be even more highly developed. European minnows can not only distinguish the odour of their own species, but can also be trained to discriminate between odours from two totally unrelated species. Furthermore, they can learn to recognize individual fish of other species as well as their own, on the basis of smell alone. Mexican blind cave fish, *Anoptichthys jordii* and *Astyanax mexicanus*, almost certainly rely on smell to school.

Electric organs, which are unique to fish, have also been implicated in schooling in some species. Electric or weakly electric fish use bio-electric fields to locate prey and for COMMUNICATION with members of their own species in turbid water (see ELECTROMAGNETIC SENSES). Members of these groups have special electro-receptors which detect variations in the environment, and research has shown that individuals can communicate with each other by highly sophisticated codes of electrical impulses.

The African mormyrid *Marcusenius cyprinodes* uses electrical communication of this sort to keep in contact with members of its species when migrating at night. Groups of up to ten individuals form lines as they move across the sea floor.

Sound, also, has been thought to play a role in schooling. Sensitive microphones can record the swimming sounds made by passing schools, and there is experimental evidence that individuals can respond to these sounds, especially in transmitting a fright reaction throughout the school.

Structure of fish schools. If the primary function of schooling is to reduce encounters with predators, one would expect a school to take up a structure of minimum surface area, that is, a sphere. Ball-like schools clearly do this, and there is some suggestion that schools swimming in open water do as well. If a school of, say, herring (*Clupea*) is near the surface or bottom, one would, of course, expect the overall shape to be flattened.

A great deal of study has been carried out on the internal structure of fish schools, especially since the hydrodynamic theories predicting advantages to schooling fish depend upon the individuals taking up characteristic positions with respect to each other. Despite this research, which has for the most part failed to demonstrate the predicted spacing, many scientists stick to the concept of an idealized school of rigid structure. In reality, school structure only exists in a dynamic sense. Fish may have preferred positions with respect to their neighbours, but there is continual movement and reorganization within the schools.

Some scientists suggest that the movement is a continual effort of the fish to arrange the school by size. Others, after the demonstration that there exists a marked difference in oxygen concentration in the centre of schools of striped mullet (*Mugil cephalus*) compared to the surrounding water, believe that movement might be due to conflicting responses to, on the one hand, low oxygen content in the centre of the school and, on the other hand, decreased safety at the periphery.   B.P.

SEARCHING may occur as an aspect of FORAG-ING or HUNTING behaviour, as a precursor to COURTSHIP, or as part of the behaviour by which a parasite finds its host (see PARASITISM). Usually a searching animal will confine its activities to a particular locality or type of HABITAT. Many animals learn to concentrate their search in areas in which they have previously been successful. Honey-bees

(*Apis mellifera*) learn that each species of food plant provides nectar only during a certain TIME of day. Carrion crows (*Corvus corone*) and herring gulls (*Larus argentatus*) arrive at particular places as if they expected to find particular types of prey there. They may overlook prey items which they associate with another place.

Having arrived in a particular locality, there are various searching strategies that a predator could employ. Random search is generally the least efficient of these, partly because it often results in the animal revisiting areas that have already been covered. Random search can be improved upon if there is greater directionality in the search path. That is, the searcher makes fewer changes of direction than would be expected on a random basis. The best searching strategy is attuned to the distribution of the prey. If the prey are widely scattered then it is better to make fewer turns and cover more distance per unit time. If the distribution of the prey is clumped into patches of high density, then it is better to make more frequent changes of direction, especially in the vicinity of a capture. The parasitoid wasp *Trichogramma evanescens* searches more widely for the eggs of its host insect when these have a low density, and is able to concentrate its search when the eggs are more densely distributed. Blackbirds (*Turdus merula*) searching for prey on open meadowland tend to concentrate in the surrounding area just after a capture, and this tendency is more marked if the prey is patchily distributed. Thus the blackbird is able to alter its search strategy according to the circumstances.

If a predator has depleted a particular area of prey, it is not in its interests to return there for some time. In some cases, however, the food is rapidly replenished, and the predator might be expected to make repeated visits. Pied wagtails (*Motacilla alba yarelli*), for example, forage along the river bank garnering insects that are washed up there. Each wagtail defends a stretch of river as a TERRITORY, and systematically exploits the food resource by searching upstream along one bank and downstream along the other. The bird usually completes a circuit over a period of time that is sufficient for the food items to be replenished.

The searching animal has to pay ATTENTION to the specific tell-tale cues that are characteristic of each type of prey. When the prey is CAMOUFLAGED, the predator will need longer to detect each prey item. Many predators increase their searching efficiency by developing a SEARCHING IMAGE, or tendency to pay attention to a particular prey type. The searching image may change as a result of lack of success, or of a chance encounter with a different prey type. It may be influenced by hunting expectations and by social aspects of PREDATORY behaviour.

28

SEARCHING IMAGE is the term used to describe the perceptual phenomenon of suddenly seeing something that was previously overlooked. A classic example is that of someone looking for one sort of object, such as a clay jug, and completely failing to see a glass carafe which is in front of his nose. It is more usually applied, however, to the ability to pick out a very cryptic object against a background which it resembles so closely as to be almost indistinguishable. Many people have probably had the experience of looking at a photograph of a CAMOUFLAGED moth, such as the underwing moth (*Catocala*), against its natural background, and at first seeing only a picture of a tree-trunk. Then, although overlooked at first, the insect itself is suddenly seen and thereafter it is perfectly obvious. They have developed a searching image for the insect.

Difficulties with the concept of the searching image begin to arise when it is applied to other species besides our own. We may know subjectively about changes in our perception of the world, and we can use language to ask other people about theirs. But how can we know about the perception of other animals when the only evidence we have available to us is that of their behaviour? To begin with, a number of anecdotes are highly suggestive of searching images in animals. Hungry toads (*Bufo bufo*) have been reported to snap at bits of moss if they have just been eating spiders (Araneae), but at long thin objects if they have just had an earthworm (e.g. *Lumbricus*). Numbers of inedible objects have been found in the stomachs of trout (Salmoninae), about the same size and shape as the real food consumed at the same time. The implication of these observations is that the prey leaves behind an 'image' in the mind of the animal as to what it should eat. Bird predators provide even more convincing evidence of these perceptual phenomena, which have become known as acquiring a searching image. Several species of insectivorous birds at first overlook the stick-like caterpillars of certain moths (Geometridae) resting among twigs. These caterpillars so closely resemble twigs in colour, shape, and in the angle that their bodies make with the branches that the birds are at first completely taken in. Behavioural observations show that the birds take a long time to find the first caterpillar, but once they have found one, some individuals are able to pick out the other insects without subsequently pecking any of the sticks, showing that they could detect the difference between sticks and caterpillars once they had 'got their eye in'. Other individuals, having once found a caterpillar, then proceeded to peck at the sticks as well; in this case, their searching image is apparently rather less specific.

Another line of evidence that animals have searching images for the prey items they are seeking comes from FIELD STUDIES of what predators are eating in the wild. A comparison between the food available to an animal in its natural environment and what the animal actually finds and eats can reveal the extent to which the animal is being selective or is overlooking some food items. This is technically a difficult thing to do, both from the point of view of knowing what food is available, and in the actual recording of what the animal manages to find. But it has been attempted in a number of studies on the great tit (*Parus major*). The food brought to the nest by parent great tits has been recorded over the course of the breeding season, and compared with the insects actually to be found in the surrounding woodland. A very interesting phenomenon has emerged from these studies. Some species of insect are not taken by the tits for some time after they first emerge, even though they would seem to be a perfectly good food source. Later in the season, however, the same species may suffer heavy PREDATION. It has been argued from this that the birds do not take new species of insect the first time they encounter them, because at that stage they quite literally cannot detect them. Only when they have had sufficient exposure to them to be able to acquire a searching image for them are they able to break the camouflage and find the insects at all readily. This conclusion has been criticized on the grounds that there is no direct evidence that the birds in these studies were getting any better at detecting their prey. A newly emerged species of insect is very likely to be found on a species of tree different from the ones where the birds are accustomed to finding food. It is quite likely, therefore, that the birds have to learn where a new species of insect is to be found, and this could account for the increase in the numbers of them eaten as the season progresses and the birds change their HUNTING grounds.

This leads us to the crux of the problem of how to demonstrate that animals have searching images, in the sense of being able to learn to see something that they initially could not see. We might observe that an animal begins to eat a particular sort of camouflaged food that it did not eat when first exposed to it. But quite obviously, not all such changes in behaviour are caused by a change in what the animal is able to see. Changes in hunger level, in where the animal spends most of its time, or in what it regards as acceptable food, are some of the many factors which could result in a change in the pattern of feeding behaviour with no change at all in the animal's ability to perceive its prey. In order to show that animals can improve their ability to detect cryptic prey, these other factors have to be eliminated. We know that birds can learn to concentrate their search in areas of abundant food. As they learn where a new food source is to be found, they will spend more time there, and eat more and more of the new food. So before concluding that they can see it any better, we have to rule out the effects of the birds LEARNING where to hunt. This can be achieved experimentally by ensuring that a new food occurs right from the start within easy striking distance of the bird. Another reason why a predator may not eat food the first time it en-

counters it is that it does not know how to attack, kill, or prepare the food in the correct way (see PREDATORY BEHAVIOUR). With an increase in the number of prey eaten by the predator, we must therefore be quite sure that this is not due to an improvement in the motor patterns of food handling, before inferring that a perceptual change has taken place. Also, palatable food which is conspicuous may sometimes be rejected when first encountered, simply because it is unfamiliar. This is true of blue jays (*Cyanocitta cristata*), which may even exhibit alarm and fear when given butterflies (Lepidoptera) of a type they have not seen before, although later they eat them readily.

In the wild, it is clearly very difficult to distinguish changes in a predator's ability to see its food (acquisition of a searching image) from these other learning processes which may be going on at the same time. In the LABORATORY, however, it has been shown that domestic chicks (*Gallus gallus domesticus*) do show a definite improvement with experience in their ability to detect camouflaged grains of rice, under conditions where these other factors can be ruled out. The chicks were observed searching for rice dyed to resemble exactly the colour of the background. When first presented with these cryptic grains, the chicks failed to eat them, even though they were almost certainly looking straight at them, and even though they readily took the same rice grains placed on a background which rendered them conspicuous. This shows that the rice grains were palatable and available to the chicks, and that their failure to eat them at first was due to their crypsis. However, after a short exposure to the camouflaged grains, the chicks began taking them readily, having, apparently, no difficulty in detecting them. They had clearly learnt to see the cryptic grains; in other words the chicks had developed a searching image, in the classical sense of the term.

The fact that predators may develop searching images has some interesting consequences as far as their prey are concerned. If a predator has already acquired a searching image for one prey type, prey which look different from that type will be at an advantage since they will tend to be overlooked. Two or more prey types may thus evolve within a single species and exist side by side, a condition known as a *polymorphism*. The banded snail *Cepaea* comes in several different colours and banding patterns, and this is thought to be due in part to the pressure of NATURAL SELECTION exerted by predators with searching images. Experiments on wild and captive birds of various species have certainly demonstrated the value of polymorphism from the prey's point of view. Carrion crows (*Corvus corone*) searching for bits of meat hidden by an experimenter under empty mussel shells (Myrtilidae) lying among stones found it much more difficult when they were searching for shells of three different colours than when searching for shells all of the same colour. They seemed to have developed a searching image for one shell colour, and to be able to find that colour efficiently, but not to be able to distinguish different colours (different searching images) all at the same time. The polymorphic population of mussel shells made up of three different colours was therefore much less vulnerable to predation because of the tendency of crows to hunt by searching image.

The concept of searching image is applied particularly to the case of predators searching for cryptic prey, but the phenomenon has much in common with studies on selective ATTENTION, which suggest that in a variety of situations animals may selectively perceive different aspects of their environments at different times. Human beings and other animals are known to switch attention from one set of environmental stimuli to another, and this process would seem to have quite a lot in common with that of acquiring a searching image. Learning to see a camouflaged object that is at first overlooked is a process in which subtle cues that enable the object to be seen are gradually attended to and used in the search. Most predators have eyes that are very good at picking out movement, and regular patterns such as bars and strong colour contrasts. Because of this, their prey have evolved to be difficult to see by these types of eyes: they remain motionless, break up their outline and their patterns, and blend in with the colour of their background. Denied obvious cues for detecting their prey, the predators in turn have to use other cues to find food at all. Despite elaborate camouflage, the prey are distinguishable: they may have shadows, shapes, or symmetries which betray them, and these their predators, baffled at first, learn to latch on to. They acquire a searching image, and the camouflage is broken. M.D.

SENSE ORGANS are organs adapted for the reception of stimuli or events outside the nervous system. The stimuli may be external to the animal as a whole, such as visual or auditory stimuli, or they may be internal to the body, such as the pressure within a tissue, or the changes in the length of a muscle. Each sense organ contains some kind of transducer which converts light-, chemical-, mechanical-energy, etc., into the type of electrical potential that can be registered by the receptor cell. The receptors are nerve cells which are part of the sense organ, and which connect to other cells of the nervous system.

In 1826, Johannes Müller noted that sensory quality depends on which nerve is stimulated, and not on how it is stimulated. This is now known as the doctrine of *specific nerve energies*. Visual sensations in man can be generated by light falling on the retina, by pressure on the eyeball, by electrical stimulation, etc. Thus it is not the stimuli that determine the gross sensory quality, but the nerves that are activated by the stimuli. The differences in sensory quality between the different sensory modalities are determined by the route the nervous messages take within the BRAIN. Any kind of acti-

vation of the optic nerve produces visual sensations because the nerve goes to the visual system of the brain. Similarly activation of the auditory nerve produces sound sensations because the nerve transports the information to the auditory system of the brain. The same general principle applies to the sub-modalities, such as the different colours or different sound frequencies. As a rule different nerve cells provide information about different qualities, and variations in a particular nerve message provide information about the intensity of stimulation.

The sense organs of different species are generally similar in their modes of transduction, but vary considerably in their accessory structures. Thus in VISION the transduction between light and electrical energy involves a photo-chemical mediator called a visual pigment. The structure of the eye, however, varies considerably from species to species. In HEARING the transduction from mechanical to electrical energy is accomplished through the movement of accessory hairs in some species, and of membranes in other species. Sometimes the accessory structures are specialized organs, such as the ears of vertebrates, and sometimes they are modifications of existing structures, such as the legs of insects. Some insects have mechano-receptors placed within a leg to detect vibration that is transmitted through the substrate. Many Orthoptera (grasshoppers, crickets, and locusts) have *tympanal organs*, consisting of a membranous drum enclosing an air sac (Fig. A). The receptors are so arranged that they are stimulated by movements of the drum.

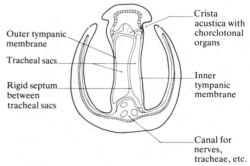

Outer tympanic membrane

Tracheal sacs

Rigid septum between tracheal sacs

Crista acustica with chorclotonal organs

Inner tympanic membrane

Canal for nerves, tracheae, etc.

Fig. A. Cross-section of the leg of an orthopteran insect, showing the tibial tympanal organ.

Animals vary considerably in their sensory requirements and there are many SENSES OF ANIMALS that involve complex and specialized sense organs. These include the CHEMICAL SENSES such as TASTE AND SMELL, the MECHANICAL SENSES, and the ELECTROMAGNETIC SENSES. There are also senses, such as those involved in PAIN, in THERMOREGULATION, and other aspects of HOMEOSTASIS, which do not involve elaborate sense organs, but rely on specialized receptors which are without accessory structures.

There are many receptors within the brain that are sensitive to chemicals, such as HORMONES, circulating in the blood. There are also numerous simple receptors within the various organs of the body.
35, 66

SENSES OF ANIMALS. The environment in which an organism lives varies over time and space. Even the most primitive organisms show some capacity to react individually to temporary or local conditions, so that the biochemical processes that underlie their life and replication can adjust and persist. Most animals have a marked ability to adapt behaviourally. However, behavioural adjustment can only be achieved if the animal can assess the state of the environment. It is the function of special mechanisms known as the senses, or more accurately as sensory systems, to provide this monitoring of the environment and to transform and process the information acquired so that it can control or steer the behaviour in an adaptive manner (see ADAPTATION).

The state of the environment is characterized by spatial and temporal variations of physical and chemical energy. It is these variations that the senses respond to. Potentially there is an infinity of environmental variables that could be sensed. Information about many of them would not contribute appreciably to the adaptiveness of behaviour, and since developing and operating a sensory system demands energy, each species has evolved responsiveness to only a selection of stimuli. Human beings, for example, can sense neither ultra-violet light nor magnetic fields.

How does one find out what stimuli animals can sense? In the first instance behavioural observations give clues. The flatworm *Planaria* is found under stones in streams. If one removes the stones they move away to hide under another stone. This suggests that they can sense light. Simple experiments using a torch with planarians kept in an aquarium in a dark room soon confirm this: they consistently avoid light and seek darkness.

Anatomical peculiarities also give clues. Animals that have eyes are likely to sense light. Indeed, if one examines *Planaria* with a magnifying lens one can see two dark, cup-like structures, one at each side of the head, that look very like eyes. Further experiments with a miniature torch soon confirm this. The light-avoidance response is released when light falls on these 'eyes', but not when it falls elsewhere on the body. Surgical removal of the eyes renders the animals completely unresponsive to light. Simple reasoning about environmental demands upon animals can also be a guide. Species that live in deep caves never encounter light and might thus be expected not to be able to sense it. And indeed, certain cave-dwelling fish (Amblyopsidae) and shrimps living in caves (Palaemonidea) have been found to be blind.

Often, however, such straightforward methods are not sufficient. Naturalists in the eighteenth century investigating the ability of bats (Chiroptera) to fly in complete darkness without colliding with obstacles were not able to deduce what sense they were relying on. Actually, they discovered that plugging the bats' ears disrupted their ORIENTATION, but as no sounds could be heard they did not know how to interpret this finding. Development of advanced acoustical equipment during the Second World War led to the solution of the riddle. The bats emit ultrasounds that are inaudible to us, but to which their own ears are exquisitely sensitive, and they make use of the ultrasound echoes to orientate.

Similarly, sophisticated equipment and formal experiments were necessary before it was possible to conclude that pigeons (*Columba livia*) are capable of perceiving the polarization plane of light, that is the vibration plane of light waves. In the LABORATORY hungry pigeons were placed in a circular chamber, and were rewarded with food grains for pecking small discs which were placed along the walls of a chamber aligned with the polarization plane of an overhead light source. If they pecked discs not so aligned they received no food, and were punished by having to sit in the dark for a period of time. From time to time the polarization plane of the overhead light source was turned to a new, unpredictable position, thus making a different set of discs the correct ones, i.e. the ones that would yield food on being pecked. The pigeons soon learned to observe the polarization plane orientation and began to choose the correct discs more often than by chance. Control experiments were necessary to prove that the increased rate of correct choices was not due to some unintentional cue other than polarization, such as brightness patterns, apparatus noise, and so forth.

Human beings are unaware of the polarization of light (except under rather special circumstances), and so it is difficult to imagine that to a pigeon, and indeed to many other animals that are sensitive to this stimulus, the sky must appear as an intricate design, since its light is polarized in a complex patterned way that is linked with the position of the sun.

Experiments such as these show that animals are responsive only to selected sets of stimuli, but that the selection can differ from species to species. Rattlesnakes (*Crotalus*) and some boas (Boinae) detect infra-red radiation with specialized pit organs, which can be seen in Fig. A as small depressions located between the nose openings and the eyes. By means of these organs they can sense small prey from a distance of 1 or 2 m and attack it. Certain fish can sense electric fields, pigeons are sensitive to changes in atmospheric pressure, many insects see ultra-violet light, some mammals are colour-blind, birds may see more colours than we do, certain moths can hear ultrasound, rodents can smell odours that are imperceptible to us, and so forth. This diversity is explained by the fact that each species has a different NICHE, and that,

Fig. A. Head of a rattlesnake showing the pit (P) which is sensitive to infra-red radiation.

accordingly, one or the other sensory capability contributes more to make its behaviour adaptive in its particular conditions of life.

A consequence of this diversity is that the often asked question of how many senses there are cannot be answered meaningfully. Even if we restrict ourselves to human beings there are difficulties. The sense of balance, for example, can arguably be subdivided into senses of angular acceleration associated with the *utricular* and *saccular maculae*, all located in the inner ear. Both are doubtlessly mechanoreceptive senses in that they are responsive to MECHANICAL forces, but then, so is the sense of touch and, strictly speaking, also the sense of HEARING, in that sounds are mechanical waves. Some arbitrariness is unavoidable, but according to the type of stimulus energy that they are sensitive to, the senses can be classified as VISION, hearing, mechanical senses, CHEMICAL SENSES, temperature sense, (see THERMOREGULATION), and ELECTROMAGNETIC SENSES. Even so, one important sense, PAIN, is difficult to fit in, because it is sensitive indiscriminately to strong thermal, chemical, and mechanical stimuli.

So far we have been considering senses that are responsive to stimuli originating in the environment. However, in higher animals most organs are not directly exposed to the outer world. Rather they exist in an environment that is internal to the organism, the *milieu intérieur*. The state of this internal environment is subject to variations. For example, with physical exertion the content of oxygen in the blood falls and that of carbon dioxide rises. If injury is to be avoided such deviations require adjustment, through either physiological or behavioural responses, in our example perhaps by an increase of the breathing depth and rate (see HOMEOSTASIS). Internal (*enteroceptive*) senses of the chemical, thermal and mechanical variety, of which human beings are only dimly aware, ensure the necessary monitoring and initiation of corrective action. The internal stimuli, such as stomach distention, blood sugar level, level of HORMONES, that play such an important role in the MOTIVATION of behaviour are all, of course, mediated by these senses. Enteroceptive senses function to coordinate physiological and be-

haviour responses into an adaptive whole. SEXUAL receptivity of a female cat (*Felis catus*), for example, is only reproductively significant if it is synchronous with ovulation; the rise in the blood of a sex hormone triggers both.

Similarly, temperature sensitive neurones located within the hypothalamus and the spinal cord (see BRAIN) of vertebrates are crucial for the regulation of the body temperature. In birds, for example, the former area appears to be involved with blood circulation and plumage adjustments while the latter controls shivering and panting.

In higher animals the coordinated action of many body structures is involved in behaviour. For example, usually a number of muscles control the flexion of a given leg joint, and a number of leg joints are involved in even such a simple behaviour as walking. Effective walking can only be achieved if the nervous system has continuous information about the state of these structures. The degree of flexion of the knee joint in one leg clearly is important for the correct adjustment of the tension of the muscles in the other leg if balance is to be maintained. Special, so-called *proprioceptive* sense organs, associated with muscles, joints, and tendons, provide just that information.

Sensory systems. We now turn to the structures that underlie the sensory processes, the sensory systems. They are made up of components subserving three different functions: *transduction*, *transmission*, and *processing*. In higher multicellular animals the structural elements of sensory systems are discernible as specialized cells or groups of cells; in lower, unicellular animals the same functions are carried out by more or less recognizable cell organelles. What follows refers mainly to vertebrate animals, but similar even though usually correspondingly simpler structures and mechanisms are found in invertebrate animals.

The stimulus transduction is achieved by receptors. These cells or *organelles* convert, or better, translate the stimulus energy impinging on them into a standard form of signal-code. In higher animals these are usually sequences of nerve cell *action potentials*, and the code transduction is usually such that the more intense the stimulus, the higher the frequency of action potentials. Receptors are specialized according to the stimuli, the type of environmental energy, that they react to. Accordingly the physico-chemical processes that mediate the transduction vary a great deal: light receptors of the visual sense, for example, contain pigments that are modified chemically by the light they absorb; mechanoreceptors mediating the sense of touch undergo electrochemical changes as a consequence of the deformation of their membranes, and so forth. It is, however, characteristic of all receptor cells that, as a stage of the translation, they convert the stimulus energy changes into a graded electrical potential, known as the *receptor* or *generator potential*, that is proportional to the stimulus intensity.

Receptors may occur singly or in clusters. The temperature receptors of mammals are distributed in a variable fashion over the body surface, with differing densities. The temperature sensitivity of a given area is related to that density. Exploration of the skin with small heating and cooling probes reveals that there are discrete sites sensitive to either cold or warmth. The snout area is often highly sensitive, and indeed it has the highest density of these warm and cold spots. The cold receptors have been identified anatomically on the skin of the nose of cats. They are thin nerve endings that are embedded in the basal layer of the epidermis (outer layer). The identity of warm receptors is uncertain.

Receptor cells can also occur in clusters, and are then often associated with accessory structures that can be of considerable complexity, and which aid the stimulus reception, or even the transduction process itself. In the eyes of vertebrates the cones (actually three different types of cones) and rods of the retina, which contain light sensitive pigments, are the only actual receptor cells. Other structures, such as the pupil, the lens, and so forth, are only accessory structures that enable the formation of an image on the retina. The receptors proper, plus such additional structures, form what is called a sensory organ.

The organization of such sensory organs often contributes much in expanding the basic capability of receptor cells, that is, in coding the relevant stimulus energy into proportional neural messages. This is illustrated by the organ of hearing, the *cochlea* situated in the inner ear (the outer ear and middle ear have largely supporting functions). The receptor cells are hair cells that react in a proportional manner to bending of their hairs. They are arranged in such a manner in relation to auxiliary structures that each hair cell is only stimulated by tones of a certain pitch, and transduces the sound energy associated with that pitch.

The conversion of generator potentials into sequences of action potentials is achieved by the receptor cells themselves in the case of primary receptors, or by follower nerve cells after mediation by synapses in the case of secondary receptors. The next stage is the transmission of the coded information. The response organs, muscles and glands, that are responsible for actual behaviour are as a rule situated some distance away from the receptor structures. The information acquired by these receptor structures has, therefore, to be conveyed to the effector organs. However the transmission is rarely direct. Rather the sensory information proceeds first to the so-called sensory projection areas, more or less discrete parts of the brain, where it is processed. Sensory nerves, which are bundles of nerve cell *axons*, form the first transmission channel. Through them the sequences of action potentials, which encode as an ensemble the qualitative, spatial, and temporal characteristics of stimulus patterning, reach the brain. There they are relayed by further neurones, whose axons make up the central sensory tracts or pathways, to the projection areas themselves.

By introducing finely sharpened needles that are insulated except for a tiny area of the tip (microelectrodes) into the sensory nerves or tracts of anaesthetized animals, we get some information about the way in which the sensory information is being transmitted. The tips of these microelectrodes pick up the electrical potentials associated with the neuronal impulses, and, when amplified, they can be made audible with a loudspeaker or visible with an oscilloscope. If we apply this electrophysiological technique to the optic nerve of a cat, and stimulate the corresponding eye with a fine pencil of light, we find that a given axon responds with an increased action potential frequency to illumination of only a particular small spot of the retina. Typically it gives off a burst of impulses at the onset of the illumination and then settles down to a steady medium rate, becomes totally silent at the ending of the stimulus, and then recovers to a low firing rate during darkness. Sensory systems are generally particularly responsive to stimulus changes, less so to steady stimuli, and often they are also spontaneously active in the absence of any stimulation. If the light being shone on the cat's retina is made to change colour we may find that the neuronal unit only responds to red light, not to green or blue. If we sample further axons in the nerve (there are approximately half a million of them) we find that each of them responds to light shone on a particular spot of the retina. The response may be to only red, green, or blue light, or it may be indiscriminate. Visual scenes are thus represented by a time-varying mosaic of action potentials travelling through individual fibres of the cross-section of the optic nerves.

Results from experiments like this can sometimes throw light on the sensory capabilities of an animal. For a long time psychologists maintained that cats were colour-blind. They could not train them to distinguish colours. Physiologists however, on the basis of the information just described (i.e. the presence of colour coding visual axons), maintained that they must have COLOUR VISION, and indeed, with improved CONDITIONING techniques, psychologists have come to the same conclusion: cats do see colours.

Finally, we must consider the information processing function of sensory systems. This takes place mainly in the sensory projection areas we mentioned earlier, although some of it occurs at each of the transmission relaying sites, such as the sensory nerve nuclei of the midbrain, or the dorsal horn of spinal grey matter, or, as in the case of vertebrate vision, in the retina itself (apart from the receptor cells, it also contains neurones forming a highly complex network).

The nerve cells of the sensory projection areas often receive the coded mosaic of nerve impulses coming from the receptors in a map-like fashion. For example, touch of the various body parts is relayed to corresponding regions of the *somatosensory* cortex of the forebrain such that there is a distorted representation of the body: the sensory *homunculus*. The face and hands of this representation are oversize, corresponding to the greater touch sensitivity of these body areas, which is related to the greater density of touch receptors in the skin. Typically for any sense there are several such representations. Touch is represented clearly twice on the forebrain cortex, and once on the cerebellar cortex, and less clearly several times more in other brain areas. The projections should not be thought of as equivalent. In the case of vision, for example, the visual cortex is concerned mainly with recognition of shapes while the visual midbrain deals with information about the location of stimuli in the visual space. Neural pathways usually interconnect such multiple projections, enabling information exchange.

How is the sensory information processed? Microelectrode recordings from neurones in these projection areas provide some insights. In the visual system a first stage is represented by line detection neurones. They only fire in response to a bar or strip of light falling onto a certain point of the retina with a particular orientation. They do so because they receive the joint input from several aligned units from the same retinal area. In the next stage neurones combine the inputs from several of these line-detecting units, and are responsive to correspondingly more complex patterns, and so forth. In the auditory system, where the processing proceeds in an analogous way, neurones have been found in squirrel monkeys (*Saimiri sciureus*) that respond specifically to the different calls that these primates make during SOCIAL encounters. They respond uniquely to quite complex sound patterns, representing stimuli that are known to influence the animal's behaviour.

Not all the information that an animal is capable of receiving is useful at all times. Accordingly, one finds quite frequently that depending on the behavioural context, sensory information that is potentially available is disregarded. The grayling butterfly (*Hipparchia semele*) reveals marked colour preferences when it visits coloured dummy flowers in search of food. This indicates a good colour DISCRIMINATION ability. None the less, in an experiment in which males pursue dummy coloured female butterflies dangled from a fishing rod they behave as if they were completely colour-blind. In this sexual context the male grayling butterfly just does not attend to colour cues, only to brightness: the darker the model the better. Similarly, touch stimuli on the snout of a resting cat have no particular consequence, at best perhaps some withdrawal or licking behaviour. But when the same cat is hunting, the snout becomes highly sensitive, and the slightest touch releases biting. It is as if the motivation of the cat directed its ATTENTION to these particular stimuli in preference to others, a sort of optional stimulus selectivity.

While sometimes the direction of attentiveness, as in the above examples, makes FUNCTIONAL sense, in other cases stimulus selection may be due to the fact that the brain simply cannot deal with the wealth of information that floods into it and has to ignore some of it. Rats (*Rattus norvegicus*) that have learned to distinguish visual patterns (say a white square from a black circle) illustrate this, for on closer examination they are found to have mastered the discrimination only on the basis of a partial stimulus feature, and to have disregarded other distinctive cues. In our example some might choose the brightness difference, others the shape difference, but none both.

Some stimuli in the environment recur repeatedly, but may prove unimportant for the adaptiveness of the behaviour of an animal Thus they need not be attended to. The process of HABITUATION represents an additional stimulus selection mechanism that deals with this situation. For example, a colony of black-headed gulls (*Larus ridibundus*) nesting near to an army gun range did not react to the firing noise. However, when a rocket was launched for the first time the birds panicked in response to the hiss. Some days later, however, they ignored this noise too.

The phenomenon of attention implies that at one or more levels of the sensory system, higher centres of the brain must be able to influence the flow and processing of sensory information at lower levels. So-called *centrifugal* or *efferent* neural pathways to virtually every relay station, sometimes even to the receptors themselves, have been described, and may in fact be those mediating such attentional control.

There remains to be considered how the highly complex networks of sensory systems are put together during the development of the individual. There is no doubt that GENETIC information plays a crucial role. An example is provided by Siamese cats which have a single gene mutation (i.e. a sudden and relatively permanent change of a gene) that produces a rearrangement of the nerve cell connections in the visual system. Because of this, information from both eyes does not converge as it does in cats that have the normal genes. The result is that Siamese cats have deficient depth vision (their well-known squinting is a by-product of the neural fault). On the other hand, it can also be shown that experience is important in organizing sensory systems. Kittens that are brought up in horizontally or vertically striped cages develop visual cortex neurones that respond exclusively to horizontally or vertically oriented line patterns respectively. In normally reared kittens these neurones, as an ensemble, respond to lines of all orientations. Of course, when animals learn to recognize very specific stimuli, such as their mate or offspring, we must in any case assume that sensory mechanisms are being modified by experience (see LEARNING).   J.D.

35, 66, 100, 132

SENSITIVE PERIODS are periods of time during which a developing animal is especially sensitive to particular types of experience. In the development of SONG in certain birds, for example, there is a period during which the young bird is capable of LEARNING the song of its parents or foster parents. Thus the chaffinch (*Fringilla coelebs*) has a sensitive period which begins soon after HATCHING, and lasts for a few months. Chaffinches caught in their first autumn and then maintained in isolation from other chaffinches develop an almost normal song, but chaffinches hand-reared in isolation from a few days of age develop only a very crude song. Chaffinches reared in isolation and exposed to tape-recordings of chaffinch song during their first autumn and winter do develop normal song. Such birds will only learn songs which are similar to normal chaffinch songs. Thus, for chaffinches, there is a sensitive period during which they will learn a particular song repertoire, provided that they are exposed to suitable auditory stimuli during this period. Once full song has developed, no further song patterns are learnt.

A chaffinch which was castrated before its first breeding season, and subsequently treated with male sex HORMONES, was capable of song-learning at 2 years of age. Thus the duration of the sensitive period seems to be a matter of both age and experience. It therefore seems likely that the sensitive period is determined by particular processes of development which include both physiological state and experience (see ONTOGENY).

Sensitive periods are especially important in IMPRINTING, the process by which SOCIAL preferences are influenced by experience. The sensitive period usually begins at birth or hatching, and an individual's social preferences become progressively narrowed to the point where they are no longer influenced by experience. The young animal is predisposed to become attached to objects which provide a particular range of stimuli, and in the natural environment these are usually provided by the parents. In the LABORATORY, the young animal can be induced to develop an attachment for unnatural objects, such as animals of other species, cardboard boxes, rubber balls, etc. The imprinting process sometimes has a dual effect, resulting in attachment to parents and in a preference for SEXUAL partners which are similar to the object of the imprinting. Thus ducks (Anatinae) of one species which are reared by ducks of another species will subsequently attempt COURTSHIP with members of the foster species. The sensitive period for filial imprinting generally terminates before that of sexual imprinting.

Sensitive periods are also important in the development of HABITAT preferences and FOOD SELECTION. Among human beings there are thought to be sensitive periods involved in the learning of LANGUAGE, and of certain movement skills.

Functional aspects of sensitive periods. At first sight it seems puzzling that an animal should be required to learn such characteristics of its parents as appearance, song, or language. And indeed, an INNATE recognition of other members of the species, and an innate ability to choose a particular habitat or sing a particular song is found in some species. Why should there be this difference between species?

One possibility is that some species have much more highly developed PARENTAL CARE than others. In some species, the parents discriminate between their own offspring and others, and may even attack young that are not their own. Shortly after she gives birth, a mother goat (*Capra*) has a brief period when she is sensitive to the smell of her kid. Five minutes of contact within the first hour of birth is sufficient, but if there is no contact during this period then the kid will be rejected. The young animal needs to be able to discriminate between the parents that care for it and other adults of the same species. To do this the young have to learn the individual characteristics of their parents.

In the case of sexual imprinting it has been suggested that the sensitive period enables animals to learn the characteristics of their kin, so that they can then choose a mate that is different, but not too different, from their parents and siblings. This arrangement makes it possible for the relative advantages of inbreeding and outbreeding to be balanced. Thus sexual imprinting may provide a standard of comparison for subsequent MATE SELECTION.

There is considerable evidence that members of the opposite sex known to each other from early life are relatively unattracted to each other sexually. For example, in one study, male and female deer mice (*Peromyscus maniculatus*) were paired at 21 days of age or at 50 days. Those that had been paired early produced fewer offspring than those paired late, and the results were the same whether or not the mates were siblings. Similar evidence has been found among human beings. Studies of children in Israeli kibbutzim, and of cultures in which marriages are arranged among children who grow up together, show that strong sexual relationships are not formed between people who have spent their childhood together. This lack of sexual attraction among girls and boys living communally is sometimes called negative imprinting, and is thought to be an important factor in preventing incest among human beings.

In birds the timing of sexual imprinting is associated with the development of adult plumage, as would be expected if a bird is to learn about the appearance of its siblings. Thus, the sensitive period for sexual imprinting in male northern mallard ducklings (*Anas platyrhynchos*) starts at about 4 weeks of age and lasts about a month. During this period the young female becomes quite adult-like. Domestic chicks (*Gallus g. domesticus*) develop adult plumage more slowly, and their sensitive period for sexual imprinting starts at about 6 weeks of age.

Sensitive periods of learning appear at various stages of development, and are relevant to different aspects of behaviour. Undoubtedly the different sensitive periods have different FUNCTIONS. Each species solves its developmental problems in its own way. In some cases this involves sensitive periods of learning, and in some it does not.

8, 9

SEXUAL BEHAVIOUR includes all behaviour leading to the fertilization of eggs by sperm. Once fertilization has taken place, there may be further sexual behaviour directed towards further fertilizations, or there may be a switch to some form of PARENTAL CARE. Sexual behaviour may involve COPULATION leading to internal or external fertilization, or copulation may be absent. In the three-spined stickleback (*Gasterosteus aculeatus*), for example, the female swims into the nest, deposits her eggs, and swims out. The male follows the female into the nest and fertilizes the eggs. In some animals, COURTSHIP precedes fertilization.

Hormones and sexual behaviour. The development and temporal organization of sexual behaviour is largely controlled by HORMONES. In some mammals the hormonal balance of the mother can have a strong influence upon the sexual development of the young. For example, if pregnant guinea-pigs (*Cavia aperea porcellus*) are treated with male sex hormone, the genitalia of female offspring may resemble those of males. In the new-born rat (*Rattus norvegicus*), the BRAIN is sexually undifferentiated and capable of producing male or female behaviour. Differentiation normally occurs during a particular SENSITIVE PERIOD, under the influence of hormones. If female rats are treated with male sex hormone at 4 days of age they are incapable of female sexual behaviour as adults, although they may show male behaviour under certain conditions. In general, male sex hormone secreted during early life imposes a masculine pattern of DEVELOPMENT, whereas feminine development occurs in the absence of hormones, or in the presence of the normal maternal hormones. In normal male infants there appears to be a short period during which male sex hormones are produced, and these influence the course of development (see ONTOGENY).

The next important period of hormonal influence is puberty. Hormones produced at puberty induce a readiness to show sexual behaviour, and initiate the pattern that lasts throughout the animal's life. In many species this pattern is seasonal, and is correlated with the CLIMATE typical of the normal HABITAT. For example, the great tits (*Parus major*) of Wytham Woods, near Oxford, England, lay their eggs in April. The laying date is strongly correlated with the average daily temperatures during March. If there is a warm spring the eggs may appear early in April, but if temperatures are low, laying may be postponed by as much as 4 weeks. A combination of increasing day-length and temperature seems to be re-

sponsible for bringing the birds into reproductive condition. Food availability during the courtship period is also of importance, because the female has to obtain sufficient resources to lay about ten eggs. The female produces her own weight in eggs within 9 or 10 days. She is aided by the COURTSHIP FEEDING of the male, but the food comes from the same source, and is likely to be more abundant if the weather is warm.

In many species the TIME of mating is determined by an internal CLOCK capable of considerable accuracy on an annual basis. In the Atlantic and Pacific palolo worms (*Eunice viridis* and *E. fucata* respectively) the mating period is confined to one or two days a year, and to a particular time of day. Their period of reproduction is closely associated with the phases of the moon. In most vertebrates, factors such as the day-length, temperature, and time of year influence the brain (possibly via the *pineal organ* of the forebrain) to stimulate *gonadotrophic hormone* secretion by the *pituitary gland*. These hormones stimulate growth and activity in the testes and ovaries, which in turn produce their characteristic sex hormones. At the end of the breeding season pituitary activity ceases, the gonads become inactive, and reproductive behaviour is no longer evident.

Some species, including man, are sexually active all the year round. This is understandable in equatorial regions where there is no optimum season for reproduction. However, some tropical sea birds, such as the sooty tern (*Sterna fuscata*), the brown booby (*Sula leucogaster*), and the lesser noddy tern (*Anous tenuirostris*), breed at 8- to 10-month intervals under equatorial conditions. Even in man there is evidence for circannual RHYTHMS in hormone levels. There appears to be a circannual change in the activity of the testes of mature male human beings, with a peak in late autumn and a trough in late spring, in the northern hemisphere. Parallel changes can be detected in women. There is also an interesting correlation between the birth rate and latitude in human populations. As the circannual peak of human gonadal activity is in November, we might expect the birth rate to be highest nine months later. This does not seem to be the case. In man, as in macaque monkeys (*Macaca*), peak birth rate is correlated with latitude, being earlier nearer to the equator. However, there are also marked differences correlating with SOCIAL practices. For instance, human populations in the northern hemisphere not using birth control devices have their birth peak around 7 January, whereas those using birth control peak in about April (July in the U.S.A.). In populations not using birth control devices and with sexual abstinence during Lent, the number of conceptions increases after Easter, which accounts for the birth peak in January. In populations using birth control devices, social factors such as tax laws may account for the annual birth peaks.

In addition to seasonal cycles of sexual activity many mammals have a much shorter cycle of sexual 'heat', called the *oestrous cycle*. Some, such as the red fox (*Vulpes vulpes*), have only one period of heat per year. The females are receptive only during a short period (1–6 days) in February. Domestic dogs (*Canis lupus familiaris*) have two periods of heat per year, one in spring and one in autumn. Some other mammals have a series of cycles at regular intervals throughout the breeding season. The oestrous cycle of the female rat lasts 4–5 days. The period of receptivity occurs when fertilization is most likely to take place, just before ovulation. Males are usually attracted to females on heat by their sensitivity to the PHEROMONES released by the female. In some species copulation is necessary for ovulation. These include the rabbit (*Oryctolagus cuniculus*), the cat (*Felis catus*), and the short-tailed shrew (*Blarina brevicauda*). In most mammals, however, ovulation occurs whether or not there is sexual activity. The oestrous cycle is under strict hormonal control, and the period of receptivity or heat is usually no more than 15% of the total period of the oestrous cycle. Some primates, including man, have a different system, called the MENSTRUAL CYCLE.

The role of experience in sexual behaviour. Although the basic aspects of sexual behaviour are INNATE in animals, experience has been shown to play an important part, particularly in birds and mammals. In birds sexual IMPRINTING can be particularly important in MATE SELECTION and subsequent reproductive success. In mammals also, early experience can have a profound effect upon subsequent sexual behaviour. Tactile contact with members of the same species during infancy is especially important in the rat, guinea-pig, cat, dog, and rhesus macaque (*Macaca mulatta*). Deprivation of tactile contact in infants and JUVENILES can have deleterious effects upon subsequent masculine sexual behaviour, and upon SOCIAL RELATIONSHIPS in general. Socially deprived males often show deficiencies in ORIENTATION towards females, and opportunity for PLAY as a juvenile seems to be particularly important for the development of proper COORDINATION in subsequent sexual behaviour.

Evolutionary aspects of sexual behaviour. Every sexually reproducing species must meet a number of criteria. These include mate selection, reproductive timing, coordination during mating, promotion of viable offspring. Although species vary greatly, each has solved the same basic problems during the course of EVOLUTION.

Mate selection is of prime importance in preventing interbreeding between species, which can have highly deleterious GENETICAL consequences. Most species have characteristic ISOLATING MECHANISMS, which include specific SIGN STIMULI and DISPLAYS. Each potential partner must be able to recognize the correct signals and respond appropriately. Different species have different MATING SYSTEMS, but mate selection is always subject to the same basic evolutionary pressure. The sexual partner should provide the best possible

benefits for the offspring, both in terms of the genetical contribution, and in terms of parental care (see SEXUAL SELECTION).

Reproductive timing is largely determined by the habitat that is characteristic of the species. Conditions must be favourable for the extra FEEDING burden imposed by the production and maintenance of offspring. In temperate latitudes daylength and temperature are the most important factors. In equatorial regions rainfall may be of prime importance. For example, the red-billed weaver bird (*Quelea quelea*) of East Africa uses green grass for NEST-BUILDING. The reproductive cycle of first-year birds is influenced by the appearance of the new grass that grows after rain. Older birds, however, respond to the rainfall itself.

Coordination during mating is partly a matter of mutual stimulation, bringing each partner to a peak of sexual MOTIVATION. In many animals the courtship of the male serves to bring the female into this condition. Female fruit flies (*Drosophila melanogaster*) are more receptive to male courtship if they are able to hear the male courtship SONG prior to their introduction to the males. Similarly, female Barbary doves (*Streptopelia risoria*) respond to the calls of the male by hormonal changes which bring them to a state of sexual readiness.

Copulation requires COOPERATIVE behaviour and sometimes provides stimuli necessary for successful fertilization. In addition to those animals (see above) in which ovulation is induced by copulation, details of the copulatory pattern may be important in the process of fertilization itself. For example, in rats a number of intromissions prior to ejaculation are necessary to induce the normal transport of sperm from the vagina to the uterus. It is also necessary that ejaculation is prolonged and is followed by a period of quiescence (the *refractory period*) on the part of the male, during which time he is unable to copulate. Maximal numbers of sperm do not reach the uterus until 6–8 min after ejaculation. If, during this period, a female rat receives copulatory intromissions, the number of sperm is reduced, and the litter size diminished. It is possible for a competing male to cancel the effects of a rival by copulating immediately after him. It is therefore important that during the refractory period a mating pair should not be disturbed by rival males.

Promotion of viable offspring is largely a matter of parental care, especially of INCUBATION behaviour in birds and of MATERNAL behaviour in mammals. However, sexual behaviour does have some influence on the viability of the offspring. In some monogamous species strong pair bonds are developed during courtship, and these may be maintained throughout the period of parental care. However, such partnerships do not depend only upon sexual attractiveness and compatibility. Among the herring gulls (*Larus argentatus*) breeding at Walney Island, Cumbria, England, individuals show SPECIALIZATION in FEEDING habits. The different types of food are usually available at different times of day, depending upon the tides, weather, etc. During incubation it is most important that the eggs should not be left unattended, as they are readily subject to PREDATION. Therefore the members of a mated pair cannot leave the nest on a FORAGING expedition simultaneously. Successful pairs seem to be made up of individuals with different feeding specializations, but it is not known how this comes about. It is possible that courtship feeding by the male influences the subsequent food preferences of the female. Alternatively, incompatible pairs may be unsuccessful in raising a brood in one year, and may 'divorce' and take other partners in subsequent years. Successful mated pairs often re-pair for a number of successive years, and tend to occupy the same TERRITORY, even though they may be absent from the breeding colony for the rest of the year.

11, 82

SEXUAL SELECTION. In 1871 Charles DARWIN published *The Descent of Man and Selection in Relation to Sex*. In this book Darwin examined and documented the complex subject of sexual selection, which he had alluded to in his *Origin of Species* (1859). Darwin regarded man as the 'greatest subject', and worked for many years collecting material for a book on the descent or origin of man. When Darwin came to consider the problem of the differentiation of the human races, he found it necessary to investigate the subject of sexual selection in detail.

The concept of sexual selection is closely related to the more general concept of NATURAL SELECTION. According to Darwin, sexual selection 'depends on the advantage which certain individuals have over others of the same sex and species solely in respect of reproduction'. In those cases where the males have acquired a particular structure or feature of behaviour, Darwin maintained that this was 'not from being better fitted to survive in the struggle for existence, but from having gained an advantage over other males, and from having transmitted this advantage to their male offspring alone'. Here Darwin is assuming that the females make a definite choice of their sexual partner. In general, this assumption is confirmed by modern studies (see MATE SELECTION).

Darwin believed that the end product of sexual selection was different from that achieved by natural selection. The end result of natural selection is a population which has greater FITNESS, in the sense of being adapted to the environment, than its predecessor. Darwin maintained that the advantage of behaviour evolved through sexual selection lies in the satisfaction of female whims. Once evolved, such behaviour may make little contribution to survival but will, nevertheless, be maintained as a result of the processes of mate selection. Present-day biologists agree with Darwin about the general principle of sexual selection, but many of Darwin's ideas about sexual selection have been found to be correct only for certain species and certain situations.

Sexual selection as envisaged by Darwin usually resulted in sexual dimorphism, that is, in a difference between males and females. In the Indian peafowl (*Pavo cristatus*), for example, the peacocks are brightly coloured and have a large distinctive tail which can be spread like a fan. The females have drab coloration and are without the distinctive tail. The males gather in an arena at mating time where they display to any female who approaches. The tail is spread, the plumes are rattled, and the wings are quivered. The display is directed mainly at the female, who, when aroused, runs to the front of the male each time he turns his back and eventually solicits him. Males display simultaneously in the arena and hence compete directly. A female can compare a number of displays and choose the one which stimulates her most. Presumably the most attractive males will fertilize the most females. Thus the characteristics of the male which attract females will be perpetuated in the species by means of sexual selection. Because the roles of the two sexes are not symmetrical, there will be no such selection favouring the development of special plumage structures in the female. Consequently a sexual dimorphism will develop through the course of EVOLUTION.

Not all sexual dimorphism, however, is a result of sexual selection. Most of the differences between the sexes are clearly the result of natural selection. Among these are characteristics having to do with PARENTAL CARE, such as the pouch of female marsupials and the mammae of female mammals. Amongst ducks (Anatinae) there is strong selection for reduced size in the females of hole-nesting species in which only the females incubate. Consequently there is a size dimorphism in these species which has nothing to do with sexual selection. The same is true for the sexual size dimorphism of many other species of animals. On the other hand biologists are agreed that there are many aspects of sexual dimorphism which are not explicable by means of natural selection alone. Sexual selection occurs mainly when males compete, and when females are coy and initially unwilling to mate. Female coyness is often related to reproductive needs: the need to avoid the wrong mate, to ovulate at some specific time, or to be guided to a mate or breeding place. Under such circumstances sexual selection will tend to be effective, but it may be tempered by opposing pressures of natural selection. For example, the male of the three-spined stickleback (*Gasterosteus aculeatus*) generally has a distinctive nuptial coloration consisting of a red belly and blue eyes. However, in a particular population of this species in North America a blackish protective coloration has evolved in response to the presence of a black predatory fish. Male sticklebacks in nuptial coloration are black-bellied in this population, and the females in the absence of other colour types spawn normally in the nests of the black-bellied males. Even though this black-bellied population has been in existence for 4000–8000 years, females tested in the laboratory still show a preference for red-bellied males. Here we can see that natural selection, in the form of predator selection, may override the effects of sexual selection.

Darwin was not correct in thinking that sexual selection was always a matter of female choice. The stag of the red deer (*Cervus elaphus*), a native of Britain, develops antlers each year for the rutting season. The males compete and fight one another for the ownership of harems. Any stag may challenge another already possessing a harem, but he usually does not do so if the possessor is larger or has larger antlers. Fights consist of clashing heads and locking antlers, after which the heads are twisted in an attempt to deal a blow to the flank, which can be fatal. Victorious males take over the females of the loser. In this case the antlers are used for fighting and threat display. The competition for females is, therefore, a direct one between the males, who use their antlers in threat and fighting and not for direct wooing of the females. In this respect, this example differs from other cases of sexual selection, but the antlers clearly confer reproductive advantage, in that those with larger antlers serve more females. Apart from this advantage, the antlers appear to contribute little to survival. The females do not possess them, nor do stags outside the breeding season. Moreover, they constitute a considerable drain on the stag's metabolism, requiring large quantities of materials such as calcium salts and phosphorus for their growth each year.

Another important factor influencing choice of sexual partner is *parental investment*. This can be defined as an investment by the parent in an individual offspring that increases the offspring's chance of surviving (and hence reproductive success) at the cost of the parent's ability to invest in other offspring. Parental investment includes any investment of time or material that benefits the young, such as feeding or guarding them. It does not include effort expended in finding a member of the opposite sex, or in competing with other members of the same sex in order to mate with a member of the opposite sex. The extent of parental investment can be measured by reference to its effect on the parent's ability to invest in other offspring. Thus a large parental investment is one that strongly decreases the parent's ability to produce other offspring. The members of a sexual partnership may not invest in their offspring to the same degree. Individuals of the sex investing less will compete among themselves to breed with members of the sex investing more, since an individual of the former can increase its reproductive success in this way. Competition for mates usually characterizes males, because males usually invest little in their offspring. In cases where male

parental investment strongly exceeds that of the female, therefore, one would expect females to compete among themselves for males, and for males to be selective about whom they accept as a mate. For example, in sea horses (*Hippocampus*), where there is high male parental investment, the females are brightly coloured and tend to dominate the courtship.

Since the time of Darwin, biologists have been concerned to find out exactly what factors influence the choice of a sexual partner (see MATE SELECTION), and how these factors prevent interbreeding between species (see ISOLATING MECHANISMS). For example, scientists have shown that experimental alteration of the colour patterning of male domestic fowl (*Gallus g. domesticus*) has a considerable influence on mate selection by females. During courtship the cock displays to the hen by a waltzing motion during which the plumage is displayed. Hens accept males by crouching in a typical 'solicitation' posture. Experiments show that females discriminate between would-be mates primarily by means of comparative male appearance, rather than by differences in the courtship displays of the males themselves. Experimental interference with the colour patterning of cocks, such as that illustrated in Fig. A, was found to have a decreasing effect on solicitation by females when they were courted by these cocks. On the whole females of different breeds, such as White Leghorn, Brown Leghorn, and Broiler, prefer to mate with cocks of their own breed,

Fig. A. A Brown Leghorn cock with white feathers attached to alter the neck colour and contour.

especially if these possess the body build and colour patterning typical of the breed (see DOMESTICATION).
22

SIGN STIMULUS. A sign stimulus is a part of a *stimulus configuration*; it is external to the animal, and relevant to a particular response.

There is often a rigid relationship between a stimulus and a REFLEX response, such as that between a moving object and the eye-blink response. This protective reflex occurs regardless of the animal's position or ORIENTATION. It also occurs irrespective of the time of day, provided that the eyes are open. In contrast, many behavioural responses depend upon an external condition in addition to the mere presence of an adequate stimulus, and an appropriate internal readiness to respond. For example, a male fish or bird in its TERRITORY shows AGGRESSION towards an interloper. If, on the other hand, the same animal is encountered outside the territory, it elicits fleeing or some other non-aggressive response. Because of their dependence upon more complex conditions of elicitation, such responses to stimulation occur less predictably than the reflex type of response. Furthermore, it has been shown that an animal responding in this non-reflex fashion utilizes only a part of the potential information contained in the stimulus situation. This relevant part is called the *sign stimulus*. Note that this definition leaves open the question of the developmental history of the response in the individual animal. For instance, a male three-spined stickleback (*Gasterosteus aculeatus*) in nuptial colours has a red belly which is an important cue for triggering FIGHTING in another male (Fig. A). Crude imitations of a male suffice to elicit aggression provided they have a red belly. In contrast, a freshly killed male lacking the red belly is virtually ineffective in evoking attack. From this, one can conclude that the many details of structure, surface texture, etc., of a natural stickleback are irrelevant in this context, whereas a particular colour cue is of prime importance. However, the seemingly simple sign stimulus, the red belly of the male stickleback, is more complicated than described so far. The red coloration must be on the underside; a model of a male with a red back is about as ineffective as a male in non-breeding dress. This spatial, *configurational* relationship is a common property of many sign stimuli. It illustrates a difficulty encountered in stimulus analysis. In order to discover which part of a situation has stimulus value that part must be experimentally altered, or even eliminated. In so doing, however, other parts may be altered as well, and in each case it must be determined to what extent a sign stimulus depends upon the total stimulus configuration.

When one part of the total situation is found not to be a sign stimulus for a particular response, it might well have sign stimulus value for another response. The herring gull (*Larus argentatus*) robs

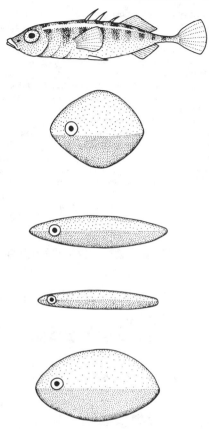

Fig. A. Models of a male stickleback that elicit fighting in a rival male. The model at the top is a dead male lacking the nuptial colours, the four lower models have a red belly but are otherwise very crude.

the eggs of other gulls nesting in the colony. An egg must be normally shaped and undamaged to act as a RELEASER. During INCUBATION, however, responses to the egg are released by quite different stimulus properties. For a herring gull approaching its own nest, factors such as colour, speckling, and size of the egg have sign stimulus value while shape does not; the 'egg' may have the shape of a cylinder, a ball, or a cone. Once the gull settles on its eggs, however, shape becomes a crucial feature. If the eggs do not have rounded smooth edges, the bird will frequently rise and resettle in a 'disturbed' manner. Thus, for three different acts, three sets of sign stimuli of one and the same stimulus object are responded to. The selection of different sign stimuli by a gull in each of the three cases permits one to reject the idea that the gull is unable to perceive the irrelevant cues in each case. Rather, it has to be assumed that the internal state of the bird determines which stimuli the bird will respond to in each context. In other words, it must have different MOTIVATION when robbing another's egg than when it incubates its own egg.

Many alterations in sign stimuli evoke a weaker or stronger response in the animal responsive to them. In stimulus analysis, therefore, it is necessary to make sure there is a reproducible stimulus property. For example, territorial pied flycatchers (*Ficedula hypoleuca*) tend to mob a male red-backed shrike (*Lanius collurio*), a potential predator. An indispensable sign stimulus eliciting the MOBBING response is the black eye bar of the shrike. Dummy experiments have shown that when the shade of the bar is varied, so as to decrease the contrast of the bar against the light grey head, the mobbing response decreases. How precisely the birds attend to the nature of the bar is indicated by the two weakest stimulus values of the series (Fig. B): the

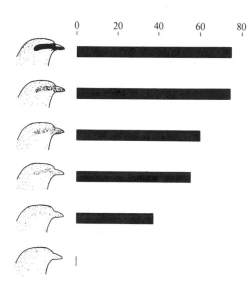

Fig. B. Influence of the strength of a sign stimulus, the eye bar of a male red-backed shrike, on the strength of a response (mobbing (calls/min) at a shrike by territorial pied flycatchers). Only the heads of the shrike models used are depicted. Top model corresponds to the natural shrike.

slightest contrast that is scarcely perceivable to our eyes still elicits mobbing, although it is not significantly different from a zero response evoked by the bar-less male. Such continuous variation of the effectiveness of sign stimuli has also been found in the body size of the female guppy (*Poecilia reticulata*) in eliciting male COURTSHIP, and in egg number as a determinant of incubation behaviour in the oystercatcher (*Haematopus ostralegus*), to cite but two further examples. In other words, sign stimuli can artificially be made SUPERNORMAL. The common occurrence of the underlying perceptual graded response raises the following question: Why has nature not tuned perceptual mechanisms more precisely to the stimuli they are designed to decode? The probable answer is that a sign

stimulus serves several functions. For example, the black eye bar in the male red-backed shrike is probably an intraspecific sign stimulus, and the advantage of intraspecific COMMUNICATION outweighs the advantage of becoming immune to mobbing by reduction of the eye bar. If constraints imposed by other FUNCTIONS keep the physical properties of a sign stimulus unchanged, there is no reason for the perceptual mechanism to be tuned so that the sign stimulus is maximally precise. It can afford to be imprecise, since in nature the presence of potentially misleading supernormal stimuli is rather rare. With no constraints upon the development of a sign stimulus any imprecision of the corresponding perceptual mechanism of the receiver may be an important selection factor in NATURAL SELECTION. It may well have led to bizarre structures and behaviour patterns with a signal function, as exemplified by the nuptial colours of many animals.

An important question is how sign stimuli of one stimulus object combine to trigger a given response. If one asks this question, one must be aware that any conclusion drawn pertains to observable behaviour that, in itself, is only the end result of many processes occurring within the animal. The simplest way in which sign stimuli combine is by algebraic summation of their individual stimulus values. For example, in the African cichlid fish *Haplochromis burtoni*, territorial males are brightly coloured. Among the various components of the complex colour pattern there is a vertical black bar running from the eye to the corner of the lips as illustrated in Fig. C. This stimulus increases aggression in other male fish. In contrast, an orange patch above the pectoral fins decreases attack readiness in others, while the rest of the whole pattern is virtually unimportant in this situation. The combined presence of both stimuli on a fish yields a total stimulus value of exactly the algebraic sum of the component values. The effects of both components are independent of each other. Note that such an additivity of cues might not be found if some other indices of attack readiness were chosen. Moreover, from the additive superimposition of the effects elicited by the vertical bar and the orange patch, it does not necessarily follow that an additive process takes place in the nervous system. Instead of adding both stimulus values directly, various multiplicative processes may combine to produce the

Fig. C. Colour pattern of territorial male *Haplochromis burtoni*.

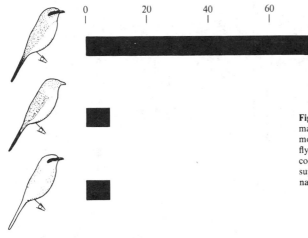

Fig. D. Stimulus values of three models of the male red-backed shrike as measured by the mobbing response (calls/min) of territorial pied flycatchers. The two lower colour patterns are complementary to each other, and after being superimposed would yield the pattern of the natural male shrike (top).

response. The various components of the whole *Haplochromis burtoni* colour pattern may appear and disappear within seconds, according to the external SOCIAL situation. This fact makes them appear ideal for functioning as *social releasers*, that is sign stimuli that have evolved to facilitate communication. The black eye bar can be 'switched on' independently of the orange patch, a fact that can be functionally understood in terms of the social behaviour of the fish.

Egg-retrieving behaviour in the herring gull is affected by a number of visual properties of an egg on the rim of the nest. Size, speckling, and background coloration (green being more effective than brown) have been shown to be important, and each of these components adds algebraically to the effect of the others. With a constant egg background coloration, the omission of speckling could be compensated for by a constant increase in size for all values of size tested. The effects of speckling and size are thus additively superimposed, and either can be substituted for the other to produce the same total stimulus value. Similar qualitative observations have led to the formulation of a *rule of heterogeneous summation*. This implies that the effectiveness of a stimulus pattern in releasing a particular behaviour increases with the number, the sign (+ or −, see above, *H. burtoni*), and the intensity (see above, pied flycatcher) of the stimulus components involved. Furthermore, these components may replace each other functionally.

In most other cases stimuli depend on the associated presence of others; that is, they show stimulus interaction. In the extreme, this may lead to the total breakdown of recognition of a pattern, even if only one characteristic of the whole is missing. Territorial pied flycatchers mob a male red-backed shrike model more strongly than any other shrike model used (Fig. D). A model possessing only the black eye bar and another showing the rest of the whole colour pattern of the male shrike are virtually ineffective. By adding up their stim-

ulus values one obtains only a minor fraction of the response to the whole pattern. Apparently, both pattern components depend on each other in their effectiveness—that is, they do not follow any pure algebraic summation paradigm as referred to above.

Components of a pattern may also interact with the background of a stimulus object, not only with each other. For example, young of the cichlid fish, *Hemihaplochromis multicolor*, are cared for by the female parent, which they follow about. A substitute for the mother that is able to elicit the following approach by the young must be somewhat darker than the background. This flexibility of the response correlates well with the nature of the physiological colour change of the adult fish. It is always somewhat darker than its background, and thereby always offers to its brood an optimum stimulus situation.

A still higher degree of complexity is involved in the evaluation by animals of other stimulus objects, inanimate or living. For instance, the bee-killing digger wasp (*Philanthus triangulum*) and, under certain circumstances, honey-bees (*Apis mellifera*), make use of landmarks to relocate their burrows or their hives, respectively. Jackdaws (*Corvus monedula*) and tits (Paridae) mob when a conspecific is threatened by the approach of a mammalian predator. Still more perplexing behaviour can be observed in hamadryas baboons (*Papio hamadryas*). A subordinate group member usually gives way to superior members, but if the dominant male is at its back, the subordinate animal will occasionally threaten one of the other more dominant members without eliciting their attack. (The dominant male would immediately interfere if the subordinate animal were attacked.) Hence, the subordinate must, in some way, be certain that it takes no risk when performing the 'safe' threat.

Sign stimuli not only have a releasing function. In addition, they direct responses in space, and they often change the readiness of the recipient to

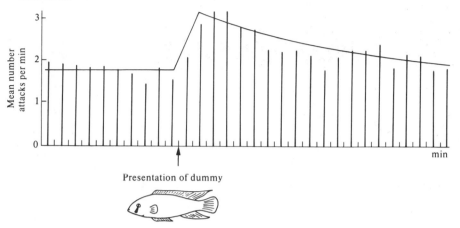

Fig. E. A brief presentation (10 s) of a model male *Haplochromis burtoni* raises attack readiness followed by a much slower decline, as measured by number of attacks/min by an experimental male directed towards juvenile test fish continuously present.

respond in a particular way. The latter function can be illustrated by the fighting behaviour of *Haplochromis burtoni*. An experimental male fish is continuously allowed to observe temporarily blinded juveniles which it attacks from time to time. A brief exposure of the male to a rather effective model of a male (without the attack-suppressing orange patch) raises its readiness to attack the young test fish (Fig. E). The stimulus model cannot be bitten because it is presented behind a glass plate. Thereafter, the raised level of responsiveness abates at a much slower rate. While the less effective blind test fish apparently elicit occasional attacks, the much more effective stimulus of the male model motivates the territorial fish to attack more often than before. The difference between both effects of a stimulus is most clearly seen in that attack readiness increases for a relatively long period compared to the period of sudden stimulation. When the model stimulus is available to the experimental fish, it receives a certain proportion of the attacks. From this one must conclude that eliciting and motivating properties are present in one and the same stimulus object.

100, 137   E.C.

SKINNER, BURRHUS FREDERIC (1904–). An American experimental psychologist who has made detailed studies of LEARNING. He rewarded animals (usually with food) if they behaved in a certain way, so the animal learned to do that rewarded activity more frequently. This is called OPERANT conditioning. Skinner could then measure the effectiveness of different methods of REWARD in causing changes in the animal's behaviour. These investigations are historically important not only because of the discoveries made, but also because they have inspired so many subsequent experimental analyses of learning.

Skinner's wide fame, however, is due not so much to his experimental research as to his popular books that argue that operant conditioning is sufficient to provide, first, a complete scientific explanation of human behaviour, and, second, a technique for the control of human behaviour. B. F. Skinner's books about human behaviour have been widely read and have stimulated impassioned controversy, though mostly in America.

Brief life. B. F. Skinner was born in Pennsylvania. After leaving school he attended Hamilton College, New York, from 1922 to 1926, where he majored in English. He was then intending to pursue a literary career. However, after reading Bertrand Russell on J. B. Watson, and H. G. Wells on PAVLOV, he changed his mind (or 'altered his behaviour' as Skinner himself would put it), and went to Harvard Graduate School to read psychology. He then worked for a Ph.D. in psychology and did 5 years of post-doctoral research at Harvard. He took his first teaching job at the University of Minnesota in 1936, then moved to the University of Indiana in 1945, and back to Harvard again in 1948. He has remained at Harvard since then.

Conditioned behaviour of animals. The philosophical context of Skinner's work is crucial for understanding the way he analyses behaviour. Indeed Skinner has sometimes defined *behaviourism* (the school of psychology in which he is categorized) as a philosophy of the science of behaviour, not the science itself. Early in the twentieth century, the philosophical tenets of logical empiricism received their most extreme formulation in the idea of *operationism*. Operationism held that the meaning of scientific terms is the public operation used to quantify them. The meaning of temperature, for example, might be defined as the movement of mercury in a thermometer. When

J. B. Watson inaugurated behaviourism in 1913 he adopted an operationist attitude to psychological terms. Watson was particularly dissatisfied with psychological introspection, the subjective analysis of one's personal consciousness. The trouble with introspection was that different psychologists rarely agreed on their observations, and it was difficult to test who was right. Watson held that the mind could only be studied through observable behaviour. B. F. Skinner was heavily influenced by the doctrines of Watson. In his early works, at least, Skinner was uncompromising in his banishment of motives, intentions, and internal states of mind from the true realm of scientific study. Operationism produced a psychology that was sternly anti-theoretical: for example, in 1950 Skinner wrote a paper entitled 'Are theories of learning necessary?' Behaviourists in general, and Skinner in particular, sought to describe, control, and predict behaviour but not to understand it, in the conventional sense.

Another important influence on Skinner's work was Pavlov's discovery of the conditional REFLEX. In Pavlov's work, a dog (*Canis lupus familiaris*) might be conditioned to salivate on hearing a bell; the salivatory response of the dog is elicited by the stimulus, which is food. Some attempts were made to explain sequences of behaviour as simple chains of stimuli and reflex responses. Skinner called the conditional reflex Pavlovian, or classical, CONDITIONING to distinguish it from operant conditioning. In a typical operant conditioning experiment the animal is fed when it does some arbitrary act, such as pressing a bar. The bar pressing is said to be *reinforced*, and the animal becomes more likely to press the bar in the future. The activity (bar pressing) persists because its consequences are effective. However, the activity itself was not initially stimulated by food; the animal first pressed the bar by accident. It is the arbitrary relation between conditioned behaviour (bar pressing) and its REINFORCEMENT (food) that makes operant conditioning so versatile in explaining behaviour. Most of Skinner's research has been on operant conditioning, usually of the pigeon (*Columba livia*) or the rat (*Rattus norvegicus*).

Skinner developed a powerful technique for studying operant conditioning: the *Skinner box*, or behavioural chamber. The Skinner box is just a box with some mechanism for an animal to operate (usually a bar for a rat to press, or a disc for a pigeon to peck at), and an automatic mechanism to reward the animal with food at the times determined by the experimenter. The presses on the bar can be recorded automatically, so it is possible for the experimenter to obtain vast quantities of data for the expenditure of little effort. Skinner boxes are now part of the standard LABORATORY apparatus of experimental psychology.

Experimental studies of animal learning have generally used many rats and then averaged the results from them all. However, Skinner's doctoral research was an intensive study of the operant conditioning of a single rat. In this and subsequent research he systematically studied the factors affecting the rate of operant behaviour. Here are some of the factors that can be varied. First, there is the time interval from the bar press to the food reward. Second, this interval can be of fixed or variable duration. Third, there is the reinforcement ratio: that is, whether the animal is rewarded once per bar press, or once per five bar presses, or whatever. Fourth, this reinforcement ratio can be fixed or variable. Skinner's doctoral research was published in his famous book *The Behavior of Organisms: An Experimental Analysis* (1938), and his later work was summarized in *Schedules of Reinforcement* (1957). If, for example, the rat is fed after a fixed interval of 1 min since the last reward, the rat learns to decrease its pressing just after a reward is delivered, and then increase it again towards the end of the 1-min interval. If the interval is made variable, the rate of bar-pressing is stabilized; the rat presses at a constant rate. The results of experiments on operant conditioning are useful in training animals because they reveal the schedule of reinforcement that is most effective for altering their behaviour. Skinner has often maintained that you can train an animal to do more or less anything by reinforcing it appropriately. A famous example is his attempt during the Second World War to train pigeons to direct missiles towards their target. The idea, however, was not taken up by the military authorities.

In later years there has been a detectable softening in Skinner's philosophy. His paper 'Behaviorism at fifty' (1963, the title alludes to the date of Watson's paper of 1913) acknowledges that psychology must account for private experiences. His book *Schedules of Reinforcement* (1957) adopts a more theoretical stance than earlier works, in which he had sought only to describe the law-like relations between the stimuli impinging on the animal and its behavioural response. And in *About Behaviorism* (1974) Skinner denies that behaviourism 'ignores consciousness, feelings, and states of mind' or that it 'neglects innate endowment'.

Although Skinner has repeatedly emphasized that all behaviour is produced by reinforcement, he does not think that this is inconsistent with the ethologists' emphasis on ADAPTATION. In *Contingencies of Reinforcement* he wrote that NATURAL SELECTION is 'responsible for the fact that men respond to stimuli, act upon the environment, and change their behavior under contingencies of reinforcement'. Furthermore, evolutionary theory can illuminate behaviourist psychology because 'the fact that operant conditioning, like all physiological processes, is a product of natural selection throws light on the question of what kind of consequences are reinforcing and why' (*About Behaviorism*).

Behaviourism as applied to man. When B. F. Skinner went to the University of Indiana in 1945, he met J. R. Kantor, a psychologist who held similar views on operationism to his own. It has been suggested that Kantor's influence was one reason why Skinner started to be more interested in applying his ideas to human beings at this time. Skinner's first book about human beings was his Utopian novel *Walden Two* (1948), which portrays a community in which behaviour is controlled from birth by reinforcement techniques. Although *Walden Two* had been a best-seller for years, it was *Beyond Freedom and Dignity* (1971) that provoked the biggest storm. In this book Skinner argued that rhetoric about the freedom and dignity of human beings is not only erroneous, because all human behaviour is determined by conditioning, but also dangerous, because it obstructs the behavioural control that is necessary for the survival of our culture. A liberal reaction was inevitable, not least because of Skinner's polemical style. For example, Skinner even used his system to give unflattering explanations of the behaviour of his critics. They were hostile because 'the scientific formulation has destroyed accustomed reinforcers', causing 'behavior previously reinforced by credit or admiration to undergo extinction. Extinction often leads to aggressive attack.' Skinner's more recent books *About Behaviorism* (1974) and *Reflections on Behaviorism and Society* (1978) are answers to his critics, together with syntheses of behaviourism.

LANGUAGE is the human faculty that is most often explained as an expression of motives or intentions. It is language, therefore, that constitutes the most serious challenge to a science of human behaviour that excludes intentions from its area of study. In *Verbal Behavior* (1957) Skinner attempted to analyse language behaviouristically. He treated language as behaviour, and so he explained verbal utterances as being due to their reinforcing consequences. If a human being says 'open the door', and someone opens it, then the speaker will find his utterance reinforcing, and will tend to emit the same noise again when he wants the door opened. The linguist Noam Chomsky wrote a famous review of *Verbal Behavior*, criticizing it for its incompatibility with both the way in which language is acquired and the grammatical structure of language.

B. F. Skinner's research on how animals adjust their behaviour to different reinforcers, and to different schedules of reinforcement, and his clear distinction between operant and Pavlovian conditioning, is his uncontroversially significant legacy to psychology. It was when he stretched these ideas, and argued that they promised a total political science of psychology, that his behaviour received a mixed reception from his readers.

126, 127, 128   M.R.

SLEEP is a behavioural phenomenon which is widely distributed throughout the animal kingdom. It is characterized mainly by prolonged periods of immobility, and an increased reluctance to respond to stimulation, although in this there is considerable variation among species. Our understanding of its nature and function has been greatly furthered by recent research along two largely separate lines of enquiry. On the one hand, ethological studies of the sleeping habits of many different vertebrate species in their natural surroundings (see ETHOLOGY) have nurtured a conception of sleep which embraces pre-sleep activities that are characteristic of each species, as well as the sleep state itself. On the other hand, physiological investigations of sleep-control mechanisms have revealed a whole range of hitherto unsuspected phenomena which have provoked considerable controversy and radical reassessment of the nature of sleep. In addition, further COMPARATIVE physiological research involving a wide range of vertebrate species has offered to bridge the gap between ethological and physiological studies, through insights into the evolution of sleep and the role it plays in the survival of individual species.

Ethological studies. Although it is normally an easy matter to agree when a given animal is asleep, a general definition of what constitutes sleep has proved elusive. This is a result of the wide variation of sleep styles among animals. The problem is further exacerbated by a general lack of agreement concerning its essential nature, its function, and its evolution. It is therefore more satisfactory, as well as easier, to work with a characterization of sleep, a list of symptoms which may or may not be present in any given species. A satisfactory definition will need to await the solution of many outstanding theoretical problems.

1. *Prolonged immobility*. Although sleep may occur for very short periods, it is most commonly associated with very long sessions lasting many hours, during which an animal remains immobile. Numerous postural adjustments, particularly among warm-blooded animals, may occur during the sleep period, but there is typically no locomotion. It appears that behaviour which is consistent with remaining in the same place is often carried out during sleep. Thus infants will suckle during sleep and ruminants ruminate. Postural adjustments involved in temperature regulation, and stereotyped grooming actions may also occur without any sign of a return to normal consciousness. On the other hand, actions which involve movement away from the sleeping site are absent, either as the result of an inability to coordinate the action, or a general reluctance to initiate such acts.

2. *Raised thresholds of responding*. Animals in so-called 'deep' sleep are less likely to respond to stimulation than they would be when awake. When a response does occur it is commonly delayed, and its execution poorly coordinated. Sleeping fish, for example, can sometimes be handled and even lifted out of the water before they begin to struggle free. Often the most timorous animals can be the most difficult to wake from

sleep, possibly because their vulnerability has led to the selection of especially secure sleeping sites in which they are rarely disturbed.

By no means all animals show such deep sleep. Many herbivorous mammals such as elephants (Elephantidae), deer (Cervidae), cattle (Bovidae), and hares (Leporidae) are reputed to be very light sleepers, reacting with great speed and vigour to the slightest danger. Observation of elephants suggests that they sleep through substantial disturbances caused by fellow herd members, but remain acutely responsive to even the quietest sounds made by an unfamiliar agent. These, and a host of other observations, indicate that a sleeping animal continues to assess the significance of environmental stimuli, but is considerably more selective in its responding than when awake.

3. *Sleep reversibility.* Unlike coma, anaesthesia, and many drugged states, sleep is readily dispelled by intense stimulation, and replaced by a state of wakefulness. It is generally believed that, among vertebrates at least, specific nervous control mechanisms exist for initiating and terminating sleep, and that these mechanisms are responsive to both internal conditions and external stimulation. The speed of reversal which is marked in warm-blooded animals may be considerably slower in cold-blooded animals, whose body temperature during sleep may have fallen well below optimal values (see THERMOREGULATION).

4. *Specific sleep posture.* Individual species differ considerably in the postures adopted during sleep, but animals within a given species usually adopt the same posture. Horses (Equidae), elephants, cattle, and deer may sleep standing up or lying down by virtue of special skeletal adaptations. The sloth (Bradypodidae) and some species of bats (Chiroptera) sleep suspended upside down. Many carnivores sleep crouched, or lie curled up to a variable degree, depending on the environmental temperature.

Unfortunately it is not possible to make any simple generalization to the effect that sleep postures permit a high degree of muscle relaxation. Physiological investigation of muscle tension during sleep shows that deep relaxation is by no means always the rule, even in man who lies fully supported by a bed. Some animals even sleep in a state of gentle but incessant activity. Moorhens (Gallinulini), for example, may sleep while swimming in tight circles well away from the lake shore. Each species has developed a posture which is best suited to its anatomy, physiology, and environmental pressures. Although such postures presumably minimize energy expenditure within these limitations, they do not necessarily reduce muscle tension to a minimum.

5. *Specific sleep site.* Once again, individual species choose widely differing places to sleep, but members of the same species very commonly select sites which are essentially similar. Moreover many individual animals sleep in the same place day after day. Typically a sleep site appears to be selected because it reduces danger from PREDATION and minimizes exposure to other inclemencies of the environment, such as extremes of heat and cold. As a result sleep sites are often in burrows underground, or high in trees along inaccessible branches. Alternatively a site may be chosen not because it is inaccessible, but because it offers optimum camouflage possibilities. Animals who have little to fear from predation and who live in relatively favourable environmental conditions may be less particular about choosing a sleep site. This appears to be true of lions (*Panthera leo*) and basking sharks (Cetorhinidae), who are often seen sleeping in unconcealed locations. Other species may rely upon VIGILANCE rather than a carefully selected sleep site. This applies to large herbivores who have special difficulties in concealing themselves in open grasslands.

Commonly the sleeping place is quite separate from the location of waking activities. Starlings (Sturnidae) who sleep in the centre of large cities disperse to the surrounding countryside to feed during the day, even though this may involve journeys of many miles. Hippopotamuses (Hippopotamidae) sleep while standing in the lake, but have to leave the water to graze. Cattle egrets (*Bubulcus ibis*) ride on the backs of grazing animals all day, feeding on insects disturbed by their hosts, but fly off at night to their special sleeping trees which may be many miles distant. However, other animals, such as the gorilla (*Gorilla gorilla*), merely make their nests where they happen to be at sunset.

6. *Timing of sleep.* Some species rest at night, some during the day, and others sleep both during the day and at night, being active only at sunset and sunrise. Not all animals can be characterized as sleeping at a particular time of day, but by far the majority can. The reasons for this *circadian* patterning are often not difficult to find (see RHYTHMS). Most birds are quite unable to fly successfully in the dark. Many reptiles are unable to maintain an adequate body temperature for effective functioning after sunset. Small warm-blooded mammals are subject to serious predation during the day, and therefore confine as much of their food-seeking activity as possible to the night time. Predators restrict their activity to times when their prey are most available (see PREDATORY BEHAVIOUR). In general, it appears that animals sleep when the environment is most unfavourable, and when the food supply is least. Among land-dwelling animals, the 24-h alternation of light and dark is clearly a dominant influence. However, in the sea, especially near the shore, tides will have an equally important influence on the rest–activity rhythms of animals which live there.

Experiments with mammals, birds, crustacea, and insects in the LABORATORY have shown that these sleep–waking rhythms are not solely determined by changes in the environment. Considerable evidence is now available to show that animals have some internal mechanism which also helps pace these rhythms. Presumably these internal CLOCKS permit the anticipation of environmental changes, and make the animal less responsive to freak fluctuations in illumination and temperature caused merely by the weather. These clocks influence sleep control mechanisms, so that sleep is induced at particular times of day without regard for how exhausting or how relaxing the previous activity period has been.

Sleep as an innate activity. Sleep has been characterized above as a state of immobility and low responsiveness. However, many authors have drawn attention to the stereotyped behaviour patterns which precede sleep and have suggested that they should be considered together as part of an INNATE sleep pattern. According to this formulation, activity preparatory to sleep is part of the APPETITIVE phase, while sleep itself or possibly 'falling asleep' is the CONSUMMATORY response (see also MOTIVATION).

The appetite phase includes the general business of disengaging from the task currently in hand before seeking out or preparing the site where the sleep will take place. Among some species this may involve certain special preparatory actions. The elephant, for example, carefully prepares a pillow of hay before lying down. The gorilla, chimpanzee (*Pan*), and possibly man build a nest each night before retiring. After dark many parrotfishes (Scaridae) retire to the bottom of the tank and secrete a mucous envelope. Once the site has been found and prepared an animal adopts the appropriate sleep posture, and suppresses all activity until sleep overcomes it.

During the appetitive phase an animal may show signs of drowsiness characterized by sluggish actions, and a reluctance to perform tasks unrelated to sleep. Drowsiness is often associated with stretching, and with yawning (found in mammals, birds, reptiles, and fish), the function of which remains a mystery. If sleep is prevented the symptoms of drowsiness are gradually accentuated, and the animal may show intensified efforts to withdraw to a place where sleep will prove possible. However, danger or excitement of any kind generally postpones sleep appetitive activity. The postponement is often for a considerable period of time, yet does no obvious harm to the animal. Sleep rarely occurs away from the security of the sleep site, however tired an animal may be.

Ethological theories emphasize the fact that the period of prolonged immobility known as the sleep state is intimately associated with, and preceded by, patterns of preparatory behaviour which are specific to the species. Sleep is therefore treated as a behaviour pattern, rather than as a state of consciousness or as a physiological state. Such views are more comprehensive and better able to provide explanations for the very wide range of sleep manifestations to be observed in the animal kingdom.

Sleep duration. The table shows the approximate sleeping times for a number of mammals which have been studied in CAPTIVITY. It may be supposed that laboratory and zoo conditions distort sleeping habits in many ways. Nevertheless the great range of sleep durations comes as a surprise to many. Some mammals spend up to 20 h a day sleeping, allowing very little time for the other activities essential for survival, whilst others, especially the large herbivores, spend less than 6 h asleep. This variation is not readily explained in terms of physiological differences which might promote a greater need for sleep among the long sleepers. However, an examination of the life-style of the individual species will often provide an answer in terms of the need to be awake.

Large herbivores typically require considerable periods of wakefulness in which to gather the large quantities of plant material which they need each day. Moreover their vulnerability to predation is such that VIGILANCE is always at a premium. It follows that their need for wakefulness is high. Opossums (Didelphidae), on the other hand, live on a nutritious diet of carrion, insects, fruit, and grain with which they can satisfy their requirements in only a few hours. Moreover they are safest from predation when asleep, by virtue of the security of their sleeping site. As a consequence their need for wakefulness is low and they sleep up to 20 h each day.

Table 1. Total sleep time per 24 h of various mammals studied in captivity

Hours	
20	two-toed sloth
19	armadillo, opossum, bat
18	
17	
16	lemur, tree shrew
15	
14	hamster, squirrel, mountain beaver
13	rat, cat, mouse, pig, phalanger
12	chinchilla, spiny anteater
11	jaguar
10	hedgehog, chimpanzee, rabbit, mole-rat
9	
8	man, mole
7	guinea-pig, cow
6	tapir, sheep
5	okapi, horse, bottle-nose dolphin, pilot-whale
4	giraffe, elephant
3	
2	
1	
0	Dall's harbour porpoise, shrew

Often there are long periods each day when some animals are unable to be active. Some birds are dependent upon vision for NAVIGATION in FLIGHT. As a consequence flying is suppressed at night, and the bird remains immobile for the duration of the dark period. For much of this time it is demonstrably asleep. For the breeding season many birds must migrate to polar regions where there are more daylight hours (see MIGRATION). Diurnal fluctuations in temperature can impose similar restrictions on the behaviour of animals, especially small terrestrial reptiles whose body temperature can be critically dependent upon environmental influences. *Liolaemus*, a common genus of South American lizard, illustrates this principle in an extreme case. This lizard lives at high altitudes which have hot sun during daylight, but freezing cold at night, with snow falling and lying on the ground. To avoid freezing, *Liolaemus* must spend the night in a burrow. During the day, however, *Liolaemus* can raise its body temperature to an effective level by basking in the intense equatorial sun.

It appears to be true that the amount of time spent sleeping that is characteristic of a species is determined on the one hand by the amount of time available for wakefulness, and on the other hand by the dangers associated with waking activities which often expose an animal to predation. It is probably true that animals indulge in wakefulness for only as long as is strictly necessary. This principle would certainly appear to apply to the sleep of young animals in all species so far studied, which is invariably more prolonged than the sleep of adults. Moreover it appears that the prolongation of sleep may be associated with the degree of maturity at birth. Among cats (Felidae) and rats (*Rattus*), whose young are born in a highly immature condition, new-born animals sleep for considerable periods. New-born guinea-pigs (*Cavia aperea porcellus*), on the other hand, show much less sleep at birth, and this has been attributed to the relatively mature condition of their nervous system. Ontogenetic studies show that older babies sleep less than younger babies, who in turn sleep less than babies born prematurely. Long periods of inactivity in young animals must contribute enormously to the manageability of offspring, and greatly lessen the strain on parents.

Evolution of sleep. The phenomenon of sleep, as we know it in man and other mammals, can be identified with a high degree of certainty in the majority of familiar vertebrates. Highly analogous phenomena can also be identified in some invertebrates, such as molluscs and insects. This suggests that sleep goes back a long way in the EVOLUTION of animals. It is possible that many of the complex aspects of sleep in vertebrates constitute specializations of a more primitive mechanism which divided an animal's TIME into periods of activity and inactivity.

Whilst little controversy attaches itself to observations of sleep in mammals and birds, sleep in reptiles is sufficiently different to warrant some discussion. Mammals and birds both show characteristic patterns of BRAIN electrical activity during sleep. These are absent from prolonged recordings made of reptiles. Such are the differences in brain morphology between reptiles on the one

hand, and birds and mammals on the other, that many physiologists accept that we have little reason to expect similar indicators of sleep. In fact brain activity during reptilian 'sleep' is similar to that of reptilian wakefulness, which is itself similar in many respects to mammalian wakefulness. Claims for the existence of sleep among reptiles have therefore been based upon behavioural considerations.

Even so, behavioural observations are not without ambiguities. Many reptiles, including snakes (Serpentes), crocodiles (Crocodylidae), and alligators (Alligatoridae), can remain immobile for many days, and show no obvious *circadian* activity rhythm which is the hallmark of avian and mammalian sleep–waking cycles. Similarly marine reptiles such as turtles (Cheloniidae) need to return to the surface regularly to breathe, and therefore cannot show prolonged immobility under water. Lizards (Lacertidae), however, have been reported to show obvious symptoms of sleep. For example, Meller's chameleon (*Camaeleo melleri*) retires to a tree branch at sunset where it adopts an unmistakable sleep posture, with its eyeballs retracted for the duration of the night. Careful observations of the crocodilian, *Caiman sclerops*, has revealed that its prolonged torpor can be subdivided into periods of vigilant rest and periods of sleep-like unresponsiveness. These states are reflected in its resting posture. When the animal lies on its belly with all four limbs lying alongside and in line with its body, it can be lifted up from the floor for a few seconds before any ESCAPE or THREAT response can be observed.

Although studies of amphibia are few, sleep is known to occur in tiger salamanders (*Ambystoma tigrinum*), Cuban tree frogs (e.g. *Hyla septentrionalis*), and toads (*Bufo arenarum*). The salamander and the toad show changes in brain electrical activity associated with sleep. The tree frog is, moreover, reluctant to jump when tickled during one of these periods of immobility. The toad sleeps with its eyes closed and its head lowered. Axolotl (*Ambystoma mexicanum*) have been observed to sleep in groups suspended in the water supported by plants. At these times they are relatively unresponsive, and have a very slow gill-stroke rate. The bull frog (*Rana catesbeiana*) has not, however, been observed to sleep.

Fish are also known to sleep at certain times, some at night and some during the day. They spend long periods immobile lying against or under rocks, on the sand or buried under the sand, or floating concealed among aquatic plants. At these times it is often possible to touch and even handle them before they wake sufficiently to struggle effectively. Some fish change colour when asleep; these changes are not fully understood, but are thought likely to be defence mechanisms.

Many people are reluctant to view the rest periods of invertebrates, especially insects, in terms of sleep, possibly because they are so very different from those of man, whose sleeping habits have formed the prototype for much analysis of the sleep concept. Nevertheless there are many examples of insect rest which meet the criteria for sleep specified above. Butterflies and moths (Lepidoptera) are well known to choose specific sleep sites which offer the best CAMOUFLAGE. It is known that, in some species at least, a specific sleep posture is adopted during these prolonged periods of immobility. In the Mediterranean flour moth (*Anagasta kuehniella*) the antennae, which are normally directed forward during locomotion, come to be laid backward close to the wings, and eventually are crossed over so that the tips are hidden beneath the wings. During sleep the moth is extremely reluctant to respond to stimuli, and it is claimed that one wing can be lifted using a fine brush and allowed to fall back without the moth showing any movement at all.

Whether or not we accept the idea that insects sleep, there is abundant evidence that they do show periods of prolonged immobility, often at particular times of day, often after adopting a posture and a sleeping site that are specific to the species, and that at these times they are often less responsive to stimulation. A similar conclusion applies to other invertebrates, especially molluscs, and this suggests that sleep has a very long evolutionary history.

Physiological studies. Advances in the study of the physiology of avian and mammalian sleep have rekindled scientific interest on a wide front, and led to a great accumulation of information, the significance of which is not fully understood. Nevertheless two major discoveries are of special interest to the student of animal behaviour. The first concerns the active control which the brain enjoys concerning the induction of sleep. The second involves the unexpected complexity and variability of the sleep state itself.

It was previously generally believed that the sleep state constituted a passive response of the brain to the accumulated deleterious effects of prolonged wakefulness, possibly as the result of an accumulation of metabolic waste products which could only be eliminated during sleep. Experimenters have since established, by the use of brain stimulation experiments, that sleep can be actively and quickly induced by passing low frequency electric currents into carefully selected brain sites. Experiments in which parts of the brain are removed have further established that the brain contains cells whose main function appears to be the initiation and control of the sleep state. The implication of this research is that sleep is unlikely to be a direct and passive response to the effects of wakefulness, but is controlled by mechanisms which can actively induce sleep at appropriate times.

The complexity of the sleep state itself remained unsuspected until the discovery that dramatic changes of state occur within a single sleep period. Among mammals and birds sleep can be diagnosed electrophysiologically by the presence of high voltage slow wave electrical activity in the brain cortex. However, at intervals, this slow wave pattern is replaced by episodes of low voltage activity, similar to the wakefulness pattern. These episodes may be of very short duration, but in larger mammals, such as man, they can last for up to 45 min. Despite the active nature of the brain's electrical pattern, the individual animal remains clearly asleep, and may even be more difficult to rouse at this time.

By international convention these two states are distinguished by the names quiet sleep (QS) and active sleep (AS). These names have been carefully chosen to replace a whole collection of alternative names which had accumulated rapidly over the years, names which often referred to characteristics which were present in some species but not in others, or in some individuals but not in others. For example, AS often contains sporadic bursts of rapid eye movements (whence the name REM sleep), but they are not continuous throughout the episode, and in some species do not occur at all (see DREAMING). Other features of AS, which include profound postural relaxation, decreased responsiveness, increased variability of certain physiological measures such as heart rate, and the presence of certain characteristic patterns of electrical activity in brain structures, have all been observed in many species, although not all are essential to the overall pattern. The only feature which is present in all cases is the characteristic electrical activity of the brain.

Although the functions of AS and QS are poorly understood, a number of interesting facts have been highlighted as relevant by researchers. It is now well established that sleep appears in the developing animal before it is born. Experiments using foetal animals *in utero* appear to show that AS matures before QS. Moreover, young animals spend a very high percentage of their sleep in AS (up to 100%), whereas in adulthood this percentage is usually low (not exceeding 25%). The suggestion has therefore been made that AS plays a crucial role during development, for example by speeding the growth of nervous tissue.

Other investigators have drawn attention to the regular alternation of QS and AS. The sleep of a normal adult mammal begins with a period of QS which later gives way to an episode of AS. This is often followed by a brief AROUSAL, and then another period of QS, and so on. The time interval between episodes of AS is normally fairly constant and appears to be related to the metabolic rate or size of the animal, since small mammals such as the rat and the mouse (*Mus musculus*) have episodes of AS only a few minutes apart, whereas AS episodes in man and the chimpanzee are separated by more than 1 h.

Some researchers have attributed this cyclic organization to an arrangement whereby some deleterious effects of prolonged QS are relieved by a short regular episode of AS. Others have focused their attention on the brief periods of arousal which often follow AS, suggesting that these may provide a vital opportunity for an animal to scan its environment for potential sources of danger. Yet others have suggested that two complementary biochemical processes take place in the two sleep stages, and that the stages need to alternate so that one process is allowed to complete the work of the other. The matter is still highly controversial.

COMPARATIVE STUDIES indicate that the alternation of two sleep states (AS and QS) is widespread among warm-blooded animals (birds and mammals). Other vertebrates (fish, amphibia, reptiles) do not appear to have these two kinds of sleep. Most reptiles, for example, show electrical activity during sleep which is very similar to that shown during wakefulness. It is not yet clear whether this pattern is similar to warm-blooded AS or QS, since reptilian sleep contains neither high voltage slow waves (QS) nor rapid eye movements (AS). However, the change from a one-state to a two-state style of sleeping appears to have accompanied the change from a cold-blooded to a warm-blooded physiology.

Function of sleep. Sleep and sleep controlling mechanisms have been studied from two points of view, ethological and physiological. The ethologist regards sleep as a mechanism which schedules behaviour so as to produce prolonged periods of immobility. The physiologist on the other hand emphasizes the role of sleep control mechanisms in producing a change in physiological state. The FUNCTION of sleep can therefore be viewed either in terms of the benefits bestowed on an animal by prolonged immobility, or in terms of the advantages of the physiological state.

Traditional views emphasize the value of sleep as a restorative process. Although scientific research has not yet managed to specify exactly which processes of restoration take place at this time, the view is still very popular and continues to dominate scientific investigation. According to such theories, the behaviour associated with sleep serves to manœuvre an animal to a safe and stress-free location, where a period of unconsciousness will lead to little harm.

One alternative view disregards the possible restorative functions of sleep, and emphasizes instead the advantages to be gained from prolonged periods of immobility, especially if they occur in well-hidden or inaccessible locations. These advantages include protection from predation and the inclemencies of the environment, such as extremes of heat and cold, as well as a reduction in overall energy expenditure. According to this theory, the state of unconsciousness characteristic of sleep in many species serves to maintain immobility by suppressing ATTENTION, and removing the will for action.

Whether function can be accounted for simply in terms of restoration or immobility, or whether both are needed, constitutes yet another controversy in this highly active area of scientific enquiry. Whatever solutions are finally adopted, it is essential that they reflect the great *phylogenetic* (evolutionary) age and very general distribution of the phenomenon of sleep in the animal kingdom. 67, 104, 142   R.M.

SOCIAL INTERACTIONS. When an animal responds to a STIMULUS situation, we speak of its reaction. In a social encounter one individual may elicit a response from another, which in its turn elicits another response from the first individual. Such a case, where each individual influences the behaviour of the other, can be described as an interaction. Interactions may, but need not, involve degrees of complexity not present in reactions to external stimuli.

TINBERGEN's (1951) classic description of the COURTSHIP of the three-spined stickleback (*Gasterosteus aculeatus*) implied that each response of the male is the stimulus for the next response of the female, and vice versa (Fig. A). However, it soon becomes apparent that such a description is accur-

Fig. A. Schematic representation of the sexual behaviour of the three-spined stickleback (*Gasterosteus aculeatus*).

ate only as an indication of the progression of response sequences. In practice each response of the male may elicit a range of responses from the female, and vice versa, with the result that the courtship sequence may progress a little way, and then drop back to an earlier stage before progressing further. Such variability in the response sequence can be ascribed to fluctuations of unknown origin in the MOTIVATION of the participants, and such an explanation may be adequate in simpler cases. However, the possibility that B will respond either with Y or with some other response is likely to affect A's initial behaviour: each act may be determined in part by the assessed probability of the possible responses of the other participant. This immediately implies a considerable degree of complexity, and the interaction cannot be described simply in terms of successive reactions.

A similar conclusion arises from consideration of FUNCTION. The function of THREAT postures, for instance, is often said to be that they reduce physical combat: they appear to be given when the displaying individual has tendencies both to approach (usually to attack) and to withdraw from its rival. Now if an individual is going to attack or is going to flee, its behaviour is most likely to be effective if it does so directly, without signalling its intention (see INTENTION MOVEMENTS). But if it may do either the one or the other, and which it will do depends on the behaviour of the other individual, COMMUNICATION is essential to assess the probabilities of the various courses of action of the opponent.

Further confirmation comes from the finding that some signals do in fact appear to have the effect of assessing the probable behaviour of another individual. Siamese fighting fish (*Betta splendens*) sometimes engage in long duels in which each fish alternately faces its opponent with its gill covers erect, and turns broadside to him (Fig. B). As the skirmish proceeds, there is a gradual increase in gill-cover erection, tail-beating, and biting. At first the displays of the two fish increase in parallel, each one matching its gill-cover erections to those of its partner. Only towards the end does one outstrip the other, which soon capitulates. It thus seems as though each skirmish involves a gradual escalation, with each fish testing out the other, until one gives up.

As this point, some caution is necessary. Although each fish behaves 'as if testing out the other', the evidence does not require us to postulate a very complex level of cognitive functioning in these cases. But at the other extreme, we know that our own behaviour in interactions can be affected by our conscious or unconscious assessment of probable responses by our partner, and that we may dissemble or act deviously to influence his course of action. That such complexities are not limited to human interactions is believed by every owner of a pet dog (*Canis lupus familiaris*), but hard data are extremely scarce. There

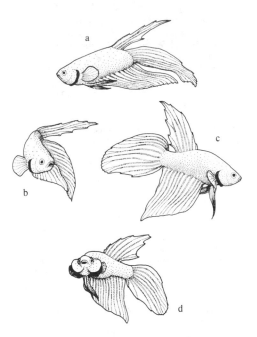

Fig. B. The Siamese fighting fish. **a** and **b** show fish when not displaying; **c** and **d** show broadside and facing displays.

is, however, a growing body of evidence for non-human primates, which suggests that they sometimes show behaviour that can be described only as involving deliberate deception.

75   R.A.H.

SOCIAL ORGANIZATION ranges from the simple cooperation between a male and a female during mating to the complex societies of ants (Formicidae) or baboons (*Papio*).

The totality of SOCIAL RELATIONSHIPS among all members of a group constitutes their social organization. This may be highly structured, in terms of DOMINANCE relationships, or it may be loosely arranged into a system of relatively independent family groups.

One of the most fundamental aspects of social organization is the MATING SYSTEM that is characteristic of the species. The two basic types of mating systems are monogamy and polygamy. In a monogamous system each breeding adult mates with only one member of the opposite sex, whereas a polygamous individual generally has two or more mates, either simultaneously or successively.

The type of mating system that is characteristic of a species develops through EVOLUTION in relation to the animal's ecological NICHE. Availability of food and degree of PARENTAL CARE are probably the most important factors. For example, many birds establish a TERRITORY during the breeding season, thus ensuring a food supply for their brood. The FORAGING efforts of both male and female are required to support the young, and so monogamy is a necessary part of this type of social organization. Over 90% of the birds of the world are monogamous, but monogamy is rare in other animal groups. For example, female mammals are specially equipped to feed the new-born, so it is inevitable that male mammals will have a lesser role to play in caring for the young, and are likely to be free to invest time and energy in more than one female. The most common type of mating system found among mammals is polygamy, in which one male mates with a number of females. Whereas a female bird is most likely to choose a mate on the basis of his apparent parental abilities, a female mammal is more likely to be influenced by social dominance and other indicators of reproductive success (see SEXUAL SELECTION).

The interplay of ecological and social factors is well illustrated by the primates, which have a wide range of types of social organization. At one extreme there are the solitary species in which individuals are widely spaced and have social contact only during COURTSHIP and COPULATION. Many prosimian species, such as lemurs (Lemuridae), loris (Lorisidae), and tarsiers (Tarsiidae), fall into this category. They are generally of small size, nocturnal, arboreal, and insectivorous. Their solitary way of life is typical of animals with a widely dispersed food supply.

Monogamous species include the white-handed gibbon (*Hylobates lar*). Like many monogamous birds, the males defend territories with fixed boundaries and small zones of overlap. The young remain with the parents until they become sexually mature.

Single-male harems, in which the group consists of a single adult male and a number of adult females plus their young, are common amongst the primates. One example is the gelada baboon (*Theropithecus gelada*) which inhabits the mountain grasslands of Ethiopia. Large AGGREGATIONS of these baboons occur in gorges with steep cliffs. The harems remain intact even when part of a larger group. During the rainy season they spend most of the day FEEDING on plants. As the dry season approaches, the herds break up into their component parts, the harems, the all-male groups, and groups of JUVENILES. The all-male groups show greater DISPERSION than do the harems, and this appears to reduce COMPETITION for food between the groups.

The social organization of gelada baboons is very similar to that of various antelope (Bovidae) species, which also have a HABITAT that is subject to marked seasonal variation in food availability. For example, among impala (*Aepyceros melampus*) there are two types of herds: bachelor herds made up of single males, and harem herds made up of females usually under the control of a single male (see HERDING).

In areas with a relatively stable food supply, social groups often contain a number of males. For example, mountain gorillas (*Gorilla g. beringei*) have a small HOME RANGE and feed on the stems and leaves of forest plants. They live in small groups each led by a silverbacked male, but often including more than one male. They do not defend territories, and their home range may overlap with that of other groups.

Troops of savannah-living baboons, such as the olive baboon (*Papio anubis*), generally contain a number of adult males and females. Typically, there will be several dominant adult males who have privileged access to food, water, and mates. These males cooperate in maintaining their social position against rivals, in protecting mothers and infants from interference by other members of the troop, and in DEFENSIVE behaviour against predators. When threatened by an enemy the huge males rush to the scene, displaying their canine teeth. They form a phalanx which cuts off any line of attack upon any other member of the troop. The advantages of this type of cooperation are considerable in rich habitats that tend to attract the large predators. In poor environments, such as the semi-desert of Ethiopia, the troops of hamadryas baboons (*Papio hamadryas*) tend to be smaller, and are dominated by a single male. It seems that the advantages of cooperation are insufficient to outweigh the tendency for a single male to become socially dominant and sexually attractive. This view is supported by the fact that the yellow baboon (*Papio cynocephalus*), which lives in a rich tropical environment, forms large troops with many cooperating males, who are not markedly distinguishable from the females. The olive baboon is intermediate in habitat, social organization, and sexual dimorphism. The hamadryas baboon, which lives in the poorest habitat and has smaller, more hierarchically organized troops, has the greatest degree of sexual dimorphism. The dominant males have an enormous cape or mane which serves to attract females and deter rivals. The most complex forms of social organization, outside the primates, are found among the insects and the colonial invertebrates.

The true social insects are from two orders (groups): (i) the Isoptera (termites), and (ii) the Hymenoptera (which includes wasps, bees, and ants). Insect society is characterized by a reproductive division of labour, with sterile individuals working on behalf of the fertile members of the society. All the members of a society have a GENETIC relationship to one another, but reproduction is generally confined to a single female, the queen. She lays many thousands of eggs, each potentially capable of developing into a member of one of the castes.

The most important factor determining caste is the diet that is fed to the larvae. Those fed on a restricted diet generally develop into workers. In the ants, bees, and wasps the workers are all sterile females, but in the termites they may be sterile males or females. The workers carry out a variety of tasks, including foraging, nest construction, rearing the young, attending the queen, and guarding the colony. In ants and termites there is sometimes a special soldier caste equipped with weapons, such as enlarged jaws. In some species division of labour is determined by age. Thus in the honey-bee (*Apis mellifera*) a worker spends the first 3 days of adult life cleaning out the cells in which the eggs are laid. She then feeds the older larvae on a mixture of pollen and honey obtained from the storage cells in the hive. Special nurse glands in the head now develop and secrete a substance called royal jelly. This is fed to all newly hatched larvae for a short period, and to larvae that are destined to become queens for the whole of their development. On about the tenth day wax secreting glands on the abdomen become active, and the worker gradually changes her behaviour to cell construction. From about the eighteenth day she may leave the hive for a few brief ORIENTATION flights, during which she learns the position of the hive in relation to the sun and to various landmarks. At this age she also guards the hive and inspects incoming bees. At about 21 days of age the worker becomes a forager, supplying the hive with nectar, pollen, and water. This role lasts until the end of her life, 2–3 weeks later. This general sequence may be modified in relation to factors such as the type of flower crop, the CLIMATE, and the age of the colony. To some extent the activities of the workers are determined by the COMMUNICATION between them (see COOPERATIVE BEHAVIOUR). Workers are also stimulated to initiate activities through their EXPLORATORY behaviour. They visit those parts of the hive where food is stored, where larvae are brooded, and where construction is under way, and they may respond to the situations that they discover. If food is short, young workers may become foragers. If the larvae are underfed, the nurse glands of an older forager may regenerate, and she may return to feeding larvae. Many scientists have been struck by the ALTRUISM of the workers in insect societies, which go about their business in an apparently unselfish manner.

Among the Hymenoptera females develop from fertilized eggs, but males develop from unfertilized eggs. The males are *haploid*, having only the mother's contribution of genetic material, while the females are *diploid*, having a contribution from both parents. Normally, the genetic affinity between parent and offspring is $\frac{1}{2}$, and between siblings it is also $\frac{1}{2}$. In the Hymenoptera, however, all sisters receive the same genetic material from their father, because all his sperm are identical. (Haploid animals have half the normal amount of genetic material, and no rearrangement takes place during sperm production.) Sisters therefore have a genetic affinity of $\frac{3}{4}$ ($\frac{1}{2}$ from father and $\frac{1}{2} \times \frac{1}{2}$ from mother). Thus the fe-

males share more genes with their sisters than they do with their mothers. In terms of EVOLUTION it is better (i.e. greater FITNESS is achieved) for hymenopteran females to help their mother produce more daughters than it is to have offspring themselves. Scientists believe that this is the explanation for the surprising fact that the social organization has evolved independently eleven times in the Hymenoptera, but only once in any of the many other insect groups (the Isoptera).

The colonial invertebrates include the corals (*Corallium* spp.), siphonophores (Siphonophora), and bryozoans (Bryozoa). The individual members, called zooids (or hydranths), are often fully subordinated to the colony as a whole. For example, in the colonial marine polyp *Campanularia flexuosa* (see RHYTHMS, Fig. B) the zooids are attached, plant-like, to a common branching stem. Each zooid has a single ring of tentacles surrounding a central projection containing the mouth. Thus each is similar to an individual hydra (*Hydra*), a free-living polyp. The zooids feed by catching small animals, e.g. brine shrimp (*Artemia salina*). They are all similar in form, but have life cycles that are independent of each other. Communication between individuals occurs, especially in relation to defence. In response to noxious stimulation of one zooid, all those in the colony may retract. It might appear that the zooids have a perfect social organization, consisting of independent members that nevertheless show close cooperation. However, some scientists doubt that such a colony can properly be called a society.

Among the bryozoans there are some species in which the zooids are morphologically specialized and have different roles, such as feeding, defence, reproduction, etc. While this type of organization may seem similar to that of some insect societies, it is also reminiscent of the various organs of an individual multi-celled animal, which are specialized for different tasks, and are incapable of independent life.

To some extent, all animals have some social organization, even if only that necessary for SEXUAL reproduction. Some species show social organization in some aspects of behaviour but not in others. Examples can be found in the defensive SCHOOLING of fishes, and the ALARM responses of many birds. More complex social organization involves some division of labour, which is often based upon sexual dimorphism. Thus we have seen that dominant male baboons may take on the responsibilities of defence. Division of labour is also often a function of age. Reproductive division of labour is genetically determined in insect societies, but may occur in vertebrate societies as a result of competition for social DOMINANCE, or for mates. In the colonial invertebrates the division of labour among semi-independent zooids may be so marked that the colony is regarded by some as a single organism, akin to an organized society.

20, 145

SOCIAL RELATIONSIIIPS. In everyday speech a relationship is said to exist between two individuals who have known each other over a period of time, and who mould some of their behaviour in accordance with each other's individual characteristics. The term is also used in a wider sense to refer to categories of individuals, as when we say 'His relationship with the police was poor', or 'The army strove to maintain a good relationship with the civilian population': it is, however, with relationships between individuals that students of ETHOLOGY are primarily concerned.

Many, and perhaps most, animal species do not form relationships. This is obviously the case with solitary species for whom mating is a transitory affair, and PARENTAL CARE non-existent. It is also probably true of many gregarious species which associate with other individuals of their kind, but not with particular other individuals. Furthermore, although extended ASSOCIATIONS between individual invertebrates have been described, it is not yet clear how far they merit the term 'relationship'. Inter-individual relationships, in the sense discussed here, occur throughout the vertebrates, but are most obvious, and have been most studied, in birds and mammals. In this article, we shall not be concerned with the nature of particular types of relationships, such as the pair bond in birds, or the relationship between a HOUSEHOLD PET, such as a dog (*Canis lupus familiaris*), and his master, but with some general issues applicable to a wide range of relationships.

Special interest attaches to the study of animal relationships because of the possibilities for cross-fertilization with the study of human relationships. Although the latter have been studied far more intensively and for much longer, their complexity makes progress slow. It is just because animals are simpler than man, lacking verbal LANGUAGE and social institutions, such as marriage, that their study can be rewarding. Their relative simplicity makes possible the elucidation of issues, and the clarification of principles, that would otherwise be lost in the complexity of the human case. For such reasons we are here concerned primarily with higher mammals, and especially primates.

Students of animal behaviour recognize that the study of relationships involves special problems. Such problems include the various ways that relationships can be described and classified, what we mean by a stable relationship, the nature of the processes involved in the development and maintenance of a relationship, and the criteria by which we can make prognoses about the course of a relationship.

Relationships involve a series of SOCIAL INTERACTIONS over time. Whereas each interaction occupies a strictly limited span of time, relationships can be discussed only in the context of a much longer time period. And in part just because relationships are extended in time, their study involves complications additional to those involved in the study of interactions. For instance, each interaction may be influenced by those that preceded it, and perhaps by expectations of those in the future. Furthermore, interactions of one type may affect interactions of another type within the same relationship. Indeed interactions within a relationship may be affected by interactions between one of the participants and a third party. Furthermore, a relationship may possess emergent qualities not present in its constituent interactions: we shall return to this point later. In spite of these complexities, progress has been possible because we now know quite a lot about the component interactions, and because many of the principles governing animal COMMUNICATION are understood.

Description of relationships. As with any other branch of science, the study of relationships requires a descriptive base before problems such as those mentioned above can be tackled effectively. So far, only a few pointers exist: these will be considered in the next few paragraphs.

Since a relationship involves a series of interactions in TIME, the first step must be to describe the constituent interactions. This could be done at many levels of precision (see CLASSIFICATION OF BEHAVIOUR). For present purposes, it would first be necessary to specify what the participants do together (i.e. do they groom each other, copulate, behave aggressively to each other, associate together, etc.?). But it would be necessary to specify not only what they do, but also how they do it. For instance, do they groom persistently or perfunctorily, fight savagely or with circumspection? Such aspects of interaction can be called 'qualities', without of course implying that they could not be subjected to quantitative treatment. To describe a relationship we must go at least one step further, describing not only the content and quality of the interactions over the specified stretch of time, but also how these interactions are patterned; that is, their absolute and relative frequencies, and how interactions of one type are related to interactions of other types.

Although the study of animal relationships must be based on the study of interactions, it is not necessarily the case that they involve nothing but their constituent interactions. For one thing, as we shall see, they may possess qualities emergent from those interactions. In addition, our own human experience suggests that inter-individual relationships may involve properties not easily accessible to the behavioural scientist. Certainly what we feel about what we do may be as important as what we do. However introspective evidence is just not available from animals, and although feelings and EMOTION may be important, we must rely on behavioural evidence concerning their nature (see ANTHROPOMORPHISM).

Description of a relationship must thus be based on the content, quality, and patterning of its component interactions. But when we describe interactions, we are initially concerned with empirical instances. Somehow we must abstract from these empirical instances to make generalizations valid for the individuals, age/sex categories, or species with which we are concerned. There is no general agreement about the procedures by which the transition from empirical instances to generalizations should be achieved, in part because the best procedure must vary with the problem in hand. The issue is, however, likely to prove important as the study of relationships proceeds.

Having described relationships in terms of their component interactions, we need some means of ordering the data. We may now consider some dimensions which seem likely to be important for understanding the dynamics of inter-individual relationships.

1. *Content of interactions.* Description of a relationship usually starts from the content of the interactions that occur within it. Thus a consort relationship in monkeys (Simiae) involves SEXUAL behaviour, GROOMING, and mutual proximity, while a mother–infant relationship involves suckling, grooming, protection, PLAY, proximity, etc. This dimension can also be applied at finer levels of analysis. Thus, in discriminating between different mother–infant relationships, whether or not play occurs might be an important index of the nature of the relationship; and we could even go further and use the dimension to distinguish between mother–infant relationships containing play according to whether the play did or did not predominantly involve contact.

2. *Diversity of interactions.* If a relationship involves only one type of interaction it can be described as single-stranded or uniplex; if many, as multi-stranded or multiplex. Here again the dimension can be applied at many levels of analysis: a mother–infant relationship could be described as uniplex, involving only MATERNAL/filial behaviour, or multiplex, involving suckling, grooming, playing, etc. The important issue is the diversity of behaviour involved, not the dichotomy.

3. *Reciprocity versus complementarity.* An issue which cuts across the previous ones concerns the extent to which the interactions are reciprocal or complementary. A reciprocal interaction is one in which the interactants show similar behaviour, either simultaneously or alternately, as in peer–peer play. In a complementary interaction the behaviour of each individual differs from, but complements, that of the other, as in mother–infant interactions. In some relationships, all interactions are reciprocal: the relationships between young monkeys approach this condition, each playing equivalent though alternating parts in each bout of play. In other relationships all interactions are complementary. The classic dominance/subordinance relationship (see DOMINANCE) is such a case: monkey A threatens monkey B, bites B, and has priority to food and water over B, whilst B avoids A and grooms A more than A grooms B. However

it must not be forgotten that dominance may vary with the content of the interaction: individual A may have priority of access to food, but not to females. It may also vary within one type of interaction: a territory-owning bird dominates its neighbours whilst on its own TERRITORY, but not when on theirs. It is thus the interactions (specified as to content and/or context) that must be described as reciprocal or complementary, and not the relationship as a whole. How consistent the different interactions within a relationship are in this respect is a further issue: in many animal relationships the consistency is considerable, and it is perhaps only in the most sophisticated of human relationships that the various types of interaction may be either reciprocal or complementary, with idiosyncratic patterns of imbalance.

An important aspect of complementarity concerns the direction of control or power. In human interpersonal relationships this has been approached through the concepts of *exchange theory*. In general, A's power over B depends on A's resources and B's dependence, which in turn depends on B's needs for A's resources and the alternative sources open to B. Furthermore, as a first approximation, A's power over B is greater than B's over A if, in their interactions, A's rewards minus his costs are greater than B's rewards minus his costs. It will be apparent that power or control is not the same as dominance. A female may be subordinate to her consort male, but have considerable control over his behaviour.

4. *Qualities of interactions.* We have seen that, in describing a relationship, we must describe not only the content of the interactions, but also their quality. For instance a mother rhesus macaque (*Macaca mulatta*) may reject her infant roughly, by hitting or pushing it away, or gently, by crossing her arms over her nipples but remaining available for the infant to cuddle against.

A quality of one of the constituent interactions of a relationship need not be applicable to others. For example, a mother may be rejecting of her infant's suckling requests but not of its requests to be groomed. At the other extreme, qualities may be applicable to all interactions in a relationship. In general, qualities which apply to reciprocal interactions may apply either to one or to both partners. Thus it is possible for both partners to show 'sensitivity' (objectively defined), or for one to behave with sensitivity and the other not. But where qualities apply to complementary interactions, they tend to apply to the interaction rather than to either partner independently: if one partner controls, the other must be controlled.

5. *Relative frequency and patterning of interactions.* Some of the judgements we make about relationships depend not on the frequency or quality of particular types of interaction, but on the relative frequencies of, or interrelations between, different types. Thus if a monkey mother frequently rejects her infant's attempts to gain ventro–ventral contact and never initiates contact, we might de-

scribe her as rejecting. If she never rejects and often initiates, we might describe her as possessive. But if she often does both, or seldom does either, we might describe her as controlling or permissive. These latter qualities can be regarded as emergent properties: they are not present in the separate interactions, but emergent from the relations between them.

6. *Multidimensional qualities.* In everyday life our readiness to apply a quality label to a relationship depends on the concurrence of a number of characteristics. This same is true with animals. For instance we would be more prone to describe a monkey mother–infant relationship as warm if the mother frequently took the initiative in ventro–ventral contact, placed her arm round the baby when in contact with it, frequently groomed the baby, and frequently played with it.

It is important to see that our judgements about relationships may depend on multidimensional qualities because, if we are to understand the bases of the judgements we make every day, we must recognize that the dimensions involved may be independent. For example we would be more prone to describe a relationship as affectionate if (i) the participants do a number of different things together, i.e. it is multiplex; (ii) it involves certain specific types of interaction, for instance if, in the absence of the other, each partner shows types of behaviour tending to restore proximity, and if the presence of the partner alleviates the anxiety produced by strange situations; (iii) if the interactions have certain qualities, for instance if the behaviour of each partner is organized in relation to the ongoing behaviour or GOALS of the other; (iv) if actions conducive to the WELFARE of the other are likely to be repeated.

It will be noted that some of these properties would characterize aggressive as well as affectionate relationships, but only in the latter case do they occur together. On the other hand, it is not suggested that all these properties are necessary before the label 'affectionate' can be applied, or indeed that there are not others that would also contribute to such a judgement. Those listed can in fact all be identified, and studied with moderate rigour, in the rhesus mother–infant relationship. To those who argue that affectionate relationships involve intangible qualities in addition to such observables it can be said that such intangible qualities may be correlated with those that we can observe, and that, if so, we may be able to use the observables to investigate conditions conducive to the formation of affectionate relationships; and that anyway the observables are all we can study with the methods at our disposal.

7. *Cognitive and moral levels.* The cognitive and moral levels of individual human subjects have important relevance for the study of human interpersonal relationships. Thus relationships involving dominance/subordinance or nurturance/dependence interactions will be markedly different in character according to whether either partner

functions at a higher moral or cognitive level, or whether both are similar. For animals, the question of moral levels is of doubtful relevance, and COGNITION can be assessed only crudely. The issue is however important in two contexts. First, as hinted earlier in this article, comparisons of relationships across species must take into account the differences in cognitive capacities of the animals concerned. Second, parent–offspring relationships inevitably involve differences in cognitive capacity, an issue of crucial importance for their understanding, as is shown below.

8. *Penetration.* The dimensions described so far, with the possible exception of that of cognitive and moral levels, have been applicable to both human and animal relationships. 'Penetration', referring to the degree of intimacy of the relationships, is more specifically human. One can picture the personality as involving central, less visible areas surrounded by more numerous peripheral ones, with the sectors corresponding to different areas of interaction (sex, family, etc.). The intimacy of a relationship can then be represented by the extent to which the personality of each participant penetrates that of the other. This can be divided into two dimensions: how many sectors are penetrated, and how deeply. The first of these is likely to be related to the second dimension discussed here, namely the number of types of interaction in the relationship.

Of course this list of dimensions is not intended to be exhaustive or definitive: its purpose is to emphasize that hard-headed descriptions of seemingly intangible relationships are not beyond the bounds of possibility.

Inter-individual relationships and social structure. The properties of a relationship may change with time and with context. Both must thus be specified when a relationship is described. Changes with time will be considered below: here we shall consider the question of the social context.

In many monkey species adult and JUVENILE females, who do not have infants of their own, covet those of others. In some species mothers readily 'loan' their babies to other females, but in other species the attentions of other females are resented by the mothers. In such cases subordinate females may have difficulty in protecting their infants from other females, and may do so only by becoming more possessive and restrictive of their baby's movements. For this reason, in CAPTIVITY, rhesus infants of group-living mothers spend more time on their mothers, and less time off their mothers at a distance from them, than do infants living alone with their mothers in similar cages. Thus the mother–infant relationship is affected by the presence of social companions.

That, however, is only the start of the matter. The extent to which a mother permits other females access to her infant depends on her relationship with them: she will be more tolerant with one of her grown-up daughters, or with one of her own grooming companions. Thus the infant's relationships with females other than its mother may be affected by their relationship with her. In some species the males often intercede in disputes, and in doing so are likely to take the side of their own favourites. Thus how much the mother threatens away another female may depend on the male's relationships with each of them. Thus the male's relationship with a female may affect the mother's relationship with that female, which may affect the female's relationship with the infant, which may affect the mother's relationship with her infant. In other words, in such a case each relationship is part of a nexus of relationships in which each relationship may be affected by many others.

This is a particular case of what must be generally true for any group of individuals of species capable of forming relationships of the type we are discussing, namely that inter-individual relationships are affected by the nexus of other relationships in which they lie. An example is illustrated in Fig. A, which summarizes some observations

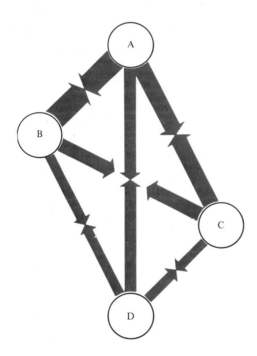

Fig. A. Diagram of relationship between dominance rank and frequency of grooming, from a study of wild-living chimpanzees (*Pan*). The males are ranked A, B, C, D. The thickness of each arrow represents the frequency with which one male grooms another. Notice that opposing arrows are of comparable thickness, indicating that there is considerable reciprocity in the grooming relationship between each pair of males. Notice also that the thickness of the arrows emanating from each male diminishes with the rank of the recipient male. This shows that the frequency of grooming is higher, the greater an individual's rank in the dominance hierarchy.

on male chimpanzees (*Pan troglodytes*) of a free-living community in Tanzania. Data on a variety of types of AGONISTIC interactions show that the males can be arranged in a straight-line HIERARCHY. The nature of the grooming interactions varies with position in the hierarchy: the higher-ranking males tend to be involved in grooming most often, and each male tends to groom high-rank partners more often than low-rank ones. It is also apparent that sessions between individuals of markedly different status are usually started by the lower-ranking ones. Furthermore, low-rank individuals tend to groom for longer sessions than high-ranking ones, and each male thus grooms a partner of higher rank for longer than that partner grooms him.

We may link this example to a more general issue. Just as a relationship is constituted by the content, quality, and patterning of its component interactions, so the social structure of a group is constituted by the content, quality, and patterning of the relationships within it (see SOCIAL ORGANIZATION). Now we may here point to an important difference between human and animal groups. To take a simple case, a dozen chicks (*Gallus g. domesticus*) reared together in a pen will form a dominance hierarchy: one alpha individual becomes able to supplant and peck all others, a beta individual to supplant and peck all except the alpha, and so on down to the omega individual who cannot supplant or peck anyone. This social structure arises from the behavioural interactions of the individuals: it may be very resistant to changes, but that is because the inter-individual relationships are stable, not because the structure has an existence independent of those relationships.

In human groups there is a sense in which the opposite is the case, the nature of the relationships between individuals being determined by the structure of the group. Thus shared beliefs about, for instance, monogamy may determine the patterning of relationships; and an elected parliament may make laws which govern the relationships between individuals. This contrast between structures which are generated by the patterning of relationships and structures which generate the pattern is of interest because of its relevance to a controversy amongst sociologists and anthropologists: some lay stress on the manner in which institutions and organizations constrain the behaviour of individuals, whilst others, focusing on the individual and the social world he creates about him, see social structure as emergent from the interactions between individuals. In the human case, both approaches are clearly valid. In animal studies, the idea of structure as the pattern of relationships is clearly of more importance than structure as a determinant of the pattern of relationships. If we make use of this difference, the study of animal societies may throw light on the more complex human case.

Stability and change. We have seen that each

interaction within a relationship may be affected by past interactions. This implies that a relationship is seldom static, and any stability it has must be essentially dynamic in nature. Indeed precisely what is meant by the 'stability' of a relationship is not easy to define, since most relationships change their properties with time; either progressively, or by moving from one temporarily stable state to another, or by merely varying within certain limits. To understand the stability of relationships we must come to terms with how relationships are preserved in the face of changes in the participants and the buffetings of the external world, and also with how they may change progressively and yet preserve their integrity. These two aspects of dynamic stability are of course interdependent.

In human relationships, stability can often be partially understood in terms of a more or less consciously perceived goal: the participants may seek to build a marriage or a parent–child relationship of a particular type. In so far as such goals are determined by CULTURAL means, comparable mechanisms are unlikely to be important in animals, though of course the separate interactions within a relationship may be directed towards shorter-term goals. Another mechanism by which stability is maintained, which probably operates in both animals and man, involves the avoidance of undesirable states, as when marriage partners retreat from states which make divorce a real possibility. This need not involve conscious recognition of the undesirable state: monkeys may be so constituted as to avoid states in which AGGRESSION between them is probable.

Within any particular species, the constitutions of particular classes of individual may permit predictions about the sorts of relationships they will form. For instance in macaque monkeys young males tend to associate together; females, even when adult, continue to associate with their mothers, but males do not; and adult females may form temporary relationships with males when in oestrus ('heat'). Of course, such statements pose further questions, such as whether the differences between the relationships formed by males and females are directly determined by differences in their HORMONE states; by differences in their earlier experience, such as the nature of their relationships with their mothers; and/or are more directly genetically determined: such issues cannot be considered here. Whilst generalizations about the propensities of particular age/sex classes are likely to be specific to particular animal species, or groups of species, comparable generalizations can be made about particular human cultures.

Given such behavioural propensities, understanding of the dynamics of relationships requires further principles which will 'explain' the various mechanisms by which interactions affect each other, and which are thus responsible for the pattern of interactions we observe. At present we are a long way from being able to specify such principles, but the following (overlapping) categories may indicate their nature.

1. *Principles concerned with learning.* Three LEARNING paradigms have been applied to the study of relationships: (i) Exposure learning: repeated exposure of an individual to a stimulus situation, including that presented by another individual, enhances his attachment to it. Related to this, and of special importance in the human case, is 'modelling', where one individual is especially likely to behave in ways that he has seen another individual to whom he is attached behaving (see IMITATION). (ii) Classical CONDITIONING, whereby the pairing of a so-called unconditional stimulus, which already elicits a response, with a (to be) conditional stimulus results in the latter coming to elicit a response similar to that initially elicited by the former. For example, if tactile stimulation elicits relaxation in an infant, it is possible that visual and auditory stimulation associated with it (e.g. voice of parents) will come to do so also. Classical conditioning has been applied to emotional states as well as to behavioural responses. (iii) Instrumental conditioning. If a stimulus event which is contingent upon a response increases the future probability of that response, the stimulus is called a reinforcer. If REINFORCEMENT occurs in the presence of a particular stimulus situation, then that situation will set the occasion for the repetition of that response. It has been suggested that individuals continue their social activities in so far as those activities produce positive reinforcement, or REWARDS from others, and do not involve excessive cost, and that many aspects of interpersonal relationships can be understood in terms of 'the laws of learning' worked out in LABORATORY situations. However, whilst it is apparent that the consequences of our social activities affect their repetition, the application of principles derived from experiments with animals to real-life social relationships is fraught with difficulties. In particular, it is much more difficult to predict ahead of time which events will and will not act as reinforcers (and thus lead to an increase in the frequency of the preceding activity).

2. *Principles concerned with degrees of compatibility between interactions.* Some types of interaction are especially likely to be associated, whilst others are incompatible. In the case of social behaviour, many such cases can be considered in terms of 'status': for instance, aggressive dominance of B by A may be incompatible with mounting of A by B. In many human societies individuals differing markedly in status are likely to interact in some ways, by gestures of condescension/obeisance for example, and unlikely to do so in others. They may, for instance, not eat at the same table.

3. *Principles concerned with the effects of one type of interaction on the probability of others.* The occurrence of one type of interaction may affect the probability of others by a wide variety of mechanisms. For example, in many species COURTSHIP

interactions enhance the likelihood of COPULA-TION, and reduce the likelihood of agonistic encounters: here the mechanisms may or may not involve hormones. Another example, in which hormonal mechanisms appear not to be involved, concerns the formation of social relationships between gelada baboons (*Theropithecus gelada*). Newly acquainted baboons first show FIGHTING, then presenting, then mounting, and then grooming as they establish a relationship. Although a stage may be omitted, the order appears to be invariant, and it seems that the sequence is usually necessary for the establishment of a relationship. And to make a human analogy, greeting gestures are often a necessary preliminary to further social intercourse. In this context, however, it must be noted that even quite insignificant gestures may play an important role in determining the future course of interactions. This has been studied especially in higher primates: in wild-living chimpanzees, for instance, a change in the dominant male may be preceded by many weeks by slight changes in patterns of interactions.

4. *Principles concerned with the operation of positive and negative feedback in dynamically changing systems.* In many relationships in which the participants actively seek each other's company, negative FEEDBACK can be said to operate: external stresses that tend to reduce their association may elicit a more active striving for association. In other cases positive feedback may occur: if one partner is unwilling to interact in a particular way, the other may show frustration-induced aggression, which may enhance the uncooperativeness of the partner.

When small changes in one of the participants threaten to disturb a relationship, the compensatory changes required in the other may depend on whether the interactions concerned are reciprocal or complementary. If they are reciprocal, then the partner must change in the same direction if the association is to continue, but if they are complementary, the partner must change in a complementary fashion. For example, if one of two rhesus infants changes in the direction of playing less, the other must change in a similar fashion. But after a rhesus monkey infant has been separated from its mother it may be both 'depressed' and more demanding, and the mother must respond to its more filial behaviour by becoming more maternal if the depression is to be relieved.

The existence of negative feedback mechanisms means that absence or excessive presence of a so-called normal aspect of behaviour may not necessarily matter in the long term: they may be compensated in other ways. However some types of relationship, although apparently well buffered in the short term, carry the seeds of their own destruction or transmutation to a different type of relationship in the very nature of the complementariness of their interactions. Parent/offspring

and teacher/pupil relationships are obvious examples.

5. *Institutionalization.* The relationship between non-human species can largely be understood in terms of the behavioural propensities of the individuals and their mutual effects on each other. In man the institutionalization of relationships raises a new dimension of complexity. This has already been mentioned briefly.

6. *Social forces.* Finally, the patterning of interactions between two individuals may be affected by outsiders. For example, if one member of a group persecutes another, other members of the group may join in on one side or the other.

The development of relationships. As stressed already, the development of relationships is to be understood in the same terms as their maintenance. Only two general points need be made here. The first of these concerns the manner in which inter-individual relationships come to be formed in young animals. The issue has perhaps been best studied in the development of the parent–young relationship in birds. A young moorhen (*Gallinula chloropus*) normally begs from its mother, obtains warmth from her, and follows her when she moves about. Thus a number of different responses are all directed towards the mother. However if eggs are hatched in an incubator the young readily learn to beg from forceps, to obtain warmth from an infra-red lamp, and to follow a wooden box. Presumably each of these objects possesses some stimulus characters in common with those aspects of the mother that normally elicit the responses in question. Under natural conditions the responses come to be directed towards the mother by a complex of learning processes known collectively as IMPRINTING, but which can be understood only with the use of all three of the paradigms discussed in the previous section. Since each response becomes associated with characteristics of the mother in addition to those that elicited it initially, some aspects of the mother will become associated with more than one filial response. Thus the mother comes to be constituted of stimulus characters specific to particular filial responses and other characters common to most or all of them.

At first, the chick's following responses can be elicited by a wide range of conspicuous objects, though some are more effective than others (see SIGN STIMULUS). As DEVELOPMENT proceeds, strange objects begin to elicit FEAR responses, and cannot elicit following until HABITUATION has occurred. 'Strangeness', however, implies a previous learning of the familiar. Thus learning the characteristics of the mother and developing a fear of strangers are two sides of the same coin. The development of an individual attachment specifically to the mother is a consequence of the fact that she mediates several types of interaction with the infant, and that the infant simultaneously becomes afraid of strangers. Similar issues arise in the development of relationships between infant and adult mammals, including man.

The second issue here arises from the first. At almost any age, strangers initially evoke fear responses. Since fear predisposes towards aggression, strangers are either attacked or actively avoided. Thus the formation of inter-individual relationships often requires first the overcoming of fear responses. For this reason it is important not to be too assertive in initial interactions with young children: often they must habituate to a stranger before interacting with him. Many human greeting ceremonies may be partially explicable in a similar manner. Particularly dramatic examples of the manner in which fear must be overcome before a relationship can be established can be seen in the pair formation ceremonies of birds (see MATE SELECTION; DISPLAYS).

In conclusion, two limitations of this survey of the study of relationships must be repeated. First, it is concerned primarily with higher mammals, and to a lesser extent birds. Whilst inter-individual relationships do occur in less complex organisms, they have been relatively little studied from the point of view with which this article is concerned. Second, the study of relationships between individual animals could not be started until a considerable amount was known about how animals behave towards their physical environment, and how individuals interact in brief encounters. The study of relationships embraces the issues that these raise, but also involves the additional problems that arise from the effects of interactions or of interactions over time. It is with the problems raised by this dynamic aspect of relationships that this article has been primarily concerned.

75   R.A.H.

SONG. The environment in which we live is almost never silent. All manner of sounds are produced, some involuntarily as a result of the movement of an animal, some by the movement of wind in the vegetation, and some as the voluntary cries or songs of animals. Though sounds may give valuable information to members of the same species, they may give potentially harmful information to predators; for this reason they tend to be highly specific, and to be produced in controlled circumstances.

Sound travels and so can be used to attract, inform, or warn. Because there may be many species singing within earshot of each other, there tends to be considerable interference and confusion in the environment, resulting in complex and stereotyped recognition songs. Because important information can be conveyed over a distance between members of the same species, songs can be used in a number of behavioural contexts, and so any individual may have a range of different songs for different situations. For our purposes, a sound may be regarded as a mechanical event, a succession of pressure waves propagated through the medium in which animals live, but song is a sound of animal origin which is neither accidental nor meaningless and thus of behavioural importance.

The nature of sound. Sounds are waves of compression which are produced locally by distortion of the surroundings, whether by displacement or compression of the medium, and which are then propagated outwards from the source as a compression wave in the elastic medium. Sound is propagated at different velocities in different media. In air this velocity is 340 m/s, so a note of 440 Hz has a wavelength of 0.77 m. In water the velocity of sound is 1500 m/s, so sounds have a shorter wavelength than in air. Because sound sources radiate most effectively if their diameter is greater than about 1/3 wavelength, small animals tend to produce higher frequency sounds than large ones; crickets (Gryllidae) which have wings about 20 mm long produce songs with dominant frequencies of 3–10 kHz or of wavelengths between 35 and 100 mm. Cows (*Bos primigenius taurus*), having mouths which open to a diameter of 0.2 m, produce sounds with frequencies of a few hundred hertz, though in principle there is nothing to prevent them from producing, as do horses (Equidae), sounds of far higher frequencies. The larger the source is, by comparison with the sound wavelength, the more directional the sound output becomes, which is why a flat wall or cliff face gives a clear echo. It may or may not be advantageous for an animal to produce a highly directed sound, and this is something which may be controlled by the animal.

Sounds may be heard at a distance because the sound wave causes both displacement of the medium and pressure changes in the medium at a distance from the source. An ear may be adapted to detect displacement and is then constructed as a vane which is waved to and fro by the sound, or it may detect pressure changes and consist of a membrane set across the opening of an enclosed space. For maximum sensitivity the receptor should be as light as possible; it can be made directional and more sensitive by increasing its area, or by extending it by a horn-shaped reflector, as is common in mammals.

Away from the sound source, the loudness of the sound falls according to the inverse square law. If the distance from the source is doubled, the sound intensity falls to a quarter. The intensity is measured in decibels (dB) above a threshold sound power of 2×10^{-12} W/m^2 which is about the quietest sound that humans can perceive. Louder sounds are expressed in terms of the sound pressure, where a 6 dB increase indicates a twofold increase of sound pressure and of displacement of the medium, and a fourfold increase in sound power. Sounds above about 120 dB are sometimes produced by animals, but tend to be painfully loud; as they are 10^{12} times as powerful as the quietest sounds, there is the possibility of communication over ranges from a few centimetres to many kilometres in species which can produce the loudest sounds, and hear over the greatest dynamic range.

The medium itself can modify the sound. Air absorbs sounds as well as propagating them; the effect is greatest with warm dry air which tends to attenuate the higher sound frequencies and to reduce their range. Reflection and refraction of sound can also occur at the interface between different media, which is why sound carries well across calm water or round the 'Whispering Gallery' of St. Paul's Cathedral. Sound tunnels of this type are found in shallow seas, in the surface layers of warm oceans, and in other such situations, resulting in greatly increased range of sound communication.

Methods of examination of sounds. The first essential in the analysis of sound is to make as faithful a record of it as possible. From this record, parameters such as the frequency–time, or the frequency–energy structure can be derived.

For recording mammal or bird song, a calibrated microphone and high quality portable tape recorder, both with a frequency response extending between 40 Hz and 20 kHz, are usually adequate. Directional microphones may improve the resolution of the song from the background noise, but tend not to have such flat response curves as non-directional types. Many mammals and insects produce sounds in the ultrasonic region between 20 kHz and 150 kHz. For recording these songs, special high speed recorders and high frequency microphones are required, but the recorded sound can be listened to or visualized on more conventional equipment by playing it back at reduced speed. For recording of water-borne sounds *hydrophones* are required; these have been extensively developed for military purposes.

The simplest type of display of a song is an *oscillogram* which shows sound pressure against time. This type of record shows the structure of simple songs very clearly (Fig. A), but is less clear for more complex songs where frequency and intensity may both change rapidly and where the notes may be rich in harmonics.

For these more complex songs, a spectral analysis such as a *sonagram* may give a characteristic analysis of a particular song pattern. In the sonagram, a short song record is passed repeatedly through a variable frequency narrow band filter to give a display of frequency against time; the plot also gives some indication (Figs. A, B), by the darkness of the bands, of the intensity of the different components. Because of the properties of the filters, rapid changes of frequency or intensity are difficult to resolve in a sonagram.

In experimental work on songs, pre-recorded songs may be played back through a tape recorder, and it is often possible to arrange a short section of song into a loop that can play repetitively. In the case of songs with a simple structure, it may be possible to synthesize songs, either by straightforward electronic methods, or by more or less complex computer programs; such synthetic songs can be used to test the biological relevance of a particular song, or to examine the critical parameters that identify a song.

Mechanisms of song production in animals. Animal sound producing mechanisms show parallels with the principal classes of musical instruments, such as violins, drums, brass, and woodwind. The analogies are not perfect, but may help in understanding the mechanisms.

The type of mechanism employed is related to the anatomical organization of the animals, so arthropods (crabs, insects, spiders, centipedes, etc.) with their hard external skeletons tend to produce songs by tapping or scraping, while land vertebrates commonly use air-powered modifications of the respiratory system. In both cases, naturally noisy biological functions are exploited, refined, and specialized to produce distinctive song frequencies and patterns.

The basic elements of song are the carrier frequency or dominant frequency, and the way in which this is modulated or structured in the song. A single sound event or pulse may be part of a short sequence of similar pulses and result in a chirp. Chirps, in turn, may be repeated or varied to form a song sequence, and a continuous series of pulses may be produced in the form of a trill. Though the carrier frequency within and between pulses may be fairly constant in the songs of insects, in the songs of birds and mammals, glissandos, arpeggios, and changes in the harmonic content of components of the song are common and indeed typical. These differences reflect the differences in the levels of complexity of nervous and anatomical organization in arthropods as opposed to vertebrates.

Among insects, the loudest and most musical songs are produced by male members of the order Orthoptera (crickets, grasshoppers, and their relatives), and by bugs of the order Homoptera such as cicadas (Cicadidae) and leafhoppers (Cicadellidae).

Grasshoppers and crickets tend to stridulate using a file, scraper, and resonator. In grasshoppers (Acridoidea) the large hind femur bears a ridge, which may be toothed, and this can be rubbed against the folded fore-wing. The rustling song so produced is structured by the way in which the leg is vibrated, and the songs tend to be composed of long sequences of chirps. Because there is no sharp resonance, the song carrier frequency is variable, and there are a lot of harmonics.

Crickets (Gryllidae) and bush crickets (Tettigoninae) sing by rubbing their fore-wings together (Fig. E). The undersides of the wings of the male have veins bearing ratchet-like teeth which may be caught by a scraper on the opposite wing. As the wing closes, a succession of the teeth are caught and released to produce a sound. Depending on whether the resonant part of the wing is set into sustained resonance, as in crickets, or excited as each tooth is passed, as in bush crickets, so the

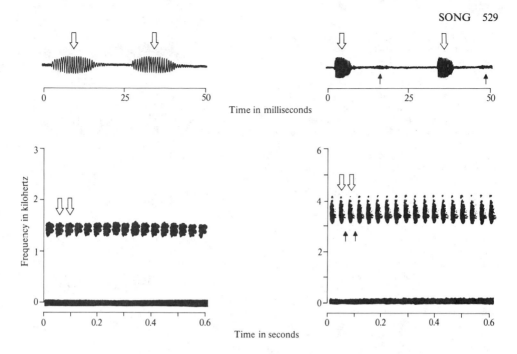

Fig. A. Songs of two species of mole cricket to show how these may be visualized. Left: song of *Gryllotalpa gryllotalpa*; right: song of *Gryllotalpa vineae*. In both cases, the upper diagram is an oscillogram showing song pressure vertically against time horizontally. A time marker of 1000 marks per second is recorded below the oscillogram of the song. The lower figures are sonagrams which show frequency vertically and time horizontally; the pulses indicated by arrows on the oscillograms are shown by arrows on the sonagrams. Note that the sonagram shows how the songs differ in their carrier frequencies, which are 1.5 and 3.5 kHz.

song is either a pure tone or a series of brief transients. Crickets tend to have a song carrier frequency of 3–10 kHz, and bush crickets a carrier frequency of 10–60 kHz; carrier frequency is determined by the dimensions and stiffness of the cuticle of the wing veins.

The muscles producing song are also used for locomotion in Orthoptera; the transition is made by changing the relative phasing of the activation of the muscles, and altering the number of muscle units that are activated, but the song tends to have similar rhythmic components to walking or flight. Some bush-crickets produce up to 250 song pulses per second using specialized wing muscles which heat up to over 35 °C during singing. The rapidly pulsating songs of cicadas are produced by specialized organs on either side of the abdomen. These have domed *tymbals* which, when pulled, click inwards making a pinging sound, and, when released, click outwards again. The muscles that cause the click are activated by the stretch that occurs as the tymbal clicks out, and so oscillate at the high rate of 200–500 Hz, which produces the pulses, each of which has a carrier frequency of 5–10 kHz. Because cicadas are large and tend to sing in concert, their songs can be painfully loud for human beings.

Many smaller insects, such as flies (Diptera), use wing flapping as a close range song. As the whole wing is beaten by the flight muscles, the carrier frequency is only a few hundred hertz. The sound power is low, so this system is only used at close range where the relatively large displacement component of the sound-wave may be received by a feathery antenna or similar vane-like structure.

While hardly classifiable as song, the squeaks and hisses produced by many insects may effectively startle predators. Such sounds may be produced by the rapid expulsion of air from the gut or respiratory system. Many other small insects communicate by sounds made by striking the substrate, which conducts sound more effectively than air. Examples include the death-watch beetle *Anobium*, and many grasshoppers and termites (Isoptera). Stridulatory mechanisms are known in many other insect, crustacean, and arachnid groups, but few of these have been well described acoustically or behaviourally. Some of the loudest sounds are those of the pistol shrimps, *Alpheus*, which, by snapping shut one of their claws, produce explosive sounds that can stun fish at close range.

Fishes, similarly, have acoustic COMMUNICATION. The sounds of fish such as haddock (*Melanogrammus aeglefinus*) are produced as rhythmic croaks by the moving of air from one part of the swim-bladder to another through a narrow orifice, and by muscles which squeeze the swim-bladder. The sounds may have effective ranges of several kilometres, as the sound power emitted is large and is confined, by acoustic reflection, between the top and bottom of the sea. Most land vertebrates produce sounds which are more readily regarded as songs, because they tend to be produced as rhythmic pulses of specific carrier frequencies. In the main, the song is produced by air movement in the respiratory system, but the mechanism of frequency control and sound emission varies from group to group.

In frogs and toads (Anura), the air is pumped between the lungs situated in the body and the air sacs situated either in the cheeks or below the lower jaw. The sound is produced by vocal cords in the throat, but is radiated by the spherical air sacs which pulsate at a frequency determined by the vocal cords. In some amphibia, such as the midwife toad (*Alytes*), the song is a very pure tone with a carrier frequency of 1.3 kHz, but in other amphibia the song may be rich in transients and harmonics, resembling that of grasshoppers. The song is structured into pulses by cyclical or interrupted movements of the air from the lungs into the air sacs, or vice versa. Because the radiating surface is the skin of the air sac, the mechanism differs fundamentally from that of the majority of birds and mammals which sing through open mouths, using the throat as an acoustic transformer between the vibrating vocal cords and the air around them.

In birds, the major vocal cords are situated not in the *larynx*, near the outer end of the trachea, but in the *syrinx* at the bifurcation of the trachea near the lungs. The syrinx is a cartilaginous box with both external and internal musculature. Membranes extend inwards from the outer wall and across the junction of the trachea and the anterior air sac. These membranes are capable of being set in vibration by air leaving the lungs or air sacs, and, because there are several vibrating membranes which can be individually modified, birds are capable of producing songs which contain simultaneously two notes or rhythms which are not harmonically related; there are instances where three or four different notes appear to be produced simultaneously (Fig. B). Birds sing with their mouths open during expiration of air, and are able to modulate their songs by: (i) altering the manner in which air is expired; (ii) altering the position of the head and neck, which alters the shape of the trachea; and (iii) by opening or closing the mouth. It is this range of possibilities that enables such a rich range of notes to be produced.

By contrast with that of birds, the larynx of mammals is highly developed as an organ of sound production. The larynx is situated at the outer end of the trachea, where it enters the mouth, and in mammals is supported by a fixed *hyoid* cartilage, on which is hinged the *thyroid* cartilage, and a pair of lateral *arytenoid* cartilages. These cartilages can be moved together by muscles to alter the shape of the laryngeal box, while the

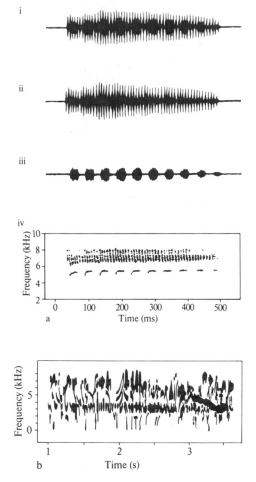

Fig. B. Complex songs of birds showing how two or more sound-producing mechanisms may act simultaneously in one individual. **a.** Song of the American wood thrush (*Hylocichla mustelina*): (i) oscillogram of the unfiltered song; (ii) oscillogram of song filtered to show only the 6–8 kHz component; (iii) filtered to show the 5 kHz component; (iv) sonagram of the song showing the two components, which have different rhythms. **b.** Sonagram of the gouldian finch (*Chloebia gouldiae*) appears to show three different song rhythms, a continuo at 3–4 kHz and an elaborate chirruping seen most clearly towards the end of the sonagram at higher frequencies.

outer opening can be closed by a muscular *epiglottis*. The vocal cords run transversely either side of the interior of the larynx, and can be tensioned by muscles to alter their stiffness. When air is expired through the larynx the vocal cords can be extended across the air stream and tensioned, and will then vibrate producing a tone which depends on the tension. In addition to the vocal cords the mammalian larynx can contain a series of air sacs which act as resonators; these are particularly well developed in monkeys and apes (Simiae), which also have air sacs in the throat that appear to radiate the sound directly in the same way as those of amphibians.

The majority of mammals produce sound through the mouth during expiration. The shape of the buccal cavity, the position of the tongue relative to the palate, and the shape of the lips are all important in controlling the temporal structure of the song, and in altering its harmonic structure. Some mammals, notably horseshoe bats (*Rhinolophus*), produce sound through the nose, which has an elaborate radiating surface which produces a narrow sound beam used in the sonar direction finding (Fig. C). Inspiratory sounds are also found to enrich the vocal repertoire; the bray of the ass *Equus A. asinus* is produced by an inspiratory 'Ee' followed by an expiratory 'Aw', a process which can be imitated in a lifelike manner by humans.

The biological functions of songs. Not surprisingly, the richness of the acoustic and behavioural repertoire of an animal is related to its degree of both functional and social specialization. Thus crickets (Gryllidae) and bees (Apidae) among the insects, and crows (*Corvus*) and finches (Fringillidae) among the birds, tend to have many specific songs for specific behavioural situations, while some of the more solitary animals, such as carnivores, tend to have a far narrower range of types of song, and may even be mute. One obvious disadvantage in developing an elaborate acoustic communication is that other species can listen and make use of the songs. Thus it may be advantageous to sing only at certain times of day, or at particular seasons of the year, or in restricted behavioural situations.

One of the most familiar types of song is the *proclamation song*. This is the song of the blackbird (*Turdus merula*) from the chimney top, the roar of the lion (*Panthera leo*) in the bush, or the chirp chirp of the house cricket (*Acheta domestica*) on the hearth. Usually these songs are loud and are designed to cover a behaviourally meaningful area, which in the case of a lion may have a radius of 7 km, or in the case of a cricket of 4–5 m. The proclamation song's prime function is to indicate to other members of the species the existence of an individual, whether in transit or in residence. One of the more important subsidiary functions of the proclamation song is to indicate the limits of a TERRITORY, and it is used in this way by many insects and birds. Two individuals may sit and sing at each other across an agreed boundary which, if traversed, either physically or in song, may provoke an attack by the defender. This situation can be seen most readily in urban birds which commonly have sites around their territory from which they can both guard and proclaim its limits.

Proclamation songs also have an important function in breeding, and so it is common to find them being produced by the males who also defend the territories, the song acting as a call to females of the species. For this reason, proclamation songs are often highly specific, which is necessary in a cluttered acoustic environment in order to avoid dangerous ambiguity. In the case of some bird and some cricket species, the nature of the proclamation song may provide the surest and fastest means of identifying a species. In the case of insects, the type of parameter that can be varied is the carrier frequency or basic tone of the song, the duration of the pulses, the number of pulses in a sequence, the use of a number of types of pulse or chirp, or the way the song varies during its duration (Fig. D). Birds, with a greater neural complexity, tend to produce not only specific songs, but songs which vary greatly from individual to individual, this can be heard easily with the great tit (*Parus major*), which produces a persistent 'teechew' which varies in rate from individual to individual, and where some individuals may duplicate the first syllable. In other examples, such as blackbirds, the song is composed of a sequence of phrases which are unlikely to be repeated in the same order in successive sequences. While elements of the song are common to many individuals, there is much embellishment of particular phrases, and repetition of phrases is rare; factors like this allow the song to be distinguished from that of the song thrush (*Turdus philomelos*), which habitually repeats rather simple phrases two to four times in succession, and then changes to a new phrase which in turn is repeated several times. By careful study it is possible for us to recognize both the species and the individual by the song.

The songs of most birds and of many mammals have carrier frequencies from 1 to 5 kHz. At these frequencies, attenuation by the air is negligible and sound absorption by vegetation is less than at higher frequencies. Insects, being far smaller, tend to be unable to produce loud low frequency sounds, and many produce ultrasonic signals which, even though attenuated and distorted by

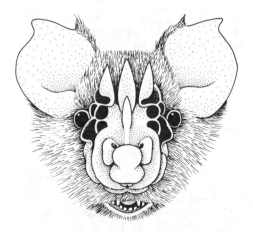

Fig. C. Head of the Persian leaf-nosed bat (*Triaenops persicus*) showing the elaborate nose leaf through which the sonar call is emitted and the horn-like ears which detect the returning echo.

a

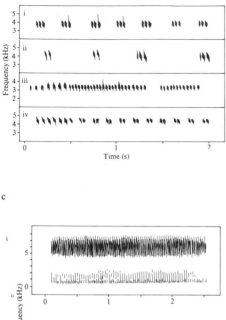

b

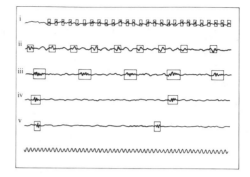

c

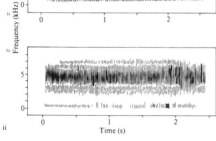

Fig. D. Specific differences in song patterns. a. Sonagrams of the songs of different cricket species: (i) *Gryllus campe stris* produces three chirps per second, each composed of three pulses; (ii) *Gryllus argentinus* produces two chirps per second, each containing two or three pulses; (iii) *Teleogryllus commodus* produces a train of pulses with large intervals and then a longer train with shorter intervals. The carrier frequency is 3.5 kHz. (iv) *Teleogryllus oceanicus* produces a similar large interval train of pulses but this is followed by a train of double-pulsed chirps, all with a carrier frequency of 4.5 kHz. b. Oscillograms of the songs of various species of fruit fly of the genus *Drosophila* showing differences in pulse form and interval. (i) *D. bipectinata*, 120 pulses per second; (ii) *D. affinis*, 36 pulses per second. (iii), (iv), (v) all produce pulses consisting of three to seven cycles of tone; (iii) *D. persimilis*, 19 pulses per second, is sympatric with (iv) the related *D. pseudo-obscura*, 6 pulses per second, but both are allopatric with (v) *D. ambigua*, 6 pulses per second. The time marker is a 200 Hz signal. c. Sonagrams of the songs of related species of warbler: (i) the grasshopper warbler (*Locustella naevia*), 31 triple pulses per second, with a carrier frequency of 6 kHz; (ii) Savi's warbler (*Locustella luscinioides*), 50 double pulses per second with carrier frequencies of 3 and 5 kHz.

the environment, may have a biologically useful range.

Associated with the proclamation song are AGGRESSION and SEXUAL behaviour. Many groups of animals have special songs for these situations: crickets, such as *Gryllus campestris*, produce a structured proclamation song consisting of a train of chirps, each of which contains three or four pulses. In the presence of another male this gives way to loud aggressive stridulation, in which the sound pulses are repeated as a continuous trill (Fig. E). In the presence of a female the song becomes quieter, the carrier frequency rises from 4.5 to 13 kHz, and the pulses are repeated at a lower rate. In all three cases a similar neuromuscular pattern is used, but at different levels and rates of excitation. In such species the aggressive song is designed to be as loud as possible, and to bluff or scare away the intruder; such songs may be less specific than the proclamation songs, and may, indeed, be used against other species. COURTSHIP songs are also less specific where there is no recognition problem, but in species which are so small that proclamation songs are physically impossible, such as the tiny fruit flies *Drosophila*, the courtship song may act as a species-specific recognition signal (Fig. D), without which the female fly is unwilling to mate.

In birds and mammals the situation is greatly complicated by PARENTAL CARE, which involves sharing of the male's territory by the female. As a consequence invitation and greeting songs are commonly observed. In other situations female birds may IMITATE the song of the juvenile to elicit feeding by the male; this occurs before egg-laying in many song-birds. Both sexes share the parental investment in the eggs, and it is suggested that this COURTSHIP FEEDING allows the female to gauge the reliability of the male. In the normal course of brood rearing, the juveniles will sing for food while following their parents and foraging for themselves. In turn, the parents have a series of calls by which they can collect straggling members of the brood or aggregate or disperse them in case of danger.

In mammals, which suckle their young, hunger of the young seems to be less important than other forms of stress in provoking songs. For example, young rodents, which are hairless and blind for the first fortnight after birth, produce ultrasonic calls when they are removed from the nest and exposed to cold. The youngest animals are fairly tolerant of cooling, but as they get older they survive less well and call more readily. The call elicits searching and retrieval by the mother (see MATERNAL BEHAVIOUR).

In SOCIAL animals songs are also used for cohesion of the flock. These parallel the conversational songs that occur between a pair which are FORAGING apart. The situation is exemplified by northern mallard ducks (*Anas platyrhynchos*) which call with increasing frequency at nightfall, gathering together before flying from the feeding place, each giving the flight call and departing. During feeding individuals of many bird species sing to attract others to the site; this is a long-range song, unlike the analogous feeding song of the honey-bee (*Apis mellifera*), which is produced during the waggle dance. The honey-bee dance has a range of only a few millimetres and gives information about distance and direction of the food supply (see COMMUNICATION).

Of the various social calls, perhaps the most obvious is that given in the presence of danger. These ALARM calls are usually loud and differ in their phrasing from the other calls of the species (Fig. F). Reaction to these calls can be species-specific or involve many species. Recordings of alarm calls of birds are used for clearing airport runways, but it has been found that birds habituate quickly to repeated presentations of the same call. Such calls are always associated with a certain risk. In some species the call is given before the flight of the threatened individual, which thereby draws off the predator by displaying itself. On the other hand, in some social birds, such as the house sparrow (*Passer domesticus*) and the common magpie (*Pica pica*), repeated calls may be given to attract other individuals, all of which then cooperate to scare off the predator. Alarm calls are not given by insects, most of which use part of the locomotor apparatus to produce song. At the approach of danger, such species tend to stop singing and rearrange the legs or wings ready for flight; this can readily be seen in the field in both grasshoppers and crickets.

Because some calls are loud, they can also be used as DEFENSIVE behaviour, by startling predators. Among birds and mammals the scream

Fig. E. Cricket song. Above, the singing posture, with the fore-wings raised. The file (F) traverses the harp (H) which resonates at about 4.5 kHz. Below, oscillograms of the different types of song: (i) proclamation song of chirps containing 4 pulses; (ii) aggressive song of a train of louder pulses, but with similar intervals to those in proclamation song; (iii), (iv) successive phases of courtship song, showing how the song becomes quieter and single pulses are produced. The time marker is a 50 Hz signal.

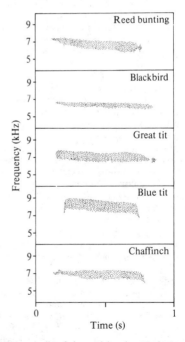

Fig. F. Alarm calls of the reed bunting (*Emberiza schoeniclus*), the blackbird (*Turdus merula*), the great tit (*Parus major*), the blue tit (*Parus caeruleus*), and the chaffinch (*Fringilla coelebs*), showing that these are sustained relatively pure notes and similar in all species.

serves this purpose. The cockroach *Gromphaderina brunneri* expires air through its spiracles when picked up or molested and the death's head hawk-moth (*Acherontia atropos*) hisses through its pro-boscis when disturbed. In some cases the sound may be so loud as to be damaging, and is used offensively; pistol shrimps, for example, produce a sound loud enough to stun fish; a level of 124 dB at 1 m has been quoted.

In another category, the echoes produced by reflections of loud sounds by objects opaque to sound are used in sonar NAVIGATION by the cave swiftlet (*Collocalia brevirostris*), bats (Chiroptera), and whales (Cetacea) (see ECHO-LOCATION). All emit loud short pulses of high frequency sound and listen to the echo. The mechanisms involved are complex and beyond the scope of this article, but it is relevant that owlet moths (Noctuidae) and lace wing flies (Chrysopidae) can hear the bats' calls while still out of range, and will then take evasive action. Tiger moths (Arctiidae) not only detect the bat calls, but then emit loud pulses of ultrasound at a rate high enough to mask the echo from their bodies.

Development of song patterns. Since so many songs, in a wide variety of animal groups, are speci-fic, it is of considerable interest to know the extent to which the song pattern is INNATE and what the role of LEARNING is.

In the case of insects, there is evidence in many species that the song pattern is innate, and that it cannot be greatly modified by experience. The pat-tern of the song of crickets, for example, is highly stereotyped, and neural impulses to the ap-propriate muscles can be detected in immature in-sects which cannot sing and which do not appear to react behaviourally to the song of adults. Ex-periments using brain stimulation in these insects have shown that stimulation of different regions, with different levels of stimulus, can initiate bursts of apparently normal song. Yet other experiments, with wingless or dwarf-winged crickets, have shown that the nervous pattern of the normal song is produced appropriately, even though the animal is silent, so FEEDBACK is relatively unimportant. In the case of a few species of bush cricket a different situation has been reported. Closely related species, which commonly occur together and sing with similar pulses but arranged in different pat-terns, have been induced to modify their songs and, in effect, to mimic the songs of another species.

In birds, learning and experience are very im-portant. It is only a few species, such as cuckoos (Cuculinae), which do not experience the songs of their parents throughout their growth. As a result, experience and practice can be of great importance in the development of the more complex types of song. One type of song that occurs in many species in immature birds, or before the normal breeding season, is termed subsong. Subsong is quieter than normal full song, tends to contain longer phrases containing less pure notes, and is generally less re-

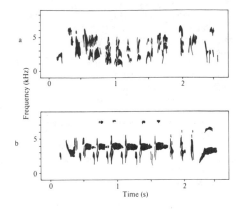

Fig. G. Development of song by the canary (*Serinus cana-ria*) **a.** Subsong, showing the ill-defined phrasing and lack of tonal purity. **b.** Full song, by contrast, shows regular phrasing and notes which are relatively pure and free of harmonics.

fined. During the practice involved in the transi-tion from subsong to full song, the song structure and tonality become refined (Fig. G). Birds reared in isolation produce normal subsong, but the full song remains simple, while birds reared in isolated groups all tend to learn a local dialect which may differ greatly from that of the wild population. Not surprisingly, deafening causes birds to sing abnormally, though the song usually contains ele-ments of the normal song.

The following model is suggested for this learn-ing situation. There is a nervous template which will accept a range of song types of the species, and which will then be refined. During subsong, this template is used to match the individual's song to that in the local dialect which has been learnt by the young bird. That there is only a limited range of dialects which can be accepted by the template has been shown by experiments in which young birds have been reared with adults of a related species, whose song they do not copy, nor do they develop the normal full song of their own species.

Of course, some birds are excellent mimics and learn and reproduce a wide variety of alien sounds (Fig. H). The reason for this is not clear, but, in contrast with the rather stereotyped songs of in-sects, the songs of birds are rarely monotonous, and the ability to incorporate new elements into the song must be important in SEXUAL SELECTION.

While learning is important in many animal groups, there is good evidence that the basic song template is inherited. Hybrids can be made of sev-eral bird, frog, and orthopteran insect genera. In the case of the bird species the hybrids appear able to accept a wider range of song models than either of the parent species, while in the case of the rather simple songs of some frogs and crickets the songs of the hybrids have shown intermediate carrier frequencies and pulse rates to those of the two parent species. The response to song by hybrids is variable. In the case of some crickets the hybrids preferred the songs of hybrid males to those of either parent species, while in some grasshoppers the hybrids accepted the song of either parent species. This suggests that inheritance may be complex and involve many genes, but, more interestingly, it suggests that the recognition template can be inherited intact, and may be similar to the song producing template; there would be strong NATURAL SELECTION for such a mechanism (see GENETICS).

Evolution of song. In the EVOLUTION of songs various factors are important and may conflict. Any behaviour of this type is subject to the pressures of natural selection. The most obvious factors involved are recognition and recognizability, economy of time and effort, and safety.

In many species the song of one or both sexes may give not only the information that the singer is conspecific, but also qualitative information about FITNESS. In crickets, grasshoppers, and fruit flies the song of the male induces either approach of the female or sexual receptivity. The songs of different members of one species tend to be closely similar, and to differ greatly from the songs of closely related species. When tested with artificial songs, or the records of songs of other species, female insects tend to ignore them, but to approach sources of 'good' imitations of their own species' song.

In encounters between species there is a real risk of mistaken identity being fatal to the female insect. Most female insects are monogamous, and so must choose the right male first time. There is little problem if only one species is found in one area, but where the ranges overlap different species may be expected to have widely differing songs; in a group of *Drosophila* species, all of which were *sympatric* (i.e. living in the same region), it was found that the most closely related species in a complex of six species had remarkably different songs, but that species which were evolutionarily rather distinct had more nearly similar songs. In a case like this, the chances of mismatching occurring between the extremes are small, as other sexual barriers tend to have evolved. An extreme example of this is the song of the grasshopper warbler (*Locustella naevia*), which is remarkably like that of some orthopteran insects. Homogeneity within a species may take a variety of forms. For example, in insects, there is little evidence that songs differ in different races of the same insect, but such *clines* and DIALECTS are common in frogs and birds. In the case of frogs, there are two Australian species which produce similar songs where each species occurs in isolation, but in the zone of overlap produce songs by which the species may be distinguished, and which differ from that normal to each parent species. Local dialects in birds probably are learnt as modifications of the general song of the species, and have the advantage that outsiders may readily be recognized by females also accustomed to a particular dialect.

The advantage of a specific song to other members of a species is clear, but it also confers a disadvantage on the singer, in so far as he advertises his presence. There is some evidence, both from birds and from insects, that songs may be designed to be hard for predators to locate. Experiments with owls (Strigiformes) have shown that they find it far harder to locate pure tone sounds than broad band noises; as songs are often pure tones, this is advantageous. Many rodent and insect species sing with carrier frequencies above 20 kHz, in a range where larger mammal and all bird predators cannot hear, while those that sing at frequencies in the audible range tend to restrict their singing to times of day when PREDATION is less likely (see also PREDATORY BEHAVIOUR).

That predators may use songs to locate their prey is known from observations on cats (Felidae) hunting both birds and crickets, and there are instances where insect parasites find their hosts by *phonotaxis* towards the song. It has been suggested, though, that the aggregation songs of cicadas, Orthoptera, etc. have the important dual role of bringing individuals together for breeding, and, because of the acoustic confusion that arises when many individuals sing together, making it hard for predators to locate any one prey. Similar functions have been suggested for *mobbing calls* in small birds which mob hawks (Accipitridae), though one hawk species is known to mimic the mobbing calls of its prey in order to aggregate them so it can make a capture. Use of song by one species to decoy another, except by man, is curiously rare,

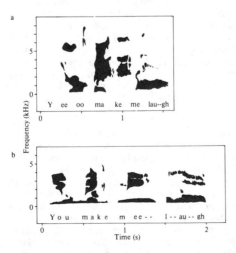

Fig. H. Sonagrams of the sentence 'you make me laugh' as spoken by (**a**) a Javan hill mynah (*Gracula religiosa*) and (**b**) an American human male, showing mimicry and close similarities of intonation, even though the methods of sound production are different.

which is surprising in view of the plasticity of the songs of some species, but there are termite species which in concert produce sounds like the rustling of a snake (Serpentes) and insects and spiders which hiss when disturbed.

Rather more difficult to understand are songs produced in STRESS situations, thus rendering the singer conspicuous. Examples of this are the alarm call of adult birds, and the distress calls of their chicks; the case of the alarm call by the alert and active adult is explained either as an example of ALTRUISM, or as ensuring a cooperative action in a group which might act individually to the detriment of the individual. The distress calls of the young are more difficult to explain because they are likely to attract the attention of predators. This behaviour, however, is normally accompanied by a parental response which silences the distress call, and can be regarded as a goad by the young to ensure closer parental care.

Conclusion. Of the various media of communication available to animals, sound offers the possibility of range, speed of communication, richness of information, and width of spectrum that is surpassed only by visual communication. It is thus not surprising that most members of the more advanced animal groups are sensitive to sounds, that many of these have evolved special sound producing organs, and that song is an important element of their behaviour.   H.C.B.C.

21, 68, 73, 100, 122, 135

SPECIALIZATION in FEEDING behaviour generally occurs as a result of COMPETITION. When Charles DARWIN visited the Galapagos Islands in 1835 he discovered a number of similar but distinct species of finch (Geospizini) living together. He developed the idea that these had all evolved from a common ancestor. It is now thought that Darwin's finches originated from a single species related to the present-day grassquits (Tiaridini) of the American mainland. As the Galapagos population grew, there would have been increasing competition for food, and NATURAL SELECTION would favour any individual that could exploit a particular food resource more efficiently than others. This pressure of natural selection would lead to increasing advantages for specialists. Some became seed eaters with conical beaks, and others insect eaters with long thin beaks, etc. The effects of interbreeding would be ameliorated to some extent by the separation of the many islands of the Galapagos, so that distinct species could eventually evolve.

The present-day finches of the Galapagos include the ground finches (*Geospiza*) which have a heavy conical beak, the insectivorous tree finches (*Camarhynchus*) which have more slender probing beaks, and the woodpecker finches (*Cactospiza*) which have taken on the role of the woodpeckers. Thus Darwin's finches have come to occupy the NICHES which are taken by other birds on the mainland.

Specialization can also occur as a form of individual ADAPTATION. For example, on Walney Island, Cumbria, there is a colony of herring gulls (*Larus argentatus*) that has been growing rapidly for a number of years. Individuals face severe competition for food. Although herring gulls are omnivorous, the individuals in this colony tend to specialize upon particular food sources. Some feed largely upon mussels (Myrtilidae), others upon earthworms, for example *Lumbricus* and *Allolobophora* spp., some feed almost exclusively upon rock crabs (*Cancer pagurus*), while others have taken up CANNIBALISM as a form of FORAGING, preying upon the eggs and chicks of other gulls in the colony.

Each crab specialist defends a feeding TERRITORY on the beach at low tide, SEARCHING for crabs that are buried under the sand. The crabs bury themselves at dawn, and if high tide coincides with sunrise they remain stranded upon the beach. The gulls search for the tell-tale cracks in the sand through which the crab breathes. Only the crab specialists seem able to find these, perhaps because they know when to look. By efficiently exploiting this food resource, the crab specialists avoid the competition for the more common types of food.

STRESS is a physiological condition that results from excessive environmental or psychological pressures. Most animals are able to cope with environmental disturbances by means of DEFENSIVE behaviour, HOMEOSTASIS, and ACCLIMATIZATION. However, there are limits to the animal's capacity to cope with very intense or prolonged disturbances.

In the natural environment stress may result from physical factors, such as sudden storms, flooding, etc., or from encounters with predators, or from SOCIAL RELATIONSHIPS. In CAPTIVITY stress can be induced by unnatural conditions. Overcrowding, for instance, may cause animals to be subjected to AGGRESSION from which they cannot ESCAPE; artificial daily routines may disrupt the animal's natural RHYTHMS. In the interests of animal WELFARE, steps are usually taken to avoid these circumstances.

Stress can be induced in LABORATORY experiments by psychological means. PAVLOV trained dogs to salivate when a circle was presented, since this was followed by food (see CONDITIONING), but to refrain from salivating when an ellipse was presented, since this was not followed by food. Once the dogs had learned the DISCRIMINATION, the ellipse was made progressively more like the circle, and as the discrimination became more and more difficult, the dogs showed increasing signs of stress. It has been shown that rhesus macaques (*Macaca mulatta*) which received mild electric shocks for making wrong decisions in a simple task developed gastric ulcers, a sign of stress, whereas control monkeys, which received the same shocks but did not have to make decisions, remained healthy.

Animals subject to stress undergo physiological changes, particularly in the balance of circulating HORMONES. Initially there is activation of the sympathetic division of the AUTONOMIC nervous system, resulting in increased rate of heartbeat, altered blood flow, and release of adrenalin hormones into the blood. The adrenal cortex, under the influence of the pituitary gland, is stimulated to produce steroid hormones, the levels of which are often taken as a physiological index of stress. Animals subjected to stress develop enlarged adrenal glands and gastric ulcers. If the stress is excessive there may be reproductive failure, heart failure, or other pathological conditions.

Stress has profound effects upon behaviour. Mild stress, such as that induced in FRUSTRATION or CONFLICT situations, gives rise to REDIRECTED and DISPLACEMENT ACTIVITY, but severe stress may induce ABNORMAL behaviour. In addition, there are many specific behavioural effects of stress. For example, there may be loss of APPETITE, reduction of lactation in domestic cattle (*Bos primigenius taurus*), and of egg laying in domestic fowl (*Gallus g. domesticus*). In social situations, stress is often related to DOMINANCE. In rhesus macaques it has been found that physiological indices of stress are related to the amount of AVOIDANCE behaviour shown by subordinate individuals. Under natural conditions such stress can be reduced by behavioural means. The subordinate individual who is frequently bullied can retire to the periphery of the social scene. In overcrowded conditions, however, stress can be severe, and can lead to reduced fertility and increased aggression. Research on people shows that physiological indices of stress are often correlated with overcrowded conditions, such as on commuter trains in large cities.

It has been suggested that overcrowding leads to stress and to changes in hormonal balance that reduce fertility, and that this is a means by which population density is kept in check, so as to prevent over-exploitation of food resources. A difficulty with this suggestion, however, is that NATURAL SELECTION acts at the level of the individual, so that any mutation which endowed an individual with a hormonal balance which was less than usually responsive to the effects of stress would immediately put the individual at a reproductive advantage, and would thus tend to spread in the population. In other words, the stress theory implies a process of EVOLUTION that appears to be unstable, and incapable of providing a means of population control that is immune to genetic cheats (see ALTRUISM).

56

SUN BATHING occurs in many animals as part of THERMOREGULATION. For example, the desert locust *Schistocerca gregaria* maintains a perpendicular ORIENTATION to the sun during the morning and evening when the air is cold, but turns parallel to the sun's rays when it reaches a temperature of about 40 °C. Similarly, after a cold night lizards often seek sunlight, basking in the sun so as to raise their body temperatures. The horned lizard (*Phrynosoma modestum*) maintains a posture that keeps its dorsal surface perpendicular to the sun's rays to absorb the maximum possible warmth.

Some sun-bathing animals, such as locusts and grasshoppers (Acridoidea), orient with respect to the heat content of sunlight, rather than the direction of light. Even blinded insects maintain an appropriate orientation to the sun. In pigeons (Columbidae), on the other hand, sun-bathing is a response to light; this has been demonstrated by experiments using a bright directional light in a temperature-controlled cage. Pigeons, and many other birds, have a typical sun-bathing posture in which they spread out their wings and ruffle their feathers. In contrast, their normal response to high temperature in the absence of bright light is to sleek the feathers rather than raise them. Thus it appears that sun-bathing in birds is not primarily thermoregulatory, but is designed to allow sunlight to reach the skin. Sunlight encourages the synthesis of vitamin D in the skin, and this may be a FUNCTION of sun-bathing.

In addition to special sun-bathing behaviour, many animals show a preference for resting in sunlight, provided that it is not too hot. This is a way of saving energy, because the solar radiation provides heat that would otherwise have to be provided by metabolic processes.

SUPERNORMAL STIMULUS. A stimulus is said to be supernormal when it surpasses a natural stimulus in its effectiveness as a RELEASER. As an illustration, let us consider the COURTSHIP of the grayling butterfly (*Hipparchia semele*). The butterfly is brown in colour, similar to tree bark. Courtship begins with a SEXUAL pursuit phase in which the male chases the female. Experiments with artificial females show that neither colour, size, nor shape has much influence on the behaviour of the male; but distance from the male, type of movement, and darkness of pigmentation of the female's wings and body have a strong influence on the male's pursuit response. Black females, which never occur in nature, are much more effective than those with natural coloration, and thus provide a supernormal stimulus for the male.

Similar experiments have been carried out on the incubating response of herring gulls (*Larus argentatus*). These experiments were designed to investigate what stimulus characteristics of the egg, such as its shape, colour, and marking, are important in eliciting INCUBATION. In other words, what characteristics of the egg are important for the bird to recognize it as an object-to-be-sat-on? During these investigations it was found that a brooding herring gull, given a choice between two artificial eggs, one of normal size and one much larger, prefers the larger egg (Fig. A). This type of response to supernormal stimuli suggests that animals often have an open-ended preference rather than a preference for a particular size or shape of object. In other words, for gulls it may be that 'the bigger the better' is the rule when it comes to eggs. It obviously makes sense for the animal to choose a larger egg under normal circumstances, because the chick is likely to be larger and have a better chance of survival in the critical early phase of life. An open-ended preference of this type is perhaps a good strategy when the choice open to the animal is normally curtailed by nature. Thus in nature herring gull eggs would never be greater than a certain size, so that there would be no deleterious consequences from exercising an open-ended preference. It is only when man intervenes by providing artificial eggs that the situation can get out of hand.

Fig. A. A herring gull attempting to roll a giant egg into its nest, in preference to a real egg.

It has been suggested that human beings have an open-ended preference for sweet-tasting substances. In nature, sugar is in short supply, being available only in honey and in fruit. Under such circumstances, an open-ended preference for sweetness may be a good thing, because sweetness generally indicates a source of quick energy. Many animals, when tested in the LABORATORY, show a preference for sweet-tasting foods, even though they would not normally find them in nature. Normally animals eat what they find palatable, and thereby supply all their nutritional needs. Most animals specialize on certain types of food and are adapted to the taste, texture, and digestibility of their food. This type of specialization to some extent prevents different species competing for the same food supply. This delicate balance of nature can easily be upset by man. So long as he did not manufacture sugar, the maxim 'a little of what you fancy does you good' held true for man as for most animals, but this situation changed with the technological revolution.

Man has a natural liking for sweet things, and

primitive man could satisfy this desire by eating fruit or honey. However, the development of the sugar industry, following on the development of the sugar plantations of the Caribbean, radically altered this situation. The average person in Western Europe or America today eats twenty times as much sugar as people did two or three hundred years ago. Ever since the natural limitation on the availability of sugar was removed by technological means, man has been free to exercise his open-ended preference for sweetness. Many scientists believe that this high consumption of sugar is deleterious for human health.

Responsiveness to supernormal stimuli is exploited by some parasites. For example, the markings inside the gaping mouth of the nestling of the European cuckoo (*Cuculus canorus*) are much more effective in eliciting a FEEDING reaction from the foster parents than are the markings of their own young. Exaggeration of stimuli may be possible in this case because the young cuckoo is normally larger than the young of its foster parents. The possibility of exaggerating natural situations has also been exploited by cartoonists. This is particularly true of the over-emphasis of baby characteristics frequently found in cartoons of children and animals. Here we see that the exaggerated features are supernormal equivalents of those features which normally distinguish a JUVENILE from an adult, such as a rounded head shape, and a small face in proportion to the size of the head. Similarly, the breasts and buttocks of women are often exaggerated in cartoons, in order to induce a SEXUAL response in the reader.

137

SURVIVAL VALUE. The survival of a trait within a population depends upon the extent to which the trait contributes to reproductive success (SEE NATURAL SELECTION). This depends upon the *selective pressures* characteristic of the environment: those features of the environment which are likely to jeopardize reproductive success. This might be jeopardized by the death of a parent from starvation, PREDATION, etc.; by failure to breed; or by failure of the young to thrive. For example, if predation was the most common cause of mortality within a population, then there would be a premium on CAMOUFLAGE, and camouflage would be said to have high survival value.

The study of the way in which characters of animals, whether structural or behavioural, contribute to survival is an important part of the biologist's approach to animal behaviour. We can think of an animal as designed by natural selection to fulfil certain FUNCTIONS. Understanding the design features of a mechanism often helps in investigating its principle of operation. For example, speculation about the survival value of the bright coloration of flowers led to the discovery that bees (Apidae) are able to distinguish colours (see COLOUR VISION). The bees are attracted by the colour patterns of the flowers, and their visits to the flowers aid in pollination (see COADAPTATION). Individual bees tend to specialize on certain types of flower, which is of advantage to the plant, because it means that the bee is more likely to bring pollen from the same species of plant. This advantage has led to the EVOLUTION of distinctive flower patterns, designed to attract bees, and to make it easier for the bee to learn the characteristics of its favourite flower. Many flowers have special markings, called *honey guides*, which lead the bee towards the nectar. These honey guides are not always visible to the human eye, but become apparent when photographed through an ultra-violet filter, as illustrated in Fig. A. Whereas human beings have high sensitivity to red, but are unable to detect ultra-violet light, bees cannot see reds but are very sensitive to ultra-violet. This profound difference in sensitivity to the wavelength of light was discovered as a result of speculation concerning the survival value of colour patterns in flowers.

Fig. A. Flowers of *Helianthus rigidus* (above) and *Oenothera biennis* (below), photographed through a yellow filter (left) or with an ultra-violet filter (right). The ultra-violet filter shows up the striking patterns which act as honey guides, and are normally unseen by man.

Ideas about the survival value of behaviour need not be purely conjectural, but can often be investigated experimentally. For example, black-headed gulls (*Larus ridibundus*) remove broken egg-shells from the area of the nest when the young have hatched. It is evident that this behaviour must have a high survival value, because in leaving the young to carry away the shell, the parent is exposing them to predation. By careful experiment, scientists have shown that an important function of egg-shell removal is that it helps to maintain the camouflage of the nest. In these experiments, sixty black-headed gull eggs were placed, singly and widely scattered, in close vegetation near to the gullery. Each egg was partially camouflaged with a few straws of grass, and half the eggs were marked by a broken egg-shell placed near by. Observations showed that predatory car-

rion crows (*Corvus corone*) and herring gulls (*Larus argentatus*) found the marked eggs much more easily than the eggs which had no broken egg-shells nearby. Thus the survival value of the egg-shell removal lies in protecting the young from predation, even though there is a slight risk of predation during the short time that the parent is not guarding the nest because it is removing the egg-shell.

Experiments on the survival value of behaviour, which are necessarily conducted in somewhat artificial conditions, can never reveal the exact way in which natural selection operates. This is because, in the complex natural situation, it is the overall effect of selection that is important. In nature there are many conflicting pressures, so that the design effected by natural selection is inevitably a compromise. For example, the black-headed gull does not remove the broken egg-shell soon after hatching, as do some other ground-nesting birds, but delays 1–2 h. This is probably due to the fact that the gull colony is closely packed, and neighbouring black-headed gulls prey on wet chicks, but not on dry ones. There may, therefore, be survival value in delaying egg-shell removal until the chicks have dried and the parents' absence does not expose them to predation from neighbours. The disadvantage of nesting close together is, in turn, counteracted by the advantage gained in protection from predators, such as crows and herring gulls. It is known that the nests near the centre of the colony suffer less from such predators than those near the edge, and it appears that this is due to the way in which black-headed gulls combine to harass larger birds that fly over the colony.

88

SYMBIOSIS is the living together of organisms of different species, to their mutual benefit. It occurs among plants, between plants and animals, and among animal species. It can occur as a relationship between individuals, between individuals and societies, and even between entire societies.

There are some instances in which one species benefits from a symbiotic relationship, while the other remains unaffected. Such *commensalism* is difficult to verify without prolonged investigation into the question of whether one participant is truly unaffected by the relationship. However, some examples have been verified, particularly in the food-finding relationships between birds and mammals. One such example is the cattle egret (*Bubulcus ibis*) which feeds upon ground-living insects disturbed by large mammals, such as elephants (Elephantidae), and buffalo (Bovidae). Similar birds are the brown-headed cowbirds (*Molothrus ater*) of North America, and the flycatcher, the cattle tyrant (*Machetornis rixosa*) of Argentina. In agricultural areas numerous birds have learned to follow the plough, obtaining food uncovered or disturbed during the ploughing of the land. An example is the lapwing (*Vanellus vanellus*) of Europe. Many sea-gulls (Laridae) follow ships to obtain food from refuse, and to catch small fish disturbed by the vessel. In all these cases, the mammalian hosts appear to be unaffected by the association.

Social commensalism is said to exist between species of ants (Formicidae), which nest in close association with each other. However, it is not clear what benefit they gain from this. More obvious cases are those of various arthropod species that inhabit ants' nests. For example, there are scavenger millepedes (Stylodesmidae), which run with army ants (Dorylinae) and seem to be accepted as nest mates by their hosts. The millepedes gain some protection by living with ants, as well as benefiting from scavenging opportunities created by them.

Mutualism, sometimes called true symbiosis, is characterized by associations from which both participants benefit. The mutual cooperation is often facilitated by simple forms of COMMUNICATION between the participants. For example, many aphids and related species derive protection by associating with ants, while the ants benefit by obtaining food. When a garden ant (*Lasius niger*) encounters a bean aphid (*Aphis fabae*) it caresses and palpates the aphid with its antennae in a characteristic way. This induces the aphid to exude from its anus a drop of sugary material, called honeydew. The ant generally eats the droplet. The honeydew is a by-product of digestion, and although the aphids produce more when stimulated by ants, the association is in no way obligatory. Other aphid species, however, are dependent upon ants, as they cannot eject honeydew without them.

An interesting symbiosis exists between various species of sea anemone of the genus *Stoichactis*, and the anemone fish *Amphiprion*. The fish are able to gain protection from predators by swimming among the tentacles of the anemone without being harmed. A fish which has been isolated from an anemone will be stung initially on making contact with the tentacles. Normally, such fish allow themselves to be stung as a result of making brief contact with the tentacles. In this way they are able to develop an immunity, and their skin becomes covered with anemone slime, a substance that inhibits the *nematocysts* (stinging organs) of the anemone. Thus the anemone and the fish become mutually adjusted through a process of ACCLIMATIZATION.

A fish acclimatized to one anemone is not stung if it makes contact with another anemone of the same species, but it may be stung by an anemone of a different species. Unacclimatized anemone fish, or fish of other species, are always stung by the anemone, even if there is an acclimatized fish living in it. Thus the behaviour of the anemone is not altered by the process of acclimatization. If an acclimatized fish is wiped with a cloth to remove the slimy covering, and then allowed to make contact with an anemone, it will be stung.

In some species, such as *Amphiprion tricinctus*, a single fish, or a mated pair, will take up residence in an anemone, and will defend it against other fish of the same species. They will also defend the anemone against members of other species, such as butterfly fish (Chaetodontidae), which have the habit of biting the ends off the anemone tentacles. Most species of anemone fish make excursions from their anemone to obtain food, and they often bring food back to the anemone, some of which it then eats. Thus the anemone benefits from the relationship by being protected from PREDATION to some extent, and by obtaining food. The anemone fish gains protection. Some species lay their eggs at the base of anemone tentacles, where they can develop undisturbed.

Another symbiosis in which one partner gains protection, while the other gains food, is found between anemones and hermit crabs. Various species of hermit crab of the genera *Dardanus* and *Pagurus* live inside mollusc shells, on which there is often a large anemone of the genus *Calliactis*. The arrangement is beneficial to the crab, because the stinging anemone gives the crab protection against enemies. It is probable that the anemone benefits by obtaining particles of food, which are dispersed into the surrounding water when the crab feeds. In some cases the association is initiated by the anemone. For example, the hermit crab *Eupagurus bernhardus* does nothing to aid the transfer of the anemone *Calliactis parasitica* to its shell. This anemone attaches itself to stones or to an empty shell in the normal manner, but it displays special alacrity in attaching itself to a shell containing a hermit crab. First the tentacles explore the surface of the shell very actively, and many of them adhere to the shell. The anemone then attaches itself to the shell with its mouth, and detaches its foot from the substratum. Then it bends double and starts to attach its pedal disc, or foot, to the shell. When this is firmly attached, the anemone relaxes the grip with its mouth and straightens up into the normal position. This remarkable behaviour is very unusual in anemones, and it is more common for the crab to take the active part in the transfer of an anemone to its shell. This situation exists in the Caribbean, between the crab *Dardanus venosus* and the anemone *Calliactis tricolor*; and in Hawaii, between *Dardanus gemmatus* and *Calliactis polypus*. When the crabs move their appendages on the surface of the anemone, the anemones relax and detach themselves. The crabs lift the anemones in their claws and place them upon the mollusc shell, where the anemones attach themselves.

Benefits of a different kind are involved in cleaning symbioses. In the marine environment, cleaning symbiosis is a relationship in which the cleaner removes parasites, diseased tissue, and food particles from cooperating fish. This mutually beneficial behaviour results in the removal of harmful materials from the host, and furnishes food for the cleaners. Cleaning symbiosis is widespread in the marine environment, and the hosts include many kinds of fish, as well as marine turtles (Cheloniidae) and lizards (Lacertidae), and some sea urchins (Echinoidea). About forty-five species of fish are known to be cleaners, as are a few species of shrimp (Natantia).

The cleaning process seems to be vitally important to many fish, and experiments have shown that if the cleaners are removed from an area, the majority of host fish move elsewhere, and those that remain become diseased. Cleaner fishes and shrimps have regular stations where they are visited by fish to be cleaned. Many fish spend as much time being cleaned as they do foraging for food. The cleaners are clearly key organisms in the COMMUNITY, and their relationships with other fish are highly developed and complex. Cleaner fish are generally much smaller than their hosts, but they are very rarely eaten by them. They are generally brightly coloured, often marked with bold stripes. They generally approach a potential host with an elaborate DISPLAY. The host responds by opening its mouth, allowing the cleaner to enter and remove parasites and food particles. The cleaner shrimps, such as *Periclimenes petersoni*, employ a similar strategy. They are usually conspicuous, and display to fish with their enlarged antennae. The fish swim up to the shrimp and present their head or gill region. If part of the body is injured, it is generally the first to be presented to the shrimp. When the fish holds itself still, the shrimp boards it, and walks rapidly over it examining injured regions and removing parasites. It enters the gills and cleans them, and it also cleans the mouth cavity. If the fish wishes to move, it first signals to the shrimp to move away, or it forcibly ejects it. The shrimps rarely fall prey to the fish, even though they may be similar to the fish's normal food.

The evolutionary origins of symbiotic relationships are a matter of speculation, but they may have developed from PARASITISM, in which the host is exploited but not usually killed by the parasite. Parasitism probably developed in situations in which free-living organisms were able to invade HABITATS provided by other species. Fleas (Siphonaptera), for example, probably arose as winged scavenging flies, whose larvae fed on the excrement in the burrows of mammals. The larvae of present-day fleas still feed on debris in the nest or den of their host. It is likely that the adult attached itself to the hair of a suitable mammal, so that it would then be taken to a suitable site in which to lay its eggs. A further small evolutionary step would then enable the adult flea to obtain food directly from its host.

In response to a parasite, the host species can either evolve DEFENSIVE mechanisms, or it can turn the situation to its own advantage. This may well be an origin of symbiosis, since an individual that can obtain benefit from the presence of a parasite will certainly be favoured by NATURAL SELECTION, over and above an animal that takes expensive counter-measures, however effective.

It is also possible that symbiosis has evolved from situations in which animals of different species are drawn together by some common factor, such as food supply. For example, numerous species of birds congregate and prey upon army ants in the tropics. It is a small step from such AGGREGATIONS to the FLOCKING of birds of different species, in which there is some social attraction between individuals and some cooperation in their behaviour. Flocking behaviour can be of advantage to members of the different participant species in locating and flushing prey, and in coordinated defensive behaviour against predators.

The tendency for members of a mixed flock to associate with each other may have evolved into more specific symbiotic associations between species. For example, the carmine bee-eater (*Merops nubicus*) accompanies the Arabian bustard (*Ardeotis arabs*) to feed on the insects flushed from the ground by the large bustard. The tendency for some animals to nest close to other species, from which they gain protection, may also have evolved into true symbiosis. Small birds, for example, sometimes nest under the edges of nests of imperial eagles (*Aquila heliaca*) and ospreys (*Pandion haliaetus*). The protection gained presumably outweighs the risk of predation by the host. Similarly, many small birds gain protection by nesting in close association with man. Some, such as the barn swallow (*Hirundo rustica*), nest in buildings, while others, such as the steppe weavers (*Textor*) of equatorial Africa, often nest close to human habitations in preference to wilder places. Some species living in close association with man have become so subject to DOMESTICATION that they have become dependent upon man in some respects. As man clearly derives benefit from the association, many cases of domestication must involve symbiosis.

18, 145

T

TASTE AND SMELL are the most important of the CHEMICAL SENSES possessed by animals. It is likely that sensitivity to chemical stimulation was one of the first senses to evolve. In vertebrates, the visual receptors have almost certainly evolved from *chemoreceptors*, and the areas of the vertebrate brain that are concerned with taste and smell are amongst the oldest, in the evolutionary sense. In the final analysis, however, whether or not something is a food for an animal depends on its chemical composition, and animals must be able to regulate the chemistry of their own bodies (see HOMEOSTASIS). Because the nervous system is used in such regulation, animals must have receptors to detect changes in body chemistry. Thus vertebrates have receptors buried deep in the BRAIN that monitor the chemical constitution of the blood, but do not contribute to an awareness of taste or smell.

The distinction between taste and smell is difficult to justify on biological grounds. Both taste and smell receptors are stimulated by the presence of certain organic molecules, and are accordingly known as chemoreceptors. In vertebrates these receptors occur on the tongue and in the nose, but many invertebrates have them distributed elsewhere on the body. Thus insects often have chemoreceptors on their antennae, on their legs and other appendages. Rather little is known about the physiology of the chemoreceptors. It is thought that different receptors respond to molecules of different shapes, but we do not know what sugar and saccharine have in common that makes them both taste sweet, not merely to ourselves but also to rats (*Rattus*), monkeys (Simiae), and blowflies (Calliphoridae). Traditionally, it has been thought that there are four basic receptors for taste in humans. The main evidence for this comes from the fact that different parts of the tongue are sensitive to different tastes. Sweet tastes affect the tip of the tongue, salty tastes affect the sides towards the front, sour tastes affect the sides towards the back, and bitter tastes affect the extreme back of the tongue. Even less is known about smell, though once again there are presumably different receptors sensitive to different molecules. Most mammals have millions of olfactory receptors, but some organisms with very few have a keen sense of smell. For example, the tomato hornworm, the caterpillar of the moth *Manduca quinquemaculata*, feeds mainly on tomato, tobacco, and potato plants, and avoids many other kinds of plants, such as geraniums. It discriminates between the vast array of plants by means of just sixteen olfactory receptors.

The most obvious use of chemoreceptors is to enable animals to select edible foods, and to regulate their intake. The blowfly has receptors on its legs that are sensitive to sweet substances, and on landing on a drop of sugar it lowers its *proboscis* (sucking tube) and feeds. Its feeding is largely determined by the stimulus, and it will continue to drink a sugar solution until the receptors in its proboscis 'adapt', i.e. cease responding. With continued stimulation most receptors reduce their rate of firing, and this adaptation is particularly noticeable in the case of taste and smell. For example, with continuous exposure to the same smell, we cease to be aware of it. Similarly rats regulate their food intake by means of taste receptors. The more hungry they are, the more tolerant they become of bitter-tasting food: bitterness is often a sign that the food is poisonous. Moreover, if a rat is deprived of salt, it will make good its intake by selecting particularly salty foods.

Rats and many other species have a subtle and interesting mechanism based on taste, which enables them to learn to avoid poisonous foods. If rats are given food with a particular taste and are then made sick, they avoid eating food with that taste. Even though the sickness may occur hours after they have eaten, they associate the sickness selectively with the taste of the food, and not, for instance, with the place in which they have eaten. They are particularly prone to avoid food with an unfamiliar taste, especially if eating it is followed by nausea. In one experiment rats were given a new food, and $\frac{1}{2}$ h later one to which they were thoroughly accustomed; they were then made sick. Afterwards they avoided the new food, but continued to eat the familiar food, although the sickness occurred closer in time to their eating of it (see FOOD SELECTION). It is for this reason that rats are so difficult to poison; if they are not killed outright by eating a poisonous food, they will thereafter avoid it.

Smell is used by many species to assist in the location of food, but it is also often used for COMMUNICATION between species, particularly as a SEXUAL attractant. The male silkmoth (*Bombyx mori*) has many thousands of smell receptors on its antennae, over 70% of which are sensitive to just one kind of molecule, that of a substance called bombykol, which is released by unmated female silkmoths. On receiving a whiff of bombykol the male flies upwind, and so approaches the female. When in the vicinity of the female the concentration of bombykol becomes stronger, and the male can actually locate the female by flying in such a way as to make the concentration always increase. If the concentration is not increasing at a given moment, the moth turns until it finds a direction of flight which results in the strength of the smell increasing. The male silkmoth can detect the odour of a receptive female from a distance of several kilometres. In fact, its receptors can respond to the presence of a single molecule of bombykol.

Volatile molecules released by a member of a species for the purpose of communicating with other members are known as PHEROMONES and their significance is not always merely sexual. For example, cats have glands in the neck that release pheromones, and when a domestic cat (*Felis catus*) rubs its neck against a surface it is often marking it with its own odour, as a TERRITORY sign (see SCENT MARKING). Dogs (*Canis l. familiaris*) leave reminders of their presence by emitting pheromones in urine. Ants (Formicidae) mark trails to food by excreting a scent characteristic of their species.

Finally, some species of fish use smell to assist in NAVIGATION. Salmon (Salmoninae) are spawned in streams and migrate hundreds of miles to feeding grounds in the open sea. At spawning time they find their way back to the stream in which they themselves were spawned, and, in the absence of other landmarks, it is almost certain that they select the river that they ascend by its characteristic smell. They must then select the right tributary, and may have to make many choices between joining streams before reaching the right spot. Salmon show a strong preference for water having the same odour as that in which they were born and bred.   N.S.S.

TAXES. A taxis is a form of ORIENTATION in which the animal heads directly towards or away from a source of stimulation. For example, the maggot larva of the house fly (*Musca domestica*) shows a negative photo-taxis. When the larva has finished feeding it leaves the food and goes to a dark place where it pupates. The fully fed larva, during the 3 or 4 days before pupation, will crawl directly away from a light source. The head of the maggot is equipped with light sensitive receptors capable of registering different light intensities, but not of forming an image or of obtaining any information about the direction of a light source. As the maggot crawls along it moves its head from side to side, testing the light intensity on each side of its body. If the intensity on the right side is greater than that on the left, then the larva is less likely to turn its head towards the right. It therefore tends to change its course and crawl more towards the left, away from the light source. The process of successive comparison of the intensity of a stimulus on one side of the body and then on the other is called *klino-taxis*. The animal increases the rate of head turning in response to an increase in the level of illumination. If the light is turned off every time the maggot turns its head to the right, and turned on every time it turns it to the left, then the animal will turn in a circle towards its right. This experiment demonstrates clearly that although the fly larvae have no direction receptors, they can perform a directional response. Klino-taxis is found in a number of other animals, including the single-celled protozoan *Euglena*.

If an animal can make a simultaneous comparison of the intensity of stimulation received from two or more receptors, and can strike a balance between them, then it can achieve *tropo-taxis*. This type of orientation is characterized by a straight course towards or away from a source of stimulation, in contrast to the wavy course characteristic of klino-taxis. The pill woodlouse (*Armadillidium vulgare*) exhibits tropo-taxis in its positive reaction to light. This animal lives under stones or decaying wood, and it shows a positive reaction to light after periods of starvation or desiccation. The animal may otherwise be photonegative, or indifferent to light. The pill woodlouse has a pair of compound eyes on its head, and one of these can easily be blacked out temporarily. With two eyes operating, the animal heads directly towards a light source, but with only one eye it describes a circle. This shows that the animal normally relies upon the balance of the stimulation from its two eyes. When presented with two light sources the woodlouse may take a course between them, or may fixate upon one. Often the animal starts off by taking a median course, but then heads towards one. If by chance the animal finds itself facing a particular light source it will head towards it, ignoring any source of side illumination. A tropo-taxic animal is likely to behave in this way, particularly if the two light directions are at a wide angle, and if the animal's eyes are shielded at the back and sides so that they are not stimulated by lateral illumination.

Many animals are able to attain directional orientation without having to compare the stimulation from two receptors. This ability is called *telotaxis*. It is shown by all animals which have eyes that are capable of providing information about the direction of light by virtue of their structure. An eye with a lens, for example, usually has some type of retina, containing a number of receptors (see VISION). The position of the image on the retina provides the directional information that is necessary for orientation.

The principles of taxis, which have here been illustrated with respect to light, also apply in many other sensory modalities, including the CHEMICAL, MECHANICAL, and ELECTROMAGNETIC senses. For example, the flatworm *Planaria alpina* exhibits tropo-taxis in relation to the direction of water currents, and many animals show klino-taxis in response to gradients of chemical stimulation.

45

TECHNIQUES OF STUDY. The study of animal behaviour can be approached in two distinct ways. In one the scientist himself is directly responsible for the observation, animal capture, recording, and data analysis. The second approach entails remote control, observation, recording, and analysis being carried out by a computer or some other electronic device. Each approach has advantages and disadvantages; the salient features of each in relation to the study of individuals and groups of animals will be described below.

Animal capture. In the early stages of FIELD STUDIES, and sometimes for LABORATORY STUDIES, animals need to be captured for marking and/or experimental treatment. The techniques used are largely dictated by the species of animal, the topography, and the number to be caught.

Single birds can be captured on the nest during INCUBATION with a walk-in trap or snare. By using a net shot from cannon, large numbers of birds may be captured while they are incubating, resting, or feeding. Passerines (Passeriformes) are readily caught with mist netting: across their flight path, at dawn or dusk, a mesh net is placed into which these small birds fly and become entangled. Extraction from the net must be done with great care to avoid damage to the bird. Some birds can be attracted by bait which has been treated with a stupefying drug, such as α-chloralose. The results are sometimes unpredictable, as a few birds are greedy and take an overdose, and others take too little making them vulnerable to predation or drowning.

Rocket nets, snares, and many variations of walk-in traps are used for capturing mammals. Commercially made traps are available and easy to operate with small mammals, such as voles (*Microtus*), mink (*Lutreola*), and ferrets (*Mustela putorius furo*). Catching larger mammals presents many problems. Large herds may have to be controlled before individuals can be captured. In general, stockades are useful for the capture of herds, and offer the possibility of further control over the animals after they are caught, for instance, for segregating males from females or juveniles from adults.

For large solitary mammals which are difficult to handle, such as bears (Ursidae), drug immobilization may be the only satisfactory technique. A shotgun can be modified to fire a hollow dart containing a stupefying drug, a plunger in the dart forcing the drug into the animal upon impact. At this point many things can go wrong; the dart may not penetrate the skin or may hit a fatty layer; the dose may be insufficient; or the animal may escape into the bush where it cannot be located, and die from exposure or predation. To overcome this last problem a radio tagged dart has been developed which transmits a signal about 1 km. The drugged animal can then be followed and retrieved with minimal danger of loss. Once caught, a blindfold keeps most animals quiet; this is particularly true for nervous mammals and larger birds.

Animal identification. Accurate study of animal behaviour often requires recognition of individuals in the field or laboratory, especially for detailed long-term studies of SOCIAL ORGANIZATION, MATING, COURTSHIP, maintenance of TERRITORY, PARENTAL CARE, and other interactions among animals. Of course there are occasions when marking of the animal is not necessary, as with individually caged laboratory animals, or animals that have distinctive natural markings, such as unique coloration, facial features, or fur pattern. Bewick's swans (*Cygnus bewickii*), for example, have varied amounts of black and yellow on both sides of the bill. Careful charting of bill colour has revealed that no two individuals are alike. Armed with a good telescope, and under reasonable weather conditions, a careful observer can consistently identify individuals from a considerable distance.

Having decided whether marking is necessary, a suitable technique must be chosen to be compatible with the animal's life style. No marking technique is useful if it severely alters the animal's behaviour. Therefore, whatever method is chosen, reasonable attempts should be made to ascertain what effects it may have on the animal's natural behaviour. Of those cases where a mark is necessary, the question must be asked whether tagging is to be temporary or permanent. Commonly practised temporary marking includes fur and toe-nail clipping, dyeing, and painting. For short-term marking of birds during incubation an apparatus may be devised which automatically deposits a small amount of paint or dye onto the bird's back while it sits on the nest. This mark may last until the next moult. Long-term marking usually requires that an object such as a coloured leg band, neck band, or ear tag be attached to the animal. There are numerous methods of permanent tagging, each usually suitable for a different type of animal. For example, most large mammals and fish may be freeze branded, producing white permanent scar tissue. Wing tagging is a commonly practised form of long-range identification for field studies on birds, along with numbered leg rings. Animals possessing shells can have metal numbered discs cemented on a conspicuous place, or a mark engraved on the shell.

Toe clipping at the first joint is routinely practised on small field rodents, but less common is the use of radioactive isotopes. Tagging through radio transmission is also a method of allowing recognition of individuals, but it requires a higher level of sophistication than just simply looking at the animal and identifying it. This technique will be explained in detail below.

The use of a number of different methods of tagging of an individual usually ensures that when one tag fails, other methods continue to allow identification. A simple case involves the marking of foxes (*Vulpes*) to allow night observation of behaviour. Radio collars are fitted around the foxes' necks. Brightly coloured tape is wrapped around each collar and reflective tape is placed at strategic locations. To facilitate spotting the animal under dim light conditions, a low energy source of β radiation is placed on top of the collar. When the fox is released, radio tracking gives approximate location, while the reflective tape and β light help visual identification of the animal. After the radio ceases to operate and the reflective tape and β light have worn off, the coloured tape on the collar is sometimes the only identification left.

Personal observation. Direct observational techniques can be applied both in the laboratory and in the field, but first a location must be selected where observations are feasible. A hide or blind is useful for observing animal behaviour without disturbing the animals. Basically, a hide is a wooden or canvas container with viewing portholes. In the laboratory the hide may take the form of a room with a one-way viewing screen. Many hides are much more elaborate because of the conditions of the study. Some researchers live for many months in a hide, and facilities for cooking, washing, etc., must be provided. In some studies, an all-terrain vehicle is fitted out with beds, monitoring equipment for radio tracking, and recording data, while also having the advantage of mobility to follow free-ranging animals. This allows observations to be made even under difficult circumstances.

The study of animal behaviour (ETHOLOGY) began with straightforward observations; and the use of pencil and paper remains a convenient and simple technique for recording and analysing the behaviour observed. One way to facilitate this method is to produce pre-printed score sheets, on which the behaviour of animals, and other aspects of the animal's situation, are recorded in tabular form. The activities selected should be unambiguous and mutually exclusive in order to make them amenable to statistical treatment (see CLASSIFICATION). For those watching animals in the field, a good pair of binoculars is often essential. High power binoculars (e.g. $8 \times$ or $10 \times$) are excellent for longer-range viewing, such as studying bird behaviour. Binoculars with large ocular measurements (40–60 mm) gather light effectively, and so are the most useful type under low light conditions, such as dusk and dawn.

For night observations infra-red binoculars are sometimes necessary. These possess an infra-red light source which the animal cannot see, but when viewed through special binoculars equipped with phosphorescent screens, the infra-red light is transformed into visible green light. Another type of night vision telescope is the image intensifier, which uses a photo-multiplier tube and does not require an external light source. However, if a light source is used, it enhances the clarity and may increase the range of night vision.

During observation, events can be counted using hand counters, or, in the case of time-dependent data, stop watches may be used to delineate time units. For night studies a small oscillator may be used to produce audible sound 'bleeps' at given time intervals, thus marking off seconds or other time units while the number of activities occurring within that time unit is recorded.

Manually controlled recording techniques. When events occur at a rapid rate, the use of pen and paper may be precluded, and the observer needs some other recording method. Vocal commentaries describing what an animal (or animals) is doing can be recorded on magnetic tape. There are two types of tape recorder, reel-to-reel and cassette. High quality sound reproduction and constant tape speed can be achieved using reel-to-reel recorders. However, they are more cumbersome and prone to damage than cassette recorders, which are simple to operate and inexpensive. Cassette recorders do not usually have high fidelity sound production, but are adequate for dictation.

Where only one channel of information is required a monaural tape recorder is sufficient, but the increasing interest in animal physiology in conjunction with behaviour necessitates the use of a stereo tape recorder, allowing simultaneous acquisition of data on more than one channel. The potential problem of accurately synchronizing behaviour with physiological events is then diminished.

Graphical display of data provides the researcher with direct access to experimental results. Such is the case when an event chart recorder is used to encode temporal data. A simple switch closure causes the event recorder to produce a mark on the chart paper, thus recording the event. The paper moves at a constant speed, thus providing a time base for the data. Activity patterns, or physiological parameters, such as heartbeat rate and respiration, may also be recorded in this way. Chart recorders traditionally have been laboratory objects, but there are several commercial models now available which will operate for several weeks from small motor-car batteries. Chart recorders also have multichannel capabilities. Up to thirty events, each on an individual channel, may be simultaneously monitored on some laboratory recorders, and up to eight events on battery operated field units.

Frequently, one encounters situations where a permanent record is needed for documentation purposes. For example, a photograph of a particular species' habitat or nest location may reveal much more about behaviour than a simple written description, especially if during later analysis something of importance is discovered relating to that area. Much animal behaviour occurs at a speed which is too high for a human observer to record by the methods described above. This difficulty can sometimes be overcome by the use of photographic methods.

Two major types of film recording are practical: still and cinematography. Still photography primarily utilizes the 35 mm format. A single lens reflex (SLR) camera body with automatic metering of the light and interchangeable lenses is essential, because there is usually insufficient time for measuring light levels, etc. by hand. The types of lenses needed are mainly dictated by interests and the questions being asked. Longer focal length (telephoto) lenses are necessary for recording bird behaviour and the activity of elusive animals. A steady tripod is indispensable for these applications. Shorter focal length (wide angle) lenses may be helpful in laboratory documentation. Macro lenses enable a 1:1 reproduction ratio or even greater magnification, and find use in studies of smaller animals, such as insects.

In circumstances where there are several interacting animals, it may be advantageous to use cinematography. Most good cine cameras have capabilities that allow variable filming speeds, thus minimizing the chance of missing rapidly occurring behavioural events. For example the total interaction time of an aggressive encounter between two herring gulls (*Larus argentatus*) may be merely 1–5 s, which can only be analysed properly by high speed cine recording. Lens considerations are similar to still photography, but particular care should be taken that the camera meters light values directly, and automatically adjusts for exposure. The ability to record sound on more sophisticated units allows vocal and postural behaviour to be recorded simultaneously.

There are three basic film types: black and white, colour, and infra-red. Black and white film is perfectly adequate for most record photography and is easily developed. Colour slide film is chosen when individual recognition of animals and habitat types is needed. Photography of nocturnal animals requires special procedures. Most animals are insensitive to infra-red light, so infra-red film and infra-red flash can be used without disturbing them. Infra-red film requires special processing, and great care needs to be taken to keep the film cool until use (this also applies to colour film).

Time and energy budget studies of animals require minute details of all the behaviour occurring throughout a particular time period. To accomplish this a time-lapse recording technique is used (taking a picture every second, minute, hour, or some other predetermined time interval). Any motor driven SLR or cine camera with an appropriate interval timer may be used. The time interval must be short enough between pictures to record on film the behaviour with the shortest duration. For instance, in studies of SLEEP patterns of animals who spend most of the day sleeping, a time interval of 1 h between frames may be adequate. Conversely, if the behaviour occurs only rarely then more frequent exposures are needed. Care must be taken to ensure that the cycle of picture shooting does not correspond directly with a cyclic behaviour pattern, and that the animal remains in the view-finder.

Electronic technology has advanced to a state where miniature video television cameras and monitors capable of immediate playback of previously recorded information (instant replay) are available. These units can be operated in the field for several hours on portable rechargeable batteries. Video cameras have much of the capability of ordinary cine cameras, and also have a vast range of lenses available for laboratory and field applications. Portable infra-red video cameras allow filming of nocturnal behaviour similar to cine techniques, but with the added feature of instant replay.

Portable video systems do sometimes have problems of visual definition, because the information is being stored line by line. Resolution of detail is limited by the number of lines per centimetre, and in practice only gross postures can be visually resolved. Where fine detail is required, cine filming is the best choice. Graininess of film is the limiting factor in resolution of still and cine photography, but does not appear to be as critical as line resolution on video recording.

Remotely controlled recording techniques. The recording of animal behaviour on a round-the-clock basis necessitates use of automatic data-gathering systems (Fig. A). For example, in a study of herring gull incubation as much information as possible was required on what an incubating gull did with its TIME. These data had to be recorded in some manner. The change in weight of the incubating bird was measured by a remotely controlled spring balance with an electrical readout. This output was connected to an automatic sampling device called a data logger. Measurements obtained in this manner provided information on weight loss during fasting, and weight gain after foraging. The data logger would operate every 5 min to sample ten balances. Output voltage was converted directly to binary codes, and punched onto paper computer tape. Certain environmental factors influencing behaviour were also measured. This involved an automatic weather station capable of recording wind speed, direction, temperature, barometric pressure, etc. The data were formulated in a computer compatible form to facilitate straightforward data entry and subsequent computer analysis.

Computer controlled experiments are also possible provided there is access to a suitable interface between the experiment and the controller. The interface transforms the language of the computer into a suitable command format. For example, a command might be to drop a pellet into a hopper within a SKINNER box. This instruction would be transformed into a suitable voltage by the interface, which in turn controls a switching relay. When the relay is energized, food is dispensed. Two-way interaction between the experiment and computer may be obtained if the interface has

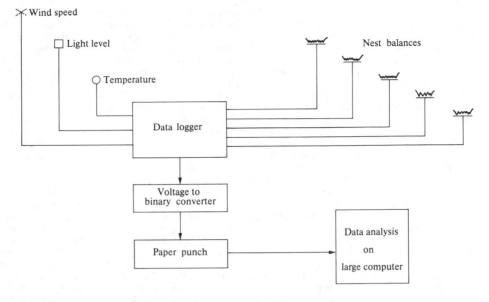

Fig. A. An automatic data-gathering system.

some type of FEEDBACK mechanism. Extending the example above, if the pellet were to fall onto a plate which was wired in such a way as to weigh food, then whenever the animal removed pellets, appropriate signals could be received and subsequent computer action could take place.

Proper design of the interface will determine the success of any project using computer control. Computers are standard off-the-shelf devices, as is much of the software (the language commands which the computer understands and obeys). The biologist is advised to seek advice from either the companies that sell equipment, or electronics engineers who design it. This saves untold hours of frustration and loss of data. Massive amounts of data can be collected in a relatively short time, as in the incubation study cited above.

Automatic sampling may be accomplished without the use of sophisticated computers. Simple and inexpensive integrated circuits which provide accurate clock timing may be employed to switch equipment on and off with variable duty cycles. Crystal clock circuits allow even more accurate timing, with precision to ± 10 s per month. These miniature circuits operate on a low voltage, and have numerous applications both in the field and laboratory. For example, in many field studies, scientists commonly need to record recurrent sounds. A systematic sampling technique may suit this experiment best; therefore, a timer is programmed to turn a tape recorder on, for example, for 5 min in every hour. It may be required that sampling be from a particular location, and this is accomplished through the use of a highly directional microphone. Omni-directional microphones

would be used if a general sample of sounds were to be made. Using this approach a long magnetic tape would last for several days, which is helpful if undisturbed behaviour patterns are being recorded.

A transducer is a device that transforms impulses, be they mechanical or electronic, into a suitable form for detection and recording. There are many forms of transducer which transform mechanical motion (for example, movement, pressure, speed, direction) into electrical energy. The potentiometer and thermistor are common transducers. Remote recording of the activity patterns of small birds in a laboratory environment may be accomplished by strategically placing perches wired to micro-switches. The number of hops on each perch may be monitored directly by a computer, and used as feedback for maintaining control over the reward rates to the animal.

Radar techniques. Information on the location of animals, on MIGRATION routes, and on some types of activity can be achieved through the use of radar. The objects viewed on a radar screen are essentially the result of high energy (40 kW to 10 MW) radio waves of very short wavelength (3–30 cm) which have scattered back after encountering, for instance, droplets of rain, birds, or insects. There are two types of radar: search, and narrow beam. Search radar is primarily for military and aviation purposes, as in air traffic control, and covers wide aerial sweeps. Narrow beam radars survey a small portion of the sky at any one time, but have the additional facility of tilting up and down which allows study of object altitude.

Relative to animal behaviour, radar range is

dependent on the ability to discriminate small objects such as insects, small birds, or bats (Chiroptera). These ranges vary from a few kilometres for low power, narrow beam radar to hundreds of kilometres for high power surveillance radar. This makes the detection of the migration routes of high-flying birds much simpler.

The viewing screen of a radar installation is called a plan position indicator (PPI). Dots or streaks of light on the PPI indicate that an object is within radar range. Much can be learned about bird and insect behaviour by personal observation of the screen and by time lapse photography of the PPI. Long streaks of light resulting from a time exposure indicate flight direction and speed of movement. Some types of radar are capable of picking up wing beat frequencies which are produced by the birds and insects modulating the radar radio waves.

Animal types can also be narrowed down to a few possibilities by noting the size of the light blip, flight speed, and wing beat frequency. FLOCKING behaviour and density can also be monitored by observing the spacing of the animals and cluster patterns. Some seabirds move in widely scattered flocks causing dot patterns to appear on the PPI. Passerine birds may move in dense flocks at night, resulting in a fine misty appearance on the radar screen. By counting the bright spots an estimate can be made of the number of animals within a flock. An individual can be identified if it carries a small metallic tag: the metal reflects radar waves better than the surrounding areas, resulting in a much brighter dot on the screen. Knowing the flight paths of birds is of interest to the student of animal behaviour, and of great importance to air traffic controllers, helping them to avoid bird strikes to aircraft which endanger human life.

Radio techniques. Telemetry is data transmission of information without direct connection between transmitter and receiver. Different modalities of transmission including light, ultrasonics, and radio have been used in the field, but radio telemetry has so far provided the most generally useful and adaptable approach to the study of behaviour of land mammals and birds. Ultrasonic telemetry is best suited for underwater use. Accomplishments in miniaturization (for example transistors) and hybrid circuitry (that is, integrated circuits containing resistors, capacitors, transistors, etc. in one small chip) have provided the means to resolve previously unanswerable questions through the use of radio telemetry. Radio signals from transmitters can be utilized for three general purposes: (i) the location of a subject by radio tracking; (ii) the transmission of details of both physiological and environmental variables; and (iii) the monitoring of activity patterns (using radio tracking and biotelemetry).

Radio tracking involves the attachment of small transmitters to animals. These devices emit pulses of radio waves on carrier frequencies between 30 and 500 MHz, which are detected by a sensitive radio receiver and directional antenna. By carefully choosing the pulse repetition rates and frequency of the carrier waves, individual recognition can be accomplished with no ambiguity. The whole concept of radio tracking is much more than merely putting location points on a large map and then determining home range. It facilitates contact with elusive animals, after which detailed observations of their behaviour may be made. In studying the behaviour of lions (*Panthera leo*) for example, a particular pride might be located once every few weeks after extensive searching. But if a single lion within the pride is radio tagged, the pride can be located every few hours, and can be followed and observed continually.

There are transmitters and harnesses suitable for most animals, be they fish, birds, or large terrestrial mammals. In addition to simple single-stage transmitters, which may weigh only 1 g and transmit over a distance of 0.5 km, there are two-stage transmitters which weigh approximately 6 g and transmit over a distance in excess of 80 km when the animal and receiver are in line of sight. In practice, most transmitters operate over a range of 1–5 km under field conditions. Power for the device is supplied by a battery, which in many cases is the limiting factor for both long-term performance and size of radio tags. Solar panels and rechargeable batteries have been incorporated into transmitters attached to diurnally active animals; under field conditions these last for several years. All components (transmitters, battery, and antenna) are encapsulated in a waterproof resinous material to which a harness is attached. Each species must have a custom-designed attachment, which must not impede or affect behaviour in any way. A common harness for mammals is a collar, for birds a backpack attachment made of loops, and for fish an implant or fin attachment.

Radio receivers are difficult to construct and can be purchased commercially. The receiving antennae may be either highly directional, or capable of reception throughout 360° (omni-directional). The omni-directional antenna is excellent for studies of nest or den attendance, where the whole system (receiver, omni-directional antenna, tape recorder, and clock switch circuit) is remotely operated in the field. With a single highly directional antenna, location can be estimated with an accuracy of 5–10°. With two antennae, radio triangulation can be accurate to ±0.5°. Measurements are affected by several parameters: (i) topographic reflection of the signal; (ii) system errors caused by antenna misalignment; and (iii) animal movement during triangulation. These can be minimized with practice, and through thorough care of equipment.

Activity monitoring can be achieved by simply changing the pulse repetition rate in the transmitter using a mercury switch. The sensitivity of this mode of measurement depends upon the ingenuity of the experimenter in placing the transducers on the animal. A distinction can be made between inactivity, sporadic activity, and continuous activity simply by listening to the pulse rate changes. This is easily recorded on magnetic tape, and transcribed into a computer compatible format, which allows an analysis of activity patterns.

Radio tracking by orbiting satellites has been carried out with larger animals. The transmitter in this case is usually bulky and large because of the high power requirement. A female polar bear (*Thalarctos maritimus*) was radio tagged in Alaska and tracked by the satellite NIMBUS 5 to the USSR. Satellite tracking is accurate to about ±1 km.

Physiological radio telemetry. Transmitters exist that are capable of sending any one (single-channel) or combination of (multi-channel) physiological variables, such as temperature, heartbeat, etc. The principal components of a simple biotelemetry system are shown in Fig. B.

As a general rule, the first single-channel physiological systems were formed by modulation of single or multi-stage radio tracking devices. These have the advantage of being smaller than multi-channel transmitters. However, when several *bits* of information are being retrieved and transmitted over a single frequency (multiplexing), or if hybrid thick and thin film electronic techniques are employed, multi-channel transmitters can be very small. There are two particular dangers in using these single transistor circuits for physiological measurements if they are not properly designed. First, frequency stability and pulse rate of the transmitter may be affected merely by changes in body posture and by temperature shifts. Secondly, as battery power decreases, the pulse repetition rate tends to drift, making the initial calibration useless. These are pitfalls for any biotelemetry transmitter, but pulsed systems are especially vulnerable.

For some physiological experiments it is desirable to obtain simultaneous recordings of multiple signals, especially when relationships between several behavioural and physiological variables are of interest. Each of the transmitted signals may be designated as a discrete channel of information. An obvious method of obtaining multiple channels of data would be to use several single-channel transmitters, but this usually necessitates redundant receivers and recording systems which require extra maintenance and may be extremely costly. When a number of signals are required, a multi-channel system (frequently called a multiplexed system) is usually the least costly and most efficient method to choose.

There are two major types of multiplexing: time-division multiplexing and frequency-division multiplexing. Time-division multiplexing requires sequential sampling of each individual transducer (electrode) input. For example, if five signals are needed, one might sample the first signal for 1 s, then the second signal for 1 s, etc. When all five signals have been sampled, a cycle is completed, after which the cycle may then repeat or turn off and repeat at a later time (thus conserving power). Frequency-division multiplexing is characterized by multiple frequencies, each related to a particular channel. Again by example, if five signals are needed, each channel may be separated by 5 kHz, that is the first channel is 5 kHz, the second is 10 kHz, and so on. Sub-carriers are used to modulate (vary or turn on and off) a main carrier frequency, which is usually the transmission frequency.

At the receiving end, special demodulators (Fig. B) are required to sort out the individual signals, and to dispatch them to recording devices. This process is usually accomplished by reversing the original encoding scheme used in the transmitter. In the case of time-division multiplexing, each part of the signal is sorted according to a particular time relationship, with a starting pulse signifying the beginning of each cycle. For frequency-division multiplexing, each channel is recognized by a special filter, which responds to its set frequency. That frequency will be equal to the corresponding sub-carrier frequency of a particular channel, in our example each filter would be respectively tuned at 5 kHz increments. Both of the techniques described above are utilized in physiological telemetry.

Despite the sophistication of electronic radio-monitoring devices, several difficulties remain. It may be difficult to record simultaneously several variables from an animal when the signal sources are far removed from the transmitter. It is par-

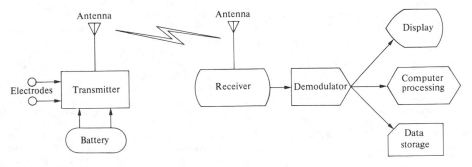

Fig. B. A simple biotelemetry system.

ticularly difficult to maintain the continuity of the electrode leads when they are long, or are attached to parts of the animal where much movement occurs. Attention should be given to the mechanical properties of wires which are subjected to bending and stress; for example, very fine stranded stainless steel or copper wire can be used. Teflon insulation is preferred because of its non-irritating and physiologically inert qualities. When wiring transducers, leads have to be as short as possible and avoid areas where excessive movement occurs (that is around the neck and limbs). Obviously this is not always possible, so each case needs separate consideration. For example, if the electrocardiogram (record of electrical activity of the heart) and respiration are to be measured, then the best location for a transmitter is near the heart. If the electroencephalogram (record of electrical activity inside the brain) were also to be recorded, leads would have to run up past the neck from the heart area, and in this situation it might be preferable to have two transmitters, one at the head and the other at the chest. Systems that do not use wire leads, such as radio pills used for measuring temperature, pressure, etc., are highly desirable, but of limited application.

In all studies in which the animal's freedom, or physiological integrity, is violated, care must be taken to avoid suffering or discomfort. This is important, not only in the interests of the WELFARE of the animal, but also in the interests of obtaining information about behaviour and physiology that is not distorted by unnatural conditions.   C.J.A.

3

TENDENCY is an everyday word widely used among students of ETHOLOGY (animal behaviour). The tendency for an animal to perform a particular activity can be loosely regarded as an index of the likelihood that the activity will be performed. This usage is adequate for casual reference, but it is not satisfactory from a theoretical point of view.

An animal may have a high tendency for FEEDING, but an even higher tendency for SEXUAL behaviour. Normally, the latter will take precedence and sexual behaviour will be observed. So long as the sexual tendency is higher than the feeding tendency, the animal will perform sexual behaviour and no feeding behaviour will be observed. In this type of situation, feeding will have a low likelihood of occurrence, even though the feeding tendency is high.

In its more academic sense the term tendency is used to express the level of underlying potential MOTIVATION to perform an activity. Thus, a high feeding tendency implies either a high degree of HUNGER, or a high APPETITE due to the presence of attractive food. The feeding tendency may be prevented from expressing itself as a result of the animal's DECISION to perform some other type of behaviour. But this does not alter the fact that the animal has a high feeding tendency.

In some cases CONFLICT occurs between tendencies of near equal level. In these, and in some other special circumstances, DISPLACEMENT ACTIVITIES, which have a relatively low tendency, are performed. Normally, however, the activity with the highest tendency is dominant over the other activities, in the sense that it determines the nature and patterning of the overt behaviour.

TERRITORY. Many animals defend patches of ground against other individuals, usually of the same species. This is most obvious when there is AGGRESSION against intruders, as in many species of birds. For example, pairs of tawny owls (*Strix aluco*) inhabit a fixed, exclusive area in a woodland for the whole of their adult lives, and they defend this against all other tawny owls by loud calling and chases. However, in many instances animals maintain exclusive areas by less overt behaviour. Even though they may seldom meet, many carnivorous mammals avoid each other by smelling the scent deposits of other individuals. Therefore the best indicator that an area is divided into territories is the finding that the occupying individuals or groups of animals are spaced out more than would be expected from a random occupation of suitable HABITATS. Territories range in size from the few millimetres that separate adjacent barnacles (Balanomorpha) on a rock to the distances of several kilometres that separate neighbouring herds of African buffalo (*Syncerus caffer*) on the Serengeti plains.

The costs and benefits of defending a territory. Territories have been described for a wide variety of animals, both vertebrates and invertebrates, and are used for many different activities. Therefore it does not make sense to search for a single SURVIVAL VALUE for territorial behaviour. Instead, we must ask the question, how does territoriality influence an animal's FITNESS? Territorial defence will have costs as well as benefits, and we would only expect an animal to defend a territory when it is economical to do so, i.e. when the benefits outweigh the costs.

COMPARATIVE STUDIES of different species, and observations of changes in behaviour within a species in response to changed ecological conditions, have shown that SOCIAL ORGANIZATION varies depending on the predictability of resources in time and space. This will influence whether or not the resource will be economically defendable. For example, spotted hyenas (*Crocuta crocuta*) live in clan territories in the Ngorongoro crater, where their food supply—game animals such as Thomson's gazelle (*Gazella thomsoni*), wildebeest (*Connochaetes*), and zebra (*Equus Hippotigris* sp.)—is predictable and abundant. In contrast, in the Serengeti, where food is very seasonal, the hyenas wander over wide ranges, and do not defend any fixed areas. Similarly, in winter, small birds such as pied wagtails (*Motacilla alba yarrellii*) defend permanent territories where food is predictable and abundant while, at the same time, other mem-

Fig. A. Pied wagtails vary their social behaviour in relation to the food supply. Some individuals defend permanent territories where food is predictably good, while others roam about in flocks exploiting transient patchy food supplies.

bers of the same species roam about in flocks and exploit transient, patchy food supplies (Fig. A). Examples will now be discussed where attempts have been made to assess the costs and benefits of territorial defence. Three types of benefit will be considered: food, mates, and predator avoidance.

Feeding territories. There is abundant indirect evidence that one benefit of territorial defence is food acquisition. Within many groups of animals, including lizards, birds, and primates, territory size increases with body weight, which is what we would expect since larger animals require more food in order to stay alive (see HOME RANGE). Diet also influences territory size. For example primates with more foliage in their diet (e.g. the black and white colobus, *Colobus geuereza*) have smaller territories than those that feed on fruit and flowers (e.g. the red colobus, *Colobus badius*), probably because the latter represents a more widely dispersed food resource. Within the same species, territory size is usually smaller where food is more dense. Ovenbirds (*Seiurus aurocapillus*) have smaller territories in areas where their insect food supply is more abundant, and limpets (*Patella*), which graze gardens of algae on rocks, defend smaller patches in areas where the algae are denser. In a LABORATORY STUDY short-term changes of a lizard's (Lacertidae) territory size were induced experimentally by manipulation of its food supply. Territory size decreased when food was added, and increased back to its original size when the food supply was removed again.

There are fewer precise quantitative studies of the costs and benefits of feeding territories. Hummingbirds (Trochilidae) and sunbirds (Nectariniidae) feed on nectar, and although the size of

their territories may vary by several hundredfold, each contains approximately the same number of flowers. It has been shown that this number is just sufficient to support an individual's daily energy requirements. Similarly in tree squirrels (*Tamiasciurus*), although the size of the territories varies between different areas, each contains approximately enough seeds to maintain a squirrel during the year. Presumably these territories contain just enough food to satisfy the owner's requirements and no more, because a larger territory would involve greater defence costs for little extra benefit.

Pied wagtails defend winter feeding territories along river banks, and in order to collect sufficient food during the short daylight hours at this time of the year they have to spend over 90% of their time searching for food. They feed by walking along the river edge and picking up insect food which is washed up onto the bank. As they walk along they deplete the food supply temporarily, so that if they visit the same stretch again soon afterwards they experience a low feeding rate. The longer they leave the stretch the more time there is for more food to wash up onto the bank, and thus the greater the wagtail's feeding rate. Each wagtail defends a territory of about the same length of river, and exploits the food systematically by walking up one bank to the territory boundary, crossing the river, and then walking back down the other side until it completes the circuit. It can be calculated that the length of territory defended is just sufficient to enable the owner to spend time walking round the territory so that when it gets back to the beginning again the food supply will have renewed to a profitable level.

The same factors probably influence the territorial behaviour of other animals that exploit a renewing food supply. Nectar feeding birds visit their flowers in sequence, so as to allow for nectar replenishment, and limpets graze their algal gardens systematically, so as to allow the algae to grow again to a profitable level in between successive feeding visits. The main benefit from maintaining exclusive use of the territory in these cases is that the owner can regulate its visits to the renewing food supply, so as to crop the food in the best way. The presence of intruders interferes with the estimate of efficient return times to patches in the territory. In addition, because intruders do not know the renewal pattern of the food, even if they do manage to land and feed on the territory the chances are that they will feed in an area that, unknown to them, was recently depleted by the owner and is therefore unprofitable. In the wagtails, even if an intruder manages to escape the owner's attention and land in the territory, it almost always experiences a low feeding rate, and after a short time it leaves of its own accord anyway. Feeding territories are therefore usually more profitable to owners than they are to intruders.

Many animals vary their defence strategies depending on the food supply. When food levels became very high, intruder pressure increases and both sunbirds and squirrels give up defence of their territories because the defence costs become uneconomical. At the other extreme, productivity may become so low that, even with territorial defence, the animal is unable to meet its daily energy requirements, and under these conditions the territory is abandoned. Therefore there is an upper and a lower threshold for economical defence.

Pied wagtails show changes in defence depending on the renewal rate of food in their territories. When the renewal rate of insects onto the river bank is very high then the owner allows other wagtails to trespass unmolested. Under these conditions the resource is being renewed so rapidly that the owner's feeding rate is unaffected, even if another wagtail has recently depleted the patch. Therefore it does not benefit the owner to waste time evicting intruders. When the renewal rate is very low, so that even long return times to patches do not give a profitable feeding rate, the owners abandon their territories and go off to feed elsewhere.

At intermediate levels of food abundance the owner may defend his territory alone or he may allow another bird (usually a juvenile) to feed on his territory. This 'satellite' bird imposes a cost on the owner because it depletes the food supply. However it also brings a benefit to the owner because it helps him to defend the territory. It can be shown that under the conditions that a satellite is allowed onto the territory, the owner is in fact benefiting from the association. His increased feeding rate arising from benefits gained by help with territory defence outweighs the costs incurred through having to share his food supply with an-

other bird. When food supplies decrease, so that the owner benefits by having the territory all to himself, he changes his behaviour and evicts the satellite wagtail.

Quantitative studies reveal that territory size and switches in defence strategies are exactly what we would expect from a cost–benefit analysis in terms of FEEDING efficiency. Animals that defend feeding territories appear to be very efficient at changing their behaviour so as to maximize their feeding rates.

Mating territories. Some animals only defend territories during the breeding season. In these cases it is usually the males who defend areas to which females come for mating. Sometimes the territories contain vital resources which the females require, such as food, nests, or good egg-laying sites. Thus the males gain access to females indirectly by controlling these resources. If the resources are patchily distributed then some males may be able to defend better areas than other males. Whenever there is a mosaic of male territories of different quality then it is those males who can command the best sites who achieve the greatest reproductive success. This situation may give rise to polygyny, when the differences among male territories are so great that females do better by mating with an already mated male on a good territory rather than with a bachelor male on a poor one (see MATING SYSTEMS).

In marshland or savannah-like habitats there is a wide range of productivity even between small adjacent areas of vegetation. In some bird species, males who can defend territories in the most productive areas attract several mates while other males in poor-quality habitats remain bachelors. In the red-winged blackbird (*Agelaius phoeniceus*) certain types of vegetation provide safer nest sites from predators, and hence result in greater fledging success. Males whose territories contain the most of this suitable vegetation attract the most females. In the lark bunting (*Calamospiza melanocorys*) the main factor influencing a male's reproductive success is the degree of shade in his territory. Territories with good shade provide better nest sites, because one of the major factors responsible for nestling mortality is exposure to strong sunlight. Males defending those territories with the best shade attract two mates, while those with poor territories fail to attract any. It is even possible to predict accurately the mating status of a male (polygynous, monogamous, or bachelor) simply from a knowledge of the available shade in his territory.

In the American bullfrog (*Rana catesbeiana*) some males achieve much greater mating success than others. These males are those that are able to defend the best egg-laying sites, namely areas which are not subject to extreme temperature variations which cause developmental abnormalities in the eggs. Females also prefer to lay their eggs in warm water because the eggs develop faster, and are therefore exposed for less time to PREDATION by leeches. As expected, males compete vigorously for these good territories, and it is the largest and strongest males who are able to command these sites, and thus achieve the greatest mating success. Similarly male dragonflies (*Plathemis lydia*) whose territories contain the best oviposition sites attract the most females.

Territory size may also affect mating success. In the three-spined stickleback (*Gasterosteus aculeatus*) males with larger territories are more successful at luring gravid females to their nests for egg laying. By preferring males on large territories females increase their reproductive success. This is because one of the major causes of egg loss is predation by other males who try to interfere with the nest; males with large territories are less susceptible to such interference. In hummingbirds, which are promiscuous, males who defend the best feeding territories attract the most females. In effect they are trading some of their food supply for the opportunity to copulate.

Not all mating territories are like this. In some situations males aggregate at traditional, communal display grounds, or LEKS, where they defend patches of ground and compete for DOMINANCE status. These territories do not contain any vital resources; they are often just tiny bare areas of ground, a few centimetres or, at the most, a few metres in diameter. Even so they are vigorously defended against other males. Lek behaviour has been reported in some birds, frogs, antelopes (Bovidae), bats (Chiroptera), and dragonflies, and may evolve in situations where neither the females themselves, nor the resources that they require, are economically defendable.

Females visit the leks in order to mate, and in all leks that have been studied almost all of the copulations are performed by a few individuals, even though all the males on the lek may be sexually responsive. The most successful males are those that occupy the central territories on the lek. For example, in a study of the sage grouse (*Centrocercus urophasianus*) over 80% of the copulations were achieved by three central males, while in the white-bearded manakin (*Manacus trinitatis*) one male alone performed over 70% of the copulations.

It is not clear whether the females are preferring to mate with particular males or whether they prefer to visit particular territories. In the leks of manakins, sage grouse, and the Uganda kob (*Adenota kob thomasi*), there is no obvious difference in the appearance of the DISPLAY of the successful and unsuccessful males. In the Uganda kob the occupancy of the territories may change, but the females still prefer the same positions on the lek for mating. Similarly, in manakins females continue to visit the central territories, even when the original males are experimentally removed and their positions are taken over by new individuals. It is possible that the females prefer the central territories, and so force the males to compete for these locations and mate with the winners of the competition. In effect they allow the inter-male rivalry to do the sorting for them. The problem is a difficult one however, and from studies of the black grouse (*Lyrurus tetrix*) and the ruff (*Philomachus pugnax*) it appears that male differences do influence female choice, with males who exhibit certain courtship tactics achieving the greatest reproductive success.

Not all of the males on a lek adopt conspicuous territorial behaviour. Some males employ silent, sneaky, mate-searching strategies. For example, in the green tree frog (*Hyla cinerea*) some males defend territories and call loudly to attract females. Other males sit quietly on the edge of a caller's territory and attempt to intercept and mate with females as they arrive. In the ruff, resident males defend territories on the lek while satellite males adopt the sneaky strategy of attempting to steal copulations while the residents are busy chasing off intruders. In this case there is a distinct plumage difference between the males who adopt the different strategies; the residents have dark head tufts and usually dark neck ruffs as well, while the satellite males are adorned with white tufts and ruffs (Fig. B). In the live-bearing fish

Fig. B. Lek display of the ruff. The only species of bird with marked individual variations in male plumage. Males with dark ruffs and tufts are territorial while the white one is a satellite male. The bird without the ruff is a female.

Poeciliopsis occidentalis some males are territorial and are black and aggressive, attracting females by elaborate courtship displays. There are also some non-territorial males which are a dull brown colour like the females, and, ignoring all the formalities of courtship, these males simply dash up to a female and attempt a quick copulation.

In theory there are two possible explanations for these differences in male behaviour at a lek. The first possibility is that some males, for example the ones owning the central territories, are fitter, and peripheral males are young and inexperienced or just poor-quality individuals. In other words, there is one best place to be to attain the greatest reproductive success, and only the strongest males are able to defend these territories. Other males have to make the best of a bad job and wait for vacancies to arise. This explanation probably applies to the black grouse leks, where it is young males who are non-territorial or who defend the peripheral territories.

The second possibility is that different behaviour results in the same reproductive success. The main idea here is that the best strategy for any one male to employ must depend on what all the other males are doing. For example, if all male frogs were callers then it would pay any individual to avoid the costs of calling, and instead put his efforts into intercepting all the arriving females for himself. On the other hand, it is obvious that not all of the males can be silent and employ this sneaky behaviour, because if they did then no females would be attracted. So we would not expect all males to be callers, nor all to be sneakers. In theory, there should be a stable equilibrium when the numbers of callers and the numbers of sneakers are such that the reproductive success of a calling frog is exactly equal to that of a sneaking frog. At this equilibrium all males have equal expectation of success, and it does not benefit any individual to change his strategy.

In the frogs it appears that the mating success of callers and non-callers is indeed about equal. However, in many cases it is difficult to measure the reproductive success of males who employ different strategies. Even if the territory owners in the centre of the lek enjoy a greater mating success in any one season this does not necessarily mean that over a whole lifetime they are more successful; for example the peripheral males may copulate less each season, but live for longer because they do not incur the costs of territorial defence.

Territories and predator avoidance. Another benefit of spacing out from neighbours is that under some circumstances predation can be decreased. In gull (Laridae) colonies and in great tits (*Parus major*) nesting in woodland, nests that are spaced further apart from nearest neighbours suffer less predation from crows (*Corvus*) and weasels (*Mustela*) respectively. The increased risk to closely spaced nests results from a successful predator intensifying its SEARCHING in the immediate vicinity of the nest it has found.

However in other cases living colonially may be a more effective means of decreasing predation. For example guillemots (*Uria aalge*) actively defend their egg sites against attacks by gulls. Eggs are better protected in dense colonies because closely packed adult guillemots form a more effective means of defence.

The examples discussed above concentrate on three main benefits arising from territorial behaviour. However, there may be other benefits involved; for example, spacing out may decrease susceptibilty to the transmission of disease. It must be emphasized that NATURAL SELECTION acts on the reproductive success of individuals relative to that of others in the population, so that benefits in terms of single factors such as food, mates, and predator avoidance will only influence selection through their effect on the overall fitness of the animal. The territorial behaviour adopted will be the best compromise between many conflicting selective pressures.

Interspecific territoriality. Animals usually defend their territories only against members of their own species. This makes sense because other species usually exploit different resources, so defence against them would involve costs for little resulting benefit. Whenever aggression is shown towards other species it can be demonstrated that this arises because of an overlap in resource requirements.

Some species of birds that exploit similar food and nest in similar places defend mutually exclusive territories, for example reed and sedge warblers (*Acrocephalus scirpaceus* and *A. schoenobaenus* respectively) in Europe (Fig. C) and the yellow-headed blackbird (*Xanthocephalus xanthocephalus*) and red-winged blackbird (*Agelaius phoeniceus*) in America. Some fish defend their territories against many other species; *Pomacentrus flavicauda* evicts no fewer than thirty-eight species from its territory, and observations show that at least thirty-five of these are food competitors. In contrast, all of the sixteen other species that are allowed to trespass unmolested exploit different food supplies.

Predator pressure may also favour interspecific territoriality, especially when two species nest in similar places so that a predator may be searching for both at the same time. In the great tit, not only does another great tit nest near by increase the chances of predation by weasels—the presence of a nearby blue tit (*Parus caeruleus*) nest probably has the same adverse effect.

Mechanisms of territorial defence. Territory owners use a wide variety of displays in the defence of their territories. For example, in songbirds there appears to be a three-tier system of defence. Song seems to act as a long-range signal that deters potential trespassers, visual displays are used at an intermediate range to repel actual trespassers, and finally, if the intruder persists, it is chased and attacked.

In some studies the keep-out signals have been

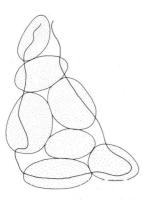

Fig. C. Interspecific territorialism in the reed and sedge warbler. Lines denote territory boundaries, lighter territories sedge warbler and darker territories reed warbler. Reed warblers arrive later and squeeze in between the sedge warbler territories.

identified experimentally. Male red-winged black-birds have a patch of bright red and yellow feathers on their upper wings which they display to intruders. If these feathers are painted black, the owners become much less successful at maintaining their territories. In another experiment territorial birds were prevented from singing by an operation which involved cutting the nerves which supply the vocal cords. These muted birds were also much less successful at keeping out intruders than were sham-operated individuals (i.e. those on which an operation was performed, but the nerves not cut).

A different experimental approach showed that song also acts as a keep-out signal in territorial defence in the great tit. Males were removed from their territories and replaced with loudspeakers that broadcast either great tit song or a control sound (a note on a tin whistle). New birds which came along to take over the vacancies settled quickly in the areas broadcasting the control sound, but took much longer to occupy the empty territories where great tit song was played. This experiment shows that the song alone can act as a signal to tell intruders that a territory is already occupied.

In mammals scent acts as a keep-out signal, and many species mark their territories with urine, faeces, or with scent produced by specialized glands (see SCENT MARKING). European badgers (*Meles meles*) live in groups and defend a communal territory, marking the boundaries of their range with droppings deposited in latrines. The home ranges of the different clans can be mapped accurately by feeding the badgers at each sett with small coloured pieces of plastic hidden in an artificial food supply. Different setts are given different colours and then the markers are recovered in the faeces deposited at the latrines round the territory

1 km

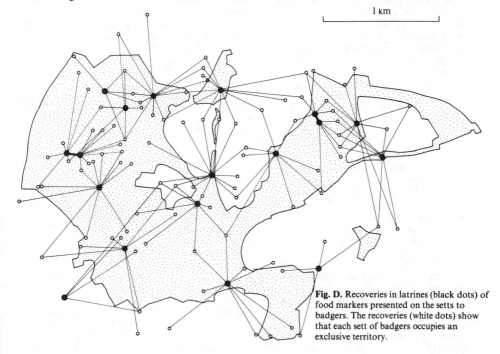

Fig. D. Recoveries in latrines (black dots) of food markers presented on the setts to badgers. The recoveries (white dots) show that each sett of badgers occupies an exclusive territory.

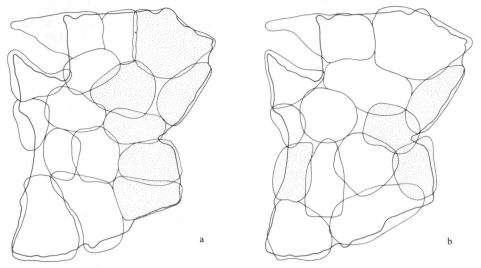

Fig. E. The replacement of removed birds. Six pairs of great tits were removed between 19 and 24 March 1969 (**a**, stippled areas). Within 3 days, four new pairs had taken up territories in the wood (**b**, stippled areas). There was some expansion of residents' territories during the removal, so that after replacement the territories again formed a complete mosaic over the wood.

boundary (Fig. D). Wolves (*Canis lupus*) mark the edges of their pack territories with urine, and in the red fox (*Vulpes vulpes*) experiments have shown that males can be inhibited from trespassing on to a territory by the smell of the owner's urine scent alone.

In ants there is a complex system of PHERO-MONES involved in territorial defence. Workers of the African weaver ant (*Oecophylla longinoda*) assemble nest mates to help in territorial defence against intruders by laying down a recruitment pheromone, and they use another pheromone as a keep-out signal to advertise their territories to other colonies.

Territories and populations. Many studies have shown that when territory owners are removed or die naturally their places are rapidly taken over by newcomers (Fig. E). This suggests that, prior to removal, potential settlers had been prevented from occupying the territories by the presence of the owners. In some cases it is known exactly where the replacements come from. In great tits they are from hedgerow territories outside the wood, which are less good territories in terms of reproductive success. In red grouse (*Lagopus lagopus scoticus*) replacements come from flocks of non-territorial birds which would not have bred, and would probably have died in the absence of a territory.

These observations show that territorial behaviour can limit population density. It is important to realize that this does not mean that a FUNCTION of territorial behaviour is to limit population density. Such reasoning would imply natural selection at the level of the population or group, and there are strong theoretical reasons for supposing that this does not normally occur. What

it does mean is that the limitation of population density can come about as a consequence of territorial behaviour that has evolved by means of individual selection. The tawny owl population in Wytham Woods near Oxford, England, has remained constant for many years even though breeding success has varied enormously over this time. This is because each pair of owls requires a certain area of woodland in which to hunt, and a consequence of this is that there is only enough room in the wood for a fixed number of territories. The best-quality individuals win territories, and the rest fail to breed and almost certainly die.

1, 20, 30   N.B.D.

THERMOREGULATION. For most animals there is an optimum body temperature, around which they function most efficiently. Below this temperature their metabolism progressively slows down, muscular activity diminishes, and the animal may become torpid. Above the optimum temperature metabolic rate rapidly increases, and this may be expensive to maintain. Moreover, there is an upper limit to the temperature at which bodily processes remain viable. For most species this limit seems to be in the region of 47 °C.

Most animals are able to influence their own body temperature to some extent, either by employing specialized physiological mechanisms, or by appropriate behaviour. In both cases it is necessary for the animal to be able to detect the environmental temperature, or its own body temperature, or both. This is done by means of various sensory processes, which are generally grouped together under the heading of *thermoreception*.

Thermoreception. Thermoreceptors probably

exist in most animals, but they have been studied in relatively few. Nerve endings sensitive to temperature are known to occur in a variety of insects. For example, in the cockroach *Periplaneta* there are thermoreceptors on the antennae which perceive air temperature, and thermoreceptors on the legs which perceive ground temperature. Fish have thermoreceptors in the skin, lateral line, and brain, and are very sensitive to temperature changes. The catfish *Ictalurus* has been shown to respond to changes in temperature of less than 0.1 °C. Many reptiles have a well-developed temperature sense, having thermoreceptors in the brain, as well as in the skin. The temperature sense is especially well developed in pit vipers (Crotalidae), which have special pits on the face (see SENSE ORGANS) that are sensitive to infra-red radiation, and are shaped so as to give the animal a directional temperature sense.

In birds there are thought to be few thermoreceptors in the skin, except on the tongue and bill of some species. In pigeons (Columbidae) it is known that there are thermoreceptors in the brain which influence behaviour and plumage adjustment, and others in the spinal cord which control shivering and panting.

In mammals, thermoreceptors occur in the skin, in deep organs, and in various parts of the central nervous system. Distinct heat and cold receptors are distributed in the skin; the heat receptors are usually deeper than the cold receptors. There are also receptors deep in the body, in veins for example, which can initiate shivering, even though the temperature at skin and brain receptors is kept constant. Thermoreceptors in the spinal cord influence shivering, panting, and blood flow, and these functions are repeated by thermoreceptors in the part of the BRAIN called the *hypothalamus*. In general, the most sophisticated forms of thermoregulation are found in mammals, and the brain receives information from many parts of the body. The integration of this information leads to appropriate activation of various mechanisms of warming and cooling.

Mechanisms of warming. The metabolic reactions of the body produce heat continuously, and the more active the animal the greater the rate of heat production. In a cold environment, however, the heat production which is merely a by-product of normal metabolism and activity may not be sufficient.

The responses made by animals to increase body temperature can be divided into two classes: (i) those which produce heat by increasing metabolic rate, and (ii) those which serve to prevent the loss of body heat. Many invertebrates are cold-blooded (*poikilothermic*) in the sense that their body temperature tends to conform with that of the environment. Because the rate of metabolic reactions is determined by the temperature at which they occur, such animals are forced to reduce their activity when the body temperature falls. However, some invertebrates, such as the common woodlouse (*Porcellio scaber*) and millepedes (Myria-

poda), are stimulated into extra activity by falling temperatures, and are thus able to maintain a body temperature which is higher than that of the environment.

In warm-blooded animals, the rate of heat production can be raised by increasing muscular activity, as in shivering, and by the direct effects of HORMONES on metabolic rate. Food intake can also serve to increase heat production, because heat is released during digestion. Many animals increase food intake in response to cold.

Loss of body heat can be reduced in a variety of ways. Birds and mammals can conserve heat by increasing the insulation at the body surface. Birds in particular can make considerable savings by ruffling their feathers, and some mammals can do the same by raising their hair. In mammals with little hair, such as man, the reduction of blood flow near the body surface serves the same purpose. In the long term, these rapid responses can be supplemented by the growth of thicker fur, and by the deposition of subcutaneous fat. In many animals such changes take place on a seasonal basis.

Many animals are able to reduce heat losses by behavioural means. The effective surface of the body can sometimes be reduced by curling up, or huddling with other members of the species. Such behaviour lowers the amount of heat lost by radiation from the body. An effective way to reduce heat loss is to move into less cold surroundings. Such behaviour may take the form of a simple taxis (see TAXES), or may involve a more complex type of HABITAT selection. For example, the bedbug *Cimex lectularius* is able to move towards a heat source by using the thermoreceptors on its antennae, which are moved from side to side and so test the way ahead.

Some animals are able to warm themselves by seeking a suitable microhabitat, not necessarily by responding directly to temperature differences, but to other clues to likely warm places. After a cold night lizards often seek sunlight, and by basking in the sun they raise their body temperature. The horned lizard (*Phrymosoma modestum*) not only does this, but also turns and tilts its body in such a way that the dorsal surface is as perpendicular as possible to the rays of the sun. The Namib desert lizard (*Aporosaura anchietae*) absorbs heat from the ground in the early morning. When the surface temperature approaches 30 °C the lizard emerges from the loose sand in which it has spent the night, and presses the ventral surface of its body against the sand. In order to achieve maximum contact the body is dished convexly, with the tail and limbs held in the air (see Fig. A). Body temperature increases rapidly under these conditions, and the lizard is soon able to move about actively.

The warming effect of sunlight can be further exploited by colour change. Absorption of heat in sunlight is greater for a dark coloured object than for a pale one, being maximal for black objects. A number of lizard species are able to change colour in accordance with their thermal requirements. Dark coloration prevails in the early morning,

Fig. A. The Namib desert lizard (*Aporosaura anchietae*) in the 'dished' warming posture (above), in which the body is pressed to the ground; and the dance-like cooling behaviour (below), in which the body is raised on the tail and one or two legs, thus minimizing contact with the hot substratum.

when body temperatures are low, and the lizard blanches when its temperature approaches the preferred level. For example, the desert iguana (*Dipsosaurus dorsalis*) reaches the half-way stage of colour change at 40 °C, and at higher temperatures becomes much paler than its surroundings. Some animals have specialized appendages which can be deployed to increase heat gain. Thus sunbathing turtles (Testudines) often extend their black feet, and the white coloured whistling swan (*Cygnus columbianus*) holds its black feet above its back, thus increasing the rate of heat gain.

Mechanisms of cooling. Overheating occurs if an animal finds itself in a particularly hot environment, or if its heat production is abnormally high, perhaps due to excessive muscular exercise, or if its heat dissipation is in some way impaired. Because the maximum body temperature that can be tolerated is close to the lethal temperature for many animals, cooling mechanisms need to be rapid and effective. This is generally the case, and animals have a wide variety of physiological and behavioural ways of dissipating heat.

Heat is lost from the body in four principal ways: (i) by *conduction* through the tissues from the interior to the surface of the body; (ii) by *convection* due to the action of the blood stream in circulating warm blood from the interior of the body to the cooler surface tissues; (iii) by *radiation* from the surface of the body to a cooler environment; (iv) by *evaporation* of water from the body surface, or from the lungs.

Conduction is involved primarily in heat flow within the body, but can also occur between the body and an external object, such as the ground. Conduction within the body depends upon temperature differences within the body, and upon the insulating properties of the tissues. The insulation of animals generally has two components: (i) a layer of tissue, fat and skin, which serves to maintain a temperature difference between the body core and the body surface, and (ii) a layer of relatively still air trapped within a coat of hair, feathers, or clothing. The insulating property of hair and feathers depends largely upon the amount of trapped air, but it is also influenced by wind and wetness. The main effect of air movement is to penetrate the coat and destroy part of its insulation by decreasing the amount of still air trapped there. Wet hair increases heat loss by conduction, and is particularly important if the body is in contact with the ground. A sitting deer (Cervidae), for example, compresses the hair on its legs and trunk, so that it holds less air. Heat is then conducted into the ground, particularly when the hair is wet.

Studies have been made of the heat loss by conduction of young pigs (*Sus scrofa domestica*) on different types of floor material. It was found that the rate of conduction is strongly affected by posture, and by the temperature difference between the body and the floor material. When the floor was cold the animals assumed a tense posture and supported their trunks off the floor. At higher temperatures the pigs relaxed and lay down with their trunks stretched on the floor. The heat loss by conduction from the body to the floor was greater when the floor was made of concrete, which has a high thermal conductivity, and much less when it was made of wood, which is a good insulator. These studies show that the place in which an animal chooses to rest, and the posture it adopts, are likely to affect heat loss from the body. Many animals take up a resting position before rain starts, thus ensuring that their contact with the ground is a dry one.

Convection, the transport of heat by movement of a warmed fluid, is important in two main respects: (i) in the transfer of heat from the body to the environment, and (ii) in the transport of heat within the body. In considering the transfer of heat by moving air around the body of an animal, it is usual to distinguish between forced and free convection. Forced convection is the transfer of heat through a boundary layer of a surface exposed to an air stream. The rate of heat loss depends upon the velocity of the flow, and upon the presence of obstacles to the free passage of air over the surface. Both the angle at which the wind strikes the body surface, and the angle in relation to the angle of hair growth are important. Thus if the wind blows 'against the grain' of an animal's hair much more heat will be lost by convection than if it strikes the animal from the other direction. Clearly, it is of thermal advantage for animals to face into the wind when resting, and many species are observed to do this.

Free convection depends upon the ascent of warm air from a heated surface. The movement of air associated with free convection from the head and limbs of an animal can be demonstrated by the technique of Schlieren photography, in which cool air appears darker than warm air. Such methods indicate that the rate at which heated air moves away from the body surface is greatly influenced by the shape of the surface and by the presence of bristles etc.

Convection within the body is primarily carried out by the movement of blood through the various parts of the body, and, to the extent that this can be controlled by the brain, it constitutes an important means of temperature regulation. For example, when the hand of an Eskimo is plunged into ice-cold water, there is an increase in the flow of blood into the hand, which serves to keep it warm and usable. The increase in flow is twice as great as that in a European. In Manchuria four groups of Mongol people were tested in this way, and a gradation of response was found, which corresponds to the climates of the regions inhabited by the people studied. The Orochoms, a nomadic tribe of northern Manchuria, who breed and hunt reindeer (*Rangifer tarandus*), had most adaptation; the Mongols and north Chinese came next; and the Japanese had the least response. This suggests that the ability to increase the flow of blood to the hand by a large amount may be a GENETIC adaptation to cold. This view is supported by the fact that the Lapp reindeer herders, a people of European origin, have a response that is no different from that of other Europeans, although they live in conditions similar to those of Eskimos. It seems that an individual cannot acquire the response in his lifetime, but that a period of evolution in a cold environment is necessary for the response to develop by NATURAL SELECTION.

Radiation is a process by which matter gives off energy into its surroundings. It is distinctly different from other modes of heat transfer in its great speed of propagation, which equals that of light, and because no material medium is required across the path of transmission. The energy is transmitted in the form of electromagnetic waves, and energy transfer therefore obeys the ordinary laws of optics. Radiated energy can thus be reflected, absorbed, transmitted, and refracted by suitable materials.

The heat loss by radiation is proportional to the area of the radiating surface, and increases as the fourth power of the temperature of the surface. However, when the temperature difference between the surface and the environment is small, the heat loss by radiation is roughly proportional to the temperature difference between the animal and its environment. However, it is the temperature of the surface which is important, and this can be affected by a number of factors. Firstly, air-flow over the body surface may, as we have seen, cool the surface by convection, so that the greater the heat loss by convection, the less the loss by radiation. The extent of this convective cooling depends upon the speed of the air-flow. Secondly, the temperature of an animal's surface rises when the animal is exposed to higher levels of environmental radiation, in the form of direct or indirect sunlight. The extent to which the surface of an animal absorbs radiant heat depends to a large extent upon its texture and colour. These together will determine the amount of radiation reflected from the surface, only the absorbed radiation affecting the temperature of the surface. As we have seen, black surfaces usually absorb more radiant heat than pale ones, and such heat gain may serve to raise the animal's body temperature. The increase in surface temperature, however, will mean that some of the heat gained will be dissipated in the form of radiant heat loss. The extent to which this occurs will depend upon the efficiency of the animal in removing heat from its surface to its interior by internal convection.

The coloration of animals probably plays no significant role in the loss of heat, only in heat gain. Thus white animals may gain less heat than black ones, due to the greater reflection of radiation, but the factors affecting heat loss remain the same. However, animals can affect their radiant heat losses in other ways, which are primarily behavioural. To escape heat many mammals seek shade, are active nocturnally in the desert, or burrow underground. Most small mammals inhabiting deserts are nocturnal, and are able to regulate their temperatures by means of well-defined activity RHYTHMS. Other animals, such as ground squirrels (*Citellus*), make sorties from a burrow in direct response to the prevailing environmental temperature. Ground squirrels are diurnal, and tend to avoid both low and high temperatures. However, it should be remembered that sorties between a warm and a cool environment may serve the purpose of heat gain, as well as heat loss. For example, certain fiddler crabs (*Uca*) living on mud flats in tropical mangrove swamps make sorties from cool burrows into the hot sunlight, where they take on a dark coloration. Upon retreating back into the burrow, the crabs blanch; then follows an interval of cooling, before they darken and return to the surface to repeat the cycle. The fact that the crabs are dark in sunlight and pale in the shade means that heat gain is maximized by exposure to sunlight.

Large animals in hot climates are often not able to seek the shade, but are able to make postural adjustments so as to minimize the area of the body that is exposed to solar radiation. Interestingly, the camel (Camelidae) seems to be the only large animal that exploits this possibility. Camels reduce radiant heat gain by continuously orienting their narrow end to the sun, but this does not seem to be a characteristic behaviour of horses (Equidae), cattle (Bovidae), or sheep (*Ovis*).

Evaporation occurs from the surface of most animals, but varies considerably with the type of body covering. The common earthworm (*Lumbricus terrestris*), for example, cannot survive for long in a dry atmosphere because of the rapid desiccation of its body surface, whereas for insect species possessing a hard wax-covered surface evaporation is a relatively insignificant factor. In general, water loss due to evaporation is greater the higher the environmental temperature, and the greater the velocity of the air-flow over the surface; but the most important factor is the environmental humidity, increases in which greatly reduce evaporative water loss. In addition to water loss from the body surface, many animals lose water and heat in respiration, where the same physical considerations apply. This form of evaporative water loss is particularly important in reptiles and birds.

Heat loss due to evaporation is very important in thermoregulation, and can have a profound effect upon an animal's way of life. For example, the common woodlouse is normally nocturnal, but it is sometimes compelled to come out into the open during the day, when a combination of high temperature and high humidity causes the body temperature to reach a dangerous level. In a saturated environment woodlice die at about 31.5 °C, a temperature which is easily reached on a sunny day under stones, etc. Although the air temperature may be greater in the open than in the animal's normal hiding place, the humidity will generally be less, and the woodlice are able to reduce their body temperature as a result of rapid evaporation from the body surface. Woodlice are not capable of physiological control of their evaporative water loss, and soon have to seek their humid hiding places again. However, some arthropods are thought to have rudimentary control in this respect. For example, the South African darkling beetle (*Onymacris bicolor*) is able to ventilate its subelytral cavities (the cavity under the hard and case-like anterior wing), and thus temporarily lower its body temperature. These diurnal beetles are thereby able to tolerate high environmental temperature for long enough to enable them to run over exposed sand from one place of shelter to another.

Reptiles and amphibians dissipate heat by evaporation from the body surface, but such evaporative cooling is obligatory, in the sense that it occurs at all body temperatures and is not under physiological control. Both cutaneous and respiratory water loss increases with temperature, but the water lost from the respiratory tract increases proportionately more than that from the skin. This is due to the increase in breathing rate at higher temperatures. Because animals in hot dry environments have to conserve water, evaporative cooling is an expensive cooling mechanism, to be used only in emergencies. In such situations, many reptiles attempt to increase their respiratory evaporation. Crocodiles (Crocodylidae) gape widely when hot, as do snakes (Serpentes) and some lizards (Lacertidae). Some sun-basking lizards, such as the desert iguana, pant somewhat like a dog. They open their mouths widely, extrude the tongue, and vibrate the lower jaw.

In birds, loss of water through the skin plays a significant role in cooling, but it is uncontrolled. Respiratory evaporation is of much greater importance, and in most birds it can be controlled to a considerable extent. Many species are able to alter their breathing pattern to facilitate respiratory evaporation, and this appears to be under the control of the parts of the BRAIN concerned with thermoregulation. BREATHING rate and depth generally increase steadily as environmental temperature rises, and vigorous panting generally occurs at about 41 °C, when environmental temperature approaches body temperature.

During FLIGHT, heat production is greatly increased, and there is a consequent requirement for heat dissipation. Some of this is achieved by convection resulting from the increased air-flow over the body surface, but evaporative heat loss must bear the main burden. Experiments with pigeons have involved fitting them with face masks containing instruments capable of measuring air-flow during breathing, and transmitting this information by radio to the scientist on the ground. It was discovered that breathing rate was very much greater than that necessary to support the muscular exercise involved in flight. This increased breathing is required to provide sufficient evaporative cooling during flight. This method of cooling involves the animal in considerable water losses, and on the long flights of MIGRATION, regular stops for drinking are obligatory.

Mammals have some degree of control over evaporative water loss, both from the skin and from the respiratory tract. The moisture evaporated from the skin is generally provided by special sweat glands present in all higher mammals, such as rats and mice (Murinae) and members of the Lagomorpha (rabbits, hares, and pikas). Different species sweat to varying extents; thus horses sweat profusely and cattle relatively little. In man, the sweat glands of the palms are emotionally controlled, while those on the rest of the body are normally thermally controlled. Sweating is controlled by thermoreceptors in the brain, and not by those on the skin. Sweating is not the only way of moistening the body surface, and some mammals, such as opossums (Didelphidae) and some rodents, salivate profusely and spread the saliva over their fur by licking. Other animals may plunge into a pool of water to wet themselves, or spray themselves with water, as do elephants (Elephantidae).

Respiratory evaporation in mammals varies in efficiency from species to species, being generally most developed in animals which sweat little. Increasing the rate of normal breathing in order to increase evaporative cooling usually interferes with the respiratory physiology. This can be avoided by specialized panting behaviour. In panting dogs (*Canis l. familiaris*), for example, most of the air enters the nose and leaves by the mouth, thus giving a unidirectional flow which does not greatly increase air-flow through the lungs. In addition, dogs extend their tongues when panting, thus increasing the evaporative surface.

Evaporative cooling is a costly form of thermoregulation in that considerable amounts of water are lost, and have to be replenished. In the case of panting, extra heat is produced by the muscular exercise involved. Desert animals, which cannot afford to lose water, tend to minimize evaporative cooling and employ other means of dissipating or avoiding heat. The dromedary (*Camelus dromedarius*) does not pant at all. It is able to store heat in the day-time by allowing its body temperature to rise, and dissipate it by radiation in the cold desert night. This enables it to economize on cooling by water evaporation during the day to a degree not possible in other species. The widely held belief that camels store water to a greater extent than other species is fallacious.

Degrees of regulation. The mechanisms of HOMEOSTASIS ensure the maintenance of a more or less constant internal environment, despite changes in the external environment. Some animals, called conformers, are not able to withstand external influences, and their bodily condition tends to conform with that of the environment. For example, the salinity of the body fluids of many marine invertebrates is identical to that of sea water. Other animals, known as regulators, maintain their bodily functions in a condition which tends to be independent of environmental fluctuations. Many fish, for instance, maintain a body fluid composition which is more dilute than sea water. There are many intermediates between the two extremes, and one species may be able to regulate one bodily function but not others.

The body temperature of fish follows the temperature of the environment much more closely than does that of many land animals. There are several reasons for this. Firstly, by virtue of living in water, fish cannot lose water from the body by evaporation, and therefore they can never achieve a body temperature that is lower than that of their surroundings. Secondly, the heat generated by muscular exercise and metabolism is rapidly removed from the body by conduction, which could be avoided only by very efficient insulation. Thirdly, heat is rapidly removed from the body surface by convection, due to the flow of water over the surface. This conformity with the environmental temperature means that fish are vulnerable to fluctuations of temperature which they are unable to avoid. There are various mechanisms by which they cope with this situation.

An important factor is that fish have evolved tissues that can function over a wide range of temperatures. That is, they have a wide thermal TOLERANCE. Although fish are tolerant of a wide temperature range, this does not mean that they are equally efficient over the range. Generally, the tissues of a given species function most efficiently

in a more restricted range of temperature, and it is within this that the fish are normally found in nature. The distribution of marine fish is known to be markedly influenced by temperature, and this suggests that they have delicate powers of temperature discrimination. Laboratory experiments, in which fish are trained to respond to temperature changes by food-seeking behaviour, show that this is so. Cod (*Gadus callarias*), for example, can detect a rise in temperature of as little as 0.05 °C. It has also been shown that many fish species have well-developed temperature preferences, and are capable of choosing water of the preferred temperature.

Although it appears that each species has a preferred temperature range over which it functions most efficiently, it is also a matter of common observation that arctic species are as active as related species living in warmer climates. What has happened is that, in many species, there has been evolutionary ADAPTATION, which ensures the maintenance of optimal activity in the particular thermal conditions that the population actually experiences in its normal life. This type of adaptation is also common in marine invertebrates. For example, the common jellyfish (*Aurelia aurita*) swims by pulsations of its bell. Individuals taken from Nova Scotian waters, where the average temperature of the surface of the sea is about 14 °C, pulsate actively over the range 2–18 °C, and cease pulsating at a lower limit of − 1 °C and an upper limit of 29 °C. In Florida waters, where the average temperature is 29 °C, members of the same species have a lower limit of 12 °C, an upper limit of 36 °C, and an optimum at about 30 °C.

In addition to genetic adaptation, many aquatic animals are able to adjust to temperature changes by ACCLIMATIZATION. For example, the brook trout (*Salvelinus fontinalis*) is a non-migratory form of salmon that lives in lakes and streams, where temperature fluctuations are greater than in the sea. This fish has considerable powers of acclimatization, enabling it to adjust to the cold conditions of winter. In other words, an individual is able to survive in winter at a temperature that would be lethal in summer. By contrast, the salmon (*Oncorhynchus keta*) is a migratory salmon with a narrow tolerance range, and poor powers of acclimatization. It is able to adjust to temperature changes behaviourally, by migration to more favourable climates.

Terrestrial animals have a much wider range of conditions to contend with than do aquatic animals. The coldest antarctic waters have a temperature no lower than − 1.5 °C, whereas the temperature on land can be as low as − 50 °C. The warmest seas are about 30 °C, while land temperatures often reach 66 °C. Within any one area the range of sea temperature is about 10° C during the year, and about 1 °C during 24 h. The high thermal capacity of water means that the proportion of water in the environment is the main factor influencing the thermal CLIMATE of an area.

Deserts cool rapidly at night, and heat up quickly when the sun rises, whereas conditions in the hot and damp tropics can be remarkably constant, with temperature fluctuations of as little as 1 °C per day.

Animals which allow their body temperature to conform with that of the environment have a difficult time on land. The necessity to avoid temperature extremes often severely restricts the range of behaviour of these animals. For example, although many lizards are able to maintain their body temperature within a narrow preferred range when they are active, this is at considerable behavioural cost. Thus the Namib desert lizard can be active on the surface of the dunes only between 27 °C and 40 °C, and this may restrict its activity time to one 2-h period in the morning, and another in the afternoon. This is not very long for the animal to forage for its daily food and water, although it is sometimes possible for the lizard to extend its activity time by means of special warming and cooling behaviour, as illustrated in Fig. A.

Although reptiles are able to regulate their body temperatures to some extent, true thermal homeostasis occurs only in birds and mammals. These are able to maintain a constant body temperature despite fluctuations in the environmental temperature. They are able to do this because of their high metabolic rate, which provides them with an internal source of heat. This, together with a high degree of insulation, makes it possible for birds and mammals to maintain a body temperature which is generally higher than that of the environment. Provided the brain receives information about the temperature of the body, and is able to exercise control over the mechanisms of warming and cooling, then it is easy to maintain a constant body temperature, so long as the environmental temperature does not approach too closely that of the body. When the brain temperature becomes greater than that of a particular *set point*, the cooling mechanisms are activated, and, if it drops below the set point, the warming mechanisms are brought into play. The principle is the same as that employed in a thermostatically controlled electric heater. Fine control of body temperature reaches a peak in man, who is able to make very sensitive adjustments, largely because of an 'early-warning' system consisting of numerous thermoreceptors in the skin. Other mammals are particularly good at maintaining a constant body temperature in the face of extreme environmental conditions. For example, an Eskimo dog can sleep in the open snow at temperatures below −40 °C, due to its high degree of insulation and its ability to increase its resting metabolic rate by 30–40%.

However, the ability to maintain a constant temperature does not mean that an animal will always do so. In man, there is a diurnal variation ranging from 36.7 °C in the early morning to 37.5 °C in the late afternoon. Other mammals show similar fluctuations, and this phenomenon is particularly marked in birds, whose body temperature may fall several degrees at night. Many small mammals and birds are able to economize on energy expenditure by employing a daily cycle of torpor. Thus the body temperature of bats (Chiroptera) and hummingbirds (Trochilidae) falls considerably when they have no need to be active. Other mammals economize on an annual basis by allowing the body temperature to drop to near environmental levels during periods of HIBERNATION. As we have seen, body temperature fluctuations are employed by the camel as a means of water conservation, and many other instances of 'deliberate' manipulation of body temperature are known. The elevation of body temperature during fever in human beings is thought to be one of these.

In general, the mechanisms of thermoregulation found within the animal kingdom vary considerably from species to species. This variation is due, in part, to the requirements of different ways of life, and in part to overall evolutionary advances. Primitive animals tend to be conformers, with some degree of behavioural thermoregulation, and some ability to adapt physiologically to local thermal conditions. More advanced animals are better able to control their own body temperatures, as can be seen by comparing true temperature-regulating mammals with more primitive mammals. The Madagascar hedgehog *Centetes ecaudatus*, one of the most primitive living mammals, can maintain its activity over a body temperature range of 24–34 °C. Similarly the three-toed sloth (*Bradypus*) shows a body temperature range of 27.7–36.8 °C at air temperatures ranging from 24.5–32.4 °C. It loses heat rapidly in cold air, and is unable to survive outside its moist and thermally equable environment. The ability to maintain a constant internal temperature in the face of environmental variation, which is characteristic of higher mammals and birds, enables animals to exploit a wide range of habitats.

143

THIRST is a state of MOTIVATION which arises primarily as a result of dehydration of the body tissues. All animals require water to maintain their metabolic processes. All animals lose water by a variety of routes, including excretion, THERMOREGULATION, and evaporation from the body surface. Lost water has to be replenished by drinking, by eating foods containing water, or, as occurs in some lizards (Lacertidae), by absorption through the skin (see DRINKING).

Many animals have well-developed mechanisms of water conservation which cut down the rate of water loss when drinking is not possible. Water loss through excretion can be reduced by reabsorption of water in the kidney, so that a more concentrated urine is produced. Water is also reabsorbed in the small intestine, so that less is lost in the faeces. Many animals eat less when thirsty, and this also helps to cut down on water loss by excretion. The waste products of digestion and food metabolism have to be excreted and some water loss is inevitable. LABORATORY STUDIES have shown that water losses in pigeons (*Columba livia*) deprived of food are only about one quarter of the normal level.

Water loss through thermoregulation can sometimes be reduced through behaviour, such as seeking a cool HABITAT or cutting down on the production of heat by eating less food or taking less exercise. Many birds lose water and heat from the lungs and air sacs, and this can be cut down by alterations in the pattern of BREATHING. When a camel (Camelidae) is short of water it stores heat in the fatty tissue of its hump in the daytime, allowing its body temperature to rise. This heat is dissipated during the cold desert night by radiation, without the loss of water that accompanies the more normal methods of thermoregulation. Camels, contrary to popular belief, do not store water, but storage of water in the intestine is known to occur in the Namib desert lizard (*Aporosaura anchietae*).

Thirst affects behaviour in two main ways: by increasing the tendency to seek water, and by altering feeding and thermoregulatory behaviour as described above. The increased tendency to drink can arise as a result of primary or of secondary thirst. Primary thirst results from dehydration of the tissues of the body, and this is monitored by the BRAIN, through changes in the salt concentration (osmotic pressure) or volume of the blood, and through the action of HORMONES, such as *angiotensin*. Secondary thirst is purely psychological, and arises as a result of events which are likely to induce dehydration in the future. For example, many birds and mammals drink at the same time as they eat, provided water is available. This prandial drinking does not arise as a result of dehydration, although food intake does in the long run cause dehydration. It arises purely as a result of the act of eating which induces a temporary secondary thirst. This arrangement has the advantage that the water taken with the meal forestalls any dehydration that might arise as a result of the food intake. Thus, the drinking is in anticipation of future primary thirst. Similarly, many birds and mammals drink in response to rises in environmental temperature (see MOTIVATION, Fig. G). This drinking forestalls the dehydration that will result from the future thermoregulation necessitated by the rise in temperature. Here the animal is drinking in order to make water available for thermoregulation.

As in other aspects of HOMEOSTASIS, thirst is sometimes a victim of compromise in the interests of the total WELFARE of the body. In hot weather it may be necessary for an animal to induce considerable dehydration in the interests of thermoregulation. Similarly, hungry animals may be obliged to run up a water debt when water is not available. Animals living in deserts are frequently faced with such problems, and their TOLERANCE of dehydration is greater than that of most animals.

THREAT behaviour is a form of COMMUNICATION that usually occurs in situations involving mild AGGRESSION, or CONFLICT between aggression and FEAR. For example, an animal may maintain its social DOMINANCE by threatening, almost as a matter of routine, when it pauses near a subordinate. A male three-spined stickleback (*Gasterosteus aculeatus*) tends to be very aggressive near to his nest, but increasingly fearful as he nears the boundary of his TERRITORY. In disputes with neighbours the fish typically adopts a head-down threat posture (Fig. A) when the tendencies to attack and escape are in equilibrium.

Fig. A. The head-down threat posture of the three-spined stickleback (*Gasterosteus aculeatus*).

Threat sometimes takes the form of INTENTION MOVEMENTS of attack. The opening of the mouth that precedes biting has evolved into a ritualized baring of the teeth that is characteristic of threat in many mammals (see FACIAL EXPRESSIONS, Fig. A). The RITUALIZATION of attack and conflict behaviour is a common form of threat. Thus the threat DISPLAYS of the American green heron (*Butorides virescens*) seem to be derived from intention attack movements (see RITUALIZATION, Fig. C).

Threat displays often involve ritualization of morphological features, such as the enlarged claws that are waved by fiddler crabs (*Uca*) during aggressive encounters (DISPLAY, Fig. A). Many snakes (Serpentes) vibrate the end of their tail during threat, and in rattlesnakes (*Crotalus*) the end of the tail is modified into a special rattling device. Similarly, the tail spines of porcupines (Hystricidae) have become modified into sound-producing organs that are used during threat. Hissing threat sounds occur in many vertebrates and can be regarded as a ritualized form of BREATHING.

The main FUNCTION of threat is to keep rivals at a distance without undue expenditure of energy or risk of injury. Many features of the behaviour of animals and of their morphology subserve this function. COLORATION, for example, may help to repel other members of the same species. Normally coloured zebra finches (*Taenopygia guttata*) maintain a characteristic INDIVIDUAL DISTANCE, but all-white zebra finches sit much closer together. The territorial SONG of birds and other animals is a means of warding off rivals. SCENT MARKING often serves a similar function. Many threat displays involve morphological features which appear to make the animal larger than it really is (CONFLICT, Fig. D). A display of weapons is also a common form of threat. For example, male walruses (*Odobenus rosmarus*) display their tusks and sticklebacks raise their spines (Fig. A).

Threat postures are often used to repel members of other species. These may be similar to those used against conspecifics, or they may take the form of *deimatic* or intimidating displays, evolved as a means of DEFENSIVE behaviour. For example, the European toad (*Bufo bufo*) inflates its lungs and appears to be much bigger than it really is. The eyed hawkmoth (*Smerinthus ocellatus*) exposes a pair of eye-like markings when it is disturbed by a predator.

75, 99, 137

TIME. Unlike human beings, animals are not able to measure time by means of artificial clocks. However, many animals do have a good sense of time, which is dependent upon internal physiological processes. For example, honey-bees (*Apis mellifera*) visit certain flowers at particular times of day, and this coincides with the secretion of nectar, which also occurs at particular times of day. Bees can be trained to search for their food at a certain time of day by offering them food for several days at exactly the same time. They will continue to look for food at this time, even when it is no longer available. That the time-sense of bees is based upon an internal CLOCK has been demonstrated by training bees to visit a conspicuous object at a certain time of day in the European time zone, and then transporting them quickly to America, where their behaviour was examined. It was found that bees visited the same conspicuous object at the correct time by the European clock.

The time-sense of human beings is less well developed than that of most animals. Curiously, time-sense seems to be better when people are asleep or hypnotized. It has been found that they are able to wake up at a pre-set time from normal sleep, and also from hypnosis, with remarkable accuracy. However, people are rather poor at estimating time when awake. In order to obtain more accurate measurements of time, the ancients devised various artificial means. These included slow-burning candles, water clocks, sundials, and simple pendulum clocks. To make these clocks accurate, it was necessary to rely on the observations of ancient astronomers, who were able to estimate the passage of time from the motions of the sun, moon, and stars.

The astronomers of antiquity studied the motions of the sun, moon, and stars from a purely terrestrial viewpoint. They had no knowledge of the rotation of the earth on its axis, or of its revolution around the sun. To them the earth seemed like a stationary platform in the centre of the universe. The viewpoint of these ancient astronomers is an important one for us to consider, because it is the same as the animal viewpoint. Although animals may be furnished with an internal clock, this cannot keep good time unless it is synchronized with external events. Just as the ancient clockmakers found it necessary to calibrate their clocks by reference to the motion of heavenly bodies, so the physiological clocks of animals are set by external factors. If maintained in a laboratory environment with constant temperature and light, the internal RHYTHMS of animals drift away from their normal 24-h periodicity. External factors, such as changes in temperature and the light–dark cycle, serve to keep these internal clocks on time. In considering the role of time in the animals' world, it is important for us to understand how these external factors vary throughout the year.

The apparent motion of the sun. The day–night alternation that we experience in everyday life is due to the rotation of the earth upon its axis. However, from a terrestrial viewpoint the sun appears to move through the sky. It rises on the eastern horizon and sets in a westerly direction. For observers in the northern hemisphere the sun moves through the southern sky, along the arc of a circle, reaching its zenith at noon. The arc appears to be lower in the sky in winter than it does in summer, as illustrated in Fig. A. The sun appears to move at a uniform speed along its daily path from horizon to horizon. The length of the arc, and therefore the time from sunrise to sunset, is

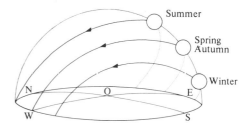

Fig. A. Paths of the sun across the sky in each of the seasons, as seen by an observer standing on the surface of the earth at O, in the northern hemisphere at about middle latitude.

less in winter than it is in summer.

The changes in the apparent motion of the sun, that are characteristic of the seasons, arise from the fact that the earth's equator is not aligned with the plane of the *ecliptic*. The ecliptic is the path of the sun in relation to the earth, and it can be seen as outlining a disc, or plane, cutting through the earth, as illustrated in Fig. B. The axis of rotation of the earth, called the *polar axis*, is tilted with respect to the plane of the ecliptic, so that the plane of the equator makes an angle of 23.5° to the plane of the ecliptic, as illustrated in Fig. B.

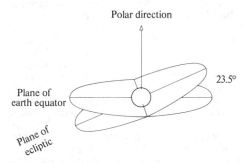

Fig. B. The plane of the earth's equator is at an angle of 23.5° to the plane of the ecliptic, which is the plane of the path of the sun in relation to the earth.

For an observer in the northern hemisphere, the sun appears to be higher in the sky in the summer than in the winter, because the polar axis is inclined towards the sun in summer, and away from it in winter, as illustrated in Fig. C. The days are longer in the summer, because the sun describes a longer arc in the sky, and summer temperatures tend to be higher than winter temperatures, both because the days are longer and because the sun is higher in the sky.

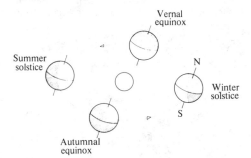

Fig. C. Diagram of the motion of the earth round the sun, showing how the North Pole is inclined towards the sun in summer and away from it in winter.

The height of the sun above the horizon can be roughly measured as the length of a shadow. The shadow cast by a stick onto a horizontal surface is shortest at noon and longest at sunrise and sunset. Because the sun travels at a constant rate along its arc through the sky, it gains height rapidly in the

morning, flattens out at midday, and loses height more and more rapidly towards evening—consequently shadows change length more rapidly before sunset and after sunrise than around noon. Changes in the length of shadow were used by the ancients as measures of the passage of time, and it is possible that similar clues are used by animals.

The direction of the sun, in relation to fixed landmarks, also changes throughout the day. The position of an observer on the surface of the earth can be specified in terms of lines of longitude, which are circles on the surface of the earth passing through both north and south poles. The line which an observer is standing on is called the local *meridian*, and the observer is looking along this line when he observes the sun at noon. The sun rises in the east and sets in the west, and when it reaches its highest point at noon its direction is due south for an observer in the northern hemisphere. If an observer knows the position of the sun at noon in relation to local landmarks he can estimate the time of day from the angle between the observed direction of the sun and its direction at noon. In the northern hemisphere the sun is to the left of its noon position during the morning, and to the right during the afternoon.

The apparent height of the sun, and its direction at any point in time, are constantly changing for an observer travelling on the earth's surface. Many animals are capable of judging the sun's position even when they are travelling over long distances. To appreciate this ability, we need to be more precise in describing the position of the sun. The points of the compass are divided into 360 *degrees*, with the north taken as the zero reference. The angle from north to a point in an eastward direction is called the *azimuth* of the point. The vertical angle from the horizon to the point is called the *altitude* of the point, as illustrated in Fig. D. Each of the 360 degrees of the compass is divided into sixty equal parts called *minutes*, and each minute is divided into sixty equal parts called *seconds*.

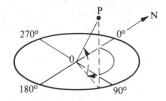

Fig. D. Calculation of the altitude and azimuth of a point, P.

The time indicated on sundials is the *local apparent solar day*, and this is defined as the interval between the successive arrivals of the sun at the local meridian. However, accurate observations show that the length of successive apparent solar days is not constant. The sun can be up to 16 min late or early in arriving at the local meridian, so that the time of sundials is sometimes fast and sometimes slow, compared to a clock. The *mean* solar day is based upon the average of the apparent solar days throughout the year, and keeps time with respect to a fictitious sun that is never early or late in arrival at the local meridian. As animals have no known way of predicting the length of the local apparent solar day, it is evident that the mean solar day is a better measure for the purpose of animal studies.

Because it is inconvenient to start the day at noon, the *local civil day* is defined as the mean solar day, starting at midnight. *Local civil time* counts the hours from midnight up to twelve, and then counts again up to twelve from noon. *Local mean solar time* starts with zero at midnight and counts up to 24 h. This 24-h clock is the one normally used for scientific purposes. Local mean solar time is different for points on the earth east or west of each other. In 1884 a system of *standard times* was adopted by international agreement. In this system the earth is divided into twenty-four time zones, which are nominally 15° wide and centred on each standard meridian, beginning with the prime meridian which runs through Greenwich. The standard time in each zone differs by 1 h from the time in either adjacent zone. Local modifications to the time zone boundaries have been made in order to avoid different times in any one political subdivision. The time zones of the world are illustrated in Fig. E. A person travelling from east to west would have to set his clock back by 1 h as he entered each successive time zone. There is an apparent loss of 1 day when this has been done twenty-four times. In order to compensate for this, it is necessary to advance by one whole day at some point on the journey. By international agreement this is always done at the *international date line*, which coincides with the 180° terrestrial meridian, apart from minor local adjustments.

In addition to indicating the time of day, the apparent motion of the sun gives a good guide to the time of year. The daily motion of the sun indicates the direction of the south for an observer in the northern hemisphere. Hence the westerly and easterly directions can be determined by simply observing the sun on any one day of the year. There are only two moments in the year when the sun rises at a point due east of an observer, and sets due west. These are the *vernal* and *autumnal equinoxes*, at which the length of the day is equal to the length of the night. The longest day of the year is generally taken as midsummer (21 June) and the shortest day as midwinter (22 December).

The apparent motion of the stars. To a terrestrial observer, the stars appear to fall on a spherical surface, as if fixed to the inside of an upturned bowl. The ancient notion of the position of the stars, called the *celestial sphere*, implies that all stars are equidistant from the earth. Although we know today that this is not so, the concept of the celestial sphere remains a useful one, particularly in considering how animals might view the stars.

At any given moment the stars appear to form a definite pattern on the celestial sphere. The whole pattern moves in a manner which suggests that the celestial sphere is spinning, although in reality it is the earth which is spinning about its polar axis. The point in the northern sky about which the celestial sphere seems to rotate is called the *north celestial pole*, and the equivalent point in the southern sky is called the *south celestial pole*. In the northern sky the star *Polaris* lies only a single degree from the north celestial pole. The effect of the earth's rotation is therefore to make Polaris move in a very small circle, so small as not to be noticed by the casual eye. For practical purposes, this star can be used for determining the observer's northerly direction. As the observer travels northwards on the surface of the earth, the altitude of Polaris increases, and stars which are close to Polaris describe a complete circle in one day, although they cannot be seen by daylight. Stars

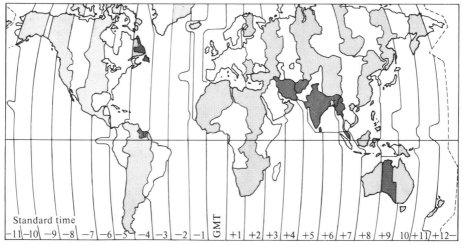

Fig. E. Time zones of the world.

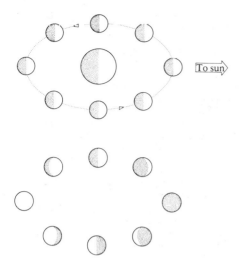

Fig. F. The illumination of the moon in its orbit round the earth, as seen from another planet (above), and from the earth (below).

which are far away from Polaris dip below the horizon for part of their path. They rise and set each night, following paths which carry them from east to west. Relative to the sun, they appear to rise and set about 4 min earlier each night. However the stars are perfectly regular in their apparent motions, unlike the sun and moon.

A clock that keeps pace with the rotation of the celestial sphere measures *siderial time*. The siderial day runs through 24 siderial hours of 60 siderial minutes each. Since the siderial day is shorter than the mean solar day, every time unit in this system is also proportionately shorter, and one siderial hour is about 10 solar seconds shorter than one solar hour.

The apparent motions of the moon. To a terrestrial observer, the moon moves along a path similar to that of the sun, but the moon rises about 50 min later each day. As a consequence, the moon is sometimes seen in the daytime, and at other times it is seen only at night. At intervals of about 29.5 days the moon appears to be crescent shaped, with the horns pointing away from the sun. At this time it appears in the evening sky, setting very soon after the sun. The moon rises and sets later each day for about 3 weeks, and during this time the crescent expands to a fully illuminated disc, rising very close to the time of sunset. Thereafter the disc shrinks again to a crescent and disappears from view, until the time that it again appears in the west at sunset.

In reality, the moon travels in an orbit around the earth, and is illuminated by the sun, as illustrated in Fig. F. The periodic phases of the moon result from the time taken for the moon to travel completely around its orbit. The changing position of the moon in the sky is a result of the earth's rotation in relation to the position of the moon in its orbit. Both these aspects of the apparent motion of the moon were used by the ancients in measuring the passage of time, and this information is also potentially available to animals.

Another important influence that the moon has on terrestrial life is its effect on the tides. The tides are an alternate rising and falling of the sea level, resulting from the gravitational pull of the moon and the sun. The moon raises a tidal bulge on the nearest side of the earth, causing high sea levels there. The gravitational pull of the moon causes the earth to move slightly nearer the moon, leaving behind the less attracted surface water on the far side of the earth. The result is two tidal bulges on opposite sides of the earth, as shown in Fig. G. The rotation of the earth causes these bulges to be displaced some distance ahead of the direction of the moon.

The sun contributes a similar, though smaller, tidal effect. When the earth, moon, and sun come into line with each other, the solar and lunar tidal forces combine to produce the high spring tides at the time of the new and full moons. When the sun and moon are at right angles with respect to the earth, their tidal forces oppose each other, and this produces the low neap tides at the time of the moon's first and third quarters.

The animal as a terrestrial observer. We cannot directly measure an animal's ability to estimate time, but we can sometimes make estimates by studying animal behaviour. In particular, studies of animal ORIENTATION have provided good evidence of the ability of animals to estimate time.

The apparent motions of the sun, stars, and moon are used by animals as indications of the passage of time. As we have seen, bees can be trained to search for food at particular times of day. This ability is known to be based upon the apparent motions of the sun. For example, honeybees native to Brazil can be trained to forage in a particular compass direction in a locality that is well known to them. When they are moved to an unknown locality, they continue to forage in the same compass direction, irrespective of the time of

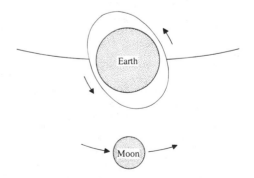

Fig. G. Diagram showing the direction of the tidal bulges on the earth in relation to the position of the moon.

day. This shows that the bees are capable of compensating for the anti-clockwise motion of the sun. However, bees whose parents were imported from the northern hemisphere, where the sun appears to move in a clockwise direction, are unable to make the appropriate compensation. It appears that there are INNATE mechanisms which compensate in northern bees for the clockwise motion of the sun, and in southern bees for its anti-clockwise motion. This change in the direction of compensation must have evolved in Brazilian bees during the years since AD 1530, when the first honey-bees were shipped from Portugal to Brazil.

Many birds are able to compensate for changes in the position of the sun with time. For example, common starlings (*Sturnus vulgaris*) placed in a cage, with a number of feeders around the edge, can be trained to feed only from the west feeder, irrespective of the direction of the sun. When the sun is obscured and replaced with an immobile artificial sun, the starling moves from one feeder to the next, so as to keep an angle to the sun that is appropriate to the time of day. This shows that the bird has an ability to tell the time by means of some internal clock, which it can use to predict the direction of the sun at any time of day. Scientists are generally agreed that many birds, including pigeons (Columbidae), can use the sun as a compass. This ability is thought to be used by homing pigeons and migratory birds. However, there is less agreement as to whether the altitude of the sun, in addition to its azimuth angle, can be used by birds as a navigational aid. The problem is complicated by the fact that some birds can navigate successfully in overcast conditions, when they must be using some factor other than the position of the sun.

Fish can use the position of the sun as a directional reference, and can maintain a line of travel in a particular direction throughout the day by making allowances for the sun's daily movement. Green sunfish (*Lepomis cyanellus*) reared from the egg stage in artificial light can compensate for the sun's daily movement at their very first exposure to the natural sun. Some other fish have to learn the direction of the sun's movement in their locality. Fish that live in the tropics are able to reverse the direction of their sun-compensating mechanism during the course of the year, in accordance with the apparent clockwise motion of the sun in the winter months, and its anti-clockwise motion in the summer.

When fish are displaced in latitude alone, no time differences are involved. If fish are transported in a southerly direction, their days become longer and the altitude of the sun is greater at any particular time of day. Experiments show that such fish make the errors that would be expected if they were responding to the sun's azimuth, but not to its altitude. These errors arise from the fact that the visible arc of the sun is greater in southerly latitudes, so that the azimuth angle is slightly different in the morning and evening. Thus it appears

that fish, like many birds and arthropods, are able to use the sun as a compass, but do not respond to its altitude.

Studies of bird MIGRATION have shown that some birds are capable of orientation by means of the stars. For example, some European warblers (*Sylvia*) are able to find their migratory direction under a clear starry sky, even on moonless nights. Under overcast conditions the birds become disoriented. In principle, no time information should be necessary to find compass direction with the help of star configuration. A terrestrial observer can always determine the north–south axis, about which the celestial sphere rotates, provided he has sufficient knowledge of star patterns. Even without such knowledge he could observe the rotation of the stars, and thus locate the celestial poles. However, it appears that warblers do compensate for the movement of the stars. Experiments conducted in a small planetarium indicate that warblers show correct orientation if the natural sky is replaced by an artificial one. The birds are correctly oriented even if only the major stars are shown, and even if the stars are not moving. This shows that the orientation must be based on the configuration of the stars. The orientation of the warblers is correct only when the position of the artificial sky roughly corresponds to that of the natural sky, at the current season and time of night. This suggests that a time-mechanism compensating for the motion of the stars must be involved. In other birds, however, no such time-mechanism appears to be present. For example, indigo buntings (*Passerina cyanea*) use the star pattern alone to locate their direction.

It is known that some animals are able to use the moon as an aid in orientation. For example, the common sand flea *Talitrus saltator* (a crustacean), in addition to being able to find compass directions with the aid of the sun during the daytime, may use the moon for the same purpose at night. Time compensation is necessary for these purposes, and it would be possible to orient to the moon using the same clock that is involved in solar direction finding, provided that both the moon's position in the sky and its shape are taken into account. From the shape of the moon, the position of the sun relative to the moon can always be derived. At full moon, for instance, the sun's position is just opposite to that of the moon. At the first quarter it is about 90° to the right of the moon, and at the last quarter it is about 90° to the left. Considering the poor vision of the sand flea, it seems unlikely that information about the shape of the moon is used by this animal.

Internal clocks. Although it is clear that many animals make use of the apparent motions of the sun, moon, and stars in telling the time, it is also known that many possess internal clocks that are capable of keeping accurate time, even when isolated from external events. For example, fiddler crabs (*Uca*) emerge from their burrows at low tide and become very active, courting, foraging, etc.

With each flood tide the crabs retreat back to their burrows. LABORATORY STUDIES show that this rhythm of activity may persist for as long as 5 weeks in isolation from tidal or lunar events.

The tendency for rhythms to persist in the apparent absence of external timing stimuli is widespread in the animal kingdom. Even some single-celled organisms exhibit this phenomenon. For example, the protozoan *Euglena* shows a rhythm of swimming activity that is synchronized with the motion of the sun. Such patterns, when they have a periodicity of about 1 day, are called *circadian* rhythms. The circadian rhythm of *Euglena* persists even when it is maintained in continuous darkness in the LABORATORY.

The persistence of rhythms in the apparent absence of synchronous periodic external stimuli is taken by some to show that the rhythms are *endogenous*; that is, originating from within the animal in a spontaneous manner. The alternative theory is that the rhythms are *exogenous*, timed by external environmental stimuli. Although it has been demonstrated many times that rhythmic activities of animals may persist when the animals are maintained in the laboratory under conditions of constant temperature and level of illumination, we have to remember that there are other characteristics of the environment which may vary in the course of the day. Factors such as barometric pressure, humidity, and cosmic radiation could possibly be used by animals to keep time. Such factors are almost impossible to exclude from laboratory experiments. Other types of experiment, however, can be used to test for endogenous clocks.

An endogenous clock is essentially a model of celestial phenomena that is embodied within the animal. Thus a circadian rhythm can be regarded as a model of the rotation of the earth upon its polar axis. The artificial clocks used by humans are essentially models of this type, which are used to predict external events, such as sunrise. If the animal clock is being triggered by some external event related to the earth's rotation, then transporting the animal along a line of latitude should result in resynchronization of the rhythm to the new local time. It has been found that circadian changes in the pigmentation of fiddler crabs are unchanged by transportation from the Atlantic coast of the USA to the Pacific coast, suggesting that the activity clock of these animals is endogenous. In another experiment, a colony of bees was placed inside a specially constructed, completely enclosed room, where they were trained to forage from artificial feeders. They were trained to obtain food at a particular time of day from a particular feeder, at a laboratory in Paris. The bees were then flown overnight to New York and tested there under identical conditions. For 3 days the bees continued to visit the empty feeder at the time they were accustomed to feed in Paris. Local conditions induced no change in their time-sense, indicating that it is endogenous.

Experiments aimed at stopping and starting the biological clock also indicate that the clock is endogenous. For example, time-trained honeybees, which have been chilled for a 5-h period, come to the food dish 5 h late. This result suggests that the bees were not relying on external timing cues, and that their internal clock had stopped during the period of chilling. The timing of rhythms is apparently inborn in many instances. Chicks (*Gallus g. domesticus*), house mice (*Mus musculus*), and lizards (Lacertidae) which have been incubated and reared under constant conditions show normal circadian rhythms at an early stage of development. In one experiment, lizards were hatched in an incubator under temperature and light periods that corresponded to an 18-h day (9 h light and 9 h dark) while other lizards were reared on a 36-h day. Both groups of lizards exhibited the normal 24-h rhythm when they were tested under constant conditions.

The shore crab (*Carcinus maenas*) has a tidal rhythm that differs from that of the fiddler crab in that activity is synchronized with high water. The rhythm persists in constant laboratory conditions for about a week, after which it fades away. The rhythm can be restored by cooling the crab at near freezing point for 6 h. This cold-shock appears to restart the tidal clock. In one experiment, shore crabs were raised in the laboratory from eggs to adulthood, under a normal 24-h day–night regime. The active period of the adult crabs was limited to the daylight hours. However, after the crabs were given a single 15-h cold-shock treatment, a tidal rhythm of activity appeared. It seems that the endogenous tidal clock had been dormant and had been restarted by the cold-shock.

Circadian rhythms in humans have been known for a long time. Rhythms of sleeping and waking, and of numerous physiological processes, such as body temperature, hormone levels, etc., follow a circadian pattern even in people carefully isolated from environmental influences in a subterranean bunker. The time-sense of humans is rather peculiar. Although most people are generally rather poor at consciously estimating time periods, some people can wake up at a pre-set time from normal sleep and from hypnosis with remarkable punctuality. It is possible that environmental clues associated with the rotation of the earth can influence this time-sense, unknown to the experimenter. However, this seems unlikely, in view of observations that have been made at the South Pole and during space flight, where the rotation of the earth cannot result in any fluctuation of the environment.

An interesting parallel is found in bees, which can be trained to perform a task at a specific time of day. It is possible to train a bee to perform different tasks at various times of day, provided that the interval between the different training times is greater than 2 h. Bees cannot be trained to perform tasks which are not based upon the 24-h clock. For example they cannot learn to visit a feeder at 17- or 19-h intervals. It is as if there is an endogenous 24-h clock in humans, bees, and other animals, which can be tapped for specific purposes.

The survival value of time-sense in animals. Regular recurring events, such as changes from day to night, and changes in the tides and seasons, are a feature of the world of most animals. By confining its activities to specific times a species is able to fill a particular ecological NICHE. An obvious example is the ecological division between nocturnal and diurnal animals. Animals specialized for daytime vision are disadvantaged at night, because they cannot forage efficiently and may be in danger from predators. Their SLEEP patterns are generally organized on a circadian basis, so that the animals are inactive, inconspicuous, and generally safe at times when it is inappropriate for them to be engaged in their normal waking activities. Nocturnal animals often SLEEP in safe places during the day, and possess specializations which enable them to be profitably active at night. Their daytime resting places are sometimes designed to avoid climatic extremes, as in many desert species, and to avoid predators.

The daily rhythm of activities in many animals is designed to exploit opportunities in food availability. For example, among the herring gulls (*Larus argentatus*) in a colony on the west coast of England there are some individuals whose pattern of feeding behaviour is synchronized with low tide, when food is exposed on the mud-flats; and other individuals whose foraging activity is timed to coincide with the arrival of refuse trucks at the rubbish tip of the nearby town. Similarly, the daily foraging of bees is often synchronized with the times at which various flowers secrete nectar.

Another important function of time-sense in animals is synchronization of reproductive activity between male and female members of the species. This is particularly important in those marine species that release their eggs and sperm into the sea. The chances of fertilization are considerably increased if release occurs at the same time in the two sexes. A famous example is the Palolo worm of the Atlantic (*Eunice fucata*) and Pacific (*Eunice viridis*) oceans. This animal reproduces only twice per year; during the neap tides of the last quarter moon in October and November. Grunion fish (*Leuresthes tenuis*) of the California coast take advantage of the spring tides. They ride on the crests of the waves until they arrive on the beach, where they deposit their eggs and sperm. The fertilized eggs develop in the warm moist sand, and the water does not reach them for the next 2 weeks. At the next spring tide the JUVENILES have become freed from their eggs, and are washed into the open sea.

As we have seen, internal clocks play an important role in direction finding. Many forms of NAVIGATION require some reference to an endogenous clock, whether solar, celestial, or magnetic factors are involved. Many birds require considerable powers of navigation to aid them in their long migration. For example, the gold plover (*Pluvialis dominica*) breeds along the northern shore of Alaska, and migrates during the autumn to Argentina via Labrador, a distance of about 11 000 km, much of which is over the open ocean from Nova Scotia to Guyana. Other groups of animals also perform long migrations. The Atlantic green turtles (*Chelonia mydas*) visit small islands to bury their eggs in the sand. Turtles which feed near the Brazilian coast migrate to Ascension Island 2000 km out into the Atlantic ocean.

Although human beings appear to have a sense of time that is of the same order as that found in many animals, man has evidently found it necessary to supplement this ability with artificial devices. In order to calibrate these instruments with a reasonable accuracy, it was necessary for people to refer to the apparent motions of the sun, moon, and stars. Time, indeed, has little meaning except in relation to the motion of the earth in space.
5, 25, 26, 115

TINBERGEN, NIKOLAAS (1907–   ). One of the founders of modern ETHOLOGY. He is particularly famous for his ingenious experimental studies of the control mechanisms and FUNCTIONS of animal behaviour in general, and of gulls (Laridae) in particular. His approach to the subject has been to seek clear solutions to simple and well-defined problems, rather than to build large explanatory systems on relatively loose tests. Even his application of the COMPARATIVE method to gulls is exceptional for its thoroughness. His book *Curious Naturalists* (1958) describes in autobiographical form some of the more important work of his school.

Brief life. Niko Tinbergen was born in The Hague, a member of an exceptionally talented family: his elder brother Jan won the Nobel Prize for economics in 1969, and his younger brother Lukas did important biological research before dying young. Tinbergen was interested in natural history from an early age, and went to Leiden University to study biology. At Leiden, he retained his enthusiasm for natural history, especially ornithology, and collaborated on a book of photographs and observations about birds, *Het Vogeleiland* (The Bird Island), 1930. He also played various sports with notable skill; he was, for instance, a hockey player of international standard. He began graduate research at Leiden, but this was cut short by an opportunity to join an expedition to Greenland in 1930–1. His thesis of only 32 pages was accepted, however. In Greenland, Tinbergen studied two species of birds, the northern phalarope (*Phalaropus lobatus*) and the snow bunting (*Plectrophenax nivalis*), as well as enjoying the local Eskimo culture; *Eskimoland* (1935) was his book about the trip. He returned from Greenland to Leiden University to continue his ethological research. He collaborated briefly with Konrad LORENZ, but the advance of ethology was soon baulked by the Second World War. Tinbergen was among several academics imprisoned by the Nazis for protesting against the treatment of Jews, and he spent most of the war in prison.

Soon after the war Tinbergen went to lecture in the United States. The negative response to Continental ethology convinced him that it was necessary to move to the English-speaking world, and he was attracted to Oxford University in 1949 by Sir Alister Hardy. Tinbergen has remained in Oxford since then. In 1973 he shared the Nobel Prize with Konrad Lorenz and Karl von FRISCH.

The aims and methods of Tinbergian ethology. In 1963 Tinbergen published an essay entitled 'On aims and methods in ethology'. In it he suggested that there are four main areas of ethological enquiry: causation (see MOTIVATION), DEVELOPMENT, SURVIVAL VALUE, and EVOLUTION. A similar conception of the content of ethology had provided the structure of his book *The Study of Instinct* (1951). Tinbergen also stressed that the ethologist's method should consist of, first, a period of observation so that the natural context and behavioural repertoire of the species under study becomes familiar, and then, second, experimental study of causation, development, and survival value, and an attempt to reconstruct the evolution of particular units of behaviour. He had emphasized the importance of the observation of behaviour in its natural surroundings (see FIELD STUDIES) largely as a criticism of those comparative psychologists who studied behaviour in artificial LABORATORY environments, and had little idea of the function of the behaviour in the animal's life. Tinbergen himself made major contributions to all areas of ethology, except the experimental study of development. This omission reflects his lifelong preference for animals in their natural environments rather than in CAPTIVITY. As a boy he watched animals in nature; in contrast to Konrad Lorenz, who kept large numbers of HOUSEHOLD PETS.

Causation of behaviour. Tinbergen's work from approximately 1930 to 1950 was on the mechanisms used by organisms in finding their way around their environment, and on the causation of behaviour. This interest, as well as his flair for simple but convincing experiments, was already apparent in his doctoral research on how the digger wasp *Philanthus triangulum* recognizes its burrow. The female digger wasp digs a burrow in the sand and lays some eggs in it; she provisions the eggs with honey-bees (*Apis mellifera*) which she hunted and caught. The female repeatedly leaves her nest to catch a bee, and then returns to the same nest. She must therefore have an ability to recognize her own nest entrance as distinct from that of the other nests near by. She might recognize her nest by some stimulus emanating from the nest entrance itself, or by the spatial arrangement of objects around the entrance. Tinbergen first found that by shuffling the objects near the entrance he could make the wasp take longer to find her nest.

In another experiment he accustomed a wasp to having a circle of pine cones around her nest entrance, and then he exactly transposed the circle to a place a little distance away. The wasp still looked for her nest in the middle of the circle of cones, so showing that it is the spatial arrangement of landmarks around the nest entrance that the wasp uses to find her burrow. Furthermore, visual cues are sufficient for the wasp; other stimuli, such as local odours, are unnecessary.

The wasp only homed in on the pine cones from close range. It must also be able to find its way home from the hunting ground, which is too far away for it to be able to see its nest. Tinbergen and his collaborators tried digging up some pine trees that were growing near the nest site, and displacing them. They found that the wasps then searched for their nests according to the displaced trees. They also did more complicated experiments in which they moved both small cues from near the nest, and larger cues from slightly further away; in this way they revealed how the wasps combined the information from the different landmarks in their environment.

Tinbergen conducted further experiments on how the digger wasp recognizes its prey, the honey-bee. The important experiments used an object hanging on a piece of string. The object could be a recently killed honey-bee, a dead, deodorized bee, a bee that had been rendered odourless and then re-odorized by shaking with some live bees, a piece of wood of about the size of a bee, or a piece of wood that had been scented by shaking with live bees. Tinbergen found that a wasp was initially attracted to any of these objects, but only if the object was moving. The wasp then hovered near by, but would only seize the object if it smelled like a honey-bee. Therefore the wasp is initially visually attracted to any of these objects, and then checks that the prey is correct by its odour. However, the wasp only completed its hunt by stinging the prey if the prey was a honey-bee (it would not sting a piece of wood that smelled like a honey-bee). So it must have used some further cues, perhaps visual or tactile, to decide whether finally to sting its prey.

The digger wasp study demonstrated the environmental cues used by a single animal to control its behaviour. Tinbergen's study of the COURTSHIP of the three-spined stickleback (*Gasterosteus aculeatus*) illustrated the SOCIAL control of an individual's behaviour. Here the cues eliciting behaviour are structures or activities of another member of the same species. Tinbergen showed that the courtship of the stickleback could be understood as a sequence of separate behavioural units in the male and the female, wherein each male activity elicits a female activity, which in turn elicits another male activity (SOCIAL INTERACTIONS, Fig. A). This analysis mirrors the stimulus–response (SR) chain theory of behavioural sequences that was a dominant part of psychology at that time (during the 1930s).

In the spring the male sticklebacks fight each other, and their attacks are specifically directed at other males. The most outstanding characteristic of the males is their red throat and belly, so Tinbergen hypothesized that this was the stimulus for attack. He tested this hypothesis by using one of his recurrently important methods: models. He built a series of models (SIGN STIMULUS, Fig. A), some of which were very crude, lacking most stickleback characteristics, but possessing a red underside; other models looked very similar to sticklebacks, but lacked red coloration on their bellies. The males attacked the crude, but red-bellied, models much more than the accurate models which lacked red coloration. The red pattern acts as a RELEASER of attack.

Tinbergen made equally famous use of models in his experiments on the bill-pecking response of newly hatched herring gull chicks (*Larus argentatus*). The chick begs for food by pecking at the tip of its parent's bill. The parent then regurgitates food. The bill is yellow with a red spot at the tip. Tinbergen tested the hypothesis that it is the red spot at the tip that stimulates the chick to peck at the bill. He presented two kinds of model of a gull's head and bill, the first of entirely natural colours, and the second identical to the first except that it lacked the red spot. The chicks pecked more at the model with the red spot (FOOD BEGGING, Fig. A). Further experiments demonstrated that it is the contrast in colour between spot and bill that stimulates the chick: the more the contrast, the more the chick pecks (see SUPERNORMAL STIMULUS).

Tinbergen's early work consisted mainly of descriptions of the natural behaviour of single species, and of studies on the internal (motivational) and external factors causing behaviour. The emphasis on the internal and external control of behaviour is also apparent in his earlier syntheses of theory, such as 'An objectivistic study of the innate behaviour of animals' (1942), and his books *The Study of Instinct* (1951) and *Social Behaviour in Animals* (1953). With the move to Oxford in 1949, the emphasis of Tinbergen's work shifted to the evolution of behaviour.

Behavioural diversity in gulls. No doubt many influences were at work in causing the change in the direction of Tinbergen's research in 1950. One important factor was his collaboration with Konrad Lorenz, who thought that ethology should emulate anatomy, and study function, evolution, and taxonomy by the comparative method. Another influence was the milieu of zoology at Oxford, which then encouraged the study of evolution; for example, Tinbergen was soon collaborating with H. B. D. Kettlewell. Kettlewell was working on a now classic example of evolution in action, in the peppered moth (*Biston betularia*). The peppered moth exists in two forms: black and peppered grey. The former is more abundant in woods that have been polluted by industrial smoke, the latter in unpolluted woods. Tinbergen collected the photographic proof that birds preferentially prey upon black forms on trees in unpolluted woods, and on grey forms in polluted woods. The more conspicuous form is eaten more in both cases, so PREDATION by birds is acting here as the agent of NATURAL SELECTION, causing changes in the proportions of the two kinds of moth.

Tinbergen chose the various species of gulls as the subject for a thorough comparative study, by himself and his many students and collaborators, of the evolution of behaviour. He was already very familiar with the behaviour of the herring gull from his earlier observations, and in 1953 published *The Herring Gull's World*. Now he embarked on a study of the evolution of the various signals of the different species. The results of this study were later made into an award-winning film, and a book with the same title, *Signals for Survival* (1967). The study of gull displays adopted the following approach. The different gull species were first observed, and the different signals described. If different species used the same signal then the signal would be given the same name in each case, just as similar organs of different species are given the same name in an anatomical study. The function of each signal was then worked out by observing what other birds do immediately after the signal is given. For example, the signal called *grass pulling* in the herring gull causes near-by birds to move away.

The evolution of the signals can in some cases be reconstructed. It is usually possible to observe in a signal some parts of other activities. The signal *upright*, for example, contains elements of the posture of a physical attack (e.g. stretching the neck, and pointing the bill downwards) and of escape (e.g. halting instead of running at the opponent, and sleeking of the plumage). The signal is thought to be evolutionarily derived from the other movements. Sometimes it is possible to arrange the signals of different species in a series of decreasing similarity to the original structure or activity from which they are thought to derive. The evolutionary process by which signals become more exaggerated and so less like their original state is called RITUALIZATION.

Tinbergen also pioneered the experimental study of function. With his collaborators, he demonstrated that a whole suite of characters are behavioural defences against predation. Colonial nesting, synchronization of egg-laying, camouflaged eggs, removal of broken egg-shells, spaced out nests within the colony, and the choice of different HABITATS by day and by night, are all mechanisms to reduce predation on eggs, chicks, or adults. If, for example, broken egg-shells are placed near a gull's nest, then that area becomes more visible to predators and any near-by chicks or unhatched eggs are more likely to be eaten.

Even in the biological study of function, which

has traditionally been notorious for being vague and inconclusive, Tinbergen showed how clear solutions can be obtained. As so often, he clearly defined the problem and was then able to solve it simply, by neat experimentation.

Human ethology. In more recent years, Tinbergen has become increasingly interested in applying ethological methods to the study of human beings. Again, he has usually avoided issuing any general pronouncements on the human predicament. He selected the problem of a form of autism, called Kanner's syndrome, studying children suffering from this illness, which is characterized by an excessive withdrawal from society. He interpreted Kanner's syndrome as resulting from a conflict between the child's fears and its frustrated desire for society. The analysis illustrates his claim that ethology can contribute methods as well as theories to the study of man.   M.R.

136, 137, 138, 139

TOLERANCE. Each species has a characteristic ability to tolerate extreme values of environmental factors, such as temperature, humidity, etc. For example, many marine invertebrate animals are affected by variation in the salinity of the water, because their body fluids generally have much the same salt concentration as sea water, and their body tissues are adapted to function efficiently in such a medium. If they become immersed in a less saline medium, water enters their tissues as a result of the physical process called *osmosis*, by which water passes from a dilute to a concentrated solution; in a more saline medium, the opposite process occurs. Animals which cannot control the passage of water into their bodies are limited in the range of conditions in which they can live by the degree of salinity which they can tolerate. Thus an animal's HABITAT is often restricted by its tolerance range. In estuaries, for example, different species of the amphipod 'shrimp' *Gammarus* are able to inhabit different parts of the estuary, due to their differing salinity tolerance ranges. As illustrated in Fig. A, *Gammarus locusta* has a high tolerance of salt water, and is found at the mouths of estuaries. *Gammarus zaddachi* has moderate tolerance of salt water, and is usually found in the stretch of river 11–16 km from the sea. *Gammarus pulex* is a true freshwater species, and does not occur at all in parts of the river showing any influence of the tide or salt water.

Although species differ in their degree of tolerance, some species have the ability to change their range of tolerance through the process of ACCLIMATIZATION. For example, the small tree lizard *Urosaurus ornatus* normally has a temperature tolerance range with a maximum at 43.1 °C. However, by maintaining these animals in the laboratory for a period of about 10 days at a temperature of 35 °C, compared with the more normal temperature of 22–26 °C, it is possible to raise their maximum temperature tolerance to 44.5 °C.

The tolerance of animals depends largely upon their physiological mechanisms and their ability to adjust to environmental changes by means of suitable behaviour. Compared to mammals and birds,

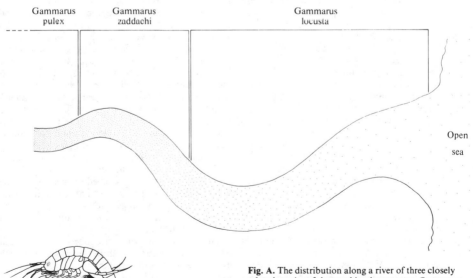

Fig. A. The distribution along a river of three closely related species of the amphipod crustacean *Gammarus*, relative to the concentration of salt water. The degree of freshness of the water is indicated by the degree of shading.

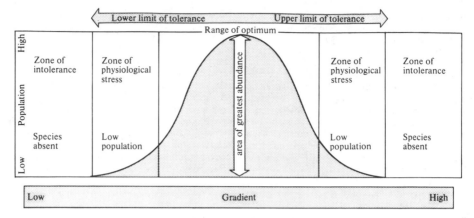

Fig. B. Idealized graph of population abundance along an ecological gradient.

which can maintain a constant body temperature when the environmental temperature changes, reptiles are much more at the mercy of the environment, because they lack the physiological mechanisms required for efficient THERMOREGULATION. Although reptiles often have a well-developed ability to adjust to changes in environmental temperature by means of suitable behaviour, the necessity for such behaviour leaves them less free to engage in other activities. In general, animals that have evolved the ability to regulate their bodily processes by largely physiological means have greater freedom of behaviour (see HOMEOSTASIS).

In the natural environment, the distribution of animals frequently reflects their tolerance along environmental gradients. Towards the high and low ends of the gradient the members of a species may be able to stay alive, but they suffer increasing physiological stress, become inefficient, and can maintain only a low population. In the middle of the gradient there is a range of optimal conditions in which a large population can be maintained. When the number of animals per unit area is plotted along an environmental gradient, a bell-shaped curve usually results, as illustrated in Fig. B. Such curves show not only the limits and range of tolerance of the species, but also the optimum value of the environmental gradient. Many animals exhibit a preference for the optimal conditions, and may try to establish themselves there through a process of habitat selection. However, it will generally be the case that some are unable to do so, because of COMPETITION from other members of their species.

Another complicating factor is that the environment of any species consists of a series of interacting gradients. Population density will be high only in those areas where the optimal ranges of the environmental gradients overlap. As an example, let us consider two species of woodlice. One species, the pill woodlouse (*Armadillidium vulgare*), is common in grasslands and scrubby woodland, both in Europe and North America. The other

species, *Venezilla arixonicus*, is rather rare and found only in arid country. The preferences of these two species have been investigated with respect to environmental gradients of temperature, humidity, and light intensity. In this type of investigation one factor at a time is varied. For example, the animals may be tested in a temperature gradient, under constant light conditions, with the whole gradient at a high humidity, or at a low humidity. At another time the animals will be tested in the temperature gradient with the hot end dry and the cold end wet, or with the hot end wet and the cold end dry. By systematic experiment it is possible to discover the interactions of the light, temperature, and humidity gradients, with respect to the animal's preferences.

As a result of such experiments, it was found that *Armadillidium* prefers temperatures around 10–15 °C, combined with high humidity and moderate light intensity. On the other hand, *Venezilla* prefers higher temperatures (20–25 °C), lower humidity, and lower light levels. These experimental results accord well with the conditions of the natural habitats of the two species. *Armadillidium* lives in cool wet places and is active during the day, whereas *Venezilla* lives in warm dry places and is active at night. Under more extreme conditions, however, the picture is not quite so clear cut. For instance, at temperatures in the region of 35–40 °C, *Armadillidium* tends to choose lower humidities irrespective of whether these are in the light or the dark. It may be that a low humidity aids efficient thermoregulation. Similarly, at high humidities *Venezilla* chooses dry conditions, even if these are in the light.

The complex physical situation in the natural environment, together with biological factors such as competition and availability of food, means that the distribution of animals is not a perfect guide to their range of tolerance. Inevitably, most animals have to compromise. Nevertheless, an animal's facility for ADAPTATION to its characteristic circumstances, over long periods of time, will tend to ensure that there is a general relationship between the animal's range of tolerance and the environment in which it habitually lives.

27

TOOL USING is usually defined by biologists as the use of an external object as a functional extension of the body, in attaining an immediate *goal*.

This rather precise definition excludes many cases of manipulation of objects by animals which bear a superficial resemblance to tool using but lack certain essential elements. For example, there are many birds that drop food items onto rock, or some other hard surface, in order to smash them open. Some gulls (Laridae) and crows (*Corvus*) drop shellfish from a height to crack them open. The raven (*Corvus corax*) and the bearded vulture (*Gypaëtus barbatus*) drop bones in order to crack them and feed on the marrow. The song thrush (*Turdus philomelos*) has the habit of smashing snails (Gastropoda) against a rock anvil. The Egyptian vulture (*Neophron percnopterus*) sometimes breaks eggs by picking one up in its beak and throwing it upon the ground. One bird was tested with plaster eggs, and after trying unsuccessfully to break one in this way, it carried it to a stone several metres away and threw the egg against it. Egyptian vultures are also known to fly above eggs in the nest of an ostrich (Struthionidae) and drop stones onto them. They may also pick up a stone in the beak and throw it at the egg, as illustrated in Fig. A.

Fig. A. Egyptian vulture (*Neophron percnopterus*) about to throw a stone at an ostrich egg.

Biologists usually distinguish two ways of breaking open food items, illustrated by these examples. When a bird drops or hits an egg onto a hard surface, such as a stone anvil, it is not using the anvil as a functional extension of its own body; it is not using a tool. But when it drops or throws the stone at the egg, it is using the stone as a functional extension of its bill, in the sense that it is manipulating an object to attain an immediate goal; it is using the stone as a tool.

Many animals scratch or rub themselves against trees, but the tree would not normally be regarded as a tool. However, elephants (Elephantidae) and horses (Equidae) have been known to pick up a stick, in the trunk or mouth respectively, and scratch themselves with it. We would normally regard the stick as a tool under these circumstances.

Tailor birds (*Orthotomus* and *Phillergates* spp.) and weaverbirds (Ploceinae) manipulate nest material in a complex manner. Tailor birds make their nests by folding a large hanging leaf and sewing the edges together with plant fibre. Weaverbirds knot and weave pieces of grass around twigs to make their basket-like nests. (See NEST-BUILDING, Fig. B.)

Some have argued that the nest could be considered as a tool for rearing the young, but this is not really a short-term goal. Moreover, to regard the nest material as a tool for making a nest would be like calling knitting-wool a tool for making a garment. Most biologists would distinguish between the material being manipulated, and the means by which the material is manipulated. Knitting-needles are normally regarded as tools, but not knitting-wool.

Evolutionary aspects of tool using. In some animals tool using is highly specialized and has probably evolved, like many aspects of INSTINCT, to meet particular ecological situations. For example, the solitary wasp *Ammophila umaria* holds a small pebble in its mandibles and uses it as a hammer to pound dirt into its nest burrow. The hermit crab *Dardanus venosus* inhabits mollusc shells upon which it places a sea anemone, often *Calliactis tricolor*. The crab induces the anemone to detach itself from a rock, and then lifts it with its claws and places it upon the mollusc shell. The anemone gives the crab some protection from predators (see SYMBIOSIS). The archer fish (*Toxotes jaculator*) forages close to the surface of the water in mangrove swamps. When it sees an insect on overhanging vegetation, it spits out an accurate stream of water which dislodges the prey so that it falls onto the surface of the water where it is eaten by the fish (Fig. B). Although only a few centimetres long, the archerfish can hit prey up to 1.2 m above the surface of the water. These are all instances of fairly stereotyped behaviour patterns that are characteristic of the species.

Some types of tool using appear to be more adaptable, and may be said to involve some degree of INTELLIGENCE. For example, the Galapagos

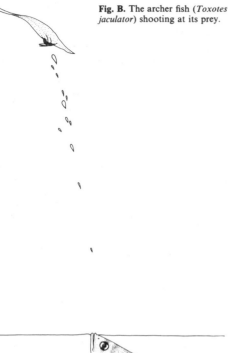

woodpecker finch (*Cactospiza pallida*) probes for insects in crevices in the bark of trees by means of a cactus spine or twig held in the beak (see INTELLIGENCE, Fig. D). The birds select spines and twigs that are appropriate to the task, and may break them to a convenient length. There is a report of a bird which, after a few unsuccessful attempts to probe with a forked twig, broke the twig above the fork and started probing with the single remaining piece. Observations of a young woodpecker finch which had been taken from the nest as a fledgling revealed that the bird manipulated twigs from an early age, but when hungry and presented with an insect in a hole, it would drop the twig and try to obtain the insect with its beak. Eventually, the bird learned to probe for insects with a twig. Thus it appears that woodpecker finches show the rudiments of tool making or modification, and are capable of LEARNING to improve their FORAGING behaviour by means of tools.

Another well-studied example of tool use is the manipulation of stones by sea otters (*Enhydra lutris*). This otter is found off the coast of California, south of San Francisco, and off the Aleutian Islands. The otters dive to the sea-bed, and bring

to the surface crabs (Brachyura), sea urchins (Echinoidea), mussels (Myrtilidae), etc. They usually eat while swimming on their backs. When FEEDING on mussels, the otter comes to the surface with a stone, about 10 cm in diameter, which it places on its chest as an anvil. The mussel is then held between the paws and banged repeatedly on the anvil until broken sufficiently to be eaten. Sometimes an otter will retain the same anvil stone for several feeding episodes, holding it in its armpit while diving for further mussels. The otter pup is dependent upon its mother for food until it is about 15 months old. The pup dives when the mother dives, but usually does not bring up any food. When the mother has broken open the food item, the pup takes or is given some of it. It is probable that the young learn to use anvil stones by IMITATION, and during manipulative PLAY.

A number of mammals use static rocks as anvils. For instance, the dwarf mongoose (*Helogale undulata rufula*) flings eggs from between its hind legs so that they hit a rock and break. Other mongooses throw eggs at the ground. The sea otter, however, is the only non-primate mammal known to use a stone as a tool as part of its normal behaviour.

It is not difficult to guess the pressures of NATURAL SELECTION that have led to the EVOLUTION of tool using in animals. Most instances of tool using confer an obvious advantage upon the animal. Eggs and shellfish are food items that are inaccessible to many animals because of their hard casing. Animals that use tools usually occupy a distinctive NICHE. Thus the woodpecker finch is the only finch to specialize in the extraction of insects from decaying wood. On isolated islands, such as the Galapagos, COMPETITION is intense, and typically leads to feeding SPECIALIZATION. Sea otters are the only otters to have entered the marine environment. They did so at a time when other mammals were already established as marine species, and were already exploiting the readily available foods. The manipulative skills of otters enabled them to develop a mode of FORAGING that was not available to other marine animals. Biologists are not agreed as to the extent to which it is necessary to account for tool-using behaviour in terms of intelligence and INSIGHT. Clearly, tool using is a form of PROBLEM SOLVING, but in some cases the problems have been solved by the process of evolution, the individual animal merely applying the solution in the form of behaviour that develops during ONTOGENY, as a result of a mixture of learning and INNATE predispositions. In other cases, however, individual animals acquire tool-using behaviour that is not characteristic of the species, and which probably originates from intelligent exploitation of opportunity, which is then sometimes passed to other members of the group through learning and imitation. This aspect of tool using is particularly common among the primates.

Tool using among the primates. Tool using has been more thoroughly studied in primates than in any other animal group. This is partly because scientists of different disciplines are interested in the problem. Students of ETHOLOGY have studied tool using as part of the natural behaviour of primates. Psychologists have studied it as an example of problem solving and intelligence. Anthropologists are particularly interested in tool using in primates for the light it can throw upon the use of tools by early man and his ancestors.

Tool using by monkeys and apes (Simiae) has been observed frequently in the wild, usually in feeding or during aggressive encounters. Sticks may be used to obtain food, stones to break open food items, and leaves or water to clean food. Chimpanzees (*Pan*) have been observed using sticks to dig up edible roots. Wild chimpanzees at the Gombe Stream Reserve in Tanzania used sticks to lever open boxes of bananas provided by scientists. After breaking a suitable branch off a tree, the chimpanzee would usually strip off the leaves and bite splinters off one edge to form a chisel-shaped edge. Sticks may also be used as levers to break open the nests of ants (*Crematogaster* spp.).

Wild chimpanzees have also been observed using sticks, twigs, and grass stems to probe for food. Sticks may be poked into the nests of bees (Apidae) to obtain honey, or into ants' nests where they are left for a few seconds and then withdrawn covered in ants. These may be eaten directly from the stick, or the stick may be swept through the free hand so that the insects are gathered there. Grass stems are frequently used by chimpanzees at the Gombe research station to probe into the mounds of giant termites (*Macrotermes*). The stems are sometimes carefully selected and prepared. If the end becomes bent it may be bitten off or a new tool selected. Infants under 2 years of age do not appear to use grass stems for termiting, although they often accompany their mothers and watch them intently. Between 1 and 2 years of age they may manipulate grass stems during play, but they do not use them in the correct context until they are between 2 and 3 years old. At this age they tend to be clumsy, and to use tools of inappropriate dimensions. By 4 years of age the JUVENILE chimpanzee exhibits a more adult-like technique. These observations show that this aspect of tool use requires a degree of skill that is not easily learned.

Wild chimpanzees sometimes use leaves as a sponge to obtain drinking water from a hole in a tree, or to wipe faeces, mud, blood, or sticky fruit juice from parts of the body. Although tool using has been most intensively studied in chimpanzees, similar behaviour has been observed in other primates. Wild capuchin monkeys (Cebinae) have been observed to use twigs to probe for insects under the bark of dead trees. Japanese macaques (*Macaca fuscata*) wash their food with water (see CULTURAL BEHAVIOUR). Baboons (*Papio*) may use sticks to probe for insects, and stones to squash scorpions (Scorpiones).

In CAPTIVITY tool use has been observed in a wider range of primates. Painting and drawing behaviour has been studied in captive capuchin monkeys, orang-utans (*Pongo pygmaeus*), gorillas (*Gorilla gorilla*), and chimpanzees. Capuchin monkeys, guenons (*Cercopithecus*), and chimpanzees have all been observed to lure ducks (Anatinae) and chickens (*Gallus g. domesticus*) with bread, either as a form of play, or to capture and kill them.

Many primates have been the subject of LABORATORY experiments designed to test tool-using capabilities. Capuchin monkeys have used rope, wire sticks, cardboard, and cloth to pull in food baits. The variety of objects suggests that the animal has some concept of the task to be performed. Experiments with object modification, or tool making, point to a similar conclusion. Chimpanzees will obtain a stick of suitable length by breaking a branch from a tree, or by fitting two poles together. However, they will only do this once they have become familiar with these objects during play, or during some other experiment. It appears that the FRUSTRATION generated by its inability to reach the food interferes with the animal's insight into the tool-making possibilities. Once familiar with the relevant aspects of object manipulation, however, chimpanzees can often achieve a solution to the problem of food that is presented out of reach. Experiments with chimpanzees show that they will uncoil lengths of wire, fit tubes together, and remove stones from boxes so that they may be stacked to reach a hanging bait. They will break boards from boxes and split them along the grain to produce a stick. In one test a chimpanzee was presented with boards that it could not split with teeth or hands. It was provided with a hand axe and shown how to use it. The chimpanzee, however, failed to make use of the axe when a stick was required to obtain food, suggesting that using a tool to make a tool is beyond the mental capabilities of chimpanzees.

Many primates use tools in situations involving AGGRESSION. Monkeys often drop fruit and sticks from trees to intimidate other animals. Chimpanzees often shake growing branches during aggressive DISPLAYS directed at other members of the troop. This may enhance the effectiveness of the display, though it can hardly be called tool using. On some occasions, however, chimpanzees, baboons, orang-utans, and various capuchin monkeys have been observed to throw stones and sticks at other animals with deliberate aim. Gorillas and chimpanzees may brandish sticks during aggressive encounters, and may occasionally use them as clubs. Sticks may also be used during play and mock aggression.

The development of tool-using behaviour. Tool use among primates is largely learned. If a monkey

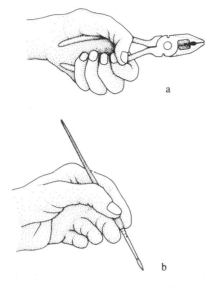

Fig. C. Power grip (**a**) and precision grip (**b**).

or ape accidentally drops a stick onto an enemy during aggressive display, and if the opponent retreats as a result, then the action of dropping the stick is likely to be rewarding (see REINFORCE-MENT). Similarly, if brandishing a stick is found to be effective in intimidating opponents, then the behaviour is likely to be repeated. Sooner or later the opponent will be hit with the stick and this will be found to be even more effective.

Apes are anatomically better adapted for throwing and hitting than monkeys. Apes are able to throw from a bipedal position, and the shape of their shoulder girdle, like that of man, enables them to throw with some force. During THREAT, chimpanzees make arm movements which are very similar to those used during throwing. These animals are thus pre-adapted to learn to use weapons during aggressive encounters.

Although much tool using may develop through TRIAL AND ERROR, some may be learned by imitation. Young chimpanzees manipulate sticks during play. They often poke them at unfamiliar objects, and then sniff the end of the stick, before touching the object with hand or mouth. Moreover, infants usually accompany their mothers during foraging, and have ample opportunity to observe them using tools. The primate hand is well suited to grasp objects. Young chimpanzees normally use the power grip (Fig. C), as do human infants, but adults use the precision grip for delicate work such as probing into a hole with a twig or straw. Many juvenile primates increase their manipulative skills during play, and make use of them in later life. Some aspects of tool use, such as probing for termites, require considerable skill to attain success, and it is doubtful that they could have developed through trial and error.

41, 92

TRIAL AND ERROR learning is a form of LEARNING in which a movement or manipulation leads to particular consequences. The animal then forms an association between its behaviour and the consequences, and the latter are said to provide REINFORCEMENT for the learning process. For example, a domestic cat (*Felis catus*) may paw at a latch so that a door swings open. Access through the door may provide positive reinforcement, and the cat will be more likely to paw at the latch on a future occasion. If opening the door had resulted in an unpleasant experience for the cat, the reinforcement would have been negative, and the cat would be less likely to paw the latch in the future.

Trial and error learning usually involves two aspects: classical CONDITIONING and OPERANT behaviour. The cat learns that the latch is associated with opening doors through a process of classical conditioning. It may show GENERALIZATION to similar latches on other doors. The cat also learns that certain paw movements (the operant) are necessary to make the door swing open.

Trial and error learning is usually contrasted with the type of learning in which the animal pays attention to particular stimuli on the basis of some hypothesis about, or INSIGHT into, the situation. It should also be distinguished from learning involving IMITATION, IMPRINTING, or any obvious involvement of INSTINCT. In the case of the learning that occurs during FOOD SELECTION, for example, the animal has a particular strategy with respect to novel foods. The essence of trial and error learning is that the animal makes a discovery as a result of an almost accidental action or situation. The learning takes place as a result of the discovery.

V

VACUUM ACTIVITIES occur in the apparent absence of the external stimuli that normally elicit the activity. For example, canaries (*Serinus canaria*) deprived of nest material will perform the movements of weaving material into a non-existent nest. Starlings (*Sturnus vulgaris*) which have not caught flies (Diptera) for some time may go through the motions of catching and eating non-existent flies.

The problem with this category of behaviour is that it is difficult to be sure that there is no relevant external stimulus which the animal can perceive although the human observer cannot. Thus the starling may be responding to a speck of floating dust as if it were a fly. Through the process of stimulus GENERALIZATION the dust may take the place of a fly, unknown to the observer. Pigeons (Columbidae) which are deprived of nest material will pick up a short piece of wire or a thin pencil, which they will repeatedly manipulate and build into a non-existent nest. This is a clear case of stimulus generalization, and not of vacuum activity. However, the border between these two categories of behaviour is a very tenuous one.

VIGILANCE, as used by students of ETHOLOGY and psychology, refers to an animal's state of readiness to detect certain specified events occurring unpredictably in the environment. We say that an animal is highly vigilant when it is very likely to detect these unpredictable events. This vigilant state is a function both of the animal's overt behaviour and of the activity of its central nervous system. It varies from the low levels associated with SLEEP to the state of heightened alertness that is provoked by an event signalling potential danger (see AROUSAL). The level of vigilance is also related to the types of event occurring in the environment, in that an animal may not be equally likely to detect all events. This is because, depending upon its MOTIVATION, it will be attending preferentially to one or another class of stimuli (see ATTENTION). A hungry finch (Fringillidae) searching for food, for example, will be in a state of readiness to detect the small seeds on which it feeds, but if it hears an ALARM call its attention will shift so that it becomes preferentially able to detect an approaching predator. We shall return later to this specific, directed aspect of vigilance.

Most of our knowledge of the relationship between vigilance and the activity of the nervous system comes from studies of man, and indeed the experimental study of vigilance has been largely concerned with the performance of human subjects in the laboratory. This may seem far removed from the situation of an animal in the wild, but the human experiments do provide information on the general properties of vigilance which is, to some extent, valid for other vertebrates.

Let us consider briefly, then, the study of the vigilance performance of human subjects. In one of the earliest experiments on this topic subjects watched a pointer move in small excursions around a white dial and their task was to detect the infrequent and irregular occurrence of double jumps of the pointer. Here a signal (double jump) appeared against a 'background' of non-signals (single jumps). Other experiments require the subject to detect the presence of a brief stimulus added to or subtracted from a monotonous environment (e.g. a momentary spot of light on a dark screen). In addition to visual stimuli, auditory and tactile signals have been used. In such experiments detection performance commonly deteriorates with time spent on the task, and the study of this vigilance decrement is of great practical importance for the efficiency of human watch-keepers, such as radar operators.

The measures of performance in such experiments are the proportion of correct detections made and the number of false alarms, i.e. the number of times a signal is reported when no signal appeared. The detection of signals, or the making of false alarms, involves two processes, DISCRIMINATION and DECISION. The first process requires the observer to be capable of distinguishing between signals and non-signals, and the latter requires him to decide whether each event is or is not a signal. These same processes of discrimination and decision must also occur in other animals, and the forces of NATURAL SELECTION will act so as to optimize their influence on behaviour. For example, prey animals may often 'decide' that sudden changes in stimulation are potentially dangerous and take avoiding action, even though some of these changes will in fact be false alarms and time and energy will be wasted in responding to them. Some compromise must be made between maximizing the response to danger signals and minimizing false alarms.

As mentioned earlier, the state of vigilance in man can be correlated with the activity of the nervous system. This is achieved by measuring the electrical activity of the BRAIN by means of the *electroencephalogram* (EEG). One component of the electrical activity of the cortex is greater just prior to omission errors (i.e. failures to detect a signal) than before correct detections, and this same component also increases in magnitude in the transition from wakefulness to sleep. This illustrates the fact that man's state of alertness can be measured in terms of the electrical activity of the brain, and shows that failures of detection are more likely to occur when the brain is in a less alert state. In cases where a correct detection is made, the frequency of the EEG trace just prior to signal detection is found to affect the speed with which a response is made to the signal. Since the EEG patterns found in man are similar to those in many other mammals, these neurophysiological correlates of vigilance are probably of quite general validity.

We have said that the level of brain activity will influence the likelihood that stimuli in the environment are detected. In addition, the detection of novel or unexpected stimuli has itself the property of alerting the animal and making subsequent detections more likely. This alerting is known as the ORIENTING RESPONSE, and involves a range of behavioural and physiological changes. Imagine the response of a drowsy dog (*Canis lupus familiaris*) to an unfamiliar, but not too intense, noise. It is likely to adopt a more alert posture, prick up its ears, look around, and dilate its pupils. These behavioural changes will be accompanied by changes in the EEG, heartbeat rate, and blood pressure, by an increase in muscle tone, and perhaps by a temporary inhibition of respiration. The function of these changes is to allow the animal to obtain more information about the nature and position of the unexpected event and to improve its chances of detecting, and of responding to, any important consequences of this alteration in its environment. The animal must first answer the question 'What is this stimulus and where is it?' before it can tackle the question 'What shall I do about it?'

We turn now to the vigilant behaviour of animals in the wild, and consider the FUNCTION of such behaviour. In general terms the function of vigilance is to enhance the animal's chance of detecting events, of extracting information from its environment. But vigilance also has more specific functions. The information to be extracted may be of many kinds, and vigilance can be seen as just one element of a behavioural complex directed towards some specific GOAL. The function of vigilance, therefore, will vary with this goal. Thus, the vigilance of a sexually motivated animal will be directed selectively towards potential mates, and in this context the goal and the function of vigilant behaviour, just as with COURTSHIP or copulation, will be to produce offspring.

The specific goals to which vigilance may be directed are many, and include the avoidance of danger, the procurement of resources, the maintenance of proximity in social groups, and the efficient receipt of COMMUNICATION signals. The most direct way that an animal can enhance its level of vigilance in relation to some goal is simply to spend more time in an alert state, actively searching the environment for the goal object, such as food, a potential mate, or a predator. Let us consider some examples of vigilant behaviour and the uses to which vigilance is put.

Eye movements are obviously direct indicators of visual scanning in those animals capable of such movements. Birds, however, have very limited powers of eye movement and in this group head movements are a behavioural indicator of visual scanning. Geese have vigilant postures ('head-up' and 'extreme head-up') which are easily distinguished from the grazing attitude (see Fig. A), and in the pink-footed goose (*Anser fabalis brachyrhynchus*) the extreme head-up posture is particularly associated with PARENTAL CARE since it is more common in those adults that are escorting goslings.

Fig. A. Grazing and vigilant postures in the pink-footed goose. **a.** Grazing posture. **b.** Head-up posture. **c.** Extreme head-up posture.

When a group of patas monkeys (*Erythrocebus patas*), in the grasslands of Africa, sense the presence of a predator, the male of the group will often climb a tree and look around for the source of the danger. If the predator is sighted the male can then take appropriate DEFENSIVE action; one of its strategies is to run past the predator, thus distracting it from the females and young of the group which lie quietly in the grass.

The male gelada baboon (*Theropithecus gelada*) of Ethiopia periodically surveys the baboon herd, monitoring the positions of the females in his harem; if they are well separated from him he stands in a particularly impressive posture, with tail erect and head high, scanning the herd. Then, on sighting one of his females, he runs after her at great speed. In this way he maintains the integrity of the harem and prevents his consorts from copulating with other males. The females, for their part, keep an eye on their male overlord, periodically moving closer to him.

When maintaining vigilance for food items animals may develop a SEARCHING IMAGE. This enables a selective search to be made for a particular type of food, and has the advantage that the desired items are more likely to be detected and that distraction by other stimuli is reduced. A great tit (*Parus major*) feeding its nestlings, for example, will repeatedly return to the nest with the same species of insect, although many different prey species are available.

Vigilance is often an important element of parental care. Young animals are often unskilled in vigilance; their perceptual and signalling systems may not have fully developed, and in many cases they will be unable to differentiate between dangerous and non-dangerous stimuli. In many cases, therefore, parents must maintain vigilance for the protection of their young. This is the case in geese, for example, where the additional vigilance burden imposed by the young is borne largely by the male. Similarly, in groups of the vervet monkey (*Cercopithecus aethiops*), the adults spend more time in look-out behaviour than do the sub-adults, the juveniles, or the infants.

Because there are always conflicting demands on an animal's time budget, and because there are physiological limits to the maintenance of a high state of alertness, animals have adopted a number of strategies to supplement active watch-keeping. GREGARIOUSNESS, for example, provides various advantages in relation to vigilance. It serves as an ANTI-PREDATOR device since an approaching predator is more likely to be detected by some member of a group than by a solitary individual. The whole group can be rapidly informed of the danger by means of an ALARM signal. This phenomenon has been illustrated experimentally with flocks of the red-billed weaver bird (*Quelea quelea*) and the common starling (*Sturnus vulgaris*). From this property of gregariousness it also follows that gregarious individuals can afford to reduce their own level of active watch-keeping and still be safer than if they lived alone. In this way they are able to save valuable time which can be spent in feeding or some other maintenance activity. Such a strategy seems to have been adopted by the white-fronted goose (*Anser albifrons*), since in larger flocks of this species individuals spend less time in the alert, head-up posture and more time grazing. Nevertheless, at any one time, larger flocks contain more individuals in the head-up posture, and are therefore more likely to detect an approaching predator.

Gregariousness also facilitates the finding of food in species of birds which feed in flocks, such as many seed-eating finches (Fringillidae). For such birds it is often far easier to detect a group of feeding conspecifics than to locate small food items directly.

Vigilance, then, is a state of readiness to detect unpredictable events. It fulfils many functions and is often aided by adaptations which, directly or indirectly, increase the probability of detection. Only when detection is successful can the animal decide what response should be made. J.L. 36

VISION is the detection of light by eyes and the behavioural response it produces. The eye signals to the BRAIN attributes of the light such as its intensity, spatial distribution, variation over time, and colour. In many the subjective experience provoked by light is called seeing, but vision and seeing are not synonymous. Human patients with damage to the visual areas of the brain may retain some responses to light, such as blinking to a bright light or following a moving pattern with their eyes, even though they deny seeing anything. When DREAMING we have all experienced the distinction between vision and seeing; in dreams the brain generates visual images without the eyes receiving light. Similarly, we have no knowledge of the subjective experiences of animals, and their vision may be quite unlike our seeing, especially if the animal has a small and primitive nervous system. Nevertheless, the word 'see' will be used as a convenient shorthand, without implying any particular conscious perception.

An eye is defined as a cell or organ specialized to absorb light by means of photosensitive pigments, and to signal that event to the nervous system. The latter is important to the definition because plants absorb light and use its energy for photosynthesis, but we would not wish to say that they have eyes or vision. The definition means that other means of detecting light, for example by its warming effect on the body, are not visual.

Historical background. Although the structure of the eye was nearly, but not quite, correctly described by the ancient Greeks, their explanations of its function were ingeniously incorrect. Beginning with Pythagoras in 532 BC, a popular theory was that invisible rays, likened by Galen to delicate threads of cobweb, emanated from the eye and touched objects in the world, which were thereby sensed like an exquisitely delicate and extended sense of touch. The emanation theory, which survived for about 1500 years, had a popular rival proposed by Democritus, namely that objects continuously emitted images of themselves, like casts or impressions, that were received by the eyes. Both theories had trouble in explaining why it became difficult to see after sunset, but Plato resolved the problem by proposing that objects emitted their impressions, or rays, by using the sun's light and that perception was only possible when the rays from the eye combined with those from the world outside. It is odd that Plato failed to acknowledge the redundancy of the inner rays, and that Aristotle's much simpler suggestion that we see solely by detecting intangible rays from illuminated objects was vigorously opposed. However, all the rival theories agreed that the lens was the principal organ of sight, and that the *retina* and *optic nerves*, which were thought to be hollow tubes, served to nourish the lens and carry its message back to the brain in visual spirit.

The next concerted attempt to explain vision was by Arab scholars, who inherited the Greek view of the structure of the eye and who provided the first recorded diagram by Hunain ibn Ishāk in the 9th century AD (Fig. A). It repeats the Greek error of having hollow optic nerves, a spherical lens, and an anterior chamber containing circulating visual spirit. The early Arab ophthalmologists also accepted the emanation theory and even devised a method of surgically removing opacities of the lens which were believed to result from excess fluid, hence 'cataract'. The main and successful opposition to the emanation theory came in the 11th century AD, from Alhazen who, reiterating Aristotle, postulated that vision was accomplished by rays entering the eye through the pupil. In placing the visual image on the back of the 'glacial sphere' it is not clear whether Alhazen was referring to the back of the lens, or the back of the vitreous humour (i.e. jelly-like material behind the lens), but if the latter he was the first to identify correctly the role of the retina. Certainly by the

16th century the Renaissance anatomists, who translated the Arabic texts, doubted the photoreceptive role of the lens, abolished the canal in the optic nerves, and suggested that the layer we now call the retina received the visual impression. Leonardo da Vinci proposed that there was an image on the retina but, believing that it must be upright like the world we see, claimed that there must be two images, the first inverted in the front of the eye and caused by the pupil acting as a pin-hole camera, and the second erect on the retina. It only remained for Maurolycus (1554) to show that light was collected and refracted by the lens to produce an image (which he nevertheless incorrectly placed on the back of the lens), and for Platter (1583) to show that vision persisted after cutting of the suspensory ligaments of the lens, along which the visual spirit was thought to flow, and the Greek theories were toppling. They were finally overturned in 1619 when Scheiner cut a small hole in the coating at the back of several animal and human eyes and observed an inverted image on the transparent retina. His drawing of the eye is shown in Fig. B.

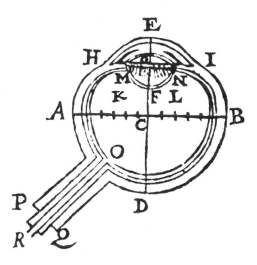

Fig. B. Horizontal section of the right human eye by Scheiner (1619). It is the first diagram in which the proportions are correct, but the iris should be tangential to the lens.

Electromagnetic radiation and light. It may seem odd that for more than 2000 years educated men failed to understand either image formation or visual sensitivity, but they lacked any knowledge of the nature of light and its action on certain animal pigments. Both are indispensable to understanding vision.

Energy is radiated by the sun at a speed of about 279 000 km/s. This electromagnetic radiation can be considered as a mixture of components, each being periodic, or wave-like (Fig. C). The distance between the peaks of the waves is the *wavelength*. The number of waves per second is the

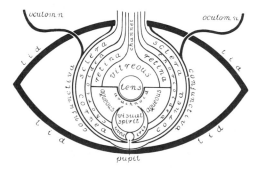

Fig. A. Earliest known diagram of the eye, prepared from the drawing by Hunain ibn Ishāk (AD 860). The Arabic labelling has been translated.

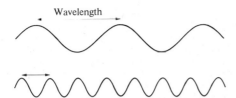

Fig. C. The wave-form nature of light. The upper wave-form has a longer wavelength but smaller frequency than the lower wave-form.

frequency. Frequency and wavelength are inversely related. Fig. D shows the enormous and continuous range of wavelengths of electromagnetic radiation. In musical terms it covers 70 octaves, yet the eye is sensitive only to radiation within one of these octaves, called the visible spectrum. In fact the existence of wavelengths outside the visible spectrum was unknown until 1800, when Herschel showed that a thermometer could be heated by infra-red radiated by a warm body in darkness. Ritter then showed that silver chloride could be photochemically altered by the invisible rays we now call ultra-violet, Maxwell predicted a continuous range of wavelengths in 1865, Röntgen discovered X-rays in 1895, and radio waves soon followed.

We must now attempt to answer two questions. The first is 'How do animals see that part of the electromagnetic spectrum we call light?' The second is 'Why do they fail to see the huge remainder?' The answer to the first question is that photosensitive pigments in receptor cells can absorb light, whose captured energy initiates a chemical change in the pigment molecules. This change alters the electrical potential across the cell membrane which in turn induces nerve impulses in the nerve cell connected to the receptor. The nerve impulses are transmitted to the brain along the optic nerve. When a photopigment has been changed (isomerized) by light it is said to be bleached, and the bleaching action of light is familiar in washing hanging out to dry. However, photoreceptor bleaching differs in being both rapid and reversible, so that the eye continuously regenerates its photopigments for further use.

So far we have described light as a wave-form, but in describing its energy it helps to consider light as discrete packets of energy called photons. The energy of one photon is inversely proportional to the wavelength of the radiation, so certain wavelengths are more effective than others in bleaching a particular photopigment. In fact one photon of the right wavelength can change (isomerize) a molecule of photopigment when the latter is maximally sensitive.

The answer to the second question is less straightforward. Why can we not see X-rays or radio waves? First, as energy is inversely proportional to wavelength of radiation, it follows that the long wavelength radio waves will contain too little energy to reliably isomerize pigments. Radio telescopes solve this energy problem by reflecting radio waves from the concave surface of an

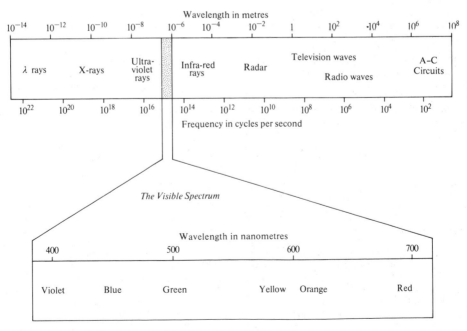

Fig. D. The electromagnetic spectrum, showing the small part that is visible.

enormous metal dish and focusing them onto an electronic collector at the centre of curvature. Conversely, the very short wavelengths of X-rays and gamma rays contain so much energy that they damage living tissue (hence the precautions in hospitals and nuclear power stations), so that even if short wavelength radiation from the sun were not filtered out by the atmosphere it is doubtful whether an eye that detected them could have survived. Eyes have therefore evolved to capture the one octave of electromagnetic radiation that has enough harmless energy.

The second reason for selecting only one octave depends on refraction. When radiation passes from one medium to another of different density, for example from air to glass, it is bent or refracted. A lens works by refracting all the rays falling on it from a point source, so that they converge on a focal point behind it, which is the image. Now the angle of refraction depends not only on the density of the lens, but also on the wavelength of radiation. The shorter the wavelength the greater the refraction. This means that no simple lens can bring all wavlengths to a common focus. We shall return below to the means by which the eye deals with the differing refraction within the visible spectrum, but it is obvious why the invisible long wavelengths are not detected. An eye that focused radio waves would have to be as large as a house, because radio waves are bent so slightly by a lens. By detecting only a narrow band of fairly short wavelengths the eye can produce a fairly crisp image and remain compact.

In summary, there is no theoretical reason why eyes could not detect wavelengths outside the visible spectrum, and indeed the visible spectrum is not identical in all animals. However, evolution has produced eyes that respond to wavelengths that penetrate the atmosphere, have about the right energy to excite photopigments, and are sufficiently refracted to form an image at a short focal length.

Types of eyes. Provided a cell can absorb light it can respond to light, by using a resulting chemical or temperature change to modify its behaviour. Many single-celled organisms such as Protozoa respond to light by moving towards or away from it. As they are generally not radially symmetrical—that is they are shaped like a Rugby ball rather than a football—most rotations of the cell will alter the amount of light absorbed and the cell can move in such a way as to maximize or minimize the absorption. However, there can be no instantaneous appreciation of the direction of illumination unless the cell can detect regional differences in absorption within itself, which is possible in organisms like the protozoan *Paramecium*, where the cilia on the more strongly illuminated side may beat more quickly in response to chemical changes beneath them. Similarly the two flagellae of *Chlamydomonas* beat asynchronously when light falls from one side on the photosensitive cyto-

plasm beneath them, and thus turn the cell, like a rowing boat when one oar is pulled hard.

In higher animals there are cells that are specialized to absorb light and to transmit information about it to the nervous system. In the simplest arrangement, seen in the earthworm *Lumbricus*, the cells are scattered on the surface of the body, and their combined pattern of activity provides crude but instantaneous information about direction as well as intensity of illumination. Photosensitive cells are more commonly found in groups, usually in a depression in the body surface. This is called an eye pit (Fig. Ea) and is the first anatomical structure that can be called an eye because an extremely crude image or shadow is formed. For example, if the light is coming mostly from the right, the left side of the pit will be more strongly illuminated than the right, which is sheltered by the rim. This regional variation can be signalled by the nerve fibres connected to the array of receptors. Another advantage of a pit is that it protects the receptors from glare which, coming from all directions, reduces the contrast of the shadow. Shielding our eyes against a bright sky has the same effect.

If the rim of a visual pit grows inwards leaving only a small hole at the centre, we have a pin-hole eye like that of the primitive cephalopod mollusc, the pearly nautilus (*Nautilus pompilius*) (Fig. Eb). It works exactly like a pin-hole camera, allowing light from any one point to reach only a small part of the interior and so form an inverted image. As no refraction and therefore no focusing are involved, the pin-hole eye is unique in producing an image whose clarity is independent of the distance of objects from the eye. But it has two major disadvantages that probably explain why it is now an evolutionary relic. The first is that light is scattered, or diffracted, when it encounters an edge so that the shadow of the edge is never completely sharp. Instead it is flanked by alternating dark and light bands whose width depends on the wavelength of the light, blue producing narrower bands than red. If white light, which contains all the visible wavelengths, enters a pin-hole, these diffraction bands for different wavelengths do not overlap perfectly and stripes of different colours occur alongside all edges in the image. The same phenomenon occurs when light is reflected from an edge, hence the beautiful reflected colours of mother-of-pearl, whose surface bears multiple minute striations, and the curious coloured fringes we see when a microgroove record is viewed obliquely. So a pin-hole eye will never produce an image of high quality because the circumference of the hole, where diffraction occurs, is so large in relation to the area of the hole.

The second disadvantage is that the pin-hole is small relative to the sensory surface behind it, so the image will always be dim, like that in a camera whose iris has been closed right down.

Both visual pits and pin-hole eyes are prone to

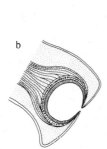

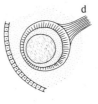

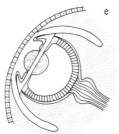

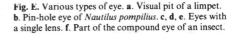

Fig. E. Various types of eye. **a**. Visual pit of a limpet. **b**. Pin-hole eye of *Nautilus pompilius*. **c, d, e**. Eyes with a single lens. **f**. Part of the compound eye of an insect.

accumulate dirt. The evolution of a protective window in front of the opening may have been the first step in the origin of a lens and the development of a single lens eye whose optical properties are so similar in all vertebrates, Arachnids (spiders), modern cephalopods, molluscs, e.g. the common octopus (*Octopus vulgaris*), and many gastropods (snails) (Fig. Ec–e). If the lens opening, or pupil, is large, the effects of diffraction are negligible because with every doubling of the circumference of the pupil, where diffraction occurs, its area increases fourfold, which means that the image is both sharp and bright.

Before considering the properties of the vertebrate eye in more detail we must look at the very different way in which insects and some crustaceans (e.g. shrimps, Natantia) have solved the problem of producing an image. The compound eye of insects consists of up to 25 000 slim light-tight tubes called *ommatidia* (Fig. Ef). Each ommatidium has two lenses at its surface, which funnel rather than focus the light onto eight elongated photoreceptor cells surrounding a central *rhabdom*. Each receptor is connected to a nerve cell beneath it which signals the intensity of light from the direction in which the ommatidium points. A single ommatidium does not produce an image, but since the entire compound eye is spherical, and adjacent ommatidia point in slightly different directions, the distribution of light over all the ommatidia mirrors the pattern of illumination in the environment and gives a picture formed of closely packed dots, like that of a newspaper photograph.

In addition to their large compound eyes many insects have a small number, typically three, of very simple eyes called *ocelli*, lying between the *compound eyes*. There is a single lens and a true image.

Structure of the vertebrate eye. Fig. F shows a horizontal section through the centre of the human eye. The *sclera* is a tough coat that protects

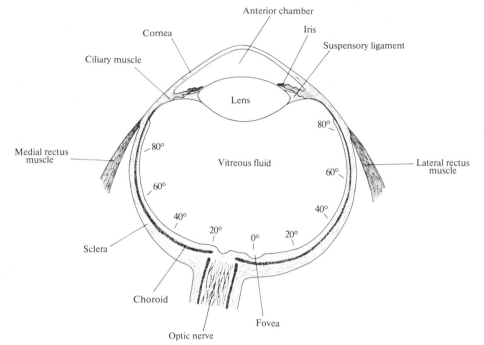

Fig. F. Section through the human eye. The numbers in degrees along the retina show where different points in space are represented.

the eye, maintains its shape by resisting the intra-ocular pressure, and provides an attachment for the eye muscles. The *choroid* is black and reduces both internal reflections and transmission of light through the side of the eye. The *retina* contains the photosensitive receptors, the nerve cells to which they are connected, and the blood vessels. The posterior chamber contains a clear jelly, the vitreous fluid, and the anterior chamber is filled with the aqueous fluid. Both aqueous and vitreous fluids have a lower density than the lens, so light is refracted at both surfaces of the lens as well as between the air and the *cornea*. The *iris* controls the size of the pupil, and therefore the amount of light entering the eye. The *ciliary muscle* surrounds the lens. When it contracts the lens becomes more convex, bringing close objects into focus on the retina; when it is fully relaxed the lens is flattened and distant objects are in focus. In possessing a *fovea* (see below) and the means of discriminating colours the human eye is not typical of all vertebrates, but its general design is fairly representative.

Image formation. The first job of the lens is to produce a sharp image. This is far from easy. Light from a source at infinity produces parallel rays. Falling on a convex lens they will be refracted and bought to a focus at one point. The distance between the centre of the lens and the image from parallel rays is the focal length of the lens. As the light source is brought closer to

the lens the light rays are no longer parallel, they strike the lens at a different angle, and are focused further behind the lens (Fig. G). This means that no lens of a constant shape and fixed distance from the retina can produce a clear image of objects at all distances, as every photographer knows. This problem is solved in the mammalian eye by *accommodation*, that is, changing the shape of the lens so that its focal length can be altered. Unfortunately the lens loses its elasticity with age, which is why we may need additional spectacle lenses for reading as we get older.

The lens is not the main refractive surface of the eye, as Kepler discovered at the end of the 16th century. The cornea provides about twice the refraction of the lens, but the importance of the latter is that it contributes all of the variable refraction for accommodation. However, the refraction at the cornea has important consequences. When we swim under water everything is blurred, because water has nearly the same refractive index as the cornea, which no longer refracts the light, and our lens cannot change sufficiently to compensate for the lost refraction. The clarity is restored by wearing goggles filled with air. How then do fish see clearly? They do so by having a nearly spherical lens of high refractive index and short focal length. But its shape is inflexible so accommodation is accomplished by moving the lens towards or away from the retina. Animals that live both in and out of water like the common frog (*Rana temporaria*), the American alligator (*Alligator mississippiensis*), and the hippopotamus (*Hippopotamus amphibius*) cannot see well in both. They have eyes near the top of the head which stick up like periscopes when the body is almost totally submerged.

The next problem facing a lens is *spherical aberration*, which occurs in all simple lenses formed of one material and having a uniform radius of curvature (Fig. H). Light from a point source passing through the edge of the lens is brought to a focus in front of that passing near the centre, which means that the image is blurred. As refraction depends on the curvature of the lens, the radius of curvature of the cornea diminishes towards the edge so that the light at the margin is less strongly refracted. In addition the density of the lens decreases at the margins with the same effect. The result is that animal eyes are as well corrected for spherical aberration as the most expensive cameras. Any remaining aberration is eliminated by constriction of the iris, so that only the centre of the cornea and lens transmits light to the retina. This is equivalent to increasing depth of focus in a camera by 'stopping' down the iris diaphragm. Hence the constriction of the iris commonly seen in people and animals when they visually attend to something.

The final obstacle to a clear image is *chromatic aberration*, which can be predicted from the earlier discussion of the relation between refraction and wavelength. A simple lens cannot bring radiation

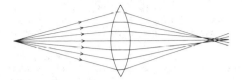

Fig. H. Diagram to show spherical aberration. The rays at the edge of the lens are brought to focus in front of those near the centre. The result is a blurred image.

of different wavelengths to a focus in the same plane (Fig. I). Blue light is in focus in front of green, and green in front of red. As white light contains all visible wavelengths it therefore produces a blurred image in which only one colour component can be in focus. The problem is solved in a camera by cementing together lenses of different shape and refractive index; the eye's solution is to discard the most troublesome but least important wavelengths. The first step is the lens,

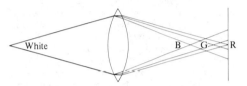

Fig. I. Chromatic aberration. When white light shines through a lens, different wavelengths are not equally refracted. Blue is focused in front of green, and green in front of red. In the diagram red is in focus, blue and green are not.

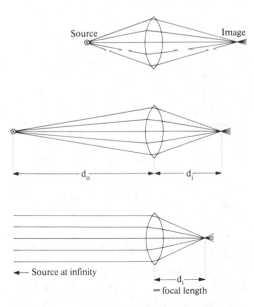

← Source at infinity

d_i = focal length

Fig. G. Diagram to show that as an object moves towards a lens its image occurs further behind the lens. If the lens cannot move backwards or forwards from the retina (as it does in fish), its shape has to be changed to keep the object focused on the retina.

which is yellowish and therefore absorbs almost all the ultra-violet light. In fact this is why we fail to see ultra-violet, and patients in whom the lens has been removed can detect the ultra-violet that now reaches the retina, just as honey-bees (*Apis mellifera*) see in ultra-violet light by having a compound eye that transmits it. The colour correction in the eyes of human beings and other primates is further accomplished by an orange screening pigment lying in front of the receptors at the *fovea* (the region of most acute vision at the centre of the retina). This filters out blue light, allowing chiefly the green and red regions of the visible spectrum to produce a tolerably clear image here. The snag is that visual acuity is poorer in blue than in green or red light, but that is probably unimportant in primates because the most salient coloured objects in their environment tend to be red or green. Apart from flowers, blue light comes from an undifferentiated sky, and we perceive it chiefly by the unscreened receptors in the periphery of the retina. We should remember that chromatic aberration is just as much a problem in colour-blind animals. Although all wavelengths of light have the same perceived hue, which a colour-blind person calls grey, the physical image and therefore the perception will still be blurred unless chromatic aberration is corrected.

The final correction for colour aberration in primates occurs only at the very centre of the fovea, occupying about half a degree of vision, where the receptors contain no pigment sensitive to blue. Any remaining aberration is therefore unnoticed because it is not even detected. This is why a tiny blue spot appears black when we look directly at it from a distance, whereas the colour of a red spot is still perceived. By glancing slightly to one side the blue reappears.

Instead of eliminating chromatic aberration, some fish and arachnids have taken advantage of it in a highly ingenious way. In the jumping spider *Metaphidippus aeneolus* the receptor cells containing photosensitive pigment lie in four layers one behind the other, and the three deepest layers correspond to the focal planes of blue, green, and red light. As fish and spiders are known to have colour vision it seems likely that each layer of receptors is responsible for detecting a different colour and precisely in the plane where that colour is focused. This proposal is strengthened by experiments on spiders in which the electrical activity of receptors in two of the layers has been recorded while the eye is stimulated with coloured light. Receptors in one layer were stimulated chiefly by blue light and in the other by green light.

Polarized light. The one remaining property of light that we have not described is its plane of vibration. The light wave shown in Fig. C is like a rope attached to a post at one end and vigorously wagged up and down at the other. But the rope could be wagged from side to side or in any other plane, and waves of ordinary light do consist of vibrations in all planes. Some substances have the property of absorbing light vibrations in one plane and transmitting those at right angles, and we use them in polarized sun-glasses. More important to vision, when ordinary light is reflected from water or a wet road the horizontal vibrations tend to be reflected and the vertical transmitted. This means that the reflected light is now plane-polarized. We can reduce such reflections, which dazzle us when driving or sailing, by wearing sun-glasses that selectively absorb horizontally polarized light. Now particles in the atmosphere also polarize light, but the proportion that is polarized and the plane in which it is polarized vary across the sky according to the position of the sun. Many animals navigate or orient by using the sun's position (see NAVIGATION, ORIENTATION). But what happens on a cloudy day? As long as part of the sky is blue the position of the sun can be determined by the amount and type of polarization of the light in the unclouded patch. Many arthropods, insects, crustaceans, and arachnids have eyes that detect this polarization. For example migrating butterflies (Lepidoptera) placed in a box will change the direction of their MIGRATION if the top of the box consists of a polarizing filter that is systematically rotated to change the plane of polarized light entering the box. Common sand fleas (*Talitrus saltator*) enclosed in a bowl jump in the direction of the seashore, but if a polarizing filter above the bowl is rotated they follow suit. But perhaps the best example is the honey-bee, which indicates the direction in which it has found food by performing a dance in a particular direction with respect to the sun. Usually the dance is performed inside the hive, but occasionally it is done outside on the alighting board. Von FRISCH showed that if the illumination on the alighting board was polarized the direction of the dance changed as he changed the plane of polarization. He then made an octagonal filter from eight pieces of triangular polaroid (Fig. J). When he looked through it at different parts of the sky he found that the pattern of brightness changed even though the sun was behind clouds.

We are not sure how the eye detects the plane of polarization, but there are clues. The retina of the jumping spider *Metaphidippus* has already been described as having four layers, three of which may respond to red, green, and blue light respectively. The fourth layer contains receptors that are too close to the lens to receive a focused image, but which are elongated and, unlike those of other layers, are arranged so that they all lie in a plane at right angles to the incident light, which may allow them to respond best to light polarized in that plane.

How the compound eye detects the plane of polarization is also uncertain, but we know that the central crystalline pillar in each ommatidium consists of eight segments, each adjacent to a receptor (Fig. Ef). Together they may function like von Frisch's octagonal polarizing filter.

Apart from navigation over long distances, what could be the purpose of detecting the polarization of light? Arthropods have low visual acuity, i.e. their detail vision is poor, and even the spider's eye with its lens and retina is very shortsighted, so that things beyond its web or its jumping distance will be blurred. Like the sand flea already described these animals may therefore use the plane of polarization to indicate the direction of home when it is out of sight. It may also prevent them from jumping onto water if they are terrestrial, or lead them to water if they breed, feed, or sometimes live there, which may explain why aquatic beetles that occasionally fly, like the great water beetle (*Dytiscus marginalis*), sometimes crash-dive on a road or a car bonnet, both of which produce plane polarized reflections like those from water.

Structure of the vertebrate retina. The image produced by the lens of the vertebrate eye lies on a layer of photoreceptor cells. When they capture light they transmit their excitation to several types of nerve cell in turn, the last type being the *ganglion cells*, whose nerve fibres leave the eye and end in the brain.

In the human eye there are about 130 million receptors and 1 million ganglion cells, which means that each ganglion cell is connected to many receptors. The vertebrate retina, unlike that

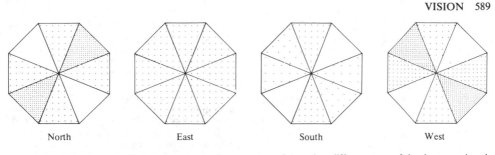

| North | East | South | West |

Fig. J. Four different brightness patterns seen at the same time of day when different parts of the sky were viewed through an octagonal filter made of eight pieces of polaroid. Depth of shading indicates brightness.

of cephalopods such as the octopus, is curious in that the light reaches the receptors after passing through all the other elements, which must impair the image slightly.

There are three features of the retina that are especially important in understanding vision. These are the distinction between rods and cones, their distribution within the retina, and the type of photopigment they contain.

Rods and *cones* are so called because of their shape. The eyes of strongly nocturnal species contain few or no cones (for example dogfish, *Scyliorhinus*, and bushbabies, *Galago*), whereas those of strongly diurnal species (for example many lizards, Lacertidae, and squirrels, Sciuridae) contain few or no rods. This led Schultze to propose (1866) that rods serve dim-light vision and cones bright-light vision, and that most animals have both to enable them to see under a variety of light intensities. But why not have only rods, which might enable us to see all lights from the very brightest to the very dimmest? The answer is that although rods are exquisitely sensitive to weak light, they are saturated and dazzled by bright light and soon cease to provide excitation about its intensity. Cones, on the other hand, begin to signal the intensity of the light when it is already bright, and continue to do so over a large intensity range. This was discovered by recording the electrical activity of single rods and cones in the amphibian mud-puppy (*Necturus maculosus*), and was the first direct confirmation of Schultze's theory over 100 years after he proposed it.

The receptors may be uniformly distributed over the retina, as in rats and mice (Murinae), but it is much more common for cones to predominate in one area, rods in another, and for both to be scarcer at the edge of the retina than in the centre. This allows different parts of the retina to be specialized for the detection of light, the resolution of detail, and the perception of colour and movement.

Sensitivity to light. After being in a totally dark room for about ½ h we can detect a flash of light so weak that only 5–10 photons reach the receptors. Moreover, if the light comes from a fairly large spot it can be shown that each photon must

have been absorbed by a different receptor and that each receptor has therefore reached the limit of perfection in sensitivity. But if so, why can we not see one photon instead of 5–10? The reason is that receptors are slightly unstable chemically, and are spontaneously active or 'noisy', sending spurious signals to the brain. By insisting that several receptors signal simultaneously the brain ignores the isolated false alarms. Like people, it is more impressed by a deputation than an isolated caller. By teaching animals to make a response whenever a light appears, it has been shown that the laboratory rat (*Rattus norvegicus*), the pigeon (*Columba livia*), the cat (*Felis catus*), and the rhesus macaque monkey (*Macaca mulatta*) are all as sensitive as we are when fully dark-adapted. They can see light so weak that the most sensitive instruments can barely detect it.

In bright sunlight we can see changes in brightness even though the overall intensity is a billion times greater than the weakest visible light in the dark. Once again experiments on monkeys and pigeons show that they are not different from us. The range of light intensities to which the eye is sensitive is therefore enormous. No camera is as versatile, for a film sufficiently sensitive to detect a few photons would be totally and equally blackened at high intensities, and one that produced an acceptable picture in sunlight would reveal nothing in near-darkness. The eye seems to adjust its sensitivity according to the prevailing illumination, and how it does so can be found by studying the *dark adaptation curve*.

When we step from sunlight into a cinema we take some time to become accustomed to the dim surroundings. This change in sensitivity can be plotted precisely by measuring the weakest light we can see at various times after entering a dark room. The intensity of the just visible light is known as the *threshold*, which falls progressively in the dark (Fig. K) until it reaches the minimum of about 5–10 photons. The most conspicuous feature of the dark adaptation curve is that it has a kink in it. If the experiment is repeated with a very rare human subject who has no cones in his retina (a rod monochromat), there is no kink and the steep curve is obtained. If it is carried out with a

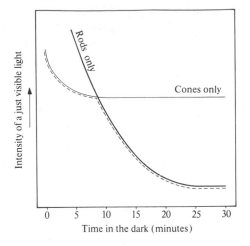

Fig. K. Graph showing how the intensity of a just visible brief light decreases as we adapt in darkness (dotted line). The steep line shows what happens in a person with only rods in the retina. The shallow line shows what happens when the light is on the fovea, where there are only cones.

normal subject and a small red target light confined to the fovea, where there are no rods, the shallow curve is obtained and again there is no kink. Together, these observations show that during the first part of the curve, while we are still light-adapted, our sensitivity is determined by cones which do not further increase their sensitivity after 10 min. The rods then take over and are entirely responsible for our detection of very weak lights. Further evidence for this is that in bright light we see best with the fovea, where there are no rods, whereas at twilight we can detect dim objects best by looking to one side and using the part of the retina where rods are commonest. A further important feature of the dark adaptation curve is that if the threshold light is coloured we only perceive its colour in that part of the curve above the kink, suggesting that cones are responsible for our colour vision.

What is the mechanism in the eye that governs the billionfold change in sensitivity during dark adaptation, and its reversal when we step out into the light again? There are five factors. First, the pupil gets bigger in darkness, but it does so in a second or two and in any case it allows no more than a sixteenfold change in the light reaching the retina. Second, it has been shown by electrical recordings from single receptors in the mud-puppy that the range of intensity over which the receptor gives a strong signal can change, according to the prevailing light level. How it does so is mysterious, but once again it will not account for more than a fraction of our adaptation because it occurs within seconds. The next method is more gradual and very important. As receptors greatly outnumber the optic fibres carrying their message to the brain,

it follows that many receptors bombard a single nerve cell, which signals the presence of light when the bombardment surpasses a certain level. One way of improving sensitivity to weak light is to increase the number of receptors bombarding each nerve cell, so that even though their signals are individually weak they are collectively strong. By recording the electrical activity of single nerve cells in the retina of anaesthetized animals it has been shown that during dark adaptation the area of retina over which each nerve cell collects signals from receptors becomes bigger. However, the increase in sensitivity is achieved at the expense of detail vision. If two lights fall on separate receptors converging onto a single nerve cell, the latter can signal nothing about the distribution of the light. But if those receptors connect with different nerve cells the lights can be separated. So whenever sensitivity is improved by pooling signals from receptors, detail vision suffers. This explains why visual acuity deteriorates at twilight, and why strongly nocturnal animals have excellent sensitivity but poor acuity.

The fourth means of adjusting sensitivity concerns the time which the brain takes to make a decision about the presence or absence of light. This is variously called *integration time, summation time*, or *perceptual moment*. If a flash of light lasts 50 ms, it looks identical to a flash lasting 1 ms but containing the same amount of light energy. This is because we are light-adapted we take about 50 ms to evaluate the nerve signals about the light, and within that period our perception of the light is a product of intensity and time. This is known as Bloch's Law. If the general illumination is dim, one way of improving sensitivity is to lengthen the time over which weak signals are summated, like lengthening the exposure time on a camera on a dull day. That the eye and brain do this can be shown by measuring the duration over which Bloch's Law holds at different stages of dark adaptation. When we are fully dark-adapted we summate signals over a period as long as $\frac{1}{2}$ s before deciding whether we have seen something. But again it is done at some cost because our reaction time is correspondingly lengthened, which is one reason why batsmen appeal against bad light when facing fast bowlers.

Changes in pupil size, the sensitivity range of receptors, and the area and time over which signals from receptors are pooled still account for less than a third of the total adaptation observed in vision. The remainder lies in photopigment regeneration. When a light is shone in the eye some of it is reflected out again, like that from an animal's eye caught in car headlights. If the photopigment is not bleached already very little is reflected out, whereas if much of the photopigment is already bleached more of the light is reflected. By shining a known amount of light in the eye and measuring how much is reflected back it is possible to assess the proportion of photopigment in the bleached condition. This technique is called *retinal densitometry*. By making the measurements at different stages of dark adaptation it has been shown that the photopigment regenerates during darkness, and that our visual threshold is proportional to the amount of unbleached pigment available to absorb light. Furthermore, by shining the measured light on the fovea or on the periphery of the retina it is possible to study the cone and rod pigments independently. Cone pigments regenerate faster, reaching full regeneration after 10 min, whereas rod pigments take much longer. This again helps to explain the kink in the dark adaptation curve of Fig. K.

It can now be seen that the eye is beautifully suited to deal with vision under different light conditions. Relatively small but rapid alterations in light intensity can be handled by the pupil and the working range of the receptors. Much larger changes in illumination are dealt with by spatial and temporal summation, and by pigment regeneration. But why is the latter so slow? Until the era of electric light, the vast majority of large changes in illumination occurred slowly, at sunrise or sunset, or when a storm was brewing. The eye has evolved to deal with these and to keep its sensitivity attuned to the heavens and not to artificial light sources.

Different visual pigments. Red glass appears red because it transmits most of the red light falling on it but absorbs most of the shorter wavelengths like blue and green. A visual pigment is like coloured glass, absorbing maximally in some part of the spectrum and much less at wavelengths to either side. This means that an eye containing only one kind of visual pigment will be most sensitive to wavelengths that are maximally absorbed by that pigment. Experiments on animals that have only one photopigment or have trivial amounts of any of the others, for example the laboratory rat, show that they are most sensitive to light with a wavelength of about 500 nm, and that when the pigment is extracted from the eye it absorbs best at this wavelength. An animal with only one type of visual pigment will be colour-blind because a single pigment provides no information about the wavelength of the light it has absorbed. All it can do is signal that light has been absorbed, and it cannot distinguish between a weak light of the optimum wavelength and an intense light of a less effective wavelength. This is known as the *principle of invariance*.

Many animals have several different photopigments in the eye, each in a separate class of receptor, and each maximally sensitive to a different wavelength. This enables the animal to be highly sensitive to light over the whole of the visible spectrum instead of in just one part of it, and it provides the basis for colour vision because different wavelengths now maximally excite different groups of receptors (see COLOUR VISION).

The commonest photopigment in vertebrates is *rhodopsin*, which occurs in rod receptors and absorbs best at about 500 nm. Cones contain the pigments responsible for colour vision, and there are typically three types, absorbing best in the blue, green, and yellow-red portions of the spectrum. However, there are many variations. The honey-bee has a visual pigment maximally sensitive to ultra-violet light, which enables it to see the nectar guides on flowers that are known to us only by photographing them with a film sensitive to ultra-violet. On the other hand honey-bees have no pigment specifically sensitive to red light, and behavioural experiments confirm that their discrimination in this part of the spectrum is poor.

Although eyes commonly contain several photopigments, they are not necessarily in use at the same time. The rods of the mammalian eye, containing rhodopsin, are only used at low levels of illumination, which is why we are most sensitive to blue-green light at twilight. In bright surroundings the rods are inactive and we see with our cones. As the cones absorbing red or green light are commoner than those absorbing blue, our sensitivity shifts towards the red end of the spectrum, and blue flowers now look paler than red ones.

Visual acuity. Visual acuity is a measure of the resolving power of vision. It is most reliably measured by presenting a stimulus of alternating black and white parallel stripes to a subject and reducing the width of the stripes until they can no longer be seen and the stimulus looks a uniform grey. Visual acuity can be measured in an animal by presenting it with stripes and a uniform grey stimulus containing the same total amount of light, and rewarding it for responding consistently to one or the other, for example by touching it, pecking it, or jumping at it. The width of the lines is then systematically varied until the animal can no longer see the difference between the two stimuli. Such experiments show that visual acuity varies in different regions of the eye, depends on illumination and visual adaptation, and shows great variation from species to species.

The variation in the distribution of cones and rods in the human eye has already been mentioned. In nearly all vertebrates the receptors are more densely packed near the centre of the retina, and in primates, birds, and reptiles there is a region where the cones are very slim and the nerve cells that normally lie on top of them are displaced to one side, leaving a pit. This region occupies about one degree of visual angle, and the visual image is sharpest here because there is nothing to hinder the passage of light to the receptors (Fig. L). Using the stripes test already mentioned, the width of the finest visible lines is 12 s of arc in the American kestrel (*Falco sparverius*), 30 s of arc in man, 36 s in the rhesus monkey, and 40 s in the blue jay (*Cynocitta cristata*) ($1°$ of arc = 60 min; 1 min of arc = 60 s). The kestrel can therefore spot a mouse from a height of 1.5 km.

The importance of these comparative measures is that in each case the finest visible line corresponds closely to the distance between adjacent receptors in the fovea, which therefore sets the

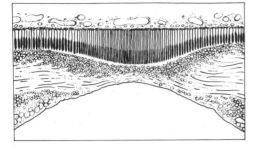

Fig. L. Diagram through the human fovea, where the cones are slim and closely packed and the nerve cells are pushed to one side. The central third of the diagram corresponds to less than 1° out of 180° of vision.

limits on acuity. In animals without a fovea, acuity is much poorer, for example it is only 4 min of arc in the domestic cat and about 1° in the rat. The rat is no better than the honey-bee, whose acuity of 1° corresponds very closely to the visual angle between adjacent ommatidia pointing in slightly different directions.

Birds are unique in that many species have two foveas in each eye, one near the centre of the retina and the other on the side closest to the back of the head. As the eyes are at the side of the head the bird has binocular vision only for a small region in front of the beak. Objects here will form images on the posterior fovea of each eye. The centre fovea is always more prominent and corresponds to monocular vision at the side of the head. Presumably this is why birds cock the head when looking intently, as did the American kestrel when its acuity was being measured.

If we examine a target with the periphery of the eye, acuity is much poorer, and its change with increasing displacement from the fovea corresponds closely with the reduction in the number of cones in the periphery (Fig. M). This indicates that our acuity in bright light is determined by the cones, because the number of rods actually increases in the periphery, yet acuity remains poor. In dim illumination we see only by means of rods, and although acuity is low it is best where the rods are densest, about 15° to the side of the fovea. However, if the retina contains mostly cones, acuity falls dramatically in poor light, as in the American kestrel.

We can now see that sensitivity and acuity are very different and each is improved at the expense of the other. Pure rod or pure cone eyes are specialized for sensitivity or acuity respectively. Most eyes have cones and rods but group them differently, so that regional specialization occurs in the retina. And probably all vertebrate eyes can adjust the balance between detecting light and resolving an image by changing the functional connections between receptors and nerve cells according to the overall level of illumination.

Visual fields and binocular vision. The eyes of many vertebrates are on the side of the head, and as each eye can cover nearly 180° vision is panoramic. The only region where the animal is blind is a small sector behind the head and as long as the head rotates slightly no predator can approach unseen. The advantages of such an arrangement are obvious, but if the two eyes never see an object simultaneously, how does the animal recognize with the left eye an object it has previously seen only with its right eye? The visual pathways in the brain make this possible. For example, in fish, amphibia, reptiles, and birds the optic fibres from each eye cross to the opposite sides of the brain, where their messages are still quite separate. However, the areas where they terminate are interconnected by *commissures* joining the two sides of the brain. In experiments where these commissures have been surgically severed it has been shown in pigeons that when the bird learns a visual DIS-CRIMINATION with one eye only, for example pecking at a particular colour or shape for food reward, it behaves as if it had never seen the stimuli when subsequently tested with its other eye. In contrast, a normal pigeon shows excellent discrimination when the untrained eye is first opened.

Panoramic vision is so useful that it may seem odd that in many other animals the eyes look forwards, causing the two visual fields to overlap. The overlap can be small, for instance about 40° in the rat, or almost complete, as in primates and in predators like cats and kestrels. The panorama has been sacrificed for stereoptic depth perception. Although the two eyes look at the same scene they see it from slightly different viewpoints, and the extent to which the view differs depends on the distance of objects in the scene: the closer they are the greater the difference. By recording the electrical activity of single cells in the visual areas of the brain in cats and monkeys it has been found that many of them are stimulated only when the difference, or disparity, in the two eyes is of a specific amount. As different cells respond best to different disparities, the brain is able to register the relative distance of objects in the binocular part of the visual field.

The smallest binocular disparity that can be seen is known as *stereoacuity*, and it is impressively small. If two needles are held vertically and at arm's length, one above the other, human observers can see when one of them is as little as 0.2 mm in front of the other. This corresponds to a disparity, or *picture difference*, of 2 s of arc. However, if one eye is closed the distance between the needles has to be increased twentyfold before we can detect it. This phenomenon is easily demonstrated by holding two fingers one above the other and at arm's length and alternately viewing them with one eye and with both eyes.

Stereoacuity has been measured in very few animals, but there is no reason to suppose that it is not just as superior to monocular depth perception as it is in men. In rhesus monkeys it has been mea-

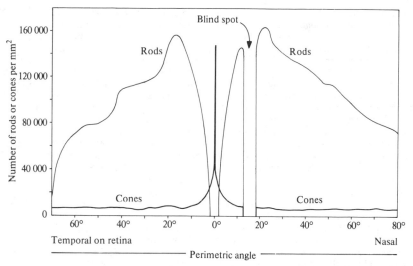

Fig. M. The graph shows the number of receptors in each square millimetre of the human retina. Cones are scarce away from the fovea, whereas the rods become more frequent. The blind spot is where the optic nerve enters the eye and there are no receptors.

sured with the needle test and is from 2 to 12 s of arc.

Why do animals need such precise distance perception? It is undoubtedly useful in jumping in trees or on rocks, and in pouncing on prey, but its principal function may be to uncover CAMOUFLAGE. A speckled moth, such as the peppered moth (*Biston betularia*), resembles the tree on which it lies, but it has to lie on top of the bark; it can never be in the same plane. A predator with good stereoacuity will detect a camouflaged moth because it is closer than the bark. This is why camouflaged animals are always more difficult to see in photographs (everything lies in the same depth plane) or with one eye closed.

Eye and brain. We have a fairly good idea how eyes detect and signal the visual scene. But the message from the nerve cells is transmitted to the nervous system, and our ignorance of how the message controls visual perception is vast, especially in insects, fish, reptiles, and amphibia. In birds and mammals the visual pathways have been traced anatomically, and their electrical activity measured whilst the eye is stimulated with various stationary and moving patterns of white and coloured light. Perhaps the most striking finding is that there is no single visual area in the brain. Different areas of the brain specialize in different aspects of vision such as the detection of pattern, colour, movement, and intensity, and although they are all interconnected, brain damage in one area can selectively disturb some aspect of vision. For example some kinds of brain damage in man produce colour-blindness, while others impair the recognition of objects.

Finally, the brain determines how an animal responds to what its eye detects, and it is wrong to assume that animals use their visual image in the way we would. The carnivorous great water beetle, for example, has an excellent compound eye, but it ignores a moving prey which is behind glass and which it therefore cannot smell, while persistently and fruitlessly attacking a muslin bag containing meat odour. A mother hen (*Gallus g. domesticus*) will ignore a chick enclosed in a glass bowl because she cannot hear its distress calls. A male robin (*Erithacus rubecula*) will attack a bunch of red feathers, which is a more provoking stimulus than a brown stuffed robin alongside it. Frogs (Anura) strike only at moving targets, and will starve even though surrounded by dead insects they can see.

It seems that the response to a visual stimulus depends on highly specific components of it (see SIGN STIMULI), and on the context in which the stimulus appears. These components cannot be predicted from the structure and function of the eye, but only from the experimental study of animal behaviour (see LABORATORY STUDIES). A.C.

66, 132, 133

VOCALIZATION, the production of sound by means of a vocal apparatus, occurs only in vertebrates, but many other animals produce sounds that have an equivalent biological role. Thus the SONG of birds is vocal, whereas that of crickets (Gryllidae) is produced by *stridulation*, involving frictional movement of modified wings.

The sounds of animals are produced in association with characteristic states of MOTIVATION. They are usually designed by NATURAL SELECTION to influence the motivational state of other animals. They form an important part of the complex systems of COMMUNICATION that occur in the animal kingdom. Even seemingly innocuous sounds have been shown to have a communicative FUNCTION. For example, the eggs of many bird species produce a regular click, which is audible to humans when the egg is held to the ear, and has long been recognized as indicating that hatching is near. The clicks are produced by the BREATHING apparatus of the embryo, and were thought to be an accidental consequence of its early operation. However, LABORATORY STUDIES have shown that the clicks have a communicative function, serving to synchronize the HATCHING of the differently aged eggs in the clutch. Hatching can be slowed down or speeded up by providing artificial clicks at an appropriate rate.

Sound production. Few invertebrates, other than arthropods, make use of auditory communication. Among the arthropods some crabs (Brachyura), shrimps (Natantia), spiders (Arachnida), and insects such as crickets communicate by sound. Among vertebrates, it is fairly common in fish, amphibians, and reptiles, and almost universal among birds and mammals.

The three main methods of producing sound are (i) beating a substrate, (ii) rubbing appendages together, and (iii) blowing air through an orifice. Beating the ground with a limb is known in fiddler crabs (*Uca*), rabbits (*Oryctolagus cuniculus*), deer (Cervidae), and kiwis (Apterygidae). Woodpeckers (Picidae) drum on wood, and beavers (*Castor fiber*) slap the water with their tails.

The rubbing together of appendages to produce sound is common in arthropods. Many insects have specially modified wings, legs, and antennae, by means of which they produce sound by stridulation. Some birds have modified feathers which produce sounds. Thus some snipe (*Gallinago*), woodcock (*Philohela*), nighthawks (*Chordeiles*), and little bustards (*Otis tetrax*) have modifications which exaggerate the sounds of FLIGHT.

When sound is produced from an air-flow the respiratory system is usually involved. Among vertebrates there are numerous specialized structures that have evolved to enhance sound production. These include the mammalian *larynx* and the avian *syrinx*. Air chambers that function as resonators are found in frogs and toads (Anura), swans (*Cygnus*), frigate-birds (Fregatidae), howler monkeys (*Alouatta*), and orang-utans (*Pongo pygmaeus*).

The different methods of sound production produce characteristic types of sound to which the HEARING mechanisms of each species are attuned. For example, all cricket stridulation sounds are composed of a series of pulses of characteristic frequency or pitch. A group of pulses, called a chirp, is separated from other groups by a pause, and a long series of pulses produced without pauses is called a trill. The cricket's auditory system has two parts, one of which responds to the characteristics of individual chirps or trills, and the other to the spacing of chirps within the song. Crickets produce several types of song, including

songs for ADVERTISEMENT, AGGRESSION, and COURTSHIP. There are also modifications induced by environmental factors, such as temperature.

The ears of frogs and toads can discriminate a wide range of frequencies, but are attuned to the characteristic frequency of the calls typical of the species. Birds and mammals, on the other hand, both produce and respond to sounds that vary considerably in frequency and amplitude. Many bats (Chiroptera) and rodents, such as rats (*Rattus*), vocalize in the ultrasonic range, which is beyond human detection. Many song-birds are able to produce two or more frequencies simultaneously, using two or more oscillating membranes in the syrinx.

Functional aspects of vocalization. As a medium of communication, sound has the advantage that it can be altered to suit the circumstances. For example, animals giving ALARM calls are in danger of attracting the attention of a predator. Sounds which provide relatively few cues to the caller's location are therefore likely to evolve through the action of natural selection. In fact, the alarm calls of different species are very similar and share features which make them difficult to locate. Most animals locate sounds by comparing the timing, intensity, and frequency of the sounds reaching the two ears. Thus ideal alarm calls should have a number of qualities: (i) They should begin and end gradually, so as not to provide useful timing cues. (ii) They should be low-pitched, so as to make intensity comparisons difficult. (iii) They should cover a narrow range of frequencies, thus making it difficult for a predator to use differences in tonal quality as a localization cue. (iv) They should not vary in pitch, because this might enable a predator to compare the time and loudness of the sounds reaching the two ears. Studies of the alarm calls of the reed bunting (*Emberiza schoeniclus*), blackbird (*Turdus merula*), great tit (*Parus major*), blue tit (*Parus caerulus*), and chaffinch (*Fringilla coelebs*) show that not only are the calls remarkably similar but they have most of the characteristics of an ideal alarm call.

Advertisement songs tend to have the opposite sound characteristics to alarm calls, thus facilitating localization of the singer. Such songs may also be tailored to suit the type of vegetation typical of the HABITAT. Birds which sing high in the trees tend to have more highly pitched songs than those which sing close to the ground. Animals inhabiting dense vegetation also tend to have low-pitched calls, probably because of the greater ability of low frequency sounds to penetrate, or carry around, objects.

The sounds used by animals are often closely tied in with their SOCIAL ORGANIZATION. Animals that maintain vocal contact over long distances, or through dense vegetation, tend to use high-pitched calls or whistles. Animals, such as baboons (*Papio*), which form close-knit groups, communicate by low-pitched grunts which do not carry far but can be modulated in complex ways through subtle movements of the mouth and tongue. Some scientists believe that the ability to modulate sounds in this way was important in the EVOLUTION of human LANGUAGE.

VOLUNTARY BEHAVIOUR is said to be controlled by the will. Some scientists and philosophers regard the will as a free agent, influenced but not dictated to by bodily processes. Others deny the existence of any such free agent governing behaviour, but this does not necessarily mean that they do not recognize a difference between voluntary and involuntary behaviour. The controversy has a long HISTORY which provides a good perspective for consideration of the problem of voluntary behaviour.

DARWIN's theory of evolution by NATURAL SELECTION, published in 1859, crystallized scientific thought into a pattern in which man is seen as an integral part of the animal kingdom. Both these developments aroused widespread opposition amongst contemporary educated people. Even today, many people are reluctant to recognize their psychological affinity with animals. The consideration of motivational questions originated as part of philosophy, and entered rather late into the development of science. In outlining the various philosophical schools of thought on the problems of voluntary behaviour, there is relatively little to be said about animal behaviour, but the stage is set for the scientific view, that the problems of human and animal MOTIVATION are not different in principle.

Rationalism. Plato, and most of the ancient Greek philosophers, regarded motivation as unimportant in the conduct of behaviour. Plato recognized certain 'forced' movements as being products of emotion or 'animal passions', but thought of them as disruptive or lawless in character, and not part of the normal and natural behaviour of man. He regarded human behaviour as the result of voluntary and rational processes, man's will being free to choose whatever course of action his reason dictates.

Rationalism became part of Christian doctrine, particularly through the writings of Thomas Aquinas (1224–74), as the following quotation from a discussion of his philosophy shows: '. . . man has sensuous desire, and rational desire or will. He is not absolutely determined in his desires and actions by sense impressions as is the brute, but possesses a faculty of self-determination, whereby he is able to act or not to act. . . . The will is determined by what intelligence conceives to be the good, by a rational purpose. This, however, is not compulsion: compulsion exists where a being is inevitably determined by an external cause. Man is free because he is rational, because he is not driven into action by an external cause without his consent, and because he can choose between the means of realising the good or the purpose which his reason conceives' (*Summa Theologica*).

This view persists amongst most Catholic philosophers and psychologists today. It is an important matter in theology, because the notion of responsibility depends upon freedom of choice. A man should not be held responsible for his behaviour if he acts under compulsion, nor should he be rewarded or punished if he does so. If salvation is to be based upon the power to choose, then the will must be free to choose, and most theologians hold this to be true axiomatically.

Thomas Aquinas clearly regarded animal behaviour as being determined by sensuous desire, though he appeared to recognize some elementary process of judgement in animals. 'Others act from some kind of choice, such as irrational animals, for the sheep flies from the wolf by a kind of judgment whereby it considers it to be hurtful to itself; such a judgment is not a free one but implanted by nature' (ibid.).

Modern rationalists assume men to be rational and to direct their behaviour towards their voluntarily chosen GOALS. At the same time they acknowledge the existence of involuntary aspects of behaviour, such as REFLEX and habitual behaviour. Few modern psychologists would deny that human motivation involves awareness of goals, and partly consists in striving towards these goals. This is thought to be true of animals to a lesser degree. Many scientists however would not agree that the notion of 'free will' is necessary in accounting for rational goal-seeking behaviour.

Materialism. A few of the ancient Greek philosophers dissented from the common rationalist view, and held that events in the mental world are caused in a similar way to events in the physical world. Democritus, a contemporary of Plato, believed that all matter, animate as well as inanimate, was composed of atoms of various sizes and shapes. Events in both the physical and the mental world occur because of the constant motion of these atoms. Democritus believed not only that physical and mental events were materially similar, but also that pleasures derived from the exercise of the 'animal passions', such as eating, drinking, and sex, were in no way inferior to pleasures of intellectual origin. He argued that all pleasures are equally good, and that men should regulate their lives so as to attain the greatest possible pleasure. He recognized that physical pleasures involved dangers, and advised men to exercise moderation in this respect. Epicurus, following Democritus, held similar views, but allowed for the possibility that volition could influence the motion of the atoms of the soul, thus making moral responsibility more plausible in materialistic terms.

Materialistic explanations of behaviour had little influence up to the time of the publication of *The Passions of the Soul* (1649) by René Descartes. Descartes maintained that all physical phenomena could be adequately explained mechanically, and that animals were merely automata. Their behaviour was held to be due to the physical forces acting upon them, some of which were external while others were internal in origin. Thus in a starved animal there would be physiological agitations which would cause the animal to eat. In animals, agitations of physical or physiological origin entered the BRAIN through SENSE ORGANS, and were reflected to the muscles, producing reflex behaviour. In the case of man, reasoning intervened to guide behaviour in accordance with his knowledge and wishes. Descartes thus proposed that human conduct was under the dual influence of mind and body, such that the mind is subject to certain agitations (passions) which emanate from the body and also from mental processes. The passions 'dispose the soul to desire those things which nature tells us are of use, and to persist in this desire, and also to bring about that same agitation of spirits which customarily causes them to dispose the body to the movement which serves for the carrying into effect of these things' (ibid., article 52). Thus although the 'rational soul' imparts a large degree of freedom to human behaviour, the mind is influenced by the physiological state of the body, and by the EMOTIONS.

A pinnacle of materialism was reached by Thomas Hobbes in his *Leviathan* (1651). For Hobbes the explanation of all things was to be found in their physical motions. He distinguished between 'vital motion', responsible for physiological processes in the body, and 'animal motion' responsible for behaviour. Pleasure and pain were attributed to physiological processes, which both gave rise to emotions and influenced behaviour. Actions were initiated by 'endeavours', which were small incipient actions. Whenever an endeavour was directed towards an object known by experience to be pleasant, an appetite was aroused which enhanced the vital motion of the body, and increased the tendency to perform the action. If an endeavour was directed towards an object known by experience to be painful, an aversion was aroused which impeded the vital motions of the body and decreased the tendency to act.

Hobbes believed that men behave in such a way as to achieve pleasure and avoid PAIN, and that, whatever we think, these are the only causes of our behaviour. Thus Hobbes stripped the pleasure–pain principle of all ethical implications, making it purely motivational. He recognized that, in deliberating on the next course of action, many ideas may come to mind. These have separate endeavours associated with them, and that which arouses the greatest pleasure determines the action. Thus Hobbes had a purely materialistic view of thinking, and he suggested a mechanism for the phenomenon of anticipation. The goal of an action influenced the act only through its endeavour, which evoked the degree of pleasure associated with the outcome of the action, on the basis of previous experience. The anticipated pleasure or pain determines the behaviour, but we think of this in retrospect as our own volition. Thus, for Hobbes, the will is simply an idea that man has about himself.

In accounting for mental events in material terms,

in seeking mechanistic explanations of purposive behaviour, and in regarding the will as an epiphenomenon, Hobbes anticipated much modern scientific thought. However, such thoroughgoing materialism was not to become acceptable for many decades.

Associationism. Associationism is based upon the premiss that there are psychological laws which describe how a man will behave in particular circumstances. These laws are entirely independent of the material and mechanisms that govern behaviour: they simply describe the behaviour as Newton's laws describe the behaviour of physical bodies. Newton ended his *Principia* (1687) with the words, 'I have not been able to discover the cause of these properties of gravity, and I make no hypothesis. ... It is enough that gravity act according to the laws which we have found.'

The associationists hoped to find similar laws governing behaviour. They joined the materialists in denying any freedom of will, but did not attempt to account for behaviour in physical or physiological terms. Materialism and associationism are logically independent, though both views could be, and were, held by the same person. The associationists, such as John Locke and David Hume, believed that the ideas of the mind arise from experience. Locke, in *An Essay concerning Human Understanding* (1690), envisaged the INNATE mind as a *tabula rasa* (cleared slate), upon which ideas are engraved as a result of experience. The contents of the mind and the laws of thought were moulded by the laws of association. Simple ideas, which occur contiguously, can become interassociated and result in a complex idea. Various laws of association were formulated and these were conceived as working automatically, without any intervention by free will. For both Locke and Hume the will was strictly determined by desires and ideas that arise from sensations of the body, and present themselves to the mind. Man's conscious will was seen simply as his impression of his own ideas.

The associationist view, that human behaviour develops entirely through experience, though very influential in the early days of psychology, is not widely accepted today. Scientists now recognize that much behaviour is influenced by the GENETIC make-up of the individual, and that behaviour develops through a combination of innate factors and individual experience (see ONTOGENY).

The scientific view. Insistence on empirical verification excludes from scientific enquiry certain types of question about behaviour. For instance, suppose that we wish to know whether animals can experience pain. We can present a stimulus which would be painful to a human being and observe the animal's response. We might find that the animal withdraws from the stimulus and avoids the stimulus situation in the future. However, as we cannot ask the animal about its feelings, we have no means of knowing whether it consciously experienced pain. Such withdrawal and AVOIDANCE behaviour could be shown by a man-made robot. Can we be sure that animals are not like such machines?

Because of the anatomical and physiological similarities between human beings and some animals, it might be argued that we have no right to assume that animals cannot feel pain. However, this is a moral rather than a scientific issue, because we have no empirical means of verifying the theory that animals do have conscious experience similar to ours. As scientists, therefore, students of ETHOLOGY are not inclined to ask whether animals consciously experience pain, love, or other emotions, or whether they enjoy freedom of the will.

Although many people in the mid-nineteenth century recognized the value of the scientific approach in the study of physics, chemistry, and physiology, few would countenance the application of deterministic principles to the mental world. This dualistic approach to life was challenged by Darwin, who maintained that the structure and behaviour of living organisms can be explained entirely in terms of random variation and natural selection. In his *Origin of Species* (1859), Darwin maintained that the SURVIVAL VALUE of any animal trait is determined by natural selection. That is, the extent to which a trait is passed from one generation to the next is determined by the breeding success of the PARENTAL generation, and the value of the trait in enabling the animals to survive natural hazards, such as food shortage, PREDATION, and SEXUAL rivals. Such environmental pressures can be looked upon as selecting those traits which fit the animal to the environment. This is what Darwin meant by the 'survival of the fittest'. Behavioural traits are just as important as morphological or physiological ones in determining the survival of the species. Thus Darwin stated that 'if it can be shown that instincts do vary ever so little, then I can see no difficulty in natural selection preserving and continually accumulating variations of instincts to any extent that may be profitable. It is thus, as I believe, that all the most complex and wonderful instincts have originated' (*Origin of Species*). Similarly, Darwin argued that it is only necessary to suppose that INTELLIGENCE in lower animal forms has survival value for EVOLUTION to proceed along this line, so that more and more intelligent members of the animal kingdom will evolve. Given the continuing adaptive value of higher intelligence, it is likely that animals as intelligent as man, or more so, will eventually evolve.

The Darwinian thesis implies that the mechanisms controlling animal behaviour have evolved continuously from simple beginnings, as part of a purely deterministic process, without any divine intervention, purpose of nature, or conscious design. It implies the continuity of man and beast, so that we should expect eventually to be able to account for human behaviour in mechanistic terms. The antithesis is that man's will must be free in order that he can be held responsible for his behaviour. This implies that there can never be a fully deterministic account of human behaviour, that there is discontinuity between man and animal.

Most ethologists believe in evolutionary continuity, and are therefore inclined to the view that the essence of voluntary behaviour in man has precursors in animals. Disagreement about the nature of voluntary behaviour remains. Some scientists take a dualistic view, others a purely deterministic one. Most recognize a distinction between reflex and habitual behaviour on the one hand, and intentional or voluntary behaviour on the other.

At the physiological level, voluntary behaviour has long been identified with the somatic nervous system which controls the skeletal muscles. The AUTONOMIC nervous system controls the muscles of the heart, intestine, etc., and is generally associated with involuntary activity. Thus we can move our fingers at will, but we cannot contract the muscles of our small intestine at will. There is some evidence that some animals can learn to exercise OPERANT control over some aspects of autonomic activity, but this work is controversial. Operant control implies that the animal is able to carry out an action in order to obtain some REWARD. Many behavioural scientists regard operant behaviour as the equivalent in animals of voluntary behaviour in human beings. If a human being could influence autonomic activity (such as alteration of the heartbeat rate) in order to obtain a reward (without mediation from the somatic nervous system), then this would be taken as evidence for voluntary control of autonomic activity. Some scientists take the view that this ability can be learnt through special training, such as yoga.

Voluntary behaviour is an aspect of motivation that may well be illuminated by further research. Improved understanding of animal COMMUNICATION and of the nature of LANGUAGE may well throw light on the nature of voluntary behaviour, and research into COGNITIVE aspects of animal behaviour may help us to understand the extent to which DECISION-MAKING in animals involves freedom of choice.

16

W

WELFARE OF ANIMALS is a subject that can be raised whenever men come into contact with one of the other animal species which have evolved on this planet. Man uses animals for food, farms them for their skins, uses them to work for him, hunts them for sport, relies on them as friends and guides, keeps them in CAPTIVITY in zoos and circuses, creates havoc with their natural HABITAT for his own designs, and uses them for research into the workings of his own body, and for making vaccines and other essentials for medical science. He has domesticated and made new varieties (see DOMESTICATION), he has set up some animals and worshipped them, he has been responsible for the extinction of whole species.

Human reliance on animals for so many essentials means that concern for the well-being and health of animals cannot be dismissed as sentimentality: indeed, it may often make sound practical and economic sense. Healthy animals work better and produce more, so that welfare considerations may arise from a form of enlightened self-interest, as well as from compassion. However, the interests of human beings and of other animals may not always coincide. For example, the need to feed the ever-increasing human population has led to the development of intensive husbandry methods for keeping farm animals, and many people believe that these impose considerable suffering on animals such as pigs (*Sus scrofa domestica*), calves (*Bos primigenius taurus*), and chickens (*Gallus g. domesticus*) that are kept under such systems. Medical and other branches of biological research, as well as the testing of new substances before they are deemed safe for human beings, frequently involve the use of live animals and the destruction of various parts of their bodies. Large numbers of animals may die when a new drug is being tested, and there are many other practices in which the lives and health of animals are sacrificed for human interests. These are just a few examples of how a conflict may arise between human and animal needs. It is difficult to think of another moral issue on which human beings differ so widely as in their attitudes as to how this conflict can best be resolved.

On the one hand, there are those who regard the human species as so important and so much the centre of the universe that there is really no conflict at all: human beings should always have priority. Thomas Aquinas, on whose teaching many orthodox Roman Catholic beliefs are based, wrote that only men have souls and are in this respect fundamentally different from all other creatures. The view that man is supreme and is the only being worthy of moral consideration is still widespread among many peoples of the world. It leads either to the view that animals are incapable of experiencing suffering, or that even if they do suffer and feel pain, no provision need be made for their welfare if this conflicts with human desires or economic considerations.

On the other hand, there have always been people who have believed that since humans share life and sentience with many other creatures, compassion should be extended beyond the boundaries of their own species. St. Francis of Assisi is renowned for preaching mercy to all God's creations, and Buddhism stresses the unity of all life. Leonardo da Vinci, Alexander Pope, and William Hogarth are among many who believed that the brutality towards animals that they witnessed as a common occurrence in their own time was not justified. The political philosophers Jeremy Bentham and John Stuart Mill both believed that moral considerations should be extended to other animals.

A feeling that, whatever the differences between man and other species, they share the capacity to suffer and feel pain has been responsible for the founding of various bodies concerned with animal welfare and the passing of legislation for the protection of animals during the 19th and 20th centuries. In England, the Royal Society for the Prevention of Cruelty to Animals (RSPCA), founded in 1824 and given its Royal Charter in 1840, has played a prominent role in promoting such legislation, and similar societies now exist in many other parts of the world. The World Federation for the Protection of Animals is an international organization based in Zürich, and has consultative status with the United Nations Organization (UN), the Food and Agriculture Organization (FAO), the United Nations Educational, Scientific, and Cultural Organization (UNESCO), and the Council of Europe. The Council of Europe has itself been responsible for the European Convention on the International Transport of Animals, and the Council is preparing similar Conventions on humane slaughter, intensive farming, and the treatment of LABORATORY animals.

Broadly speaking, animal welfare legislation has aimed at ensuring the physical health of animals under various conditions, including measures to control various diseases, and at the avoidance of unnecessary suffering. It is the 'avoidance of unnecessary suffering' which causes the greatest problems of definition and enactment. Even if it were possible to agree on how much suffering is necessary, there would still be the difficulty of recognizing when an animal is suffering. PAIN and its avoidance are clearly central issues here. Many mammals are known to have the nervous apparatus which in human beings mediates the sensation of pain. Animals also squeal, struggle, and give behavioural evidence which is generally regarded as the accompaniment of painful feelings. It seems reasonable to conclude, therefore, that many animals suffer pain in much the same way as do human beings. Rather more difficult to define are the less acute forms of suffering which generally come under the headings of 'distress' or STRESS. The term 'stress' was originally used to refer to physiological changes which were found to occur to a wide variety of stimuli such as cold, electric shocks, etc. These changes involved heightened activity of certain glands, development of gastric ulcers, and a number of related physiological features. In addition to factors known to lead to the classic stress syndrome, other 'stressors' have been described, such as overcrowding, which lead to ABNORMAL behaviour, even in the absence of some or even all of the above physiological changes. It seems reasonable to allow that animals may be distressed by being unable to feed and drink, to move their limbs, to SLEEP, and to have social interaction with their fellows, but the difficulty of defining distress in an objective and convincing way has been a stumbling-block in the formation of animal welfare legislation even in countries where there is widespread public interest in the way that animals are treated.

Animal welfare legislation varies greatly from country to country, some parts of the world having no legal safeguards for animals at all, others having many different laws governing various aspects of the treatment of animals. The following account should in no way be taken as comprehensive: it is merely intended to give some idea of the variety of laws that have been enacted and of the main areas where it has been felt that animals need legal protection.

General animal protection. In many countries (e.g. in all the countries of the European Economic Community) acts of cruelty or excessive ill-treatment of animals are punishable by fines or imprisonment or both. Sometimes this is the only law under which animals are protected, but some countries have other more specific legislation as well. One of the most far-reaching laws was passed by the German Federal Republic in 1972. This law has as its basis that it 'shall serve to protect the well-being of the animal' in many different spheres. There are sometimes curious anomalies, however. In Britain, a captive fox (*Vulpes*), along with other animals in captivity, is protected by the Protection of Animals Act of 1911, but the hunting of wild foxes is perfectly legal.

The word 'animal' is usually taken to mean vertebrate animal and often 'warm-blooded animal'. Invertebrates generally have no legal protection.

Farm animals. The World Federation for the Protection of Animals estimates that, excluding poultry, 1000 million animals are killed each year to provide meat. The husbandry, transportation, and ultimate slaughter of food animals is a primary concern of animal welfare.

Husbandry and management. Modern extreme forms of intensive farming, such as keeping hens in battery cages and the intensive rearing of pigs and calves, have given rise to public disquiet. The idea of severely restricting the movements and possible behaviour patterns of animals, and keeping them on restricted regimes of light and food, is repugnant to some people, but the economic advantages of such systems over more traditional methods are generally accepted. Some traditional practices such as branding also cause concern. In the United Kingdom, the Agriculture (Miscellaneous Provisions) Act, 1968, made it an offence to cause or permit unnecessary suffering, pain, or distress to agricultural animals, and codes for recommended treatment of farm animals are issued by the government, although these are non-statutory. Denmark, although having no special legislation on farm animals, outlaws the keeping of laying birds in cages.

Transport. Meat animals often travel many hundreds of miles on hoof, road, rail, or even by air. The possibility of distress, injury, and even death is clearly very great during transportation, and because of this the European Convention for the Protection of Animals during International Transport was drawn up. However, not all these journeys may be essential, and attempts have been made in several countries to have the transport of live animals replaced by that of carcass meat.

Slaughter. Methods of slaughter used in the 20th century are little changed in most areas of the world from medieval times. Although many European countries insist on pre-slaughter stunning (with certain exceptions with regard to Jewish and Muslim requirements), most animals in the rest of the world are made into meat by being attacked with a knife or hammer, and may be fully aware of what is happening to them.

Experiments on animals. To find out about the workings of the human body and also about human behaviour, other animals are studied because they are sufficiently similar to ourselves to act as 'models' for the human case, without, in many people's minds, raising quite such acute ethical problems as would experimentation on humans. Just how far this position is tenable is another contentious issue raised by animal welfare. Some scientists believe that it is perfectly acceptable to insert electrodes into the brains of animals, to deliberately damage their brains or sense organs, or to bring up baby monkeys in complete social isolation. To other people, these and other practices are immoral.

The World Federation for the Protection of Animals estimates that, throughout the world, some 140 million animals per year are used for research purposes, these being mostly rats and mice (Murinae), hamsters (Cricetini), and guinea-pigs (*Cavia aperea porcellus*), but including also large numbers of cats (*Felis catus*), dogs (*Canis lupus familiaris*), and monkeys (Simiae). Many are used for medical and biological research, but, in addition, the mandatory testing of drugs, cosmetics, and food additives in some countries leads to a very large consumption of mammals for these purposes.

The use of live animals for research became a regular practice in Europe in the 19th century. Public outcry in Britain against operations performed in Britain, France, and Italy led to the setting up of a Royal Commission in 1875 to look into vivisection. The result of this was the Cruelty to Animals Act of 1876 which ostensibly gave a great deal of protection to animals by restricting what could be done to a living creature. However, in practice, cruel and painful experiments are allowed in the United Kingdom as the Home Office is empowered to issue licences which exempt the holder from the restrictions of the Act. Many practices, such as using animals for the production of vaccine, fall outside the Act as they are not counted as experiments. In other countries, there is even less protection for experimental animals. Only Denmark, Ireland, India, and Zimbabwe have any legislation comparable to the United Kingdom Act of 1876. Many countries, such as Chile, France, Israel, and Japan, have no laws at all relevant to animal experimentation. In New Zealand, vivisection is specifically excluded from the Animal Protection Act of 1960.

One school of thought is that live animals should be replaced in research with non-sentient material, for example by the use of tissue culture methods instead of whole animals, but the feasibility of this on a large scale has still to be explored.

Wild animals. There are many welfare organizations concerned about the killing of seals (Phocidae), and kangaroos (Macropodidae), about the cruelty involved in killing whales (Cetacea) and porpoises (Phocaenidae), and with blood sports generally. Wild animals as a whole enjoy little legal protection, although there are certain specific exceptions and some countries have laws governing which methods of killing are permissible.

Pets. The World Federation for the Protection of Animals estimates that throughout Europe some 5 million dogs are destroyed each year because they are unwanted HOUSEHOLD PETS. It is considered ironic by some that this slaughter is often done by 'animal welfare' organizations.

Conclusion. Animal welfare raises moral as well as biological and economic problems. The balance must somehow be struck between extreme ANTHROPOMORPHISM, which sees animals as humans in another skin, and its equally misleading opposite, the refusal to allow that animals share with us any of the features that would warrant our moral concern for them.

The range of feeling that humans have towards other animals is enormous, as expressed in the practices that they allow or outlaw. Very great cruelty is certainly inflicted by some people, and yet there are many others who care deeply for animals. Concern for animal welfare should not be seen as an alternative to concern for human beings; rather, both are manifestations of a caring attitude towards others.   M.D.

32

WILDLIFE MANAGEMENT embraces all the diverse topics that are involved in man's various relationships with wildlife. The aim of management can range from CONSERVATION of endangered species, through exploitation of a resource species, to extermination of a pest. In practice topics of this sort are not so conveniently compartmentalized; the control of a pest may involve the destruction of other aspects of the COMMUNITY; the exploitation of a resource species results in the condemnation of other species which compete with man for that resource, such as predators of game birds which compete with sportsmen for prey. This complexity obviously arises from the intricate nature of natural communities of animals and plants. If man manipulates one element of the community the consequences reverberate throughout. The resolution of the optimal solutions to such complex problems is the *raison d'être* of wildlife management, but exactly what is optimal depends ultimately on subjective rather than scientific criteria, and for this reason wildlife management is almost as much a sphere of politics as of biology. One person's Utopia might be a countryside devoid of predators and thronged with pheasants (Phasianinae), another's ideal could be a self-regulating balance between wild predators and their prey. Here we will assume that the most robust single criterion for good wildlife management is a concern for minimizing disruptive side-effects on natural communities of animals and plants; what has been called an 'ecological conscience'.

There are three major categories of problem involved in wildlife management. First, there is the case of the endangered species: how is the rapid decline in numbers of the Javan rhinoceros (*Rhinoceros sondaicus*), or the puma (*Puma concolor*), to be halted? This kind of problem may be difficult to solve in practice, especially when it leads to discussion of the necessity or desirability of struggling to snatch the obscure creature from the jaws of extinction. After all, many more species have become extinct in the past than are extant today, and like as not the salvation of the one in question will always be against somebody's interests, for instance the man whose crops it eats. Argument often revolves around the academically appealing but often inapplicable (in the case of already very rare species) principle of maintaining the diversity of animal communities. In the end debate rests on the power of those who like rhinoceroses or pumas versus those who do not.

The second category of problem involves many more lives. In these cases a pest, a fur-bearing or game species, may be managed in one of a variety of ways, each with different consequences for the involved parties. Rabbits (*Oryctolagus cuniculus*), for example, may be infected with the normally fatal disease of myxomatosis, or they may be poisoned, shot, or left alone. The merits of these different possibilities depend on whether one's principal concern rests with farmers, areas of grassland, foxes (*Vulpes*), individual rabbits, or populations of rabbits. To take another example, seamen have dumped unwanted cats (Felidae) on many oceanic islands around the world. In some cases these islands harboured unique faunas, hopelessly susceptible to the hunting talents of a newly introduced predator. Two different instances illustrate the nuances of the problem. On Aldabra in the Indian Ocean there lived the last remnants of a population of flightless rails (Rallidae). The atoll is comprised of three adjoining islets which together form a ring, with each of the three being separated from its neighbours by narrow channels. Before the arrival of feral cats the flightless rails thrived on all three islets, but now cats have colonized two of these islands and exterminated the rails. The biologists studying the remaining population watch with dread as cats meander along the shores of the channel of water which insulates the surviving rails. One piece of driftwood is all that is needed to obliterate the last of these birds. As it happens this particular species is of unusually great interest to biologists, but even if it were not its demise would be irreparable. A work of art that is destroyed can never be replaced, and this is the gravity of the decision which can face the wildlife manager.

The second case is slightly different. On Marion Island in the sub-Antarctic the original introduction of a handful of cats several decades ago has exploded into thousands now. These cats are severely reducing the sea-bird breeding colonies along the shores. None of these birds are in immediate danger of extinction, and even the birds on Marion Island itself may survive the onslaught, but in much reduced numbers. So here the danger may not be one of extinction, but one part of a process of attrition affecting sea-bird refuges on a wide scale. What is to be done? The simple answer is to kill the cats. (That, sadly, has been the traditional and unimaginative answer to most wildlife management problems in the past; in this case it may be the right one.) However, this is easier said than done: on Aldabra the cats live at such low density that they cannot easily be found to be killed, on Marion there are so many of them that the task would be hopeless by most conventional means. On both islands the rocky, crevice-filled terrain stacks the odds heavily in the cats' favour.

One recent suggestion has been to introduce the disease cat-flu amongst the cats. This might be effective, but not enough is known about its epidemic properties to be certain. What is certain is that it is a most unpleasant disease, and so the argument also involves WELFARE considerations.

The resolution of this particular problem is still a long way off, but the example does serve to illustrate several general points. First, there is a real problem and something has to be done if undesirable consequences are to be averted. Second, the problem is man-made, fishermen having introduced the cats in the first place. There is a school of thought which argues that the best thing that people can do for wildlife, and nature in general, is to leave it alone. Clearly leaving it alone will not solve this problem, at least not in any desirable sense, and as man caused it he will have to solve it. Indeed, even where we have not so directly precipitated a problem, there seem to be rather few cases where simply leaving things alone is a realistic solution. Human beings have so influenced almost every part of the world that, like it or not, we carry the responsibility of managers. A further point is that any thoughtful solution to the cat problem will clearly require a combination of skills. It will require a knowledge of the biology, especially the behaviour and ecology, of both cats and sea-birds, of the principles of predator–prey interactions, of the characteristics of transmission of infectious diseases, of virology and immunology. In addition attention has to be paid to the quite different disciplines of ethics and economics. It is this necessity of an amalgam of expertise that makes wildlife management so challenging.

The island cats are just one isolated problem, but this general type of dilemma is commonplace. It includes the control of agricultural pests the world over, the control of carriers of diseases, such as rats (*Rattus*) and plague, foxes and rabies, badgers (Melinae) and bovine tuberculosis. The fact that control (too often a euphemism for killing) is so frequently linked with wildlife management should not create the impression that it is a purely destructive subject. This is simply a consequence of the fact that almost everything is a pest in somebody's eyes. In spite of this, the control can be designed to do everything possible to minimize the disadvantage to both individuals and populations of the species involved.

The third major category of wildlife management embraces both of the first two in practice, and ultimately must be the most important. This is HABITAT management and conservation. The most pervasive effects of man's actions on wildlife are seen as consequences of his treatment of habitats. Destroy the rain forest and you can forget about conserving the Javan rhinoceros. Habitats are threatened everywhere, all ultimately as a consequence of the alarming increase in human population, and the concomitant demands for ever more resources. The decisions which will really determine the possibilities for conserving, say, tropical rain forest lie not with conservationists, but with silviculturists and the economists of the timber trade. For this reason it seems possible that the next generation may never see rain forest, nor many of the creatures that inhabit it.

Turning now to some examples which illustrate these different facets of wildlife management, we look first at the problems of conserving wolves (*Canis lupus*). Wolves pose quite different problems in different areas, and the definition of a wildlife problem often depends largely on scale. Thus, on a worldwide scale wolves are probably not in immediate danger of extinction, but in some countries they are hovering on the edge of oblivion. In Europe, for example, the only viable populations that remain are in Romania, Yugoslavia, Greece, and Albania. Wolves are also probably holding their own in Russia, although the number of wolf-fur coats on the market precludes complacency anywhere. But over the remainder of Europe the wolf has either been exterminated already, or soon will be. Similarly, the geographic range of wolves in North America is a fraction of its former extent. In the USA the only state with any number is Minnesota. Americans by and large think of the wolf as desperately endangered; but Minnesotan farmers are campaigning for the reintroduction of a wolf bounty.

In Europe the range of the wolf is fragmented and the species is endangered in some areas. Italy provides an example. In the early 1970s surveys revealed that about one hundred wolves survived in the Apennines, of which the greatest concentration was in the Abruzzo Mountains. The census involved many techniques, but one particularly relied on the wolves' behaviour. An important element of wolf COMMUNICATION is howling, and this is often a communal behaviour; when one wolf howls others join in. The biologists conducting the survey were sufficiently conversant with wolf howls to mimic them, and to add to their survey material on the basis of the number of wolves replying. They found that the biggest concentration was no more than twenty to twenty-five animals scattered within the Abruzzo at a density of one wolf per 65 km². The survey seemed to confirm fears that without immediate intervention the Italian wolf would disappear. The first practical problem was that nothing was known either of the wolves' behaviour or of the reasons for their decline. The latter could easily be guessed at: most of their prey species, such as red deer (*Cervus elephus*) and roe deer (*Capreolus capreolus*), have been either greatly reduced or hunted to extinction in Italy. In addition the remaining wolves had turned to domestic stock, as well as household garbage, and so come into conflict with shepherds. In order to discover how the wolves were behaving and why they continued to decline they were studied intensively. It would not have been legitimate to generalize from the exhaustive studies of wolves in North America, since behaviour, especially SOCIAL

behaviour, is very flexible amongst carnivores and seems to be adapted to ecological circumstances. The wolves in mountainous Italy might be quite different in their life-style from those in wooded Minnesota. In fact, as the study eventually demonstrated, they were.

Apart from revealing more about the biology of these wolves, their relatively large HOME RANGES, their small group sizes, and greater tendency to travel alone instead of in packs, the study brought to light several general points relevant to their management. First there was the problem of how to study the wolves. They were very elusive, and although a lot could be learnt by traditional FIELD STUDIES, following tracks and signs, examining faeces, etc., the major questions about their lifestyle remained unassailable. The best TECHNIQUE for tackling such problems is radiotracking. But in order to attach a collar with a radio transmitter to a wolf, the animal must be trapped, tranquillized, and handled, all of which are dangerous and traumatic for the animal. The decision had to be taken that the information which such a study could yield merited the risks involved. Another problem to be faced was who should be responsible for these wolves. This was an economic problem, since the wolves killed sheep (*Ovis*) and shepherds demanded reimbursement. For a trial period the regional government of the Abruzzo compensated farmers for wolf damage. The claims were very large, but at least 50% of them were obviously fraudulent. Shepherds began to incite their dogs (*Canis lupus familiaris*) to kill their own sheep in order to claim compensation. Nevertheless, some of the wolf damage was serious, especially when they engaged in killing surplus prey. To the individual farmer who suffers such a kill, in which dozens of sheep are lost in a single night, the effect is disastrous, even if the overall impact of wolves on the economics of sheep farming is negligible. Wolves could only do so much damage because of the archaic shepherding practices, but equally they could only survive at all in the absence of their former prey because of this same inefficiency.

The compensation to shepherds raised another unexpected dilemma: a female from one remnant pack paired with a large shepherd dog and raised a family of hybrid puppies. To make matters worse, while all the pups behaved like wolves, some looked like wolves while others were black with blazes of white. First, this raised a new problem for the wolves: with only one such litter of these hybrids the gene pool (genes in the population; see EVOLUTION) had been significantly altered because the whole population was so tiny. Of course, this must have been happening to all wolf populations since man first tamed them, and began the process of DOMESTICATION. Nevertheless, with the wild population so small this hybridization might have been the 'last straw' for the wolves, diluting them to extinction. The second problem was that it became tricky to know what actually constituted a wolf. When a shepherd claimed for losses caused by a hybrid, was he to be reimbursed? Of course it might be a wolf-like or a dog-like hybrid. Furthermore, where did the responsibility of the biologists lie? Should they make every effort to kill the hybrids, or should they study them? In practice this last question was resolved by the wolf-dogs, as nobody could catch them.

The next problem revealed by the wolf study was the way in which species other than the one in question influence the outcome of management and conservation. It was originally thought that one amongst several factors that might be contributing to the wolf's decline was ecological competition with foxes. This notion became all the more plausible when diet studies showed that both species were eating a lot of garbage, and that both obtained most of this from commune garbage dumps on the outskirts of every mountain hamlet. Study of the foxes eventually showed that, while their ranges were overlapped by those of the wolves, and although the two species did meet at garbage tips, the competition between them was not significant, at least for the wolves. However, the fox did turn out, quite inadvertently, to be the wolves' biggest enemy. In every village commune each hunter was issued with poison in order to kill foxes (the motive being that foxes were presumed to reduce the crop of game available to the hunters). Since the hunting system was such that each hunter had access to all of the commune's land, the effect was that each piece of land was poisoned by many hunters, each acting as if he was the only one there. The result was thousands of dead foxes every year and enough wolves accidentally poisoned to threaten their complete extinction. The solution, which may anyway be too late, was legislation which banned the use of poison for fox control.

Turning to the other type of wolf problem, in North America, where the wolf is either endangered or too numerous depending on which State is considered, the eventual management plan contains a common irony: the solution which seemed to best ensure the wolves' survival involved permitting some to be killed. The conclusion was that if farmers were to tolerate complete protection for the wolf in wilderness areas, they would have to be permitted at least limited control in agricultural regions. In the long run such a compromise again rests on habitat conservation; if the area of wilderness dwindles then the refuge will be ineffectual.

In the long run the solution to any problem concerning a predator leads to its prey. Wolf conservation in Italy will rest on the effectiveness of schemes now under way to reintroduce the red deer that used to support the wolves. Similarly the Abruzzo region is an enclave of another prey species, the chamois (*Rupicapra rupicapra*), and attempts are being made to boost their numbers too, not just for the benefit of wolves, but as a conservation exercise in its own right. As with the wolves the decline of the Abruzzo chamois seems to result from human interference. During the last glaciation chamois occurred in both the Alps and the Apennines; today they survive quite well in the Alps, perhaps because they are less accessible there, while only four to five hundred survived in the Abruzzo. But this is a success story, for in 1915 only thirty to forty survived. These were all inside the Abruzzo National Park, the last chamois outside the park having been shot at the end of the last century. So, in this case, the existence of the park and ever more stringent wardening seem to have saved these creatures. However, a population in a single herd is very fragile, susceptible to disease and interference. The chamois do not seem to colonize elsewhere, and attempts to reintroduce them to former parts of their range have failed because of attacks by feral dogs and hunters. Until these two threats are removed the conservation of these antelope cannot proceed further.

Game ranching. The objective of game ranching is to use species for meat production which are resistant to local diseases, which can tolerate excessive heat and drought, and which can therefore be ranched to increase meat production on land of otherwise marginal agricultural potential. Ideally this idea would not simply be a means of boosting food production, but also of minimizing ecological damage caused by introduced FARM ANIMALS, and of reducing COMPETITION between wild game and domestic stock. Of course, to achieve these ambitions requires careful planning, but provisional results indicate that this type of approach will prove worthwhile.

At the Galana Game Ranch in Kenya the project involves comparison between wild species, such as eland (*Taurotragus oryx*), oryx (*Oryx gazella*), and buffalo (*Syncerus caffer*), and breeds of domestic species of, for example, cattle (*Bos primigenius taurus*) and sheep (*Ovis ammon aries*). The comparisons involve everything from their different diets, FORAGING behaviour, and energy assimilation, to their ranging behaviour, tractability in CAPTIVITY, disease tolerance, and the taste of their meat. Some problems soon become clear. For example, bull buffalo are so dangerous as to be impractical to keep. For the most part the management considerations require very sophisticated measurements. For instance, scientists have made detailed studies of the feeding behaviour and metabolism of cattle, sheep, and oryx. This involves following individuals of each species, noting what they eat at 3-min intervals, day and night. Slight, but possibly critical differences emerge: 19.2% of the sheep's diet is obtained by browsing, 4.7% of the cow's, and none of the oryx's. The oryx eats slightly more herbs than either of the other two, and all three species feed predominantly by grazing on grasses. In considering which species is the most efficient converter of grass to meat the first consideration is the theoretical prediction that smaller animals lose relatively more body heat because of their larger surface area:volume ratio compared with larger animals (see THERMOREGULATION). Smaller animals consequently have to eat relatively more than larger ones to 'refuel'. The sheep weighs about 44 kg, the oryx about 142 kg, and the cow about 320 kg. Indeed, weighing the food consumed by each species and weighing the resulting faeces showed that the rule held good, and indicated that the oryx did not have any special ADAPTATION to live on less food than predicted. Next the efficiency of the digestion of the three species on different diets was compared. All were much less able to digest protein deficient diets than protein rich diets, and there was no significant difference in their overall ability to digest grasses. The significant difference comes in connection with their relative tolerance of drought conditions. Using the requirements of cattle as a standard, the requirements of other species are: oryx 25%, sheep 45%, eland 60%. Oryx can survive without DRINKING for up to 25 days, while the other species normally drink every other day. This difference could have important implications for their ranching. Oryx could be grazed at much greater distances from water-holes, and marginal land could be used for their production that would require expensive irrigation if cattle were to be ranched there. The next step in this study is to discover the extent to which each species eats different species of plant and different parts of the leaf. This will enable the biologists to decide whether the oryx are complementary to the sheep and cattle in their feeding habits, or whether they are competitive. The answer to this question will allow the optimal combination of herbivores for given grazing land to be chosen, hopefully to the advantage of both the habitat and food production.

Another attempt at wildlife ranching involves rodents in South America. The capybara (*Hydrochaerus hydrochaerus*) is a large relative of the guinea-pig (*Cavia aperea porcellus*), and inhabits swamplands. The capybara is a heavy animal, weighing over 70 kg, and in some areas lives in large social groups of as many as eighty animals. The swamplands are threatened by drainage schemes designed to increase the grazing land available to cattle, and sometimes with military objectives too. Swamplands are an endangered habitat throughout the world, so extensive drainage is undesirable. Furthermore, in some cases it is arguable that the new pastures so created would be very fragile, with little topsoil and easily devastated by wind erosion. A possible alternative is to crop the capybaras and leave the habitat unscathed. In Venezuela considerable work has already been done on the behaviour of these unusual animals with this aim in mind. Information about their behaviour incorporated with knowledge of their population dynamics has been used to estimate optimal cropping rates. The difference between effective, responsible management and mere exploitation lies in the thoroughness of theoretical exercises of this sort. Whether the aim is to

boost the numbers of desirable species, or reduce the damage caused by a pest, the principle holds: the results of interference should be predicted as accurately as possible beforehand, and the consequences considered. This can only be achieved on the basis of adequate biological knowledge and often expensive research.

Limiting animal numbers. Two of the examples above have been concerned with boosting numbers of certain species with the long-term aim of habitat protection. The opposite case is exemplified by the damage done to woodlands by elephants (Elephantidae) in some national parks in Africa. For example, in Ruaha, Tanzania, scientists have examined the age distributions of the populations of various tree species and have also looked at all the available information about trends in elephant numbers. The results of turning these two sets of figures into mathematical predictions about the future impact of elephants on trees in the park are devastating. By the end of the century the whole landscape (and the ecology of the area) will have been changed if something is not done. In this case the acceleration in damage is so fast that a timorous cull of elephants would be insignificant, and therefore large-scale massacres have to be considered. Incidentally, it seems that the elephants knock over trees to eat merely the leaves of the crown.

Much wildlife management is concerned with large-scale attempts to reduce numbers of wild animals. A case which illustrates a wide variety of important principles is the control of rabies. In North America rabies is transmitted by a variety of species, including foxes, skunks (Mephitinae), racoons (*Procyon*), and bats (Chiroptera), but in Western Europe the red fox (*Vulpes vulpes*) is the main vector and victim. Waves of rabies have swept across Europe periodically throughout recorded history. In the past rabies has been a serious threat to human life. This risk has been reduced to a minimal level by modern vaccines and prophylaxis, but the disease remains a serious blemish on people's peace of mind and an enormous economic burden. Throughout the world it remains a major threat to human life, and is variously important to agrarian economics.

As successive European governments react to the invasion of rabies with vaccination schemes for domestic animals the proportion of foxes in the total of reported animal victims increases. For example, in Poland in 1966 29.3% of rabies infections diagnosed were in foxes, but by 1971 the figure had risen to 70.6%. The result of the foxes' undeniable involvement has been a series of campaigns aimed at killing as many foxes as possible. There is evidence that if fox populations can be sufficiently reduced then the chances of infection being passed on fall below the critical level at which the disease spreads or is maintained. In practice this figure is often thought to be about 0.2 foxes/km^2. Yet in spite of extensive campaigns involving shooting, poisoning, and, most importantly, gassing of earths, the disease has spread at a remarkably constant rate (of about 60 km/year). This has led to increased efforts to understand better the mechanics of a rabies outbreak in terms of fox behaviour. Perhaps such an understanding could explain not only why the killing schemes fail, but also what alternatives might succeed.

At one level there is an obvious reason why the killing has not stopped the disease, nor completely annihilated foxes. Foxes are very difficult to kill and very resilient to control. In addition, human interest and the availability of money for control soon run out. Most authorities believe that they cannot maintain more than an annual kill of 70% of the population, and fox populations are known to be able to withstand about 60–70% reduction while still being able to replace the losses within one breeding season. Most control has traditionally assumed that foxes behave rather similarly everywhere, and live at rather constant densities. However, more and more studies have shown that mammalian social behaviour is very flexible, and that societies of one species vary from habitat to habitat in an adaptive way. These intraspecific variations are often traceable to differences in the abundance and distribution of food in different habitats. This would seem to be one of the important factors which determines fox population densities, and is reflected in the sizes of their home ranges. For instance, studies of radiotracked foxes have shown that some maintain small home ranges of little more than 10 ha, while others have enormous ranges of well over 1000 ha. In general, home ranges in a given habitat are of comparable size. So, on the outskirts of Oxford, England, they average about 45 ha, whilst in the town centre they are bigger (about 250 ha). On the hill country of the north of England home ranges of 1300 ha are the average. So this variation in fox numbers from one district to another must be considered in formulating their control. Furthermore, there is evidence that these home ranges are territories, each occupied by a group of foxes which may number as many as five or six adults, depending on the habitat. One drawback to killing schemes as a way of reducing fox numbers and so combating rabies is the possible operation of a *vacuum effect*, by which one TERRITORY holder is removed and another quickly takes his place. The more one looks at fox society the more complex becomes the task of controlling them, and of understanding the consequences of that control. For example, studies both in the wild and in captivity indicate that only some vixens within a group of foxes breed, and this is probably the consequence of social forces determining that the more dominant vixens reproduce. Breeding vixens are more susceptible to control by gassing as they lie with the cubs, and the effects on other members of the group of the removal of these breeding animals is quite unknown.

These problems have led to a search for alternative methods. Various ideas have been suggested, including the distribution of contraceptive baits. Perhaps the most exciting involves oral vaccination of foxes against rabies. The notion is that if a sufficiently high proportion of foxes could be persuaded to vaccinate themselves against the disease by eating baits impregnated with vaccine, then they could form a far more effective buffer against the spread of the disease than control through killing could ever achieve. The biggest drawback to this attractive idea is that it is very difficult to make an appropriate vaccine. There are two sorts of vaccine, the so-called killed vaccine, which is made from dead virus and is quite safe but must be injected to be really effective, and the live vaccines, made of live virus whose virulence has been attenuated artificially. Live vaccines can give effective immunity when absorbed through the wall of the intestine, but they are less stable in outside weather conditions, and more dangerous in the sense that there is a chance that while they immunize foxes they may risk giving the disease to other species, such as some rodents. Nevertheless, in spite of the great complexities of vaccine production, there is considerable room for optimism that a suitable vaccine will be available soon. To make such a policy effective, information would be required on all aspects of the behaviour of foxes. How far do they travel in a given habitat, and hence how far do baits need to be distributed? What are their food preferences and hence what sort of bait should the vaccine be placed in? How can chemical attractants be used to lure foxes to a bait? How long is a vaccinated fox likely to survive? Scientists in Canada have produced some ingenious findings on the best way to distribute baits. They have been able to monitor the effectiveness of different methods of baiting by loading each bait with a dose of the antibiotic, tetracycline. When a fox eats this drug in the bait, traces of it are absorbed into growth rings in the teeth, and fluoresce under an ultra-violet microscope. Thus by examination of the teeth of foxes killed by hunters it is possible to see how many of them took the bait, and hence to gauge how many would have been immunized if a vaccine had been used. When sausage baits in plastic bags were dropped by aeroplanes as many as 74% of foxes in the study area ate the bait. This is certainly as effective as any killing scheme, and is likely to be preferred on every count. Not only would such a high rate of vaccination probably be more effective in eliminating the disease, but it would satisfy humanitarian and economic considerations. The economics involved have two facets: first the baits are so cheap that the immunization of each fox would cost very little. Second, since the early days of Canada's pioneers foxes have been an important resource within the fur trade, and rabies is currently seen mainly as an enemy of the fur trappers.

The research into the correct way to distribute bait is exhaustive in its attention to detail. The scientists have discovered a number of factors which make a particular plastic bag the best vehicle for the bait. The bag gathers dew in the early morning sun, glistening to attract the fox's attention, but the dew evaporates by the time children emerge and the bag is almost invisible then. The bait ferments quickly within the bag, increasing the attractive smells, and the open top acts as a funnel out of which these aromas are channelled. When an animal tears at the bag to eat the bait it leaves teeth marks in the plastic which are diagnostic of the species. In fact a field trial of oral vaccination has recently been completed in Switzerland, using attenuated live vaccine, and the spread of rabies through the trial valleys was successfully halted.

Another facet of rabies control concerns the spread of the disease amongst mongooses (Herpestinae) on the West Indian island of Grenada. The fact that rabies exists in mongooses on this island is a result of a typical example of improperly conceived wildlife management. The mongooses were introduced to control rats in the sugar plantations; they failed to do this, but rabies soon spread amongst them. Research on the behaviour of the mongooses and the transmission of disease amongst their communities revealed a significant factor that differs greatly from the fox case: mongooses can recover from the disease, and are thereafter immune to it. By studying the numbers of mongooses carrying rabies serum neutralizing antibodies (indicative of immunity to the disease), it was found that between 1971 and 1974 the proportion of immune mongooses had increased from 20.8% to 43.2%. It may well be that attempts to kill mongooses as a way of eliminating the disease should be abandoned, and that instead attempts should be made to boost the proportion with immunity artificially, and so to raise it over the threshold at which rabies will disappear.

A similar case where wildlife managers face a disease problem is that of bovine tuberculosis (TB) in badgers in the south-west of England. This disease used to be rife amongst cattle, but in the mid-1930s voluntary testing and slaughter of reactors to tuberculin tests started a process which had effectively vanquished the disease by 1960. It was at this point that the south-west was found to be mysteriously a last stronghold of bovine TB, and in June 1971 the disease was diagnosed in a badger from an infected area. Subsequent testing of many badgers has shown an overall link between the distribution of infected badgers and cattle, and this has led to government action to kill badgers on infected farms. The merits or otherwise of the policy of the British government on this matter have received critical attention, but it is again the subtleties of badger behaviour and TB transmission that presently hold most interest for the wildlife manager, and pose the greatest paradoxes to control authorities. The major difficulty is that, superimposed on the general association between badger and bovine TB, is a peculiar patchwork of anomalies. On some farms where the cattle contract the disease the badgers are quite free of it, and vice versa. There is no explanation for this,

nor is the process of transmission satisfactorily explained. It is assumed that the cattle eat the bacteria in the faeces and urine which are excreted by a tuberculous badger in the later stages of the disease. Certainly, by confining badgers in immediate proximity to cattle it is possible to achieve infection (although apparently with some difficulty), but this sheds little light on what actually happens in nature, which presents an unsolved problem. It seems that the only hope of solving problems like this is to study the behaviour of badgers in their natural surroundings in very great detail (see FIELD STUDIES). In particular, it is important to know about their relative population densities in different areas, and in this context scientists have devised an ingenious technique. Badgers live in social groups which inhabit (at least in some habitats) a main communal sett, centrally placed within a group territory. The borders of the territory are delineated by middens where faeces and glandular deposits accumulate. By feeding the badgers in each sett with coloured markers mixed with peanuts and syrup and then collecting the colour-marked faeces from the border latrines, it is possible to discover quickly the size of each territory. If one badger from each sett is injected with a weak (and harmless) dose of a radioactive substance, a sensitive Geiger counter can be used to identify its droppings. By comparing the proportion of radioactive to unmarked droppings the number of badgers in the sett can be calculated. The two techniques together give an elegant method of estimating population densities. As more information is gathered on badger sociology by techniques like these, the bovine TB problem should become assailable. Without adequate information the management of such complicated animal societies cannot hope to succeed.

Carnivores—like the fox and the badger—are involved in many wildlife management problems. In particular, carnivores are persecuted for depredations on domestic stock or game and are trapped for their furs. Coyotes (*Canis latrans*) are accused of killing sheep, and are widely harassed. Recently attempts have been made to stop them doing this through aversive CONDITIONING. The idea was to fit sheep with collars containing an emetic. The collar should be so constructed that the chemical (such as lithium chloride) is eaten when the coyote devours the sheep. Preliminary tests with rabbits in pens seemed promising, but of course the practical problems in the field are immense. Not least, enormous numbers of sheep would have to be equipped with deterrent collars. Second, the emetic would have to take effect quickly enough for the coyote to associate its sickness with the appropriate meal (see FOOD SELECTION). One possible advantage of the method is that there may be IMITATION; for example, a coyote that avoided sheep as food might pass the habit to its pups. However, how such a conditioned animal would fare in competition or in SOCIAL RELATIONSHIPS with others of its kind is unknown. As an alternative, attention has focused on a collar equipped with a poison. Toxic collars made with rubber pouches containing sodium monofluoroacetate have been tested and have worked in trials. The problem is that a poison which acts quickly enough to kill the coyote before it can do more damage and yet is suitable for use in the collar seems hard to find. The philosophy of a toxic or aversive collar is attractive in that it should select for habitual sheep killers rather than slaughter indiscriminately. The application of this idea may seem to have immense practical complications, but it must be considered in the context of hundreds of years of work on other methods which have failed to dissuade coyotes from killing sheep. Certainly, any method which is selective for a rogue animal is preferable to blanket control, though the problem with coyotes in sheep country may well be that they are all potentially rogue. This however would certainly not apply to foxes in Britain, where losses of lambs are so small and scattered that it simply must be the case that only the minority of foxes do any damage. A more promising application of the aversive conditioning approach might be the protection of hand-reared pheasants around their release pen.

Some efforts have gone into making artificial products which could be used either to attract or repel animals. One product looked as if it could fulfil both purposes. A group of American biologists developed a so-called 'fermented egg product' (FEP). The chemical composition of this material was analysed in detail and it was found that it could readily be manufactured in bulk. FEP was reported to repel deer, and thus gave the possibility of protecting forestry plantations from deer damage. In contrast, it apparently attracted coyotes and thus was potentially useful as a lure to bring coyotes to traps or to *scent stations*. SCENT stations have been used widely as a tool for estimating coyote numbers: an attractive odour is put in the middle of a patch of raked sand and coyote numbers are estimated on the basis of how many footprints are counted in the sand each morning. In conclusion, though, it seems that there are indeed many problems attached to devising new ways of coyote control, and for many people killing remains the only realistic solution.

The second main area of management of predatory animals is in their role as fur-bearers. In Britain fur-trapping has not been a traditional pursuit in the countryside, but with a recent upsurge in demand for fox fur this has suddenly changed. In 1975 the average price for a British fox pelt was under £5; by early 1979 it was almost £30. Monitoring and controlling sudden changes in relationship with wildlife like this are part of wildlife management.

Amongst the mammals the most notorious mismanagement has been directed towards marine creatures, and it is here that one may look forward to an increasing contribution from sound scientific considerations. The case of the abortive cull of grey seals (*Halichoerus grypus*) is particularly interesting. In 1977 the Department of Agriculture and Fisheries for Scotland (DAFS) proposed a 6-year plan to cut grey seal numbers back to their population density of the mid-1960s. This was equivalent to a reduction of up to a third, and was prompted by concern that the expanding seal populations were detrimental to the fishing industry. In 1978 the cull was scheduled to remove nine hundred adult breeding females and their associated pups, and a further four thousand moulted pups, in Orkney and North Rona. However, the cull had to be abandoned as a result of public pressure. In addition to a widely held belief that not enough forethought had been put into the cull, the public outcry exposed just how little was known about the critical issues, and how hard it might be to get the necessary answers. The central question of whether grey seals really do compete with fishermen is, almost unbelievably, still only poorly answered. More than five hundred grey seal stomachs were examined, and the majority did contain commercially important fish. However, 58% of the seals had been killed near some sort of fishing net, so it was clearly not an unbiased sample. Neither is there adequate information on the quantities of commercially important fish that seals consume; the DAFS apparently based their calculations on an extrapolation from a different species made in 1934. The calculations involved in much of the debate about culling these seals not only rest on surprisingly shaky evidence, but incline to the assumption that what the seals do not eat the fishermen will catch. Since the fish suffer other forms of PREDATION, most notably from other fish, this seems most improbable. This is clearly a case where much more groundwork is needed before satisfactory decisions can be taken.

Part of the uproar over the grey seal cull was stimulated not by a concern for their conservation as a population but by revulsion at the methods used to kill them. Indeed, the brutality involved in the killing of many marine mammals is causing international concern and, sadly, exposing many double standards. For example, not only are many species of whales (Cetacea) endangered by overexploitation, but the methods used to kill them are undeniably cruel. Whales are killed with explosive hand harpoons which spread fragments of metal through the body, lacerating blood vessels and organs. Alternatively, they are impaled by electric harpoons, or exploding carbon dioxide harpoons. With none of the available techniques could death be thought to be instantaneous. The extent of inhumanity should be a consideration in wildlife management, although it has often not been so. Recently more attention has been given not only to the efficiency of traps of all sorts but also to the cruelty they involve. A balance should be sought between concern for populations and for individuals.

A different sort of fish problem which illustrates a further facet of wildlife management, or again, lack of it, concerns the pet trade. There is a gross inequality in the distribution of legislation to protect different HOUSEHOLD PETS, and fish are amongst those with the least protection. For example, some 40% of aquarium fish are caught in the wild. In the south of England alone there are about half a dozen large importers, each handling four to five million fish every year, and about twenty smaller importers, perhaps handling a little over a million fish every year. About 10% of the fish die in transit, and losses before that are unknown. Anyone can apply for an import licence for an unlimited number of fish.

The wildlife manager often has to strike a compromise between many conflicting interests. These may include the future viability of the habitat; the population size of a species within a particular area; the welfare of individual animals; the economic impact of management practices; the traditional interests of the local human population, etc. Too often, management decisions have been taken without adequate knowledge of these various factors. Research into ETHOLOGY and ecology can do much to provide this knowledge, and to safeguard the interests of animal and plant communities around the world.   D.W.M.

107, 117

BIBLIOGRAPHY

1. Alcock, J. *Animal Behaviour*. 1975
2. Alexander, R. McN., Goldspink, G. *Mechanics and Energetics of Animal Locomotion*. 1977
3. Amlaner, C. J., Macdonald, D. W. *A Handbook on Biotelemetry and Radio Tracking*. 1980
4. Anderson, R. S. *Pet Animals and Society*. 1975
5. Aschoff, J. *Circadian Clocks*. 1965
6. Babkin, B. P. *Pavlov: A Biography*. 1949
7. Baker, R. *The Evolutionary Ecology of Animal Migration*. 1978
8. Bateson, P. P. G. How Do Sensitive Periods Arise and What are they for? *Animal Behaviour*, 27. 470–86. 1979
9. Bateson, P. P. G. How Does Behaviour Develop? In Bateson, P. P. G., and Klopfer, P. H. (eds.), *Perspectives in Ethology*, Vol. 3. 1978
10. Bastock, M. *Courtship: a Zoological Study*. 1967
11. Beach, F. A. *Sex and Behaviour*. 1965
12. Beadle, M. *The Cat: History, Biology and Behaviour*. 1977
13. Birch, M. C. *Pheromones*. 1974
14. Blurton-Jones, N., Reynolds, V. *Human Behaviour and Adaptation*. 1978
15. Bodenheimer, F. S. *The History of Biology: an Introduction*. 1958
16. Bolles, R. C. *Theory of Motivation*. 1975
17. Bonner, J. T. *The Evolution of Culture in Animals*. 1980
18. Bossema, I. *Jays and Oaks: an Eco-ethological Study of a Symbiosis*. 1979
19. Breder, C. M., Jr. *Modes of Reproduction in Fishes*. 1966
20. Brown, J. L. *The Evolution of Behaviour*. 1975
21. Busnel, R. G. *Acoustic Behaviour of Animals*. 1963
22. Campbell, B. *Sexual Selection and the Descent of Man*. 1972
23. Carlsoo, S. *How Man Moves*. 1972
24. Cloudsley-Thompson, J. *Animal Migration*. 1978
25. Cloudsley-Thompson, J. L. *Rhythmic Activity in Animal Physiology and Behaviour*. 1961
26. Cole, F. W. *Fundamental Astronomy*. 1974
27. Cox, C. B., Healey, I. N., Moore, P.D. *Biogeography: an Ecological and Evolutionary Approach*. 1976
28. Curio, E. *The Ethology of Predation*. 1976
29. Darwin, C. *On the Origin of Species by Means of Natural Selection, or the Preservation of Favoured Races in the Struggle for Life*. 1859
30. Davies, N. B. Ecological Questions about Territorial Behaviour. In Krebs, J. R., and Davies, N.B. (eds.), *Behavioural Ecology*. 1978
31. Davis, M. Imitation: a Review and Critique. In Bateson, P. P. G., and Klopfer, P. H. (eds.), *Perspectives in Ethology*. 1973.
32. Dawkins, M. S. *Animal Suffering: the Science of Animal Welfare*. 1980
33. Dawkins, R. Hierarchical Organisation: a Candidate Principle for Ethology. In Bateson, P. P. G., and Hinde, R. A. (eds.), *Growing Points in Ethology*. 1976
34. Dawkins, R. *The Selfish Gene*. 1976
35. Dethier, V. G. *The Physiology of Insect Senses*. 1963
36. Dimond, S., Lazarus, J. *The Problem of Vigilance in Animal Life*. 1974
37. Dorst. J. *The Migration of Birds*. 1962
38. Drent, R. *Incubation*. 1975
39. Edmunds, M. *Defense in Animals*. 1974
40. Ehrman, L., Parsons, P. *The Genetics of Behaviour*. 1976
41. Eibl-Eibesfeldt, I. *Ethology. The Biology of Behaviour*. 1970
42. Elton, C. *Animal Ecology*. 1927
43. Ewer, R. T. *Ethology of Mammals*. 1968
44. Fox, M. W. *Understanding Your Dog*. 1972
45. Fraenkel, G. S., Gunn, D. L. *The Orientation of Animals*. 1961
46. Fraser, A. F. *Farm Animal Behaviour*. 1974
47. von Frisch, K. R. *A Biologist Remembers*. 1967
48. Frith, H. J. *The Mallee-Fowl: the Bird that Builds an Incubator*. 1962
49. Gambaryan, P. P. *How Mammals Run*. 1974
50. Ghiselin, M. T. *The Triumph of the Darwinian Method*. 1969
51. Gladwin, T. *East is a Big Bird: Navigation and Logic on Puluwat Atoll*. 1970
52. Gottlieb, G. *Comparative Psychology and Ethology*. 1979
53. Gould, J. L. *An Introduction to Ethology*. 1980
54. Gould, S. J. *Ontogeny and Phylogeny*. 1977
55. Gray, J. *Animal Locomotion*. 1968
56. Gray, J. A. *The Psychology of Fear and Stress*. 1971
57. Gray, J. A. *Pavlov*. 1979
58. Gray, P. H. *The Early Animal Behaviourists: Prolegomenon to Ethology*. 1969
59. Gregory, R. L. *Eye and Brain*. 1966
60. Griffin, D. R. *Listening in the Dark*. 1958
61. Griffin, D. R. *The Question of Animal Awareness*. 1976
62. Grigorian, N. A. *Pavlov, Ivan Petrovich*. 1974
63. Gruber, H. E. *Darwin on Man*. 1974
64. Hafez, E. S. E. *The Behaviour of Domestic Animals*. 1975
65. Hailman, J. P. *Optical Signals*. 1977
66. Harrison-Matthews, L., Knight, M. *The Senses of Animals*. 1963
67. Hartmann, E. L. *The Functions of Sleep*. 1973
68. Hartshorne, C. *Born to Sing*. 1973
69. Hasler, A. D. *Olfactory Imprinting in Coho Salmon*. 1978
70. Hearnshaw, L. S. *A Short History of British Psychology (1840–1940)*. 1964
71. Hediger, H. *Wild Animals in Captivity*. 1964
72. Heinroth, K. *History of Ethology*. 1977
73. Hinde, R. A. *Bird Vocalizations*. 1969
74. Hinde, R. A. *Animal Behaviour* (2nd edn.). 1970
75. Hinde, R. A. *Biological Bases of Human Social Behaviour*. 1974
76. Hinde, R. A. (ed.). *Non-Verbal Communication*. 1972
77. Hirsch, J. *Behaviour Genetic Analysis*. 1967
78. Honig, W.K., James, P. H. R. (eds.). *Animal Memory*. 1971
79. Honig, W. K., Staddon, J. E. R. (eds.). *Handbook of Operant Behaviour*. 1977

80. Horn, G., Hinde, R. A. *Short-term Changes in Neural Activity and Behaviour.* 1970
81. Hulse, S. H., Fowler, H., Honig. W. K. (eds.). *Cognitive Processes in Animal Behaviour.* 1978
82. Hutchison, J. B. (ed.). *Biological Determinants of Sexual Behaviour.* 1978
83. Jenkins, F. A. *Primate Locomotion.* 1974
84. Keeton, W. T. *Avian Orientation & Navigation: a Brief Overview.* 1979
85. Kiley-Worthington, M. *Behaviour Patterns of Farm Animals.* 1977
86. Kimble, G. A. *Hilgard and Marquis' Conditioning and Learning.* 1961
87. Kleiman, D. G. *Monogamy in Mammals.* 1977
88. Klopfer, P. H., Hailman, J. P. *Function and Evolution of Behaviour.* 1972
89. Kohler, W. *The Mentality of Apes.* 1957
90. Krebs, J. R., Davies, N. B. (eds.). *Behavioural Ecology.* 1978
91. Kummer, H. *Social Organization of Hamadryas Baboons.* 1968
92. Lawick-Goodall, J. van. *Tool-using in Primates and other Vertebrates.* 1970
93. Leshner, A. I. *An Introduction to Behavioural Endocrinology.* 1978
94. Lewis, D. *We, the Navigators: the Ancient Art of Land-finding in the Pacific.* 1972
95. Lorenz, K. *Evolution and Modification of Behaviour.* 1966
96. Lorenz, K. *On Aggression.* 1966
97. Lynn, R. *Attention, Arousal and the Orientation Reaction.* 1966
98. Mackintosh, N. J. *The Psychology of Animal Learning.* 1974
99. Manning, A. *An Introduction to Animal Behaviour* (3rd edn.). 1979
100. Marler, P. R., Hamilton III, William J. *Mechanisms of Animal Behaviour.* 1966
101. Marshall, A. J. *Bower-Birds: their Displays and Breeding Cycles.* 1954
102. Masterton, R. B., Hodos, W., Jerison, H. (eds.). *Evolution, Brain & Behaviour: Persistent Problems.* 1976
103. McFarland, D. J. *Feedback Mechanisms in Animal Behaviour.* 1971
104. Meddis, R. *The Sleep Instinct.* 1977
105. Millar, S. *The Psychology of Play.* 1968
106. Milner, P. M. *Physiological Psychology.* 1970
107. Moen, A. N. *Wildlife Ecology.* 1973
108. Muntz, W. R. A. *Comparative Aspects in Behavioural Studies of Vertebrate Vision.* 1974
109. Nachtigall, W. *Insects in Flight.* 1974
110. Nisbett, A. *Konrad Lorenz.* 1976
111. Noble, G. K. *The Biology of the Amphibia.* 1931
112. Oatley, K. *Brain Mechanisms and Mind.* 1972
113. Oppenheim, R. W. *Prehatching and Hatching Behaviour: a Comparative & Physiological Consideration.* 1973
114. Passmore, J., *Man's Responsibility for Nature.* 1974
115. Pengelley, E. *Circannual Clocks.* 1974
116. Pennycuick, C. J. *Animal Flight.* 1972
117. Pratt, D. J., Gwynne, M. D. *Rangeland Management and Ecology in East Africa.* 1977
118. Ridley, M. Paternal Care. *Animal Behaviour,* 26. 904–33. 1978
119. Riopelle, A. J. (ed.). *Animal Problem Solving.* 1967
120. Rothschild, M., Clay, T. *Fleas, Flukes, and Cuckoos.* 1957
121. Ruse, M. *The Darwinian Revolution.* 1979
122. Sales, G. *Ultrasonic Communication by Animals.* 1974
123. Schmidt-Koenig, K. *Migration and Homing in Animals.* 1975
124. Schmidt-Koenig, K. *Avian Orientation & Navigation.* 1979
125. Silverman, P. *Animal Behaviour in the Laboratory.* 1978
126. Skinner, B. F. *Cumulative Record.* 1961
127. Skinner, B. F. *Particulars of My Life.* 1976
128. Skinner, B. F. *The Shaping of a Behaviourist.* 1979
129. Skutch, A. F. *Parent Birds and Their Young.* 1976
130. Sluckin, W. *Imprinting and Early Learning.* 1964
131. Smith, N. E. *Visual Isolation in Gulls.* 1967
132. Smythe, R. H. *Vision in the Animal World.* 1975
133. Tansley, K. *Vision in Vertebrates.* 1965
134. Thompson, R. F. *Foundations of Physiological Psychology.* 1967
135. Thorpe, W. H. *Bird-Song.* 1961
136. Thorpe, W. H. *The Origins and Rise of Ethology.* 1979
137. Tinbergen, N. *The Study of Instinct.* 1951, 1969
138. Tinbergen, N. *The Animal in its World.* 1972
139. Tinbergen, N. *Curious Naturalists.* 1958
140. Trueman, E. R. *The Locomotion of Soft-bodied Animals.* 1975
141. Twitty, V. C. *Of Scientists and Salamanders.* 1966
142. Webb, W. B. *Sleep—the Gentle Tyrant.* 1975
143. Whittow, G. C. *Comparative Physiology of Thermoregulation.* 1970
144. Williams, C. B. *Insect Migration.* 1958
145. Wilson, E. O. *Sociobiology: the New Synthesis.* 1975
146. Wright, W. D. *The Measurement of Colour* (3rd edn.). 1964

INDEX OF ENGLISH NAMES OF ANIMALS

References are to titles of articles

INDEX OF SCIENTIFIC NAMES OF ANIMALS

References are to titles of articles

CONFLICT, DOMINANCE, IMITATION, NEST-BUILDING
Branta leucopsis (barnacle goose),
  FIELD STUDIES
Bryozoa (bryozoa),
  SOCIAL ORGANIZATION
Bubulcus ibis (cattle egret),
  FLOCKING, HUNTING, INTERACTIONS AMONG ANIMALS,
  PREDATORY, SYMBIOSIS
Bucephala clangula (common golden-eye duck),
  INTENTION MOVEMENTS, RITUALIZATION
Bucerotidae (hornbills),
  INCUBATION, PREDATORY
Buconidae (puffbirds),
  NEST-BUILDING
Bucorvus abyssinicus (Abyssinian ground hornbill),
  PREDATORY
Bufo arenarum (toad),
  SLEEP
Bufo bufo (European toad),
  COPULATION, COURTSHIP, DEFENSIVE, HABITUATION,
  ORIENTATION, SEARCHING IMAGE, THREAT
Bufo fowleri (Fowler's toad),
  MIGRATION
Buphagus erythrorhynchus (tick bird),
  DEFENSIVE, INTERACTIONS AMONG ANIMALS
Burhinus oedicnemus (common stone curlew),
  NEST-BUILDING
Buteo (buzzards),
  COLORATION
Butorides virescens (American green heron),
  INTENTION MOVEMENTS, RITUALIZATION, THREAT

Cacicus (caciques),
  PARASITISM
Cacicus cela (yellow-rumped cacique),
  DEFENSIVE
Cactospiza (woodpecker finches),
  SPECIALIZATION
Cactospiza pallida (woodpecker finch),
  INTELLIGENCE, TOOL USING
Caiman sclerops (caiman),
  SLEEP
Calamospiza melanocorys (lark bunting),
  TERRITORY
Calandra granaria (beetle),
  RHYTHMS
Calliactis parasitica (hermit crab anemone),
  DEFENSIVE, SYMBIOSIS
Calliactis polypus (hermit crab anemone),
  SYMBIOSIS
Calliactis tricolor (hermit crab anemone),
  SYMBIOSIS, TOOL USING
Callichthys (mudsucker),
  BREATHING
Callimorpha jacobaeae (cinnabar moth),
  MIMICRY
Calliphora (bluebottle),
  FLIGHT
Calliphoridae (blowflies),
  PARASITISM, TASTE AND SMELL
Callithrix jacchus (common marmoset),
  BINOCULAR VISION, FUNCTION, SCENT MARKING
Callorhinus ursinus (northern fur seal),
  DOMINANCE
Camarhynchus (insectivorous tree finches),
  FIELD STUDIES, SPECIALIZATION
Camelidae (camels),
  DRINKING, HISTORY, LOCOMOTION, THERMOREGULATION,
  THIRST
Camelus dromedarius (dromedary),
  THERMOREGULATION

Campaea perlata (moth),
  CAMOUFLAGE
Campanularia flexuosa (marine polyp),
  RHYTHMS, SOCIAL ORGANIZATION
Cancer pagurus (rock crab),
  MIGRATION, SPECIALIZATION
Canidae (canids),
  COMPARATIVE STUDIES, HEARING, PARASITISM
Canis (dogs, wolves, jackals),
  BRAIN, COMMUNICATION, COPULATION, DOMESTICATION,
  LOCOMOTION, PARASITISM
Canis familiaris dingo (dingo),
  EVOLUTION
Canis latrans (coyote),
  WILDLIFE MANAGEMENT
Canis lupus (wolf),
  COMMUNICATION, COOPERATIVE, FACIAL EXPRESSIONS,
  FEEDING, HERDING, HISTORY, HUNTING, MATERNAL,
  PREDATION, SCHOOLING, TERRITORY, WILDLIFE
  MANAGEMENT
Canis lupus familiaris (domestic dog),
  ABNORMAL, AGGRESSION, AROUSAL, ASSOCIATION,
  AVOIDANCE, COLOUR VISION, COMFORT, DARWIN,
  DEFENSIVE, DRUGS, EMOTION, EXTINCTION, FEAR,
  FRUSTRATION, GENERALIZATION, GENETICS, HISTORY,
  HORMONES, HOUSEHOLD PETS, HUNGER, IMITATION,
  IMPRINTING, INTELLIGENCE, LEARNING, MATERNAL,
  MEMORY, MOTIVATION, ORIENTATION, PLAY, PROBLEM
  SOLVING, REINFORCEMENT, SEXUAL, SKINNER, SOCIAL
  INTERACTIONS, TASTE AND SMELL, THERMOREGULATION,
  VIGILANCE, WELFARE, WILDLIFE MANAGEMENT
Canis mesomelas (jackal),
  HUNTING, PREDATION
Capra (goats)
  HUNTING, IMPRINTING, MATERNAL, PHEROMONES,
  SENSITIVE PERIODS
Capra hircus (domestic goat),
  FARM ANIMAL, ONTOGENY
Capra ibex (ibex),
  HISTORY
Capra ibex nubiana (Nubian ibex),
  CAPTIVITY
Capreolus capreolus (roe deer),
  CAPTIVITY, WILDLIFE MANAGEMENT
Caprimulgidae (goatsuckers),
  HIBERNATION
Caprimulgus europaeus (nightjar),
  COURTSHIP, FLIGHT, HIBERNATION
Caranx hippos (carangid),
  SCHOOLING
Carassius auratus (goldfish),
  ANTHROPOMORPHISM, BRAIN, COLOUR VISION,
  INTELLIGENCE
Carcinus maenas (shore crab),
  BREATHING, RHYTHMS, TIME
Cardinalis cardinalis (cardinal),
  ADVERTISEMENT, ANTHROPOMORPHISM
Carduelis carduelis (goldfinch),
  ISOLATING MECHANISMS
Carduelis chloris (greenfinch),
  FOOD SELECTION, ISOLATING MECHANISMS
Carduelis flammea (redpoll finch),
  IMITATION
Carparachne alba (white lady spider),
  COMMUNITY
Castor fiber (beaver),
  SCENT MARKING, VOCALIZATION
Casuariidae (cassowaries),
  PARENTAL CARE
Cathartidae (New World vultures),
  CHEMICAL SENSES, SCHOOLING

COMPARATIVE STUDIES
Urosaurus ornatus (tree lizard),
    HABITAT, TOLERANCE
Ursidae (bears),
    FOOD SELECTION, HISTORY, HOARDING, SCENT MARKING,
    TECHNIQUES OF STUDY
Ursus americanus (black bear),
    HIBERNATION, PLAY
Ursus arctos (brown bear),
    HIBERNATION, HUNGER

Vanellus vanellus (lapwing),
    COURTSHIP, SYMBIOSIS
Vanessa atalanta (red admiral),
    MIGRATION
Vanessa cardui (painted lady),
    FLIGHT, MIGRATION
Venezilla arixonicus (woodlouse),
    TOLERANCE
Vermileo vermileo (snipe fly),
    FEEDING
Vespertilionidae (vespertilionid bats),
    ECHO-LOCATION
Vespinae (paper wasps),
    COMPARATIVE STUDIES
Vespoidea (wasps),
    FLIGHT
Vespula (wasp),
    DEFENSIVE, MIMICRY
Vespula vulgaris (wasp),
    SCENT MARKING
Viduinae (whydahs),
    FOOD BEGGING
Vipera xanthina (mountain viper),
    HUNTING
Viperidae (vipers or adders),
    AGGRESSION
Vonones sayi (harvestman),
    DEFENSIVE

Vulpes (fox),
    AGGRESSION, CAPTIVITY, DEFENSIVE, DISPLAYS, HISTORY,
    TECHNIQUES OF STUDY, WELFARE OF ANIMALS, WILDLIFE
    MANAGEMENT
Vulpes vulpes (red fox),
    DISPERSION, HOARDING, SEXUAL, TERRITORY, WILDLIFE
    MANAGEMENT
Vulpes vulpes fulva (American red fox),
    DEATH FEIGNING

Wallabia (wallabies),
    NICHE

Xanthocephalus xanthocephalus (yellow-headed
    blackbird),
    TERRITORY
Xantusia vigilus (desert night lizard),
    PARENTAL CARE
Xenopus (clawed toads),
    MECHANICAL SENSES
Xenopus laevis (African clawed toad),
    HATCHING, HORMONES
Xerus erythropus (African ground squirrel),
    ESCAPE

Zabilius aridus (long-horned grasshopper),
    CAMOUFLAGE
Zarhynchus (oropendola),
    PARASITISM
Zarhynchus wagleri (Wagler's oropendola),
    DEFENSIVE
Zenaidura macroura (mourning dove),
    NEST-BUILDING
Zonotrichia leucophrys (white-crowned sparrow),
    CULTURAL, DIALECT, FIXED ACTION PATTERNS, GOAL,
    IMITATION, ONTOGENY
Zygoptera (damselflies),
    BREATHING, FLIGHT

ACKNOWLEDGEMENTS FOR ILLUSTRATIONS

The publishers are grateful to the following for their permission to make use of illustration material. Every effort has been made to contact copyright holders, but we apologize to anyone who may have been omitted.

ADVERTISEMENT Figs. A & B: After Figs. 7-1, & 7-5 from J. P. Hailman, 'Optical signals', in *How Animals Communicate* (ed. Thomas A. Sebeok 1977). By permission of Indiana University Press.

AGGRESSION Fig. B: After F. R. Walther (1958), *Zeitschrift Tierpsychologie*, 15, 340–80. By permission of Verlag Paul Parey.

Fig. C: Redrawn from Eibl-Eibesfeldt, *Ethology*, Fig. 199 after C. E. Shaw in *Herpetologica*, 4, 137–45. By permission of the Editor.

ANTHROPOMORPHISM Fig. A: Redrawn from a photograph by Paul Lemmon in *Animal Behaviour* (N. Tinbergen, Time-Life Books, 1966). By permission of Frank W. Lane.

APPEASEMENT Fig. A: Redrawn from a photograph by W. Wickler (1965), in *Naturwiss.*, 52, 335–41. By permission of Springer-Verlag, Inc.

CAMOUFLAGE Fig. A (top) and Fig. C: After J. Alcock, *Animal Behaviour: An Evolutionary Approach* (2nd edn. 1979). By permission of the author and Sinauer Associates.

Fig. B: After Fig. 2.10, p. 17, and Fig. D: After Fig. 2.28, p. 39, from M. Edmunds, *Defence in Animals* (Longman 1974).

CLIMATE Fig. A: After Fig. 11, p. 33; Fig. B: After Fig. 16, pp. 40 & 41; Fig. C: After Fig. 26, p. 57, from *Biogeography*, by C. B. Cox, I. N. Healey, and P. D. Moore (1976) By permission of Blackwell Scientific Publ. Ltd.

COMMUNITY Fig. A: After Beard (1955), *Ecology* (Duke University Press, Durham, NC 27708).

CONFLICT Fig. B: After N. Tinbergen (1959), 'Comparative studies of the behaviour of gulls (Laridae)', in *Behaviour*, 15, 1–70. By permission of the Editor.

Fig. D: After Fig. 1, p. 13, M. J. A. Simpson (1968), in *Animal Behaviour Monographs* 1. By permission of Ballière Tindall and the author.

COORDINATION Fig. A: After Fig. 3.7, p. 35, R. A. Hinde, *Animal Behaviour* (1970). By permission of McGraw-Hill Book Co.

COURTSHIP Fig. D: After Fig. 23 (from Morris [192]), in M. Bastock, *Courtship: A Zoological Study* (1967). By permission of Heinemann Educational Books.

Fig. E: After Fig. 16.4, from R. A. Hinde, *Animal Behaviour* (1970). Copyright © 1970, R. A. Hinde. By permission of R. A. Hinde and McGraw-Hill Book Co.

DEFENSIVE BEHAVIOUR Fig. B: After Fig. 2.20, p. 30; Fig. E: After Fig. 93, p. 178; Fig. G: After Fig. 10.6, p. 129, from M. Edmunds, *Defence in Animals* (Longman 1974).

DISPLAYS Fig. A: After Figs. 20 & 21 (after Warren, from Crane [68]), in M. Bastock, *Courtship: A Zoological Study* (1967). By permission of Heinemann Educational Books.

Fig. B: After Fig. 6.6, p. 62; Fig. C: After Fig. 6.13, p. 75, in R. A. Hinde, *Biological Bases of Human Social Behaviour* (1974). By permission of McGraw-Hill Book Co.

Fig. D: After Fig. 7.13, p. 169, in M. Edmunds, *Defence in Animals* (Longman 1974).

EXTINCTION Fig. A: From J. Konorski, *Integrative Activity of the Brain* (1967), p. 318. By permission of The University of Chicago Press.

Fig. C: From R. A. Boakes and M. S. Halliday, *Journal of Comparative and Physiological Psychology*, 88 (1975).

Fig. D: From R. A. Boakes and M. S. Halliday, *Inhibition and Learning* (1972), p. 214. By permission of Academic Press Inc. (London) Ltd.

FEAR Fig. A: After N. Tinbergen (1948), in *Wilson Bulletin*, 60, 6–52. By permission of the Editor.

FIXED ACTION PATTERNS Fig. A: After Lorenz and Tinbergen (1938), in *Zeitschrift Tierpsychologie*, 2, 1–29. By permission of Verlag Paul Parey.

FLIGHT Fig. E: Modified from Fig. 3.3 in C. Pennycuick, *Animal Flight* (Studies in Biology No. 33), Edward Arnold, 1972. By permission of the publisher.

Fig. G: After T. Weis Fogl, *Jnl. Experimental Biology*, 59, 169–230 (1973).

Fig. H(a): After Dr. H. Oehme, in *Scale Effects in Animal Locomotion*, p. 491 (edited by T. J. Pedley, Academic Press 1977). By permission of T. J. Pedley and Prof. Dr Hans Oehme.

Fig. I: After Magnan, *La Locomotion chez les animaux*, Vol. I, 'Le Vol des insectes', Hermann, Paris, 1934. By permission of the publisher.

Fig. J: From Knut Schmidt-Nielsen, *Animal Physiology: Adaptation and Environment* (1979). By permission of Cambridge University Press.

Fig. K(c): Modified from Fig. 7.3 in C. Pennycuick, *Animal Flight* (Studies in Biology No. 33), Edward Arnold, 1972. By permission of the publisher.

Fig. M: After C. D. Bramwell & G. R. Whitfield, *Philosophical Transactions of The Royal Society*, 267, 503–92 (1974). By permission of The Royal Society and the authors.

Fig. N: After G. Heilman (1926), *The Origin of Birds* (H. F. & G. Witherby Ltd.).

Fig. O (b, c & d): After Fig. 1.4, in C. Pennycuick, *Animal Flight* (Studies in Biology No. 33), Edward Arnold, 1972. By permission of the publisher.

GENERALIZATION Fig. B: After N. Guttmann & H. I. Kalish, *Journal of Experimental Psychology*, 51 (1956).

HATCHING Figs. A–G: After Figs. 7, 8b, 9, 9b, 10, and 13 from Oppenheim and Foelix in *Studies in the Development of Behavior and the Nervous System*, Vol. I, *Behavioral Embryology* (ed. G. Gottlieb, 1973). By permission of Academic Press, New York, Inc.

HISTORY OF THE STUDY OF ANIMAL BEHAVIOUR Fig. A: After A. Sieveking, p. 95 in *The Cave Artists* (Thames & Hudson 1979).

Fig. B: After A. Leroi-Gourhan, *The Art of Pre-*

historic Man in Western Europe (Thames & Hudson 1968).

Fig. C(a) from p. 260 and Fig. C(b) from p. 256, T. H. White, *The Book of Beasts* (Cape 1956). By permission of David Higham Associates Ltd.

HORMONES Fig. A: After S. A. Temple (1974), *General and Comparative Endocrinology*, 22. By permission of Academic Press, New York, Inc., and the author.

HYPNOSIS Fig. C: Redrawn from the fig. on p. 514 of G. G. Gallup *et al.*, 'Immobility Response: A Predator-Induced Reaction in Chickens', *The Psychological Record*, 1971, Vol. 21, pp. 513–19. Reprinted by permission of the Editor.

INTELLIGENCE Figs. B & C: Redrawn from Figs. 1 & 3, M. R. A. Chance, *Man* (Old Series), Vol. 60, 1960. Reprinted by permission of The Royal Anthropological Institute of Great Britain and Ireland.

LOCOMOTION Fig. J: Redrawn from P. P. Gambaryan [After Muybridge], in *How Mammals Run* (Wiley 1974).

MIMICRY Fig. A: Redrawn from a combination of Figs. 4.10 & 4.12, pp. 101 & 103, in M. Edmunds, *Defence in Animals* (Longman 1974).

Figs. B & C: Redrawn from photographs in *Mimicry*, by Wolfgang Wickler (1968). By permission of Weidenfeld & Nicolson and the author.

MOTIVATION Figs. E & F: After P. Leyhausen (1956), *Zeitschrift Tierpsychologie*, 2–1. By permission of Verlag Paul Parey.

NAVIGATION Fig. A: After S. T. Emlen (1975) in *Avian Biology* (eds. D. S. Farner and J. R. King). By permission of Academic Press, New York, Inc., and the author.

NEST-BUILDING Figs. A, B, & C: After N. E. and E. C. Collias, *The Auk*, Vol. 79 (1962). By permission of the American Ornithologists' Union.

NICHE Fig. A. (a) & (b): Redrawn from Fig. 8.16 ('G' p. 301) [After Fisher & Peterson]; and Fig. A (c) & (d): Redrawn from Fig. 8.16 ('E' p. 301) [After Salthe]; from *Evolutionary Ecology*, 2nd edn. by Eric R. Pianka. Copyright © 1978 by Eric R. Pianka. By permission of Harper & Row, Publ. Inc.

ONTOGENY Fig. B: From J. Kear, *Wildfowl Trust 18th Annual Report* (1967), pp. 122–4. By permission of The Director, The Wildfowl Trust.

Fig. C: After W. W. Cruze, *Journal of Comparative Psychology*, 19 (1935), 371–408.

Fig. D: After P. P. G. Bateson, *Advances in the Study of Behaviour*, Vol. 6 (1976), 1–20. By permission of the author and Academic Press, New York, Inc.

Fig. E: After R. A. McCance, *Lancet*, 2 (1962), 671–6. By permission of the author and the Editor.

Fig. F: After Fig. 1, P. Klopfer and M. Klopfer, *Animal Behaviour*, 25 (1977), 286–91. By permission of Ballière Tindall and the authors.

ORIENTATION Figs. A, B, C, & E: After Herman Schöne. By permission of Wissenschaftliche Verlag.

Fig. D: After Asch & Witkin, *Journal of Experimental Psychology*, Vol. 38.

PERCEPTUAL LEARNING Fig. A: Redrawn from *Personality Development* by Jerome Kagan (1970). By permission of Harcourt Brace Jovanovich, Inc.

PREDATORY BEHAVIOUR Figs. A & C: After Fig. 49 (p. 141), & Fig. 50 (p. 143) from E. Curio, *The Ethology of Predation* (1976). By permission of Springer-Verlag, New York, Inc.

REINFORCEMENT Fig. A: After R. M. Yerkes & S. Morgulis, *Psychological Bulletin*, 6 (1909).

Fig. B: After G. V. Anrep (1920), 'Pitch discrimination in the dog', *Journal of Physiology*, 53, 1920:367–85. By permission of the Editorial Board.

RHYTHMS Fig. B: After M. A. Brock, from *Circannual Clocks* (ed. E. T. Pengelley, 1974). By permission of Academic Press, New York, Inc., and the author.

Fig. C: After Fig. 108, p. 117, from E. Bunning, *The Physiological Clock* (1967). By permission of Springer-Verlag, New York, Inc.

Fig. D: After Fig. 6–8 (after Bangert 1960), from P. R. Marler & W. J. Hamilton, *Mechanisms of Animal Behaviour* (1966). By permission of John Wiley & Sons, Inc.

SCENT MARKING Fig. A: After Gosling (1972), *Zeitschrift Tierpsychologie*, 30, 271–6. By permission of Verlag Paul Parey.

SENSE ORGANS Fig. A: After Fig. 11.2 (from Haskell 1961 after Schwabe), from P. R. Marler and W. J. Hamilton, *Mechanisms of Animal Behaviour* (1966). By permission of John Wiley & Sons, Inc.

SEXUAL SELECTION Fig. A: After A. Lill (1968), *Behaviour*, 30, 127–45. By permission of the Editor.

SIGN STIMULUS Fig. A: After Fig. 20, p. 28 from N. Tinbergen, *The Study of Instinct*. By permission of Oxford University Press.

Fig. B: After Fig. 24, p. 55; and Fig. D: After Fig. 21, p. 52 from E. Curio, *Animal Behaviour*, 23 (1975). By permission of Ballière Tindall and the author.

SOCIAL INTERACTIONS Fig. B: After M. J. A. Simpson (1968), *Animal Behaviour Monographs* 1, Fig. 1, p. 13. By permission of Ballière Tindall and the author.

SONG Fig. B(a): Reproduction of Fig. 53: Woodthrush by Crawford H. Greenwalt, 1968. Reproduced by permission of The Smithsonian Institution Press from Crawford H. Greenwalt's *Bird Song: Acoustics and Physiology*, p. 67, Washington D.C. 1968 © Smithsonian Institution.

Fig. B(b): After Fig. 61, from W. H. Thorpe, *Birdsong* (Cambridge Monographs in Experimental Biology 12). By permission of Cambridge University Press.

Fig. D(a): After Y. Leroy, 'Signaux acoustiques. Comportement et systématique de quelques espèces de gryllides (Orthoptera Ensiferes)', from *Bulletin Biologique de la France et de la Belgique*, Vol. 100 (1966).

Fig. D(b): After the figure on p. 87 of H. C. Bennet-Clark and A. W. Ewing, 'The Love Song of the Fruit Fly', *Scientific American*, July 1970, Vol. 223, No. 1. By permission of Dr. A. W. Ewing.

Fig. D(c): After Fig. 51, from W. H. Thorpe, *Birdsong* (Cambridge Monographs in Experimental Biology 12). By permission of Cambridge University Press.

Fig. D(e): Reproduced from Abb. 2, from F. Huber in *Naturwissenschaftliche Rundschau*, 18, No. 4, pp. 143–56. By permission of the Editor and Prof. Dr. Franz Huber.

Fig. F: After P. Marler, 'Development in the study of animal communication', Ch. 4 in *Darwin's Biological Work: Some Aspects Reconsidered* (ed. P. R. Bell 1959). By permission of Cambridge University Press.

Fig. G: After Fig. 34; and Fig. H: After Fig. 63, from W. H. Thorpe, *Birdsong* (Cambridge Monographs in Experimental Biology 12). By permission of Cambridge University Press.

SURVIVAL VALUE Fig. A: After K. Daumer (1958), 'Blumenfarben, wie sie die Bienen Sehen', from *Zeitschrift für Vergleichende Physiologie*, 41, 49–110. By permission of Springer-Verlag, New York, Inc.

TERRITORY Fig. B: After Robert Gillmor, from *Ecological Adaptation for Breeding in Birds* (ed. David Lack, Methuen 1968). By permission of Associated Book Publ. Ltd.

Fig. C: After Fig. 2 from C. K. Catchpole (1972), 'A comparative study of territory in the Reed warbler (*Acrocephalus scirpaceus*) and Sedge warbler (*A.*

schoenobaenus)', *J. Zool., Lond.* 166: 213–31. By permission of The Zoological Society of London.

Fig. D: After Fig. 6 from H. Kruuk (1976), 'Spatial organization and territorial behaviour of the European badger *Meles meles*', *J. Zool., Lond.* 184: 1–19.

Fig. E: After Fig. 31, p. 111, from J. R. Krebs, 'Territory and breeding density in the great tit Parus Major L', *Ecology*, 52: 2–22. Copyright © 1971, The Ecological Society of America. By permission of Duke University Press.

TIME Fig. A: After Fig. 2–1; and Fig. D: After Fig. 2–9, from F. W. Cole, *Fundamental Astronomy* (1974). By permission of John Wiley & Sons, Inc.

TOLERANCE Fig. B: After Fig. 8, p. 27, from *Biogeography*, by C. B. Cox, I. N. Healey, and P. D. Moore (1976). By permission of Blackwell Scientific Publ. Ltd.

TOOL USING Fig. C: After Fig. 5, p. 437 from J. Alcock, *Animal Behaviour* (1979). By permission of the author and Sinauer Associates.

VISION Figs. A & B: After S. Polyak, *The Vertebrate Visual System* (1957), pp. 17 & 35. By permission of The University of Chicago Press.

Fig. D: After Fig. 1 (I. Abramov and J. Gordon), p. 328 in *Handbook of Perception*, Vol. 3 (1973). By permission of Academic Press, New York, Inc., and the authors.

Fig. E: Compiled from Fig. 3.1, p. 24 & Fig. 3.3, p. 29 from R. L. Gregory, *Eye and Brain* (1966). By permission of Weidenfeld & Nicolson.

Fig. H: After the figure on p. 31; and Fig. K: After Fig. 74, p. 140, from *Visual Perception* (ed. T. N. Cornsweet 1970). By permission of Academic Press, New York, Inc.

Fig. L: After S. Polyak, *The Vertebrate Visual System* (1957), p. 276. By permission of The University of Chicago Press.

Fig. N: After M. H. Pirenne, *Vision and the Eye* (Chapman & Hall 1967). By permission of Associated Book Publ. Ltd.